T0137169

Lecture Notes in Computer Science 5769

Commenced Publication in 1973
Founding and Former Series Editors:
Gerhard Goos, Juris Hartmanis, and Jan van Leeuwen

Cesare Alippi
Marios Polycarpou
Christos Panayiotou
Georgios Ellinas (Eds.)

Artificial Neural Networks – ICANN 2009

19th International Conference
Limassol, Cyprus, September 14-17, 2009
Proceedings, Part II

Volume Editors

Cesare Alippi
Politecnico di Milano, Dipartimento di Elettronica
Piazza L. da Vinci 32, 20133 Milano, Italy
E-mail: alippi@elet.polimi.it

Marios Polycarpou
Christos Panayiotou
Georgios Ellinas
University of Cyprus, Department of Electrical and Computer Engineering
75 Kallipoleos Street 1678 Nicosia, Cyprus
E-mail: {mpolycar, christosp, gellinas}@ucy.ac.cy

Library of Congress Control Number: 2009933887

CR Subject Classification (1998): F.1, I.2, I.4, I.5, I.6, J.3, C.1.3, C.2

LNCS Sublibrary: SL 1 – Theoretical Computer Science and General Issues

ISSN 0302-9743
ISBN-10 3-642-04276-7 Springer Berlin Heidelberg New York
ISBN-13 978-3-642-04276-8 Springer Berlin Heidelberg New York

springer.com

Printed in Germany

Typesetting: Camera-ready by author, data conversion by Scientific Publishing Services, Chennai, India
Printed on acid-free paper SPIN: 12757459 06/3180 5 4 3 2 1 0

Preface

This volume is part of the two-volume proceedings of the 19th International Conference on Artificial Neural Networks (ICANN 2009), which was held in Cyprus during September 14–17, 2009. The ICANN conference is an annual meeting sponsored by the European Neural Network Society (ENNS), in cooperation with the International Neural Network Society (INNS) and the Japanese Neural Network Society (JNNS). ICANN 2009 was technically sponsored by the IEEE Computational Intelligence Society. This series of conferences has been held annually since 1991 in various European countries and covers the field of neurocomputing, learning systems and related areas.

Artificial neural networks provide an information-processing structure inspired by biological nervous systems. They consist of a large number of highly interconnected processing elements, with the capability of learning by example. The field of artificial neural networks has evolved significantly in the last two decades, with active participation from diverse fields, such as engineering, computer science, mathematics, artificial intelligence, system theory, biology, operations research, and neuroscience. Artificial neural networks have been widely applied for pattern recognition, control, optimization, image processing, classification, signal processing, etc.

In 2009, the ICANN conference was organized by the KIOS Research Center for Intelligent Systems and Networks and the Department of Electrical and Computer Engineering of the University of Cyprus. The conference was held at the seaside city of Limassol, which is the second largest city in Cyprus. The participants had the opportunity to enjoy the technical program, as well as the rich cultural heritage of Cyprus, whose 9,000-year cultural legacy has been at the crossroads of world history. Currently, Cyprus is a full member of the European Union that combines European culture with ancient enchantment.

Out of approximately 300 paper submissions to ICANN 2009, the Program Committee selected about 200 papers, which are published in the two volumes of these proceedings. The selection of the accepted papers was made after a thorough peer-review process, where each submission was evaluated by at least three reviewers. The submitted papers were authored by peer scholars coming from 47 countries, which geographically cover the whole planet (Europe, Middle East, Africa 69%; Asia/Pacific 18%; Americas 13%). The large number of accepted papers, variety of topics and high quality of submitted papers reflect the vitality of the field of artificial neural networks. In addition to the regular papers, the technical program featured keynote plenary lectures by worldwide renowned scholars, two tutorials on exciting new topics, two competitions on immunology and environmental toxicology prediction, and two workshops. One of the workshops was supported by the EU-sponsored COST Action "Intelligent Monitoring, Control and Security of Critical Infrastructure Systems" (IntelliCIS).

The two-volume proceedings contain papers on the following topics: Learning Algorithms; Computational Neuroscience; Hardware Implementations and Embedded Systems; Self Organization; Intelligent Control and Adaptive Systems; Neural and Hybrid

Architectures; Support Vector Machines; Recurrent Neural Networks; Neuro-informatics and Bioinformatics; Cognitive Machines; Data Analysis and Pattern Recognition; Signal and Time Series Processing; Applications; Neural Dynamics and Complex Systems; Vision and Image Processing; Neuro-evolution and Hybrid Techniques for Mobile Agents Control; Neural Control, Planning and Robotics Applications; Intelligent Tools and Methods for Multimedia Annotation; Critical Infrastructure Systems.

It is our pleasure to express our gratitude to everybody that contributed to the success of ICANN 2009. In particular, we thank the members of the Board of the European Neural Networks Society for entrusting us with the organization of the conference, as well as for their assistance during the preparation of ICANN 2009. Special thanks to the President of ENNS, Wlodzislaw Duch, who helped significantly toward the success of the conference. We would like to express our sincere gratitude to the members of the Program Committee and all the reviewers, who did a tremendous job under strict time limitations during the reviewing process. We thank the members of the Organizing Committee for the great effort in the organization of the conference and the members of the Local Organizing Committee for their assistance. We are grateful to the University of Cyprus (ECE Department) and the Cyprus Tourist Organization for their financial support. We thank the conference secretariat, Top Kinisis, and especially Christina Distra, for their excellent and timely support in the organization of the conference. We are grateful to several researchers at the University of Cyprus and the Politecnico di Milano, who assisted in various ways in the organization of ICANN 2009, and especially Alexandros Kyriakides and Andreas Kartakoullis, who spent several days working on the formatting of the final proceedings, and Manuel Roveri, who addressed several software-related and procedural problems raised during the review process. We would also like to thank the publisher, Springer, for their cooperation in publishing the proceedings in the prestigious series of *Lecture Notes in Computer Science*. Finally, we thank all the authors who contributed to this volume for sharing their new ideas and results with the community. We hope that these ideas will generate further new ideas and innovations for the benefit of society and the environment.

July 2009

Cesare Alippi
Marios Polycarpou
Christos Panayiotou
Georgios Ellinas

Organization

Conference Chairs

General Co-chairs	Marios Polycarpou, University of Cyprus, Nicosia, Cyprus Christos Panayiotou, University of Cyprus, Nicosia, Cyprus
Program Chair	Cesare Alippi, Politecnico di Milano, Italy
Publications Chair	Georgios Ellinas, University of Cyprus, Nicosia, Cyprus
Finance Chair	Maria Michael, University of Cyprus, Nicosia, Cyprus
Local Arrangment Chair	Elias Kyriakides, University of Cyprus, Nicosia, Cyprus
Registration Chairs	Constantinos Pitris, University of Cyprus, Nicosia, Cyprus Manuel Roveri, Politecnico di Milano, Italy
Publicity Chairs	Derong Liu, University of Illinois at Chicago, USA Ouejdane Mejri, Politecnico di Milano, Italy

Program Committee

Angelo Alessandro	Angelo Università di Genova, Italy
Luís Alexandre	University of Beira Interior, Portugal
Bruno Apolloni	University of Milan, Italy
Amir Atiya	Cairo University, Egypt
Bartlomiej Beliczynski	Warsaw University of Technology, Poland
Monica Bianchini	Università degli Studi di Siena, Italy
Ivo Bukovsky	Czech Technical University in Prague, Czech Republic
Ke Chen	The University of Manchester, UK
Andrej Dobnikar	FRI, Slovenia
Gerard Dreyfus	ESPCI – Paristech, France
Wlodek Duch	Nicolaus Copernicus University, Poland
David A. Elizondo	De Monfort University, UK
Peter Erdi	Hungarian Academy of Sciences, Hungary
Pablo Estévez	University of Chile, Chile
Jay Farrell	University of California, Riverside, USA
Erol Gelenbe	Imperial College London, UK
Mark Girolami	University of Glasgow, UK
Barbara Hammer	Clausthal University of Technology, Germany
Tom Heskes	Radboud University Nijmegen, The Netherlands
Timo Honkela	Helsinki University of Technology, Finland
Amir Hussain	University of Stirling, UK
Giacomo Indiveri	University of Zurich - ETH Zurich, Switzerland

Sylvain Jaume	Massachusetts Institute of Technology, USA
Yaochu Jin	Honda Research Institute Europe, Germany
Stefanos Kollias	National Technical University of Athens, Greece
Jan Koutník	IDSIA, Switzerland
Vera Kurkova	Academy of Sciences of the Czech Republic
Aristidis Likas	University of Ioannina, Greece
Derong Liu	University of Illinois at Chicago, USA
Francesco Marcelloni	University of Pisa, Italy
Thomas Martinetz	University of Lubeck, Germany
Matteo Matteucci	Politecnico di Milano, Italy
Ali Minai	University of Cincinnati, USA
Roman Neruda	Institute of Computer Science, Czech Republic
Erkki Oja	Helsinki University of Technology, Finland
Stanislaw Osowski	Warsaw University of Technology, Poland
Guenther Palm	University of Ulm, Germany
Thomas Parisini	University of Trieste, Italy
Constantinos Pattichis	University of Cyprus, Cyprus
Vincenzo Piuri	University of Milan, Italy
Mark Plumbley	Queen Mary University of London, UK
Manuel Roveri	Politecnico di Milano, Italy
George Rovithakis	Aristotle University of Thessaloniki, Greece
Leszek Rutkowski	Technical University of Czestochowa, Poland
Karl Sammut	Flinders University, Australia
Edgar Sanchez	Cinvestav, Mexico
Marcello Sanguineti	University of Genova, Italy
Fabio Scotti	University of Milan, Italy
Olli Simula	Helsinki University of Technology, Finland
Alessandro Sperduti	Università degli Studi di Padova, Italy
Andreas Stafylopatis	National Technical University of Athens, Greece
Johan Suykens	KULeuven, Belgium
John Taylor	King's College London, UK
Sergios Theodoridis	University of Athens, Greece
Marc Van Hulle	KU Leuven, Belgium
Marley Vellasco	Pontificia Universidade Catolica do Rio de Janeiro, Brazil
Michel Verleysen	Université Catholique de Louvain, Belgium
Alessandro E.P. Villa	Neuroheuristic, France
Jun Wang	The Chinese University of Hong Kong, Hong Kong
Zidong Wang	Brunel University, UK
Stefan Wermter	University of Sunderland, UK
Hujun Yin	The University of Manchester, UK
Michalis Zervakis	Technical University of Crete, Greece
Xiaodong Zhang	Wright State University, USA
Pedro Zufiria	Universidad Politécnica de Madrid, Spain
Rodolfo Zunino	University of Genoa, Italy

Local Organizing Committee

Constantinos Pattichis	University of Cyprus
Christos Schizas	University of Cyprus
Julius Georgiou	University of Cyprus
Vasilios Promponas	University of Cyprus
Elpida Keravnou	University of Cyprus
Petros Ioannou	Cyprus University of Technology
Chris Christodoulou	University of Cyprus
Andreas Pitsillides	University of Cyprus
Spyros Martzoukos	University of Cyprus
Theocharis Theocharides	University of Cyprus
Christoforos Hadjicostis	University of Cyprus
Nicolas Tsapatsoulis	Cyprus University of Technology

Referees

M. Adankon
L. Aijun
F. Aiolli
T. Aksenova
A. Al-Ani
A. Alanis
A. Alessandri
L. Alexandre
E. Alhoniemi
C. Alippi
H. Allende
C. Alzate
F. Amigoni
S. An
D. Andina
G. Andreoni
M. Antonelli
B. Apolloni
A. Araujo
B. Arnonkijpanich
Y. Asai
S. Asteriadis
M. Atencia
A. Atiya
Y. Avrithis
D. Bacciu
A. Bagirov
A. Bahramisharif
S. Bamford
A. Bartczuk
S. Bassis
G. Battistelli
F. Bazso
C. Bechlioulis
B. Beliczynski
M. Bianchini
M. Bicego
M. Biehl
K. Blekas
S. Bohte
G. Boracchi
C. Borgelt
A. Borghese
A. Bors
P. Bouboulis
Z. Buk
I. Bukovsky
K. Bunte
M. Bursa
K. Cameron
R. Camplani
A. Cangelosi
A. Carvalho
M. Casey

I. Castelli
A. Castillo
B. Castillo-Toledo
D. Cavouras
M. Celebi
M. Cepek
C. Cervellera
J. Chambers
A. Chaves
K. Chen
D. Chen
L. Chen
T. Chen
Z. Chengwen
M. Chetouani
S. Choi
N. Chumerin
R. Cierniak
M. Cimino
J. Cires
J.C. Claussen
M. Cococcioni
V. Colla
S. Coniglio
C. Constantinopoulos
S. Contassot-Vivier
E. Corchado
O. Cordon
F. Corona
J.A. Costa
A. Cruz
E. Cuadros-Vargas
E. Culurciello
M. Cuneo
W. Czajewski
G. Da San Martino
A. d'Avila Garcez
Z. Dai
S. Damas
T. Damoulas
J. Dauwels
T. De Bie
A. De Paola
S. Decherchi
Y. Demiris
B. Denby
M. Di Francesco
E. Di Iorio
V. Di Massa
D. Dias
M. Diesmann
A. Dobnikar
S. Doboli
E. Donchin
H. Dong
J. Drchal
G. Dreyfus
N. Drosopoulos
P. D'Souza
P. Ducange
W. Duch
D. Durackova
P. Dziwnski
M. Elgendi
D. Eliades
D. Elizondo
M. Elshaw
M. El-Telbany
P. Erdi
M. Ernandes
H. Erwin
P. Estevez
A. Esuli
R. Ewerth
P. Földesi
J. Farrell
H. Fayed
J. Feng
G. Fenu
S. Ferrari
K. Figueiredo
J. Fraile-Ardanuy
D. François
L. Franco
G. Fumera
C. Fyfe
M. Gabryel
M. Gaggero
X-Z. Gao
E. Gelenbe

X. Georgiou
A. Ghio
G. Gigante
G. Gini
M. Girolami
M. Giulioni
G. Gnecco
E. Gomez-Ramirez
V. Gomez-Verdejo
P. Gopych
L. Graening
M. Hagenbuchner
B. Hammer
K. Harris
F. He
X. He
M. Hernanez
A. Herrero
T. Heskes
L. Hey
H. Hidalgo
M. Hilairet
T. Honkela
L. Hontoria
P. Hoyer
E. Hruschka
J. Hu
T. Huang
G.-B. Huang
H. Huang
C. Hung
A. Hussain
A. Ilin
G. Indiveri
B. Indurkhya
S. Jaume
J. Jerez
Y. Jin
M. Jirina
M. Jirina
A. Kaban
M. Kaipainen
J. Karhunen
K. Karpouzis
N. Kasabov
A. Keyhani
F. Khan
Y. Kim
T. Kimura
A. Klami
A. Knoblauch
O. Kohonen
P. Koikkalainen
S. Kollias
M. Korytkowski
M. Koskela
D. Kosmopoulos
C. Kotropoulos
J. Koutnìk
K. Koutroumbas
O. Kovarik
C. Krier
D. Krusienski
F. Kurfess
V. Kurkova
G. La Camera
K. Lagus
J. Lampinen
C. Laoudias
A. Lazaric
L. Y. Lean Yu
J. Lee
A. Lendasse
L. Li
W. Li
A. Likas
B. Linares-Barranco
T. Lindh-Knuutila
M. Lippi
D. Liu
J. Liu
W. Liu
Y. Liu
D. Loiacono
P.-Y. Louis
G. Loukas
J. Lücke
L. Ma
D. Macciò
M. MacIver

M. Mahfouf
M. Manry
N. Manyakov
A. Marakakis
F. Marcelloni
R. Marek
M. Markou
U. Markowska-Kaczmar
T. Martinetz
T. Martinez-Marin
A. Marullo
P. Masci
S. Maskell
P. Massobrio
F. Masulli
G. Matsopoulos
M. Matteucci
E. Merenyi
M. Michaelides
A. Micheli
A. Minai
S. Mitra
N. Mitrakis
U. Modigliani
F. Morabito
G. Morgavi
S. Moustakidis
J. Murray
P. Mylonas
S. Nara
M. Nelson
A. Neme
R. Neruda
M. Niazi
K. Nikita
C. Nikou
D. Novak
R. Nowicki
K. Ntalianis
D. Oberhoff
E. Oja
K. Okada
K. Okuma
C. Orovas
J. Ortiz de Lazcano Lobato
M. Ortolani
S. Osowski
Y. Oussar
N. Oza
T. Pahikkala
A. Paiva
G. Palm
P. Panagi
D. Panagiotopoulos
C. Panayiotou
T. Papini
T. Parisini
K. Parsopoulos
E. Pasero
C. Pattichis
G. Pazienza
F. Pellegrino
S. Petridis
A. Petrosino
B. Phillips
F. Piazza
V. Piuri
M. Plumbley
T. Poggi
M. Polycarpou
F. Prieto
A. Pucci
K. Raftopoulos
K. Raivio
F. Ramos-Corchado
M.A. Ranzato
K. Rapantzikos
M. Restelli
A. Revonsuo
R. Riaza
B. Ribeiro
L. Ricalde
S. Ridella
L. Rigutini
F. Rivas
T. Rodemann
R. Rodriguez
S. Rogers
J. Ronda
F. Rossi

M. Roveri
S. Rovetta
G. Rovithakis
G. Ruthenbeck
L. Rutkowski
J. Ruz
D. Saez
K. Saito
M. Salomon
K. Sammut
E. Sanchez
M. Sanguineti
J. Santos
L. Sarti
S. Sayeed
G. Schaefer
R. Scherer
F.-M. Schleif
P. Schneider
F. Schwenker
F. Scotti
G. Sen Gupta
R. Setiono
E. Severin
I.-U.-H. Shah
I. Shah
V. Shaposhnyk
B. Shen
Y. Shen
H. Shimazu
N. Siebel
L. Silva
O. Simula
G. Siolas
M. Skrbek
K. Slavakis
S. Slusny
L. Smetana
Z. Somogyvari
Q. Song
A. Sorjamaa
U. Spagnolini
A. Sperduti
E. Spyrou
S. Squartini
A. Stafylopatis
G. Stamou
J. Starczewski
J. Stastny
V. Stathopoulos
D. Stavrakoudis
N. Steele
M. Strickert
J. Suykens
K. Swingler
F. Talantzis
Y. Tang
R. Tanscheit
H. Tarik
J. Taylor
A. Tefas
S. Theodoridis
T. Theoharis
C. Thiel
S. Timotheou
P. Tino
M. Titsias
J. Togelius
E. Trentin
N. Tsapatsoulis
S. Tzikopoulos
J. Vayrynen
M. Valle
H. Valpola
M. Van Hulle
M. Vellasco
M. Verleysen
J. Vesanto
A. Villa
T. Villmann
G. von der Emde
B. Vrusias
M. Wallace
J. Wang
Z. Wang
X. Wang
Y. Wang
Y.X. Wang
Z.-W. Wang
C. Weber

Q. Wei
G. Wei
H. Wei
T. Wennekers
S. Wermter
D. Wierstra
T. Windeatt
T. Xiang
T. Yagi
Z. Yang
C. Yang
Y. Yang
H. Yin
S. Zafeiriou
P. Zegers
S. Zeng
M. Zervakis
X. Zhang
Q. Zhao
Z. Zhao
M. Zhong
S. Zhou
J. Zhou
Z.-H. Zhou
P. Zufiria
R. Zunino

Table of Contents – Part II

Neuroinformatics and Bioinformatics

Epileptic Seizure Prediction and the Dimensionality Reduction Problem 1
André Ventura, João M. Franco, João P. Ramos, Bruno Direito, and António Dourado

Discovering Diagnostic Gene Targets and Early Diagnosis of Acute GVHD Using Methods of Computational Intelligence over Gene Expression Data 10
Maurizio Fiasché, Anju Verma, Maria Cuzzola, Pasquale Iacopino, Nikola Kasabov, and Francesco C. Morabito

Mining Rules for the Automatic Selection Process of Clustering Methods Applied to Cancer Gene Expression Data 20
André C.A. Nascimento, Ricardo B.C. Prudêncio, Marcilio C.P. de Souto, and Ivan G. Costa

A Computational Retina Model and Its Self-adjustment Property 30
Hui Wei and XuDong Guan

Cognitive Machines

Mental Simulation, Attention and Creativity 40
Matthew Hartley and John G. Taylor

BSDT Atom of Consciousness Model, AOCM: The Unity and Modularity of Consciousness 54
Petro Gopych

Generalized Simulated Annealing and Memory Functioning in Psychopathology 65
Roseli S. Wedemann, Luís Alfredo V. de Carvalho, and Raul Donangelo

Algorithms for Structural and Dynamical Polychronous Groups Detection 75
Régis Martinez and Hélène Paugam-Moisy

Logics and Networks for Human Reasoning 85
Steffen Hölldobler and Carroline Dewi Puspa Kencana Ramli

Data Analysis and Pattern Recognition

Simbed: Similarity-Based Embedding 95
John A. Lee and Michel Verleysen

PCA-Based Representations of Graphs for Prediction in QSAR Studies 105
Riccardo Cardin, Lisa Michielan, Stefano Moro, and Alessandro Sperduti

Feature Extraction Using Linear and Non-linear Subspace Techniques 115
Ana R. Teixeira, Ana Maria Tomé, and E.W. Lang

Classification Based on Combination of Kernel Density Estimators 125
Mateusz Kobos and Jacek Mańdziuk

Joint Approximate Diagonalization Utilizing AIC-Based Decision in the Jacobi Method 135
Yoshitatsu Matsuda and Kazunori Yamaguchi

Newtonian Spectral Clustering 145
Konstantinos Blekas, K. Christodoulidou, and I.E. Lagaris

Bidirectional Clustering of MLP Weights for Finding Nominally Conditioned Polynomials 155
Yusuke Tanahashi and Ryohei Nakano

Recognition of Properties by Probabilistic Neural Networks 165
Jiří Grim and Jan Hora

On the Use of the Adjusted Rand Index as a Metric for Evaluating Supervised Classification 175
Jorge M. Santos and Mark Embrechts

Profiling of Mass Spectrometry Data for Ovarian Cancer Detection Using Negative Correlation Learning 185
Shan He, Huanhuan Chen, Xiaoli Li, and Xin Yao

Kernel Alignment k-NN for Human Cancer Classification Using the Gene Expression Profiles 195
Manuel Martín-Merino and Javier de las Rivas

Convex Mixture Models for Multi-view Clustering 205
Grigorios Tzortzis and Aristidis Likas

Strengthening the Forward Variable Selection Stopping Criterion 215
Luis Javier Herrera, G. Rubio, H. Pomares, B. Paechter, A. Guillén, and I. Rojas

Features and Metric from a Classifier Improve Visualizations with Dimension Reduction 225
Elina Parviainen and Aki Vehtari

Fuzzy Cluster Validation Using the Partition Negentropy Criterion 235
Luis F. Lago-Fernández, Manuel Sánchez-Montañés, and Fernando Corbacho

Bayesian Estimation of Kernel Bandwidth for Nonparametric Modelling 245
Adrian G. Bors and Nikolaos Nasios

Using Kernel Basis with Relevance Vector Machine for Feature Selection 255
Frédéric Suard and David Mercier

Acquiring and Classifying Signals from Nanopores and Ion-Channels ... 265
Bharatan Konnanath, Prasanna Sattigeri, Trupthi Mathew, Andreas Spanias, Shalini Prasad, Michael Goryll, Trevor Thornton, and Peter Knee

Hand-Drawn Shape Recognition Using the SVM'ed Kernel 275
Khaled S. Refaat and Amir F. Atiya

Selective Attention Improves Learning 285
Antti Yli-Krekola, Jaakko Särelä, and Harri Valpola

Signal and Time Series Processing

Multi-stage Algorithm Based on Neural Network Committee for Prediction and Search for Precursors in Multi-dimensional Time Series 295
Sergey Dolenko, Alexander Guzhva, Igor Persiantsev, and Julia Shugai

Adaptive Ensemble Models of Extreme Learning Machines for Time Series Prediction 305
Mark van Heeswijk, Yoan Miche, Tiina Lindh-Knuutila, Peter A.J. Hilbers, Timo Honkela, Erkki Oja, and Amaury Lendasse

Identifying Customer Profiles in Power Load Time Series Using Spectral Clustering 315
Carlos Alzate, Marcelo Espinoza, Bart De Moor, and Johan A.K. Suykens

Transformation from Complex Networks to Time Series Using Classical Multidimensional Scaling 325
Yuta Haraguchi, Yutaka Shimada, Tohru Ikeguchi, and Kazuyuki Aihara

Predicting the Occupancy of the HF Amateur Service with Neural Network Ensembles ... 335
Harris Papadopoulos and Haris Haralambous

An Associated-Memory-Based Stock Price Predictor ... 345
Shigeki Nagaya, Zhang Chenli, and Osamu Hasegawa

A Case Study of ICA with Multi-scale PCA of Simulated Traffic Data ... 358
Shengkun Xie, Pietro Lió, and Anna T. Lawniczak

Decomposition Methods for Detailed Analysis of Content in ERP Recordings ... 368
Vasiliki Iordanidou, Kostas Michalopoulos, Vangelis Sakkalis, and Michalis Zervakis

Outlier Analysis in BP/RP Spectral Bands ... 378
Diego Ordóñez, Carlos Dafonte, Minia Manteiga, and Bernardino Arcay

ANNs and Other Machine Learning Techniques in Modelling Models' Uncertainty ... 387
Durga Lal Shrestha, Nagendra Kayastha, and Dimitri P. Solomatine

Comparison of Adaptive Algorithms for Significant Feature Selection in Neural Network Based Solution of the Inverse Problem of Electrical Prospecting ... 397
Sergey Dolenko, Alexander Guzhva, Eugeny Obornev, Igor Persiantsev, and Mikhail Shimelevich

Efficient Optimization of the Parameters of LS-SVM for Regression versus Cross-Validation Error ... 406
Ginés Rubio, Héctor Pomares, Ignacio Rojas, Luis Javier Herrera, and Alberto Guillén

Applications

Noiseless Independent Factor Analysis with Mixing Constraints in a Semi-supervised Framework. Application to Railway Device Fault Diagnosis ... 416
Etienne Côme, Latifa Oukhellou, Thierry Denœux, and Patrice Aknin

Speech Hashing Algorithm Based on Short-Time Stability ... 426
Ning Chen and Wang-Gen Wan

A New Method for Complexity Reduction of Neuro-fuzzy Systems with Application to Differential Stroke Diagnosis ... 435
Krzysztof Cpałka, Olga Rebrova, and Leszek Rutkowski

LS Footwear Database - Evaluating Automated Footwear Pattern Analysis 445
Maria Pavlou and Nigel M. Allinson

Advanced Integration of Neural Networks for Characterizing Voids in Welded Strips 455
Matteo Cacciola, Salvatore Calcagno, Filippo Laganá, Giuseppe Megali, Diego Pellicanó, Mario Versaci, and Francesco Carlo Morabito

Connectionist Models for Formal Knowledge Adaptation 465
Ilianna Kollia, Nikolaos Simou, Giorgos Stamou, and Andreas Stafylopatis

Modeling Human Operator Controlling Process in Different Environments 475
Darko Kovacevic, Nikica Pribacic, Mate Jovic, Radovan Antonic, and Asja Kovacevic

Discriminating between V and N Beats from ECGs Introducing an Integrated Reduced Representation along with a Neural Network Classifier 485
Vaclav Chudacek, George Georgoulas, Michal Huptych, Chrysostomos Stylios, and Lenka Lhotska

Mental Tasks Classification for a Noninvasive BCI Application 495
Alexandre Ormiga G. Barbosa, David Ronald A. Diaz, Marley Maria B.R. Vellasco, Marco Antonio Meggiolaro, and Ricardo Tanscheit

Municipal Creditworthiness Modelling by Radial Basis Function Neural Networks and Sensitive Analysis of Their Input Parameters 505
Vladimir Olej and Petr Hajek

A Comparison of Three Methods with Implicit Features for Automatic Identification of P300s in a BCI 515
Luigi Sportiello, Bernardo Dal Seno, and Matteo Matteucci

Neural Dynamics and Complex Systems

Computing with Probabilistic Cellular Automata 525
Martin Schüle, Thomas Ott, and Ruedi Stoop

Delay-Induced Hopf Bifurcation and Periodic Solution in a BAM Network with Two Delays 534
Jian Xu, Kwok Wai Chung, Ju Hong Ge, and Yu Huang

Response Properties to Inputs of Memory Pattern Fragments in Three Types of Chaotic Neural Network Models 544
Hamada Toshiyuki, Jousuke Kuroiwa, Hisakazu Ogura, Tomohiro Odaka, Haruhiko Shirai, and Yuko Kato

Partial Differential Equations Numerical Modeling Using Dynamic Neural Networks 552
Rita Fuentes, Alexander Poznyak, Isaac Chairez, and Tatyana Poznyak

The Lin-Kernighan Algorithm Driven by Chaotic Neurodynamics for Large Scale Traveling Salesman Problems 563
Shun Motohashi, Takafumi Matsuura, Tohru Ikeguchi, and Kazuyuki Aihara

Quadratic Assignment Problems for Chaotic Neural Networks with Dynamical Noise 573
Takayuki Suzuki, Shun Motohashi, Takafumi Matsuura, Tohru Ikeguchi, and Kazuyuki Aihara

Global Exponential Stability of Recurrent Neural Networks with Time-Dependent Switching Dynamics 583
Zhigang Zeng, Jun Wang, and Tingwen Huang

Approximation Capability of Continuous Time Recurrent Neural Networks for Non-autonomous Dynamical Systems 593
Yuichi Nakamura and Masahiro Nakagawa

Spectra of the Spike Flow Graphs of Recurrent Neural Networks 603
Filip Piekniewski

Activation Dynamics in Excitable Maps: Limits to Communication Can Facilitate the Spread of Activity 613
Andreas Loengarov and Valery Tereshko

Vision and Image Processing

Learning Features by Contrasting Natural Images with Noise 623
Michael Gutmann and Aapo Hyvärinen

Feature Selection for Neural-Network Based No-Reference Video Quality Assessment 633
Dubravko Ćulibrk, Dragan Kukolj, Petar Vasiljević, Maja Pokrić, and Vladimir Zlokolica

Learning from Examples to Generalize over Pose and Illumination 643
Marco K. Müller and Rolf P. Würtz

Semi–supervised Learning with Constraints for Multi–view Object Recognition 653
Stefano Melacci, Marco Maggini, and Marco Gori

Large-Scale Real-Time Object Identification Based on Analytic Features 663
Stephan Hasler, Heiko Wersing, Stephan Kirstein, and Edgar Körner

Estimation Method of Motion Fields from Images by Model Inclusive Learning of Neural Networks 673
Yasuaki Kuroe and Hajimu Kawakami

Hybrid Neural Systems for Reduced-Reference Image Quality Assessment 684
Judith Redi, Paolo Gastaldo, and Rodolfo Zunino

Representing Images with χ^2 Distance Based Histograms of SIFT Descriptors 694
Ville Viitaniemi and Jorma Laaksonen

Modelling Image Complexity by Independent Component Analysis, with Application to Content-Based Image Retrieval 704
Jukka Perkiö and Aapo Hyvärinen

Adaptable Neural Networks for Objects' Tracking Re-initialization 715
Anastasios Doulamis

Lattice Independent Component Analysis for fMRI Analysis 725
Manuel Graña, Maite García-Sebastián, and Carmen Hernández

Adaptive Feature Transformation for Image Data from Non-stationary Processes 735
Erik Schaffernicht, Volker Stephan, and Horst-Michael Gross

Bio-inspired Connectionist Architecture for Visual Detection and Refinement of Shapes 745
Pedro L. Sánchez Orellana and Claudio Castellanos Sánchez

Neuro-Evolution and Hybrid Techniques for Mobile Agents Control

Evolving Memory Cell Structures for Sequence Learning 755
Justin Bayer, Daan Wierstra, Julian Togelius, and Jürgen Schmidhuber

Measuring and Optimizing Behavioral Complexity for Evolutionary Reinforcement Learning 765
Faustino J. Gomez, Julian Togelius, and Juergen Schmidhuber

Combining Multiple Inputs in HyperNEAT Mobile Agent Controller 775
Jan Drchal, Ondrej Kapral, Jan Koutník, and Miroslav Šnorek

Evolving Spiking Neural Parameters for Behavioral Sequences 784
Thomas M. Poulsen and Roger K. Moore

Robospike Sensory Processing for a Mobile Robot Using Spiking Neural Networks 794
Michael F. Mc Bride, T.M. McGinnity, and Liam P. Maguire

Neural Control, Planning and Robotics Applications

Basis Decomposition of Motion Trajectories Using Spatio-temporal NMF 804
Sven Hellbach, Julian P. Eggert, Edgar Körner, and Horst-Michael Gross

An Adaptive NN Controller with Second Order SMC-Based NN Weight Update Law for Asymptotic Tracking 815
Haris Psillakis

Optimizing Control by Robustly Feasible Model Predictive Control and Application to Drinking Water Distribution Systems 823
Vu Nam Tran and Mietek A. Brdys

Distributed Control over Networks Using Smoothing Techniques 835
Ion Necoara

Trajectory Tracking of a Nonholonomic Mobile Robot Considering the Actuator Dynamics: Design of a Neural Dynamic Controller Based on Sliding Mode Theory 845
Nardênio A. Martins, Douglas W. Bertol, and Edson R. De Pieri

Tracking with Multiple Prediction Models 855
Chen Zhang and Julian Eggert

Sliding Mode Control for Trajectory Tracking Problem - Performance Evaluation 865
Razvan Solea and Daniela Cernega

Bilinear Adaptive Parameter Estimation in Fuzzy Cognitive Networks 875
Thodoris Kottas, Yiannis Boutalis, and Manolis Christodoulou

Intelligent Tools and Methods for Multimedia Annotation

AM-FM Texture Image Analysis of the Intima and Media Layers of the Carotid Artery 885
Christos P. Loizou, Victor Murray, Marios S. Pattichis, Christodoulos S. Christodoulou, Marios Pantziaris, Andrew Nicolaides, and Constantinos S. Pattichis

Unsupervised Clustering of Clickthrough Data for Automatic Annotation of Multimedia Content 895
Klimis Ntalianis, Anastasios Doulamis, Nicolas Tsapatsoulis, and Nikolaos Doulamis

Object Classification Using the MPEG-7 Visual Descriptors: An Experimental Evaluation Using State of the Art Data Classifiers 905
Nicolas Tsapatsoulis and Zenonas Theodosiou

MuLVAT: A Video Annotation Tool Based on XML-Dictionaries and Shot Clustering 913
Zenonas Theodosiou, Anastasis Kounoudes, Nicolas Tsapatsoulis, and Marios Milis

Multimodal Sparse Features for Object Detection 923
Martin Haker, Thomas Martinetz, and Erhardt Barth

Critical Infrastructure Systems

Multiple Kernel Learning of Environmental Data. Case Study: Analysis and Mapping of Wind Fields 933
Loris Foresti, Devis Tuia, Alexei Pozdnoukhov, and Mikhail Kanevski

Contributor Diagnostics for Anomaly Detection 944
Alexander Borisov, George Runger, and Eugene Tuv

Indoor Localization Using Neural Networks with Location Fingerprints 954
Christos Laoudias, Demetrios G. Eliades, Paul Kemppi, Christos G. Panayiotou, and Marios M. Polycarpou

Distributed Faulty Sensor Detection in Sensor Networks 964
Xuanwen Luo and Ming Dong

Detection of Failures in Civil Structures Using Artificial Neural Networks 976
Zhan Wei Lim, Colin Keng-Yan Tan, Winston Khoon-Guan Seah, and Guan-Hong Tan

Congestion Control in Autonomous Decentralized Networks Based on the Lotka-Volterra Competition Model 986
Pavlos Antoniou and Andreas Pitsillides

Author Index 997

Table of Contents – Part I

Learning Algorithms

Mutual Information Based Initialization of Forward-Backward Search for Feature Selection in Regression Problems 1
Alberto Guillén, Antti Sorjamaa, Gines Rubio, Amaury Lendasse, and Ignacio Rojas

Kernel Learning for Local Learning Based Clustering 10
Hong Zeng and Yiu-ming Cheung

Projective Nonnegative Matrix Factorization with α-Divergence 20
Zhirong Yang and Erkki Oja

Active Generation of Training Examples in Meta-Regression 30
Ricardo B.C. Prudêncio and Teresa B. Ludermir

A Maximum-Likelihood Connectionist Model for Unsupervised Learning over Graphical Domains 40
Edmondo Trentin and Leonardo Rigutini

Local Feature Selection for the Relevance Vector Machine Using Adaptive Kernel Learning 50
Dimitris Tzikas, Aristidis Likas, and Nikolaos Galatsanos

MINLIP: Efficient Learning of Transformation Models 60
Vanya Van Belle, Kristiaan Pelckmans, Johan A.K. Suykens, and Sabine Van Huffel

Efficient Uncertainty Propagation for Reinforcement Learning with Limited Data 70
Alexander Hans and Steffen Udluft

Optimal Training Sequences for Locally Recurrent Neural Networks 80
Krzysztof Patan and Maciej Patan

Statistical Instance-Based Ensemble Pruning for Multi-class Problems 90
Gonzalo Martínez-Muñoz, Daniel Hernández-Lobato, and Alberto Suárez

Robustness of Kernel Based Regression: A Comparison of Iterative Weighting Schemes 100
Kris De Brabanter, Kristiaan Pelckmans, Jos De Brabanter, Michiel Debruyne, Johan A.K. Suykens, Mia Hubert, and Bart De Moor

Mixing Different Search Biases in Evolutionary Learning Algorithms ... 111
Kristina Davoian and Wolfram-M. Lippe

Semi-supervised Learning for Regression with Co-training by Committee ... 121
Mohamed Farouk Abdel Hady, Friedhelm Schwenker, and Günther Palm

An Analysis of Meta-learning Techniques for Ranking Clustering Algorithms Applied to Artificial Data ... 131
Rodrigo G.F. Soares, Teresa B. Ludermir, and Francisco A.T. De Carvalho

Probability-Based Distance Function for Distance-Based Classifiers ... 141
Cezary Dendek and Jacek Mańdziuk

Constrained Learning Vector Quantization or Relaxed k-Separability ... 151
Marek Grochowski and Włodzisław Duch

Minimization of Quadratic Binary Functional with Additive Connection Matrix ... 161
Leonid Litinskii

Mutual Learning with Many Linear Perceptrons: On-Line Learning Theory ... 171
Kazuyuki Hara, Yoichi Nakayama, Seiji Miyoshi, and Masato Okada

Computational Neuroscience

Synchrony State Generation in Artificial Neural Networks with Stochastic Synapses ... 181
Karim El-Laithy and Martin Bogdan

Coexistence of Cell Assemblies and STDP ... 191
Florian Hauser, David Bouchain, and Günther Palm

Controlled and Automatic Processing in Animals and Machines with Application to Autonomous Vehicle Control ... 198
Kevin Gurney, Amir Hussain, Jon Chambers, and Rudwan Abdullah

Multiple Sound Source Localisation in Reverberant Environments Inspired by the Auditory Midbrain ... 208
Jindong Liu, David Perez-Gonzalez, Adrian Rees, Harry Erwin, and Stefan Wermter

A Model of Neuronal Specialization Using Hebbian Policy-Gradient with “Slow” Noise ... 218
Emmanuel Daucé

How Bursts Shape the STDP Curve in the Presence/Absence of GABAergic Inhibition 229
Vassilis Cutsuridis, Stuart Cobb, and Bruce P. Graham

Optimizing Generic Neural Microcircuits through Reward Modulated STDP 239
Prashant Joshi and Jochen Triesch

Calcium Responses Model in Striatum Dependent on Timed Input Sources 249
Takashi Nakano, Junichiro Yoshimoto, Jeff Wickens, and Kenji Doya

Independent Component Analysis Aided Diagnosis of Cuban Spino Cerebellar Ataxia 2 259
Rodolfo V. García, Fernando Rojas, Jesús González, Belén San Román, Olga Valenzuela, Alberto Prieto, Luis Velázquez, and Roberto Rodríguez

Hippocampus, Amygdala and Basal Ganglia Based Navigation Control 267
Ansgar Koene and Tony J. Prescott

A Framework for Simulation and Analysis of Dynamically Organized Distributed Neural Networks 277
Vladyslav Shaposhnyk, Pierre Dutoit, Victor Contreras-Lámus, Stephen Perrig, and Alessandro E.P. Villa

Continuous Attractors of Lotka-Volterra Recurrent Neural Networks 287
Haixian Zhang, Jiali Yu, and Zhang Yi

Learning Complex Population-Coded Sequences 296
Kiran V. Byadarhaly, Mithun Perdoor, Suresh Vasa, Emmanuel Fernandez, and Ali A. Minai

Structural Analysis on STDP Neural Networks Using Complex Network Theory 306
Hideyuki Kato, Tohru Ikeguchi, and Kazuyuki Aihara

Time Coding of Input Strength Is Intrinsic to Synapses with Short Term Plasticity 315
Márton A. Hajnal

Information Processing and Timing Mechanisms in Vision 325
Andrea Guazzini, Pietro Lió, Andrea Passarella, and Marco Conti

Review of Neuron Types in the Retina: Information Models for Neuroengineering 335
German D. Valderrama-Gonzalez, T.M. McGinnity, Liam Maguire, and QingXiang Wu

Brain Electric Microstate and Perception of Simultaneously Audiovisual Presentation 345
Wichian Sittiprapaporn and Jun Soo Kwon

A Model for Neuronal Signal Representation by Stimulus-Dependent Receptive Fields 356
José R.A. Torreão, João L. Fernandes, and Silvia M.C. Victer

Hardware Implementations and Embedded Systems

Area Chip Consumption by a Novel Digital CNN Architecture for Pattern Recognition 363
Emil Raschman and Daniela Ďuračková

Multifold Acceleration of Neural Network Computations Using GPU ... 373
Alexander Guzhva, Sergey Dolenko, and Igor Persiantsev

Training Recurrent Neural Network Using Multistream Extended Kalman Filter on Multicore Processor and Cuda Enabled Graphic Processor Unit 381
Michal Čerňanský

A Non-subtraction Configuration of Self-similitude Architecture for Multiple-Resolution Edge-Filtering CMOS Image Sensor 391
Norihiro Takahashi and Tadashi Shibata

Current-Mode Computation with Noise in a Scalable and Programmable Probabilistic Neural VLSI System 401
Chih-Cheng Lu and H. Chen

Minimising Contrastive Divergence with Dynamic Current Mirrors 410
Chih-Cheng Lu and H. Chen

Spiking Neural Network Self-configuration for Temporal Pattern Recognition Analysis 421
Josep L. Rosselló, Ivan de Paúl, Vincent Canals, and Antoni Morro

Image Recognition in Analog VLSI with On-Chip Learning 429
Gonzalo Carvajal, Waldo Valenzuela, and Miguel Figueroa

Behavior Modeling by Neural Networks 439
Lambert Spaanenburg, Mona Akbarniai Tehrani, Richard Kleihorst, and Peter B.L. Meijer

Statistical Parameter Identification of Analog Integrated Circuit Reverse Models 449
Bruno Apolloni, Simone Bassis, Cristian Mesiano, Salvatore Rinaudo, Angelo Ciccazzo, and Angelo Marotta

A New FGMOST Euclidean Distance Computational Circuit Based on Algebraic Mean of the Input Potentials 459
Cosmin Radu Popa

FPGA Implementation of Support Vector Machines for 3D Object Identification 467
Marta Ruiz-Llata and Mar Yébenes-Calvino

Reconfigurable MAC-Based Architecture for Parallel Hardware Implementation on FPGAs of Artificial Neural Networks Using Fractional Fixed Point Representation 475
Rodrigo Martins da Silva, Nadia Nedjah, and Luiza de Macedo Mourelle

Self Organization

A Two Stage Clustering Method Combining Self-Organizing Maps and Ant K-Means 485
Jefferson R. Souza, Teresa B. Ludermir, and Leandro M. Almeida

Image Theft Detection with Self-Organising Maps 495
Philip Prentis, Mats Sjöberg, Markus Koskela, and Jorma Laaksonen

Improved Kohonen Feature Map Associative Memory with Area Representation for Sequential Analog Patterns 505
Tomonori Shirotori and Yuko Osana

Surface Reconstruction Method Based on a Growing Self-Organizing Map 515
Renata L.M.E. do Rego, Hansenclever F. Bassani, Daniel Filgueiras, and Aluizio F.R. Araujo

Micro-SOM: A Linear-Time Multivariate Microaggregation Algorithm Based on Self-Organizing Maps 525
Agusti Solanas, Arnau Gavalda, and Robert Rallo

Identifying Clusters Using Growing Neural Gas: First Results 536
Riccardo Rizzo and Alfonso Urso

Hierarchical Architecture with Modular Network SOM and Modular Reinforcement Learning 546
Masumi Ishikawa and Kosuke Ueno

Hybrid Systems for River Flood Forecasting Using MLP, SOM and Fuzzy Systems 557
Ivna Valença and Teresa Ludermir

Topographic Mapping of Astronomical Light Curves via a Physically Inspired Probabilistic Model 567
Nikolaos Gianniotis, Peter Tiňo, Steve Spreckley, and Somak Raychaudhury

Generalized Self-Organizing Mixture Autoregressive Model for Modeling Financial Time Series 577
Hujun Yin and He Ni

Self-Organizing Map Simulations Confirm Similarity of Spatial Correlation Structure in Natural Images and Cortical Representations 587
A. Ravishankar Rao and Guillermo Cecchi

Intelligent Control and Adaptive Systems

Height Defuzzification Method on L^∞ Space 598
Takashi Mitsuishi and Yasunari Shidama

An Additive Reinforcement Learning 608
Takeshi Mori and Shin Ishii

Neural Spike Suppression by Adaptive Control of an Unknown Steady State 618
Arūnas Tamaševičius, Elena Tamaševičiūtė, Gytis Mykolaitis, Skaidra Bumelienė, Raimundas Kirvaitis, and Ruedi Stoop

Combined Mechanisms of Internal Model Control and Impedance Control under Force Fields 628
Naoki Tomi, Manabu Gouko, and Koji Ito

Neural Network Control of Unknown Nonlinear Systems with Efficient Transient Performance 638
Elias B. Kosmatopoulos, Diamantis Manolis, and M. Papageorgiou

High-Order Fuzzy Switching Neural Networks: Application to the Tracking Control of a Class of Uncertain SISO Nonlinear Systems 648
Haris E. Psillakis

Neural and Hybrid Architectures

A Guide for the Upper Bound on the Number of Continuous-Valued Hidden Nodes of a Feed-Forward Network 658
Rua-Huan Tsaih and Yat-wah Wan

Comparative Study of the CG and HBF ODEs Used in the Global Minimization of Nonconvex Functions 668
Amit Bhaya, Fernando A. Pazos, and Eugenius Kaszkurewicz

On the Knowledge Organization in Concept Formation: An Exploratory Cognitive Modeling Study 678
Toshihiko Matsuka, Hidehito Honda, Arieta Chouchourelou, and Sachiko Kiyokawa

Dynamics of Incremental Learning by VSF-Network 688
Yoshitsugu Kakemoto and Shinchi Nakasuka

Kernel CMAC with Reduced Memory Complexity 698
Gábor Horváth and Kristóf Gáti

Model Complexity of Neural Networks and Integral Transforms 708
Věra Kůrková

Function Decomposition Network 718
Yevgeniy Bodyanskiy, Sergiy Popov, and Mykola Titov

Improved Storage Capacity in Correlation Matrix Memories Storing Fixed Weight Codes 728
Stephen Hobson and Jim Austin

Multiagent Reinforcement Learning with Spiking and Non-Spiking Agents in the Iterated Prisoner's Dilemma 737
Vassilis Vassiliades, Aristodemos Cleanthous, and Chris Christodoulou

Unsupervised Learning in Reservoir Computing: Modeling Hippocampal Place Cells for Small Mobile Robots 747
Eric A. Antonelo and Benjamin Schrauwen

Switching Hidden Markov Models for Learning of Motion Patterns in Videos 757
Matthias Höffken, Daniel Oberhoff, and Marina Kolesnik

Multimodal Belief Integration by HMM/SVM-Embedded Bayesian Network: Applications to Ambulating PC Operation by Body Motions and Brain Signals 767
Yasuo Matsuyama, Fumiya Matsushima, Youichi Nishida, Takashi Hatakeyama, Nimiko Ochiai, and Shogo Aida

A Neural Network Model of Metaphor Generation with Dynamic Interaction 779
Asuka Terai and Masanori Nakagawa

Almost Random Projection Machine 789
Włodzisław Duch and Tomasz Maszczyk

Optimized Learning Vector Quantization Classifier with an Adaptive Euclidean Distance 799
Renata M.C.R. de Souza and Telmo de M. Silva Filho

Efficient Parametric Adjustment of Fuzzy Inference System Using Error Backpropagation Method 807
Ivan da Silva and Rogerio Flauzino

Neuro-fuzzy Rough Classifier Ensemble 817
Marcin Korytkowski, Robert Nowicki, and Rafał Scherer

Combining Feature Selection and Local Modelling in the KDD Cup 99 Dataset 824
Iago Porto-Díaz, David Martínez-Rego, Amparo Alonso-Betanzos, and Oscar Fontenla-Romero

An Automatic Parameter Adjustment Method of Pulse Coupled Neural Network for Image Segmentation 834
Masato Yonekawa and Hiroaki Kurokawa

Pattern Identification by Committee of Potts Perceptrons 844
Vladimir Kryzhanovsky

Support Vector Machine

Is Primal Better Than Dual 854
Shigeo Abe

A Fast BMU Search for Support Vector Machine 864
Wataru Kasai, Yutaro Tobe, and Osamu Hasegawa

European Option Pricing by Using the Support Vector Regression Approach 874
Panayiotis C. Andreou, Chris Charalambous, and Spiros H. Martzoukos

Learning SVMs from Sloppily Labeled Data 884
Guillaume Stempfel and Liva Ralaivola

The GMM-SVM Supervector Approach for the Recognition of the Emotional Status from Speech 894
Friedhelm Schwenker, Stefan Scherer, Yasmine M. Magdi, and Günther Palm

A Simple Proof of the Convergence of the SMO Algorithm for Linearly Separable Problems 904
Jorge López and José R. Dorronsoro

Spanning SVM Tree for Personalized Transductive Learning 913
Shaoning Pang, Tao Ban, Youki Kadobayashi, and Nik Kasabov

Improving Text Classification Performance with Incremental Background Knowledge 923
Catarina Silva and Bernardete Ribeiro

Empirical Study of the Universum SVM Learning for High-Dimensional Data 932
Vladimir Cherkassky and Wuyang Dai

Relevance Feedback for Content-Based Image Retrieval Using Support Vector Machines and Feature Selection 942
Apostolos Marakakis, Nikolaos Galatsanos, Aristidis Likas, and Andreas Stafylopatis

Recurrent Neural Network

Understanding the Principles of Recursive Neural Networks: A Generative Approach to Tackle Model Complexity 952
Alejandro Chinea

An EM Based Training Algorithm for Recurrent Neural Networks 964
Jan Unkelbach, Sun Yi, and Jürgen Schmidhuber

Modeling D_{st} with Recurrent EM Neural Networks 975
Derrick Takeshi Mirikitani and Lahcen Ouarbya

On the Quantification of Dynamics in Reservoir Computing 985
David Verstraeten and Benjamin Schrauwen

Solving the CLM Problem by Discrete-Time Linear Threshold Recurrent Neural Networks 995
Lei Zhang, Pheng Ann Heng, and Zhang Yi

Scalable Neural Networks for Board Games 1005
Tom Schaul and Jürgen Schmidhuber

Reservoir Size, Spectral Radius and Connectivity in Static Classification Problems 1015
Luís A. Alexandre and Mark J. Embrechts

Author Index 1025

Epileptic Seizure Prediction and the Dimensionality Reduction Problem

André Ventura, João M. Franco, João P. Ramos,
Bruno Direito, and António Dourado

Centro de Informática e Sistemas da Universidade de Coimbra
Department of Informatics Engineering
Pólo II University of Coimbra
P 3030-290 Coimbra Portugal
{anrod,jmfranco,jpramos}@student.dei.uc.pt,
{bmleitao,dourado}@dei.uc.pt
http://www.dei.uc.pt

Abstract. Seizures prediction may substantially improve the quality of life of epileptic patients. Processing EEG signals, by extracting a convenient set of features, is the most promising way to classify the brain state and to predict with some antecedence its evolution to a seizure condition. In this work neural networks are proposed as effective classifiers of brain state among 4 classes: interictal, preictal, ictal and postictal. A two channels set of 26 features is extracted. By correlation analysis and by extracting the principal components, a reduced features space is obtained where, by an appropriate neural network, over 90% successful classifications are achieved, for dataset with several patients from the Freiburg database.

Keywords: Classification, Neural Networks, Feature Selection, PCA, Correlation, Epilepsy, Seizure Prediction, EEG Processing.

1 Introduction

Epilepsy is one of the most common neurological disorders, affecting people of all ages. This disease is characterized by recurrent abnormal electrical discharge of a group of neurons. Seizures, according with their location can produce strange sensations, emotions, convulsions, muscle spasms and loss of consciousness, decreasing social and professional capacities of the patient [1]. About 60 million people in the world have epilepsy [2] and approximately 75% of them can be controlled by medications or curable by surgery. Unfortunately, the remaining 25% do not respond to available therapies and cannot control the disease, which cause a risk of serious injury and an intense feeling of helplessness that has a strong impact on the patient's everyday life [3].

During the past decades, several studies have evidenced that seizures do not begin abruptly but develop several minutes before clinical symptoms which lead to the possibility of their prediction. Therefore, patients can be forewarned to

C. Alippi et al. (Eds.): ICANN 2009, Part II, LNCS 5769, pp. 1–9, 2009.

take timely safety and preventive steps and thus substantially improve their quality of life [4]. With EEG it is possible to record local voltage potentials corresponding to large neural populations. Then the seizure prediction can be summarized into extracting features from the EEG and classify into interictal (normal state), preictal (approach of a crisis), ictal (elapse of a crisis) or postictal (after the crisis) [5]. The great question is to find a good group of features to obtain a classifier with high sensitivity (being able to predict seizures) and high specificity (avoiding false alarms) [6].

When developing classifiers, such as neural networks, one ought to consider several aspects. Besides the architecture of the neural network, the selection of a better subset of features plays an important role. The performance of a classifier actually improves when redundant features are removed from the original set.

Assessing the accuracy of neural networks employing feature selection methods for epilepsy prediction has been the subject of a few studies [7,8,9], but not in a classifier oriented approach.

In the present work different classifiers with different feature selection are compared in order to get insights into the best strategy to built an alarming system.

The organization of the paper is as follows: the next section describes the extraction of the original set of features and the methods used to select a better subset; Section 3 introduces the neural networks used as classifiers; In section 4 detailed information about the experiments and their results are presented and finally in section 5 the conclusions are discussed.

2 Patient's Dataset

The data used in this study was collected from three patients from the Freiburg Center for Data Analysis and Modeling Database [10]. Patients 8 and 19 suffer from a frontal lobe epilepsy and patient 12 from a temporal lobe epilepsy.

The EEG data was acquired using a Neurofile NT digital video EEG system with 128 channels, 256Hz sampling rate, and a 16 bit analogue-to-digital converter.

2.1 Two Channels Features Set

Two channels were considered to build the original set of features, being one located in the epileptic focus. By this way it is possible to consider the dynamics in both channels, hoping that its differences will contribute to the class separability. Applying energy concepts, wavelet transform, nonlinear dynamics, 26 features have been extracted, listed in Table 1. The methods were developed in Matlab and its appropriate toolboxes [11], and other freely available software, such as the nonlinear time series analysis TSTOOL) [12].

Energy analysis is based on the algorithm presented in [13]. EEG signal is processed through two windows with different length to analyze energy patterns: a short-term energy observation window and a long-term energy observation window. The short term window is a later subset of the long term window, aiming

Table 1. The 13 extracted features from EEG to be used in classification of the brain state

Concept	Feature
Nonlinear System Dynamics	Correlation dimension
	Max Lyapunov Exponent
Wavelet Transform coefficient energy	short term energy band (0Hz - 12.5Hz)
	long term energy band (0Hz - 12.5Hz)
	short term energy band (12.5Hz - 25Hz)
	long term energy band (12.5Hz - 25Hz)
	short term energy band (25Hz - 50Hz)
	long term energy band (25Hz - 50Hz)
	short term energy band (50Hz - 100Hz)
	long term energy band (50Hz - 100Hz)
Signal Energy	Energy level
	Energy variation (short term energy)
	Energy variation (long term energy)

to compute a rate of energy growth since one of the major characteristics of a seizure is a dramatic increase in electric energy in EEG signals. The main objective is to observe energy patterns before epileptic seizures, confirming eventually the increase of energy bursts in the periods that precede seizures. A similar displacement is applied to both windows and both end at the same time point.

Wavelet coefficients have been considered in the same approach as the energy signal, to a similar energy analysis, allowing identification of rate variations in the different frequency bands that constitute the EEG signal. The mother wavelet used in the presented study was daubechies-4 and the decomposition was completed with four levels.

Nonlinear analysis faces the EEG as trajectories of a nonlinear system. Two nonlinear dynamic features, the maximum Lyapunov exponent and the correlation dimension, through a sliding window, are computed using [12]. The construction of the attractor, after the determination of the parameters time delay and the embedding dimension, allows the calculation of the maximum Lyapunov exponent and the correlation dimension. The estimation of the maximum Lyapunov exponents consists in the quantification of the exponential growth of the average distance between two nearby trajectories of the attractor. In our case the estimation is performed by error aproximation (TSTOOL). Correlation dimension is determined by takens estimator method [12].

The joint analysis of the extracted features created a 13-dimension space which represents the EEG signal in several components (energy signal, frequency and system dynamics). The objective of the study is to investigate the eventual occurrence of hidden characteristics in data such that clusters can be discovered allowing an acceptable classification of EEG data into 4 classes:

- interictal (normal EEG pattern)
- preictal (two minutes prior to the seizure onset)
- ictal (the seizure onset)
- postictal (two minutes subsequent to seizure end)

The extracted features are sampled with a constant 5s interval. One cycle is composed by series of these 4 classes.

2.2 Feature Selection and Feature Space Reduction

Correlation analysis. Correlation is widely used in statistics as it describes the degree of relationship between two variables.

Analysing the 26x26 correlation matrix, one of those with correlation above 90% has been eliminated. Moreover, the correlation above the threshold between the two features should exist in all patients. By this procedure, a subset of the original features with 18 features has been obtained, listed in table 2.

Table 2. New Set of Features

Extraction Technique	Features	
	Channel 1	Channel 2
Nonlinear Dynamics	Correlation dimension Max Lyapunov Exponent	Correlation dimension Max Lyapunov Exponent
Wavelet Transform	Energy STE 1 Energy STE 2 Energy STE 4 Energy LTE 2 Energy LTE 4	Energy STE 1 Energy STE 3 Energy LTE 1 Energy LTE 2 Energy LTE 3 Energy LTE 4
Signal Energy	Energy LTE	Energy STE Energy LTE

Independent Correlation. Some studies point out that a trained neural network's performance decreases when tested with different patients [7,14,15], leading to the conclusion that for each patient a proper neural network, with a proper set of features, must be engineered. By this perspective, the need for the same number of features for every patient becomes dispensable. Furthermore, as we remove more features highly correlated, the performance of the neural network for that patient may actually improve.

Keeping this in mind, at this step we continued using the correlation, only this time we remove all the features with a correlation above the threshold and regardless of any other patient, hence the use of "independent" in the term.

Principal Component Analysis. Principal Component Analysis (PCA) has been called one of the most valuable results from applied linear algebra. PCA is used abundantly in all forms of analysis, from neuroscience to computer graphics. With minimal additional effort PCA provides a roadmap for how to reduce a complex data set to a lower dimension to reveal the sometimes hidden, simplified structure that often underlie it. A more detailed explanation about PCA can be found in [16].

To proceed with this method we used Matlab (princomp), to obtain the principal components coefficients and their variances. To project into a new dimension we considered the coefficients which retain 99% of the variance. With this, we achieved a reduction of about 50% in the number of original features, and were able to keep 13 features from patient 8; also 13 features from patient 12; and finally 11 features from patient 19.

Once more, like independent correlation, by using PCA, which may return different subsets of features for each patient, we are not interested in applying the same neural network to different patients.

3 Neural Networks (NN)

In the present work three neural networks (Feed-Forward Backpropagation Network, Layered-Recurrent Network and Radial Basis Network) have been selected. Although we have tested for many others types of NN, these three presented better results in a computational time considered reasonable.

3.1 Feed-Forward Backpropagation Networks (FF) / Layered-Recurrent Networks (LRN)

These neural networks consist of two layers' network, with the hidden layer composed by ten neurons with *logsig* or *tansig* as the transfer function and the output layer composed by four linear neurons. LRN also have a feedback loop around the hidden layer, providing a single delay to the network. Gradient descent weight and bias and Levenberg-Marquardt backpropagation were used as learning and training functions, respectively.

3.2 Radial Basis Function (RBF)

Radial Basis Networks are two layers networks, with the first layer composed by radial basis transfer functions and the second layer with linear neurons. Radial Basis Network adds neurons to the hidden layer until it meets the mean squared error goal. The spread constant used was 1.0.

4 Experiments

4.1 Evaluation

In order to compare across the different methods and cases, three performance criteria were used: (1) accuracy (the closeness by which a set of measurements approaches the true value), (2) sensitivity (the ability to classify positive cases (preictal)) and (3) specificity (the ability to classify negative cases (non-preictal)). A high sensibility and a high specificity are required to be considered as useful in a clinical environment. The datasets were previously classified by a neurologist into the four stages of an epileptic cycle.

$$Accuracy(\%) = \frac{CorrectCases}{TotalCases} \times 100 \quad (1)$$

$$Sensitivity(\%) = \frac{TruePositives}{TruePositives + FalseNegatives} \times 100 \quad (2)$$

$$Specificity(\%) = \frac{TrueNegatives}{TrueNegatives + FalsePositives} \times 100 \quad (3)$$

Table 3. Results Table, *LogSig* as Transfer Function SS - Sensitivity, SP - Specificity, AC - Accuracy

Trained	Tested	FF		
		SS	SP	AC
Pat 8	Pat 12	0	100	42,5
	Pat 19	0	100	73,5
Pat 12	Pat 8	0	90,9	25,6
	Pat 19	0	94,1	9,9
Pat 19	Pat 8	0	83,5	51,5
	Pat 12	50,4	15,9	11

Table 4. Results Table, *LogSig* as Transfer Function SS - Sensitivity, SP - Specificity, AC - Accuracy PCA - Principal Component Analysis None - The original Dataset ICorr - Independent Correlation

		FF			LRN		
		SS	SP	AC	SS	SP	AC
Correlation	Pat 8	93,8	99,8	96,9	93,8	98,8	96,5
	Pat 12	92,5	99,0	96,7	92,5	99,5	97,6
	Pat 19	87,5	98,6	96,5	92,5	99,5	96,7
PCA	Pat 8	90,6	99,8	95,6	100	97,4	94,1
	Pat 12	97,5	99,3	97,1	100	99,5	97,4
	Pat 19	92,5	98,3	95,4	90,0	97,6	95,6
None	Pat 8	87,5	96,7	94,7	96,9	99,8	95,6
	Pat 12	92,5	99,8	96,9	90,0	99,5	96,9
	Pat 19	95,0	98,8	97,1	92,5	99,3	96,9
ICorr	Pat 8	96,9	98,1	91,6	84,4	98,3	93,8
	Pat 12	100	98,1	94,5	97,5	98,8	95,2
	Pat 19	45,0	99,3	92,1	47,5	98,3	92,5

4.2 Results

Several studies have shown that each patient should have his/her own epileptic prediction mechanism, each neural network is tested with a single patient, the one whose dataset was applied for training. Table 3 demonstrates the decrease of performance results, when the neural network is tested with different patients's dataset of those the network was trained.

To build the train and test set the 70% empirical rule was followed. This rule states that 70% of the dataset should be used as a train set and the rest (30%) as a test set. To carry out this rule, from each subset of 3 entries, 2 were taken as train data and the other one as test data.

The results obtained using Feed-Forward Backpropagation Network (FF) and Layered-Recurrent Network (LRN) are shown in tables 4 and 5. For each table we tested with different transfer functions: *logsig* for table 4 and *tansig* for table 5. The results are very similar, but some improvement is noticed in table 4. In table 6 we used Radial Basis Function as neural network. The results presented in this table are worse than the results obtained with other neural networks (table 4 and 5).

Table 5. Results Table, *Tansig* as Transfer Function SS - Sensitivity, SP - Specificity, AC - Accuracy PCA - Principal Component Analysis None - The original Dataset ICorr - Independent Correlation

		FF			LRN		
		SS	SP	AC	SS	SP	AC
Correlation	Pat 8	87,5	99,3	96,5	81,3	99,3	95,6
	Pat 12	100,0	99,3	98,2	92,5	98,8	96,9
	Pat 19	97,5	97,8	96,7	95,0	98,6	97,1
PCA	Pat 8	93,8	99,3	96,9	90,6	100	97,1
	Pat 12	97,5	99,5	97,1	100	99	97,4
	Pat 19	82,5	99	96,9	90	99,3	97,4
None	Pat 8	87,5	98,1	94,5	87,5	99,8	95,6
	Pat 12	100	99,3	98	95	99,8	97,8
	Pat 19	77,5	98,3	93,8	85	98,1	94,9
ICorr	Pat 8	87,5	97,9	91,9	46,9	99,1	90,5
	Pat 12	97,5	98,8	95,8	97,5	99,3	95,8
	Pat 19	47,5	99	93	75	97,3	93

Table 6. RBF Results Table SS - Sensitivity, SP - Specificity, AC - Accuracy PCA - Principal Component Analysis None - The original Dataset ICorr - Independent Correlation

		RBF		
		SS	SP	AC
Correlation	Pat 8	90,6	92,4	85,7
	Pat 12	62,5	99,0	73,0
	Pat 19	62,5	61,7	45,1
PCA	Pat 8	78,1	90,3	83,1
	Pat 12	77,5	96,1	84,0
	Pat 19	60,0	70,1	62,2
None	Pat 8	90,6	92,2	88,6
	Pat 12	72,5	99,8	86,2
	Pat 19	55	75,9	68,8
ICorr	Pat 8	78,1	91,7	82,4
	Pat 12	60	94	78,9
	Pat 19	35	65,1	53,2

5 Conclusions

The quality of a set of features used in a neural network has a direct effect on the neural network's performance. A set of features highly correlated between them may lead the neural network to a local minimum error during the training epochs and, consequently, to a poor performance of the network. Furthermore, as the number of features increases, the computational cost of using a neural network also augments. Therefore, an optimal subset of features should be extracted before applying them to a neural network.

In this paper different approaches were studied to reduce the dimensionality of the original set of features extracted from an EEG signal. The first approach (hereafter "correlation") considers the correlation between features of each patient and removes it if the same feature has a correlation above 90% in every patient. This experience may be useful when applying the same classifier mechanism to all patients. However, some studies have revealed that each patient should have his/her own classifier. Our second approach (hereafter "independent correlation") follows the latter idea by removing the feature in each patient regardless the other patient's features' correlation. The last approach uses the Principal Components Analysis (hereafter "PCA") technique selecting the vectors that retain 99% of the total variance between the data.

The experiments show very similar results between approaches. There is a slight degradation of results using the "independent correlation". This degradation may be caused by the loss of information that had been removed along the reduction process, albeit the fact that the same feature has a correlation above 90% with another. The "correlation" has some good results. However, the problem of the wrongly removed features, mentioned before, may arise once more. Moreover, the study involved in the process of feature selection between all the patients increases with the number of patients, becoming unattainable at a certain point. Finally, PCA exhibited the best results in the experiment, proving to be a good method for the selection of a subset of features for each individual.

Several approaches have been studied to overcome the problem of feature extraction/selection and it is difficult to say which one should be used. Therefore, further work comparing different methods of feature extraction/selection, like Independent Component Analysis, Kruskal-Wallis, Genetic Algorithms , multidimensional scaling, or others , should be considered.

Acknowledgments. This research is supported by European Union FP7 Framework Program, EPILEPSIAE Grant 211713 and by CISUC (FCT Unit 326). The authors express their gratitude to Freiburg Center for Data Analysis and Modeling (FDM) of Albert Ludwigs University of Freiburg, for the access to the epilepsy database (`http://www.fdm.uni-freiburg.de/groups/timeseries/epi/EEGData/download/infos.txt`).

References

1. National Institute of Neurological Disorders and Stroke, `http://www.ninds.nih.gov/disorders/epilepsy/epilepsy.htm`
2. Wyllie, E.: Epilepsy - Information for You and Those Who Care About You. Cleveland Press (2008)
3. People against childhood epilepsy, `http://www.paceusa.org/indexframe5.html`
4. Swiderski, B., Osowski, S., Cichocki, A., Rysz, A.: Epileptic Seizure Prediction Using Lyapunov Exponents and Support Vector Machine. In: Beliczynski, B., Dzielinski, A., Iwanowski, M., Ribeiro, B. (eds.) ICANNGA 2007. LNCS, vol. 4432, pp. 373–381. Springer, Heidelberg (2007)
5. Direito, B., Dourado, A., Vieira, M., Sales, M.: Combining energy and wavelet transform for epilectic seizure prediction in an advanced computational system. In: Proc. of the 2008 International Conference on Biomedical Engineering and Informatics BMEI2008, vol. 2, pp. 380–385, International Conference on Biomedical Engineering and informatics, Hainan, PR China (2008)

6. Mormann, F., Kreuz, T., Rieke, C., Andrzejak, R.G., Kraskov, A., David, P., Elger, C.E., Lehnertz, K.: On the predictability of epilectic seizures. Clinical Neurophysiology 116(3), 569–587 (2005)
7. Jerger, K.K., Netoff, T.I., Francis, J.T., Sauer, T., Pecora, L., Weinstein, S.L., Schiff, S.J.: Early Seizure Detection. Journal of Cinical Neurophysiology 18, 259–269 (2001)
8. Ahmad, T., Fairuz, R.A., Zakaria, F., Isa, H.: Selection of a Subset of EEG Channels of Epileptic Patient During Seizure Using PCA. In: Mathematics and Computers in Science, Proceedings of the 7th WSEAS International Conference on Signal Processing, Robotics and Automation, Cambridge, UK, pp. 270–273 (2008) ISBN ISSN: 1790-5117, 978-960-6766-44-2
9. Hu, S., Stead, M., Worrel, G.: Removal of Scalp Reference Signal and Line Noise for Intracranial EEGs. In: IEEE International Conference on Networking, Sensing and Control, Sanya, China, pp. 1486–1491 (2008) ISBN: 978-1-4244-1685-1
10. Schelter, B., Winterhalder, M., Maiwald, T., Brandt, A., Schad, A., Timmer, J., Schulze-Bonhage, A.: Do false predictions of seizures depend on the state of vigilance? A report from two seizure prediction methods and proposed remedies. Epilepsia 47(12), 2058–2070 (2006)
11. The Mathworks, Inc. Natick, MA
12. Merkwirth, C., Parlitz, U., Wedekind, I., Lauterborn, W.: TSTOOL User Manual, Version 1.11., `http://www.dpi.physik.uni-goettingen.de/tstool/HTML/index.html`
13. Esteller, R., Echauz, J., DAlessandro, M., Worrell, G., et al.: Continuous Energy Variation During the Seizure Cycle: Towards an On-line Accumulated Energy. Clinical Neurophysiology 116(3), 517–526 (2005)
14. Mormann, F., Andrzejak, R.G., Elger, C.E., Lehnertz, K.: Seizure prediction: the long and winding road. Brain 130(2), 314–333 (2007)
15. D'Alessandro, M., Vachtsevanos, G., Hinson, A., Esteller, R., Echauz, J., Litt, B.: A Genetic Approach to Selecting the Optimal Feature for Epileptic Seizure Prediction. In: IEEE International Conference on Engineering, Medicine and Biology, vol. 2, pp. 1703–1706 (2001) ISSN: 1094-687X
16. Schlens, J.: A Tutorial on Principal Component Analysis, `http://www.cs.cmu.edu/~elaw/papers/pca.pdf`

Discovering Diagnostic Gene Targets and Early Diagnosis of Acute GVHD Using Methods of Computational Intelligence over Gene Expression Data

Maurizio Fiasché[1], Anju Verma[2], Maria Cuzzola[3], Pasquale Iacopino[3], Nikola Kasabov[2], and Francesco C. Morabito[1]

[1] University "Mediterranea" of Reggio Calabria, DIMET
Via Graziella Feo di Vito, I-89100 Reggio Calabria, Italy
{maurizio.fiasche,morabito}@unirc.it

[2] KEDRI, Auckland University of Technology
350 Queen Street, Auckland, New Zealand
{averma,nkasabov}@aut.ac.nz

[3] Regional Center of Stem Cells and Cellular Therapy, "A. Neri"
Via Petrara,11 , 89100 Reggio Calabria, Italy
{everest1@libero.it,iacopinoctmo@tin.it}

Abstract. This is an application paper of applying standard methods of computational intelligence to identify gene diagnostic targets and to use them for a successful diagnosis of a medical problem - acute graft-versus-host disease (aGVHD). This is the major complication after allogeneic haematopoietic stem cell transplantation (HSCT) in which functional immune cells of donor recognize the recipient as "foreign" and mount an immunologic attack. In this paper we analyzed gene-expression profiles of 47 genes associated with allo-reactivity in 59 patients submitted to HSCT. We have applied 2 feature selection algorithms combined with 2 different classifiers to detect the aGVHD at on-set of clinical signs. This is a preliminary study and the first paper which tackles both computational and biological evidence for the involvement of a limited number of genes for diagnosis of aGVHD. Directions for further studies are outlined.

Keywords: Neural Networks, Feature Selection, GEP, GVHD, Gene selection, Machine Learning.

1 Introduction

With the completion of the first draft of the human genome the task is now to be able to process this vast amount of ever growing dynamic information and to create intelligent systems for detection, prediction and knowledge discoveries about human pathology and disease. When genes are in action, the dynamics of the processes in which a single gene is involved are very complex, as this gene interacts with many other genes and mediators, and is influenced by many environmental factors. The genes in an individual may mutate, change slightly their code, and

C. Alippi et al. (Eds.): ICANN 2009, Part II, LNCS 5769, pp. 10–19, 2009.

may therefore express differently at a next time. Modeling these events, learning about them and extracting knowledge, is major goal for bioinformatics [1,2].The branch of information sciences for the analysis, modeling and knowledge discovery of biological phenomenons such as genetic processes is bioinformatics. The potential applications of microarray technology are numerous and include identifying markers for classification, diagnosis, disease outcome prediction, target identification and therapeutic responsiveness [1,2]. However microarray analysis might not identify unique markers (e.g. a single gene) of clinical utility for same diseases. Indeed, diagnosis and prediction of the biological state/disease is likely to be more accurate by identifying clusters of gene expression profiles (GEPs) performed by macroarray analysis. Based on profile, it 's possible to set a diagnostic test, so a sample can be taken from a patient, the data related to the sample processed, and a profile related to the sample obtained [2]. This profile can be matched against existing gene profiles and based on similarity, it can be confirmed with a certain probability a diagnosis of disease or if the patient is at risk of developing it in the future. We apply this approach to detect acute graft-versus-host disease (aGVHD) in allogeneic hematopoietic stem cell transplantation (HSCT), a curative therapy for several malignant and non malignant disorders [3]. Acute GVHD remains the major complication and the principal cause of mortality and morbility following HSCT [4,5]. At present, the diagnosis of aGVHD is merely based on clinical criteria and may be confirmed by biopsy of one of the 3 target organs (skin, gastrointestinal tract, or liver) [6]. The severity of aGVHD is graded clinically from I to IV using a standardized system, with increased mortality rates associated with significant aGVHD (grades II-IV), [7]. There is no definitive diagnostic blood test for aGVHD, although a lot of blood proteins have been described as potential biomarkers in small studies [8,9]. A recent report indicates a preliminary molecular signature of aGVHD in allogeneic HSCT patients [10]. In the current project, our primary objective was to validate a novel and not invasive method to confirm the diagnosis of aGVHD in HSCT patients at onset of clinical symptoms. For this purpose, a database has been built by pre-processing experimental measures, and features were selected to enable a good class separation without using all features and facing the "curse of dimensionality" problem, i.e., an excessive number of training inputs that increases the system complexity without remarkable advantages in terms of prediction performances. This problem can be considered as a typical inverse problem of pattern classification starting from experimental database. The proposed approach, exploits a Correlation-based Feature Selection (CFS) algorithm combined with a neural network (ANN) classifier and also a wrapper method combined with the Naive Bayesian classifier to select the most important features (genes) for the diagnosis. This is the first paper which discusses both computational and biological evidence to confirm the early statement of aGVHD based on selected genetic diagnostic markers. The organization of the rest of the paper is as follows: in section 2 the data analyzed and feature subset selection techniques in order to reduce the number of variables are described; in section 3 the Neural Network classifier is described; finally, in section 4 and 5 the results of the diagnostic method are discussed and conclusions are inferred with some possible future applications.

2 Methodology

In this paper we consider two general approaches to feature subset selection, more specifically, wrapper and filter approaches, for gene selection. Wrappers and filters differ in the evaluation of feature subsets. Filter approaches remove irrelevant features according to general characteristics of the data. Wrapper approaches, by contrast, apply machine learning algorithms to feature subsets and use cross-validation to evaluate the score of feature subsets. In theory, wrappers should provide more accurate classification results than filters [11]. Wrappers use classifiers to estimate the usefulness of feature subsets. The use of "tailor-made" feature subsets should provide a better classification accuracy for the corresponding classifiers, since the features are selected according to their contribution to the classification accuracy of the classifiers. The disadvantage of the wrapper approach is its computational requirement when combined with sophisticated algorithms such as support vector machines. As a filter approach, CFS was proposed by Hall [12]. The rationale behind this algorithm is "a good feature subset is one that contains features highly correlated with the class, yet uncorrelated with each other". It has been shown in Hall [12] that CFS gave comparable results to the wrapper and executes many times faster. It will be shown later in this paper that combining CFS with a suitable ANN, provides a good classification accuracy for diagnosis of aGVHD.

2.1 Experimental Data

Fifty-nine HSCT patients were enrolled in our study between March 2006 and July 2008 in Transplants Regional Center of Stem Cells and Cellular Therapy "A. Neri" Reggio Calabria, Italy, during a Governative Research Program: "*Project of Integrated Program: Allogeneic Hemopoietic Stem Cells Transplantation in Malignant Hemopathy and Solid Neoplasia Therapy - Predictive and prognostic value for graft vs. host disease of chimerism and gene expression*". Because experimental design plays a crucial role in a successful biomarker search, the first step in our design was to choose the most informative specimens and achieve adequate matching between positive cases aGVHD (YES) and negative controls aGVHD (NO) to avoid bias. This goal is best achieved through a database containing high-quality samples linked to quality controlled clinical information. Patients with clinical signs of aGVHD (YES) were selected, and in more than 95% of them aGvHD was confirmed by biopsy including those with grade I. We used 26 samples from aGVHD (YES) patients that were taken at the time of diagnosis and we selected 33 samples from patients that didn't experienced aGVHD (NO). All together YES/NO patient groups comprised a validation set. Total RNA was extracted from whole peripheral blood samples using a RNA easy Mini Kit (Qiagen) according to the manufacturer's instructions. Reverse transcription of the purified RNA was performed using Superscript III Reverse Transcriptase (Invitrogen). A multigene expression assay to test occurrence of aGVHD were carried out with TaqMan® Low Density Array Fluidic (LDA-macroarray card) based on Applied Biosystems 7900HT comparative dd CT method, according to manufacturer's instructions. Expression of each gene was measured in triplicate and then normalized to the reference gene 18S mRNA, who was included in macroarray card. About the project of macroarray card, we selected 47 candidate genes from the published literature, genomic databases, pathway analysis. The 47 candidate genes were involved in immune network and inflammation pathogenesis.

2.2 Feature Subset Selection

Feature Selection is a technique used in machine learning of selecting a subset of relevant features to build robust learning models. The assumption here is that not all genes measured by a macroarray method are related to aGVHD classification. Some genes are irrelevant and some are redundant from the machine learning point of view [13]. It is well-known that the inclusion of irrelevant and redundant information may harm performance of some machine learning algorithms. Feature subset selection can be seen as a search through the space of feature subsets. CFS evaluates a subset of features by considering the individual detector ability of each feature along with the degree of redundancy between them.

$$CFS_s = \frac{k \cdot \overline{r}_{cf}}{\sqrt{k + k \cdot (k-1) \cdot \overline{r}_{ff}}} \tag{1}$$

where:

- CFS_S is the score of a feature subset S containing k features
- $\overline{r}_{cf}$ is the average feature to class correlation ($f \in S$)
- $\overline{r}_{ff}$ is the average feature to feature correlation

The distinction between normal filter algorithms and CFS is that while normal filters provide scores for each feature independently, CFS presents a heuristic "merit" of a feature subset and reports the best subset it finds. To select the genes with CFS, we have:

a. Choose a search algorithm,
b. Perform the search, keeping track of the best subset encountered according to CFS_S,
c. Output the best subset encountered.

The search algorithm we used was best-first with forward selection, which starts with the empty set of genes. The search for the best subset is based on the training data only. Once the best subset has been determined, and a classifier

Table 1. The 13 genes selected from CFS and their name and meaning

Gene Name	Official full name	Immune function
BCL2A1	BCL2-related protein A1	Anti- and pro-apoptotic regulator.
CASP1	Caspase 1, apoptosis-related cysteine peptidase	Central role in the execution-phase of cell apoptosis.
CCL7	chemokine (C-C motif) ligand 7	Substrate of matrix metalloproteinase 2
CD83	CD83 molecule	Dendritic cells regulation.
CXCL10	chemokine (C-X-C motif) ligand 10	Pleiotropic effects, including stimulation of monocytes, natural killer and T-cell migration, and modulation of adhesion molecule expression.
EGR2	Early growth response 2	transcription factor with three tandem C2H2-type zinc fingers.
FAS	TNF receptor superfamily, member 6)	Central role in the physiological regulation of programmed cell death.
ICOS	Inducible T-cell co-stimulator	Plays an important role in cell-cell signaling, immune responses, and regulation of cell proliferation.
IL4	Interleukin 4	Immune regulation.
IL10	Interleukin 10	Immune regulation.
SELP	selectin P	Correlation with endothelial cells.
SLPI	Stomatin (EPB72)-like 1	Elemental activities such as catalysis.
STAT6	transducer and activator of transcription 6, interleukin-4 induced	Regulation of IL4- mediated biological responses.

has been built from the training data (reduced to the best features found), the performance of that classifier is evaluated on the test data. The 13 genes selected by CFS are reported in Table 1. A leave-one-out cross validation procedure was performed to investigate the robustness of the feature selection procedures. In 29 runs, the subset of 13 genes was selected 28 times (96%) by CFS. Now it is possible to use a classifier to estimate the usefulness of feature subset.

2.3 Wrapper Method

While CFS assigns a score to subset of features, wrapper approaches take biases of machine learning algorithms into account when selecting features. *Wrapper* apply a machine learning algorithm to feature subsets and use cross-validation to compute a score for them. In general, filters are much faster than wrappers. However, as far as the final classification accuracy is concerned, wrappers normally provide better results. The general argument is that the classifier that will be built from the feature subset should provide a better estimate of accuracy than other methods. The main disadvantage of wrapper approaches is that during the feature selection process, the classifier must be repeatedly called to evaluate a subset. For some computationally expensive algorithms such as SVMs or artificial neural networks, wrappers is very heavy. To select the genes using a wrapper method, we have to:

(a) Choose a machine learning algorithm to evaluate the score of a feature subset.
(b) Choose a search algorithm.
(c) Perform the search, keeping track of the best subset encountered.
(d) Ouput the best subset encountered.

As machine learning algorithm we used simple Bayesian classifier naïve Bayes, it assumes that features are independent given the class. Its performance on data sets with redundant features can be improved by removing such features. A forward search strategy is normally used with naïve Bayes as it should immediately detect dependencies when harmful redundant features are added.

Also here the search algorithm was best-first with forward selection, starting with the empty set of genes. We reported accuracy estimates for classifiers built from the best subset found during the search. The search for the best subset is based on the training data only. Once the best subset has been determined, and a classifier has been built from the training data (reduced to the best features found), the performance of that classifier is evaluated on the test data. Genes selected from wrapper method are in table 2. Most of the genes selected are

Table 2. The 7 genes selected from wrapper and their name and meaning

Gene Name	Official full name	Immune function
CASP1	Caspase 1, apoptosis-related cysteine peptidase	Central role in the execution-phase of cell apoptosis.
EGR2	Early growth response 2	transcription factor with three tandem C2H2-type zinc fingers.
CD52	CD52 antigen	B-cell activation.
SLPI	Stomatin (EPB72)-like 1	Elemental activities such as catalysis.
ICOS	Inducible T-cell co-stimulator	Plays an important role in cell-cell signaling, immune responses, and regulation of cell proliferation.
IL10	Interleukin 10	Immune regulation.
CXCL10	chemokine (C-X-C motif) ligand 10	Pleiotropic effects, including stimulation of monocytes, natural killer and T-cell migration, and modulation of adhesion molecule expression.

also part of the 13 genes selected using the CFS method and the only two genes that are different are actually correlated to other genes from the set of 13 genes. A leave-one-out cross validation procedure was performed to investigate the robustness of the method over the training set: in 29 runs, the subset of 7 genes was selected 26 times (90%) by the naïve Bayes wrapper. In section 4 it has been show the performance of this technique estimated on the testing data.

3 Neural Network Model for Early Diagnosis Using the Selected Gene Diagnostic Markers

Artificial neural networks (ANNs) are commonly known as biologically inspired, highly sophisticated analytical techniques, capable of modeling extremely complex non-linear functions. Formally defined, ANNs are analytic techniques modeled after the processes of learning in the cognitive system and the neurological functions of the brain and capable of predicting new observations (on specific variables) from other observations (on the same or other variables) after executing a process of so-called learning from existing data [14]. Here we have used the selected 13 genes via the filtering CFS method and a popular ANN architecture called MLP with back-propagation (a supervised learning algorithm). The MLP is known to be a robust function approximator for prediction/classification problems. The training data set had 29 patient samples (13 aGVHD(Yes) and 16 aGVHD(No)). The test data set consisted of 30 patient samples (13 aGVHD(Yes) and 17 aGVHD(No)). After the step of test, final results has been obtained according Fig. 2. The ANN's outputs were:

- 0, if aGVHD diagnosis was Yes;
- 1, if aGVHD diagnosis was No.

The ANN based system was trained with adaptive rate of learning during a period of 500 epochs. The ANN, according to a consequence of the Kolmogorov's theorem [15], has a hidden layer with 27 neurons; activation functions are: tan-sigmoid between input and hidden layer, and pure linear between hidden and output layer (Fig. 1). After the training phase, the ANN has been tested; final results are shown in section 4.

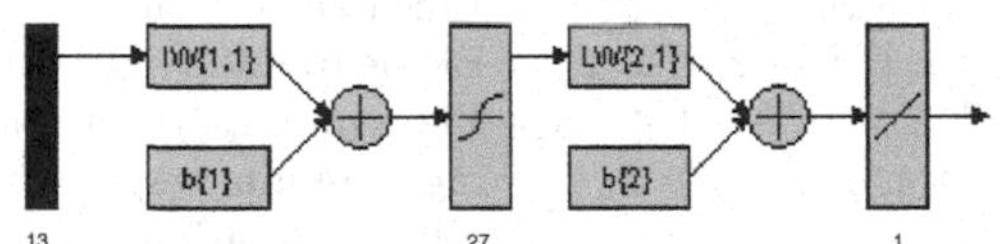

Fig. 1. Structure of the implemented ANN: $b\{1\}$ and $b\{2\}$ represent the biases of input and hidden layers respectively; $IW\{1,1\}$ and $LW\{2,1\}$ represent the weights for the input and the hidden layers respectively. 13 is the number of ANN-inputs (i.e., neurons in the input layer), 27 is the number of hidden neurons and 1 is the number of the ANN-outputs (i.e., output neuron).

4 Results

In this ANN, the 13 genes were inputs and the evaluation of syndrome was output. We have explored different kinds of ANN [16], have compared them to

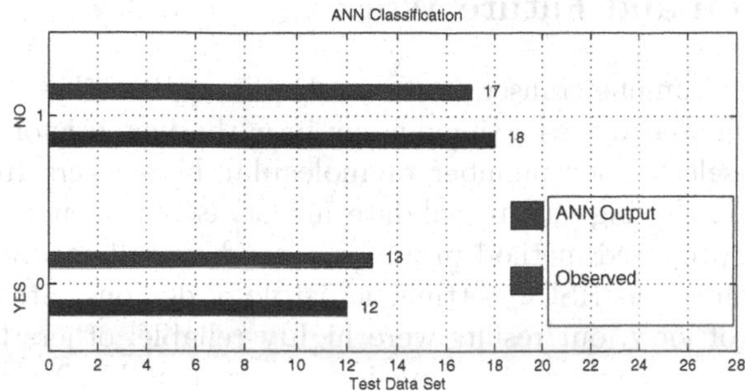

Fig. 2. Observed classes of patients and results obtained by ANN

Table 3. Experimental results of a CFS with ANN classifier and a wrapper method combined with the nave Bayesian classifier. The starting set has been divided in training set and test set, a leave one-out cross-validation has been calculated for the two subsets.

Method	Training set	Test set
CFS-ANN	28(29)	29(30)
Wrapper-nave Bayes	26(29)	29(30)

improve results and our experimental runs also proved the notion that for this type of classification problems MLP performs better than other ANN architectures such as radial basis function (RBF), recurrent neural network (RNN), and self-organizing map (SOM). The final obtained results were good, and tell us that it was possible to diagnose the aGVHD using a restrict number of variables. Only 1 case escaped our classification model (Fig. 2), which achieves 96% accuracy in a leave one-out cross-validation on the training set and 97% on the test data set. In this section we want to report also the results of the classification system based on Wrapper and naïve Bayesian approach and to compare it with the ANN classification results. In table 3 it's showed that Wrapper approach is less robust of CFS approach with an accuracy of 90% on the training data set, but for the classification aim, Wrapper with naïve Bayes classification technique gave the same results of ANN, 97% of accuracy. In patients with aGVHD (YES), level expression of immune gene pattern showed a different behaviour: BCL2A1, CASP1, CCL7, CD83 were up-expressed than reference normal value (it's assumed to be = 1). For these genes it's very important to establish the cut-off expression value correlating with event. In contrast, CXCL10, EGR2, FAS, ICOS, IL-4, IL-10, SELP, SLP1, STAT6 was always down-regulated during aGvHD and before of pharmacological treatment. When clinical manifestation was resulted, expression level of all significant genes was strongly increased. In aGVHD (NO) group, for all genes the transcriptosome expression showed a very high value.

5 Conclusion and Future Work

We examined the immune transcripts to study the applicability of gene expression profiling (macroarray) as a single assay in early diagnosis of aGVHD. Our interest was to select a low number of molecular biomarkers from an initial gene panel and exploiting this to validate a fast, easy and non-invasive diagnostic tool. The proposed method provides a good overall accuracy to confirm aGVHD development in HSCT setting, as istology demonstrated. Concerning biological point of view, our results were highly reliable: others have reasoned that Th2 cell therapy could rapidly ameliorate severe aGVHD via IL-4 and IL-10 mediated mechanisms [17]. It's noteworthy that in our study a set of genes, indicated by computational analysis, included same mediators of Th2 response such as IL10, and signal transducer and activator of transcription 6, interleukin-4 induced (STAT6). All these were strongly down-regulated in aGVHD (YES) setting suggesting an absent control mediated by Th2 cells. Therefore, we highlight the fact that defective expression of ICOS impaired the immune protective effectors during clinical aGVHD. This evidence is in according to previous reported about ICOS as regulatory molecule for T cell responses during aGVHD. It has been showed that ICOS signal inhibits aGVHD development mediated by CD8 positive effector cells in HSCT [18]. According to previous reports, mediators of apoptosis cells and dendritic cell activators were involved.All together our results strongly outlined the importance and utility of non-invasive tool for aGVHD diagnosis based on GEP. We believe that to achieve an advantage from GEP performance, it's very important known: **a)** the transcript levels of immune effector cells in early time post-engraftment to better understand polarization of Th2 cell, **b)** the CD8 positive cell action. In conclusion, in current practice, tissue biopsies are performed to confirm this diagnosis and our molecular tool may obviate the need for an invasive procedure.This study demonstrated, for the first time, that with the use of our computational intelligence approach to select gene diagnostic targets and use them for an early diagnosis of aGVHD with 97% accuracy in the test data set of HSCT population. We plan to extend the system as a personalized model to capture peculiarity of patients through an optimization method [19,20]. A further approach to feature selection and model creation is the so called integrated approach [21], where features and model parameters are optimised together for a better accuracy of the model, which is an extension of the wrapper approach. The authors are engaged in this direction.

References

1. Kasabov, N.: Evolving Connectionist Systems: The Knowledge Engineering Approach, 2nd edn. Springer, London (2007)
2. Kasabov, N., Sidorov, I.A., Dimitrov, D.S.: Computational Intelligence, Bioinformatics and Computational Biology: A Brief Overview of Methods, Problems and Perspectives. J. Comp. and Theor. Nanosc. 2(4), 473–491 (2005)
3. Appelbaum, F.R.: Haematopoietic cell transplantation as immunotherapy. Nature 411, 385–389 (2001)

4. Weisdorf, D.: Graft vs. Host disease: pathology, prophylaxis and therapy: GVHD overview, Best Pr. & Res. Cl. Haematology 21(2), 99–100 (2008)
5. Lewalle, P., Rouas, R., Martiat, P.: Allogeneic hematopoietic stem cell transplantation for malignant disease: How to prevent graft-versus-host disease without jeopardizing the graft-versus-tumor effect? Drug Discovery Today: Therapeutic Strategies — Immunological disorders and autoimmunity 3(1) (2006)
6. Ferrara, J.L.: Advances in the clinical management of GVHD, Best Pr. & Res. Cl. Haematology 21(4), 677–682 (2008)
7. Przepiorka, D., Weisdorf, D., Martin, P.: Consensus Conference on acute GVHD grading. Bone Marrow Transplanation 15, 825–828 (1995)
8. Paczesny, S., Levine, J.E., Braun, T.M., Ferrara, J.L.: Plasma biomarkers in Graft-versus-Host Disease: a new era? Biology of Blood and Marrow Transplantation 15, 33–38 (2009)
9. Paczesny, S., Oleg, I.K., Thomas, M.: A biomarker panel for acute graft-versus-host disease. Blood 113, 273–278 (2009)
10. Buzzeo, M.P., Yang, J., Casella, G., Reddy, V.: A preliminary gene expression profile of acute graft-versus-host disease. Cell Transplantation 17(5), 489–494 (2008)
11. Langley, P.: Selection of relevant features in machine learning. In: Proceedings of AAAI Fall Symposium on Relevance, pp. 140–144 (1994)
12. Hall, M.A.: Correlation-based feature selection for machine learning. Ph.D. Thesis. Department of Computer Science, University of Waikato (1999)
13. Wang, Y., Tetko, I.V., Hall, M.A., Frank, E., Facius, A., Mayer, K.F.X., Mewes, H.W.: Computational Biology and Chemistry 29(1), 37–46 (2005)
14. Bishop, C.: Neural Networks for Pattern Recognition. Calderon-Press, Oxford (1995)
15. Kurkova, V.: Kolmogorov's theorem and multilayer neural networks. N. Net 5, 501–506 (1992)
16. Fogel, D.B.: An information criterion for optimal neural network selection. IEEE Tran. N.N. 490–497 (1991)
17. Foley Jason, J.E., Mariotti, J., Ryan, K., Eckhaus, M., Fowler, D.H.: The cell therapy of established acute graft-versus-host disease requires IL-4 and IL-10 and is abrogated by IL-2 or host-type antigen-presenting cells. Biology of Blood and Marrow Transplantation 14, 959–972 (2008)
18. Yu, X.-Z., Liang, Y., Nurieva, R.I., Guo, F., Anasetti, C., Dong, C.: Opposing effects of ICOS on graft-versus-host disease mediated by CD4 and CD8 T cells1. The Journal of Immunology 176, 7394–7401 (2006)
19. Hu, Y., Song, Q., Kasabov, N.: Personalized Modeling based Gene Selection for Microarray Data Analysis. In: The 15th Int. Conf. on Neuro-Information Processing, ICONIP, Auckland, New Zealand. LNCS, vol. 5506/5507. Springer, Heidelberg (2009)
20. Kasabov, N.: Global, local and personalised modelling and profile discovery in Bioinformatics: An integrated approach. Pattern Recognition Letters 28(6), 673–685 (2007)
21. Schliebs, S., Defoin-Platel, M., Kasabov, N.: Integrated Feature and Parameter Optimization for an Evolving Spiking Neural Network. In: Proc. of ICONIP 2008, Auckland, NZ. LNCS, vol. 5506/5507. Springer, Heidelberg (2009)

Mining Rules for the Automatic Selection Process of Clustering Methods Applied to Cancer Gene Expression Data

André C.A. Nascimento[1], Ricardo B.C. Prudêncio[1], Marcilio C.P. de Souto[2], and Ivan G. Costa[1]

[1] Center of Informatics, Federal University of Pernambuco, Recife, Brazil
{acan,rbcp,igcf}@cin.ufpe.br

[2] Dept. of Informatics and Applied Mathematics, Fed. Univ. of Rio Grande do Norte, Natal, Brazil
marcilio@dimap.ufrn.br

Abstract. Different algorithms have been proposed in the literature to cluster gene expression data, however there is no single algorithm that can be considered the best one independently on the data. In this work, we applied the concepts of *Meta-Learning* to relate features of gene expression data sets to the performance of clustering algorithms. In our context, each *meta-example* represents descriptive features of a gene expression data set and a label indicating the best clustering algorithm when applied to the data. A set of such meta-examples is given as input to a learning technique (the *meta-learner*) which is responsible to acquire knowledge relating the descriptive features and the best algorithms. In our work, we performed experiments on a case study in which a meta-learner was applied to discriminate among three competing algorithms for clustering gene expression data of cancer. In this case study, a set of meta-examples was generated from the application of the algorithms to 30 different cancer data sets. The knowledge extracted by the meta-learner was useful to understanding the suitability of each clustering algorithm for specific problems.

1 Introduction

New biotechnology methodologies, such as microrrays, allow the measurement of the expression of all genes of a cell sample. Medical researchers can use such methodologies to measure the expression of cancer cell samples of several patients with distinct cancer types. With these data, machine learning methods can be applied to perform computational diagnosis, i.e., to classify the type of a cancer cell based only on the gene expression profile. Another analysis of particular interest is the application of clustering to search for cancer tissues sharing similar molecular signatures. As demonstrated in [1] and [2], this kind of analysis does not only allows to distinguish between distinct cancer types, but also it has lead to the discovery of new cancer sub-types. Such gene expression data sets impose

C. Alippi et al. (Eds.): ICANN 2009, Part II, LNCS 5769, pp. 20–29, 2009.

several challenges to clustering methods, as they usually have a small number of observations (<100 cancer tissues), high dimensionality (> 1,000 of genes), the distribution of cancer types is unbalanced and there is a high level of noise [3].

While several clustering methods have been proposed in the bioinformatics literature, there is no consensus in the community on which method should be preferably used [4,5,6]. Recently, [7] performed a large scale evaluation of classical clustering methods over 35 data sets of cancer gene expression, which showed that k-means and mixture of multivariate Gaussians had best clustering performance for most of the data sets. That work also showed that hierarchical methods perform poorly for the majority of the sets. Despite of these experimental evidences, medical researchers are still faced with the question on which is the most appropriate method for a particular data set. As in other Machine Learning domains, there is a large variety of clustering algorithms considered suitable to be employed in the cluster analysis of given gene expression data sets. The selection of such algorithms requires empirical knowledge that is not easy to acquire. In general, the choice of algorithms is basically driven by the familiarity of biological experts to the algorithm, rather than the characteristics of the algorithms themselves and of the data [6].

This work is a first attempt to investigate the performance of clustering algorithms on gene expression data, by extracting rules that relate the characteristics of the data sets of gene expression to the performance achieved by the algorithms. The proposed work is directly derived from the Meta-Learning framework [8,9], originally proposed to support algorithm selection for classification and regression problems. According to [10], Meta-Learning can be defined by considering four aspects: (a) the problem space, P, representing the set of instances of a given problem class (usually classification and regression problems); (b) the meta-attribute space, M, that contains characteristics used to describe the problems (e.g., number of training examples, correlations between attributes, among others); (c) the algorithm space, A, that is the set of candidate algorithms to solve the problems in P; (d) a performance metric, Y, that measures the performance of an algorithm on a problem (e.g., classification accuracy estimated by cross-validation).

In this framework, Meta-Learning receives as input a set of meta-examples, in which each meta-example is derived from the empirical evaluation of the algorithms in A on a given problem in P. More specifically, each meta-example stores: (1) the values of the meta-attributes M extracted from a problem; and (2) the best candidate algorithm, considering the performance information Y. Hence, the *meta-learner* is only another learning technique that relates a set of predictor attributes (the meta-attributes) to a target attribute (the best algorithm).

The concepts of Meta-Learning have been extensively applied to select algorithms for classification and regression tasks (e.g., [11,12]). In recent years, Meta-Learning has been extended to other domains of application, as reported in [10]. In [13,14], for instance, the authors proposed the use of Meta-Learning to select algorithms for time series forecasting. In [15], the authors applied Meta-Learning to support the design of planning systems. In [16], Meta-Learning is employed to analyze the performance of meta-heuristics for optimization problems. Considering these applications, Meta-Learning can be viewed as a more general framework to algorithm selection. Hence, one would expect it to be useful in analyzing experiments in clustering of gene expression data.

In the current work, we applied a Meta-Learning procedure to analyze the experiments performed with three clustering algorithms (k-means, finite mixture of Gaussians and spectral clustering), since they were the winners among the seven clustering methods considered initially, on 30 data sets of cancer gene expression. Each data set was described by 13 descriptive meta-attributes and associated to a class label, which indicates the best clustering algorithm among the three candidates. In order to verify the viability of our proposal, different learning techniques (including Support Vector Machines, k-NN and two ensemble techniques) were used as meta-learners. We also applied the MLRules ensemble algorithm to extract interpretable knowledge, which provided useful insights on what makes an algorithm to perform better than another.

Section 2 describes the generation of meta-examples in our domain, as well as the techniques used for Meta-Learning. Section 3 introduces the experiments that evaluate the Meta-Learning process and discusses the obtained results. Finally, Section 4 presents some final remarks and future work.

2 Experimental Work

This research is directly derived from a previous work [7], in which we performed an empirical evaluation of clustering methods on different data sets of cancer gene expression. In the present work, we applied Meta-Learning to analyze the results of our clustering experiments, aiming to extract useful knowledge for selecting clustering methods. In this section, we briefly describe the experiments performed in [7], followed by the description of how the meta-examples were produced in the current work.

In [7], seven distinct clustering algorithms were analyzed: single linkage (SL), complete linkage (CL), average linkage (AL), k-means (KM), finite mixture of Gaussians (FMG), spectral clustering (SP), and Shared Nearest Neighbors algorithm (SNN). Also, four different proximity measures were employed, when applicable: Pearson's Correlation coefficient, Cosine, Spearman's correlation coefficient and Euclidean Distance. In the case of the Euclidean Distance, four different versions were applied: original (Z0), standardized (Z1), scaled (Z2) and ranked (Z3). The algorithms were evaluated in [7] over a set of 35 microarray datasets (See Table 1). These data sets present different values for characteristics such as type of microarray chip (second column), number of samples (third column), number of classes (fourth column) and distribution of samples within the classes (fifth column). In terms of the data sets, it is important to point out that microarray technology is usually available in two different platforms, cDNA and Affymetrix.

2.1 Meta-data

For each gene expression data set, we generated a meta-example composed by features (meta-attributes) that describe the data set and a label indicating the algorithm that obtained the best results. The criterion used for this labeling process and the meta-attributes considered are described in this section.

Table 1. Gene expression data sets considered

Dataset Name	Array Type	N	k	Samples per class	Class label
Armstrong-2002-v1	Affy	72	2	24, 48	FMG
Armstrong-2002-v2	Affy	72	3	24, 20, 28	FMG
Bhattacharjee-2001	Affy	203	5	139, 17, 6, 21, 20	FMG
Chowdary-2006	Affy	104	2	62, 42	-
Dyrskjot-2003	Affy	40	3	9, 20, 11	FMG
Golub-1999-v1	Affy	72	2	47, 25	KM
Golub-1999-v2	Affy	72	3	38, 9, 25	-
Gordon-2002	Affy	181	2	31, 150	FMG
Laiho-2007	Affy	37	2	8, 29	SP
Nutt-2003-v1	Affy	50	4	14, 7, 14, 15	FMG
Nutt-2003-v2	Affy	28	2	14,14	FMG
Nutt-2003-v3	Affy	22	2	7,15	-
Pomeroy-2002-v1	Affy	34	2	25,9	FMG
Pomeroy-2002-v2	Affy	42	5	10, 10, 10, 4, 8	SP
Ramaswamy-2001	Affy	190	14	11, 10, 11, 11, 22, 10, 11, 10, 30, 11, 11,11, 11, 20	KM
Shipp-2002-v1	Affy	77	2	58,19	SP
Singh-2002	Affy	102	2	50, 52	SP
Su-2001	Affy	174	10	26, 8, 26, 23,12, 11, 7, 27, 6, 28	KM
West-2001	Affy	49	2	25,24	FMG
Yeoh-2002-v1	Affy	248	2	43, 205	FMG
Yeoh-2002-v2	Affy	248	6	15, 27, 64, 20, 79, 43	KM
Alizadeh-2000-v1	cDNA	42	2	21, 21	KM
Alizadeh-2000-v2	cDNA	62	3	42, 9, 11	FMG
Alizadeh-2000-v3	cDNA	62	4	21, 21, 9, 11	FMG
Bittner-2000	cDNA	38	2	19, 19	KM
Bredel-2005	cDNA	50	3	31, 14, 5	FMG
Chen-2002	cDNA	179	2	104, 75	-
Garber-2001	cDNA	66	4	17, 40,4, 5	FMG
Khan-2001	cDNA	83	4	29, 11, 18, 25	-
Lapointe-2004-v1	cDNA	69	3	11, 39, 19	FMG
Lapointe-2004-v2	cDNA	110	4	11, 39, 19, 41	KM
Liang-2005	cDNA	37	3	28, 6, 3	FMG
Risinger-2003	cDNA	42	4	13, 3, 19, 7	KM
Tomlins-2006-v1	cDNA	104	5	27, 20, 32, 13, 12	KM
Tomlins-2006-v2	cDNA	92	4	27, 20, 32, 13	KM

Performance evaluation. In order to evaluate the performance of each combination of algorithm and proximity measure considered, an external validation index was used, the corrected Rand (cR) index [17]. The corrected Rand index takes values from -1 to 1, with 1 indicating a perfect agreement between the partitions generated by the clustering algorithm and the true classes known a priori, and values near 0 or negatives corresponding to cluster agreement found by chance. Unlike the majority of other indices, the cR is not biased towards a given algorithm or number of clusters in the partition [17].

The labeling of each meta-example was done according to the following procedure: at first, for each clustering algorithm, we selected the proximity measure that achieved the best results, i.e., the largest cR indices. In order to do so, we took into account only the partition with the number of clusters equal to the number of actual classes in the respective data set [7]. Finally, in order to detect the best algorithm for each data set, a ranking of the algorithms was made.

Only three algorithms were selected as class labels: FMG, KM and SP, since they were the only winners. In case of ties among these three algorithms, the data set on which it happened was excluded for generating a meta-example. This occurred in five data sets, indicated in Table 1 by a "-" at the last column. Hence, an actual number of 30 meta-examples were produced.

Meta-attributes. For the construction of the meta-dataset we used a set of 14 descriptive attributes (meta-attributes). Some of them were first proposed for the case of supervised learning tasks [9]. Recently, they have been also employed in the non-supervised learning context [18]. The samples (examples) considered in our study are labeled, i.e., they have a class label vector $Y = \{y_i\}$, $y_i \in \{1, ..., k\}$, where k is the number of classes for each data set. The class distribution among examples can be defined as $C = \{c_1, ..., c_k\}$, $c_j = \sum_j^N 1(y_i = j)$. Based on this and in other statistics, we define our set of meta-attributes as:

1. LgE: $\log_{10}$ of the number of examples. A raw indication of the available amount of training data.
2. LgREA: $\log_{10}$ of the ratio of the number of examples by the number of attributes. A rough indicator of the number of examples available to the number of attributes.
3. PMV: percentage of missing values. An indication of the quality of the data.
4. MN: multivariate normality, which is the proportion of examples transformed via T^2 that are within 50% of a Chi-squared distribution (degree of freedom equal to the number of attributes describing the example). A rough indicator of the approximation of the data distribution to a normal distribution.
5. SK: skewness of the T^2 vector. Same as the previous item.
6. Chip: type of microarray technology used (either cDNA or Affymetrix).
7. PFA: percentage of the attributes that were kept after the application of the attribute selection filter.
8. PO: percentage of outliers. In this case, the value stands for the proportion of T^2 distant more than two standard deviations from the mean. Another indicator of the quality of the data.
9. NRE: normalized relative entropy. An indicator of how uniformly examples are distributed among classes, i.e. the divergence between the actual class distribution and an uniform distribution. Its calculation is made using the Kullback-Leibler divergence equation, normalized by $2 \log k$, where k is the total number of classes. Let $P(c_j) = \frac{c_j}{N}$ be the probability of the uniform class distribution, the normalized entropy is given by the equation:

$$NE = \frac{\sum_{j=1}^{k} P(c_j) \log(\frac{P(c_j)}{1/k})}{2 \log k} \tag{1}$$

10. SC_{10}: "small" clusters. A measure of the number of classes with size inferior to the threshold $\theta = 10$. Its value is given by: $SC_\theta = \sum_{j=1}^{k} 1(c_j < \theta)/k$.
11. SC_{15}: same measure of previous item, but with its threshold set to $\theta = 15$.
12. BC: "big" clusters. A measure of the number of classes with size superior to the threshold $\theta = 50$, given by: $BC_\theta = \sum_{j=1}^{k} 1(c_j > \theta)/k$.
13. k-NN outliers: classification error obtained by the k-NN algorithm ($k = 3$) [19]. Another indicator of the quality of the data.

2.2 Meta-learner

We evaluated six algorithms as meta-learners: J48, PART, MLRules [20], Random Forest, k-Nearest Neighbors (k-NN) and also Support Vector Machines (SVM). With the exception of the SVM experiments, which were performed using the libSVM[1] package, all experiments were executed within the WEKA framework [2].

The J48 algorithm is the WEKA implementation of the C4.5 decision tree algorithm. The varied parameters were the confidence factor (from 2^{-15} to 2^{15}) and the minimum number of instances per leaf (from 2 to 7). Another method considered is the PART algorithm, which actually builds a partial C4.5 decision tree in each iteration, making the "best" leaf into a rule. For the experiments with PART, we used the same parameter values evaluated for the J48 algorithm.

The random forest algorithm belongs to the so-called "ensemble methods", a combination of various methods that generate many classifiers (in this case, decision trees) aggregating their results. This method has several features, which include the possibility to be used on a mixture of discrete and continuous descriptors, to classify binary or multi-class data sets and work with data sets where there are more variables than observations. The algorithm also presents good performance even when most predictive variables are noise [21]. For this work, we fixed the number of trees parameter to 100 and varied the number of attributes to be selected in each tree from 1 to 15. Another ensemble method evaluated was the Maximum likelihood rule ensembles (MLRules), which is a recent rule induction algorithm for solving classification problems via probability estimation. The ensemble is built minimizing the negative loglikelihood to estimate the class conditional probability distribution. We varied the minimization technique (Newton and gradient) and the shrinkage parameter in $\{0.1, 0.2, ..., 1\}$.

SVMs are supervised learning methods that construct a separating hyperplane in an n-dimensional space, trying to maximize the margin between the classes. We executed the experiments considering polynomial and RBF kernels. For polynomial kernel, we varied the cost and the degree parameters in the intervals $[2^{-15}, 2^{15}]$ and $[2, 6]$ respectively. For RBF kernel we varied both the cost and gamma parameters in the interval $[2^{-15}, 2^{15}]$. For the k-NN algorithm, we varied k in the interval $[1, 20]$.

Classification experiments were developed according to the leave-one-out procedure, with some remarks: as the class distribution of the meta-examples was

[1] http://www.csie.ntu.edu.tw/~cjlin/libsvm/
[2] http://www.cs.waikato.ac.nz/ml/weka/

unbalanced (class distribution is, respectively, 16 examples for FMG, 10 for KM and 4 for SP), this could led to overfitting towards the classes with larger number of examples (FMG and KM). So, in order to make class distribution more uniform, each example from the FMG, KM and SP classes was replicated 2, 3 and 8 times, respectively. This replication process was performed only in the training data. Thus, by doing so, an example would never be at the same time on the training and test sets. Rule extraction experiments were developed employing the same balanced data, except by the fact that we did not evaluate the accuracy of the obtained model using the leave one out procedure: instead, we utilized the full training set.

3 Results and Discussion

3.1 Meta-learner

The average test accuracy of the leave-one-out experiments realized with the five methods compared can be seen in Table ??. According to this table, Random Forest obtained the best classification accuracy followed by MLRules. All other methods had a cross-validation accuracy equal or lower than the base line error (taking the majority class as reference). This is probably a consequence of the difficulty of the classification problem, as there are very few samples to classify (30 samples) and one of the classes (SP) has only four samples. Ensemble methods, like Random Forest, are often expected to have a better performance on such difficult classification scenarios, which is confirmed in our study.

Table 2. Accuracy rates - runs over balanced meta-data

Method	Accuracy
PART	40.00%
J48	30.00%
MLRules	56.67%
k-NN	53.33%
SVM	53.33%
Random Forest	63.33%
Base Line Error	53.33%

3.2 Rule Mining

The next step in our analysis was to extract interpretable knowledge from the meta-learning learning analysis of the data. Our goal is to discover explanatory (partial) models of performance of clustering algorithms on cancer gene expression data. Observing the generated rules, one can notice the suitability of clustering algorithms studied, as well as the actual relations to characteristics (meta-attributes), with respect to the underlying structure in the data sets.

In order to do so, we used the MLRules algorithm with all data as training set and Newton steps as minimization technique and shrinkage to 0.5. A total of 100 rules were generated, but we selected only the ten rules with biggest weights

```
Rule 1 (12/3):
  If PO ≥ 0.059 then
    Suggest MFG with weight 0.18179

Rule 2 (13/3):
  If LgREA entre [-1.604,-0.982]
   and PO ≥ 0.03
   and SK ≥ 0.397 then
    Suggest MFG with weight 0.16655

Rule 3 (4/0):
  If chip = cDNA
   and ERN ≤ 0.011 then
    Suggest KM with weight 0.16259

Rule 4 (11/4):
  If knn-outliers ≥ 0.17
   and NM ≤ 0.724
   and PO ≥ 0.026 then
    Suggest KM with weight 0.14597

Rule 5 (10/4):
  If LgE ≥ 1.926 then
    Suggest KM with weight 0.14176

Rule 6 (4/0):
  If LgE ≥ 2.025
   and ERN ≤ 0.026 then
    Suggest KM with weight 0.13758

Rule 7 (18/5):
  If PAR ≥ 0.806
   and LgREA ≤ -0.868
   and PVF ≤ 0.121 then
    Suggest MFG with weight 0.13560

Rule 8 (7/3):
  If PVF ≤ 0.011
   and PO ≥ 0.046
   and LgE in [1.55,2.072]
   and NM ≥ 0.419 then
    Suggest SP with weight 0.13339

Rule 9 (12/3):
  If PO ≥ 0.059 then
    Suggest MFG with weight 0.13006

Rule 10 (12/1):
  If LgE ≤ 1.872
   and PAR ≥ 0.809 then
    Suggest MFG with weight 0.12998
```

Fig. 1. Rules induced by the MLRules algorithm

for analysis. The produced rules can be seen in Figure 1. They are listed in a pseudo-code like structure to ease readability.

Here, at each rule one can find, respectively, the method indicated (KM, FMG and SP), the number of meta-examples classified by the node and how many are misclassified (in parenthesis), as well as the rule weight. Interestingly to notice that, in general, rules that suggest the KM method involves the LgE and the ERN meta-attributes. This agrees with literature information, in which this method tends to find equal sized clusters (low ERN, rules 3 and 6) and is very sensitive to a small number of training patterns (low LgE, rules 5 and 6). We can also observe the presence of the meta-attribute PO (percentual of outliers) requiring bigger values in most of the rules that suggest the MFG method, an indication that this method presents good tolerance to datasets with a high number of outliers (rules 1, 2 and 9). Only one rule related to the Spectral algorithm was generated (rule 8), possibly due to the small number of examples labeled with this class available. The Spectral method employed in [18] is based in a Gaussian similarity function, which matches the requirements of data normality. This fact agrees with the assertive $MN > 0.419$ on rule 8. Furthermore, the rule suggests the use of the Spectral method in the presence of outliers.

Another interesting fact is the presence of chip type in rule 3. It is well known in the microarray literature that cDNA and Affymetrix chips generate expression values with distinct characteristics [22]. The cDNA arrays are based on log-ratios of the expression between the reference cell (tumor) and a control cell (healthy cell), whereas Affymetrix data is based only on the tumor cell and expression values should reflect the absolute count of transcripts in that cell. As a result, the log-ratios used in cDNA measurements make the expression values to have a normal distribution. Differently, Affymetrix expression values are positive and have a distribution skewed towards lower expression values. Furthermore, measurements of cDNA chips are less susceptible to probe problems in a specific chip, as a problematic probe will have the same effects to both control and reference values [22]. While there is no consensus in the microarray literature regarding the data quality and microarray platform, the cDNA chip type verification on rule 3 is another indication that data from cDNA microarrays are less sensitive to noise, suggesting the k-means method in this case.

The rules induced by MLRules could be susceptible to overfitting, as there are very few examples in the data sets. Nevertheless, as discussed in previous paragraphs, the rules extracted are in accordance to general knowledge in the clustering literature. Thus, rather than proposing the use of the rules and the attribute thresholds in their own, we interpret them as "soft" guidelines to the choice of a clustering method given a certain cancer gene expression data set.

4 Final Remarks

In this paper, we presented a preliminary study that explores the ability to automatically generate rules to guide the choice of clustering algorithms for gene expression data. One of the our main contributions is to show that two rule-based ensemble classifiers — random forests and MLRules — on average, presented the most accuracy rates in predicting the best clustering algorithm for gene expression data sets. We emphasize that the classification problem analyzed here is a difficult one, as there are very few meta-examples. Thus, no classification method had a high classification accuracy.

Another contribution of this work was to extract rules for the selection of clustering algorithms, by using an rule ensemble algorithm. Overall, the rules extracted give us some interesting guidelines for choosing the method. For instance, in the case of gene expression data from cDNA microarrays, k-means method should not be used when class size distribution is not uniform. Although, when a large number of samples is present, the method is preferred. Finite mixture of Gaussians should be used when there are few samples and a non-uniform class distribution. In cases where the data follows a normal distribution and there's a large amount of outliers, Spectral clustering is adequate. Such guidelines, based on meta-attributes of data sets, had not been empirically demonstrated before in the gene expression literature. As a future work, we will try to increase the number of meta-examples, as well as investigate other meta-attributes.

Acknowledgments. The authors would like to thank CNPq (Brazilian Agency) for its financial support.

References

1. Golub, T., et al.: Molecular classification of cancer: class discovery and class prediction by gene expression monitoring. Science 286(5439), 531–537 (1999)
2. Alizadeh, A.A., et al.: Distinct types of diffuse large b-cell lymphoma identified by gene expression profiling. Nature 403(6769), 503–511 (2000)
3. Spang, R.: Diagnostic signatures from microarrays: a bioinformatics concept for personalized medicine. Biosilico 1(2), 64–68 (2003)
4. Costa, I.G., et al.: Comparative analysis of clustering methods for gene expression time course data. Genetics and Molecular Biology 27(4), 623–631 (2004)
5. Datta, S., Datta, S.: Methods for evaluating clustering algorithms for gene expression data using a reference set of functional classes. BMC Bioinformatics 7, 397 (2006)
6. D'haeseleer, P.: How does gene expression clustering work? Nature Biotechnology 23(12), 1499–1501 (2005)
7. de Souto, M.C., et al.: Clustering cancer gene expression data: a comparative study. BMC Bioinformatics 9, 497 (2008)
8. Vilalta, R., et al.: Using meta-learning to support data- mining. Intern. Journal of Computer Science Application 1(31), 31–45 (2004)
9. Giraud-Carrier, C., et al.: Introduction to the special issue on meta-learning. Machine Learning 54(3), 187–193 (2004)
10. Smith-Miles, K.: Towards insightful algorithm selection for optimisation using meta-learning concepts. In: Proceedings of the IEEE International Joint Conference on Neural Networks 2008, pp. 4118–4124 (2008)
11. Brazdil, P., et al.: Ranking learning algorithms: Using IBL and meta-learning on accuracy and time results. Machine Learning 50(3), 251–277 (2003)
12. Kalousis, A., Gama, J., Hilario, M.: On data and algorithms - understanding inductive performance. Machine Learning 54(3), 275–312 (2004)
13. Prudĕncio, R.B.C., Ludermir, T.B.: Meta-learning approaches to selecting time series models. Neurocomputing 61, 121–137 (2004)
14. Wang, X., et al.: Rule induction for forecasting method selection: Meta-learning the characteristics of univariate time series. Neurocomputing (2008) (to appear)
15. Tsoumakas, G., et al.: Lazy adaptive multicriteria planning. In: Proceedings of the 16th European Conference on Artificial Intelligence, ECAI 2004, pp. 693–697 (2004)
16. Smith-Miles, K.: Cross-disciplinary perspectives on meta-learning for algorithm selection. ACM Computing Surveys 41(1), 1–25 (2008)
17. Milligan, G., Cooper, M.: A study of standardization of variables in cluster analysis. Journal of Classification 5, 181–204 (1988)
18. de Souto, M.C.P., et al.: Ranking and selecting clustering algorithms using a meta-learning approach. In: Proceedings of the International Joint Conference on Neural Networks. IEEE Computer Society, Los Alamitos (2008)
19. Wilson, D.L.: Asymptotic properties of nearest neighbor rules using edited data. IEEE Transactions on Systems, Man and Cybernetics (3) (1972)
20. Dembczyński, K., Kotłowski, W., Słowiński, R.: Maximum likelihood rule ensembles. In: Proceedings of the 25th International Conference on Machine Learning, ICML, pp. 224–231 (2008)
21. Breiman, L.: Random forests. Machine Learning 45, 5–32 (2001)
22. Cauton, H.C., Quackenbush, J., Brazma, A.: Microarray Gene Expression Data Analysis: A Beginner's Guide. Blackwell Publishing, Malden (2003)

A Computational Retina Model and Its Self-adjustment Property

Hui Wei and XuDong Guan

Department of Computer Science, Fudan University, Shanghai 200433, P.R. China

Abstract. The results of neurophysiology and biology about early vision inspire us to establish an effective and economical computational model. As the most important part of the early vision, retina is a complex but orderly structure, in which the first essential information is processed preliminarily. In this paper, according to anatomic structure, a multi-layer digital retina model is presented to simulate biological retina and it is placed in a physical visual field of a reduced eye to analyze why the retina can be capable of fulfilling all tasks. The model tries to reach a kind of feasible balance among hardware complexity, computing load and performance as the retina does. This research also contributes to the design and implementation of artificial retina chips to improve perception of visually impaired patients.

Keywords: retina, early visual information processing, neural network.

1 Introduction

Target tracking, Robot navigation, autopilot, and the traffic control etc. enhance the requirement for digital image processing, especially for real-time image processing technology. The traditional methods of digital image processing mainly deal with images via spatial domain processing or via frequency domain processing. These methods seem to take a long time but act to be low efficiency. These real-time image processing applications demand higher processing rates (10-1000 Gops/s) [1] than can be provided by commercial microprocessors (1-5 Gops/s). The mechanism of the higher mammals' vision, which is precise and complex, can immediately sense the environment information and then does rapid judgment. Now, even the most advanced artificial vision system is still nowhere near of the human retina on capacity of information processing. Therefore, it is very useful to develop a new way of machine vision from the perspective of simulation of neurophysiology structure and psychological process. In the design of artificial retina chips, it's more instructive to explore how the physical retina coordinates the information processing requirement and neural structure. Is there possible that we might find a way not only to save energy but also maintain good performance like our retina does? Therefore, it's necessary to analyze the self-adjustment property of physical retina to improve the design and implementation of the artificial retina. This paper presents the retinal early visual model integrating the physiological structure and computer model closely, which not

C. Alippi et al. (Eds.): ICANN 2009, Part II, LNCS 5769, pp. 30–39, 2009.

only simulates the multi-layer structure of the retina and the ganglion cell receptive field, but also simulates the precise retinal cells distribution, so the model is more accurate and very similar to the real retina mechanism. Meanwhile, a large numbers of photographs of the real world have been tested, and then the self-adjustment properties of ganglion cells were verified to achieve a feasible balance among hardware complexity, processing time and accuracy.

2 Model Design on Simulating Information Processing of Biological Retina

2.1 Computation Model of Retina Structure

In the field of vision, retina is crucial for both biology research and computer modeling. As the most important part of the early vision, retina is a complex but orderly structure, in which the first essential information is processed preliminarily. The intricate structures of retina are corresponding to its various functions. Retina is composed by the multi - layer cells, the most important of three layers from the outside to inside are the photoreceptor cell layer, bipolar cell layer and the ganglion cell layer. Each layer contains more than one type of cell. Figure 1 illustrates the retina visual information processing flow. Based on

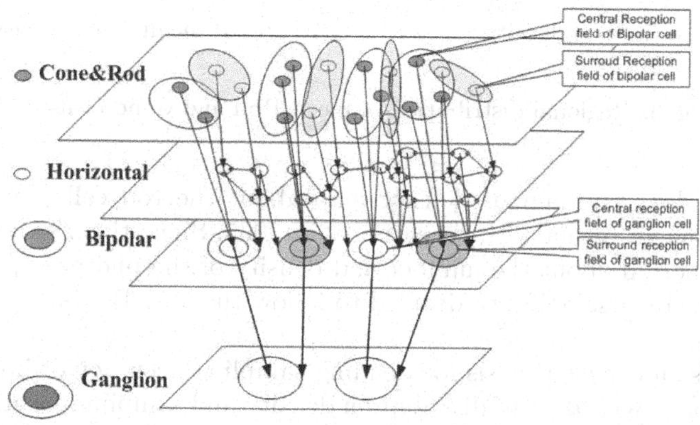

Fig. 1. Retina information process flow. The model contains three main cell layers. It shows the cells network and the vision information process flow.

the simplified structure, a reduced eye model was used in our model. According to the physiological size of retina, the physical model projects simulated point to the retina with similar triangles. The algorithm diagram is shown in Figure2. It shows the mapping between the point of outer scene and the receptor cell position of retina. $Rmax = 27.5mm.\Theta_{max} = 70^o, d = Rmax * tan(\Theta) = 10mm$,D is adjustable.$\Theta \leq \Theta_{max}, (x, y) = (X * 10/D, Y * 10/D)$.

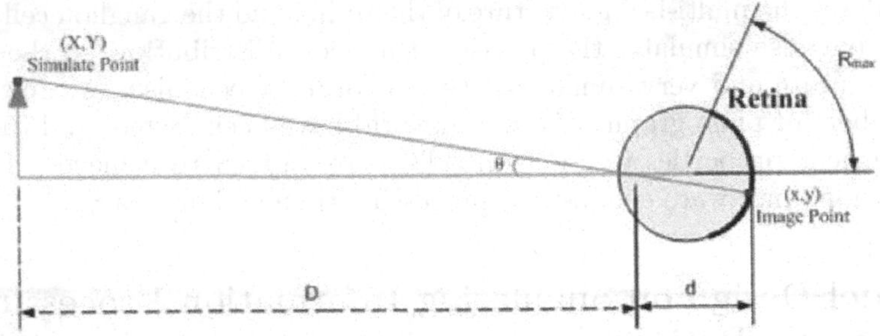

Fig. 2. Receptor cell's position point mapping model

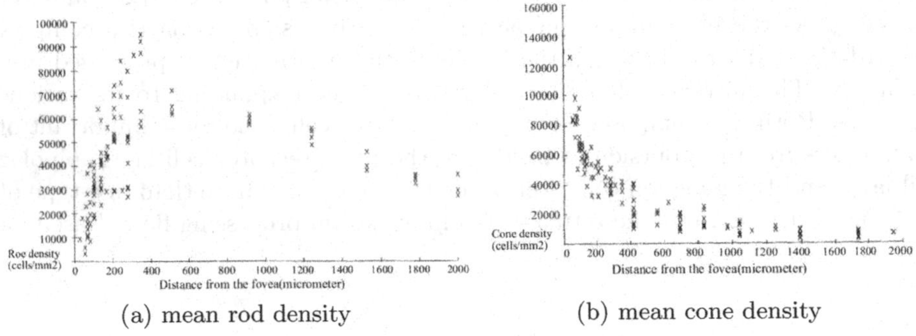

(a) mean rod density

(b) mean cone density

Fig. 3. Regional distribution of mean Rod and Cone Density

Osterberg drew the famous density map about the rod cells and the cone cells of the human retina with eccentricity in 1935. Since then, several studies have been reported about the number and density of the photoreceptors. After comparing the diverse data, we decided to follow the Jost B. Jonas's data[2] as Fig. 3.

It provides an excellent basis for defining sampling accuracy of image in different locations. According to distribution density and sampling locations of the rod cells and cone cells, we consider retina as the disk which is composed of a center in fovea and several rings. Rod cells and cone cells in the different bands have different density. We simplify the distribution of cone cells and rod cells in the retina as Fig. 4.

The following pseudo-code describes the retina cells generation in the first quadrant.

1) Calculate the side length of the inscribed square and circumscribed square as L1, L2, and simulate every ring.

2) Generating ring cell matrix according to the density D of ring cells and describing these with four two-dimensional matrixes and (Figure 5).'m' means

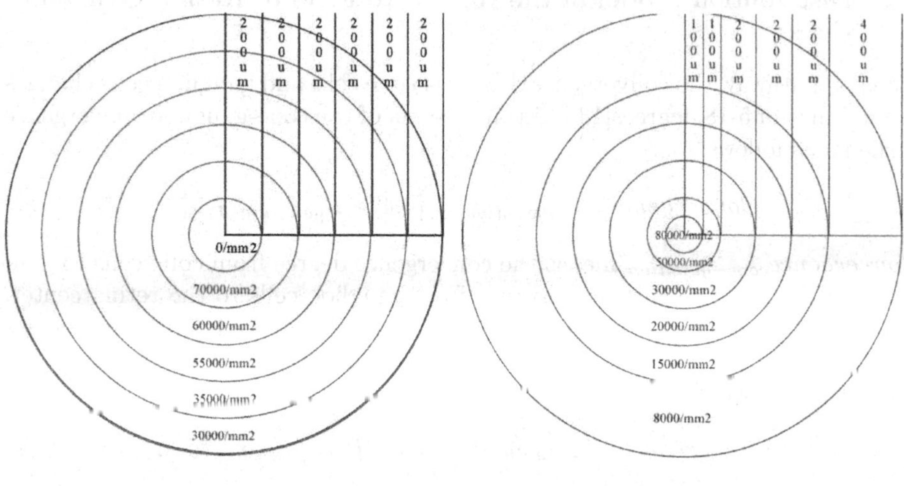

(a) Different rod rings density (b) Different cone rings density

Fig. 4. Density of Rod and Cone rings

the number of cells in the horizontal direction, and 'n' is the number of cells in the vertical direction.$m = (L2 - L1) * \sqrt{D}, n = L1 * \sqrt{D}$

3) Generating the position of cell point in each matrix respectively. For matrix A_{m*n}:$(x, y)_{i,j} = (L1 + i * 1/\sqrt{D}, j * 1/\sqrt{D})$;for matrix $B_{m,m}$,$(x, y)_{i,j} = (L1 + i * 1/\sqrt{D}, L1 + j * 1/\sqrt{D})$, for matrix C_{n*m},$(x, y)_{i,j} = (i * 1/\sqrt{D}, L1 + j * 1/\sqrt{D}), i \in [0, m-1], j \in [0, n-1]$.

To improve the performance of program, we calculate each point in the square area. But in the final output, we remove the points outside the gray areas.

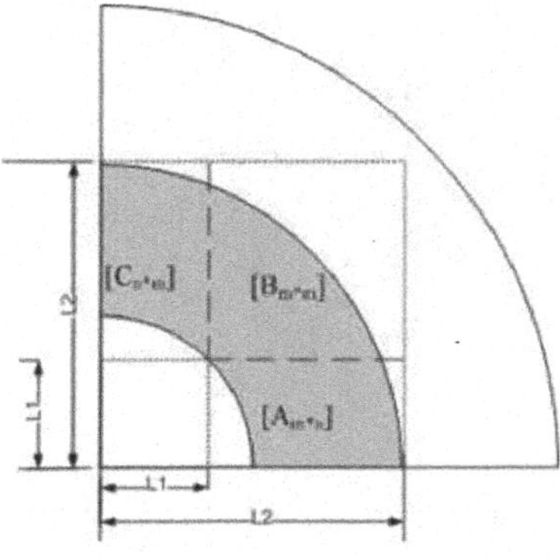

Fig. 5. The cell matrix of retina

2.2 Distribution Model of the Receptive Field of Retina Ganglion Cells

For cone pathway, the convergence degree of cone cells and ganglion cells changes in the range of 5-18 degrees[13]. Linear formula of the cone-ganglion convergence is shown as follows:

$$Convergence_{cone-ganglion} = \lceil 0.47 * R_{ganglion} + 5 \rceil . \tag{1}$$

$Convergence_{cone-ganglion}$ means the convergence degree from cone cells to ganglion cell.$R_{ganglion}$ means the distance from ganglion cells to the retina center. For rod pathway, the convergence degree of rod cells and ganglion cells changes in the range of 100-1500 degrees[14][15]. Linear formula of the rod-ganglion convergence is shown as follows:

$$Convergence_{rod-ganglion} = \lceil 53.8 * R_{ganglion} + 27.5 \rceil . \tag{2}$$

According to the formula (1) and (2), we can get the convergence degree of any ganglion cells in any position. And then, we can get the receptor cells of certain ganglion cells through gradually expanding scan radius of receptive field from inner to outer.

2.3 Algorithm Model of Photoreceptors

The cone cell algorithm model. According to sensitive curves of photoreceptor cells drawn by Dowling in the 1987, the cone cells can be divided into three types in the light-visible area from 400nm to 700nm, which are respectively sensitive to the wavelength of 560nm, 530nm and 430nm, as Fig.6 show: In the RGB color model, the three wavelengths of light correspond to R, G and B color values. We design the similar sensitivity curve based on the RGB color model, firstly we make a transformation for the R, G, B among the value of

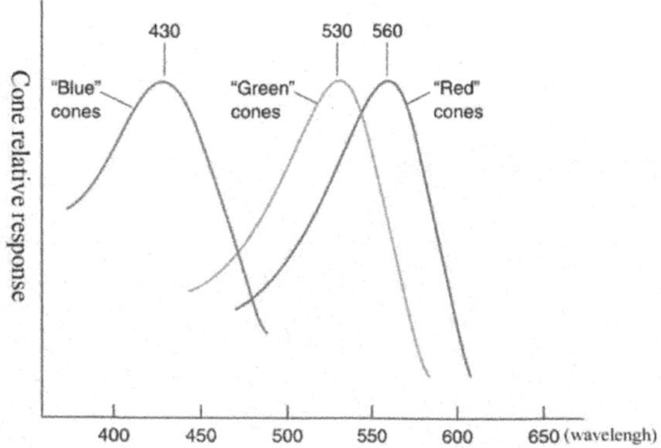

Fig. 6. Different cone response to different wavelength light

0 255: $B \in [0, 0xFF], G \in [0x100, 0x1FF], R \in [0x200, 0x2FF]$, then translate the RGB value for standardization based on the formula$R = R/767, G = G/767, B = B/767$. We use quadratic curve (3) to simulate cone sensitive curve approximately.

$$R(\lambda) = -a * (\lambda - F)^2 + c. \tag{3}$$

$R(\lambda)$ means the output of a single cell, λ is the input of a single cone cell, F as sensitive color constants,a,c are the coefficients,$a \succ 0, 0 \prec c \leq 1$. When F=1/6,F=1/2 and F=5/6, the cone cell is respectively sensitive to blue, green or red light, but different response curves overlap in the edge. To simulate the physiological responsive characteristic of photoreceptor cells, we limit response value of the photoreceptor cells in the range 0-1, so we set a=16,c=1 .

In this experiment, we did not care how to represent and transport color information. What we focus on is the number of activated cells and response intensity of every cell, so we use average weighted formula (4) to simulate the final output value of cone cells. The quantities of three types cone cells are equal. In the same density region, they are cross-distributing uniformly.

$$R_{cone}(r, g, b) = \frac{R(r)^2 + R(g)^2 + R(b)^2}{R(r) + R(g) + R(b)}. \tag{4}$$

The rod cell algorithm model. Rod cell is the carrier of luminance information responding to the light intensity. In traditional image processing, we use gray value to represent the luminance of a pixel, so it's reasonable that we use the gray value to map the response intensity. We translate the color RGB value into the gray value as the input of rod cell. According to ITU-R BT.601a standard formula recommended by International Telecommunication Union to transform a colorful image to gray imageGray = 0.299*R + 0.587*G + 0.114*B. We just choose one behind the decimal, and then we get the function (5).

$$R_{rod}(r, g, b) = (0.6 * g + 0.3 * r + 0.1 * b)/255. \tag{5}$$

2.4 Algorithm Model of Ganglion Cells

Ganglion cells have the center-surrounding receptive field structure. We use model (6) to imitate the output of ganglion cells, satisfying the standards of DOG model.

$$R^{(x,y)}_{ganglion}(r, g, b) = R^{(x,y)}_{center}(r, g, b) - R^{(x,y)}_{surround}(r, g, b) \tag{6}$$

In which,$R^{(x,y)}_{ganglion}(r, g, b)$ means the output of ganglion cells in the retina position (x,y);$R^{(x,y)}_{center}(r, g, b)$ means the output from the central receptive field of ganglion cells in the retina position (x, y);$R^{(x,y)}_{surround}(r, g, b)$is the output of the surrounding ganglion cells located in the retina location(x, y). For the cone cells,$R^{(x,y)}_{center}(r, g, b) = \sum_{center} R_{cone}(r, g, b)$,
$R^{(x,y)}_{surround}(r, g, b) = \sum_{surround} R_{cone}(r, g, b)$;

for the rod cells, $R^{(x,y)}_{center}(r,g,b) = \sum_{center} R_{rod}(r,g,b)$, $R^{(x,y)}_{surround}(r,g,b) = \sum_{surround} R_{rod}(r,g,b)$;

3 Experiment and Results Analysis

3.1 Experimental Results and Analysis for Real Scene Photos

We tested and verified 63 real scene photos with different complexity to calculate the activation rate of ganglion cells (Fig.7), and we classified these complex photos by their degree of complexity to analyze the relationship between the output characteristics and input complexity when retina cells handle the different complexity of visual information. As results of experiments show, for real

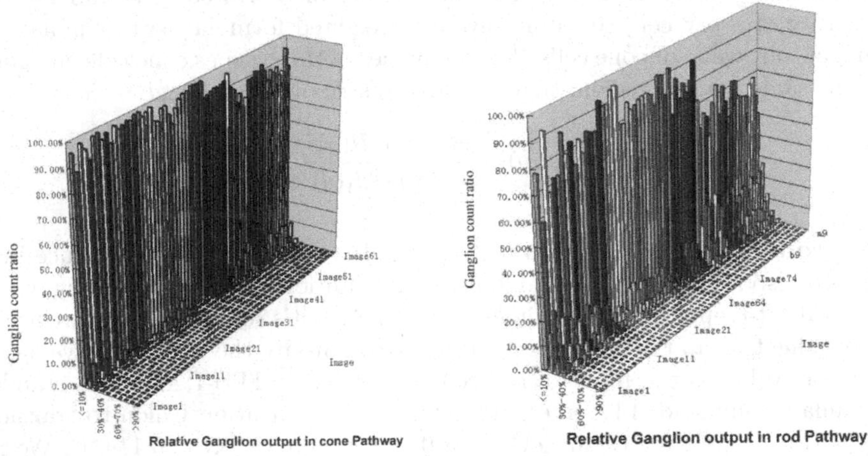

(a) Relative Ganglion output in cone Pathway (b) Relative Ganglion output in rod Pathway

Fig. 7. Rod and Cone Pathway

stationary scenes, about 80%ganglion cells are activated, but the ganglion cells' outputs are mainly in the range of 10%-20% of the max output. It indicates ganglion cells are in a state of low energy consumption when human's retina is receiving static visual information. Most ganglion cells are not at work under this kind of status. They are possibly in a state of rest or waiting for other information input. The retina will balance the input information processing and the complexity of hardware: sufficiently receiving outside stimulation as well as completely processing visual information under the limited hardware condition.

3.2 Analysis for All Activated Ganglion Cell Position

To clearly show position and output of activated ganglion cells in cone-ganglion and rod-ganglion pathway, the paper combined two pathways into one output.

In our experiment, we only choose one quarter–the upper-right quarter of the retina because four quarters of retian are basically symmetrical with each other. we depicted the pixels which activated ganglion cells in the image, then we set gray value of each pixel according to ganglion cells output.

The experiment takes the real world image (Figure11a) as input and the experiment results is shown (Figure11b). From results of experiment, we can

(a) input image (b) result image

Fig. 8. Input image and result image

learn that the fovea has high visual acuity because cone cells are intensive in this region. The resolution is also high in fovea just like the lower left of the image. When the distance between photoreceptor and fovea center increases, the visual acuity decreases gradually. The results of experiment do verify these characteristics of retina in information processing.

3.3 Verification on Self-adjustment Property of Retina

Generally, all characteristics of the retina network are finally exhibited in the receptive field of ganglion cells. Changing different background image or changing moving object velocity, receptive field size would change correspondingly. This dynamic characteristic is extremely important in physiology. For example, in a dark environment, receptive field would become bigger to acquire more light in a larger area to sense the general information. At the same time, visual acuity has to adjust to be low. When we need to distinguish the slight differences, receptive field would turn to be smaller so as to improve space acuity.

We take cone-path as an example to verify this amazing characteristic, we adjust the convergence to change the size of receptive field, and then we record the corresponding complexity, scanning efficiency, computing load and accuracy. The model complexity is defined as the total connections between photoreceptors and ganglion cells. Scanning efficiency is a measure that estimates how fast the center area may scan all visual field by a way like oculomotor phenomenon. Computing load is defined as the computational intensity. Accuracy is defined as a ratio of edge coverage. Obviously, all characteristics have direct relations with receptive field size. The results of experiment is shown below Fig.9.

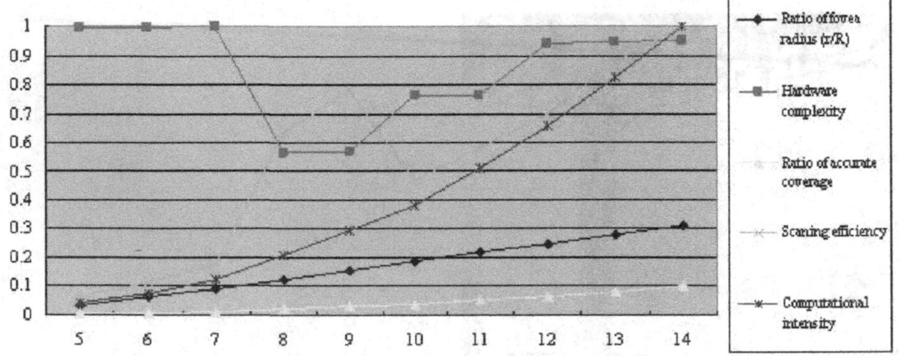

Fig. 9. The balance of parameters in different retina designs

Fig.9 is an illustration of retina designs with different parameters. With the growth of convergence from 5 to 14, the ratio of accurate coverage and computational intensity increase, but the later increases more steeply. And other two measures of hardware complexity and scanning efficiency are nonmonotonic. It is obvious that when the convergence is 9 the hardware complexity reduces to 55% of maximum complexity and scanning efficiency can stay at near 80% of the best efficiency, meanwhile the computational intensity is no more than 30% of maximum intensity. This means that the size-changeable receptive fields model can balance the hardware resource, processing precision, computational intensity very well.

4 Discussions

This model verified the research of physiology and psychology from the perspective of visual information processing, and brought up the new field and direction for the physiology and psychology. This model has not yet simulated the dynamic information input. We will take the real-time visual information processing model as the next target for our study.

Acknowledgments. Our research is supported by National Natural Science Foundation of P. R. China (60303007) and Shanghai Science and Technology Development Fund (08511501703) Thanks for their support.

References

1. Wills, D.S., Baker, J.M., Cat, H.H., Chai, S.M., Codrescu, L., Cruz-Rivera, J.L., Eble, J.C., Gentile, A., Hopper, M.A., Lacy, W.S., Lpez-Lagunas, A., May, P., Smith, S., Taha, T.: Processing Architectures for Smart Pixel Systems. IEEE Journal of Selected Topics in Quantum Electronics 2(1), 24–34 (1996)
2. Jonas, J.B., Schneider, U., Naumann, G.O.H.: Count and density of human retinal photoreceptors. Graefe's Arch. Clin. Exp. Ophthalmol 230, 505–510 (1992)
3. Shah, S., Martin, D.: Visual information processing in primate cone pathways. I.A model. IEEE Transactions On Systems, Man, And Cybernetics-Part B: Cybernetics 26(2), 259–274 (1996)
4. Shah, S., Martin, D.: Visual information processing in primate cone pathways. II. Experiments. IEEE Transactions On Systems, Man, And Cybernetics-Part B: Cybernetics 26(2), 275–289 (1996)
5. Mertoguno, S., Bourbakis, N.: A digital retina-like low level vision processor. Digital Object Identifier 33(5), 782–788 (2003)
6. Herault, J.: A model of colour processing in the retina of vertebrates from photoreceptors to colour opposition and colour constancy phenomena. Neurocomputing 12, 113–129 (1996)
7. Andersen, J.D.: Methods for modeling the first layers of the retina. In: Neuroinformatics and Neurocomputers, vol. 1, pp. 179–186. IEEE, Los Alamitos (1992)
8. Lee, J.-W.: A moving detectable retina model considering the mechanism of an amacrine cell for vision. In: Proceedings Industrial Electronics, vol. 1, pp. 106–109. IEEE, Los Alamitos (2001)
9. Fangtu, Q., Chaoyi, L.: Mathematical simulation of disinhibitory properties of concentric receptive field. ACTA biophysica sinca 16(2), 214–220 (2000)
10. Zang, L., Cheng, Q.Z.: A new computational model of retinal ganglion cell receptive fields I: a model of ganglion cell receptive fields with extended disinhibitory area. ACTA biophysica sinca 16(2), 288–295 (2000)
11. Zang, L., Cheng, Q.Z.: A new computational model of retinal ganglion cell receptive fields II: modeling center - surround interactions in orientation selectivity of a ganglion cell receptive field with extended disinhibitory area. ACTA biophysica sinca 16(2), 296–302 (2000)
12. Tombran-Tink, J., Barnstable, C.J., Rizzo III, J.F.: Visual Prosthesis and Ophthalmic Devices, pp. 135–158. Humana Press (2007)
13. Sterling, P., Freed, M.A., Smith, R.G.: Architecture of rod and cone circuits to the On-beta ganglion cell. J. Neurosci. 8, 623–642 (1988)
14. Kolb, H., Famiglietti, E.V.: Rod and cone pathways in the inner plexiform layer of the cat retina, vol. 186, pp. 47–49 (1974)
15. Nelson, R., Kolb, H.: A17: a broad-field amacrine cell of the rod system in the retina of the cat. J. Neurophysiol. 54, 592–614 (1985)

Mental Simulation, Attention and Creativity

Matthew Hartley and John G. Taylor

Department of Mathematics, King's College, Strand, London UK
mhartley@mth.kcl.ac.uk, john.g.taylor@kcl.ac.uk

Abstract. We extend an attention-and brain-based based architecture for mental simulation to allow for subliminal or creative thinking as part of a thinking sequence. In particular a mechanism for switching between attended and subliminal thinking is proposed. A simulation of this architecture is then presented, specifically for the task of 'Unusual uses of a cardboard box'.

Keywords: thinking, reasoning, cognition, observational learning, consciousness.

1 Introduction

Numerous cases of successful thinking at an unconscious level are recounted in a recent book by Gladwell [5], where a 'thin slicing' (rapid) method of assessment of a situation leads to optimal recognition, as compared to a slower and more deliberate perusal. He quotes the psychologist [12], who says we toggle back and forth between our unconscious and conscious modes of thinking, depending on the situation.

How can we begin to understand at a brain and model level how such unconscious thinking is achieved? Here we regard the crucial part of thinking as that of mental simulation, where trains of thought can be regarded as sequences of mental states which lead from one initial (present) state to a final (goal) state. This latter state may be well-defined beforehand, or it may only be determined by the application of a success criterion, such as through some form of template indicating that the final state satisfies a further set of conditions. This is what occurs in mathematical thinking towards proving a theorem, when the steps in the proof of the theorem will, as they approach a crucial such step, lead to the satisfaction of some internally created criterion.

We conclude from this that in order to create a cognitive machine, one that can 'think' we need to allow for processing at two levels: one being the conscious one, the other being unconscious. These two levels would seem to occur sequentially, with the unconscious steps guided by well-used rules or meeting some essential criterion, but lying outside attention control. However they can emerge into consciousness if some important criterion is met in one of the states of the unconscious sequence.

Unconscious processing is crucial in creative activity: in painting, musical composition, photography, writing and so on. Such subliminal creativity has

C. Alippi et al. (Eds.): ICANN 2009, Part II, LNCS 5769, pp. 40–53, 2009.

even been proposed as antithetic to consciousness and advocates have proposed to strongly play down the role of conscious processing [6]. Here we consider that both components – the subliminal and the consciously controlled forms – of thought should be considered together in order to understand them. To do that we need to understand, through architecture building and simulation, of what each of the two components consists and how they complement each other.

In order to develop a suitable architecture to achieve such two-level processing, we will therefore present and analyse a neural (brain-guided) architecture that is able to support mental simulation as a process of 'thinking about the world' by means of the pairs of forward and inverse internal models it contains. This model allows for cycling through a set of neural loops to emulate the manner in which actions on objects in the external world will lead from a given initial state to a valuable final one. In this way the mental simulation process of the 'world model in the head' produces a limited version of thinking, one without the necessary conscious component.

We extend this attention-controlled model of thinking to include an alternate route outside the control of attention. It is that route, we suggest, that involves creative processing, and leads to solutions of seemingly impossible problems through a period of subliminal processing.

In the next section we present the architecture we regard as at the basis of cognitive processing and consider the manner in which mental simulation can occur as part of this processing. We have already used this architecture in modelling results from a complex paradigm used in studying observational learning on infants, to which we refer for a fuller account [11]. In section 3 we consider in more detail the manner in which two levels of thinking processing – conscious and subliminal - can be achieved by modifications to the overall mental simulation architecture. This requires an additional attention control system that can lead to the switching between conscious and unconscious process in thinking, which as noted earlier occurs sequentially in creative thinking. The following section 4 looks at the simulation of learning in the well known creative task of thinking of unusual uses for a specific object. The final section concludes the paper.

2 The Architecture of Mental Simulation

The brain basis for the architecture from which we start is that originally proposed for mental simulation [11]. With attention added, this architecture is that shown in figure 1. Each box represents a component processing area that receives input from one or more other areas and passes output to other processes.

Here we briefly describe the function of each module. Details of the model neuron used are given in full in the appendix, and where coding differs from the use of dedicated nodes to represent individual elements, this is specified.

Vision - This represents the basic visual input available to the model. We model this in an extremely simple manner - the region has dedicated nodes each of which holds a possible view available to the model, and they are activated depending on the simulation setup.

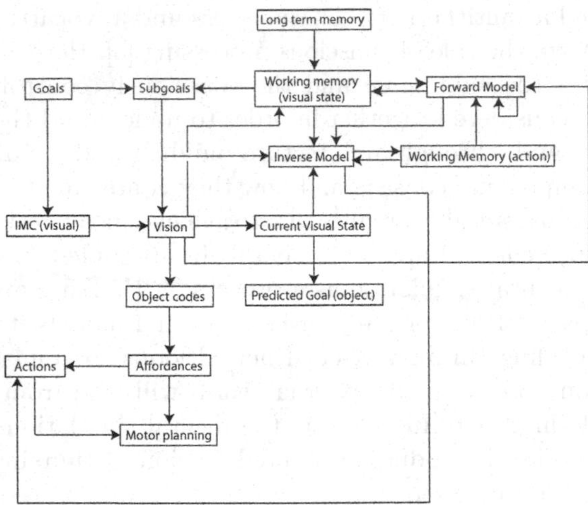

Fig. 1. Details of the Full Mental Simulation Loop with Visual Attention. The functions of the various modules present in the figure are described in the text.

Object codes - Here objects already known to the system are stored, such that they can be activated by the object representation module. Each object has a corresponding node.

Goals - Here the simulation's current goal is stored and used to influence behaviour. Goals are represented by dedicated nodes. The module stores higher level end goals (such as opening a box) without specific details of how those goals are to be accomplished.

Subgoals - This region codes for lower level goals that form the components of a higher level directive - for example when opening a box subgoals might include unlatching the lid, or removing a cover. The subgoals are coded as dedicated nodes.

Affordances - This area relates objects to their uses. We suggest that object uses are coded as the affordances offered by those objects, which can then be realised through actions used on those affordances. These are coded by nodes representing these affordances (such as the affordance of opening primed by a box object).

Actions – Here the actions necessary to achieve affordances have representations. These representations are primed by the affordances module, so the affordance of opening would activate an action representing a specific opening movement.

Inverse model – The IM is used to determine what action will achieve the goal supplied to it given the current working visual state. In a simple simulation, this output is produced by predetermined connections between a set of possible visual states and goal nodes, and output actions. These can also be learned through Hebbian style plasticity. The IM outputs the result coded as an action the system can perform

Working memory (action) – This area acts as a buffer for the actions produced by the IM, as one of a set of precoded action representations.

Forward model – The FM calculates the effect of a proposed action on the current state of the real or imagined world. This produces a new state. Similarly to the IM the module contains state nodes and action nodes as inputs, which are connected to state nodes as outputs, with connectivity that can either be pre-wired or learned.

Working memory (visual state) – This area holds the currently imagined visual state (as one of a set of possible states represented by dedicated nodes) produced by the forward model in a buffer such that it can be supplied to the inverse model to reach the next step towards achieving the simulated goal.

Long term memory – This allows information to be supplied to the forward model about the consequences of actions that require extra knowledge, such as imagining the room that lies behind a door and would be visible as a result of opening that door.

The attention control modules and connections in the architecture of figure 1 are as follows:

The Visual Attention Inverse Model Controller (denoted IMC (visual) in the figure): This module sends an attention feedback signal to the vision modules. Only one such module is shown in figure 1 but sensory attention can also be directed to object codes and possibly to affordances, so these modules should also be connected to the IMC(visual). The function of this attention signal is to amplify neuron activity relevant to the goal being pursued by attention (and reduce that caused by distracters).

The connection from the Goals module to the IMC (visual): This connection biases the competition assumed to be occurring on the IMC (visual) so as to allow for the attainment of attention amplification satisfying the focussing of attention on the goal object.

The connection from the Vision module to the Working Memory (visual state) module: This allows for the attention-amplified lower level activations representing the attended object to attain the visual state working memory so as to be available for report (so be in consciousness).

The connection from the Inverse model to the Action module: This connection enables the action **u** generated by the inverse model to be used to bias the Actions module representations. This causes as output from the Action module that action representation corresponding to **u**. Such a biasing process corresponds to the model of motor attention of [10], in which the Actions module acts as the plant in a control model in which the action IMC is the inverse model of figure 1. We see that the Taylor-Fragopanagos model is extended, in figure 1, by the presence of the forward model, thereby allowing more flexibility by possible imagined manipulations of the visual state to check for modifications a given action could produce. We emphasise that the visual state influence on the action, in both the original model and our present one, is that of the attended state in consciousness. We should add that this is not the only influence on the Actions module: there is also input from Affordances, as is expected also to influence the action being generated. This affordance may have been in consciousness (with suitable further connections to allow the affordance values to be attended to and the resultant activation attain a relevant working memory site) or not in consciousness (as in the connectivity of figure 1).

The architecture of figure 1, without the visual attention components mentioned above, was used in [11] to model the results of observational learning by infants in a paradigm requiring a sequence of movements to open a hierarchy of boxes contained in boxes. This was achieved in the simulation by the infant performing a mental simulation of each stage of the box openings, and then performing such action in reality. There was no need for use of the visual attention components of the architecture of figure 1 since there was only one object in the field of view and also there were only the simplest forms of internal models (the FM and IMC). At no point was there need to filter out distracters, nor to allow lateral spreading of activity in object maps in order to solve difficult problems by analogy. How this latter feature of attention and its removal allow for creative thinking will be considered in the next section.

3 The Mechanism of Conscious Attention in Thinking

We consider the architecture of figure 1 as supporting the process of thinking at the two levels we described in the introduction: at conscious and at unconscious levels. In order to switch between these levels it is necessary to consider in more detail than heretofore the visual attention components in the architecture of figure 1, especially the visual attention inverse model controller and the further attention connections included in figure 1. It is through these, in concert with the other modules already present and some additional ones to be mentioned, that it will be possible to see how two levels of processing, conscious and unconscious, will be possible with the architecture.

We described briefly in the previous section the manner in which visual activity can be used as part of the motor control system. In addition, and as seen from the architecture of figure 1, it is also possible to see how the position of the Working Memory (visual state) module as sandwiched between the forward and inverse models allows there to be consciousness of the set of visual states in a mental simulation loop. Numerous neural models of consciousness have been proposed ([11,2] and references therein), but few have used attention as the gateway to consciousness. One such model is the CODAM model ([9] and references therein), where CODAM stands for Corollary Discharge of Attention Model; we refer interested readers to that reference.

We need now to consider how the visual states in a mental simulation loop can be taken out of consciousness but yet be part of the mental simulation loop. This can be achieved by the insertion of a switching device to allow output from the forward model to avoid the Working Memory (visual state) module, so totally avoid conscious report. This switching device is based on an error module, as in the CODAM model of attention [9]; we propose it is part of the cingulate activity in the brain.

The mechanism to achieve mental simulation at a non-conscious level is by means of the connection lines in figure 1 described in the previous section, which avoid the Working Memory (visual state) module:

1. The direct connection from the forward model to the inverse model. This enables the inverse model to produce the next action to achieve the sub-goal.
2. The direct connection from the visual state module to the forward model. This will allow generation of the next state brought about by the new output of the inverse model and the visual state.
3. Recurrent connection of the FM to itself if there is a sequence of virtual states to be traversed.

Let us turn to the example mentioned in section 1, of giving unusual uses for an object: we take a cardboard box as an example. We can say "As a hat" as one such unusual use. That could arise from the flow of information in our brains:

Cardboard box (in picture or as words) → input processing → box nodes in object map → hat nodes in object map (by learnt lateral connections) → hat nodes in affordance map (by direct connections from the hat node in the object map and by lateral connections from the box representation of affordances to the hat representation there) → test of viability of putting on the box as a hat.

If the test of viability works, then the 'putting on hat' action becomes attended to and there is report, either by putting on the box as a hat or saying "As a hat". If the box is too large to fit stably on our head then we put it on our head and keep our hands on it to steady it; if the box is too small then we may desist from saying it could be used as a hat, or try it on as a little 'pillar box' hat.

These various responses indicate that we try out subliminally what happens if we try to put the box on our head, using the simulation loop. If successful and the action is viable then we attend to it, and hence report it. If it is not we move on to another subliminally-analysed use.

To achieve the subliminal processing stage as well as the final report there must be an attention switch, generated as part of the IMC(visual), so that when there is an attention control signal output there is normal transmission from the forward and inverse models to their relevant working memory modules shown in figure 1. When there is no attention then the mental simulation loop circuit functions without the relevant working memory modules. It thus functions in a subliminal or unconscious manner. There will need to be an extra module for assessing the relevance of states achieved during this unconscious activation of the loop; that will be fed by the forward model in parallel with the self-recurrence (or external running of the FM) and the signal to the inverse model. Given an error-based output from this assessment module then its output would be used to bring attention to the final state and the sequence of intermediate states (assumedly not many) so as to attain the sub-goal more explicitly.

The reason for the presence of the switch itself is that of allowing the reasoning process to go 'underground' when an apparently insuperable obstacle is met by the conscious reasoning system. This may be seen as part of the extended reasoning system discussed, for example, in [1]. However such a switching process plays a crucial role in the truly creative cognitive process. When a blockage is met in 'simpler' logical reasoning then the attention control of processing has to loosen its iron grip on what is allowed to follow what in the processing, with increased reasoning and recall efficiency by subliminal-level processing. This feature is well known, for example, in answering quiz questions and solving puzzles of a variety of sorts. So the switch into the subliminal mode may be achieved in the case of quizzes or creative processes such as painting or other artistic acts from the start of the search or creation process. In more general reasoning, the creative and subliminal component need only be used at points where logic gives out and more general 'extended' and creative reasoning has to step in.

In the case of our example of unusual uses of the cardboard box, the attention switch is assumed to be turned off by the goal 'unusual uses', since we know that going logically (and consciously) through a list of all possible uses of anything will not get us there, nor any other logically-based search approach. We have learnt that we need to speak 'off the top of our head', in an unattended manner. So we can regard, in a simulation of this task, that we are not using attention at all after the switch has turned it elsewhere, or reduced it to a very broad focus.

From this point of view there may well be access by the internal models during this creative phase to a considerable range of neural modules for memory of both episodic and semantic form right across the cortex. The best approach to model this would thus be to have these connections develop as part of earlier learning processes, but such they can function initially in an attentive phase and then be useable in a subliminal one. But the presence of the unattended learning of the required lateral connections may also be possible and need to be considered.

4 Simulation Results for Unusual Uses of a Cardboard Box

To simulate the paradigm involving imagining unusual uses of a cardboard box, we emphasise certain aspects of the model described in section 1. In particular we need to allow the use of lateral spreading with object and affordance codes and look at the more specific effects of attention. We can see the architecture of the model to be used here in figure 2 (as an extension of parts of figure 1 to handle the switch between attended and unattended processing).

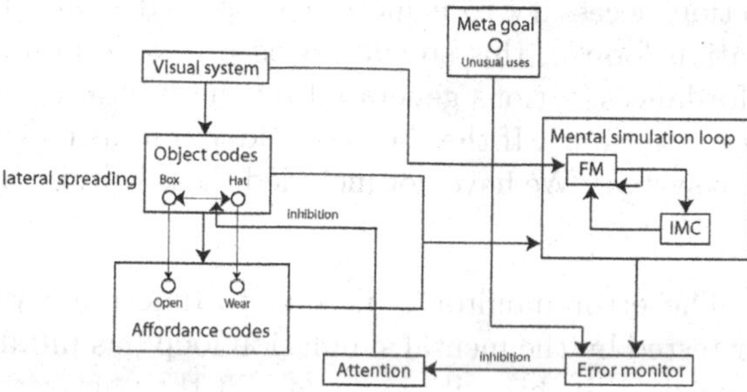

Fig. 2. Simplified Brain-Based Architecture for Creativity: Solving the 'Unusual Uses Task'. The functions of the specific modules are specified in the text

Here we explain the overall function of the specific modules used, giving details as far as the scope of the paper permits:

Meta goal – The overall goal of the simulation is to find unusual uses of the cardboard box. As part of this process, simple action/object goal pairs are created, so we need to code both the overall goal of imagining unusual uses and the immediate goals that are tested to see if they are unusual. We have not specified the immediate goals, since we are unclear if these are used in the creative, unattended lateralisation processing. If the processing is automatic from the affordance/action codes module to the action being taken, and then used by the mental simulation loop, then no such immediate goals module is needed; that is what has been used in our architecture and simulation. If needed it can be included between the affordance codes and the mental simulation loop without any expected change in the results we report below. The meta goal is coded as a single dedicated node (with the possibility of adding more nodes for expansion, either as a distributed representation or to include other meta goals.).

Object codes - Here objects are represented by single nodes. Lateral excitatory connections between the nodes allow similar objects to be activated by this spreading, in addition to visual stimulation. It is these lateral connections which allow analogy.

Affordance codes - The affordance module contains nodes representing specific affordant actions that can be used on objects (such as the action of opening a box). These are primed by the object code module using pre-selected connections.

Mental simulation loop – In the full model shown in figure 1, the mental simulation loop incorporates a forward model (FM), inverse model (IM) and buffer working memories. In unattended mental simulation we suggest that these working memories are not active, such that activity passes straight between the FM and IM. The forward model generates an expected result of carrying out the action (these are pre-coded in this simplified model) while the inverse model determines the action necessary to achieve a suggested state. The function of the mental simulation loop in this simulation is to test subliminally the pairs of objects and affordances/actions generated by the lateral spreading to see if they are considered "unusual". If the use is considered unusual then attention is brought back to the system. We have not included these working memory buffers in figure 2.

Error monitor – The error monitor is needed to determine whether a given object/action pair tested by the mental simulation loop has fulfilled the goal criterion of being "unusual". If this criterion is met, it then activates the attention control module such that attention is restored to the goal of finding an unusual use for the box. In this simulation the error monitor compares the selected action result (passed on from the mental simulation loop) against an internally maintained list of those considered novel.

Attention – Here we use a more specific property of the attention control system than that used so far. In particular we now require the attention system to control lateral spreading in both object and affordance modules by inhibition of lateral connections. In our model, this occurs by output from the attention module stimulating the inhibitory connections present in the object code module. When attention is focussed, representations will be activated singly in each region, while after the removal of attention activity can spread to similar representations (we assume that the organisation of the module is such that similar objects are laterally connected). How this attentional attenuation of lateral connection takes place at the neurobiological level is indicated to some extent by studies of visual attention [3,4]. We have not included the working memory buffers, present in figure 1, in figure 2, so as to keep the architecture as simple as possible, although they should be there; they play no direct role in our simple simulation.

We can see the flow of activations of the simulation areas in the following chart. Activation can be split into two phases, where the first activates the goal of finding an unusual use and tries the action of opening the box which is found not to be unusual. The second, after attention is relaxed, spreads activity such that the extra object (the hat) and its affordances become involved.

The flow paths in the upper diagram carry attention-controlled processing. That in the lower diagram have no attention focussed on them, so allowing more lateral spreading between concepts, as shown in the first line of that flow.

We then simulate the model using a neural network simulation package designed for simple timestep simulations and record the activations for key nodes which we can then plot using a spreadsheet based graph tool. We look at activations of specific nodes from the model in the following figure 5.

Here we see that initially attention is active. Presentation of the box activates the corresponding object representation – the action of opening. This pair of object/action (open box) is tested by the mental simulation loop (see next figure 5). This fails to achieve the overall goal of finding an unusual use of the box and the

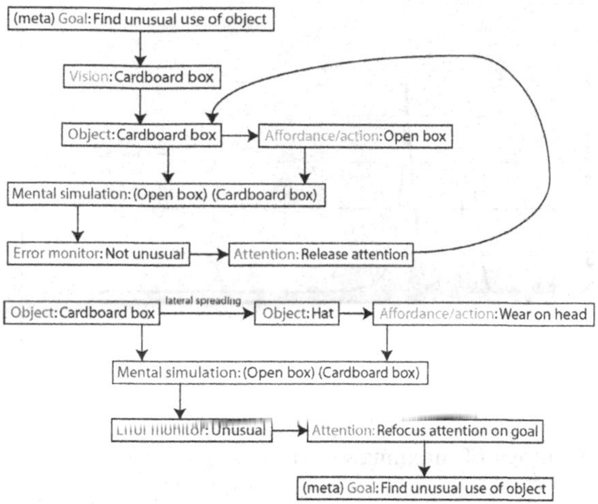

Fig. 3. The Information Flow during the Creative Act

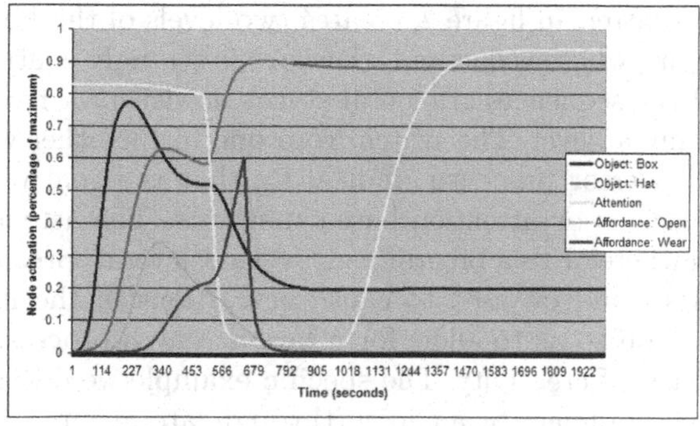

Fig. 4. The Dynamical Activations of the Neurons Involved in the Simulation I, activation level is a percentage of maximum output

error monitor then turns off attention. This allows lateral spreading to activate the hat object through lateral spreading from the box object. The hat object then activates the affordant action of wearing.

Figure 5 gives the activation of the mental simulation area as well as the goal and attention areas. We see two peaks in activation for the MS loop, the first being when the object/action pair "open box" is considered and the second when the pair "wear box" is tested. After this second pair is seen to achieve the goal, attention is reactivated. This reactivation of attention allows reporting of the goal achievement to occur, and also stabilises the representation of the selected action and object by inhibiting the lateral spreading.

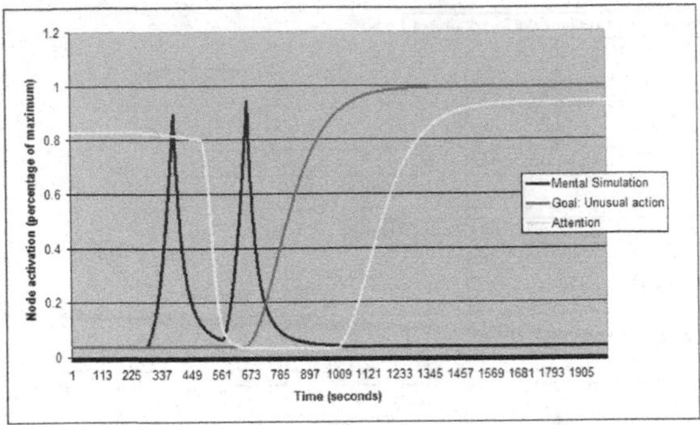

Fig. 5. The Dynamical Activations of the Neurons Involved in the Simulation II, activation level is a percentage of maximum output

5 Conclusions

We have presented a complex architecture for mental simulation and more general thinking. This latter, in figure 2, creates two levels of thinking. One is that of conscious thought, whose sequential structure is strongly controlled by attention on goals and the sequences of mental states needed to arrive at them. The other is at a subliminal level. The switch from one to the other was determined in our model by an error processor, indicating that the goal was not directly achievable by conscious or attention-based methods. The attention focus was then switched elsewhere or to a broad focus, so that lateral spreading in concept and other modules could be used to cause activations in the internal models of the simulation loop so as to allow for a larger search process. This latter is regarded as the basis of creativity. The specific example we developed (unusual uses for a hat) used argument by analogy (through lateral spreading), which has been argued for by some as basic to all creativity [7]. It is clearly so for surrealist artists (consider Duchamp's famous urinal, used unusually in an art gallery as an exhibition), and can be seen to be so for a much broader range of creativity in art and literature.

An important component in the success of such subliminal thinking is clearly the size and storage capacity of the modules involved in the process. The larger such sizes (developed through attention-controlled learning) then the more effective will be the creative process in attaining a solution to the task. This may help explain why the various factors which have been explored for creativity by such as the Torrance Creativity Tests (TTCT) [12], including fluency, flexibility, adaptability and perseverance, appear to be correlated. These creativity metrics, in our approach, are expected to be correlated, since they will depend on the capacity of the network of modules most crucially involved in the creative act; this network depends crucially on the subset involved in lateral spreading in an attention-free manner.

There are numerous open questions posed by this work. Firstly they involve the details of the brain processes, as predicted by our model architecture (this may be regarded as the ground truth of the model). Is the activation of cingulate especially concerned with the switching process, and acting as an error monitor? If it were then there should be a correlation between such activation and the resulting spreading of activity in the object and affordance modules (temporal lobe and tempero-parietal junction). There are further predictions made by the model, such as the further on- or off- switches during such creative reasoning expected in parietal sites of the visual IMC for attention; these switches would be expected to be absent in attention switching per se. Other than these unanswered questions arising from the model there is considerable ground truth supporting the underlying attention model of thinking of figure 1 [11] as well as its extension to figure 2 (part of the cingulate as acting as an error corrector, for example).

We make use of the concept of a "meta goal", which is an overall goal (in this case to find unusual objects) that relies on testing combinations of actions and objects that also in some sense represent goals (such as "open box" or "wear box"). How and where these different types of goal are coded is not fully understood at present. It can be conjectured to arise from a hierarchy of self-organising maps in prefrontal cortex, with the highest level corresponding to that of meta-goals. Additionally, the mechanism by which removal of attention allows lateral spreading of activation within object codes is unclear – while similar mechanisms are known to exist for spatial attention this process takes place in a higher level of representational space. With topography in the relevant concept space, then removal of attention (acting in a topographic manner) would remove inhibition of distracters, seen as nearby concepts; in this manner there would be lateral spreading.

The way in which the lateral spreading might work also has significant effect on the types of creative reasoning possible. For example a small cardboard box may be considered similar to a hat, while a large cardboard box is like a car. However a car is unlike a hat, and two cardboard boxes are like each other. Whether object coding contains both semantic and physical properties and how factors such as object scale are coded is an interesting topic for further research.

There are also many questions about the developmental aspects of the modules of figures 1 and 2: are they learnt under attention or is there also unattended learning occurring during visual processing? This could be tested by detecting the ability of an infant to creatively handle a new object subliminally even though it had only been exposed to it without any attention given to it. This and numerous other questions are open and of interest to study.

Acknowledgments. The authors would like to thank the EU Cognitive Systems Unit for financial support during the project MATHESIS during which part of this work was done.

References

1. Clarke, A.: Being There. MIT Press, Cambridge (2004)
2. Edelman, G.M.: Naturalising consciousness: A theoretical framework. PNAS 100(9), 5520–5524 (2003)
3. Fang, F., Boyaci, H., Kersten, D., Murray, S.O.: Attention-dependent representations of a size illusion in human V1. Curr. Biol. 18(21), 1707–1712 (2008)
4. Friedman-Hill, S.R., Robertson, L.C., Desimone, R., Ungerleider, L.G.: Posterior parietal cortex and the filtering of distracters. Proc. Natl. Acad. Sci. USA 199(7), 4263–4268 (2003)
5. Gladwell, M.: Blink, The power of thinking without thinking. Penguin Books, London (2005)
6. Norretranders, T.: The User Illusion: Cutting Consciousness Down to Size. Penguin, London (1997)
7. Sowa, J.F., Majumdar, A.K.: Analogical Reasoning, in Proceedings of the International Conference on Conceptual Structures (Dresden). In: Ganter, B., de Moor, A., Lex, W. (eds.) ICCS 2003, vol. 2746, pp. 16–36. Springer, Heidelberg (2003)
8. Taylor, J.G., Freeman, W., Cleeremans, A.(eds.): Brain and Consciousness. Neural Networks 20(9), 929–1060 (2007)
9. Taylor, J.G.: CODAM: A Model of Attention leading to the Creation of Consciousness. Scholarpedia 2(11), 1598 (2007)
10. Taylor, J.G., Fragopanagos, N.: Simulations of attention control models in sensory and motor paradigms. In: Wunsch II, D.C., Hasselmo, M., Venayagamoorthy, Wang, D. (eds.) Proc Int Conf Artificial Neural Networks (ICANN 2003), pp. 298–303. IEEE Press, Los Alamitos (2003)
11. Taylor, J.G., Hartley, M.: Towards a neural model of mental simulation. In: Kůrková, V., Neruda, R., Koutník, J. (eds.) ICANN 2008,, Part II. LNCS, vol. 5164, pp. 969–980. Springer, Heidelberg (2008)
12. Torrance, E.P.: The Manifesto: A Guide to Developing a Creative Career. In: Strangers to Ourselves, Ablex Publ. Co. Wilson TD, Westport. MIT Press, Cambridge (2002)

Appendix - Model Details

The model is implemented using a simple neural simulation programming language.

Modules in the model contain graded nodes, the parameters of which are described below. More complex modules (such as the IMC and FM) use dedicated non neural processing structures. The dedicated nodes consist of graded neurons, the membrane potential of which obeys the equation:

$$C\frac{dV}{dt} = g_{leak}(V - V_{leak}) \tag{1}$$

Where C is the capacitance of the neuron, g_{leak} its leak conductance, V_{leak} its equilibrium potential and I its input current. The output of these graded neurons follows the form:

$$I_{out} = \frac{I_{base}}{1 + e^{\frac{V}{V_{scale}}}} \tag{2}$$

Here I_{base} and V_{scale} are constants controlling the maximum neuron output and its scaling. Parameters have the following default values:

Table 1. Neuron Parameter Values

Parameter	Value	Units
C	25	nF
g_{leak}	0.025	μS
V_{leak}	-70	mV
I_{base}	1	μA
V_{scale}	0.001	mV

Table 2. Neuron Connection Weights

Connection	Weight
Meta goal to Error monitor	0.5
Visual system to Object codes	1
Visual system to FM	1
Object codes (lateral spreading)	0.2
Object codes to Affordance codes	1
Object codes to MS loop	0.5
Affordance codes to MS loop	0.5
FM to IMC	1
IMC to FM	0.5
MS loop to Error monitor	0.5
Error monitor to Attention	-1
Attention to Object codes lateral spreading	-2

These parameter values are chosen to give suitable neuronal response times, changing the parameters significantly would alter the timing response of the neurons and prevent proper operation of the model.

Connections between modules are subject to the following weights (normalised such that relative connection strengths are emphasised):

These connection strengths are chosen through testing the model through a range of parameters. The values are important to within about 30%; beyond this the simulation results change significantly.

BSDT Atom of Consciousness Model, AOCM: The Unity and Modularity of Consciousness

Petro Gopych

Universal Power Systems USA-Ukraine LLC,
3 Kotsarskaya Street, Kharkiv 61012 Ukraine
pmg@kharkov.com

Abstract. At the ground of most brain computations may be minimal abstract selectional machines (ASMs) implementing optimal algorithms of recent binary signal detection theory (BSDT). Using the BSDT ASMs, such fundamental cognitive notions as subjectivity and the meaning of a message have already been defined mathematically. BSDT neural network assembly memory model provides strict and biologically plausible definition of optimal assembly memory units (AMUs, implementations of ASMs) which may be considered as 'atoms' of consciousness (AOCs). The idea of an AOC is here developed into an 'atom' of consciousness model (AOCM) — a mathematical theory of consciousness. Neuronal computational structures leading to the emergence of subjective experience or a 'quale' (a formal solution of the 'hard problem' of consciousness) are presented as complex dynamical hierarchical associations of AMUs/AOCs of infinite prehistory. Within the AOCM framework some cognitive phenomena are explained and it has been demonstrated that unified and modular biological models of consciousness are not antithetical.

Keywords: Neural networks, memory, semantic information, context, binding, feeling, thinking, qualia, neural Darwinism, micro-consciousness.

1 Introduction

Human consciousness is now studied by natural sciences [1,2] and, as any natural phenomenon, it demands to be described on a strict mathematical background which, in spite of many efforts, is absent until now. The main obstacle is the inability of available theories to introduce a mathematical definition of subjectivity — fundamental human cognitive faculty to experience internal feelings, thoughts and other private events available to the experiencing subject only (that is the so-called 'hard problem' of consciousness [3]).

In present work, assuming that the brain is a selectional device [1], a model of consciousness defined in mathematical terms — an 'atom' of consciousness model (AOCM) — is built on the basis of recent binary signal detection theory (BSDT, [4]), BSDT neural network assembly memory model (NNAMM, [5]) and BSDT minimal abstract selectional machines (ASMs, [6]). ASM theory gives a rigorous definition of subjectivity and, together with mentioned and other BSDT results [7,8], underlies mathematically the AOCM discussed below.

C. Alippi et al. (Eds.): ICANN 2009, Part II, LNCS 5769, pp. 54–64, 2009.

2 ASMs and AMUs/AOCs of the BSDT

The BSDT ([4] and references therein) defines in an N-dimensional binary vector space 2^N different N-dimensional binary vectors x with spin-like components $x^i = \pm 1$, reference vector $x = x_0$ representing the information stored or that should be stored in a neural network (NN), and noise $x = x_r$. Vectors $x(d)$ (x_0 damaged to the damage degree, d) are introduced by using a '*replacing*' coding rule:

$$x_i(d) = \begin{cases} x_0^i \text{ if } u_i = 0 \\ x_r^i \text{ if } u_i = 1 \end{cases}, \; d = \sum u_i/N, \; i = 1, \ldots, N, \tag{1}$$

where u_i are marks, 0 or 1 (if $u_i = 1$ then in x_0 its ith component, x_0^i, is replaced by the ith component of noise, x_r^i; otherwise x_0^i remains intact). If m is the number of $u_i = 1$ then $d = m/N$, $0 \leq d \leq 1$; $q = 1 - d$ is a fraction of intact components of x_0 in $x(d)$ or an intensity of cue, $0 \leq q \leq 1$. As the total number of vectors $x(d)$ is 3^N, each x can be presented by at least one of $x(d)$ [4,8].

Binary data coded as described are decoded by a two-layer NN with N model neurons in its entrance and exit layers. For perfectly learned intact NN storing one reference pattern only, x_0, its synapse matrix elements are $w_{ij} = x_0^i x_0^j = \pm 1$. The NN's input $x = x_{\text{in}}$ is decoded (x_0 is identified in x_{in}) successfully if it is transformed into the NN's output $x_{\text{out}} = x_0$ (such an x_{in} is called successful input, x_{succ}); additional 'grandmother' neuron checks this fact. If $x_{\text{out}} = x_0$, x_{in} is interpreted as x_0 damaged by noise; otherwise as a sample of noise, x_r. BSDT NN decoding algorithm can also be presented in functionally equivalent convolutional and Hamming distance forms each of which is optimal (the best) in the sense of pattern recognition quality [4].

2.1 ASMs, Meaning, Subjectivity, Qualia and Awareness

ASM theory [6] applies the BSDT to optimal selections. The ASMs (minimal abstract selectional machines) are hypercomplex super-Turing goal-specific learnable selectional devices whose properties are defined by their prehistories having evolutionary and developmental components. *ASM prehistory* defines both the ASM itself (complete *process* of its creation including causes, forces, rules, etc) and the *meaning* of x_0 selected by it, $M(x_0)$. $M(x_0)$ is the content of prehistory of ASM selected the x_0 and may be in either implicit, $M_{\text{impl}}(x_0)$, or explicit, $M_{\text{expl}}(x_0)$, form. An ASM's internal architecture, the set of its parameters, its relations to the environment, etc in their primary intrinsic physical (material, truly continuous, implicit) form constitute jointly the $M_{\text{impl}}(x_0)$ which, due to its objective material essence, is always true but never available outside the ASM. In that sense, $M_{\text{impl}}(x_0)$ is fundamentally *subjective* and may be interpreted as a *quale* [1-3] of the x_0 and all others $x_{\text{in}} = x_{\text{succ}}$ recognized as x_0 by this ASM. In other words, $M_{\text{impl}}(x_0)$ is the prehistory recorded in the form of intrinsic physical properties of the ASM selected the x_0 and has no symbolic form even in principle.

To communicate the $M_{\text{impl}}(x_0)$ to a distant ASM, the ASM-sender has initially to transform its $M_{\text{impl}}(x_0)$ into a *symbolic* form. The description obtained is explicit meaning of x_0, $M_{\mathrm{expl}}(x_0)$, that can already be available for an ASM-receiver. Because separate ASMs have no tools to watch themselves, without help they cannot read out their own physical states in order to translate their implicit meanings into an explicit (symbolic/sign) form. For this reason, ASMs can exchange by symbolic information if they are inherent parts of an ASM society[1] — a set of interacting ASMs equipped additionally by sensory and executive devices [6]. As ASM prehistory spans infinitely back in time [6], $M_{\mathrm{expl}}(x_0)$ (its complete text written by using, e.g., -1s and $+1$s) has in general *infinite*[2] length or Kolmogorov complexity [9] and, consequently, cannot be communicated.

This paradox can be solved supposing that all ASMs share common initial parts of their prehistories interpreted as their common *context* (or *semantic* cue, c) of *infinite* length: $M_{\mathrm{expl}}(x_0) = x_0 + c$ where '+' means an adjunction of strings (in Kolmogorov complexity theory, x_0 and c are called prefix and cylinder [9, p.13-14]). If for the ASM-sender (ASM$_i$) and ASM-receiver (ASM$_j$) their meanings are respectively $M_{\mathrm{expl}}(x_0)_i = x_0 + c_i$ and $M_{\mathrm{expl}}(x_0)_j = x_0 + c_j$ then their 'difference,' excluding common infinite initial parts of c_i and c_j, is $\Delta M_{\mathrm{expl}}(x_0)_{ij} = |c_i - c_j| = \Delta c_{ij}$. The smaller the Δc_{ij}, the greater the *similarity*[3] between ASM$_i$ and ASM$_j$ is; for ASMs of the same *category*, $\Delta c_{ij} = 0$. For correct decoding ('understanding') the $M_{\mathrm{expl}}(x_0)_i$ by ASM$_j$, given their common infinite context, it is enough to have a finite amount, $x_0 + \Delta c_{ij}$, of bits of semantic information. Hence, for communicated ASMs, it is their common context that makes possible conventional computations over infinite strings $M_{\mathrm{expl}}(x_0)$.

Cosmological models predict that total number of bits available in the universe has to be either $\sim 10^{120}$-10^{122} or infinity [10]. All extant life on the Earth has a universal ancestor — the basis of the universal phylogenetic tree [11]. The later the tree's peripheral branching occurs, the longer the common evolution prehistory of related animal species is and the shorter their individual evolution stories are. These findings support assumptions on both the one-side infinity of $M_{\mathrm{expl}}(x_0)$ and the identity of initial infinite parts of $M_{\mathrm{expl}}(x_0)$, for respective ASMs, in animals of cognate species.

Passive ASMs are devoted to a routine feedforward classification of their inputs, x_{in}. Active ASMs are designed to search, among their inputs x_{in}, successful ones, x_{succ}, corresponding to a given x_0. An active ASM runs in a cyclic manner while an x_{succ} will be found or while the number of consecutive failure attempts achieves a limit defined beforehand. In latter case, active ASM requests for an explicit advice from ASM society whether or not to continue the search [6].

---

[1] The size of ASM society, though not specified here, must be larger than a threshold.

[2] Finite-length $M_{\mathrm{expl}}(x_0)$ may represent pure mathematics. In this special case, $M_{\mathrm{expl}}(x_0)$ has no corresponding $M_{\mathrm{impl}}(x_0)$ and defines e.g. a calculus' vocabulary, axioms, formation and inference rules and, as a result, its *meaningless* signs and mathematical truths [13]. They become *meaningful* when the calculus will be embedded into a real-world model whose meaning, as the meaning of any real-world notion, is specified by its $M_{\mathrm{impl}}(x_0)$ and/or by its $M_{\mathrm{expl}}(x_0)$ of already infinite length.

[3] In algorithmic/Kolmogorov complexity theory, similarity is defined in another way — as the length of the shortest algorithm for transforming the comparing strings one into the other [9].

Operating over qualia, $M_{\mathrm{impl}}(x_0)$, or/and over end-side finite fractions, $x_0+\Delta c_{ij}$, of infinite meaningful strings $M_{\mathrm{exmpl}}(x_0)$ that have common infinite initial parts (context), ASM society generates the required advice. We interpret this essentially nontrivial act as an act of awareness of the meaning of x_0, $M(x_0)$, in an 'aha moment' [12]; otherwise $M(x_0)$ remains unconscious.

2.2 AMUs/AOCs for Memory and Consciousness

The NNAMM [4,5] applies the BSDT to solving memory-related problems. According to it, a cell assembly or a cell 'coalition' [2], responding to a perceptual stimulus and corresponding to a particular AMU, is allocated and activated by a dynamic spatiotemporal synchrony mechanism. This AMU (assembly memory unit, it stores an x_0) is associated with AMUs representing other elements of this stimulus, they may in turn be associated with AMUs representing features/attributes of these elements, etc (in such a way a hierarchy of overlapping AMUs is built, (2) and Fig. 1). The NNAMM implies also a specific stage of memory running — memory activation — when target AMU is allocated and activated together with AMUs connected to it (they constitute its contextual surround, 'fringe' or 'penumbra' [2]). Target AMU is an AMU responding to focal stimulus/feature — stimulus in the focus of attention. If it is activated then other AMUs closely associated with it are also activated. Within particular set of interacting AMUs, constituting working, 'fleeting' or 'iconic' memory, the trace x_0 from one AMU may be used as a cue by another one.

Each BSDT AMU is constructed as specific to its memory trace complicated hierarchy of BSDT universal NN units (UNNUs) [5,7]. The hierarchy's lower-level UNNUs and its apex UNNU are real-neuron implementations of passive and active ASMs, respectively; each of them is learned to remember its specific reference pattern x_0 and is feeding by vectors x_{in} produced, in a feedforward fashion, by lower-level UNNUs for higher-level ones. Thus, particular AMU stores its memory item in a *semi-representational* way: as representational code x_0 remembered in an apex UNNU and, simultaneously, as specific to this memory record hierarchy of learned UNNUs generating, for the apex UNNU, its inputs x_{in} from continuous input sensory signals. We refer to particular AMU or its complete learned UNNU-hierarchy as a neural subspace [7] which is an implementation of a partial ASM society. The whole memory (neural space), a set of interactive AMUs understanding as their complete hierarchies of UNNUs, is a counterpart to the whole ASM society.

In an AMU, its low-level UNNUs run routinely and meanings of their x_0 never become conscious. But its apex UNNU (and, consequently, its complete hierarchy) cannot run without advices from ASM/AMU society and every time, when the AMU requests for such an advice, the meaning of its x_0 becomes aware (conscious). For this reason, any AMU may be interpreted as an 'atom' of consciousness (AOC) [5] where 'atom' means 'the smallest entity having that (emergent) property.' An AMU/AOC is a brain-scale object because its most probable biological implementation includes, though widely distributed but individually specified, neural circuits related to the cortex (e.g., here the networks housing NN memories themselves may be stored [5]) and brain subcortical areas (e.g., thalamus, serving as an ASM's N-channel scanner [6], or hippocampus and amygdala where 'grandmother' neurons may be placed [7]). It is important that in the construction of real AMUs/AOCs (their hierarchies of UNNUs) not only the brain but also body peripheral neural circuits, serving senses and motion, may be involved. That is one of the reasons why BSDT AOC idea is consistent with popular concept of embodied cognition [14].

AMU is an implementation of ASMs for solving memory-related problems, AOC is an AMU applied to solving consciousness-related problems. Hence, in relevant context, these terms can next be used either differently or as synonyms.

3 Dynamic Hierarchical Associations of AMUs/AOCs

Let AMU is a real-brain assembly memory unit in its primary physical form while the same in italic, AMU, is its complete symbolic description: $AMU = M_{\mathrm{expl}}(x_0)$. In further equations (including Sect. 5) we shall use AMU and $M_{\mathrm{expl}}(x_0)$ because, as any other formulae, these ones can have a sign form while AMU implicit meaning, $M_{\mathrm{impl}}(x_0)$, has not. Thus, AMU_i is, though of infinite length but complete, symbolic description of an AMU_i representing a working, 'fleeting' or 'iconic' memory for a scene/stimulus at the ith instant. The AMU_i in its active state is

$$AMU_i = AMU_i^t + \sum_{j \neq t} AMU_i^j, \tag{2}$$

where AMU_i^j ($j = 1, ..., J$) represents the jth feature/attribute of the ith scene and J is the number of these features. Equation (2) defines also an AMU hierarchy (cf. Fig. 1) where AMU_i and AMU_i^j correspond to a higher-level (apex or dominant) AMU and to one of its lower-level AMUs, respectively (dominant AMU defines its relation to lower-level AMUs and their relations to each other). The sign '=' means 'includes dominant AMU and,' the sign '+' means 'associations with dominant and other AMUs.' In words: an active memory state with the dominant AMU_i includes AMU_i itself and J memories AMU_i^j interacting with the AMU_i and each other with strengths that are not specified so far. Each AMU_i^j could in turn be written as $AMU_i^j = \sum AMU_i^{jk}$ where an AMU_i^{jk} ($k = 1, ..., K$) is the kth feature of the AMU_i^j, each AMU_i^{jk} could in turn be written as $AMU_i^{jk} = \sum AMU_i^{jkl}$ where an AMU_i^{jkl} ($l = 1, ..., L$) is the lth feature of the AMU_i^{jk} and so forth. The depth of hierarchy (maximal number of AMU upper indices) and the size of it (total number of AMUs constituting it) are not limited in theory; in practice, AMUs with small activities are excluded from the consideration. For the depth and the size of a hierarchy, 3-5 useful levels [15] and magical short-term memory capacity for 4 or 7 ± 2 items [16] may respectively be considered as their empirical upper limits. In right-hand side of (2), AMU_i^t and separate 'sum' of AMUs are respectively focal AMU and its context or 'fringe.' For an acute attentive state, activation degree of the AMU_i^t (target AMU) is larger than that for its 'fringe'; for a diffuse attentive state, activities of constituting it AMUs have rather similar values.

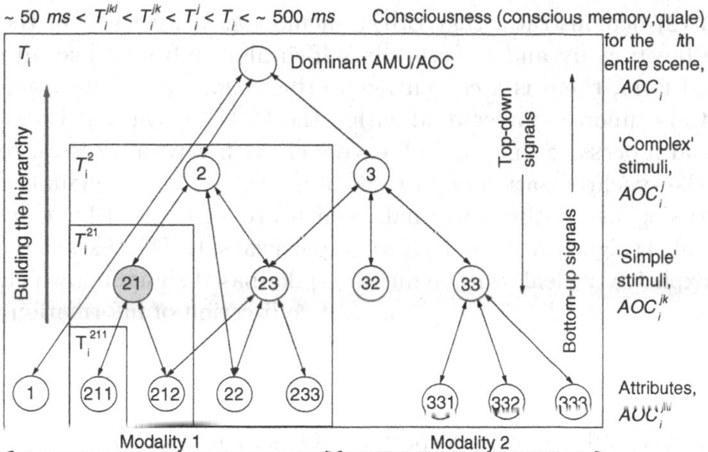

Fig. 1. An AMU/AOC hierarchy representing a percept of an entire scene and its microgenesis [17]; two unspecified modalities are taken into account. Circles, dominant AMUs specified on the right (their UNNU hierarchies are not shown); numbers inside the circles, superscripts designating the AMUs; empty circle, dominant AMU of the scene; shadowed circle, focal AMU (AMU_i^{21}, here the scene's dominant AMU and focal AMU are not coincide). For Modality 1, rectangles encompass subhierarchies dynamically synchronized during time periods indicated in their left-hand top corners (the subhierarchy's dominant AMU and the time of its binding share their indices); binding periods for particular subhierarchies are related on the top (this relation shed also light on progressive *microgenesis* [17] of the AMU hierarchy — time asynchrony of neural activity leading to the emergence of unconscious percept of entire scene from unconscious percepts of its parts); T_i, the scene's time of binding or *psychological refractory period* [18]; as locations of circles 1 and 32 are ambiguous, their synchronization periods, T_i^1 and T_i^{32}, may be from ranges $T_i^{jkl} \leq T_i^1 \leq T_i^j$ and $T_i^{jkl} \leq T_i^{32} \leq T_i^{jk}$, respectively (circles 1 and 32 are actually shown at their lowest and highest possible levels). Bidirectional arrows connect higher-level AMUs (left-hand side of (2)) with lower-level AMUs (right-hand side of (2)) and reflect *recurrent processing* of sensory signals, e.g. [19]. For the hierarchy presented, fractal-like 'sum' of its AMUs is as follows (for a given level of the hierarchy, the signs '=' are typed with a fixed indention):

$$\begin{aligned}
AMU_i &= AMU_i^1 + AMU_i^2 + AMU_i^3,\\
AMU_i^2 &= AMU_i^{21} + AMU_i^{22} + AMU_i^{23},\\
AMU_i^3 &= AMU_i^{31} + AMU_i^{32} + AMU_i^{33},\\
AMU_i^{21} &= AMU_i^{211} + AMU_i^{212},\\
AMU_i^{22} &= AMU_i^{232},\\
AMU_i^{23} &= AMU_i^{231} + AMU_i^{232} + AMU_i^{233},\\
AMU_i^{31} &= AMU_i^{23},\\
AMU_i^{33} &= AMU_i^{331} + AMU_i^{332} + AMU_i^{333},\\
AMU_i^{212} &= AMU_i^{231}\ .
\end{aligned}$$

AMUs are fed by bottom-up signals (from sense organs, upward arrows) and top-down signals (from higher-order brain areas, downward arrows) though the whole hierarchy is mainly built in bottom-up direction. The tight relations between memory and consciousness (i.e., AOC = AMU) reflect the *intentionality* [1] of consciousness (i.e., that consciousness is in general about objects or events); memory and attention systems, though distinct, are rather common.

Equation (2) demonstrates that BSDT memories, at all levels of their hierarchies, are functionally and structurally self-similar or fractal-like. In (2), despite its final form, there is a content (reflecting, e.g., a machinery generating AMUs/AOCs) cannot be understood within the BSDT because it is devoted to discrete signal processing only and all extremely rich wave processes, responsible e.g. for the synchronization of patterns of neural activity, remain beyond its scope [8]. Analogous/continuous signals, which are well suited for transferring information along rigidly defined/hardwired pathways, BSDT takes into account implicitly; explicitly it deals with symbolic, spike-based signaling, which is better if in a system (e.g., the brain) permanent redirection of information flows is required to respond to permanently changing environment.

Almost everything we said of AMUs concerns AOCs including their biological ground, structure and even the magic number of simultaneously aware stimuli. If in (2) to supersede all AMUs by respective AOCs then it will describe a percept of a stimulus/scene and its attributes to be aware (see Fig. 1 for some details).

4 The 'Hard Problem' and Complex Scene Awareness

We follow the view [1] according to which a theory has only to explain (not to reproduce) subjective phenomena. Our consideration is also constrained to primary (sensory) consciousness [1]. Higher-order consciousness is in general outside of the story.

Figure 1 shows at a given moment a dynamically constructed AMU/AOC hierarchy for the percept of a complex scene consisting of some elements or features represented by their AMU subhierarchies. Each AMU is connected to its specific UNNU hierarchy either directly or through its lower-level AMUs. Generation of apex AMU inputs, x_{in}, is time-consuming; the time required depends on sense organ and particular UNNU hierarchy. Dominant AMU of each (sub)hierarchy receives its inputs after a time period needed for the binding of their lower-level AMUs taken together with their UNNU hierarchies. This progressive specification and stabilization of the scene's percept is known as microgenesis [17] (see also Sect. 5.2). The scene's AMU hierarchy is created for $T_i \sim$ 250-500 ms (that is psychological refractory period [18]). During the T_i, parameters of AMU hierarchy of interest are tuned to current pattern of sensory signals for processing it *routinely*. If this pattern (the scene or its viewpoint) changes then, for the new one, previous parameter tuning becomes inappropriate, all attempts to recognize new pattern with old parameter tuning end in failure, the scene's dominant AMU requests for AMU society advice and previous scene's unconscious quale *becomes conscious* (see Sects. 2.1, 2.2) while the AMU hierarchy is tuning to new pattern of activity. New sensory pattern (scene, its quale) becomes conscious when it will be replaced by the next one, etc. Hence, the scene's feeling/quale gradually emerges *before* and *independently* of the act of its awareness.

As the meaning of x_0 selected/recalled/recognized by the scene's AMU is defined as a process of selection, perceptual memory trace x_0 is jointly specified by all consistently activated links of the chain 'sense organ – sensory pathways

(UNNU hierarchy) – their apex AMU storing the x_0.' This structure and its ongoing activity lead to the emergence of unconscious *feeling* or *quale* of the current x_0-specific sensory input (e.g., the feeling of the color red, $M_{\text{impl}}(x_0)$). *Unconscious* quale of the scene (which may then become conscious, see above) is *hierarchically* constructed of *unconscious* qualia of its components (Fig. 1) in the process of implicit (with no appeal to AMU/AOC society) *perceptual thinking* implying that all qualia are from a very-high-dimensional qualia space [1].

If perceptual chain is disrupted then its vivid quale disappears/mitigates but neural activity of the chain's last link, apex AMU, can be interpreted as an *abstract thought* of this quale which can also be generated by top-down signals, without any activity in sensory pathways. A thought of x_0 could participate in (serial/logical) associations with other (abstract) thoughts or in the process of *abstract* ('higher order') *thinking*.

Additionally to thalamocortical (and cortico-thalamo-cortical) activity needed for running each AMU, feeling (a counterpart to *phenomenal*, verbally unreportable, consciousness [20]) requires the activity in sensory pathways while abstract thinking (a counterpart to *access*, verbally reportable, consciousness [20]) does not. These two patterns of neural activity may correspond to two hypothetical patterns of neural correlates of consciousness (NCC), phenomenal NCC and access NCC [20], and may respectively substantiate implicit (automatic) and explicit (voluntary controlled) styles of learning, memory and behavior.

5 The AOCM and Other Consciousness Theories

5.1 AOCM and Unified Models of Consciousness

One of central ideas of popular theory of neuronal group selection (TNGS or neural Darwinism) is *dynamic core* hypothesis [1]. The dynamic core is a temporally ordered, serial and changeable process of neuron brain activity underlying a person's conscious experience. Current dynamic core pattern or *functional cluster* is associated with a current conscious state from a repertoire [1,21] of such possible states which is 'as large as one's experience and imagination.' Functional cluster is an instant implementation of the dynamic core and in that sense they are equivalent *at a moment* (during the T_i, cf. Fig. 1). Each functional cluster consists in turn of many *neuronal groups* which 'are more strongly interactive among themselves than with the rest of the brain.' Their interaction 'is achieved through the process of reentry — the ongoing, reversal, highly paralleled signaling within and among brain areas' [21].

By analogy with Sect. 3 and (2), we can write $FC_i = \sum NG_i^j$ where at the ith time moment FC_i and NG_i^j are infinite symbolic descriptions of the ith functional cluster and the jth neuronal group of the FC_i, respectively (dynamic core is a sequence of FC_i). This equation reflects also other suggestion of the TNGS: within a short time period the distributive neuronal processes underlying the unified consciousness are highly differentiated (represented by different rather separate neuronal groups and functional clusters) and, simultaneously, highly integrated (neuronal groups are strongly interactive among themselves constituting the current functional cluster).

If to posit additionally that functional clusters and their neuronal groups are structurally and functionally similar organized (i.e., the forms of their formal sign representations are equivalent, $FC = NG$; this assumption is absent from the TNGS and makes up our *expanded* TNGS), then the analogy between the AOCM and the (expanded) TNGS becomes practically complete and we could believe that $FC = NG = AOC$. If it is, then (cf. (2))

$$FC_i = FC_i^t + \sum_{j \neq t} FC_i^j, \; NG_i = NG_i^t + \sum_{j \neq t} NG_i^j, \; FC_i = NG_i^t + \sum_{j \neq t} NG_i^j, \quad (3)$$

where FC_i^t (or NG_i^t) represents the scene's 'target' or focal functional cluster (or neuronal group) while separate sum represents its context or 'fringe.'

The AOCM and the TNGS share also the assumption on the role, for consciousness, of evolution and development, of brain thalamocortical system, of synchrony and memory. Unlike the TNGS, the AOCM allows the existence of 'grandmother' neurons.

5.2 AOCM and Modular Models of Consciousness

The idea of modularity of consciousness was inspired by neurophysiological and neurological data, highlighting multiple spatially separate and functionally different brain areas for processing perceptual inputs of different modalities or submodalities, and by psychophysical experiments, revealing a temporal asynchrony in perceiving different visual scene attributes. Additionally, 'activation of specific neural regions might not only correlate with specific perceptual experiences but could be sufficient to cause them' [23] or 'processing sites are simultaneously perceptual sites' [22].

It is supposed [22], perceptions of separate attributes (their *micro*-consciousnesses) appear in brain modules processing respective sensory data. They are distributed in time and space and organized hierarchically because submodalities differ in their processing times. *Macro*-consciousness is introduced for binding the micro-consciousnesses. *Unified* consciousness is hypothesized as consciousness of 'a perceiving person,' it is consciousness of micro- and macro-consciousnesses and 'requires communication with others and, especially, the use of language.'

By analogy with Sect. 3 and (2), at the ith time moment, T_i, we can write $MC_i = \sum mC_i^j$ where MC_i and mC_i^j are the ith macro-consciousness and the jth micro-consciousness of the MC_i, respectively; the sum reflects the binding of micro-consciousnesses into a macro-consciousness. In this case T_i is explicitly treated as consisting of the ijth time periods, T_i^j, distributed according to their temporal hierarchy (Fig. 1) and representing time intervals needed to perceive the scene's separate ijth attributes.

If to posit additionally that macro-consciousnesses and their micro-consciousnesses are structurally and functionally similar organized (i.e., the forms of their formal symbolic descriptions are equivalent, $mC = MC$; this assumption is absent from original micro-consciousness theory and makes up its *expanded* version we introduce) then the analogy between the AOCM and (expanded) micro-consciousness theory becomes practically complete and we could believe that $mC = MC = AOC$. If it is, then (cf. (2))

$$MC_i = MC_i^t + \sum_{j \neq t} MC_i^j, \; mC_i = mC_i^t + \sum_{j \neq t} mC_i^j, \; MC_i = mC_i^t + \sum_{j \neq t} mC_i^j, \quad (4)$$

where MC_i^t (or mC_i^t) represents the scene's 'target' or focal macro- (or micro-) consciousness while separate sum represents its context or 'fringe.'

The micro-consciousness theory's three-level hierarchy is consistent with the AOCM if to accept the upper limits (see Sect. 3) for the number of hierarchical levels, 3-5 [15], and working memory capacity, 4 or 7±2 [16]. The AOCM implies (Sect. 4) the binding of unconscious qualia of the scene's attributes into its entire unconscious quale which may then become conscious. The micro-consciousness theory assumes conversely that conscious quale of a scene is produced by the binding of already conscious qualia of its attributes (that is a 'post-conscious' binding [22]). This confusion disappears if to remember that the AOCM predicts directly (see Fig. 1) a temporal asynchrony in perceiving different the scene's attributes in experiments designed [22] to separately observe these attributes by compulsory disrupting the entire scene percept's microgenesis.

Thus, expanded TNGS (Sect. 5.1) and expanded micro-consciousness theory (Sect. 5.2) are practically equivalent in the sense of their formal structural, functional and computational organization. Of this, taking into account that the former and the latter are respectively unified and modular models, follows that, in contrast to [21], they and their original versions are actually not antithetical. Fractal-like self-similarity of these theories says that in the brain there is no higher-order executive system or 'homunculus' (but see [2]).

6 Conclusion

On the basis of BSDT formal definition of subjectivity, and without any appeal to non-physical or unknown physical forces, a biologically plausible mathematical theory of consciousness — an 'atom' of consciousness model (AOCM) — has been introduced and a solution of the 'hard problem' of consciousness has been proposed. Within the AOCM, the unity and possible modularity of consciousness as well as some particular cognitive phenomena are explained. Simple computational examples providing, e.g., performance of conscious face recognition can be found in ref. 7.

References

1. Edelman, G.M.: Naturalizing Consciousness: A Theoretical Framework. Proc. Natl. Acad. Sci. 100, 5520–5524 (2003)
2. Crick, F., Koch, C.: A Framework for Consciousness. Nature Neurosci. 6, 119–126 (2003)

3. Chalmers, D.: The Conscious Mind. Oxford Uni. Press, Oxford (1996)
4. Gopych, P.M.: Elements of the Binary Signal Detection Theory, BSDT. In: Yoshida, M., Sato, H. (eds.) New Research in Neural Networks, pp. 55–63. Nova Science, New York (2008)
5. Gopych, P.M.: Foundations of the Neural Network Assembly Memory Model. In: Shannon, S. (ed.) Leading Edge Computer Sciences, pp. 21-84. Nova Science, New York (2006)
6. Gopych, P.: Minimal BSDT abstract selectional machines and their selectional and computational performance. In: Yin, H., Tino, P., Corchado, E., Byrne, W., Yao, X. (eds.) IDEAL 2007. LNCS, vol. 4881, pp. 198–208. Springer, Heidelberg (2007)
7. Gopych, P.: Biologically Plausible BSDT Recognition of Complex Images: The Case of Human Faces. Int. J. Neural Systems 18, 527–545 (2008)
8. Gopych, P.: BSDT Multi-valued Coding in Discrete Spaces. In: Corchado, E., Zunino, R., Gastaldo, P., Herrero, Á. (eds.) CISIS 2008. ASC 53, pp. 258–265. Springer, Heidelberg (2009)
9. Li, M., Vitányi, P.: An Introduction to Kolmogorov Complexity and its Applications, 2nd edn. Springer, Heidelberg (1997)
10. Davies, P.C.W.: Emergent Biological Principles and the Computational Properties of the Universe. Complexity 10, 11–15 (2004)
11. Woes, C.: The Universal Ancestor. Proc. Natl. Acad. Sci., USA 95, 6854–6859 (2003)
12. Bar, M., Kassam, K.S., Ghuman, A.S., et al.: Top-Down Facilitation of Visual Recognition. Proc. Natl. Acad. Sci., USA 103, 449–454 (2006)
13. Nagel, E., Newman, J.R.: Gödel's Proof, rev. edn. New York Uni. Press, New York (2001)
14. Clark, A.: Supersizing the Mind. Oxford Uni. Press, Oxford (2008)
15. Ullman, S.: Object Recognition and Segmentation by a Fragment-based Hierarchy. Trends Cogn. Sci. 11, 58–64 (2007)
16. Glassman, R.B.: Topology and Graph Theory Applied to Cortical Anatomy May Help Explain Working Memory Capacity for Four or Three Simultaneous Items. Brain Res. Bull. 60, 25–42 (2003)
17. Bachmann, T.: Microgenetic Approach to the Cognitive Mind. John Bendjamins Publ., Amsterdam (2000)
18. Pashler, H.: Dual-Task Interference in Simple Tasks: Data and Theory. Psych. Bull. 116, 220–244 (1994)
19. Lamme, V., Roelfsema, P.: The Distinct Models of Vision Offered by Feedforward and Recurrent Processing. Trends Neurosci. 23, 571–579 (2000)
20. Block, N.: Two Neural Correlates of Consciousness. Trends Cogn. Sci. 9, 46–52 (2005)
21. Tononi, G., Edelman, G.M.: Consciousness and Complexity. Science 282, 1846–1851 (1998)
22. Zeki, S.: The Disunity of Consciousness. Trends Cogn. Sci. 7, 214–218 (2003)
23. Cooney, J.W., Gazzaniga, M.S.: Neurological Disorders and the Structure of Human Consciousness. Trends Cogn. Sci. 7, 161–165 (2003)

Generalized Simulated Annealing and Memory Functioning in Psychopathology

Roseli S. Wedemann[1], Luís Alfredo V. de Carvalho[2], and Raul Donangelo[3]

[1] Instituto de Matemática e Estatística, Universidade do Estado do Rio de Janeiro
Rua São Francisco Xavier, 524, 20550-900, Rio de Janeiro, RJ, Brazil
roseli@ime.uerj.br

[2] Eng. Sistemas e Computação, COPPE - Universidade Federal do Rio de Janeiro,
Caixa Postal 68511, 21945-970, Rio de Janeiro, RJ, Brazil
LuisAlfredo@ufrj.br

[3] Instituto de Física, Universidade Federal do Rio de Janeiro
Caixa Postal 68528, 21941-972, Rio de Janeiro, RJ, Brazil
donangel@if.ufrj.br

Abstract. We compare the use of Generalized Simulated Annealing (GSA) to the traditional Boltzmann Machine (BM), to model memory functioning, in a neural network model that describes conscious and unconscious processes involved in neurosis, which we proposed in earlier work. Modules corresponding to sensorial and symbolic memories interact, representing unconscious and conscious mental activity. We previously developed an algorithm, based on known microscopic mechanisms that control synaptic properties, and showed that the network self-organizes to a hierarchical, clustered structure. Some properties of the complex networks which result from this self-organization indicate that the use of GSA may be more appropriate than the BM, to model memory access mechanisms. We illustrate the model with simulations.

1 Introduction

Freud observed that neurotic patients systematically repeated symptoms in the form of ideas and impulses and called this tendency a *compulsion to repeat* [1], which he related to repressed or traumatic memory traces [2]. These traumatic and repressed memories are knowledge which is present in the subject, but which is not accessible to him through symbolical representation. It forms the unconscious mind and as it cannot be expressed symbolically, it does so through other body response mechanisms, in the form of neurotic (unconscious) symptoms, similar to reflexes. With the term *symbolic expression* we refer to the association of symbols to meaning as in language and also other nonlinguistic forms of expressing thought and emotions, such as artistic representations (e.g., a painting or musical composition) and remembrance of dreams. Neurotics have obtained relief and cure of painful symptoms through a psychoanalytic method called *working-through*, which aims at developing knowledge regarding the symptoms by accessing unconscious memories and understanding and changing the analysand's compulsion to repeat [1]. It involves mainly analyzing free associative talking, symptoms, parapraxes (slips of the tongue and pen, misreading, forgetting, etc.), dreams and also that which is acted out in transference.

C. Alippi et al. (Eds.): ICANN 2009, Part II, LNCS 5769, pp. 65–74, 2009.

We have described a model in [3,4,5], where we proposed that the neuroses manifest themselves as an associative memory process, where the network returns a stored pattern when it is shown another input pattern sufficiently similar to the stored one [6]. We modeled the compulsion to repeat neurotic symptoms by supposing that such a symptom is acted when the subject is presented with a stimulus which resembles a repressed or traumatic memory trace. The stimulus causes a stabilization of the neural net onto a minimal energy state, corresponding to the memory trace that synthesizes the original repressed experience, which in turn generates a neurotic response (an *act*). The neurotic act is not a result of the stimulus as a new situation but a response to the repressed memory trace. We mapped the linguistic, symbolic, associative process involved in psychoanalytic working-through into a corresponding process of reinforcing synapses among memory traces in the brain.

These connections should involve declarative memory, leading to at least partial transformation of repressed memory to consciousness. This has a relation to the importance of language in psychoanalytic sessions and the idea that unconscious memories are those that cannot be expressed symbolically. A model emphasizing brain mechanisms for attention, which is essential for conscious activity and which we have not modeled, is discussed in [7]. We propose that as the analysand symbolically elaborates manifestations of unconscious material through transference in psychoanalysis, he reconfigures the topology of his neural net, by creating new connections and reinforcing or inhibiting older ones. The network topology which results from this reconfiguration process will stabilize onto new energy minima, associated with new acts.

Memory functioning was originally modeled by a Boltzmann Machine (BM). However, the power-law and generalized q-exponential behavior we have found, for the node-degree distributions of the network topologies generated by our model, indicate that they may not be well described by Boltzmann-Gibbs (BG) statistical mechanics, but rather by Nonextensive Statistical Mechanics (NSM) [8,9,10]. We have thus modeled memory by a generalization of the BM called Generalized Simulated Annealing (GSA) [9]. In GSA, the probability distribution of the system's microscopic configurations is not the BG distribution, assumed in the BM, and this should affect the chain of associations of ideas which we are modeling.

We review in Section 2 the main features of our associative memory model for the neuroses. In Section 3, we discuss the use of the BM and GSA as memory access mechanisms and in Section 4, we show computer simulation results with properties of these mechanisms. We then draw some conclusions and perspectives for future work.

2 Hierarchical Memory Model for the Neuroses

We proposed a memory organization [4,5], where neurons belong to two hierarchically structured modules corresponding to *sensorial* and *symbolic memories* (see Fig. 1). Traces stored in sensorial memory represent mental images of stimuli received by sensory receptors. Symbolic memory stores higher level representations of traces in sensorial memory, *i.e. symbols*, and represents brain structures associated with symbolic processing, language and consciousness. Sensorial and symbolic memories interact, producing unconscious and conscious mental activity. If the retrieval of a sensorial

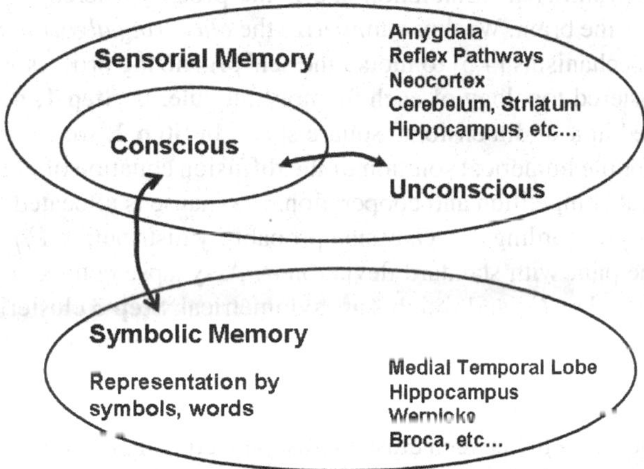

Fig. 1. Memory modules which represent storage of sensorial input and symbolic representations, and also memory traces which can or cannot become conscious

memory trace can activate retrieval of a pattern in symbolic memory, it can become conscious. We refer to the work of Edelman [11] for a neurophysiological discussion of these issues.

Memory functioning was initially modeled in [3,4,5] by a BM with N neurons, where node states take binary values and connections have symmetrical weights [6] $w_{ij} = w_{ji}$. The states S_i of the units n_i, take binary output values in $\{0, 1\}$. Because of the symmetry of the connections, there is an energy functional

$$H(\{S_i\}) = -\frac{1}{2} \sum_{ij} w_{ij} S_i S_j \,, \tag{1}$$

which allows us to define the BG distribution function for network states

$$P_{BG}(\{S_i\}) = \exp\left[-\frac{H(\{S_i\})}{T}\right] / \sum_{\{S_i\}} \exp\left[-\frac{H(\{S_i\})}{T}\right] , \tag{2}$$

where T is the network temperature parameter. The corresponding transition probability (acceptance probability) from state $S \equiv \{S_i\}$ to S', if $H(S') \geq H(S)$, is given by

$$P_{BG}(S \to S') = \exp\left[\frac{H(S) - H(S')}{T}\right] . \tag{3}$$

Pattern retrieval on the net is achieved by a standard simulated annealing process, in which T is gradually lowered by a factor α, according to the BG distribution, given by Eq. (2). A detailed treatment of the BM may be found in [6].

Brain neural topology is structured by cooperative and competitive mechanisms, controlled by neurosubstances, where neurons interact mainly with nearby neighbors, having fewer long-range synaptic connections to distant neurons [11,12]. This is started and controlled by environmental stimulation and is the process whereby the environment represents itself in the brain. We thus summarize the *clustering algorithm* we developed based on these mechanisms [4,5] to model the self-organizing process which results in a structured, clustered topology of each memory module. In **Step 1**, neurons are uniformly distributed in a bi-dimensional square sheet. In **Step 2**, we assume a Gaussian approximation for the numerical solution of the diffusion equation of neural substances that control neural competition and cooperation. A synapse is allocated to connect two neurons n_i and n_j, according to a Gaussian probability distribution P_{ij} of the distance that separates the pair, with standard deviation σ. A synapse connecting n_i to n_j has strength proportional to P_{ij} and weights are symmetrical. **Step 3** clusterizes neurons in the memory sheets, based on mechanisms for forming cortical maps [4,5,6,13], where a group of neurons spatially close to each other represents a sensorial stimulus or an idea. **Step 4** regulates synaptic intensities by strengthening synapses within a cluster and reducing synaptic strength between clusters, disconnecting them. Neurons that have received stronger sensorial stimulation and are more strongly connected, stimulate their neighborhoods and promote still stronger connections.

We represent the association of ideas or symbols (such as in culture and language) by long-range synapses, which should connect clusters by considering the basic Hebbian learning mechanism [6,11,12], where synaptic growth between two neurons is promoted by simultaneous stimulation of the pair. Since we are still not aware of synaptic distributions which result in such topologies, as a first approximation, we allocated synapses randomly among clusters. If the synapse connects clusters in different memory sheets (sensorial and symbolic memories), its randomly chosen weight is multiplied by a real number ζ in the interval $(0, 1]$, reflecting the fact that, in neurotic patterns, sensorial information is weakly accessible to consciousness, *i.e.* repressed.

A full description of the algorithms in the model, with memory storage and retrieval and working-through simulation, can be found in [3,4,5].

3 Generalized Simulated Annealing

In neural network modeling, temperature is inspired from the fact that real neurons fire with variable strength, and there are delays in synapses, random fluctuations from the release of neurotransmitters, and so on. These are effects that we can loosely think of as noise [6,12], and we may thus consider that temperature in BMs controls noise. In our model, temperature allows associativity among memory configurations, lowering synaptic inhibition, in an analogy with the idea that freely talking in analytic sessions and stimulation from the analyst lower resistances and allow greater associativity.

The BM differs from a gradient descent minimization scheme, in that it allows the system to change state with an increase in energy, depending on the temperature value, according to Eq. (3). The Boltzmann distribution function favors changes of states with small increases in energy, so that the machine will strongly prefer visiting state space in a nearby energy neighborhood from the starting point.

The topologies we have generated with the algorithm reviewed in Section 2 are hierarchically clustered [5], containing synapses that connect neurons that are nearest neighbors in spatial coordinates, and also long-range synapses. In Fig. 2 [14], we show the average node degree (k) distributions for 10000 different initial topologies, generated with the clustering algorithm [4,5], for different values of N, such that $N_{sens} = N_{symb} = N/2$ of them belong to sensorial and symbolic memories respectively and $\sigma = 0.58$. Other model parameters are also described in [4,5].

Figure 2 shows an asymptotic power-law behavior with exponent $\gamma \approx -3.2$, which indicates scale independence [10]. It is known that random graphs follow the Poisson distribution of node degrees [10]. For $N = 4000$, the deviation from the fit by the Poisson distribution, $P_\lambda(k) = \lambda^k \exp(-\lambda)/k!$, for $k > 10$ is quite evident. The deviation from Poisson for higher values of k may be attributed to the cooperative-competitive biological mechanisms mentioned earlier, which introduce structure. Smaller values of k correspond to neurons that did not participate significantly in the competition-cooperation process and hence, distributions for small k are approximately fitted by Poisson forms.

Figure 2 also shows a fit by a generalization of the q-exponential function [10,15] given by

$$P_q(k) = p_0 k^\delta \frac{1}{\left[1 - \frac{\tau}{\mu} + \frac{\tau}{\mu} e^{(q-1)\mu k}\right]^{\frac{1}{q-1}}}, \tag{4}$$

where p_0, δ, τ, μ and q are additional adjustable parameters. The curves indicate that asymptotically, the power-law and generalized q-exponential fits are appropriate, with $q \approx 1.113$. This is a common feature of many biological systems and indicates that they may not be well described by BG statistical mechanics [16].

There is no theoretical indication of the exact relation between network topology and memory dynamics. There have been some indications that complex systems which present a power-law behavior (are asymptotically scale invariant) may be better described by the NSM formalism [8,9,10,16]. Since the neural systems we are studying do not have only local interactions and present the scale-free topology characteristic, we have begun to investigate memory dynamics with a generalized acceptance probability distribution function [9] for a transition from state S to S', if $H(S') \geq H(S)$, given by

$$P_{GSA}(S \to S') = \frac{1}{[1 + (q_A - 1)(H(S') - H(S))/T]^{1/(q_A-1)}}, \tag{5}$$

where q_A is a model parameter and other variables and parameters are the same as defined in Section 2.

If one substitutes Eq. (5) for Eq. (3) in the simulated annealing algorithm of the BM, the resulting procedure is called GSA [9]. The GSA procedure presented in [9] also proposes a *visiting distribution function* for generating the possible states S', which we have not yet studied. In the $q_A \to 1$ limit, GSA recovers the BM. The acceptance probability distribution given by Eq. (5) should allow more associativity among memory states, with transitions to more distant minima in the energy functional H than Eq. (3), during the annealing process. This implies that the GSA machine will tend to make many local associations (state transitions) and, more often than the BM, will also make looser, more distant associations. This should correspond to a more flexible and creative memory dynamics in the brain.

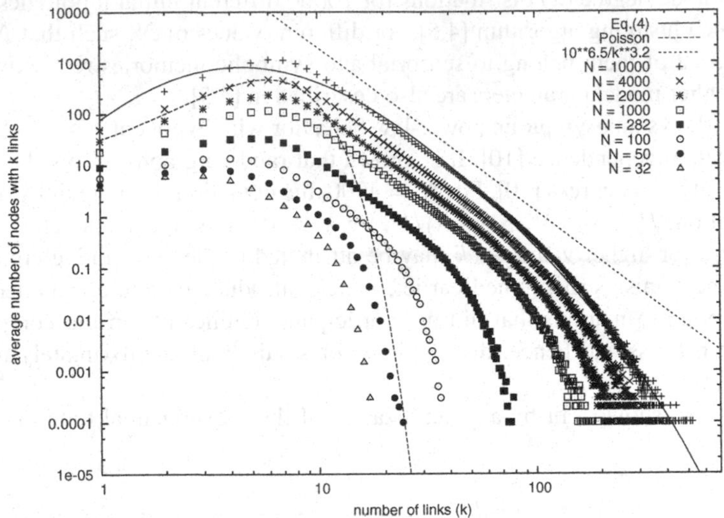

Fig. 2. Average node degree distributions for various N [14]. The fit by Eq. (4) corresponds to $q = 1.113$, $p_0 = 610$, $\delta = 4.82$, $\tau = 2.34$ and $\mu = 0.014$. The Poisson fit corresponds to $N = 2950$ and $\lambda = 6.4$. For large k, there is an exponential finite size effect.

4 Simulation Illustration and Network Properties

We have thus constructed a generalization of the BM, employed in our previous work, and modeled memory access functioning with GSA [9], derived from the just mentioned nonextensive formalism. In GSA, the probability distribution of the system's microscopic configurations is not the BG distribution, assumed in the BM, and this should affect the chain of associations of ideas which we are modeling. To illustrate this, we compare the energies of the patterns accessed by the BM and GSA at two different initial temperatures. States activated by stronger synaptic connections have lower energies (see Eq.(1)) and, although there is degeneracy (different states corresponding to the same energy value), since the acceptance probability distributions of the BM and GSA depend on the energies of states, the frequency of visits to energy states reflects the efficiency of the methods in exploring possible memory configurations. Since we are searching for local minima, we use lower initial temperature values and higher values of the annealing schedule α.

Simulation of memory access is very time consuming and thus, in the following simulations, we have analyzed small networks with $N = 32$ and $N_{sens} = N_{symb} = 16$. Memory sheets have size 1.5 X 1.5, $\sigma = 0.58$ and $\zeta = 0.5$, as before. Although this network size is extremely small, it has allowed a preliminary illustration of the model. We are considering parallelizing the algorithms, in order to consider larger networks. The simulation experiment followed was to perform up to 10000 minimization procedures, starting each one from a different random network configuration, which was presented to both the BM and GSA. When a new minimum energy pattern is found, it is stored and the procedure is repeated from other random starting configurations, otherwise the search stops. We note in Figs. 3(a) and 3(b) that, for $T = 0.2$, there are patterns found by GSA that are not found using the BM, while the opposite takes place at $T = 0.1$ (Figs. 3(c) and 3(d)). For the experiment described above, GSA appears to visit state space more loosely at higher temperatures, while the traditional BM visits state space more uniformly at lower temperatures. For lower temperatures, the BM functions more like a gradient descent method, and randomly generated patterns will stabilize at the closest local minima. We see in Figs. 4(a) and 4(b) that, for $T = 0.2$, when we lower the value of q_A, GSA becomes less associative and finds less patterns.

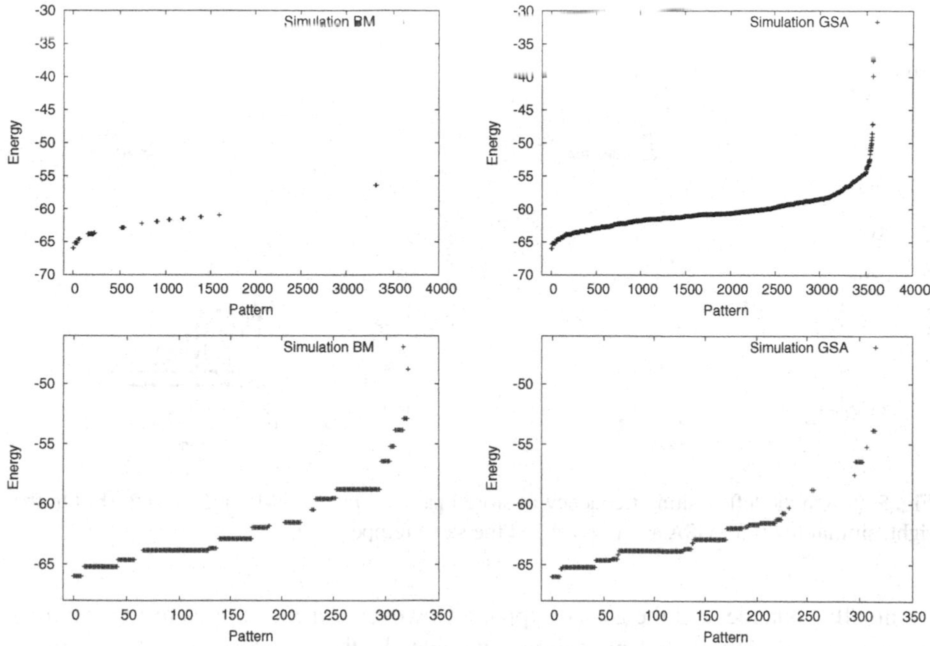

Fig. 3. The numbers taken as abcissas in (a) and (b), and in (c) and (d), identify the same patterns. (a) Upper left, energy of stored patterns visited by the BM for $T = 0.2$. (b) Upper right, energy of stored patterns visited by GSA for $q_A = 1.4$ and $T = 0.2$. (c) Lower left, similar to (a) for $T = 0.1$. (d) Lower right, similar to (b) for $q_A = 1.4$ and $T = 0.1$.

In order to understand the features of GSA that led to the results presented in Fig. 3, we compare in Fig. 5 the frequency with which the different minimum energy states corresponding to patterns are found, with the BM and with GSA, for $q_A = 1.4$. Both calculations were performed at $T = 0.2$, which corresponds to the upper row of Fig. 3. In the case of GSA, the frequency with which the hardest to detect patterns are found is much larger than the corresponding ones in the BM. In particular, several patterns that are not found by the BM are detected employing GSA. This corresponds to the gaps encountered in the spectrum shown in Fig. 3(a). If the number of iterations allowed in the experiment were increased, the patterns detected through both procedures should

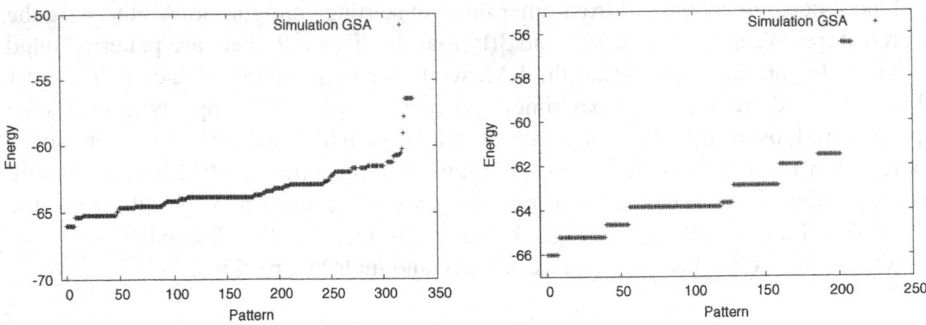

Fig. 4. (a) On the left, energy of stored patterns visited by GSA for $q_A = 1.3$. (b) On the right, similar to (a), for $q_A = 0.7$. In both cases $T = 0.2$.

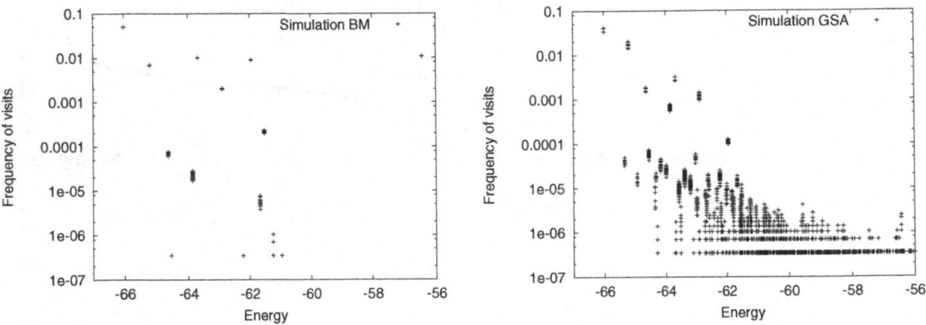

Fig. 5. (a) On the left, visiting frequency to stored patterns by the BM for $T = 0.2$. (b) On the right, similar to (a) for GSA at $q_A = 1.4$ and the same temperature.

eventually coincide, and the gaps disappear. However, our intention is to find minima without an exhaustive search procedure, but guided by the probability distribution function for network states. GSA tends to prefer the lower energy states, but will also find, with low probability, higher energy states. One can observe an exponential upper limit for the frequency of visits, as a function of energy for GSA. The BM tends to visit states with a more uniform distribution of frequencies, as is expected from the characteristic of the locality of visits of state space, which we mentioned in the end of Section 3.

5 Conclusions

We have previously proposed a memory organization, where two hierarchically structured modules corresponding to sensorial and symbolic memories interact, producing sensorial and symbolic activity, representing unconscious and conscious mental processes. This memory structure and functioning along with an adaptive learning process is used to explain a possible mechanism for neurotic behavior and psychoanalytical working-through. The model emphasizes that symbolic processing, language and meaning are important for consciousness [4,5].

The complex network topologies which result from the self-organizing processes, based on biological mechanisms, which we have modeled have properties that are common features of many biological systems and indicate that they may not be well described by BG statistical mechanics, but rather by NSM. These mechanisms are characteristic of much of the brain's functioning and suggest the use of GSA to model memory functioning and the way we associate ideas in thought. The study of network quantities such as node degree distributions and clustering coefficients may indicate possible experiments, that would validate models such as the one presented here.

Temperature and noise in the simulated annealing process that occurs in the model for memory activity should be related to associativity. Very high temperatures allow the production of logically disorganized thought, because they allow associations of excessively distant, usually uncorrelated ideas. In the model we have presented, temperature and q_A-values regulate associativity among memory configurations, in an analogy with the idea that freely talking in analytic sessions and stimulation from the analyst lower resistances and allow greater associativity. We are continuing systematic study of the parameter dependency of the model and its interpretation and also expanding and generalizing it to treat new features and phenomena, as well as larger networks.

Our main contribution in recent work has been to propose a neuronal model, based on known microscopical, biological brain mechanisms, that describes conscious and unconscious memory activity involved in neurotic behavior, as described by Freud. The model emphasizes that symbolic processing, language and meaning are important for consciousness. Although biologically plausible, in accordance with many aspects described by psychoanalytic theory and clinical experience, and based on simulations, the model is very schematic and we do not sustain or prove that this is the actual mechanism that occurs in the human brain. It nevertheless illustrates and seems to be a good metaphorical view of facets of mental phenomena, for which we seek a neuronal substratum, and suggests directions of search.

Acknowledgments. We are grateful for fruitful discussions regarding basic concepts of psychoanalytic theory, with psychoanalyst and professor Angela Bernardes of the Escola Brasileira de Psicanálise of Rio de Janeiro and the Department of Psychology of the Universidade Federal Fluminense. Suggestions and discussions with Sumiyoshi Abe, Evaldo M. F. Curado and Constantino Tsallis regarding NSM were also helpful and enlightening. This research was developed with grants from the Brazilian National Research Council (CNPq), the Rio de Janeiro State Research Foundation (FAPERJ) and the Brazilian agency which funds graduate studies (CAPES).

References

1. Freud, S.: Beyond the Pleasure Principle, Standard Edition. The Hogarth Press, London (1974); First German edition (1920)
2. Freud, S.: Introductory Lectures on Psycho-Analysis, Standard edn. W. W. Norton and Company, New York (1966), First German edition (1917)

3. Wedemann, R.S., Donangelo, R., de Carvalho, L.A.V., Martins, I.H.: Memory Functioning in Psychopathology. In: Sloot, P.M.A., Tan, C.J.K., Dongarra, J., Hoekstra, A.G. (eds.) ICCS-ComputSci 2002. LNCS, vol. 2329, pp. 236–245. Springer, Heidelberg (2002)
4. Wedemann, R.S., de Carvalho, L.A.V., Donangelo, R.: A Complex Neural Network Model for Memory Functioning in Psychopathology. In: Kollias, S.D., Stafylopatis, A., Duch, W., Oja, E. (eds.) ICANN 2006. LNCS, vol. 4131, pp. 543–552. Springer, Heidelberg (2006)
5. Wedemann, R.S., Carvalho, L.A.V., Donangelo, R.: Network Properties of a Model for Conscious and Unconscious Mental Processes. Neurocomputing 71, 3367–3371 (2008)
6. Hertz, J.A., Krogh, A., Palmer, R.G. (eds.): Introduction to the Theory of Neural Computation. Lecture Notes, vol. I. Perseus Books, Cambridge (1991)
7. Taylor, J.G.: Mind and consciousness: Towards a final answer? Physics of Life Reviews 2, 1–45 (2005)
8. Abe, S., Okamoto, Y. (eds.): Nonextensive Statistical Mechanics and Its Applications. Lecture Notes in Physics. Springer, Heidelberg (2001)
9. Tsallis, C., Stariolo, D.: Generalized Simulated Annealing. Physica A 233, 395–406 (1996)
10. Thurner, S.: Nonextensive Statistical Mechanics and Complex Scale-free Networks. Europhysicsnews 36(6), 218–220 (2005)
11. Edelman, G.M.: Wider than the Sky, a Revolutionary View of Consciousness. Penguin Books, London (2005)
12. Kandel, E.R., Schwartz, J.H., Jessel, T.M. (eds.): Principles of Neural Science. MacGraw Hill, USA (2000)
13. Carvalho, L.A.V., Mendes, D.Q., Wedemann, R.S.: Creativity and Delusions: The Dopaminergic Modulation of Cortical Maps. In: Sloot, P.M.A., Abramson, D., Bogdanov, A.V., Gorbachev, Y.E., Dongarra, J., Zomaya, A.Y. (eds.) ICCS 2003. LNCS, vol. 2657, pp. 511–520. Springer, Heidelberg (2003)
14. Wedemann, R.S., Donangelo, R., Carvalho, L.A.V.: Properties of a memory network in psychology. In: Abe, S., et al. (eds.) Complexity, Metastability and Nonextensivity-CTNEXT 2007. CP965, pp. 342–345. American Institute of Physics, New York (2007)
15. Tsallis, C., Bemski, G., Mendes, R.S.: Is re-association in folded proteins a case of nonextensivity? Physics Letters A 257, 93–98 (1999)
16. Boon, J.P., Tsallis, C.: Special Issue Overview: Nonextensive Statistical Mechanics: New Trends, New Perspectives. Europhysics News 36(6), 185 (2005)

Algorithms for Structural and Dynamical Polychronous Groups Detection

Régis Martinez[1] and Hélène Paugam-Moisy[2]

[1] LIRIS, UMR CNRS 5205, Université de Lyon, F-69676 Bron, France
[2] TAO - INRIA, LRI - Université Paris-Sud 11, F-91405 Orsay, France
regis.martinez@liris.cnrs.fr, hpaugam@lri.fr

Abstract. Polychronization has been proposed as a possible way to investigate the notion of cell assemblies and to understand their role as memory supports for information coding. In a spiking neuron network, polychronous groups (PGs) are small subsets of neurons that can be activated in a chain reaction according to a specific time-locked pattern. PGs can be detected in a neural network with known connection delays and visualized on a spike raster plot. In this paper, we specify the definition of PGs, making a distinction between structural and dynamical polychronous groups. We propose two algortihms to scan for structural PGs supported by a given network topology, one based on the distribution of connection delays and the other taking into account the synaptic weight values. At last, we propose a third algorithm to scan for the PGs that are actually activated in the network dynamics during a given time window.

1 Introduction

One of the main challenges in cognitive science is to understand how knowledge is represented and processed in the brain. From the early notions of cell assemblies [5] and "grand-mother cells" (see [4]), many open questions are still debated. What is the support of memory? How and where information is coded in the brain activity? The recent hypothesis that information could be encoded by precise spike timings gives arguments for the thesis of temporal rather than spatial cell assemblies. Pointing out the fact that connection delays have non uniform values between neurons in the brain, Izhikevich proposed the concept of polychronization [7], which is far richer than the current concepts of synchronization and synfire chains [1]. Also in computer science the concept of polychronization yields valuable tracks for defining new learning rules in spiking neuron networks [10] and polychronous groups have been confirmed to play the role of dynamical cell assemblies in a classification task [9]. Studying PGs and understanding their role in information coding could help improving both the network structure and the effectiveness of learning rules acting on delays.

So far, no formal definition has been given for a polychronous group, and only specific methods to inventory them have been proposed [6,8]. In section 2 we give a precise definition of a polychronous group (PG), making a distinction between structural and dynamical PGs. In section 3 we present three algorithms and data structures to inventory all polychronous groups of a given network topology and to detect which groups

C. Alippi et al. (Eds.): ICANN 2009, Part II, LNCS 5769, pp. 75–84, 2009.

are triggered in spike activity. In section 4 we give a complexity analysis of the algorithms, and we present experimental measurements: Number of PGs (mean values, from several experiments) when varying different parameters.

2 Definition of Polychronous Groups

2.1 PG Definition

In the founding paper [7] polychronization denotes the fact that several neurons can be activated in a chain reaction according to a specific time-locked pattern of firings, not only in pure synchrony. A polychronous group is characterized by a precise temporal pattern of firings, in a subset of neurons, that is more likely to happen than just by chance in neural activity. Such patterns are intrinsically based on the network topology: connectivity, synaptic weights and especially conduction delays.

In recent work [8] a polychronous group, referred to as a *polygroup*[1], is defined as the set of neurons that supports the time-locked pattern. We state that the set of neurons involved in the temporal pattern is not enough to characterize the PG. Indeed, if one neuron can appear in more than one PG, it is clear that a given set of neurons could also fire with several different timings, and support or participate to more than one PG. The chain reaction that defines a PG is a series of causal interactions between neurons, such as $[N_1, t_1], [N_2, t_2], ..., [N_i, t_i] \Rightarrow [N_j, t_j]$, where a pair $[N_k, t_k]$ denotes a spike fired by neuron N_k at time t_k. The chain is started by a specific firing pattern of a small number s of neurons, the triggering neurons, further named *triggers*. Hence we propose to define a PG by a list of triggers associated to a temporal firing pattern:

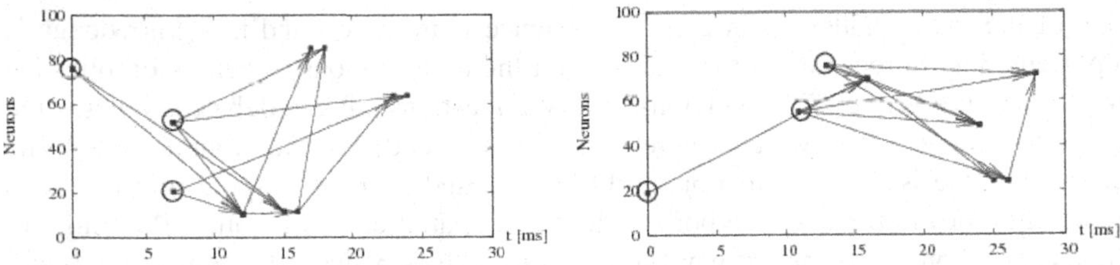

Fig. 1. Graphs for two polychronous groups: the $21-52-76$ $(7,7,0)$ (left PG) and the $19-55-76$ $(0,11,13)$ (right PG). They share neuron 76 among their respective sets of triggers.

Definition 1. *A s-triggered* **polychronous group** *refers to the set of neurons that can be activated by a chain reaction whenever the triggers $N_k (1 \leq k \leq s)$ fire according to the timing pattern $t_k (1 \leq k \leq s)$. The PG is denoted by: $N_1 - N_2 - ... - N_s(t_1, ..., t_s)$ where the firing times t_k are listed in the same order as the corresponding triggers N_k.*

[1] We do not use the term *polygroup* because it might bring confusion with other uses in Physics.

To decide whether a neuron could be activated by counting the number of spikes it recieves simultaneoulsy, we follow Izhikevich [7] by extending the notion of simultaneous arrival to a short time range denoted $jitter$.

A graphical representation of a PG can be plotted on a spike raster plot of the network activity, focusing attention on a subset of neurons, and drawing additional links for representing the causality of interactions (see Figure 1). Such a representation is a subgraph of the neural activity, where a vertex is a pair $[N_k, t_k]$, and a directed edge denotes the causal influence from a pre-synaptic neuron to a post-synaptic one.

2.2 Structural PGs vs. Dynamical PGs

Definition 1 allows to inventory all the possible s-triggered polychronous groups supported by a given spiking neuron network with known connectivity and conduction delays, disregarding weight values. Since such PGs depend on the network architecture only, they are structural and we call them the **supported PGs**.

However, synaptic weights are usually subject to learning rules and their values change through time, which can influence the decision whether a certain amount of spikes simultaneoulsy incoming to a neuron N_j is sufficient or not for triggering a spike fired by N_j. Then we define **adapted PGs** by applying Definition 1 in taking into account the membrane potential dynamics of neurons and the values of synaptic weights for finding the causal relations yielding a neuron N_j to actually fire a spike at time t_j. Since activated PGs do not depend on the network dynamics (e.g. under the influence of a given input), they are also structural PGs.

Another question is to find which PGs actually appear in the spike activity of a neuron network during a given time window. These PGs are a subset of the structural PGs and we call them **activated PGs**. Unlike structural PGs, activated polychronous groups are dynamical PGs, since they depend on the network dynamics.

3 Scanning for Polychronous Groups

Algorithms 1 and 2 are designed to scan for supported PGs and adapted PGs respectively. They are based on a given network topology, with known connectivity, conduction delays and synaptic weights. In Algorithm 1, all the combinations of s neurons are tested as possible triggers, under the hypothesis that $NbSpikesNeeded$ are required to generate a causal relation making a neuron N_j to spike (s and $NbSpikesNeeded$ are set by the user). Actual post-synaptic potentials (PSPs) are not taken into account in Algorithm 1 whereas in Algorithm 2, the decision to let a neuron N_j spike requires the computation of the weigthed PSPs recieved by N_j. In Algorithm 2, the amount of incoming spikes can differ from $NbSpikesNeeded$, a parameter which is no longer useful (s is still required and set by the user). Algorithm 3 is written to scan for the actual appearance of previously inventoried adapted polychronous groups in the network activity during a time slice of simulation data (with varying inputs, for instance).

All three algorithms are suited for any model of neuron that can be run in event-driven mode. Algorithm 2 uses the neuron model equation to decide whether the neuron spikes or not, on the basis of its recent PSP history. This constraint may be relaxed for

Algorithm 1 that can be straightforward adapted to more general neuron models, since it is only based on spike events, and do not take care of the intrinsic dynamics of neurons.

The model neuron we used for experiments has the following characteristics (based on SRM_0 [2]): The firing threshold ϑ is set to a fixed negative value (e.g. $\vartheta = -50\,mV$) and the threshold kernel simulates an absolute refractory period τ_{abs}, when the neuron cannot fire again, followed by a reset to the resting potential u_{rest}, lower than ϑ (e.g. $u_{rest} = -65\ mV$). We assume that the simulation is computed in discrete time (with $0.1\ ms$ time steps). The variables of each neuron are updated at each new incoming spike (event-driven programming), which is sufficient for computational purpose.

The network structure is typical of Reservoir Computing methods: Connections between neurons are drawn randomly, according to a given connectivity; Weights start from 0.5 as initial value and can vary under STDP, a temporal Hebbian rule of synaptic plasticity; Delays are fixed but random, between 1 and 20 ms.

3.1 Definitions

Notations

n : number of neurons in the network
N_i : neuron numbered i
N_{T_k} : triggering neuron numbered T_k, with k from 1 to s
t : current time (virtual biological time, in simulations)
d_{ij} : conduction delay on the connection from N_i to N_j
w_{ij} : synaptic weight from N_i to N_j, equals 0 if connection does not exist.

Data structures

Post-Synaptic Potentials. A PSP is denoted PSP_{t_{psp},N_l,N_m} : Post-Synaptic Potential evoked at time t_{psp} by a pre-synaptic neuron N_l on a post-synaptic neuron N_m.

Event queue. The queue $EventQueue$ is the structure for processing simulation events (evoked PSPs). It contains the events ordered by their chronological occurence in the future.

PSP list stored in neurons. For the purpose of our algorithms, we need to store, for each neuron, the list of the spikes it recieved during a given time course elapsed, the $Jitter$ (see Section 2.1). This list is called $PSPList_i$, for neuron N_i.

Data structure for a PG. When a PG is calculated, informations have to be stored :

- timings $t_{T_1}, ..., t_{T_s}$ of the triggers N_{T_1} to N_{T_s};
- for each PSP, PSP_{t_{psp},N_l,N_m}, the ID of N_m and its spike-firing time, the ID of N_l and the time of evocation of the PSP.

Variables

$NbTriggeringConnections_i$: number of connections received by N_i from triggers
$MaxPotential_i$: maximum membrane potential that neuron N_i might reach under the action of triggers
$tLastSpike_i$: time of most recent firing of neuron N_i, initialized to $-RefractoryPeriode$, so that previous spikes are already out of refractory periode

Parameters

s : number of triggers, fixed for every polychronous groups
$NbSpikesNeeded$: number of simultaneous impinging spikes necessary to trigger a new spike in any neuron
$Jitter$: time range for spikes to be considered as "simultaneous"
$NbSpikesMax$: maximum number of spikes in a polychronous group
$NbSpikeMin$: minimum numer of spikes in a polychronous group for it to be saved
$MaxTimeSpan$: maximum time span of the polychronous group
$MinTimeSpan$: minimun time span of the polychronous group
$PSPStrength$: amplitude of a Post Synaptic Potential (PSP)
$RestingPotential$: default membrane potential of a neuron when it has no input
$Threshold$: membrane potential value above which the neuron spikes
$RefractoryPeriode$: time after a spike, during which a neuron cannot fire again

3.2 Algorithms

Algorithm 1. In order to list the supported PGs, we first look at all the combinations of a given number s of neurons. We check each combination, looking for neurons that might be excited enough to fire in turn, because they recieve more than a certain amount of spikes, $NbSpikesNeeded$. If such neurons exist, then the combination becomes a set of triggers. We simulate the firing of the triggers with the right starting timing and record the propagation of the neural activity, until it dies or it reaches an upper limit $MaxTimeSpan$ set for the time span of a PG. Moreover, we limit the record to a given number of neurons $NbSpikesMax$ in order to truncate the possible cyclic PGs.

In this algorithm, the propagation of the activity is based on the number of spikes recieved by each neuron in a time window. For instance, the neurons may be parameterized so that they fire whenever they are impinged by at least three spikes within a millisecond. A full description, in pseudo-language, is given in Annex.

Algorithm 2. The principle of this algorithm is very similar to the previous one, except that the decision of firing or not is based on the level of the membrane potential, which depends on (a) the weights of the incoming connections that will modulate the increase of the membrane potential and thus the probability to generate a new spike and (b) the elapsed time since the previous PSP.

As in Algorithm 1, we look at all the possible combinations of s triggers, except that neuron activity is calculated upon its membrane potential exceeding or not the firing threshold. In this algorithm, the propagation of the activity is based on the fact that the mambrane potential exceedes the threshold. See Annex.

Algorithm 3. Algorithm 3 is written to scan for the appearance of known polychronous groups (already detected by Algorithm 1 or 2) in the activity recorded from a simulation, during a given time range.

In order to detect the activation of a PG in a particular time window in the recorded activity of a simulation (Algorithm 3), it would be ideal to check if the whole group is activated. For sake of computational time, we only look for the firing of the triggers of the PG, with the good timing pattern within a precision of $Jitter$. We based this algorithmic simplification on the assumption that the activation of the triggers will activate the tail of the polychronous group with little change, which is likely if the known PG has been previously detected to be an adapted PG, but could fail in case of supported PGs.

// We look for the actual activation of known PGs in a temporal range $[Start; End]$. For each neuron N_i, the list of spikes fired by this neuron between times $Start$ and End is stored already (i.e. we assume that the spike raster has been recorded).

Let $Spike_{mi}$ be the mth spike of the list of spikes fired by N_i, at time $t_{Spike_{mi}}$

for all PG triggered by $\{N_{T_1}, N_{T_2}, \dots, N_{T_s}\}$ with timing $\{t_{T_1}, t_{T_2}, \dots, t_{T_s}\}$ **do**
    $\forall Spike_{mT_1}$ at time $t_{Spike_{mT_1}}$
    if $\forall k \in [T_2; T_s], \exists\ Spike_{mT_k}$ at time $t_{Spike_{mT_1}} + (t_{T_k} - t_{T_1})$ **then**
        Save activation of PGs at time $t_{Spike_{mT_1}} - t_{T_1}$
    end if
end for

4 Complexity Study and Experiments

4.1 Computational Complexity

First, the complexity estimation for **Algorithm 1** is developed below. The complexity of Algorithm 2 and Algorithm 3 will be discussed afterwards.

Let c be the connectivity of a network, i.e. the probability that one neuron projects a connection to another. Let $AC = (n-1) \times c$ be the average number of connections recieved by any neuron in the network. Remind that s is the number of triggers.

The number of combinations to parse is $C_n^s = \frac{n!}{s!(n-s)!}$. For each combination, we search for neurons that recieve concurrent connections from triggers and count such connections. This step is computed with complexity $O(n + s \times AC)$.

The probability that a neuron recieves $k = NbSpikesNeeded$ connections from other neurons, with $k \leq s$, is c^k. We watch which neurons have enough input connections to trigger a new spike. For each excited neuron, we initialise the calculation of the corresponding PG and enqueue the spike events, i.e. $s \times AC$ operations.

The `while` block is the most difficult to evaluate because of the various parameters it envolves. In worst case, all n neurons will recieve $NbSpikesNeeded$ spikes, and spike in turn. Then there would be $n \times k$ events to process, and $n \times AC$ new PSP to enqueue. Neurons would spike again as soon as they can, in regard to their refractory period. There should be at most $M/R \times n \times AC$ events to process in the whole calculation, where $M = MaxTimeSpan$ and $R = RefractoryPeriode$.

Hence, the overall complexity X_{algo1} of Algorithm 1, in worst case, is :

$$\underbrace{C_n^s}_{combin.} \times [\ \underbrace{(n + s \times AC)}_{search\ triggers} + \underbrace{(s \times AC) \times (c^k)}_{nb\ triggers} \times (\ \underbrace{s \times AC}_{init.\ spikes} + \underbrace{M/R \times n \times AC)}_{worst\ case\ events}\ ]$$

Since $s \ll n$, the number of combinations C_n^s can be approximated by $\left(\frac{n}{s}\right)^s$ and $AC = (n-1) \times c$ replaced by $c \times n$, which yields an upper bound for X_{algo1}:

$$X_{algo1} \leq \frac{1 + s \times c}{s^s} n^{s+1} + \frac{c^{k+2}}{s^{s-2}} n^{s+2} + \frac{M}{R} \times \frac{c^{k+2}}{s^{s-1}} n^{s+3} \quad (1)$$

It results that the complexity X_{algo1} is of order $O(n^{s+3})$ in worst case. In practical cases, spiking neuron networks usually have a sparse connectivity. Fixing the order of magnitude of the connectivity c to $1/\sqrt{n}$ looks like a realistic estimation, both for artificial networks (e.g. $c = 0.1$ in a network of 100 neurons, when computing with spiking neuron networks) and biological networks (around 10^5 connections per neuron between 10^{11} neurons in the human brain). Replacing c by $n^{-1/2}$ in Equation (1) gives:

$$X_{algo1} \leq \frac{1}{s^s} n^{s+1} + \frac{1}{s^{s-1}} n^{s+1/2} + \frac{1}{s^{s+2}} n^s + \frac{M}{R} \times \frac{1}{s^{s-1}} n^{s+1} \quad \text{since } k \geq 2 \quad (2)$$

Finally, in practical cases, the complexity of Algorithm 1 is of order $O(n^{s+1})$.

The complexity X_{algo2} of **Algorithm 2** is similar to X_{algo1} because both algorithms have the same control structure. Running Algorithm 2 should be slightly more time consuming if the neuron model is complex.

The complexity X_{algo3} of **Algorithm 3** is of order $O(P * S/n)$, where P is the number of known polychronous groups (computed by Algorithm 1 or 2), and S the total number of spikes in the time slice chosen for scanning the network activity.

4.2 Experiments

A direct comparison with other algorithms is not straightforward since the Izhikevich's code available on [6] starts from cutting off the weights under an arbitrary value of 0.95, and does not exactly compute any of the PG categories we defined. The Maier & Miller's method [8] (noted "MM algo" in Table 1) is close to Izhikevich's one: both are based on a $n \times t_{max}$ matrix of spike arrival counts, with the risk of consuming a huge amount of memory, since t_{max} is the maximum time to which the simulation is run.

In Table 1, $NbSpikesNeeded = s = 3$. The first two lines show the strong influence of the connectivity c on the number of PGs. In the next three lines, the network size is varied with an adapted value of c in order to keep fixed the degree of the connection graph. Different parameters of the neuron model (with coherent value of $Jitter$) highly

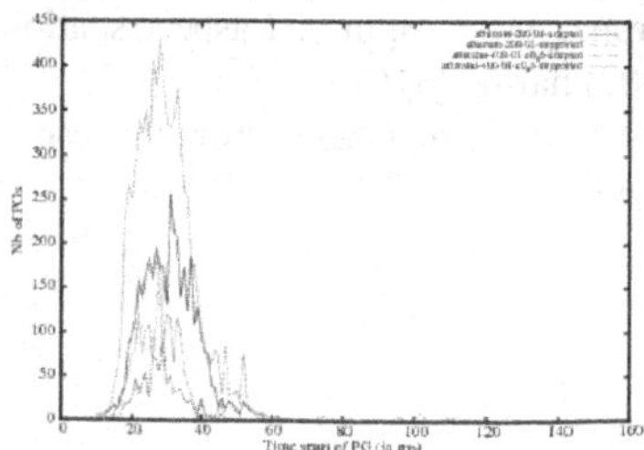

Fig. 2. PGs timespan comparison

Table 1. Experimental measurements

n	c	Jitter	supported PGs	adapted PGs	MM algo.
100	0.1	1ms	13.6	13.2	13.6
100	0.2	1ms	1295	1308	1300
100	0.18	1ms / 0.4	697 / 697	702 / 72	732
200	0.09	1ms / 0.4	295 / 295	274 / 16	289
500	0.036	1ms / 0.4	103 / 103	112 / 6	107
200	0.09	1.2ms	431		434
200	0.09	1.0ms	295		289
200	0.09	0.7ms	176		167
200	0.09	0.5ms	79		67
200	0.09	0.1ms	0		0

influence the number of adapted PGs (but not sructured PGs). The last line shows that an absence of tolerance ($Jitter = 0.1ms$ = simulation time step) could lead to an absence of PGs. With $NbSpikesNeeded = 2$, $n = 100$ and $c = 0.18$ (not in Table 1), there are 387 supported PGs with $s = 3$ triggers, but only 4 with $s = 2$.

Figure 2 shows that the distribution of the PGs timespans (time from the spike of the earliest trigger to the spike of the latest neuron belonging to the PG) is similar for supported and adapted PGs, for different network sizes. Moreover, such a time span distribution is comparable to Izhikevich's observation ([7], p.127).

5 Discussion

We have proposed to clarify the definition of a polychronous group and set a standard notation, taking into account both the set of triggers and their specific firing pattern. We make a distinction between three categories of PGs, whether they are scanned from the network architecture only, or they take into account the variations of weights under a learning algorithm, or they are scanned for reflecting the dynamical activity inside the network in response to input data. Though designed for the analysis of simulations with spiking neuron network models, the algorithms could also be applied to real data since multi-neuron activities appear to be recordable in natural neuron networks [3].

References

1. Abeles, M.: Corticonics: Neural Circuits of the Cerebral Cortex. Cambridge Press, New York (1991)
2. Gerstner, W., Kistler, W.: Spiking Neuron Models: Single Neurons, Populations, Plasticity. Cambridge University Press, Cambridge (2002)
3. Gourévitch, B., Eggermont, J.: Maximum decoding abilities of temporal patterns and synchronized firings. In: NeuroComp. 2008 (2008) hal-00331583
4. Gross, C.G.: Genealogy of the "grandmother cell". The Neuroscientist (2002)
5. Hebb, D.O.: The Organization of Behaviour. Wiley, New York (1949)
6. Izhikevich, E.M.: `http://vesicle.nsi.edu/users/izhikevich/publications/spnet.htm` (2006)
7. Izhikevich, E.M.: Polychronization: Computation with spikes. Neural Computation 18(2), 245–282 (2006)
8. Maier, W.L., Miller, B.N.: A minimal model for the study of polychronous groups. arXiv:0806.1070v1 [cond-mat.dis-nn] (2008); (presented at the TS4CF08 Meeting of The American Physical Society)
9. Martinez, R., Paugam-Moisy, H.: Les groupes polychrones pour capturer l'aspect spatio-temporel de la mémorisation. In: NeuroComp. 2008 (2008) hal-00331613
10. Paugam-Moisy, H., Martinez, R., Bengio, S.: Delay learning and polychronization for reservoir computing. Neurocomputing 71(7-9), 1143–1158 (2008)

A Algorithm 1

```
// Lines starting with // are comments.

for all combination of s neurons out of n neurons of the network do
    // Look for PGs triggered by this combination
```

for all i from 1 to n **do**
  $NbTriggeringConnections_i = 0$
end for
// Count connections comming from triggers, to find common triggers output neurons
for all p from 1 to s **do**
  for all i with $w_{iT_p} \neq 0$ **do**
    $NbTriggeringConnections_i = NbTriggeringConnections_i + 1$
  end for
end for
for all p from 1 to s **do**
  for all i with $w_{iT_p} \neq 0$ **do**
    $NbPSP_i = 0$ // Reset count of PSP evoked in N_i in the last $Jitter$ ms
    if $NbTriggeringConnections_i \geqslant NbSpikesNeeded$ **then**
      // Reset $NbTriggeringConnections_i$
      $NbTriggeringConnections_i = 0$
      // A spike from N_i is triggered.
      We will calculate the PG with trigger neurons $\{N_{T_1}, N_{T_2}, \ldots, N_{T_s}\}$ firing at N_i
      firing with timing $\{d_{max} - d_{T_1 i}, d_{max} - d_{T_2 i}, \ldots, d_{max} - d_{T_s i}\}$
      with $d_{max} = max(d_{T_1 i}, d_{T_2 i}, \ldots, d_{T_s i})$
      thus triggering neuron N_i.
      Add triggering spikes to PG. // Store triggering spikes to the PG data structure
      $PGSpikeCount = s$ // Count of spikes in this PG
      $t = 0$ // Initialise clock
      // Enqueue PSPs from triggers starting at $t = 0$.
      for all neuron N_h recieving a connection from N_{T_k}, $\forall k$ from 1 to s **do**
        Enqueue the new upcoming PSP evoked in N_h at time $t + (d_{max} - d_{T_k i}) + d_{T_k h}$ by the spike from N_{T_k}
      end for
      while ($PGSpikeCount < NbSpikesMax$) and (PSP queue not empty) and ($t < MaxTimeSpan$) **do**
        Consider next upcoming PSP PSP_{t_{psp}, N_l, N_m} evoked at time t_{psp} with
        N_l : firing pre-synaptic neuron of the spike that evoked PSP_{t_{psp}, N_l, N_m}
        N_m : post-synaptic neuron in which the PSP is evoked
        $t = t_{psp}$
        $NbPSP_m = NbPSP_m + 1$
        for all PSP_{t_{psp}, N_o, N_m} with $t - t_psp > Jitter$ **do**
          Erase PSP // Erase PSPs evoked in N_m older than $t - Jitter$ ms.
        end for
        if ($t - tLastSpike_m > RefractoryPeriode$) and ($NbPSP_m \geqslant NbSpikesNeeded$) **then**
          // N_m fires a spike
          $tLastSpike_m = t$
          Add a spike from N_m at time t, to PG
          $PGSpikeCount = PGSpikeCount + 1$
          for all neuron N_m recieving a connection from N_l **do**
            Enqueue an upcoming PSP evoked in N_m at time $t + d_{lm}$ by the spike from N_l
          end for
        end if
      end while
      if $PGSpikeCount > NbSpikeMin$ **then**
        Save the PG
      end if
    end if
  end for
end for
end for

B Algorithm 2

N.B. Red printing is for the lines that differ from Algorithm 1.

```
for all combination of $s$ neurons out of $n$ neurons of the network do
  // Look for PGs triggered by this combination
  for all Neuron $N_i$, output of a triggering neuron do
    $NbPSP_i = 0$ // Count of PSP evoked in $N_i$ in the last $Jitter$ ms
    $Potential_i = RestingPotential$ // Set initial membrane potential for $N_i$
    $MaxPotential_i = Potential_i + \sum_{T_j} w_{T_j i} PSPStrength$
    if $MaxPotential_i \geqslant Threshold$ then
      // A spike from $N_i$ is triggered.
      We will calculate PG with trigger neurons $\{N_{T_1}, N_{T_2}, ..., N_{T_s}\}$ firing at $N_i$
       firing with timing $\{d_{max} - d_{T_1 i}, d_{max} - d_{T_2 i}, ..., d_{max} - d_{T_s i}\}$
       with $d_{max} = max(d_{T_1 i}, d_{T_2 i}, ..., d_{T_s i})$
       thus triggering neuron $N_i$.
       Add triggering spikes to PG. // Store triggering spikes to PG data structure
       $PGSpikeCount = 0$ // Count of spikes in this PG
       $t = 0$
       // Enqueue PSPs from triggers
       for all neuron $N_h$ recieving a connection from $N_{T_k}$, $\forall k$ from 1 to $s$ do
         Enqueue the new upcoming PSP evoked in $N_h$ at time $t + (d_{max} - d_{T_k i}) + d_{T_k h}$ by the spike from $N_{T_k}$
       end for
       while ($PGSpikeCount < NbSpikesMax$) and (PSP queue not empty) and ($t < MaxTimeSpan$) do
         Consider next upcoming PSP $PSP_{t_{psp}, N_l, N_m}$ evoked at time $t_{psp}$ with
         $N_l$ : firing pre-synaptic neuron of the spike that evoked $PSP_{t_{psp}, N_l, N_m}$
         $N_m$ : post-synaptic neuron in which the PSP is evoked
         $t = t_{psp}$
         for all $PSP_{t_{psp}, N_o, N_m}$ with $t - t_p sp > Jitter$ do
           Erase PSP // Erase PSPs evoked in $N_m$ older than $t - Jitter$ ms.
         end for
         // Re-evaluate decreasing membrane potential, with regard to last spike impact recieved $t^f$
         $Potential_m = \eta(Potential_l, t^f)$
         $Potential_m = Potential_m + w_{lm} \times PSPStrength$
         if ($Potential_m \geqslant Threshold$) then
           // $N_m$ fires a spike
           Add a spike to from $N_m$ at time $t$, to PG
           $PGSpikeCount = PGSpikeCount + 1$
           for all neuron $N_m$ recieving a connection from $N_l$ do
             Enqueue an upcoming PSP evoked in $N_m$ at time $t + d_{lm}$ by the spike from $N_l$
           end for
         end if
       end while
       if $PGSpikeCount > NbSpikeMin$ then
         Save the PG
       end if
    end if
  end for
end for
```

Logics and Networks for Human Reasoning

Steffen Hölldobler and Carroline Dewi Puspa Kencana Ramli

International Center for Computational Logic,
TU Dresden, 01062 Dresden, Germany
sh@iccl.tu-dresden.de
http://www.computational-logic.org/~sh/

Abstract. We propose to model human reasoning tasks using completed logic programs interpreted under the three-valued Łukasiewicz semantics. Given an appropriate immediate consequence operator, completed logic programs admit a least model, which can be computed by iterating the consequence operator. Reasoning is then performed with respect to the least model. The approach is realized in a connectionist setting.

Keywords: Human Reasoning, Logic Programs, Connectionist Models.

1 Introduction

It has been widely argued in the field of Cognitive Science that logic is inadequate for modelling human reasoning (see e.g. [3]). In this context, "logic" is meant to be classical logic and, indeed, classical logic fails to capture some well-documented forms of human reasoning. However, in the field of Artificial Intelligence many non-classical logics have been studied and widely used. These logics try to capture many assumptions or features that occur in commonsense reasoning like, for example, the closed world assumption or non-monotonicity.

Recently, in [19] Stenning and van Lambalgen have suggested that completed logic programs under the three-valued Fitting semantics [9] can adequately model many human reasoning tasks. In addition, they propose a connectionist realization of their approach.

While trying to understand Stenning and van Lambalgen's approach we made the following observations: *(i)* Łukasiewicz semantics [17] seems to be better suited for the approach as the law of equivalence holds under this semantics, whereas it does not hold under Fitting semantics. *(ii)* [19] contains some (minor) errors. *(iii)* The immediate consequence operator introduced in [19] differs in a subtle way from the one in [9] and it turned out, that the latter is inadequate for human reasoning. *(iv)* The core method, a connectionist model generator for logic programs first presented in [12], can easily be adapted to handle Stenning and van Lambalgen's approach.

The paper discusses these observation by presenting three-valued logics, logic programs, their completion semantics as well as their immediate consequence operators in Section 2, by specifying an algorithm for mapping Stenning and van Lambalgen's immediate consequence operator onto a recurrent neural network

C. Alippi et al. (Eds.): ICANN 2009, Part II, LNCS 5769, pp. 85–94, 2009.

Table 1. Truth tables for 3-valued logics

	$\neg$
$\top$	$\bot$
$\bot$	$\top$
U	U

		$\wedge$	$\vee$	$\rightarrow_K$	$\rightarrow_L$	$\leftrightarrow_K$	$\leftrightarrow_L$	$\leftrightarrow_C$
$\top$	$\top$	$\top$	$\top$	$\top$	$\top$	$\top$	$\top$	$\top$
$\bot$	$\top$	$\bot$	$\top$	$\top$	$\top$	$\bot$	$\bot$	$\bot$
U	$\top$	U	$\top$	$\top$	$\top$	U	U	$\bot$
$\top$	$\bot$	$\bot$	$\top$	$\bot$	$\bot$	$\bot$	$\bot$	$\bot$
$\bot$	$\bot$	$\bot$	$\bot$	$\top$	$\top$	$\top$	$\top$	$\top$
U	$\bot$	$\bot$	U	U	U	U	U	$\bot$
$\top$	U	U	$\top$	U	U	U	U	$\bot$
$\bot$	U	$\bot$	U	$\top$	$\top$	U	U	$\bot$
U	U	U	U	U	$\top$	U	$\top$	$\top$

with feed-forward core in Section 3, by showing how some human reasoning task can be adequately modelled in the proposed logic and its connectionist realization in Section 4, and by discussing our findings in Section 5.

2 Logics, Programs and Consequence Operators

2.1 Three-Valued Logics

In this paper we consider (propositional logic) languages over an alphabet consisting of (propositional) variables, the connectives $\{\neg, \wedge, \vee, \rightarrow, \leftrightarrow\}$ and paranthesis. We will consider various three-valued logics based on the semantics of their connectives (see Table 1): Łukasiewicz has proposed $\{\neg, \wedge, \vee, \rightarrow_L, \leftrightarrow_L\}$ [17], Kleene uses $\{\neg, \wedge, \vee, \rightarrow_K, \leftrightarrow_K\}$ in his *strong three-valued logic* [16] and $\{\neg, \wedge, \vee, \rightarrow_K, \leftrightarrow_C\}$ is suggested by Fitting for logic programming [9] and used by Stenning and van Lambalgen to model human reasoning [19].

An *interpretation* is a mapping from the language to the set of truth values $\{\top, \bot, \mathsf{U}\}$. Using Table 1 an interpretation for a given formula is uniquely determined by specifying the values for the propositional variables occurring in it. We will represent interpretations by pairs $\langle I^\top, I^\bot \rangle$, where $I^\top \cap I^\bot = \emptyset$, $I^\top$ contains all variables which are mapped to $\top$, $I^\bot$ contains all variables mapped to $\bot$, and all variables which occur neither in $I^\top$ nor in $I^\bot$ are mapped to U. We use I_F and I_L to denote that an interpretation I uses Fitting or Łukasiewicz semantics, respectively. Furthermore, let $\mathcal{I}$ denote the set of all interpretation. $(\mathcal{I}, \subseteq)$ is a complete semilattice (see [9]). Finally, an interpretation I is said to be a *model* for a formula G iff $I(G) = \top$.

One should observe, that the law of equivalence (for all interpretations I: $I(F \leftrightarrow G) = I((F \rightarrow G) \wedge (G \rightarrow F))$) holds under Łukasiewicz semantics, but not under Fitting semantics.

2.2 Programs

A *(program) clause* is an expression of the form $A \leftarrow B_1 \wedge \ldots \wedge B_n$, $n \geq 1$, where A is an atom and each B_i, $1 \leq i \leq n$, is either a literal (i.e., atom or negated atom), $\top$ or $\bot$. A is called *head* and $B_1 \wedge \ldots \wedge B_n$ *body* of the program clause. $\top$ is a valid formula, whereas $\bot$ is an unsatisfiable one. One should observe that the body of each clause is non-empty. A clause of the form $A \leftarrow \top$ is called *positive fact.* A clause of the form $A \leftarrow \bot$ is called *negative fact.* A *(logic) program* is a finite set of clauses. Two examples are $\mathcal{P}_1 = \{p \leftarrow q\}$ and $\mathcal{P}_2 = \{p \leftarrow q, q \leftarrow \bot\}$.

2.3 Completion

Let $\mathcal{P}$ be a program. It is turned into a single formula in the following way:

1. All clauses with the same head $A \leftarrow Body_1, \ldots, A \leftarrow Body_n$ are replaced by the single formula $A \leftarrow Body_1 \vee \ldots \vee Body_n$.
2. The resulting set is replaced by its conjunction.
3. If A is a variable occurring in $\mathcal{P}$ with no clause in $\mathcal{P}$ of the form $A \leftarrow Body$, then add $A \leftarrow \bot$.
4. All occurrences of $\leftarrow$ are replaced by $\leftrightarrow$.

The resulting formula is called *completion of* $\mathcal{P}$ and is denoted by *comp*($\mathcal{P}$). If the third step is omitted, then the resulting formula is called *weak completion of* $\mathcal{P}$ and is denoted by *wcomp*($\mathcal{P}$). For example, $wcomp(\mathcal{P}_1) = (p \leftrightarrow q) \neq comp(\mathcal{P}_1) = (p \leftrightarrow q) \wedge (q \leftrightarrow \bot) = comp(\mathcal{P}_2) = wcomp(\mathcal{P}_2)$.

Let $I = \langle \{p, q\}, \emptyset \rangle$. Then, $I_F(\mathcal{P}_2) = I_{\text{L}}(\mathcal{P}_2) = \top$, whereas $I_F(comp(\mathcal{P}_2)) = I_{\text{L}}(comp(\mathcal{P}_2)) = \bot$. The completion forces all models to map q to $\bot$, and the same holds for the weak completion.

Let $I = \langle \emptyset, \emptyset \rangle$. Then, $I_F(p \leftrightarrow p) = \top$, whereas $I_F(p \leftarrow p) = \mathsf{U}$. In other words, I is a model for the completed program $p \leftrightarrow p$, but it is not a model for $p \leftarrow p$ under the Fitting semantics. On the other hand, $I_{\text{L}}(p \leftrightarrow p) = I_{\text{L}}(p \leftarrow p) = \top$.

2.4 Consequence Operators

In [9] Fitting has defined an immediate consequence operator $\Phi_{F,\mathcal{P}}(I) = \langle J^\top, J^\bot \rangle$, where $J^\top = \{A \mid A \leftarrow Body \in \mathcal{P} \text{ and } I(Body) = \top\}$ and $J^\bot = \{A \mid \text{for all } A \leftarrow Body \in \mathcal{P} : \ I(Body) = \bot\}$. One should note that $I_F(Body) = I_{\text{L}}(Body)$ because the body of a clause is a conjunction of literals. Fitting has shown that $\Phi_{F,\mathcal{P}}$ is monotone on $(\mathcal{I}, \subseteq)$. We call I is a *fixed point for Fitting operator* if and only if $\Phi_{F,\mathcal{P}}(I) = I$.

Proposition 1. *$\Phi_{F,\mathcal{P}}$ is continuous and admits a least fixed point lfp($\Phi_{F,\mathcal{P}}$).*

Proof. Because our programs are finite and, thus, the underlying alphabet and all directed subsets of $\mathcal{I}$ are finite, we find that a monotone $\Phi_{F,\mathcal{P}}$ is also continuous (see e.g. [20]). Hence, $\Phi_{F,\mathcal{P}}$ admits a least fixed point, which can be computed by iterating $\Phi_{F,\mathcal{P}}$ starting with the empty interpretation. □

lfp($\Phi_{F,\mathcal{P}}$) is equal to the least model of *comp*($\mathcal{P}$) under Fitting semantics. For example, $lfp(\Phi_{F,\mathcal{P}_1}) = lfp(\Phi_{F,\mathcal{P}_2}) = \langle \emptyset, \{p, q\} \rangle$.

In [19] Stenning and van Lambalgen have defined a slightly different immediate consequence operator: $\Phi_{SvL,\mathcal{P}}(I) = \langle J^\top, J^\bot \rangle$, where $J^\top = \{A \mid A \leftarrow Body \in$

$\mathcal{P}$ and $I(\mathit{Body}) = \top\}$ and $J^{\perp} = \{A \mid$ there exists $A \leftarrow \mathit{Body} \in \mathcal{P}$ and for all $A \leftarrow \mathit{Body} \in \mathcal{P}: \; I(\mathit{Body}) = \perp\}$. They showed that this operator is also monotone. Similarly, I is a *fixed point for Stenning and van Lambalgen operator* if and only if $\Phi_{SvL,\mathcal{P}}(I) = I$

Proposition 2. *$\Phi_{SvL,\mathcal{P}}$ is continuous and admits a least fixed point $\mathit{lfp}(\Phi_{SvL,\mathcal{P}})$.*

This result can be proven along the lines of the proof of Propostion 1. For example, $\mathit{lfp}(\Phi_{SvL,\mathcal{P}_1}) = \langle\emptyset,\emptyset\rangle$ and $\mathit{lfp}(\Phi_{SvL,\mathcal{P}_2}) = \langle\emptyset,\{p,q\}\rangle$. In addition, Stenning and van Lambalgen make the following claims, where they assume that models are defined with respect to Fitting semantics:

A. The least fixed point of $\Phi_{SvL,\mathcal{P}}$ can be shown to be the minimal model of $\mathcal{P}$ (Lemma 4(1.) in [19]).
B. All models of $\mathit{comp}(\mathcal{P})$ are fixed points of $\Phi_{SvL,\mathcal{P}}$ (Lemma 4(3.) in [19]).

Both claims are false. Consider $\mathcal{P}_1 = \{p \leftarrow q\}$ and let $I = \langle\emptyset,\emptyset\rangle$. As discussed before, $\mathit{lfp}(\Phi_{SvL,\mathcal{P}_1}) = I$, but $I_F(\mathcal{P}_1) = \mathsf{U}$; thus, we have obtained a counter example for A. One should observe that the minimal models of $\mathcal{P}_1$ under Fitting semantics are $\langle\{p\},\emptyset\rangle$ and $\langle\emptyset,\{q\}\rangle$, both of which are not fixed points of $\Phi_{SvL,\mathcal{P}_1}$. Now let $I'' = \langle\emptyset,\{p,q\}\rangle$ and $I' = \langle\emptyset,\{p\}\rangle$. $I''_F(\mathit{comp}(\mathcal{P}_1)) = \top$, but $\Phi_{SvL,\mathcal{P}_1}(I'') = I'$, $\Phi_{Svl,\mathcal{P}_1}(I') = I$, and $\Phi_{SvL,\mathcal{P}_1}(I) = I$; thus, we have obtained a counter example for B.

Problem A. can be overcome if we interprete programs under Łukasiewicz semantics. For the discussed example we find $I_{\text{Ł}}(\mathcal{P}_1) = \top$, which is no coincidence as we will show in the sequel.

Proposition 3. *(i) If $I_{\text{Ł}}(\mathrm{wcomp}(\mathcal{P})) = \top$ then $\Phi_{SvL,\mathcal{P}}(I) \subseteq I$.*
(ii) If $\Phi_{SvL,\mathcal{P}}(I) = I$ then $I_{\text{Ł}}(\mathrm{wcomp}(\mathcal{P})) = \top$.

Proof Sketch. *(i)* If $I_{\text{Ł}}(\mathit{wcomp}(\mathcal{P})) = \top$ then for for each equivalence $A \leftrightarrow \mathit{Body}_1 \vee \ldots \vee \mathit{Body}_n$ occurring in $\mathit{wcomp}(\mathcal{P})$ we find that

$$I_{\text{Ł}}(A) = I_{\text{Ł}}(\mathit{Body}_1 \vee \ldots \vee \mathit{Body}_n). \tag{1}$$

Now let $I = \langle I^{\top}, I^{\perp}\rangle$ and $\langle J^{\top}, J^{\perp}\rangle = \Phi_{SvL,\mathcal{P}}(\langle I^{\top}, I^{\perp}\rangle)$. By definition of $\Phi_{SvL,\mathcal{P}}$ we find $J^{\top} \subseteq I^{\top}$ and $J^{\perp} \subseteq I^{\perp}$ given (1). Hence, $\Phi_{SvL,\mathcal{P}}(I) \subseteq I$.

(ii) Suppose $\Phi_{SvL,\mathcal{P}}(I) = I$. Let $F := A \leftrightarrow \mathit{Body}_1 \vee \ldots \vee \mathit{Body}_n$ be an arbitrary but fixed conjunct occurring in $\mathit{wcomp}(\mathcal{P})$. If $I_{\text{Ł}}(A) = \top$, then there exists $A \leftarrow \mathit{Body}_i \in \mathcal{P}$ such that $I_{\text{Ł}}(\mathit{Body}_i) = \top$. Hence, $I_{\text{Ł}}(\mathit{Body}_1 \vee \ldots \vee \mathit{Body}_n) = \top$ and, consequently, $I_{\text{Ł}}(F) = \top$. The cases $I_{\text{Ł}}(A) = \perp$ and $I_{\text{Ł}}(A) = \mathsf{U}$ follow similarly. Because F was arbitrary but fixed we conclude $I_{\text{Ł}}(\mathit{wcomp}(\mathcal{P})) = \top$. □

Proposition 4. *If $I = \mathit{lfp}(\Phi_{SvL,\mathcal{P}})$ then $I_{\text{Ł}}(\mathrm{wcomp}(\mathcal{P})) = I_{\text{Ł}}(\mathcal{P}) = \top$.*

Proof Sketch. From Proposition 3*(ii)* we learn that $I = \mathit{lfp}(\Phi_{SvL,\mathcal{P}})$ entails $I_{\text{Ł}}(\mathit{wcomp}(\mathcal{P})) = \top$. Moreover, $I_{\text{Ł}}(\mathit{wcomp}(\mathcal{P})) = \top$ iff for each equivalence $A \leftrightarrow \mathit{Body}_1 \vee \ldots \vee \mathit{Body}_n$ occurring in $\mathit{wcomp}(\mathcal{P})$ we find that $I_{\text{Ł}}(A) = I_{\text{Ł}}(\mathit{Body}_1 \vee \ldots \vee \mathit{Body}_n)$. A careful case analysis reveals that $I_{\text{Ł}}(\mathcal{P}) = \top$ holds. □

3 The Core Method

In [12] a connectionist model generator for propositional logic programs using recurrent networks with feed-forward core was presented. It was later called the *core method* [2]. The core method has been extended and applied to a variety of programs including modal (see e.g. [7]) and first-order logic programs [1]. It is based on the idea that feed-forward connectionist networks can approximate almost all functions arbitrarily well [14,11] and, hence, they can also approximate – and in some cases compute – the immediate consequence operators associated with logic programs. Moreover, if such an operator is a contraction mapping on a complete metric space, then Banach's contraction mapping theorem ensures that a unique fixed point exists such that the sequence constructed from applying the operator iteratively to any element of the metric space converges to the fixed point [10]. Turning the feed-forward core into a recurrent network allows to compute or approximate the least model of a logic program [13].

Kalinke has applied the core method to logic programs under the Fitting semantics presented in Section 2 [15]. In particular, her feed-forward cores compute $\Phi_{F,\mathcal{P}}$ for any given program $\mathcal{P}$. Seda and Lane showed that the core method can be extended to many-valued logic programs [18]. Restricted to three-valued logic programs considered here, their cores also compute $\Phi_{F,\mathcal{P}}$. In the sequel, these approaches are modified in order to compute $\Phi_{SvL,\mathcal{P}}$.

Given a program $\mathcal{P}$, the following algorithm translates $\mathcal{P}$ into a feed-forward core. Let m be the number of propositional variables occurring in $\mathcal{P}$. Without loss of generality, we may assume that the variables are denoted by natural numbers from $[1, m]$. Let $\omega \in \mathbb{R}^+$.

1. The input and output layer is a vector of binary threshold units of length $2m$ representing interpretations. The $2i - 1$-st unit in the layers, denoted by $i^\top$, is active iff the i-th variable is mapped to $\top$. The $2i$-th unit in the layers, denoted by $i^\bot$, is active iff the i-th variable is mapped to $\bot$. Both, the $2i - 1$-st and the $2i$-th unit, are passive iff the i-th variable is mapped to U. The case where the $2i - 1$-st and the $2i$-th unit are active is not allowed. The threshold of each unit occurring in the input layer is set to $\frac{1}{2}$. The threshold of each $2i - 1$-st unit occurring in the output layer is set to $\frac{\omega}{2}$. The threshold of each $2i$-th unit occurring in the output layer is set to $max\{\frac{\omega}{2}, l - \frac{\omega}{2}\}$, where l is the number of clauses with head i in $\mathcal{P}$.
   In addition, two units representing $\top$ and $\bot$ are added to the input layer. The threshold of these units is set to $-\frac{1}{2}$.
2. For each clause of the form $A \leftarrow B_1 \wedge \ldots \wedge B_k$ occurring in $\mathcal{P}$, do the following.
   (a) Add two binary threshold units $h^\top$ and $h^\bot$ to the hidden layer.
   (b) Connect $h^\top$ to the unit $A^\top$ in the output layer. Connect $h^\bot$ to the unit $A^\bot$ in the output layer.
   (c) For each B_j, $1 \leq j \leq k$, do the following.
      i. If B_j is an atom, then connect the units $B_j^\top$ and $B_j^\bot$ in the input layer to $h^\top$ and $h^\bot$, respectively.

   ii. If B_j is the literal $\neg B$, then connect the units $B^\perp$ and $B^\top$ in the input layer to $h^\top$ and $h^\perp$, respectively.
   iii. If B_j is $\top$, then connect the unit $\top$ in the input layer to $h^\top$.
   iv. If B_j is $\perp$, then connect the unit $\perp$ in the input layer to $h^\perp$.
 (d) Set the threshold of $h^\top$ to $k - \frac{\omega}{2}$, and the threshold of $h^\perp$ to $\frac{\omega}{2}$.
3. Set the weights associated with all connections to ω.

Proposition 5. *For each program $\mathcal{P}$, there exists a core of binary threshold units computing $\Phi_{SvL,\mathcal{P}}$.*

Proof. Assume that the input layer is actived at time t such that it represents an interpretation I. Then, at time $t+1$ an $h^\top$-unit representing $A \leftarrow B_1 \wedge \ldots \wedge B_k$ in the hidden layer becomes active iff all units representing $B_1 \wedge \ldots \wedge B_k$ in the input layer are active, i.e., if $I(B_1) = \ldots = I(B_k) = \top$. Likewise, at time $t+1$ an $h^\perp$-unit representing $A \leftarrow B_1 \wedge \ldots \wedge B_k$ in the hidden layer becomes active iff one unit representing the negation of $B_1 \wedge \ldots \wedge B_k$ in the input layer is active, i.e., if $I(\neg B_1) \vee \ldots \vee I(\neg B_k) = \top$. At time $t+2$ a unit representing A in the output later becomes active iff there is an active $h^\top$-unit representing $A \leftarrow B_1 \wedge \ldots \wedge B_k$ at time $t+1$. Likewise, at time $t+2$ a unit representing $\neg A$ in the ouput layer becomes active iff all $h^\perp$ units reprenting rules with head A are active at time $t+1$. Thus, the core is a direct encoding of $\Phi_{SvL,\mathcal{P}}$. □

Given a program $\mathcal{P}$ and its core, a recurrent network can be constructed by connecting each unit in the output layer to its corresponding unit in the input layer with weight 1. In Figure 1 the construction is illustrated.

Proposition 6. *For each program $\mathcal{P}$, the corresponding recurrent network initialized by the empty interpretation will converge to a stable state which corresponds to the least fixed point of $\Phi_{SvL,\mathcal{P}}$.*

Proof. The result follows immediately from the construction of the recurrent network using Propositions 4 and 5. □

4 Human Reasoning

In this section we will discuss some examples taken from [3]. These examples were used by Byrne to show that classical logic cannot appropriately model human reasoning. Stenning and van Lambalgen argue that a three-valued logic programs under a completion semantics can well model human reasoning [19]. Moreover, as we will see, the core method presented in Section 3 serves as a connectionist model generator in these cases.

Consider the following sentences: *If Marian has an essay to write, she will study late in the library. She has an essay to write.* In [3] 96% of all subjects conclude that *Marian will study late in the library.* The two sentences can be represented by the program $\mathcal{P}_3 = \{l \leftarrow e \wedge \neg ab,\ e \leftarrow \top,\ ab \leftarrow \perp\}$. The first sentence is interpreted as a licence for a conditional and the atom ab is used to

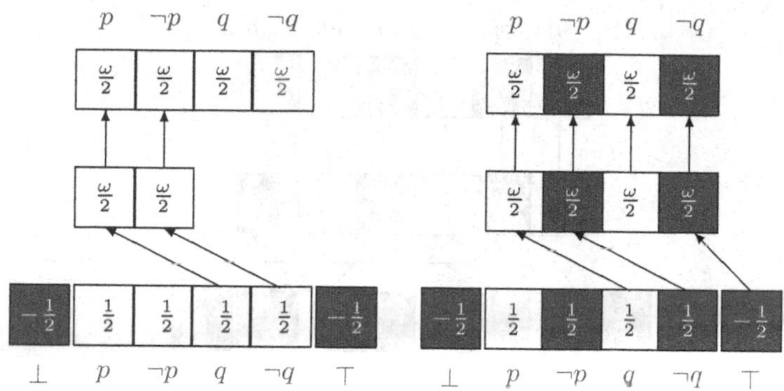

Fig. 1. The stable states of the feed-forward cores for $\mathcal{P}_1$ (left) and $\mathcal{P}_2$ (right), where all connections have weight ω, active units are shown in grey and passive units in white. The recurrent connections between corresponding units in the output and input layer are not shown.

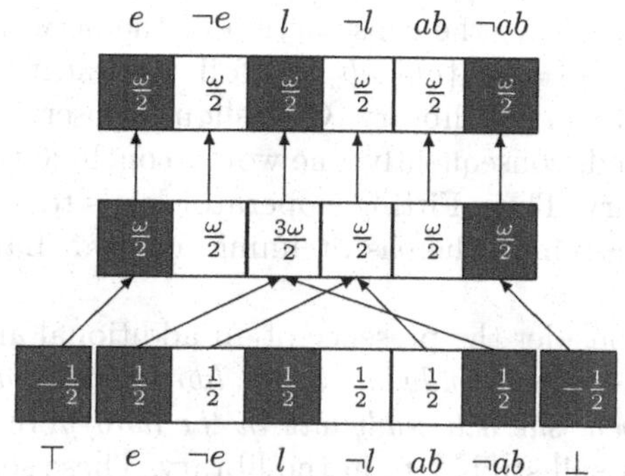

Fig. 2. The stable state of the feed-forward core for $\mathcal{P}_3$

cover all additional preconditions that we may be unaware of. As we know of no such preconditions, the rule $ab \leftarrow \bot$ is added. The corresponding network as well as its stable state are shown in Figure 2. From $lfp(\Phi_{SvL,\mathcal{P}_3}) = \langle\{l, e\}, \{ab\}\rangle$ follows that Marian will study late in the library.

Suppose now that the antecedent is denied: *If Marian has an essay to write, she will study late in the library. She does not have an essay to write.* In [3] 46% of subjects conclude that Marian will not study late in the library. These subject err with respect to classical logic. But they do not err with respect to the non-classical logic considered here. The two sentences can be represented by the program $\mathcal{P}_4 = \{l \leftarrow e \wedge \neg ab,\ e \leftarrow \bot, ab \leftarrow \bot\}$. The corresponding network as well as its stable state are shown in Figure 3. From $lfp(\Phi_{SvL,\mathcal{P}_4}) = \langle\emptyset, \{ab, e, l\}\rangle$ follows that Marian will not study late in the library.

Now consider an alternative argument: *If Marian has an essay to write, she will study late in the library. She does not have an essay to write. If she has textbooks to read, she will study late in the library.* In [3] 4% of subjects conclude

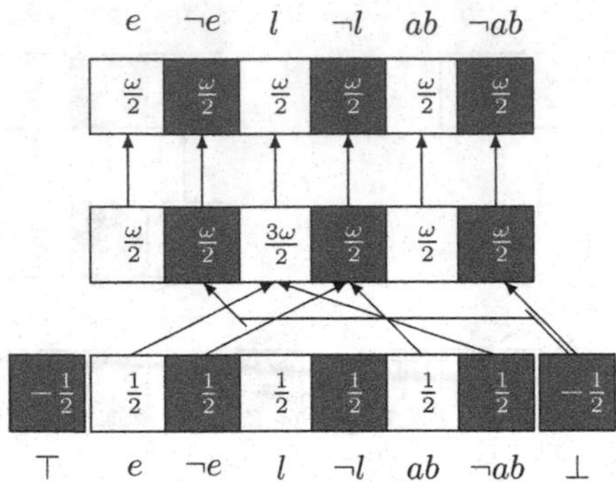

Fig. 3. The stable state of feed-forward core for $\mathcal{P}_4$

that Marian will not study late in the library. These sentences can be represented by $\mathcal{P}_5 = \{l \leftarrow e \wedge \neg ab_1,\ e \leftarrow \bot,\ ab_1 \leftarrow \bot,\ l \leftarrow t \wedge \neg ab_2, ab_2 \leftarrow \bot\}$. Due to lack of space we leave the construction of the network to the interested reader. From $lfp(\Phi_{SvL,\mathcal{P}_5}) = \langle\emptyset, \{ab_1, ab_2, e\}\rangle$ follows that it is unknown whether Marian will study late in the library. One should observe that $lfp(\Phi_{F,\mathcal{P}_5}) = \langle\emptyset, \{ab_1, ab_2, e, t, l\}\rangle$ and, consequently, one would conclude that Marian will not study late in the library. Thus, Fitting's operator leads to a wrong answer with respect to human reasoning, whereas Stenning and van Lambalgen's operator does not.

As final example consider the presence of an additional argument: *If Marian has an essay to write, she will study late in the library. She has an essay to write. If the library stays open, she will study late in the library.* In [3] 38% of subjects conclude that Marian will study late in the library. These sentences can be represented by $\mathcal{P}_6 = \{l \leftarrow e \wedge \neg ab_1,\ e \leftarrow \top,\ l \leftarrow o \wedge \neg ab_2,\ ab_1 \leftarrow \neg o,\ ab_2 \leftarrow \neg e, \}$. As argued in [19] the third sentence gives rise to an additional argument for

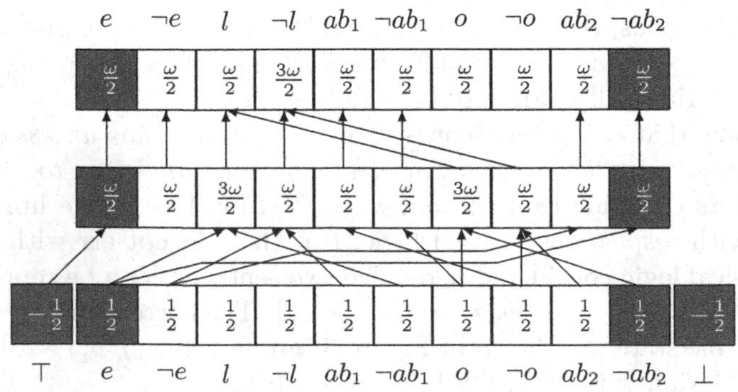

Fig. 4. The stable state of feed-forward core for $\mathcal{P}_6$

studying in the library, viz. that the library is open. Likewise, there must be a reason for going to the library like, for example, writing an essay. The corresponding network as well as its stable state are shown in Figure 4. From $lfp(\Phi_{SvL,\mathcal{P}_6}) = \langle\{e\}, \{ab_2\}\rangle$ follows that it is unknown whether Marian will study late in the library.

5 Discussion

We propose to use the Łukasiewicz semantics for three-valued logic programs in the area of human reasoning. Returning to our last example, $lfp(\Phi_{SvL,\mathcal{P}_6}) = \langle\{e\}, \{ab_2\}\rangle$ is a model for both, the weak completion of $\mathcal{P}_6$ and $\mathcal{P}_6$ itself, under the Łukasiewicz semantics, whereas it is – somewhat surprisingly – a model for the weak completion of $\mathcal{P}_6$ but not for $\mathcal{P}_6$ under the Fitting semantics. Under the Łukasiewicz semantics the completion of a three-valued logic program is exactly what completion was originally thought of, viz., the addition of the only-if halves to a program specifying the if-halves [4]. Nevertheless, whether Łukasiewicz semantics is adequate for human reasoning in a broader sense remains to be seen.

We showed that the core method can be adapted to implement the revised immediate consequence operator of Stenning and van Lambalgen. We presented an algorithm for constructing the networks and proved that the networks settle down in a state encoding the least fixed point of the operator. Although our networks consist of logical threshold units, they can be replaced by bipolar sigmoidal ones while preserving the relationship to logic programs by applying the method first presented in [8] (see also [5]). The modified networks can then be trained using backpropagation or related techniques, rule extraction methods can be applied to the trained networks and the neural-symbolic cycle can be closed. On the other hand, to the best of our knowledge there is no evidence that backpropagation is neurally plausible.

In our networks units come in pairs, where the first (second) element represents the fact that the corresponding variable or formula is mapped to true (false). In [19] similar networks are proposed – albeit in a three-dimensional setting – and, in addition, the elements of each pair inhibit each other. From a logical point of view such an inhibition is unnecessary – it can never be the case that both elements are active – as long as the units of the input layer are not externally activited. In a general setting, however, where context information is used to activate the input units, such inhibitory connections establishing a winner-take-all behaviour are very useful. Unfortunately, the presented results concerning the core networks and their relation to logic programs do not apply in this case anymore unless we would be able to show that $\Phi_{SvL,\mathcal{P}}$ is a contraction.

In [19] integrity constraints and abduction are suggested to handle additional human reasoning tasks. We would like to investigate whether the techniques developed in [6] can be applied to model the tasks in a connectionist setting.

In the field of Logic Programming Fitting's three-valued (first-order) logic has been used in termination analysis. It remains to be seen whether these results carry over to Łukasiewicz semantics. How important is Stenning and van Lambalgen's operator from a logic programming perspective?

References

1. Bader, S., Hitzler, P., Hölldobler, S., Witzel, A.: A fully connectionist model generator for covered first-order logic programs. In: Veloso, M.M. (ed.) Proceedings of the Twentieth International Joint Conference on Artificial Intelligence, January 2007, pp. 666–671. AAAI Press, Menlo Park (2007)
2. Bader, S., Hölldobler, S.: The core method: Connectionist model generation. In: Kollias, S.D., Stafylopatis, A., Duch, W., Oja, E. (eds.) ICANN 2006. LNCS, vol. 4132, pp. 1–13. Springer, Heidelberg (2006)
3. Byrne, R.M.J.: Suppressing valid inferences with conditionals. Cognition 31, 61–83 (1989)
4. Clark, K.L.: Negation as failure. In: Gallaire, H., Minker, J. (eds.) Logic and Databases, pp. 293–322. Plenum, New York (1978)
5. d'Avila Garcez, A.S., Broda, K., Gabbay, D.M.: Neural-Symbolic Learning Systems: Foundations and Applications. Springer, Heidelberg (2002)
6. d'Avila Garcez, A.S., Gabbay, D.M., Ray, O., Woods, J.: Abductive reasoning in neural-symbolic learning systems. TOPOI 26, 37–49 (2007)
7. d'Avila Garcez, A.S., Lamb, L.C., Gabbay, D.M.: A connectionist inductive learning system for modal logic programming. In: Proceedings of the IEEE International Conference on Neural Information Processing (ICONIP), Singapore (2002)
8. d'Avila Garcez, A.S., Zaverucha, G., de Carvalho, L.A.V.: Logic programming and inductive learning in artificial neural networks. In: Herrmann, C., Reine, F., Strohmaier, A. (eds.) Knowledge Representation in Neural Networks, pp. 33–46. Logos Verlag, Berlin (1997)
9. Fitting, M.: A Kripke–Kleene semantics for logic programs. Journal of Logic Programming 2(4), 295–312 (1985)
10. Fitting, M.: Metric methods – three examples and a theorem. Journal of Logic Programming 21(3), 113–127 (1994)
11. Funahashi, K.-I.: On the approximate realization of continuous mappings by neural networks. Neural Networks 2, 183–192 (1989)
12. Hölldobler, S., Kalinke, Y.: Towards a massively parallel computational model for logic programming. In: Proceedings of the ECAI 1994 Workshop on Combining Symbolic and Connectionist Processing, pp. 68–77, ECCAI (1994)
13. Hölldobler, S., Kalinke, Y., Störr, H.-P.: Approximating the semantics of logic programs by recurrent neural networks. Applied Intelligence 11, 45–59 (1999)
14. Hornik, K., Stinchcombe, M., White, H.: Multilayer feedforward networks are universal approximators. Neural Networks 2, 359–366 (1989)
15. Kalinke, Y.: Ein massiv paralleles Berechnungsmodell für normale logische Programme. Master's thesis, TU Dresden, Fakultät Informatik (1994) (in German)
16. Kleene, S.C.: Introduction to Metamathematics. North-Holland, Amsterdam (1952)
17. Łukasiewicz, J.: O logice trójwartósciowej. Ruch Filozoficzny 5, 169–171 (1920); English translation: On Three-Valued Logic. In: Borkowski, L. (ed.) Jan Łukasiewicz Selected Works, pp. 87–88. North Holland, Amsterdam (1990)
18. Seda, A.K., Lane, M.: Some aspects of the integration of connectionist and logic-based systems. In: Proceedings of the Third International Conference on Information, pp. 297–300. International Information Institute, Tokyo (2004)
19. Stenning, K., van Lambalgen, M.: Human Reasoning and Cognitive Science. MIT Press, Cambridge (2008)
20. Stoy, J.E.: Denotational Semantics. MIT Press, Cambridge (1977)

Simbed: Similarity-Based Embedding

John A. Lee[1,⋆] and Michel Verleysen[2]

[1] Imagerie Moléculaire et Radiothérapie Expérimentale,
Avenue Hippocrate, 54, B-1200 Bruxelles
[2] Machine Learning Group,
Place du Levant, 3, B-1348 Louvain-la-Neuve, Belgium
{john.lee,michel.verleysen}@uclouvain.be
http://www.ucl.ac.be/mlg

Abstract. Simbed, standing for similarity-based embedding, is a new method of embedding high-dimensional data. It relies on the preservation of pairwise similarities rather than distances. In this respect, Simbed can be related to other techniques such as stochastic neighbor embedding and its variants. A connection with curvilinear component analysis is also pointed out. Simbed differs from these methods by the way similarities are defined and compared in both the data and embedding spaces. In particular, similarities in Simbed can account for the phenomenon of norm concentration that occurs in high-dimensional spaces. This feature is shown to reinforce the advantage of Simbed over other embedding techniques in experiments with a face database.

Keywords: Nonlinear dimensionality reduction, similarity measure, manifold learning, stochastic gradient, multiscale optimization.

1 Introduction

Dimensionality reduction is the task of finding faithful, low-dimensional representations of high-dimensional data. Although the case of clustered data can be considered, it usually relies on the assumption that the data are sampled from a smooth manifold. For instance, if the underlying manifold is a linear subspace, then methods such as principal component analysis (PCA) [12] or classical metric multidimensional scaling [21] can be successfully applied. However, these techniques are not optimal if the manifold is heavily curved or folded [19]. This issue can be addressed by using methods of nonlinear dimensionality reduction [10] (NLDR) instead of a linear projection. The development of nonlinear variants of MDS lead in the early sixties to many techniques that are based on the principle of distance preservation. Nonmetric MDS [18,8] and Sammon's nonlinear mapping [15] (SNLM) are the best known methods in this family. The eighties and early nineties saw the advent of methods related to artificial neural networks and soft-computing. Auto-encoders with multilayer perceptrons [7] and

⋆ J.A.L. is a Postdoctoral Researcher with the Belgian National Fund for Scientific Research (FNRS).

C. Alippi et al. (Eds.): ICANN 2009, Part II, LNCS 5769, pp. 95–104, 2009.

Kohonen's self-organizing maps [6] (SOMs) are the most prominent examples in this trend. Since the late nineties and the seminal paper describing kernel PCA [17], many recent develoments in NLDR have targeted spectral embedding [16]. Isomap [19] and locally linear embedding [14] are probably the most representative methods in this branch. Spectral methods provide the guarantee of finding the global optimum of their cost function. In contrast, methods based on other optimization techniques generally do not offer this advantage. However, they usually compensate for this drawback by the capability of handling a broader range of cost functions. Successful nonspectral methods are for instance curvilinear component analysis [1,4] (CCA), stochastic neighbor embedding [5] (SNE), and its variant t-SNE [20].

This paper introduces Simbed, a new NLDR method that relies on similarity matching in order to embed data in a low-dimensional space. Simbed's most prominent feature is its principled way of computing pairwise similarities that accounts for the phenomenon of norm concentration [3]. Briefly put, this term refers to the fact that the norm of high-dimensional vectors tends to have a low variance/expectation ratio. Hence, Simbed owns a decisive advantage when it comes to real-life data that combine non-negligible noise with a high dimensionality. The paper also weaves connections with other methods that involve similarities, such as SNE and t-SNE. An unexpected relationship with CCA is also pointed out, which shows that CCA is closer to similarity matching than to distance preserving techniques such as SNLM.

The rest of this paper is organized as follows. Section 2 introduces notations for distances and gives a principled definition of pairwise similarities in high-dimensional spaces. Section 3 describes a cost function that assesses the similarity matching, along with an algorithm that optimizes it. Section 4 points out some connections with other techniques, such as CCA and t-SNE. Section 5 gathers some experimental results with both artificial and real data. Finally, Section 6 draws the conclusions.

2 Distances and Similarities

Let $\mathbf{\Xi} = [\boldsymbol{\xi}_i]_{1\leq i\leq N}$ denote a data set of N vectors picked in an M dimensional space. The symbol δ_{ij} denotes the pairwise Euclidean distance $\|\boldsymbol{\xi}_i - \boldsymbol{\xi}_j\|_2$.

In order to define a similarity measure, let us consider vector $\boldsymbol{\xi}_i$ and an isotropic k-dimensional normal distribution centered on it, with $k \leq M$. We define the similarity of $\boldsymbol{\xi}_j$ with respect to $\boldsymbol{\xi}_i$ to be the probability of the event $\|\boldsymbol{\xi} - \boldsymbol{\xi}_i\|_2 \geq \delta_{ij}$ where $\boldsymbol{\xi} \sim \mathcal{N}(\boldsymbol{\xi}_i, \lambda\mathbf{I})$. In other words, the similarity is the probability of observing a larger distance than the measured value and thus varies between 0 and 1. The normality assumption allows us to write $\|\boldsymbol{\xi} - \boldsymbol{\xi}_i\|_2/\lambda \sim \chi_k$, where χ_k denotes a chi distribution with k degrees of freedom [2]. The probability density function of $c \sim \chi_k$ is given by

$$p(c,k) = \frac{\sqrt{2}}{\Gamma(k/2)}\left(\frac{c}{2^{1/2}}\right)^{k-1} \exp(-c^2/2) \ , \tag{1}$$

where Γ is the Gamma function. Therefore, the similarity between $\boldsymbol{\xi}_i$ and $\boldsymbol{\xi}_j$ can be defined by

$$\sigma_{ij}(\lambda, k) \doteq \text{Prob}\left[\delta_{ij} \geq \lambda c\right] = \int_{\delta_{ij}}^{\infty} \frac{p(z/\lambda, k)}{\lambda}\, dz = Q\left(\frac{\delta_{ij}^2}{2\lambda^2}, \frac{k}{2}\right) , \tag{2}$$

where $c \sim \chi_k$ and Q is the regularized upper incomplete gamma function. This definition contains two free parameters, namely λ and k.

Standard deviation λ reflects the fact that the proposed similarity measure is a scale-dependent concept. Hence, this parameter basically sets up the threshold between the 'local' neighborhood of some vector and the rest of the space. Its value can then be arbitrarily fixed by the user. If we consider that Ξ contains noisy vectors sampled from some manifold, λ should not go below the (local) noise standard deviation. Similarly, larger values than $\max_{1\leq i,j\leq N} \delta_{ij}$ make little sense. As will be shown later on, a multiscale or multiresolution approach that explores several values of λ can be useful.

As to k, which specifies the number of degrees of freedom, its optimal value depends on λ. If λ is close to the standard deviation of the noise, k should be equal to M, since noise indifferently spans all dimensions of space. For slightly larger values of λ, noise becomes negligible and the manifold can be locally approximated by a linear subspace with as many degrees of freedom as the manifold intrinsic dimensionality. For larger values of λ, k should be chosen according to the global shape of the manifold, which is difficult to investigate.

It is noteworthy that for appropriate values of k and λ the proposed similarity can account for the phenomenon of norm concentration [3]. Defining the similarity with a Gaussian kernel (this turns out to be the case $k = 2$) does not offer this possibility. The model behind the proposed similarity can be refined, for instance by using anisotropic normal distributions, but this introduces additional parameters.

3 Matching the Pairwise Similarities

NLDR aims at finding a low-dimensional representation of data set Ξ. Let $\mathbf{X} = [\mathbf{x}_i]_{1\leq i\leq N}$ denote this representation and let P be its dimensionality. Symbol d_{ij} refers to the pairwise Euclidean distance $\|\mathbf{x}_i - \mathbf{x}_j\|_2$. As in the previous section, we can define the similarity between $\mathbf{x}_i$ and $\mathbf{x}_j$ in the same way, i.e.

$$s_{ij}(\lambda, l) \doteq \text{Prob}\left[d_{ij} \geq \lambda c\right] = Q\left(\frac{d_{ij}^2}{2\lambda^2}, \frac{l}{2}\right) , \tag{3}$$

where $c \sim \chi_l$. Scale parameter λ can be reused from the previous definition, since we look at the same scale in both high- and low-dimensional spaces. The key difference lies in l, whose value is simply equal to embedding dimensionality P in this case.

Having defined the similarity measures in both spaces, one can try to match them by minimizing the mean square error

$$E(\mathbf{X}, \Xi, \lambda, k, l) = \frac{1}{N^2} \sum_{i=1}^{N} \sum_{j=1}^{N} (\sigma_{ij}(\lambda, k) - s_{ij}(\lambda, l))^2 \tag{4}$$

with respect to $\mathbf{X}$. The intuition behind this cost function is similar to that behind distance preservation: building a low-dimensional representation that preserves pairwise distances or similarities hopefully keeps neighboring data items close to each other, while maintaining the gap between dissimilar ones. Simbed —which stands for similarity-based embedding— minimizes (4) by performing a stochastic gradient descent such as in [1]. For this purpose, we separately consider all terms of the outer sum in the cost function. They can be written as

$$E_i(\mathbf{X}, \Xi, \lambda, k, l) = \frac{1}{N^2} \sum_{j=1}^{N} (\sigma_{ij}(\lambda, k) - s_{ij}(\lambda, l))^2 \quad . \tag{5}$$

The partial derivative of $E_i(\mathbf{X}, \Xi, \lambda, k, l)$ with respect to $\mathbf{x}_j$ is

$$\frac{\partial E_i(\mathbf{X}, \Xi, \lambda, k, l)}{\partial \mathbf{x}_j} = \frac{2}{N^2} (\sigma_{ij}(\lambda, k) - s_{ij}(\lambda, l)) \frac{p(d_{ij}/\lambda, l)}{\lambda} \frac{\mathbf{x}_j - \mathbf{x}_i}{d_{ij}} \quad . \tag{6}$$

This leads to the iterative update $\mathbf{x}_j := \mathbf{x}_j + \Delta \mathbf{x}_j^{(t,i)}$ for $1 \leq j \leq N$, where

$$\Delta \mathbf{x}_j^{(t,i)} = -\alpha^{(t)} \frac{\partial E_i(\mathbf{X}, \Xi, \lambda^{(t)}, k^{(t)}, l)}{\partial \mathbf{x}_j} \quad . \tag{7}$$

Index t denotes the current iteration (or 'epoch') and $\alpha^{(t)}$ is the step size (or learning rate). At each iteration, index i runs over the whole data set. Each iteration thus performs N^2 updates for a time complexity of $\mathcal{O}(N^2)$.

In the spirit of a pseudo-Newton optimization scheme, an appropriate value of $\alpha^{(t)}$ should be related to the second-order derivative. A single value of $\alpha^{(t)}$ that fits for all i and j should satisfy the second inequality in

$$\left\| \frac{\partial^2 E_i(\mathbf{X}, \Xi, \lambda^{(t)}, k^{(t)}, l)}{\partial \mathbf{x}_j^2} \right\|_2 \leq \frac{2}{(\lambda^{(t)} N)^2} \leq \frac{1}{\alpha^{(t)}} \quad . \tag{8}$$

The first inequality gives an upper bound of the second partial derivative. Therefore, $\alpha^{(t)}$ should be lower than $(N\lambda^{(t)})^2/2$. In a classical stochastic gradient descent, $\alpha^{(t)}$ should slowly decay [13]. This can be indirectly achieved by progressively reducing $\lambda^{(t)}$ in a multiscale approach. The principle of a multiscale optimization is rather simple and consists in 'blurring' the cost function in the early part of the process. Doing so smooths out narrow pits and peaks, thus putting the emphasis on the widest and deepest basins. Next, the amount of blur is slowly reduced as the optimization progresses. Hence, a multiscale strategy increases the probability of finding the global optimum. As a matter of fact,

Simbed's cost function can be optimized with a multiscale scheme because it can be written as a continuous function of smooth and rapidly decaying kernels. The parameters to be optimized, namely the coordinates in $\mathbf{X}$, exclusively appear in the arguments of these similarity kernels.

The efficiency of a multiscale optimization increases if scale parameter λ slowly decays. In the case of a stochastic gradient descent, we can combine the decay of λ with that of the step size. Simbed relies on the schedules given by $\lambda^{(t)} = 4\max_{1\leq i,j\leq N} \delta_{ij}/t$ and $\alpha^{(t)} = 0.1(N\lambda^{(t)})^2$. The latter fulfills the inequality mentioned in the previous section and ensures that $\|\Delta \mathbf{x}_j^{(t,i)}\|_2$ slowly vanishes as t grows to infinity. In practice, Simbed runs for at most T iterations but an early stop is possible when the criterion

$$\sum_{i=1}^{N}\sum_{j=1}^{N}\left|\left(d_{ij}^{(t)}\right)^2 - \left(d_{ij}^{(t-1)}\right)^2\right| \leq \epsilon \sum_{i=1}^{N}\sum_{j=1}^{N}\left(\left(d_{ij}^{(t)}\right)^2 + \left(d_{ij}^{(t-1)}\right)^2\right) , \quad (9)$$

is met, in which $d_{ij}^{(t)}$ refers to the pairwise distance at the end of iteration t. As to dimensionality parameter k, the constant value P can be used for noisefree data, whereas a schedule such as $k^{(t)} = P + (M - P)t/T$ can be adopted for noisy data.

The initialization of the algorithm can be achieved with a P-dimensional PCA projection. During the stochastic gradient descent, it is advised to randomly permute the order of the updates with respect to index i in (7). A reinitialization of the random number generator with always the same seed makes the optimization fully deterministic, provided the data vectors are not permuted in Ξ from run to run. These permutations can be avoided by computing $c = \arg\min_i \sum_{j=1}^{N} \delta_{ij}^2$ and by sorting the vectors in Ξ according to δ_{cj}.

4 Connection with Other Techniques

Simbed can be related to several other methods described in the literature, such as SNE, t-SNE, CCA, and SOMs. SNE and t-SNE follow the same paradigm as Simbed, that is, similarity matching with a stochastic optimization scheme. However several important differences can be pointed out.

First, the pairwise similarities used in Simbed involve cumulative distribution functions whose value depends only on the corresponding distances. In contrast, SNE and t-SNE rely on empirical probability density functions defined as

$$\sigma_{ij}(\lambda) \doteq \frac{g(\delta_{ij}/\lambda)}{\sum_{m\neq n} g(\delta_{mn})} \quad \text{and} \quad s_{ij} \doteq \frac{h(d_{ij})}{\sum_{m\neq n} h(d_{mn})} , \quad (10)$$

where $g(u) = \exp(-u^2/2)$ and either $h(u) = g(u)$ for SNE or $h(u) = 1/(1 + u^2)$ for t-SNE ($h(u)$ is proportional to a Student's t pdf with one degree of freedom). These similarity functions involve a softmax denominator; it ensures that $\sum_{i\neq j} \sigma_{ij}(\lambda) = \sum_{i\neq j} s_{ij} = 1$. Such a normalization makes all similarities interdependent and leads to paradoxical situations. For instance, equalities $g = h$

and $\delta_{ij} = \lambda d_{ij}$ do not imply $\sigma_{ij}(\lambda) = s_{ij}$. On the other hand, similarities in Simbed are such that $\delta_{ij} = d_{ij} \Leftrightarrow \sigma_{ij}(\lambda) = s_{ij}$, provided $k = l$ or $\delta_{ij} = 0$.

The second difference between Simbed and SNE/t-SNE results from the first one. Simbed estimates the similarity matching between the high- and low-dimensional spaces with a mean square error. In comparison, similarities in SNE and t-SNE are pdfs and their matching is thus computed by sums of Kullback-Leibler divergences in the cost function (see [20] for details and variants). This is a key point knowing that the particular choice of the similarity measure deeply impacts the form of the stochastic gradient update. The Gaussian functions in SNE lead to simplifications in the KL divergences whereas this does not happen in t-SNE. In (symmetric) SNE, the stochastic gradient update is proportional to $(\sigma_{ij}(\lambda) - s_{ij})(\mathbf{x}_j - \mathbf{x}_i)$ whereas it is proportional to $(\sigma_{ij}(\lambda) - s_{ij})(1/(1 + d_{ij}))(\mathbf{x}_j - \mathbf{x}_i)$ in t-SNE. This additional factor in t-SNE explains why it outperforms SNE [20]. Such a damping factor that decreases with respect to d_{ij} can be also found in Simbed as well as in CCA. This factor accounts for the capability of these methods to 'tear' manifolds [1,4,9].

A third difference concerns the absence of multiscale approach in SNE and t-SNE, although such a strategy has proved to be useful in methods such as CCA, SOMs and their variants.

The connection between Simbed and CCA can be investigated by looking at the terms of CCA's cost function, which are written as $E_i(\mathbf{X}, \boldsymbol{\Xi}, \lambda) = \sum_{j=1}^{N} (\delta_{ij} - d_{ij})^2 H(\lambda - d_{ij})$, where H denotes Heaviside's step function. The stochastic gradient update is proportional to $(1 - \delta_{ij}/d_{ij})H(\lambda - d_{ij})(\mathbf{x}_j - \mathbf{x}_i)$. Although CCA is often related to Sammon's nonlinear mapping and other methods based on distance preservation, it can easily be cast within the framework of similarity preservation. For this purpose, we can equivalently rewrite CCA's cost function as $E_i(\mathbf{X}, \boldsymbol{\Xi}, \lambda) = \sum_{j=1}^{N} (\sigma_{ij}(\lambda) - s_{ij}(\lambda))^2$, where $\sigma_{ij}(\lambda) \doteq (\lambda - \delta_{ij})H(\lambda - d_{ij})$ and $s_{ij}(\lambda) \doteq (\lambda - d_{ij})H(\lambda - d_{ij})$. While $s_{ij}(\lambda)$ satisfies all conditions to be a similarity function, positivity of $\sigma_{ij}(\lambda)$ can be enforced by the approximation $\sigma_{ij}(\lambda) \approx (\lambda - \delta_{ij})H(\lambda - \delta_{ij})$, which leads to

$$\frac{\partial E_i(\mathbf{X}, \boldsymbol{\Xi}, \lambda)}{\partial \mathbf{x}_j} = \left(1 - \frac{\delta_{ij}}{d_{ij}} - \frac{(\delta_{ij} - \lambda)H(\delta_{ij} - \lambda)}{d_{ij}}\right) H(\lambda - d_{ij})(\mathbf{x}_j - \mathbf{x}_i) \ . \quad (11)$$

The only difference with CCA's genuine gradient is the additional term in the first factor, which is non-zero only if $d_{ij} < \lambda < \delta_{ij}$. This shows that CCA is closely related to similarity-based embedding and that the multiplication by a Heaviside function in its cost function plays a much more important role than a simple weighting of the cost function terms, such as in SNLM.

5 Experiments

The experiments involve several data sets as well as several NLDR techniques. The first data set contains 750 vectors that sample a Swiss roll with uniform distribution. Its equation is written as $\boldsymbol{\xi} = [\sqrt{u}\cos(3\pi\sqrt{u}),\ \sqrt{u}\sin(3\pi\sqrt{u}),\ v]^T$, where random parameters u and v have uniform distributions between 0 and 1. The second data set stems from the same manifold and includes 750 vectors as well, but these have three additional coordinates that are kept constant. Gaussian noise with standard deviation 0.05 is added to all six dimensions. The third data set contains 1965 pictures of B.J. Frey's face [14]. Each face is 20 pixels wide and 28 pixels high. After concatenation into 560-dimensional vectors, PCA achieves a first dimensionality reduction that leaves 20 coordinates.

Simbed is compared to t-SNE, CCA, SNLM, and PCA, whose result serves as baseline. Two versions of Simbed are used, one with constant and equal values for k and l, the second with the adaptive schedule for k. The implementation of t-SNE is provided by the authors of [20]; the 'perplexity' (i.e. the scale parameter) is left to its default value. CCA is implemented as in [4], with a constant step size that is equal to 0.2. The scale parameter follows a similar schedule as in Simbed, except that $\lambda^{(1)}$ is doubled; the stopping criterion is the same. SNLM is implemented as in [15] with a step size equal to 0.3. All methods are fed with pairwise Euclidean distances, no geodesic distances are used.

Performance assessment is achieved by means of the criteria proposed in [11]. These criteria look at K-ary neighborhoods around each vector in the data space as well as in the embedding space. The first criterion is denoted by $Q_{\mathrm{NX}}(K)$ and reflects the overall quality of the embedding; its value corresponds to the average percentage of identical neighbors in both spaces. The second criterion is denoted by $B_{\mathrm{NX}}(K)$ and reveals the 'behavior' of a NLDR method. A positive value indicates that distant points are embedded close to each other whereas a negative one indicates that close neighbors are embedded far away. Results for the three data sets are shown in Figs. 1 to 3. Each figure includes three panels; the first one spans the interval $1 \leq K \leq N-1$, whereas the small ones on the right focus on small values of K, for each criterion separately.

As to the noisefree Swiss roll, CCA slightly outperforms all other methods for small values of K. Simbed closely follows, whereas t-SNE comes third and precedes both SNLM and PCA. On the other hand, the global shape of the manifold is best preserved by PCA and SNLM, followed by Simbed, CCA, and t-SNE. Simbed thus reaches the best 'global-local' tradeoff. The multiscale optimization of CCA and Simbed actually leads to flatter curves than for the other methods. For this noisefree data set, Simbed with $k = l = 2$ performs slightly better than the variant with an increasing value of k, as expected.

The situation gets reversed in Fig. 2 for the noisy Swiss roll. Thanks to its use of similarity kernels with heavier tails in the embedding space than in the data space, t-SNE unfolds the Swiss roll better than the other methods and achieves the best performance for small values of K. Simbed is second and its version with an increasing k takes advantage of its more appropriate noise model. CCA comes next and precedes both SNLM and PCA.

The results for the face bank are shown in Fig. 3. Simbed with increasing values of k clearly outperforms the version with constant k. CCA climbs on the third step, followed by SNLM and PCA. Despite numerous attempts, t-SNE has never converged. As can be seen, similarity matching proves to be very

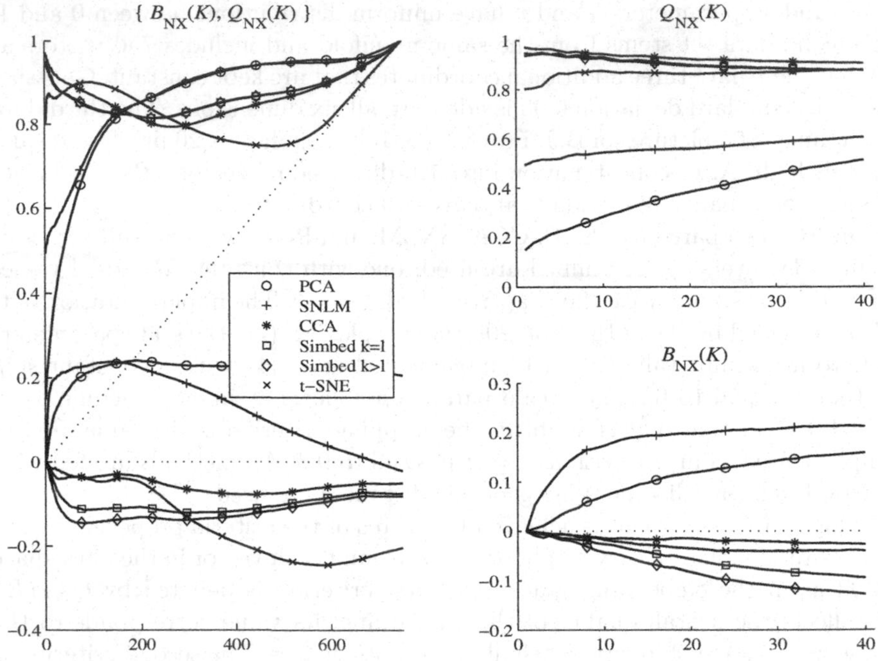

Fig. 1. Quality assessment for the embeddings of the noisefree Swiss roll

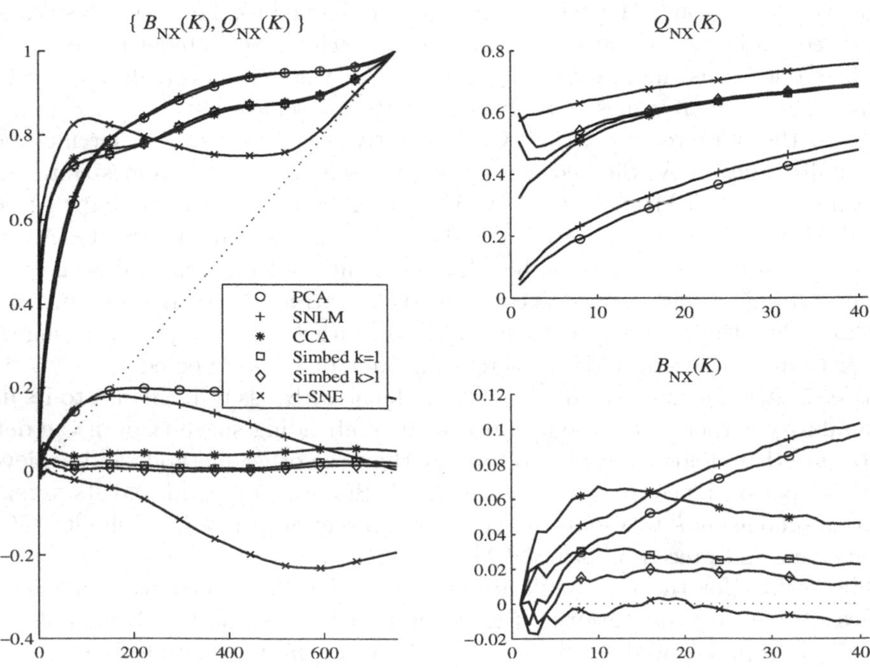

Fig. 2. Quality assessment for the embeddings of the noisy Swiss roll

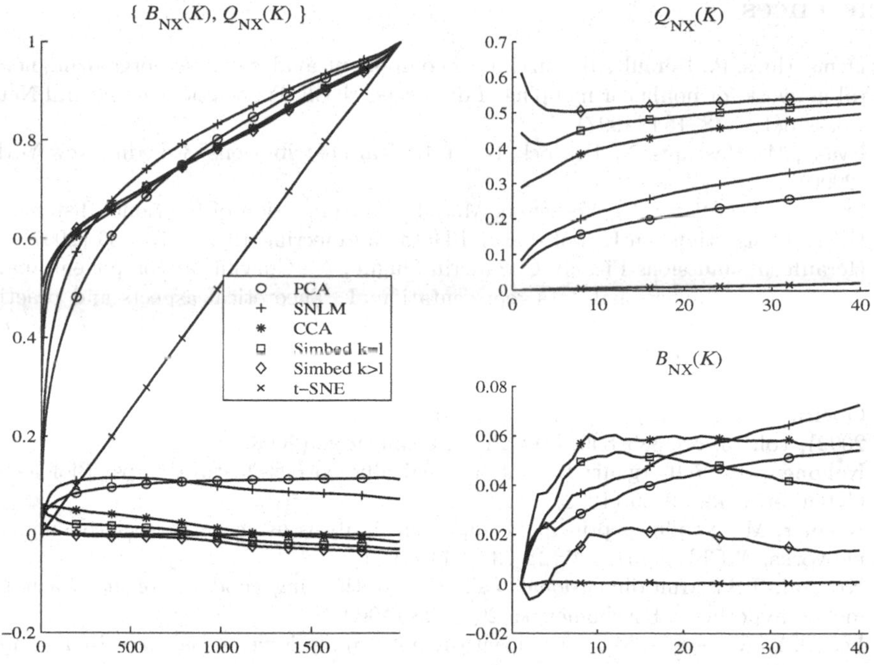

Fig. 3. Quality assessment for the embeddings of the face bank

efficient for this very high-dimensional data set. More importantly, the successive performance leaps between CCA and the two versions of Simbed indicate that the definition of the similarity kernel plays a key role as to the quality of the results. In the case of this data set, shifting from the piecewise linear kernel of CCA to smooth kernels that take into account the properties of high-dimensional spaces proves to be decisive.

6 Conclusion

Simbed is a new method of nonlinear dimensionality reduction that relies on similarity matching. It has two prominent features. First, it involves a principled similarity measure that can cope with the phenomenon of norm concentration in high-dimensional spaces. Second, its cost function can be optimized with a multiscale approach, which diminishes the probability of getting stuck in a local optimum. Simbed can be related to other methods such as SNE and t-SNE, and it also extends CCA.

Experiments with both artificial and real data show that Simbed compares to some of the best NLDR methods. It can provide excellent quantitative results in the case of noisy and very high-dimensional data, such as face images.

References

1. Demartines, P., Hérault, J.: Curvilinear component analysis: A self-organizing neural network for nonlinear mapping of data sets. IEEE Transactions on Neural Networks 8(1), 148–154 (1997)
2. Evans, M., Hastings, N., Peacock, B.: Statistical Distributions, 3rd edn., New York (2000)
3. François, D., Wertz, V., Verleysen, M.: The concentration of fractional distances. IEEE Transactions on Knwoledge and Data Engineering 19(7), 873–886 (2007)
4. Hérault, J., Jaussions-Picaud, C., Guérin-Dugué, A.: Curvilinear component analysis for high dimensional data representation: I. Theoretical aspects and practical use in the presence of noise. In: Mira, J., Sánchez, J.V. (eds.) Proceedings of IWANN 1999, vol. II, pp. 635–644. Springer, Alicante (1999)
5. Hinton, G., Roweis, S.T.: Stochastic neighbor embedding. In: Becker, S., Thrun, S., Obermayer, K. (eds.) Advances in Neural Information Processing Systems (NIPS 2002), vol. 15, pp. 833–840. MIT Press, Cambridge (2003)
6. Kohonen, T.: Self-organization of topologically correct feature maps. Biological Cybernetics 43, 59–69 (1982)
7. Kramer, M.: Nonlinear principal component analysis using autoassociative neural networks. AIChE Journal 37(2), 233–243 (1991)
8. Kruskal, J.B.: Multidimensional scaling by optimizing goodness of fit to a nonmetric hypothesis. Psychometrika 29, 1–28 (1964)
9. Lee, J.A., Verleysen, M.: Curvilinear distance analysis versus isomap. Neurocomputing 57, 49–76 (2004)
10. Lee, J.A., Verleysen, M.: Nonlinear dimensionality reduction. Springer, Heidelberg (2007)
11. Lee, J.A., Verleysen, M.: Quality assessment of dimensionality reduction: Rank-based criteria. Neurocomputing (2009)
12. Pearson, K.: On lines and planes of closest fit to systems of points in space. Philosophical Magazine 2, 559–572 (1901)
13. Robbins, H., Monro, S.: A stochastic approximation method. Annals of Mathematical Statistics 22, 400–407 (1951)
14. Roweis, S.T., Saul, L.K.: Nonlinear dimensionality reduction by locally linear embedding. Science 290(5500), 2323–2326 (2000)
15. Sammon, J.W.: A nonlinear mapping algorithm for data structure analysis. IEEE Transactions on Computers CC-18(5), 401–409 (1969)
16. Saul, L.K., Weinberger, K.Q., Ham, J.H., Sha, F., Lee, D.D.: Spectral methods for dimensionality reduction. In: Chapelle, O., Schoelkopf, B., Zien, A. (eds.) Semisupervised Learning. MIT Press, Cambridge (2006)
17. Schölkopf, B., Smola, A., Müller, K.-R.: Nonlinear component analysis as a kernel eigenvalue problem. Neural Computation 10, 1299–1319 (1998)
18. Shepard, R.N.: The analysis of proximities: Multidimensional scaling with an unknown distance function (parts 1 and 2). Psychometrika 27, 125–140, 219–249 (1962)
19. Tenenbaum, J.B., de Silva, V., Langford, J.C.: A global geometric framework for nonlinear dimensionality reduction. Science 290(5500), 2319–2323 (2000)
20. van der Maaten, L., Hinton, G.: Visualizing data using t-SNE. Journal of Machine Learning Research 9, 2579–2605 (2008)
21. Young, G., Householder, A.S.: Discussion of a set of points in terms of their mutual distances. Psychometrika 3, 19–22 (1938)

PCA-Based Representations of Graphs for Prediction in QSAR Studies

Riccardo Cardin[1], Lisa Michielan[2], Stefano Moro[2], and Alessandro Sperduti[3]

[1]Engeneering Ingegneria Informatica
[2]Dipartimento di Scienze Farmaceutiche
[3]Dipartimento di Matematica Pura ed Applicata, Università di Padova

Abstract. In recent years, more and more attention has been paid on learning in structured domains, e.g. Chemistry. Both Neural Networks and Kernel Methods for structured data have been proposed. Here, we show that a recently developed technique for structured domains, i.e. PCA for structures, permits to generate representations of graphs (specifically, molecular graphs) which are quite effective when used for prediction tasks (QSAR studies). The advantage of these representations is that they can be generated automatically and exploited by any traditional predictor (e.g., Support Vector Regression with linear kernel).

Keywords: PCA for graphs, prediction on structured domains, supervised learning.

1 Introduction

Principal Component Analysis (PCA) ([3]) constitutes one of the oldest and best known tools in Pattern Recognition and Machine Learning. It is a powerful technique for dimensionality reduction, while preserving much of the relevant information conveyed by a set of variables. It is theoretically well founded and reduces to the solution of an eigenvalue problem involving the covariance (or correlation) matrix of the available data. More recently, exploiting the well known kernel trick, Kernel PCA ([7]), which is a nonlinear form of PCA, has been proposed. Through the kernel trick, it is possible to implicitly project data into high-dimensional feature spaces. PCA is then performed in feature space, discovering principal directions that correspond to principal curves in the original data space. By defining a kernel on structured data, such as sequences, trees, and graphs, it is also possible to apply PCA to application domains where it is natural to represent data in a structured form; just to name a few, Chemistry, Bioinformatics, and Natural Language Processing. One problem with using KPCA in structured domains is that kernels are usually defined a priori and because of that they may not capture the structural features that are relevant for the task at hand. Explicit common structural features can be captured by data mining for graphs, however this is usually a time consuming approach that has no guarantee to return relevant structural features.

C. Alippi et al. (Eds.): ICANN 2009, Part II, LNCS 5769, pp. 105–114, 2009.

In this paper, we investigate how the recent definition of PCA for structures [9,5,10] can be used to define informative vectorial representations for graphs amenable to treatment by state of the art classification and prediction techniques. Specifically, we have considered the prediction of biological activity of chemical compounds (Quantitative Structure-Activity Relationship, or QSAR for short) on two datasets involving chemical compounds. On this datasets, we observed that Support Vector Regression using a linear kernel is much more effective when applied to representations of molecules based on PCA for graphs than using representations that are standard in QSAR studies, i.e. autocorrelation MEP vectors [1].

2 Principal Components Analysis for Vectors and Structures

One of the aims of standard PCA ([3]) is to reduce the dimensionality of a data set, while preserving as much as possible the information present in it. This is achieved by looking for orthogonal directions of maximum variance within the data set. The principal components are sorted according to the amount of variance they explain, so that the first few retain most of the variation present in all of the original variables. It turns out that the qth principal component is given by the projection of the data onto the eigenvector of the (sample) covariance matrix $\mathbf{C}$ of the data corresponding to the qth largest eigenvalue.

From a mathematical point of view, PCA can be understood as given by an orthogonal linear transformation of the given set of variables (i.e., the coordinates of the vectorial space in which data is embedded):

$$\mathbf{y}_i = \mathbf{A}\mathbf{x}_i \tag{1}$$

where $\mathbf{x}_i \in \mathbb{R}^k$ are the vectors belonging to the data set, and $\mathbf{A} \in \mathbb{R}^{r \times k}$ is the orthogonal matrix whose qth row is the qth eigenvector of the covariance matrix. Typically, larger variances are associated with the first $p < k$ principal components. Thus one can conclude that most relevant information occur only in the first p dimensions. Given a fixed value for p, principal components allow also to minimize the reconstruction error, i.e.

$$\mathbf{A}^{(p)} = \arg\min_{\mathbf{M} \in \mathbb{R}^{p \times k}} \sum_i \|\mathbf{x}_i - \mathbf{M}^\mathsf{T}\mathbf{M}\mathbf{x}_i\|^2$$

where the rows of $\mathbf{A}^{(p)} \in \mathbb{R}^{p \times k}$ corresponds to the first p eigenvectors of $\mathbf{C}$. In fact, let $\mathbf{X} = [\mathbf{x}_1, \mathbf{x}_2, \mathbf{x}_3, \ldots, \mathbf{x}_n]^\mathsf{T}$ then, by imposing with no loss in generality that $\|\mathbf{w}\| = 1$, the direction of maximum variance can be computed as

$$\mathbf{w}^* = \arg\max_{\|\mathbf{w}\|=1} \mathbf{w}^\mathsf{T}\mathbf{X}^\mathsf{T}\mathbf{X}\mathbf{w} = \arg\max_{\|\mathbf{w}\|=1} \mathbf{w}^\mathsf{T}\mathbf{C}\mathbf{w}.$$

This is a constrained optimization problem that can be solved by optimizing the Lagrangian

$$\mathcal{L}(\mathbf{w}, \lambda) = \mathbf{w}^\mathsf{T}\mathbf{C}\mathbf{w} - \lambda(\mathbf{w}^\mathsf{T}\mathbf{w} - 1).$$

By differentiating the Lagrangian with respect to $\mathbf{w}$ and equating to zero leads to the following symmetric eigenvalue problem $\mathbf{C}\mathbf{w} = \lambda\mathbf{w}$. Thus the first principal direction corresponds to the eigenvector with maximum eigenvalue, while the other principal directions correspond to the other eigenvectors (pairwise orthogonal by definition), sorted according to the corresponding eigenvalues.

2.1 Sequences and Graphs

Given a temporal sequence $\mathbf{x}_1, \mathbf{x}_2, \ldots, \mathbf{x}_t, \ldots$ of input vectors[1], where t is a discrete time index, in [9] it is proposed to model the sequence through the following linear dynamical system:

$$\mathbf{y}_t = \mathbf{A}\mathbf{x}_t + \mathbf{B}\mathbf{y}_{t-1} \tag{2}$$

so to extend the linear transformation defined in eq. (1) by introducing a *memory term* involving a non null matrix $\mathbf{B} \in \mathbb{R}^{r\times r}$. The basic idea is to look for the minimum value of r such that the input sequence can be fully reconstructed starting from the final state vector $\mathbf{y}_n \in \mathbb{R}^r$. We say that the input sequence can be fully reconstructed if, starting from the state vector $\mathbf{y}_n$ obtained applying eq. (2) (sequence *encoding*), all the $\mathbf{x}_t$, $t = 1, \ldots, n$, can be generated by applying recursively the two linear transformations $\mathbf{x}_t = \mathbf{A}^\mathsf{T}\mathbf{y}_t$ and $\mathbf{y}_{t-1} = \mathbf{B}^\mathsf{T}\mathbf{y}_t$ (sequence *decoding*).

By using the initial condition $\mathbf{y}_0 = \mathbf{0}$ (the null vector), and the dynamical linear system (2), the state vectors generated by a single sequence $\mathbf{x}_1, \mathbf{x}_2, \ldots, \mathbf{x}_t, \ldots, \mathbf{x}_n$, can be collected as rows of a matrix, which can be written as

$$\mathbf{Y} = \begin{bmatrix} \mathbf{y}_1^\mathsf{T} \\ \mathbf{y}_2^\mathsf{T} \\ \mathbf{y}_3^\mathsf{T} \\ \vdots \\ \mathbf{y}_n^\mathsf{T} \end{bmatrix} = \underbrace{\begin{bmatrix} \mathbf{x}_1^\mathsf{T} & \mathbf{0} & \mathbf{0} & \mathbf{0} & \cdots & \mathbf{0} \\ \mathbf{x}_2^\mathsf{T} & \mathbf{x}_1^\mathsf{T} & \mathbf{0} & \mathbf{0} & \cdots & \mathbf{0} \\ \mathbf{x}_3^\mathsf{T} & \mathbf{x}_2^\mathsf{T} & \mathbf{x}_1^\mathsf{T} & \mathbf{0} & \cdots & \mathbf{0} \\ \vdots & \vdots & \vdots & \vdots & \vdots & \vdots \\ \mathbf{x}_n^\mathsf{T} & \mathbf{x}_{n-1}^\mathsf{T} & \mathbf{x}_{n-2}^\mathsf{T} & \cdots & \mathbf{x}_2^\mathsf{T} & \mathbf{x}_1^\mathsf{T} \end{bmatrix}}_{\boldsymbol{\Xi}} \underbrace{\begin{bmatrix} \mathbf{A}^\mathsf{T} \\ \mathbf{A}^\mathsf{T}\mathbf{B}^\mathsf{T} \\ \mathbf{A}^\mathsf{T}\mathbf{B}^{2^\mathsf{T}} \\ \vdots \\ \mathbf{A}^\mathsf{T}\mathbf{B}^{n-1^\mathsf{T}} \end{bmatrix}}_{\boldsymbol{\Omega}},$$

where $\boldsymbol{\Xi}$ is a data matrix collecting all the (inverted) input subsequences (including the whole sequence) as rows, and $\boldsymbol{\Omega}$ is the parameter matrix of the dynamical system. Notice that rows of $\boldsymbol{\Xi}$ can be considered as snapshots of a stack after each reading of a new input, i.e. the t-th row represents the status of the stack after that the input vectors $\mathbf{x}_1, \mathbf{x}_2, \ldots, \mathbf{x}_t$ have been read.

In [9] it is observed that the sequence can be encoded with no loss of information by a dynamical system with optimal matrices $\mathbf{A}$ and $\mathbf{B}$, if all the eigenvectors of the covariance matrix $\frac{1}{n}\boldsymbol{\Xi}^\mathsf{T}\boldsymbol{\Xi}$ corresponding to non-null eigenvalues are used. Specifically, given the tiny spectral decomposition of $\frac{1}{n}\boldsymbol{\Xi}^\mathsf{T}\boldsymbol{\Xi} = \mathbf{U}^{(p)}\boldsymbol{\Lambda}\mathbf{U}^{(p)^\mathsf{T}}$ where the columns of $\mathbf{U}^{(p)}$ are the p eigenvectors corresponding to non-null eigenvalues, let $s = nk$, and

$$\mathbf{P}_{k,s} \equiv \begin{bmatrix} \mathbf{I}_{k\times k} \\ \mathbf{0}_{(s-k)\times k} \end{bmatrix}, \quad \text{and} \quad \mathbf{R}_{k,s} \equiv \begin{bmatrix} \mathbf{0}_{k\times(s-k)} & \mathbf{0}_{k\times k} \\ \mathbf{I}_{(s-k)\times(s-k)} & \mathbf{0}_{(s-k)\times k} \end{bmatrix},$$

[1] In the following, for the sake of presentation, we assume $\mathbb{E}[\mathbf{x}_t] = 0$.

then we have

$$\mathbf{A} = \mathbf{U}^{(p)^{\mathsf{T}}} \mathbf{P}_{k,s} \in \mathbb{R}^{p \times k}, \quad \text{and} \quad \mathbf{B} = \mathbf{U}^{(p)^{\mathsf{T}}} \mathbf{R}_{k,s} \mathbf{U}^{(p)} \in \mathbb{R}^{p \times p}. \tag{3}$$

Matrix $\mathbf{R}_{k,s}$ implements a shift operator into the dynamical system memory stack, while $\mathbf{P}_{k,s}$ implements a push operator on the same stack. In fact, if we consider rows $\boldsymbol{\xi}_t$ and $\boldsymbol{\xi}_{t-1}$ of Ξ, we have $\boldsymbol{\xi}_t^{\mathsf{T}} = \mathbf{P}_{k,s}\mathbf{x}_t + \mathbf{R}_{k,s}\boldsymbol{\xi}_{t-1}^{\mathsf{T}}$.

The same result can be obtained by considering more sequences. For example, let consider two sequences $(\mathbf{x}_1^a, \mathbf{x}_2^a, \mathbf{x}_3^a, \mathbf{x}_4^a)$ and $(\mathbf{x}^b{}_1, \mathbf{x}^b{}_2)$, coded into the matrices $\Xi_\mathbf{a}$ and $\Xi_\mathbf{b}$, respectively. Then, these can be collected together into the matrix $\Xi = \left[\begin{array}{c} \Xi_\mathbf{a} \\ \hline \Xi_\mathbf{b}\ \mathbf{0}_{2\times 2} \end{array}\right]$ and treated as described above.

When considering the possibility to extend the dynamical system to graphs, either with directed or undirected edges, two problems need to be faced: *i)* how to deal with cycles during the encoding; *ii)* how to identify the origin and destination of an edge during decoding.

In [5], these two problems are solved through a coding trick. The basic idea is to enumerate the set of vertexes following a given convention and representing a (directed or undirected) graph as an (inverted) ordered list of vertex's labels associated with a list of edges for which the vertex is origin and where the position in the associated list is referring to the destination vertex. The idea is that the list is used by the linear dynamical system during encoding to read one by one the information about each vertex and associated edges, pushing the read information into the internal stack. Decoding is obtained by popping from the internal stack, one by one, the information about vertexes and associated edges.

The proposed linear dynamical system supporting the above idea is defined as

$$\mathbf{y}_i = \mathbf{W}_\mathbf{v}[\mathbf{v}_{label}^{\mathsf{T}}, \mathbf{v}_{edges}^{\mathsf{T}}]^{\mathsf{T}} + \mathbf{W}_\mathbf{y}\mathbf{y}_{i-1} \tag{4}$$

where i ranges over the enumeration of the vertexes, i.e. positions in the list representing the graph, $\mathbf{v}_{label} \in \mathbb{R}^k$ is the numerical encoding of the current label, $\mathbf{v}_{edges} \in \mathbb{R}^N$ is the vector representing the information about the edges entering the current vertex where N is the maximum number of vertexes that the system can manage for a single input graph, and $\mathbf{y}_0$ is the null vector. Thus $[\mathbf{v}_{label}^{\mathsf{T}}, \mathbf{v}_{edges}^{\mathsf{T}}]^{\mathsf{T}} \in \mathbb{R}^d$, where $d = k + N$.

In Figure 1 we have reported an example of how vertexes of the molecular graph of Aziridine can be coded. Assuming that the Nitrogen atom is coded by label [001], the Carbon atom by label [010], and the Hydrogen atom by [100], and assuming vertex enumeration shown on the left side of the figure, on the right side of the figure we have reported how the information about each vertex is coded. Specifically, the Nitrogen atom is inserted first (n_1) in the stack. Because of that, no edge is represented. Then a Carbon atom is inserted (n_2), and it is connected with the Nitrogen atom already in the stack. The insertion of the second Carbon atom (n_3) comes also with the edge connecting it to the already inserted Carbon atom and the edge connecting it to the already inserted

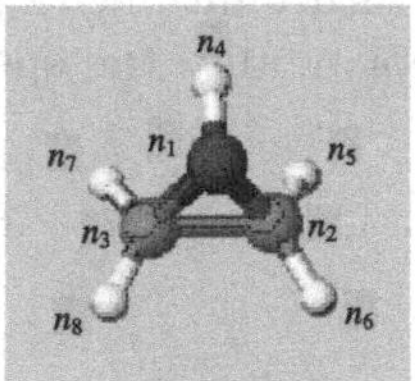

vertex (atom)	label	undirected edge with n_1	n_2	n_3	n_4	n_5	n_6	n_7	n_8
n_8 (H)	1 0 0	0	0	1	0	0	0	0	0
n_7 (H)	1 0 0	0	0	1	0	0	0	0	0
n_6 (H)	1 0 0	0	1	0	0	0	0	0	0
n_5 (H)	1 0 0	0	1	0	0	0	0	0	0
n_4 (H)	1 0 0	1	0	0	0	0	0	0	0
n_3 (C)	0 1 0	1	1	0	0	0	0	0	0
n_2 (C)	0 1 0	1	0	0	0	0	0	0	0
n_1 (N)	0 0 1	0	0	0	0	0	0	0	0

input sequence coding the full Aziridine molecular graph ($[\mathbf{x}_8,\mathbf{x}_7,\mathbf{x}_6,\mathbf{x}_5,\mathbf{x}_4,\mathbf{x}_3,\mathbf{x}_2,\mathbf{x}_1]$)
[*100*0010000,*100*0010000,*100*0100000,*100*0100000,*100*1000000,*010*1100000,*010*1000000,*001*0000000]

Fig. 1. Examples of vertexes' coding for the molecular graph of Azeridine. Here $k = 3$ and $N = 8$. The input sequence representing the molecular graph, where the label information is shown in italics, is shown as well. Notice that, in the sequence the last column of the matrix is not reported since in chemical compounds no self-connection on vertexes is allowed.

Nitrogen atom, and so on[2]. In Figure 1, the input sequence coding the full Aziridine molecule is shown as well.

The space embedding the explicit representation of the stack is Nd since no more than N vertexes can be inserted. It should be noted that this size of the stack is needed only if the input graphs are directed, and the above system is basically equivalent to system (2) for sequences. However, if undirected graphs are considered, a specific state space optimization can be performed. In fact, when inserting the first vertex into the internal stack only the first entry of the vector $\mathbf{v}_{edges}$ may be non null (the one encoding the self-connection), since no other vertex has already been presented to the system. In general, if vertex i is being inserted, only the first i components of $\mathbf{v}_{edges}$ may be non null. Because of that, the shift operator embedded into matrix $\mathbf{W_y}$ may "forget" the last component of each field into which the internal stack is organized. Formally, the shift operator described above can be implemented by the following matrix

$$\mathbf{S} \equiv \begin{bmatrix} \mathbf{0}_{d\times s} & & \\ \mathbf{I}_{(d-1)\times(d-1)} & \mathbf{0}_{(d-1)\times(s-d+1)} & \\ \mathbf{0}_{(d-2)\times d} & \mathbf{I}_{(d-2)\times(d-2)} & \mathbf{0}_{(d-2)\times(s-2(d-1))} \\ \mathbf{0}_{(d-3)\times(2d-1)} & \mathbf{I}_{(d-3)\times(d-3)} & \mathbf{0}_{(d-3)\times(s-3(d-1))} \\ & \cdots & \\ & \mathbf{0}_{(k+1)\times(s-k-1)} & \mathbf{I}_{(k+1)\times(k+1)} \end{bmatrix} \quad (5)$$

and the solution matrices defined as $\mathbf{W_v} \equiv \mathbf{U}^{(p)^\mathsf{T}}\mathbf{P}_{d,s}$ and $\mathbf{W_y} \equiv \mathbf{U}^{(p)^\mathsf{T}}\mathbf{S}\mathbf{U}^{(p)}$.

In [10] different strategies to speed-up the spectral decomposition of the data matrix $\mathbf{\Xi}$, which can be quite large, in the case of discrete labels have been suggested. One of the most effective strategies to reduce the size of $\mathbf{\Xi}$, after

[2] For chemical compounds, atoms' occurrence order in the canonical SMILE representation can be used as enumeration of vertexes. Moreover, the entry for an edge can code the type of bond, e.g. for a double bond the number 2 can be used.

removal of null columns, is the substitution of rows $\mathbf{r}_j$ with multiplicity $\mu_j > 1$ by a single occurrence of the row $\sqrt{\mu_j}\mathbf{r}_j$, which does not modify the spectrum of the matrix.

3 Prediction for Graphs

The idea explored in this paper is to use the dynamical system described by eq. (4), which is a linear recurrent neural network, to project the graphs belonging to the dataset into a vectorial space of dimension p. The obtained vectorial representations, or just some of its components, i.e. the ones corresponding to the first q principal components, can then be used to train any state of the art classifier or predictor. It must be observed that the proposed approach is different by an approach where an extended representation of each graph is generated for each graph in the dataset and then PCA computed. In fact, in this latter case, the substructures belonging to the graphs are not considered and consequently, the representation space obtained by PCA is much smaller, i.e., considering n graphs, no more than n principal directions can be obtained. On the contrary, by considering all the m substructures belonging to the graphs, as generated by the encoding procedure[3], we have $m \gg n$ and a richer representation space is generated. In addition, if a specific substructure is very frequent, the principal direction accounting for that substructure will be more important than principal directions that account for less frequent substructures.

In order to assess the usefulness of the proposed approach, we have performed some experiments involving chemical compounds, represented through the associated molecular graph. We have selected two datasets which are representative of the type and size of data typically involved in QSAR studies for drug design. Then, on a prediction task, we have compared the results obtained using Support Vector Regression [8] on two different vectorial representations. The first representation is typically used in QSAR studies and is based on the use of autocorrelated molecular descriptors encoding for the Molecular Electrostatic Potential (autoMEP) [1]. These descriptors are derived from a classical point charge model, and their computation is quite involved. The second representation is obtained by the proposed approach, i.e. principal components of graphs.

We start the description by first presenting the datasets, then the prediction tasks and related evaluation measure, and finally the obtained results.

3.1 Datasets

An important task in medicinal chemistry is the definition of computational models for the prediction of properties of interest for drug design, such as the quantitative prediction of relative binding affinity of chemical compounds with respect to a given receptor. The human adenosine A_{2A} receptor ($hA_{2A}R$) has been discovered to be involved in some neurological disorders, such as Parkinson's

[3] Notice that which substructure is generated by the encoding procedure depends on the definition of the associated dynamical system.

Table 1. Occurrences of atoms symbols in the chemical datasets and some of their statistical properties

Chemical Symbol	C	N	O	S	F	Cl	Br	H
Frequency in A_{2A}	2444	913	317	16	16	25	25	2224
Frequency in A_3	2115	808	256	6	16	32	13	1822

Dataset	Num. of molecules	Max. number atoms	Max. number bonds	Avg. number atoms (bonds)	Tot. number items (atoms+bonds)
A_{2A}	127	61	66	47.09 (50.89)	12,443
A_3	104	69	73	48.73 (52.80)	10,559

or Huntington's diseases, and for this reason it is important to develop predictive models for this receptor [2]. Dataset A_{2A} involves a collection of 127 known human A_{2A} antagonists, i.e. pyrazolo-triazolo-pyrimidine and triazolo-pyridine analogues [4]. In this dataset, 8 distinct atoms (C, N, O, S, F, Cl, Br, H) occur. In Table 1 we report the frequencies of such atoms through the compounds as well as some general statistics. A second dataset consists of a group of 104 pyrazolo-triazolo-pyrimidines, which are antagonists of the human adenosine A_3 receptor (hA_3R), that have a potential application in the tumor growth inhibition and in the treatment of glaucoma. The set of occurring atoms into the A_3 dataset is the same as the A_{2A} dataset, with of course different frequencies, as reported in Table 1, where other general statistics about the compounds in the dataset are presented. Two different types of vectorial representations for the molecules of the datasets have been considered. The first representation type is obtained by resorting to a standard molecule representation in QSAR studies, i.e. autocorrelation MEP vectors. This representation has been introduced by Gasteiger and collaborators [1] as molecular descriptors computed on the molecular surface. Ligands and proteins interact through molecular surfaces and, therefore, clearly, the representations of molecular surfaces have to be sought in the endeavour to understand the biological activity. In our case, twelve autocorrelation coefficients were calculated. This transformation produces a unique fingerprint of each molecule under consideration. The autocorrelation vectors were calculated by Surface module of the Adriana suite[4]. The second representation type has been obtained by computing the first q principal components of the molecular graphs.

3.2 Prediction Task and Evaluation Measure

The computational task for both datasets is to accurately predict the (A_{2A} or A_3) receptor-binding affinity, as measured by pK_i. Specifically, the target for each molecule is represented by the negative logarithmic value of the corresponding $hA_{2A}R$ or hA_3R binding constant K_i ($pk_i = -\log K_i$). The binding constant (K_i) is a concentration value, expressed in nanomolar (nM) units. It is a measure

[4] ADRIANACode; version 2.0; Molecular Networks GmbH: Erlangen, Germany, 2005.

of the strength of the binding affinity between the ligands (in our case antagonists) and the receptor, that represents the biological target. In general lower K_i values (and higher pK_i values) correspond to more potent antagonists. The target values for the A_{2A} dataset range over the interval $[-3.14, 0.92]$, while the target values for the A_3 dataset range over the interval $[-3.65, 0.85]$. To address the regression tasks, we used Support Vector Regression [8] over either the vectorial representations of molecules obtained by autoMEP or PCA for graphs as outlined in the paper.

The quality of the models was assessed by using a standard measure in QSAR studies, i.e. the correlation coefficient R between the predicted and the experimental values for pKi. R ranges from -1 to 1, with value 1 indicating the highest possible quality for the model. Due to the small size of the two datasets, we decided to evaluate the merits of the two different representation schemes by using a leave-one-out approach. Consequently, denoting with t_i the target associated with the i-th input vector $\mathbf{x}_i$ and with $SVR_i(\mathbf{x}_i)$ the output to $\mathbf{x}_i$ of the SVR trained by using all the training examples except for the i-th example, we used the following definition of leave-one-out R

$$R_{loo} = \frac{\frac{1}{n-1}\sum_{i=1}^{n}\left(t_i - \bar{t}\right)\left(SVR_i\left(\mathbf{x}_i\right) - \overline{SVR}\right)}{\sqrt{\frac{\sum_{i=1}^{n}\left(t_i-\bar{t}\right)^2}{n-1}}\sqrt{\frac{\sum_{i=1}^{n}\left(SVR_i(\mathbf{x}_i)-\overline{SVR}\right)^2}{n-1}}} \tag{6}$$

where

$$\bar{t} = \frac{\sum_{i=1}^{n} t_i}{n} \quad \text{and} \quad \overline{SVR} = \frac{\sum_{i=1}^{n} SVR_i\left(\mathbf{x}_i\right)}{n}.$$

3.3 Results

AutoMEP representations of molecules are constituted by twelve indicators, regardless of the dataset where the molecules belong to. Concerning molecules's representations obtained by PCA for graphs, in the second column of Table 2 we have reported, for each dataset, the size of the full data matrix Ξ. In the third column of Table 2 we have reported the size of Ξ after removal of null columns and redefinition of the rows as described in Section 2.1. It can be observed that there is a significant reduction of the size of Ξ with a computational

Table 2. Sizes of the full and reduced data matrix Ξ for the datasets. Since nodes in molecular graphs cannot have self-connections, the number of columns for the full data matrix Ξ is given by $mk + \frac{1}{2}m(m-1)$, where m is the maximum number of nodes in a single graph for the dataset. The size of the data embedding space is reported as well.

Dataset	size full Ξ	size reduced Ξ	size embedding space
A_{2A}	5980×2318	3752×1544	1438
A_3	5068×2898	3130×1759	1459

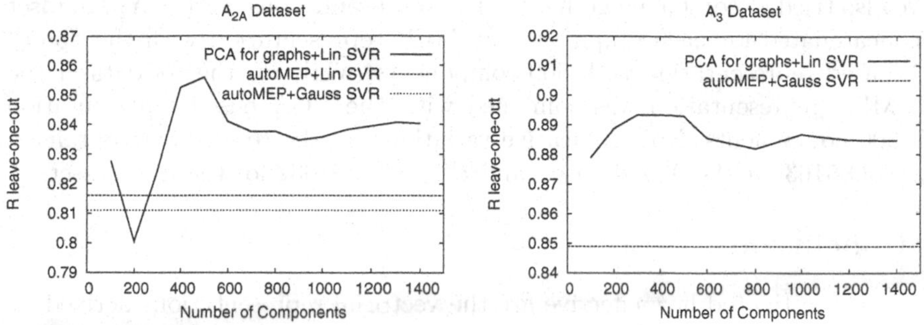

Fig. 2. R_{loo} values obtained by SVR, using a linear kernel (Lin) or a gaussian kernel (Gauss) for the A_{2A} and A_3 datasets. Correlation values obtained by using PCA for graphs are plotted versus the number of principal components used to represent the molecules. Correlation values obtained by autoMEP (where each molecule is represented by 12 numerical features) are constant (since they do not depend on the number of principal components). For dataset A_3, the SVR with linear kernel using autoMEP representations got a very low figure ($R_{loo} = 0.716$).

effort which is $O(r \log(r) + c)$, where r is the number of rows and c the number of columns. Principal directions are then computed starting from the reduced Ξ matrix. In the fourth column of Table 2, for each dataset, we have reported the number of principal directions, i.e. eigenvectors with non-null eigenvalues of the correlation matrix, we obtained. This number corresponds to the size of the space embedding all the data. It can be observed that both datasets have an embedding space of similar dimension.

For the PCA based representation of molecules, we have considered different numbers of components. Specifically, we have considered the first $q = 100w$ principal components where $w = 1, 2, \ldots, 14$.

Concerning SVR using linear kernel, we have considered $C = 10^a$ with $a \in \{-4, -3, -2, -1, 0, 1, 2, 3\}$ and $\epsilon = 10^z v$ with $v \in \{0.01, 0.03, 0.06\}$ and $z \in \{0, -1, -2\}$. For autoMEP representations, since they are constituted by just 12 features, we considered also a gaussian kernel with parameter γ. In this case, we run a preliminary set of experiments to determine the "best" range for the hyper-parameters for each dataset. This lead to the determination of the following set of values for dataset A_{2A}: C as above, $\epsilon \in \{0.2, 0.22, 0.26, 0.4, 0.42, 0.45, 0.47, 0.52, 0.6, 0.63, 0.67, 0.7, 0.78\}$, $\gamma = 10^a$ with $a \in \{-4, -3, -2, -1, 0, 1\}$. For dataset A_3, we used $C = 50 + 10a$ with $a = 1, \ldots, 15$, $\epsilon \in \{0.1, 0.2, 0.3, 0.4, 0.5, 0.6, 0.7, 0.8, 0.9\}$, and $\gamma = 0.005 + 0.01z$ with $z = 0, 1, \ldots, 5$.

In Figure 2 we have reported the R_{loo} results obtained by SVR for the different representation methods, number of principal components for PCA for graphs, kernels, and datasets. It can be observed that best results for R_{loo} most of the times are obtained by PCA based representations. Specifically, the best results for the A_{2A} dataset are obtained with $k = 500$ ($R_{loo} = 0.855929$) and for the A_3 dataset are obtained with $k = 300$ ($R_{loo} = 0.893401$). For dataset A_3, the SVR with linear kernel using autoMEP representations got a very low figure ($R_{loo} = 0.716$).

We also tried to combine together the two representations: for the A_{2A} dataset we concatenated for each graph the autoMEP representation with the "best" PCA based representation with 500 components, while for the A_3 dataset the auto MEP representation was combined with the PCA based repersentation with 300 components. No significative variations on the results were obtained: $R_{loo} = 0.85163$ for the A_{2A} dataset, and $R_{loo} = 0.897032$ for the A_3 dataset.

4 Conclusion

We have investigated how effective are the vectorial representations derived by PCA for graphs in prediction tasks involving chemical compounds. Specifically, we have compared the results obtained by Support Vector Regression using the representations of molecules based on PCA for graphs versus ad hoc representations that are standard in QSAR studies, i.e. autocorrelation MEP vectors. The PCA for graphs based representations clearly outperform autocorrelation MEP vectors.

For the future, we plan to investigate whether the proposed approach is competitive also with respect to more sophisticated approaches, such as graph kernels defined for Chemoinformatics (e.g. [6]). We also need to assess the performance of Recurrent Neural Network models trained on the sequential representations of molecules used by PCA for graphs.

References

1. Gasteiger, J., Li, X., Rudolph, C., Sadowski, J., Zupan, J.: Representation of molecular electrostatic potentials by topological feature maps. Am. Chem. Soc. 116, 4608–4620 (1994)
2. Jacobson, K.A., Gao, Z.-G.: Adenosine receptors as therapeutic targets. Nat. Rev. Drug Discov. 5, 247–264 (2006)
3. Jolliffe, I.T.: Principal Component Analysis. Springer, Heidelberg (2002)
4. Schiesaro, A., Bolcato, C., Pastorin, G., Spalluto, G., Cacciari, B., Klotz, K.N., Kaseda, C., Michielan, L., Bacilieri, M., Moro, S.: Linear and non-linear 3d-qsar approaches in tandem with ligand-based homology modeling as computational strategy to depict the pyrazolo-triazolo-pyrimidine antagonists binding site of the human adenosine a_2a receptor. J. Comput. Inf. Model 48, 350–363 (2008)
5. Micheli, A., Sperduti, A.: Recursive principal component analysis of graphs. In: ICANN (2), pp. 826–835 (2007)
6. Ralaivola, L., Swamidass, S.J., Saigo, H., Baldi, P.: Graph kernels for chemical informatics. Neural Networks 18(8), 1093–1110 (2005)
7. Schölkopf, B., Smola, A.J., Müller, K.-R.: Kernel principal component analysis. In: Gerstner, W., Hasler, M., Germond, A., Nicoud, J.-D. (eds.) ICANN 1997. LNCS, vol. 1327, pp. 583–588. Springer, Heidelberg (1997)
8. Smola, A.J., Schölkopf, B.: A tutorial on support vector regression. Statistics and Computing 14(3), 199–222 (2004)
9. Sperduti, A.: Exact solutions for recursive principal components analysis of sequences and trees. In: ICANN (1), pp. 349–356 (2006)
10. Sperduti, A.: Efficient computation of recursive principal component analysis for structured input. In: ECML, pp. 335–346 (2007)

Feature Extraction Using Linear and Non-linear Subspace Techniques

Ana R. Teixeira[1], Ana Maria Tomé[1], and E.W. Lang[2]

[1] DETI/IEETA, Universidade de Aveiro, 3810-193 Aveiro, Portugal
ana@ieeta.pt

[2] Institute of Biophysics, University of Regensburg 93040 Regensburg, Germany
elmar.lang@biologie.uni-regensburg.de

Abstract. This paper provides a new insight into unsupervised feature extraction techniques based on subspace models. In this work the subspace models are described exploiting the dual form of the basis vectors. In what concerns the kernel based model, a computationally less demanding model based on incomplete Cholesky decomposition is also introduced. An online benchmark data set allows the evaluation of the feature extraction methods comparing the performance of two classifiers having as input the raw data and the new representations.

1 Introduction

Finding better representations of a given set of data with more informative features is sometimes fundamental to improve the performance of a classifier. Often not all original features are appropriate, and even the number of features might be too large to conduct an efficient training. Subspace techniques can be applied as unsupervised feature generators simultaneously providing dimension reduction and more suitable representations.

Principal Component Analysis (PCA) is a subspace technique widely used in many fields like face recognition [1] and related computer vision tasks [2]. In this application a new representation of a given data set is formed by a linear combination of the original features whereby the data is projected onto orthogonal basis vectors. These projections represent new features which are non-correlated and even can be of smaller number. This model also implies that the original features are linear combinations of these projections. This assumption is a limitation if it is to model highly complex data. Kernel PCA methods are well suited in such cases to find the non-linear principal components. And in a classification task, the new representation provided by the non-linear kernel methods belongs to a new space (called feature space) where the data most probably become linearly separable [3]. The main characteristic of kernel methods is that the non-linear components (in input space) are computed via a transformation to a space of higher dimension. In this feature space, the main steps of the PCA are formulated using dot products. However, the non-linear mapping and the dot product are performed simultaneously using kernel functions [4],[3]. The parameters of these functions are the features in the input space, thus avoiding an

C. Alippi et al. (Eds.): ICANN 2009, Part II, LNCS 5769, pp. 115–124, 2009.

explicit mapping to the higher dimensional space. Kernel methods are computationally demanding whenever it is needed to manipulate huge training data sets. If the kernel (dot product) matrices are large, then their storage as well as their eigendecomposition might be prohibitive in any practical application because of memory limitations. Besides this, the basis vectors of the subspace have to be written in their dual form, i.e., as a linear combination of the training set. So, in a classification task, the training set has also to be available even during the testing phase of the classifier, i.e. , even when the parameters of the subspace model do not change.

In this work we show different strategies to perform feature extraction either in input or feature space. In input space the features are calculated by using the PCA decomposition. In feature space KPCA and greedy KPCA are applied. The latter is based on an incomplete Cholesky approach to compute the new representation of the training data set in feature space. Then, the dual form of the kernel subspace model is computed, formed by a subset of the training set which turns the model less demanding, during the application to new data. Another issue to be discussed is the influence of centering the data on the models. The proposal is to perform the centering and simultaneously maintain the new representation of the training data set non-correlated. The numerical simulations compare the performance of classifiers using kernel features, principal component features and a direct classification of the raw data using two classifiers: the nearest neighbor (NN) and linear discriminant function (RL). Furthermore, to evaluate the impact of the projective techniques, a comparative study with the best results published in [5] is presented and discussed.

2 Subspace and Classification

With subspace methods, denoising or classification is achieved by projecting the data onto basis vectors ($\mathbf{U}$), i. e. by computing products between the data vectors and basis vectors. The projections constitute the new representation of the data which can be a simple linear combination of the input data (PCA projections) or it can represent non-linear components of the data (KPCA projections). In a classification task the projections are then the input to the classifier. During the training phase the basis vectors are computed and the projections are used to adapt the parameters of classifiers. Afterwards, the performance of the classifier is evaluated with the projections of new data (test data) onto the basis vectors computed using the training set. These steps are the same either using PCA or KPCA, the differences are only concerned with the introduction of a kernel to replace the dot products.

2.1 Subspace Model

Using the dual form to describe the basis vector matrix $\mathbf{U}$, the basis vectors (columns of the matrix) are obtained as a linear combination of the training data set, either in input space or after a non-linear mapping. Considering that the mapped training data set is $\mathbf{\Phi} = [\phi(\mathbf{x}_1), \phi(\mathbf{x}_2) \ldots \phi(\mathbf{x}_N)]$, the basis vector matrix reads

$$\mathbf{U} = \mathbf{\Phi}\mathbf{V}\mathbf{D}^{-1/2} \tag{1}$$

In this form the matrices $\mathbf{V}$ and $\mathbf{D}$ are obtained by computing the eigendecomposition of the kernel matrix $\mathbf{K} = \mathbf{\Phi}^T\mathbf{\Phi}$. Usually the eigenvectors, i.e. the columns of $\mathbf{V}$, are ordered according to the value of the corresponding eigenvalues, the diagonal of matrix $\mathbf{D}$. Assuming that the eigenvalues are ordered in decreasing order,$\lambda_1 >, \lambda_2, \ldots > \lambda_L \ldots > \lambda_{last}$,the number ($L$) basis vectors can be chosen according to the percentage of variance of the data to be kept in the new representation. Afterwards the mapped training data set $\mathbf{\Phi}$ projected into the basis vector leads to new representation,

$$\mathbf{Z} = \mathbf{D}^{-1/2}\mathbf{V}^T\mathbf{\Phi}^T\mathbf{\Phi} \tag{2}$$

where each column (j) of $\mathbf{Z}$, of dimension L, is the representation of $\mathbf{x}_j$ in the feature space. The substitution of the kernel matrix by its eigendecomposition in previous equation leads to $\mathbf{Z} = \mathbf{D}^{1/2}\mathbf{V}^T$. Then, the new representation of the training data set is non-correlated, i.e. , $\mathbf{Z}\mathbf{Z}^T$ is a diagonal matrix. Also notice that a low-rank approximation for the kernel matrix $\mathbf{K}$ can be obtained by computing $\mathbf{Z}^T\mathbf{Z}$. If the dimension of $\mathbf{Z}$ is L, the approximation corresponds to the L leading eigenvalues and related eigenvectors. Furthermore, note that the Principal Component Analysis model and its corresponding projections can be obtained by substituting the mapped data set ($\mathbf{\Phi}$) by the raw data set $\mathbf{X} = [\mathbf{x}_1, ..., \mathbf{x}_N]$ in the previous equations. Due to lack of space those descriptions are omitted and it can also be verified that the properties discussed above are also accomplished.

2.2 Basis in Input Space

The common approach to compute the matrix of basis vectors $\mathbf{U} = [\mathbf{u}_1, ..., \mathbf{u}_L]$ is to perform the eigendecomposition of the covariance matrix (or the scatter matrix $\mathbf{S} = \mathbf{X}\mathbf{X}^T$). However, instead of computing $\mathbf{S}$, a matrix of dot products, the kernel matrix ($\mathbf{K} = \mathbf{X}^T\mathbf{X}$) can be an alternative. This strategy is often used when the dimension of the data D , as in case of face recognition applications, is larger than the size of the training set N to avoid the eigendecomposition of the scatter matrix. Taking the singular value decomposition (SVD) of the data, we can establish the relations between eigenvectors of both matrices and the non-zero eigenvalues of both matrices, that are identical [6].

2.3 Basis in Feature Space

In feature space, the dot products are evaluated by kernel functions using the data in input space. For example, with the radial basis function (RBF), the dot product between a pair of points is

$$\phi^T(\mathbf{x}_i)\phi(\mathbf{x}_j) = k(\mathbf{x}_i, \mathbf{x}_j) = k_{i,j} = \exp\left(-\frac{\|\mathbf{x}_i - \mathbf{x}_j\|^2}{2\sigma^2}\right) \tag{3}$$

where σ is a value to be assigned according to the range of values of the data set. Thus the matrix of dot products $\mathbf{K} = \mathbf{\Phi}^T\mathbf{\Phi}$ can be calculated easily this way. Each entry $k_{i,j}$ is the result of the dot product between a pair (i,j) of mapped examples of the training set. As referred before the drawbacks of these methods are

- the size of $\mathbf{K}$. The eigendecomposition of large matrices can be unfeasible in practical applications requiring the manipulation of large data sets.
- the dual form of the model (eqn. 1). This form in kernel methods needs that the training set must be stored, even during the test phase, to compute the projections of any new point $\phi(\mathbf{y})$ into the model. This is because the mapping is never explicitly computed but is simultaneously obtained with dot product (eqn.3)- the so called kernel trick.

In the next section a method is suggested to deal with both problems. The kernel matrix of the complete training set is not computed and the eigendecomposition is performed with matrices of smaller size ($R < N$). The description of the model is then also based on a subset of the training data set.

2.4 Basis Vectors and Cholesky Decomposition

The Cholesky decomposition is a decomposition of an $N \times N$ symmetric positive matrix into the product of a $N \times N$ triangular matrix by the transpose of the triangular matrix. The incomplete approach with symmetric pivoting leads to $R \times N$ matrix, $\mathbf{C}$, which allows to compute an approximation of the original matrix controlling the error of the approximation. In [7] and [8] an algorithm is proposed which allows the computation of the decomposition of the kernel matrix having as input the training data set $\mathbf{X}$. The outcomes of the algorithm are the indexes of the pivoting and the matrix is

$$\mathbf{C} = \left[\mathbf{L}\ \ \mathbf{L}^{-T}\mathbf{K}_{rs}\right] \tag{4}$$

The pivoting scheme leads to the division of the training set into two subsets which can identified also using a block notation $\mathbf{X} = [\mathbf{X}_r\ \mathbf{X}_s]$. Then, the approximation of the kernel matrix $\tilde{\mathbf{K}} = \mathbf{C}^T\mathbf{C}$ can be expressed with four blocks: the upper left block matrix $\mathbf{K}_r = \mathbf{L}^T\mathbf{L}$ has dimension $R \times R$, the upper right block matrix $\mathbf{K}_{rs}$ has dimension $R \times (N - R)$, the lower left block is $\mathbf{K}_{rs}^T$ and the lower right block matrix $\mathbf{K}_{rs}^T\mathbf{K}_r^{-1}\mathbf{K}_{rs}$ has dimension $S \times S$ where $S = N - R$. The block $\mathbf{K}_r$ is the kernel matrix of the subset $\mathbf{\Phi}_r \equiv \phi(\mathbf{X}_r)$ and $\mathbf{K}_{rs}$ corresponds to the kernel matrix between the subsets. It can be easily shown that the last block represents an approximation of the corresponding block of the original matrix which should be $\mathbf{K}_s = \mathbf{\Phi}_s^T\mathbf{\Phi}_s \equiv \phi^T(\mathbf{X}_s)\phi(\mathbf{X}_s)$ [9]. The minimization of $trace(\mathbf{K}_s - \mathbf{K}_{rs}^T\mathbf{K}^{-1}\mathbf{K}_{rs}) = trace(\mathbf{\Delta}_s)$ is used as criterion to the pivoting scheme of the incomplete Cholesky algorithm [8]. The goal is to iteratively construct $\mathbf{C}$ so that $trace(\mathbf{\Delta}_s) < \varepsilon$, is an user defined threshold. The pivoting is the choice of the index of the elements in $\mathbf{X}_s$ that is moved to $\mathbf{X}_r$ in each iteration [6],[9].

The matrix $\mathbf{C}$ can also be used to form an orthogonal representation of the training data set in the feature space

$$\mathbf{Z} = \mathbf{V}_q^T \mathbf{C} \tag{5}$$

where $\mathbf{V}_q$ is the eigenvector matrix of the matrix $\mathbf{Q} = \mathbf{C}\mathbf{C}^T$. Note that $L \leq R$ projections can be considered by choosing L eigenvectors that correspond to the largest eigenvalues. Furthermore note the new representation is always non-correlated, i.e. $\mathbf{Z}\mathbf{Z}^T$ is diagonal whatever the value of L ($< R$). A simple manipulation of $\mathbf{Z}$ leads to the description of the subspace model in its dual form

$$\mathbf{Z} = \mathbf{V}_q^T \mathbf{C} = \mathbf{V}_q^T \mathbf{L}^{-T} \boldsymbol{\Phi}_r^T \left[\boldsymbol{\Phi}_r \ \boldsymbol{\Phi}_s \right] \tag{6}$$

Consequently, the basis vector matrix can be written as

$$\mathbf{U} = \boldsymbol{\Phi}_r \mathbf{L}^{-1} \mathbf{V}_q \tag{7}$$

So, the dual form of the basis vector matrix is written using only a subset of the training data set thus reducing the storage and computational requirements during testing phases of the classifier. It has to be noticed that the vectors form an orthogonal basis in the feature space, i.e., $\mathbf{U}^T\mathbf{U} = \mathbf{I}$. Several algorithms [10], [11], [12] lead to similar solutions by exploiting the idea that there are samples in the training set that can be expressed as a linear combination of others.

2.5 Centering the Data

All previous deductions were conducted assuming that the data is centered. In the input space this can considered a pre-processing step that must be accomplished before computing the scatter or kernel matrix and before projecting any new data vector. So computing the mean of the training set $\mathbf{x}_{mean}$, the mean must be subtracted from every data vector whether it belongs to the training set or not.

KPCA and a complete training set: In feature space centering the mapped data is a more elaborate procedure that must performed mostly during the computation of the projections. To facilitate the exposition, let's consider a vector $\mathbf{m}$ with N elements all of which equal $1/N$, and a matrix $\mathbf{M}$ filled with N column vectors $\mathbf{m}$. Therefore to project a new data point $\phi(\mathbf{y})$ and to take into account the centered training data set, the following operations need to be integrated into the dot product

$$(\boldsymbol{\Phi} - \boldsymbol{\Phi}\mathbf{M})^T(\phi(\mathbf{y}) - \boldsymbol{\Phi}\mathbf{m}) \tag{8}$$

The first term removes the mean to the training data set, the second subtracts the mean of the training set from the new data. Then the manipulation of the previous expression results into four terms that contribute to the L projections in the feature space of the input data point y as

$$\mathbf{z}_y = \mathbf{D}^{-1/2}\mathbf{V}^T(\boldsymbol{\Phi}^T\phi(\mathbf{y}) - \mathbf{M}^T\boldsymbol{\Phi}^T\phi(\mathbf{y}) - \boldsymbol{\Phi}^T\boldsymbol{\Phi}\mathbf{m} + \mathbf{M}^T\boldsymbol{\Phi}^T\boldsymbol{\Phi}\mathbf{m}) \tag{9}$$

The last two terms only depend on the training set and they are present in every data point projected into $\mathbf{U}$, so they can be stored in advance and constitute a bias term that is present in every projection. It can be easily shown that projecting the complete training set $\mathbf{\Phi}$ to obtain $\mathbf{Z}$, the last three terms within parenthesis arise from the centered kernel matrix $\mathbf{K}_c = (\mathbf{I} - \mathbf{M})\mathbf{\Phi}^T\mathbf{\Phi}(\mathbf{I} - \mathbf{M})$, where $\mathbf{I}$ is an $N \times N$ identity matrix. Then, to accomplish non-correlated projections for the training data set the matrices $\mathbf{V}$ and $\mathbf{D}$ should be obtained from the eigendecomposition of $\mathbf{K}_c$. It should also be noticed that with an RBF kernel the dot products in feature space are always less than the unit (see eqn. 3). And in particular the contribution of the last two terms depends on the parameter σ of the kernel function.

KPCA and a reduced training set: The starting point of a Cholesky approach is the incomplete Cholesky decomposition of the full matrix. The projections can be written (see eqn.5), in order to turn the projections related to the centered data, the low rank approximation of the kernel matrix can be centered $\tilde{\mathbf{K}}_c = (\mathbf{I} - \mathbf{M})\mathbf{C}^T\mathbf{C}(\mathbf{I} - \mathbf{M})$ where the mean $\mathbf{Cm}$ is subtracted from every column of $\mathbf{C}$. In that case the eigenvectors $\mathbf{V}_q$ must be computed with $\mathbf{Q}$ after centering the matrix $\mathbf{C}$. Then, the term $\mathbf{b} = \mathbf{V}_q^T\mathbf{Cm}$ should also be subtracted from every data projected onto the model (see eqn.7).

$$\mathbf{z}_y = \mathbf{U}^T\phi(\mathbf{y}) - \mathbf{b} \tag{10}$$

3 Numerical Simulations

The effectiveness of the subspace feature extraction methods discussed above is evaluated by comparing the performance of the classifiers. For that we carried out experiments on thirteen artificial and real world data sets available on Gunnar Ratsch's web site (accessible at http://ida.first.fraunhofer.de/projects/bench). The data sets represent benchmarks and several algorithms [5], [13], [14], in wich it has been been applied to these data sets. In this work generalization error rates of the classifiers are presented using as input: the raw data, the PCA, the KPCA, and greedy KPCA projections.

Data sets. Table 1 resumes the information of the 13 data sets. All data sets have 100 random partitions of pairs training/test sets, except *Splice* and *Image* which have 20 partitions. On each partition data sets, different classification algorithms were used, the second column shows the average and the standard deviation of the generalization error published in [5]. Furthermore, it is possible to download from the web page the generalization errors for every partition of data.

Evaluation. The performance of classifiers is used to illustrate the influence of projecting the data into the different models. Two classifiers were considered: one-nearest neighbor (NN) and the linear discriminant function (RL). With an NN classifier, each element of the test set is classified according to the nearest

Table 1. Data set description. The results of t-test (95%): Best results of [5] versus raw data classification (column I1) and Raw data versus PCA projections (column I2), where $\oplus$ accepts $H0$ and $\ominus$ rejects $H0$.

		Data description			Projections PCA				t-test	
		D	N	Benchs [5]	L	NN	L	RL	I1	I2
Group 1	***B. Cancer (BC)***	9	200	25.9 ± 4.6	9	32.5 ± 4.8	2	26.2 ± 2.3	$\oplus$	$\oplus$
	Diabetis (Di)	8	468	23.5 ± 1.7	8	30.1 ± 2.0	8	23.4 ± 1.7	$\oplus$	$\oplus$
	German (GR)	20	700	23.6 ± 2.1	20	29.4 ± 2.4	17	23.9 ± 2.1	$\oplus$	$\oplus$
	Heart (Hr)	13	170	16.0 ± 3.3	9	22.0 ± 3.1	10	15.9 ± 3.1	$\oplus$	$\oplus$
	F. Solar (FS)	9	400	32.4 ± 1.8	9	39.0 ± 4.0	6	32.9 ± 1.8	$\oplus$	$\ominus$
	Thyroid (Ty)	5	140	4.4 ± 2.2	3	3.9 ± 2.1	5	14.7 ± 3.2	$\ominus$	$\oplus$
	Titanic (Ti)	3	150	22.4 ± 1.0	1	33.0 ± 11.0	3	22.0 ± 1.0	$\oplus$	$\oplus$
	Twonorm (Tn)	20	400	2.7 ± 0.2	1	3.4 ± 0.5	1	2.3 ± 0.1	$\ominus$	$\ominus$
Group 2	***Image (Im)***	18	1010	2.7 ± 0.7	13	3.30 ± 0.5	18	16.5 ± 0.98	$\ominus$	$\oplus$
	Ringnorm (Rg)	20	400	1.6 ± 0.1	6	21.3 ± 1.2	19	24.6 ± 0.7	$\ominus$	$\ominus$
	Splice (Sp)	60	1000	9.5 ± 0.7	6	22.4 ± 1.4	60	16.31 ± 0.6	$\ominus$	$\oplus$
	Waveform(Wv)	21	400	9.8 ± 0.8	2	11.7 ± 0.7	2	12.6 ± 0.7	$\ominus$	$\ominus$
	Banana (Ba)	2	400	10.7 ± 0.4	2	13.6 ± 7.0	1	46.9 ± 7.0	$\ominus$	$\oplus$

neighbor in the training set. In case of linear discriminant functions, the weight vectors are computed within the training data set by using the mean-square-error criterion [15]. In spite of having two classes, the multi-class strategy is followed and each element in the test set is assigned to the class whose discriminant function has the largest value. The averages (and standard deviation) of the generalization error rates are presented. The difference between error percentages achieved by pairs of methods are also compared using t-test with 95% significance. The statistical test has been carried out in an attempt to reject $\ominus$ or accept $\oplus$ the null hypothesis ($H0$), i.e, the equality of mean performance of both methods.

3.1 Linear Features

Both classifiers were applied to the raw data. The results of the classification leads us to organize the data sets into two groups (see table 2). In the first group (group 1), at least one of the classifiers achieves an error rate comparable to the ones published in [5]. In the remaining five data sets (the group 2) the performance was far from the results presented in [5]. A t-test is used to compare the best result with the ones published, and the $H0$ is rejected in the group 2 and it is accepted in group 1 except for two data sets (see column I1 of table 1). The data is pre-processed using PCA and its L projections are used as input to the classifiers. The number of projections was varied from 1 to D (the number of features of raw data). In table 1, for each data set the average minimum error rate and the corresponding dimension (L) of the new data representation are shown. Globally we verified that the error rate of the linear discriminant classifier is similar to the one achieved with the raw data. Using a t-test to compare the best results of raw data versus PCA projections, the $H0$ hypothesis is accepted for most of the data sets, except in four data sets (see column I2 of table 1). However, also to be noticed is that in some data sets a considerable dimension reduction is reached. The most significant occurs in the *twonorm* data set where $D = 20$ and $L = 1$ for both classifiers and in both cases the performance improves when compared with the raw data version. In the case of the RL classifier the result is even better than the result published in [5]. In this case $H0$ hypothesis is rejected (see column I1 and I2 of table 1).

3.2 Non-linear Features

Both versions of KPCA are used to compute the model: the first depends on the full training set (KPCA) and the second depends on a subset with R elements (greedy KPCA). In both cases the number of projections varied from $L = 1$ up to R. Both models are computed using centered versions of the kernel matrices as proposed before. Using the RBF kernel function to evaluate the dot products, a value must be assigned to σ. This parameter is often a variable of the experimental studies [16] or it is optimized using a cross validation strategy using the training data [13]. In this study, the value of σ^2 is chosen as the average of euclidian squared distances of each training vector to the center of training set. Notice that the choice of sigma also interferes with the size R of the subset to be included in the basis vector model of greedy KPCA. If the decay of eigenvalues is too smooth the complete training set will be chosen in the incomplete Cholesky decomposition using a fixed threshold for the approximation error. The threshold $0.01N$ was considered, because with the RBF kernel the trace of the kernel matrix is always the size of training data set.

In table 2 the average error rates and standard deviation are shown. Column $I1$ shows the result of the t-test between the results of [5] and either the NN or the RL classifier . And we can see that the H0 hypothesis is accepted in all but the German data set. We can see that in group 1 there is no significant improvement in the performance of the classifiers using the non-linear features of the data sets. In most cases the minimum error rate is achieved using a number ($L > D$) of projections higher than the dimension (D) of the raw data. One of the exceptions is the *twonorm* where $L = 1$ as in input space. Another t-test was performed to see how the number of projections can affect the results. The error rates were compared to the error rates achieved when the number of projections is $L = D$, KPCA$_D$. The column $I2$ of table 2 shows the result and it can be verified that the $H0$ hypothesis is rejected in group 2 and accepted in group 1. Furthermore, notice that with the linear discriminant function the best performance is achieved with $L > D$ with the exception of the *waveform* set.

Table 2 also shows the performance of classifiers using the greedy KPCA to compute projections and we can verify that the results are similar to the ones computed with KPCA. The null hypothesis is accepted for every data set (see column I3). The number of projections used in both methods was not always the same, but considering that the second method is obtained using an approximation of the kernel matrix, some variations had to be expected. In what concerns

Table 2. Error rate (%) using KPCA and greedy KPCA. Results of t-test: Best versus KPCA (column I1), KPCA versus KPCA$_D$ (column I2) and greedy KPCA versus KPCA (column I3) where $\oplus$ accept $H0$ and $\ominus$ reject $H0$.

	KPCA				$H0$		greedy KPCA					$H0$
	L	NN	L	RL	I1	I2	R	L	NN	L	RL	I3
BC	7	32.5 ± 4.8	21	25.2 ± 4.5	$\oplus$	$\oplus$	90	7	32.5 ± 4.8	22	25.2 ± 4.5	$\oplus$
Di	17	25.3 ± 1.8	10	23.2 ± 1.6	$\oplus$	$\oplus$	140	61	30.2 ± 1.9	10	23.1 ± 1.6	$\oplus$
Gr	12	30.0 ± 2.5	12	23.3 ± 2.1	$\ominus$	$\oplus$	400	13	29.1 ± 2.4	12	23.4 ± 2.3	$\oplus$
Hr	8	22.7 ± 3.4	12	15.8 ± 3.0	$\oplus$	$\oplus$	110	48	22.79 ± 2.9	11	15.8 ± 3.1	$\oplus$
FS	55	32.2 ± 0.5	25	32.1 ± 0.6	$\oplus$	$\oplus$	74	70	35.3 ± 0.7	48	33.8 ± 0.6	$\oplus$
Ty	6	4.0 ± 2.2	15	5.8 ± 2.4	$\oplus$	$\oplus$	25	6	3.9 ± 2.2	25	5.3 ± 2.3	$\oplus$
Ti	9	32.3 ± 1.1	10	22.3 ± 1.0	$\oplus$	$\oplus$	10	10	31.1 ± 1.4	6	21.8 ± 1.0	$\oplus$
Tn	1	3.4 ± 0.4	1	2.3 ± 0.1	$\oplus$	$\oplus$	205	1	3.5 ± 0.6	1	2.3 ± 0.1	$\oplus$
Im	23	2.8 ± 0.6	75	7.9 ± 1.3	$\oplus$	$\ominus$	120	21	2.9 ± 0.7	80	8.1 ± 1.2	$\oplus$
Rg	40	3.5 ± 0.4	25	1.6 ± 0.1	$\oplus$	$\ominus$	262	45	3.8 ± 0.4	31	1.7 ± 0.1	$\oplus$
Sp	600	7.5 ± 2.6	720	4.3 ± 2.1	$\oplus$	$\ominus$	874	620	7.7 ± 2.6	764	4.4 ± 2.1	$\oplus$
Wv	29	9.7 ± 0.7	2	12.0 ± 0.8	$\oplus$	$\ominus$	258	30	9.8 ± 0.3	2	12.0 ± 0.7	$\oplus$
Ba	5	13.6 ± 0.4	34	10.7 ± 0.4	$\oplus$	$\ominus$	15	15	13.6 ± 0.7	5	10.8 ± 1.8	$\oplus$

the approximation, we see that, using the same threshold, the relative decrease (N/R) of the number of examples to describe the model is very heterogenous, it ranges from 1.1 to 26.6, but in 7 data sets is higher than 2.

4 Concluding Remarks

In this work we introduce projective subspace techniques and cast them in a concise presentation by using the dual form for the models. Besides that an algorithm to compute the KPCA model using a subset of the training data set using a greedy approach is also presented. We further consider the centering problem and adapt the model description to remove the mean of the data. We verify that these techniques have a different impact on the performance of the classifiers. The reason is mostly related to the data characteristics. We showed that for some data sets the performance achieved using raw data is similar to the results published [5]. In these data sets a dimension reduction by PCA yields similar results. It can also be verified that for these data sets, using KPCA projections, the generalization errors rate remains roughly constant. The other group are nonlinear data sets. The performance on the high-dimensional feature space clearly improves, and is comparable to the one described in [5]. Another aspect to point out is that with KPCA projections the linear discriminant function classifier performs better than the nearest neighbor one, in 10 of the 13 data sets. Confirming that having decision functions based on an hyperplane is possible but the dimension has to increase like in the case of *banana* data sets. The numerical simulations corroborate that the greedy approach to KPCA does not harm the performance.

Acknowledgment

A.R. Teixeira received a PhD Scholarship (SFRH/BD/28404/2006) supported by the Portuguese Foundation for Science and Technology (FCT).

References

1. Yang, M.-H., Kriegman, D.J., Ahuja, N.: Detecting faces in images: a survey. IEEE Transactions on Pattern Analysis and Machine Intelligence 24(1), 34–58 (2002)
2. Moghaddam, B.: Principal manifolds and probabilistic subspace for visual recognition. IEEE Transactions on Pattern Analysis and Machine Intelligence 24(6), 780–788 (2002)
3. Schölkopf, B., Mika, S., Barges, C.J., Knirsch, P., Müller, K.-R., Ratsch, G., Smola, A.J.: Input space versus feature space in kernel-based methods. IEEE Transactions on Neural Networks 10(5), 1000–1016 (1999)
4. Müller, K.-R., Mika, S., Rätsch, G., Tsuda, K., Schölkopf, B.: An introduction to kernel-based algorithms. IEEE Transactions on Neural Networks 12(2), 181–202 (2001)
5. Rätsch, G., Onoda, T., Müller, K.R.: Soft margins for adaboost. Machine Learning 42(3), 287–320 (2001)
6. Teixeira, A.R., Tomé, A.M., Lang, E.W.: Exploiting low-rank approximations of kernel matrices in denoising applications. In: IEEE International Workshop on Machine Learning for Signal Processing (MLSP 2007), Thessaloniki, Greece (2007)
7. Bach, F.R.: Kernel independent component analysis (2003), `http://www.di.ens.fr/~fbach/kernel-ica/index.htm`
8. Fine, S., Scheinberg, K.: Efficient SVM training using low-rank kernel representations. Journal of Machine Learning Research 2, 243–264 (2001)
9. Teixeira, A.R., Tomé, A.M., Lang, E.W.: Feature extraction using low-rank approximations of the kernel matrix. In: Campilho, A., Kamel, M.S. (eds.) ICIAR 2008. LNCS, vol. 5112, pp. 404–412. Springer, Heidelberg (2008)
10. Franc, V., Hlaváč, V.: Greedy algorithm for a training set reduction in the kernel methods. In: 10th International Conference on Computer Analysis of Images and Patterns, pp. 426–433. Springer, Holland (2003)
11. Cawley, G.C., Talbot, N.L.C.: Efficient formation of a basis in a kernel induced feature space. In: Verleysen, M. (ed.) European Symposium on Artificial Neural Networks, pp. 1–6. d-side, Belgium (2002)
12. Baudat, G., Anouar, F.: Feature vector selection and projection using kernels. Neurocomputing 55, 21–38 (2003)
13. Xu, Y., Zhang, D., Song, F., Yang, J.-Y., Jing, Z., Li, M.: A method for speeding up feature extraction based on kpca. Neurocomputing 70(4-6), 1056–1061 (2007)
14. Cawley, G.C., Talbot, N.L.: Efficient leave-one-out cross-validation of kernel Fisher discriminant classifiers. Pattern Recognition 36, 2585–2592 (2003)
15. Duda, R., Hart, P., Stork, D.G.: Pattern Classification. John Wiley & Sons, Chichester (2001)
16. Mika, S., Schölkopf, B., Smola, A., Müller, K.R., Scholz, M., Rätsch, G.: Kernel pca and de-noising in feature spaces. In: Advances in Neural Information Processing 11, pp. 536–542. MIT Press, Cambridge (1999)

Classification Based on Combination of Kernel Density Estimators

Mateusz Kobos and Jacek Mańdziuk

Warsaw University of Technology,
Faculty of Mathematics and Information Science,
Plac Politechniki 1, 00-661 Warsaw, Poland
{M.Kobos,J.Mandziuk}@mini.pw.edu.pl

Abstract. A new classification algorithm based on combination of kernel density estimators is introduced. The method combines the estimators with different bandwidths what can be interpreted as looking at the data with different "resolutions" which, in turn, potentially gives the algorithm an insight into the structure of the data. The bandwidths are adjusted automatically to decrease the classification error. Results of the experiments using benchmark data sets show promising performance of the proposed approach when compared to classical algorithms.

Keywords: kernel density estimation, classification, density estimators combination.

1 Introduction

Classification based on density estimators is one of the basic methods used in machine learning (see e.g. [1] for an introduction to the subject). Among the non-parametric density estimation methods, the most popular are the Gaussian Mixture Model (GMM) and the Kernel Density Estimator (KDE), the latter also called the Parzen windows method. In KDE, in order to get the density estimate in a given point, a distance-based influence of all points from the training set on that point is calculated. A "kernel function" is used to put a relatively greater emphasis on points that are closer than on those which are placed further. Typically, kernel's definition includes a parameter called "bandwidth" which determines how much emphasis is put on the closest points.

A method to estimate the density (but not to make a classification) using a linear combination of predefined GMMs and KDEs was introduced in [2]. The parameters of the combination are computed using the stacking meta-learning method with the EM algorithm. In [3], another fusion of GMM and KDE is proposed. The GMM algorithm is used to assign a weight to each of the predefined KDE models. Yet another meta-learning approach – boosting – is proposed in [4]. The base boosted classifiers are simple algorithms based on KDE; each of the training points gets a different weight in each algorithm's iteration. A different meta-learning approach is proposed in [5]. Authors use ensemble averaging method based on GMMs to estimate the density. Boosting and bagging

C. Alippi et al. (Eds.): ICANN 2009, Part II, LNCS 5769, pp. 125–134, 2009.

meta-learning algorithms are also used in [6] for the density estimation. The EM algorithm is used to maximize training data likelihood but the classification error is not directly optimized. In [7], authors describe an algorithm which uses "Gaussian product kernel estimators" where the bandwidths are chosen independently for each class-dimension combination. What is more, the bandwidths can vary depending on the localization in the feature space.

In this paper, we propose a new classification method which is based on a combination of KDEs. The algorithm is significantly different from each of the above mentioned methods, since it optimizes directly the classification error and does not use explicitly any meta-learning algorithm. The combined kernels bandwidth is not predefined but adjusted to the data.

The method exhibits some similarities to the Ghosh et al. approach [8], where the authors introduced a method which is at heart a binary classifier. They search for an optimal bandwidth using the cross-validation method for each of the 2 classes independently. As a result, they obtain a pair of bandwidth values which can be interpreted as a point in a 2-dimensional bandwidth space. In order to classify a given test point, a couple of density estimations are made. Each of them corresponds to a pair from the neighborhood of the optimal bandwidth values pair in the bandwidth space. The estimation results are transformed in a certain way and their weighted sum is computed. The sum yields the final classes probabilities. The main difference between the method presented in [8] and our approach is that the latter one is simpler (and possibly faster) because the parameter space that is searched is one-dimensional instead of two-dimensional.

The paper is organized as follows: Sect. 2 contains a description of the proposed algorithm, Sect. 3 presents results of the tests on benchmark data sets and comparison with the literature results, Sect. 4 concludes the paper.

2 Algorithm Description

Every classification machine learning algorithm has two modes of operation: training phase and classification/recall phase. In Sections 2.1, 2.2, 2.3 we describe the classification phase and Sect. 2.4 contains a description of the training phase.

2.1 Introduction

The problem of classification is to create a decision rule $d(\mathbf{x}) : \mathbb{R}^d \to \{\omega_1, \omega_2, \ldots, \omega_c\}$ to classify a d-dimensional observation (point) $\mathbf{x}$ into one of c classes ω_i. The rule is usually built using the observations from a training set, its robustness is tested on the observations from a testing set. One possibility to create such a rule is to employ the Bayesian classifier of the form

$$d_B(\mathbf{x}) = \arg\max_{w_i} \hat{P}(\omega_i|\mathbf{x}) = \arg\max_{w_i} \frac{\hat{p}(\mathbf{x}|\omega_i)\hat{P}(\omega_i)}{\hat{p}(\mathbf{x})} \quad , \tag{1}$$

where $\hat{P}(\omega_i|\mathbf{x})$ is a posterior probability estimator of class ω_i, $\hat{P}(\omega_i)$ is a prior estimator of class ω_i (in practice, it is equal to the fraction of observations from a given class in the training set), $\hat{p}(\mathbf{x}|\omega_i)$ is an estimated probability density function of class ω_i, and $\hat{p}(\mathbf{x}) = \sum_{i=1}^{c} \hat{p}(\mathbf{x}|\omega_i)\hat{P}(\omega_i)$ is a normalization factor. All of the quantities in this formula are simple to compute except for the class probability density estimate $\hat{p}(\mathbf{x}|\omega_i)$.

One of the most popular density estimators is the Kernel Density Estimator. When applied in the Bayesian classifier, it has the form of

$$\hat{p}_h(\mathbf{x}|\omega_i) = \frac{1}{|\mathcal{D}_i|} \sum_{\mathbf{x}' \in \mathcal{D}_i} \frac{1}{h^d} \phi\left(\frac{\mathbf{x} - \mathbf{x}'}{h}\right), \tag{2}$$

where $\mathcal{D}_i$ is the set of observations belonging to class ω_i, the function $\phi(\mathbf{x}) : \mathbb{R}^d \to [0, \infty)$ is a density function called "kernel function" and h is a smoothing factor called "bandwidth".

One of the most popular choices for the kernel function is the Gaussian kernel. In the general case, it has the form of

$$\phi(\mathbf{x}) = \frac{1}{(2\pi)^{d/2}|\mathbf{\Sigma}|^{1/2}} \exp\left[-\frac{1}{2}(\mathbf{x} - \mu)^T \mathbf{\Sigma}^{-1}(\mathbf{x} - \mu)\right], \tag{3}$$

where the covariance matrix $\mathbf{\Sigma}$ is responsible for the hyper-ellipsoidal shape of the kernel (cf. [1, Sect. 2.5.2]). Generally, the shape of the kernel should be adjusted to match the layout of the training points in the feature space. There are two approaches to reach this goal. The first one is to set an appropriate shape of the kernel i.e. to adjust the covariance matrix to match the data. The second one is to let the covariance matrix be fixed and equal to the identity matrix $\mathbf{\Sigma} = \mathbf{I}$ (the shape of the kernel will be circular in this case) and to transform the feature space instead (this method was used e.g. in [7]). We have chosen the second approach because it makes the algorithm simpler to analyze and is less computationally intensive. During the experiments, two different transformations were used: standardization and whitening, the latter implemented with Principal Component Analysis (PCA).

2.2 Combination of Estimators

The kernel's bandwidth parameter specifies how smooth the resulting estimation of density function will be. The larger the factor, the smoother and less concentrated on the training points the estimation of the density function. We can also interpret the bandwidth as a "resolution" of the data view – the larger the bandwidth, the smaller resolution and the more general view on the data (i.e. the density assumes similar values even in distant points of the space, which makes them difficult to distinguish).

The main idea presented in this paper is to combine a certain number of KDEs with different bandwidths which are selected to match the analyzed data set. Such an approach of looking at the data with different "resolutions" should give a better insight into the data structure and result in a better classification than a method using a single "resolution". In this "multi-resolution" case, the formula for posterior probability estimator will be an average of different-bandwidth

KDEs: $\hat{p}(\mathbf{x}|\omega_i) = \frac{1}{E}\sum_{j=1}^{E} \hat{p}_{h_j}(\mathbf{x}|\omega_i)$, where E is the number of KDEs, and h_j is individual bandwidth of j-th KDE.

The other underlying idea is to make the estimators' bandwidths related in some way. It is proposed that bandwidths of different estimators decrease in an exponential manner. This way it is possible to combine vastly different data view "resolutions". As a result, bandwidth is a function of the following form:

$$h_j(a) = h_{\min} + a^j(h_{\max} - h_{\min}) \ , \tag{4}$$

where $j \in \{1, \dots, E\}$ is an estimator number, $a \in [0, 1]$ is a parameter determining how fast the bandwidths decrease, $[h_{\min}, h_{\max}]$ is a range of bandwidth values.

2.3 Bandwidth Range

The first question to be answered is what is a "sensible" bandwidth range in (4). If one uses a very small bandwidth, numerical problems occur. If a given testing point is far from any other point in the training set, then every class-conditional density in the given point will be close to zero and, as a result, $\hat{p}(\mathbf{x})$ in the denominator of the Bayes classifier formula (1) will be close to zero. For a sufficiently small bandwidth, the value will be smaller than the computer's machine precision and, as a consequence, assumed to be equal to zero. This, in turn, will make the formula impossible to evaluate. The next problem is that for small bandwidths one can get unreliable and possibly misleading information for classification [8, Sect. 2.2]. To deal with these problems, we decided to take an approach similar, but not the same, to the one presented in [8]. We chose a small ratio $\frac{1}{\xi}$ of the smallest non-zero percentile of the pairwise distances of the transformed data points from the training set as the lower limit. The ξ is a radius of a sphere which contains 99% of kernel's probability mass – it is assumed that outside this sphere, the kernel's influence on the overall density is negligible. In the presented algorithm, the Gaussian kernel is used, and for such a kernel it can be shown that $\xi = \sqrt{F^{-1}_{\chi^2(d)}(0.99)}$, i.e. it is a square root of inverted χ^2 distribution function with d degrees of freedom in point 0.99.

The proposed solution is not always sufficient to solve the above mentioned problems. In extreme situations, an outlier point can be situated far away from all the training points, and in such case the algorithm would not be able to evaluate the formula (1) even for larger bandwidths. We can note that in such situations (i.e. outlier point or, equivalently, a very small bandwidth) the KDE classification result mimics the result of Nearest Neighbor algorithm [9, p.251]. Thus, if the denominator of the formula (1) is equal to zero (i.e. the value is smaller than the computer's machine precision), the probabilities that would be yielded by the Nearest Neighbor algorithm are returned as the classification result.

The upper limit of the bandwidth range, on the other hand, is set to be the 99-th percentile of pairwise distances of the transformed data points in the training set. As can be seen, when calculating the lower and the upper limits, small and large percentiles are used instead of simply using the minimum and the maximum. The reason is making the calculations resistant to outlier points, which could unnecessarily widen the bandwidth range. Apart from that, the estimation of both lower and upper limits is rather conservative.

2.4 Algorithm Training

In this section the training phase in which the parameters of KDEs combination (especially a) are computed is described. Parameter a in (4) is selected to minimize the classification error on the training set. The classification error that is minimized is the Mean Squared Error (MSE) defined as

$$\mathrm{MSE}(\hat{P}(\cdot), \mathcal{D}) = \frac{1}{|\mathcal{D}|} \sum_{\mathbf{x} \in \mathcal{D}} \sum_{i=1}^{c} (\hat{P}(\omega_i|\mathbf{x}) - \mathbf{t}_i(\mathbf{x}))^2 \ , \tag{5}$$

where $\mathcal{D}$ is the data set on which the error is computed, $\hat{P}(\omega_i|\mathbf{x})$ is the algorithm's posterior probability estimation for class ω_i, and $\mathbf{t}_i(\mathbf{x})$ is a vector whose i-th component, where i corresponds to the actual class ω_i, is 1 and all other components are 0.

The training phase of the algorithm consists of several steps. 1) As the first step, the sequence of the training instances is randomly permuted. The randomization is required for the cross-validation folds (which are created from the data in later steps) to be independent as much as possible (e.g. we do not want to cumulate all of the samples from one class in one fold) which is a standard requirement in the cross-validation method. It is worth noting that, apart from this step, the algorithm is completely deterministic. 2) In the next step, the data is transformed. The transformation parameters (e.g. for standardization transformation: sample expected values and sample standard deviations) are saved – they will be used later to build the classification error function. 3) Next, using the transformed data, a "sensible" bandwidth range (as it was described in Sect. 2.3) is calculated. The range will be searched for the optimal bandwidth value. 4) Finally, in the last step, the value of a that minimizes the 10-fold stratified cross-validation estimator of a classification error function is searched for. The minimization is performed in a exhaustive way with a grid-search method. In this method, we compute the value of the cross-validation estimator function in 100 equidistant points from range $[0, 1]$ (see Fig. 1 for an example of examined error values and location of the optimal a). As a result of the training phase, the optimal a is obtained along with the transformation parameters, transformed data and bandwidth range. These values will be used later in the classification phase.

The construction of 10-fold stratified cross-validation estimator of the classification error function needs some further explanation. First, 10 splits of the training data are created. Each split consists of two disjoint data sets: the fitting set $\mathcal{D}$ which is used to train the classifier, and the validation set $\mathcal{D}^v$ which is used to compute the classification error of the trained algorithm. For each split, the classification MSE (5) defined as $\mathrm{Error}(a) = \mathrm{MSE}(\hat{P}(\cdot; a), \mathcal{D}^v)$ is calculated,

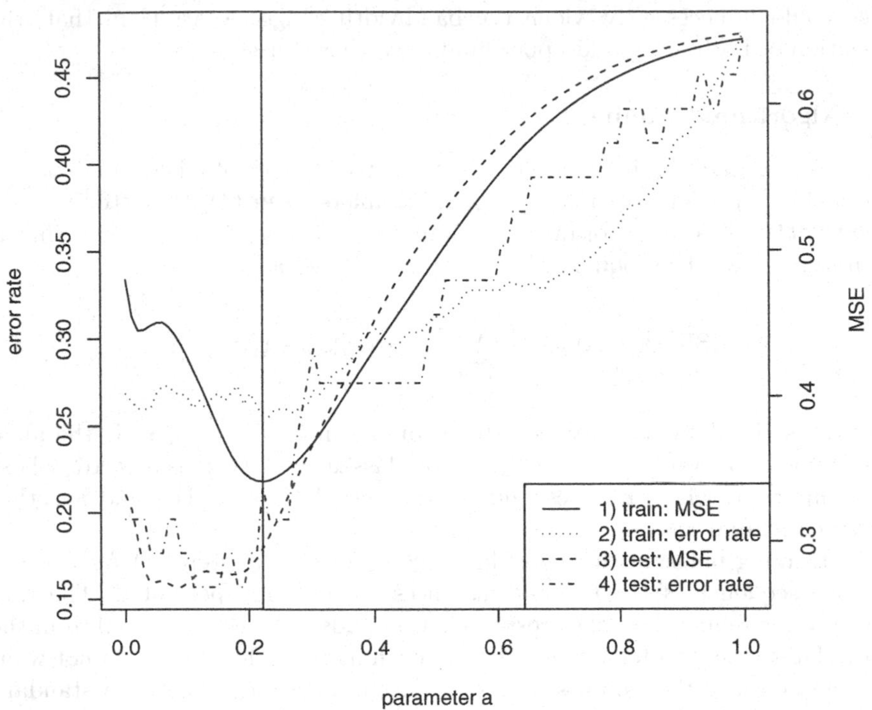

Fig. 1. An example of classification errors for different parameter a values. The algorithm with 2 KDEs and standardization transformation is tested on the *Boston housing* data set. Examined error types: cross-validation MSE on the training set (*line 1*), cross-validation error rate on the training set (*line 2*), MSE on the test set (*line 3*), error rate on the test set (*line 4*). The optimal a value found by the algorithm is equal to the global minimum of the cross-validation MSE error on the training set (*vertical line*).

where a is the bandwidths' decrease parameter from (4), $\hat{P}(\omega_i|\mathbf{x}; a)$ is the posterior probability estimator from (1) dependent on a. The MSE function is used instead of direct use of the error rate function (i.e. misclassification ratio) because it seems to be less affected by the random dependencies in the data (see Fig. 1). The function uses the training set transformation computed in step 2 of the training phase to transform the data in each of the cross-validation splits. The reason the transformation is not computed for every split's fitting set independently, as would a standard cross-validation procedure suggest, is that we want the bandwidths calculated for every split to correspond to the same values in the original non-transformed space. If the transformations were calculated independently, the same bandwidth value would correspond to different bandwidth values in each split in the original space. The priors used in each split are also estimated on the whole training set (for similar reasons).

3 Experiments

The efficacy of the proposed algorithm was compared with the results published in [10] and [8]. In [10], the authors consider 33 classification algorithms and verify them using different data sets, which establishes a broad comparison base for the proposed algorithm. Among the data sets examined in [10], the ones that matched the proposed algorithm (i.e. sets with numerical attributes only) were chosen. The raw error rates used for comparison were retrieved from the article's appendix available at one of the author's website. In [8], on the other hand, the authors compare the algorithm they introduced with literature results. The same data is used in this paper to compare our method with that of [8].

The following data sets were used in the experiments: *Boston housing* (Boston housing, used in [10]), *breast cancer* (Wisconsin breast cancer data set, collected at the University of Wisconsin by W.H. Wolberg [11], used in [10]), *glass* (forensic glass data, used in [8]), *Indian diabetes* (PIMA Indian diabetes, used in [10]), *liver disorders* (BUPA liver disorders, used in [10]), *Ripley's synthetic* (Ripley's synthetic data, used in [8]), *satellite image* (StatLog satellite image, used in [10]), *sonar* (sonar data, used in [8]), *vehicle silhouette* (StatLog vehicle silhouette, used in [10]). All of the data sets except for *Ripley's synthetic* were downloaded from the UCI Machine Learning Repository [12]; the *Ripley's synthetic* data set was downloaded from [13]. In the cited articles, some of the original data sets were preprocessed and we executed the same preprocessing steps. When testing our algorithm, we followed the methodology used in adequate articles with an exception for the holdout experiments (for data sets with a selected testing set). The holdout experiments were executed 10 times instead of once, because our algorithm is non-deterministic and repeating the experiment several times results in a more unbiased efficacy estimation.

3.1 Results

During the experiment, the algorithm was tested with data standardization transformation and number of estimators equal to: 1 (*E=1*), 2 (*E=2*), 5 (*E=5*), or the number of classes in the data set (*E=cl.no.*). For comparison, the algorithm was also tested with data whitening transformation and the number of estimators equal to: 1 (*PCA E=1*), or the number of classes in the data set (*PCA E=cl.no.*).

Four of the algorithms yielded results better than the literature ones on at least one data set (see Table 1), and one of them (*E=2*) yielded results better than the literature ones on 2 data sets. On average, the results yielded by *E=2*, *E=5*, *E=cl.no.* were better than the results of the simplest version *E=1* (average differences equal to .0035, .0018, .002, resp.); the results yielded by *PCA E=1*, *PCA E=cl.no.* were worse (average differences equal to -.01, -.012, resp.).

When comparing the classification results, it might be helpful to check at which quantile is a given result situated among the literature results (Fig. 2) (the quantiles were calculated using `ecdf` function in the R environment [14]). This method potentially allows to assess if the improvement of the result is

Table 1. Comparison of experimental results. Error rate (misclassification error) for different algorithm versions and different data sets was measured. The results that are better than the best literature result (*best lit.*) are marked with an asterisk. All of the results are given with the same number of decimal places as in the respective literature source.

data set	best lit.	E=1	E=2	E=5	E=cl.no.	PCA E=1	PCA E=cl.no.
Boston housing	.221	.239	.247	.249	.245	.243	.247
breast cancer	.0278	.0323	.0323	.0323	.0323	.0661	.0661
glass	.236	.252	.233*	.247	.247	.271	.308
Indian diabetes	.221	.251	.256	.259	.256	.264	.269
liver disorders	.279	.405	.365	.365	.365	.321	.310
Ripley's synthetic	.090	.105	.094	.097	.094	.105	.094
satellite image	.098	.097*	.095*	.095*	.096*	.292	.288
sonar	.135	.194	.222	.216	.222	.181	.181
vehicle silhouette	.145	.286	.285	.284	.286	.208	.205

meaningful (e.g. the same error rate improvement for a difficult data set can be more important than for a simple one). As can be seen in Fig. 2, the results of all of the algorithm's versions were situated among the top-50% literature results in 4 out of 9 examined data sets (except for *PCA E=1* where the ratio was 5 out of 9). On average, the quantile results yielded by *E=2*, *E=cl.no.* were better than the results of the simplest version *E=1* (average differences equal to .0168, .0101, resp.); the results yielded by *E=5*, *PCA E=1*, *PCA E=cl.no.* were worse (average differences equal to -.0034, -.11, -.0797, resp.).

In summary, the results yielded by all of the algorithm's versions are promising when compared to the literature results. Although the error rate on some of the data sets was high, it can be argued that according to the no free lunch theorem [1, Sect. 9.2.1] no single classifier can achieve great results on all of the problems. Furthermore, the algorithms that used the whitening transformation generally yielded worse results than the others. On the other hand, the whitening transformation improved the results on some of the data sets (see e.g. results on data set *liver disorders* in Fig. 2), so we can conclude that this transformation can improve or worsen the results depending on the data. What is more, although various algorithm's versions excelled in the classification of various data sets, the version which seemed to be generally the best is the one which uses two KDEs with standardization transformation (*E=2*). This version yielded the results that were better than the best literature results on two data sets: *glass* and *satellite image*. The results were also better, on average, than the results of the simplest version which uses one KDE (*E=1*); this observation justifies application of 2-element KDEs combination instead of a simpler version with a single KDE.

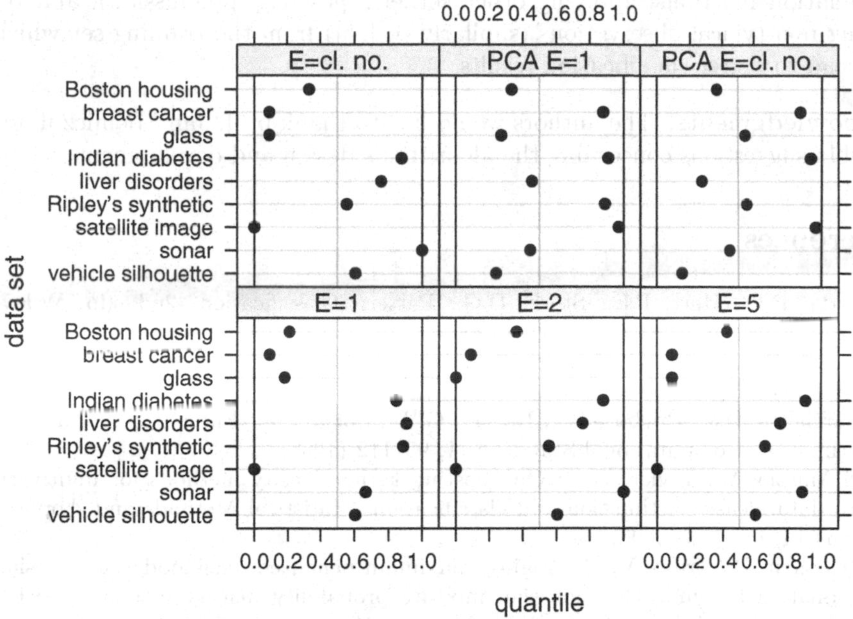

Fig. 2. Comparison of the experimental results with the literature results. Each panel contains relative error rates of a certain algorithm version on different data sets. Each point corresponds to a quantile position of the experiment result among the literature results. For example, in the panel *E=1*, 4 results are among 50% best literature results and 5 results are among 50% worst literature results. Points at quantile 0 correspond to the results that are as good as the best literature result or better; points at quantile 1 correspond to the results that are as bad as the worst literature result or worse.

4 Conclusions and Future Work

A new classification algorithm based on a combination of Kernel Density Estimators, where the classification error is minimized directly is presented in the paper. The algorithm performs well on the benchmark data sets when compared to the literature results; one of the algorithm's versions yields the results which are better on two data sets than the best ones reported in the literature. These results confirm the algorithm's potential, especially in the domains related to the examined data sets. It will be the object of further research to examine the detailed characteristics of these data sets, and, as a result, to determine which algorithm version matches best each data set domain.

The next steps of the algorithm's development involve applying a numerical optimization method (e.g. one of the pseudo-Newton algorithms) instead of the exhaustive optimization currently employed, which should result in significantly shorter training times. Other extension paths concern testing other KDEs combination functions, applying other kernel types (e.g. p-Gaussian), and removing non-typical observations (similarly to [15]) from the training set which may lead to better classification results.

Acknowledgments. The authors would like to thank prof. Jan Mielniczuk for valuable suggestions concerning the algorithm's design and development.

References

1. Duda, R.O., Hart, P.E., Stork, D.G.: Pattern Classification, 2nd edn. Wiley-Interscience Publication, Hoboken (2000)
2. Smyth, P., Wolpert, D.: Linearly combining density estimators via stacking. Machine Learning 36, 59–83 (1999)
3. Marchette, D.J., Priebe, C.E., Rogers, G.W., Solka, J.L.: Filtered kernel density estimation. Computational Statistics 11, 95–112 (1996)
4. Di Marzio, M., Taylor, C.C.: On boosting kernel density methods for multivariate data: density estimation and classification. Statistical Methods and Applications 14, 163–178 (2005)
5. Ormoneit, D., Tresp, V.: Averaging, maximum penalized likelihood and bayesian estimation for improving gaussian mixture probability density estimates. IEEE Transactions on Neural Networks 9, 639–650 (1998)
6. Ridgeway, G.: Looking for lumps: boosting and bagging for density estimation. Computational Statsistics and Data Analysis 38, 379–392 (2002)
7. Cooley, C.A., MacEachern, S.N.: Classification via kernel product estimators. Biometrika 85, 823–833 (1998)
8. Ghosh, A.K., Chaudhuri, P., Sengupta, D.: Classification using kernel density estimates: Multiscale analysis and visualization. Technometrics 48, 120–132 (2006)
9. Scott, D.W.: Multivariate Density Estimation: Theory, Practice, and Visualization. Wiley, New York (1992)
10. Lim, T.S., Loh, W.Y., Shih, Y.S.: A comparison of prediction accuracy, complexity, and training time of thirty-three old and new classification algorithms. Machine Learning 40, 203–228 (2000)
11. Mangasarian, O., Wolberg, W.: Cancer diagnosis via linear programming. Siam News 23, 1–18 (1990)
12. Asuncion, A., Newman, D.: UCI Machine Learning Repository (2007)
13. Ripley, B.: Pattern recognition and neural networks datasets collection (1996), http://www.stats.ox.ac.uk/pub/PRNN/
14. R Development Core Team: R: A Language and Environment for Statistical Computing. R Foundation for Statistical Computing, Vienna, Austria (2008)
15. Dendek, C., Mańdziuk, J.: Improving performance of a binary classifier by training set selection. In: Kůrková, V., Neruda, R., Koutník, J. (eds.) ICANN 2008, Part I. LNCS, vol. 5163, pp. 128–135. Springer, Heidelberg (2008)

Joint Approximate Diagonalization Utilizing AIC-Based Decision in the Jacobi Method

Yoshitatsu Matsuda[1] and Kazunori Yamaguchi[2]

[1] Department of Integrated Information Technology,
Aoyama Gakuin University,
5-10-1 Fuchinobe, Sagamihara-shi, Kanagawa, 229-8558, Japan
matsuda@it.aoyama.ac.jp
http://www-haradalb.it.aoyama.ac.jp/~matsuda

[2] Department of General Systems Studies,
Graduate School of Arts and Sciences, The University of Tokyo,
3-8-1, Komaba, Meguro-ku, Tokyo, 153-8902, Japan
yamaguch@graco.c.u-tokyo.ac.jp

Abstract. Joint approximate diagonalization is one of well-known methods for solving independent component analysis and blind source separation. It calculates an orthonormal separating matrix which diagonalizes many cumulant matrices of given observed signals as accurately as possible. It has been known that such diagonalization can be carried out efficiently by the Jacobi method, where the optimization for each pair of signals is repeated until the convergence of the whole separating matrix. Generally, the Jacobi method decides whether the optimization is actually applied to a given pair by a convergence decision condition. Then, the whole convergence is achieved when no pair is actually optimized any more. Though this decision condition is crucial for accelerating the speed of the whole optimization, many previous works have employed simple conditions based on an arbitrarily selected threshold. In this paper, we propose a novel decision condition which is based on Akaike information criterion (AIC). It is derived by assuming each cumulant matrix to be a sample generated independently. In each pair optimization, the condition compares the reduction rate of the objective function with a constant depending on the number of cumulant matrices. It involves no thresholds (and no parameters) to be set manually. Numerical experiments verify that the proposed decision condition can accelerate the optimization speed for artificial data.

1 Introduction

Independent component analysis (ICA) is a widely-used method in signal processing [1,2]. It solves blind source separation problems under the assumption that source signals are statistically independent of each other. In the linear model (given as $\boldsymbol{x} = \boldsymbol{W}\boldsymbol{s}$), it estimates the mixing matrix $\boldsymbol{W}$ and the source signals $\boldsymbol{s}$ from only the observed signals $\boldsymbol{x}$. The dimension of $\boldsymbol{x}$ corresponds to the number of signals N. Joint approximate diagonalization [3,4] is one of efficient methods

C. Alippi et al. (Eds.): ICANN 2009, Part II, LNCS 5769, pp. 135–144, 2009.

for estimating $\boldsymbol{W}$. Now, $\boldsymbol{\Delta}_{pq}$ is defined as an $N \times N$ matrix whose (i,j) element is κ_{ijpq}. Here, κ_{ijpq} is the 4-th order cumulants of $\boldsymbol{x}$. It is easily proved that $\tilde{\boldsymbol{\Delta}}_{pq} = \boldsymbol{V}\boldsymbol{\Delta}_{pq}\boldsymbol{V}'$ is a diagonal matrix for any p and q if $\boldsymbol{V}$ is the accurate separating matrix. Therefore, $\boldsymbol{W}$ can be estimated as $\boldsymbol{V}$ which diagonalizes $\boldsymbol{\Delta}_{pq}$ as accurately as possible for many p's and q's. The most well-known joint diagonalization algorithm is JADE [3], where q is given as the same value of p. Besides, because $\boldsymbol{x}$ is assumed to be pre-whitened, $\boldsymbol{V}$ is constrained to an orthonormal matrix. Then, the estimated separating matrix $\hat{\boldsymbol{V}}$ is given as

$$\hat{\boldsymbol{V}} = \mathrm{argmin}_{\boldsymbol{V}} \sum_{i,j \neq i,k} \left(\tilde{\kappa}_{ijkk}\right)^2 \tag{1}$$

where $\tilde{\boldsymbol{\Delta}}_{kk} = (\tilde{\kappa}_{ijkk})$. Because it is relatively difficult to calculate $\hat{\boldsymbol{V}}$ directly, the Jacobi method is often used. The method optimizes the objective function $\psi = \sum_{i,j \neq i,k} \left(\tilde{\kappa}_{ijkk}\right)^2$ only for each pair (i,j). By sweeping the optimizations over all the pairs repeatedly, the whole $\boldsymbol{V}$ can be estimated. Because $\boldsymbol{V}$ is an orthonormal matrix, each pair optimization is given as a 2×2 rotation matrix $(\cos\phi, \sin\phi; -\sin\phi, \cos\phi)$ which has only a single parameter ϕ. Because each pair optimization can be solved analytically and efficiently, JADE is known to be quite efficient.

In this paper, we focus on the convergence decision in each pair optimization. Generally, the Jacobi method decides whether the convergence is achieved for each pair, then actually optimizes the pair only if the convergence is not achieved. If the convergence of every pair is achieved, the convergence of the whole estimated matrix is declared. If the decision method can remove adequately unnecessary pair optimizations, the computational costs can be reduced. So, the decision method is crucial for the efficiency and the convergence rate of the Jacobi method. However, most of previous works have employed simple decision methods, for example, whether the estimated $\hat{\phi}$ is over a threshold ϵ. Here, we propose a new decision condition based on Akaike information criterion (AIC). AIC is a measure of the "goodness" of a probabilistic model for given samples and is often used in the model selection [5,6]. AIC is based on a mathematical framework explaining "Occam's razor," and can suppress the increase of unnecessary parameters of the model. This property is suitable for the decision on each pair optimization.

This paper is organized as follows. In Section 2, the JADE algorithm and AIC are briefly explained. In Section 3, a new decision condition for each pair optimization and an extension of JADE are proposed. In Section 4, some numerical experiments show that the proposed method improves the convergence rate of JADE. Lastly, this paper is concluded in Section 5.

2 Background

2.1 JADE Algorithm

As shown in the above, JADE [3] is a joint approximate diagonalization algorithm which minimizes $\psi = \sum_{i,j \neq i,k} \left(\tilde{\kappa}_{ijkk}\right)^2$ by the Jacobi method. Now, the pair optimization for (i, j) is given as follows. By picking up the terms depending on i and j, the objective function is simplified into

$$\psi_{(i,j)} = \sum_k (\tilde{\kappa}_{ijkk})^2. \tag{2}$$

By using the rotation angle ϕ and the original κ_{ijkk} of $\boldsymbol{x}$, $\tilde{\kappa}_{ijkk}$ is given as

$$\tilde{\kappa}_{ijkk} = -\sin\phi\left(\cos\phi \cdot \kappa_{iikk} + \sin\phi \cdot \kappa_{jikk}\right) + \cos\phi\left(\cos\phi \cdot \kappa_{ijkk} + \sin\phi \cdot \kappa_{jjkk}\right). \tag{3}$$

It has been shown that there exists the analytical solution of $\hat{\phi}$ minimizing $\psi_{(i,j)}$ (see [3] for the details). In consequence, $\left(\cos\hat{\phi}, \sin\hat{\phi}\right)$ is given as the principal eigenvector of a 2×2 matrix $\boldsymbol{G}\boldsymbol{G}^t$ where

$$\boldsymbol{G} = \begin{pmatrix} g_{11} \; g_{12} \cdots g_{1N} \\ g_{21} \; g_{22} \cdots g_{2N} \end{pmatrix}, \tag{4}$$

$g_{1k} = \kappa_{iikk} - \kappa_{jjkk}$, and $g_{2k} = \kappa_{ijkk} + \kappa_{jikk}$

Then, the complete description of JADE is given as follows:

1. *Initialization.* Whiten $\boldsymbol{x}$ and calculate the cumulant matrices $\tilde{\boldsymbol{\Delta}}_{kk} = (\tilde{\kappa}_{ijkk})$ for $k = 1, \ldots, N$. Besides, $\boldsymbol{V} = \boldsymbol{I}_N$ (where $\boldsymbol{I}_N$ is the $N \times N$ identity matrix).
2. *Sweep.* For every pair (i, j),
   (a) Calculate $\boldsymbol{G}\boldsymbol{G}^t$ and $\hat{\phi}$,
   (b) If $\hat{\phi} > \epsilon$ (a small threshold), do the actual rotation of $\boldsymbol{V}$ by $\hat{\phi}$ and update $\boldsymbol{\Delta}_{kk} := \tilde{\boldsymbol{\Delta}}_{kk}$ for every k.
3. *Convergence decision of the whole matrix.* If no pair has been actually rotated in the current sweep (in other words, $\hat{\phi} \leq \epsilon$ holds for every pair), end. Otherwise, go to the next sweep.

It is valuable to note that the computational costs of $\hat{\phi}$ is much less than those of the update of the cumulant matrices $\boldsymbol{\Delta}_{kk} := \tilde{\boldsymbol{\Delta}}_{kk}$ if N is large. In the former, because the calculation of $\boldsymbol{G}\boldsymbol{G}^t$ is dominant and $\boldsymbol{G}$ includes only $O(N)$ elements, the cost is $O(N)$. On the other hand, because the latter has to update all the cumulants depending on i *or* j (including κ_{ilkk} $(l \neq i, j)$), its cost is $O\left(N^2\right)$. Therefore, the computational costs of JADE can be reduced by employing an adequate convergence condition depending on only $\hat{\phi}$.

2.2 AIC

Akaike information criterion (AIC) is a measure of the "goodness" of a probabilistic model $g(\boldsymbol{z}|\boldsymbol{\theta})$ for given samples $\boldsymbol{z}$'s, where $\boldsymbol{\theta}$ denotes the adjustable parameters of the model. Though the usual likelihood is give as $\sum_{\boldsymbol{z}} \log g(\boldsymbol{z}|\boldsymbol{\theta})$, it can be increased without restriction by employing more complex models including more number of parameters. On the other hand, AIC takes into consideration the distance between the given model $g(\boldsymbol{z}|\boldsymbol{\theta})$ and the "true" model. In consequence, AIC is given as

$$\mathrm{AIC} = -2\sum_{\boldsymbol{z}} \log g\left(\boldsymbol{z}|\hat{\boldsymbol{\theta}}\right) + 2K \tag{5}$$

where $\hat{\boldsymbol{\theta}}$ is the optimized $\boldsymbol{\theta}$ maximizing the likelihood and K is the number of parameters in $\boldsymbol{\theta}$. It can be proved under some assumptions and approximations that a probabilistic model with less AIC is nearer to the true model (see [6] for the details). The term $2K$ can be regarded as a penalty for increasing the number of parameters. So, it is expected to suppress the maximization of the objective function with unnecessary (and often harmful) parameters. Besides, if the model $g(\boldsymbol{z}|\boldsymbol{\theta})$ assumes Gaussian-distributed errors with a constant variance, Eq. (5) is simplified further into

$$\mathrm{AIC} = N \log \hat{\sigma}^2 + 2K. \tag{6}$$

Here, $\hat{\sigma}^2$ is the estimated variance of errors $\frac{\sum_k \hat{\delta}_k^2}{N}$, where $\hat{\delta}_k$ is a residual error for each sample k under $\hat{\boldsymbol{\theta}}$.

3 Method

3.1 Decision Method Based on AIC

The essential conception is to regard the set of the four variables (κ_{iikk}, κ_{jjkk}, κ_{ijkk}, and $\tilde{\kappa}_{ijkk}$) as a sample $\boldsymbol{z}$ in the optimization of a pair (i, j). Here, each $k = 1, \ldots, N$ is regarded as a trial in sampling. Then, AIC can be calculated by $\hat{\delta}_k = \tilde{\kappa}_{ijkk}$ for $\hat{\phi}$ and the number of parameters $K = 2$ where $\boldsymbol{\theta}$ consists of $\hat{\phi}$ and $\hat{\sigma}^2$. The change of AIC by a pair optimization (denoted by δAIC) is given by

$$\begin{aligned}
&\delta\mathrm{AIC} \\
&= \left(N \log \frac{\sum_k \left(\tilde{\kappa}_{ijkk}\right)^2}{N} + 2\left(K_{\mathrm{cur}} - 1 + 2\right) \right) - \left(N \log \frac{\sum_k \left(\kappa_{ijkk}\right)^2}{N} + 2K_{\mathrm{cur}} \right) \\
&= N \log \frac{\sum_k \left(\tilde{\kappa}_{ijkk}\right)^2}{\sum_k \left(\kappa_{ijkk}\right)^2} + 2
\end{aligned} \tag{7}$$

where K_{cur} is the number of parameters which has been optimized so far. The term "-1" in $(K_{\mathrm{cur}} - 1 + 2)$ is caused by the elimination of one parameter $\sigma^2 = \frac{\sum_k \kappa_{ijkk}^2}{N}$ (the variance in the current state). Because AIC has to be reduced ($\delta\mathrm{AIC} < 0$), the decision condition in a pair optimization is given by

$$\frac{\sum_k \left(\kappa_{ijkk}\right)^2 - \sum_k \left(\tilde{\kappa}_{ijkk}\right)^2}{\sum_k \left(\kappa_{ijkk}\right)^2} > 1 - \exp\left(-\frac{2}{N}\right). \tag{8}$$

The left and right terms can be regarded as the reduction rate of the objective function $\sum_k \left(\kappa_{ijkk}\right)^2$ and the threshold, respectively. If N is large, the threshold is close to $\frac{2}{N}$.

Remarks. Because the above derivation is based on some approximations and assumptions whose validity is not theoretically verifiable, only numerical experiments can verify the effectiveness of this decision condition (see Section 4).

Nevertheless, we now focus on the essential assumption that κ_{ijkk} (including κ_{iikk}, κ_{jjkk}, and $\tilde{\kappa}_{ijkk}$ for fixed $\hat{\phi}$) is regarded as a sample at the k-th trial. In other words, it asserts that κ_{ijkk} is statistically independent of κ_{ijll} for $l \neq k$. In the following, we show that this assumption approximately holds if the initial separating matrix is given randomly. For simplicity, the cumulants are assumed to be accurately estimated by the large number of $\boldsymbol{x}$. By $\boldsymbol{x} = \boldsymbol{W}\boldsymbol{s}$ ($\boldsymbol{W} = (w_{ij})$ is orthonormal), κ_{ijkk} $(i \neq j)$ is given as

$$\kappa_{ijkk} = \sum_{p} w_{ip} w_{jp} w_{kp}^{2} \kappa_{p}^{\mathrm{source}} \tag{9}$$

where $\kappa_{p}^{\mathrm{source}}$ is the kurtosis of the p-th source. Then, $\kappa_{ijkk} = \sum_{p} \alpha_p w_{kp}^2$ and $\kappa_{ijll} = \sum_{p} \alpha_p w_{lp}^2$ where the variable α_p does not depend on k and l. Thus, if w_{kp}^2 and w_{lp}^2 are statistically independent, κ_{ijkk} and κ_{ijll} are independent similarly. The independence is not destructed by any rotations of the cumulant matrices. Therefore, if the initial $\boldsymbol{W}$ is given randomly, the essential assumption always holds approximately.

3.2 Extension of JADE with AIC-Based Decision

A simple extension of JADE can be constructed by replacing the decision condition $\hat{\phi} > \epsilon$ with Eq. (8). But, as will be shown in Section 4, the condition is not suitable for the convergence phase. Therefore, the following algorithm of two stages is constructed:

1. Do the JADE algorithm with the decision condition Eq. (8) for each pair optimization until the convergence.
2. Continuously, do the usual JADE algorithm with the condition $\hat{\phi} > \epsilon$.

This algorithm is called JADE-AIC hereafter.

Remarks. It will be observed in Section 4 that Eq. (8) rejects necessary rotations at the convergence phase. It is probably one of the significant reasons that the derivation of Eq. (8) implicitly assumes that every κ_{ijkk} converges to 0. Because the number of sample is limited in practice, the optimum of κ_{ijkk} (denoted by $\hat{\kappa}_{ijkk}$) is not equal to 0 generally. Therefore, the following condition is preferable at the convergence phase:

$$\frac{\sum_k (\kappa_{ijkk})^2 - \sum_k (\tilde{\kappa}_{ijkk})^2 - \sum_k (\hat{\kappa}_{ijkk})^2}{\sum_k (\kappa_{ijkk})^2 - \sum_k (\hat{\kappa}_{ijkk})^2} > 1 - \exp\left(-\frac{2}{N}\right). \tag{10}$$

Because it is difficult to estimate $\hat{\kappa}_{ijkk}$ in advance, the condition $\hat{\phi} > \epsilon$ is employed at the convergence phase in this paper.

3.3 Extension by BIC

BIC (Bayesian information criterion) is another well-known criterion measuring the "goodness" of a model [7,6]. BIC is given as

$$\mathrm{BIC} = -2 \sum_{\boldsymbol{z}} \log g\left(\boldsymbol{z}|\hat{\boldsymbol{\theta}}\right) + \log(N) K \tag{11}$$

where the constant factor 2 in the penalty term of AIC (Eq. (5)) is replaced by a factor $\log(N)$ depending on the number of signals N. Because $\log(N) > 2$ in almost all cases, BIC rejects the increase of the number of parameters more strongly than AIC. BIC can be utilized also as the decision condition in our proposed method. In this case, Eq. (8) is replaced with

$$\frac{\sum_k (\kappa_{ijkk})^2 - \sum_k (\tilde{\kappa}_{ijkk})^2}{\sum_k (\kappa_{ijkk})^2} > 1 - \exp\left(-\frac{\log(N)}{N}\right). \quad (12)$$

By this condition, the JADE-BIC algorithm can be constructed in the same way as in Section 3.2.

4 Results

Here, the proposed JADE-AIC and JADE-BIC algorithms are compared with usual JADE in blind source separation of artificial data and an image separation problem. Regarding artificial data, three experiments were carried out by 10, 20, and 30 sources ($N = 10$, 20, and 30). In each experiment, a half of the sources were generated by the Laplace distribution (super-Gaussian) and the other half by the uniform distribution (sub-Gaussian). The number of samples was set to 100000, and the mixing matrix $\boldsymbol{W}$ was randomly generated. Regarding the image separation, the sources were 12 grayscale images of 256×256 pixels from SIDBA and a 12×12 mixing matrix was given randomly, where $N = 12$ and the number of samples is 65536. In JADE, the decision condition $\hat{\phi} > \epsilon$ were used ($\epsilon = 10^{-6}$). In addition, the following condition was also used:

$$\frac{N^2 \sum_k \left((\kappa_{ijkk})^2 - (\tilde{\kappa}_{ijkk})^2\right)}{\sum_k (\tilde{\kappa}_{kkkk})^2} > \epsilon_2 \quad (13)$$

where ϵ_2 was set to 10^{-3}. Eq. (13) has been implemented in a public JADE package at `http://www.tsi.enst.fr/~cardoso/Algo/Jade/jadeR.m`. It is an ad-hoc condition measuring the effect of the pair optimization to the objective function ψ. Eq. (13) was also employed as the decision condition for comparison. JADE with Eq. (13) is called JADE2 here. Besides, JADE-AIC without the second stage (usual JADE) is also used for comparison at the convergence phase. It is called JADE-AIC-raw. Similarly, JADE-BIC-raw was also used.

Fig. 1 shows the decreasing curves of Amari's separating errors [8] along the number of the actual rotations by JADE-AIC, JADE-BIC, and JADE. The separating error is defined as the sum of normalized non-diagonal elements of the product of the estimated separating matrix and the given mixing one. If the error is equal to 0, the estimated separating matrix is equivalent to the inverse of the mixing one except for scaling factors. They were averaged over 10 runs. JADE-AIC-raw, JADE-BIC-raw, and JADE2 were omitted in Fig. 1. It is because the distinct differences about those methods were observed only in the convergence phase (in Fig. 2). It shows that JADE-AIC and JADE-BIC were

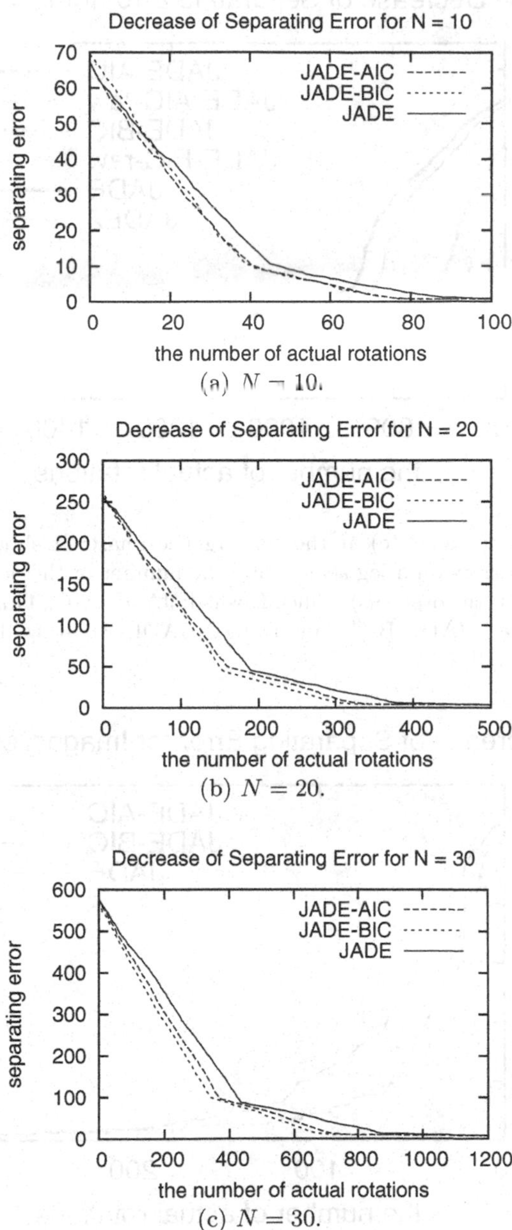

(a) $N = 10$.

(b) $N = 20$.

(c) $N = 30$.

Fig. 1. Decreasing curves of separating errors along the number of the actual rotations ($N = 10$, 20, and 30) for artificial data. Dashed: JADE-AIC. Dotted: JADE-BIC. Solid curves: JADE.

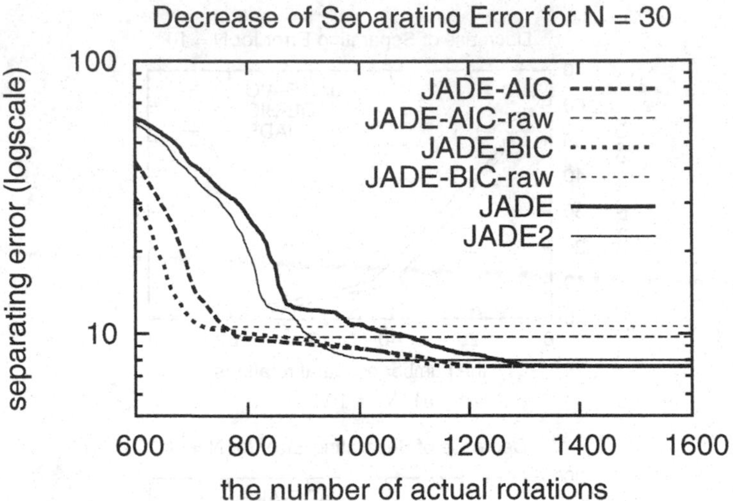

Fig. 2. Enlargement of Fig. 1-(c) at the convergence phase: It shows the decreasing curves of separating errors on a log scale along the number of the actual rotations ($N = 30$ and from the 600-th rotations). Thick Dashed: JADE-AIC. Thin Dashed: JADE-AIC-raw. Thick Dotted: JADE-BIC. Thin Dotted: JADE-BIC-raw. Thick solid curves: JADE. Thin solid: JADE2.

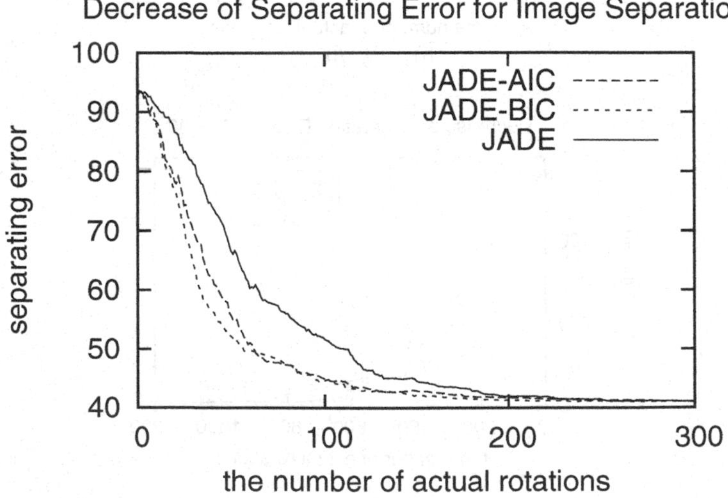

Fig. 3. Decreasing curves of separating errors in image separation. Dashed: JADE-AIC. Dotted: JADE-BIC. Solid curves: JADE.

always superior to JADE after the same number of actual rotations. Though the threshold in Eq. (8) depends on N, the superiority of JADE-AIC and JADE-BIC were always observed irrespective of N. Regarding the comparison of AIC with BIC, JADE-BIC seemed to be slightly superior to JADE-AIC for larger N. But, the difference was not distinct. Fig. 3 shows the results in image separation. It shows that JADE-AIC and JADE-BIC were superior to JADE even for an actual application. Fig. 2 is an enlargement of Fig. 1-(c) at the convergence phase. It shows the decreasing curves of the separating errors on a log scale from the 600-th rotations by JADE-AIC, JADE-AIC-raw, JADE-BIC, JADE-BIC-raw, JADE, and JADE2. JADE-AIC and JADE-AIC-raw (using only the AIC-based decision condition) branched off around the 800-th rotation. Similarly, JADE-BIC and JADE-BIC-raw did around the 700-th one. It shows that the information-criteria-based conditions Eqs. (8) and (12) were no longer effective at the convergence phase. Except for JADE-AIC-raw and JADE-BIC-raw, almost the same optimum was achieved. The curves of JADE-AIC and JADE-BIC were almost the same at the convergence phase. Though JADE-AIC and JADE-BIC were slightly inferior to JADE2 around the 1000-th rotations, the final optima of JADE-AIC and JADE-BIC were slightly superior to JADE2. It shows that JADE-AIC and JADE-BIC were effective even at the convergence phase but some further improvements may be needed. In summary, those results verified that the utilization of the information criteria such as AIC and BIC in JADE is effective especially at the early phase. Besides, the information-criteria-based methods are effective even at the convergence phase.

5 Conclusion

In this paper, we proposed a new decision condition of the Jacobi method in joint approximate diagonalization. The condition is naturally derived from AIC and does not involve any threshold to be set manually. Besides, an extension of JADE with the condition is proposed. The numerical experiments verified the effectiveness of the proposed method.

Though the information criteria such as AIC and BIC are widely used in order to estimate the number of sources [9], they give a criterion for estimating the "goodness" of the whole separating matrix. On the other hand, the proposed method focuses on each pair optimization and the information criterion is estimated for only a pair. The numerical results suggest that such pair-focused utilization of the information criteria is effective in signal processing.

Currently, though numerical results suggest the superiority of the proposed method, the theoretical foundation is not sufficient. For example, the validity of the application of AIC to each pair optimization is not theoretically guaranteed yet. We are now constructing a more rigorous theoretical framework for this method. Besides, the current condition is useful only at the early phase. We are now planning to construct a new condition suitable for the convergence phase by assuming different probabilistic models of cumulants and estimating their optima at the convergence in advance. We are also planning to extend our approach to general ICA frameworks such as information theoretic ones and to utilize other information criteria. In addition, we are planning to apply the proposed AIC-based condition to our previously-proposed ICA algorithm named LMICA [10]. LMICA is a quite efficient Jacobi method which optimizes only the significant pairs. So, the proposed condition seems suitable for the method. This work is partially supported by Grant-in-Aid for Young Scientists (KAKENHI) 19700267.

References

1. Hyvärinen, A., Karhunen, J., Oja, E.: Independent Component Analysis. Wiley, Chichester (2001)
2. Cichocki, A., Amari, S.: Adaptive Blind Signal and Image Processing: Learning Algorithms and Applications. Wiley, Chichester (2002)
3. Cardoso, J.F., Souloumiac, A.: Blind beamforming for non Gaussian signals. IEEE Proceedings-F 140(6), 362–370 (1993)
4. Cardoso, J.F.: High-order contrasts for independent component analysis. Neural Computation 11(1), 157–192 (1999)
5. Akaike, H.: A new look at the statistical model identification. IEEE Transactions on Automatic Control 19(6), 716–723 (1974)
6. Burnham, K.P., Anderson, D.R.: Model selection and multimodel inference: A practical-theoretic approach, 2nd edn. Springer, Berlin (2002)
7. Schwarz, G.: Estimating the dimension of a model. The Annals of Statistics 6(2), 461–464 (1978)
8. Amari, S., Cichocki, A.: A new learning algorithm for blind signal separation. In: Touretzky, D., Mozer, M., Hasselmo, M. (eds.) Advances in Neural Information Processing Systems, vol. 8, pp. 757–763. MIT Press, Cambridge (1996)
9. Wax, M., Kailath, T.: Detection of signals by information theoretic criteria. IEEE Transactions on Acoustics Speech and Signal Processing 33, 387–392 (1985)
10. Matsuda, Y., Yamaguchi, K.: Linear multilayer ICA generating hierarchical edge detectors. Neural Computation 19, 218–230 (2007)

Newtonian Spectral Clustering

Konstantinos Blekas, K. Christodoulidou, and I.E. Lagaris

Dept. of Computer Science, University of Ioannina,
P.O. Box. 1186, 45110 Ioannina, Greece
{kblekas,kchristo,lagaris}@cs.uoi.gr

Abstract. In this study we propose a systematic methodology for constructing a sparse affinity matrix to be used in an advantageous spectral clustering approach. Newton's equations of motion are employed to concentrate the data points around their cluster centers, using an appropriate potential. During this process possibly overlapping clusters are separated, and simultaneously, useful similarity information is gained leading to the enrichment of the affinity matrix. The method was further developed to treat high-dimensional data with application to document clustering. We have tested the method on several benchmark data sets and we witness a superior performance in comparison with the standard approach.

1 Introduction

Given a set of data points, the problem of clustering is to discover a number of subsets, called *clusters*, that contain points with similar properties. In the literature there is a plethora of clustering approaches that have been proposed rather recently. In this work we concentrate on the class of methods which are based on spectral clustering [1], [2]. Spectral clustering has become increasingly popular during the last decade. Such algorithms are based on similarity information between data points. That is, similar data points (or points with high affinity) are more likely to belong to the same cluster than points with low affinity. These kind of algorithms have proved to be quite successful in numerous application domains, such as computer vision [3], [4], [5], speech recognition [6], bioinformatics [7], [8], text mining [9], etc.

Spectral clustering techniques make use of information obtained from an appropriately defined affinity matrix. Their primary strength is their ability to treat complex data shapes where other well-known methods (such as k-means) either cannot be directly applied, or fail. The similarity matrix must be built in such a way so as to reflect the topological characteristics of the data set. In addition *sparsity* is another desired property, since it offers computational advantages [2], [10]. In applications of computer vision and related problems, the similarity matrix is naturally sparse due to the local character of the similarities.

Methodologies leading to sparse affinity matrices have been proposed in the past [2]. For instance, the ϵ-neighborhood technique connects only points whose pairwise distances are smaller than ϵ. Another similar method is the (mutual)

C. Alippi et al. (Eds.): ICANN 2009, Part II, LNCS 5769, pp. 145–154, 2009.

k-nearest neighbor, where every point is connected only with its k nearest neighbors. However, these methods heavily depend on the choice of the control parameter (ϵ or k) that acts as a threshold for cutting some edges of the associated graph.

We present here an alternative spectral clustering method that consists of two phases. The data points are initially manipulated in a way suggested in the Newtonian clustering [11], where the original data set is transformed and the cluster appearance becomes more prominent. This is done via a dynamic procedure based on Newton's equation of motion using a properly constructed potential function. During the next phase, the affinity matrix is calculated not in the usual way, but with extra information embedded that was gained in the previous phase. At the same time this information has a sparsifying effect, and hence our affinity matrix is both sparser and richer. We further modified our method in order to treat problems of high dimensionality, such as those appearing in document clustering. The modification is carried out by choosing a different potential function and likewise a slightly different equation of motion. We have tested our method on a suite of well known benchmarks ranging from continuous feature data to image segmentation and document clustering problems. We compare to the standard spectral clustering method and the classical k-means algorithm.

In section 2 we lay out an algorithmic description of the proposed Newtonian spectral clustering while in section 3 we report experimental results for several data sets. Finally in section 4 we summarize and conclude with some remarks.

2 The Proposed Method

2.1 Spectral Clustering with a Dynamic Procedure

Let the set $X = \{x_1, \ldots, x_N\}$ denote the input set of N observations that we want to partition into K groups. We consider that the data points correspond to particles of unit mass, interacting via a two-body attractive, short-range potential. Let V_{ij} be the potential between particles located at points x_i and x_j. In this section we will consider a simple potential of Gaussian form given by:

$$V_{ij} = -\exp(-\frac{||x_i - x_j||^2}{2\sigma^2}) \ , \tag{1}$$

where the scale parameter (σ) determines the range of the potential. The value of σ is important since it affects the dynamic procedure that shrinks the clusters, as well as the performance of the subsequent spectral clustering application. The determination of this parameter will be detailed later on.

Under this consideration, the data points move under the influence of a *force*. Data that move toward different clusters either repel each other, or they are too far to interact. We expect that after an ample number of steps in time, points belong to the same cluster will come together forming to shrank clusters. The proposed dynamic procedure is governed by the Newton's equations of motion, which are:

$$\frac{d^2x_i(t)}{dt^2} = -\nabla_i \sum_{\substack{j=1 \\ j \neq i}}^{N} V_{ij} \equiv F_i, \ \forall i = 1, 2, \cdots, N \ . \qquad (2)$$

The initial positions are taken to be the original data points, i.e. $x_i(t=0) = x_i$ ($\forall i = 1, \ldots, N$), while the initial velocities ($v_i \equiv \frac{dx_i}{dt}$) are set to zero. We integrate the equations of motion in small time steps δt, considering that the forces F_i remain constant during this short time interval. At each step we reset the velocities to zero in order to avoid artifacts due to "heating". Hence we obtain the following motion scheme:

$$x_i(t + \delta t) = x_i(t) + \frac{1}{2}\delta t^2 F_i \ . \qquad (3)$$

Since the interaction is attractive, after a time period T the particles belonging to the same neighborhood-cluster will concentrate around its center. So an initially spread–out cluster is being shrunk as a result of the dynamic procedure. The simulation terminates, after a certain number of steps or when the steps become too small and further iterations hardly make any difference. Two typical examples are presented in Fig. 1 (a) and (c), where the initial data points (red) are concentrated (black) after 100 steps.

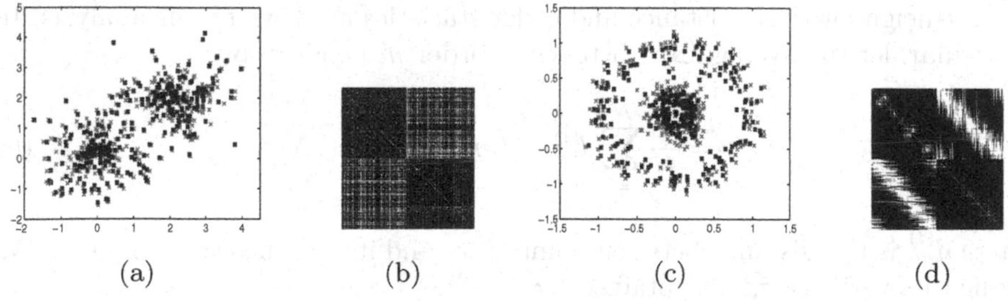

(a) (b) (c) (d)

Fig. 1. Two typical examples of the effect of the dynamic procedure

The path traveled by each particle can offer useful information. In particular, at the end of the dynamic process every point $x_i = x_i(0)$ has been moved into a new position $x_i(T)$. Let $dist_{ij}(t)$ denote the distance between two points x_i and x_j at time t. The elements of the affinity matrix A are then given by:

$$A_{ij} = b_{ij} \exp(-\frac{dist_{ij}^2(T)}{2\sigma^2}) \ , \qquad (4)$$

where $b_{ij} = 0$ if $dist_{ij}(T) > dist_{ij}(0)$ and $b_{ij} = 1$ otherwise. The above rule denotes that when two points move apart, they belong to different cluster and hence have zero affinity. Points that cluster together have a prominent affinity since $dist_{ij}(T) < dist_{ij}(0)$. Figure 1 (b) and (d) shows the sparsity of the affinity matrix in the case of the two artificial data sets of Figure 1 (a) and (c), respectively. More than 20% of the Affinity matrix elements are discarded (white pixels) due to the shrinking effect.

Spectral clustering is based on the data set's affinity matrix. In the literature there are several variations of the standard methodology described in [1], which we follow in our study. After having calculated the affinity matrix A, the *Laplacian* matrix L is then given by

$$L = D^{-1/2} A D^{-1/2} , \tag{5}$$

where D is a diagonal matrix with elements $D_{ii} = \sum_{j=1}^{N} A_{ij}$. The Laplacian matrix is known to be symmetric and positive semi-definite. Next, the K normalized eigenvectors $u_1, \ldots, u_K$ of matrix L (where K is the desired number of clusters) that correspond to the largest eigenvalues are computed and eventually fed into the k-means algorithm in order to estimate the final clustering solution.

Estimating the scale parameter σ^2. As mentioned before, the determination of the scale parameter σ is crucial and has to be chosen carefully. Sparse data sets require a longer range than dense data sets. Hence σ depends on the data set. An automatic determination of its value was suggested in [1] by running the clustering algorithm repeatedly for a number of values of σ and selecting the one which provides the least distorted clustering solution.

In this direction, we present here a more systematic methodology. The average nearest-neighbor (NN) distance and order statistics are keys to our analysis. In particular, let the average NN distance of order m be given by

$$< d_m >= \frac{1}{N} \sum_{i=1}^{N} d_m^{(i)}, \quad \forall\, m = 1, 2, \cdots, N-1 , \tag{6}$$

where $d_m^{(i)}$ is the distance between point at x_i and its mth nearest neighbor. We studied its variance $\tilde{\sigma}_m^2$ as obtained by

$$\tilde{\sigma}_m^2 = \frac{1}{m} \sum_{k=1}^{m} (< d_k^2 > - < d_k >^2) , \tag{7}$$

using order statistics. It was found in [11] that in the case of a single cluster the functional form of $\tilde{\sigma}_m^2$ is given as

$$\tilde{\sigma}_m^2 = \alpha(m+1)^2 + \beta(m+1) . \tag{8}$$

When there are more than one clusters within the data set, $< d_m >$ acquires discontinuities and the cumulative quantity $\tilde{\sigma}_m^2$ is given by a superposition of translated quadratics. Then, the value for the range of the potential is estimated by finding the number of neighbors m^* for which the second difference of $\frac{\tilde{\sigma}_m^2}{m+1}$

(with respect to m) vanishes. Figure 2 illustrates this behavior by plotting the quantity $\frac{\tilde{\sigma}_m^2}{m+1}$ versus m, in the case of two typical examples of Fig. 1. As our experiments have shown, there is a wide stability region around m^* for estimating the range value ($\sigma^2 = \sigma_{m^*}^2$), where the performance of our approach was identical. A detailed description of the above method for estimating the proper value of σ can be found in [11].

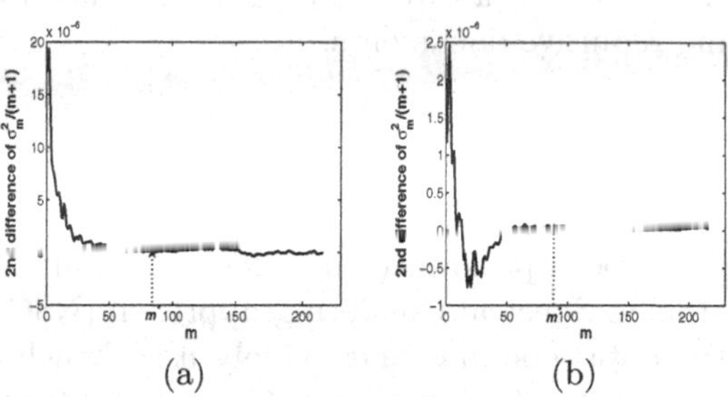

Fig. 2. Plots of second difference of quantity $\frac{\tilde{\sigma}_{M,m}^2}{m+1}$ with respect to m where the number m^* is estimated

2.2 Extension to Document Clustering

An important issue in clustering is treating high-dimensional data. Since spectral clustering is a common technique used for this purpose, we have tried to adjust the proposed method to deal with such problems. Document clustering is a very interesting application in text mining and information retrieval, aiming to the division of a collection of documents into groups based on their similarity.

In our study, each input document is transformed into a feature vector $x_i \in R^M$, where M is the size of the corpus vocabulary, such that every feature denotes the weight of the corresponding term. We have applied the TF-IDF (term frequency, inverse document frequency) weighting scheme for creating feature vectors. Moreover, the proximity between each pair of documents is computed used the cosine similarity metric. Since documents are normalized vectors, the similarity measure is reduced to the following simple rule:

$$V_{ij} = x_i^T x_j \ . \tag{9}$$

The above metric is also used as the potential function V_{ij} during the Newtonian dynamic procedure (see Eq. 1 - Eq. 3). The introduction of such kind of potential requires an alternative motion scheme of data points. Now, the interaction is not always attractive. Naturally, any particle is influenced positively from similar documents (that belong to the same cluster) and their interaction is attractive (positive force). In the opposite case, dissimilar document vectors have a repulsive effect to the particle and thus offering a negative sign force within its motion update rule. It can be easily found that the formulation of the force F_i now becomes as:

$$F_i = \sum_{\substack{j=1 \\ j \neq i}}^{N} c_{ij}(t) x_j \text{ , where } c_{ij}(t) = \begin{cases} +1 & \text{if } V_{ij} > \overline{V}_{ij}/2 \\ -1 & \text{otherwise} \end{cases} . \tag{10}$$

In fact the quantity $\overline{V}_{ij}/2$ acts as a threshold similarity value for distinguishing between attractive and repulsive documents.

3 Experimental Results

Several experiments have been performed in order to examine the effectiveness of the proposed Newtonian Spectral Clustering approach (NSC). We have considered both simulated data sets and other widely used benchmarks. We compare with the standard Spectral Clustering (SC) and the traditional k-means algorithm. During all experiments the number of Newtonian steps was fixed at $T = 100$, while the value of time step was set to $\delta t = 10^{-5}$. Moreover, both approaches, NSC and SC, used the same value of σ in the Gaussian similarity function as estimated by the proposed method. Finally, since we were aware of the true class label of data, all clustering methods were evaluated using the purity metric (classification accuracy), by assuming that all objects of a cluster are assigned to its dominant class.

The first series of experiments was performed on two simulated datasets (150 points per class) with two class ($K = 2$) presented in Fig. 1 (a) and 1 (c). By considering different levels of noise, we performed 50 experiments for each noise value and kept record of the mean accuracy for every method. The depicted comparative results are illustrated in the two diagrams of Fig. 3 in terms of different noise values. As it was expected, in the first data set with two spheres all three methods displayed identical behavior, since data were generated by sampling from two Gaussian densities that have the same spherical-type covariance matrix of the form $\sigma^2 I$. In the second data set (Fig. 1 (c)) which is more complex with two concentric clusters, our method performs better than the standard SC method especially in high-level noisy environments. The traditional k-means algorithm fails in situations with non-spherical data shapes.

Additional experiments were made using four known benchmarks (Fig. 4). The first one Fig. 4(a) is a two-class problem with a moon and a sun shape, while the next Fig. 4(b) is the CRAB data set of Ripley [12], that contains $N = 200$ data belonging to four clusters ($K = 4$). Here, we have created a 2-dimensional data set by projecting the data on the plane defined by the second and third principal components. We have also studied two UCI benchmarks [13] the renowned Fisher-IRIS data set Fig. 4(b) with $N = 150$ points belonging to three clusters ($K = 3$) (projected on the plane using the first two principal components), and the wine set consisted of $N = 178$ $K = 3$-classes data with 13 features (were we have applied zero-mean normalization). Table 1 summarizes

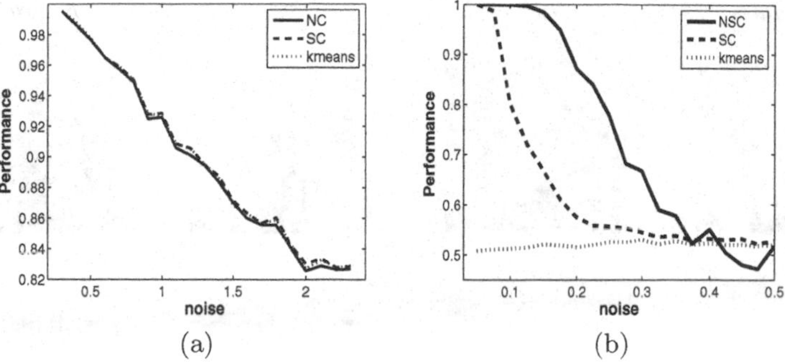

Fig. 3. Comparative results of NSC and its peers in terms of noise value using the two data sets of Fig. 1

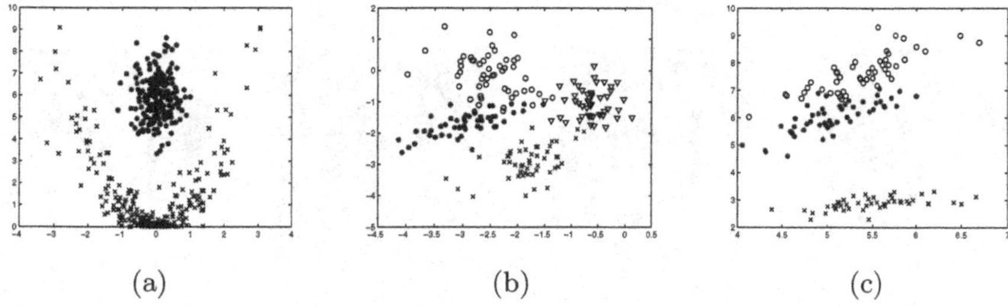

Fig. 4. Three known benchmarks used in our experiments: (a) the moon & sun, (b) the crabs and (c) the iris data set

the results obtained by the application of the three comparative approaches to the above mentioned data sets. In these cases the performance of NSC and the SC yielded comparable results, while being superior to k-means.

We have also applied our method to tackle the problem of image segmentation. For this purpose, we have selected six colored images from the Berkeley segmentation database[1] presented in Fig. 5, all with resolution around 150×150. We note here that in this series of experiments, since the number of input data is large, we have followed the Nyström method [14] for finding a numerical approximation to eigendecomposition. Fig. 5 illustrates the segmentation results of each method, where in the reconstructed images every pixel takes the intensity value of the cluster center that belongs. It is interesting to notice here that the NSC creates much smoother regions in comparison with the standard SC. We believe that if we take into account additional information, such as spatial, texture, etc. the resulting segmentation will be improved.

Finally, we have studied the performance of our method when dealing with high-dimensional spaces. For this purpose we have selected sets of documents

[1] http://www.eecs.berkeley.edu/Research/Projects/CS/vision/grouping/segbench/

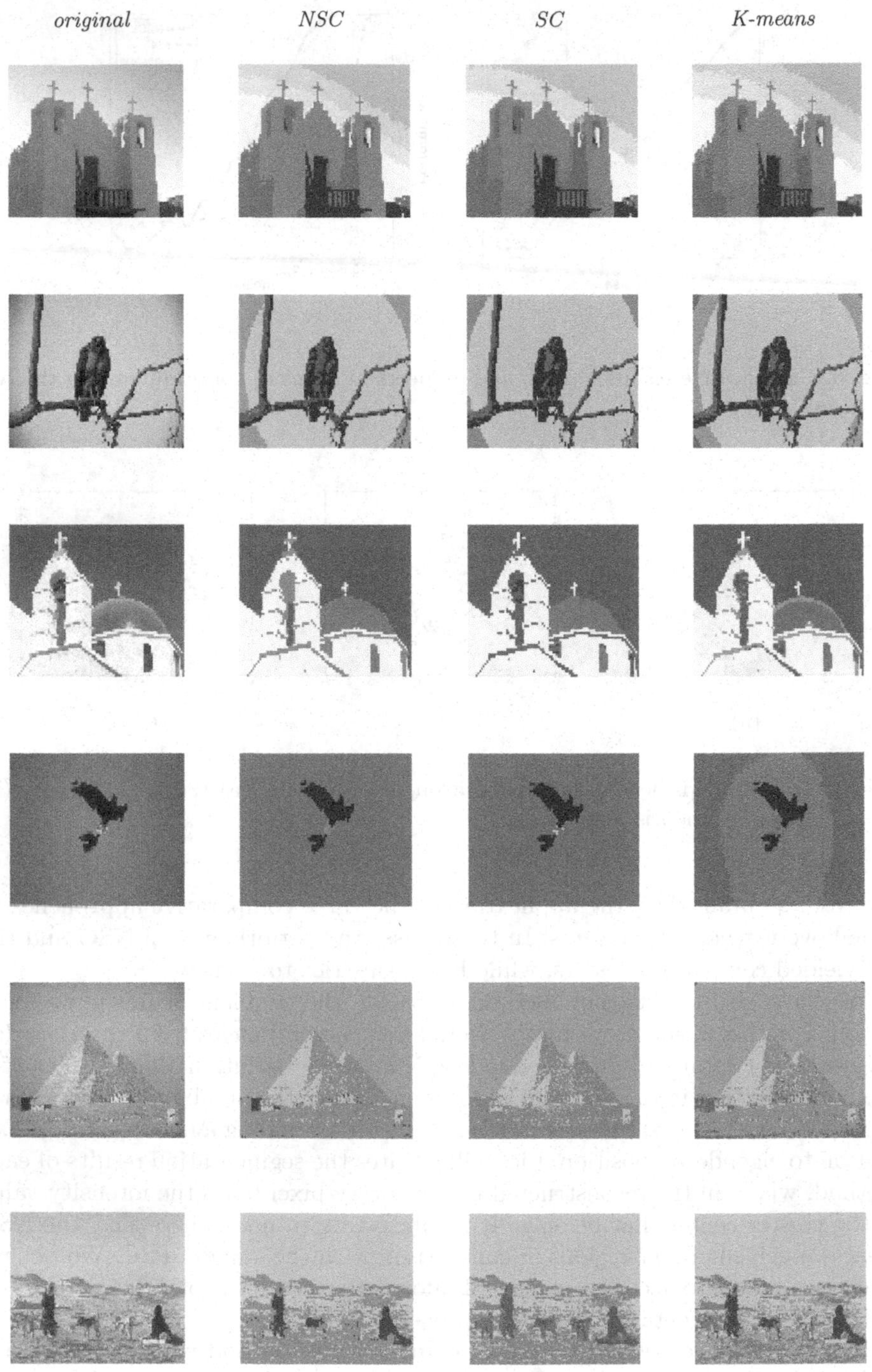

Fig. 5. Segmentation results obtained by three comparative clustering methods in six real colored images. NSC creates smoother regions.

Table 1. Comparative results using four known experimental data sets

Experimental dataset	*Performance of*		
	NSC	*SC*	*k-means*
moon & sun (Fig. 4(a))	0.94	0.94	0.92
crabs (Fig. 4(b))	0.94	0.93	0.93
iris (Fig. 4(c))	0.93	0.91	0.89
wine	0.98	0.98	0.97

and in particular four subsets of the popular 20-Newsgroup collection[2]. Their characteristics are presented in Table 2. The first set Talk_3 consists of documents of the talk subjects (politics.guns, politics.mideast, politics.misc), the next two of scientific documents (crypt, electronics, med, space), and the fourth set has documents from five newsgroups (comp.graphics, rec.motorcycles, rec.sport.baseball, sci.space, talk.politics.mideast). Table 2 shows also the results from the above data obtained by both approaches, NSC and SC. As it can be observed, the performance of our method is significantly better showing that the proposed way of constructing affinity matrix is worthwhile in high-dimensional data. Several other experiments were made with other subsets from the same data collection with similar results.

Table 2. Document data used in our experiments and the accuracy results obtained by both NSC and SC methods

Document dataset		*Performance of*	
name	*description*	*NSC*	*SC*
Talk_3	$N = 300, K = 3, M = 4515$	0.78	0.71
Science_4-400	$N = 400, K = 4, M = 4855$	0.71	0.62
Science_4-2000	$N = 2000, K = 4, M = 10250$	0.74	0.73
Multi_5	$N = 500, K = 5, M = 5589$	0.75	0.63

4 Conclusions

In this study we presented a novel method, the Newtonian spectral clustering, that inherits from Newtonian clustering information such that renders possible the formation of a proper affinity matrix that is sparse and contains enriched information. An extension of this approach has also been presented in order to deal with high-dimensional data such as documents. We have applied the method to several benchmark problems and we noticed performance superior to the standard spectral clustering approach. It is our intention to further pursue and develop the method to handle different problems with complex type of data such as time-series, multimedia data, discrete sequences, etc. Finally, the persistent

[2] http://www.cs.cmu.edu/afs/cs.cmu.edu/project/theo-20/www/data/news20.html

issue of discovering the optimal number of clusters may be examined in the framework of this method as well.

References

1. Ng, A., Jordan, M.I., Weiss, Y.: On spectral clustering: Analysis and an algorithm. In: Advances in Neural Information Processing Systems, vol. 14, pp. 849–864 (2001)
2. Luxburg, U.: A tutorial on spectral clustering. Statistics and Computing 17, 395–416 (2007)
3. Shi, J., Malik, J.: Normalized cuts and image segmentation. IEEE Trans. Pattern Recognition and Machine Intelligence 22, 888–905 (2000)
4. Park, J., Zha, H., Kasturi, R.: Spectral clustering for robust motion segmentation. In: Pajdla, T., Matas, J(G.) (eds.) ECCV 2004. LNCS, vol. 3024, pp. 390–401. Springer, Heidelberg (2004)
5. Chang, H., Yeung, D.: Robust path-based spectral clustering with application to image segmentation. In: Proc. Intern. Conf. on Computer Vision, pp. 278–285 (2005)
6. Bach, F., Jordan, M.I.: Learning spectral clustering, with application to speech seperation. Journal of Machine Learning Research 7, 1962–2001 (2006)
7. Pentney, W., Meila, M.: Spectral clustering fof Biological sequence data. In: Proc. of the 25th Annual Conference of AAAI, pp. 845–850 (2005)
8. Higham, D.J., Kalna, G., Kibble, M.: Spectral clustering, and its use in bioinformatics. Journal of Computational and Applied Mathematics 204, 25–37 (2007)
9. Dhillon, I.S.: Co-clustering documents and words using bipartite spectral graph partitioning. In: Proc. seventh ACM SIGKDD Intern. Conf. on Knowledge Discovery and Data mining (KDD), pp. 269–274 (2001)
10. Chen, W., Song, Y., Bai, H., Lin, C., Chang, E.Y.: Parallel spectral clustering. In: Daelemans, W., Goethals, B., Morik, K. (eds.) ECML PKDD 2008, Part II. LNCS (LNAI), vol. 5212, pp. 374–389. Springer, Heidelberg (2008)
11. Blekas, K., Lagaris, I.E.: Newtonian clustering: an approach based on molecular dynamics and global optimization. Pattern Recognition 40, 1734–1744 (2007)
12. Ripley, B.D.: Pattern Recognition and Neural Networks. Cambridge Univ. Press Inc., Cambridge (1996)
13. Merz, C.J., Murphy, P.M.: UCI repository of machine learning databases. Irvine, CA (1998), `http://www.ics.uci.edu/~mlearn/MLRepository.html`
14. Fowlkes, C.S., Belongie, F., Chung, F., Malik, J.: Spectral grouping using the Nyström method. IEEE Trans. on Pattern Analysis and Machine Intelligence 26, 214–225 (2004)

Bidirectional Clustering of MLP Weights for Finding Nominally Conditioned Polynomials

Yusuke Tanahashi[1] and Ryohei Nakano[2]

[1] Nagoya Institute of Technology
Gokiso-cho, Showa-ku, Nagoya 466-8555 Japan
tanahasi@ics.nitech.ac.jp
[2] Chubu University
1200 Matsumoto-cho, Kasugai 487-8501 Japan
nakano@cs.chubu.ac.jp

Abstract. We present a method for finding nominally conditioned polynomials to fit multivariate data containing both numeric and nominal variables. Here a polynomial is accompanied with a nominal condition stating when the polynomial is applied. Our method employs a four-layer perceptron (MLP) having shared weights. To get succinct polynomials, we employ weight sharing method called BCW, where each weight is allowed to be one of common weights, and a near-zero common weight can be eliminated. BCW performs bidirectional search to obtain an excellent set of common weights. Moreover, we employ the Bayesian Information Criterion (BIC) to efficiently select the optimal model parameters. In our experiments the proposed method successfully restored the original polynomials for artificial data, and found succinct polynomials for real data sets, showing excellent generalization.

Keywords: polynomial regression, multi-layer perceptron, weight sharing, rule extraction, information criteria.

1 Introduction

Discovering understandable numeric relationships such as polynomials from data is one of the key issues of data mining. Given multivariate data containing both numeric and nominal variables, we consider finding nominally conditioned polynomials, where each polynomial is accompanied with a nominal condition stating when the polynomial is applied.

To find nominally conditioned polynomials, there has been a combinatorial approach, such as ABACUS [1] and GMDH [4]. However, such a search-based approach intrinsically has the limited scalability due to combinatorial explosion. As an alternative, a connectionist numeric approach, such as RF6.4 [12] and REFANN [11], has been investigated. RF6.4 can solve this problem by using a four-layer perceptron and rule restoring, free from combinatorial explosion.

To find succinct results from data in the context of neural networks, we focus on *weight sharing* [3], where weights are divided into clusters, and weights within

C. Alippi et al. (Eds.): ICANN 2009, Part II, LNCS 5769, pp. 155–164, 2009.

the same cluster have the same value called a *common weight.* A common weight very close to zero can be removed. Thus, a weight sharing problem is to find both a set of adequate common weights and their mapping onto neural network weights.

As for weight sharing, there have been basic ideas, such as OBS [2] and soft weight sharing. OBS iteratively prunes the least influential weight using Hessian-based computation. We employ weight sharing called *BCW (bidirectional clustering of weights)* [13], which uses Hessian-based computation. In a merging phase, BCW iteratively merges common weights, and in a splitting phase it does the reverse operation. These two phases are repeated in turn to find a globally excellent set of common weights.

When we apply weight sharing BCW to connectionist polynomial regression RF6.4, we should determine vital model parameters. The existing method [13] uses cross-validation for model selection, requiring heavy computation. This paper employs the Bayesian Information Criterion (BIC) [10] instead. Since BIC doesn't need repetitive learning, we can select the optimal model very fast.

Section 2 explains connectionist polynomial regression RF6.4, and Section 3 explains weight sharing method BCW. Section 4 explains how to apply BCW to RF6.4 using BIC, and Section 5 evaluates the performance of RF6.4+BCW+BIC using artificial and real data sets.

2 Connectionist Polynomial Regression: RF6.4

Basic Framework. Let $(q_1, ..., q_{K_1}, x_1, ..., x_{K_2}, y)$ or $(\boldsymbol{q}, \boldsymbol{x}, y)$ be a vector of variables, where q_k and x_k are nominal and numeric explanatory variables, and y is a numeric dependent variable. For each q_k we introduce a *dummy variable* $q_{k\ell}$ defined as follows: $q_{k\ell} = 1$ if q_k matches the ℓ-th category, and $q_{k\ell} = 0$ otherwise. Here $\ell = 1, ..., L_k$, and L_k is the number of distinct categories appearing in q_k.

As a true model governing data, we consider the following set of *regression rules.* A regression rule represents a nominally conditioned polynomial.

$$if \bigwedge_{k} \bigvee_{q_{k\ell} \in Q_k^i} q_{k\ell} \qquad then\ \ y = \phi(\boldsymbol{x}; \boldsymbol{w}^i), \quad i = 1, ..., I^*. \tag{1}$$

Here Q_k^i and $\boldsymbol{w}^i$ denote a set of $q_{k\ell}$ and a weight vector respectively used in the i-th rule. As a regression function $\phi(\boldsymbol{x}; \boldsymbol{w}^i)$, we consider the following multivariate polynomial, whose power values are not restricted to integers. A weight vector $\boldsymbol{w}^i$ is composed of weights w_0^i, w_j^i and w_{jk}^i.

$$\phi(\boldsymbol{x}; \boldsymbol{w}^i) = w_0^i + \sum_{j=1}^{J^i} w_j^i \prod_{k=1}^{K_2} x_k^{w_{jk}^i} = w_0^i + \sum_{j=1}^{J^i} w_j^i \exp\left(\sum_{k=1}^{K_2} w_{jk}^i \ln x_k\right). \tag{2}$$

To express a nominal condition numerically, we introduce the following.

$$c(\boldsymbol{q}; \boldsymbol{v}^i) = \sigma\left(\sum_{k=1}^{K_1} \sum_{\ell=1}^{L_k} v_{k\ell}^i q_{k\ell}\right), \tag{3}$$

where $\boldsymbol{v}^i$ is composed of $v_{k\ell}^i$ and $\sigma(h) = 1/(1 + e^{-h})$. When we consider

$$v_{k\ell}^i = \begin{cases} \beta_2 & if\ q_{k\ell} \in Q_k^i, \\ -\beta_1 & if\ q_{k\ell} \notin Q_k^i\ and\ q_{k\ell'} \in Q_k^i\ for\ some\ \ell' \neq \ell, \\ 0 & if\ q_{k\ell'} \notin Q_k^i\ for\ any\ \ell', \end{cases} \tag{4}$$

where $\beta_1 \gg \beta_2 \gg 0$, we see the numeric function $c(\boldsymbol{q}; \boldsymbol{v}^i)$ can approximate well the truth value (1 or 0) of the nominal condition of the i-th rule.

Hence a total set of regression rules shown in Eq. (1) can be represented by a single numeric function shown in Eq. (5). The function can be learned by using a single *four-layer perceptron* as shown in Fig. 1. Here $\boldsymbol{\theta}$ is composed of weights v_{0r}, v_{jr}, $v_{rk\ell}$, and w_{jk}. How to find the optimal numbers of hidden units, J^* and R^*, will be described in Section 4.

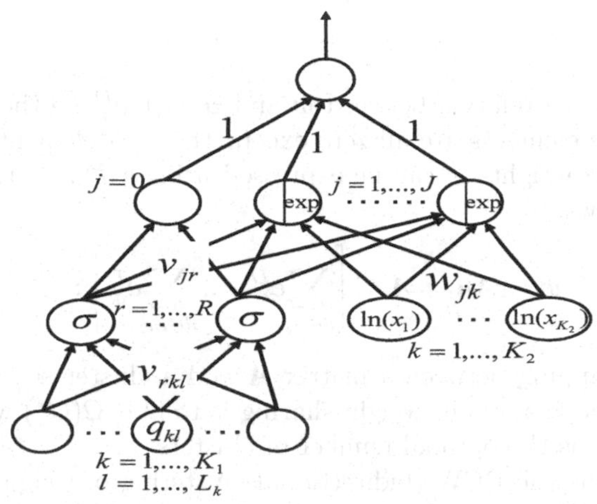

Fig. 1. Four-layer perceptron for RF6.4

$$f(\boldsymbol{q}, \boldsymbol{x}; \boldsymbol{\theta}) = c_0 + \sum_{j=1}^{J} c_j \exp\left(\sum_{k=1}^{K_2} w_{jk} \ln x_k\right), \tag{5}$$

$$c_0 = \sum_{r=1}^{R} v_{0r}\sigma_r, \quad c_j = \sum_{r=1}^{R} v_{jr}\sigma_r, \quad \sigma_r = \sigma\left(\sum_{k=1}^{K_1}\sum_{\ell=1}^{L_k} v_{rk\ell}\, q_{k\ell}\right), \tag{6}$$

Rule Restoring. Rule restoring [9] extracts a set of regression rules from a learned four-layer perceptron. It goes in two steps. In the first step, coefficient vectors $\boldsymbol{c}^\mu = (c_0^\mu, c_1^\mu, ..., c_{J^*}^\mu)$ for all the data points $\mu = 1, ..., N$ are quantized into I representatives $\{\boldsymbol{a}^i = (a_0^i, a_1^i, ..., a_{J^*}^i) : i = 1, ..., I\}$. For vector quantization (VQ) we employ the K-means [6] due to its simplicity, to obtain such I disjoint subsets $\{G_i : i = 1, ..., I\}$ that the distortion d_{VQ} is minimized. How to find the optimal number of regression rules, I^*, will be mentioned in Section 4.

$$d_{VQ} = \sum_{i=1}^{I} \sum_{\mu \in G_i} \|\boldsymbol{c}^\mu - \boldsymbol{a}^i\|^2, \qquad \boldsymbol{a}^i = \frac{1}{N_i} \sum_{\mu \in G_i} \boldsymbol{c}^\mu. \tag{7}$$

In the second step, the final rules are obtained by solving a simple classification problem whose training data are $\{(\boldsymbol{q}^\mu, i(\boldsymbol{q}^\mu)) : \mu = 1, ..., N\}$, where $i(\boldsymbol{q}^\mu)$ indicates the representative label of the μ-th data point. Here we employ the C4.5 decision tree generation program [8] due to its wide availability.

3 Weight Sharing Method: BCW

Basic Framework. Let $E(\boldsymbol{w})$ be an error function to minimize, where $\boldsymbol{w}$ denotes a vector of weights $(w_1, ..., w_d, ..., w_D)$. Then, we define a set of disjoint clusters $\Omega(G) = \{S_1, ..., S_g, ..., S_G\}$, where S_g denotes a set of weight numbers such that $S_1 \cup ... \cup S_G = \{1, ..., D\}$. Also, we define a vector of common weights $\boldsymbol{u} = (u_1, ..., u_g, ..., u_G)^T$ associated with a cluster set $\Omega(G)$ such that $w_d = u_g$ if $d \in S_g$. Let $\widehat{\boldsymbol{u}}$ be a vector obtained by training a neural network whose structure is defined by $\Omega(G)$.

Now we consider a relation between $\boldsymbol{w}$ and $\boldsymbol{u}$. Let $\boldsymbol{e}_d^D$ be the D-dimensional unit vector whose elements are all zero except the d-th element which is equal to unity. Then the weights $\boldsymbol{w}$ can be expressed using a $D \times G$ transformational matrix $\boldsymbol{A}$ as follows.

$$\boldsymbol{w} = \boldsymbol{A}\boldsymbol{u}, \quad \boldsymbol{A} = \left[\sum_{d \in S_1} \boldsymbol{e}_d^D, ..., \sum_{d \in S_G} \boldsymbol{e}_d^D \right]. \tag{8}$$

Note that the mapping between a matrix $\boldsymbol{A}$ and a cluster set $\Omega(G)$ is one-to-one. Therefore, the goal of our weight sharing is to find $\Omega(G^*)$ which minimizes $E(\boldsymbol{A}\boldsymbol{u})$, where G^* is the optimal number of clusters.

Below we outline the BCW (bidirectional clustering of weights) [13]. Since a weight sharing problem may have many local optima, BCW repeats merge and split operations in turn to find a globally excellent solution.

Bottom-up Clustering. A one-step bottom-up clustering transforms $\Omega(G)$ into $\Omega(G-1)$ by a merge operation; i.e., clusters S_g and $S_{g'}$ are merged into a cluster $\widetilde{S}_g = S_g \cup S_{g'}$. We want to select a suitable pair so as to minimize the increase of $E(\boldsymbol{w})$ as defined below. Here $\boldsymbol{H}(\boldsymbol{w})$ denotes the Hessian of $E(\boldsymbol{w})$.

$$DisSim(S_g, S_{g'}) = \frac{(\widehat{u}_g - \widehat{u}_{g'})^2}{(\boldsymbol{e}_g^G - \boldsymbol{e}_{g'}^G)^T (\boldsymbol{A}^T \boldsymbol{H}(\widehat{\boldsymbol{w}})\boldsymbol{A})^{-1} (\boldsymbol{e}_g^G - \boldsymbol{e}_{g'}^G)}. \tag{9}$$

This is the second-order criterion for merging S_g and $S_{g'}$, called the *dissimilarity*. We select a pair of clusters which minimizes $DisSim(S_g, S_{g'})$ and merge the two clusters. After the merge, the network with $\Omega(G-1)$ is retrained. This is the *one-step bottom-up clustering with retraining*.

Top-down Clustering. A one-step top-down clustering transforms $\Omega(G)$ into $\Omega(G+1)$ by a split operation; i.e., a cluster S_g is split into two clusters S_g' and S_{G+1} where $S_g = S_g' \cup S_{G+1}$. In this case, we want to select a suitable cluster and its partition so as to maximize the decrease of the error function.

Just after the splitting, we have a $(G+1)$-dimensional common weight vector $\widetilde{\boldsymbol{v}} = (\widehat{\boldsymbol{u}}^T, \hat{u}_g)^T$, and a new $D \times (G+1)$ transformational matrix $\boldsymbol{B}$ defined as

$$\boldsymbol{B} = \left[\sum_{d\in S_1} \boldsymbol{e}_d^D, \ldots, \sum_{d\in S_g'} \boldsymbol{e}_d^D, \ldots, \sum_{d\in S_G} \boldsymbol{e}_d^D, \sum_{d\in S_{G+1}} \boldsymbol{e}_d^D \right]. \tag{10}$$

Then, we define the general *utility* as follows. The utility values will be positive, and the larger the better. Here $\boldsymbol{g}(\boldsymbol{w})$ denotes the gradient of $E(\boldsymbol{w})$.

$$GenUtil(S_g, S_{G+1}) = \kappa^2 \boldsymbol{f}^T (\boldsymbol{B}^T \boldsymbol{H}(\boldsymbol{B}\widetilde{\boldsymbol{v}})\boldsymbol{B})^{-1} \boldsymbol{f}. \tag{11}$$

$$\kappa = \boldsymbol{g}(\boldsymbol{B}\widetilde{\boldsymbol{v}})^T \sum_{d\in S_{G+1}} \boldsymbol{e}_d^D, \quad \boldsymbol{f} = \boldsymbol{e}_{G+1}^{G+1} - \boldsymbol{e}_g^{G+1}, \tag{12}$$

When a cluster g to split is unchanged, $\boldsymbol{f}^T(\boldsymbol{B}^T\boldsymbol{H}(\boldsymbol{B}\widetilde{\boldsymbol{v}})\boldsymbol{B})^{-1}\boldsymbol{f}$ won't be significantly changed. Since κ is the summation of gradients over the members of a cluster $G+1$, the gradients to add together should have the same sign if you want a larger κ^2. Thus, the gradients of the cluster are sorted in ascending order and examined is only splitting into smaller-gradients and larger-gradients. Examining all such candidates, we select the cluster to split and its splitting which maximize the criterion (11). After the splitting, the network with $\Omega(G+1)$ is retrained. This is the *one-step top-down clustering with retraining.*

BCW Procedure. The procedure of BCW is shown below, where h denotes the width of search. It always converges since the number of different $\boldsymbol{A}$ is finite.

Step 1: Get the initial set $\Omega(D)$ through learning. Perform scalar quantization for $\Omega(D)$ to get $\Omega_1(2)$. Remember the matrix $\boldsymbol{A}^{(0)}$ at $\Omega_1(2)$. $t \leftarrow 1$.

Step 2: Perform repeatedly the one-step top-down clustering with retraining from $\Omega_1(2)$ to $\Omega(2+h)$. Update the best performance for each G if necessary.

Step 3: Perform repeatedly the one-step bottom-up clustering with retraining from $\Omega(2+h)$ to $\Omega_2(2)$. Update the best performance for each G if necessary. Remember $\boldsymbol{A}^{(t)}$ at $\Omega_2(2)$.

Step 4: If $\boldsymbol{A}^{(t)}$ is equal to one of the previous ones $\boldsymbol{A}^{(t-1)}, \ldots, \boldsymbol{A}^{(0)}$, then stop and output the best performance of $\Omega(G)$ for each G as the final result. Otherwise, $t \leftarrow t+1$, $\Omega_1(2) \leftarrow \Omega_2(2)$ and go to step 2.

4 Applying BCW to Connectionist Polynomial Regression Using BIC

This section explains how to apply BCW to connectionist polynomial regression RF6.4 together with the following model selection. Given data, we don't know in advance the optimal numbers J^* and R^* of hidden units, the optimal number I^* of regression rules, or the optimal number G^* of BCW clusters. Therefore, we need criteria suitable for selecting them. As such criteria, we employ the

Bayesian Information Criterion (BIC) [10]. Compared to other means such as cross-validation or bootstrap method which need heavy repetitive learning, BIC can select the optimal model in much less time because we need only one learning for each value of J, R, I, and G.

The whole procedure of **RF6.4+BCW+BIC** goes as follows.

(step 1). Train the four-layer perceptron without weight sharing for each combination of $J = 1, 2, ...$ and $R = 1, 2, ...$, and select the J and R which minimizes the following equation as J^* and R^*. Here $\widehat{\boldsymbol{\theta}}_{J,R}$ denotes an estimate for the model having J and R hidden units. Since M is the number of weights, we have $M = JK_2 + R(K_{all} + J + 1)$, where K_{all} is the total number of dummy variables.

$$\mathrm{BIC}(J, R) = \frac{N}{2} \log \left(\frac{1}{N} \sum_{\mu=1}^{N} \left(f(\boldsymbol{x}^\mu; \widehat{\boldsymbol{\theta}}_{J,R}) - y^\mu \right)^2 \right) + \frac{M}{2} \log N \tag{13}$$

(step 2). Train the perceptron of J^* and R^* with weight sharing BCW, and select the G which minimizes the following equation as G^*. Here $\widehat{\boldsymbol{\theta}}_G$ denotes an estimate for the model having G clusters.

$$\mathrm{BIC}(G) = \frac{N}{2} \log \left(\frac{1}{N} \sum_{\mu=1}^{N} \left(f(\boldsymbol{x}^\mu; \widehat{\boldsymbol{\theta}}_G) - y^\mu \right)^2 \right) + \frac{G}{2} \log N \tag{14}$$

(step 3). Train the perceptron of J^* and R^* under the condition of G^* clusters with a near-zero common weight pruned.

(step 4). Quantize the coefficient vectors using the K-means algorithm for each $I = 1, 2, ...$, and select the I which minimizes the following equation as I^*. As for BIC(I), we adopt Pelleg's X-means approach [7].

$$\mathrm{BIC}(I) = -\sum_{i=1}^{I} N_i \log N_i + \frac{N}{2} \log |\widehat{\boldsymbol{\Sigma}}| + \frac{1}{2} \sum_{\mu=1}^{N} (\boldsymbol{c}^\mu - \boldsymbol{a}^{(\mu)})^T \widehat{\boldsymbol{\Sigma}}^{-1} (\boldsymbol{c}^\mu - \boldsymbol{a}^{(\mu)})$$

$$+ \frac{I(J^* + 2)}{2} \log N, \text{ where } \widehat{\boldsymbol{\Sigma}} = \frac{1}{N - I} \sum_{i=1}^{I} \sum_{\mu \in G_i} (\boldsymbol{c}^\mu - \boldsymbol{a}^{(\mu)})(\boldsymbol{c}^\mu - \boldsymbol{a}^{(\mu)})^T \tag{15}$$

(step 5). Solve the classification problem using the C4.5 program [8] to get the final result.

5 Experimental Evaluation

Artificial Data. We consider the following set of regression rules.

$$\begin{cases} \text{if } q_{21} \wedge q_{31} \text{ then } & y = 1 + 2x_1^2 x_2 x_3^{0.5} + 3x_2 x_3^{0.5} x_4^2 x_5, \\ \text{if } q_{22} \wedge q_{31} \text{ then } & y = 2 - x_1^2 x_2 x_3^{0.5} + x_2 x_3^{0.5} x_4^2 x_5, \\ \text{if } (q_{12} \vee q_{13}) \wedge q_{32} \text{ then } & y = -3 + x_1^2 x_2 x_3^{0.5} - x_2 x_3^{0.5} x_4^2 x_5, \\ \text{if } q_{11} \wedge q_{32} \text{ then } & y = 2 + 3x_1^2 x_2 x_3^{0.5} + 4x_2 x_3^{0.5} x_4^2 x_5, \end{cases} \tag{16}$$

Here we have 15 numeric and 4 nominal explanatory variables with L_1=3, L_2 = L_3=2, and L_4=5. Note that variables $q_4, x_6, ..., x_{15}$ are irrelevant. For each data point, values of x_k and q_k were randomly taken from the range (1, 2) and from its categories respectively, while the corresponding y was calculated using Eq. (16) with Gaussian noise $\mathcal{N}(0, 0.3)$ added. Training data size was 500 (N=500).

The initial weight values were randomly generated from the range $(-1, +1)$. The learning was terminated when each element of the gradient got less than 10^{-5}. BCW was applied only to weights w_{jk} since their simplicity is directly linked to the readability. We set the width of BCW as h=10. The numbers of hidden units were changed from one (J=1, R=1) to five (J=5, R=5), and the number of rules was change from one (I=1) to ten (I=10). Note that J^*=2, G^*=4, and I^*=4 for our artificial data.

Table 1 compares the frequency with which BIC(J, R) was minimized out of 100 runs. From the table we have J^*=2 and R^*=3. For any run we have J^*=2, which is correct. As for R, most runs supported R=3, while 14 runs supported R=4, which is no problem since we can restore the original rules even if R=4.

Table 1. Frequency with which BIC(J, R) is minimized (artificial data)

model	J=1	J=2	J=3	J=4	J=5
R=1	0	0	0	0	0
R=2	0	0	0	0	0
R=3	0	**86**	0	0	0
R=4	0	14	0	0	0
R=5	0	0	0	0	0

Table 2 compares the frequency with which BIC(G) was minimized out of 100 runs. Note that each run was performed under the model where J^*=2 and R^*=3. From the table we can select the optimal number of common weights, G^*=4, which is correct.

Table 2. Frequency with which BIC(G) is minimized (artificial data)

G	**2**	**3**	**4**	**5**	**6**	**7**	**8**	**9**	**10**	**11**	**12**	total
Freq.	0	0	**83**	12	5	0	0	0	0	0	0	100

Figure 2 shows how training error MSE and BIC(G) changed through BCW learning under the model where J^*=2 and R^*=3. The bidirectional clustering was repeated twice until convergence. The figure shows G^*=4 since BIC(G) was minimized at G=4, which is located at iteration No. 19.

At this point we have J^*=2, R^*=3, and G^*=4. Then, we pruned the near-zero common weight, and retrained the network. Table 3 compares the frequency with which BIC(I) was minimized out of 100 K-means runs. From the table we have the optimal number of rules, I^*=4, which is correct.

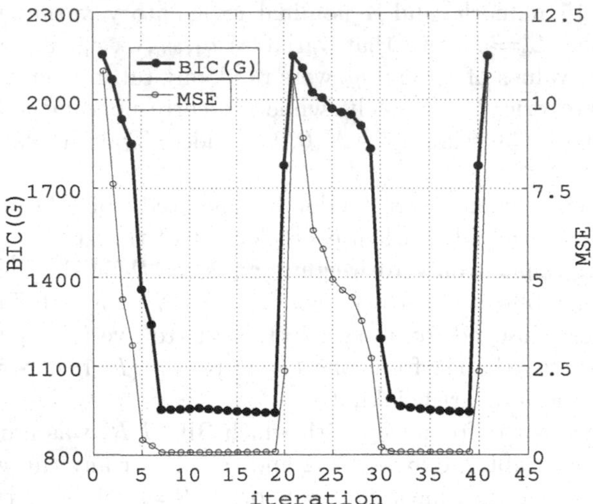

Fig. 2. Bidirectional clustering for artificial data

Table 3. Frequency with which BIC(I) is minimized (artificial data)

I	**1**	**2**	**3**	**4**	**5**	**6**	**7**	**8**	**9**	**10**	total
Freq.	0	0	0	**82**	10	5	0	0	0	3	100

By applying the C4.5 program, we have the following set, almost equivalent to the original Eq. (16). Note that the rule set is free from any irrelevant variables. The CPU time required for RF6.4+BCW+BIC was about 11 min.

$$\begin{cases} \text{if} \;\; q_{21} \wedge q_{31} \;\; \text{then} \\ \quad y = 1.008 + 1.982 x_1^{2.011} x_2^{1.052} x_3^{0.498} + 2.983 x_2^{1.052} x_3^{0.498} x_4^{2.011} x_5^{1.052} \\ \text{if} \;\; q_{22} \wedge q_{31} \;\; \text{then} \\ \quad y = 2.012 - 1.000 x_1^{2.011} x_2^{1.052} x_3^{0.498} + 0.970 x_2^{1.052} x_3^{0.498} x_4^{2.011} x_5^{1.052} \\ \text{if} \;\; (q_{12} \vee q_{13}) \wedge q_{32} \;\; \text{then} \\ \quad y = -2.980 + 1.128 x_1^{2.011} x_2^{1.052} x_3^{0.498} - 0.991 x_2^{1.052} x_3^{0.498} x_4^{2.011} x_5^{1.052} \\ \text{if} \;\; q_{11} \wedge q_{32} \;\; \text{then} \\ \quad y = 1.975 + 3.065 x_1^{2.011} x_2^{1.052} x_3^{0.498} + 4.042 x_2^{1.052} x_3^{0.498} x_4^{2.011} x_5^{1.052} \end{cases}$$

Real Data Sets. We evaluated the performance of RF6.4+BCW+BIC using 10 real data sets[1]. Table 4 shows specs of the data sets, and the optimal model parameters obtained using RF6.4+BCW+BIC. Here K_2 and K_{all} denote the

[1] Cpu and Boston data sets were taken from UCI Repository of ML Databases. College, cholesterol, b-carotene, mpg, lung-cancer and wage data sets were from StatLib. Yokohama data set was from Web page of Kanagawa Pref.,Japan. Baseball data set was taken from the directory of Japanese professional baseball players in 2006.

total numbers of numeric explanatory variables and dummy variables respectively. Before processing, all numeric variables are normalized as follows.

$$\widetilde{x}_k = \frac{x_k}{max(x_k)}, \quad \widetilde{y} = \frac{y - mean(y)}{std(y)} \tag{17}$$

The initial weight values and stopping criteria are set in the same way as described previously. We varied model parameters as follows: J=1,..,5, R=1,...,5, G=2,...,12, and I=1,...,10. These ranges were determined considering computational cost and our earlier experiments. These model parameters were optimized sequentially in the same way as described before. Table 4 shows we have a small number of common weights implying the succinctness of the resulting rules.

Table 4. Specs and optimal parameters (real data sets)

data set	contents	N	K_2	K_{all}	J^*	R^*	G^*	I^*
cpu	performance of cpu	205	6	26	3	3	6	6
Boston	housing prices in Boston	486	12	76	2	1	12	10
college	instructional expenditure of colleges	1,129	11	51	2	2	10	9
cholesterol	amount of cholesterol	297	5	13	2	3	6	8
b-carotene	amount of beta-carotene	315	10	6	1	4	8	6
mpg	fuel cost of cars	388	5	28	2	3	6	5
lung-cancer	survival time of lung cancer patients	126	3	6	3	5	4	4
wage	wages of workers per hour	534	2	14	4	3	5	7
Yokohama	housing prices in Yokohama	558	4	23	5	2	6	5
baseball	annual salaries of baseball players	219	10	14	3	4	8	6

We compared the generalization performance of RF6.4+BCW+BIC with other regression methods. The performance was measured by 10-fold cross-validation. As other methods we considered the following five: linear multiple

Table 5. Comparison of cross-validation errors (real data sets)

data set	**MR**	**QT**	**HME**	**MLP**	**RF6.4**	**RF6.4+BCW**
cpu	4.609E+3	4.301E+3	1.453E+3	1.566E+3	2.021E+3	**1.363E+3**
Boston	2.355E+1	1.350E+1	1.209E+1	9.450E+0	9.262E+0	**7.533E+0**
college	9.570E+6	9.703E+6	8.661E+6	6.910E+6	**6.330E+6**	8.283E+6
cholesterol	2.641E+3	2.584E+3	2.502E+3	2.577E+3	2.446E+3	**2.418E+3**
b-carotene	2.989E+4	2.928E+4	2.895E+4	2.893E+4	2.958E+4	**2.646E+4**
mpg	1.809E+1	1.622E+1	1.392E+1	1.373E+1	**1.283E+1**	1.495E+1
lung-cancer	2.284E+4	2.136E+4	2.154E+4	2.156E+4	2.172E+4	**1.857E+4**
wage	2.124E+1	1.884E+1	1.925E+1	1.864E+1	1.849E+1	**1.830E+1**
Yokohama	1.232E+9	4.151E+8	3.790E+8	**3.462E+8**	3.911E+8	4.667E+8
baseball	1.755E+7	1.849E+7	**1.406E+7**	1.670E+7	1.732E+7	1.620E+7

regression (MR), linear quantification theory type 1 (QT) [14], HME [5], multi-layer perceptron (MLP), RF6.4 without BCW.

Table 5 compares the generalization performances. RF6.4+BCW+BIC showed the best performance for 6 data sets out of 10. When we compare its performance with that of plain RF6.4, we see BCW+BIC surely contributed to the improvement in generalization.

6 Conclusion

We applied weight sharing BCW to connectionist polynomial regression using BIC for model selection. In our experiments the combined method finds a set of very succinct polynomials having excellent generalization.

References

1. Falkenhainer, B.C., Michalski, R.S.: Integrating quantitative and qualitative discovery in the ABACUS system. Machine Learning 3, 153–190 (1990)
2. Hassibi, B., Stork, D.G., Wolf, G.: Optimal brain surgeon and general network pruning. In: Proc. IEEE Int. Conf. on Neural Networks 1992, pp. 293–299 (1992)
3. Haykin, S.: Neural Networks, 2nd edn. Prentice-Hall, Englewood Cliffs (1999)
4. Ivakhnenko, A.G.: Polynomial theory of complex systems. IEEE Trans. Syst. Man Cybern SMC-1(4), 364–378 (1971)
5. Jordan, M.I., Jacobs, R.A.: Hierarchical mixtures of experts and the EM algorithms. Neural Computation 6(2), 181–214 (1994)
6. Lloyd, S.P.: Least squares quantization in PCM. IEEE Trans. Information Theory 28(2), 129–137 (1982)
7. Pelleg, D., Moore, A.: X-means: extracting K-means with efficient estimation of the number of clusters. In: Proc. 17th Int. Conf. on Machine Learning, pp. 727–734 (2000)
8. Quinlan, J.R.: C4.5: Programs for Machine Learning. Morgan Kaufmann, San Francisco (1993)
9. Saito, K., Nakano, R.: Extracting regression rules from neural networks. Neural Networks 15(10), 1279–1288 (2002)
10. Schwarz, G.: Estimating the dimension of a model. Annals of Statistics 6, 461–464 (1987)
11. Setiono, R., et al.: Extraction of rules from artificial neural networks for nonlinear regression. IEEE Trans. Neural Networks 13(3), 564–577 (2002)
12. Tanahashi, Y., Saito, K., Nakano, R.: Piecewise multivariate polynomials using a four-layer perceptron. In: Negoita, M.G., Howlett, R.J., Jain, L.C. (eds.) KES 2004. LNCS (LNAI), vol. 3214, pp. 602–608. Springer, Heidelberg (2004)
13. Tanahashi, Y., Saito, K., Nakano, R.: Model selection and weight sharing of multi-layer perceptrons. In: Khosla, R., Howlett, R.J., Jain, L.C. (eds.) KES 2005. LNCS (LNAI), vol. 3684, pp. 716–722. Springer, Heidelberg (2005)
14. Tanaka, Y.: Review of the methods of quantification. Environmental Health Perspectives 32, 113–123 (1979)

Recognition of Properties by Probabilistic Neural Networks

Jiří Grim[1] and Jan Hora[2]

[1] Institute of Information Theory and Automation
P.O. BOX 18, 18208 PRAGUE 8, Czech Republic
[2] Faculty of Nuclear Science and Physical Engineering
Trojanova 13, CZ-120 00 Prague 2, Czech Republic
grim@utia.cas.cz, hora@utia.cas.cz

Abstract. The statistical pattern recognition based on Bayes formula implies the concept of mutually exclusive classes. This assumption is not applicable when we have to identify some non-exclusive properties and therefore it is unnatural in biological neural networks. Considering the framework of probabilistic neural networks we propose statistical identification of non-exclusive properties by using one-class classifiers.

Keywords: Probabilistic neural networks, Non-exclusive classes, One-class classifiers, Biological compatibility.

1 Introduction

The statistical approach is known to enable general and theoretically well justified decision making in pattern recognition. Given the probabilistic description of the problem in terms of class-conditional probability distributions, we can classify objects described by discrete or continuous variables. The Bayes formula provides full classification information in terms of *a posteriori* probabilities of a finite number of classes. A unique decision, if desirable, can be obtained by means of Bayes decision function which minimizes the probability of error. We recall that the classification information contained in the *a posteriori* probabilities is partly lost if only a unique decision is available [5].

On the other hand, introducing Bayes formula, we assume that the unconditional distribution of the recognized data vectors can be expressed as a weighted sum of class-conditional distributions, according to the formula of complete probability. In this way we implicitly assume the classes to be mutually exclusive. Nevertheless, the probabilistic classes may overlap in the sample space, they are mutually exclusive just in the sense of the complete probability formula.

The abstract statistical concept of mutually exclusive classes is rather unnatural in biological systems since most real life categories are non-exclusive. In this respect the multiclass Bayes decision scheme is unsuitable as a theoretical background of neural network models. In the following we use the term *property* to emphasize the fact that the recognized object may have several different properties simultaneously. In order to avoid the strict assumption of mutually

C. Alippi et al. (Eds.): ICANN 2009, Part II, LNCS 5769, pp. 165–174, 2009.

exclusive classes we propose recognition of properties by probabilistic neural networks based on one-class classifiers. We assume that for each property there is a single training data set. To identify a property we evaluate the log-likelihood ratio of the related conditional probability distribution and of the product of unconditional univariate marginals. Hence the only information about alternative properties is assumed to be given in the form of global marginal distributions of all involved variables. The proposed recognition of properties has two qualitative advantages from the point of view of biological compatibility: a) it is applicable both to the non-exclusive and exclusive properties (cf. Sec. 5) and b) provides a unified approach to recognition of properties and feature extraction (cf. [4]).

The concept of probabilistic neural networks (PNNs) relates to the early work of Specht [10] who proposed a neural network model closely related to the non-parametric Parzen estimates of probability density functions. In comparison with other neural network models the PNN of Specht may save training time essentially but, according to Parzen formula, one neuron is required for each training pattern. Moreover, there is a crucial problem of the optimal smoothing of Parzen estimates in multidimensional spaces. The PNN approach of Specht has been modified by other authors and, in some cases, simplified by introducing finite mixtures [9]. In this paper we refer mainly to our results on PNNs published in the last years (cf. [3] - [6]). Unlike previous authors we approximate the class-conditional probability distributions by finite mixtures of product components. The product-mixture-based PNNs do not provide a new technique of pattern recognition but they may contribute to better understanding of the functional principles of biological neural networks [3], [4].

In the following we first discuss the theoretical differences between multiclass classifiers (Sec. 2) and one-class identification of properties (Sec. 3). In Sec. 4 we summarize basic features of probabilistic neural networks and their application to identification of properties. In Sec. 5 we compare both schemes in application to recognition of handwritten numerals.

2 Multiclass Bayes Decision Scheme

Considering the statistical pattern recognition we assume that some multivariate observations have to be classified with respect to a finite set of mutually exclusive classes $\Omega = \{\omega_1, \ldots, \omega_K\}$. The observation vectors $\boldsymbol{x} = (x_1, x_2, \ldots, x_N) \in \mathcal{X}$ from the N-dimensional space $\mathcal{X}$ (which may be real, discrete or binary) are supposed to occur randomly according to some class-conditional distributions $P(\boldsymbol{x}|\omega)$ with *a priori* probabilities $p(\omega), \omega \in \Omega$. Recall that, given an observation $\boldsymbol{x} \in \mathcal{X}$, all statistical information about the set of classes Ω is expressed by the Bayes formula for a posteriori probabilities

$$p(\omega|\boldsymbol{x}) = \frac{P(\boldsymbol{x}|\omega)p(\omega)}{P(\boldsymbol{x})}, \quad P(\boldsymbol{x}) = \sum_{\omega \in \Omega} P(\boldsymbol{x}|\omega)p(\omega), \quad \boldsymbol{x} \in \mathcal{X} \tag{1}$$

where $P(\boldsymbol{x})$ is the joint unconditional probability distribution of $\boldsymbol{x}$.

The posterior distribution $p(\omega|\boldsymbol{x})$ may be used to define a unique final decision by means of the Bayes decision function [1]

$$d : \mathcal{X} \to \Omega, \qquad d(\boldsymbol{x}) = \arg\max_{\omega \in \Omega}\{p(\omega|\boldsymbol{x})\}, \quad \boldsymbol{x} \in \mathcal{X} \tag{2}$$

which is known to minimize the probability of classification error.

Remark 2.1. The Bayes decision function is a typical multiclass classifier directly identifying one of a finite number of classes on output. A weak point of Bayes classification is the unknown probabilistic description to be estimated from training data. In multidimensional spaces the number of training samples is usually insufficient to estimate the underlying distributions reliably and, in case of large data sets, even the computational complexity may become prohibitive.

Alternatively there are numerous non-statistical methods like support vector machines, AdaBoost, back-propagation perceptron and others having proved to yield excellent results in different practical problems. Unlike Bayes formula they are typically based on complex separating surfaces suitable to distinguish between two classes. In such a case the multiclass problems have to be reduced to multiple binary problems. It is possible to construct individual binary classifier for each class (one-against-all approach), to distinguish each pair of classes (all-pairs method) [8], or to use a more general method of error-correcting output codes which can utilize binary classifiers for all possible partitions of the set of classes [2]. We recall that, by nature of the underlying separating planes, multiclass solutions based on binary classifiers are discrete and therefore the *a posteriori* probabilities, if desirable, have to be approximated by heuristic means. There is no exact relation to the probability of classification error, usually the learning algorithm minimizes some heuristic criterion (e.g. a margin-based loss function [1]). As it can be seen, from the point of view of "binary" approximating multiclass decision functions, the concept of properties is basically irrelevant. □

3 Identification of Properties

In case of non-exclusive classes we assume that the multivariate observations may have some properties from a finite set $\Theta = \{\theta_1, \ldots, \theta_K\}$. Considering a single property $\theta \in \Theta$ we are faced with a two-class (binary) decision problem. For any given sample $\boldsymbol{x} \in \mathcal{X}$ we have to decide if the property is present or not. In other words, the decision is *positive*, (θ) if the property has been identified and *negative*, $(\bar{\theta})$ if it has not been identified. Since both alternatives are mutually exclusive, we can solve the binary classification problem in a standard statistical way. In full generality we denote $p(\theta)$ the *a priori* probability that the property θ occurs and $p(\bar{\theta}) = 1 - p(\theta)$ denotes the complementary *a priori* probability that the property is missing. Analogously, we denote $P(\boldsymbol{x}|\theta)$ and $P(\boldsymbol{x}|\bar{\theta})$ the conditional probability distributions of $\boldsymbol{x} \in \mathcal{X}$ given the property θ and $\bar{\theta}$ respectively. Thus, given the probabilistic description of a binary problem $\{\theta, \bar{\theta}\}$ we can write

$$P(\boldsymbol{x}) = P(\boldsymbol{x}|\theta)p(\theta) + P(\boldsymbol{x}|\bar{\theta})p(\bar{\theta}), \quad \boldsymbol{x} \in \mathcal{X}, \quad (p(\bar{\theta}) = 1 - p(\theta)) \tag{3}$$

[1] Here and in the following we assume that possible ties are uniquely decided.

and by using Bayes formula

$$p(\theta|\boldsymbol{x}) = \frac{P(\boldsymbol{x}|\theta)p(\theta)}{P(\boldsymbol{x})}, \quad p(\bar{\theta}|\boldsymbol{x}) = \frac{P(\boldsymbol{x}|\bar{\theta})p(\bar{\theta})}{P(\boldsymbol{x})}. \tag{4}$$

we obtain the related decision function in the form

$$\Delta : \mathcal{X} \to \{\theta, \bar{\theta}\}, \quad \Delta(\boldsymbol{x}) = \begin{cases} \theta, & p(\theta|\boldsymbol{x}) \geq p(\bar{\theta}|\boldsymbol{x}), \\ \bar{\theta}, & p(\theta|\boldsymbol{x}) < p(\bar{\theta}|\boldsymbol{x}), \end{cases} \quad \boldsymbol{x} \in \mathcal{X}. \tag{5}$$

Remark 3.1. As mentioned earlier (cf. Remark 2.1), the problem of mutually exclusive classes can be formally decomposed into a set of "one-against-all" binary classification problems. If we define for each class $\omega_k \in \Omega$ the property $\theta = \{\omega_k\}$ and the opposite property, $\bar{\theta} = \Omega \setminus \{\omega_k\}$, then we can construct the two corresponding components $P(\boldsymbol{x}|\theta)p(\theta)$ and $P(\boldsymbol{x}|\bar{\theta})p(\bar{\theta})$ in terms of the class conditional distributions $P(\boldsymbol{x}|\omega)$. In particular, we can write

$$P(\boldsymbol{x}|\theta)p(\theta) = P(\boldsymbol{x}|\omega_k)p(\omega_k), \quad p(\theta) = p(\omega_k), \tag{6}$$

$$P(\boldsymbol{x}|\bar{\theta})p(\bar{\theta}) = \sum_{\omega \in \Omega_k} P(\boldsymbol{x}|\omega)p(\omega), \quad p(\bar{\theta}) = \sum_{\omega \in \Omega_k} p(\omega), \quad \Omega_k = \Omega \setminus \{\omega_k\}. \tag{7}$$

Expectedly, the classification accuracy of the multi-class decision function (2) may be different from that of the binary decision functions (5) based on the distributions (6) and (7). We discuss the problem in Sec. 4 in detail. □

Let us recall that by introducing binary classifiers we assume the training data to be available both for the property θ and for its opposite $\bar{\theta}$. Unfortunately, in real life situations it is often difficult to characterize the negative property $\bar{\theta}$ and to get the corresponding representative training data. In such cases the binary classifier (5) cannot be used since the probabilistic description of the negative property is missing. In this sense the identification of properties is more naturally related to one-class classifiers [11], [12] when only a single training data set for the "target" class is available.

Given some training data set $\mathcal{S}_\theta$ for a property $\theta \in \Theta$, we can estimate the conditional distribution $P(\boldsymbol{x}|\theta)$ and the property θ could be identified by simple thresholding. Nevertheless, the choice of a suitable threshold value is difficult if any information about the opposite property $\bar{\theta}$ is missing [12]. In the following we assume a general "background" information in the form of unconditional marginal distributions $P_n(x_n)$ of the variables x_n which is applicable to all properties $\theta \in \Theta$. This approach is motivated by the underlying PNN framework since the information about the unconditional marginal probabilities $P_n(x_n)$ may always be assumed to be available at the level of a single neuron.

In order to identify a property $\theta \in \Theta$ we propose to use a one-class-classifier condition based on the log-likelihood ratio

$$\pi_\theta(\boldsymbol{x}) = \log \frac{P(\boldsymbol{x}|\theta)}{\prod_{n \in \mathcal{N}} P_n(x_n)} \geq \epsilon. \tag{8}$$

Note that asymptotically the mean value of the criterion $\pi_\theta(\boldsymbol{x})$ converges to the Kullback-Leibler discrimination information between the two distributions $P^*(\boldsymbol{x}|\theta)$ and $\prod_{n\in\mathcal{N}} P_n(x_n)$

$$\bar{\pi}_\theta = \frac{1}{|\mathcal{S}_\theta|}\sum_{x\in\mathcal{S}_\theta}\log\frac{P(\boldsymbol{x}|\theta)}{\prod_{n\in\mathcal{N}} P_n(x_n)} \quad\rightarrow\quad \sum_{x\in\mathcal{X}} P^*(\boldsymbol{x}|\theta)\log\frac{P^*(\boldsymbol{x}|\theta)}{\prod_{n\in\mathcal{N}} P_n(x_n)}. \tag{9}$$

The last expression is nonnegative and for independent variables it is zero. In this sense it can be interpreted as a measure of dependence of the involved variables distributed by $P^*(\boldsymbol{x}|\theta)$ (cf. [3]).

We recall that the criterion $\pi_\theta(\boldsymbol{x})$ does not depend on the a priori probability $p(\theta)$. The property $\theta \in \Theta$ is identified if the probability $P(\boldsymbol{x}|\theta)$ is significantly higher than the corresponding product probability $\prod_{n\in\mathcal{N}} P_n(x_n)$. The threshold value ϵ in (8) can be related to the log-likelihood function of the estimated distribution $P(\boldsymbol{x}|\theta)$ (cf. (21)). Thus the only information about the negative properties $\bar{\theta}$ is contained in the unconditional product distribution $\prod_{n\in\mathcal{N}} P_n(x_n)$ which implies the assumption of independence of the variables x_n.

4 Probabilistic Neural Networks

Considering PNNs we approximate the class-conditional distributions $P(\boldsymbol{x}|\omega)$ by finite mixtures of product components

$$P(\boldsymbol{x}|\omega) = \sum_{m\in\mathcal{M}_\omega} F(\boldsymbol{x}|m)f(m), \quad F(\boldsymbol{x}|m) = \prod_{n\in\mathcal{N}} f_n(x_n|m). \tag{10}$$

Here $\mathcal{M}_\omega$ are the component index sets of different classes, $\mathcal{N} = \{1,\ldots,N\}$ is the index set of variables, $f(m)$ are probabilistic weights and $F(\boldsymbol{x}|m)$ are the products of component specific univariate distributions $f_n(x_n|m)$.

In order to avoid the biologically unnatural complete interconnection of neurons we have introduced the structural mixture model [5], [6]. In particular, considering binary variables $x_n \in \{0,1\}$, we define

$$F(\boldsymbol{x}|m) = F(\boldsymbol{x}|0)G(\boldsymbol{x}|m,\phi_m)f(m), \; m \in \mathcal{M}_\omega \tag{11}$$

where $F(\boldsymbol{x}|0)$ is a "background" probability distribution defined as a fixed product of global marginals

$$F(\boldsymbol{x}|0) = \prod_{n\in\mathcal{N}} f_n(x_n|0) = \prod_{n\in\mathcal{N}} \vartheta_{0n}^{x_n}(1-\vartheta_{0n})^{1-x_n}, \quad (\vartheta_{0n} = \mathcal{P}\{x_n = 1\}) \tag{12}$$

and the component functions $G(\boldsymbol{x}|m,\phi_m)$ include additional binary structural parameters $\phi_{mn} \in \{0,1\}$

$$G(\boldsymbol{x}|m,\phi_m) = \prod_{n\in\mathcal{N}}\left[\frac{f_n(x_n|m)}{f_n(x_n|0)}\right]^{\phi_{mn}} = \prod_{n\in\mathcal{N}}\left[\left(\frac{\vartheta_{mn}}{\vartheta_{0n}}\right)^{x_n}\left(\frac{1-\vartheta_{mn}}{1-\vartheta_{0n}}\right)^{1-x_n}\right]^{\phi_{mn}}. \tag{13}$$

The main advantage of the structural mixture model is the possibility to confine the decision making only to "relevant" variables. Making substitution (11) in (6), we can express the probability distributions $P(\boldsymbol{x}|\omega), P(\boldsymbol{x})$ in the form

$$P(\boldsymbol{x}|\omega) = \sum_{m\in\mathcal{M}_\omega} F(\boldsymbol{x}|m)f(m) = F(\boldsymbol{x}|0) \sum_{m\in\mathcal{M}_\omega} G(\boldsymbol{x}|m,\phi_m)f(m), \tag{14}$$

$$P(\boldsymbol{x}) = \sum_{\omega\in\Omega} p(\omega)P(\boldsymbol{x}|\omega) = F(\boldsymbol{x}|0)\sum_{m\in\mathcal{M}} G(\boldsymbol{x}|m,\phi_m)w_m, \quad w_m = p(\omega)f(m).$$

As the background distribution $F(\boldsymbol{x}|0)$ cancels in the Bayes formula we obtain

$$p(\omega|\boldsymbol{x}) = \frac{\sum_{m\in\mathcal{M}_\omega} G(\boldsymbol{x}|m,\phi_m)w_m}{\sum_{j\in\mathcal{M}} G(\boldsymbol{x}|j,\phi_j)w_j} = \sum_{m\in\mathcal{M}_\omega} q(m|\boldsymbol{x}), \quad \omega\in\Omega. \tag{15}$$

$$q(m|\boldsymbol{x}) = \frac{w_m G(\boldsymbol{x}|m,\phi_m)}{\sum_{j\in\mathcal{M}} w_j G(\boldsymbol{x}|j,\phi_j)}, \quad \boldsymbol{x}\in\mathcal{X}. \tag{16}$$

Thus the posterior probability $p(\omega|\boldsymbol{x})$ becomes proportional to a weighted sum of the component functions $G(\boldsymbol{x}|m,\phi_m)$ each of which can be defined on a different subspace. In other words the input connections of a neuron can be confined to an arbitrary subset of input nodes. The structural mixtures (14) can be optimized by means of EM algorithm in full generality (cf. [3] -[6]).

In view of Eq. (15) the structural mixture model provides a statistically correct subspace approach to Bayesian decision-making. In particular, considering Eq. (15), we can write the decision function (2) equivalently in the form

$$d(\boldsymbol{x}) = \omega_k : \quad \sum_{m\in\mathcal{M}_{\omega_k}} q(m|\boldsymbol{x}) \geq \sum_{m\in\mathcal{M}_\omega} q(m|\boldsymbol{x}), \ \forall\omega\in\Omega_k, \ \boldsymbol{x}\in\mathcal{X}. \tag{17}$$

Remark 4.1. Applying binary classifier (5) to the multiclass problem of Remark 3.1, we can write (cf. (4), (6), (7))

$$\Delta(\boldsymbol{x}) = \{\omega_k\} : \quad p(\omega_k|\boldsymbol{x}) \geq p(\Omega_k|\boldsymbol{x}) = 1 - p(\omega_k|\boldsymbol{x}), \ \ \boldsymbol{x}\in\mathcal{X} \tag{18}$$

and, after substitution (15), we obtain the following equivalent form of Eq. (18)

$$\Delta(\boldsymbol{x}) = \{\omega_k\} : \quad \sum_{m\in\mathcal{M}_{\omega_k}} q(m|\boldsymbol{x}) \geq \frac{1}{2}, \ \ \boldsymbol{x}\in\mathcal{X}. \tag{19}$$

Condition (19) is stronger than (17) and, unlike the multiclass decision function (17), it may happen that no class will be identified by the binary classifiers (18) for a given $\boldsymbol{x}\in\mathcal{X}$. However, the two different decision functions (17) and (19) will perform comparably in multidimensional problems. In high dimensional spaces the mixture components $F(\boldsymbol{x}|m)$ in (14) are almost non-overlapping and therefore the conditional weights $q(m|\boldsymbol{x})$ have nearly binary properties by taking values near zero or one. It can be seen that if for some $m_0\in\mathcal{M}_{\omega_k}$ the value

$q(m_0|\boldsymbol{x})$ is near to one then both the multiclass and binary classifiers (17) and (19) will decide equally $d(\boldsymbol{x}) = \Delta(\boldsymbol{x}) = \omega_k$. In numerical experiments we have obtained false positive and false negative frequencies differing in both schemes in several units only. □

The structural mixture model (14) is particularly useful to identify the properties by means of the one-class-classifier condition (8). In view of definition of the background distribution $F(\boldsymbol{x}|0)$ and considering the properties $\theta = \{\omega\}, \omega \in \Omega$ defined by numerals, we can write (cf. (8), (12), (14)):

$$\pi_\omega(\boldsymbol{x}) = \log \left[\frac{P(\boldsymbol{x}|\omega)}{\prod_{n \in \mathcal{N}} P_n(x_n)} \right] = \log \sum_{m \in \mathcal{M}_\omega} G(\boldsymbol{x}|m, \phi_m) f(m) \geq \epsilon_\omega. \quad (20)$$

The mean value of the criterion $\pi_\omega(\boldsymbol{x})$ is actually maximized by EM algorithm since the background distribution $F(\boldsymbol{x}|0)$ is fixed and *a priori* chosen. Thus, having estimated the conditional distributions $P(\boldsymbol{x}|\omega)$ we can derive the threshold values ϵ_ω for each property $\omega \in \Omega$ from the related log-likelihood function:

$$\epsilon_\omega = \frac{c_0}{|\mathcal{S}_\omega|} \sum_{x \in \mathcal{S}_\omega} \pi_\omega(\boldsymbol{x}) = \frac{c_0}{|\mathcal{S}_\omega|} \sum_{x \in \mathcal{S}_\omega} \log \sum_{m \in \mathcal{M}_\omega} G(\boldsymbol{x}|m, \phi_m) f(m). \quad (21)$$

Here the coefficient c_0 can be used to control the general trade-off between the false positive and false negative decisions.

The proposed statistical recognition of properties based on the threshold condition (20) is closely related to the mutually exclusive Bayesian decision making. Note that by choosing $\omega \in \Omega$ which maximizes $\pi_\omega(\boldsymbol{x})$ we obtain Bayes decision function very similar to (17). The only difference is the missing *a priori* probability $p(\omega)$ which implies the latent assumption of equiprobable classes.

5 Numerical Example

To illustrate the problem of recognition of properties we have applied PNNs to the widely used benchmark NIST Special Database 19 (SD19) containing about 400000 handwritten numerals. It is one of the few sufficiently large databases to test statistical classifiers in multidimensional spaces. The SD19 digit database consists of 7 different parts - each of about 60000 digits. They were written by Census Bureau field personnel stationed throughout the United States, except for one part (denoted as hsf_4) written by high school students in Bethesda, Maryland. Thus different parts of the SD19 database differ in origin and also in the quality. In particular, the digits written by students are known to be more difficult to recognize. Unfortunately, in the report [7] there is no unique recommendation concerning the choice of the training and test set respectively, except that the hsf_4 data should not be used as a test set. A frequent choice of the test set is to use numerals written by "independent" persons - not involved in preparation of training data. However, the use of "writer independent" test data is incorrect from the statistical point of view. The purpose of any benchmark data

Table 1. Recognition of numerals from the NIST SD19 database by differently complex structural mixtures. The classification error is given in the last row.

Experiment No.	I	II	III	IV	V	VI	VII	VIII
Number of Components	10	40	80	100	299	858	1288	1571
Number of Parameters	10240	38758	77677	89973	290442	696537	1131246	1462373
Classification error in %	**11.93**	**6.23**	**4.81**	**4.28**	**2.93**	**2.31**	**1.95**	**1.84**

Table 2. Classification error matrix obtained in the Experiment VIII. Each row contains frequencies of decisions for test data from a given class. The last column contains percentage of false negative decisions. The last row contains the total frequencies of false positive decisions in percent of all test patterns.

CLASS	0	1	2	3	4	5	6	7	8	9	false n.
0	19950	8	43	19	39	32	36	0	38	17	1.1 %
1	2	22162	30	4	35	7	18	56	32	6	0.9 %
2	32	37	19742	43	30	9	8	29	90	16	1.5 %
3	20	17	62	20021	4	137	2	28	210	55	2.6 %
4	11	6	19	1	19170	11	31	51	30	247	2.1 %
5	25	11	9	154	4	17925	39	6	96	34	2.1 %
6	63	10	17	6	23	140	19652	1	54	3	1.6 %
7	7	12	73	10	73	4	0	20497	22	249	2.1 %
8	22	25	53	97	30	100	11	11	19369	72	2.1 %
9	15	13	25	62	114	22	3	146	93	19274	2.5 %
false p.	0.09%	0.07%	0.27%	0.20%	0.17%	0.23%	0.07%	0.16%	0.33%	0.35%	**1.84%**

is to test the statistical performance of classifiers regardless of any "practically useful" aspects. For this reason the statistical properties of the training- and test data should be identical since otherwise we test how the classifier "overcomes" the particular differences between both sets. In order to guarantee the same statistical properties of both training and test data sets we have used the odd samples of each class for training and the even samples for testing. We have normalized all digit patterns (about 40000 for each numeral) to a 32x32 binary raster. In order to increase the natural variability of data we have extended the training data sets four times by making three differently rotated variants of each pattern (by -10,-5,+5 degrees) - with the resulting sets of about 80000 patterns for each numeral. The same procedure has been applied to the test data, too.

First, considering the problem of mutually exclusive classes, we have used the training data to estimate the class-conditional mixtures for all numerals separately by means of EM algorithm. Each digit pattern has been classified by the *a posteriori* probabilities computed for the most probable variant of the four rotated test patterns, i.e. for the variant with the maximum probability $P(\boldsymbol{x})$.

Table 3. Identification of properties (numerals) by means of one-class classifier. Each row corresponds to one test class, the columns contain frequencies of the identified numerals respectively.

CLASS	**0**	**1**	**2**	**3**	**4**	**5**	**6**	**7**	**8**	**9**	false n.
0	18815	2	954	23	30	292	76	0	406	103	6.8 %
1	6	21857	55	46	2756	111	52	4436	5039	410	2.2 %
2	4	9	18660	105	5	2	6	6	207	3	6.9 %
3	6	2	43	18971	1	1733	0	12	3177	373	7.7 %
4	1	0	6	1	18494	5	5	83	265	3229	5.5 %
5	7	2	4	918	0	17211	35	0	1246	282	6.0 %
6	50	10	30	0	60	888	18833	0	360	1	5.7 %
7	1	5	601	924	209	4	0	19817	242	6735	5.4 %
8	9	13	22	620	19	289	6	5	18201	154	8.0 %
9	3	4	6	70	1722	90	2	1060	1266	18667	5.6 %
false p.	0.0%	0.0%	0.9%	1.0%	2.4%	1.7%	0.1%	2.8%	6.1%	5.6%	**6.0%**

This approach simulates a biological analyzer choosing the best position of view. Table 1 shows the classification accuracy of differently complex mixture models, as estimated in eight independent experiments. The total numbers of mixture components and of the component specific parameters ($\sum_m \sum_n \phi_{mn}$) are given in the second and third row of Tab. 1 respectively. The last row contains the classification error in percent. It can be seen that the underlying mixture model is rather resistant against overfitting.

Table 2 comprises detailed classification results of the decision function from the Experiment VIII (cf. Tab. 1, last column) in terms of error frequency matrix. Each row contains frequencies of different decisions for the respective class with the correct classifications on diagonal. The last "false negative" column contains the error frequencies in percent. Similarly, the last "false positive" row of the table contains frequencies of incorrectly classified numerals in percent.

Table 3 shows how the properties (numerals) can be identified by means of one-class classifier (20). The threshold values have been specified according to the Eq. (21) by setting $c_0 = 0.75$ after some experiments[2]. Each column contains frequencies of decisions obtained by the one-class classifier (20) for the respective numeral (first row). Hence, the numbers on the diagonal correspond to correctly identified numerals. The last column contains percentage of false negative decisions defined as complement of the correctly identified patterns. The last row contains percentage of the false positive decisions which correspond to incorrectly identified numerals. Note that the only difference between Tables 2 and 3 is the information about the mutual exclusivity of the recognized numerals which is not available in case of properties. As the a priori probabilities of numerals are nearly identical, the resulting tables reflect the net gain provided by the Bayes formula.

[2] A validation set would be necessary to optimize the trade-off between the false negative and false positive decisions and also the underlying mixture complexity.

6 Concluding Remark

We propose to identify properties by means of one-class-classifier based on the log-likelihood ratio which compares the conditional probability distribution of the "target" property with the product of univariate marginals of the unconditional background distribution. The only information about the "negative" properties is contained in the global univariate marginals of involved variables. In the numerical example we compare the problem of identification of properties with the standard "multiclass" Bayes decision function. The proposed identification of properties performs slightly worse than Bayes rule because of the ignored mutual exclusivity of classes. On the other hand recognition of properties should be more advantageous in case of non-exclusive classes. The method is applicable both to the non-exclusive and exclusive properties and provides a unified approach to recognition of properties and feature extraction in the framework of probabilistic neural networks.

Acknowledgement. Supported by the project GAČR No. 102/07/1594 of Czech Grant Agency and by the projects 2C06019 ZIMOLEZ and MŠMT 1M0572 DAR.

References

1. Allwein, E.L., Schapire, R.E., Singer, Y.: Reducing multiclass to binary: a unifying approach for margin classifiers. Journal of Machine Learning Research 1, 113–141 (2001)
2. Dietterich, T.G., Bakiri, G.: Solving Multiclass Learning Problems via Error-Correcting Output Codes. Journal of Artif. Intell. Res. 2, 263–286 (1995)
3. Grim, J.: Neuromorphic features of probabilistic neural networks. Kybernetika 43(5), 697–712 (2007)
4. Grim, J., Hora, J.: Iterative principles of recognition in probabilistic neural networks. Neural Networks, Special Issue 21(6), 838–846 (2008)
5. Grim, J., Kittler, J., Pudil, P., Somol, P.: Information analysis of multiple classifier fusion. In: Kittler, J., Roli, F. (eds.) MCS 2001. LNCS, vol. 2096, pp. 168–177. Springer, Heidelberg (2001)
6. Grim, J.: Extraction of binary features by probabilistic neural networks. In: Kůrková, V., Neruda, R., Koutník, J. (eds.) ICANN 2008,, Part II. LNCS, vol. 5164, pp. 52–61. Springer, Heidelberg (2008)
7. Grother, P.J.: NIST special database 19: handprinted forms and characters database, Technical Report and CD ROM (1995)
8. Hastie, T., Tibshirani, R.: Classification by pairwise coupling. The Annals of Statistics 2(26), 451–471 (1998)
9. Haykin, S.: Neural Networks: a comprehensive foundation. Morgan Kaufman, San Francisco (1993)
10. Specht, D.F.: Probabilistic neural networks for classification, mapping or associative memory. Proc. IEEE Int. Conf. on Neural Networks I, 525–532 (1988)
11. Tax, D.M.J.: One-class classification. PhD thesis, Delft University of Technology, The Netherlands (2001)
12. Tax, D.M.J., Duin, R.P.W.: Combining one-class classifiers. In: Kittler, J., Roli, F. (eds.) MCS 2001. LNCS, vol. 2096, pp. 299–308. Springer, Heidelberg (2001)

On the Use of the Adjusted Rand Index as a Metric for Evaluating Supervised Classification

Jorge M. Santos[1] and Mark Embrechts[2]

[1] ISEP - Instituto Superior de Engenharia do Porto, Portugal
[2] Rensselaer Polytechnic Institute, Troy, New York, USA
jms@isep.ipp.pt, embrem@rpi.edu

Abstract. The Adjusted Rand Index (ARI) is frequently used in cluster validation since it is a measure of agreement between two partitions: one given by the clustering process and the other defined by external criteria. In this paper we investigate the usability of this clustering validation measure in supervised classification problems by two different approaches: as a performance measure and in feature selection. Since ARI measures the relation between pairs of dataset elements not using information from classes (labels) it can be used to detect problems with the classification algorithm specially when combined with *conventional* performance measures. Instead, if we use the class information, we can apply ARI also to perform feature selection. We present the results of several experiments where we have applied ARI both as a performance measure and for feature selection showing the validity of this index for the given tasks.

1 Introduction

One of the main difficulties in classification problems consists on the correct evaluation of the classifier performance. This is usually done by applying a common performance measure like the Mean Squared Error (MSE) or the Classification Correct Rate. Other measures like AUC (area in percentage under the Receiver Operating Characteristic (ROC) curve), Sensitivity and Specificity, are also used specially for two class problems like those involving medical applications. All these measures *compare* the labeled outcome of the supervised classification algorithm with the known labeled targets. By doing this they evaluate how good the algorithm has labeled the input data according to the required target labels. This can lead to poor results derived only by the fact that the output labels could be switched even if the classes are well identified. In these cases we deemed useful the introduction of a measure that can evaluate how well the algorithm split the input data in different classes by looking at the relation between elements of each class and not to the given labels. This is the main reason for our proposal of using a clustering validation measure in supervised classification problems.

Usually, as we will show on the experiments, the ARI performs in a similar way as other common measures. Lower values for bad classification results and higher values for good classification results. We advise to include ARI in the set of

C. Alippi et al. (Eds.): ICANN 2009, Part II, LNCS 5769, pp. 175–184, 2009.

performance measures usually used on the evaluation of supervised classification algorithms.

Since ARI is a measure of agreement between partitions and the target data is partitioned by means of the labeling we can also use ARI to perform feature selection if we split each feature in non-overlapping equal intervals and compare the partition derived from the split with the one given by the targets. By doing this we are evaluating each feature's discriminant power and we can rank the features according to the computed ARI value. We can then select the most discriminant features to apply in our classification algorithm.

This work is organized as follows: the next section introduces the Adjusted Rand Index; Section 3 explains how we intend to use ARI as a performance measure for supervised classification problems and for feature selection; Section 4 presents several experiments that show the applicability of the proposed measure with results detailed in Section 5. In the final section we draw some conclusions about the paper.

2 The Adjusted Rand Index

There are several performance indices for cluster evaluation. Indices are measures of correspondence between two partitions of the same data and are based on how pairs of objects are classified in a contingency table.

Consider a set of n objects $S = \{O_1, O_2, ..., O_n\}$ and suppose that $U = \{u_1, u_2, ..., u_R\}$ and $V = \{v_1, v_2, ..., v_C\}$ represent two different partitions of the objects in S such that $\cup_{i=1}^{R} u_i = S = \cup_{j=1}^{C} v_j$ and $u_i \cap u_{i'} = \emptyset = v_j \cap v_{j'}$ for $1 \leq i \neq i' \leq R$ and $1 \leq j \neq j' \leq C$. Given two partitions, U and V, with R and C subsets, respectively, the contingency Table 1 can be formed to indicate group overlap between U and V.

Table 1. Contingency Table for Comparing Partitions U and V

Partition		V				
	Group	v_1	v_2	$\cdots$	v_C	Total
	u_1	t_{11}	t_{12}	$\cdots$	t_{1C}	$t_{1.}$
U	u_2	t_{21}	t_{22}	$\cdots$	t_{2C}	$t_{2.}$
	$\vdots$	$\vdots$	$\vdots$	$\ddots$	$\vdots$	$\vdots$
	u_R	t_{R1}	t_{R2}	$\cdots$	t_{RC}	$t_{R.}$
Total		$t_{.1}$	$t_{.2}$	$\cdots$	$t_{.C}$	$t_{..} = n$

In Table 1, a generic entry, t_{rc}, represents the number of objects that were classified in the rth subset of partition R and in the cth subset of partition C. From the total number of possible combinations of pairs $\binom{n}{2}$ from a given set we can represent the results in four different types of pairs:

a - objects in a pair are placed in the same group in U and in the same group in V;
b - objects in a pair are placed in the same group in U and in different groups in V;
c - objects in a pair are placed in the same group in V and in different groups in U and;
d - objects in a pair are placed in different groups in U and in different groups in V.

This leads to an alternative representation of Table 1 as a 2×2 contingency table (Table 2) based on a, b, c, and d.

Table 2. Simplified 2×2 Contingency Table for Comparing Partitions U and V

Partition	V	
U	Pair in same group	Pair in different groups
Pair in same group	a	b
Pair in different groups	c	d

The values of the four cells in Table 2 can be calculated using the values of Table 1 by:

$$a = \sum_{r=1}^{R}\sum_{c=1}^{C}\binom{t_{rc}}{2} = \left(\sum_{r=1}^{R}\sum_{c=1}^{C} t_{rc}^2 - n\right)/2 \tag{1}$$

$$b = \sum_{r=1}^{R}\binom{t_{r.}}{2} - a = \left(\sum_{r=1}^{R} t_{r.}^2 - \sum_{r=1}^{R}\sum_{c=1}^{C} t_{rc}^2\right)/2 \tag{2}$$

$$c = \sum_{c=1}^{C}\binom{t_{.c}}{2} - a = \left(\sum_{c=1}^{C} t_{.c}^2 - \sum_{r=1}^{R}\sum_{c=1}^{C} t_{rc}^2\right)/2 \tag{3}$$

$$\begin{aligned} d = \binom{n}{2} - a - b - c &= \binom{n}{2} - \sum_{r=1}^{R}\binom{t_{r.}}{2} - \sum_{c=1}^{C}\binom{t_{.c}}{2} + a \\ &= \left(\sum_{r=1}^{R}\sum_{c=1}^{C} t_{rc}^2 + n^2 - \sum_{r=1}^{R} t_{r.}^2 - \sum_{c=1}^{C} t_{.c}^2\right)/2 \end{aligned} \tag{4}$$

where t_{rc} represents each element of the $R \times C$ matrix of Table 1.

Using these four values we are able to compute several performance indices that we will present in the following paragraphs.

Together with the well known Jaccard Index [1], the Rand Index (RI), proposed by Rand [2], was, and still is, a popular index and probably the most used for cluster validation. Rand Index can be easily computed by:

$$RI = \frac{a + d}{a + b + c + d} \tag{5}$$

and it basically weights those objects that were classified together and apart in both U and V. There are some known problems with RI such as the fact that the expected value of the RI of two random partitions does not take a constant value (say zero) or that the Rand statistic approaches its upper limit of unity as the number of clusters increases. With the intention to overcame these limitations researchers have created several different measures. Examples are the Fowlkes-Mallows [3] Index $(a/\sqrt{(a+b)(a+c)})$ or the Adjusted Rand Index (ARI) proposed by Hubert and Arabie [4] as an improvement of RI. In fact ARI became one of the most successful cluster validation indices and in [5] it is recommended as the index of choice for measuring agreement between two partitions in clustering analysis with different numbers of clusters. ARI can be computed by

$$ARI = \frac{\binom{n}{2}(a+d) - [(a+b)(a+c) + (c+d)(b+d)]}{\binom{n}{2}^2 - [(a+b)(a+c) + (c+d)(b+d)]} \tag{6}$$

or

$$ARI = \frac{\binom{n}{2}\sum_{r=1}^{R}\sum_{c=1}^{C}\binom{t_{rc}}{2} - \left[\sum_{r=1}^{R}\binom{t_{r.}}{2}\sum_{c=1}^{C}\binom{t_{.c}}{2}\right]}{\frac{1}{2}\binom{n}{2}\left[\sum_{r=1}^{R}\binom{t_{r.}}{2} + \sum_{c=1}^{C}\binom{t_{.c}}{2}\right] - \left[\sum_{r=1}^{R}\binom{t_{r.}}{2}\sum_{c=1}^{C}\binom{t_{.c}}{2}\right]} \tag{7}$$

with expected value zero and maximum value 1.

3 Using ARI as a Performance Measure and for Feature Selection

When using classification algorithms one must use performance measures to evaluate the classification results. There are some well known performance measures with their inherent advantages and drawbacks. For a detailed comparison of performance measures for classification please refer to [6].

The simple use of the classification correct rate in percentage (COR) may lead to erroneous conclusions specially if we are dealing with unbalanced data sets. Consider the case of a two-class problem with one class having 90% of the cases. If all the outputs of the classification algorithm are from the majority class we will get a COR value of 90 that can be misleading specially if one intends to detect and classify the minority class (e.g. medical applications), therefore one should be aware that special care must be taken when using COR in problems with low representative classes.

There are some performance measures specially suited for two-class problems that one must definitely use when working with these kind of datasets. Examples of these measures are:

- AUC: The area in percentage under the Receiver Operating Characteristic (ROC) curve, which measures the trade-off between sensitivity and specificity in two-class decision tables [7]. The higher the area the better is the decision rule.

– BCR: The balanced correct rate defined as $50\frac{a}{a+b} + 50\frac{d}{c+d}$ in percentage.

These two measures are based on the resulting 2×2 decision table, considering as abnormal class the one with lesser cases. They are specially suitable for unbalanced datasets where an optimistically high COR could arise from a too high sensitivity or specificity. AUC and BCR give an adequate picture in those situations.

The same way we use BCR or AUC for two-class unbalanced datasets we can also use ARI for unbalanced datasets with any number of classes. By analyzing each pair of elements ARI will measure not only the correct separation of elements belonging to different classes but also the relation between elements of the same class. In a certain way this measure pays more attention to the relation between elements than to the relation between each element and its target label. We can say that ARI evaluates the capability of the algorithm to separate the elements belonging to different classes.

Consider we have a two-class problem with half of the data belonging to each class and we apply a classification algorithm. Suppose that the result of the classification algorithm is a classification matrix (confusion matrix) with half of the elements as False Positives and the other half as False Negatives. In this case the COR is 0% meaning that the algorithm is a total disaster in terms of classification goal but, the ARI value is 1 (maximum) meaning that the algorithm is doing the correct distinction between classes but the problem is only with the data labeling. The elements are well separated but the given labels are incorrect or there is some problem in the implementation of the algorithm (we could be facing the perfect *lying machine*!). By combining ARI with other measures we can gain valuable information about the performance of our classification algorithm.

We also used ARI to perform feature selection. Since ARI gives a measure of the agreement between partitions and in classification problems the training data is partitioned by means of the given labels we can make a partition for each feature and compare it with the one given by the labels. To do this we rank the feature values by splitting them in non-overlapping equal intervals (categories) that could be as many as the number of classes. These intervals will define the partition to use, together with the class partition, in the computation of ARI index. Let us consider a simple example just to clarify this concept. Table 3 represents the values of two features from a given dataset with 12 elements with the respective class labels. By computing the ARI value for features 1 and 2 using the partition defined by the class labels $P_c = \{\{a, b, c, d\}, \{e, f, g, h\}, \{i, j, k, l\}\}$ and the partition defined for each feature $P_{feat1} = \{\{a, b, c, e\}, \{d, f, h, l\}, \{g, i, j, k\}\}$, $P_{feat2} = \{\{e, i, j, k, l\}, \{f, g, h\}, \{a, b, c, d\}\}$ we can rank the features according to their ARI value. In the presented case the feature with highest ARI is feat2 and therefore is the most discriminant feature.

ARI will give us the feature's discriminant power. Having ranked the existent features we select a certain number of the most discriminant ones to use in our classification algorithm. This approach is suitable for datasets with an extremely large number of features like those related with gene expression.

Table 3. A simple example to illustrate the use of ARI for feature selection

Element	a	b	c	d	e	f	g	h	i	j	k	l
Class label	1	1	1	1	2	2	2	2	3	3	3	3
feat1	0	0.3	0.1	0.5	0.2	0.4	0.7	0.5	0.9	1	0.7	0.4
feat2	1	0.8	0.9	0.7	0.2	0.4	0.4	0.5	0	0.1	0.1	0.2

4 Experiments

In the context of using ARI as a performance measure we have performed some experiments in artificial and real-world datasets. As artificial datasets we used checkerboard datasets such as the one shown in Figure 1. Checkerboard datasets are complex, controllable and unbalanced datasets. We used two different configurations: 2×2 and 4×4 checkerboards. For each one of the configurations we built three datasets with different numbers of elements (points) but with a common characteristic: a fixed number of elements belonging to the minority class (100). The percentage of elements of this minority class is 50, 25 and 10% of the total number of elements. The names of these datasets in Table 5 have the following meaning: CheckN×N(T, p) means "checkerboard N×N dataset with a total of T elements, p% of them of the minority class".

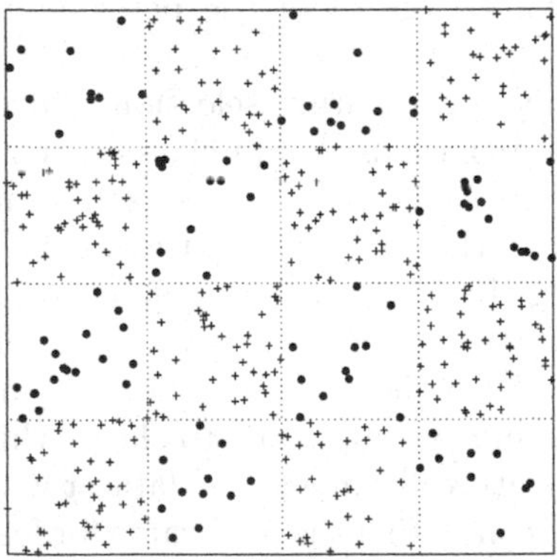

Fig. 1. An example of the 4×4 checkerboard dataset with 400 points (100 elements in the minority class: dots). Dotted lines are for visualization purpose only.

The real-world datasets are summarized in Table 4, with the top ones being the two-class datasets, the middle ones the datasets with more than two classes (multi-class problems) and the bottom ones the datasets used for feature selection. Almost every datasets can be found in the UCI repository [8] with the exception

Table 4. The real-world datasets

Data set	number of elements	number of features	number of classes	number of elem. per class
Clev. Heart Disease 2	297	13	2	160-137
Diabetes	768	8	2	500-268
Ionosphere	351	34	2	225-126
Liver	345	6	2	200-145
Sonar	208	60	2	111-97
Wdbc	569	30	2	357-212
Breast Tissue	106	9	6	21-15-18-16-14-22
Clev. Heart Disease 5	297	13	5	160-54-35-35-13
Glass	214	9	6	70-76-17-13-9-29
Iris	150	4	3	50-50-50
Wine	178	13	3	59-71-48
Leukemia	72	7129	2	47-25
Arcene	100	10000	2	44-56

of Olive [9], Breast Tissue [10] and Leukemia [14]. The datasets differ a lot among them specially in what concerns the number of features and their topology.

We used neural networks (MLP's) as classification algorithms in all problems and for the two-class problems we also used Support Vector Machines [11] that are known to be an excellent classifier for these kind of problems. In the experiments with MLP's we used the following architectures: as many inputs as the number of features, one hidden layer and one output layer for the two-class problems and as many outputs as the number of classes for multi-class problems. The number of hidden neurons, n_h, was chosen in order to assure a not too complex network with acceptable generalization. For that purpose we took into account the minimum number of lines needed to separate the checkerboard classes and the well-known rule of thumb $n_h = w/\epsilon$ (based on a formula given in [12]), where w is the number of weights and the expected error rate. Other MLP characteristics were chosen following [13]: all neurons use the hyperbolic tangent as activation function; as risk functional we used the MSE and as learning algorithm the backpropagation (BP) of the errors. The inputs were all pre-processed in order to standardize them to zero mean and unit variance. In all experiments we used the 2-fold cross validation method. In this method in each run half of the data set is randomly chosen for training and the other half for testing, then they are used with reverse roles (the original training set becomes the test set and vice-versa). Each experiment consisted of 20 runs of the algorithm. After the 20 runs the mean and standard deviation of the following performance measures were computed: AUC, COR, BCR, and ARI for the two-class problems; COR and ARI for the multi-class problems.

In the context of using ARI for feature selection we performed exploratory experiments in two data sets: a Mass-spectrometric Data for detecting cancer and; a Microarray Gene Expression Data for detecting Leukemia referred in Table 4 as Arcene and Leukemia respectively. In both experiments we used several different values for the number of intervals (categories) to split each feature and we find better results when choosing values for the number of intervals around the double of the number of classes. We selected 50 features from the 10000 of Arcene and 15 features from the 7129 of Leukemia. We have applied a Naive Bayes classifier in both cases.

5 Results

In Table 5 we show the mean and standard deviation (in brackets) of the several performance measures for the performed experiments with two-class and multi-class problems and the results for the feature selection data sets. In multi-class problems we only computed the COR, BCR and ARI performance measures since AUC is mainly for two-class problems (we also compute AUC in our daily experiments with multi-class problems since it can be obtained from the confusion matrix, however in that kind of problems it has a different meaning, reason for not showing AUC in the results because it's not appropriate for the presented comparison).

The results for the multi-class problems show a straight correlation between ARI and the *traditional* indices, specially BCR, a more reliable performance measure. The results for the Glass dataset deserve a special attention. We can see that the ARI value is more related with BCR than with COR. This is due to the characteristics of this dataset. This is a highly unbalanced dataset and by analyzing the confusion matrices (due to lack of space we do not show here the confusion matrices) we can see that the predictions are mainly restricted to 3 classes (classes 1,2 and 6) reason for the different ARI value. The results show that ARI is a good performance measure for multi-class problems.

The results of the two-class problems clearly show that ARI also gives valuable information regarding the performance of the classification algorithms. Higher values of ARI are related with higher values of the other indices. The extremely small ARI values for Liver dataset clearly points to a very complex dataset with extremely overlapping classes. When analyzing the confusion matrices we see that there are an extremely high number of misclassified elements (almost 40% of the data). These are the situations where the ARI values are smaller. Results for Diabetes also present some of this behavior.

We also can see that the ARI results for the SVM are always lower than the ones for MLP. We do not have an explanation for this, specially considering that the other performance measures do not show this same behavior.

In the feature selection problems the results for Leukemia are better than those published in [14] and for Arcene the results are not as good as those reported but we were not able to get access to all the data to perform a fair comparison. However we think that these results are very promising.

Table 5. The results with real-world and artificial datasets

Dataset		Performance Measures			
Two-class		AUC	COR	BCR	ARI
Cleaveland HD 2	MLP	0.89 (0.01)	82.42 (1.08)	82.12 (1.03)	0.36 (0.03)
	SVM	0.90 (0.01)	83.31 (0.97)	82.90 (0.96)	0.18 (0.02)
Diabetes	MLP	0.83 (0.01)	76.58 (0.88)	72.44 (0.92)	0.20 (0.01)
	SVM	0.82 (0.01)	75.34 (2.01)	67.85 (2.91)	0.15 (0.01)
Ionosphere	MLP	0.90 (0.02)	87.81 (1.21)	84.05 (1.58)	0.56 (0.04)
	SVM	0.98 (0.01)	94.26 (0.75)	93.11 (0.85)	0.45 (0.03)
Liver	MLP	0.72 (0.02)	68.52 (1.97)	67.00 (1.98)	0.07(0.01)
	SVM	0.73 (0.02)	70.23 (2.49)	67.61 (2.76)	0.05 (0.02))
Sonar	MLP	0.89 (0.03)	78.82 (2.51)	78.59 (2.56)	0.33 (0.06)
	SVM	0.93 (0.02)	84.71 (2.01)	84.34 (2.05)	0.23 (0.03)
Wdbc	MLP	0.99 (0.001)	97.39 (0.67)	97.04 (0.71)	0.87 (0.02)
	SVM	0.99 (0.002)	96.79 (0.62)	96.43 (0.66)	0.28 (0.02)
Check2×2(1000,10)	MLP	0.63 (0.16)	95.32 (1.26)	76.97 (6.51)	0.61 (0.10)
	SVM	0.99 (0.01)	98.39 (0.49)	92.55 (2.43)	0.53 (0.03)
Check2×2(400,25)	MLP	0.96 (0.08)	95.57 (2.53)	92.47 (4.73)	0.75 (0.09)
	SVM	0.99 (0.004)	95.11 (1.13)	92.22 (1.83)	0.65 (0.03)
Check2×2(200,50)	MLP	0.98 (0.01)	92.85 (2.48)	92.87 (2.48)	0.67 (0.06)
	SVM	0.98 (0.01)	92.40 (2.09)	92.42 (2.10)	0.45 (0.05)
Check4×4(1000,10)	MLP	0.71 (0.05)	93.66 (0.76)	70.77 (3.63)	0.48 (0.06)
	SVM	0.98 (0.01)	96.04 (0.55)	82.04 (2.60)	0.27 (0.02)
Check4×4(400,25)	MLP	0.88 (0.03)	86.22 (1.61)	77.96 (3.80)	0.48 (0.06)
	SVM	0.96 (0.01)	89.70 (1.37)	84.06 (2.07)	0.35 (0.02)
Check4×4(200,50)	MLP	0.83 (0.05)	78.54 (3.80)	78.51 (3.79)	0.30 (0.07)
	SVM	0.91 (0.02)	83.18 (2.56)	83.12 (2.54)	0.24 (0.04)
Multi-class					
Breast Tissue			64.01 (3.47)	62.40 (3.47)	0.46 (0.04)
Cleaveland HD 5			58.55 (1.43)	58.55 (1.43)	0.42 (0.03)
Glass			63.13 (3.56)	53.02 (3.56)	0.29 (0.04)
Iris			96.47 (1.17)	96.17 (1.47)	0.90 (0.03)
Olive			94.19 (0.73)	94.19 (0.72)	0.90 (0.01)
Thyroid			95.19 (3.09)	92.64 (3.09)	0.84 (0.09)
Wine			97.42 (1.31)	97.42 (1.30)	0.92 (0.04)
Arcene		0.76	74.00	74.00	0.22
Leukemia		0.98	91.18	92.50	0.67

6 Conclusions

We presented and proposed in this work the use of an unsupervised classification performance measure in supervised classification problems. We have presented several experiments that show the validity of ARI index as a performance measure in classification both in two-class and multi-class datasets. We have showed that ARI is especially good for multi-class classification. By analyzing the relations between pairs of elements belonging to each predicted class and the correspondent label ARI gives valuable information about the correct separability of the classes.

We also presented two preliminary experiments that show that ARI can also be used for feature selection specially for datasets with a high number of features but we are conscious that this issue deserves a more detailed study particularly to evaluate the influence of the number of intervals (categories) in the final results.

Finally, we must say that we use this index in our daily experiments and it shows to be useful in some of them, therefore we advise all the researchers to include this index as a measure of performance of their classification algorithms.

References

1. Jaccard, P.: Étude comparative de la distribution florale dans une portion des alpes et des jura. Bulletin del la Société Vaudoise des Sciences Naturelles 37, 547–579 (1901)
2. Rand, W.M.: Objective criteria for the evaluation of clustering methods. Journal of the American Statistical Association 66, 846–850 (1971)
3. Fowlkes, E., Mallows, C.: A method for comparing two hierarchical clusterings. Journal of the American Statistical Association 78, 553–569 (1983)
4. Hubert, L., Arabie, P.: Comparing partitions. Journal of Classification 2(1), 193–218 (1985)
5. Milligan, G., Cooper, M.: A study of the comparability of external criteria for hierarchical cluster analysis. Multivariate Behavioral Research 21, 441–458 (1986)
6. Ferri, C., Hernández-Orallo, J., Modroiu, R.: An experimental comparison of performance measures for classification. Pattern Recognition Letters 30(1), 27–38 (2009)
7. Metz, C.E.: Basic principles of ROC analysis. Seminars in Nuclear Medicine 8(4), 283–298 (1978)
8. Blake, C., Keogh, E., Merz, C.: UCI repository of machine learning databases (1998), `http://www.ics.uci.edu/~mlearn/MLRepository.html`
9. Forina, M., Armanino, C.: Eigenvector projection and simplified non-linear mapping of fatty acid content of italian olive oils. Ann. Chim. (Rome) 72, 127–155 (1981)
10. de Sá, J.M.: Pattern Recognition: Concepts, Methods ans Applications. Springer, Heidelberg (2001)
11. Cortes, C., Vapnik, V.: Support-vector networks. Machine Learning 20(3), 273–297 (1995)
12. Baum, E., Haussler, D.: What size net gives valid generalization? Neural Computation 1(1), 151–160 (1990)
13. Bishop, C.M.: Neural Networks for Pattern Recognition. Oxford University Press, N.Y. (1996)
14. Golub, T., Slonim, D., Tamayo, P., Huard, C., Gaasenbeek, M., Mesirov, J., Coller, H., Loh, M., Downing, J., Caligiuri, M., Bloomfield, C., Lander, E.: Molecular classification of cancer: Class discovery and class prediction by gene expression monitoring. Science 286(5439), 531–537 (1999)

Profiling of Mass Spectrometry Data for Ovarian Cancer Detection Using Negative Correlation Learning

Shan He, Huanhuan Chen, Xiaoli Li, and Xin Yao

The Centre of Excellence for Research in Computational Intelligence and Application (Cercia)
School of Computer Science
University of Birmingham
Birmingham, B15 2TT, United Kingdom

Abstract. This paper proposes a novel Mass Spectrometry data profiling method for ovarian cancer detection based on negative correlation learning (NCL). A modified Smoothed Nonlinear Energy Operator (SNEO) and correlation-based peak selection were applied to detected informative peaks for NCL to build a prediction model. In order to evaluate the performance of this novel method without bias, we employed randomization techniques by dividing the data set into testing set and training set to test the whole procedure for many times over. The classification performance of the proposed approach compared favorably with six machine learning algorithms.

Keywords: negative correlation learning, bioinformatics, proteomics, data mining.

1 Introduction

Ovarian cancer is aggressive: it is rarely detected in early stage and when detected in late stages, e.g., stage III and beyond, the 5-year survival rate is only approximately 15% [6]. Detection of early-stage ovarian cancer can reduce the death rate significantly. For example, the reported 5-year survival rate is about 90% for those women detected in stage I. Cancer antigen 125 (CA125) has been introduced for cancer diagnosis [21]. However, the accuracy for early-stage cancer diagnosis is very low (about 10%) and is prone to large false positive rate.

In recent year, Mass Spectrometry (MS) as a proteomics tool is applied for early-stage ovarian cancer diagnosis. This new proteomics tool is simple, inexpensive and minimally invasive [20]. The first application of MS to the early-stage ovarian cancer diagnosis was done by Petricoin [17]. The author employed genetic algorithms (GAs) coupled with clustering analysis to generate diagnosis rule sets to predict ovarian cancer. The study was based on the SELDI-TOF (Surface-enhanced Laser Desorption/Ionzation Time-Of-Flight) low-resolution MS data. With the advance of the mass spectrometry technology, high-resolution SELDI-TOF was employed and studied by the same authors to discriminate ovarian cancer from normal tissue. This dataset is collected with extensive quality control and assurance (QC/QA) analysis which are supposed to have superior classification patters when compared to those collected with low-resolution instrumentation [6]. In their paper, the sensitivity and specificity were

C. Alippi et al. (Eds.): ICANN 2009, Part II, LNCS 5769, pp. 185–194, 2009.

claimed to be both almost 100%. However, a reproducing study done by Jerries [12] shows that the performance of the best prediction model generated by their GA only achieved 88% accuracy at 25th percentile and 93% accuracy at 75th percentile.

Recently, in attempt to improve the accuracy of identifying cancer on the high-resolution SELDI-TOF ovarian cancer data, Yu et.al. [19] proposed a method that consists of Kolmogorov-Smirnov (KS) test, wavelet analysis and Support Vector Machine (SVM). The average sensitivity and specificity are 97.38% and 93.30%. Before the classification using SVM, the proposed method selected 8094 m/z values via KS test and further compressed to a 3382-dimensional vector of approximation coefficients with Discrete Wavelet Transformation (DWT). Although the accuracy achieved by the procedure was improved, the biological interpretability was greatly sacrificed since the 3382-dimensional DWT coefficient vector for classification is not biologically meaningful.

In this paper, we propose a novel MS data profiling method based on a novelensemble learning technique, Negative Correlation Learning (NCL) for ovarian cancer detection, which can generate more accurate and biologically meaningful results. We firstly employed MS data preprocessing techniques for signal denoising, peak detection and selection proposed in [10]. The selected peaks will be used for NCL to build a prediction model. We compared the classification performance of NCL with Support Vector Machines (SVM), Prediction Analysis for Micro-arrays (PAM), Bagging, and Random Forests (RF). We also compared the proposed method with Linear Discriminant Analysis (LDA) and Quadratic Discriminant Analysis (QDA).

The paper is organized as follows: Section 2 describes details of the proposed methods, followed by the detailed setting of our experiments, control parameter selection and experimental results in Section 3. Finally, Section 4 concludes the paper.

2 Peak Detection and Classification Algorithms

The proposed profile method consists of the two major steps: data preprocessing and NCL classification model. In the data preprocessing step, there are five components: 1). data preprocessing; 2). SNEO based peak detection; 3). peak calibration; 4). correlation-based peak selection; 5). peak qualification. In the following subsections, we give details of each step.

2.1 Data Preprocessing

Data preparation. As the *m/z* data points of each original MS spectrum are different, in order to compare different spectra under the same reference and at the same resolution, we homogenize the *m/z* vector using a resampling algorithm in the MATLAB Bioinformatics Toolbox.

We correct the baseline caused by the chemical noise in the matrix or by ion overloading using the following procedure: 1). estimated the baseline by calculating the minimum value within the width of 50 *m/z* points for the shifting window and a step size of 50 *m/z* points; 2). regresses the varying baseline to the window points using a spline approximation; and 3). subtract the resulting baseline from the spectrum. Finally, each spectrum was normalized by standardizing the area under the curve (AUC) to the median of the whole set of spectrum. The dataset is split for training and testing as detailed in Section 3.

Modified SNEO based Peak Detection Algorithm. Smoothed Non-linear Energy Operator (SNEO), or also known as the Smoothed Teager Energy Operator, has been used to detected hidden spikes in EEG and ECG biomedical signal. The method is sensitive to any discontinuity in the signal. It was shown by [16] that the output of SNEO is the instantaneous energy of the high-pass filtered version of a signal. For MS data, true peaks can be regarded as instantaneous changes in the signal. Therefore, the SNEO is ideal for the detection peaks in MS data because of its instantaneous nature. The generalized SNEO Ψ_s is defined as [16]:

$$\Psi_s[x(n)] = \Psi[x(n)] \otimes w(n) \tag{1}$$

$$\Psi[x(n)] = x^2(n) - x(n+j)x(n-j) \tag{2}$$

where $\otimes$ is the convolution operator and $w(n)$ is a smoothing window function; in this study, bartlett window function is used. Usually, the step size j is set to be 1 which gives us a standard SNEO. For the high-solution MS data, we selected the step size $j = 3$, which gives the best classification results.

Conventionally after applying SNEO to pre-emphasize peaks in the signal, potential peaks are detected using a threshold approach [16]. However, our research indicated that the threshold method is not suitable for MS peak detection since the background noise, e.g., false peaks, is non-stationary and there is no precise knowledge on the energy distribution of the true peaks and background noise. In this study, we replace the threshold detection method by a naive peak finding algorithm: we firstly calculate the first derivatives in the SNEO emphasized signals, then those local maximum will be detected as peaks. Obviously, this naive peak finding algorithm detects true peaks as well as a large number of false peaks. We therefore employ a filter-based peak selection algorithm as detailed in Section 2.1 to filter out false peaks.

Peak calibration. After the peak detection, it is necessary to calibrate the peaks in order to alleviate the impact of the *m/z* axis shifting problem. We divide each spectrum into windows with an equal number of *m/z* values. The selection of an optimal number of *m/z* values is done by experiments as detailed in 3.3. At each *m/z* point in each window, the total number of peaks across all the spectra is calculated. The *m/z* point that has the highest number of peaks within the window is set as the calibration *m/z* value. Then the peaks in all spectra within the window are moved to this calibrated *m/z* point. For each spectrum, if there is more than one peak in the *m/z* window, only the highest peak will be moved to the calibrated *m/z* point, all the other lower intensity peaks will be removed. We plot a MS spectrum of a cancer sample and peaks detected and calibrated by our method in Figure 1.

Correlation-based Peak Selection. The correlation-based feature selection [8] uses a correlation based heuristic to determine the usefulness of feature subsets. The usefulness is determined by measuring the "merit" of each individual feature for predicting

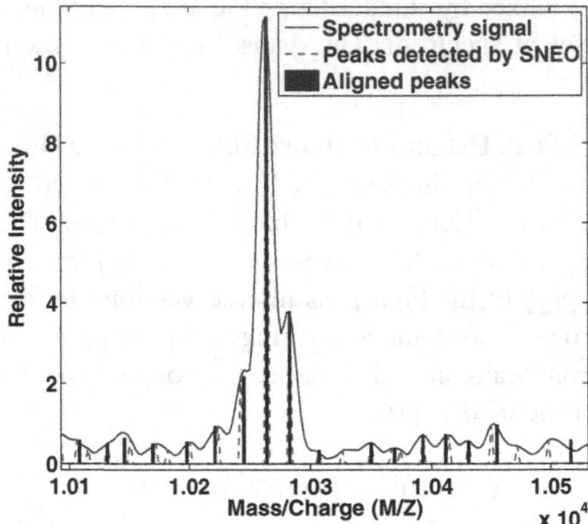

Fig. 1. MS spectrum of an ovarian cancer sample, peaks detected and calibrated by our SNEO peak detection and calibration method

the class label as well as the level of inter-correlation among them. First, an evaluation function is defined as:

$$G_s = \frac{k\overline{r_{ci}}}{\sqrt{k + k(k-1)\overline{r_{ii}}}} \tag{3}$$

where k is the number of features in the subset; $\overline{r_{ci}}$ is the mean feature correlation with the class, and $\overline{r_{ii}}$ is the average feature intercorrelation.

Equation (3) is the core of the feature selection algorithm. With this evaluation function, heuristic search algorithm then can be applied to search the feature subset with the best merit as measured in Equation (3). In the implementation, a best first heuristic search strategy was used to search the feature subset space in reasonable time. The peak selection algorithm starts from the empty set of features and uses a forward best first search to search an optimal subset. The stopping criterion of five consecutive fully expanded non-improving subsets was used [8].

There are two approaches to measure the correlation between features and the class (r_{ci}), and between features (r_{ii}). One is based on classical linear correlation and the other is based on information theory. The correlation-based feature selection employed here used the information theory based approach since it can capture correlations that are either linear or nonlinear. For details of implementation, please refer to [8].

Peak Qualification. After applying the correlation-based peak selection to the detected peaks from the training set, a small set of peaks then can be generated. This set of peaks will be used as inputs for NCL to build a prediction model. Based on the selected peak set from training set, we construct *m/z* windows using the same width as used in the calibration step. For each spectrum in test set, peaks detected by SNEO peak detection algorithm is qualified by the constructed *m/z* windows, that is, only those peaks within the *m/z* windows will be used as inputs in testing. If there are more than one peak in a *m/z* window in a spectrum, only the highest peak will be retained.

2.2 Negative Correlation Learning

Ensemble of multiple learning machines, i.e. a group of learners that work together as a committee, has attracted a lot of research interests in the machine learning community because it is considered as a good approach to improve the generalization ability [9].

Negative Correlation Learning (NCL) [14,13] is a successful ensemble technique and it has shown a huge number of empirical applications [15,11,3,7,5]. NCL introduces a correlation penalty term into the cost function of each individual network so that each neural network minimizes its MSE error together with the correlation of the ensemble.

Given the training sets $\{\mathbf{x}_n, y_n\}_{n=1}^N$, NCL combines M neural networks $f_i(\mathbf{x})$ to constitute the ensemble.

$$\bar{f}(\mathbf{x}_n) = \frac{1}{M}\sum_{i=1}^{M} f_i(\mathbf{x}_n). \tag{4}$$

In training network f_i, the cost function e_i for network i is defined by

$$e_i = \sum_{n=1}^{N}(f_i(\mathbf{x}_n) - y_n)^2 + \lambda p_i, \tag{5}$$

Algorithm Negative Correlation Learning (NCL)

Input: the training set $\mathbf{D} = \{\mathbf{x}_n, y_n\}_{n=1}^N$, integer M specifying size of ensemble, the learning rate η in backpropagation (BP) algorithm and integer T specifying the number of iterations.

For $t = 1, \cdots, T$ **do**:

1. Calculate $f_{ens}(\mathbf{x}_n) = \frac{1}{M}\sum_{i=1}^{M} f_i(\mathbf{x}_n)$.
2. For each network from $i = 1$ to M do: for each weight $w_{i,j}$ in network i, perform a desired number of updates,

$$\begin{aligned}
e_i &= \sum_{n=1}^{N}(f_i(\mathbf{x}_n) - y_n)^2 - \lambda \sum_{n=1}^{N}(f_i(\mathbf{x}_n) - f_{ens}(\mathbf{x}_n))^2, \\
\frac{\partial e_i}{\partial w_{i,j}} &= 2\sum_{n=1}^{N}(f_i(\mathbf{x}_n) - y_n)\frac{\partial f_i}{\partial w_{i,j}} - 2\lambda \sum_{n=1}^{N}(f_i(\mathbf{x}_n) - f_{ens}(\mathbf{x}_n))(1 - \frac{1}{M})\frac{\partial f_i}{\partial w_{i,j}}, \\
\Delta w_{i,j} &= -2\eta \left\{ \sum_{n=1}^{N}(f_i(\mathbf{x}_n) - y_n)\frac{\partial f_i}{\partial w_{i,j}} - \lambda \sum_{n=1}^{N}(f_i(\mathbf{x}_n) - f_{ens}(\mathbf{x}_n))(1 - \frac{1}{M})\frac{\partial f_i}{\partial w_{i,j}} \right\}.
\end{aligned}$$

Output: NCL ensemble

$$f(\mathbf{x}) = \frac{1}{M}\sum_{i} f_i(\mathbf{x}).$$

Fig. 2. Negative Correlation Learning Algorithm

where λ is a weighting parameter on the penalty term p_i:

$$p_i = \sum_{n=1}^{N} \left\{ (f_i(\mathbf{x}_n) - \bar{f}(\mathbf{x}_n)) \sum_{j \neq i} (f_j(\mathbf{x}_n) - \bar{f}(\mathbf{x}_n)) \right\}$$
$$= - \sum_{n=1}^{N} (f_i(\mathbf{x}_n) - f_{ens}(\mathbf{x}_n))^2 . \tag{6}$$

The first term in the right-hand side of (5) is the empirical training error of network i. The second term p_i is a correlation penalty function. The purpose of minimizing p_i is to negatively correlate each network's error with errors for the rest of the ensemble. The λ parameter controls a trade-off between the training error term and the penalty term. With $\lambda = 0$, we would have an ensemble with each network training with plain back propagation, exactly equivalent to training a set of networks independently of one another. If λ is increased, more and more emphasis would be placed on minimizing the penalty. The algorithm is summarized in Figure 2.

For the NCL model, we used radial basis function (RBF) networks as base classifiers. The training of RBF network is separated into two steps. In the first step, the means μ_k are initialized with randomly selected data points from the training set and the variances σ_k are determined as the Euclidean distance between μ_k and the closest $\mu_i (i \neq k, i \in \{1, \cdots, K\})$. Then in the second step we perform gradient descent in the regularized error function (weight decay)

$$\min e = \frac{1}{2} \sum_{n=1}^{N} (y_n - f(\mathbf{x}_n))^2 + \alpha \sum_{k=1}^{K} w_k^2 . \tag{7}$$

In order to fine-tune the centers and widths, we simultaneously adjust the output weights, the RBF centers and variances. Taking the derivative of Equation (7) with respect to RBF means μ_k and variances σ_k^2 we obtain

$$\frac{\partial e}{\partial \mu_k} = \sum_{n=1}^{N} (f(\mathbf{x}_n) - y_n) \frac{\partial f(\mathbf{x}_n)}{\partial \mu_k}, \tag{8}$$

with $\frac{\partial f(\mathbf{x}_n)}{\partial \mu_k} = w_k \frac{\mathbf{x}_n - \mu_k}{\sigma_k^2} \phi_k(\mathbf{x}_n)$ and

$$\frac{\partial e}{\partial \sigma_k} = \sum_{n=1}^{N} (f(\mathbf{x}_n) - y_n) \frac{\partial f(\mathbf{x}_n)}{\partial \sigma_k}, \tag{9}$$

with $\frac{\partial f(\mathbf{x}_n)}{\partial \sigma_k} = w_k \frac{\|\mathbf{x} - \mu_k\|^2}{\sigma_k^3} \phi_k(\mathbf{x}_n)$. These two derivatives are employed in the minimization of Equation (7) by scaled conjugate gradient descent, where we always compute the optimal output weights in every evaluation of the error function. The optimal output weights $\mathbf{w}$ can be computed in closed form by

$$\mathbf{w} = (\Phi^T \Phi + \alpha I)^{-1} \Phi^T \mathbf{y}, \tag{10}$$

where $\mathbf{y} = (y_1, \cdots, y_n)^T$ denotes the output vector, and I an identity matrix.

3 Numerical Experiments

3.1 Datasets

The SELDI-TOF high resolution ovarian cancer dataset OC-WCX2-HR was collected by NCI-FDA using a hybrid quadruple time-of-flight spectrometer with extensive quality control and assurance (QC/QA) analysis, which are supposed to have superior classification patters when compared to those collected with low-resolution instrumentation. The dataset consists of 216 samples, of which 95 control and 121 cancer. Each spectrum contains 350,000 *m/z* values. We condensed the spectrum into 7064 *m/z* values following [6].

3.2 Experimental Setting

The objective of our experiments is to assess the classification performance of our proposed method. In order to evaluate the classification performance of the proposed method with minimal bias, we employed the same experimental setting used in [12], which is also similar to the "proportional validation" in [19] by randomly splitting all the datasets into a training set and a test set. 52 control samples and 53 cancer samples from the OC-WCX2-HR dataset were selected for training data, the rest 43 control samples and 68 cancer samples were set aside for evaluation as a blind dataset. These same settings were used in [6] and [12].

For comparison purpose, we ran experiments on the dataset with Bagging [1], RF [2], SVM [4], PAM [18], LDA and QDA. We investigated the classification performance of these machine learning algorithms in comparison with NCL on the same peak set extracted by SNEO peak detection and correlation based peak selection algorithms.

We first carried out preliminary experiments to select optimal parameters for each algorithm as detailed in Section 3.3. Then based on the optimal parameters, we executed the experiments to evaluate the classification performance.

3.3 Parameter Selection

For the SNEO peak detection method, there is no tunable parameter. However, in the peak calibration step, the calibration window width is adjustable. Following [10], we selected calibration window width of 10, which generated the best results.

For Random Forests classifier, we tuned the following parameters: the number of candidate variables for each split and the minimum size of terminal nodes. We search the grid $\{1, 2, \cdots, 8\} \times \{100, 200, \cdots, 500\}$ on the training data.

The shrinkage parameter of PAM was selected by 10-fold cross-validation as suggested by [18].

For SVM we used C-classification with RBF kernel. In order to select optimal parameters for SVM, e.g., kernel parameter σ and cost C, we executed grid search on $\{2^{-10}, 2^{-9}, \cdots, 2^5\} \times \{2^{-5}, 2^{-4}, \cdots, 2^{10}\}$ by 10-fold cross-validation.

In the Bagging algorithm, 100 classification and regression trees are grown in each Bagging ensemble.

Table 1. Test set accuracy (%) percentiles of the dataset OC-WCX2-HR from 50 runs of NCL and 6 other machine learning algorithms. The 6 machine learning algorithms used the peaks selected by the proposed SNEO peak detection and correlation based selection method.

Algorithm	Test set accuracy 25th			Test set accuracy 75th		
	Overall	Sensitivity	Specificity	Overall	Sensitivity	Specificity
NCL	**93.69**	**93.83**	92.48	**97.39**	98.30	97.39
PAM	90.99	86.78	**94.67**	94.60	92.30	97.99
RF	91.89	89.28	91.07	96.40	94.64	98.27
SVM	92.79	94.83	87.75	95.50	**98.31**	96.15
Bagging	90.54	93.10	87.23	95.04	96.28	94.23
LDA	88.29	82.76	92.45	93.68	95.00	**98.21**
QDA	91.89	89.47	90.90	95.49	96.61	97.92

We use the traditional Linear Discriminant Analysis (LDA) and Quadratic Discriminant Analysis (QDA). In matlab, the LDA and QDA are called by the classify function.

The number of hidden nodes in RBFs of the NCL model is randomly selected but restricted in the range of 5 to 15. The ensemble consists of 100 RBF networks.

3.4 Experimental Results

In total, 128 peaks were detected and selected as biomarkers. These peaks were then used for NCL to build a prediction model.

The overall average accuracy obtained by our method from 50 runs is 95.16% with a standard deviation of 2.75%. The average sensitivity and specificity are 96.21% with a standard deviation of 2.83% and 93.96% with a standard deviation of 3.9%, respectively.

The results obtained from 6 other machine learning methods on the same training set and test set are presented in Table 1. It can be seen from the table that the SVM and RF algorithms generated better results than the other machine learning algorithms but still worse than the results generated by NCL.

Jeffries et al. [12] employed GA coupled with clustering analysis on the same dataset, the overall average accuracy from 50 runs of the GA was only 88% and 93% at 25 and 75 percentiles, respectively, which is far worse than the results obtained by our proposed method. Apart from its poorer accuracy, the GA based method actually fall into the whole-spectrum method since the method treated each *m/z* point in the spectrum as a separate test. The outputted biomarkers were a set of significant *m/z* values but were not necessarily a peak set. Therefore, the biological interpretation of their results is not guaranteed.

In [19], the average sensitivity and specificity were improved to 97.47% and 93.35% in 1000 independent 2-fold proportional validation using SVM. These results are slightly better than the results generated by our method. However, in their study, KS test and Discrete Wavelet Transform (DWT) were employed to reduce the dimensionality of the data. The biological interpretability of this method was greatly sacrificed since the features used for classification were a set of coefficients of DWT, which are even less biologically meaningful than a set of *m/z* values.

4 Conclusion

In this study, we have propose a novel MS data profiling method for ovarian cancer detection based on a novel ensemble method, Negative Correlation Learning (NCL). To our best knowledge, it is the first time NCL have been applied to proteomics.

In order to assess the classification performance of the proposed method, we evaluated our method on one high resolution ovarian cancer dataset using the same experimental settings used in [6] and [12]. We compared the proposed profiling method with six machine learning algorithms. The experimental results show that the proposed method can generate excellent classification accuracy. Results from our method are also better than most of the results in the literature, even some whole-spectrum methods. The most notable merit of our proposed method is that, besides its excellent classification performance, it obtains more biologically meaningful results, e.g., a parsimonious set of peaks, for further study and validation.

Acknowledgment

This work is supported by the Leverhulme Trust Early Career Fellowship (ECF/2007/0433) awarded to Dr. Shan He.

References

1. Breiman, L.: Bagging predictors. Machine Learning 24, 123–140 (1996)
2. Breiman, L.: Random forests. Machine Learning 45, 5–32 (2001)
3. Brown, G., Wyatt, J., Tiňo, P.: Managing diversity in regression ensembles. Journal of Machine Learning Research 6, 1621–1650 (2005)
4. Burges, C.: A tutorial on support vector machines for pattern recognition. Data Mining And Knowledge Discovery 2, 121–167 (1998)
5. Chen, H., Yao, X.: Evolutionary random neural ensemble based on negative correlation learning. In: Proceedings of the 2007 IEEE Congress on Evolutionary Computation (CEC 2007), pp. 1468–1474 (2007)
6. Conrads, T.P., Fusaro, V.A., Ross, S., Johann, D., Rajapakse, V., Hitt, B.A., Steinberg, S.M., Kohn, E.C., Fishman, D.A.: High-resolution serum proteomic features for ovarian cancer detection. Endocr. Relat. Cancer 11(2), 163–178 (2004)
7. García, N., Hervás, C., Ortiz, D.: Cooperative coevolution of artificial neural network ensembles for pattern classification. IEEE Transactions on Evolutionary Computation 9(3), 271–302 (2005)
8. Hall, M.A.: Correlation-based Feature Selection for Machine Learning. PhD thesis, The University of Waikato (1999)
9. Hansen, L.K., Salamon, P.: Neural network ensembles. IEEE Transactions on Pattern Analysis and Machine Intelligence 12(10), 993–1001 (1990)
10. He, S., Li, X.: Profiling of high-throughput mass spectrometry data for ovarian cancer detection. In: Yin, H., Tino, P., Corchado, E., Byrne, W., Yao, X. (eds.) IDEAL 2007. LNCS, vol. 4881, pp. 860–869. Springer, Heidelberg (2007)

11. Islam, M.M., Yao, X., Murase, K.: A constructive algorithm for training cooperative neural network ensembles. IEEE Transaction on Neural Networks 14(4), 820–834 (2003)
12. Jeffries, N.O.: Performance of a genetic algorithm for mass spectrometry proteomics. BMC Bioinformatics 5(1), 180 (2004)
13. Liu, Y., Yao, X.: Ensemble learning via negative correlation. Neural Networks 12(10), 1399–1404 (1999)
14. Liu, Y., Yao, X.: Simultaneous training of negatively correlated neural networks in an ensemble. IEEE Transactions on Systems, Man, and Cybernetics, Part B: Cybernetics 29(6), 716–725 (1999)
15. Liu, Y., Yao, X., Higuchi, T.: Evolutionary ensembles with negative correlation learning. IEEE Transaction on Evolutionary Computation 4(4), 380–387 (2000)
16. Mukhopadhyay, S., Ray, G.C.: A new interpretation of nonlinear energy operator and its efficacy in spike detection. IEEE Transactions Biomedical Engineering 45(2), 180–187 (1998)
17. Petricoin, E.F., Ardekani, A., Hitt, B., Levine, P., Fusaro, V., Steinberg, S., Mills, G., Simone, C., Fishman, D., Kohn, E.: Use of proteomic patterns in serum to identify ovarian cancer. The Lancet 359, 572–577 (2002)
18. Tibshirani, R., Hastie, T., Narasimhan, B., Chu, G.: Diagnosis of multiple cancer types by shrunken centroids of gene expression. Proc. Natl. Acad. Sci. USA 99, 6567–6572 (2002)
19. Yu, J.S., Ongarello, S., Fieldler, R., Chen, X.W., Toffolo, G., Cobelli, C., Trajanoski, Z.: Ovarian cancer identification based on dimensionality reduction for high-throughput mass spectrometry data. Bioinformatics 21(10), 2200–2209 (2005)
20. Zhang, X., Wei, D., Yap, Y., Li, L., Guo, S., Chen, F.: Mass spectrometry-based "omics" technologies in cancer diagnostics. Mass Spectrometry Reviews 26, 403–431 (2007)
21. Zurawski, V.R., Orjaseter, H., Andersen, A., Jellum, E.: Elevated serum ca 125 levels prior to diagnosis of ovarian neoplasia: relevance for early detection of ovarian cancer. Int. J. Cancer 42(5), 677–680 (1988)

Kernel Alignment *k*-NN for Human Cancer Classification Using the Gene Expression Profiles

Manuel Martín-Merino[1] and Javier de las Rivas[2]

[1] Universidad Pontificia de Salamanca
C/Compañía 5, 37002, Salamanca, Spain
mmartinmac@upsa.es

[2] Cancer Research Center (CIC-IBMCC, CSIC/USAL)
Salamanca, Spain
jrivas@usal.es

Abstract. The k Nearest Neighbor classifier has been applied to the identification of cancer samples using the gene expression profiles with encouraging results. However, the performance of k-NN depends strongly on the distance considered to evaluate the sample proximities. Besides, the choice of a good dissimilarity is a difficult task and depends on the problem at hand.

In this paper, we learn a linear combination of dissimilarities using a regularized version of the kernel alignment algorithm. The error function can be optimized using a semi-definite programming approach and incorporates a term that penalizes the complexity of the family of distances avoiding overfitting.

The method proposed has been applied to the challenging problem of cancer identification using the gene expression profiles. Kernel alignment k-NN outperforms other metric learning strategies and improves the classical k-NN based on a single dissimilarity.

1 Introduction

DNA microarrays provide rich profiles that are used in cancer prediction considering the gene expression levels across a collection of related samples. This technology has been applied to the identification of cancer samples with encouraging results [2].

The k Nearest Neighbor (k-NN) classifier has been widely applied to the identification of cancer samples using the gene expression profiles. However, k-NN relies strongly on the distance considered to evaluate the object proximities. The choice of a dissimilarity that reflects accurately the proximities among the sample profiles is a difficult task and depends on the problem at hand [4]. Moreover, there is no optimal dissimilarity in the sense that each dissimilarity reflects different features of the data and misclassifies frequently a different subset of patterns [3]. Therefore, different dissimilarities should be integrated in order to reduce the misclassification errors.

Several authors have proposed techniques to learn the metric from the data [17,18]. Some of them, are based on linear transformations of the Euclidean

C. Alippi et al. (Eds.): ICANN 2009, Part II, LNCS 5769, pp. 195–204, 2009.

metric [15,17] that fail often to reflect the proximities among the sample profiles. Other approaches such as [18] are more general, but are prone to overfitting when the sample size is small because they learn the metric without taking into account the generalization ability of the classifier. Besides, they rely on complex non-linear optimization algorithms.

Our approach integrates a set of heterogeneous dissimilarities that reflect different features of the data in order to reduce the classification errors. To this aim, a linear combination of dissimilarities is learnt considering the relation between kernels and distances. Each dissimilarity is embedded in a feature space using the Empirical Kernel Map [13]. Next, learning the dissimilarity is equivalent to optimize the weights of the linear combination of kernels. The combination of kernels is learnt in the literature [1,5] maximizing the alignment between the input kernel and an idealized kernel. However, this error function does not take into account the generalization ability of the classifier and is prone to overfitting.

In this paper, we consider a regularized version of the kernel alignment proposed by [1]. The linear combination of kernels is learnt in a HRKHS (Hyper Reproducing Kernel Hilbert Space) following the approach of hyperkernels proposed in [11]. This formalism exhibits a strong theoretical foundation, is less sensitive to overfitting and allow us to work with infinite families of distances.

The algorithm has been applied to the identification of human cancer samples using the gene expression profiles with remarkable results.

This paper is organized as follows: Section 2 introduces briefly the idea of Kernel Alignment, section 3 presents the algorithms considered to learn a linear combination of dissimilarities. Section 4 illustrates the performance of the algorithm in the challenging problem of gene expression data analysis. Finally, Section 5 gets conclusions and outlines future research trends.

2 Kernel Target Alignment

Let $\mathcal{X}$ be a compact subset of $\mathbb{R}^d$ where d is the space dimensionality. The function $k : \mathcal{X} \times \mathcal{X} \to \mathbb{R}$ is a kernel if it is symmetric and semi-definite positive (see [14] for more details). Given two kernels k_1 and k_2 and a sample $\mathcal{S}$, the empirical alignment evaluates the similarity between the corresponding kernel matrices. Mathematically it is defined as:

$$A(\mathcal{S}, K_1, K_2) = \frac{\langle K_1, K_2 \rangle_F}{\sqrt{\langle K_1, K_1 \rangle_F \langle K_2, K_2 \rangle_F}}, \tag{1}$$

where K_1 denotes the kernel matrix for the kernel k_1, and $\langle K_1, K_2 \rangle_F = \sum_{ij} K^1_{ij} K^2_{ij} = Tr(K_1 K_2)$ is the Frobenius product between matrices. If the kernel matrices K_1 and K_2 are considered as bidimensional vectors, the alignment evaluates the cosine of the angle and is a similarity measure.

For classification purposes we can define an ideal target matrix kernel as $K_2 = yy^T$, where y is the vector of labels for the sample $\mathcal{S}$. $K_2(x_i, x_j) = 1$ if $y(x_i) = y(x_j)$ and -1 otherwise. Substituting K_2 in equation (1) the empirical alignment between the matrix kernel K_1 and the target labels for the sample $\mathcal{S}$ can be written as:

$$A(\mathcal{S}, K_1, yy^T) = \frac{y^T K_1 y}{m\|K_1\|_F}, \tag{2}$$

where m is the size of the training set $\mathcal{S}$.

It has been shown in [1] that the empirical alignment is stable with respect of different splits of the data and that larger values for the alignment increase the separability among the classes.

3 Learning the Metric in a HRKHS Using Kernel Alignment

In order to incorporate a linear combination of dissimilarities into k-NN, we follow the approach of Hyperkernels developed by [11]. To this aim, each distance is embedded in a RKHS via the Empirical Kernel Map (see [13,7] for details). Next, a regularized version of the alignment is introduced that incorporates a L_2-penalty over the complexity of the family of distances considered. The solution to this regularized quality functional is searched in a Hyper Reproducing Kernel Hilbert Space. This allows to minimize the quality functional using a semidefinite programming approach (SDP).

Let $X = \{x_1, x_2, \ldots, x_m\}$ and $Y = \{y_1, y_2, \ldots, y_m\}$ be a finite sample of training patterns where $y_i \in \{-1, +1\}$. Let $\mathcal{K}$ be a family of semidefinite positive kernels. Our goal is to learn a kernel of dissimilarities [7] $k \in \mathcal{K}$ that represents the combination of dissimilarities and that minimizes the empirical quality functional defined by:

$$Q_{emp}^{align}(K, X, Y) = 1 - A(K, X, Y) = 1 - \frac{y^T K y}{m\|K\|_F}, \tag{3}$$

where K is the matrix kernel of k. However, if the family of kernels $\mathcal{K}$ is complex enough it is possible to find a kernel ($k^* = y^T y$) that achieves training error equal to zero overfitting the data. To avoid this problem, we introduce a term that penalizes the kernel complexity in a Hyper Reproducing Kernel Hilbert Space (HRKHS) [11]. This HRKHS is generated by a hyperkernel defined as follows. Let $\mathcal{X}$ be a non-empty set and $\underline{\mathcal{X}} = \mathcal{X} \times \mathcal{X}$ the compounded index set. Then $\underline{k} : \underline{\mathcal{X}} \times \underline{\mathcal{X}} \to \mathbb{R}$ is a hyperkernel if it is symmetric and positive semi-definite.

Thus, the quality functional optimized by the regularized kernel alignment can be written as:

$$Q_{reg}(k, X, Y) = Q_{emp}^{align}(K, X, Y) + \frac{\lambda_Q}{2}\|k\|_{\underline{H}}^2, \tag{4}$$

where $\| \|_{\underline{\mathcal{H}}}$ is the L_2 norm defined in the Hyper Reproducing Kernel Hilbert space generated by the hyperkernel $\underline{k}$. As we will see next, the kernel k that minimizes (4) belongs to this HRKHS and therefore can be written as a linear combination of hyperkernels. λ_Q is a regularization parameter that controls the complexity of the resulting kernel. For a rigorous definition of the HRKHS the reader is referred to [11].

The following theorem allows us to write the solution to the minimization of this regularized quality functional as a linear combination of hyperkernels in a HRKHS.

Theorem 1 (Representer theorem for Hyper-RKHS [11]). *Let X, Y be the combined training and test set, then each minimizer $k \in \underline{\mathcal{H}}$ of the regularized quality functional $Q_{reg}(k, X, Y)$ admits a representation of the form:*

$$k(x, x') = \sum_{i,j=1}^{m} \beta_{ij} \underline{k}((x_i, x_j), (x, x')) \tag{5}$$

for all x, $x' \in X$, where $\beta_{ij} \in \mathbb{R}$, for each $1 \leq i, j \leq m$.

However, we are only interested in solutions that give rise to positive semidefinite kernels. The following condition over the hyperkernels [11] allow us to guarantee that the solution is a positive semidefinite kernel.

Property 1. Given a hyperkernel $\underline{k}$ with elements such that for any fixed $\underline{x} \in \underline{X}$, the function $k(x_p, x_q) = \underline{k}(\underline{x}, (x_p, x_q))$, with x_p,$x_q \in \mathcal{X}$, is a positive semidefinite kernel, and $\beta_{ij} \geq 0$ for all $i, j = 1, \ldots, m$, then the kernel

$$k(x_p, x_q) = \sum_{i,j=1}^{m} \beta_{ij} \underline{k}(x_i, x_j, x_p, x_q) \tag{6}$$

is positive semidefinite.

Now, we address the problem of combining a finite set of dissimilarities. As we mentioned earlier, each dissimilarity can be represented by a kernel using the Empirical Kernel Map. Next, the hyperkernel is defined as:

$$\underline{k}(\underline{x}, \underline{x}') = \sum_{i=1}^{n} c_i k_i(\underline{x}) k_i(\underline{x}') \tag{7}$$

where each k_i is a positive semidefinite kernel of dissimilarities, c_i is a constant ≥ 0 and n is the number of dissimilarities.

Now, we show that $\underline{k}$ is a valid hyperkernel: First, $\underline{k}$ is a kernel because it can be written as a dot product $\langle \underline{\Phi}(\underline{x}), \underline{\Phi}(\underline{x}') \rangle$ where

$$\underline{\Phi}(\underline{x}) = (\sqrt{c_1}\, k_1(\underline{x}), \sqrt{c_2}\, k_2(\underline{x}), \ldots, \sqrt{c_n}\, k_n(\underline{x})) \tag{8}$$

Next, the resulting kernel (6) is positive semidefinite because for all $\underline{x}$, $\underline{k}(\underline{x}, (x_p, x_q))$ is a positive semidefinite kernel and β_{ij} can be constrained to be ≥ 0. Besides, the linear combination of kernels is a kernel and therefore is positive semidefinite. Notice that $\underline{k}(\underline{x}, (x_p, x_q))$ is positive semidefinite if $c_i \geq 0$ and

k_i are pointwise positive for training data. Both Laplacian and multiquadratic kernels verify this condition.

Finally, we show that the resulting kernel is a linear combination of the original k_i. Substituting the expression of the hyperkernel (7) in equation (6), the kernel is written as:

$$k(x_p, x_q) = \sum_{i,j=1}^{m} \beta_{ij} \sum_{l=1}^{n} c_l k_l(x_i, x_j) k_l(x_p, x_q) \tag{9}$$

Now the kernel can be expressed as a linear combination of base kernels.

$$k(x_p, x_q) = \sum_{l=1}^{n} \left[c_l \sum_{i,j=1}^{m} \beta_{ij} k_l(x_i, x_j) \right] k_l(x_p, x_q) \tag{10}$$

Therefore, the above kernel introduces into the k-NN a linear combination of base dissimilarities represented by k_l with coefficients $\gamma_l = c_l \sum_{i,j=1}^{m} \beta_{ij} k_l(x_i, x_j)$.

The previous approach can be extended to an infinite family of distances. In this case, the space that generates the kernel is infinite dimensional. Therefore, in order to work in this space, it is necessary to define a hyperkernel and to optimize it using a HRKHS. Let k be a kernel of dissimilarities. The hyperkernel is defined as follows [11]:

$$\underline{k}(\underline{x}, \underline{x'}) = \sum_{i=0}^{\infty} c_i (k(\underline{x}) k(\underline{x'}))^i \tag{11}$$

where $c_i \geq 0$ and $i = 0, \ldots, \infty$. In this case, the non-linear transformation to feature space is infinite dimensional. Particularly, we are considering all powers of the original kernels which is equivalent to transform non-linearly the original dissimilarities.

$$\underline{\Phi}(\underline{x}) = (\sqrt{c_1}\, k(\underline{x}), \sqrt{c_2}\, k^2(\underline{x}), \ldots, \sqrt{c_n}\, k^n(\underline{x})) \tag{12}$$

where n is the dimensionality of the space which is infinite in this case.

As for the finite family, it can be easily shown that $\underline{k}$ is a valid hyperkernel provided that the kernels considered are pointwise positive. The inverse multiquadratic and Laplacian kernels satisfy this condition. The following proposition allows us to derive the hyperkernel expression for any base kernel.

Proposition 1 (Harmonic Hyperkernel). *Suppose k is a kernel with range $[0, 1]$ and $c_i = (1 - \lambda_h)\lambda_h^i$, $i \in \mathbb{N}$, $0 < \lambda_h < 1$. Then, computing the infinite sum in equation (11), we have the following expression for the harmonic hyperkernel:*

$$\underline{k}(\underline{x}, \underline{x'}) = (1 - \lambda_h) \sum_{i=0}^{\infty} (\lambda_h k(\underline{x}) k(\underline{x'}))^i = \frac{1 - \lambda_h}{1 - \lambda_h k(\underline{x}) k(\underline{x'})}, \tag{13}$$

λ_h is a regularization term that controls the complexity of the resulting kernel. Particularly, larger values for λ_h give more weight to strongly non-linear kernels while smaller values give coverage for wider kernels.

3.1 Kernel Alignment k-NN in a HRKHS

We start with some notation that is used in the kernel alignment algorithm. For p,q,r $\in \mathbb{R}^n$, n $\in \mathbb{N}$ let $r = p \circ q$ be defined as element by element multiplication, $r_i = p_i \times q_i$. Define the hyperkernel Gram matrix $\underline{K}$ by $\underline{K}_{ijpq} = \underline{k}((x_i, x_j), (x_p, x_q))$, the kernel matrix $K = reshape(\underline{K}\beta)$(reshaping an m^2 by 1 vector, $\underline{K}\beta$, to an $m \times m$ matrix), where β are the linear coefficients in equation (5) that allow us to compute the kernel as a linear combination of hyperkernels. Finally, $\mathbf{1}$ a vector of ones.

The optimization of the regularized quality functional (3) for the kernel alignment in a HRKHS can be written as:

$$\max_{k \in \underline{H}} \quad tr(Kyy^T) + \frac{\lambda_Q}{2} \|k\|^2_{\underline{H}} \tag{14}$$

$$\text{subject to} \quad \|K\|^2_F = C \tag{15}$$

$$\tag{16}$$

where λ_Q is a parameter that penalizes the complexity of the family of kernels considered, $\|K\|^2_F = tr(KK^T) = \sum_{ij}(K_{ij})^2$ is the Frobenius norm of the kernel and C is a constant such that the denominator in equation (2) is restricted to be constant while the numerator is maximized.

The minimization of the previous equation leads to the following SDP optimization problem [6].

$$\min_{\beta} \quad \frac{1}{2}t_1 + \frac{\lambda_Q}{2}t_2 \tag{17}$$

$$\text{subject to} \quad \beta \geq 0 \tag{18}$$

$$\|\underline{K}^{\frac{1}{2}}\beta\| \leq t_2, \ \mathbf{1}^T\beta = 1 \tag{19}$$

$$\begin{bmatrix} K & y \\ y^T & t_1 \end{bmatrix} \succeq 0 \tag{20}$$

Once the kernel is learnt, the first k nearest neighbors are identified considering that the Euclidean distance in feature space can be written exclusively in terms of kernel evaluations:

$$d_e^2(x_i, x_j) = k(x_i, x_i) + k(x_j, x_j) - 2k(x_i, x_j) \tag{21}$$

where k is the kernel of dissimilarities learnt by the regularized kernel alignment algorithm introduced previously.

The computational complexity of the algorithm is high because we have to estimate m^2 coefficients β_{ij}. However, it can be significantly reduced if the Hyperkernel $\{\underline{k}((x_i, x_j), .)|1 \leq i, j \leq m^2\}$ is approximated by a small fraction of terms, $p \ll m^2$ for a given error using the incomplete Cholesky factorization method.

4 Experimental Results

The algorithms proposed have been applied to the identification of several cancer human samples using microarray gene expression data.

We have chosen problems with a broad range of signal to noise ratio, different number of samples and varying priors for the larger category. All the datasets are available from the Broad Institute of MIT and Harvard. Next we detail the features and preprocessing applied to each dataset.

The first dataset consists of frozen tumors specimens from newly diagnosed, previously untreated MLBCL patients (34 samples) and DLBCL patients (176 samples). They were hybridized to Affymetrix *hgu133b* gene chip containing probes for 44000 genes [9]. The raw intensities have been normalized using the rma algorithm [3]. The second problem we address concerns the clinically important issue of metastatic spread of the tumor. The determination of the extent of lymph node involvement in primary breast cancer is the single most important risk factor in disease outcome and here the analysis compares primary cancers that have not spread beyond the breast to ones that have metastasized to axillary lymph nodes at the time of diagnosis. We identified tumors as 'reported negative' (24) when no positive lymph nodes were discovered and 'reported positive' (25) for tumors with at least three identifiably positive nodes [16]. All assays used the human HuGeneFL Genechip microarray containing probes for 7129 genes. The third dataset [8] addresses the clinical challenge concerning medulloblastoma due to the variable response of patients to therapy. Whereas some patients are cured by chemotherapy and radiation, others have progressive disease. The dataset consists of 60 samples containing 39 medulloblastoma survivors and 21 treatment failures. Samples were hybridized to Affymetrix HuGeneFL arrays containing 5920 known genes and 897 expressed sequence tags.

All the datasets have been standardised subtracting the median and dividing by the Inter-quantile range. The rescaling were performed based only on the training set to avoid bias.

In order to assure a honest evaluation of all the classifiers we have performed a double loop of crossvalidation [12]. The outer loop is based on stratified ten fold cross-validation that iteratively splits the data in ten sets, one for testing and the others for training. The inner loop perform stratified nine fold cross-validation over the training set and is used to estimate the optimal parameters avoiding overfitting. The stratified variant of cross-validation keeps the same proportion of patterns for each class in training and test sets. This is necessary in our problem because the class proportions are not equal. Finally, in order to evaluate the accuracy of the classifiers the misclassification rate is reported. This metric computes the proportion of samples misclassified, it is easy to interpret and allow us to compare with the results obtained by previously published studies.

Regarding the value of the parameters, $c_i = 1/n$ for the finite family of distances where n is the number of dissimilarities which is fixed to 6 in this paper. The regularization parameter $\lambda_Q = 1$ which gives good experimental results for all the problems considered. Finally, for the infinite family of dissimilarities, the regularization parameter λ_h in the Harmonic hyperkernel (13) has been set up to 0.6 which gives an adequate coverage of various kernel widths. Smaller values emphasize only wide kernels. All the base kernel of dissimilarities have been normalized so that all ones have the same scale. Three different kernels have been considered, linear, inverse multiquadratic and Laplacian.

The number of genes has been reduced using an standard f-statistics [3]. The optimal values for the kernel parameters, the number of genes and the nearest neighbors considered have been set up by crossvalidation and using a grid search strategy.

We have compared with the Lanckriet formalism [6] that allow us to incorporate a linear combination of dissimilarities into the SVM considering the connection between kernels and dissimilarities, the Large Margin Nearest Neighbor algorithm [15] that learns a Mahalanobis metric maximizing the k-NN margin in input space and the classical k-NN with the best dissimilarity for a subset of six measures widely used in the Microarray literature.

Table 1. Empirical results for the k-NN classifier considering different distances. The ν-SVM based on coordinates and the best dissimilarity have also been considered.

Technique	DLBCL-MLBCL	Breast LN	Medulloblastoma
ν-SVM (Coordinates)	16%	8.16%	16.6%
ν-SVM (Best Distance)	11%	8.16%	13.3%
k-NN Euclidean	10%	10%	10%
k-NN Cosine	15.1%	6%	10%
k-NN Manhattan	10%	12%	16.6%
k-NN Correlation	23%	18%	15%
k-NN χ^2	16%	6%	10%
k-NN Spearman	31%	28%	23.3%

From the analysis of tables 1 and 2, the following conclusions can be drawn:

- Kernel alignment k-NN outperforms two widely used strategies to learn the metric such as Large Margin NN and Lanckriet SVM. The first one is prone to overfitting and does not help to reduce the error of k-NN based on the best dissimilarity. Similarly, our method improves the Lanckriet formalism particularly for Breast LN problem in which the sample size is smaller.
- Kernel alignment k-NN is quite insensitive to the kind of non-linear kernel employed.
- Kernel alignment k-NN considering an infinite family of distances outperforms k-NN with the best distance and the ν-SVM, particularly for breast cancer and Leukemia DLBCL-MLBCL. The infinite family of dissimilarities helps to reduce the errors of the finite counterpart particularly for breast cancer. This suggests that for certain complex non-linear problems, the non-linear transformation of the original dissimilarities helps to improve the

Table 2. Empirical results for the kernel alignment k-NN based on a combination of dissimilarities. For comparison we have included two learning metric strategies proposed in the literature.

Technique	DLBCL-MLBCL	Breast LN	Medulloblastoma
Kernel align. k-NN (Finite family, linear kernel)	10%	6%	11.66%
Kernel align. k-NN (Infinite family, linear kernel)	10%	4%	10%
Kernel align. k-NN (Finite family, inverse kernel)	10%	8%	10%
Kernel align. k-NN (Infinite family, inverse kernel)	9%	4%	10%
Kernel align. k-NN (Finite family, laplacian kernel)	9%	6%	8.33%
Kernel align. k-NN (Infinite family, laplacian kernel)	9%	4%	10%
Lanckriet SVM	11%	8.16%	11.66%
Large Margin NN	17%	8.50%	13.3%

classifier accuracy. We report, that only for the Medulloblastoma and with Laplacian base kernel the error is slightly larger for the infinite family. This suggests that the regularization term controls appropriately the complexity of the resulting dissimilarity.

– Table 1 shows that the best distance depends on the dataset considered and that the performance of k-NN depends strongly on the particular measure employed to evaluate the sample proximities. Finally, an interesting result is that k-NN outperforms the ν-SVM algorithm for all the datasets.

5 Conclusions

In this paper, we propose two methods to incorporate in the k-NN algorithm a linear combination of non-Euclidean dissimilarities. The family of distances is learnt in a HRKHS (Hyper Reproducing Kernel Hilbert Space) using a Semidefinite Programming approach. A penalty term has been added to avoid the overfitting of the data. The algorithm has been applied to the classification of complex cancer human samples.

The experimental results suggest that the combination of dissimilarities in a Hyper Reproducing Kernel Hilbert Space improves the accuracy of classifiers based on a single distance particularly for non-linear problems. Besides, this approach outperforms other learning metric strategies widely used in the literature and is robust to overfitting.

Future research trends will apply this formalism to integrate heterogeneous data sources.

References

1. Cristianini, N., Kandola, J., Elisseeff, J., Shawe-Taylor, A.: On the kernel target alignment. Journal of Machine Learning Research 1, 1–31 (2002)
2. Dudoit, S., Fridlyand, J., Speed, T.P.: Comparison of Discrimination Methods for the Classification of Tumors Using Gene Expression Data. Journal of the American Statistical Association 97(457), 77–87 (2002)
3. Gentleman, R., Carey, V., Huber, W., Irizarry, R., Dudoit, S.: Bioinformatics and Computational Biology Solutions Using R and Bioconductor. Springer, Berlin (2006)
4. Jiang, D., Tang, C., Zhang, A.: Cluster Analysis for Gene Expression Data: A Survey. IEEE Transactions on Knowledge and Data Engineering 16(11), 1370–1386 (2004)
5. Kandola, J., Shawe-Taylor, J., Cristianini, N.: Optimizing kernel alignment over combinations of kernels, NeuroCOLT, Tech. Rep (2002)
6. Lanckriet, G., Cristianini, N., Barlett, P., El Ghaoui, L., Jordan, M.: Learning the kernel matrix with semidefinite programming. Journal of Machine Learning Research 3, 27–72 (2004)
7. Pekalska, E., Paclick, P., Duin, R.: A generalized kernel approach to dissimilarity-based classification. Journal of Machine Learning Research 2, 175–211 (2001)
8. Pomeroy, S.E.A.: Prediction of central nervous system embryonal tumour outcome based on gene expression. Nature 415 (2002)
9. Savage, K., et al.: The molecular signature of mediastinal large B-cell lymphoma differs from that of other diffuse large B-cell lymphomas and shares features with classical hodgkin lymphoma. Blood 102(12) (December 2003)
10. Scholkopf, B., Tsuda, K., Vert, J.: Kernel Methods in Computational Biology. MIT Press, Cambridge (2004)
11. Soon Ong, C., Smola, A., Williamson, R.: Learning the kernel with hyperkernels. Journal of Machine Learning Research 6, 1043–1071 (2005)
12. Statnikov, A.: A comprehensive evaluation of multicategory classification methods for microarray gene expression cancer diagnosis. Bioinformatics 21(5), 631–643 (2004)
13. Tsuda, K.: Support Vector Classifier with Assymetric Kernel Function. In: Proceedings of ESANN, Bruges, pp. 183–188 (1999)
14. Vapnik, V.: Statistical Learning Theory. John Wiley & Sons, New York (1998)
15. Weinberger, K.Q., Saul, L.K.: Distance Metric Learning for Large Margin Nearest Neighbor Classification. J. Machine Learning Research 10, 207–244 (2009)
16. West, M., et al.: Predicting the clinical status of human breast cancer by using gene expression profiles. PNAS 98(20) (2001)
17. Wu, G., Chang, E.Y., Panda, N.: Formulating distance functions via the kernel trick. In: ACM SIGKDD, Chicago, pp. 703–709 (2005)
18. Xiong, H., Chen, X.-W.: Kernel-Based Distance Metric Learning for Microarray Data Classification. BMC Bioinformatics 7(299), 1–11 (2006)

Convex Mixture Models for Multi-view Clustering

Grigorios Tzortzis and Aristidis Likas

Department of Computer Science, University of Ioannina,
GR 45110, Ioannina, Greece
{gtzortzi,arly}@cs.uoi.gr

Abstract. Data with multiple representations (views) arise naturally in many applications and multi-view algorithms can substantially improve the classification and clustering results. In this work, we study the problem of multi-view clustering and propose a multi-view convex mixture model that locates exemplars (cluster representatives) in the dataset by simultaneously considering all views. Convex mixture models are simplified mixture models that exhibit several attractive characteristics. The proposed algorithm extends the single view convex mixture models so as to handle data with any number of representations, taking into account the diversity of the views while preserving their good properties. Empirical evaluations on synthetic and real data demonstrate the effectiveness and potential of our method.

Keywords: clustering, mixture models, multi-view learning.

1 Introduction

The most common approach for the machine learning setting, is to assume that data are represented in a single vector or graph space. However, in many real-life problems multi-view data arise naturally. Multi-view data are instances that have multiple representations (views) from different feature spaces. Usually these multiple views are from different vector spaces or different graph spaces or a combination of vector and graph spaces. The most typical example are web pages. Web pages can be represented with a term vector for the words in the web page text, another term vector for the words in the anchor text and a hyper-link graph.

The natural and frequent occurrence of multi-view data has raised interest in the so called *multi-view learning*. The main challenge of multi-view learning is to develop algorithms that use multiple views simultaneously, given the diversity of the views. Most studies on this topic address the semi-supervised classification problem and multi-view classification algorithms have often proven to utilize unlabeled data effectively and improve classification accuracy (e.g. [1,2,3]).

This work focuses on multi-view unsupervised learning and particularly in *multi-view clustering*. Multi-view clustering explores and exploits multiple representations simultaneously in order to produce a more accurate and robust

C. Alippi et al. (Eds.): ICANN 2009, Part II, LNCS 5769, pp. 205–214, 2009.

partitioning of the data than single view clustering. The available literature for this topic (e.g. [4,5,6,7]) is still limited, with encouraging results though. Borrowing the terminology of [7], there exist two approaches in multi-view clustering: *centralized* and *distributed*. Centralized algorithms simultaneously use all available views to cluster the dataset, while distributed algorithms first cluster each view independently from the others, using an appropriate single view algorithm, and then combine the individual clusterings to produce a final partitioning.

Most studies in multi-view clustering follow the centralized approach and extend well-known clustering algorithms to the multi-view setting. Bickel and Scheffer [4] developed a two-view EM and a two-view k-means algorithm under the assumption that the views are independent. They also studied the problem of mixture model estimation with more than two views and showed that co-EM [8] is a special case of their formulation [9]. De Sa [5] proposed a two-view spectral clustering algorithm that creates a bipartite graph of the views and is based on the "minimizing-disagreement" idea. This method also assumes that the views are independent. An algorithm that generalizes the single view normalized cut to the multi-view case and can be applied to more than two views was introduced by Zhou and Burges [6]. Following the distributed approach, Long *et al.* [7] proposed a general model for multi-view unsupervised learning which handles more than two views and representations from both vector and graph spaces.

In this paper we follow the centralized approach and present a multi-view clustering algorithm based on the *convex mixture model* of Lashkari and Golland [10]. Convex mixture models are a special case of mixture models that identify exemplars in the dataset by optimizing a convex criterion and have shown promising results in [10]. One of many attractive features is their applicability when only the dataset pairwise distance matrix is available and not the data points. The proposed *multi-view convex mixture model* finds exemplars *based on all views* and handles any number of views. The experiments with our algorithm demonstrate a considerable improvement on the clustering results compared to i) single view convex mixture models applied on the individual views and ii) single view convex mixture models that use the concatenation of the views.

The rest of this paper is organized as follows. Section 2 reviews the single view convex mixture model, while the proposed multi-view algorithm is presented in section 3. The experimental evaluation on artificial data and linked documents is discussed in section 4 and section 5 concludes this work.

2 Convex Mixture Models

Exemplar-based mixture models [10], also called *convex mixture models* (*CMM*), result in soft assignments of data points to clusters and in the extraction of representative exemplars from the dataset. They are simplified mixture models whose components equal in number the dataset size, the components' distributions are centered at the dataset points, thus representing all data points as cluster center candidates (candidate exemplars), and the only adjustable parameters are the components' priors.

Given a dataset $\mathcal{X} = \{\mathbf{x}_1, \mathbf{x}_2, \ldots, \mathbf{x}_N\}$, $\mathbf{x}_i \in \Re^d$ the convex mixture model distribution is $Q(\mathbf{x}) = \sum_{j=1}^{N} q_j f_j(\mathbf{x})$, $\mathbf{x} \in \Re^d$, where q_j denotes the prior probability of the j-th component, satisfying the constraint $\sum_{j=1}^{N} q_j = 1$, and $f_j(\mathbf{x})$ is an exponential family distribution on random variable $\mathbf{X}$ with its expectation parameter equal to the j-th data point. Taking into account the bijection between regular exponential families and Bregman divergences [11], we write $f_j(\mathbf{x}) = C(\mathbf{x}) \exp(-\beta d_\varphi(\mathbf{x}, \mathbf{x}_j))$ with d_φ denoting the Bregman divergence corresponding to the components' distributions.

A clustering is produced by maximizing the log-likelihood $L\left(\{q_j\}_{j=1}^{N}; \mathcal{X}\right)$, shown in (1), over $\{q_j\}_{j=1}^{N}$ s.t. $\sum_{j=1}^{N} q_j = 1$. The constant β controlls the sharpness of the components and also *the number of clusters* identified by the convex mixture model when the soft assignments are turned into hard ones. Higher β values result in more clusters in the final solution.

$$L\left(\{q_j\}_{j=1}^{N}; \mathcal{X}\right) = \frac{1}{N}\sum_{i=1}^{N} \log\left[\sum_{j=1}^{N} q_j f_j(\mathbf{x}_i)\right] = \frac{1}{N}\sum_{i=1}^{N} \log\left[\sum_{j=1}^{N} q_j e^{-\beta d_\varphi(\mathbf{x}_i, \mathbf{x}_j)}\right] + \text{const.} \tag{1}$$

The log-likelihood function (1) can be expressed in terms of the Kullback-Leibler (KL) divergence if we define $\hat{P}(\mathbf{x}) = 1/N, \mathbf{x} \in \mathcal{X}$ to be the empirical distribution of the dataset and by noting that

$$D(\hat{P}\|Q) = -\sum_{\mathbf{x}\in\mathcal{X}} \hat{P}(\mathbf{x}) \log Q(\mathbf{x}) - \mathbb{H}(\hat{P}) = -L\left(\{q_j\}_{j=1}^{N}; \mathcal{X}\right) + \text{const.}, \tag{2}$$

where $\mathbb{H}(\hat{P})$ is the entropy of the empirical distribution that does not depend on the parameters of the convex mixture model. Now the maximization of (1) is equivalent to the minimization of (2). This minimization problem is *convex* and can be solved with an efficient-iterative algorithm. As proved in [12], the updates on the components' prior probabilities are given by

$$q_j^{(t+1)} = q_j^{(t)} \sum_{\mathbf{x}\in\mathcal{X}} \frac{\hat{P}(\mathbf{x}) f_j(\mathbf{x})}{\sum_{j'=1}^{N} q_{j'}^{(t)} f_{j'}(\mathbf{x})} \tag{3}$$

and the algorithm is *guaranteed to converge to the global minimum* as long as $q_j^{(0)} > 0, \forall j$. The prior probability q_j associated with data point $\mathbf{x}_j$ is a measure of *how likely this point is to be an exemplar* and will be of great importance when we present our multi-view algorithm in section 3.

Clustering with a convex mixture model requires to select a value for the parameter β ($0 < \beta < \infty$). It is possible to identify a reasonable range of β values by determining a reference value β_0. In [10], the following empirical value (4) has been proposed, achieving good results in their experiments.

$$\beta_0 = N^2 \log N / \sum_{i,j} d_\varphi(\mathbf{x}_i, \mathbf{x}_j) \tag{4}$$

Convex mixture models showed their potential when a Gaussian convex mixture model, i.e. with Euclidean distance as the Bregman divergence, outperformed a fully parametrized Gaussian mixture model in [10]. This proved that the smaller flexibility of convex mixture models, as $\{q_j\}_{j=1}^N$ are the only parameters, is well compensated by their ability to avoid the initialization problem and always locate the global optimum. Another important feature is that only the pairwise data distances take part in the calculation of the priors, thus the values of the data points are not required. As stated in [10], the method can be extended to any proximity data as long as the distance matrix $\mathbf{D}$ is available, by simply replacing $d_\varphi(\mathbf{x}_i, \mathbf{x}_j)$ with D_{ij} in (1) and the convexity is not affected.

3 Multi-view Convex Mixture Models

Motivated by the potential and the advantages of the convex mixture models of section 2, in this work we extend them to data with multiple representations. Following the centralized approach, exemplars are identified by defining for each view a convex mixture model distribution, *with common priors q_j across all views*, as well as the corresponding empirical distribution and minimizing the KL divergence between those two distributions summed over all views.

3.1 Model Description

Suppose we are given a dataset with N instances, $\mathfrak{X} = \{\mathbf{x}_1, \mathbf{x}_2, \ldots, \mathbf{x}_N\}$, and for each instance V views are available. Let us define $\mathcal{X} = \{\mathcal{X}^1, \mathcal{X}^2, \ldots, \mathcal{X}^V\}$, such that $\mathcal{X}^v$ contains the representations of the instances in the v-th view, i.e. $\mathcal{X}^v = \{\mathbf{x}_1^v, \mathbf{x}_2^v, \ldots, \mathbf{x}_N^v\}$, $\mathbf{x}_i^v \in \Re^{d^v}$. Also, assuming that no prior information for the data is available in any view, define for each view a uniform empirical dataset distribution (5), as in [10], and also a convex mixture model distribution (6). Note that all distributions $\{Q^v(\mathbf{x})\}_{v=1}^V$ *share the same prior probabilities* $\{q_j\}_{j=1}^N$, but have different component distributions $f_j^v(\mathbf{x})$.

$$\hat{P}^v(\mathbf{x}) = \begin{cases} \frac{1}{N}, & \mathbf{x} \in \mathcal{X}^v \\ 0, & \text{otherwise} \end{cases} \tag{5}$$

$$Q^v(\mathbf{x}) = \sum_{j=1}^N q_j f_j^v(\mathbf{x}) = C^v(\mathbf{x}) \sum_{j=1}^N q_j e^{-\beta^v d_\varphi^v(\mathbf{x}, \mathbf{x}_j^v)}, \; \mathbf{x} \in \Re^{d^v} \tag{6}$$

Our aim is to locate high quality exemplars (cluster centroids) in the dataset, by considering all views simultaneously, around which the remaining instances will cluster. To achieve this, the proposed *multi-view convex mixture model* minimizes the sum of the KL divergences between the empirical distribution and the convex mixture distribution of each view, given by the following equation:

$$\min_{\substack{q_1,\ldots,q_N \\ \text{s.t. } \sum_{j=1}^N q_j = 1}} \left\{ \sum_{v=1}^V D(\hat{P}^v \| Q^v) = -\sum_{v=1}^V \sum_{\mathbf{x} \in \mathcal{X}^v} \hat{P}^v(\mathbf{x}) \log Q^v(\mathbf{x}) - \sum_{v=1}^V \mathbb{H}(\hat{P}^v) \right\}, \tag{7}$$

where $\mathbb{H}(\hat{P}^v)$ is the entropy of the empirical distribution of the v-th view that does not depend on the parameters of the multi-view convex mixture model.

It is well known that the sum of convex functions is also a convex function. Therefore, the above optimization problem, which is a generalization of the single view case, is also *convex*, since its objective function is the sum of the single view objectives which are convex functions. To solve (7) the same efficient-iterative algorithm as in (2) can be used. It can be shown that the updates on the components' prior probabilities are given by

$$q_j^{(t+1)} = \frac{q_j^{(t)}}{V} \sum_{v=1}^{V} \sum_{\mathbf{x} \in \mathcal{X}^v} \frac{\hat{P}^v(\mathbf{x}) f_j^v(\mathbf{x})}{\sum_{j'=1}^{N} q_{j'}^{(t)} f_{j'}^v(\mathbf{x})} \tag{8}$$

and the algorithm is *guaranteed to converge to the global minimum* as long as $q_j^{(0)} > 0, \forall j$. The prior q_j associated with instance $\mathbf{x}_j$ is again a measure of *how likely this instance is to be an exemplar* and takes into account all views.

In the derivation of the above multi-view convex mixture model the following facts were considered. Different views can have very different statistical properties, therefore we allow the convex mixture model distribution (6) of each view to have its own β value and Bregman divergence, i.e. different component distributions. For example, for one view we can use a Gaussian CMM and for another a Bernoulli CMM. An important property of the single view convex mixture model is convexity and we wish to preserve this property in the multi-view setting. As a result, summing the single view objectives to construct the multi-view objective is a natural choice. Finally, since our target is to extract representative exemplars from the dataset based on all views, we require all convex mixture model distributions to have common priors q_j. Intuitively, this means that an instance whose corresponding prior probability has a high value, is more or less a good exemplar in all views.

3.2 Algorithm Implementation

We follow the same steps as in [10] to implement the algorithm that optimizes (7). Letting $s_{ij}^v = \exp(-\beta^v d_\varphi^v(\mathbf{x}_i^v, \mathbf{x}_j^v))$ and using an auxiliary matrix $\mathbf{Z}$ and an auxiliary vector $\mathbf{n}$ we update the prior probabilities q_j as follows:

$$Z_{iv}^{(t)} = \sum_{j=1}^{N} s_{ij}^v q_j^{(t)}, \quad n_j^{(t)} = \frac{1}{V} \sum_{v=1}^{V} \sum_{i=1}^{N} \frac{\hat{P}^v(\mathbf{x}_i^v) s_{ij}^v}{Z_{iv}^{(t)}}, \quad q_j^{(t+1)} = n_j^{(t)} q_j^{(t)}, \tag{9}$$

where $q_j^{(0)} > 0$ for all instances we want to consider as possible exemplars. Obviously, our formulation requires only the pairwise distances in each view and not the instances themselves in order to calculate the priors. Thus it can be extended to use proximity values, analogously to the single view case.

Suppose we wish to partition a multi-view dataset into M disjoint clusters $C_1, C_2, \ldots, C_M$ using the multi-view convex mixture model. To identify M exemplars (cluster centroids), the instances with the M highest q_j values are determined. Specifically, we run the algorithm until the M highest q_j values correspond to the same instances for a number of consecutive iterations. Moreover, we require that the order among the M highest q_j values remains the same during these iterations. This convergence criterion differs from that in [10]. After finding the M exemplars, we assign each of the remaining $N - M$ instances to cluster C_k, associated with the k-th exemplar, that has the largest posterior probability over all views, according to (10). Note that we refer to the exemplar instances as $\mathfrak{X}^E = \{\mathbf{x}_1^E, \mathbf{x}_2^E, \ldots, \mathbf{x}_M^E\} \subset \mathfrak{X}$ and their prior probabilities and component distributions in the v-th view as q_k^E and $f_k^{vE}(\mathbf{x})$, $k = 1, \ldots, M$ respectively.

$$C_k = \{\mathbf{x}_k^E\} \cup \left\{ \mathbf{x}_i \left| \sum_{v=1}^{V} \frac{q_k^E f_k^{vE}(\mathbf{x}_i^v)}{\sum_{j=1}^{N} q_j f_j^v(\mathbf{x}_i^v)} > \sum_{v=1}^{V} \frac{q_l^E f_l^{vE}(\mathbf{x}_i^v)}{\sum_{j=1}^{N} q_j f_j^v(\mathbf{x}_i^v)}, \forall l \neq k, \mathbf{x}_i \notin \mathfrak{X}^E \right. \right\} \tag{10}$$

A final issue on the implementation of the multi-view convex mixture model is the choice of appropriate values for the β^v parameters. Since a separate single view convex mixture model is defined for each view, we can identify a reasonable range of β^v values in the same way as in the single view case. Following the ideas of the single view setting the following empirical β_0^v value is derived:

$$\beta_0^v = N^2 \log N / \sum_{i,j} d_\varphi^v(\mathbf{x}_i^v, \mathbf{x}_j^v) \; . \tag{11}$$

As for the complexity of the algorithm, calculation of the auxiliary quantities and the update of the priors costs $O(N^2 V)$ scalar operations per iteration. If the distance matrices of the views are not given, computing the s_{ij}^v quantities usually costs $O(N^2 V d)$, $d = \max\{d^1, d^2 \ldots, d^V\}$. Assuming τ iterations are required until convergence, the overall cost becomes $O(N^2 V(\tau + d))$ scalar operations.

4 Experimental Evaluation

We aim to examine whether simultaneously considering all views helps to improve the clustering results obtained from the individual views, i.e. compare single view clustering to multi-view clustering. Since a very common approach to cluster multiple represented data is to concatenate all the views and then apply a single view algorithm on the concatenated view, we wish to investigate whether a multi-view algorithm provides any gains compared to a single view algorithm applied on the concatenated view. To answer these questions, we study the performance of the single view and multi-view convex mixture model on multi-view artificial data and two collections of linked documents, where multiple representations for the data occur naturally.

In all experiments we use Gaussian convex mixture models (Gaussian CMM), i.e. $d_\varphi^v(\mathbf{x}_i^v, \mathbf{x}_j^v) = \|\mathbf{x}_i^v - \mathbf{x}_j^v\|^2, \forall v$ and a uniform empirical dataset distribution (5). We report clustering results i) separately for each view, ii) for the concatenated view and iii) for the multiple views. To assess the clustering quality we use the *average entropy* metric, as in [4,5,9], which measures the impurity of the returned clusters. Average entropy is given by (12), where N is the dataset size, M the

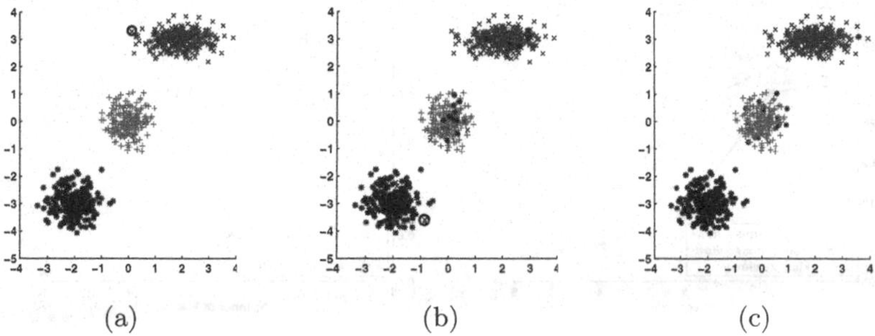

Fig. 1. Examples of the artificial dataset: (a) the original dataset generated from three Gaussian distributions belonging to three classes; (b) one of the five views for $\omega = 50$ and zero translation; (c) clustering into three clusters with the three-view dataset ($\omega = 50$, $\beta^v = \beta_0^v$) using a Gaussian multi-view CMM. Only 25 instances are misplaced.

number of clusters, c the number of classes, n_i^j the number of points in cluster i from class j and n_i the size of the i-th cluster. Lower average entropy values indicate that each cluster consists of instances belonging to the same class.

$$H = \sum_{i=1}^{M} \frac{n_i}{N} \left(- \sum_{j=1}^{c} \frac{n_i^j}{n_i} \log \frac{n_i^j}{n_i} \right) \tag{12}$$

4.1 Artificial Dataset

As a first step towards evaluating the performance of the multi-view convex mixture model, we generated a synthetic dataset, illustrated in Fig. 1(a) and consisting of 700 instances, from three two-dimensional Gaussian distributions. Each distribution represents a different class. Views were constructed with the following mechanism: for each view, we equally translated all instances of the original dataset and randomly selected ω of them to be moved to a different class. For example, assume that instance $\mathbf{x}_i$ had been selected, shown in circle in Fig. 1(a), that was generated by the first distribution (first class). We randomly picked one of the other two classes and generated a new point from the corresponding Gaussian distribution. This point, shown in circle in Fig. 1(b), is the representation of instance $\mathbf{x}_i$ in the view. Hence, an instance of the first class is now wrongly represented as an instance belonging to another class.

The above view generation mechanism will help us discover if simultaneously considering multiple views can correct some of the errors of the individual views and approach the optimum of $H = 0$ achieved by a convex mixture model on the well separated original dataset, since a convex mixture model on a single view will most probably misclassify all ω instances. For the experiments $\omega = 50$ and five views were generated, one of which is illustrated in Fig. 1(b). Five multi-view datasets were created, containing $1, \ldots, 5$ of the five views respectively. Results for these datasets are reported in Fig. 2(a) for three clusters and $\beta^v = \beta_0^v, \forall v$.

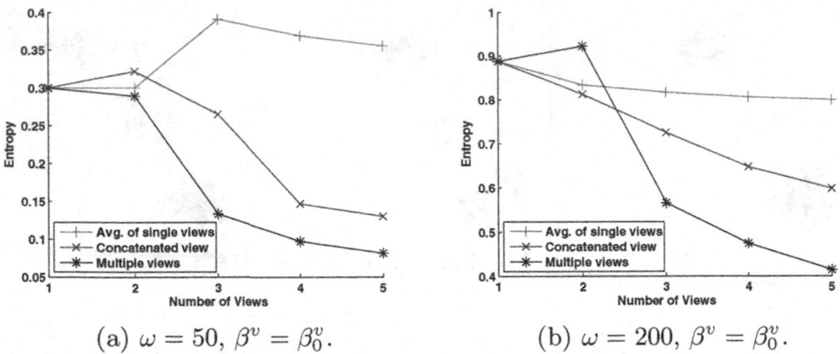

(a) $\omega = 50$, $\beta^v = \beta_0^v$.

(b) $\omega = 200$, $\beta^v = \beta_0^v$.

Fig. 2. Artificial dataset results with Gaussian CMMs in terms of entropy for different number of views and three clusters

The multi-view convex mixture model constantly achieves the lowest entropy and it always outperforms the model that uses the concatenated view. Four of the five individual views have an entropy around 0.3 while one has $H = 0.57$. This view is included in the three-view dataset and explains the peak in the graph for the single views average. The corresponding clustering is illustrated in Fig. 1(c). At the same time our method achieves $H = 0.08$ with five views, confirming that it can considerably boost the clustering performance. Finally, the multi-view convex mixture model takes advantage of every available view as the entropy constantly falls as the number of views increases.

We also executed the same experiments as above, but with views for which $\omega = 200$. Fig. 2(b) depicts the results for this case. The multi-view convex mixture model is again the best algorithm and for five views it achieves $H = 0.41$, which is almost half the entropy of the individual views average and 33% less than the entropy of the concatenated view.

4.2 Document Archives

We selected two archives of linked documents. The *WebKB* dataset is a collection of academic web pages from computer science departments of various universities, while the *Citeseer* dataset is a collection of scientific publications. Both are very popular datasets for evaluating multi-view clustering algorithms [4,5,9] and multi-view classifiers [1,2]. We used the Bickel and Scheffer [9] version in which both collections have six classes and two or three views respectively. The first view of web pages is their text and the second the anchor text of the inbound links. Publications are represented in terms of a text view, consisting of the title and abstract of each paper, and two link views, made up of the inbound and outbound references. For some of the web pages no inbound links with anchor text exist, while some papers do not have inbound or outbound references. Such instances were removed, resulting in 2076 web pages and 742 papers.

For each view we generated tfidf vectors and normalized them to unit length (normalized tfidf), so that square Euclidean distances reflect the commonly used

Table 1. *WebKB* results with Gaussian CMMs in terms of entropy and six clusters

Method-View	***WebKB* Entropy**	
	$\beta^v = \beta_0^v$	$\beta^v = \alpha\beta_0^v$
Single view CMM-text	1.54	1.49 ($\alpha = 1.5$)
Single view CMM-anchor text	1.55	1.44 ($\alpha = 3.5$)
Single view CMM-concat. text & anchor text	1.56	1.48 ($\alpha = 1.7$)
Multi-view CMM-text & anchor text	**1.5**	**1.4** ($\alpha = 1.5$)

Table 2. *Citeseer* results with Gaussian CMMs in terms of entropy and six clusters

Method-View	***Citeseer* Entropy**	
	$\beta^v = \beta_0^v$	$\beta^v = \alpha\beta_0^v$
Single view CMM-text	1.61	1.56 ($\alpha = 0.5$)
Single view CMM-inbound references	1.65	1.65 ($\alpha = 1$)
Single view CMM-outbound references	1.57	1.56 ($\alpha = 1.5$)
Single view CMM-concat. text & two link views	1.6	1.54 ($\alpha = 1.5$)
Multi-view CMM-text & two link views	**1.5**	**1.5** ($\alpha = 1$)

cosine similarity. Both datasets were partitioned into six clusters. Tables 1 and 2 report results for the *WebKB* and *Citeseer* collections respectively, where the multi-view convex mixture model is compared to its single view counterpart.

In a first series of experiments we set $\beta^v = \beta_0^v, \forall v$. As can be seen, the multi-view algorithm improves the clustering of the individual and concatenated views, making once again apparent the potential of our method and the advantages of using simultaneously multiple views. Remarkably, for both datasets the concatenated view's performance is even worse than that of some of the single views. This result explains the need to develop multi-view algorithms and not resort to tricks that allow single view algorithms to handle multiple represented data.

We also investigated the impact of the β^v parameter by searching around the range of values defined by β_0^v and selecting the fraction α of β_0^v that yields the smallest entropy (shown in parentheses in Tables 1, 2). A decrease in entropy for the two collections can be observed and again the multi-view convex mixture model is the best performer. Note that setting $\beta^v = \beta_0^v$ is the best choice for the inbound references view and the multi-view setting ($\alpha = 1$) of the *Citeseer* dataset, indicating that β_0^v provides a good range of values for the β^v parameter.

5 Conclusions and Future Work

We have proposed the multi-view convex mixture model, a method that extends convex mixture models [10] to the multi-view case and identifies exemplars in the dataset by simultaneously considering all available views. The main advantages of our method are the convexity of the optimized objective, the ability to handle views with different statistical properties and its applicability when only pairwise distances are available and not the data points. Our empirical evaluation with multi-view artificial data and two popular document collections, showed that the presented algorithm can considerably improve the results of a single view convex mixture model based either on the individual views or the concatenated view.

As for future work, we plan to compare our algorithm to other multi-view methods and experiment using additional datasets so as to thoroughly investigate the potential of the multi-view convex mixture model. We also aim to use multi-view convex mixture models in conjunction with other clustering algorithms which will treat the exemplars as a good initialization. Finally, another interesting research direction is the assignment of different weights to different views and the ability to learn those weights automatically under our framework.

Acknowledgments. We would like to thank Steffen Bickel and Tobias Scheffer for kindly providing their processed datasets.

References

1. Blum, A., Mitchell, T.: Combining labeled and unlabeled data with co-training. In: Proceedings of the 11th Annual Conference on Computational Learning Theory, pp. 92–100 (1998)
2. Muslea, I., Minton, S., Knoblock, C.A.: Active + semi-supervised learning = robust multi-view learning. In: Proceedings of the 19th International Conference on Machine Learning, pp. 435–442 (2002)
3. Brefeld, U., Scheffer, T.: Co-em support vector learning. In: Proceedings of the 21st International Conference on Machine Learning (2004)
4. Bickel, S., Scheffer, T.: Multi-view clustering. In: Proceedings of the 4th IEEE International Conference on Data Mining, pp. 19–26 (2004)
5. de Sa, V.R.: Spectral clustering with two views. In: Proceedings of the 22nd International Conference on Machine Learning Workshop on Learning with Multiple Views, pp. 20–27 (2005)
6. Zhou, D., Burges, C.J.C.: Spectral clustering and transductive learning with multiple views. In: Proceedings of the 24th International Conference on Machine Learning, pp. 1159–1166 (2007)
7. Long, B., Yu, P.S., Zhang, Z.M.: A general model for multiple view unsupervised learning. In: Proceedings of the 2008 SIAM International Conference on Data Mining, pp. 822–833 (2008)
8. Nigam, K., Ghani, R.: Analyzing the effectiveness and applicability of co-training. In: Proceedings of the 9th International Conference on Information and Knowledge Management, pp. 86–93 (2000)
9. Bickel, S., Scheffer, T.: Estimation of mixture models using co-em. In: Proceedings of the 16th European Conference on Machine Learning, pp. 35–46 (2005)
10. Lashkari, D., Golland, P.: Convex clustering with exemplar-based models. In: Advances in Neural Information Processing Systems, vol. 20, pp. 825–832 (2008)
11. Banerjee, A., Merugu, S., Dhillon, I.S., Ghosh, J.: Clustering with Bregman divergences. J. Machine Learning Research 6, 1705–1749 (2005)
12. Csiszár, I., Shields, P.C.: Information theory and statistics: A tutorial. Communications and Information Theory 1(4), 417–528 (2004)

Strengthening the Forward Variable Selection Stopping Criterion

Luis Javier Herrera, G. Rubio, H. Pomares, B. Paechter, A. Guillén, and I. Rojas

Department of Computer Architecture and Technology
University of Granada, Spain
jherrera@atc.ugr.es
http:\\atc.ugr.es

Abstract. Given any modeling problem, variable selection is a preprocess step that selects the most relevant variables with respect to the output variable. Forward selection is the most straightforward strategy for variable selection; its application using the mutual information is simple, intuitive and effective, and is commonly used in the machine learning literature. However the problem of when to stop the forward process doesn't have a direct satisfactory solution due to the inaccuracies of the Mutual Information estimation, specially as the number of variables considered increases. This work proposes a modified stopping criterion for this variable selection methodology that uses the Markov blanket concept. As it will be shown, this approach can increase the performance and applicability of the stopping criterion of a forward selection process using mutual information.

Keywords: Variable Selection, Mutual Information, Function Approximation.

1 Introduction

Selecting the most relevant features in a problem before building up a learning machine is a key preprocess step. The identification of redundant and irrelevant variables provide better generalization capabilities, a better interpretability of the constructed model and diminishes the computational cost of the learning process. Given a modeling case in which a set of input/output data D is given with input variables $X = \{x_1, x_2, \ldots, x_n\}$ and output variable $Y = y$, the objective of a variable selection process is to select the variables x_i that are most relevant to predict the value of Y. This problem is very complex, specially in function approximation problems in which the variables involved are continuous. Among the different strategies of variable selection applied to function approximation problems, forward selection using the Mutual Information (MI) criterion is a commonly used approach [1] [2].

Forward selection is a straightforward variable selection strategy, and is based on the following: starting with an empty subset of variables $X_G = \{\}$, the

C. Alippi et al. (Eds.): ICANN 2009, Part II, LNCS 5769, pp. 215–224, 2009.

variable that added to the current X_G provides the largest amount of mutual information with respect to the output variable Y will be added to it. It is an iterative process that is repeated until the mutual information of the current subset X_G with respect to Y, i.e. $I(X_G, Y)$, stops increasing. Other variable selection strategies include backward selection, forward-backward selection, subset selection and block-addition and block-deletion selection; however those strategies are normally more complex, and they haven't showed to obtain better results comparing to forward selection.

The Mutual Information criterion is traditionally a well known and used criterion that has a strong theoretical background based on Shannon's Information Theory. A new estimator for continuous variables based on the k-nearest neighbors [3] has shown to provide more robust results in comparison with other histogram or kernel-based MI estimators, and has received increasing attention in the recent literature. The problem with this estimator (and in general of any other estimator or criterion) is that it suffers from the curse of dimensionality, i.e. the problem caused by the exponential increase in volume associated with adding extra dimensions to a (mathematical) space [4]. Laterally, although in theory the MI should not decrease when additional variables are taken into account, in practice this happens for this estimator under some conditions in a forward selection process, showing a decrease trend of the MI estimation as the number of variables considered increases [1]. As this last work showed, the immediate stopping criterion of forward selection isn't robust as it might stop too early due to those effects. This work presents a modified stopping criterion for the forward selection strategy using Mutual Information, which is based on a heuristic derived from the Markov blanket concept [5][6]. In the simulations section it will be shown that this novel stopping criterion can improve the performance of the traditional forward selection strategy.

The rest of the paper is organized as follows. Section 2 reviews the concepts of mutual information, forward selection and Markov blanket. Section 3 presents the proposed forward selection with modified stopping criterion. Section 4 reviews the Least Squares Support Vector Machine paradigm using efficient learning. Section 5 presents comparative results on three well known data sets. The main conclusions of the work are drawn in section 6.

2 Background

2.1 Mutual Information

Shannon's mutual information (also called cross-entropy) between X and Y can be defined as the amount of information that the group of variables X provide about Y, and can be expressed as

$$I(X, Y) = H(Y) - H(Y|X), \tag{1}$$

where $H(Y)$ is the entropy (measure of uncertainty) of the variable Y, and $H(Y|X)$ is the conditional entropy of Y given X. The mutual information

$I(X,Y)$ thus represents the decrease of uncertainty on Y once we know X. Due to the mutual information and entropy properties, the mutual information can also be defined as

$$I(X,Y) = H(X) + H(Y) - H(X,Y), \tag{2}$$

from which we can easily see that $I(X,Y) = I(Y,X)$. To estimate the mutual information, only the estimate of the joint probability density function (PDF) between X and Y is needed [2]. For continuous variables, this estimation is very complex. This work will use a mutual information estimator based on the k-nearest neighbors technique [3]. The curse of dimensionality and the consequent possible systematic decrease of the MI with increasing size of the groups of variables involved [1] are the problems that this estimator presents.

Therefore no matter which is the strategy used to design a variable selection methodology, if comparisons are made between mutual information estimations among medium or large groups of variables, the performance of the method can be affected. The higher the number of input dimensions, the higher the number of samples needed to adequately cover the input space and obtain a reliable estimation.

2.2 Forward Selection Using Mutual Information

Given the above definitions, the objective of a forward variable selection process can be defined as finding a subset $X_G \subset X$ such that

$$I(X,Y) \approx I(X_G,Y). \tag{3}$$

A forward selection approach would start with an empty subset $X_G = \{\}$ and add input variables to the selected set as the mutual information with respect to the output variable of the selected subset increases. The iterative process would add a new variable x_i to the current subset if

$$I(\{X_G \cup x_i\},Y) > I(X_G,Y). \tag{4}$$

The x_i would be selected as the variable that makes $I(\{X_G \cup x_i\},Y)$ highest. This procedure is fast and straightforward comparing to the other strategies used in the literature.

The stopping criterion of this iterative strategy is the case in which the condition in equation 4 does not hold. The evaluation of this condition requires the comparison of two mutual information estimations. Both the curse of dimensionality problem and the possible systematic decrease of the MI estimations as the number of variables considered increases affect this stopping criterion. The performance of this strategy is therefore not guaranteed. In [1], it is suggested the use of the permutation test to calculate a threshold to stop the forward procedure. This technique however can presents a missleading operation when the input variables are interrelated among each other. The k-nn approach using Euclidean distance, on which the MI estimator is based, is very sensitive to the addition of a random variable to a set of closely valued variables, which is the normal case in for example spectrometric and time series prediction problems.

2.3 Markov Blankets

Markov blankets were first introduced in the variable selection literature in the work by Koller and Sahami [5]. Markov blankets are defined in a set of variables Z as:
Definition: Let M be a subset of variables taken from Z that does not contain x_i. We say that M is a Markov blanket for x_i if $I(\{M \cup x_i\}, Z - \{M \cup x_i\}) \approx I(M, Z - \{M \cup x_i\})$.
That is, a Markov blanket M_i of a variable x_i in a problem with variables Z, is a subset of variables that contains all the information that x_i has of Z. This concept provides a different point of view in order to deal with problems with a high number of variables. From the previous definition, the following corollary is immediate:
Corollary: given a modeling problem with X as set of input variables and Y the output variable. Given a subset X_G of X, and an input variable $x_i \in \{X - X_G\}$. Assume that some subset M_i of X_G is a Markov blanket of x_i in $Z = \{X, Y\}$. Then $I(X_G, Y) \approx I(X_G \cup \{x_i\}, Y)$.
This corollary is immediate since from the definition it is straightforward that $I(M_i, Y) \approx I(M_i \cup \{x_i\}, Y)$. The corollary states that when evaluating the relevance of a variable x_i with respect to the output variable, given a certain subset X_G, if there is a Markov blanket M_i of x_i in X_G, its added relevance will be null. Thus, intuitively, variables for which we find Markov blankets in X_G will not be added to the current X_G in a forward selection procedure.

However, finding either a true or even an approximate Markov blanket of a variable in a set of variables might be very hard; it is a task similar to a variable selection process, in the sense that it is intended to find a subset of variables according to a certain minimization criterion. In the literature [5][6], this theoretical approach is used to design an heuristic to estimate Markov blankets for the variables, to deal with the given variable selection problem.

3 Modified Stopping Criterion for Forward Selection

The most straightforward heuristic for the estimation of the Markov blankets of a variable x_i in a problem with variables X is to select the subset of variables most interrelated with it in X. It is in principle expectable (although not necessarily), that the variables in the Markov blanket M_i of a variable x_i will be quite strongly correlated with it. Then the set of p variables in X most interrelated with x_i will be taken as the approximation of Markov blankets of a variable x_i in a problem. In order to identify the variables most interrelated with a variable x_i, the mutual information with respect to the rest of variables will be calculated $I(x_i, x_j)$, $\forall i, j = 1 \ldots n$.

The idea under the application of the Markov blanket concept for the design of a modified stopping criterion is the following: a candidate Markov blanket M_i will be found in X_G for the newly added variable x_i. If M_i is in fact a good Markov blanket approximation, then the condition $I(M_i, Y) \approx I(M_i \cup \{x_i\}, Y)$ will hold, and therefore for the previous corollary, $I(X_G, Y) \approx I(X_G \cup \{x_i\}, Y)$, i.e. the

Algorithm 1. Forward selection with Markov-blanket based stopping criterion

Calculate the MI between every two input variables $Interrelation(i,j) = I(x_i, x_j)$
Starting with an empty subset of variables $X_G = \{\}$
repeat
  select the variable x_i that added to the current X_G obtains highest $I(\{x_G \cup x_i\}, Y)$
  let the candidate Markov blanket M_i be the set of p variables in X_G for which $Interrelation(i,j)$ is highest.
  Compute the additional information that x_i provides with respect to Y given M_i, i.e., compute $I(\{M_i \cup x_i\}, Y)$ and $I(M_i, Y)$
until $I(\{M_i \cup x_i\}, Y) \approx I(M_i, Y)$ showing that there is not additional information provided by x_i with respect to that already present in X_G.

stopping criterion holds. Otherwise, if $I(M_i \cup \{x_i\}, Y) > I(M_i, Y)$, then M_i is not a good Markov blanket approximation of x_i, and then it will be supposed that there is not a Markov blanket of x_i in X_G; then x_i is considered to be relevant with respect to Y.

Therefore, according to the forward selection strategy and the calculation of the approximate Markov blankets of the variables described before, the proposed forward selection procedure with modified stopping criterion stays as shown in algorithm 1.

The advantage of using Markov blankets candidates in the variable selection process is the following. As seen from the corollary in the previous section, it is equivalent to perform the comparison $I(X_G, Y) \approx I(\{X_G \cup x_i\}, Y)$ than $I(M_i, Y) \approx I(\{M_i \cup x_i\}, Y)$. However there is a great advantage in evaluating the mutual information in the second case comparing to the first one: the size of the set of variables is lower than for the first possibility, thus diminishing the problems related to the dimensionality of the sets of variables in the mutual information estimators, and increasing its robustness. The estimation of the difference in mutual information of the selected set of variables when adding a new one (see equation 4), is translated into a similar estimation but using a lower-sized set of variables. If the Markov blanket estimation for the variable is correct, this approach leads to the same theoretical result, but avoiding the estimation of the mutual information among large-sized groups of variables.

In order to evaluate the effectiveness of the proposed forward filtering approach, it is necessary to evaluate its behavior using a certain learning methodology and compare their performance. The following section shortly reviews the paradigm chosen: the Least Squares Support-Vector-Machines learning methodology for function approximation problems. It also describes an optimized technique to improve their computational cost in training.

4 Efficient Training of Least-Squares Support Vector Machines

Least Squares Support Vector Machines (LS-SVMs) [7], also known as the Kernel Ridge Regression method (KRR) [8], are a kernel-based paradigm specially well suited for function approximation problems. The main advantages of the LS-SVMs for regression with respect to traditional Support Vector Regression (SVR), is their easier mathematical resolution, and that the parameter ε from the SVR disappears and number of Lagrange multipliers is reduced to half. The LS-SVMs on the other hand have the disadvantage that they don't generate sparse models.

In case we consider Gaussian kernels, σ is the width of the kernel, that together with the regularization parameter γ, are the hyper-parameters of the problem. Note that in the case in which Gaussian kernels are used, the models obtained resemble Radial Basis Function Networks (RBFN); with the particularities that there is an RBF node per data point, and that overfitting is controlled by a regularization parameter instead of by reducing the number of kernels [2].

In LS-SVM, the hyper-parameters of the model can be optimized by cross-validation. Nevertheless, in order to speed-up the optimization, a special formulation for a reduced cost evaluation of the cross-validation error of order l (l-fold CV) taken from the work [9] was used. With this formulation, the error evaluation cost of cross-derivation does not depend on the order l, but on the number of data points of the problem, since in fact the computational cost is dominated by the inversion of the kernels K activation matrix. Such inversion is performed through a Cholesky decomposition; the most efficient exact algorithm for this case is $O(N^3)$ where N is the number of samples.

In order to perform the evaluation of the performance of the stopping criteria in the forward selection strategy, it is necessary to learn a number of LS-SVMs, each one considering the eventual state of the selected subset in the iterative process X_G of the variable selection process. This requires therefore the training of a considerable number of LS-SVM, depending on the problem. In this work, this process was distributed in a computer cluster, so that each training process of a LS-SVM was sent to a different node. This way the computational time was reduced in a factor of N (considering a computer with N nodes, ignoring communication delays), supposing that every execution takes the same amount of computational time. The executions were performed in the Ness supercomputer at the EPCC in the University of Edinburgh. The system runs linux and it has two back-end X4600 SMP nodes, both containing 16 processor-cores with 2GB of memory per core. The programming interface used for the simulations was MPIMEX [10] under MATLAB.

5 Simulations

This section presents a comparative to evaluate the performance of the proposed approach for forward selection with modified stopping criterion. A number of significant function approximation examples typically used in the machine learning literature as benchmarks has been considered, including spectrometric problems and time series prediction problems.

In the simulations performed in this work the parameter p, that defines the size of the Markov blanket approximations was assigned a medium value $p = 2$ for the tradeoff between effectiveness and avoidance of the dimensionality problem.

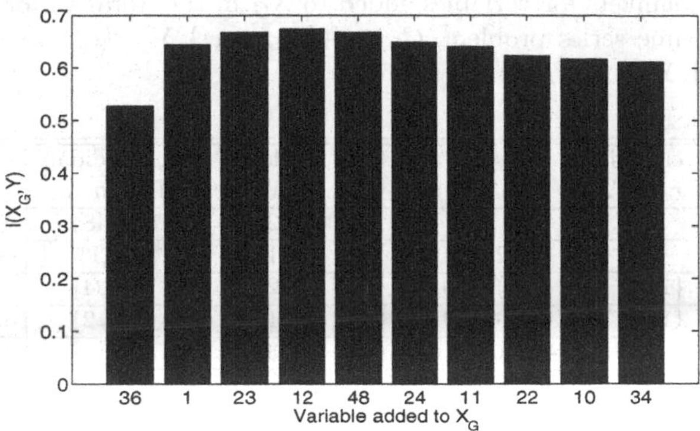

Fig. 1. Evolution of the mutual information of X_G with respect to Y, $I(X_G, Y)$ in the iterative forward variable selection approach

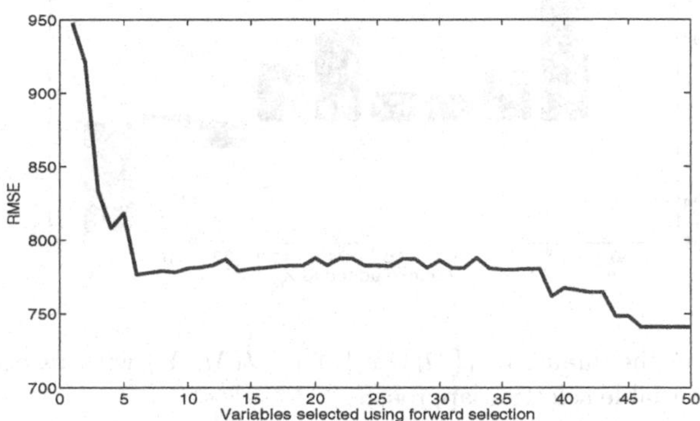

Fig. 2. Evolution of the test error in the iterative forward variable selection process

5.1 Results

This subsection will first show the evolution of the designed algorithm for a river flow time series prediction problem: it measures the monthly river flow in 1000 acre-feet of the Snake River near Moran, Wyoming among 1904-1994 [11]. For this example, 500 data samples were considered for training, and 500 for testing [12]. The objective is to obtain a prediction model to estimate

$$\hat{y}(t+1) = F(y(t), y(t-1), \ldots, y(t-49)) \tag{5}$$

It is intended to use a forward variable selection approach to select the most relevant variables (among the previous 50 values, $X = \{y(t) \ldots y(t-49)\}$) to predict $Y = \hat{y}(t+1)$ using the training data set. A code for the k-nearest neighbors mutual information estimator can be found in [13]. The mutual information estimation was improved using L-fold resampling, and the k parameter in the MI estimator was obtained according to the work in [1]. Figure 1 shows the

Table 1. Markov blankets for variables added to X_G in the iterative forward process, for the river flow time series problem. $C_1 = I(\{X_G \cup x_i\}, Y)$ -$I(X_G, Y)$ see figure 1; $C2 = I(\{M_I \cup x_i\}, Y)$ - $I(M_I, Y)$ see figure 2

It.	added variable x_i	M_i candidate of x_i in X_G	C_1	C_2	It.	added variable x_i	M_i candidate of x_i in X_G	C_1	C_2
1	36	-	-	-	6	24	{23,48}	-0.0187	0.0449
2	1	36	0.12	0.12	7	11	{12,1}	-0.0088	0.0283
3	23	{1,36}	0.023	0.023	8	22	{23,24}	-0.0173	-0.0126
4	12	{1,36}	0.0063	0.012	9	10	{11,12}	-0.0055	0.0031
5	48	{36,1}	-0.0064	0.0140	10	34	{36,12}	-0.0054	-0.0543

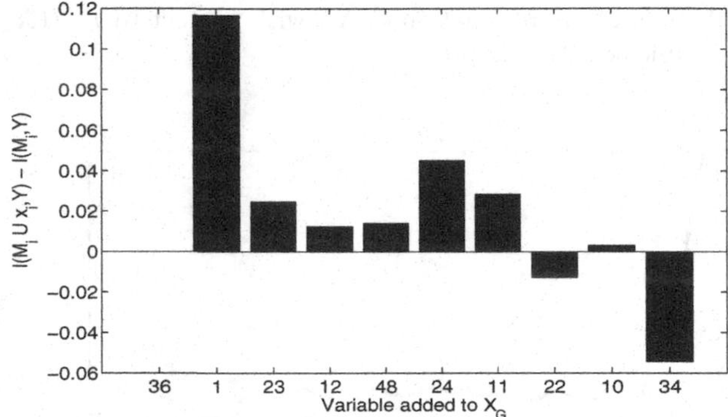

Fig. 3. Evolution of the quantity $I(\{M_i \cup x_i\}, Y) - I(M_i, Y)$ with respect to Y in the iterative forward variable selection approach

evolution of the mutual information of the current selected subset of variables X_G with respect to the output variable Y in the iterative forward process. It is observed that the maximum of the quantity $I(x_G, Y)$ is obtained when X_G includes 4 variables; from this point as the size of X_G increases, the value of $I(x_G, Y)$ decreases.

Figure 2 shows the evolution of the test error in the iterative forward selection process. As it is observed, the mutual information forward stopping criterion does not identify correctly the correspondence with the optimal performance in the iterative process, as it stops too early. The best subset of variables should include at least 6 variables to obtain the best performance (RMSE = 776, with four variables RMSE = 810).

Table 2 shows the sequence of Markov blanket candidates selected for each variable added to X_G according to algorithm 1. Figure 3 shows the evolution of the quantity $I(\{M_i \cup x_i\}, Y) - I(M_i, Y)$, that controls the modified stopping of the algorithm for that sequence. The proposed criterion would stop when checking the addition of variable $y(t-22)$, as the Markov blanket condition provides a negative value, showing that $y(t-22)$ provides no additional information

Table 2. Evaluation of the traditional and the proposed stopping criteria for the forward selection in the Mackey-Glass and the Tecator Meat function approximation problems

	Forward criterion		Proposed approach		First optimal found	
	Size of X_G	Test Error	Size of X_G	Test Error	Size of X_G	Test Error
Mackey Glass	3	4.5e-03	7	1.6e-05	8	1.0e-05
Tecator Meat	3	1.0e-01	6	5.9e-02	6	5.9e-02

from what variables $\{y(t-23), y(t-24)\}$ present in X_G already provide. As it is observed, the criterion based on the candidate Markov blankets presents a better correspondence with the real optimal performance in the given problem. According to this criterion, the selected subset of variables X_G has size equal to 7, that corresponds to the best local minimum area seen in figure 2.

The following table shows the results and performance for other two well known problems: the Mackey-Glass time series, including 50 previous regressors to predict $y(t+1)$, having 500 training data samples and 500 test data samples; the Tecator meat data set is a spectrometric data set consisting of 100 spectral input variables and one output variable (as in [2], the spectra are reduced to zero mean and unit variance, keeping the original mean and standard deviation as two additional variables to avoid loss of information).

6 Conclusions

Forward selection with mutual information is a straightforward and widely used strategy for variable selection. However it presents the problem that establishing an adequate stopping criterion is an unsolved question. This work has presented an alternative stopping criterion based on Markov blankets for forward selection using the k-nearest neighbors mutual information estimator for function approximation problems. In this context, the traditional forward stopping criterion can be far from optimality due to the deficiencies of the mutual information estimator. The proposed approach shows a better approach to optimality in the examples used. For the evaluation of the performance of the alternatives in the forward process, Least Squares Support Vector Machines have been used, using an efficient training to avoid the large computational cost.

Acknowledgments. This work was carried out using the facilities of the Edinburgh Parallel Computing Centre[1] (EPCC) and under the HPC-EUROPA project (RII3-CT-2003-506079), with the support of the European Community - Research Infrastructure Action under the FP6 "Structuring the European Research Area" Programme. This work has been partially supported by the Spanish CICYT Project TIN2007-60587 and Junta Andalucia Project P07-TIC-02768.

[1] `http://www.epcc.ed.ac.uk/`

References

1. François, D., Rossi, F., Wertz, V., Verleysen, M.: Resampling methods for parameter-free and robust feature selection with mutual information. Neurocomputing 70, 1276–1288 (2007)
2. Rossi, F., Lendasse, A., François, D., Wertz, V., Verleysen, M.: Mutual information for the selection of relevant variables in spectrometric nonlinear modelling. Chem. and Int. Lab. Syst. 80, 215–226 (2006)
3. Kraskov, A., Stogbauer, H., Grassberger, P.: Estimating mutual information. Phys.Rev. E 69, 66138 (2004)
4. Bellman, R.: Adaptive Control Processes: A Guided Tour. Princeton University Press, Princeton (1961)
5. Koller, D., Sahami, M.: Toward optimal feature selection. In: Proc. Int. Conf. on Machine Learning, pp. 284–292 (1996)
6. Herrera, L., Pomares, H., Rojas, I., Verleysen, M., Guillén, A.: Effective input variable selection for function approximation. In: Kollias, S.D., Stafylopatis, A., Duch, W., Oja, E. (eds.) ICANN 2006. LNCS, vol. 4131, pp. 41–50. Springer, Heidelberg (2006)
7. Suykens, J., Gestel, T.V., Brabanter, J.D., Moor, J.D., Vandewalle, B.: Least Squares Support Vector Machines. World Scientific, Singapore (2002)
8. Saunders, C., Gammerman, A., Vovk, V.: Ridge regression learning algorithm in dual variables. In: Proceedings of the 15th International Conference on Machine Learning, pp. 515–521. Morgan Kaufmann, San Francisco (1998)
9. An, S., Liu, W., Venkatesh, S.: Fast cross-validation algorithms for least squares support vector machine and kernel ridge regression. Pattern Recogn. 40(8), 2154–2162 (2007)
10. Guillen, A., Rojas, I., Rubio, G., Pomares, H., Herrera, L., Gonzalez, J.: A new interface for mpi in matlab and its application over a genetic algorithm. In: ESTSP 2008: Proceedings of the European Symposium on Time Series Prediction, pp. 37–46 (2008)
11. Hyndman, R.: Time series data library (1994), `http://www-personal.buseco.monash.edu.au/~hyndman/TSDL/hydrology.html`
12. Herrera, L., Pomares, H., Rojas, I., Guillén, A., Prieto, A., Valenzuela, O.: Recursive prediction for long term time series forecasting using advanced models. Neurocomputing 70, 2870–2880 (2007)
13. Astakhov, S., Grassberger, P., Kraskov, A., Stögbauer, H.: Mutual information least dependent component analysis (2004), `http://www.klab.caltech.edu/~kraskov/MILCA/`

Features and Metric from a Classifier Improve Visualizations with Dimension Reduction

Elina Parviainen and Aki Vehtari

Helsinki University of Technology,
Department of Biomedical Engineering and Computational Science

Abstract. The goal of this work is to improve visualizations by using a task-related metric in dimension reduction. In supervised setting, metric can be learned directly from data or extracted from a model fitted to data. Here, two model-based approaches are tried: extracting a global metric from classifier parameters, and doing dimension reduction in feature space of a classifier. Both approaches are tested using four dimension reduction methods and four real data sets. Both approaches are found to improve visualization results. Especially working in classifier feature space is beneficial for showing possible cluster structure of the data.

Keywords: dimension reduction, visualization, classification, metric, feature space.

1 Introduction

A successful visualization makes discussions and presentations about data more concrete and easy to follow; a bad one may just add to the cognitive load of a listener. Visualizations should not only be readable but also be related to the application at hand. If we are dealing with a classification task, we would like also our visualizations to bring forth factors having a great impact on classification. However, if visualization is done without regard to the classification task, variables having biggest effect on visualization results may not be those most relevant for classification.

In this paper, we show how we can exploit information a classifier model has learned about the covariance structure of the data, and use it in dimension reduction for more task-related visualizations. Very little extra work is needed; we don't need to learn a metric separately, or to do separate feature extraction, but just use the knowledge we get as a side-product of model fitting.

Our paper is motivated by having worked with a health care application. In health care, a black-box solution is not enough, but the results and their explanations need to be studied in more detail. Especially groups of patients with any specific characteristics are of interest. Therefore, we would like the visualizations clearly show any cluster structure in the data. Clustering algorithms are known to be affected by the metric used, and we expect also the visualizations, especially on data with some cluster structure, to be affected by a metric.

C. Alippi et al. (Eds.): ICANN 2009, Part II, LNCS 5769, pp. 225–234, 2009.

We approach the problem from two different angles. First, we can see the classifier as having learned a (global) metric from the data. The metric information is contained in the model parameters, and we show how to extract a metric in the form of a similarity matrix. We term this *metric approach*. Secondly, we can use a familiar euclidean metric, but work in classifier feature space. As the model has been trained to perform a certain task, the latent features learned contain information relevant to that task, and should therefore give a more task-related visualization than using the raw data. To this we refer as *feature approach*.

In this work we use Gaussian processes (GP), [1] and multilayer perceptron networks (MLP) [2] . It should be possible to handle other kernel based models similarly to GP and models, which can be interpreted as feature extractors, like MLP. Our examples are from binary classification tasks, but the same technique should be usable for regression models or multiclass classifiers.

2 Related Work

Using a good distance metric, whether as a metric or as a kernel, has been found important for performance of classifiers and clustering algorithms. Some of these metrics and kernels could probably be used for better visualizations as well. Learning metrics and kernels has been done both for unsupervised case [3,4], unsupervised specifically for dimension reduction in [5], and in supervised context for better classification [6,7,8]. Using similarity information from an existing probability model has been studied e.g. in [9] and [10]. These works start with a generative model and build a kernel for classification.

Metric information from a predictive model has been used for improving visualizations with self-organizing maps [11], and same technique could probably be used for other dimension reduction methods. Of the papers mentioned here, this is probably closest to our work. The approach of [11] uses local metric based on Fisher information. Local metric captures well the covariance structure of data, but determining the distance between two points becomes computationally intensive since it requires integration (or an approximation thereof) over the metric. We use both a global metric, which is computationally cheap, and features, which capture local behavior without requiring much computation. We reduce ourselves to two specific classes of models, whereas [11] uses a generic model.

3 Metric Approach

We extract metric information from GP and MLP models in the form of a similarity matrix S with elements s_{ij}. If needed, it is converted into a dissimilarity matrix D of elements d_{ij} by

$$d_{ij}^2 = s_{ii} + s_{jj} - 2s_{ij} \ ,$$

which produces a valid dissimilarity even for similarities produced by nonstationary covariance functions (like NN below).

3.1 Metric from GP

Gaussian processes are a kernel method, in which prior assumptions about function to be fitted are presented by choosing a covariance function. Optimal values for its parameters are found during training. Covariance function evaluated at data points using these parameters gives a similarity matrix which we use in dimension reduction. Here we use neural network (NN) covariance [12]

$$k_{NN}(\mathbf{x}, \mathbf{x}') = \frac{2}{\pi} \sin^{-1} \frac{2\tilde{\mathbf{x}}^T \Sigma \tilde{\mathbf{x}}'}{\sqrt{(1 + 2\tilde{\mathbf{x}}^T \Sigma \tilde{\mathbf{x}})(1 + 2\tilde{\mathbf{x}}'^T \Sigma \tilde{\mathbf{x}}')}} \ . \tag{1}$$

where $\tilde{\mathbf{x}} = [1\ \mathbf{x}]$ is an augmented input vector and Σ is a diagonal matrix with variances of inputs. We chose NN covariance both because it has performed well in our applications and for its connection to extracting features from an MLP model (below).

3.2 Metric from MLP

NN covariance (1) is derived by integrating over the weights of an MLP network with an infinite number of hidden units. This leaves variances of inputs as parameters. To extract a metric from MLP, we can use our trained classifier network as an approximation to the hypothetical MLP of NN derivation. Even though 10 or 20 hidden units may seem like a poor substitute for an infinite number of units, our results show that the approximation may be good enough for practical purposes.

As an approximating MLP, we use a network of one hidden layer of H units, working on C-dimensional input $\mathbf{x} = [x_1, x_2, \ldots, x_C]$. Weights w_{ch} are from input x_c to hidden unit h, and hidden units have biases b_h. In the output layer, weights w'_h are for hidden unit h, and bias is b'.

The output of hidden unit h for data point $\mathbf{x}$ is given by

$$\mathbf{z}_h(\mathbf{x}) = \sigma(\sum_{c=1}^{C} w_{ch} x_c + b_h) \ , \tag{2}$$

where $\sigma(\cdot)$ is a sigmoid function used in hidden units, and hidden unit outputs are combined into the output of the network as

$$y = \sum_{h=1}^{H} w'_h \mathbf{z}_h(\mathbf{x}) + b' \ . \tag{3}$$

After training the network, we have estimates for all weights and biases available. To use MLP weights to approximate the NN covariance (1), we set

$$\Sigma = \mathrm{diag}(\mathrm{var}(\{b_*\}), \mathrm{var}(\{w_{1*}\}), \mathrm{var}(\{w_{2*}\}), \ldots, \mathrm{var}(\{w_{C*}\})) \ ,$$

where variance is taken over the hidden units ($*$ denotes numbers from 1 to H).

4 Feature Approach

We extract features f corresponding to data points as described below. To get a dissimilarity matrix D, we compute squared pairwise euclidean distances of the features and normalize them to range $[0, 1]$. For similarity-based methods we transform this matrix into a similarity matrix S of elements

$$s_{ij} = \sqrt{1 - d_{ij}^2} \ .$$

4.1 GP as Feature Extractor

A covariance function computes a mapping from data space to model feature space. In feature space, each data point is represented by its similarities to all other data points. If data set size N is big, it is also possible to use only a subset of it for GP training, and compute similarities to training points only. Features for data point $\mathbf{x}_i$ are simply the values in the ith column of the covariance matrix,

$$f = [k(\mathbf{x}_i, \mathbf{x}_1), k(\mathbf{x}_i, \mathbf{x}_2), \ldots, k(\mathbf{x}_i, \mathbf{x}_N)] \ ,$$

with $k(\cdot, \cdot)$ the covariance used in the GP model, evaluated using the parameter values found during model training.

4.2 MLP as Feature Extractor

An MLP network with one hidden layer can be considered as a feature extractor. The hidden units form different features from the data, and the output layer does classification or other computation in the feature space. The outputs of the MLP hidden layer (2) become the features for data point $\mathbf{x}_i$,

$$f = [\mathbf{z}_1(\mathbf{x}_i), \mathbf{z}_2(\mathbf{x}_i), \ldots, \mathbf{z}_H(\mathbf{x}_i)] \ .$$

5 Dimension Reduction Methods

Dimension reduction is a problem of finding a low-dimensional (for visualization purposes usually 2D or 3D) target space presentation for points lying in a high-dimensional data space. For this work we have chosen four dimension reduction methods that use either similarities (kernel PCA) or dissimilarities (t-SNE, Sammon mapping, Isomap) as inputs. Such a method can be easily adapted to use a new (dis)similarity measure by using a similarity derived from the model as kernel matrix (kernel PCA) or replacing pairwise distances (other methods) by a dissimilarity matrix. As a base case to which we compare our method we use pairwise euclidean distances of original data points, transformed into similarities when needed.

T-SNE [13] is developed for 2D and 3D visualizations. It tries to ensure both target space proximity of nearby data points and large target space distance of distant data points. Distances are transformed into conditional probabilities, which tell how probably a point has another point as its neighbor. Locations for target space points are found by minimizing the Kullback-Leibler divergence between neighborhood probabilities in data space and target space.

Sammon mapping [14] is a version of metric multidimensional scaling (MDS). It finds locations for target space points such that interpoint distances match, as closely as possible, the distances between original data points. Discrepancy between target distances and data space distances can be measured using various goodness-of-fit functions, of which the Sammon criterion is one often used in data visualization.

Isomap [15] is designed for finding lower-dimensional manifolds embedded in a high-dimensional space. It performs a preprocessing step for MDS. Where plain MDS uses euclidean distances, measured in data space, Isomap calculates an approximate geodesic distance along a low-dimensional manifold, and then uses MDS.

Kernel PCA [16] is one of nonlinear generalizations of principal component analysis (PCA). It performs PCA in a feature space determined by a kernel. We use a similarity measure derived from a classifier as a kernel.

6 Experiments

We have tried both metric and feature approaches using the four dimension reduction methods introduced above, using four different data sets. All data sets are associated with a binary classification task. Data sets d1 (19 covariates, continuous and binary) and d2 (17 covariates, continuous and binary) are from health care applications we work with [1] , and d3 (9 covariates, continuous) and d4 (16 covariates, binary) are publicly available [2]. Binary classifiers (MLP and GP) were trained on all data sets. Classification results are presented as probability p of belonging to class 1.

Similarity and dissimilarity matrices representing global metric information were built for each data set/model combination (as explained in Sect. 3) and these were used in dimension reduction methods. This forms the cases for the metric approach, presented in figures as 'GP metric' and 'MLP metric'.

[1] d1 is data on institutionalization of elderly, analyzed in co-operation with Dr. Matti Mäkelä, City of Vantaa and National Institute for Health and Welfare, and d2 is data on rehabilitation of hip fracture patients [17,18], from collaboration project with Dr. Reijo Sund, National Institute for Health and Welfare.

[2] d3 is Wisconsin Breast Cancer Database (from Dr. William H. Wolberg of University of Wisconsin Hospitals, Madison) and d4 is 1984 United States Congressional Voting Records Database; both available at `http://archive.ics.uci.edu/ml/`.

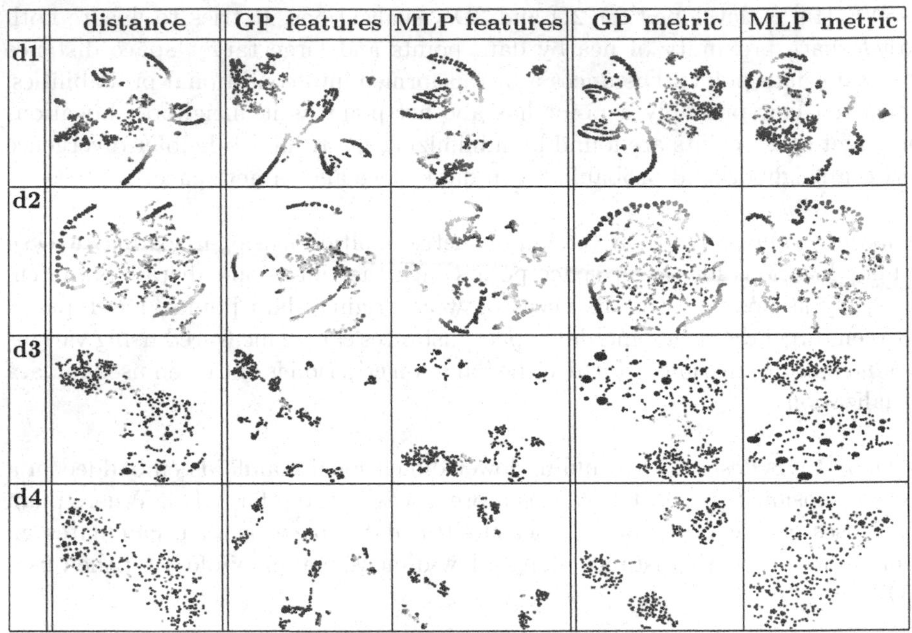

Fig. 1. Visualizations using t-SNE (perplexity=30)

Fig. 2. Visualizations using Sammon mapping

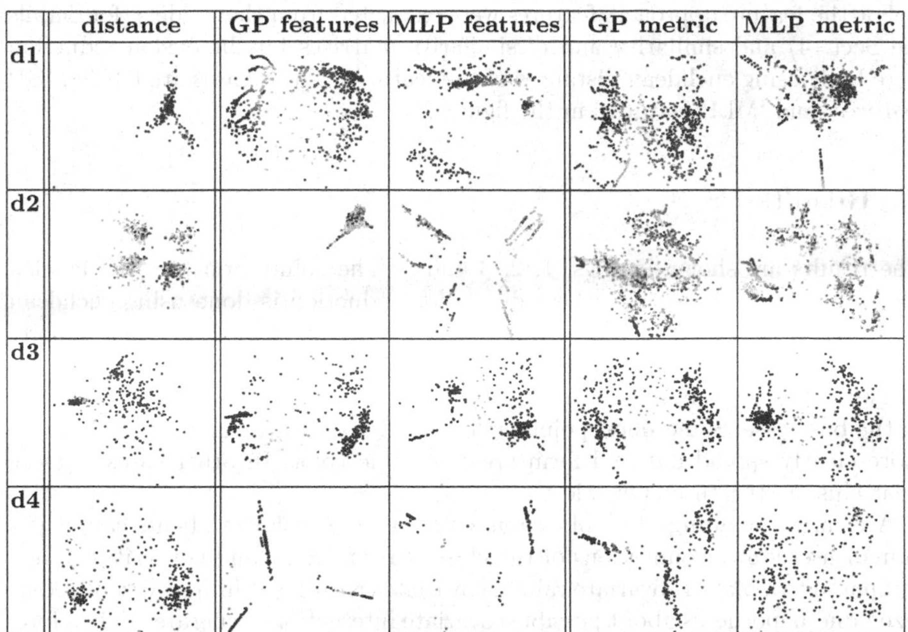

Fig. 3. Visualizations using Isomap (k=15)

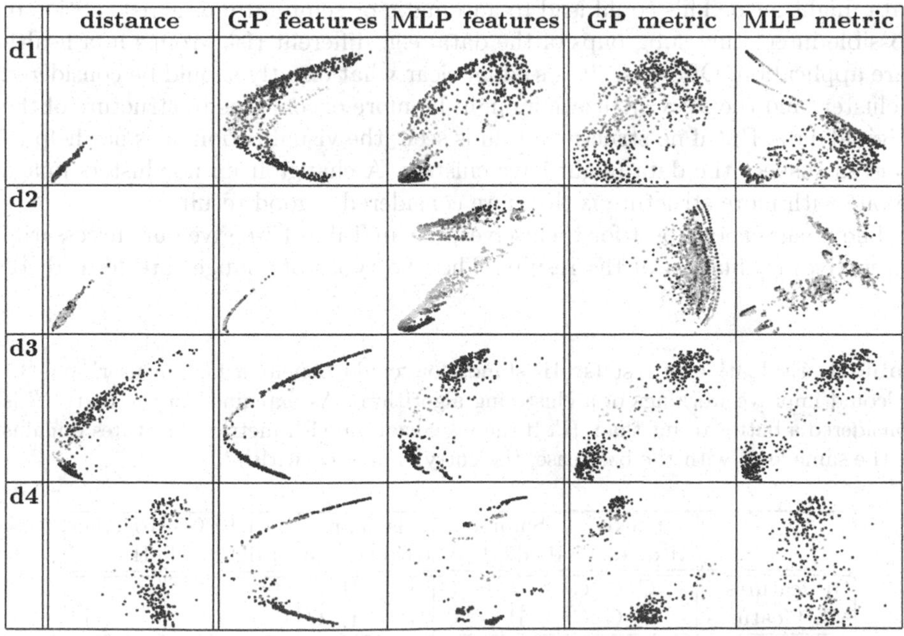

Fig. 4. Visualizations using kernel PCA

For the feature approach, features were extracted from the models (for details see Sect. 4) and similarity and dissimilarity matrices for dimension reduction were built using euclidean distances of the features. These cases are labeled 'GP features' and 'MLP features' in the figures.

7 Results

The results are shown in Figs. 1, 2, 3 and 4. The column on the left, labeled "distance", is the base case where dimension reduction is done using euclidean distances of the original data points. Colors show the class probability, as predicted by the classifier from which the metric or features were extracted.

Two types of changes are seen when metric or feature approach is compared to the base case. Sometimes, points which nearly co-located in the base case are more clearly spread out and form areas of same color. In other cases, a more clear cluster structure is visible.

A clearer spreading of points, same color areas or color gradients can help a human user study connections of the classification result and covariates. When compared to a plot of covariate values, covariates with great impact can be recognized and hypotheses about possible covariate interactions generated. Therefore, we consider this kind of result to be an improvement, if the base case does not show clear areas.

What we by choice would like to see, however, is any cluster structure the data might have. This could lead to automatic or semi-automatic recognition of possible interesting subgroups of the data, e.g. different risk groups in a health care application. Of course, it is seldom clear what exactly should be considered a cluster, and the 2D visualization might be more or less true to structure of the original data. But if no structure at all is seen, the visualization does not help us to even suspect the data might have clusters. A change from no clusters image to one with more structure is therefore considered a good result.

The images speak best for themselves, but in Table 1 we give our (necessarily subjective) evaluation of the results. The two types of changes are termed 'H'

Table 1. Evaluation of results. 'H' stands for 'could benefit a human user' and 'C' is 'could improve workings of a clustering algorithm'. As explained in the text, 'C' is considered a better result than 'H'. If the visualization with metric or features remains at the same level with the base case, the entry is marked with '-'.

	t-SNE				Sammon				Isomap				kernel PCA				together	
	d1	d2	d3	d4	d1	d2	d3	d4	d1	d2	d3	d4	d1	d2	d3	d4	H	C
GP features	-	C	C	C	-	-	C	C	H	-	H	C	H	-	-	C		
MLP features	-	C	C	C	-	H	-	C	H	-	H	C	H	H	C	C	8	14
GP metric	-	-	-	C	H	-	H	H	H	-	H	C	H	H	C	C		
MLP metric	-	-	-	-	-	-	H	-	H	-	H	-	H	H	C	-	12	5

(could benefit a human user) and 'C' (could improve workings of a clustering algorithm). Both conditions were judged visually. Using features improves results in 22 cases of the 32 tried; 14 of these show cluster structure more clearly than the base case. The metric approach improves results in 17 cases, in 5 of them clusters are seen more clearly.

8 Discussion and Conclusions

In this work, we experimented with exploiting the metric and feature information a classifier learns from the data, for visualizing the data using dimension reduction to 2D space. We used a binary classifier, but the same technique could be used with multiclass classifiers or regression models. For classification we used Gaussian processes and multilayer perceptron networks, and for dimension reduction, t-SNE, Sammon mapping, Isomap and kernel PCA.

We tried two different approaches, using a global metric and using features. We compared these approaches to visualizing the data as such, and also assessed the relative performance of the two approaches. Both approaches improved the quality of visualizations, but metric approach fell second to feature approach in most cases. When dimension reduction was done using features, readability of the visualizations was improved in 2/3 of the cases tried, and with metric approach, in about half of the cases. Especially cluster structure of the data was seen more clearly with features than when using a global metric.

In this work, we demonstrated usefulness of our method in rather informal way, and the next step will be a more rigorous analysis of the results. Using a single quantitative criterion will hardly be enough, since evaluation of the results intertwines questions about quality of visualization (e.g. preservation of distances or trustworthiness of mapping), goodness of clustering, and assessing the effect of the chosen metric.

We assume the feature approach worked better because the features, unlike a global metric, are able to capture local variations of data. With this assumption, it would be interesting to see how features would compare to learning a local metric from the data or extracting it from a model, e.g. as in [11].

As conclusion of this work, we recommend doing 2D visualization with dimension reduction using features learned by a model, instead of using the data as such. No separate metric learning of feature extraction phase is needed, but features are easily obtained from GP or MLP model. If modeling and dimension reduction are to be done anyway, doing the visualization in feature space can improve the results with negligible implementation and running time costs.

Division of work. Ideas and experiments in this paper are those of the first author. The second author has supervised the work, commenting on text and ideas.

References

1. Rasmussen, C.E., Williams, C.K.I.: Gaussian processes for machine learning. MIT Press, Cambridge (2006)
2. Lampinen, J., Vehtari, A.: Bayesian approach for neural networks – review and case studies. Neural Networks 14(3), 7–24 (2001)
3. Ye, J., Zhao, Z., Liu, H.: Adaptive distance metric learning for clustering. In: Proc. of the IEEE Computer Society Conference on Computer Vision and Pattern Recognition, pp. 1–7 (2007)
4. Xing, E.P., Ng, A.Y., Jordan, M.I., Russell, S.: Distance metric learning, with application to clustering with side-information. In: NIPS 15, pp. 505–512 (2002)
5. Weinberger, K.Q., Sha, F., Saul, L.K.: Learning a kernel matrix for nonlinear dimensionality reduction. In: Proc. 21st International Conference on Machine Learning, pp. 839–846 (2004)
6. Strickert, M., Schneider, P., Keilwagen, J., Villmann, T., Biehl, M., Hammer, B.H.: Discriminatory data mapping by matrix-based supervised learning metrics. In: Prevost, L., Marinai, S., Schwenker, F. (eds.) ANNPR 2008. LNCS (LNAI), vol. 5064, pp. 78–89. Springer, Heidelberg (2008)
7. Globerson, A., Roweis, S.: Metric learning by collapsing classes. In: NIPS 18, pp. 451–458 (2005)
8. Lanckriet, G.R.G., Cristianini, N., Bartlett, P., El Ghaoui, L., Jordan, M.I.: Learning the kernel matrix with semidefinite programming. JMLR 5, 27–72 (2004)
9. Jaakkola, T.S., Haussler, D.: Exploiting generative models in discriminative classifiers. In: NIPS 11 (1998)
10. Seeger, M.: Covariance kernels from Bayesian generative models. In: NIPS 14, pp. 905–912 (2002)
11. Peltonen, J., Klami, A., Kaski, S.: Learning more accurate metrics for self-organizing maps. In: Dorronsoro, J.R. (ed.) ICANN 2002. LNCS, vol. 2415, pp. 999–1004. Springer, Heidelberg (2002)
12. Williams, C.K.I.: Computation with infinite neural networks. Neural Computation 10, 1203–1216 (1998)
13. van der Maaten, L., Hinton, G.: Visualizing data using t-SNE. JMLR 9, 2579–2605 (2008)
14. Sammon, J.W.: A nonlinear mapping for data structure analysis. IEEE Transactions on Computers C-18(5), 401–409 (1969)
15. Tenenbaum, J.B., de Silva, V., Langford: A global geometric framework for nonlinear dimensionality reduction. Science 290, 2319–2322 (2000)
16. Schölkopf, B., Smola, A., Müller, K.: Kernel principal components analysis. In: Advances in kernel methods: support vector learning, pp. 327–352. MIT Press, Cambridge (1999)
17. Sund, R.: Methodological perspectives for register-based health system performance assessment. Developing a hip fracture monitoring system in Finland. Technical Report Stakes Research Report 174, National Research and Development Centre for Welfare and Health, Helsinki, Finland (2008)
18. Sund, R., Riihimäki, J., Mäkelä, M., Vehtari, A., Lüthje, P., Huusko, T., Häkkinen, U.: Modeling the length of the care episode after hip fracture: does the type of fracture matter? Scandinavian Journal of Surgery (in press, 2009)

Fuzzy Cluster Validation Using the Partition Negentropy Criterion

Luis F. Lago-Fernández[1], Manuel Sánchez-Montañés[1], and Fernando Corbacho[2]

[1] Departamento de Ingeniería Informática, Escuela Politécnica Superior, Universidad Autónoma de Madrid, 28049 Madrid, Spain
[2] Cognodata Consulting, Calle Caracas 23, 28010 Madrid, Spain

Abstract. We introduce the Partition Negentropy Criterion (PNC) for cluster validation. It is a cluster validity index that rewards the average normality of the clusters, measured by means of the negentropy, and penalizes the overlap, measured by the partition entropy. The PNC is aimed at finding well separated clusters whose shape is approximately Gaussian. We use the new index to validate fuzzy partitions in a set of synthetic clustering problems, and compare the results to those obtained by the AIC, BIC and ICL criteria. The partitions are obtained by fitting a Gaussian Mixture Model to the data using the EM algorithm. We show that, when the real clusters are normally distributed, all the criteria are able to correctly assess the number of components, with AIC and BIC allowing a higher cluster overlap. However, when the real cluster distributions are not Gaussian (i.e. the distribution assumed by the mixture model) the PNC outperforms the other indices, being able to correctly evaluate the number of clusters while the other criteria (specially AIC and BIC) tend to overestimate it.

Keywords: Clustering, Cluster validation, Mixture Model, Negentropy, EM algorithm.

1 Introduction

Cluster analysis [1] deals with the automatic partition of a data set into a finite number of natural structures, or clusters. The elements inside a cluster must be similar, while those belonging to different clusters must not. Clustering algorithms are usually divided into crisp and fuzzy. In crisp clustering, each data point is uniquely assigned to a single cluster. On the contrary, fuzzy clustering allows each point to belong to any of the clusters with a certain degree of membership.

A standard approach in fuzzy clustering is model clustering, where it is assumed that the observed data are generated from a mixture of probability distributions (clusters or components). The mathematical structure of these distributions is assumed to be of a certain type (usually Gaussians) but the specific parameters (e.g. means and covariances) must be found. Once the number of

C. Alippi et al. (Eds.): ICANN 2009, Part II, LNCS 5769, pp. 235–244, 2009.

components and their parameters have been selected following some strategy, the degree of membership of the point x with respect to the cluster c is usually related to the probability $p(c|x)$.

There exist different methodologies to select the parameters of the mixture model, the most popular being the Expectation-Maximization (EM) algorithm [2]. However, a common problem is how to determine the correct number of components in the mixture. A different but closely related problem is how to measure the validity of the outcomes of a particular fuzzy clustering method. This is the subject of cluster validation [3], whose objective is to provide a quality measure, or validity index, that allows to evaluate the results obtained by a clustering algorithm. Many cluster validity indices have been proposed in the literature, including geometric [4,5,6], probabilistic [7,8,9], graph theoretic [10], and visual [11,12] approaches.

In the context of density estimation using mixture models, different strategies have been explored to automatically select the number of mixture components [13,14,15,16], the most popular being the Akaike's Information Criterion (AIC) [17] and the Bayesian Inference Criterion (BIC) [18,19]. The last two approaches are based on the maximization of criteria that combine a term based on the likelihood of the observations and a term that penalizes the complexity of the model. In principle, they could also be used as validity indices in model clustering, to compare models with different number of components and thus automatically select the number of clusters. This possibility has been deeply explored in the literature, and different indices based on these and similar criteria have been proposed to validate clustering partitions and assess the number of components in clustering problems using mixture models [7,8,20]. In particular, the Information Completed Likelihood (ICL) [21,22], which is essentially the BIC criterion penalized by subtraction of the estimated partition mean entropy, has been shown to outperform AIC and BIC when the focus is clustering rather than density estimation. The AIC and BIC criteria do not explicitly penalize the overlap amongst the clusters, and so they tend to overestimate the number of cluster components when the kind of distribution followed by the real clusters does not match the distribution assumed by the mixture model [21].

In this article we present the *Partition Negentropy Criterion* (PNC), a new cluster validity index which combines a term that rewards the average normality of the clusters and a term that penalizes the average overlap. The normality of a cluster is computed using the negentropy, while the average overlap is measured by the partition entropy. Here we use the new index to validate Gaussian mixtures that are fitted with the EM algorithm, and compare its performance to the AIC, BIC and ICL criteria in a set of synthetic clustering problems. We first check that the PNC is able to assess the correct number of components in problems where the underlying cluster distributions are Gaussian. When the Gaussian clusters are highly overlapped the AIC and BIC criteria obtain a slightly better detection rate, that is they are able to assess the number of components under higher overlap, than the ICL and the PNC. However, in situations where the underlying cluster distributions are not purely Gaussian, the AIC and BIC criteria systematically overestimate the number of components, regardless of the separation amongst the clusters. In these cases only the ICL and the PNC obtain satisfactory results. In particular, the proposed PNC index provided a high detection rate and the lowest overestimation rate in all the tests performed.

2 The Partition Negentropy Criterion

In this section we develop the Partition Negentropy Criterion, a cluster validity index whose aim is to find well separated clusters as normally distributed as possible. The normality of a cluster is characterized by means of its negentropy, a standard measure of distance to normality which computes the difference between the cluster's entropy and the entropy of a Gaussian distribution with the same covariance matrix [23]. The negentropy of a continuous random variable $\mathbf{X}$ is defined as:

$$J(\mathbf{X}) = \hat{H}(\mathbf{X}) - H(\mathbf{X}) \tag{1}$$

where $H(\mathbf{X})$ is the differential entropy of $\mathbf{X}$ and $\hat{H}(\mathbf{X})$ is the differential entropy of a normal distribution with the same covariance matrix. The Gaussian distribution maximizes the differential entropy for a given covariance matrix [24], so the negentropy is always non-negative, being zero if and only if $\mathbf{X}$ is normally distributed. The maximum entropy property associated to the normal distribution also provides a hint on why normality is a desired property of any cluster. Maximum entropy, or equivalently maximum uncertainty, implies minimum structure, and so a normally distributed cluster can not be expected to contain other substructures.

Let us consider a set of data points $\{\mathbf{x}_i\}$ and a fuzzy partition $p(c|\mathbf{x}_i)$ into a set of clusters C. Our goal is to obtain a cluster validity index that rewards partitions into well separated Gaussian clusters. We will measure the quality of the partition by:

$$H(C|\mathbf{X}) + J(\mathbf{X}|C) \tag{2}$$

The first term measures the average degree of overlap amongst the clusters, while the second corresponds to the average negentropy of the clusters, which measures how distant they are from a Gaussian distribution. So minimization of this expression will favour partitions that consist of well separated normally distributed clusters. We can write $J(\mathbf{X}|C)$ as:

$$J(\mathbf{X}|C) = \hat{H}(\mathbf{X}|C) - H(\mathbf{X}|C) \tag{3}$$

And, using basic properties of the conditional entropy [24], rewrite it as:

$$J(\mathbf{X}|C) = \hat{H}(\mathbf{X}|C) + H(C) - H(\mathbf{X}) - H(C|\mathbf{X}) \tag{4}$$

Note that the term $H(\mathbf{X})$ is constant for the problem, and so it can be ignored when minimizing the expression in 2. The term $\hat{H}(\mathbf{X}|C)$ can be expressed as:

$$\hat{H}(\mathbf{X}|C) = \sum_{c=1}^{n_c} p(c)\hat{H}(\mathbf{X}|c) \tag{5}$$

where the sum extends to all the n_c clusters in C, $p(c)$ is the a-priori probability of cluster c, and $\hat{H}(\mathbf{X}|c)$ is the differential entropy of the cluster c assuming normality, that is:

$$\hat{H}(\mathbf{X}|c) = \frac{1}{2}\log|\Sigma_c| + \frac{d}{2}\log 2\pi e \tag{6}$$

where Σ_c is the covariance matrix of cluster c and d is the dimension of $\mathbf{X}$. If we substitute equations 4, 5 and 6 into expression 2, and neglect terms that do not depend on the partition C, we obtain the Partition Negentropy Criterion as:

$$PNC(C) = \frac{1}{2}\sum_{c=1}^{n_c} p(c)\log|\Sigma_c| - \sum_{c=1}^{n_c} p(c)\log p(c) \tag{7}$$

Given different partitions of a data set, we will select that with a lower PNC.

3 Evaluation of the PNC

To test the new cluster validity criterion we use the PNC to validate fuzzy partitions resulting from the application of the EM algorithm to a set of synthetic problems, and compare the results to those obtained with the AIC, the BIC, and the ICL criteria. For every problem we follow the same approach. First, the EM algorithm is used to fit a set of Gaussian mixtures with different number of components. Then, for each index (AIC, BIC, ICL, PNC) we select the mixture which provides the best index value. We compute the PNC by directly substituting the covariance matrices and the prior probabilities given by the EM algorithm into equation 7.

3.1 Two Simple Examples

We will first illustrate the PNC with two simple examples consisting of three well separated clusters in two dimensions. In the first case each cluster consists of 1000 points drawn from a normal distribution with covariance matrix equal to the identity matrix, $\Sigma = \mathbf{I}$, and centered at $\mu_1 = (0,0)$, $\mu_2 = (5,0)$, and $\mu_3 = (5,5)$ respectively (see figure 1). For this problem we have run the EM algorithm to fit a mixture of n_c Gaussians, with $n_c \in \{1,2,3,4,5\}$. The algorithm has been run 10 times for each n_c. The best partitions according to each of the four validity criteria are shown in figures 1A (AIC), 1B (BIC), 1C (ICL), and 1D (PNC). The solid lines represent the contours of the Gaussian components. Note that for well separated normal clusters all the criteria select partitions with the correct number of clusters ($n_c = 3$).

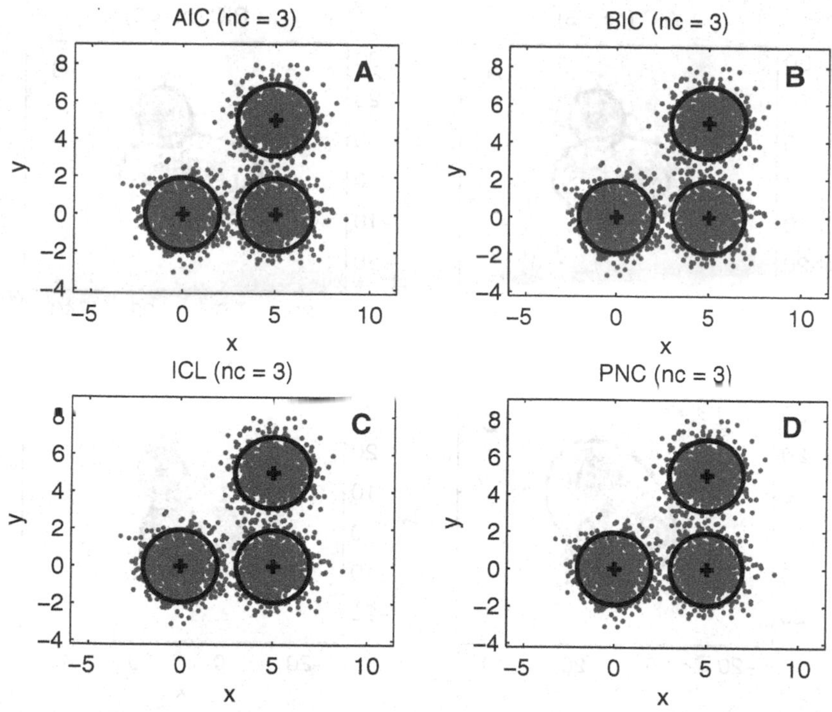

Fig. 1. Assessment of the number of clusters in a problem consisting of a mixture of three Gaussian distributions in 2D. The EM algorithm has been run 10 times for each number of components (ranging from 1 to 5), and the best solution according to the four criteria is selected. **A.** Akaike's criterion (AIC). **B.** Bayesian inference criterion (BIC). **C.** Information Completed Likelihood (ICL). **D.** Partition Negentropy Criterion (PNC).

The second problem consists of three non-Gaussian clusters of 1000 points each. In polar coordinates, the clusters follow a gamma distribution in the radius and a uniform distribution in the angle. The gamma distribution has a shape parameter $k = 2$ and a scale parameter $\theta = 1.5$. The three clusters are centered at $\mu_1 = (0, 0)$, $\mu_2 = (15, 0)$, and $\mu_3 = (15, 15)$ respectively. As before, we have run the EM algorithm 10 times for each n_c, and we have selected the best partitions according to the four validity criteria. The results are shown in figures 2A (AIC), 2B (BIC), 2C (ICL), and 2D (PNC). Note that, although the clusters are easily separable, only the PNC is able to correctly assess the number of clusters ($n_c = 3$). The other criteria overestimate the number of components. The AIC and BIC select partitions with $n_c = 5$ clusters, while the ICL selects a partition with $n_c = 4$ clusters.

These examples show that standard criteria such as AIC, BIC or ICL can perform poorly as cluster validity indices when the underlying data distribution does not match the assumed mixture model. On the other hand, the PNC shows a good performance even when the cluster distributions are not pure Gaussians.

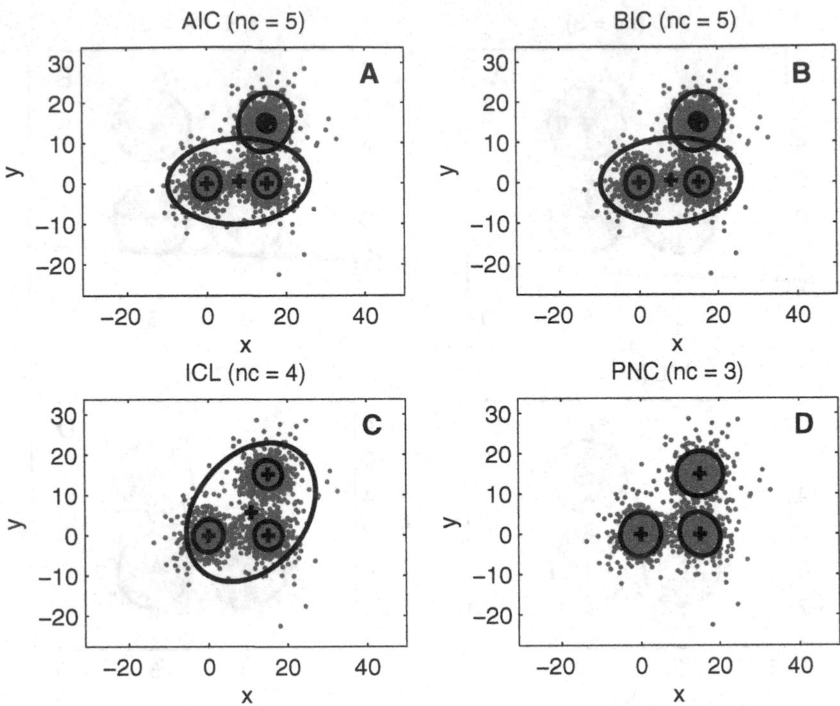

Fig. 2. Assessment of the number of clusters in a problem consisting of a mixture of three non-Gaussian distributions in 2D. The EM algorithm has been run 10 times for each number of components (ranging from 1 to 5), and the best solution according to the four criteria is selected. **A.** Akaike's criterion (AIC). **B.** Bayesian inference criterion (BIC). **C.** Information Completed Likelihood (ICL). **D.** Partition Negentropy Criterion (PNC).

3.2 Number of Detected Clusters Versus Inter-cluster Distance

As a second test we consider the assessment of the number of components in problems consisting of two spherical clusters with covariance matrices equal to the identity matrix, $\Sigma = \mathbf{I}$, and 1000 points each in two dimensions. The distance between the cluster centers is varied in order to obtain problems with different degree of overlap. We want to study the performance of the PNC and the other criteria as a function of the overlap for different cluster shapes. The shape of the clusters is modeled according to the following four kinds of probability distributions:

1. *Pure normal distribution.* Its covariance matrix is equal to the identity matrix, $\Sigma = \mathbf{I}$.
2. *Truncated normal distribution.* Points are generated using a pure normal distribution, and those whose distance to the mean exceeds 1.8 times the standard deviation are discarded. A scaling factor is applied to ensure that the resulting distribution has a covariance matrix $\Sigma = \mathbf{I}$.

3. *Uniform distribution inside a circle.* The circle radius is selected such that the covariance matrix is $\Sigma = \mathbf{I}$.
4. *Gamma-uniform distribution.* The same distribution as in the second example of section 3.1 is used. The shape parameter of the gamma is $k = 2$, and the scale parameter is selected to ensure that the resulting distribution has a covariance matrix $\Sigma = \mathbf{I}$.

In all the cases the first cluster is centered at $\mu_1 = (0, 0)$ and the second one at $\mu_2 = (d, 0)$, where d is the inter-cluster distance. For each problem and for values of d ranging between 0 and 5, we use the EM algorithm to fit a Gaussian mixture with a number of components $n_c \in \{1, 2, 3, 4, 5\}$. As before, the EM algorithm is run 10 times for each n_c and the best partition according to each of the four criteria is selected. A total of 40 different problems are generated for each d in order to average. In figure 3 we plot the average number of clusters in the best partition selected by each criterion versus the inter-cluster distance. When the clusters follow a Gaussian distribution (figure 3A), the average number of clusters selected by all the criteria is between 1 and 2. As we can see AIC and BIC can stand a higher overlap than the other indices. On the other hand, when the underlying cluster distribution is not Gaussian (figures 3B, 3C and 3D) AIC and BIC tend to overestimate the number of clusters for any degree of overlap. In these cases only the ICL and PNC criteria provide satisfactory results. The ICL criterion admits a slightly higher overlap, but it also overestimates the number of components for gamma-uniform distributed clusters at high d. At the price of a higher tendency to merge overlapping clusters, the proposed PNC is the only index that correctly assesses the number of clusters for all the considered cluster shapes when the clusters are well separated.

3.3 Number of Clusters in Randomly Generated Problems

Finally, we make the last test using 1000 randomly generated problems. Each problem contains three clusters in two dimensions, but the shape, orientation, position and scale of the clusters are selected randomly. The 4 cluster shapes of section 3.2 are considered. As for previous tests, the EM algorithm is used to fit a Gaussian mixture model to each problem. We try different number of components, $n_c \in \{1, 2, 3, 4, 5\}$, and the algorithm is run 5 times for each n_c. Then the four validity criteria are used to select a preferred partition. In table 1 we show the percentage of problems for which each criterion selects a partition with 1, 2, 3, 4 or 5 clusters. Note that the PNC is the one which selects the correct number of clusters ($n_c = 3$) with a higher probability (76% of the problems). In addition, it only overestimates the number of clusters for the 9% of the problems. The ICL follows closely (72.4% of correct partitions and 15.5% of overestimation), but AIC and BIC perform very poorly and they only select the correct number of clusters for the 11.6% and the 26% of the problems respectively.

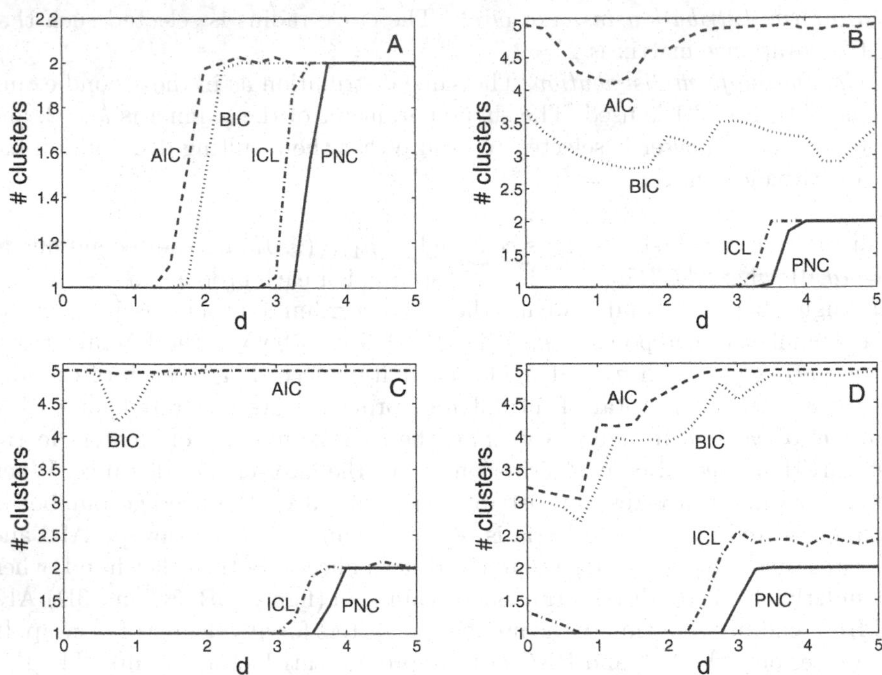

Fig. 3. Average number of clusters selected by the four considered validity criteria versus inter-cluster distance. **A.** The two clusters are normally distributed. **B.** The clusters are Gaussians without tails. **C.** The clusters are uniformly distributed inside a circle. **D.** The clusters follow a gamma-uniform distribution.

4 Discussion

In this article we have presented the Partition Negentropy Criterion (PNC), a cluster validity index whose aim is to find clearly separated clusters whose shape is as Gaussian as possible. The index measures the normality of the clusters using the negentropy, and it measures the cluster separation using the partition entropy. We investigated the performance of the PNC using synthetic clustering problems, where the ability to assess the number of clusters was compared to the AIC, BIC and ICL criteria. We first checked that the PNC is able to determine the correct number of clusters in problems whose data points are generated from a mixture of well separated Gaussian distributions. We observed, however, that the performance degrades when the Gaussians are overlapped. In these cases the PNC obtains a similar performance to ICL, but slightly worst than AIC and BIC. On the other hand, we showed that the PNC provides good results even when the data are generated from mixtures of non-Gaussian distributions. In such situations the PNC is still able to detect the correct number of clusters, outperforming the other criteria, which are more prone to overestimate this number.

Table 1. Percentage of data sets for which each criterion predicts 1, 2, 3, 4 or 5 clusters

n_c	AIC	BIC	ICL	PNC
1	0.0	0.0	0.2	0.6
2	0.3	0.5	11.9	14.4
3	11.6	26.0	72.4	76.0
4	27.7	38.7	14.1	5.2
5	60.4	34.8	1.4	3.8

Although the results here presented are promising, future work using real datasets is needed in order to validate the proposed criterion. On the other hand, the mathematical simplicity of the PNC, whose evaluation involves just the computation of the determinants of the covariance matrices of each cluster, may permit an analytical study of its performance. This is specially interesting as it would allow to compare the PNC to other indices at the mathematical level.

In the present work, the calculation of the mixture model parameters via EM is conceptually separated from the validation step, in which the validity indices are used to evaluate the outcomes of a particular run of the algorithm. Given the mathematical simplicity of the PNC, we believe that it could be possible to integrate it into an EM algorithm, thus obtaining an iterative procedure that simultaneously searches for the number of clusters and the mixture model parameters.

Finally, we will investigate how the PNC could be extended to deal with variables which are not continuous.

Acknowledgments. This work has been partially supported with funds from MEC BFU2006-07902/BFI, CAM S-SEM-0255-2006 and CAM/UAM project CCG08-UAM/TIC-4428.

References

1. Everitt, B., Landau, S., Leese, M.: Cluster Analysis. Hodder Arnold, London (2001)
2. Dempster, A.P., Laird, N.M., Rubin, D.B.: Maximum Likelihood from Incomplete Data via the EM Algorithm. J. Royal Statistical Soc. B 39, 1–38 (1977)
3. Gordon, A.D.: Cluster Validation. In: Hayashi, C., Ohsumi, N., Yajima, K., Tanaka, Y., Bock, H.H., Baba, Y. (eds.) Data Science, Classification and Related Methods, pp. 22–39. Springer, New York (1998)
4. Bezdek, J.C., Pal, R.N.: Some New Indexes of Cluster Validity. IEEE Trans. Systems, Man and Cybernetics B 28(3), 301–315 (1998)
5. Pakhira, M.K., Bandyopadhyay, S., Maulik, U.: Validity Index for Crisp and Fuzzy Clusters. Pattern Recognition 37(3), 487–501 (2004)

6. Bouguessa, M., Wang, S., Sun, H.: An Objective Approach to Cluster Validation. Pattern Recognition Letters 27(13), 1419–1430 (2006)
7. Bozdogan, H.: Choosing the Number of Component Clusters in the Mixture-Model Using a New Information Complexity Criterion of the Inverse-Fisher Information Matrix. In: Opitz, O., Lausen, B., Klar, R. (eds.) Data Analysis and Knowledge Organization, pp. 40–54. Springer, Heidelberg (1993)
8. Biernacki, C., Celeux, G., Govaert, G.: An Improvement of the NEC Criterion for Assessing the Number of Clusters in a Mixture Model. Pattern Recognition Letters 20(3), 267–272 (1999)
9. Geva, A.B., Steinberg, Y., Bruckmair, S., Nahum, G.: A Comparison of Cluster Validity Criteria for a Mixture of Normal Distributed Data. Pattern Recognition Letters 21(6-7), 511–529 (2000)
10. Pal, N.R., Biswas, J.: Cluster Validation Using Graph Theoretic Concepts. Pattern Recognition 30(6), 847–857 (1997)
11. Hathaway, R.J., Bezdek, J.C.: Visual Cluster Validity for Prototype Generator Clustering Models. Pattern Recognition Letters 24(9-10), 1563–1569 (2003)
12. Ding, Y., Harrison, R.F.: Relational Visual Cluster Validity (RVCV). Pattern Recognition Letters 28(15), 2071–2079 (2007)
13. Richardson, S., Green, P.: On Bayesian Analysis of Mixtures with Unknown Number of Components. J. Royal Statistical Soc. 59, 731–792 (1997)
14. Rasmussen, C.: The Infinite Gaussian Mixture Model. In: Solla, S., Leen, T., Müller, K.-R. (eds.) Advances in Neural Information Processing Systems, vol. 12, pp. 554–560. MIT Press, Cambridge (2000)
15. Neal, R.M.: Markov Chain Sampling Methods for Dirichlet Process Mixture Models. J. Computational and Graphical Statistics 9(2), 249–265 (2000)
16. Figueiredo, M.A.T., Jain, A.K.: Unsupervised Learning of Finite Mixture Models. IEEE Trans. Pattern Analysis and Machine Intelligence 24(3), 381–396 (2002)
17. Akaike, H.: A new look at the statistical model identification. IEEE Trans. Automatic Control 19, 716–723 (1974)
18. Schwartz, G.: Estimating the Dimension of a Model. Annals of Statistics 6, 461–464 (1978)
19. Fraley, C., Raftery, A.: How Many Clusters? Which Clustering Method? Answers Via Model-Based Cluster Analysis. Technical Report 329, Dept. Statistics, Univ. Washington, Seattle, WA (1998)
20. Bezdek, J.C., Li, W.Q., Attikiouzel, Y., Windham, M.: A Geometric Approach to Cluster Validity for Normal Mixtures. Soft Computing 1, 166–179 (1997)
21. Biernacki, C., Celeux, G., Govaert, G.: Assessing a Mixture Model for Clustering with the Integrated Completed Likelihood. IEEE Trans. Pattern Analysis Machine Intelligence 22(7), 719–725 (2000)
22. Samé, A., Ambroise, C., Govaert, G.: An Online Classification EM Algorithm Based on the Mixture Model. Stat. Comput. 17, 209–218 (2007)
23. Comon, P.: Independent Component Analysis, a New Concept? Signal Processing 36(3), 287–314 (1994)
24. Cover, T.M., Thomas, J.A.: Elements of Information Theory. John Wiley, New York (1991)

Bayesian Estimation of Kernel Bandwidth for Nonparametric Modelling

Adrian G. Bors and Nikolaos Nasios

Department of Computer Science, University of York, York YO10 5DD, UK
adrian.bors@cs.york.ac.uk

Abstract. Kernel density estimation (KDE) has been used in many computational intelligence and computer vision applications. In this paper we propose a Bayesian estimation method for finding the bandwidth in KDE applications. A Gamma density function is fitted to distributions of variances of K-nearest neighbours data populations while uniform distribution priors are assumed for K. A maximum log-likelihood approach is used to estimate the parameters of the Gamma distribution when fitted to the local data variance. The proposed methodology is applied in three different KDE approaches: kernel sum, mean shift and quantum clustering. The third method relies on the Schrödinger partial differential equation and uses the analogy between the potential function that manifests around particles, as defined in quantum physics, and the probability density function corresponding to data. The proposed algorithm is applied to artificial data and to segment terrain images.

Keywords: Kernel density estimation, bandwidth, quantum clustering.

1 Introduction

There are two approaches for data modelling in statistics: parametric and non-parametric, depending on whether there is a model assumption or not. Kernel density estimation (KDE) is a non-parametric approach in which the kernel function is centered at each data sample location while exerting an influence in the region around it. A scale parameter, also called bandwidth or window width controls the kernel function smoothing over the surrounding space. In this study we consider three different nonparametric modelling methods. The first one represents the probability density function (pdf) by simply summing the kernel functions for all data [1]. The mean-shift is an updating algorithm which employs the local gradient for finding the maxima of the pdf representation [2]. A method which relies on the analogy between the pdf representation and the quantum potential of physical particles, was proposed in [3] and analysed in [4].

The performance of KDE methods does not depend on the actual kernel function [5] but on the value of the kernel's bandwidth [6,7,8]. The bandwidth is responsible for smoothing the resulting KDE representation as well as for defining an appropriate mode localization. The algorithms used for finding the bandwidth can be classified into two categories: quality-of-fit and plug-in methods. The first category uses cross-validation by leaving certain data samples out

C. Alippi et al. (Eds.): ICANN 2009, Part II, LNCS 5769, pp. 245–254, 2009.

while approximating the pdf with the sum of kernels located at the remaining data. The plug-in methods calculate the bias in the pdf approximation such that it minimizes the mean integrated square error (MISE) between the real density and its kernel-based approximation [5,8,10,11]. However, the plug-in algorithms require an initial pilot estimate of the bandwidth for the iterative processing [9].

In this paper we consider that the kernel bandwidth corresponds to a local data variance estimate. Such variances are calculated from randomly sampled K-nearest neighbours, when assuming a prior distribution for K, in a Bayesian framework. The proposed bandwidth estimation method is employed in three KDE methods: kernel sum, mean shift and quantum clustering. These methods are applied for segmenting modulated signals and terrain topography estimated from radar images [12]. Quantum clustering is described in the context of KDE in Section 2 while the inferrence of the bandwidth parameter is described in Section 3. Experimental results are provided in Section 4 and the conclusions are drawn in Section 5.

2 Kernel Density Estimation Using Quantum Mechanics

Various kernel density estimation methods have been used in pattern recognition and computer vision including the mean shift [2] and the classical approach of kernel sums [1]. In the following we describe in detail a KDE method called quantum clustering [3]. In quantum clustering each data sample is associated with a particle that is part of a quantum mechanical system. The activation field at a location $\mathbf{X}$, calculated from N data samples $\{\mathbf{X}_i, i = 1, \ldots, N\}$ is given as in classical KDE by the sum of kernels centered at the data samples:

$$\psi(\mathbf{X}) = \sum_{i=1}^{N} \exp\left[-\frac{(\mathbf{X}-\mathbf{X}_i)^2}{2\sigma^2}\right] \tag{1}$$

where σ represents the bandwidth parameter.

According to the fifth postulate of quantum mechanics, a quantum system evolves according to the Schrödinger differential equation. The time-independent Schrödinger equation is given by:

$$\mathcal{H}\psi(\mathbf{X}) \equiv \left(-\frac{\sigma^2}{2}\nabla^2 + V(\mathbf{X})\right)\psi(\mathbf{X}) = E \cdot \psi(\mathbf{X}) \tag{2}$$

where $\mathcal{H}$ is the Hamiltonian operator, E is the eigenvalue energy level associated with a specific particle orbit, $\psi(\mathbf{X})$ corresponds to the state of the given quantum system, $V(\mathbf{X})$ is the Schrödinger potential and ∇^2 is the Laplacian. In quantum mechanics, the potential $V(\mathbf{X})$ is given and equation (2) is solved in order to find solutions $\psi(\mathbf{X})$. In this case, $\psi(\mathbf{X})$ describes the probability of locating a particle on a specific orbit. From the computational intelligence perspective we consider the inverse problem by assuming known the location of data samples and their state as given by equation (1). This location is considered as a solution for (2), subject to the calculation of a set of constants. In equation (2) the potential is always positive, $V(\mathbf{X}) > 0$. After replacing $\psi(\mathbf{X})$ from (1) into (2), we calculate the Schrödinger potential for the given data set as, [3,4]:

$$V(\mathbf{X}) = E - \frac{d}{2} + \frac{1}{2\sigma^2\psi(\mathbf{X})}\sum_{i=1}^{N}\|\mathbf{X}-\mathbf{X}_i\|^2 \exp\left(-\frac{\|\mathbf{X}-\mathbf{X}_i\|^2}{2\sigma^2}\right) \quad (3)$$

From the statistics point of view, the quantum potential formulation can be written as:

$$V(\mathbf{X}) = E - \frac{d}{2} + \frac{\sum_{i=1}^{N}\|\mathbf{X}-\mathbf{X}_i\|^2 P(\mathbf{X}|\mathbf{X}_i)}{2\sigma^2} \quad (4)$$

where $P(\mathbf{X}|\mathbf{X}_i)$ is the *a posteriori* probability for $\mathbf{X}$, given the data samples $\{\mathbf{X}_i, i = 1, \ldots, N\}$. This expression represents the weighted Euclidean distance from $\mathbf{X}$ to a set of given data samples, where the weights are represented by its *a posteriori* probabilities. This resulting function models the hypersurface of the potential function produced by the quantum clustering algorithm that depends on the bandwidth σ. Data clusters are indicated by maxima in (1) and by minima in the quantum potential from (3).

3 Bayesian Estimation of the Bandwidth

Estimating the bandwidth in KDE is very important, particularly in applications where we need to detect the local extrema of the pdf. Classical statistics estimation methods such as those using the quality-of-fit and plug-in are known to provide biased estimates for the bandwidth leading to either spurious bumpiness or oversmoothing in the resulting pdf approximation [6,7,9]. In this study we propose a Bayesian approach for the estimation of the scale σ.

3.1 Defining Local Neighbourhoods

The proposed approach considers that the bandwidth σ can be associated with the local data spread, which is statistically characterized by the variance. Let us consider that the bandwidth is modelled by the *a posteriori* probability density function $p(s|\mathbf{X})$, where s is the statistical variable associated with the bandwidth, and $\mathbf{X}$ represents a data sample. Because the bandwidth determines the local smoothness in the resulting pdf approximation, it should be calculated from a data subset defined locally. Let us consider K, the number of nearest neighbours to a specific data sample $\mathbf{X}_i$. All the other data samples are ordered according to their Euclidean distance to $\mathbf{X}_i$. The proposed approach estimates the kernel bandwidth using the statistics of the variances corresponding to K-nearest neighbours (KNN) data sample populations [13], where $K < N$.

Let us assume n neighbourhoods of various sizes $\{K_j, j = 1, \ldots, n\}$. We define a probability density function of the bandwidth $p(s|\mathbf{X})$ that is expressed as a pseudo-likelihood of the form:

$$P(s|\mathbf{X}) = \prod_{j=1}^{n} P(s|\mathbf{X}_{K_j}) \quad (5)$$

where $P(s|\mathbf{X}_{K_j})$ represents the probability of the bandwidth depending on K_j nearest neighbourhood data samples to $\mathbf{X}_{K_j}$. These probabilities can be evaluated over an entire range of K_j:

$$P(s|\mathbf{X}_{K_j}) = \int P(s|K_j, \mathbf{X}_{K_j})P(K_j|\mathbf{X}_{K_j})dK_j \tag{6}$$

and after using the Bayes rule we obtain:

$$P(K_j|\mathbf{X}_{K_j}) = \frac{P(\mathbf{X}_{K_j}|K_j)P(K_j)}{P(\mathbf{X}_{K_j})} \tag{7}$$

where $P(\mathbf{X}_{K_j}|K_j)$ is the probability of the data sample population depending on the specific neighbourhood size. In the following we consider $P(K_j)$ as a uniform distribution limited to the range $[K_1, K_2]$.

The bandwidth estimation approach consists of three steps which are outlined in the graphical model from Figure 1. In our case, the only necessary elements are the bounds of the uniform distribution characterizing $P(K_j)$, as given by $K_j \in [K_1, K_2]$, $j = 1, \ldots, n$.

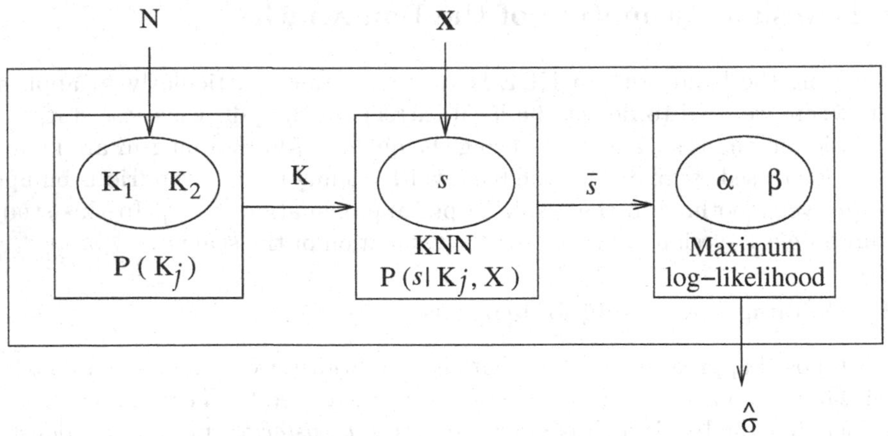

Fig. 1. Graphical model describing the stages for estimating the bandwidth $\hat{\sigma}$

3.2 Statistic Modelling of the Local Variance

After sampling K_j from the uniform distribution $P(K_j)$, we randomly sample the given data set $\{\mathbf{X}_i, i = 1, \ldots, N\}$ and consider these samples as centers for evaluating their K_j-nearest neighbours (KNN). The K_j-nearest neighbours are selected according to their Euclidean distances to sampled $\mathbf{X}_{i'}$. This results in data sets $\mathbf{X}_{K_j}$ for which we calculate the variance as:

$$s_i = \frac{\sum_{k=1}^{K_j} \|\mathbf{X}_{i,(k)} - \mathbf{X}_{i'}\|^2}{K_j - 1} \tag{8}$$

where $\{\mathbf{X}_{i,(k)}, k = 1, \ldots, K_j\}$ are the nearest neighbours to the sampled data $\mathbf{X}_{i'}$. The estimate s_i from (8) is considered as a random variable, whose statistics can be used for inferring the bandwidth $\hat{\sigma}$. The main idea behind the proposed methodology is that the bandwidth should be a measure depending on the distribution of local data variance.

The distribution of variances for data samples that are generated independently from the Normal distribution with mean 0 and variance 1 are modelled by the Chi-square distribution. Gamma distribution is a generalization of the Chi-square distribution, and is suitable for modelling distributions of data sample variances in the case when we have no knowledge about the underlying data distribution. Due to its generality, Gamma distribution can model the bandwidth distribution for kernels that are not necessarily Gaussian. The Gamma probability density function is given by the following expression:

$$P(s|\alpha, \beta) = \frac{\beta^{\alpha} s^{\alpha-1}}{\Gamma(\alpha)} e^{-\beta s} \tag{9}$$

where $s \geq 0$, $\alpha > 0$ is the shape parameter and $\beta > 0$ is the scale parameter of the Gamma distribution. $\Gamma(\cdot)$ represents the Gamma function:

$$\Gamma(t) = \int_0^{\infty} r^{t-1} e^{-r} dr \tag{10}$$

After calculating the local variance for each of the data subsets describing $P(s|\mathbf{X}_{K_j})$ and after taking into account $P(K_j)$, when varying uniformly $K_j \in [K_1, K_2]$, we form a data set corresponding to the random variable s calculated as in (8). The proposed approach relies on fitting the Gamma pdf from (9) to the resulting empirical distribution of s_i.

3.3 Estimating Gamma Distribution Parameters

Two methods have been proposed for estimating the parameters of a Gamma distribution: data moments and the maximum likelihood estimation [14]. The data moments method lacks the efficiency required for estimating the shape parameter of the Gamma distribution. The likelihood function corresponding to the distribution from (9) is:

$$\mathcal{L}(\alpha, \beta) = \prod_{i=1}^{M} p(s|\alpha, \beta) = \frac{\beta^{\alpha M}}{\Gamma^{M}(\alpha)} \left[\prod_{i=1}^{M} s_i \right]^{\alpha-1} e^{-\beta \sum_{i=1}^{M} s_i} \tag{11}$$

where we consider M data samples $\{s_i | i = 1, \ldots, M\}$, each representing the variance of a local neighbourhood, calculated according to equation (8). We estimate the parameters α and β by equating the likelihood derivatives to zero resulting in the following system of equations:

$$\begin{cases} \ln(\hat{\alpha}) - \dfrac{\Gamma'(\hat{\alpha})}{\Gamma(\hat{\alpha})} + \ln \left[\dfrac{(\prod_{i=1}^{M} s_i)^{1/M}}{\overline{s}} \right] = 0 \\ \hat{\beta} = \dfrac{\hat{\alpha}}{\overline{s}} \end{cases} \tag{12}$$

where $\overline{s}$ is the sample mean for the variable s:

$$\overline{s} = \frac{\sum_{i=1}^{M} s_i}{M} \tag{13}$$

The third term from the first equation of the system provided in (12) is the logarithm of the ratio between the geometric and arithmetic means. The first nonlinear equation from (12) is solved using the Newton-Raphson iterative algorithm with respect to $\hat{\alpha}$, [14]. This results in the following updating equation:

$$\hat{\alpha}_{t+1} = \hat{\alpha}_t - \frac{\ln(\hat{\alpha}_t) - \Psi(\hat{\alpha}_t) + \ln\left[\frac{(\prod_{i=1}^{M} s_i)^{1/M}}{\sum_{i=1}^{M} s_i}\right]}{1/\hat{\alpha}_t - \Psi(\hat{\alpha}_t)} \tag{14}$$

where $\Psi(\hat{\alpha}_t)$ is the Digamma function, which represents the logarithmic derivative of the Gamma function:

$$\Psi(\hat{\alpha}) = \frac{\Gamma'(\hat{\alpha})}{\Gamma(\hat{\alpha})} \tag{15}$$

where $\Gamma(\cdot)$ is provided in (10) and $\Gamma'(\cdot)$ represents its derivative. After estimating $\hat{\alpha}$ at the convergence of (14), we replace it in the second equation from (12) and estimate $\hat{\beta}$. A good initialization for the Newton-Raphson optimization is achieved when using the moments method estimate for $\hat{\alpha}$, [4]. The Gamma distribution defined by the parameters α and β is used afterwards for inferring the kernel bandwidth.

4 Experimental Results

The proposed methodology of bandwidth selection is embedded into three different KDE based methods: classical kernel sum, mean shift and quantum clustering. In the first example we consider phase-shifting-key modulated signals (8-PSK) which correspond to 8 clusters, whose centers are located radially at equal angles from each other. The perturbation channel equations for 8-PSK signals assuming interference are :

$$x_I(t) = I(t) + 0.2I(t-1) - 0.2Q(t) - 0.04Q(t-1) + \mathcal{N}(0, 0.11)$$
$$x_Q(t) = Q(t) + 0.2Q(t-1) + 0.2I(t) + 0.04I(t-1) + \mathcal{N}(0, 0.11)$$

where $(x_I(t), x_Q(t))$ makes up the in-phase and in-quadrature signal components at time t on the communication line, and $I(t)$ and $Q(t)$ correspond to the signal symbols and where we also consider additive Gaussian noise with SNR = 22 dB. We have generated 960 signals, by assuming equal probabilities for all inter-symbol combinations. After modelling the local variance using the Gamma distribution we use its mean as a bandwith and embed it into the three KDE based methods. Consequently, we segment the resulting potential function into the component clusters, each associated with a signal. The kernel sum and quantum potentials, calculated according to (1) and (4), for the kernel sum and

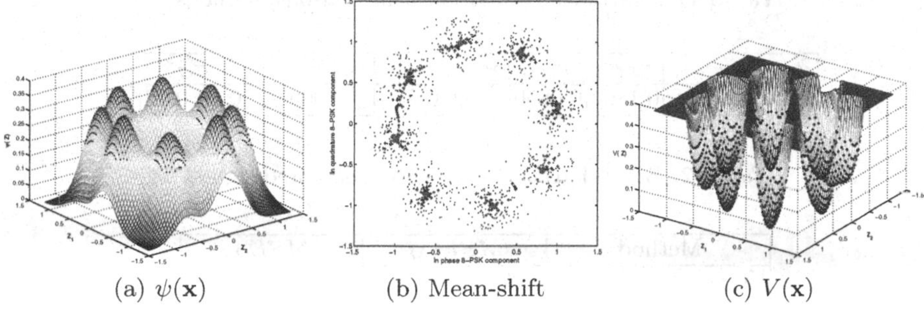

(a) $\psi(\mathbf{x})$ (b) Mean-shift (c) $V(\mathbf{x})$

Fig. 2. Blind detection of 8-PSK modulated signals using various methods. The intermediate mean shifts are shown by "+" while the mean shifts obtained at convergence are indicated with larger circles in (b).

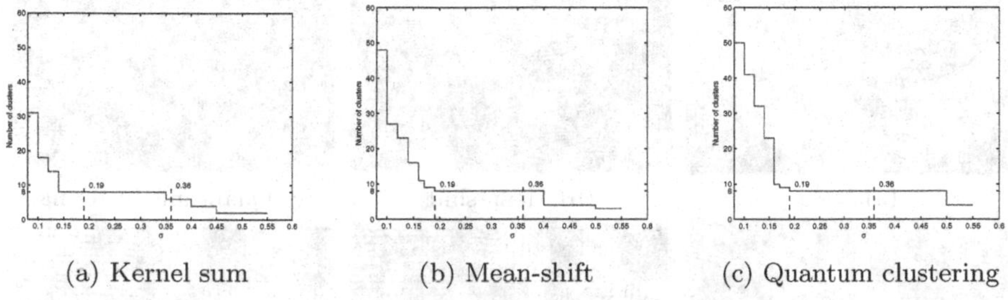

(a) Kernel sum (b) Mean-shift (c) Quantum clustering

Fig. 3. Finding the number of modes in corrupted 8-PSK data

quantum clustering, respectively, are displayed in Figs. 2(a) and 2(c), while the mean-shift intermediary and final results are provided in Fig. 2(b).

By looking to local maxima in the potential function surfaces we identify the modes, each corresponding to a source signal when using all three computational intelligence approaches described in Section 2, and when varying the bandwidth σ. The number of detected modes for the corrupted 8-PSK data are shown in Fig. 3. The range of $\hat{\sigma}$, estimated according to various intervals of K, is shown in between two dashed vertical lines in each of the plots from Fig. 3. A series of bandwidth estimators are considered for the given data and the resulting estimates are provided in Table 1. The bandwidth estimates include the classical rule-of-thumb bandwidth $\hat{\sigma}_{ROT} = 1.06S/N^{1/5}$, where S is the standard deviation of the data set and N is the number of data, and Silverman estimate $\hat{\sigma}_{SROT}$ [7]. Other comparative bandwidth estimators are the Terrell bandwidth $\hat{\sigma}_{TER}$ from [5] and Sheather and Jones from [10] which is denoted as $\hat{\sigma}_{SJ}$. The "Rule-of-Thumb" and "Direct Plug-in" methods from [11] are denoted as $\hat{\sigma}_{RSW-ROT}$ and $\hat{\sigma}_{RSW-DPI}$. The bandwidth estimated by using the proposed methodology is denoted as $\hat{\sigma}_G$. When looking to the valid mode range 8-PSK there are three methods which provide suitable bandwidth for the classical kernel sum approach but none for the other two KDE-based methods. It can be observed that the

Table 1. Bandwidth estimation using various methods

$\hat{\sigma}_{ROT}$	$\hat{\sigma}_{SROT}$	$\hat{\sigma}_{TER}$	$\hat{\sigma}_{SJ}$	$\hat{\sigma}_{RSW-ROT}$	$\hat{\sigma}_{RSW-DPI}$	$\hat{\sigma}_{G}$
0.1729	0.1469	0.1867	0.0850	0.0470	0.0696	0.19-0.36

Table 2. Blind detection of 8-PSK modulated signals

Method	MSE_{SM}	MSE_{O}
Kernel sum	0.0024 ± 0.0007	0.0041 ± 0.0007
Mean shift	0.0168 ± 0.0097	0.0200 ± 0.0109
Quantum clustering	0.0009 ± 0.0003	0.0022 ± 0.0004

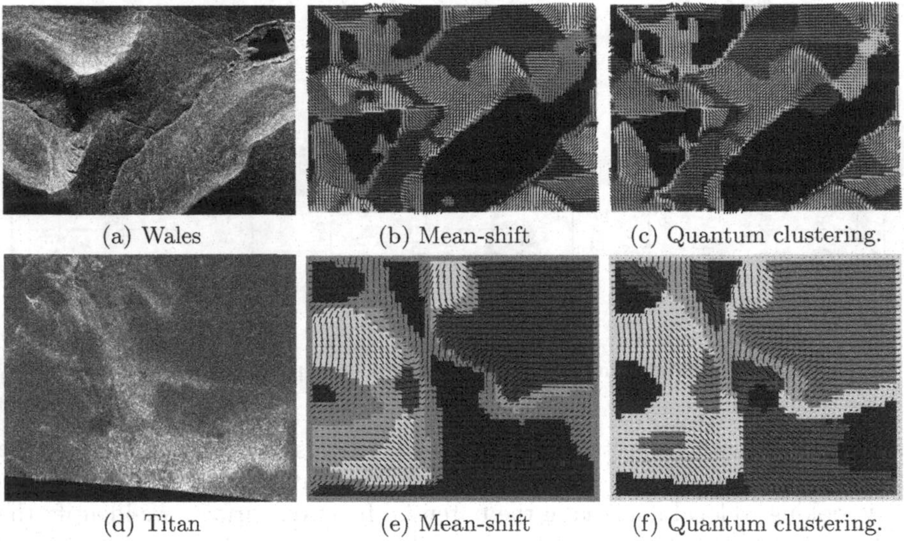

(a) Wales (b) Mean-shift (c) Quantum clustering.

(d) Titan (e) Mean-shift (f) Quantum clustering.

Fig. 4. Topographical segmentation of SAR images from Wales (a) and Titan (d) using the mean shift in (b) and (e), and quantum clustering in (c) and (f)

results provided by the proposed bandwidth estimation method is almost always included in the right range of values for all three KDE methods according to the plots from Fig. 3. Table 2 provides the bias and confidence intervals in blind detection of modulated signals for 8-PSK when estimating locations of cluster centers. We calculate the mean square error between the detected centers and the sample mean (MSE_{SM}) as well as the mean square error between the cluster centers and the ground truth centers (MSE_{O}). The number of clusters was always estimated correctly as 8. As it can be observed from Table 2, quantum clustering provides better results than the other methods.

In another application we consider Synthetic Aperture Radar (SAR) images representing terrain information from Wales and Titan, a moon of Saturn, as

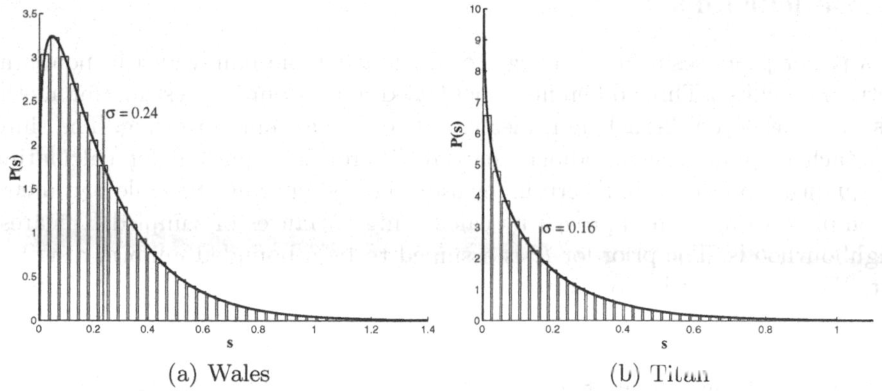

(a) Wales (b) Titan

Fig. 5. Gamma distribution fitting and bandwidth estimation

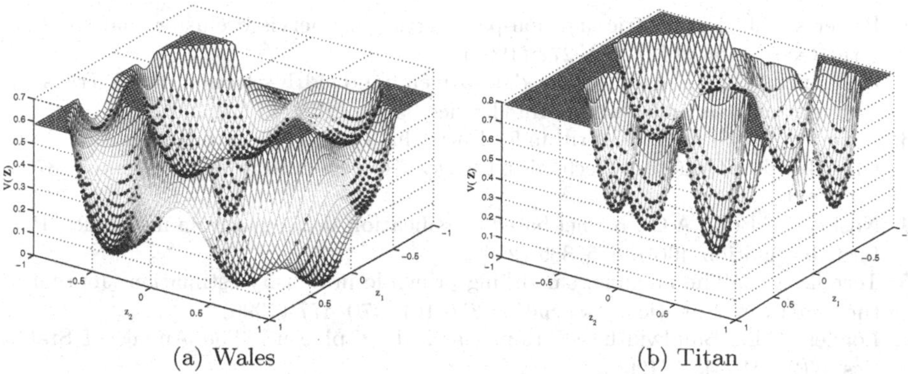

(a) Wales (b) Titan

Fig. 6. Quantum potential for the topography orientaion from the SAR images

shown in Figs. 4(a) and 4(d), respectively. We aim to identify various topographic regions in these images according to the local surface orientation clustering. In [12] a surface orientation estimation method was used for SAR images of terrain. The histograms of local neighbourhood distances for $K_j \in [N/20, N/2]$, where N is the data size, are shown when fitted to Gamma distributions in Figs. 5(a) and 5(b) for Wales and Titan data, respectively. The quantum potentials $\psi(\mathbf{x})$ from (1) are shown in Figs. 6(a) and 6(c) for the two data sets. The vector field of surface normals is segmented based on the vector orientation similarity. Each segmented region corresponds to a local maxima in $\psi(\mathbf{x})$ or to a local minima in $V(\mathbf{x})$. The segmentation of SAR images in topographic regions based on the orientation of the local surface normals is shown in Figs. 4(b) and (e), when using the mean shift method, and in Figs. 4(d) and (f), for the quantum potential. Both methods use the estimated kernel bandwidth as shown in the plots from Fig 5 for each of the images, respectively.

5 Conclusions

This paper proposes a methodology for estimating the bandwidth in nonparametric modelling. Three different kernel-based non-parametric estimation methods are considered: kernel sum, mean shift and quantum clustering. The third approach employs the Shrödinger partial differential equation for calculating the quantum potential in a certain location. For estimating the scale parameter we employ a Bayesian approach by calculating variances of sampled K-nearest neighbourhoods. The prior for K is assumed to be a bounded uniform distribution. We fit the distribution of variances with a Gamma function and evaluate its parameters using maximum log-likelihood estimation. The proposed algorithm is applied in artificial data and for segmenting vector fields of surface normals extracted from radar images of terrain.

References

1. Roberts, S.J.: Parametric and non-parametric unsupervised cluster analysis. Pattern Recognition 30(2), 261–272 (1997)
2. Comaniciu, D.: An algorithm for data-driven bandwidth selection. IEEE Trans. on Pattern Analysis and Machine Intelligence 25(2), 281–288 (2003)
3. Horn, D., Gottlieb, A.: Algorithm for Data Clustering in Pattern Recognition Problems Based on Quantum Mechanics. Physical Review Letters 88(1), art. no. 018702, 1–4 (2002)
4. Nasios, N., Bors, A.G.: Kernel-based classification using quantum mechanics. Pattern Recognition 40(3), 875–889 (2007)
5. Terrell, G.R.: The maximal smoothing principle in density estimation. Journal of the American Statistical Association 85(410), 470–477 (1990)
6. Loader, C.L.: Bandwidth selection: classical or plug-in? The Annals of Statistics 27(2), 415–438 (1999)
7. Silverman, B.W.: Density estimation for statistics and data analysis. Chapman & Hall, Boca Raton (1986)
8. Yang, L., Tschernig, R.: Multivariate bandwidth selection for local linear regression. Journal of Royal Stat. Soc., Series B, 61, part. 4, 793–815 (1999)
9. Jones, M.C., Marron, J.S., Sheather, S.J.: A brief survey of bandwidth selection for density estimation. Journal of the American Statical Association 91(433), 401–407 (1996)
10. Sheather, S.J., Jones, M.C.: A reliable data-based bandwidth selection method for kernel density estimation. Journal of the Royal Statistical Society, Series B 53(3), 683-690 (1991)
11. Ruppert, D., Sheather, S.J., Wand, M.P.: An effective bandwidth selector for local least square regression. Journal of the American Statical Association 90(432), 1257–1270 (1995)
12. Bors, A.G., Hancock, E.R., Wilson, R.C.: Terrain analysis using radar shape-from-shading. IEEE Trans. on Pattern Analysis and Machine Intelligence 25(8), 974–992 (2003)
13. Duda, P.O., Hart, P.E., Stork, D.G.: Pattern Classification. J. Wiley, Chichester (2000)
14. Choi, S.C., Wette, R.: Maximum likelihood estimation of the parameters of the Gamma distribution and their bias. Technometrics 11(4), 683–690 (1969)

Using Kernel Basis with Relevance Vector Machine for Feature Selection

Frédéric Suard and David Mercier

CEA, LIST, Laboratoire Intelligence Multi-capteurs et Apprentissage,
F-91191 Gif sur Yvette, France
{frederic.suard,david.mercier}@cea.fr

Abstract. This paper presents an application of multiple kernels like Kernel Basis to the Relevance Vector Machine algorithm. The framework of kernel machines has been a source of many works concerning the merge of various kernels to build the solution. Within these approaches, Kernel Basis is able to combine both local and global kernels. The interest of such approach resides in the ability to deal with a large kind of tasks in the field of model selection, for example the feature selection. We propose here an application of RVM-KB to a feature selection problem, for which all data are decomposed into a set of kernels so that all points of the learning set correspond to a single feature of one data. The final result is the selection of the main features through the relevance vectors selection.

Keywords: Relevance Vector Machine, Multiple Kernel, Kernel Basis, Feature Selection.

1 Introduction

During the last decade, Kernel Machines have been developed significantly, because of their ability to deal with a large variety of data, for example Support Vector Machines [12], kernel PCA [9] or kernel Fisher discriminant [7]. Approach involved by many kernel machines aims at defining a prediction function according to a weighted linear combination of kernel functions : $f(\mathbf{x}) = \sum_{i=1}^{n} w_i \cdot K(\mathbf{x}, \mathbf{x}_i) + w_0$. Among the cited algorithms, the SVM are particularly well known for their performance and their ability to face a large variety of dataset problems.

However, using a single kernel can be a limitation for some tasks, since all features are merged into a single kernel. To face this limitation, multiple kernel framework aims at using a set of kernels, instead of a single one, [5], by using a set of kernels. Lanckriet and Bach [1] have adapted this framework has been particularly to SVM thanks to composite kernel, by proposing to build a linear combination of kernels coming from various descriptors or a subset of data or a set of kernels with different parameters. The prediction function is then of the form : $f(\mathbf{x}) = \sum_{i=1}^{n} \alpha_i \sum_{j=1}^{k} \beta_j \cdot K_j(\mathbf{x}, \mathbf{x}_i) + b$.

A probabilistic model has also been developped to deal with such composite kernel [3], which shows the interest of such framework.

C. Alippi et al. (Eds.): ICANN 2009, Part II, LNCS 5769, pp. 255–264, 2009.

However, using a composite kernel have some restrictions, since the same set of kernels is used for all vectors of the solution, so that we can not merge global and local kernels to define the solution.

We consider here an other definition of multiple kernel, namely the Kernel Basis, which has been adapted recently to the *Least Angle Regression Stepwise* algorithm [2] by Guigue et al. [13]. The aim is to define the original kernel by concatenating a set of kernels, so that the algorithm can adapt a specific subset of kernels for each vector of the solution. Compared with the composite kernel, the kernel basis is able to deal both with local and global kernel. The prediction function of the kernel basis is then defined by : $f(\mathbf{x}) = \sum_{i=1}^{n} \sum_{j=1}^{k} w_{i,j} \cdot K_j(\mathbf{x}, \mathbf{x}_i) + w_0$.

But the LARS algorithm imposes a regularization parameter which helps to tune the complexity of the solution. The interest of such parameter is to limit the number of vectors, but can also impose a cross validation step.

In this paper we are looking at the Relevance Vector Machine algorithm proposed by Tipping et al. [11]. This method is based on a Bayesian approach which expects to maximize the distribution probability according to the linear combination of the relevance vectors. We propose here to define a multiple kernel like Kernel Basis, and to adapt it to the RVM algorithm. The interest is to go deeper in the model selection task, by applying a specific kernel for each relevance vector, so that we can modeling the data distribution with more efficiency.

To show the interest of our approach, we will present some preliminary results obtained for a feature selection task. The set of kernels is composed here by computing one kernel for each feature of each data. The feature selection is then operated by selecting the best kernels. We will compare the performance obtained on three different benchmark datasets with the state of the art. Results will show that the RVM-Kernel Basis is very promising, since it performs as well as SVM composite kernel but implies also a more sparse solution. The sparsity of the solution is a major advantage in such field of real time computation, for example. A sparse solution has also a higher generalization capacity.

2 Relevance Vector Machine

In the first part, we will present briefly the RVM algorithm and the way to apply a multiple kernel strategy. The relevance vector machine is a probabilistic sparse kernel model that has been introduced by M. Tipping in 2000 [11,14].

The aim is to reveal the underlying distribution of a set of data $\{\mathbf{x}_i, y_i\}_{i=1...n}$, where $\mathbf{x} \in \mathbb{R}^d$. Each data is associated with a label y defined by $p(y|\mathbf{x}) \sim \mathcal{N}(f(\mathbf{x}), \sigma^2)$, with the standard deviation coming from the addition of a gaussian noise : $\epsilon \sim \mathcal{N}(0, \sigma^2)$. The decision function is of the form :

$$f(\mathbf{x}) = \sum_{i=1}^{n} w_i \cdot \Phi(\mathbf{x}, \mathbf{x}_i) + w_0 \quad (1)$$

with Φ a kernel function, $\mathbf{w}$ the coefficient associated to each support vector.

So we can rewrite the probability of the data according to the parameters :

$$p(\mathbf{y}|\mathbf{w},\sigma^2) = \frac{1}{(\sigma\sqrt{2\pi})^n} \exp^{-\frac{1}{2\sigma^2}\|\mathbf{y}-\Phi\mathbf{w}\|^2} \tag{2}$$

with Φ a $n \times (n+1)$ matrix containing the kernel and a bias : $\Phi = \begin{bmatrix} 1 & \\ \vdots & K \\ 1 & \end{bmatrix}$.

The key of this approach is to define a prior on each coefficient w_i. According to the *Automatic Relevance Determination*[6] mechanism, all coefficient which are unecessary are pruned. This mechanism can explained the sparsity of the solution, since it prunes all parameters that add complexity to the probabilistic model. By pruning coefficients, the likelihood is then maximised regarding the input data. For further information the reader can refer to [11], this paper also presents the adaptation of this algorithm for the classification case.

3 Extension of RVM to Multiple Kernel - Kernel Basis

A recent approach in SVM pushed by Lankriet and Bach [1] , introduced the notion of multiple kernel learning, more precisely a composite kernel, by constituting a set of kernels, where each kernel has been obtained with different kernel formulation or parameters. The prediction function is then written :

$$f(\mathbf{x}) = \sum_{i=1}^{n} \alpha_i \sum_{j=1}^{k} \beta_j \cdot K_j(\mathbf{x}, \mathbf{x}_i) + b. \tag{3}$$

The solution is build around a composite kernel, that is to say that only one kernel is applied to all vectors. This approach has been proposed to tackle the descriptor fusion problem, by merging in a single kernel a set of kernels coming from different descriptors. It can also solve the problem of kernel parameter setting, since we can propose kernels from the same data but with a different parametrization.

However, this method assigns the same kernel for all support vectors without taking into account the impact of local kernels. Our problematic is to be able to assign specific kernel for each vector, which can be written:

$$f(\mathbf{x}) = \sum_{i=1}^{n} \sum_{j=1}^{k} w_{i,j} \cdot \Phi_j(\mathbf{x}, \mathbf{x}_i) + w_0. \tag{4}$$

Instead of using a couple of weighting coefficients ($\mathbf{w} \in \mathbb{R}^n$ and $\beta \in \mathbb{R}^d$), this formulation is based on a weighting coefficient $w \in \mathbb{R}^{n \times d}$. We can then apply for each vector a specific set of kernels.

The difference between composite kernel and kernel basis is illustrated on figure 1, where one can see that the kernel basis is able to assign for each vector a specific kernel which considers only one feature (RV 1 and 2) or two features

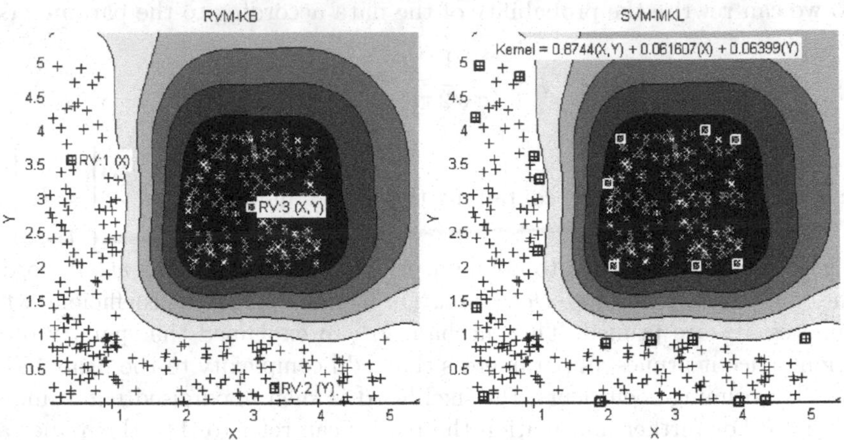

Fig. 1. This example shows the difference between RVM Kernel Basis (left) and SVM Composite Kernel (right). The Kernel Basis is able to adapt a specific kernel for each vector, namely a kernel based on 'x', a second on 'y' and the third combines 'x' and 'y'. The composite kernel aims at defining a unique kernel for all vectors.

(RV3), whereas a composite kernel can only consider the same kernel for all vectors.

One possible way to write this function is to define a kernel basis as described in [4,13]. This definition aims at decompose the kernel Φ into different blocks. The multiple kernel is then composed like a Kernel Basis :

$$\Phi = \begin{bmatrix} 1 & & & & \\ \vdots & K_1 & K_2 & \dots & K_k \\ 1 & & & & \end{bmatrix}, \tag{5}$$

for k kernels when a bias is added. This matrix has a size of $n \times (n \times k)$, so we can associate a weight w_i with each column. If we consider that all columns are independant, we can finally write the prediction function with :

$$f(\mathbf{x}) = \sum_{i=1}^{n*k} w_i \cdot \phi_i(\mathbf{x}) + w_0. \tag{6}$$

This formulation shows that we can deal with relevance vectors, each one defined by at least one kernel function. Since the formulation of Tipping can deal with non-Mercer kernels, the extension of RVM to multiple kernel is well adapted with a Kernel Basis approach.

4 Experimental Results

In the last section, we will present some results to illustrate our approach. We propose here to use the RVM-Kernel Basis for a task of feature selection. Starting from a dataset $\{\mathbf{x}_i, y_i\}_{i=1..n}$ of n data $\mathbf{x} \in \mathbb{R}^d$, the aim is to build a kernel Φ which has a size of n rows and $n \times d$ columns. Applying the RVM algorithm helps us to select the vectors to build the solution.

The more pertinent features are those corresponding to the selected vectors, so that we can simply count the number of times a features has been selected.

We compare the performance with a composite kernel approach, namely the SVM-MKL. The kernel set is constituted with d kernels of size $n \times n$. The feature selection is operated here by considering the value of the weighting coefficients β (see eq. 3) associated with each kernel in the linear combination. The weight of a feature is directly linked with the β value. We are using the SVM-MKL implementation for Matlab proposed by Rakotomamonjy et al. [8], that we already used for a task of image descriptors fusion [10].

We are using some datasets available on the UCI website (`http://archive.ics.uci.edu/ml/`) which are frequently used for machine learning benchmark : Boston Housing, Auto Mpg and Blood transfusion.

For each dataset, we first evaluated the performance of a single kernel, that is to say when only one kernel is used for all features. We tested different parameters for the kernel, namely a gaussian kernel for different bandwidths and a polynomial kernel of different orders. So that we obtained a reference to assess the multiple kernel approach. We carried a 4 fold cross validation for each kernel parameter.

In order to make our results comparable, we organized randomly the cross validation but used the same data for all tests. Concerning the multiple kernel, we set the same kernel parameter for all features, which can be considered as sub-optimal, but since our data are normalized to obtain a mean of 0 and a standard deviation of 1, this compromise is justified. At the moment, we are limited by the memory aspect of such algorithms, but this aspect is a perspective for our future works, by integrating the kernel parameter choice during the optimization.

Since the SVM algorithm involves some parameters within the optimization procedure, we have to set the weight of the misclassified points of the learning set (C), and the width of the regression tube (ϵ). We tried an exhaustive list of values for each parameter and retained the parameters corresponding to the best performance.

To estimate the performance, we have computed the mean squared error, for the regression case, and the AUC value, that is to say the area under the roc curve for the classification case.

The multiple kernel set is built differently in SVM and RVM. For the SVM, we define one kernel for each feature (which produces a kernel of size $n \times n \times d$), so that the SVM will select a subset of kernels and a set of support vectors. For the RVM the kernel basis has an original size of $n \times (n \times d)$, and the algorithm selects a subset of vectors among the $n \times d$ available. This can explained the fact that the percentage of relevance vectors have no correspondence between the single and kernel basis cases.

Auto MPG. The first dataset : AUTO MPG estimates the pollution emission of a car, by considering some mechanical characteristics of a vehicle. The database contains 398 cars, each one described by 7 features.

We have reported in table 1 the performance obtained for RVM and SVM approach for both single and multiple kernel. We report only the performance corresponding to the best parameter set (C, ϵ) for the SVM.

One can see that when using a single kernel, the SVM obtained a better score with an error of 2.74 against 2.84 for the RVM. When we use a multiple kernel, both algorithms improved their performance, with a significant improvement for the RVM-KB, which reached an error of 2.7.

However, we have to note that, due to its sparse solution, the RVM requires only 101 vectors, that is to say 101 scalars, to build the solution. The SVM requires a larger solution that implies 6 kernels and 292 points, that is to say 1752 scalars. In terms of percentage, the RVM retains 4.2% of the learning set against 97% for the SVM. The RVM percentage is computed as follows : $\frac{\#RV}{\#Data \times \%CrossValidation \times \#kernels} = \frac{101}{398 \times 3/4 \times (7+1)}$.

Table 1. Performance obtained for AUTO dataset. We report here for each algorithm, for both single and multiple kernel the error and the size of the solution, according to different kernel parameter setting.

AUTO	RVM				SVM				
	Single Kernel		Kernel Basis		Single Kernel $\epsilon = 0.2$, $C = 100$		Composite Kernel $\epsilon = 0.1$, $C = 10$		
Kernel	error	#RV (%)	error	#RV (%)	error	#SV (%)	error	#SV (%)	#kernels
gaussian 0.1	24.69	82 (27.55%)	3.14	1128 (51.4)	7.68	293 (98.16)	3.11	293 (98.2)	7
gaussian 0.5	8.99	174 (58.29%)	3.06	549 (23)	4.00	279 (93.4)	2.84	291 (97.4)	7
gaussian 1	4.47	64 (20.85%)	2.74	352 (14.8)	3.34	273 (91.46)	**2.70**	292 (97.8)	6
gaussian 2	**2.84**	20 (6.70%)	2.74	222 (9.3)	2.78	272 (91.12)	2.8	293 (98.4)	1
gaussian 3	2.86	18 (6.20%)	2.77	186 (7.8)	**2.74**	272 (91.12)	2.82	294 (98.6)	1
gaussian 5	2.97	13 (4.27%)	2.9	664 (27.8)	2.79	268 (89.78)	2.9	294 (98.6)	1
gaussian 10	3.20	18 (6.20%)	2.86	175 (7.3)	2.89	277 (92.80)	3.25	292 (97.8)	1
gaussian 20	5.20	8 (2.76%)	2.84	254 (10.6)	3.27	283 (94.81)	3.78	294 (98.6)	1
poly 1	23.80	1 (0.34%)	23.8	0.75 (0.03)	3.46	280 (93.80)	3.47	295 (98.8)	1
poly 2	2.93	18 (6.20%)	2.83	118 (4.9)	2.82	273 (91.46)	2.81	294 (98.6)	1
poly 3	3.40	44 (14.74%)	**2.70**	101 (4.2)	5.48	275 (92.13)	4.96	294 (98.6)	1

We have detailed on table 2 the mutiple kernel solution for RVM and SVM. We have reported the number of relevance vectors for the RVM and the total weight according to each feature. We simply added all the weights for all vectors using a given feature. Concerning the SVM, we report the weighting coefficient of each kernel of the composite kernel. This coefficient is directly linked to the feature pertinence. As we can see, SVM put a more important weight to the last kernel which contains all features ($\beta = 0.75$), but we can observe some similarities by comparing RVM-KB and SVM-MKL. For example, kernels 5 and 7 are neglicted in both cases, whereas kernels 1, 3 and 4 have a major weight. Considering the features 5 and 7 : acceleration and origin, this selection seems

Table 2. Auto MPG : details of the RVM-KB and SVM-MKL solution. For each feature, we report the number of relevance vectors and the cumulated weights, and for the SVM-MKL we detail the weight of each kernel for the linear combination of kernels.

		RVM-KB		SVM-MKL
kernel	Feature(s)	#RV	$\sum_i \lvert w_i \rvert$	β
1	# cylinders	18.2500	0.1886	0.0081
2	displacement	13.2500	0.0596	0.0323
3	horsepower	10.2500	0.2772	0.0289
4	weight	7.7500	0.2242	0.1053
5	acceleration	1.0000	0.0029	0
6	model year	38.2500	0.2147	0.0721
7	origin	0	0	0
8	[1-7]	12.2500	0.0328	0.7530

to be valuable, since the objective is to estimate the consumption of the car, so the other attributes seem to be more significant, namely the weight, horsepower or cylinders.

Boston Housing. The second dataset : Boston Housing, has been originally performed in the RVM article of Tipping. This regression task aims at predicting the value of a residential house by considering a set of features describing the environment, local supplies or roads. 506 prices are given, each one described by 13 features.

As we said above, we tried different kernel parameters to estimate the performance with a single kernel. We then accomplished the same procedure with the multiple kernel approach.

We report the results on the table 3 for both single and multiple kernel. Using only a single kernel, SVM performs better than RVM with an error of 3.27 against 3.81. For the multiple kernel approach, both SVM and RVM increase their performance, but the RVM has a major gain and reached an error of 3.22. Another point is the sparsity of the solution, which is also in favor of the RVM due to the lower number of points necessary : 322 scalars for the RVM against 375×7 for the SVM.

We have also reported the details of the multiple kernel solution in table 4, which clearly shows that both algorithms are able to select the more significant

Table 3. Best performance for single and multiple kernel for both RVM and SVM. For each case, we report the kernel parameter, the performance (error) and the size of the solution.

BOSTON	Kernel	error	#RV (%)
RVM	Gaussian 5	3.81	30 (7.8)
RVM-KB	Gaussian 2	3.22	322 (6.1)

BOSTON	Parameters	Kernel	error	#SV (%)
SVM	$\epsilon = 0.05, C = 100$	Gaussian,3	3.27	373 (98)
SVM-MKL	$\epsilon = 0.01, C = 10$	Gaussian,1	3.26	375 (98)

Table 4. Details of the RVM-KB and SVM-MKL solution for the Boston dataset. For each feature, we report the number of relevant vectors and the cumulated weights, and for the SVM-MKL we detail the weight of each kernel for the linear combination of kernels.

		RVM-KB		SVM-MKL
Kernel	Feature(s)	#RV	$\sum_i \lvert w_i \rvert$	β
1	Criminal rate	5.7500	0.0050	0.0040
2	Residential %	74.2500	0.0005	0
3	Industrial %	13.0000	0.0014	0
4	River distance	11.7500	0.0006	0
5	% NOx	16.0000	0.0037	0.0104
6	# rooms	15.0000	0.0035	0.1271
7	Age	3.5000	0.0008	0
8	Distance	0	0	0
9	Road	61.0000	0.0200	0
10	Tax	21.2500	0.0027	0.0181
11	Teachers	10.7500	0.0016	0.0185
12	% blacks	1.2500	0.0003	0
13	Lower salary	14.7500	0.0085	0.0939
14	[1-13]	73.7500	0.9513	0.7281

features. To assess efficiently the multiple kernel, we added to the set of kernels a kernel computed from all data. We have to precise that the bandwidth of this kernel is applied for each element of the vector, so that the bandwidth is comparable between single feature kernels and all features kernels. It is interesting to note that SVM and RVM obtained some similarities beyond their selection, since the last kernel (14 containing all features) obtained an important weight, so that it can be assimilated as a single kernel, but the performance grows up because of adding some single features.

In terms of feature selection, the SVM-MKL selected 7 kernels out of the 14 originals, and the RVM retained 9 main kernels. The 5 last kernels got a minor weight (namely the kernels 2, 4, 7, 8 and 12). On top of that, SVM totally removes kernels 2, 3, 4, 7, 8, 9 and 12. We can then notice the interest of RVM-KB, which is able to add minor kernels, that is to say local kernels, in opposition to the SVM which generally prefers to suppress this kind of kernel ($\beta = 0$).

Blood transfusion. The last experiment is a classification task based on the Blood Transfusion dataset. The aim of such database is to predict if a person has donated or not its blood in March 2007. The set contains 748 persons which are characterized by 4 features, concerning the frequency and the volume.

Since it is a classification problem, we give the performance by considering the area under the roc curve (AUC). We also report the size of the solution.

As shown on table 5, the RVM algorithm performs better than the SVM, in both single and multiple kernel case. We have to note that the SVM-MKL does not really takes benefit from the multiple kernel approach, since the performance is not increasing at all. This can be explained by the fact that all features are

Table 5. Performance obtained on the dataset BLOOD. We give the AUC value for the best kernel parameter.

BLOOD	Kernel	AUC	#RV (%)
RVM	gaussian 10	0.744	3 (0.54)
RVM-KB	poly 3	0.75	247 (11.02)

BLOOD	Parameter	Kernel	AUC	#SV (%)
SVM	$C = 10$	gaussian 10	0.718	299 (55)
SVM-MKL	$C = 100$	linear	0.71	378(67)

necessary, as shown on table 6. On top of that, the RVM is sparse, since for a single kernel only 3 vectors are necessary. However, the multiple kernel lost this sparsity, since 247 vectors are used. It should be noticed that in this case, it means 247 scalars, since we have originally 2241 vectors in the learning set, one vector beeing associated with one feature only. The sparsity can be explained by the fact that this set seems to have few overlap, considering also the SVM that retains around 50% of the learning set.

Table 6. Details of the RVM-KB and SVM-MKL solution for the Blood dataset. For each feature, we report the number of relevant vectors and the cumulated weights, and for the SVM-MKL we detail the weight of each kernel for the linear combination of kernels.

BLOOD		RVM-KB		SVM-MKL
Kernel	Feature	#RV	$\sum_i \lvert w_i \rvert$	β
1	Recency	22	0.6678	0.2484
2	Frequency	104	0.7228	0.1983
3	Monetary	104	0.7228	0.2500
4	Time	16	0.6656	0.2499

Talking about feature selection, this experiment is interesting because both RVM-KB and SVM-MKL made the same selection. As shown on table 6, all features are necessary to build the solution, and both give a similar weight to all features. It means that all features are relevant for this task.

This last experiment also point out the fact that one have to take care when building the kernel set. We have to pay attention to the fact that computing each kernel from a single feature can fail, since we loose dependancy between features. At the moment, we have to constitute the set of kernels for each feature combination. For example, if we have two features x_1 and x_2, the kernel set is composed of $K(x_1, .)$, $K(x_2, .)$ and $K(x_{12}, .)$. This point is a particular perspective, that we propose to face in a near future.

5 Conclusion

In this paper, we proposed to extend the Relevance Vector Machine to multiple kernel as a Kernel Basis. The interest is to build a solution which combines both local and global kernels, so that each vector uses a specific set of kernels, whereas a composite kernel can only deal with the same kernel for all vectors.

This approach is very useful for model selection tasks, like kernel parameter setting or feature selection. The results we obtained on such problem have shown that this approach is very promising since we obtained equivalent and sometimes best performance compared to SVM using a composite kernel, with a major advantage concerning the size of the solution. Actually, due to the sparsity of RVM, the solution is really smaller, which is very interesting by considering real time application, for example. The sparse aspect has also other interests that we propose to exploit in a near future concerning the explanation of the solution.

The main drawback of the RVM-KB, resides in the computational complexity, since we have to face a large matrix inversion, so that we can have a limitation. However, due to the parametrization of the SVM, learning time is comparable since we have to test different values for the SVM parameters. A last perspective is then linked with the large dataset, that we will have to face.

References

1. Bach, F.R., Lanckriet, G.R.G., Jordan, M.I.: Multiple kernel learning, conic duality, and the smo algorithm. In: ICML 2004: Proceedings of the twenty-first international conference on Machine learning, p. 6. ACM Press, New York (2004)
2. Efron, B., Hastie, T., Johnstone, I., Tibshirani, R.: Least angle regression, pp. 407–499 (January 2003)
3. Girolami, M., Rogers, S.: Hierarchic bayesian models for kernel learning. In: 22nd International Conference on Machine Learning, pp. 241–248 (2005)
4. Guigue, V., Rakotomamonjy, A., Canu, S.: Kernel basis pursuit. In: CAP, pp. 93–106 (2005)
5. Gunn, S., Kandola, J.: Structural modelling with sparse kernels. Machine Learning 48, 137–163 (2002)
6. Mackay, D.J.: Probable networks and plausible predictions - a review of pratictal bayesian methods for supervised neural networks, vol. 6, pp. 469–505 (1995)
7. Mika, S., Rätsch, G., Weston, J., Schölkopf, B., Smola, A.J., Mueller, K.-R.: Constructing descriptive and discriminative non-linear features: Rayleigh coefficients in kernel feature spaces. IEEE Transactions on Pattern Analysis and Machine Intelligence (2004)
8. Rakotomamonjy, A., Bach, F., Canu, S., Grandvalet, Y.: Simple MKL. Journal of Machine Learning Research (2008)
9. Schölkopf, B., Smola, A.J.: Learning with Kernels. MIT Press, Cambridge (2002)
10. Suard, F., Rakotomamonjy, A., Bensrhair, A.: Model selection in pedestrian detection using multiple kernel learning. In: Intelligent Vehicles Symposium 2007, Istanbul (June 2007)
11. Tipping, M.: The relevance vector machine. In: Solla, T.K.L.S.A., Müller, K.-R. (eds.) Advances in Neural Information Processing Systems, vol. 12. MIT Press, Cambridge (2000)
12. Vapnik, V.: The Nature of Statistical Learning Theory. Springer, N.Y. (1995)
13. Vincent, P., Bengio, Y.: Kernel matching pursuit. Mach. Learn. 48(1-3), 165–187 (2002)
14. Wu, L., Schölkopf, B., Bakir, G.: A direct method for building sparse kernel learning algorithms. Journal of Machine Learning Research 7, 603–624 (2006)

Acquiring and Classifying Signals from Nanopores and Ion-Channels

Bharatan Konnanath*, Prasanna Sattigeri, Trupthi Mathew, Andreas Spanias, Shalini Prasad, Michael Goryll, Trevor Thornton, and Peter Knee

SenSIP Center and CSSER,
Department of Electrical Engineering,
Arizona State University,
Tempe, AZ 85287-5706,
USA
{bkonnana,psattige,tmathew2,spanias,sprasad5,
mgoryll,t.thornton,paknee}@asu.edu

Abstract. The use of engineered nanopores as sensing elements for chemical and biological agents is a rapidly developing area. The distinct signatures of nanopore-nanoparticle lend themselves to statistical analysis. As a result, processing of signals from these sensors is attracting a lot of attention. In this paper we demonstrate a neural network approach to classify and interpret nanopore and ion-channel signals.

Keywords: Nanopore devices, Ion-channel sensors, Denoising using wavelets, PCA, WHT, Sensing using nanopores and neural networks.

1 Introduction

Resistive pulse sensing or Coulter counting [1] is a wide research area centered on nanopores. Though originally developed for counting particles suspended in a fluid using micrometer sized pores, Coulter counting has recently been applied at the nanoscale level [2]. Through the reduction of device aperture size to the nanometer range, Coulter counting experiments of small particles such as DNA molecules [3,4] and bovine serum albumin (BSA) [5] through solid state devices as well as virus particles [6] and DNA [7] through polymer materials have been demonstrated. In the Coulter counting experiments, individual molecules are constrained to pass through a small constrained electric path in a suspending fluid as shown in Fig 1(left panel). As the molecule passes through the orifice, it causes an increase in resistance which leads to a drop in the current as shown in Fig 1(right panel). By observing the curvature of these spikes, the size, type and the concentration of the particles can be determined [8,9].

Ion channel proteins are naturally occurring nanopores that mediate the flow of ions and molecules across membranes. The utility of ion channels for stochastic sensing has been pioneered by Bayley and several of his collaborators [10].

* This work was supported through the NSF EXP Award 0730810 program award.

C. Alippi et al. (Eds.): ICANN 2009, Part II, LNCS 5769, pp. 265–274, 2009.

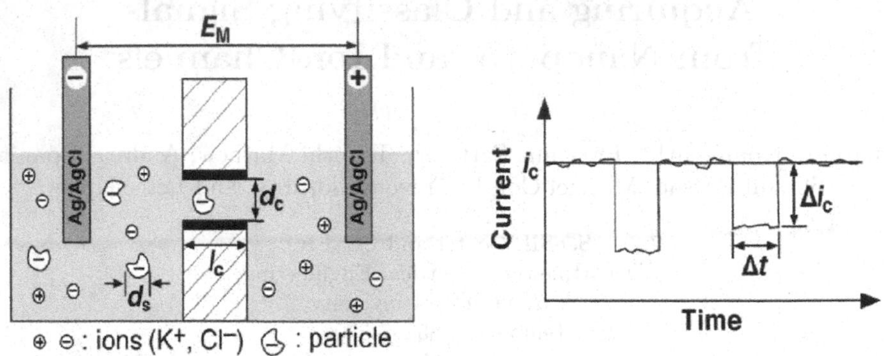

Fig. 1. Graphical Rendering of the Coulter counter (Reproduced from [8])

Ion-channels can be engineered to act as biosensors that can detect metal ions and organic molecules such as proteins [10,11,12]. Potential applications of ion-channel sensing include detection of reactive molecules in pharmaceutical products, chemical weapons, pesticides and foodstuffs [13].

An example of an engineered pore is shown in Fig 2. An applied potential to the pore creates a small current flow. A binding site for an analyte is engineered into each pore. An analyte binding event to the pore causes the current to be modulated as shown in the trace below the figure. The signature of the signal through the pore is generally different for distinct analytes. The frequency of occurrence of the binding events was shown to correlate with the concentration of the analyte while parameters of the current, such as the mean duration and amplitude correlate with the type of the analyte. Features of interest fall into two categories: switching and non-switching [12,13,14], both of which contain information that may be important in detecting an agent. The conventional modeling procedures used to classify the ion-channel signals are dwell-time analysis

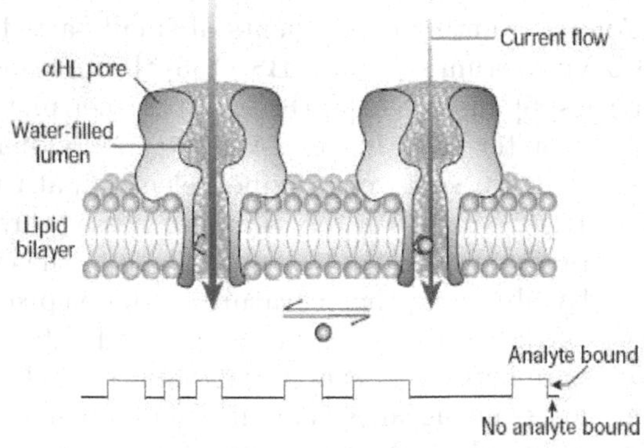

Fig. 2. Single Engineered Pore (Reproduced from [10])

[15,16,17] and Hidden Markov Models (HMMs) [18,19,20,21]. Feature extraction for ion-channel signals have been explored in [22,23].

In this paper, we present a wavelet transform based approach for denoising both nanopore and ion-channel signals. A two-step feature extraction process using Walsh Transforms and Principal Component Analysis (PCA) is also tested for ion-channel signals. Robust analyte classification is carried out for both cases using Support Vector Machines (SVMs).

2 Data Processing for Nanopore Signals

2.1 Data Generation

Nanopore data was generated using a Coulter counting element which was constructed using a Teflon chamber with two baths surrounding the nanopore. The two baths were filled with 0.1M KCl electrolyte solution. The nanopore used for the Coulter counting experiments was patterned to a diameter of 300nm but the measured diameter was 212 nm. The recordings were taken using the Axopatch 200B. Voltage traces were incremented from 0-200 mV in steps of 20 mV and each trace lasted 1s. The input signal was filtered with a 5 kHz low pass filter before the A/D conversion stage. The sampling rate was 50 kHz.

2.2 Wavelet Transform Based Signal Denoising

In the Coulter counting experiment considered in this paper, the baseline current is at the pA level. Due to the low signal-to-noise (SNR) ratio, the signal peaks, which indicate the transition events, can be easily corrupted. This creates difficulties in measurement of peak parameters such as peak height, width and shape. Hence signal denoising is essential to improve the sensitivity and accuracy of the Coulter counters. Wavelet based denoising techniques for Coulter counting experiments have been discussed in [24]. Denoising using the Discrete Wavelet Transform (DWT) is a nonlinear operation and involves the following steps:

- Use a suitable wavelet transform on the noisy data to produce the wavelet coefficients.
- Select an appropriate threshold depending on the noise variance and perform a thresholding operation of the wavelet coefficients to remove the noise
- Zero-pad the signal appropriately and perform the inverse DWT on the thresholded coefficients obtained from the previous step to get the signal estimate.

While traditional linear filtering techniques provide a trade-off for noise suppression against a broadening of signal features, denoising using the DWT preserves the sharpness of features in the original signal. The type of wavelet function, the threshold limits and the level of decomposition is determined on a case by case basis. In our simulations, we determined, using cross-validation, that the biorthogonal wavelet gave the best performance. To capture most of the features in the signal, the level of decomposition was chosen to be 3. Fig 3 shows the reproduced signal for a sample frame.

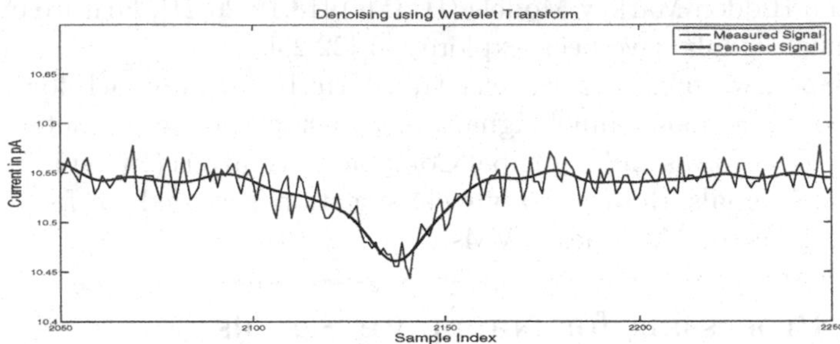

Fig. 3. Nanopore signal Denoising using the Discrete Wavelet Transform (DWT)

2.3 Feature Extraction: Baseline Current, Peak Height and Peak Width

The useful features to be extracted from the nanopore signals are the baseline current, peak height and peak width. The combination of baseline current and height of the peak indicates whether a bead has passed through the nanopore completely or not. The peak amplitude is proportional to the baseline current; i.e. greater the baseline current I, greater will be the drop in current ΔI for beads of the same diameter [9]. The width of the peak is proportional to the diameter of the bead [9]. Fig 4 shows a sample event, where the bead collided with the pore but bounced back (A) and a few milliseconds later passed through the pore (B).

2.4 Event Classification Using Neural Networks

Support vector machines (SVMs) are widely used for solving binary classification problems [25]. SVMs are decision machines that rely on transforming lower-dimensional data into higher dimensional patterns, so that data from two categories can always be separated by a hyperplane, in accordance with Cover's Theorem [26].

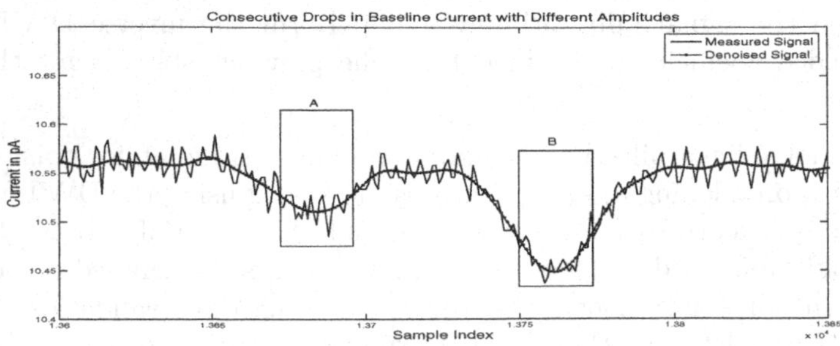

Fig. 4. Sample Nanopore Events

The SVM uses the concept of the margin, which is defined to be the smallest distance between the decision boundary and any of the samples [27]. The support vectors are the training samples that are closest to the decision boundary and thus define the optimal separating hyperplane. In support vector machines the decision boundary is chosen to be the one for which the margin is maximized. It can be shown that the larger margin minimizes the total generalization error [26]. The choice of the nonlinear function that maps the input into a higher-dimensional space is usually dependent on the problem domain. Usually polynomial or radial basis functions are used to perform the mapping.

Experimental data for eight different bias voltages, ranging from 0-200mV, with 40,512 samples at each bias voltage are available. A rectangular window of size 1000 samples with no overlap was used to segment the data. In each segment, peaks were extracted using a gradient method. Each peak was labeled either as an event or a non-event. An event indicates that a bead passed through the nanopore completely whereas a non-event indicates either: (i) a bead bounced back instead of passing through the nanopore or (ii) a spike due to noise. In the given dataset, 3979 peaks were extracted from the signals, out of which 75 peaks indicated events. Peak width, mean baseline current and drop in current amplitude were chosen as features. The dataset was partitioned into a training set (containing 34 events and 1923 non-events) and a test set (containing 41 events and 2056 non-events).

3 Data Processing for Ion-Channel Signals

3.1 Data Generation

Multiple recordings of OmpF ion channels of E. coli in a lipid bilayer across a 50 μm wide pore in silicon, sandwiched between reservoirs containing bathed in a 1M KCl solution are used for generating experimental data. Each recording is generated using a sampling rate of 10 kHz for 4 seconds and an applied voltage of 200 mV. The current amplifier employed was a HEKA EPC-8, operating at

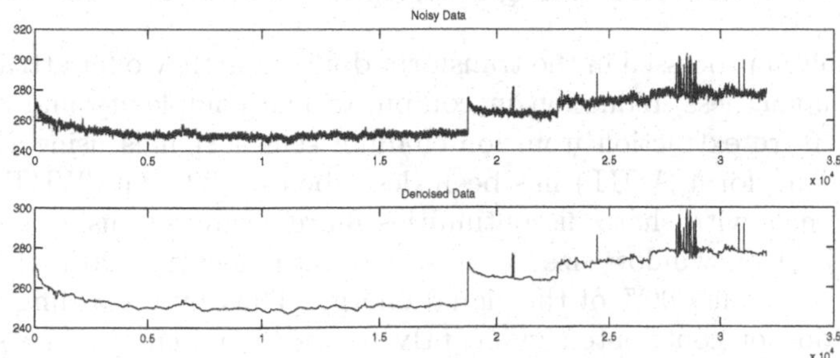

Fig. 5. Top panel: Current Recording for the OmpF ion channel in a membrane across a 50 μm wide pore, bathed in 1M KCL solution. Bottom panel: Ion-channel Signal denoising using the Discrete Wavelet Transform (DWT).

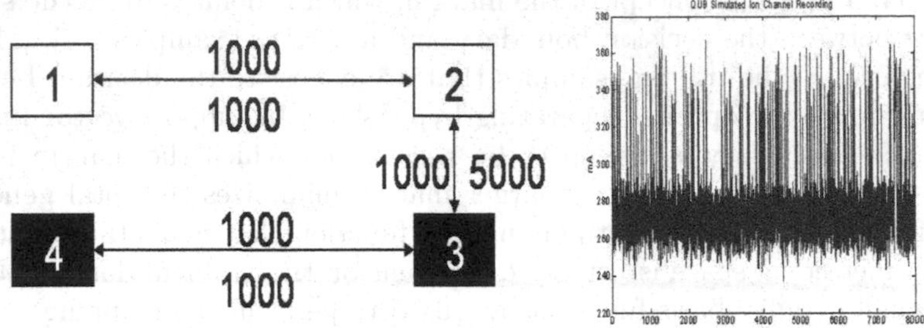

Fig. 6. Left Panel: Basic two class, four state model used by QUB to simulate two analyte ion channel. Right Panel: 8000 ms simulation using QUB.

a gain of 1mV/pA, using a resistive feedback headstage. The input signal was filtered using an 8-pole Bessel filter with a corner frequency of 1kHz before the A/D conversion. The command voltage was generated by a National Instruments 6251 DAQ board.

As shown in Fig 5, we demonstrate that the Discrete Wavelet Transform can be used for denoising ion-channel signals also using the same technique discussed in Sec 2.2. The Haar wavelet, with the level of decomposition set to 8, was found to give the best performance.

Since experimental data for two-analyte simulations are not yet available, we have used the QUB scientific package to generate synthetic data [28]. Fig 6(left panel) shows a sample 4-state Markov model used for generating data and a sample trace is shown in Fig 6(right panel). We constructed models to simulate responses of two highly similar analytes which closely resemble the authentic data. Utilizing multiple recordings, an input data matrix of dimension 400×10000 is formed by extracting four 10,000 point sequences of one second duration from each recording for a total of 50 input files for each analyte.

3.2 Feature Extraction Using Principal Component Analysis

Signals are often processed in the transform domain as they offer attractive benefits like compactness, reduction in computational complexity and robustness to noise. Feature extraction from ion-channel sensor signals using the Walsh-Hadamard Transform (WHT) has been described in [22]. The WHT is able to represent signals with sharp discontinuities more accurately using fewer coefficients. For a given window size N, it was determined that 20% of the WHT coefficients represents 90% of the signal energy. Thus by discarding the coefficients that do not contribute significantly to the signal energy, the size of the dataset was reduced by 80%. Even after WHT is performed, further dimensionality reduction is required on the dataset. For example, $N = 4096$, the size of the transformed dataset is 400×819. It is likely that many of the selected coefficients are highly correlated and there is scope for further compaction.

Principal components analysis (PCA) is a commonly used linear technique for dimensionality reduction. It performs a linear mapping of multidimensional data to a lower dimensional space while retaining as much as possible of the data variability. It was determined that that the first 10 components account for 99% of the total variance. Thus we project the data on the bases represented by the 10 principal components. Now the dimension of the dataset is 400×10. This dataset is used as input to the pattern classification algorithms.

3.3 Analyte Classification

Since we are dealing with a binary classification problem, support vector machines (SVMs) were used in this case also. As mentioned earlier, the dataset consists of 400 vectors. The transformed dataset is randomly permuted and partitioned into a training set of 200 vectors and test set of 200 vectors. To compensate for the small size of the dataset, m-fold cross-validation was used for model selection [29].

4 Results

All simulations were run on MATLAB version 7.5. The Spider toolbox [30] was used for classification using SVMs.

4.1 Nanopore Signals

The classification performance using SVMs are shown in Table 1. The best performance was obtained using RBF kernels using a kernel width of 5. All the events were captured correctly and only one non-event was wrongly labeled as an event. The best performance using a polynomial kernel function (of order 5) is also given in Table 1.

Table 1. Classification Performance on the Test Set

Kernel Used	Classification (%)
RBF	97.56
Polynomial	95.13

4.2 Ion-Channel Signals

The goal of the SVM is to classify input data as quickly as possible and therefore a smaller window length would be preferable. However, there has to be sufficient transition data contained in the input window in order to be able to characterize the signal. For this reason, for each scenario, three different window lengths, $N =$ 4096, 2048 and 1024, were considered. The results of our simulations are shown in Table 2. Polynomial kernels (of order 8) and RBF kernels (of width 6) were found to yield the best results. Our results indicate that as the window length decreases, the error rate increases for all classifiers. This is due to the fact that not only are less coefficient values being used to characterize the signal, but fewer binding events are occurring giving rise to the possibility that there is not enough signal data contained in the windowed segment.

Table 2. Classification Performance on the Test Set

Kernel Used	Classification (%)		
	$N = 1024$	$N = 2048$	$N = 4096$
RBF	69.0	74.5	80.5
Polynomial	66.5	72.5	80.0

5 Conclusion

Denoising signals using DWT was demonstrated for nanopore signals. Three features extracted from the peaks that occur in the signal - peak width, peak amplitude and mean baseline current were used to detect the passage of a bead through the nanopore. Classification was carried out using SVMs with 96 % accuracy. Denoising using DWT was demonstrated for experimental data. Feature extraction and pattern classification for discriminating between two highly similar analytes was carried out for ion-channel signals. Two-stage feature extraction using WHT and PCA provided feature vectors that could be used for classification using the four algorithms. Classification accuracy is at the 80th percentile for a frame length, $N = 4096$. We plan to improve the accuracy of the classifiers using real data generated from experiments.

References

1. Coulter, W.H.: Means for Counting Particles Suspended in a Fluid, U.S. Patent Number 2656508 (1953)
2. Henriquez, R.R., Ito, T., Sun, L., Crooks, R.M.: The resurgence of Coulter counting for analyzing nanoscale objects. Analyst 129, 478–482 (2004)
3. Heng, J.B., Ho, C., Kim, T., Timp, R., Aksimentiev, A., Grinkova, Y.V., Sligar, S., Schulten, K., Timp, G.: Sizing DNA using a nanometer-diameter pore. Biophysical Journal 87, 2905–2911 (2004)
4. Heng, J.B., Aksimentiev, A., Ho, C., Marks, P., Grinkova, Y.V., Sligar, S., Schulten, K., Timp, G.: Stretching DNA using the electric field in a synthetic nanopore. Nano Letters 5, 1883–1888 (2005)
5. Han, A.P., Schurmann, G., Mondin, G., Bitterli, R.A., Hegelbach, N.G., de Rooij, N.F., Staufer, U.: Sensing protein molecules using nanofabricated pores. Applied Physics Letters 88 (2006)

6. Deblois, R.W., Bean, C.P., Wesley, R.K.A.: Electrokinetic Measurements with Submicron Particles and Pores by Resistive Pulse Technique. Journal of Colloid and Interface Science 61, 323–335 (1977)
7. Mara, A., Siwy, Z., Trautmann, C., Wan, J., Kamme, F.: An asymmetric polymer nanopore for single molecule detection. Nano Letters 4, 497–501 (2004)
8. Ito, T., Sun, L., Henriquez, R.R., Crooks, R.M.: A Carbon Nanotube-Based Coulter Nanoparticle Counter. Acc. Chem. Res. 37(2), 937–945 (2004)
9. Petrossian, L.: Cylindrical Solid State Nanopores, Ph. D Thesis, Arizona State University, Tempe, AZ-85287
10. Bayley, H., Martin, C.R.: Resistive-pulse sensing-From microbes to molecules. Chemical Rev. 100, 2575–2594 (2000)
11. Braha, O., Gu, L.Q., Zhou, L., Lu, X.F., Cheley, S., Bayley, H.: Simultaneous stochastic sensing of divalent metal ions. Nature Biotechnology 18, 1005–1007 (2000)
12. Gu, L.Q., Braha, O., Conlan, S., Cheley, S., Bayley, H.: Stochastic sensing of organic analytes by a pore-forming protein Containing a molecular adapter. Nature 398, 686–690 (1999)
13. Luchian, T., Shin, S.H., Bayley, H.: Single-molecule covalent chemistry with spatially separated reactants. Angewandte Chemie-International Edition 42, 3766–3771 (2003)
14. Wilk, S.J., Goryll, M., Laws, G.M., Goodnick, S.M., Thornton, T.J., Saraniti, M., Tang, J., Eisenberg, R.S.: Teflon (TM)-coated silicon apertures for supported lipid bilayer membranes. Appl. Phys. Lett. 85(15), 3307–3309 (2004)
15. McManus, O.B., Blatz, A.L., Magleby, K.L.: Sampling, log binning, fitting, and plotting durations of open and shut intervals from single channels and the effects of noise. Pflugers Arch. 410, 530–553 (1987)
16. Sigworth, F.J., Sine, S.M.: Data transformations for improved display and fitting of single-channel dwell time histograms. Biophysical Journal 52, 1047–1054 (1987)
17. Ball, F.G., Kerry, C.J., Ramsey, R.L., Sansom, M.S.P., Usherwood, P.N.R.: The use of dwell time cross-correlation functions to study single-ion channel gating kinetics. Biophysical Journal 54, 309–320 (1988)
18. Venkataramanan, L., Sigworth, F.J.: Applying Hidden Markov Models to the analysis of single ion channel activity. Biophysical Journal 82, 1930–1942 (2002)
19. Venkataramanan, L., Walsh, J., Kuc, R., Sigworth, F.: Identification of HMM for ion channel currents-part I: colored noise. IEEE Trans. on Sig. Proc. 46(7), 1901–1915 (1998)
20. Qin, F., Auerbach, A., Sachs, F.: Hidden Markov modeling for single channel kinetics with filtering and correlated noise. Biophysical Journal 79, 1928–1944 (2000)
21. Spanias, A., Goodnick, S., Thornton, T., Phillips, S., Wilk, S., Kwon, H.: Signal processing for silicon ion-channel sensors. In: Proc. IEEE SAFE 2007 (2007)
22. Kwon, H., Knee, P., Spanias, A., Goodnick, S., Thornton, T., Phillips, S.: Transform-domain features for ion-channel sensors. In: Proc. IASTED SPPRA 2008, Paper 599-104 (2008)
23. Konnanath, B., Knee, P., Spanias, A., Wichern, G.: Classification of Ion-Channel Signals using Neural Networks. In: Proc. IASTED SPPRA 2009, Paper 643-075 (2009)
24. Jagtiani, A.V., Sawant, R., Carletta, J., Zhe, J.: Wavelet transform-based methods for denoising of Coulter counter signals. Meas. Sci. Technol. 19(6), 1–15 (2008)
25. Bishop, C.M.: Pattern Recognition for Machine Learning. Springer, Heidelberg (2006)

26. Haykin, S.: Neural Networks: A Comprehensive Foundation. Prentice Hall, Englewood Cliffs (1999)
27. Duda, R.O., Hart, P.E., Stork, D.G.: Pattern Classification. Wiley, Chichester (2001)
28. QuB: A software package for Markov analysis of single-molecule kinetics, `http://www.qub.buffalo.edu/wiki/index.php/Main_Page`
29. Ripley, B.D.: Pattern Recognition and Neural Networks. Cambridge University Press, Cambridge (1996)
30. The Spider (SVM Toolbox) version 1.71 for MATLAB, `http://www.kyb.mpg.de/bs/people/spider/main.html`

Hand-Drawn Shape Recognition Using the SVM'ed Kernel

Khaled S. Refaat and Amir F. Atiya

Computer Department, Faculty of Engineering, Cairo University, Egypt
khaled.saeed84@gmail.com, amir@alumni.caltech.edu

Abstract. We describe an application of the novel Support Vector Machined Kernel (SVM'ed Kernel) to the Recognition of hand-drawn shapes. The SVM'ed kernel function is itself a support vector machine classifier that is learned statistically from data using an automatically generated training set. We show that the new kernel manages to change the classical methodology of defining a feature vector for each pattern. One will only need to define features representing the similarity between two patterns allowing many details to be captured in a concise way. In addition, we illustrate that features describing a single pattern could also be used in this new framework. In this paper we show how the SVM'ed Kernel is defined and trained for the multiclass shape recognition problem. Simulation results show that the SVM'ed Kernel outperforms all other classical kernels and is more robust to hard test sets.

Keywords: Shape recognition, Support Vector Machine, Kernel, Similarity.

1 Introduction

Structured diagrams are very prevalent in many document types. Most people who need to create such diagrams use structured graphics editors such as Microsoft Visio [16]. Structured graphics editors are extremely powerful and expressive but they can be cumbersome to use [17]. It was shown through extensive timing experiments that structured diagrams drawn by hand takes only about 10% of the time it takes to draw one using a tool like Visio [10]. This indicates the value of automated recognition of hand-drawn diagrams.

One of the main steps in the problem of diagram understanding is the recognition of individual hand-drawn shapes. The input to the shape recognition system is a geometric hand-drawn shape. Whereas, the output is its classification to one of predefined classes.

In the classical classification framework, shapes are first converted to feature vectors which are then used to train the classifier. At the point of replacing the shape by a feature vector representing it, a significant amount of information is lost. This could be easily noticed when we discover that we could not recover the shape pattern once converted to a feature vector.

In other problems, sometimes representing the pattern by a feature vector could be problematic. For example if we would like to represent a document by

C. Alippi et al. (Eds.): ICANN 2009, Part II, LNCS 5769, pp. 275–284, 2009.

a feature vector, we could use a dictionary to create a feature vector of word. This could lead to a huge number of features. An alternative and seemingly more efficient way is to define a feature vector that represents the similarity between a pair of documents. In such case we could just define a vector that consists of a few simple and effective high level features. This similarity vector could, for example, consist of the number of common stemmed words between the pair of documents, the number of common named entities, the number of common semantic relations, and finally a binary feature showing whether the two documents were extracted from the same source or not. This suggests that a significant achievement could be acquired, if we could change the classification framework to using feature vectors that represent the similarity between a pair of patterns rather than using feature vectors that represent single patterns.

SVM is a suitable classifier for applying this new framework. In SVM, the classical kernels take two feature vectors as input (each feature vector represents a pattern) and return a real number representing the similarity between them [1]. In order to make use of high level similarity features as stated previously, a domain expert is required to invent a user defined kernel which is an algorithm that measures the similarity between two patterns without converting them to feature vectors. The domain expert is required in order to determine the contribution of each component similarity feature to the final similarity measure. This is a time consuming task since it has to be done for each problem. Moreover, a hard quantitative approach would lead to more consistent performance, and allows the use of cutting edge optimization methods.

We propose a novel kernel function that is extracted from data through a statistical learning procedure. The input of this new kernel function will be only one feature vector representing the similarity between the two input patterns. We name our new proposed kernel the SVM'ed-Kernel.

We propose a method to automatically generate a recreated training set from the original training set. The recreated training set is then used to learn the SVM'ed-Kernel. Interestingly, the SVM'ed-Kernel will be learned as a separate SVM classification problem. Once trained, the SVM'ed-Kernel will then be used as a kernel function in the original classification SVM problem.

Using the SVM'ed-Kernel, we do not need to define features to represent a single pattern. We will only need to define features that represent the similarity between a pair of patterns. This allows novel features to be defined that could not have been defined using the classical feature definition framework.

Moreover, a simple similarity feature between a pair of patterns could eliminate a large number of features representing a single pattern as it was shown in the example of representing a document by a feature vector. This contributes to dimensionality reduction.

In the SVM'ed-Kernel, the contribution of each similarity feature to the final similarity measure is learned statistically from the recreated training set. This eliminates the need for a domain expert, allows the definition of novel high level similarity features, and leads to optimizing the contributions of the different similarities.

In this paper, we describe the application of the SVM'ed Kernel to the Recognition of hand-drawn shapes. The SVM'ed Kernel will allow adding a chain code similarity feature representing the similarity between a pair of patterns that will boost the accuracy significantly.

This paper is organized as follows: Section 2 describes the related work. In Section 3 we introduce preliminaries of SVM as a classifier. The SVM'ed Kernel will be presented in Section 4. Section 5 describes the shape recognition problem. Finally we introduce the experimental results in section 6. The paper ends with a conclusion and future work in section 7.

2 Related Work

Many kernels have been proposed in the SVM literature. We divide the related work into general kernels and specific user-defined kernels. The general kernels are not defined for a specific problem. On the other hand, the user defined kernels are domain dependent and they are defined specifically for the problem at hand.

From among the general proposed kernels, Thadani *et al* [2] creates a kernel function suitable for the training data using a genetic algorithm mechanism. They showed that their genetic kernel has good generalization abilities when compared with the polynomial and the radial basis kernel functions. Kong *et al* [3] proposed the autocorrelation kernel by borrowing this concept from signal processing. The autocorrelation functions give comparable results to the RBF kernel when used to classify some UCI datasets. George *et al* [4] proposed a Sinc-Cauchy hybrid wavelet kernel and shows that it is admissible which means that it is positive definite [1]. They used it for the classification of Cardiac Single Photon Emission Computed Tomography images and Cardiac Arrhythmia signals. Their experimental results showed that promising generalization can be achieved with the hybrid kernel compared to conventional kernels. Wang *et al* [5] proposed the Weighted Mahalanobis Distance Kernels. They first find the data structure for each class in the input space via agglomerative hierarchical clustering and then construct the weighted Mahalanobis distance kernels which are affected by the size of clusters they reside in. They showed that, although WDM kernels are not guaranteed to be positive definite or conditionally positive definite, satisfactorily classification results can still be achieved because regularizes in SVMs with WDM kernels are empirically positive in pseudo-Euclidean spaces.

From among the specific user-defined kernels, XU *et al* [6] proposed using the weighted Levenshtein distance as a kernel function for strings. They used the UCI splice site recognition dataset for testing their proposed specific kernel which got the best results in this problem. Wu *et al* [7] proposed a new user-defined kernel for RNA classification. They showed that the new kernel takes advantage of both global and local structural information in RNAs. Their experimental results showed that the new kernel outperforms existing kernels when used to classify non-coding RNA sequences. Yan *et al* [8] proposed the position weight subsequences kernel (PWSK) that could be used for identifying gene sequences. This kernel was used for splice site identification and the performance was better than that of the string subsequences kernel (SSK). Cuturi *et al* [9] proposed a mutual information kernel for strings which borrows techniques from information theory and data compression. They showed that their kernel reported encouraging classification results on a standard protein homology detection experiment.

Our proposed kernel falls in the general kernels class while having the ability of defining similarity features which have been only used in specific user defined kernels. Moreover, it does not need a domain expert to determine the contribution of each similarity feature to the similarity measure since the kernel is learned statistically from data.

For the problem of shape recognition, Valveny and Marti discussed a method for recognizing hand-drawn architectural symbols [13] using deformable template matching. They achieved recognition rates around 85%, but did not discuss how the user might correct an incorrect recognition. Notowidigo and Miller [14] presented a novel approach to creating structured diagrams. There system aims to provide drawing freedom by allowing the user to sketch entirely off-line using a pure pen-and paper interface. The system can infer multiple interpretations for a given sketch to aid during the user's polishing stage. The UDSI program uses a novel recognition architecture that combines low-level recognizers with domain-specific heuristic filters and a greedy algorithm that eliminates incompatible interpretations. Refaat *et al* (2008) [10] has proposed a new approach for context-independent hand-written diagram recognition using support vector machines achieving an acceptable segmentation accuracy and approaching 90% recognition accuracy.

3 Preliminaries

The basic idea of SVM classifiers is to map a given data set from input space into higher dimensional feature space , called dot product space, via a map function ϕ , where

$$\phi : R^N \longrightarrow F \tag{1}$$

Then, it performs a linear classification in the higher dimensional space . This requires the evaluation of dot products:

$$k(x, y) = (\phi(x), \phi(y)) \tag{2}$$

Where k is called the kernel function. If F is high dimensional, the right hand side of equation (2) will be very expensive to compute [1]. Therefore, kernel functions are used to compute the dot product in the feature space using the input parameters which means that the mapping to is done implicitly. A kernel function returns a real number representing the similarity of its two input patterns. There are many types of kernels such as the RBF kernel, given by:

$$k(x, x_i) = e^{-\|x-x_i\|^2/2\sigma^2} \tag{3}$$

Other similar kernels are also widely used.

The function used for the assignment of new objects to one of the two classes is called the decision function which takes the form:

$$f(x) = \begin{cases} +1 & \text{if } \sum_{i=1}^{l} \alpha_i y_i k(x, x_i) + b > 0, \\ -1 & \text{if otherwise.} \end{cases} \quad (4)$$

Where, l denotes the number of training patterns
x denotes the unseen pattern vector
x_i denotes training pattern vector
y_i denotes the label of the training pattern
b denotes constant offset (or threshold)
1 and -1 are the labels of the decision classes
The parameters α_i's are computed as the solution of a quadratic programming problem.

4 The SVM'ed Kernel

The SVM'ed-Kernel could be used in any machine learning task that requires a kernel function. In this paper we illustrate its use as a kernel function for support vector machine classification problems.

Internally, the SVM'ed-Kernel is constructed as a support vector machine classification problem. Therefore we have two SVMs; the first one is the original SVM classification problem which we will call it the original SVM, while the other is the one used as a kernel function which we call it the SVM'ed-Kernel.

This SVM'ed-Kernel will be trained using a recreated training set extracted from the original one. The steps to create and use the SVM'ed-Kernel are: *A*. Define a feature vector representing the similarity between a pair of patterns, *B*. Automatically generate the recreated training set from the original one. *C*. Train the SVM'ed-Kernel as a normal classification problem using the recreated training set in *B*. *D*. Use the trained SVM'ed-kernel as a kernel function in the original SVM problem. *E*. Train the original SVM using the SVM'ed-Kernel.

We now explain each step in details. In step *A*, we define a feature vector that represents the similarity between two patterns. For example in a text categorization classification problem where we need to classify a document according to whether it is related to either sport or politics. One could define a similarity feature vector of two features. The first feature could be the number of common words after stemming, while the second one could be the number of common semantic relations.

In step *B*, assume that we have an original simple training set similar to that in Table 1. To create the recreated training set that will be used to train the SVM'ed-Kernel, we select every pair of patterns from the original training set (order is not important). So we have pattern 1 and pattern 2, pattern 1 and pattern3, pattern 2 and pattern 3, pattern 2 and pattern 4, and so on. We label each pair as being matching (1) if the two patterns have the same label in the original training set or not matching (-1) if they have different labels. Table 2

Table 1. The original training set

patterns	class label 1 or -1
pattern 1	1
pattern 2	-1
pattern 3	1
pattern 4	-1

Table 2. The recreated training set

patterns	class label 1 or -1
pattern 1 and pattern 2	-1
pattern 1 and pattern 3	1
pattern 1 and pattern 4	-1
pattern 2 and pattern 3	-1
pattern 2 and pattern 4	1
pattern 3 and pattern 4	-1

illustrates the recreated training set. One can see here that the recreated training set is of larger size than the original training set.

In step C, we first convert each pair in the recreated training set, created in step B, to a feature vector using the similarity feature vector definition we have defined in step A. After that, we train the SVM'ed-Kernel as a normal SVM classification problem using the recreated training set and any arbitrary kernel. The output of this SVM classifier will be a label indicating whether the two input patterns are matching or not. After training, we save the SVM trained as our SVM'ed-Kernel after removing the decision component from the function in equation (4), to be in the form

$$f(x) = \sum_{i=1}^{l} \alpha_i y_i k(x, x_i) + b \quad (5)$$

The decision component was removed because we are interested in the real value returned from (5), which represents how confident we are in the match. A larger returned value represents a better match (high similarity) and vice versa. Figure 1 shows three pairs, from the recreated training set, and their locations from the decision boundary of the SVM'ed-Kernel. When substituting in equation (5), pair one's similarity feature vector will return a positive number indicating high similarity between this pair. On the other hand, pair two's similarity feature vector will return a smaller positive number indicating that this pair is less similar than pair one. Finally, pair three's vector will return a negative number indicating low similarity between this pair.

The training patterns in the recreated training set were used to determine the maximal margin classifier. The distance of a new pair from the maximal margin classifier decision boundary is a direct measure of the similarity between the two

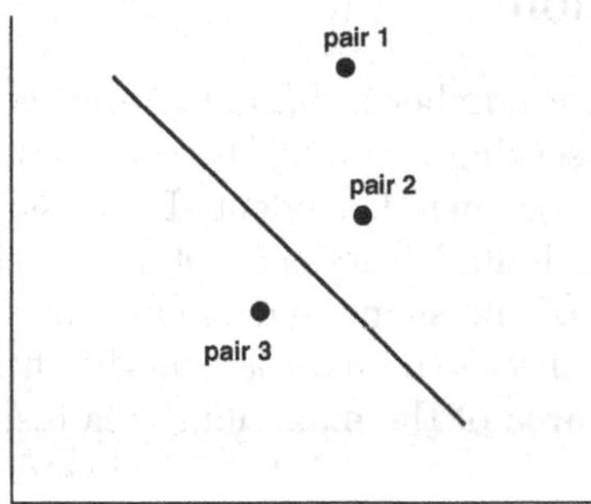

Fig. 1. Three pairs and their locations from the decision boundary of the SVM'ed-Kernel

patterns of this pair due to the way we formed the recreated training set in B and the definition of the similarity feature vector in A.

In step D, we use our trained SVM in C as our kernel function in the original SVM problem. In our work we have added an offset to the output of the SVM'ed-Kernel so that the similarity measure returned becomes always positive.

In step E, we train our original SVM using our SVM'ed-Kernel and the original training set. After training the original SVM, the original SVM model is ready for the real operation phase. We conclude here that the original SVM classifier component does not require the pattern to be extracted to a feature vector on its own. It takes the form:

$$f(pattern) = \sum_{i=1}^{l} \alpha_i y_i SVK(pattern, pattern_i) + b \tag{6}$$

where SVK is the SVM'ed-Kernel in the form:

$$SVK(pattern_x, pattern_y) = \sum_{j=1}^{l'} \alpha_j y_j k(m(x,y), m_j) + b' \tag{7}$$

Where, m_j denotes a similarity feature vector of a support vector pattern for the SVM'ed-Kernel.
$m(x,y)$ denotes the feature vector representing similarity between $pattern_x$ and $pattern_y$.
y_i denotes the label of the ith pattern in the original training set.
y_j denotes the label of the jth pair in the recreated training set.
b and b' denote constant offsets (or thresholds).
k denotes an arbitrary kernel function.
l and l' denote the number of training patterns in the original and the recreated training sets respectively.
The parameters $\alpha_i's$ and $\alpha_j's$ are computed as the solutions of quadratic programming problems.

5 Shape Recognition

In our previous work [10], a number of basic features were extracted from each shape for the purpose of serving as inputs to the classifier. The features were selected to be size and orientation independent. The basic shape features used are Ach, the area of the convex hull;Pch, its perimeter;Aer, the area of the rectangle enclosing the convex hull of the shape and having the minimum area;Alq, the area of the maximum area inscribed quadrilateral that fits inside the convex hull of the shape [12];Alt, the area of the maximum area inscribed triangle that fits inside the convex hull of the shape [12];Plt, its perimeter;Her, the height of the rectangle enclosing the convex hull of the shape and having the minimum area; and Wer, its width.

The features used in the SVM model were: Pch^2/Ach, Ach/Aer, Alq/Aer, Alq/Ach, Alt/Alq, Alt/Ach, Plt/Pch,and Her/Wer. For the SVM'ed Kernel, we decided to make use of the features already designed before together while adding a novel high level similarity feature to show the power of the novel proposed kernel.

The similarity feature vector definition that is defined for the sake of the SVM'ed-Kernel was defined simply to be the subtraction of the two original feature vectors of the two input patterns. However, we have added a single similarity feature that represents the similarity between a pair of patterns. This feature was a modified chain code distance measure [15].

6 Testing Results

In our experiments, we trained the system using hand-drawn circles, triangles, rectangles, diamonds and ellipses from Refaat *et al* data set (2008) [10], we have added some new shapes to the set to increase its size. We divided the new extended data set into 750 patterns for training and we kept two test sets unseen. The first test set is a normal one of 236 patterns while the other one consists of 234 hard patterns. In the hard test set the shapes may be drawn similar to more than one shape class and it is the responsibility of the model to discover its true class. The hard test set was created in order to measure our models' robustness to hard patterns or shapes drawn carelessly. We used the SVM'ed Kernel with the pairwise classification method [1] to handle the multiclass problem. A recreated set was generated from the training set of each pair of classes. Each recreated set was then used to train an SVM'ed Kernel which was used subsequently as a kernel function for the corresponding binary classifier. All SVM'ed Kernels use the rbf kernel with the gamma parameter set to 2. We did not perform any tuning for the gamma of the rbf kernel used by the SVM'ed Kernels.

We compared the test accuracy of the SVM'ed Kernel to that achieved by Refaat *et al* (2008) SVM model and also to that achieved by using the rbf kernel with the pairwise classification method. In the last case, by trial and error, the gamma parameter was chosen to be set to 2. We used SVMlight [11] in all our simulations. Table 3 shows the testing accuracies of the three models for both the normal and the hard sets.

Table 3. Testing Accuracies

Test Set-Model	RBF gamma = 2 *pairwise*	Refaat 2008	SVM'ed Kernel *pairwise*
normal	81.355%	88.135%	92.796%
hard	61.96%	69.23%	85.89%

The testing results showed that the SVM'ed kernel outperforms Refaat *et al* 2008 in both the normal and the hard sets by about 4.6% and 16.5% respectively. The reason of this significant gain was that the SVM'ed kernel used the modified chain code similarity measure. This mutual feature could not have been used by the classical kernels because it represents a pair of patterns rather than only one. In addition, the SVM'ed kernel did not neglect the predefined classical features which made it act as a statistical integrator of all information about the task of shape recognition.

7 Conclusion and Future Work

In this paper, we proposed the application of the SVM'ed-Kernel function to the problem of shape recognition. We showed how the SVM'ed Kernel allows defining features of similarity between a pair of patterns. In addition, we showed that the old feature definitions for single patterns could also be used by just subtracting each two corresponding features. In the shape recognition problem, the enhancement was about 4.5 % in the normal test set and interestingly about 16.5% in the hard test set. In our future work, we are going to use the SVM'ed-Kernel in various real world applications in both Natural Language Processing and Bioinformatics.

References

1. Scholkopf, B., Smola, A.J.: Learning with Kernels. The MIT Press, Cambridge (2002)
2. Thadani, K., Jayaraman, A., Sundararajan, V.: Evolutionary Selection of Kernels in Support Vector Machines. In: International Conference on Advanced Computing and Communication (2006)
3. Kong, R., Zhang, B.: Autocorrelation Kernel Functions for Support Vector Machines. In: Third International Conference on Natural Computation (2007)
4. George, J., Rajeev, K.: SINC-CAUCHY Hybrid Wavelet Kernel for Support Vector Machines. In: IEEE Workshop on Machine Learning for Signal Processing. IEEE Press, Los Alamitos (2008)
5. Wang, D., Yeung, D., Eric, C.: Weighted Mahalanobis Distance Kernels for Support Vector Machines. IEEE Transaction on Neural Networks 18, 1453–1462 (2007)
6. Xu, J., Zhang, X.: Kernels Based on Weighted Levenshtein Distance. In: International Joint Conference on Neural Networks, pp. 3015–3018. IEEE Press, Budapest (2004)

7. Wu, X., Wang, J., Herbet, K.: A New Kernel Method for RNA Classification. In: Sixth IEEE International Symposium on BioInformatics and BioEngineering, Virginia, pp. 201–208 (2006)
8. Yan, C., Wang, Z., Gao, Q., Du, Y.: A Novel Kernel for Sequences Classification. In: IEEE International Conference on Natural Language Processing and Knowledge Engineering, Wuhan, pp. 769–773 (2005)
9. Cuturi, M., Vert, J.: A Mutual Information Kernel for Sequences. In: International Joint Conference on Neural Networks, pp. 1905–1910. IEEE Press, Budapest (2004)
10. Refaat, K., Helmy, W., Ali, A., Abdelghany, M., Atiya, A.: A New Approach for Context-Independent Handwritten Offline Diagram Recognition using Support Vector Machines. In: International Joint Conference on Neural Networks, pp. 177–182. IEEE Press, Hong Kong (2008)
11. Jaochims, T.: SVMlight is an implementation of Support Vector Machines in C
12. Boyce, J., Dobkin, D., Drysdale, R., Guibas, L.: Finding External Polygons. In: Annual Symposium on the Theory of Computing (1982)
13. Valveny, E., Martí, E.: Deformable template matching within a bayesian framework for hand-written graphic symbol recognition. In: Chhabra, A.K., Dori, D. (eds.) GREC 1999. LNCS, vol. 1941, pp. 193–208. Springer, Heidelberg (2000)
14. Notowidigo, M., Miller, C.: Offline sketch interpretation. In: AAAI Fall Symposium on Making Pen-Based Interaction Intelligent and Natural. Washington (2004)
15. Ahmad, M.B., Park, J.-A., Chang, M.H., Shim, Y.-S., Choi, T.-S.: Shape registration based on modified chain codes. In: Zhou, X., Xu, M., Jähnichen, S., Cao, J. (eds.) APPT 2003. LNCS, vol. 2834, pp. 600–607. Springer, Heidelberg (2003)
16. Microsoft software product for creating a wide variety of business and technical drawings, `http://www.office.microsoft.com`
17. Notowidigdo, M.: User-Directed Sketch Interpretation. MEng thesis, Massachusetts Institute of Technology (2004)

Selective Attention Improves Learning

Antti Yli-Krekola, Jaakko Särelä, and Harri Valpola

Department of Biomedical Engineering and Computational Science,
Aalto University, Helsinki, Finland
{antti.yli-krekola,jaakko.sarela,harri.valpola}@tkk.fi
http://www.becs.tkk.fi/en/

Abstract. We demonstrate that selective attention can improve learning. Considerably fewer samples are needed to learn a source separation problem when the inputs are pre-segmented by the proposed model. The model combines biased-competition model for attention with a habituation mechanism which allows the focus of attention to switch from one object to another. The criteria for segmenting objects are estimated from data and are shown to generalise to new objects.

Keywords: Selective attention, perceptual learning, segmentation.

1 Introduction

Learning task-relevant feature and object representations is a crucial problem for an autonomous agent trying to cope in a real-world environment. Sometimes the problem can be facilitated by collecting data from controlled environments, leading for instance to reduced noise and fewer objects present simultaneously. Such simplifications allow even fairly difficult problems to be solved with the current machine learning methods.

In many situations, however, these controlled environments cannot be provided due to cost, infeasibility of human intervention or other reasons. In those cases, the system should be able to learn feature and object representations autonomously. Furthermore, the learnt representations should be relevant for the tasks the agent faces. For these really difficult cases, machine learning research has provided us with painfully few methods.

The key problem is that the relevant associations and relations are complex and dynamic. As an example, let us consider the interplay between the visual and the motor system in picking up an object. There are many degrees of freedom in the task: the object can be in several places with respect to the hand and the head, the eyes can be viewing in several directions and the hand can be in several orientations, just to name a few. Yet, the autonomous agent should be able to learn the associations that are needed to perform the task of picking up the object. In any particular context of hand, eye and object positions, there exist many simple correlations between the needed motor output and the visual input. However, averaged over all the contexts, the correlations cancel each other out. Thus the agent needs a representational system that can learn and use dynamic

C. Alippi et al. (Eds.): ICANN 2009, Part II, LNCS 5769, pp. 285–294, 2009.

associations and relations that describe the short-lasting correlations between the different modalities.

The best example of a system that has been able to solve the above problems is the human brain. In neuroscience, it is known that attention plays a key role in perceptual learning [1]. The purpose of this paper is to discuss the information processing mechanisms of attention and to show that it can facilitate learning of feature and object representations.

2 Attention and Learning

From psychophysical experiments it has become clear that attention plays a significant role in learning. For instance, Ahissar et al. [1] showed that attention guides low-level perceptual learning by focusing the representational capacity (low-level perceptual discriminations) to features that are relevant for the task at hand.

There is experimental evidence to support the idea that attention is realised by a competitive binding process that forms functional networks dynamically [2]. This dynamical binding has been shown to gate the coherence between cortical areas, thereby affecting the associations learnt between these areas [3].

Taken together, it seems plausible that selective attention and the formation of dynamical bindings are the necessary ingredients by which a large learning system can deliver training signals from distant areas, such as from motor cortex to visual cortex [4].

In order to use attention for perceptual learning in machine learning context, it is necessary to 1) implement a model which gives rise to attention, 2) learn the parameters of the model, making attention adaptive, and 3) use it successfully to facilitate perceptual learning. Although each of these three aspects have been studied independently and in pairs, to our knowledge the model presented in this article is the first to combine all three into a functional model.

2.1 Gestalt Principles

When we humans see a new object, we may not know its identity but we can nevertheless tell what is part of the object and what is not. In other words, we are able to segment out an object without having seen it before.

In perceptual psychology, the rules of the organisation of perceptual scenes are called Gestalt principles [5]. Psychologists have identified several principles, such as proximity, common fate, similarity, continuity, closure, symmetry and convexity. The Gestalt principle of continuity is illustrated in Figure 1a, where the human visual system groups some of the line segments to form a circle.

What makes the Gestalt principles interesting in the current context is that they can be learnt from data. In neural terms, the Gestalt principles can be implemented by giving positive connections between certain neurons in one area and some other neurons in an adjacent area. Learning the connections can be

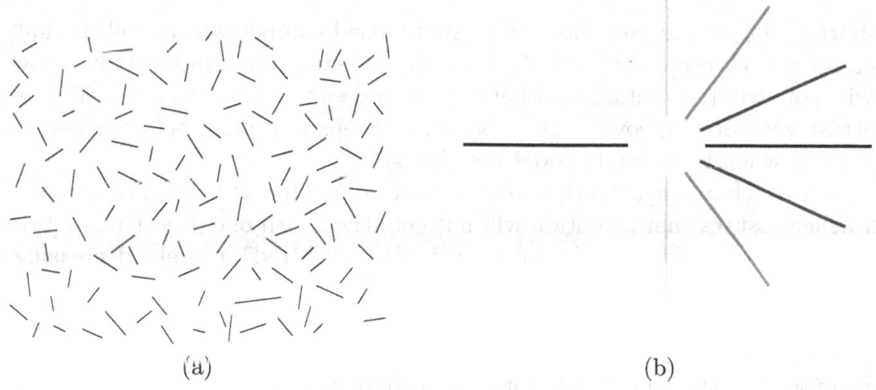

Fig. 1. a) Because of the Gestalt principles, a circle is perceived rather than some other grouping of the lines. b) Gestalt grouping of the neurons. The lines are features coded by different neurons. The shades of gray illustrate the connection strengths between the neurons on the right and the neuron on the left, darker meaning stronger. The lateral connections are stronger when the Gestalt principle is better fulfilled.

based on simple correlations found in the data. For example, features responding to lines of certain orientation in one part of the visual field are more probably co-activated with features of similar orientation in some other part of the visual field. This mechanism is illustrated in Fig. 1b.

These "neural" Gestalt rules can be learnt from the data and they operate on the level on individual feature-coding neurons. The principle is therefore applicable to any modality and also between modalities unlike, for example, many segmentation procedures that make use of the spatial structure of visual images. Moreover, the neural Gestalt rules can be learnt locally and in parallel. In the visual domain this means that the local correlations found in familiar objects generalise to new objects which have different overall shapes but nevertheless obey the same local correlations.

2.2 Biased-Competition Model for Attention

Contextual (predominantly top-down) biasing of local lateral competition had been proposed as a model of covert attention in humans [6]. Usher and Nierbur [7] then suggested a computational model for biased competition that has been shown to replicate many attentional phenomena, for instance both bottom-up and top-down aspects of attention [8].

Deco and Rolls [8] also showed that it is possible to learn the weights for contextual biasing by the mechanism outlined in Sec. 2.1. In other words, the neural Gestalt rules can be applied in a relatively straight-forward manner to implement selective attention.

2.3 Suggested Model

One shortcoming of the previously suggested biased-competition models is that they converge to a representation of one of the objects present in the inputs and then will not switch attention to other objects unless the input changes. This is in contrast with human covert attention which keeps switching between salient objects even when the stimulus does not change.

Models with changing attention usually have some kind of habituation mechanism which assures that attention will not get stuck with one object (e.g., [9]). Habituation means that active neurons gradually get "tired", thereby decreasing the stability of the currently active population of neurons. After the support for a population erodes, another population of recovered neurons takes over and the original tired population starts recovering.

Taken together the model has four key mechanisms:

1. Bottom-up input which mostly determines the activation level of the neuron,
2. Contextual (lateral or top-down) input which reflects learnt Gestalt principles,
3. Local competition which is biased by the contextual input and
4. Habituation which ensures that the winning population gradually gets tired and makes room for the winning population.

A more detailed description of the implementation is given in Sec. 3. However, it should be emphasised that the exact details of these mechanisms are not important although they of course need to fit together.

2.4 Relation to Previous Work

Several systems have been suggested that segment objects and represent them sequentially. Many of them are based on weakly coupled oscillators or other related mechanisms (e.g., [10,11,12,13]). Biased competition has the added benefit that it not only groups objects but can also select among them. This will be important when scaling up the system.

There are only a few examples of tackling the problem of using attention to improve learning. Selective attention was used for improving learning by Walther et al. [14] but their selective attention specialised in the visual domain and did not use learnable Gestalt rules which could be applied in any modality and even across modalities. Learning associations between different features has also shown to improve with attention by Kruschke [15], but his model has an external teacher controlling the attention.

3 Experiments

In this section, we use artificially generated data to demonstrate that it is possible and useful to combine attentional mechanisms and learning of feature representations in a single scheme. The Gestalt principles are first learnt on one data set. The resulting lateral connections are then used for segmenting new objects. We show that this greatly improves learning at the next stage, for which we used FastICA [16]. The MATLAB scripts for performing the experiments can be downloaded at `http://www.lce.hut.fi/research/eas/compneuro/repository/attention_learning.zip`.

3.1 The Data

We generated artificial data which had "objects" analogous to closed contours. For instance, the closed contour in Fig. 1a (circle) consists of 12 line segments which follow the local Gestalt rule of continuation. Our objects had five 100-dimensional patches (analogous to line segments) that were connected cyclically as shown in Fig. 2. Each object had one active element on each patch. In other words, an object was a 500-dimensional binary vector with five ones and 495 zeros.

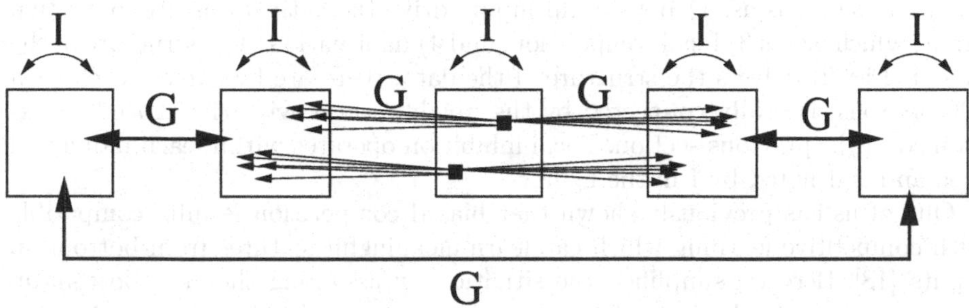

Fig. 2. The structure of the data is an idealisation of the Gestalt rules for closed contours (Fig. 1). Each of the five patches consists of 100 elements. According to the Gestalt rules (G), each element has five permissible neighbours in the adjacent patch. The model structure is similar, with local inhibitory connection (I) and excitatory lateral connections (G).

The objects were generated as follows. First the Gestalt rules (G in Fig. 2), which hold for all objects, were chosen randomly. Each element had five randomly selected permissible neighbours in both the adjacent patches. The five active elements of each object were selected in stages: 1) select one of 100 elements on the first patch, 2) select one of the five permissible elements (out of 100) on the second patch, 3) repeat for all the patches and finally 4) accept or reject the object depending on whether the element on the last patch is a permissible neighbour of the selected element on the first patch. On average there are 3,125 different objects that fulfil our continuity rules. The exact number depends on the Gestalt rules which were randomised.

We used these objects to generate noisy data which follow a linear independent component analysis (ICA [17]) model. Each 500-dimensional sample vector was a sum of five randomly selected objects and additive binary noise with 25 ones and 475 zeros. A noisy sample vector together with the five constituent objects are depicted in Fig. 3a.

3.2 Learning the Gestalt Principles

We selected 20 objects which were reserved as "new objects" for the testing phase. We generated a data set with 10,000 samples using the remaining objects (3,105 objects on average). The lateral connections where then set to the values corresponding to the covariances between the input elements. Note that it would be difficult to learn reliably any correlations between 3,105 objects from such a small data set but it is perfectly feasible to estimate the correlations of the constituent elements. The estimated covariances are noisy but good enough for the next stage.

3.3 Biased-Competition Model with Habituation

As explained in Sec. 2.3, the biased-competition model used for segmenting data has four mechanisms: 1) bottom-up inputs drive the activations, 2) contextual input, which biases 3) local competition, and 4) habituation. The structure of the model (Fig. 2) reflects the structure of the data: there are five areas (each with 100 neurons) laterally connected by the weights learnt with the procedure explained in the previous section. Local inhibition operates within each individual area and is denoted by I in the figure.

One of us has previously shown that biased competition is fully compatible with competitive learning which can learn meaningful features from bottom-up inputs [18]. Here we simplified the situation by assuming that the bottom-up inputs are already the input features to be represented. The neurons thus get bottom-up activations $\mathbf{x}$ which are simply the data samples.

Contextual lateral input from previous activations $\mathbf{y}(t-1)$ modulates the bottom-up activations as follows:

$$y_i^*(t) = [(g_i(t) + \alpha \mathbf{a}_i \mathbf{y}(t-1))\, x_i]_+ \ , \tag{1}$$

where $\mathbf{a}_i$ is a row vector of lateral connections implementing the estimated Gestalt rules and $\alpha = 0.1$. The term $g_i(t)$ is a gain which implements the habituation and will be explained shortly. The activations $y_i^*(t)$ are restricted to be positive.

After this, lateral competition selects the final activations

$$y_i(t) = [y_i^*(t) - I_{\text{area}}]_+ \ , \tag{2}$$

where I_{area} is a function of $y_i^*(t)$ within an area. All the neurons in one area have the same I_{area}. This inhibitory term is adapted with a fast time-constant such that the target sparseness would be reached. We use the following sparseness measure for a local activation pattern $\mathbf{y}_{\text{area}}$:

$$s(\mathbf{y}_{\text{area}}) = \frac{1}{||\mathbf{y}_{\text{area}}||} \sum_{i \in area} y_i \ . \tag{3}$$

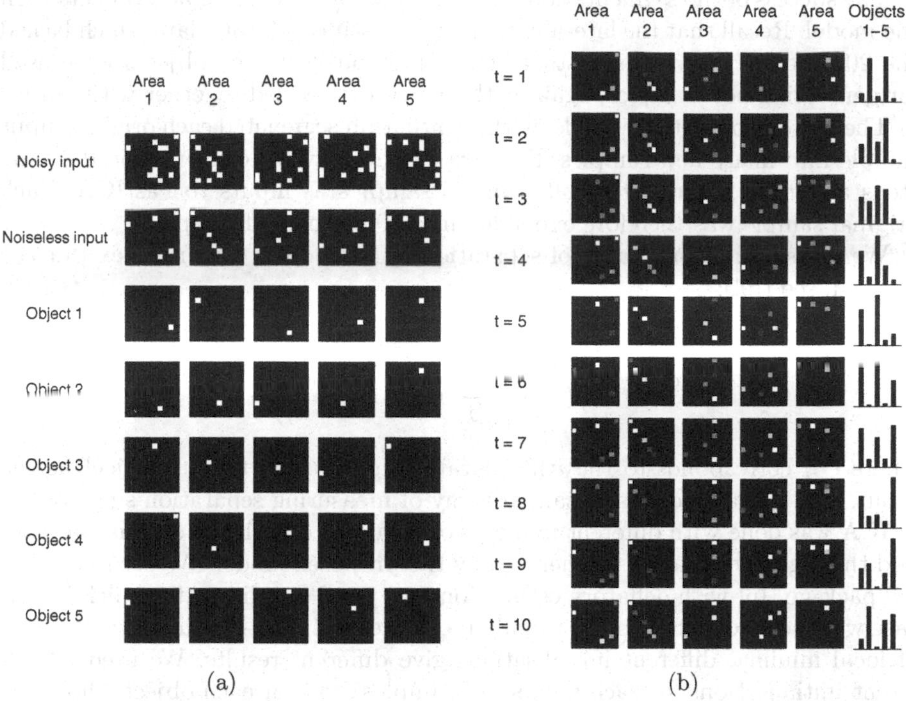

Fig. 3. a) A sample input, on the top row. The next row is the same input without noise. This noiseless input is used to produce the segmentation on the right (b). This input consists of the five shown objects. b) An example of segmentation of the noiseless input on the left (a). The 10 consequent time steps are taken after 50 steps after introducing the input. Two examples: In the first row, the 2nd and the 4th objects are seen. The 3rd object is growing from $t = 1$ until $t = 4$, and then starts to disappear.

On each time step, the local inhibition I_{area} is adjusted to make the pattern closer to the right sparseness level. We chose it to be the sparseness of the vector in which there are three ones and 97 zeros.

Habituation was implemented as follows. The gains are adjusted on each time step with a slower time-constant than the inhibition. The updates try to match the average activity with the original input: $E\{y_i\} \approx x_i$. On each time step, g_i is increased (decreased) a little if the moving average of y_i is below (above) x_i.

3.4 Results

An example of the segmentation dynamics is shown in Figure 3b. In the segmented representations, individual objects can be seen to appear and disappear more or less coherently. Note that for the sake of visual clarity, the segmentation dynamics is shown for the noiseless input from Figure 3a although all data used in the learning experiments contained noise.

The success of the segmentation was measured by separating new objects with the model. Recall that the lateral weights were estimated from data which lacked the 20 objects reserved for testing. These previously unseen objects were used for generating new samples, again with five objects added together with noise.

The biased-competition model with habituation segmented each original input sample into many new samples. First we let the network converge for 100 time steps and then we used the following 30 samples as inputs to FastICA. Each original sample was therefore expanded into 30 segmented samples.

We measured the accuracy of separation by a modified Amari index (for the original, see [19]):

$$a(C) = \frac{1}{N} \sum_{i}^{N} \left(\sum_{j}^{N} \frac{C_{ij}^2}{\max_k C_{kj}^2} - 1 \right) , \tag{4}$$

where C_{ij} corresponds to the ith separated signal using the jth object as the input. The Amari index is a standard way of measuring separation success.

ICA was done with different numbers of samples to both the original samples and the segmented samples generated by the proposed model. We used FastICA 2.5 package [16] with deflatory estimation and pow3 non-linearity, which in this case was more robust than the usually recommended tanh-nonlinearity. Because of local minima, different initialisations give different results. We used 30 different initialisations for each number of samples, and for each object, chose the component that gives the smallest Amari index.

The results are shown in Figure 4. ICA for the original non-segmented data needs about a hundred times as much samples as does ICA for the segmented data. For fine-tuning though, the non-segmented case seems to be better. The segmentation gives rough guesses about what the objects could be, but can also sometimes break them, and move the fixed points of the FastICA algorithm. The segmented case Amari index saturated to 1 milliAmaris at about 200 samples. The non-segmented case got better results with $N > 7000$.

4 Discussion

In this paper we demonstrated that selective attention can improve learning. We concentrated on showing that, with pre-segmentation, considerably fewer samples are needed to learn meaningful features. The segmentation was based on lateral connections whose strengths were estimated from another data set. The setting thus mimicked a situation where local Gestalt rules have already been learnt from past experience, allowing new objects to be segmented and thus greatly reducing the number of samples needed for learning about new objects. In this paper the learning task was chosen to be independent component analysis but reduced learning time should generalise to other types of associative learning as well due to reduced amount of distractors.

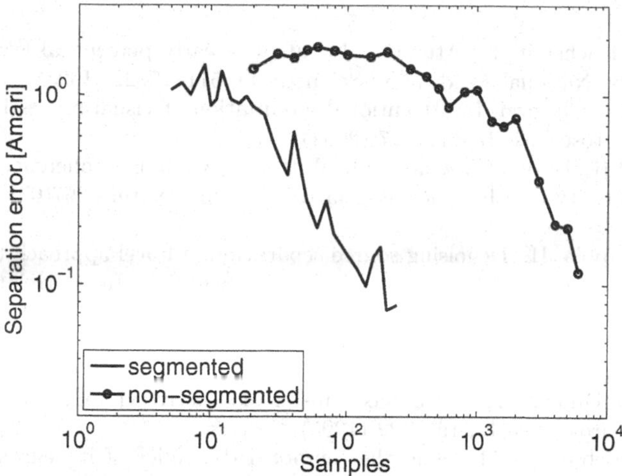

Fig. 4. Separation results with segmented and original data are shown. The separation quality is measured with a modified Amari index which measures the deviation of the unmixing matrix from optimal. Segmenting the data reduces the number of samples needed for reaching a given value of Amari index by roughly a factor of 100.

In the reported experiment, the segmentation principles were learnt offline. In actual use, it would be more useful to learn the object representations and the segmentation principles at the same time in a feedforward-feedback loop. This would, for instance, allow selective attention to guide the learning by discarding some structure in the data and focusing the representational capacity to relevant features. However, it will also be necessary to take into account the danger of run-away learning of self-induced correlations. Similar problems arise in learning any non-directed graphs, such as Markov fields [20]. A popular solution is to have two separate learning stages: one driven by the input and another, sleep-like, driven by expectations. The idea is to forget the unwanted representations during the sleep stage. When learning and forgetting balance each other, the learnt weights have captured the statistics of the input.

The model proposed in this article is based on biased-competition model which has been shown to be able to implement attention in large hierarchical networks. We have previously shown that the model is compatible with competitive learning and thus can learn meaningful bottom-up features under the guidance of selective attention [18]. In this paper, we added a mechanism for habituation which allows the focus of attention to change from one object to another and then showed that the resulting segmentation greatly improves associative learning. We believe that this work provides a fruitful starting point for future efforts in building a representational system flexible and powerful enough for an autonomous agent to survive in a complex real-world environment.

References

1. Ahissar, M., Hochstein, S.: Attentional control of early perceptual learning. Proceedings of the National Academy of Sciences 90, 5718–5722 (1993)
2. Reynolds, J.H., Chelazzi, L.: Attentional modulation of visual processing. Annual Review of Neuroscience 27, 611–647 (2004)
3. Miltner, W.H.R., Braun, C., Arnold, M., Witte, H., Taub, E.: Coherence of gamma-band EEG activity as a basis for associative learning. Nature 397(6718), 434–436 (1999)
4. Särelä, J., Valpola, H.: Denoising source separation: a novel approach to ICA and feature extraction using denoising and Hebbian learning. In: Correlation Learning Workshop in the Eighteenth Canadian Conference on Artificial Intelligence, Victoria, Canada, May 2005, pp. 45–56 (2005)
5. Todorovic, D.: Gestalt principles. Scholarpedia 3(12), 5345 (2008)
6. Desimone, R., Duncan, J.: Neural mechanisms of selective visual attention. Annual Review of Neuroscience 18, 193–222 (1995)
7. Usher, M., Niebur, E.: Modeling the temporal dynamics of it neurons in visual search: A mechanism for top-down selective attention. Journal of cognitive neuroscience 8, 311–327 (1996)
8. Deco, G., Rolls, E.T.: A neurodynamical cortical model of visual attention and invariant object recognition. Vision research 44, 621–642 (2004)
9. Itti, L., Koch, C.: A saliency-based search mechanism for overt and covert shifts of visual attention. Vision Research 40, 1489–1506 (2000)
10. Wang, D.L., Terman, D.: Locally excitatory globally inhibitory oscillator networks. IEEE Trans. Neural Net. 6, 283–286 (1995)
11. Choe, Y., Miikkulainen, R.: Self-organization and segmentation in a laterally connected orientation map of spiking neurons. Neurocomputing 21(1-3), 139–158 (1998)
12. Weng, S., Wersing, H., Steil, J., Ritter, H.: Learning lateral interactions for feature binding and sensory segmentation from prototypic basis functions. IEEE Transactions Neural Networks 17(4), 843–862 (2006)
13. Lessmann, M., Würtz, R.P.: Image segmentation by a network of cortical macrocolumns with learned connection weights. In: Proceedings of Biologically Inspired Cooperative Computing (BICC). Springer, Heidelberg (2008)
14. Walther, D., Rutishauser, U., Koch, C., Perona, P.: Selective visual attention enables learning and recognition of multiple objects in cluttered scenes. Computer Vision and Image Understanding 100, 41–63 (2005)
15. Kruschke, J.K.: Toward a unified model of attention in associative learning. Journal of Mathematical Psychology 45, 812–863 (2001)
16. FastICA: The FastICA MATLAB package (1998), `http://www.cis.hut.fi/projects/ica/fastica/`
17. Hyvärinen, A., Karhunen, J., Oja, E.: Independent component analysis. Wiley, Chichester (2001)
18. Yli-Krekola, A.: A bio-inspired computational model of covert attention and learning. Master's thesis, Helsinki University of Technology, Finland (2007)
19. Amari, S., Cichocki, A., Yang, H.H.: A new learning algorithm for blind source separation. In: Advances in Neural Information Processing 8 (Proc. NIPS 1995), pp. 757–763. MIT Press, Cambridge (1996)
20. Ackley, D.H., Hinton, G.E., Sejnowski, T.J.: A learning algorithm for boltzmann machines. Cognitive Science 9(1), 147–169 (1985)

Multi-stage Algorithm Based on Neural Network Committee for Prediction and Search for Precursors in Multi-dimensional Time Series

Sergey Dolenko, Alexander Guzhva, Igor Persiantsev, and Julia Shugai

D.V. Skobeltsyn Institute of Nuclear Physics, M.V.Lomonosov Moscow State University, Leninskie Gory, Moscow, 119991 Russia
dolenko@srd.sinp.msu.ru

Abstract. The studied problem is prediction of time series based on preceding values of several time series (a multi-dimensional time series). Besides prediction itself, the task is finding precursors, i.e. determination of a set of the most significant input features in coordinates "initial time series – lag". A four-stage prediction algorithm based on neural network committee has been suggested, implemented and studied. The algorithm has been successfully tested on one model problem and on one real world problem.

Keywords: time series prediction, neural network committee, precursor, multi-dimensional time series.

1 Introduction

Consider a multi-dimensional time series (TS) being an aggregate of several single-dimensional TS, each describing time changes of the value of some physical feature ϕ_i characterizing the object of study.

In the process of prediction, one should take into consideration the values of the physical features not only in a single point in time, but within some interval in the past. Therefore, for each TS, delay embedding is performed, i.e. each physical feature gives rise to a set of *input features* (input variables of the problem), which are the values of this physical feature in adjacent time moments in the past.

One of the shortcomings of such approach is significant increase in the total number of the input features of the problem, which becomes equal to the product of the number of physical features and the embedding depth (embedding window width). Decreasing the embedding depth for some physical features based on *a priori* information may somewhat simplify the problem, but it does not eliminate it. Large number of input features hampers the work of the prediction system, increases the building time of the prediction and deteriorates its quality. Besides that, sense analysis of the interconnections between the input features and the predicted variable becomes even more difficult.

C. Alippi et al. (Eds.): ICANN 2009, Part II, LNCS 5769, pp. 295–304, 2009.

In connection with that, besides prediction itself, one more very important task is finding precursors, i.e. determination of a set of the most significant input features in coordinates "physical feature – lag". The four-stage algorithm based on neural network (NN) committee, which is considered in this study, has been intended to solve both problems – prediction and search for precursors in multi-dimensional time series.

This study is the development of preceding studies [1,2,3,4].

2 Problem Statement and Description of the Algorithm

Assume that occurrence of the event that is interesting for the researcher is preceded by some (unknown) combination of the values of the input features, which we shall call *phenomenon.* Assume that the delay between emergence of the phenomenon and occurrence of the event is fixed for this type of phenomenon (yet also unknown). Say in this case that this phenomenon initiates this event. Assume also that the *search interval* for the delay between the phenomenon and the event occurrence is given, and that maximum duration of the event set *a priori* (*initiation interval* T_{init}) is much less than the search interval. Call *precursor* a combination of only those input features that are significantly connected with the occurrence of the event, and that can be used for its prediction. The main notions are illustrated in Fig. 1. Note that all the same considerations and the same approach are applicable to the solution of the prediction problem not for binary events, but for continuous valued *sought-for quantity* (SQ), i.e. the value of the predicted variable changing in a continuous range.

The general problem of prediction of event occurrence (of the SQ value) and analysis of the TS is split within the described context into the following stages:

1. Forming of the *initial feature set* (multi-dimensional TS) describing the studied object. Assume that the researcher has manually formed a *preliminary feature set*, i.e. he has selected a number of TS that may in his opinion refer to genesis of the predicted events. This number may turn out to be quite large, making the input dimensionality of the next stage extremely high, taking into account that this input dimensionality (the number of input features) is equal to the product of the number of features in the preliminary set and the number of time steps in the initiation interval. Therefore it seems reasonable to perform an adaptive estimation of significance of the physical features in the preliminary set in order to exclude the least significant ones from the following consideration. To perform such an estimation, the quickest of the methods of significance analysis of the input variables can be used (again, because of high dimensionality of data). The spectrum of methods of significance analysis of the input variables, which can be used, can be found in [4]. In this study, the cross-correlation method was used at this stage.
2. Finding within the search interval the most probable *phenomenon* causing the event, determination of the duration of the phenomenon (initiation interval), and determination of the delay between the phenomenon

(event initiation) and the event itself. This task can be solved by the original NN algorithm for analysis of multi-dimensional TS, developed by the authors [1,2,3] and based on use of an NN committee. A "by-product" of this stage of analysis is creation of a system that is already capable to predict the event (the SQ value), as the criterion of correct finding of the phenomenon (correct determination of the delay) is the precision of the prediction. The quality of prediction at this stage can be improved by using a hierarchical NN structure based on the stacked generalization principle [5]. Unfortunately, at this stage it is possible only to make a prediction of the event (the SQ value) with some precision, but it is impossible to understand why it has happened and by what combination of features (precursor) it has been initiated.

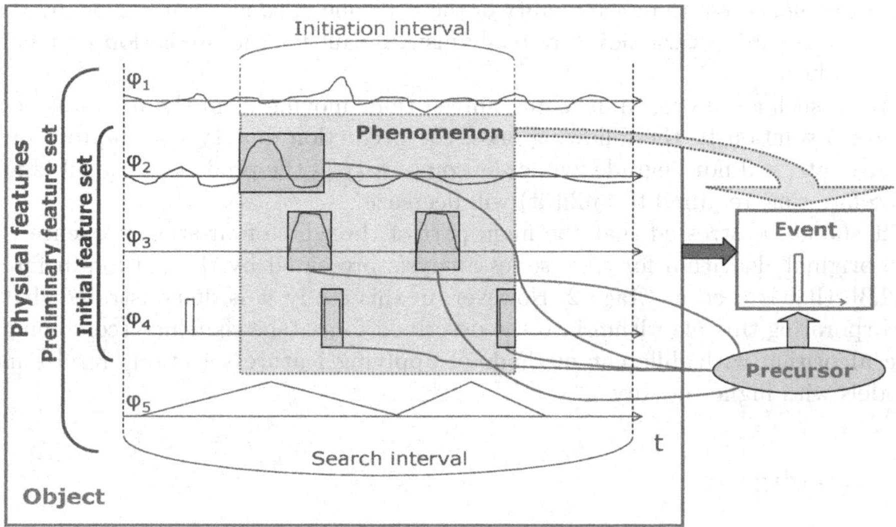

Fig. 1. Illustration of the main notions

3. To answer the question "why", i.e. to extract the *precursor* of the event, it is necessary to perform additional analysis of the phenomenon with the purpose to determine significance of separate input features making up the phenomenon, from the point of view of predicting the event. To solve this problem, one may use the full spectrum of the methods of analysis of input variable significance [4]. The combination of the most significant input features thus extracted makes up the precursor, whose dimensionality is much lower than that of the initial problem. This fact makes sense analysis of the results easier, and it may allow one to understand, what the precursor means in the terms of the initial problem.
4. At the final stage, one may put the question of solving the event (the SQ value) prediction problem again, using as input variables only the features

making up the precursor. To improve the quality of the prediction, here one can apply a committee of neural network based algorithms.

It should be stressed that at both stages where determination of significance of features is performed, the final criterion of the necessity of including this or that variable in the model is influence of taking this variable into account on the performance quality of the base predicting model. For example, a perceptron with one hidden layer and fixed settings can be used as such model. Prior to any experiments, the base model over all the available input variables is built. Then the available algorithms of analysis of feature significance are used to determine one or several least significant features, and a new model is built that is completely like the base one except that it does not take into account the selected low-significant features. If the prediction quality of the new model turns out to be not worse than the quality of the base one, a conclusion can be drawn that the discarded variables were really insignificant for the prediction problem being solved.

With such an approach, it is guaranteed that, moving from the first stage of problem solution to the following ones, the prediction quality will improve or, at any rate, will not degrade, while the complexity of the predicting model (and therefore time required to build it) will decrease.

It should be stressed that the main part of the whole four-stage sequence is the original algorithm for time series analysis presented by the authors before [1,2,3], that is used at Stage 2. However, in this study it is demonstrated that incorporating this algorithm into the described four-stage sequence (combining the algorithm with different methods of applying feature selection) results in models with higher quality.

3 Results

3.1 Criterion of Model Quality

Throughout Section 3, the main statistics that is used to estimate the quality of the predicting model is the multiple determination coefficient R squared. In this study, the following definition of R squared was used:

$$R^2 = 1 - \frac{\sum (y - \widetilde{y})^2}{\sum (y - \overline{y})^2} \quad (1)$$

Here y is the true value of the predicted TS for a sample;

$\widetilde{y}$ is the value predicted by the studied model for the same sample;

$\overline{y}$ is the average value of y over all samples of a data set.

Thus, R squared compares the error of the studied model to that of the trivial benchmark model that is a simple average over the whole data set. R squared equal to 1 describes a model with zero prediction error, while negative value of R squared corresponds to a very poor model whose prediction ability is worse than that of the trivial benchmark model.

3.2 Model Problem

An attempt of predicting values of the following functional dependence has been made:

$$y = 1.5 * x_1^{(-21)} + 3 * x_2^{(-23)} - 2 * x_3^{(-43)} - 4.5 * x_4^{(-70)} + 2 * x_5^{(-66)} \quad (2)$$

Here $x_i^{(-n)}$ is the value of TS x_i n steps ago;

$x_1(t)$ are the values of a TS consisting of random numbers in the range from 0 to 1, smoothed by averaging over 100 preceding values and then normalized into [0,1] range;

$x_2(t)$ are the values of another completely similar TS calculated from another set of random numbers,

$x_3(t)$ are the values of a TS calculated by the formula $z = \sin(t/300) + cos(t/150) + sin(t/200 + 1)$ and then normalized into [0,1] range;

$x_4(t)$ are the values of a binary TS, equal to 1, if the tens digit in the decimal representation of t is greater than the hundreds digit, and 0 otherwise;

$x_5(t)$ are the values of the TS $x_4(t)$ smoothed by averaging over 100 preceding values and then normalized into [0,1] range.

Such time series was suggested to model common behavior of real world problems, combining time series of different nature (in this case, random, harmonic and binary) to form predictors for the value of the studied SQ or for the studied event.

The preliminary feature set, except the TS $x_1 \ldots x_5$, included also TS rnd_1 and rnd_2, pseudorandom sequences of numbers in the range from 0 to 1.

All TS $x_1 \ldots x_5, rnd_1, rnd_2, y$ were represented by sequences of 65336 samples.

To solve the problem, delay embedding has been performed to the depth of 100 (search interval was equal to 100); then the data was sequentially divided into training, test and examination sets with 70:20:10 ratio.

First stage. Correlations of all the 700 input features with output variable y have been calculated. Table 1 lists the maximum values over the whole search interval of the correlation coefficient r with the output variable for all the features corresponding to each physical feature of the preliminary feature set. One can see that the physical features that are pseudorandom sequences significantly differ from all others by small values of r maximum and of its standard deviation, allowing one to discard them from future consideration.

Second stage. A NN committee (perceptrons with 8 neurons in the single hidden layer) has been trained first for initiation interval equal to 50 and overlapping interval for neighboring networks in the committee equal to 49. Note that due to use of test set, the results were practically independent of the number of neurons in the hidden layer in a quite wide range; the same refers to all the other experiments described below. Therefore, in all cases we performed test experiments with number of neurons in the hidden layer equal to powers of 2 (4, 8, 16, 32, and 64) and selected the number that gave the best results.

Table 1. Results of the first stage for the model problem

Variable	x_1	x_2	x_3	x_4	x_5	rnd_1	rnd_2
r max	0.078	0.170	0.208	0.923	0.349	0.011	0.008
St. deviation of r	0.022	0.037	0.002	0.243	0.111	0.002	0.002

The result presented in Table 2 (the value of multiple determination coefficient R^2 on examination set close to the absolute one) has been obtained in the delay range from 21 to 70; as can be seen from formula (2), this is exactly the case. Attempts of window narrowing demonstrate degradation of the results, which evidences the fact that the event duration is from 46 to 50. Thus, the event turned out to be limited with physical features $x_1 \ldots x_5$ and the delay range from 21 to 70.

Table 2. Results of the second stage for the model problem

Initiation interval	25	40	45	50
R^2 on examination set	0.928	0.952	0.957	1.000

Third stage. To extract the precursor, the method of progressive build-up of neural networks has been applied. Perceptrons with 8 neurons in the single hidden layer were used. First, $50 * 5 = 250$ perceptrons with one input were trained; the best result has been obtained when the feature corresponding to x_4 variable with delay equal to 70 was used as the precursor (Table 3). At the next iteration, each of the remaining 249 features was in turn added to the selected one; the best result has been demonstrated by the network that had the precursor composed of the features $x_4^{(-70)}$ and $x_5^{(-59)}$. In the same way, 6 iterations were performed. The precursor composed of the 6 features listed in Table 3, provided the value of multiple determination coefficient close to the ideal one. Note that 4 of the 6 extracted features participate in formula (2), and they were determined correctly; the fifth one differs from the remaining true feature only by a small delay of 2, which is not significant, as the corresponding time series x_3 is a slowly changing one.

Provided the fact that the NN model obtained at the third stage performed prediction with a very high precision, the *fourth stage* for this problem was not necessary.

Another model problem differed from the described one by the following. Prediction of the value of continuous variable y has been changed to prediction of events. The event was defined based on the value of variable y; the event was considered to happen in the current moment if the modulus of the value of y variable in this moment exceeded 2, and in the preceding moment it was less than

Table 3. Results of the third stage for the model problem

Iteration	1	2	3	4	5	6
Feature added	$x_4^{(-70)}$	$x_5^{(-59)}$	$x_3^{(-45)}$	$x_2^{(-23)}$	$x_1^{(-21)}$	$x_1^{(-66)}$
R^2 on exam. set	0.861	0.916	0.954	0.990	0.999	1.000

2. The number of events defined in this way was 1225 for 65360 data patterns, i.e. the events presented less than 1.9% of the data set. Unfavorable ratio of the determined classes is typical for most problems connected with prediction of events.

The results of solving this problem with the current version of the algorithms turned out to be unsatisfactory. "False alarms", i.e. erroneously predicted events (type I errors), were practically absent (on the examination set, 5 patterns of 12778, or 0.04%); however, the rate of misses (events not predicted by the system, type II errors) turned out to be unacceptably high (195 of 269, or more than 72% of misses). Note that earlier tests on other model binary valued problems performed by the authors [6] showed that even the single algorithm of Stage 2 with no feature selection could be efficient for binary valued problems, with low rates of both Type I and Type II errors. The poor result obtained in this study for the discussed problem may be explained by the pointed low fraction of events in the total number of patterns. Therefore, the algorithm needs additional methods to be included, that would allow the system to work successfully in the conditions of strong misbalance of classes.

3.3 Real World Problem

The considered problem was analysis of the influence of solar wind (SW) on the magnetic field of Earth. For quantitative description of perturbations of the magnetic field of Earth, different geomagnetic indexes are introduced. In this study, the so called Dst index [7] was used; its hourly values were provided by world data center WDC-C2 Kyoto [8]. Physical investigations showed [7] that the value of Dst index is most influenced by such variables as the components of interplanetary magnetic field (IMF), SW velocity and proton density in SW. As input data for prediction of Dst index value, the values of the following SW parameters and IMF parameters, recorded by ACE satellite [9], were used: the values of IMF components in two different coordinate systems Bx, By_GSE, By_GSM, Bz_GSE, Bz_GSM, IMF modulus B_magn, proton density in SW n, SW velocity v and temperature T. Gaps present in the data, not numerous, were removed by linear interpolation. Besides physical features mentioned above, the following timing variables were used: *year*, month of year (*month*), day of month (*day*), hour of day (*hour*), sine and cosine of hour of year, reduced to year period (*sin_t*, *cos_t*). The thus obtained set of 15 physical features was expanded by three time series of the same length from completely different problems, without any matching in time, taken from [10] (*CompGen*, *Physio*, *Astro*), and also two pseudorandom sequences of numbers (*rnd*1, *rnd*2). Thus, the preliminary set of physical features consisted of 20 time series. Each of them was normalized into [-1,1] range. The total number of patterns in this problem was about 59000.

To solve the problem, the time series were embedded to the depth of 48 hours (the search interval was equal to 48), then the data was sequentially divided into training, test and examination sets with 70:20:10 ratio.

First stage. Correlations of all the $20 * 48 = 960$ input features with output variable y have been calculated. For each physical feature of the preliminary set, the maximum value of the correlation coefficient r with the output variable y over the whole search interval has been calculated. As the threshold value for discarding physical features, the maximum value of correlation of y variable with five sets of pseudorandom numbers with length equal to that of the initiation interval, was accepted; this value turned out to be equal to 0.0123. Using this threshold, the following physical features were discarded: *hour* (0.0086), *rnd*2 (0.0059), and *rnd*1 (0.0055). All the other variables divided by the level of r into two groups: significant *B_magn* (0.528), v (0.451), *Bz_GSM* (0.406), *Bz_GSE* (0.341), T (0.310), *year* (0.240), and n (0.237), and less significant – all others, with r lying in the range from 0.0706 (*sin_t*) to 0.0145 (*CompGen*). At this stage, one should prefer a more conservative strategy, discarding only the features with r below the threshold.

Second stage. A NN committee (perceptrons with 16 neurons in the single hidden layer) has been trained for four values of initiation interval length: 4, 8, 16, and 24 hours. The results (the values of multiple determination coefficient R^2 on training, test, and examination sets, averaged over three experiments) are presented in Table 4, where the results obtained for initiation interval equal to search interval are also given for comparison. In all cases, the optimal position of the initiation interval was that with minimal delays; the upper-bound estimation of the phenomenon duration was 16 hours, based on the results on all the three sets. Significant difference of the results on different sets is connected with the fact that the sets were extracted sequentially, while the whole period of observation corresponded to gradual reduction of the solar activity at the fall of the 11-year cycle.

Third stage. To extract the precursor, the method of progressive build-up of neural networks has been applied. Perceptrons with 8 neurons in the single hidden layer were used. Network build-up was performed in the same way as for the model problem, except for the facts that at each iteration 3 best variable sets from the preceding iteration were used as basis for build-up, and that each network was trained 3 times (with different initial weights). As the result of network build-up, the precursor that was found to be optimal was a set of 12 most significant features; the values of R^2 are presented in Table 5. The list of the selected features included: *Bz_GSM* lagged for 2, 3, 6, 9, 11, and 15 hours; *B_magn* lagged for 2, 8 and 14 hours; n lagged for 1 hour; v lagged for 14 hours,

Table 4. Results of the second stage for the real world problem

Initiation interval	4	8	16	24	48
R^2 on training set	0.659	0.713	0.757	0.778	0.739
R^2 on test set	0.509	0.588	0.631	0.667	0.456
R^2 on examination set	0.375	0.436	0.400	0.360	-0.091

and *sin_t* lagged for 7 hours. In the whole, the selected features match physical notions about the studied problem [7].

Fourth stage. At this stage, the prediction problem was solved based on the set of features selected at stage 3, with the help of a perceptron with three hidden layers containing 24, 16, and 8 neurons. As can be seen from Table 5, the results have been improved for all the three sets of data.

Table 5. Comparison of the results of the second, third, and fourth stages for the real world problem

Stage	2	3	4
R^2 on training set	0.757	0.686	0.693
R^2 on test set	0.631	0.652	0.660
R^2 on examination set	0.400	0.361	0.374

4 Conclusion

A four-stage algorithm based on neural network committee for prediction and search for precursors in multi-dimensional time series has been suggested, implemented, and studied. The algorithm allows increasing efficiency of prediction and extracting the precursor (a combination of the most significant input features, defined in coordinates "initial time series – lag"). The suggested algorithm has been successfully tested on two problems with continuous output, and demonstrated its efficiency.

Testing of the algorithm on one model problem with binary output revealed a very high rate of false alarms, due to strong misbalance of classes. Therefore, while the general algorithm outline may remain the same for both continuous valued and binary output problems, supplementary algorithms need to be included in the method to handle misbalanced classes.

The program of future studies will also include: expansion of the range of methods used for selection of significant variables; use of NN committee at the last stage of the algorithm; use of the stacked generalization approach to work with NN committees at the second and the fourth stages of the algorithm; comparison of the results to those produced with other standard methodologies, such as recurrent NN or similar.

Acknowledgments. This study has been conducted under partial financial support of the Russian Foundation for Basic Research (RFBR), grant no. 07-01-00651-a.

References

1. Dolenko, S.A., Orlov, Y.V., Persiantsev, I.G., Shugai, J.S.: Neural network algorithm for events forecasting and its application to space physics data. In: Duch, W., Kacprzyk, J., Oja, E., Zadrożny, S. (eds.) ICANN 2005. LNCS, vol. 3697, pp. 527–532. Springer, Heidelberg (2005)
2. Dolenko, S.A., Orlov, Y. V., Persiantsev, I.G., Shugai, J. S.: Neural Network Algorithms for Analyzing Multidimensional Time Series for Predicting Events and Their Application to Study of Sun-Earth Relations. Pattern Recognition and Image Analysis 17, 584–591 (2007)
3. Shugai, J.S., Guzhva, A.G., Dolenko, S.A., Persiantsev, I.G.: An Algorithm for Construction of a Hierarchical Neural Network Complex for Time Series Analysis and its Application for Studying Sun-Earth Relations. In: 8th International Conference, Pattern Recognition and Image Analysis: New Information Technologies (PRIA-8-2007), vol. 2, pp. 355–358, Yoshkar-Ola (2007)
4. Guzhva, A.G., Dolenko, S.A., Persiantsev, I.G.: Comparative Analysis of Methods for Determination of Significance of Input Variables in Neural Network Based Modeling: Method of Comparison and Its Application to Known Real World Problems. In: Neuroinformatics-2008. IX All-Russian Scientific-Technical Conference, part 2, pp. 216–225. MEPhI Publishers, Moscow (2008) (in Russian)
5. Wolpert, D.H.: Stacked Generalization. Neural Networks 5, 241–259 (1992)
6. Dolenko, S.A., Orlov, Y. V., Persiantsev, I.G., Shugai, J. S.: A Search for Correlations in Time Series by Using Neural Networks. Pattern Recognition and Image Analysis 13, 441–446 (2003)
7. Gleisner, H., Lundstedt, H., Wintoft, P.: Predicting geomagnetic storms from solar-wind data using time-delay neural networks. Annales Geophysicae 14, 679–686 (1996)
8. World Data Center for Geomagnetism, Kyoto – Real-time (Quicklook) Dst index, `http://swdcwww.kugi.kyoto-u.ac.jp/dst_realtime/index.html`
9. ACE (Advanced Commission Explorer) Browse Data Information, `http://www.srl.caltech.edu/ACE/ASC/browse/view_browse_data.html`
10. `http://www-psych.stanford.edu/~andreas/Time-Series/SantaFe.html`

Adaptive Ensemble Models of Extreme Learning Machines for Time Series Prediction*

Mark van Heeswijk[1,3,**], Yoan Miche[1,2], Tiina Lindh-Knuutila[1,4], Peter A.J. Hilbers[3], Timo Honkela[1], Erkki Oja[1], and Amaury Lendasse[1]

[1] Adaptive Informatics Research Centre, Helsinki University of Technology P.O. Box 5400, 02015 TKK - Finland
heeswijk@cis.hut.fi
[2] INPG Grenoble - Gipsa-Lab, UMR 5216
961 rue de la Houille Blanche, Domaine Universitaire, 38402 Grenoble - France
[3] Eindhoven University of Technology
Den Dolech 2, P.O. Box 513, 5600 MB Eindhoven - The Netherlands
[4] International Computer Science Institute of University of California, 947 Center Street, Suite 600, Berkeley, CA 94704 - USA

Abstract. In this paper, we investigate the application of adaptive ensemble models of Extreme Learning Machines (ELMs) to the problem of one-step ahead prediction in (non)stationary time series. We verify that the method works on stationary time series and test the adaptivity of the ensemble model on a nonstationary time series. In the experiments, we show that the adaptive ensemble model achieves a test error comparable to the best methods, while keeping adaptivity. Moreover, it has low computational cost.

Keywords: time series prediction, sliding window, extreme learning machine, ensemble models, nonstationarity, adaptivity.

1 Introduction

Time series prediction is a challenge in many fields. In finance, experts predict stock exchange courses or stock market indices; data processing specialists predict the flow of information on their networks; producers of electricity predict the load of the following day [1,2]. The common question in these problems is how one can analyze and use the past to predict the future.

A common assumption in the field of time series prediction is that the underlying process generating the data is stationary and that the data points are independent and identically distributed (IID). Under this assumption, the training data is generally a good indication for what data to expect in the test phase.

However, a large number of application areas of prediction involve nonstationary phenomena. In these systems, the IID assumption does not hold since the

* This work was supported by the Academy of Finland Centre of Excellence, Adaptive Informatics Research Centre.

** Corresponding author.

C. Alippi et al. (Eds.): ICANN 2009, Part II, LNCS 5769, pp. 305–314, 2009.

system generating the time series changes over time. Therefore, contrary to the stationary case, one cannot assume that one can use what has been learned from past data and one has to keep learning and adapting the model as new samples arrive. Possible ways of doing this include: 1) retraining the model repeatedly on a finite window of past values and 2) using a combination of different models, each of which is specialized on part of the state space.

Besides the need to deal with nonstationarity, another motivation for such an approach is that one can drop stationarity requirements on the time series. This is very useful, since often we cannot assume anything about whether or not a time series is stationary.

To construct the ensemble model presented in this paper, a number of Extreme Learning Machines (ELMs) [3] of varying complexity are generated, each of which is individually trained on the data. After training, these individual models are combined in an ensemble model. The output of the ensemble model is a weighted linear combination of the outputs of the individual models. During the test phase, the ensemble model adapts this linear combination over time with the goal of minimizing the prediction error: whenever a particular model has bad prediction performance (relative to the other models) its weight in the ensemble is decreased, and vice versa. A detailed description can be found in Section 2.3.

In the first experiment, we test the performance of this adaptive ensemble model in repeated one-step ahead prediction on a time series that is known to be stationary (the Santa Fe A Laser series [4]). The main goal of this experiment is to test the robustness of the model and to investigate the different parameters influencing the performance of the model. In the second experiment, the model is applied to another time series (Quebec Births [5]) which is nonstationary and more noisy than the Santa Fe time series.

Ensemble methods have been applied in various forms (and under various names) to time series prediction, regression and classification. A non-exhaustive list of literature that discusses the combination of different models into a single model includes *bagging* [6], *boosting* [7], *committees* [8], *mixture of experts* [9], *multi-agent systems for prediction* [10], *classifier ensembles* [11], among others. Out of these examples, our work is most closely related to [10], which describes a multi-agent system prediction of financial time series and recasts prediction as a classification problem. Other related work includes [11], which deals with classification under concept drift (nonstationarity of classes). The difference is that both papers deal with classification under nonstationarity, while we deal with regression under nonstationarity.

In Section 2, the theory of ensemble models and the ELM are presented, as well as how we combine both of them in the adaptive ensemble method. Section 3 describes the experiments, the datasets used and discusses the results.

2 Methodology

2.1 Ensemble Models

In ensemble methods, several individual models are combined to form a single new model. Commonly, this is done by taking the average or a weighted average of the individual models, but other combination schemes are also possible [11]. For example, one could take the best n models and take a linear combination of those. For an overview of ensemble methods, see [8].

Ensemble methods rely on having multiple good models with sufficiently uncorrelated error. The individual models are typically combined into a single ensemble model as follows:

$$\hat{y}_{ens}(t) = \frac{1}{m}\sum_{i=1}^{m}\hat{y}_i(t)), \tag{1}$$

where $\hat{y}_{ens}(t)$ is the output of the ensemble model, $\hat{y}_i(t)$ are the outputs of the individual models and m is the number of models.

Following [8], it can be shown that the variance of the ensemble model is lower than the average variance of all the individual models:

Let $y(t)$ denote the true output that we are trying to predict and $\hat{y}_i(t)$ the estimation for this value of model i. Then, we can write the output $\hat{y}_i(t)$ of model i as the true value $y(t)$ plus some error term $\epsilon_i(t)$:

$$\hat{y}_i(t) = y(t) + \epsilon_i(t). \tag{2}$$

Then the expected square error of a model becomes

$$\mathbb{E}[\{\hat{y}_i(t) - y(t)\}^2] = \mathbb{E}[\epsilon_i(t)^2]. \tag{3}$$

The average error made by a number of models is given by

$$E_{avg} = \frac{1}{m}\sum_{i=1}^{m}\mathbb{E}[\epsilon_i(t)^2]. \tag{4}$$

Similarly, the expected error of the ensemble as defined in Equation 1 is given by

$$E_{ens} = \mathbb{E}\Big[\Big\{\frac{1}{m}\sum_{i=1}^{m}\hat{y}_i(t) - y(t)\Big\}^2\Big] = \mathbb{E}\Big[\Big\{\frac{1}{m}\sum_{i=1}^{m}\epsilon_i(t)\Big\}^2\Big]. \tag{5}$$

Assuming the errors $\epsilon_i(t)$ are uncorrelated (i.e. $E[\epsilon_i(t)\epsilon_j(t)] = 0$) and have zero mean ($E[\epsilon_i(t)] = 0$), we get

$$E_{ens} = \frac{1}{m}E_{avg}. \tag{6}$$

Note that these equations assume completely uncorrelated errors between the models, while in practice errors tend to be highly correlated. Therefore, errors are often not reduced as much as suggested by these equations, but can be improved by using ensemble models. It can be shown that $E_{ens} < E_{avg}$ always holds. Note that this only tells us that the test error of the ensemble is smaller than the average test error of the models, and that it is not necessarily better than the best model in the ensemble. Therefore, the models used in the ensemble should be sufficiently accurate.

2.2 ELM

The ELM algorithm is proposed by Guang-Bin Huang *et al.* in [3] and makes use of Single-Layer Feedforward Neural Networks (SLFN). The main concept behind ELM lies in the random initialization of the SLFN weights and biases. Under the condition that the transfer functions in the hidden layer are infinitely differentiable, the optimal output weights for a given training set can be determined analytically. The obtained output weights minimize the square training error. The trained network is thus obtained in very few steps and is very fast to train, which is why we use them in the adaptive ensemble model.

Below, we review the main concepts of ELM as presented in [3]. Consider a set of M distinct samples $(\mathbf{x}_i, y_i)$ with $\mathbf{x}_i \in \mathbb{R}^d$ and $y_i \in \mathbb{R}$; then, a SLFN with N hidden neurons is modeled as the following sum

$$\sum_{i=1}^{N} \beta_i f(\mathbf{w}_i \mathbf{x}_j + b_i), j \in [1, M], \tag{7}$$

with f being the activation function, $\mathbf{w}_i$ the input weights to the ith neuron in the hidden layer, b_i the biases and β_i the output weights.

In the case where the SLFN would perfectly approximate the data (meaning the error between the output $\hat{y_i}$ and the actual value y_i is zero), the relation is

$$\sum_{i=1}^{N} \beta_i f(\mathbf{w}_i \mathbf{x}_j + b_i) = y_j, j \in [1, M], \tag{8}$$

which can be written compactly as

$$\mathbf{H}\beta = \mathbf{Y}, \tag{9}$$

where $\mathbf{H}$ is the hidden layer output matrix defined as

$$\mathbf{H} = \begin{pmatrix} f(\mathbf{w}_1 \mathbf{x}_1 + b_1) & \cdots & f(\mathbf{w}_N \mathbf{x}_1 + b_N) \\ \vdots & \ddots & \vdots \\ f(\mathbf{w}_1 \mathbf{x}_M + b_1) & \cdots & f(\mathbf{w}_N \mathbf{x}_M + b_N) \end{pmatrix}, \tag{10}$$

and $\beta = (\beta_1 \ldots \beta_N)^T$ and $\mathbf{Y} = (y_1 \ldots y_N)^T$.

Given the randomly initialized first layer of the ELM and the training inputs $\mathbf{x}_i \in \mathbb{R}^d$, the hidden layer output matrix $\mathbf{H}$ can be computed. Now, given $\mathbf{H}$ and the target outputs $y_i \in \mathbb{R}$ (i.e. $\mathbf{Y}$), the output weights β can be solved from the linear system defined by Equation 9. This solution is given by $\beta = \mathbf{H}^\dagger \mathbf{Y}$, where $\mathbf{H}^\dagger$ is the Moore-Penrose generalized inverse of the matrix $\mathbf{H}$ [12]. This solution for β is the unique least-squares solution to the equation $\mathbf{H}\beta = \mathbf{Y}$. Overall, the ELM algorithm then is:

Algorithm 1. ELM

Given a training set $(\mathbf{x}_i, y_i), \mathbf{x}_i \in \mathbb{R}^d, y_i \in \mathbb{R}$, an activation function $f : \mathbb{R} \mapsto \mathbb{R}$ and N the number of hidden nodes,

1: - Randomly assign input weights $\mathbf{w}_i$ and biases b_i, $i \in [1, N]$;
2: - Calculate the hidden layer output matrix $\mathbf{H}$;
3: - Calculate output weights matrix $\beta = \mathbf{H}^\dagger \mathbf{Y}$.

Theoretical proofs and a more thorough presentation of the ELM algorithm are detailed in the original paper [3].

2.3 Adaptive Ensemble Model of ELMs

When creating a model to solve a certain regression or classification problem, it is unknown in advance what the optimal model complexity and architecture is. Also, we cannot always assume stationarity of the process generating the data (i.e. in cases where the IID assumption does not hold). Therefore, since the information that has been gathered from past samples can become inaccurate, it is needed to keep learning and keep adapting the model once new samples become available. Possible ways of doing this include: 1) retraining the model repeatedly on a finite window into the past and 2) use a combination of different models, each of which is specialized on part of the state space. In this paper, we employ both strategies in repeated one-step ahead prediction on (non)stationary time series. On the one hand, we use diverse models and adapt the weights with which these models contribute to the ensemble. On the other hand, we retrain the individual models on a limited number of past values (sliding window) or on all known values (growing window).

Adaptation of the Ensemble. The ensemble model consists of a number of randomly initialized ELMs, which each have their own parameters (details are discussed in the next subsection). The model ELM_i has an associated weight w_i which determines its contribution to the prediction of the ensemble. Each ELM is individually trained on the training data and the outputs of the ELMs contribute to the output $\hat{y}_{ens}$ of the ensemble as follows: $\hat{y}_{ens}(t+1) = \sum_i w_i \hat{y}_i(t+1)$.

Once initial training of the models on the training set is done, repeated one-step ahead prediction on the 'test' set starts. After each time step, the previous predictions $\hat{y}_i(t-1)$ are compared with the real value $y(t-1)$. If the square error $\epsilon_i(t-1)^2$ of ELM_i is larger than the average square error of all models at time step $t-1$, then the associated ensemble weight w_i is decreased, and vice versa. The rate of change can be scaled with a parameter α, called the learning rate. Furthermore, the rate of change is normalized by the number of models and the variance of the time series, such that we can expect similar behaviour on time series with different variance and ensembles with a different number of models. The full algorithm can be found in Algorithm 2.

Adaptation of the Models. As described above, ELMs are used in the ensemble model. Each ELM has a random number of input neurons, random number of hidden neurons, and random variables of the regressor as input.

Besides changing the ensemble weights w_i as a function of the errors of the individual models at every time step, the models themselves are also retrained. Before making a prediction at time step t, each model is either retrained on a past window of n values $(\mathbf{x}_i, y_i)_{t-n}^{t-1}$ (sliding window), or on all values known so far $(\mathbf{x}_i, y_i)_1^{t-1}$ (growing window). Details on how this retraining fits in with the rest of the ensemble can be found in Algorithm 2.

As mentioned in Section 2.2, ELMs are very fast to train. In order to further speed up the retraining of the ELMs, we make use of PRESS statistics, which allow you to add and remove samples from the training set of a linear model and give you the linear model that you would have obtained, had you trained it on the modified training set. Since an ELM is essentially a linear model of the responses of the hidden layer, PRESS statistics can be applied to (re)train the ELM in an incremental way. A detailed discussion of incremental training of ELMs with PRESS statistics falls outside the scope of this paper, but details on PRESS statistics can be found in [13].

Algorithm 2. Adaptive Ensemble of ELMs

Given a set $(\mathbf{x}(t), y(t)), \mathbf{x}(t) \in \mathbb{R}^d, y(t) \in \mathbb{R}$, and m models,
Create m random ELMs: $(ELM_1 \ldots ELM_m)$
Train each of the ELMs individually on the training data
Initialize each w_i to $\frac{1}{m}$
while $t < t_{end}$ **do**
  generate predictions $\hat{y}_i(t+1)$
  $\hat{y}_{ens}(t+1) = \sum_i w_i \hat{y}_i(t+1)$
  $\mathrm{t} = \mathrm{t} + 1$
  compute all errors $\rightarrow \epsilon_i(t-1) = \hat{y}_i(t-1) - y(t-1)$
  for i = 1 to #models **do**
    $\Delta w_i = -\epsilon_i(t-1)^2 + mean(\epsilon(t-1)^2)$
    $\Delta w_i = \Delta w_i \cdot \alpha / (\#models \cdot \mathbf{var}(y))$
    $w_i = \max(0, w_i + \Delta w_i)$
    Retrain ELM_i
  end for
  renormalize weights $\rightarrow \mathbf{w} = \mathbf{w} / \|\mathbf{w}\|$
end while

3 Experiments

3.1 Experiment 1: Stationary Time Series

The Santa Fe Laser Data time series [4] has been obtained from a far-infrared-laser in a chaotic state. This time series has become a well-known benchmark

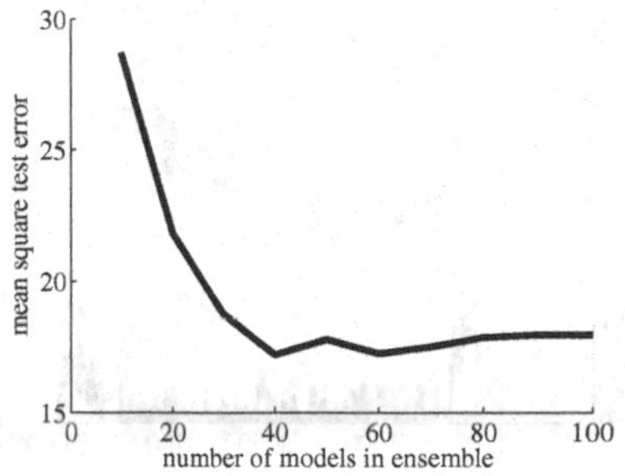

Fig. 1. MSE_{test} of ensemble on laser time series for varying number of models (no window retraining, learning rate 0.1)

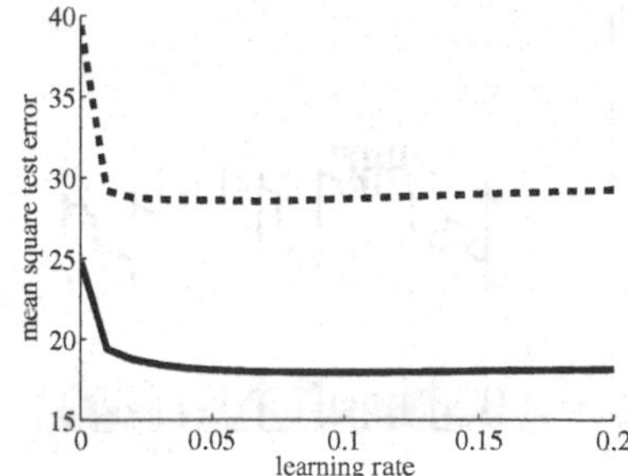

Fig. 2. MSE_{test} of ensemble on laser time series as a function of learning rate (no window retraining), for 10 models (*dotted line*) and 100 models (*solid line*)

in time series prediction since the Santa Fe competition in 1991. It consists of approximately 10000 points and the time series is known to be stationary.

The adaptive ensemble model is trained on the first 1000 values of the time series, after which sequential one-step ahead prediction is performed on the following 9000 values. This experiment is repeated for various combinations of learning rate α and number of models in the ensemble. Each ELM has a regressor size of 8 (of which 5 to 8 variables are randomly selected) and between 150 and 200 hidden neurons with a sigmoid transfer function.

Figure 1 shows the effect of the number of models on the prediction accuracy. It can be seen that the number of models strongly influences the prediction accuracy and that at least 60 models are needed to get good prediction accuracy. Figure 2 shows the effect of the learning rate on the prediction accuracy. The influence of the various (re)training strategies can be found in Table 1. This table also shows that the method is able to achieve a prediction accuracy comparable to the best methods [14].

Table 1. MSE_{test} of ensemble for laser (training window size 1000)

		retraining		
learning rate	#models	none	sliding	growing
0.0	10	39.39	58.56	34.16
0.1	10	28.70	37.93	18.42
0.0	100	24.80	33.85	20.99
0.1	100	17.96	27.30	**14.64**

Figures 3 and 4 show the adaptation of some of the ensemble weights over the length of the entire prediction task.

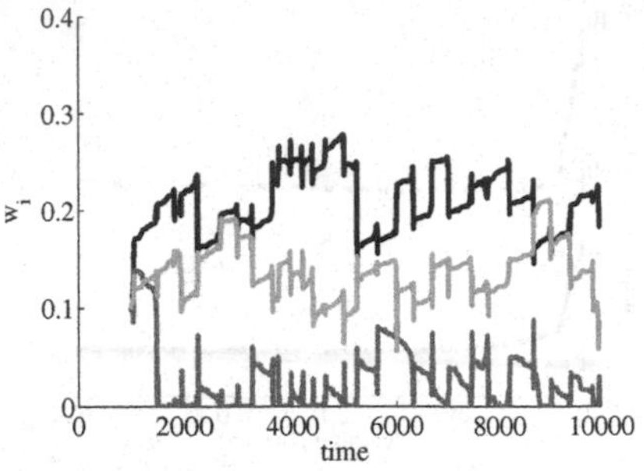

Fig. 3. plot showing part of the ensemble weights w_i adapting over time during sequential prediction on laser time series (#models=10, learning rate=0.1, no window retraining)

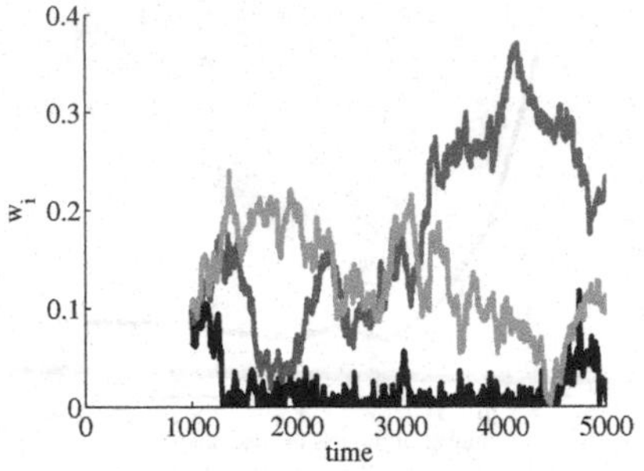

Fig. 4. plot showing part of the ensemble weights w_i adapting over time during sequential prediction on qbirths time series (#models=10, learning rate=0.1, no window retraining)

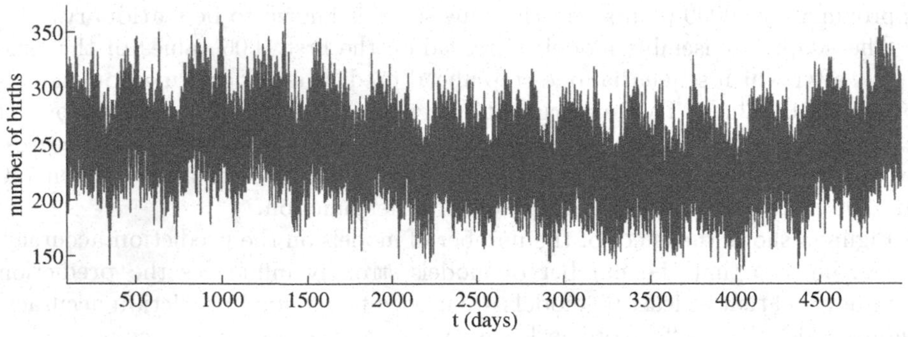

Fig. 5. The Quebec Births time series

3.2 Experiment 2: Nonstationary Time Series

The Quebec Births time series [5] consists of the number of daily births in Quebec over the period of January 1, 1977 to December 31, 1990. It consists of approximately 5000 points, is nonstationary and more noisy than the Santa Fe Laser Data.

The adaptive ensemble model is trained on the first 1000 values of the time series, after which sequential one-step ahead prediction is performed on the following 5000 values. This experiment is repeated for varying learning rates α and numbers of models. Each ELM has a regressor size of 14 (of which 12 to 14 variables are randomly selected) and between 150 and 200 hidden neurons.

Figure 6 shows the effect of the number of models on the prediction accuracy. It can be seen that the number of models strongly influences the prediction accuracy, as was the case with the Santa Fe time series. However, we need more

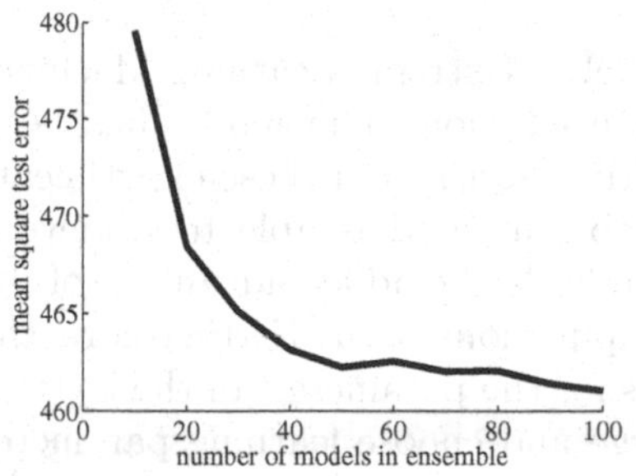

Fig. 6. MSE_{test} of ensemble on Quebec Births time series for varying number of models (sliding window retraining, learning rate 0.1)

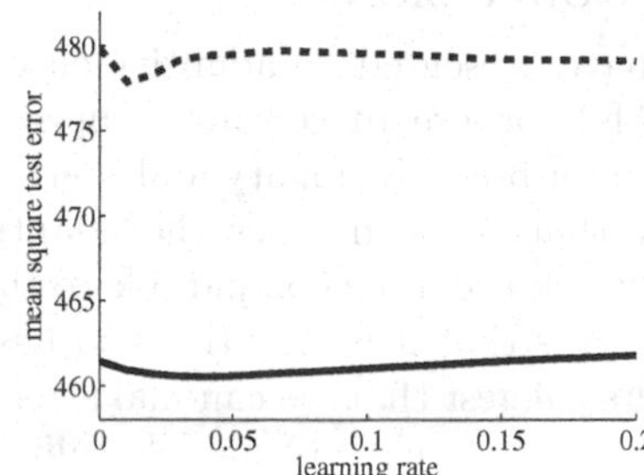

Fig. 7. MSE_{test} of ensemble on Quebec Births time series as a function of learning rate (retraining with sliding window of size 1000), for 10 models (*dotted line*) and 100 models (*solid line*)

Table 2. MSE_{test} of ensemble for Quebec Births (training windows size 1000)

		retraining		
learning rate	#models	none	sliding	growing
0.0	10	594.04	479.84	480.97
0.1	10	582.09	479.58	476.87
0.0	100	585.53	**461.44**	469.79
0.1	100	567.62	**461.04**	468.51

models in order to get a good prediction accuracy. Figure 7 shows the effect of the learning rate on the prediction accuracy. The influence of the various (re)training strategies can be found in Table 2.

3.3 Discussion

The experiments clearly show that it is important to have a sufficient number of models (more is generally better). Furthermore, the shape of the learning rate graph is independent of the number of models, which means that these parameters can probably be optimized independently from each other. We are currently performing a more thorough statistical analysis for determining the best strategy for optimizing the parameters. However, the results suggest that choosing the number of models high and choosing a sufficiently large learning rate (i.e. $\alpha = 0.1$) is a good and robust strategy.

The results also show that the proposed adaptive ensemble method is able to achieve a prediction accuracy comparable to the best methods [14], and is able to do so for both stationary and nonstationary series. An added advantage of the method is that it is adaptive and has low computational cost.

4 Conclusions

We have presented an adaptive ensemble model of Extreme Learning Machines (ELMs) for use in repeated one-step ahead prediction. The model has been tested on both stationary and nonstationary time series, and these experiments show that in both cases the adaptive ensemble method is able to achieve a prediction accuracy comparable to the best methods. An added advantage of the method is that it is adaptive and has low computational cost. Furthermore, the results suggest that we can make good guesses for the parameters of the method (i.e. choose number of models sufficiently large and choose learning parameter $\alpha = 0.1$). We are currently performing more thorough statistical analysis of the model, in order to determine the best strategy for optimizing the parameters. In addition, we would like to extend the model to include other models like OP-ELM [15] and investigate how we can guide adding new models to the ensemble in an online fashion, in order to introduce an extra degrees of adaptivity.

References

1. Sorjamaa, A., Hao, J., Reyhani, N., Ji, Y., Lendasse, A.: Methodology for long-term prediction of time series. Neurocomputing 70(16-18), 2861–2869 (2007)
2. Simon, G., Lendasse, A., Cottrell, M., Fort, J.-C., Verleysen, M.: Time series forecasting: Obtaining long term trends with self-organizing maps. Pattern Recognition Letters 26(12), 1795–1808 (2005)
3. Huang, G.-B., Zhu, Q.-Y., Siew, C.-K.: Extreme learning machine: Theory and applications. Neurocomputing 70(1-3), 489–501 (2006)
4. Weigend, A., Gershenfeld, N.: Time Series Prediction: Forecasting the Future and Understanding the Past. Addison-Wesley, Reading (1993)
5. Quebec Births Data, `http://www-personal.buseco.monash.edu.au/~hyndman/TSDL/misc/qbirths.dat`
6. Breiman, L.: Bagging predictors. In: Machine Learning, pp. 123–140 (1996)
7. Schapire, R.E., Freund, Y., Bartlett, P., Lee, W.S.: Boosting the margin: a new explanation for the effectiveness of voting methods. The Annals of Statistics 26, 322–330 (1998)
8. Bishop, C.M.: Pattern Recognition and Machine Learning (Information Science and Statistics). Springer, Secaucus (2006)
9. Jacobs, R.A., Jordan, M.I., Nowlan, S.J., Hinton, G.E.: Adaptive mixtures of local experts. Neural Computation 3, 79–87 (1991)
10. Raudys, S., Zliobaite, I.: The multi-agent system for prediction of financial time series. In: Rutkowski, L., Tadeusiewicz, R., Zadeh, L.A., Żurada, J.M. (eds.) ICAISC 2006. LNCS (LNAI), vol. 4029, pp. 653–662. Springer, Heidelberg (2006)
11. Kuncheva, L.I.: Classifier ensembles for changing environments. MCS, 1–15 (2004)
12. Rao, C.R., Mitra, S.K.: Generalized Inverse of Matrices and Its Applications. John Wiley & Sons Inc., Chichester (1972)
13. Myers, R.H.: Classical and Modern Regression with Applications, 2nd edn. Duxbury, Pacific Grove (1990)
14. Suykens, J.A.K., Gestel, T.V., Brabanter, J.D., Moor, B.D., Vandewalle, J.: Least Squares Support Vector Machines. World Scientific, Singapore (2002)
15. Miche, Y., Sorjamaa, A., Lendasse, A.: OP-ELM: Theory, experiments and a toolbox. In: Kůrková, V., Neruda, R., Koutník, J. (eds.) ICANN 2008, Part I. LNCS, vol. 5163, pp. 145–154. Springer, Heidelberg (2008)

Identifying Customer Profiles in Power Load Time Series Using Spectral Clustering

Carlos Alzate, Marcelo Espinoza, Bart De Moor, and Johan A.K. Suykens

Department of Electrical Engineering, ESAT-SCD-SISTA
Katholieke Universiteit Leuven
Kasteelpark Arenberg 10, B-3001 Leuven, Belgium
{carlos.alzate,johan.suykens}@esat.kuleuven.be

Abstract. An application of multiway spectral clustering with out-of-sample extensions towards clustering time series is presented. The data correspond to power load time series acquired from substations in the Belgian grid for a period of 5 years. Spectral clustering methods are a class of unsupervised learning algorithms where the solutions can be obtained from the eigenvectors of a Laplacian matrix derived from the data. Nonlinearity can easily be added to the analysis by the use of non-linear similarity functions that can be regarded as Mercer kernels. In this paper, a weighted kernel PCA formulation to spectral clustering is used to find interpretable customer profiles underlying the power consumption load time series. The main advantage of the interpretation as kernel PCA is the extension of the clustering model to out-of-sample points. The clustering model can be trained, validated and tested in a learning framework working directly with the data and without the use of pre-modeling steps. The experimental results with real-life data demonstrate the applicability of the multiway spectral clustering method compared to an existing method pre-modeling the data based on periodic autoregressions (PAR).

1 Introduction

Spectral clustering provides a powerful unsupervised learning tool to find similar patterns underlying the data. These methods are typically expressed as relaxations of NP-hard graph partitioning problems. Although several spectral clustering methods have been proposed in the literature [1,2,3], they all share the use of information contained in the eigenspectrum of Laplacian matrices derived from the data. Spectral clustering can be also interpreted as a kernel method when the chosen similarity gives rise to pairwise similarity matrices with positive eigenvalues. Kernel-based modeling is a powerful and elegant way of introducing nonlinearity into the analysis. However, in practice, the models can become too general for the application at hand with a high risk of overfitting. A careful and systematic parameter selection is thus needed in order to obtain relevant and informative groupings among the data. A recent approach linking spectral clustering with a form of weighted kernel PCA has been proposed in [4]. This

C. Alippi et al. (Eds.): ICANN 2009, Part II, LNCS 5769, pp. 315–324, 2009.

method fits into the least squares support vector machine (LS-SVM) [5] framework for kernel-based learning. One of the main advantages of this approach is the possibility to extend the clustering model to out-of-sample points without the need of approximation or ad-hoc schemes.

This paper contains an application of the multiway spectral clustering with out-of-sample extensions introduced in [4] towards clustering time series. Time series data pose a challenge to standard clustering methods due mainly to the high-dimensionality. The particular dataset corresponds to readings of the power load measure at 245 different substations in the Belgian grid for a period of 5 years. The objective is to identify types of costumers underlying the data. This task is particularly important for short, mid and long-term planning in the electricity sector. An approach to cluster this type of data was proposed in [6]. This approach uses a pre-modeling scheme based on periodic autoregressions (PAR) in order to reduce the dimensionality together with representing the time series independent of seasonal and calendar variations. In a later stage, k-means was used to find the clusters. Pre-modeling can be useful for interpretability and dimensionality reduction but there might be a loss of information while trying to pre-model the complete set of load time series with the same template. The proposed approach aims at obtaining interpretable customer profiles of power load consumption by clustering the data directly in a learning framework with model selection and without the use of pre-modeling steps. This paper is organized as follows. Section 2 contains a short description of the dataset used in the sequel. Section 3 summarizes classical spectral clustering. In Section 4, the multiway formulation to spectral clustering introduced in [4] is described. Section 5 reviews a model selection method. Section 6 contains the clustering results of an existing methodology and the proposed approach and in Section 7, we give conclusions.

2 Power Load Time Series

A time series dataset containing readings of the power load measured at 245 different high voltage - low voltage substations in the Belgian grid is used in this paper. The power load was measured every hour for a period of 5 years. The data characterize different profiles of load consumption such as business, residential and industrial. Each time series was normalized to have zero mean and unitary standard deviation. Figure 1 shows one particular time series of the dataset. The yearly cycles are visible together with the daily cycles. From the daily cycle it is also possible to visualize morning, noon and evening peaks.

3 Classical Spectral Clustering

3.1 Spectral Graph Theory

Given a set of N data points $\{x_i\}_{i=1}^N, x_i \in \mathbb{R}^d$ and some similarity measure $s_{ij} \geq 0$ between each pair of points x_i and x_j, an intuitive form of representing

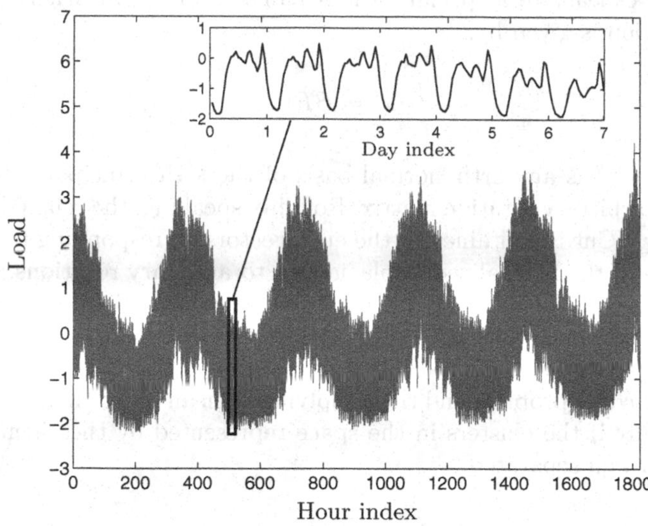

Fig. 1. Example of time series from the power load dataset. Yearly and daily cycles are visible. It is also possible to visualize morning, noon and evening peaks.

the data is using a graph $G = (\mathcal{V}, \mathcal{E})$. The vertices $\mathcal{V}$ represent the data points and the edge $e_{ij} \in \mathcal{E}$ between vertices v_i, v_j has a weight determined by s_{ij}. The affinity matrix of the graph is the matrix S with ij-entry $S_{ij} = s_{ij}$. The degree of vertex v_i is defined as $\deg_i = \sum_{j=1}^{N} s_{ij}$, the degree matrix D is the matrix $D = \mathrm{diag}([\deg_1, \ldots, \deg_N])$ on the diagonal.

A basic problem in graph theory is the graph bipartitioning problem, that is to separate the graph G into two disjoint sets $\mathcal{A}, \mathcal{B}$ based on a cut criterion. The resulting sets should be disjoint: $\mathcal{A} \cap \mathcal{B} = \emptyset$ and $\mathcal{A} \cup \mathcal{B} = \mathcal{V}$. This problem has been extensively studied [7] and several cut criteria have been proposed [8,1]. The normalized cut *NCut* [1] is a common bipartitioning criterion and is defined as $\mathrm{NCut}(\mathcal{A}, \mathcal{B}) = \mathrm{cut}(\mathcal{A}, \mathcal{B})/\mathrm{Vol}(\mathcal{A}) + \mathrm{cut}(\mathcal{A}, \mathcal{B})/\mathrm{Vol}(\mathcal{B})$, where $\mathrm{Vol}(\mathcal{A}) \equiv \sum_{i \in \mathcal{A}} d_i$ denotes the volume of the set $\mathcal{A}$.

3.2 Multiway Cut Criteria

A more general problem in graph theory is the so called multiway cut. In this case, the aim is to partition the graph into $k > 2$ disjoint parts $\mathcal{A}_1, \ldots, \mathcal{A}_k$, with $\mathcal{A}_1 \cap \ldots \cap \mathcal{A}_k = \emptyset$. The multiway NCut problem as introduced in [9] is $\min_G k - \mathrm{trace}(G^T \hat{L} G)$ such that $G^T G = I_k$, where $\hat{L} = D^{-1/2} L D^{-1/2}$ is the normalized Laplacian and $G \in \{0, c_L\}^{N \times k}$ is a cluster indicator matrix with one non-zero entry per row (indicating cluster membership) and c_L is a normalization constant. Minimizing the multiway Ncut is NP-hard. Relaxing G and allowing to take real values leads to a special form of Fan's theorem [10] where the relaxed optimal solution is given by:

$$\tilde{G}^{\star} = BR_L,$$

where $B \in \mathbb{R}^{N \times k}$ is any orthonormal basis of the k-th principal subspace of $\tilde{L}$ and R_L is an arbitrary rotation matrix. Roughly speaking, the relaxed solution of the multiway NCut is contained in the eigenvectors corresponding to the largest k eigenvalues of the normalized Laplacian up to arbitrary rotations.

Several methods to convert the partitioning information contained in the eigenvectors to clusters have been proposed [1,11]. A popular method introduced in [1] consists of *reclustering* and consists of computing the top k eigenvectors of the multiway *NCut* problem and then applying k-means onto them. Reclustering works well only if the clusters in the space represented by the eigenvectors are spherical and well separated.

3.3 Markov Random Walks

A Markov random walks view of spectral clustering was discussed in [12]. A random walk on a graph consists of random jumps from vertex to vertex. The ij-th entry of the stochastic transition matrix $P = D^{-1}S$ represents the probability of moving from node i to node j in one step. The corresponding eigenvalue problem becomes $Pr = \xi r$. Maximizing the random walks model is equivalent to minimizing the binary *NCut* and can be interpreted as finding a partition of the graph in such a way that the random walk remains most of the time in the same cluster with few jumps to other clusters.

In the ideal case of a graph with k disconnected components, the transition matrix P will have k eigenvalues equal to 1. The eigenvectors corresponding to these unitary eigenvalues are the indicator functions to the respective disconnected components and the clustering is trivial because each cluster is represented as a single point in $\mathbb{R}^k$. This also corresponds to the piecewise constant property of eigenvectors discussed in [12].

4 Spectral Clustering via Weighted Kernel PCA

In this Section, we summarize the multiway formulation for spectral clustering introduced in [4]. This algorithm considers weighted kernel PCA with D^{-1} as the weighting matrix to perform spectral clustering. The formulation fits into a constrained optimization framework typical of least squares support vector machines (LS-SVM). The choice of D^{-1} as the weighting matrix is motivated by the random walks model of spectral clustering. Moreover, [4] can be interpreted as a weighted centered version of the random walks algorithm with bias terms and out-of-sample extensions.

4.1 Multiway Formulation

Given training data $\{x_i\}_{i=1}^N, x_i \in \mathbb{R}^d$, the primal problem is formulated as follows:

$$\min_{w^{(l)},e^{(l)},b_l} J_P = \frac{1}{2N}\sum_{l=1}^{k-1}\gamma_l e^{(l)^T} D^{-1} e^{(l)} - \frac{1}{2}\sum_{l=1}^{k-1} w^{(l)^T} w^{(l)} \tag{1}$$
$$\text{such that } e^{(l)} = \Phi w^{(l)} + b_l 1_N, l = 1,\ldots,k-1$$

where k is the number of desired clusters, $\gamma_l \in \mathbb{R}^+$ are regularization constants, $e^{(l)} = [e_1^{(l)};\ldots;e_N^{(l)}]$ is the l-th score variables vector, b_l are bias terms, Φ is the $N \times n_h$ feature matrix $\Phi = [\varphi(x_1)^T;\ldots;\varphi(x_N)^T]$, $\varphi(\,) : \mathbb{R}^d \to \mathbb{R}^{n_h}$ is the mapping to a high dimensional feature space $\mathcal{F}$ of dimension n_h (which can be infinite dimensional) and $l = 1,\ldots,k-1$. Thus, each training data point $x_i \in \mathbb{R}^d$ is represented as point in the score variables space $e_i \in \mathbb{R}^{k-1}$, $e_i = [e_i^{(1)},\ldots,e_i^{(k-1)}]$. After binarization, the score variables become cluster indicators $q_i^{(l)} = \text{sign}(e_i^{(l)}), i,\ldots,N, l = 1,\ldots,k-1$. The Lagrangian of the constrained optimization problem (1) is expressed as $\mathcal{L}(w^{(l)}, e^{(l)}, b_l; \alpha^{(l)}) = J_P - \sum_{l=1}^{n_e} \alpha^{(l)^T}(e^{(l)} - \Phi^{(l)} w^{(l)} - b_l 1_N)$ where $\alpha^{(l)} \in \mathbb{R}^N$ are Lagrange multiplier vectors. Using the KKT optimality conditions $\partial\mathcal{L}/\partial e^{(l)} = 0, \partial\mathcal{L}/\partial w^{(l)} = 0, \partial\mathcal{L}/\partial \alpha^{(l)} = 0, l = 1,\ldots,k-1$, eliminating $w^{(l)}, e^{(l)}, b_l$ leads to $(\gamma_l/N)D^{-1}M_D\Phi\Phi^T\alpha^{(l)} = \alpha^{(l)}$ where

$$M_D = I_N - \frac{1}{1_N^T D^{-1} 1_N} 1_N 1_N^T D^{-1}. \tag{2}$$

Applying the kernel trick $K(x_i, x_j) = \varphi(x_i)^T\varphi(x_j)$, and defining $\lambda_l = N/\gamma_l$ leads to the following eigenvalue problem:

$$D^{-1}M_D\Omega\alpha^{(l)} = \lambda_l\alpha^{(l)}, j = 1,\ldots,k-1 \tag{3}$$

where Ω is the kernel matrix with ij-th entry $\Omega_{ij} = K(x_i, x_j)$. Note that the matrix $D^{-1}M_D\Omega$ is not symmetric but it is the product of two symmetric positive semidefinite matrices therefore its eigenvalues are real [13]. The bias terms become $b_l = -\frac{1}{1_N^T D^{-1} 1_N} 1_N^T D^{-1}\Omega\alpha^{(l)}, l = 1,\ldots,k-1$, and the projections of the training points onto the eigenvectors can be expressed in terms of the dual variables: $e_i^{(l)} = w^{(l)^T}\varphi(x_i) + b_l = \sum_{j=1}^N \alpha_j^{(l)} K(x_i, x_j) + b_l$.

4.2 Out-of-Sample Extension and Encoding/Decoding

In classical spectral clustering, the cluster indicators obtained from the eigenvectors are defined only for training data. Extensions to compute cluster indicators for unseen points are typically done in an ad-hoc manner using the Nyström approximation [14]. However, due to the primal-dual nature of the weighted kernel PCA framework, it is possible to extend the clustering model to out-of-sample points in an straightforward way using the projections onto the eigenvectors.

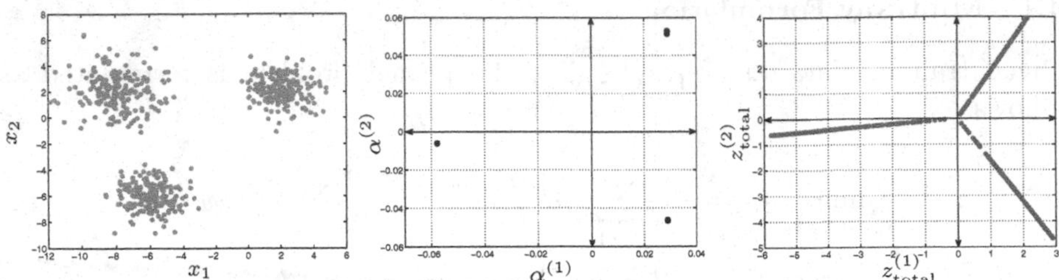

Fig. 2. Toy dataset illustrating piecewise constant eigenvectors and projections. **Left:** Training data consisting of 3 Gaussian clouds and $N = 600$ points. **Center:** Piecewise constant eigenvectors $\alpha^{(1)}, \alpha^{(2)}$ obtained by solving (3) with an RBF kernel, $\sigma = 0.5$. **Right:** Corresponding projections of the full dataset $N_{\text{total}} = 1,200$. Note that the clusters are represented in the eigenvectors and the corresponding projections in distinct quadrants.

Given a set of N_{test} test points $\{x_t^{\text{test}}\}_{t=1}^{N_{\text{test}}}$, the score variables for these points become: $z_t^{(l)} = w^{(l)^T} \varphi(x_t^{\text{test}}) + b_l = \sum_{j=1}^{N} \alpha_j^{(l)} K(x_t^{\text{test}}, x_j) + b_l$ and the cluster indicators are given by $\check{q}_t^{(l)} = \text{sign}(z_t^{(l)}), l = 1, \ldots, k-1, t = 1, \ldots, N_{\text{test}}$.

The final clustering is obtained through encoding the $k-1$ binary indicators. Each training point $x_i \in \mathbb{R}^d$ has an associated binary encoding $q_i \in \mathbb{R}^{k-1}$. From the clustering indicators q_i a codebook is formed with the k binary encodings with most occurrences [4]. Figure 2 illustrates a simple example with 3 Gaussian clouds in a $2D$ space. Note that, the eigenvectors are piecewise constant thus representing each cluster as a single point in a different quadrant. The out-of-sample extension induced by the projections displays a line structure which can be used for model selection as will be discussed in the sequel. The decoding step consists of comparing the cluster indicators for out-of-sample data with each codeword in the codebook and selecting the encoding that minimizes the Hamming distance.

5 Model Selection

Model selection is a critical issue in unsupervised learning methods such as spectral clustering. Typically, the number of clusters k and the kernel parameters are chosen in a heuristic way. However, by means of the out-of-sample extensions, it is possible to perform model selection in an intuitive and straightforward way. A model selection criterion called the Balanced Line Fit (BLF) was introduced in [4]. The clusters are represented as lines in the score variables space for well-chosen parameters. Thus, it becomes important to characterize the fitness of a particular cluster to a line. The BLF as defined in [4] is a weighted sum of a collinearity measure called the linefit and a measure of the balance of the obtained clusters: $\text{BLF}(k) = \eta \text{linefit}(k) + (1-\eta) \text{balance}(k)$, where η controls the importance given to the collinearity measure with respect to the balance, $0 \leq \eta \leq 1$, $\text{linefit}(k) = \frac{1}{k}\sum_{p=1}^{k} \frac{k-1}{k-2}\left(\frac{\zeta_1^{(p)}}{\sum_l \zeta_l^{(p)}} - \frac{1}{k-1}\right)$ where $\zeta_1^{(p)} \geq \ldots \geq \zeta_{k-1}^{(p)}$ are the ordered eigenvalues of the sample covariance matrix $C_{\tilde{Z}}^{(p)} = (1/|\mathcal{A}_p|)\tilde{Z}^{(p)^T}\tilde{Z}^{(p)}, p = 1, \ldots, k$, $\tilde{Z}^{(p)} \in \mathbb{R}^{|\mathcal{A}_p| \times (k-1)}$ is the matrix representing the zero mean score variables for out-of-sample data assigned to the p-th cluster. The term $\zeta_1^{(p)} / \sum_l \zeta_l^{(p)}$ indicates how much of the total variance is contained on the top eigenvector of $C_{\tilde{Z}}^{(p)}$. In this way, if the validation score variables for the p-th cluster are collinear, then $\zeta_1^{(p)} / \sum_l \zeta_l^{(p)}$ equals 1. The additional terms in the linefit normalize the criterion between 0 and 1. The balance measure is defined as $\text{balance}(k) = \min\{|\mathcal{A}_1|, \ldots, |\mathcal{A}_k|\}/\max\{|\mathcal{A}_1|, \ldots, |\mathcal{A}_k|\}$. The balance index equals 1 when the clusters have the same number of elements and tends to 0 in extremely unbalanced cases. Note that, the linefit is defined only for $k > 2$. For $k = 2$, a slight modification of the BLF was discussed in [4].

6 Experimental Results

6.1 Existing Methodology

An approach to cluster the load time series was proposed in [6]. This method computes a Typical Daily Profile (TDP) from the identified parameters in a Periodic Autoregression (PAR) system estimated for each times series, later performing k-means on the TDPs. PAR models allow the autoregression parameters to vary according to cyclic patterns (such as hours, days, years, seasons). These models have been used for forecasting in economical modeling and electricity price modeling. The method proposed in [6] aims at representing the time series as an hourly PAR model of order 48 with additional exogenous variables accounting for monthly and weekly cyclic variations together with temperature effects. The PAR parameters can be obtained by solving an Ordinary Least Squares (OLS) system. A TDP was defined in [6] as a convergence vector of the PAR(48) model after extracting all exogenous information. The main advantage of a TDP is that it represents a daily load profile independent of the seasonal and calendar variations. In summary, [6] creates a TDP for each time series which implies a dimensionality reduction from 43, 824 to 24. PCA is then applied over the TDP to capture 99% of the information explained by the variance. This corresponds to a further reduction to 9 dimensions. The last step consists of applying k-means over the 9 principal components of the typical daily profile. Figure 3 shows the clustering results. Several types of customers can be identified from the results, particularly clusters 1 and 2 which corresponds to *residential* profiles, clusters 6 and 7 which display a *commercial* profile, clusters 4, 5, 8 show a profile typical of street lighting and certain industrial activities taking place during the night.

6.2 Proposed Clustering Scheme

The proposed scheme consists of clustering directing the time series in a learning framework with model selection without any premodeling steps. The training

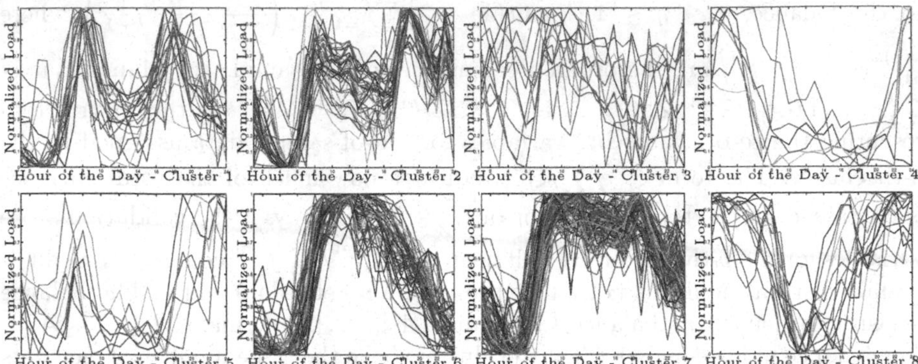

Fig. 3. Typical daily profile (TDP) clustering results using the method proposed in [6] which corresponds to preprocessing the TDP with PCA to reduce the dimensionality to 9 and then applying k-means

scenario consists of 123 time series for training and 122 times series for validation. Each time series is a 43,824-dimensional vector with zero mean. The training / validation sets were randomized 10 times and the model selection results report average values. We used the BLF for tuning the number of clusters and the RBF kernel parameter σ. The η parameter was set to 0.5 hence, giving equal weight to the linefit and to the balance. The tuning range for the RBF kernel parameter was varied from 500 to 2,000 in 100 steps. The number of clusters k ranged from 2 to 12. The choice for the 500 as the minimum value of the range is motivated by the fact that with lower values the RBF kernel matrix was tending to the identity matrix which has trivial eigenspace with no discriminatory information about the clusters. On the other hand, values greater than 2,000 led to a almost constant-valued kernel matrix independent of the data. The maximum of the BLF gives the optimal number of clusters and the optimal value for σ. Figure 4 shows the model selection results. The BLF detected 7 clusters and the optimal σ value was 200. For visualization purposes, each time series is transformed into a mean daily pattern by averaging the corresponding 1,824 24-hour non-overlapping windows. This mean daily pattern is similar to the TDP but differs in the sense that we did not model any periodicity or exogenous variables. Moreover, the mean daily pattern is computed *after* the clustering has been performed and serves only for visualization and clustering interpretation purposes. Figure 5 shows the mean daily patterns for the 245 time series using the multiway spectral clustering with parameters $k = 7, \sigma = 200$ obtained through model selection with the BLF. The obtained clusters also display typical customer profile information. Namely, cluster 1 suggests *residential* behavior with morning and evening peaks. Cluster 5 characterizes low consumption during the day increasing during the night. Clusters 2, 6 and 7 suggest *commercial* or *business* profiles while cluster 3 and 4 display load behavior with no particular associated profile.

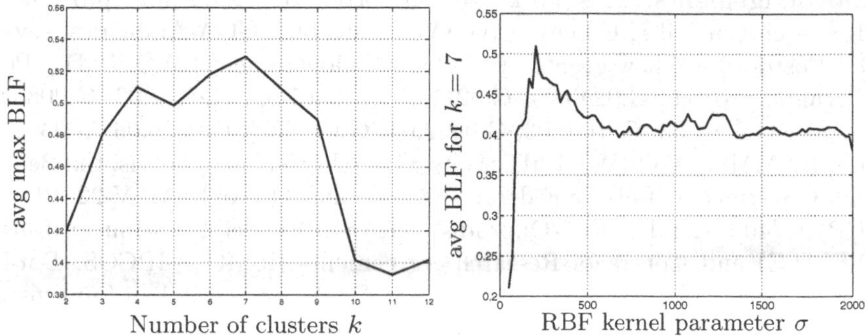

Fig. 4. Model selection using the BLF. **Left:** Average maximal BLF to determine the number of clusters k. **Right:** Average BLF for $k = 7$ to determine the optimal value of the RBF kernel parameter σ. Tuned parameters are $k = 7$ and $\sigma = 200$. The plots show average BLF values after 10 randomizations of the training and validation sets.

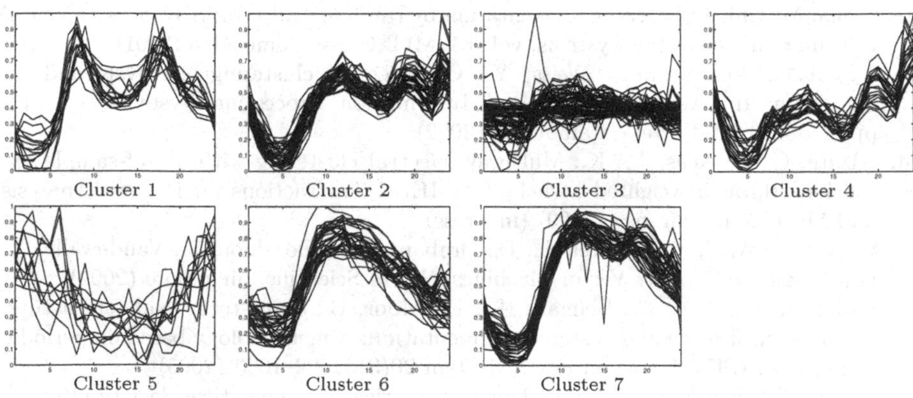

Fig. 5. Clustering results using the proposed methodology. The results show typical customer profiles while clustering the data directly without pre-processing. The x-axis corresponds to the hour of the day while the y-axis is the normalized load.

7 Conclusions

We have presented an application of the multiway spectral clustering method with out-of-sample extensions towards grouping types of customers in power load time series. The proposed scheme can be applied directly to the raw data without the need of pre-modeling steps. The parameters of the clustering model are obtained in a learning framework using training and validation scenarios. The experimental results show the applicability of the proposed methodology using a real-life high-dimensional dataset compared to an existing approach using periodic autoregressions.

Acknowledgements. This work was supported by grants and projects for the Research Council K.U.Leuven (GOA-Mefisto 666, GOA-Ambiorics, several PhD / Postdocs & fellow grants), the Flemish Government FWO: PhD / Postdocs grants, projects G.0240.99, G.0211.05, G.0407.02, G.0197.02, G.0080.01, G.0141.03, G.0491.03, G.0120.03, G.0452.04, G.0499.04, G.0226.06, G.0302.07, ICCoS, ANMMM; AWI;IWT:PhD grants, GBOU (McKnow) Soft4s, the Belgian Federal Government (Belgian Federal Science Policy Office: IUAP V-22; PODO-II (CP/01/40), the EU(FP5-Quprodis, ERNSI, Eureka 2063-Impact;Eureka 2419-FLiTE) and Contracts Research/Agreements (ISMC / IPCOS, Data4s, TML,Elia, LMS, IPCOS, Mastercard). Bart De Moor and Johan Suykens are professors at the K.U.Leuven, Belgium. The scientific responsibility is assumed by its authors.

References

1. Shi, J., Malik, J.: Normalized cuts and image segmentation. IEEE Transactions on Pattern Analysis and Machine Intelligence 22(8), 888–905 (2000)
2. Meila, M., Shi, J.: Learning segmentation by random walks. In: Advances in Neural Information Processing Systems, vol. 13. MIT Press, Cambridge (2001)
3. Ng, A.Y., Jordan, M.I., Weiss, Y.: On spectral clustering: Analysis and an algorithm. In: Advances in Neural Information Processing Systems, vol. 14, pp. 849–856. MIT Press, Cambridge (2002)
4. Alzate, C., Suykens, J.A.K.: Multiway spectral clustering with out-of-sample extensions through weighted kernel PCA. IEEE Transactions on Pattern Analysis and Machine Intelligence (2009) (in Press)
5. Suykens, J.A.K., Van Gestel, T., De Brabanter, J., De Moor, B., Vandewalle, J.: Least Squares Support Vector Machines. World Scientific, Singapore (2002)
6. Espinoza, M., Joye, C., Belmans, R., De Moor, B.: Short-term load forecasting, profile identification and customer segmentation: A methodology based on periodic time series. IEEE Trans. on Power System 20(3), 1622–1630 (2005)
7. Chung, F.R.K.: Spectral Graph Theory. American Mathematical Society (1997)
8. Fiedler, M.: A property of eigenvectors of nonnegative symmetric matrices and its applications to graph theory. Czech. Math. J. 25(100), 619–633 (1975)
9. Gu, M., Zha, H., Ding, C., He, X., Simon, H.: Spectral relaxation models and structure analysis for k-way graph clustering and bi-clustering. Tech. report, Penn. State Univ, Computer Science and Engineering (2001)
10. Fan, K.: On a theorem of Weyl concerning eigenvalues of linear transformations. Proc. Natl. Acad. Sci., USA 35(11), 652–655 (1949)
11. Bach, F.R., Jordan, M.I.: Learning spectral clustering. In: Advances in Neural Information Processing Systems, vol. 16. MIT Press, Cambridge (2004)
12. Meila, M., Shi, J.: A random walks view of spectral segmentation. In: Artificial Intelligence and Statistics AISTATS (2001)
13. Horn, R.A., Johnson, C.R.: Matrix Analysis. Cambridge University Press, Cambridge (1990)
14. Fowlkes, C., Belongie, S., Chung, F., Malik, J.: Spectral grouping using the Nyström method. IEEE Transactions on Pattern Analysis and Machine Intelligence 26(2), 214–225 (2004)

Transformation from Complex Networks to Time Series Using Classical Multidimensional Scaling

Yuta Haraguchi[1,2], Yutaka Shimada[1,3], Tohru Ikeguchi[1,4,7], and Kazuyuki Aihara[5,6,7]

[1] Graduate school of Science and Engineering, Saitama University, 255 Shimo-Ohkubo Saitama 338–8570, Japan
[2] haraguchi@nls.ics.saitama-u.ac.jp
[3] sima@nls.ics.saitama-u.ac.jp
[4] tohru@ics.saitama-u.ac.jp
[5] Graduate School of Information Science and Technology, The University of Tokyo, 4-6-1 Komaba, Meguro-ku Tokyo, 153-8505, Japan
[6] aihara@sat.t.u-tokyo.ac.jp
[7] Aihara Complexity Modelling Project, ERATO, JST, 4-6-1 Komaba, Meguro-ku, Tokyo, 153-8505, Japan

Abstract. Various complex phenomena exist in the real world. Then, many methods have already been proposed to analyze the complex phenomena. Recently, novel methods have been proposed to analyze the deterministic nonlinear, possibly chaotic, dynamics using the complex network theory [1, 2, 3]. These methods evaluate the chaotic dynamics by transforming an attractor of nonlinear dynamical systems to a network. In this paper, we investigate the opposite direction: we transform complex networks to a time series. To realize the transformation from complex networks to time series, we use the classical multidimensional scaling. To justify the proposed method, we reconstruct networks from the time series and compare the reconstructed network with its original network. We confirm that the time series transformed from the networks by the proposed method completely preserves the adjacency information of the networks. Then, we applied the proposed method to a mathematical model of the small-world network (the WS model). The results show that the regular network in the WS model is transformed to a periodic time series, and the random network in the WS model is transformed to a random time series. The small-world network in the WS model is transformed to a noisy periodic time series. We also applied the proposed method to the real networks - the power grid network and the neural network of C. elegans - which are recognized to have small-world property. The results indicate that these two real networks could be characterized by a hidden property that the WS model cannot reproduce.

1 Introduction

Various networks exist in the real world, for example, the Internet, WWW, neural networks, power grid networks, and so on. These real networks often have

C. Alippi et al. (Eds.): ICANN 2009, Part II, LNCS 5769, pp. 325–334, 2009.

complex structures. The complex network theory has emerged in 1998 as a new theory to analyze such complex structures. The complex networks widely distribute in several fields such as biology, sociology, physics, and so on. In the last decade, researches on the real networks have been drastically advanced and many methods to analyze the complex networks have been proposed. The complex network theory clarifies that complicated network structures in the real world could be described by a common and hidden rule [4, 5, 6].

On the other hand, to analyze complex phenomena in the real world, for example, air temperature, brain wave, heartbeat, and so on, the nonlinear dynamical system theory has been used. The nonlinear dynamical system theory has an impact that a low dimensional nonlinear dynamical system could produce complicated behavior. Even if the deterministic nonlinear dynamical systems produce the complex phenomena, the time series observed from the complex phenomena can be analyzed by using time series analysis method based on the nonlinear dynamical system theory.

Recently, novel methods have been proposed to analyze the nonlinear dynamical systems using the complex network theory [1, 2, 3]. In Refs. [1, 2], networks are constructed from attractors of the nonlinear dynamical systems and an observed time series. The networks are analyzed by the methods of the complex network theory. In Refs. [1, 2], it is clarified that the networks transformed from chaotic time series have several characteristic features, such as the small-world property [1] and the fit-get-rich property [2]. These features in the transformed networks reflect important properties of the chaotic dynamical systems, for example, orbital instability, self-similarity, and stretching and folding mechanism. In this sense, to analyze the networks through the nonlinear dynamical system theory is an interesting field which expresses the nonlinear dynamical system from different viewpoints.

In this paper, we take an opposite direction: we transform a network to a time series. We realize the transformation of the network to the time series by the classical multidimensional scaling [7]. To show that the proposed method is an invertible transformation, we reconstruct the original network from the time series transformed from the networks. The time series transformed from the networks by the proposed method can completely preserve the adjacency relationship between nodes in the networks. Then, we apply the proposed method to the famous Watts-Strogatz model that can produce the small-world network. In addition, we investigate the properties of the time series generated from real networks. As a result, the time series transformed from different real networks show different spectral structures, although the original networks of the time series have the same property, called small-world property [6] in the complex network theory. These results indicate that the proposed method reveals hidden structure of the complex networks which cannot be clarified yet.

2 Transformation from Networks to Time Series

2.1 Classical Multidimensional Scaling

We use the classical multidimensional scaling (CMDS) to transform a network to a time series. Multidimensional scaling (MDS) is a set of mathematical techniques. It enables us to generate representation of objects which preserves given dissimilarities between the objects. The CMDS is one of the MDS. The CMDS might find multi-dimensional coordinates for many points which preserve distance information between any two points in Euclidean space. Let the ith point in an m-dimensional Euclidean space be $\boldsymbol{x}_i$ $(i = 1, \ldots, n)$ where $\boldsymbol{x}_i = (x_{i1}, x_{i2}, \ldots, x_{im})^\mathsf{T}$. Then, the coordinate matrix X is represented as follows:

$$X = \{\boldsymbol{x}_1, \boldsymbol{x}_2, \ldots, \boldsymbol{x}_n\}^\mathsf{T}. \tag{1}$$

The distance between the ith point and the jth point is given by

$$d_{ij}^2 = |\boldsymbol{x}_i - \boldsymbol{x}_j|^2 = (\boldsymbol{x}_i - \boldsymbol{x}_j)^\mathsf{T}(\boldsymbol{x}_i - \boldsymbol{x}_j) = |\boldsymbol{x}_i|^2 + |\boldsymbol{x}_j|^2 - 2\boldsymbol{x}_i^\mathsf{T}\boldsymbol{x}_j. \tag{2}$$

Then, the matrix $D^{(2)} = \{d_{ij}^2\}$ can be represented as the following equation:

$$D^{(2)} = \mathrm{diag}(XX^\mathsf{T})\mathbf{1}_n\mathbf{1}_n^\mathsf{T} + \mathbf{1}_n\mathbf{1}_n^\mathsf{T}\mathrm{diag}(XX^\mathsf{T}) - 2XX^\mathsf{T}, \tag{3}$$

where $\mathbf{1}_n = (1, \ldots, 1)^\mathsf{T}$ is a vector of n ones and $\mathrm{diag}(XX^\mathsf{T})$ is a square matrix in which the elements are all zero except the diagonal element. Even if we only have distance information between the points, we can obtain the coordinate matrix X by using Eq. (3) and a centering matrix J. Let an inner product matrix be $P = \{p_{ij}\}$, where $p_{ij} = \boldsymbol{x}_i^\mathsf{T}\boldsymbol{x}_j$. Then, P is written by the following equation:

$$P = -\frac{1}{2}JD^{(2)}J, \tag{4}$$

where the centering matrix, $J = \boldsymbol{I}_n - n^{-1}\mathbf{1}_n\mathbf{1}_n^\mathsf{T}$, where $\boldsymbol{I}_n$ is the identity matrix of size n. Using Eqs. (3) and (4), the matrix P can be rewritten by the following equation:

$$\begin{aligned} P &= -\frac{1}{2}J[\mathbf{1}_n\mathbf{1}_n^\mathsf{T}\mathrm{diag}(XX^\mathsf{T}) + \mathrm{diag}(XX^\mathsf{T})\mathbf{1}_n{\mathbf{1}_n}^\mathsf{T} - 2XX^\mathsf{T}]J \\ &= JXX^\mathsf{T}J \\ &= X_c{X_c}^\mathsf{T} \end{aligned} \tag{5}$$

where $X_c = JX$. Here, P is a symmetric and positive semi-definite matrix whose rank is m. Because P has m positive eigenvalues and $n - m$ zero eigenvalues, the spectral decomposition of the matrix P is written by

$$P = S_m\Lambda_m S_m^\mathsf{T} = (S_m\Lambda_m^{1/2})(S_m\Lambda_m^{1/2})^\mathsf{T}, \tag{6}$$

where $\Lambda_m^{1/2}$ is the diagonal matrix of the square root of eigenvalues $\{\lambda_i\}$ of P and S_m is the matrix of eigenvectors. The eigenvalues satisfy $\lambda_1 \geq \lambda_2 \geq \ldots \geq \lambda_m > 0$ and $\lambda_{m+1} = \lambda_{m+2} = \ldots = \lambda_n = 0$. Then, $\Lambda_m^{1/2}$ is given by

$$\Lambda_m^{1/2} = \begin{pmatrix} \sqrt{\lambda_1} & & & O \\ & \sqrt{\lambda_2} & & \\ & & \ddots & \\ O & & & \sqrt{\lambda_m} \end{pmatrix}. \tag{7}$$

From Eqs. (5) and (6), the coordinate matrix $X_c(= JX)$ is equal to $S\Lambda_m^{1/2}$.

In the proposed method, we define the distance between any two nodes in a network using the adjacency relationship among the nodes and generate the coordinate vectors from the distance matrix. Considering the coordinate vectors as a time series, we can generate the time series from the networks.

2.2 The Proposed Method

Now, our purpose is to generate a time series which preserves the adjacency relationship of nodes in a network. Thus, it is necessary to generate the distance matrix which preserves the adjacency relationship of the nodes in the network. So we generate the distance matrix $D = \{d_{ij}\}$ from the adjacency matrix $A = \{a_{ij}\}$ according to the following rules:

$$d_{ij} = \begin{cases} 0 & (i = j), \\ w & (a_{ij} = 0,\ i \neq j), \\ 1 & (a_{ij} = 1), \end{cases} \tag{8}$$

where $w(> 1)$ is the weight between two disconnected nodes. Equation (8) comes from a simple idea that the distances between disconnected nodes should be larger than the distances between connected nodes. The distance of Eq. (8) satisfies the distance axiom. Applying the CMDS to the distance matrix D produced by the above procedure, CMDS arranges the nodes of the network in such a way that their adjacency relationship is preserved. Then, we can obtain the coordinate vectors which completely preserve the distance relationship between each node (Eq. (8)). Here, we now only consider undirected and unweighted networks, then the distance matrix D is symmetric. Namely, the existence of the solution of CMDS is guaranteed, then we obtain non-negative eigenvalues of the matrix D. Although the coordinate obtained by using CMDS has a rotational freedom around the origin, it does not affect the construction process of the time series from the network. Even if the coordinate vectors rotate, the adjacency information is preserved. If the coordinate vectors preserve the adjacency relationship, or the structural feature of the network, then the coordinate vectors represent the structural properties of the network. Thus, considering the coordinate vectors as a time series and analyzing the time series by the method of the nonlinear time series analysis, we can evaluate the networks through the nonlinear time series analysis.

3 Experiments

We apply our proposed method to a mathematical model and real networks. We introduced the WS model [6] that can reproduce small-world property as the mathematical model. We used the power grid network and the neural network of C. elegans [6] as the real network.

According to the procedure described in Ref. [6], we start from the regular network which has 1,000 nodes and 100 degree. Then, we rewire all the edges with the rewiring probability $p = 0.0$, 0.1, and 1.0. We transform these networks to time series and calculate their power spectra. Here, we use the eigenvector of the maximum (first) eigenvalue as the time series because the first eigenvector most represent the adjacency relationship of the network.

In addition, to check the validity of the proposed method, we re-transform the transformed time series by the proposed method to a network. We take the following procedure:

1. Set $R(i,j) = 0$ $(1 \leq i \leq N,\ 1 \leq j \leq N)$ where R represents the adjacency matrix of the reconstructed network and N is the number of nodes in the network.
2. Calculate m-dimensional coordinate values $\boldsymbol{x}_i$ from the transformed time series $s_1(t), s_2(t), \ldots, s_m(t)$ according to the following equation:

$$\boldsymbol{x}_i = (s_1(i), s_2(i), \ldots, s_m(i)). \tag{9}$$

3. Calculate the distances between two points $d_{ij} = |\boldsymbol{x}_i - \boldsymbol{x}_j|$ for all i and j. If the distance $d_{ij} < \theta$, set $R(i,j) = 1$. We decide the threshold θ which satisfies the condition that the reconstructed network has as many edges as the original network.

After the above-mentioned procedure, we compare R with the adjacency matrix of the original network O and evaluate how the proposed method preserves the adjacency information by the following value:

$$C(O,R) = \frac{|K(O) \cap K(R)|}{|K(O)|}, \tag{10}$$

where $K(A)$ is a set of elements which satisfies $a_{ij} = 1$ in the adjacency matrix $A = \{a_{ij}\}$, and is defined by Eq. (11):

$$K(A) = \{i,j|\ a_{ij} = 1,\ 1 \leq i \leq n,\ 1 \leq j \leq n\}. \tag{11}$$

In Eq. (10), $|K(O)|$ represents the number of elements in $K(O)$, and $|K(O) \cap K(R)|$ represents the number of elements in $K(O) \cap K(R)$. If $C(O,R) = 1$, R is perfectly the same as O, namely, our method completely preserves the adjacency information of the original network in the transformed time series.

4 Results

4.1 Network Reconstruction from Transformed Time Series

At first, we show that the original network can be reconstructed from the time series generated by the proposed method. From Fig. 1, the value of $C(O, R)$ is close to 0 if $w = 1$ because it is impossible to distinguish the adjacent nodes and the nonadjacent nodes. On the other hand, $C(O, R)$ becomes slightly lower than unity if w increases. The reason is that the points which satisfy the distance relationship given by the matrix D cannot be arranged in Euclidean space. For example, assuming the case of three nodes as shown in Fig. 2, when $w > 2$, we cannot arrange the points with $d_{12} = w$, $d_{13} = 1$, and $d_{23} = 1$ in Euclidean space. This simple example clearly explains that we cannot take a large value of w.

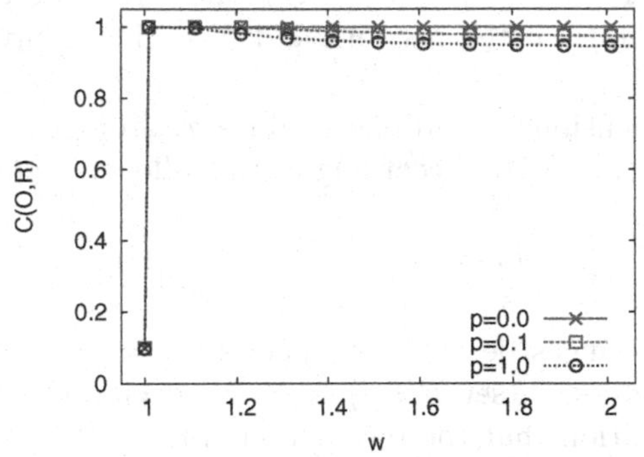

(a) The WS model: $p = 0.0, 0.1$, and 1.0

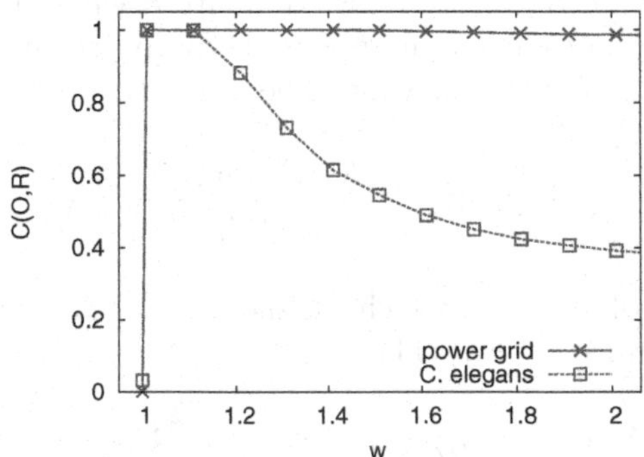

(b) Real networks: power grid and C. elegans

Fig. 1. Relation between the weight w and $C(O, R)$ between the original network and the reconstructed network. The result of (a) the WS model with $p = 0.0$ ($\times$), $p = 0.1$ (), and $p = 1.0$ () and (b) the power grid ($\times$) and the C. elegans ().

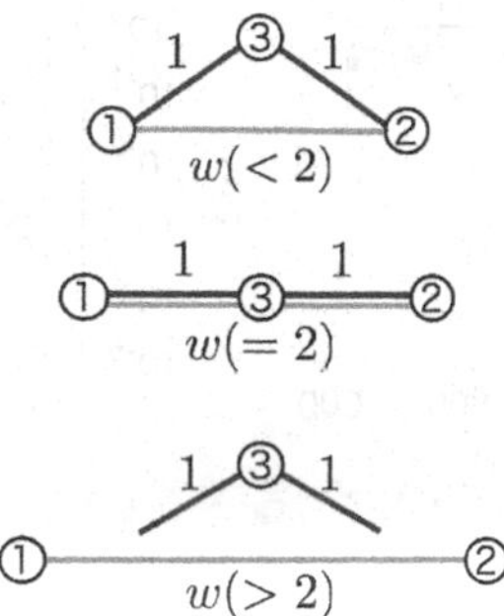

Fig. 2. The example for the case that the coordinate which preserve the distance relationship does not exist in Euclidean space

The results indicate that depending on the network structure, the weight value w is so constrained that we cannot get the coordinate which completely preserve the adjacency information of D. When $1 < w \leq 1.01$, for all the networks, the value of $C(O, R)$ takes 1. Namely, we can obtain the time series which completely preserve the adjacency relationship of the networks with the value w selected in the range $1 < w \leq 1.01$. In the following analyses, we fix $w = 1.01$.

4.2 Time Series Generated by the Proposed Method

The time series transformed from the WS model and its power spectrum are shown in Figs. 3, 4, and 5. Figures 6 and 7 are the results of real networks (power grid and C. elegans) for the same experiments. From Figs. 3 and 5, our method transforms the regular network ($p = 0.0$) and the random network ($p = 1.0$) in the WS model to periodic and random time series. In addition, the time series generated from the small-world network ($p = 0.1$) exhibits a noisy periodic time series. These results indicate that randomness in the WS model corresponds to the noise intensity of the transformed time series.

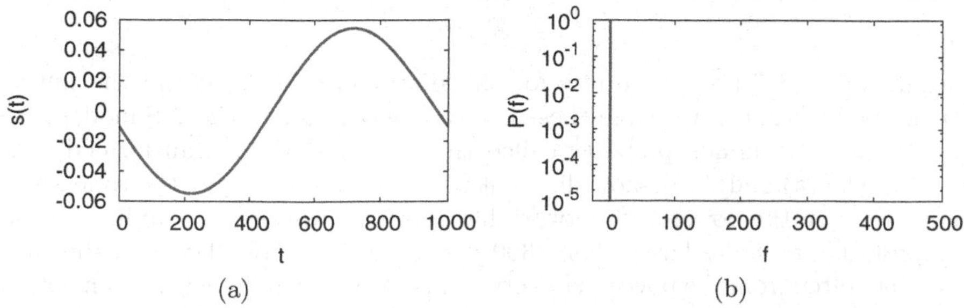

Fig. 3. (a) Time series of the WS model ($p = 0.0$) and (b) its power spectrum

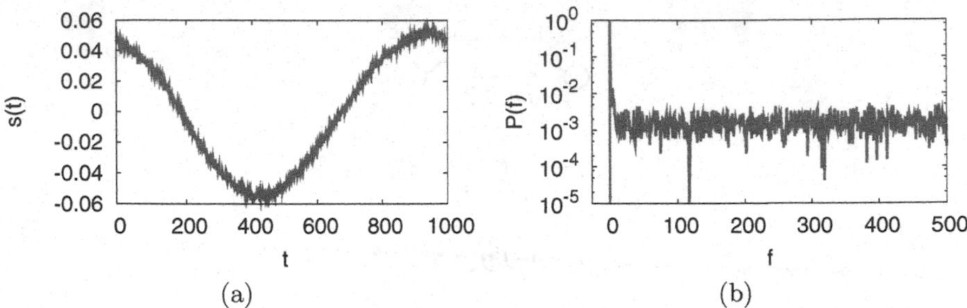

Fig. 4. (a) Time series of the WS model ($p = 0.1$) and (b) its power spectrum

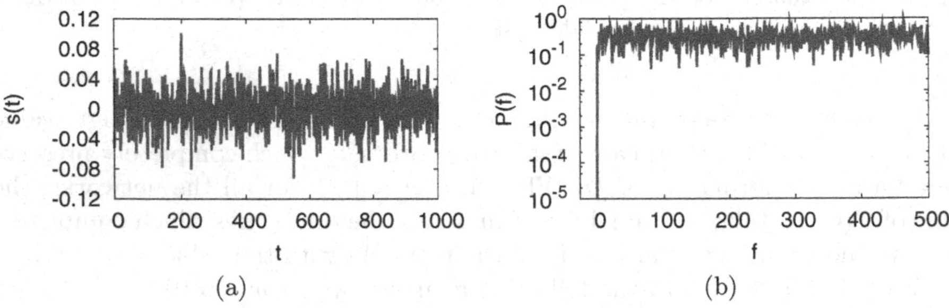

Fig. 5. (a) Time series of the WS model ($p = 1.0$) and (b) its power spectrum

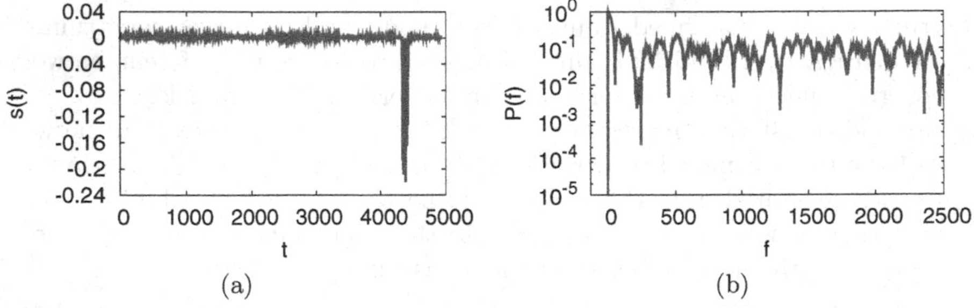

Fig. 6. (a) Time series of the power grid network and (b) its power spectrum

Figures 6 and 7 are the results for the power grid network and the neural network of C. elegans. Although these two real networks and the WS model with $p = 0.1$ have a common property called the small-world [6], transformed time series (Figs. 6(a) and 7(a)) show different property. For example, the time series generated from the power grid network has almost always small amplitudes but the amplitude becomes large when $4300 < t < 4500$. The oscillation of the time series generated from the power grid network is obviously different from the other time series for the small-world network, the WS model ($p = 0.1$) and the neural network of C.elegans. The time series generated from the neural network of C.

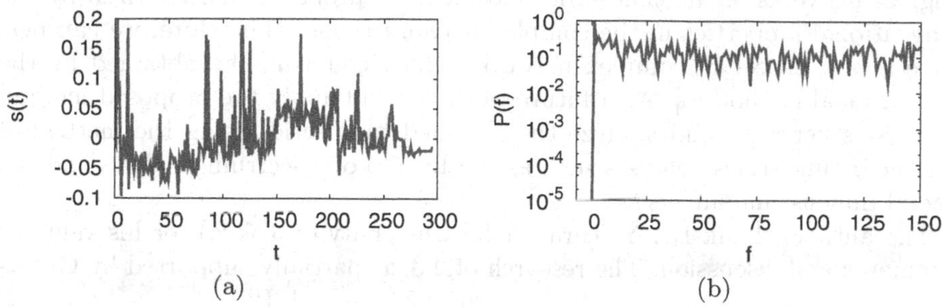

Fig. 7. (a) Time series of the neural network of C. elengans and (b) its power spectrum

elegans exhibits noisy periodic time series like the time series for the WS model ($p = 0.1$). However, the noise in the time series transformed from the neural network of C. elegans is not uniform and the noise intensity of the time series is larger than that for the WS model. Thus, the time series transformed from the neural network of C. elegans exhibits different property from that of the WS model. From these results, our method can reveal a hidden complex structure that the real networks have. The proposed method evaluates complexity in the network from different viewpoints.

5 Conclusions

In this paper, we proposed a method for transforming a network to a time series. In the proposed method, we decided the distances between any two nodes in the network using the information of adjacency relationship of the nodes. Applying classical multidimensional scaling to the distance matrix of the network, the coordinate vectors were generated. We analyzed the coordinate vectors as the time series. In numerical simulations, we generated the time series from the networks of the WS model and two real networks: the power grid network and the neural network of C. elegans. As a result, a periodic time series and a random time series were generated from the regular network and the random network of the WS model, respectively. On the other hand, the small-world network of the WS model was transformed to the noisy periodic time series. From the result for the WS model, noise intensity of the time series transformed from the networks depends on the rewiring probability p. It can be said that the time series transformed by the proposed method reflects the randomness of the networks. In addition, we reconstructed the network from the time series transformed from an original networks and compared the reconstructed network with the original networks. As a result, we can make the time series which preserves the adjacency information of the original networks by the proposed method because the original networks and the time series transformed from the networks are invertible. From these results, the time series transformed from the real networks by the proposed method shows different spectral structure, although the original networks of the time series have a same property characterized by the conventional properties in the complex network theory. Therefore, we can perceive a new aspect of complex networks which could not be obtained by the conventional method [6]. As a future work, we will apply the proposed method to various networks and analyze the obtained time series using the method of nonlinear time series analyses such as the Lyapunov spectrum analysis and the fractal dimensional analysis.

The authors thank Dr. Y. Hirata (the University of Tokyo) for his valuable comments and discussion. The research of T.I. is partially supported by Grant-in-Aid for Exploratory Research (No. 20650032) from JSPS.

References

[1] Shimada, Y., Kimura, T., Ikeguchi, T.: Analysis method of chaotic dynamics using measures of complex networks. In: Kůrková, V., Neruda, R., Koutník, J. (eds.) ICANN 2008, Part I. LNCS, vol. 5163, pp. 61–70. Springer, Heidelberg (2008)
[2] Shimada, Y., Ikeguchi, T.: Analysis of chaotic attractors by measures of complex networks. In: The International Symposium on Nonlinear Theory and its Applications, pp. 140–143 (2008)
[3] Zhang, J., Small, M.: Complex network from pseudoperiodic time series: Topology versus dynamics. Phys. Rev. Lett. 96, 238701 (2008)
[4] Newman, M.E.J.: The structure and function of complex networks. SIAM Review 45, 167–256 (2003)
[5] Costa, L.D.F., Rodrigues, F.A., Travieso, G., Boas, P.R.V.: Characterization of complex networks: A survey of measurements. Advances in Physics 56, 167–242 (2007)
[6] Watts, D., Strogatz, S.: Collective dynamics of 'small-world' networks. Nature 393, 440–442 (1998)
[7] Hirata, Y., Horai, S., Aihara, K.: Reproduction of distance matrices and original time series from recurrence plots and their applications. Eur. Phys. J. Special Topics 164, 13–22 (2008)

Predicting the Occupancy of the HF Amateur Service with Neural Network Ensembles

Harris Papadopoulos and Haris Haralambous

Computer Science and Engineering Department, Frederick University,
7 Y. Frederickou St., Palouriotisa, Nicosia 1036, Cyprus
{h.papadopoulos,h.haralambous}@frederick.ac.cy

Abstract. The Amateur Service is allocated approximately 3 MHz of spectrum in the HF band (3-30MHz) which is primarily used for long range communications via the ionosphere. However only a fraction of this resource is usually available due to unfavourable propagation conditions in the ionosphere imposed by solar activity on the HF channel. In this respect interference is considered a significant problem to overcome, in order to establish viable links at low transmission power. This paper presents the development of a set of Neural Network ensembles which can serve as a tool for predicting the likelihood of interference in the frequency allocations utilized by amateur users. The proposed approach successfully captures the temporal and long-term solar dependent variability of congestion, formally defined as the fraction of channels within a certain frequency allocation with signals exceeding a given threshold.

1 Introduction

The importance of the amateur radio service is recognized worldwide as it serves as a significant spectrum resource at an international level. In particular, during major disasters, public protection, humanitarian and disaster relief operations amateur radio frequencies have supported auxiliary or emergency communications when communications infrastructures have suffered considerable damage. The number and variety of transmission modes used by radio amateurs are significantly expanding, creating internal pressures within the amateur services for their accommodation at the expense of users of established modes such as single-sideband telephony. These new modes include digital voice, data and image. Their use improves the efficiency of amateur operations, but also increases the popularity of Amateur Radio and therefore the level of interference which gives rise to increased spectral congestion. This challenge is in addition to the high variability of the HF channel communication properties which vary on a 24-hour, seasonal, and long-term time scales reducing the useful frequency range of the HF spectrum. It is also subjected to extreme solar driven ionospheric

C. Alippi et al. (Eds.): ICANN 2009, Part II, LNCS 5769, pp. 335–344, 2009.

Table 1. Amateur user allocation frequency ranges

Code	Frequency Range (MHz)
2	1.810 - 1.850
26	7.000 - 7.100
50	14.000 - 14.350
62	18.068 - 18.168
71	21.000 - 21.450
82	24.890 - 25.000
92	28.000 - 28.500
93	28.500 - 29.000
94	29.000 - 29.700

phenomena leading to a variety of propagation impairments such as high attenuation, multipath, doppler spread and deep fading. Under these circumstances a measure of the level of interference experienced by an amateur user would be an important decision support tool that could serve as guide for adjustments of appropriate communication parameters like transmission frequency and power.

This paper presents the development of an ensemble (committee) of Neural Networks (NNs) for each amateur frequency allocation, see table 1, which can predict the likelihood of spectral congestion in that allocation. This study is a continuation of our previous work [1,2,3], in which single NNs were shown to be very successful in modelling the variation of congestion. In this paper we focus on a different user type (Amateur users) and extend our original approach to ensembles of NNs in an effort to obtain more accurate predictions.

Neural Network ensemble is a learning approach where a number of NNs are trained on the same task and their predictions are then combined to form the final prediction of the system. This approach leads to the significant improvement of the generalisation ability of a NN system [4] and has been successfully applied to many different areas such as medical diagnosis [5], face recognition [6], and optical character recognition [7].

The interference measurements used for the development of the ensembles are part of a dataset recorded at Linkoping (Sweden) in the frames of a long-term project being undertaken jointly by the University of Manchester and by the Swedish Defence Research Establishment, to measure systematically and to analyse the occupancy of the entire HF spectrum.

The rest of this paper is structured as follows. Section 2 describes the procedure followed for the measurement of the congestion values and Section 3 discusses the special characteristics that Amateur bands exhibit with respect to the measured parameter. Section 4 first details the parameters that were used as inputs to the ensembles, while it then describes their development. Finally, Section 5 presents the procedure followed in the experiments and the obtained results, while Section 6 gives the concluding remarks of the paper.

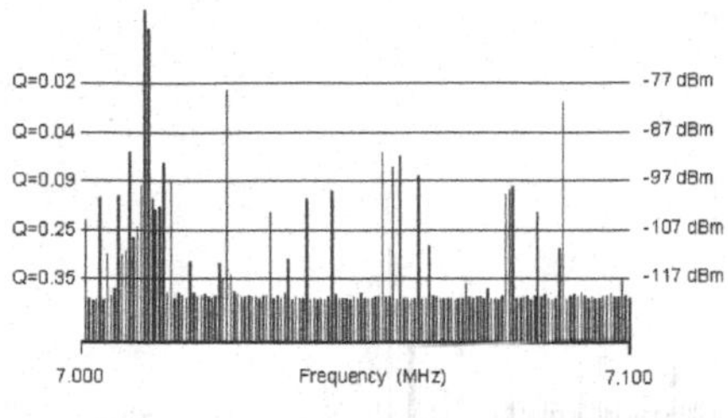

Fig. 1. Congestion measurement in an amateur user allocation

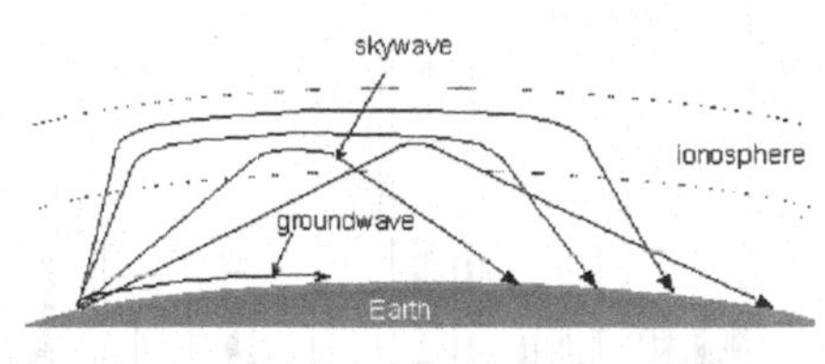

Fig. 2. Communication via the ionosphere

2 Measurement of HF Spectral Occupancy

The measure of occupancy used is congestion (Q), which is defined to be the probability that the RMS value of the output signal produced by a bandpass filter of a given bandwidth, placed at random in a given ITU defined frequency allocation, exceeds a predefined threshold level [8,9,10]. For the purpose of taking occupancy measurements the HF spectrum was divided into 95 allocations, which are shared by twelve different types of user [9]. The dataset of 24-hour occupancy measurements used for the model development was recorded over a period of six years (April 1994 to January 2000). 24-hour measurements of occupancy were obtained once a week by stepping a filter of 1 kHz bandwidth through each of the 95 ITU user defined allocations, spending 90ms at each increment. The fraction of the allocation spectral width for which the RMS signal level at each step exceeded a defined field-strength threshold level was determined (see Figure 1). A value of zero represents an empty band and a value of one a fully congested band at that particular threshold level. This defines the congestion (Q) for that allocation, for the corresponding threshold level [8,9]. A single congestion value represents an ITU frequency user allocation occupancy; thus the same congestion level will apply to contiguous 1 kHz channels within an allocation. For all 95 user allocations, a complete measurement of the HF spectrum resulted in 95(allocations) $\times$ 5(thresholds) $\times$ 24(hours) = 11400 experimental congestion values, which constitutes a complete 24-hour measurement. However in this paper only allocations that are dedicated to Amateur users (see Table 1) are considered in the modelling process so the congestion values per measurement taken into account are 1080. A total of 197 measurement sessions were carried out corresponding to a total of 23640 congestion values for each Amateur allocation.

3 HF Channel Propagation and Amateur User Occupancy Characteristics

The ionosphere is the medium which supports long-distance communication in the HF spectrum. By exploiting the ability of the ionosphere to reflect HF radio

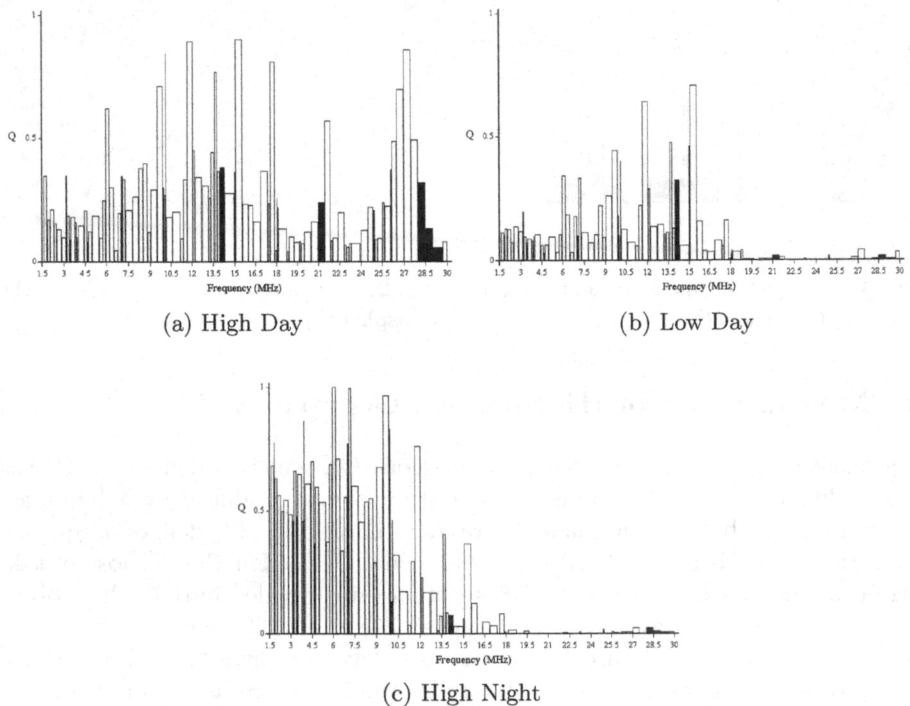

(a) High Day (b) Low Day

(c) High Night

Fig. 3. Congestion of amateur user allocations in the HF spectrum

waves (skywaves) radiating upward at some angle from the antenna we can establish long-distance communication links at relatively low cost (figure 2). The ionosphere is defined as a region of the earth's upper atmosphere where sufficient ionisation can exist to affect the propagation of radio waves in the frequency range 1 to 30 MHz. It ranges in height above the surface of the earth from approximately 50 km to 600 km. The upper atmosphere is partially ionised, the level of ionisation at various altitudes being governed by the intensity of the solar radiation and the ionisation efficiency of the neutral gases in the atmosphere. The influence of this region on radio waves is accredited to the presence of free electrons. The density of free electrons at a given height in the ionosphere depends upon the strength of the solar ionising radiation and is therefore a function of time of day, season, geographical location and solar activity [10].

At night-time and also at low solar activity periods the usable frequencies that can be supported for ionospheric propagation are significantly lower because the ionisation of the ionosphere diminishes. As a consequence the available spectrum is limited and gives rise to overcrowded low frequency allocations. This is evident in figure 3 where the usage of amateur frequency allocations (shown in black colour) at high solar activity period (figure 3a) adjusts to the available spectrum at a low solar activity daytime period (figure 3b) and also at night-time period (figure 3c).

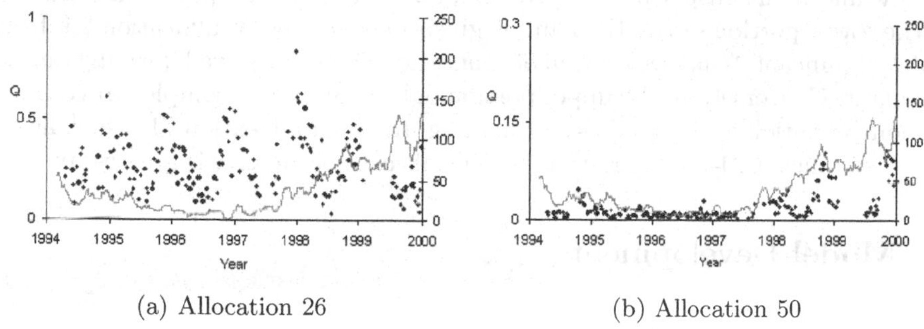

(a) Allocation 26 (b) Allocation 50

Fig. 4. Long-term and seasonal variation of congestion in the lower (a) and upper (b) part of the HF spectrum

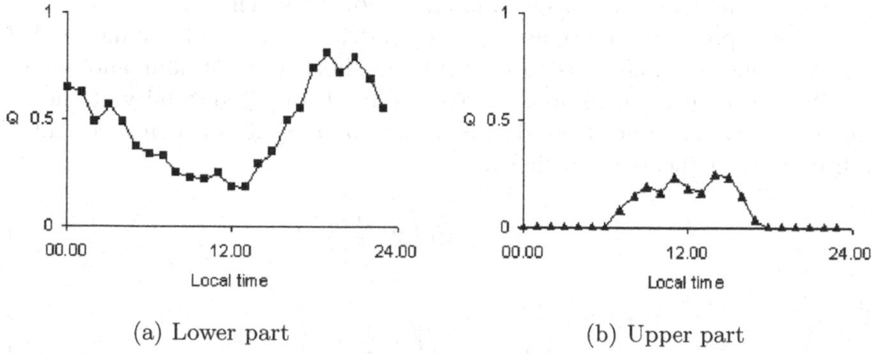

(a) Lower part (b) Upper part

Fig. 5. Typical diurnal variation of congestion in the lower (a) and upper (b) part of the HF spectrum

In figure 4 the measured congestion is plotted (with dots) for two amateur frequency allocations for a specific time together with the 50-day running mean of the daily sunspot number (an index of solar activity depicted as a continuous line in the background). This figure shows that congestion decreases for high sunspot number periods in allocation 26 (7.000 - 7.100 MHz) at the lower part of the spectrum and increases in allocation 50 (14.000 - 14.350 MHz) at the higher part of the spectrum. This is due to the fact that as solar activity increases, amateur users tend to move form lower to higher frequency allocations in an effort to take advantage of the favourable propagation conditions and avoid interference from other amateur users at the lower frequency allocations. The same figure also demonstrates the seasonal variation of congestion in each allocation. This is a consequence of the seasonal variation of the usable frequencies supported by the ionosphere, in response to seasonal change in extreme ultraviolet (EUV) radiation from the Sun [10].

Examples of typical variation of congestion with time of-day are given in figure 5 for amateur allocations in the lower and upper part of the HF spectrum for low and high sunspot activity. An example of typical occupancy encountered in the lower portion of the HF band is given in figure 5a for allocation 26, from which significant 24-hour variation of congestion can be observed, peaking during the night. Conversely, in the upper portion of the HF band a complete reversal of 24-hour variation is observed as shown in figure 5b for allocation 71, which again shows significant 24-hour variation, but in this case occupancy is greatest by day.

4 Model Development

4.1 Input Parameters

As the variation of occupancy of the HF spectrum is primarily dependant on prevailing ionospheric conditions, the input parameters used were selected so as to represent the known variations that give rise to the most characteristic properties of the ionosphere as a communications channel.

The first input parameters used correspond to the hour of the day and day of the year information so as to capture the short-term 24-hour and seasonal variability of congestion. In order to avoid unrealistic discontinuity at the midnight boundary the hour of the day, $hour, 0 \leq hour \leq 23$ was converted into its quadrature components according to:

$$sinhour = sin\left(2\pi\frac{hour}{24}\right) \tag{1}$$

and

$$coshour = cos\left(2\pi\frac{hour}{24}\right). \tag{2}$$

Similarly the day of the year $daynum, 1 \leq daynum \leq 365$ was converted to:

$$sinday = sin\left(2\pi\frac{daynum}{365}\right) \tag{3}$$

and

$$cosday = cos\left(2\pi\frac{daynum}{365}\right). \tag{4}$$

The fifth input parameter captures the solar long-term variation, as the running mean value of the daily sunspot number (R) which is a well established index of solar activity. A 50-day running mean value of the daily sunspot number ($R50$) was found to be the optimum parameter to represent the long-term variation of congestion [3]. The sixth and final parameter was the signal threshold ST that was used for the measurement.

4.2 Neural Network Ensembles

A Neural Network Ensemble was developed for each one of the 9 amateur user allocations (see table 1). The Neural Networks (NNs) composing each ensemble were feed forward networks with one input, one hidden and one output layers. Their input layer consisted of six units, one for each of the parameters described in Subsection 4.1: cosday (CD), sinday (SD), coshour (CH), sinhour (SH), 50-day running mean of the daily sunspot number ($R50$) and signal threshold (ST). Their output layer consisted of a single unit whose target output was the congestion for the corresponding input.

All units of the networks had a logistic sigmoid activation function whose outputs lie in the range [0,+1]. One advantage of this activation function is that its range of outputs coincides with the range of possible congestion values so no transformation of the network outputs was needed.

The development of the NNs was performed using the MATLAB Neural Network Toolbox [11]. The NNs were trained using the Levenberg-Marquardt backpropagation algorithm with early stopping based on a set of validation examples. This was considered as the best choice of algorithm to be used as it appears to be the fastest method for training moderate-sized feed-forward networks [11].

The diversity in the ensemble was introduced from three different sources. The first of these was the use of a fold cross validation process, as suggested in [12]. More specifically the examples available for training were split in three parts and three sets of NNs were trained, each using one of the three parts of examples as a validation set and the two remaining parts as training set. The two other sources of diversity were the variation of the number of hidden units of the NNs and the random initialization of their weights. The ensembles consisted of networks with five different numbers of hidden units {6, 9, 12, 15, 18}, while three networks with different initial weights were trained for each one. Therefore, each ensemble consisted of $3 \times 5 \times 3 = 45$ NNs in total.

The fusion of the predictions produced by each member of the ensemble was performed by computing their average. The prediction of the ensemble for a new instance x_j was

$$\hat{f}(x_j) = \frac{1}{45} \sum_{i=1}^{45} \hat{f}_i(x_j), \tag{5}$$

where $\hat{f}_i(x_j)$ is the prediction of the ith NN of the ensemble for x_j. It is worth to mention that alternative ways of combining the individual NN predictions based on their performance on the corresponding validation sets were also atempted. Their results however, were more or less the same with those obtained using (5), which was much simpler.

5 Experiments and Results

The experiments performed followed a 10-fold cross validation procedure. More specifically the 197 measurement sessions were split into 10 parts of almost equal size (7 parts of 20 sessions and 3 of 19 sessions) and the predictions for each part were obtained using an ensemble trained on the examples of the other 9 parts. Thus, the results reported here are over all 23640 congestion values for each allocation. Note that all 120 congestion values of each measurement session

Table 2. Performance comparison of the single Neural Network and Ensemble Models

Allocation	RMSE			CC	
	Single NN	Ensemble	Improvement	Single NN	Ensemble
2	0.0863	0.0825	4.4%	0.892	0.902
26	0.0762	0.0736	3.5%	0.930	0.935
50	0.0397	0.0382	3.9%	0.891	0.900
62	0.0293	0.0289	1.3%	0.520	0.534
71	0.0146	0.0141	3.3%	0.737	0.755
82	0.0117	0.0112	4.1%	0.694	0.719
92	0.0114	0.0110	3.6%	0.768	0.780
93	0.0066	0.0064	2.0%	0.757	0.760
94	0.0056	0.0054	3.7%	0.513	0.549

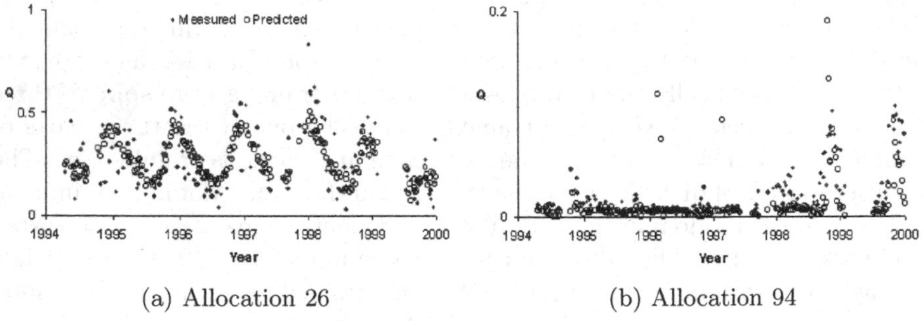

(a) Allocation 26 (b) Allocation 94

Fig. 6. Measured and predicted long term variation of congestion in the lower and upper parts of the HF spectrum

were treated as a group so that there was no temporal correlation between the examples in the training and test parts. Of course the examples were randomized before each experiment and the predictions obtained were then mapped back to their original order. Furthermore, before conducting our experiments all input parameters were normalized setting the mean value of each input to 0 and its standard deviation to 1.

For comparison reasons the same experiments were also performed with single NNs. Exactly the same type of NNs were used and the same 10-fold coross validation process was followed. In this case the validation set of each NN was formed by randomly selecting 1/10th of the training examples, while in an effort to avoid local minima the training of each NN was repeated 10 times and the trained NN that gave the best performance on the validation set was selected for application to the test examples. The choice of the number of hidden units was also based on the performance of the networks on their validation sets. NNs with the five different numbers of hidden units {6, 9, 12, 15, 18} used in the ensembles were developed and the one with the best validation set performance was selected for each allocation.

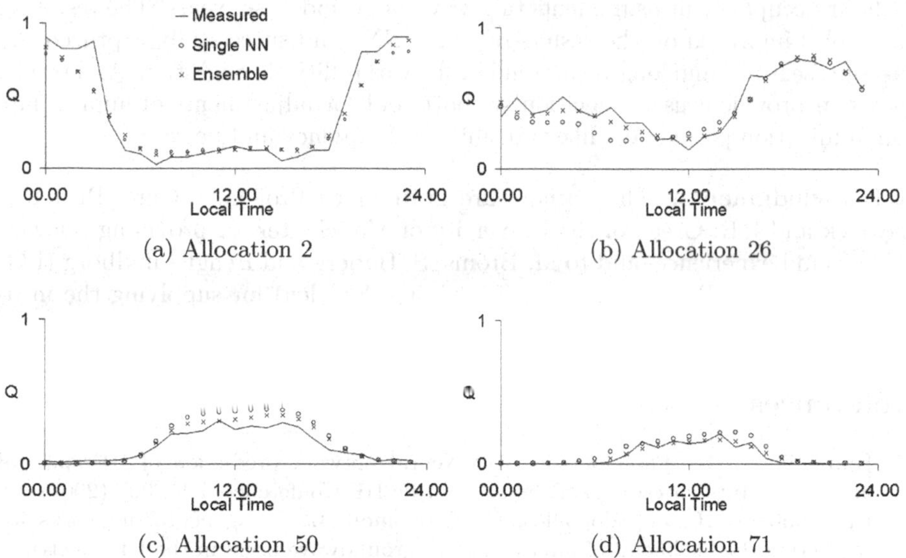

(a) Allocation 2

(b) Allocation 26

(c) Allocation 50

(d) Allocation 71

Fig. 7. Examples of 24-hour measured and predicted congestion

Table 2 presents the results of both the sigle NN and the Ensemble on each amateur allocation, in terms of their Root Mean Squared Error (RMSE) and of the correlation coefficient (CC) between their predictions and the measured congestion values. In addition, it gives the percentage of improvement in RMSE that each ensemble achieved over the corresponding single NN. The values reported in this table show that the ensembles outperform the corresponding single NNs in all 9 allocations. It is important to note that this improvement in performance, although not very big, was obtained at no extra cost. For each of the five numbers of hidden units that were tried the corresponding single NN was trained 10 times, so a total of $5 \times 10 = 50$ NNs were trained and evaluated on their validation set in order to select the one that was used to obtain the final results reported in table 2, while each ensemble consisted of only 45 NNs.

Figure 6 shows the long-term seasonal measured and predicted congestion for allocations 26 and 94 at 12:00. This figure demonstrates that the different occupancy characteristics are captured by the ensembles both in the lower and upper parts of the HF spectrum. The successful performance of the ensembles is also supported by the plots of measured and predicted 24-hour congestion values in figure 7. The improvement in performance of the ensembles over single NNs is also evident in these plots.

6 Conclusions

In this study we have developed a set of NN ensembles for the prediction of occupancy in HF Amateur allocations. The ensembles were trained on extensive

24-hour occupancy measurements taken over a period of six years. The resulting ensembles improved on the results of single NNs and successfully captured the 24-hour, seasonal and long-term trend in the variability of congestion. As a result, they can provide a useful decision support tool for adjustments of appropriate communication parameters like transmission frequency and power.

Acknowledgments. The authors are grateful to Prof G.F. Gott, Prof P.J. Laycock and P.R. Green of the University of Manchester for providing relevant background experience, and to M. Bröms, S. Boberg and Bengt Lundborg (FOI Swedish Defence Research Agency, Linköping, Sweden) for supplying the measurement data.

References

1. Haralambous, H., Papadopoulos, H.: Neural network prediction of HF spectral occupancy. In: Proceedings of the 8th Nordic HF Conference, HF 2007 (2007)
2. Haralambous, H., Papadopoulos, H., Economou, L.: Using neural networks for predicting the likelihood of interference to groundwave users in the HF spectrum. In: Proceedings of the 10th International Conference on Engineering Applications of Neural Networks (EANN 2007), pp. 200–209 (2007)
3. Haralambous, H., Papadopoulos, H.: 24-hour neural network congestion models for high-frequency broadcast users. IEEE Transactions on Broadcasting 55(1), 145–154 (2009)
4. Hansen, L.K., Salamon, P.: Neural network ensembles. IEEE Transactions on Pattern Analysis and Machine Intelligence 12(10), 993–1001 (1990)
5. Zhou, Z.H., Jiang, Y., Yang, Y.B., Chen, S.F.: Lung cancer cell identification based on artificial neural network ensembles. Artificial Intelligence in Medicine 24(1), 25–36 (2002)
6. Huang, F.J., Zhou, Z.H., Zhang, H.J., Chen, T.H.: Pose invariant face recognition. In: Proceedings of the 4th IEEE International Conference on Automatic Face and Gesture Recognition, pp. 245–250. IEEE Computer Society Press, Los Alamitos
7. Mao, J.: A case study on bagging, boosting and basic ensembles of neural networks for OCR. In: Proceedings of the 1998 IEEE International Joint Conference on Neural Networks, vol. 3, pp. 1828–1833. IEEE Computer Society Press, Los Alamitos
8. Chan, S.K., Gott, G.F., Laycock, P.J., Poole, C.R.: Hf spectral occupancy - a joint british/swedish experiment. In: Proceedings of HF 1992, Nordic Shortwave Conference (August 1992)
9. Economou, L., Haralambous, H., Green, P., Gott, G., Laycock, P., Broms, M., Boberg, S.: Aspects of hf spectral occupancy. In: Proceedings of the Eighth International Conference on hf Radio Systems and Techniques (IEE Conf. Publ. no. 474), pp. 367–372 (2000)
10. Maslin, N.: HF Communications, a Systems Approach. Pitman (1987)
11. Demuth, H., Beale, M.: Neural Network Toolbox User's Guide: For use with MATLAB. The MathWorks (1998)
12. Krogh, A., Vedelsby, J.: Neural network ensembles, cross validation, and active learning. In: Advances in Neural Information Processing Systems, pp. 231–238. MIT Press, Cambridge (1995)

An Associated-Memory-Based Stock Price Predictor

Shigeki Nagaya, Zhang Chenli, and Osamu Hasegawa

Imaging Science and Engineering Laboratory, Tokyo Institute of Technology,
4259 Nagatsuta, Midori-ku, Yokohama, 226-8503 Japan
{nagaya.s.aa,zhang.c.ad,hasegawa.o.aa}@m.titech.ac.jp
http://www.isl.tittech.ac./~hasegawalab

Abstract. We propose a novel method to predict stock price based on the Neural Associative Memory with Self-Organizing and Incremental Neural Networks (SOINN-AM). Our method has two advantages: 1) the predictor can determine its inner state space by the input training patterns automatically, 2) the predictor can modify itself by online-learning. Consequently, the predictor is more flexible for real world data than previous prediciton approaches. We demonstrate effectiveness of our approach with experiment result on real stock price data from the US and Japan market in 2002 - 2004.

Keywords: Associated-memory, Self-Organized Incremantal Neural Network (SOINN), Singal prediction.

1 Introduction

In recent years, the algorithmic trading is widely spreading the financial markets. And the stock price prediction technology is attracting more and more attention from the financial industry and the research community. Several machine learning methods, such as neural networks and reinforcement learning, have been employed to solve this problem. However, most of these approaches require complicated design for the learning model which turns out to be impractical in the ever-changing real-world market.

Neural network essentially defines a mapping M: X→Y. It is relatively straight-forward to be used to forecast market movement as long as we assign price-related information to the input and output layer. For example, we can feed the input layer with stock price of today and moving average of the last 2 weeks, and forecast the price of tomorrow at the output layer. Although some successful experimental results have been reported by using neural network, the difficulty lying here is that we usually do not know how to implement the layer structure of the network to efficiently reflect the nature of the stock price, and the best optimized network structure may vary from stock to stock, which prevented the neural network strategy from being widely applied in the real-world market.

Reinforcement learning is a kind of learning algorithm about how an agent can take action in an environment to maximize its long-term reward. During

C. Alippi et al. (Eds.): ICANN 2009, Part II, LNCS 5769, pp. 345–357, 2009.

the training process, the agent initially takes random actions and rewards will be given as a measure of how "profitable" these actions are, and the agent will optimize its action on a step-by-step basis. When the training process is over, the agent will map the states of the environment to the actions that it should take in those states.

Table 1 shows a typical learning result (inner state space) that we will likely have after employing reinforcement learning on forecasting stock price. In table 1, the state space is represented by different ranges of the stock price, and possible actions in these states are buy, sell and hold. After training the agent with historical price information, each action of each state will have a potential reward value, and in our decision making process, the agent will tend to take the action which maximizes its reward in that state. It is very obvious that strategy as simple as the example above is not profitable in the real market. But it does give us a glimpse on how reinforcement learning works in the financial market.

Although reinforcement learning is widely used in forecasting stock price in the research community, it also has its own critical flaw, that is, it is usually highly heuristic on determining the structure of the state space, and the proper state space could vary from stock to stock, which means that the user of the reinforcement learning strategy may end up struggling to design a proper state space for his portfolio. This is exactly the problem we will try to address in our proposed method.

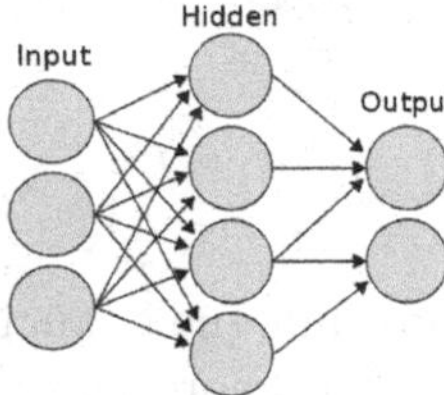

Fig. 1. Basic structure of neural network

Table 1. Example of reinforcement learning result

State(Stock Price)	Action		
	Buy	Sell	Hold
0 - 10	5	0	1
11 - 50	3	-2	2
51 - 100	1	4	2
100 <	2	3	4

2 Proposed Method

In this section, we will introduce SOINN and SOINN-AM, as they are the basis of the proposed method, and end up with details of our predictor.

2.1 Self-Organized Incremantal Neural Network (SOINN)

SOINN [1] is a two-layered neural network used for unsupervised data clustering and topology learning. It is flexible in terms of online learning while being robust against noise at the same time. Fig. 2 shows the basic structure of SOINN. There are 2 layers in the SOINN structure, both of which share very similar clustering algorithm. The input patterns are initially clustered at layer 1, and the topology

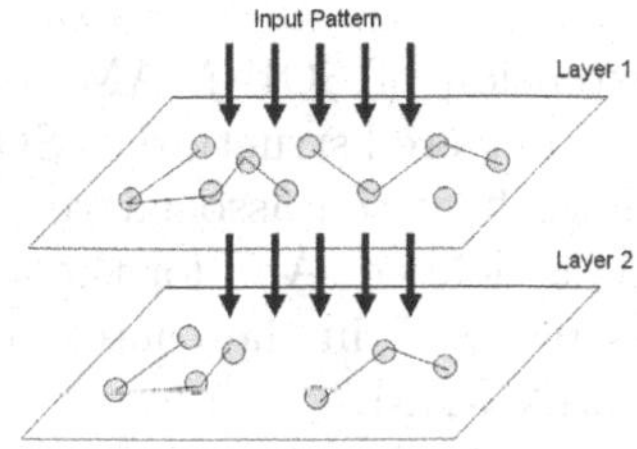

Fig. 2. Basic structure of SOINN

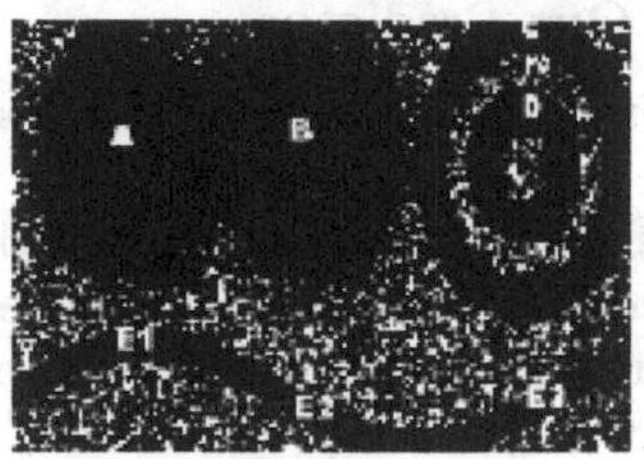

Fig. 3. Input Pattern

is then obtained at layer 2. For instance, noise-tainted patterns in Fig. 3, where there are A, B, C, D, E, 5 regions, each of which represents a unique distribution of data, are input into SOINN. The clustering result at layer 1 and the topology obtained at layer 2 are shown in Fig. 4 and Fig. 5, respectively.

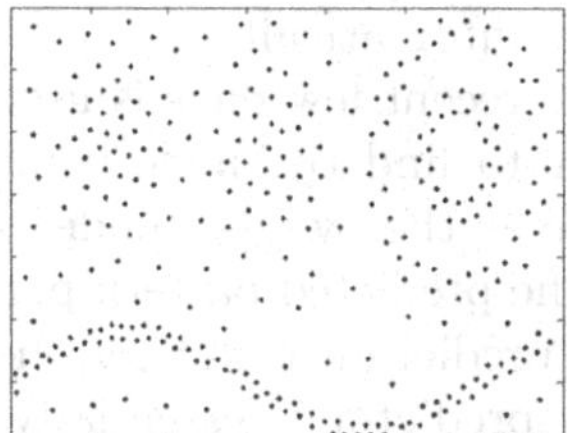

Fig. 4. Clustering Result

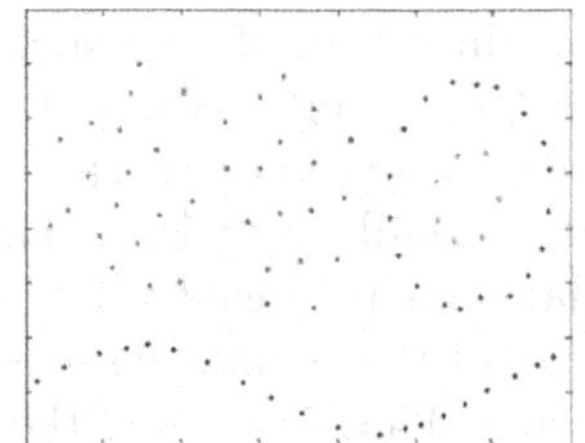

Fig. 5. Obtained Topology

2.2 Associated Memory with SOINN (SOINN-AM)

SOINN-AM [2] is an extension of SOINN to associative memory. It is also composed of 2 layers, but the first layer is only an "input layer", where the "key pattern" and the "recalled pattern" are combined as an associative pair to be later clustered by SOINN, while the second layer is called "competitive layer", which is essentially the first layer of SOINN. The basic structure of SOINN-AM is shown in Fig. 6.

There are 2 phases working sequentially in the framework of SOINN-AM. The first one is called "training phase", and the second one is called "recalling phase". In the training phase, key pattern and recalled pattern are obtained by the input layer and combined as an associative pair, which will then be rendered to the competitive layer to be clustered. The recalling phase occurs after the associative pairs have been properly clustered. The key pattern given by the user will go to the SOINN layer directly, to find the associative pair which bears the highest similarity with it. The recalled pattern part of that associative pair will be output as the answer to the given key pattern. Fig. 7 gives an example of the recalled result of SOINN-AM.

2.3 Our Predictor

Overview. Our predictor is essentially an extension of SOINN-AM to time series data. The predictor has an exactly same two-layered structure as SOINN-AM shown in Fig. 6, but it varies in the fashion how the associative pair is formed and how the similarity between patterns is defined. And for the sake of avoiding confusion, we use the term "predicted pattern" in the context of our predictor to refer to the "recalled pattern" in SOINN-AM.

In the training phase, a piece of stock prices is broken into 2 consecutive parts, the key pattern part and the predicted pattern part, and treated together as an associative pair, which will later be input into competitive layer to be clustered. At the competitive layer, associative pairs are called as "nodes", which are connected with one another by edges or not, according to their relative distances. All nodes that are directly or indirectly connected by edges form a "class" and class is represented by a prototype node, whose weight is the average weight of all of the nodes that belong to that class. Besides, the concept of the age of node and edge is introduced as our measure against noise. Nodes and edges are considered to be results of random noise if not having been updated for a certain period of time and will be removed automatically.

In the predicting phase, the stock price of the recent few days is used as our "key pattern". It goes to the competitive layer to find the node which bears the highest similarity with it and label the node as the "winner node" and the class that node belongs to the "winner class". The predicted pattern part of the prototype of the winner class is output as our prediction result for the future stock price. That is to say, the strategy of our predictor is essentially to find the "historically most possible" upcoming pattern. Note that we output the prototype of the winner class instead of outputting the winner node directly, as a bid to eliminate the influence of noises.

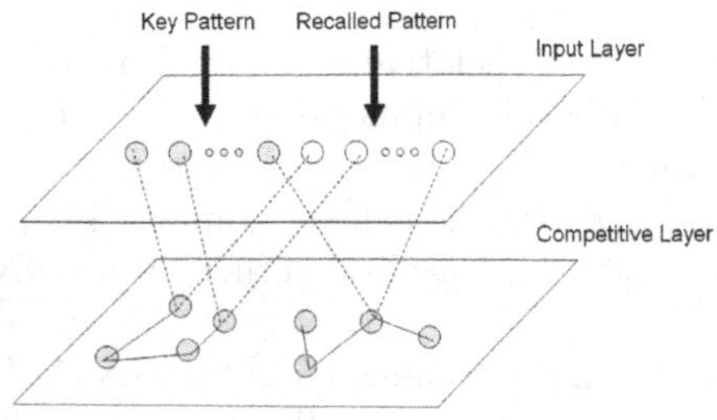

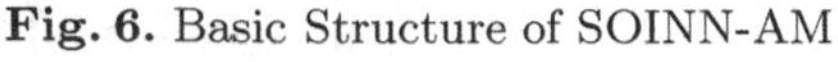

Fig. 6. Basic Structure of SOINN-AM

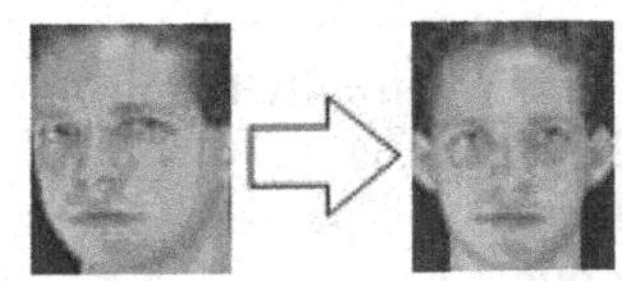

Fig. 7. Recalled result of SOINN-AM

Preparation. Notations used in the following explanation are followings. Note that each node represents one "associative pair", and each edge connects two nodes.

M	dimension of the key pattern
N	dimension of the predicted pattern
F	key pattern of the associative pair
R	predicted pattern of the associative pair
I_c	associative pair
A	set of generated nodes
W_i	weight of the i-th node
N_i	set of neighbors of the i-th node
n_p	the number of times that new patterns are presented after latest removing nodes
d_i	similarity threshold of the i-th node
Λ_{edge}	the lifetime of edges
χ_i	the number of times that the i-th node becomes the winner node
δ_r	similarity threshold at the predicting phase
λ	frequency of node removal
$\|a-b\|$	similarity between pattern $a$and pattern b

Training Algorithm (1)

1) Initialize A as an empty set, and setn_p to 0.
2) Randomly select price data of M consecutive days as the key pattern F, and the following N consecutive days as the predicted pattern R
3) Input F and R to the input layer
4) Combine F and R as $I_c = (F,R)$
5) Render I_c to the SOINN layer
6) If there are no less than 2 nodes at the SOINN layer, go to 7). Otherwise, generate a new node with weight of I_c, and set χ_i to 1, then go to 15)
7) Find the first winner node r whose weight W_r is the nearest to I_c and the second winner node q whose weight W_q is the second nearest toI_c.
8) Verify if $\|I_c - W_r\| < d_r$, $\|I_c - W_q\| < d_q$
where:

$$d_i = \begin{cases} \max\limits_{Kthnode \in N_i} \|W_i - W_k\| & (if\ N_i = \phi) \\ \min\limits_{Kthnode \in A} \|W_i - W_k\| & (if\ N_i = \phi) \end{cases}$$

If true, go to 9). Otherwise, generate a new node with the weight of I_c, and set χ_i to 1, then go to 12)
9) If there is no edge between r and q, create one between them.
10) Set the age of the newly created edge to 0.
11) Increase the age of all edges emanating from r by 1, then remove edges with ages greater than Λ_{edge}.
12) Increase χ_r by 1 and add ΔW_r and ΔW_i to the weights of r and its neighbors where

$$\Delta W_r = \frac{1}{\chi_r}(I_c - W_i)$$
$$\Delta W_i = \frac{1}{100\chi_r}(I_c - W_i) \qquad (i \in N_r)$$

13) If $n_p \| \lambda$, remove nodes with less than 1 neighbors; reset n_p to 0.
14) Go to 2) if there are remaining training data.
15) Classify the nodes.
16) Generate/Update prototype node for each cluster.
17) Training phase ends.

In step 8), we check if the winner node r and the second winner node q should belong to the same class. If the distance between I_c and W_r, and the distance between I_c andW_q are both less than a certain similarity threshold d_i, a new edge will be created between r and q, which means that they are of the same class from now on. The similarity threshold of any node is defined as the farthest distance between the node and any other neighboring nodes (nodes that belong to the same class), or the nearest distance between the node and any other nodes in case the node does not have any neighboring nodes. If the requirements of the threshold are not met, the newly input node is considered to be an independent node and it will be added to A directly.

The purpose of step 12) is to reflect the influence of the newly-input node to the current SOINN network. The winner node and the nodes in the winner class will be moved towards the newly-input node by a certain distance. In order to achieve a balance between the increment of the network and the storage of already-learnt knowledge, the distance to be moved is divided by χ_i, which is the number of times that the i-th node becomes the winner node, to ensure that node with more "winning history" is less liable to the influence of new nodes.

Training Algorithm (2): Classifying the Nodes. Here we describe how to classify the nodes, which has not yet been explained in step 15 of the training phase algorithm. The classification of nodes is conducted by the following steps:

1) Initially, none of the nodes are labeled as a member of any class.
2) Randomly select node i. Label i as a member of a new class if i has at least one neighboring node.
3) All nodes that are linked directly or indirectly with node i are labeled as members of the same class.
4) Back to 2) if there is any node with at least one neighboring node.
5) Classification ends.

Predicting Algorithm

1) Input pattern K as associative key
2) Calculate the similarity between K and each node i according to:

$$d_{ki} = \|K - W_i^F\| \qquad Where: \quad W_i = \begin{bmatrix} W_i^F \\ W_i^R \end{bmatrix}$$

3) Find the node $i = c$, which minimizes d_{ki}.
4) Find the prototype node of the class that node c belongs to.
5) The predicted pattern part of the prototype is output as the prediction result.

3 Experiment

3.1 The Prediction of the US Market

The objective of this experiment is to demonstrate the effectiveness of our predictor on the US Stock Market. The experimental result will be compared with the method proposed by H. Li, C.H. Dagli and D. Enke [9], which is an implementation of reinforcement learning.

A. Experimental Dataset. The dataset used in this experiment is collected from Yahoo! Finance, which has already been plot in Fig 10 and Fig 11. For the sake of comparison, stock indexes of S&P500 and NASDAQ from 1998 to 2002, inclusive, are used as the training dataset, and those of 2003 are used as the test dataset. As stock splits, dividends, distributions and rights offerings may happen in the real market, daily adjusted closing price is used instead of the real closing price to ensure the continuity of the experimental data.

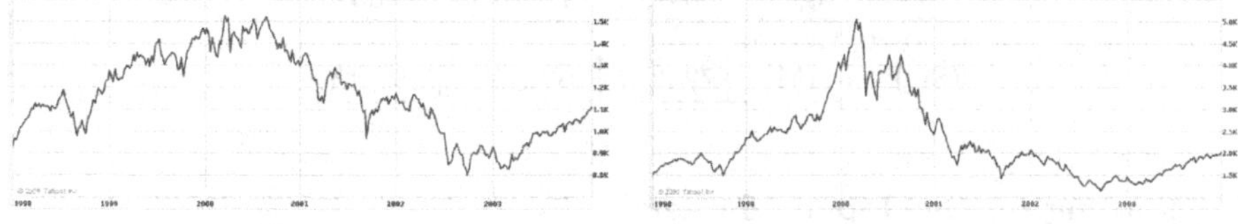

Fig. 8. Historical Price of S&P500 (1998-2003)

Fig. 9. Historical Price of NASDAQ (1998-2003)

B. Procedure of Prediction. First of all, the dimension of the key pattern is set to 9, and that of the predicted pattern is set to 1, that is to say, we will use price information of 9 consecutive days to forecast the coming 1 day.

Besides, there are two parameters to be set in SOINN, the Dead-Edge Time and Remove-Node Time, which are the parameters used by SOINN to remove unnecessary edges and nodes to eliminate the influence of noises. In this case, they are set to 50 and 200, respectively, which are large enough values so that SOINN virtually does not conduct any reduction of edges, and only remove unnecessary nodes very occasionally (More details will be given in section 5).

After having finished setting the parameters, training data are randomly picked up from the training dataset and input into to the predictor for 5000 times, and test data are also randomly selected and input for 1000 times to check if the predicted price matches the real price:

$$(P_o > P_i \ and \ R_o > R_i) \ \ or \ \ (P_o < P_i \ \ and \ \ R_o < R_i)$$

Where:

- P_i average of the key pattern part of the predicted price
- P_o average of the predicted pattern part of the predicted price
- R_i average of the key pattern part of the real price
- R_o average of the predicted pattern part of the real price

This is to say that the prediction result is regarded as correct as long as the trend of the predicted price matches that of the real price.

C. Experimental Result. The experimental result is shown in Table 2. As we can see from the table, the proposed method outperformed the reinforcement learning approach significantly in terms of the S&P500 index. Although the conventional reinforcement learning approach showed better performance for the NASDAQ index, our predictor still forecasts the indexes more precisely on an average basis. Besides, we want to point out once more that there is no need to design the state space for our predictor while it is quite a challenge in the case of reinforcement learning.

Table 2. Prediction result

	S&P500	NASDAQ	Average
Reinforcement Learning	53.30%	79.76%	66.53%
Proposed Method (SOINN-AM)	72.30%	63.90%	68.10%

3.2 Trading on the Japan Market

The objective of this experiment is to demonstrate the profitability of our predictor on the Japan Stock Market. Unlike the former experiment, in which we only predicted the trend of the price movement, in this experiment, a trading strategy will be adopted so that our predictor is able to make order to the market based on its prediction result. To begin with, a naïve strategy is employed in this experiment:

1. If the prediction result shows that the stock price will rise the next day, use all available cash to take long position at the current market price, and clear the position at the end of the next day.
2. If the prediction result shows that the stock price will fall the next day, use all available cash to take short position at the market price, and clear the position at the end of the next day.

In this strategy, we take both long and short position to ensure that the experimental result is "market neutral". And for the sake of simplicity, trading cost is not included in this experiment, and we assume that we are able to execute all of our orders, which is reasonable for we only order at the current market price. Finally, the experimental result will be compared with the approach proposed by T. Matsui and H. Ohwada [8], which is an implementation of reinforcement learning and pair trading.

A. Experimental Dataset. The dataset of this experiment is collected from Yahoo! Japan Finance, which has already been plot in Fig 12 and Fig 13.

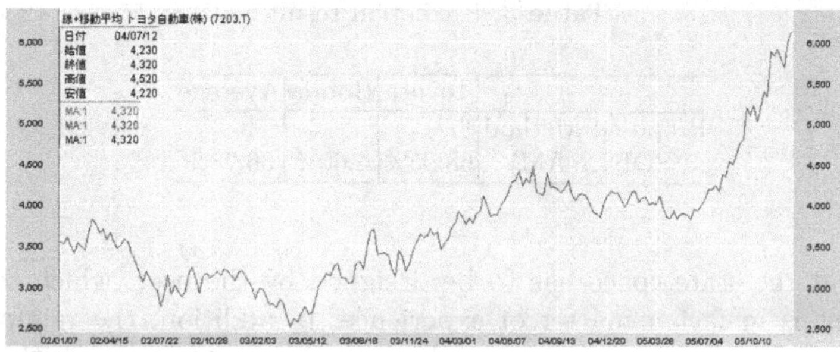

Fig. 10. Stock Price of Toyota (2002-2005)

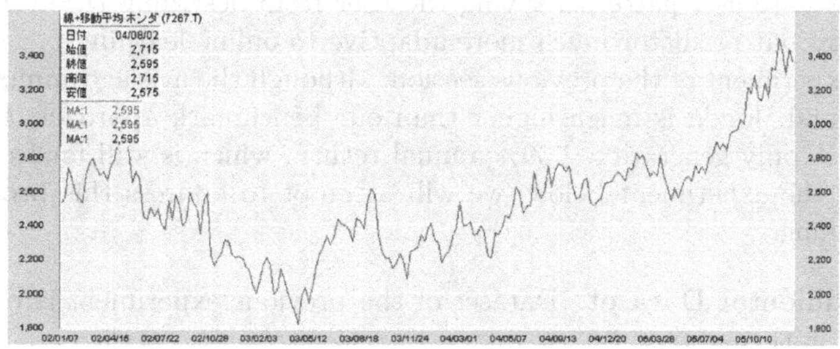

Fig. 11. Stock Price of Honda (2002-2005)

For the same reason in experiment 4.1, daily adjusted closing prices of Toyota and Honda from 2002 to 2004, inclusive, are used as the training dataset, and those of 2005 are used as the test dataset.

B. Procedure of Prediction. The selection of parameters is same as that of experiment 4.1, except that the test data starting from 1/1/2005 are input into the predictor sequentially, instead of being randomly picked up from the test dataset. Since our predictor forecasts the future price at the end of every day, order will be made to the market once a day.

C. Experimental Result. Table 3 shows the result of this experiment. The annual return of our predictor is 36.78%, which is significantly higher than the 25.60% profit of the reinforcement learning approach.

3.3 Experiment of Online Learning

In this section, we will experimentally demonstrate the online learning ability of the proposed method. As has been pointed out at the beginning of this paper, one of the biggest problems of reinforcement learning in forecasting the stock

Table 3. Prediction result

	Toyota	Honda	Average
Proposed Method (SOINN-AM)	68.95%	4.60%	36.78%

price is that the state space has to be designed by the user, which to a very large extent is in fact a matter of experience. In addition, the reinforcement learning model usually is unable to adapt to new state space in the middle of the training process. Therefore, reinforcement learning is considered to be inflexible in terms of online learning. However, in our approach, the training patterns are clustered automatically and new cluster is automatically created to accommodate new patterns, saving the user from designing the state space, which makes our predictor much more adaptive to online learning.

In the experiment of the previous section, although the average annual return of Toyota and Honda is much higher than our benchmark approach, trades of Honda stock only generate a 4.60% annual return, which is still far from satisfactory. In the experiment below, we will attempt to address this problem by online learning.

A. Experimental Dataset. Dataset of the previous experiment is continued to be used in this section.

B. Procedure of Prediction. First of all, we complete the training process of Honda as we did in section 4.2. As has been proved already, the predictor at this moment is not yet very profitable on the market of year 2005. Therefore, we additionally use data of the first half of 2005 to adapt the predictor and test its performance with the data of the second half of 2005.

C. Experimental Result. The annual return of Honda after online learning is presented in Table 4. The predictor generates an annual return of 12.66% after online learning, compared with the 4.60% annual return before it. It is clear that the profitability of our predictor is significantly improved by online learning.

Table 4. Prediction result

	Offline Learning	Online Learning
Annual Return of Honda	4.60%	12.66%

4 Consideration

In this section, we will consider about several aspects of the proposed approach, including the selection of parameters and the problem of overfitting.

4.1 The Parameters of SOINN

Λ_{edge}, or dead-edge time, is one of the most important parameters in SOINN. SOINN periodically removes edges with ages greater than Λ_{edge}, so as to prevent itself from being flooded with noises or any other useless information.

However, to find the proper Λ_{edge} is a tricky problem and there is no golden rule guiding us to the best value. We use the training dataset of Toyota of section 4.1 to conduct cross validation, as an attempt to shed some light on the best optimized value of Λ_{edge}. The experimental result is shown in Table 5.

Table 5. The profitability of the predictor after conducting cross validation

Λ_{edge}	Training by '03-'04 Tested by '02	Training by '02 and '04 Tested by '03	Training by '03+'04 Tested by '02 (Reverse order)	Average
5	2.36%	1.72%	11.96%	5.35%
10	-8.95%	7.25%	25.84%	8.05%
20	-6.38%	31.49%	1.27%	8.79%
30	11.08%	33.06%	-7.20%	12.31%
50	11.08%	33.06%	4.80%	16.31%

In our first attempt, the predictor is trained by the data of 2003 and 2004, and then tested by those of 2002. The result shows that the most profitable model can be constructed at Λ_{edge}=30 or 50. We want to notify the reader here that it is reasonable for the predictor to have multiple optimized values for Λ_{edge}, for in this case Λ_{edge}=30 is in fact a large enough value that no edge can be removed during the training process. Therefore, Λ_{edge}=50 will have the same effect and as a matter of fact, in this specific case, any value greater than 30 can be used as the optimized Λ_{edge}.

Furthermore, we use data of 2004 and 2002 as the training set and those of 2003 as the test set. Similar result is obtained here that the predictor shows best performance at Λ_{edge}=30 or 50.

Lastly, the data of 2004 and 2003 are used as the training set while those of 2002 are used as the test set. Note that data of 2004 and 2003 are input into the predictor sequentially, which is in reverse order with what we did in the first attempt. This time the result is a bit more complicated than the first two. The best optimized value appears at the point that Λ_{edge}=10.

Although it is difficult to find a constant Λ_{edge} which suits all kinds of data, it is fairly obvious that the average profitability of the predictor is still proportional to Λ_{edge}. Thus, in our practice, we fix Λ_{edge} at 50 in spite that it is not always the best optimized value, which means, in our predictor, the removal of dead edges will never be conducted. In fact, setting Λ_{edge} in this manner is common in applications based on SOINN, such as SOINN-DP [3].

4.2 The Dimension of Train/Test Data

The major parameters of the input layer of the proposed method are the dimensions of the key pattern and the predicted pattern, which in our previous experiments, were set to 9 and 1. That is to say, we would predict the upcoming 1 day price based on the prices of last 9 consecutive days.

Table 6 shows the precision rate of the predictor as we gradually adjust the dimensions of the key pattern and the predicted pattern. The result shows that the precision rate of prediction improves as we increase the dimension of the key pattern. This is in line with our instinctive perception that key pattern with more detailed information will associate more accurate prediction result.

Table 6. Annual returns for different combinations of dimensions

Key:Predicted ratio	6:4	7:3	8:2	9:1
Toyota	64.70%	68.20%	69.70%	66.10%
Honda	63.20%	63.40%	65.00%	63.10%

5 Conclusion

In this paper we proposed a novel Associated-Memory based predictor for stock price. Our SOINN-AM predictor estimates future stock price by recalling learning result from input stock price time series. With experiment results on several stock price data of the US and Japan market in 2002-2004, we demonstrate the effectiveness of our approach that SOINN-AM based predictor can gain more profitable than the conventional approaches like reinforcement learning.

Our predictor can automatically determine its inner state space according to the training patterns without any help, which the conventional approaches needed. Additionally our method can modify itself by online learning. Consequently our method is more flexible and more applicable in the ever-changing real-world financial market. In future, we will evaluate our predictor on a large scale stock price data and improve its prediction accuracy.

References

1. Shen, F., Hasegawa, O.: An incremental network for on-line unsupervised classification and topology learning. Neural Networks 19(1), 90–106 (2006)
2. Sudo, A., Sato, A., Hasegawa, O.: Associative Memory for Online Incremental Learning in Noisy Environments. In: International Joint Conference on Neural Networks, vol. 12(17), pp. 619–624 (2007)
3. Okada, S., Hasegawa, O.: Incremental Learning, Recognition,and Generation of Time-Series Patterns Based on Self-Organizing Segmentation. Journal of Advanced Computational Intelligence and Intelligent Informatics 10(3), 395 (2006)

4. Yoo, P.D., Kim, M.H., Jan, T.: Machine Learning Techniques and Use of Event Information for Stock Market Prediction: A Survey and Evaluation. In: International Conference on International Conference on Computational Intelligence for Modelling, Control and Automation and Intelligent Agents, Web Technologies and Internet Commerce, vol. 2, pp. 835–841 (2005)
5. Schierholt, K., Dagli, C.H.: Stock market prediction using different neural network classification architectures. In: IEEE/IAFE Conference on Computational Intelligence for Financial Engineering, pp. 72–78 (1996)
6. Phua, P.K.H., Xiaotian, Z., Chung, H.K.: Forecasting stock index increments using neural networks with trust region methods. In: International Joint Conference on Neural Networks, vol. 1, pp. 260–265 (2003)
7. Lee, J.W.: Stock Price Prediction Using Reinforcement Learning. In: IEEE International Symposium on Industrial Electronics, vol. 1, pp. 690–695 (2001)
8. Matsui, T., Ohwada, H.: A Reinforcement Learning Agent for Stock Trading: An Evaluation. In: The 20th Annual Conference of the Japanese Society for Artificial Intelligence, vol. 20, 3C1-6 (2006)
9. Li, H., Dagli, C.H., Enke, D.: Forecasting Series-based Stock Price Data Using Direct Reinforcement Learning. In: International Joint Conference on Neural Networks, vol. 2, pp. 1103–1108 (2004)
10. Li, H., Dagli, C.H., Enke, D.: Short-term Stock Market Timing Prediction under Reinforcement Learning Schemes. In: IEEE International Symposium on Approximate Dynamic Programming and Reinforcement Learning, pp. 233–240 (2007)

A Case Study of ICA with Multi-scale PCA of Simulated Traffic Data

Shengkun Xie[1], Pietro Lió[2], and Anna T. Lawniczak[1]

[1]Department of Mathematics and Statistics, University of Guelph, Guelph, Ont N1G 2W1, Canada
[2]The Computer Laboratory, University of Cambridge, 15 JJ Thomson Avenue, Cambridge CB3 0FD, UK

Abstract. Often packet traffic data is non-stationary and non-gaussian. These data complexity causes difficulties in its analysis by standard techniques and new methods must be employed. Recent theoretical and applied works have demonstrated the appropriateness of wavelets for analyzing multivariate signals containing non-stationarity and non-gaussianity. This paper presents a new pre-processing method, a multi-scale PCA that combines a wavelet filtering method with principal component analysis (PCA), for a noise free independent component analysis (ICA) model. By applying the proposed method to a set of test data coming from simulations of a packet switching network (PSN) model we see improvements of data analysis results.

Keywords: Independent Component Analysis, Multi-scale Principal Component Analysis, Wavelet Transform, De-nosing.

1 Introduction

Independent component analysis (ICA) [1] is an important method for extracting useful information from mixture data [2]. It belongs to a class of blind source separation methods for separating data into informational components. When ICA model reflects well the relationship between the underlying source signals and data mixtures, the obtained components are statistically independent and tend to be more non-gaussian than data mixtures. As a multivariate statistical method, ICA is strongly related to principal component analysis (PCA) and factor analysis (FA) [3]. Instead of finding independent components from a set of data mixtures, PCA searches for uncorrelated components that have the largest variance and FA looks for correlated components that lead to minimum values of independent residuals among the sets of data mixtures. There are several definitions of ICA including noise free and noisy ICA models ([1], [4]). Most of the research focuses on a noise free ICA model because it is less computationally intensive than the noisy ICA model ([1], [4]). In order to improve the result of data analysis some current works in the field modify the noise free ICA model. Noise free ICA mixture was first studied by Lee [5] to capture potentially different class of signals for the purpose of better representation of data through a mixture

C. Alippi et al. (Eds.): ICANN 2009, Part II, LNCS 5769, pp. 358–367, 2009.

model. Zhou [6] used ICA mixture with the hidden Markov model (HMM) to investigate the usefulness of the method in video content analysis. From the application domain point of view, the noise free ICA model has been successfully used in studies of biomedical signal processing. In biomedical applications, ICA has been used for removing artifacts from the ECG data [7] or for detection of Functional MRI activation data [8]. Because of the importance and frequent use of a noise free ICA model and the fact that data usually contains noise, a denoising procedure is required. In practice when ICA is used, a PCA whitening approach is usually applied to remove the effect from a Gaussian noise before a final procedure of ICA is implemented. Application of PCA whitening procedure to mixture data seems to be beneficial when a signal contains a Gaussian white noise. However, various types of data coming from complex systems have locally high frequency noise or corrupted irregular noise which is not of interest and has to be removed. The PCA whitening procedure may not work well for this type of noise. In biomedical data, such as ECG data, the corrupted or irregular component in the noisy signal is part of the signal that is of importance. In this case, an elimination of such irregularity may have serious implications for the validity of ECG test and the use of ICA has to be treated carefully. This rises up an important practical issue when for the sake of analysis is beneficial to use ICA, because, the preprocessing step in an ICA algorithm may affect the performance of ICA. In many practical applications, the observed data are non-stationary time series. Application of wavelet filtering method to this type of data is advisable because of good localization properties of wavelet method. Thus, wavelet filtering combined with PCA whitening as a preprocessing procedure in ICA may become useful approach in dealing with the noisy mixtures of data, in particular, the signals with important corrupted components.

With these motivations in mind we organize the paper as follows. In Section 2 we provide a brief description of wavelet method combined with PCA, namely multi-scale PCA ([10], [11]). Section 3 discusses the proposed preprocessing method for ICA. Section 4 provides a justification for the appropriateness of using the proposed method in our study of network traffic data; and sections 5 reports on conclusions. The software used for the paper is available upon request from the first author.

2 Combining Wavelet and PCA

Wavelet transform decomposes an original signal into elements of multi-scale subspaces that are spanned by the dilated and translated version of a mother wavelet [9]. In discrete wavelet transform (DWT), the mother wavelet functions provide different wavelet scaling filters $\mathbf{H}$ and wavelet details filters $\mathbf{G}$. DWT method recently has been used in PCA, called often, multi-scale PCA. In multi-scale PCA approach, first, a $n \times n$ DWT matrix $\mathcal{W}$ is applied to a $n \times p$ processed data matrix of mixtures Φ, which transforms Φ to $\mathcal{W}\Phi \equiv \Phi^*$. For DWT with level L, $\mathcal{W}$ can be defined as, $\mathcal{W} = [\mathbf{H}_L, \mathbf{G}_L, \mathbf{G}_{L-1}, \cdots, \mathbf{G}_1]^T$, where $\mathbf{H}_L$ and $\mathbf{G}_i$, $1 \leq i \leq L$ are row vectors. In the above, if $\mathbf{H}$ represents projection on the wavelet scaling function and $\mathbf{G}$ is projection on the wavelet

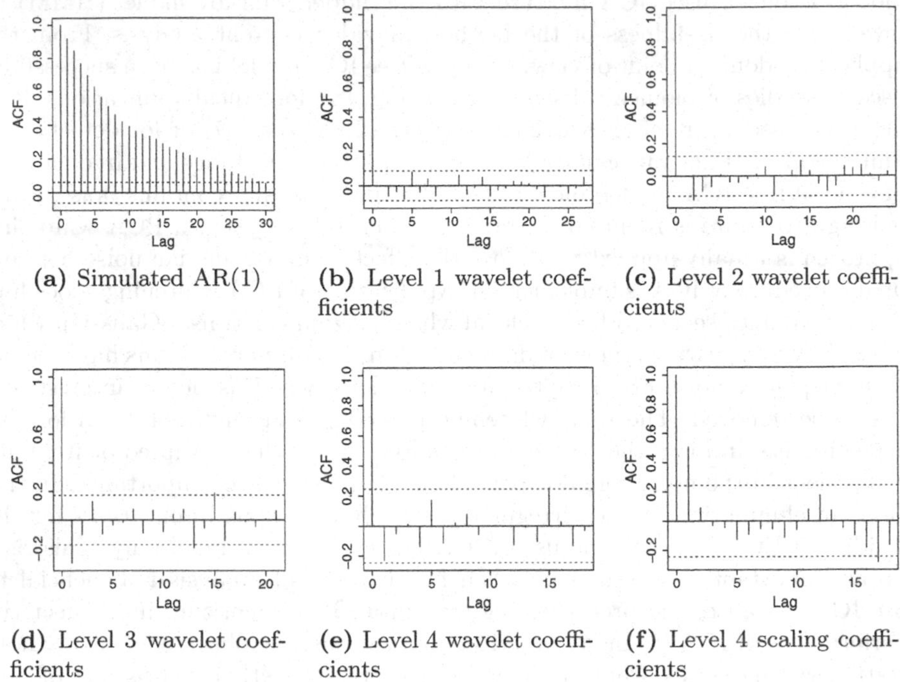

(a) Simulated AR(1)

(b) Level 1 wavelet coefficients

(c) Level 2 wavelet coefficients

(d) Level 3 wavelet coefficients

(e) Level 4 wavelet coefficients

(f) Level 4 scaling coefficients

Fig. 1. The plot (a) is the autocorrelation function of AR(1) process with parameter $\phi_1 = 0.9$ and the plots (b)-(f) are the autocorrelation functions of the wavelet coefficients at each level when a level 4 DWT with the Haar wavelet basis was applied to the original data

details, then $\mathbf{H}_L$ is obtained by applying L times the wavelet scaling filter $\mathbf{H}$ and $\mathbf{G}_i$ is obtained by applying $\mathbf{H}$ filter $i-1$ times and the wavelet details filter $\mathbf{G}$ once. The filtering operations $\mathbf{H}$ and $\mathbf{G}$ depend on the wavelet basis function, that is, different wavelet basis leads to different forms of filtering operation. The wavelet coefficients at each level i can be obtained by the following iterative procedure of applying filtering on the wavelet coefficients at level $i-1$. For $i = 1$, $\mathbf{A}_1 = \mathbf{H}\Phi$, $\mathbf{D}_1 = \mathbf{G}\Phi$, where $\mathbf{A}_1$ and $\mathbf{D}_1$ are the wavelet scaling coefficients and the wavelet details coefficients at level 1, respectively. For $2 \leq i \leq L$, this procedure is defined as $\mathbf{A}_i = \mathbf{H}\mathbf{A}_{i-1}$, $\mathbf{D}_i = \mathbf{G}\mathbf{D}_{i-1}$, where $\mathbf{A}_i$ and $\mathbf{D}_i$ are the wavelet scaling coefficients and the wavelet details coefficients at level i, respectively. After taking DWT of Φ, PCA is applied at each level of the wavelet coefficients matrix. This procedure eliminates the principal components loading and their scores [3] that correspond to smaller eigenvalues and it reconstructs the wavelet coefficients by using the selected significant components $\hat{\mathbf{P}}$ and their associated scores $\hat{\mathbf{L}}$ at each level i, that is, $\hat{\mathbf{A}}_L = \hat{\mathbf{L}}_L \hat{\mathbf{P}}_L^T$, $\hat{\mathbf{D}}_i = \hat{\mathbf{L}}_i \hat{\mathbf{P}}_i^T$, for $1 \leq i \leq L$. The reconstructed signal is then obtained by taking the PCA of wavelet approximations plus principal components of the wavelet details at each level. Therefore, the wavelet coefficient matrix Φ^* may be reconstructed by

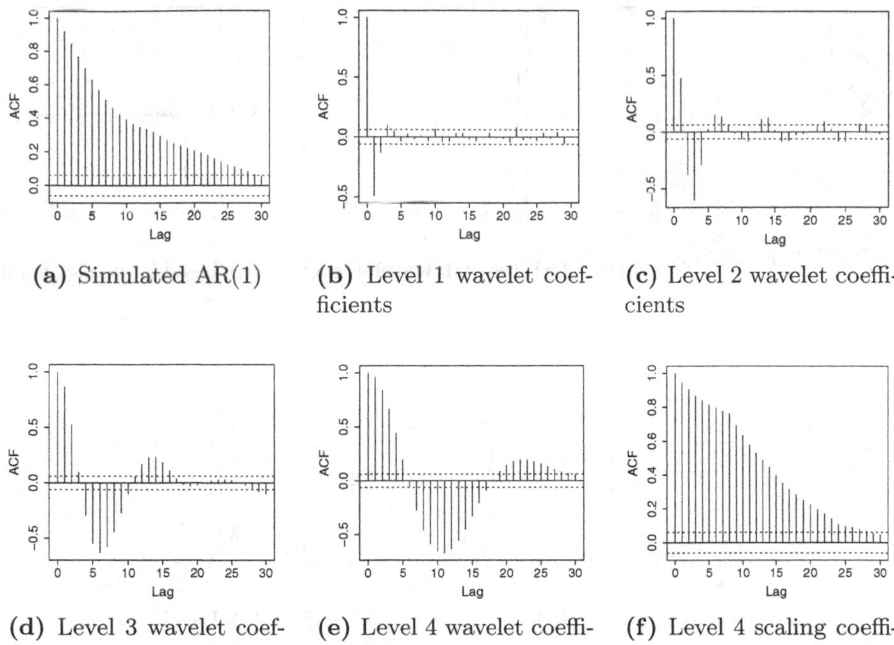

(a) Simulated AR(1) (b) Level 1 wavelet coefficients (c) Level 2 wavelet coefficients

(d) Level 3 wavelet coefficients (e) Level 4 wavelet coefficients (f) Level 4 scaling coefficients

Fig. 2. The plot (a) is the autocorrelation function of AR(1) process with parameter $\phi_1 = 0.9$ and the plots (b)-(f) are the autocorrelation functions of the wavelet coefficients at each level when a level 4 SDWT with Haar wavelet basis was applied to the original data

$\hat{\Phi}^* \equiv [\hat{\mathbf{A}}_L, \hat{\mathbf{D}}_L, \cdots, \hat{\mathbf{D}}_1]^T$ and the de-noised data matrix $\hat{\Phi}$ is obtained by taking inverse wavelet transform. That is, $\hat{\Phi} = \mathcal{W}^* \hat{\Phi}^*$, where $\mathcal{W}^*$ is the inverse wavelet transform operator associated with the DWT operator $\mathcal{W}$.

One of the important aspects of this methodology is the fact that orthogonality of the wavelet transform maintains the variance-covariance structure of the processed data [3]. This means that the variance-covariance matrix of the processed data and the wavelet coefficients matrix of DWT of the processed data are the same. For the wavelet transform one may use a non-orthogonal wavelet basis or even a stationary DWT (SDWT) [9]. However, using SDWT may create potential problems coming from altering the variance-covariance matrix because the SDWT makes the signals more autocorrelated due to the redundancy of the wavelet coefficients.

3 Independent Component Analysis with Wavelet and PCA

Noise free ICA model is important and popular in applications. Wavelet filtering and PCA are two popular preprocessing methods in data analysis and

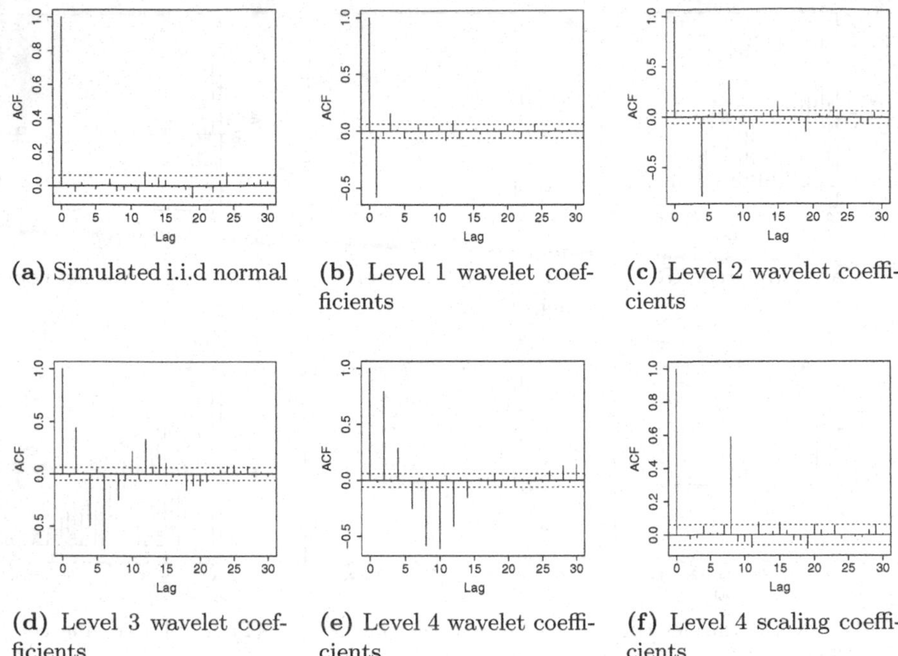

(a) Simulated i.i.d normal

(b) Level 1 wavelet coefficients

(c) Level 2 wavelet coefficients

(d) Level 3 wavelet coefficients

(e) Level 4 wavelet coefficients

(f) Level 4 scaling coefficients

Fig. 3. The plot (a) is the autocorrelation function of time series simulated from a standard normal distribution and the plots (b)-(f) are the autocorrelation functions of the wavelet coefficients at each level when a level 4 SDWT with Haar wavelet basis was applied to the original data

often used as preprocessing methods for ICA. The use of wavelet filtering as a preprocessing method may cause lost of some important and significant information at some wavelet details and, PCA may not work well for non-stationary data. This work focuses on the improvement of data analysis using noise free ICA by combining wavelet method and PCA as a preprocessing approach to avoid the potential problems caused by wavelet filtering or PCA preprocessing method. For signal mixtures $\mathcal{X} \equiv (\mathbf{X}_1(t), \mathbf{X}_2(t) \cdots, \mathbf{X}_p(t))$ being a set of observations of random variables with t as a time index or a sample index, and for a $p \times p$ mixing matrix $\mathbf{A}$, a noise free ICA model can be described as $\mathcal{X} = \mathcal{S}\mathbf{A}$, $\mathcal{S} \equiv (\mathbf{S}_1(t), \mathbf{S}_2(t) \cdots, \mathbf{S}_p(t))$, where $\mathbf{S}_i(t)$, for $1 \leq i \leq p$, are called independent components. In practice, the number of source signals may not be equal to the number of mixtures and also only two or three independent components may be of interest. To achieve a dimension reduction for multivariate data $\mathcal{X}$, PCA is applied to reduce the dimension of data mixtures before ICA is used. Before we discuss the proposed preprocessing method, we first introduce the concept of wavelet ICA which is related to ICA with multi-scale PCA. In wavelet ICA, DWT multiplies $\mathcal{X}$ from the left by wavelet decomposition matrix $\mathcal{W}$ for a column vector $\mathbf{X}_i(t)$ with index i. This gives $\mathcal{X}^* = \mathcal{W}\mathcal{X} = \mathcal{W}\mathcal{S}\mathbf{A} = \mathcal{S}^*\mathbf{A}$, where $\mathcal{S}^*$ is the wavelet transform of source signals $\mathcal{S}$. By taking the inverse wavelet

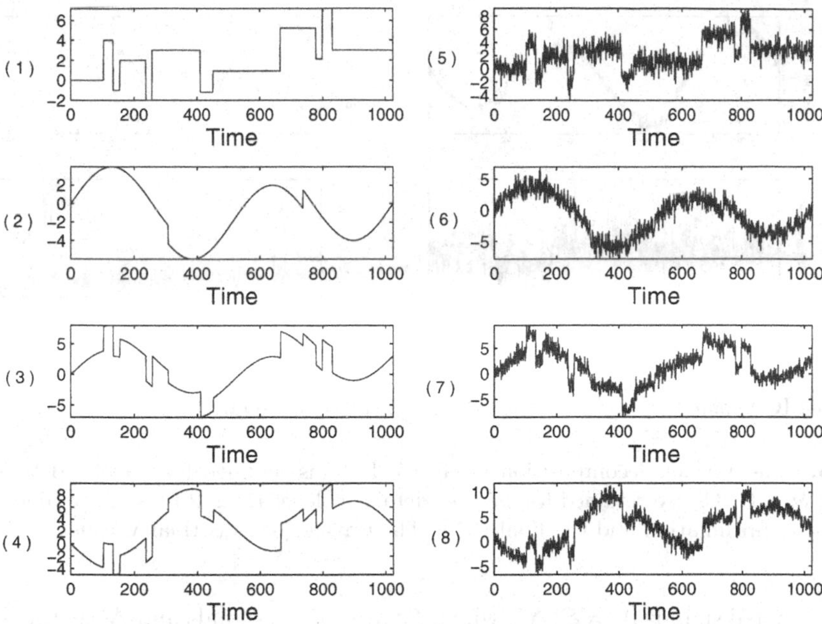

Fig. 4. The plot (1) is a blocky function; the plot (2) is a heavy Sine, which is regular except around time 300 and 750. The plot (3) and (4) are the sum and the difference between the blocky function and the heavy Sine, respectively. Multivariate Gaussian white noise exhibiting strong spatial correlation is added to the resulting four signals to produce plot (5)-(8) in the second column. The plots (5)-(8) are the input signals for ICA.

transformation, one is able to reconstruct the original data mixtures $\mathcal{X}$ and source signals $\mathcal{S}$ as the discrete wavelet transformation does not influence the estimate of the mixing matrix $\mathbf{A}$. However, the benefit of conducting analysis in wavelet domain is a multi-resolution of ICA, because ICA is applied to the wavelet coefficients at each level. In an application to signal reconstruction, a wavelet ICA is applied only to wavelet details at the selected level. The wavelet approximation coefficients should be retained at the same form and the lower level of wavelet detailed should be eliminated. The main idea of wavelet ICA is to reconstruct data mixtures $\mathcal{X}$ that are less noisy. The main drawback of this method is that wavelet ICA is not able to separate source signals $\mathcal{S}$ from mixtures. One can only obtain the source signals at each selected level. Because of this, we propose an ICA with a multi-scale PCA method that can give better result in separating the source signals from the mixtures after preprocessing data in time domain. The procedure is described as follows: first one obtains de-noised mixtures using a multi-scale PCA approach according to the goal of the quality of signal reconstruction, then one applies ICA to reconstructed signals to separate the source signals. Mathematically, $\hat{\mathcal{X}} = \mathcal{W}^* \hat{\mathcal{X}}^*$, where $\hat{\mathcal{X}}$ is reconstructed data matrix from the multi-scale PCA of $\mathcal{X}$ and $\hat{\mathcal{X}}^*$ is the matrix of significant wavelet coefficients after applying PCA. Then one finds source signals for the

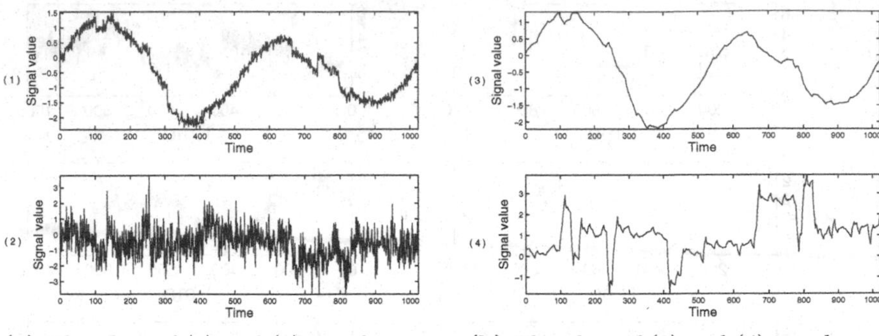

(a) The plots of (1) and (2) are the estimated source signals under the noise free ICA model.

(b) The plots of (3) and (4) are the estimated source signals under ICA with the orthogonal DWT and PCA.

Fig. 5. The wavelet decomposition level is 5; 1 PC is retained for wavelet details at level 3 & 4; 0 PCs are retained for wavelet details at level 1& 2; 2 PCs are retained for wavelet approximation and the final PCA. The wavelet basis is Haar wavelet.

reconstructed signals $\hat{\mathcal{X}} = \mathcal{S}^*\mathbf{A}^*$, where $\mathcal{S}^*$ are source signals and $\mathbf{A}^*$ is the mixing matrix. The important aspect of this approach is that one is able to separate source signals by combining wavelet and PCA because signals are de-noised in wavelet domain using PCA and wavelets and source signals are separated in the original domain. An ICA with a multi-scale PCA becomes a conventional ICA if all principal components of all wavelet coefficients at all levels are kept. If the wavelet coefficients at smaller levels are eliminated and all principal components of remaining coefficients are kept, this proposed method becomes ICA on a low-pass filtered mixtures. An ICA with a multi-scale PCA can eliminate both stationary and non-stationary noise by controlling the number of principal components of wavelet coefficients at each level.

4 A Case Study of ICA with Multi-scale PCA

In this section we study the performance of the proposed method in obtaining a set of source signals from data mixtures with known structure and from data mixtures with unknown structure. We first show the effect of the application of DWT and SDWT on the autocorrelation of signals. From Figure 1-3 we observe that the orthogonal DWT appears to perform well in de-correlating the autocorrelated data at each wavelet decomposition level, but the application of SDWT creates more autocorrelation for both autocorrelated and i.i.d data at each wavelet decomposition level. This indicates that the orthogonality of preprocessing method for ICA is an important feature as the non-orthogonal DWT not only has a problem in de-correlating a correlated signal but also creates autocorrelation for i.i.d data. Next, we investigate the goodness of the proposed method and compare it with the standard procedure of ICA when applied to

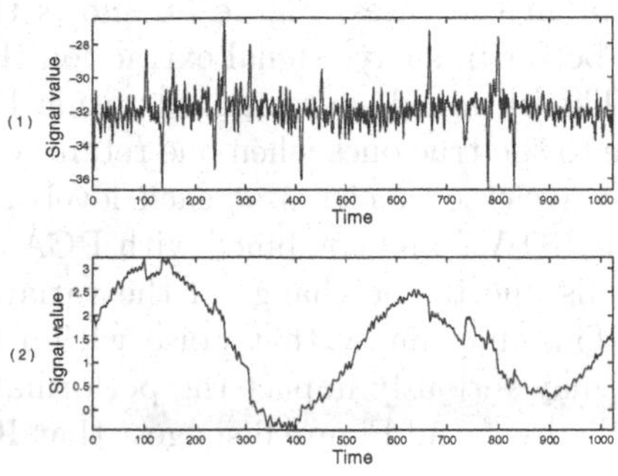

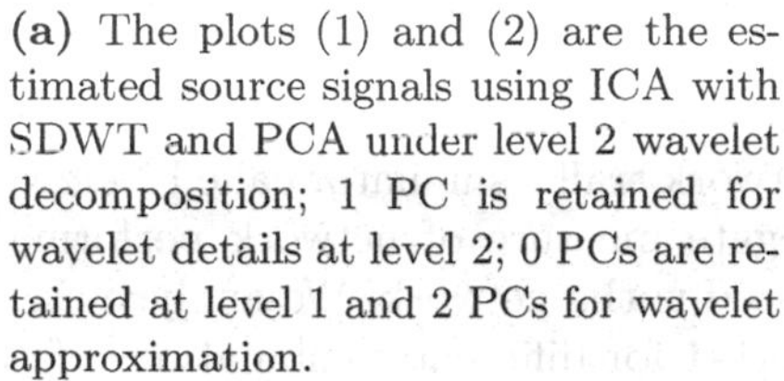

(a) The plots (1) and (2) are the estimated source signals using ICA with SDWT and PCA under level 2 wavelet decomposition; 1 PC is retained for wavelet details at level 2; 0 PCs are retained at level 1 and 2 PCs for wavelet approximation.

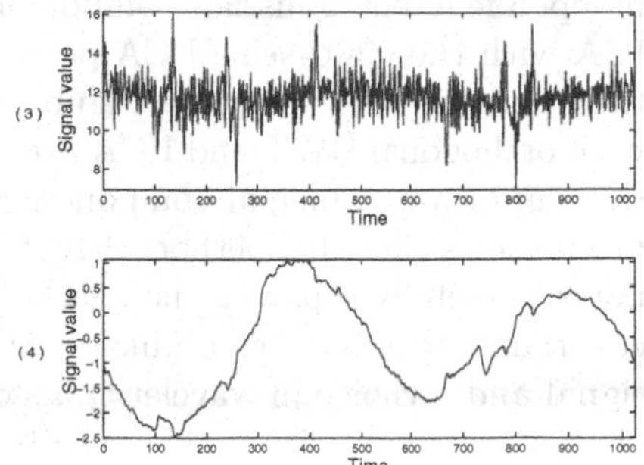

(b) The plots (3) and (4) are the estimated source signals using ICA with SDWT and PCA under level 3 wavelet decomposition; 1 PC is retained for wavelet details at level 2 & 3; 0 PCs are retained at level 1; 2 PCs are retained for wavelet approximation.

Fig. 6. Plots (a) and (b) are under the method of ICA with SDWT and PCA with different wavelet decomposition levels. The wavelet basis is Haar wavelet.

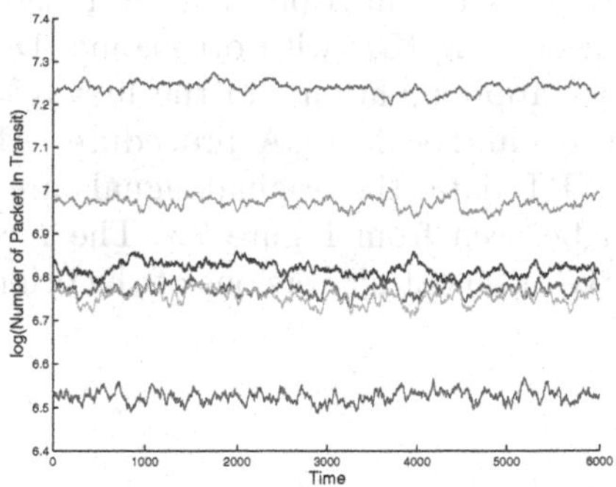

(a) The figure shows 6 graphs of log transformed number of packets in transit signals obtained form simulations of network model set-up for various combinations of edge cost functions and source loads.

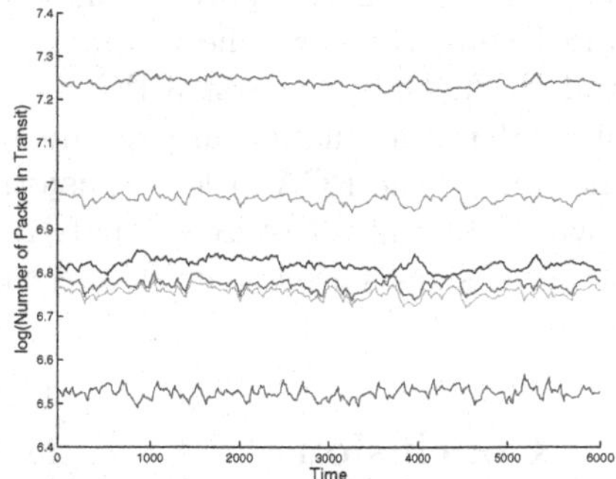

(b) The figure displays the log transformed number of packets in transit signals reconstructed using ICA with orthogonal DWT with Haar wavelet basis and PCA. The wavelet decomposition level is 5.

Fig. 7. ICA with orthogonal DWT and PCA is used to reconstruct signals

the mixtures of two well known source signals a blocky function and a heavy Sine shown in Figure 4. We use ICA and ICA with the multi-scale PCA to extract the underlying source signals. In this paper, we present results when for WT we used the Haar wavelet basis. The results obtained using other types of wavelet basis will be discussed elsewhere. The FastICA algorithm is used to find independent non-gaussian components for in the cases. Figure 5b shows that ICA with the multi-scale PCA performs better in source signal extraction than the standard ICA shown in Figure 5a. The obtained source signals using ICA with orthogonal DWT and PCA are close to the true ones when one retains only the significant principal components in wavelet coefficients at each level. Figure 6 shows that the method of ICA using SDWT and combined with PCA can not successfully separate the source signals due to the change of the variance-covariance matrix after taking SDWT. This may imply that noise within the signal and a choice of wavelet transform may seriously impact the performance of ICA search algorithm. The results in Figure 5 and Figure 6 suggest that ICA using orthogonal DWT and combined with PCA should be considered for the case study of network traffic data.

In this paper, we analyze number of packets in transit (NPT) signals. They are simulation traffic data generated by a network traffic simulator called Netzwerk-1[12]. The NPT is an important aggregate measure of network performance providing information about how congested is the network. We analyze six different NPT signals generated by Netzwerk-1 for different combinations of edge cost functions and network source loads [12] used in the simulator set-up. Each signal is the mean function calculated from a total of 24 samples that are independently simulated using Netzwerk-1. The length of each signal is 6000. The natural logarithm is applied to NPT time series to speed up the convergence of ICA algorithm. Figure 7a shows the log transformed original NPT signals and Figure 7b shows the reconstructed ones using ICA with orthogonal DWT and PCA. Only the first 3 PCs of wavelet approximation and the first 3 PCs of final reconstruction are retained in the multi-scale PCA procedure. After the multi-scale PCA of log transformed NPT data, the original signals are recovered by only 3 source signals as can be seen from Figure 7b. The reconstructed signals are less noisy and can serve as smoother versions of the original signals.

5 Conclusions

In this paper, we present a new preprocessing method for the standard ICA, namely, a multi-scale PCA to handle a noise contained in mixture data. This preprocessing method potentially achieves data dimension reduction through PCA procedure. The results of the case study show that the proposed method outperforms the standard ICA and is promising in improving the performance of extractions of non-gaussian components when orthogonal DWT is used. Our study suggests that orthogonal DWT should be chosen when the multi-scale PCA is used in preprocessing for ICA. Thus, it appears that for some types of network traffic data ICA method with the multi-scale PCA preprocessing could be useful in reconstructing the original data from a set of source signals with fewer number of signals than the set of the original signals.

Acknowledgement

S. Xie acknowledges financial support from the Univ. of Guelph and SHARCNET (through graduate research fellowship). P. Liò thanks EC IST SOCIALNETS - Grant agreement number 217141. A.T.L. acknowledges partial financial support from SHARCNET and NSERC of Canada.

References

1. Hyvarinen, A., Karhunen, J., Oja, E.: Independent Component Analysis. John Wiley & Sons Inc., New York (2001)
2. Valenzuela, W., Carvajal, G., Figueroa, M.: Blind source separation in mixed-signal VLSI using the InfoMax algorithm. In: Kůrková, V., Neruda, R., Koutník, J. (eds.) ICANN 2008,, Part II. LNCS, vol. 5164, pp. 208–217. Springer, Heidelberg (2008)
3. Jolliffe, I.T.: Principal Component Analysis. Springer Science+Bussiness Media, Inc., New York (2004)
4. Hyvarinen, A.: Survey on independent component analysis. Neural Computing Surveys 2, 94–128 (1999e)
5. Lee, T.-W., Lewicki, M.S.: Unsupervised image classification, segmentation, and enhancement using ICA mixture models. IEEE Trans. Image Process 11(3), 270–279 (2002)
6. Zhou, J., Zhang, X.P.: An ICA mixture hidden Markov model for video content analysis. IEEE Trans. on Circuit and Systems for Video Technology, Special Issue on Event Analysis in Videos 18(11), 1576–1586 (2008)
7. He, T., Clifford, G., Tarassenko, L.: Application of independent component analysis in removing artefacts from the electrocardiogram. Neural Computing and Application 15, 105–116 (2006)
8. Shen, M.F.: Application ICA method for detecting functional MRI activation data. In: Yin, F.-L., Wang, J., Guo, C. (eds.) ISNN 2004. LNCS, vol. 3173, pp. 726–731. Springer, Heidelberg (2004)
9. Percival, D.B., Walden, A.T.: Wavelet Methods for Time Series Analysis. Cambridge University Press, Cambridge (2000)
10. Aminghafari, M., Cheze, N., Poggi, J.-M.: Multivariate de-noising using wavelets and principal component analysis. Computational Statistics & Data Analysis 50, 2381–2398 (2006)
11. Bakshi, B.: Multiscale analysis and modeling using wavelets. Journal of chemometrics 13(3-4), 415–434 (1999)
12. Lawniczak, A.T., Gerisch, A., Di Stefano, B.: OSI Network-layer Abstraction: Analysis of Simulation Dynamics and Performance Indicators, Science of Complex Networks. In: Mendes, J.F. (ed.) AIP Conference Proc., vol. 776, pp. 166–200 (2005)

Decomposition Methods for Detailed Analysis of Content in ERP Recordings

Vasiliki Iordanidou[1], Kostas Michalopoulos[1], Vangelis Sakkalis[2], and Michalis Zervakis[1]

[1] Department of Electronic and Computer Engineering, Technical University of Crete, Chania 73100, Greece
[2] Institute of Computer Science, Foundation for Research and Technology, Heraklion 71110, Greece
{vasiliki_han@yahoo.com,kosmixal@gmail.com,sakkalis@ics.forth.gr, michalis@display.tuc.gr}

Abstract. The processes giving rise to an event related potential engage several evoked and induced oscillatory components, which reflect phase or non-phase locked activity throughout the multiple trials. The separation and identification of such components could not only serve diagnostic purposes, but also facilitate the design of brain-computer interface systems. However, the effective analysis of components is hindered by many factors including the complexity of the EEG signal and its variation over the trials. In this paper we study several measures for the identification of the nature of independent components and address the means for efficient decomposition of the rich information content embedded in the multi-channel EEG recordings associated with the multiple trials of an event-related experiment. The efficiency of the proposed methodology is demonstrated through simulated and real experiments.

Keywords: EEG, ICA, ERP, time-frequency measures, PCA.

1 Introduction

Event related or event locked activity induced by an external or internal stimulus involves both phase locked and non-phase locked rhythmic oscillations. Event related potentials (ERP) encompass the phase-locked (evoked) activity at different frequency bands. Recent studies have also revealed EEG responses non-phase locked to the event occurrence (induced), which vary with stimulus and interact with the ERP. Traditional ERP analysis considers the averaged signal over trials as to increase signal-to-noise ratio. The process of averaging, however, suppresses any other induced activity of non phase-lock nature associated with the event or stimulus. Such activity is often measured by the power of the AM demodulated signal at specific frequency bands, after the subtraction of the evoked activity [1,2]. Furthermore, time-frequency (TF) analysis has become important for assessing both evoked and induced brain activity from event-related EEG

C. Alippi et al. (Eds.): ICANN 2009, Part II, LNCS 5769, pp. 368–377, 2009.

recordings. In this paper we use the term ERP to denote all processes in an event-related experiment. The P300 ERP waveform, as a response to oddball experiments, is perhaps the most widely studied response due to the variety of activations it produces. Through TF analysis, theta and delta-band activities have been shown to underlie its formation, whereas alpha is also induced during the P300 response.

Several neurophysiological studies indicate that the evoked processes possibly originate from stable phase-locking transient synchronization of brain regions, with different signal peaks being evoked from specific brain regions at distinct frequency bands. Furthermore, the induced activity has been attributed to phase-resetting of ongoing EEG activity at various topological areas. Because of their different neurophysiological origin of evoked and induced activity, the analysis of both types of signal waveforms is useful in the analysis of event related recordings [1]. Results on real data recordings demonstrated that the number of independent components that correspond to event-related activity, phase-locked or not, is between five to fifteen independent components for a 31 electrodes montage [3]. The interpretation and analysis of distinct content of the EEG recordings becomes difficult not only by the complexity of information messages but also by the unavoidable signal mixing at the electrodes, produced by volume conduction effects [3].

In the above context it is quite important to provide efficient means of decomposing the multichannel EEG signal into meaningful components. Prominent methods that have been proposed for signal and energy-content decomposition include the independent component analysis (ICA) of EEG channels and the principal component analysis (PCA) of the TF energy spectra of EEG channels. In this paper we propose an improved decomposition of ERP information content through pre-filtering of the EEG signal. The proposed scheme employs ICA decomposition in order to select specific components for EEG filtering and then exploits PCA decomposition of the TF representation for efficient analysis of the overall EEG content over all trials. It is shown that the filtered signal preserves the relevant information and allows the separation and interpretation of information content more clearly than the original EEG signal, either for phase or non-phase locked activity. Thus, the proposed scheme can be effectively used for the detailed analysis and synthesis of ERP responses, for diagnostic purposes or for design of BCI systems.

Overall, the contribution of the paper is identified in the following areas. 1) It provides distinct interpretations on the usefulness of ICA decomposition of EEG vs. the PCA decomposition of its TF map. The former is mainly used for the decomposition of EEG into meaningful sub-components that can be directly related to brain source activity. The latter is primarily used for a detailed analysis of the content of EEG channels or ICA components. It is most appropriate for the analysis of summarized information content of the EEG over all channels and trials. 2) It provides measures for the interpretation and quantization of non-phase locked synchronization over trials. 3) It establishes a rigorous scheme for considering significant ICs through their time-space-frequency distribution along with their repeatability along multiple trials. 4) It defines an algorithmic scheme for filtering EEG based on only the important components and then analyzing the TF content of the filtered EEG in terms of principal components identified over all channels and repeated trials. These innovative aspects are established in the methodological and demonstrated in the experimental sections.

2 Methodology of Signal and Content Decomposition

Considering the wealth of information embedded in EEG recordings, it is quite important to provide efficient means of decomposing the multi-channel EEG signal into meaningful components. Towards this direction, the method of independent component analysis (ICA) provides a tool of EEG decomposition into spatially fixed, timely localized, maximally independent components. This decomposition is compliant with the neurophysiological attributes of brain sources and has received significant attention in ERP analysis [6,7]. Studies have shown that ICA applied to EEG datasets can separate data into physiologically and functionally distinct sources, while separating non-brain artifact signals, as eye movement, line noise and muscle activities. Furthermore, under the assumption of spatially consistent sources, the ICA decomposition can be performed in a concatenated trials scheme, with the EEG signal extended by one trial following the other, in the same way for each channel. Besides its increased stability and generalization capabilities, the concatenated trials approach has the add-on advantage of preserving the correspondence of components throughout the trials, while it is effective in recovering the inter-trial variability of sources (derived components) [5]. Thus, the content of each ICA component can be subsequently analyzed in several perspectives including its topological origin, the time and frequency distribution of its energy, as well as its coherence over trials.

Focusing on the analysis of the content of EEG signal rather than its concrete components, a more detailed decomposition scheme has been proposed on the basis of principal component analysis (PCA) of the TF energy spectra of all EEG channels [8,9]. This analysis decomposes the energy content of the entire set of EEG signals into orthogonal, spatially localized components, which are consistently induced by all channels. In this form, the PCA decomposition acts as a well established data reduction scheme in order to extract the major characteristics composing the entire ERP data, from the wealth of information embedded in its multichannel TF representation. In this paper we implement TF decompositions by means of the Wavelet transform using the complex Morlet wavelet functions. Brief details of the ICA and PCA decompositions are provided in Sections 2.1 and 2.2, respectively.

In order to simultaneously cope with the content of multi-trial EEG recordings, which is often the case in the analysis of evoked response experiments due to the low SNR at individual trials, PCA decomposition has been attempted on summary TF maps of all channels. Specific forms of content summarization schemes include the TF energy maps of the average signal for each electrode, as well as on the TF inter-trial coherence maps, which are obtained form the TF signal maps of the individual trials and reflect the coherence among trials at each time and frequency bin. Inter-trial coherence metrics that can be employed in the construction of summarization TF maps are presented in Section 2.3. In this form, which is extensively studied in our work, the PCA analysis reveals major components that are consistent with all trials and all signal channels. Notice that the potential of summary TF maps in compacting the detailed content of a time-signal in multiple trials can be very useful for content analysis and visualization of EEG channels, but also for content analysis and characterization of independent components, in order to assess their significance in the formation of the recorded EEG signal. Thus, we propose to filter the EEG recordings by back-projecting only the significant component identified by the TF content analysis, as described in Section 2.4.

2.1 Independent Component Analysis on EEG Data

Let the n EEG channels be arranged as rows of a matrix X with dimensions nxt, where t denotes the number of signal samples. Independent component analysis performs blind separation of the observed data X using the restriction that the resulting components arranged in a similar form in a component matrix S are maximally independent. Alternatively, ICA computes an unmixing matrix W, which multiplied with the observed data X results in a matrix S of independent components. Mapping the weights of W-1 on the electrodes provides a scalp topography of the projection of each component. This presumes that the source locations are spatially fixed and the independent components reveal the time-course activation of each source. In the examples section we utilize the scalp topography of each components in order to infer the brain area of its origin. Another fundamental assumption in ICA decomposition is that the number of sources is the same as the number of electrodes, which is questionable given the wealth of information encoded into the EEG signal. Applying ICA decomposition to few data channels should, thus, result in some or all extracted components being mixture of sources, summing up the activity from more than one neuronal assembly. Even in this case, however, ICA should efficiently arrange for these mixtures to have minimal common or mutual information [4]. In this paper, we attempt to provide a further unmixing of information sources by filtering the EEG channels from potential noise sources and preserving only relevant ICA components in the filtered EEG signal.

2.2 Principal Component Analysis on Time-Frequency Data

The PCA approach employed here is a general data reduction technique for TF signal representations. Methods developed for this purpose are often simplistic, considering the entire surface of TF representation as a collection of time-series signals each filtered to a certain range of frequencies. The PCA method employed here was recently developed [7], offering a data driven method for decomposing a dataset of TF surfaces. The application of PCA to timefrequency energy is much the same as its application to signals specified in the time or frequency domain.

Each timefrequency surface is rearranged into a vector, recasting the timefrequency energy into concatenated time segments each of different frequency content. In this form, the PCA data is formulated into a matrix of trials in rows and different points of activity (different timefrequency point) in columns. This arrangement is still amendable to decomposition, since PCA makes no assumption about the ordering of the columns for decomposition.

Overall, starting from the time-frequency surface of each channel, we form a three-dimensional matrix of channel x time x frequency. Then, we concatenate the time and frequency dimensions into a single dimension, obtaining the representation of the two-dimensional data matrix X (in the dimensions of: channels x time-frequency). The PCA analysis is performed on this domain, resulting in the principal-components matrix S. Finally, by folding this matrix back to three dimensions, we obtain the time-frequency surfaces of the principal components. The number of principal components can be decided in terms of the singular values of the decomposition.

2.3 Coherence Metrics

In order to quantify phase locked coherence along the trials, we can utilize the intertrial coherence TF maps for all channels [8]. This measure, referred to as phase intertrial coherence (PIC), reflects the phase-locked consistency among trials and is derived from the analysis of TF maps of individual trials at each specific channel. It is defined as

$$c_{pic}[k] = \frac{|\sum_i x_i[k]|}{\sum_i |x_i[k]|} \leq 1 \tag{1}$$

where $X_i[k]$ denotes the frequency coefficient at the i-th trial and the k-th frequency tick. Equality holds if and only if all trials involve the same signal with the same phase. This metric is expanded to the time-frequency representation of a signal, with k and t indicating the frequency and time ticks, respectively.

For the quantification of event related but not phase-locked activity, we propose a related measure for the analysis of non-phase locked activity based on the energy distribution over the TF domain for all different trials of the signal. More specifically, we introduce the so-called phase-shift intertrial coherence (PsIC), which is defined as

$$c_{PsIC}[k,t] = \frac{\sum_i |x_i[k]|^2}{max_{k,t} \sum_i |x_i[k]|^2} \leq 1 \tag{2}$$

where equality implies the same magnitude of X[k,t], even with different shifts at each trial, so that it highlights frequency bands of increased energy in all trials. Recall that these measures can be applied to summarize the information content within a single electrode or a single ICA component over all trials.

These maps, along with the TF energy spectrum, will be used for the characterization of relevant content, as each one emphasizes on different aspects of synchronous activity. Notice that both the phase and the shift-phase coherence factors can be utilized as global metrics on a multi-trial signal (channel or component), measuring its overall intertrial coherence (preferably at specific bands). In this form, they can be effectively used for significance ranking of components in each band. Alternatively, they can be computed for each tick in the time-frequency representation, in order to provide timely localized maps of the coherence over trials.

2.4 Selection of ICA Components for EEG Filtering

In this paper the analysis of information content is attempted on both the original EEG signal and its filtered version, which engages only relevant ICA components. Recall that these components are obtained from a concatenated trials ICA decomposition, so that they can easily be split into the corresponding trials. The selection of relevant components is of particular interest in this work. We propose and test two different selection schemes, guided by specific assumptions on the properties of underlying brain sources. In the first scheme we exploit the fact that the P300 waveform has a specific form in both its time structure and its frequency content, which should be exemplified in all relevant components of the ERP signal. Thus, for each ICA component we consider its back-projection into channels and we form the average (over all trials) back-projected component on channel Cz. We select those components that express high correlation with the average EEG signal on Cz and also reflect high frequency energy at the frequency bands of interest. In the second scheme, we exploit the coherence metrics on the TF decomposition of each component over all trials, so that we select components with maximally coherent activation (phase or non-phase locked) on the particular frequency bands of interest (mainly delta, theta and alpha).

Both schemes allow for the separate study of each ICA component at different frequency bands. This study of components is in accordance with the nature of sources comprising the EEG signal, since the neuronal assemblies organize and operate at specific frequency bands. Furthermore, through the separate consideration of frequencies, we can allow for the preservation of non-phase locked activities, which would have been lost in considerations of the time-domain signal (e.g. averaging, which suppresses the energy of individual frequency components occurring at the same time interval) [2]. Notice that even though the two schemes originate from different considerations, they both share similar attributes. The first scheme is essentially based on the time and frequency content of components, whereas the second one considers their consistency at specific frequency bands throughout the trials. Their conceptual similarity is further verified in the examples section, where the selection of ICA components is discussed.

3 Experimental Results

3.1 Experiments on Simulation Data

The simulated dataset is used in order to demonstrate the effects of spatial mixing and the need for ICA preprocessing. Toward this direction, we created a dataset consisting of five sources, each sampled at 1024Hz, which are mixed to only four channels using a 5x4 mixing matrix. The mixing weights for each channel were calculated as to reflect sources arriving from different origins (different topographies). The first four sources simulate signal peaks at different time locations and at 3, 6, 8 and 9HZ, respectively, whereas the fifth source simulates noise with ongoing EEG power spectrum. The TF energy maps of the four mixed channels are depicted in figure 1a. Following PCA decomposition of the TF maps, the resulting principal components are depicted in figure 1b, where we observe that the information content cannot be efficiently unmixed; the principal components form a mixture of the different sources in the TF surface. In the sequel, we apply ICA decomposition on the dataset. The independent components can separate the EEG-like noise, but the other components are mixture of the initial sources. By removing the noise-like component and back-projecting the remaining components to the channels, we obtain a filtered dataset, whose TF maps are depicted in figure 1c. Despite the remaining effects, the channels appear as much simpler mixtures. Applying PCA decomposition on these TF surfaces provides the results of figure 1d, which separate well each single source utilized in the mixture. The color-maps for all representations range from minimum to maximum values individually for each component; the actual values of color-bars are not important, since we only consider the content of each component and do not compare components themselves.

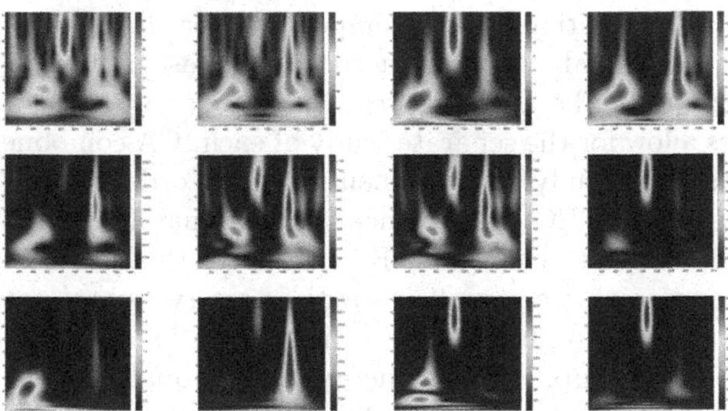

Fig. 1. Time frequency measures for the simulated data. 1a) First row: TF energy maps of the four channels; 1b) second row: four PCA components of original TF energy maps; 1c) third row: four PCA components of filtered TF maps.

3.2 Experiments on Real EEG Data

We applied the proposed scheme for improving content identification on 27-channel recordings from an auditory oddball experiment. The dataset was provided by the Ecological University of Bucharest, Romania and was obtained after an approved ethics protocol. Recordings were captured from 9 healthy participants (3 females and 6 males), who had no history of neurological or psychiatric disorder. Signals were digitally sampled at 1024Hz, with a high pass filter of cut-off frequency 0.016Hz. A stimulator provided 40 2kHz target tones (20%) and 160 1kHz non-target tones (80%). The inter-stimulus interval was 1.29s. The records used for analysis last 683ms and contain 700 samples after the stimulus. The auditory oddball experimental set-up is expected to produce both phase-locked oscillations, especially in the theta and delta bands related to P300 activity (including P3a and P3b components [10]), and non phase-locked (induced) oscillatory activity, particularly related to alpha-range event related desynchronization (ERD).

In order to filter the EEG recordings, we applied ICA on the concatenated trials dataset of each subject. For the 27 resulting independent components, we attempted an evaluation of their significance in the original signal, based on the two selection approaches aiming at discriminating event related activity from irrelevant brain and artifact activations. Recall that the first scheme relies on the similarity of the average back-projected component with the form of the recorded average signal, whereas the second scheme utilizes the intertrial coherence measures as to assess the relative consistency of components throughout the trials. A good subset of components selected by the two methods is common, whereas other components are structurally different. More specifically, the common components have frequency content primarily in the delta and theta bands. A closer inspection revealed that these components reflected phase-locked activations. This result was expected, since the method based on the average ERP waveform is biased towards phase-locked activity, which is primarily expressed in these specific bands. On the other hand, for the method using coherence measures the results are more balanced with components expressing phase-locked theta and delta bands as well alpha non-phase locked activity. Some selected components are displayed for comparison in figure 2.

Regarding the information content of the original and filtered EEG with the proposed approach, the results are illustrated in figures 3 and 4, respectively. From the measures in figure 3 regarding the original EEG recordings, we can observe that the principal components have mixed activations in frequency content, which obscures the evaluation of these findings. Alternatively, for the filtered EEG in figure 4, the results reveal more clear information regarding the underlying frequency activities. In particular, the energy of the fourth component in Fig.3 (fourth column) has faded out for all three measures considered, indicating that no useful information has been allocated to this component. Fig. 4 presents a different image, where all four components bear useful information. Furthermore, the PCs of the two coherence measures depicted in rows 2 (PIC measure) and 3 (PsIC measure) reflect better frequency concentration in the filtered compared to original signal. In particular, the second component (2nd column) of Fig.3 appears to be distributed into multiple frequency bands, whereas its counterpart of Fig.4 reflects good localization, i.e. theta phase-locked and alpha non-phase locked activity.

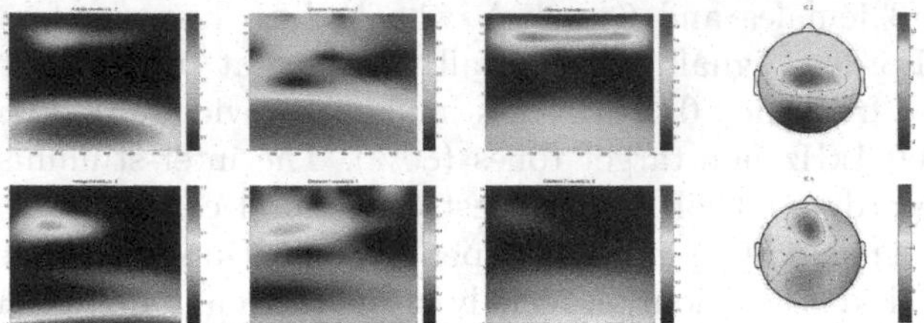

Fig. 2. Time frequency measures for selected (two) ICA components; each row depicts one component. First column displays the average TF energy, second column: phase-locked coherence, third column: non-phase-locked coherence, fourth column: brain topography of component.

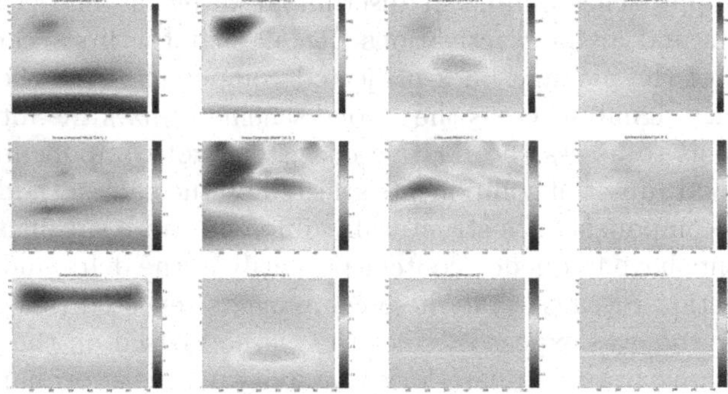

Fig. 3. First four principal components of coherence measures (PCA applied on time-frequency surfaces). Original data decomposition: 3a) First row displays the PCs of average TF energy, 3b) second row: PCs of phase-locked coherence, 3c) third row: PCs of non-phase-locked coherence.

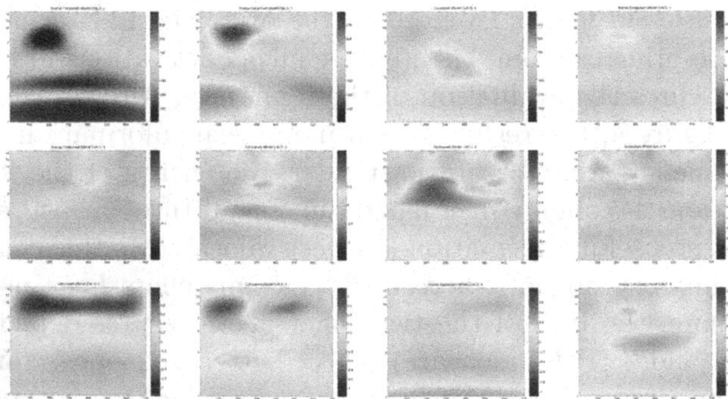

Fig. 4. First four principal components of coherence measures (PCA applied on time-frequency surfaces). Filtered data decomposition: 4a) First row displays the PCs of average TF energy, 4b) second row: PCs of phase-locked coherence, 4c) third row: PCs of non-phase-locked coherence.

4 Conclusions

The methodology developed in this paper addresses several concepts useful in the analysis of event-related EEG recordings. First, it provides measures for identifying and separating phase from non-phase locked activity and facilitates the rejection of noise activity and artifacts. Furthermore, our analysis provides the means of summarizing the extensive time-frequency information content embedded into a multi-trial, multi-channel EEG signal by means of coherence measures. Our methodology makes a clear distinction between signal and content decomposition for complex multi-trial, multi-channel EEG signals, the first using ICA on concatenated trials and the second using PCA on the summary TF maps for all channels. Finally, it demonstrates the benefits of pre-filtering the EEG signal as to remove the effects of irrelevant sources in the analysis of the relevant content.

Acknowledgments

Present work was supported by Biopattern, IST EU funded project, Contract no: 508803. The authors would like to thank Prof. Cristin Bigan at the Ecological University of Bucharest, Romania for kindly providing the EEG dataset.

References

1. Onton, J., Makeig, S.: Information-based modeling of event-related brain dynamics. Elsevier Progress in Brain Research book series (February 2006)
2. Tallon-Baudry, C., Bertrand, O.: Oscillatory gamma activity in humans and its role in object representation. Trends in Cognitive Sciences 3(4), 151–162 (1999)
3. Makeig, S., Debener, S., Onton, J., Delorme, A.: Mining Event-Related Brain Dynamics. Trends Cogn. Sci. 8, 204–210 (2004a)
4. Jung, T.P., Makeig, S., et al.: Imaging Brain Dynamics Using Independent Component Analysis. Proc. of the IEEE 89(7), 1107–1122 (2001)
5. Tsai, A.C., Liou, M., Jung, T., Onton, J.A., Cheng, P.E., Huang, C., Duann, J., Makeig, S.: Mapping single trial EEG records on the cortical surface through a spatiotemporal modality. NeuroImage 32, 195–207 (2006)
6. Bell, A.J., Sejnowski, T.J.: An Information-Maximization Approach to Blind Separation and Blind Deconvolution. Neural Comput. 7, 1129–1159 (1995)
7. Bernat, E.M., Williams, W.J., Gehring, W.J.: Decomposing ERP Time-Frequency Energy using PCA. Clinical Neurophysiology 116, 1314–1334 (2005)
8. Tallon-Baudry, C., Bertrand, O., Delpuech, C., Pernier, J.: Stimulus Specificity of Phase-Locked and Non-Phase-Locked 40 Hz Visual Responses in Human. The Journal of Neuroscience 16(13), 4240–4249 (1996)
9. Makeig, S., Delorme, A., Westerfield, M., Jung, T., Townsend, J., Courchesne, E., Sejnowski, T.J.: Electroencephalographic Brain Dynamics following manually responded visual targets. Plos Biology 2, 747–762 (2004)
10. Polich, J.: Updating P300: An integrative theory of P3a and P3b. Clinical Neurophysiology 118, 2128–2148 (2007)

Outlier Analysis in BP/RP Spectral Bands

Diego Ordóñez[1,*], Carlos Dafonte[1], Minia Manteiga[2], and Bernardino Arcay[1]

[1] Department of Information and Communications Technologies, University of A Coruña, 15071, A Coruña, Spain
{dordonez,dafonte,cibarcay}@udc.es
http://fic.udc.es

[2] Department of Navigation and Earth Sciences, University of A Coruña, 15071, A Coruña, Spain
manteiga@udc.es

Abstract. Most astronomic databases include a certain amount of exceptional values that are generally called outliers. Isolating and analysing these "outlying objects" is important to improve the quality of the original dataset, to reduce the impact of anomalous observations, and most importantly, to discover new types of objects that were hitherto unknown because of their low frequency or short lifespan. We propose an unsupervised technique, based on artificial neural networks and combined with a specific study of the trained network, to treat the problem of outliers management. This work is an integrating part of the GAIA mission of the European Space Agency.

1 Introduction

This study is an integrating part of the European Space Agency's mission GAIA, and is based on a previously defined taxonomy of celestial bodies that will be observed by means of a spectrophotometer on board. Other groups that work on the same project will be in charge of classifying the sources into each of the predefined object types. But there remains a certain group of rare objects that, even though they undoubtedly belong to one of the predefined classes, cannot be identified to belong to a concrete type with an adequate level of reliability. On the one hand, the proposed outliers analysis tries to identify clusters in this particular group of objects so as to be able to analyse them more efficiently. On the other hand, our analysis will also provide useful information to other work groups in the shape of a study of residual examples that were not classified appropriately.

Spectral classification is a well-known problem in the field of astrophysics. Its object of study is the electromagnetic radiation of the light spectrum that is emitted by stars and other objects. At first spectral classification was carried out by hand, but technological evolution and new data sources have put at

* Spanish MEC project ESP2006-13855-CO2-02.

C. Alippi et al. (Eds.): ICANN 2009, Part II, LNCS 5769, pp. 378–386, 2009.

our disposal such large amounts of information, that the manual procedures became too slow and subjective and research started focusing on automatized techniques. In that respect, Artificial Intelligence techniques (especially Artificial Neural Networks and Expert Systems) have proven their usefulness in various spectral classification tasks ([3], [7], [2],[4], [6], [1], [5]).

Modern telescopes are equipped with spectrometers that cover a significant number of objects per "frame". Not only current and planned data sources, but also spatial missions such as the GAIA mission (whose launch is foreseen for 2011), will provide large amounts of spectra belonging to various components of our galaxy. This type of information must be handled automatically. A long and complex data analysis will allow scientists to transform GAIA's signal into parameters that can be used to define the nature of the observed objects and, finally, classify and parameterize them.

In order to manage the tasks and human resources, the GAIA team has divided the work into areas or Coordination Units (CU); each CU is subdivided into work groups with specific tasks or Working Packages (WP). Our research team is part of CU 8 and in charge of WP 36, called "Outlier Analysis". Our work consists in analysing the objects that other teams, i.c. WP 21 (Discrete Source Classifier) and WP 24 (Object Clustering Analysis), are unable to classify with an acceptable degree of reliability and have identified as "outliers".

2 Objectives

Automatic information processing techniques are not flawless, and for certain objects it will be impossible to obtain a classification with an adequate degree of reliability. This is due to various reasons: because their nature is different from that of the known objects, because of measurement errors, because the objects (e.g. a star) are very distant, or simply because a certain type of noise accompanies the information. These objects are called "outliers" and constitute our object of study as an integrating part of the GAIA research consortium.

Therefore the main objective of the system consists in performing an initial approach to data analysis on outliers detected from RP/BP spectra. Outliers are objects that cannot be identified by DSC (GWP-S-8310) or OCA (GWP-S-8350) as belonging to any known GAIA-class. The objective of the Outliers Analysis WP is to analyse the data on outliers, in particular by submitting them to non-supervised cluster analysis and establishing the natural classes both by statistical methods and Artificial Intelligence techniques. A secondary objective of this WP is to label such classes in order to check whether any of them are misclassified known objects.

Initially, all the resulting groups of the outlier analysis are unknown, meaning that we are able to identify them but unable to give them a name (or label), since we do not, in principle, know what they are. Subsequently, each group identified by module Outlier Analysis can reflect two different realities:

1. The objects that are classified under this identifier cannot be assigned to a label
2. Some of the already grouped outliers could be assigned a label, either by identifying them in the position error boxes of selected astronomical databases and surveys, or with the help of a human expert who has identified the common properties among objects populating that outlier class. As soon as this assignation has taken place, the data that are classified under this identifier will be named according to the explanatory label.

The label is only a more or less appropriate name that a user can give to the cluster. The really important data are the objects that are grouped together in a cluster, because they are related in some way. The objective of this work is not to discover what the relationship is, but to carry out the classification.

3 Material and Methods

3.1 Data

Our object of study is the electromagnetic radiation of the objects that are observed by the GAIA telescopes. At the present moment, the data of which we dispose are simulated and compiled in spectral libraries. Objects in the spectral libraries BP/RP (Blue Photometric / Red Photometric) are simulated, as are the parallax and the proper motion, including random error. GAIA observed every part of the sky between 40 and 200 times in the course of five years. GAIA prism spectra are obtained in two channels, one for the blue channel (BP) and one for the red channel (RP), with cut-offs defined by the (silver) mirror response in the bluechannel, CCD QE in the red channel, and bandpass filters in the middle.

The outlier spectra identification was performed in the DSC WP, as mentioned above, and the resulting spectra set is the input of the algorithm that will be described in this work. The method considered for the outlier detection was to select all the sources with maximum probability for a single class of 0.6 or less, i.e. the type of objects that are considered not well classified and are therefore outliers for us.

Table 1. Outlier data distribution

Run 1	Object type	Number of objects	Percentage
	STAR	37845	(73%)
	GALAXY	1780	(4%)
	QUASAR	916	(2%)
	PHYSBINARY	11046	(21%)
		Total 51588	

Table 1 shows the distribution of the objects in types. It is not an optimal distribution, but it is the best that could be obtained until now to carry out the tests, whose results will be shown in section (5).

3.2 Clustering Algorithm

Our purpose is to group the outliers in such a way that the more their spectra resemble each other, the more likely they are to be classified into the same category. We also wish to know the relationships that exist between the identified groupings, with a view to subsequent analyses, such as the identification of spectra that are mixtures of objects.

We have opted for classification by means of Artificial Neural Networks, more concretely with unsupervised training algorithms. given the fact that we do not know the correct classification of the "outliers", we do not dispose of sufficient information to apply a supervised technique. We can nevertheless dispose of this information to carry out tests with the already trained algorithm, because at this point in the project we are working with simulated data. In any case, the unsupervised perspective is the adequate one for this work, because when working with real data, we have no knowledge on the concrete classification values.

In order to group the spectra that represent outliers, we have designed an algorithm that uses neural networks of the "SOM" type, i.e. the "Self Organizing Maps" ([8]) type, to carry out clustering tasks. This neural network architecture is a competitive architecture that applies unsupervised training and the Kohonen training algorithm ([8]) to identify the relationships of similarity between data, and allows us to carry out the groupings. A SOM network consists of two layers: the input layer and the output layer. Whereas the number of process elements of the input layer is determined by the dimension of the spectrum, for the output layer there is no concrete reference with which to determine the correct number to carry out the grouping. We have therefore tested various configurations in this layer, as can be seen in section 5.

3.3 Hardware and Software Tools

Our experiment required a rack of four servers equipped with two Intel Xeon Quad Core processors and 16 GB RAM each. This hardware architecture allowed us to launch a total of 32 parallel trainings (eight per computer) without any significant impact on the equipments productivity.

The neural networks and the processing algorithms were implemented in JAVA (requirement of the GAIA project). The neural networks were defined and trained with a tool that was developed by our research group: XOANE ([9]), *eXtensible Object Oriented Artificial Neural Networks Engine*, is a tool that allows us to arbitrarily shape network architectures, training algorithms, and tests. We use this framework instead of other, more popular ones (e.g: Matlab), because its execution times are shorter and it enables us to obtain intermediate results that allow us to select the best intermediate network to perform the clustering for each experiment.

4 Analysis Tasks

The Kohonen algorithm provides us with an unsupervised technique to train the self-organized maps. This means that we need measurements to quantify, a posteriori, the result of the learning task. We do not know a priori whether the training has resulted in a network that is well adjusted to the problem determined by the training patterns. To make this measurement, we focused the result analysis on trying to quantify the quality of the training by identifying the nature of the clusters that took shape at the network output (groupings of objects in determined process elements). For this analysis, we used the same training patterns, but taking into account their correct classification - this being an information that was not considered for the network learning because in a real situation it will not be available.

The experiment is marked by a series of problems inherent to the data. On the one hand, as could be observed in section 3.1, the training set is not well distributed between the different classes, due to the fact that we are treating outliers, i.c. spectra of objects that other groups could not assign to one of the existing object classes; this implies that we have no control over this distribution. On the other hand, some of the object spectra that we treat as outliers are a mixture of other objects, which only increases the confusion. It is only normal that the training is conditioned by these factors.

The proposed analysis method will try to measure these aspects on the basis of the way in which the Kohonen algorithm tries to group the output objects. The application of the learning algorithm has taught us that the process elements near the output represent object groups that, in turn, are similar and therefore more likely to be of the same type than those that were classified into distant process elements. This constitutes the basis of the entire analysis development: identify areas of process elements, at the output of the neural network, that represent the same object type.

4.1 Frequency Maps

Once the learning is accomplished, we analyse the resulting network: for each example that we know to be correctly classified, we register the result of its classification (process element that was activated); subsequently we carry out a recount for each individual process element to know how many objects of each type it was able to classify. In order to visualize this result we have reflected it in an image according to the following method:

- We select one of the known object classes that constitute the set of classes of the training set patterns.
- Each pixel of the resulting image represents a process element of the network.
- For the selected object type, we establish the pixel color according to the number of times that the selected process element was activated for this type of object, and represent this as follows: F_{ij}^{k}, i.e. we represent the number of times that the process element located in position (i, j) on the map is activated (remembering that the map is rectangular) for object type K.

– When all the examples have gone through the previous step, we scale the resulting values so that they can be drawn. The result is an image in grey scales, where 0 represents black and 1 represents white, based on the following formula: $f_{i,j}^k = \frac{f_{i,j}^k}{N_k}$, i.e. we calculate the number of times that the process element is activated compared to the number of elements of that class of which we dispose.

Figure 1 shows several images of this type. It gives us an idea of how the objects of a certain type are distributed among the network output and, hence, if the map can be segmented to define the areas that classify each object type.

5 Results

Section 4 has defined a method to carry out an a posteriori analysis of the result of a network training. This analysis is based on the way in which the training is performed and gives us a general idea of the dispersion of objects of a specific type on the map. At the same time, it is a tool that helps us to visually compare the existing confusion between certain objects and estimate how well the training has gone. We have converted this intuitive idea into something more fomal by elaborating a numeric comparative based on the analysis of a confusion matrix that informs us, for each object type, on the number and detail of badly classified objects. Figure 1 shows a classification example that visualizes how the training has distributed the classes between the different process elements of the network output layer.

When observing Figure 1, we can extract useful information. We can conclude, for instance, that galaxies constitute a clearly distinguishable group of objects, because when we compare their image with that of other objects, we find that they belong to areas where the other images show void spaces. On the other hand, quasars are the least numerous objects in the training set, and then there are the physical binaries and stars, whose high confusion level (upper part of map) may appear to be a bad result, but we must remember that physical binaries are also made of stars, and it is therefore to be expected that the spectra of both show certain similarities.

Since size restrictions of this article do not allow us to show the numerical analysis for a map of 2500 elements, we prepared the same experiment with a more reduced and manageable map of 10x10 process elements, shown in Figure 2, that assembles the objects of interest. We can observe the number of times a process element is activated after having fed the network with the training set. The grey tones represent the map segmentation, and the colours represent, going from light to dark tones, Physical Binaries, Quasars, Stars, and Galaxies.

Table 2 shows the confusion matrix for the classification example shown in Figure 2; the percentages refer to the map areas that represent the object types mentioned in the previous section.

When looking closely at the result, we can observe that the areas were selected so as to minimize (as much as possible, depending on the nature of the treated

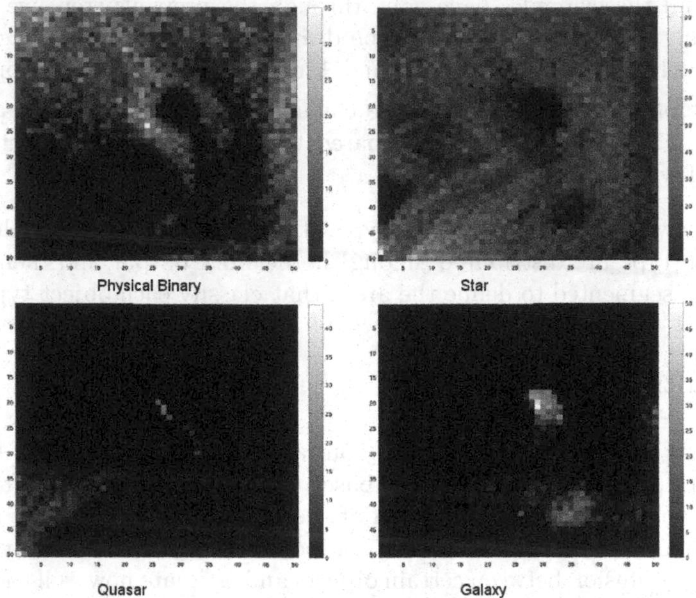

Fig. 1. Analysis of each object type with frequency maps

1549	1399	787	010	006	410	688	989	907	1207
1205	364	838	011	004	471	507	554	418	460
1197	870	693	005	005	677	560	243	359	381
1011	894	651	013	011	675	683	317	285	311
941	870	667	049	010	697	751	424	271	234
874	868	561	168	052	294	558	602	186	249
785	872	478	317	232	288	364	323	357	559
699	637	518	481	452	439	319	328	372	399
550	422	564	644	547	626	562	329	233	431
433	493	629	770	833	662	570	556	361	202

Fig. 2. SOM activation distribution

information) the false positives: the area selected for the galaxies represents only 66% of the galaxies, but since the percentage of other objects in the same area is very low, an activated process element in this area is very likely to be a galaxy.

The map shown in Figure 1 has the highest resolution considered in this context: it was configured with a distribution of 50x50 process elements. We have also tried to expand the map and observe its evolution: Figure 3 shows an example of a training experiment with increasing map sizes.

Table 2. Percentage of the distributions of the object types in the selected areas for each object type in the map (confusion matrix)

%	Phys. Bin.	Star	Quasar	Galaxy
Phys. Bin.	64	28	0	8
Star	31	52	8	9
Quasar	4	13	78	5
Galaxy	13	20	1	66

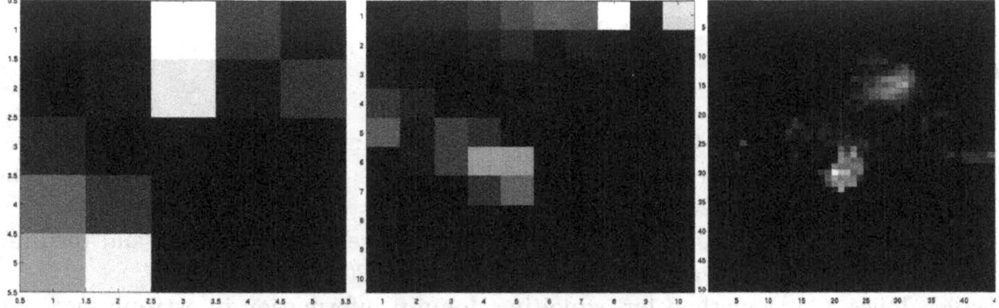

Fig. 3. Different SOM output layer dimensions: 5x5, 10x10, and 50x50 process elements

6 Conclusions

An inherent problem of information that proceeds from spectra such as the ones described in section 3.1 is that the objects strongly resemble each other in shape, which complicates their classification. In this respect, the size of the map is fundamental when segmenting the areas that represent certain object types: a map of reduced dimensions creates enormous confusion, because one and the same process element will be activated indistinctively for different object types and make the analysis useless. To counter this problem, we developed a strategy based on progressively increasing the map size until areas appear that identify, as reliably as possible, a certain type of object. However, the arbitrary increase of map sizes may also produce negative effects: the training time of a very large network increases exponentially, and after the learning there may remain a high percentage of process elements that, instead of representing one of the predefined classes, represent a misleading mixture of various, inconcrete object types. This is due to the fact that during the learning they were dragged towards an area of the input space under the influence of the neighbouring process elements.

The map configurations that we have tested are the following: 5x5, 10x10, and 50x50 in a rectangular map. This represents 25, 100, and 2500 process elements respectively at the network output. As can be observed in the Results section, more concretely in Figure 3, the definition of the map is a highly relevant factor, because it allows us to segment objects that in a more reduced map would be classified into one and the same process element.

We must consider the fact that we are working with outliers, i.e. objects that were already classified previously with another technique and identified as rare, "outlying" objects. The nature of the treated information, as well as the way of obtaining this information, result in training sets that are not well distributed among the different object types.

Acknowledgments

Spanish MEC project ESP2006-13855-CO2-02.

References

1. Allende, C., Rebolo, R., Garcia, R., Serra-Ricart, M.: The int search for metal-poor stars. spectroscopic observations and classification via artificial neural networks. The Astronomical Journal 120, 1516–1531 (2000)
2. Bailer-Jones, C.: Stellar parameters from very low resolution spectra and medium band filters. Astronomy and Astrophysics 357, 197–205 (2000)
3. Bailer-Jones, C.: A method for exploiting domain information in astrophysical parameter estimation. In: Astronomical Data Analysis Software and Systems XVII. ASP Conference Series, vol. XXX (2008)
4. Christlieb, N., Wisotzki, L., Graßhoff, G.: Statistical methods of automatic spectral classification and their application to the hamburg/eso sourvey. Astronomy and Astrophysics 391, 397–406 (2002)
5. Fiorentin, P., Bailer-Jones, C., Lee, Y., Beers, T., Sivarani, T., Wilhelm, R., Allende, C., Norris, J.: Estimation of stellar atmospheric parameters from sdss/segue spectra. Astronomy and Astrophysics 467, 1373–1387 (2007)
6. Hippel, T.V., Allende, C., Sneden, C.: Automated stellar spectral classification and parameterization for the masses. In: The Garrison Festschrift conference proceedings (June 2002)
7. Kaempf, T., Willemsen, P., Bailer-Jones, C., de Boer, K.: Parameterisation of rvs spectra with artificial neural networks first steps. In: 10th RVS workshop, Cambridge (September 2005)
8. Kohonen, T.: Self-organizing Maps. Springer Series in Information Sciences, Heidelberg (1995)
9. Ordonez, D., Dafonte, C., Arcay, B., Manteiga, M.: A canonical integrator environment for the development of connectionist systems. Dynamics of Continuous, Discrete and Impulsive Systems 14, 580–585 (2007)

ANNs and Other Machine Learning Techniques in Modelling Models' Uncertainty

Durga Lal Shrestha, Nagendra Kayastha, and Dimitri P. Solomatine

UNESCO-IHE Institute for Water Education,
Westvest 7, P.O.Box 3015, Delft, The Netherlands
{d.shrestha,kayas2,d.solomatine}@unesco-ihe.org
http://www.unesco-ihe.org/hi

Abstract. The paper presents examples of using ANNs and other machine learning (ML) techniques to assess uncertainty of a mathematical (computer-based) model M. Two approaches have been developed to estimate parametric and residual uncertainty, and they were tested on process based hydrological models. One approach emulates computationally expensive Monte Carlo simulations, and the second one uses residuals of a calibrated model M outputs to assess the remaining uncertainty of this model. ML models are trained to approximate the functional relationships between the input (and state) variables of the model M and the uncertainty descriptors. ML model, being trained, encapsulates the information about the model M errors specific for different conditions in the past, and is used to estimate the probability distribution of the model M error for the new model runs. Methods are tested to estimate uncertainty of a conceptual rainfall-runoff model of a catchment in UK.

Keywords: artificial neural networks, machine learning techniques, modelling uncertainty, clustering.

1 Introduction

Since a model in only an abstraction of reality, there is a good share of simplifications and idealisations. Models predictions are far from being exact and subjected to different degree of uncertainty. Uncertainty of the model predictions are resulting mainly from the inadequate model structure, input and parameter uncertainty. A deterministic model M of a real-world system predicting a system output variable y^* given input vector $\mathbf{x}$, initial condition of the state variables s_o and the vector of the parameters θ is considered. Let y be the measurement of an unknown true value y^*, made with error ε_y. Various types of errors propagate through the model M while predicting the observed output y and have the following form:

$$y = M(\mathbf{x}, s, \theta) + \varepsilon_s + \varepsilon_\theta + \varepsilon_x + \varepsilon_y. \tag{1}$$

where ε_s , ε_θ and ε_x are the errors associated with the model structure M, parameter θ and input vector $\mathbf{x}$, respectively. In most practical cases, it is difficult

C. Alippi et al. (Eds.): ICANN 2009, Part II, LNCS 5769, pp. 387–396, 2009.

to estimate the error components of Eq (1) unless some important assumptions are made. Thus, the different components that contribute to the total model error are generally treated as a single lumped variable and Eq (1) can be reformulated as:

$$y = M(\mathbf{x}, s, \theta) + \varepsilon. \tag{2}$$

where ε is the total remaining (or residual) error between the observed response y and the corresponding model response $\hat{y}$. Before running the model M, the components of the model, i.e. input data vector $\mathbf{x}$, initial conditions s_o, parameters vector θ and the model structure itself have to be specified, while the output or model response $\hat{y}$ and the state variable s are computed by running the model. These components may be uncertain in various ways to various degrees; the consequences of these uncertainties will be propagated into the model states and the outputs.

A number of methods were proposed to estimate model uncertainty of model M. In general, these methods can be grouped into six categories: (a) analytical methods (b) approximation methods, (c) simulation and sampling-based methods (e.g., [1]), (d) Bayesian methods (e.g., "generalized likelihood uncertainty estimation" (GLUE) [2]), (e) methods based on the analysis of model errors (e.g., [3]) and (f) methods based on fuzzy set theory (see, e.g., [4]).

Most of the existing methods (e.g., categories (c) and (d)) analyze the uncertainty of the uncertain input variables by propagating it through the model M to the outputs, and hence requires the assumption of their distributions. Most of the approaches based on the analysis of the model M errors require certain assumptions regarding the residuals (e.g., normality and homoscedasticity). Obviously, the relevance and accuracy of such approaches depend on the validity of these assumptions. The fuzzy theory-based approach requires knowledge of the membership function of the quantity subject to the uncertainty which could be very subjective. Furthermore, the majority of the uncertainty methods deal only with a single source of uncertainty. For instance, Monte Carlo (MC) based methods analyze the propagation of uncertainty of parameters θ (measured by the probability distribution function, pdf) to the pdf of the output. Similar types of analysis are performed for the input or structural uncertainty independently. Note, the methods based on analysis of model errors typically compute the uncertainty of the optimal model (with the calibrated parameters and the fixed structure), and not of the class of models (i.e. a group of models with the same structure but parameterized differently) as, for example, MC methods do.

MC based method for uncertainty analysis of the outputs of such models is straightforward, but becomes impractical in real time applications when there is no time to perform the uncertainty analysis because the large number of model runs is required. Shrestha et al. [5] presented a novel methodology to replicate computationally expensive MC simulations using machine learning techniques. In a separate line of research, Shrestha and Solomatine [6] presented the basis of a novel method to estimate the uncertainty of the optimal model i.e. residual uncertainty that takes into account all sources of errors without attempting to disaggregate the contribution given by their individual sources. The approach is referred to as an "uncertainty estimation based on local errors and clustering" (UNEEC). The method uses clustering and machine learning techniques to estimate the uncertainty of a process model by analyzing its residuals (errors). This paper presents some preliminary results of the experiments for both parameteric uncertainty and residual uncertainty of the output of the process based hydrological model with application to the Brue catchment in UK.

2 Methodology

2.1 Model of Residual Uncertainty

The UNEEC method estimates the residual uncertainty associated with the given model structure M, and parameter set θ by analyzing historical model residuals ε which is an aggregate effect of all sources of error. The historical model residuals (errors) between the model prediction $\hat{y}$ and the observed data y are the best available quantitative indicators of the discrepancy between the model and the real-world system or process, and they provide valuable information that can be used to assess the predictive uncertainty. The residuals and their distribution are often the functions of the model input variables and can be predicted by building separate model mapping of the input space to the model residuals.

Uncertainty estimated with the UNEEC method is consistent only for the given model structure and the parameter set θ. It does not mean that the model structure and parameter uncertainty are ignored, but it is assumed that the uncertainty associated with the wrong model structure, inaccurate parameter values, and observational errors (if any) are manifested implicitly in the model residuals. This type of uncertainty analysis based on the model residuals is different from the classical uncertainty analysis methods where uncertainty of parameters, input data (presented by pdf) or plausible model structures are propagated to the pdf of the output.

The UNEEC method consists of three main steps which are described briefly in this subsection.

Clustering the data. Clustering of data is an important step of the UNEEC method. Its goal is to partition the data into several natural groups that can be interpreted. By data we understand here the vectors of some variable (input) space, and the input space here means not only input variables of the process model, but also all the relevant state variables which characterizes different mechanism of the modeled process. It is assumed that the input data belonging to the same cluster will have similar characteristics and correspond to similar real-life situations. Furthermore, the distributions of the model errors within different clusters have different characteristics. This assumption would be reasonable to test before using this method. In hydrological modeling, for exampe, this assumption seems to be quite natural: a hydrological model is often inaccurate in simulating extreme events (high consecutive rainfalls) which can be identified in one group by the process of clustering resulting in high model residuals (wide error distribution). When data in each cluster belongs to a certain class (in this case, a hydrological situation), local error models can be built: they will be more robust and accurate than the global model which is fitted on the whole data.

Estimating probability distribution of the process model error. Real-life models are typically non-linear, complex and contains many parameters. This will hinder the analytical estimation of the pdf of the model error. Thus the empirical pdf of the model error for each cluster is independently estimated by analyzing historical model residuals on the calibration data. In order to avoid a biased estimate of pdf or its quantiles of the model error, it is important to check if there is any over-fitting by the process model on the calibration data. Note that when dealing with limited calibration data, the empirical distribution might be a very poor approximation of the theoretical distribution, so the reliability of such a method depends on the availability of data.

Since the pdf of the model error is estimated for each cluster, it depends on the clustering method used. For example, in the case of K-means clustering where each instance of data belongs to only one cluster, the quantiles are taken from the empirical error distribution for each cluster independently. However, in the case of fuzzy clustering method (FCM) where each instance belongs to more than one cluster, and is associated with several membership functions, the computation of the quantiles should take this into account. The following expression gives the pth [0, 1] quantile of the model error for cluster i:

$$ec_i^p = \varepsilon_t \quad t : \sum_{k=1}^{t} \mu_{i,k} < p \sum_{t=1}^{n} \mu_{i,t} \tag{3}$$

where t is the maximum integer value running from unity that satisfies the above inequality, ε_t is the residual associated with the tth data (data are sorted with respect to the associated residual), and $\mu_{i,t}$ is the membership function of the tth data to cluster i. This is not the only way of calculating quantiles for fuzzy clusters, and we tested several of them before choosing the one presented; unfortunately the space available does not allow for providing the details.

Building a model for probability distribution of the process model error. In order to predict the quantiles of the process model error for the unseen input vector, a machine learning model was built which will have predictive power after being trained using the calibration data. This model is referred to as an uncertainty model U. In order to train the model U, the quantiles of the model error has to be estimated for the individual input data vector from the historical data. The estimation of quantiles for the individual input data vector depends on the types of clustering techniques employed. In the case of fuzzy clustering an approach that can be termed fuzzy committee is used to compute the quantiles for each individual input data vector and given by:

$$e_t^p = \sum_{i=1}^{c} \mu_{i,t}^{2/m} ec_i^p / \sum_{i=1}^{c} \mu_{i,t}^{2/m} \tag{4}$$

where e_t^p is the pth quantile of the error for tth input data, ec_i^p is the pth quantile of the error for cluster i, and m is the smoothing exponential coefficient. Once the quantiles of the model error for each example in the training data are obtained, machine learning model U is constructed:

$$e^p = U^p(\mathbf{x}) \tag{5}$$

Model U, after being trained on input data X, encapsulates the pdf of the model error and maps the input vector $\mathbf{x}$ to the pdf or quantiles of the process model error. It is worthwhile noting that the model U can take any form, from linear to non-linear regression function such as an artificial neural network (ANN). The choice of the model depends on the complexity of the problem to be handled and the availability of data. Once the model U is trained on the calibration data X, it can be employed to estimate the quantiles or the pdf of the model error for the new data input.

The quantile of the predictive uncertainty of the model output can be estimated as:

$$y^p = \hat{y} + e^p \tag{6}$$

where y^p is the pth quantile of the model output. In order to estimate, for example, 90% prediction interval, it is necessary to build two models, U^5 and U^{95}, that will predict 5% and 95% quantiles, respectively. Details of the methodology can be found in Shrestha and Solomatine [6].

2.2 Model of Parametric Uncertainty

The MC simulation is performed by running the model M multiple times either changing the input data $\mathbf{x}$ or parameters vectors or even the structure of the model or combination of them. For assessing parametric uncertainty we assume that the model structure and the input data is certain (correct), so mathematically this can be expressed as:

$$\hat{y}_{t,i} = M(\mathbf{x}, \theta_i); \quad t = 1, 2, ..., n; \quad i = 1, 2, ..., s \tag{7}$$

where θ_i is the set of parameters sampled for ith run of MC simulation, $\hat{y}_{t,i}$ is the model output of the tth time step for ith run, n is the number of time steps and s is the number of simulations. Having large enough realizations of MC simulations, the statistical properties such as moments or quantiles or even cumulative distribution function of the model prediction at any time step can be estimated. If the model predictions are weighted by likelihood as done in GLUE methodology, then the prediction quantile at any time step is given by

$$P(\hat{y}_t < \hat{Q}(p)) = \sum_{i=1}^{n} w_i | \hat{y}_{i,t} < \hat{Q}(p) \tag{8}$$

where $\hat{y}_t$ is the model output at time step t, $\hat{y}_{t,i}$, is the value of model outputs at time t simulated by the model $M(\mathbf{x}, \theta_i)$ at simulation i, $\hat{Q}(p)$ is p% quantile,

w_i is the likelihood weight given to the model output at simulation i. Quantiles obtained in this way are conditioned on the inputs to the model, model structure, and the weight vector w_i.

Once the quantiles are estimated in the calibration period, the next step is to build the machine learning models that can learn the complex relationship between the quantiles and the input variables. Instead of building the model for quantiles directly, we build the model for the transferred variable $\Delta\hat{Q}(p)$:

$$\Delta\hat{Q}(p) = \hat{Q}(p) - \bar{y}. \tag{9}$$

where $\bar{y}$ is the calibrated model output. Model V encapsulating the functional relationship between the input data $\mathbf{x}$ and the variable $\Delta\hat{Q}(p)$ will take the following form:

$$\Delta\hat{Q}(p) = V(\mathbf{x}) + \xi. \tag{10}$$

where ξ is residual error in estimating quantiles. Once the model V is trained using the input data from the calibration period, the trained model is used to estimate the variable $\Delta\hat{Q}(p)$. Then the prediction quantile is computed by substituting the variable to the Eq (10):

$$\hat{Q}(p) = V(\mathbf{x}) + \bar{y}. \tag{11}$$

More details of the methodology can be found in Shrestha et al. [5].

3 Case Study

To test the two methods, a hydrological model is used in the role of model M. The Brue catchment, located in UK, is selected as the case study. The catchment has a drainage area of 135 km^2 with the average annual rainfall of 867 mm and the average river flow of 1.92 m^3/s, for the period from 1961 to 1990. Splitting of available data set is done as follows: one year hourly data from 1994/06/24 05:00 to 1995/06/24 04:00 was selected for calibration and data from 1995/06/24 05:00 to 1996/05/31 13:00 was used for the verification (testing).

A simplified version of the HBV-96 model was used as the process model to simulate river flows for the case study. The HBV model [7] is a rainfall-runoff model, which includes conceptual numerical descriptions of hydrological processes at the catchment scale.

4 Results and Discussions

A version of the HBV model with 9 parameters (4 parameters for soil, and 5 for the response routine) was used. The model was first calibrated using the adaptive cluster covering algorithm, ACCO [8], part of GLOBE tool [http://www.data-machine.com]. Nash and Sutcliffe efficiency (CE) was the objective function.

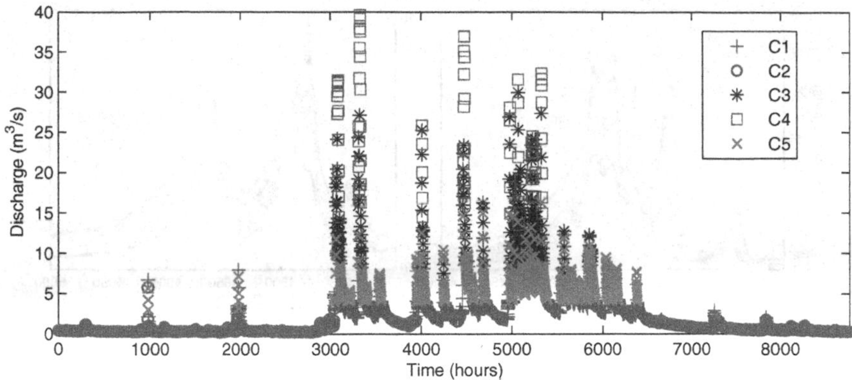

Fig. 1. Fuzzy clustering of the input data in the calibration period (from 1994/06/24 05:00 to 1995/06/24 04:00)

For the calibration period CE was 0.96 . The model was validated by simulating the flows for the independent validation data set, and CE was 0.83. The analysis of the model residuals in the calibration period shows that model residuals are highly correlated with the observed flows. The most of the high flows have relatively high residuals whereas low flows have small residuals. The presence of heteroscedasticity in the residuals is observed as well.

4.1 Residual Uncertainty

Fuzzy clustering of the input data for the calibration period is shown on Figure 1. The input data (in this figure discharge data) is attributed to one of those clusters (cluster C1 through C5) to which the data has the highest degree of membership. Cluster C2 contains input examples with very low runoff, whereas cluster C4 is associated with very high values of runoff. Fuzzy C-means clustering is able to identify clusters corresponding to various mechanisms of runoff generation process such as high flow, low flow, medium flow etc.

90% prediction interval was estimated by building two independent models for 5% and 95% quantiles of the model errors. Figure 2 shows the estimated 90% prediction intervals in the validation period. M5 model tree [9] was used as a regression model. It was found that 90.8% of the observed data points are enclosed within the computed 90% uncertainty bounds. Detail analysis reveals that 6.8% of the validation data points fall below lower bound whereas only 2.4% data points fall above upper bound. This difference is consistent with the fact that the simulation model M is biased and overestimates the flow on the validation data. In order to check if the percentage of the bracketed data points are more or less similar for any range of river flows, we have compared the histogram of the observed river flows which are outside the bounds with the observed ones. The results, which was not presented here, reveals that the distribution of the observed flows which are outside the uncertainty bounds is relatively consistent

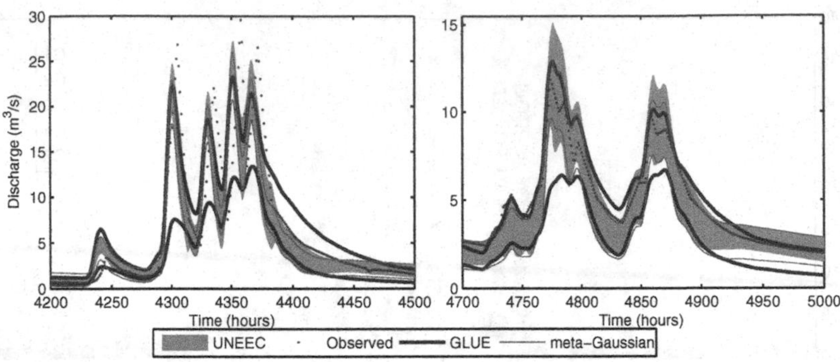

Fig. 2. Comparison of 90% prediction bounds estimated with the UNEEC, GLUE, and Meta-Gaussian approach in the validation period

with the observed ones. The average values of uncertainty bounds is 1.50 m^3/s. This value is reasonable if compared with the order of magnitude of model error in test data (root mean squared error was 0.97 m^3/s).

We compared the prediction bounds estimated by the UNEEC method with those computed by the GLUE (Beven and Binley, 1992) and meta-Gaussian methods [3]. The GLUE method is setup as follows: (a) Prior feasible ranges of parameter values are set to be the same as those used in automatic calibration of the HBV model; (b) Uniform sampling was used to sample the parameter set from the feasible parameter ranges; (c) Likelihood measure was based on the coefficient of model efficiency, CE criterion; (d) Rejection threshold values was set to 0.7; (e) The number of behavioral parameter sets was set to 3000. In meta-Gaussian method, the pdf of the model error, conditioned by the contemporary value of the simulation is carried out by using a meta-Gaussian model. The space does not allow to provide the detailed description of the experiments with these two methods and readers are referred to above citations.

The comparison results are presented in Figure 2. It is noticed that 62% and 84.8% of the observed discharge in the validation data fall inside the 90% prediction bounds estimated by the GLUE and Meta-Gaussian method, respectively. The widths of the prediction bounds estimated by GLUE and Meta-Gaussian methods are 1.28 m^3/s and 1.31 m^3/s, respectively.

4.2 Parametric Uncertainty

ANN-based uncertainty model V trained to replicate the MC simulations, and it was tested on the verification data set. It is worth mentioning that theoretically we can use othe machine learning methods instead of ANN, however in many experiemtns ANNs appeared to be the preferred option. We have also done experiments with other machine learning techniques, but the space does not allow to report these results.

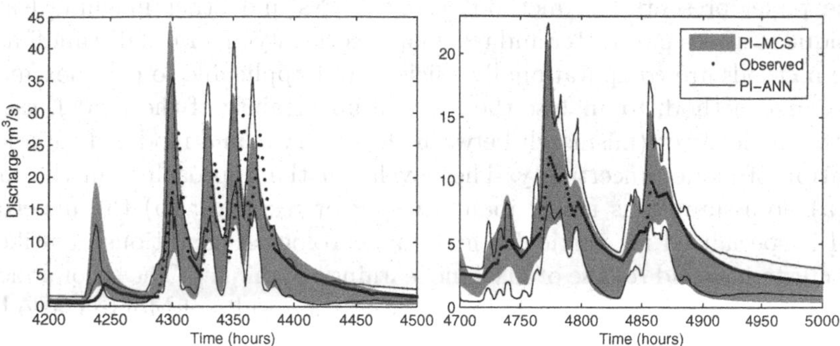

Fig. 3. Hydrograph of 90% prediction bounds estimated by MC simulation and ANN in verification period

The same data set used for calibration and verification of HBV model were used for training and verification of model V, respectively. However, for proper training of the ANN model the calibration data set is segmented into two sets; 15% of data sets for cross validation (CV) and 85% for training. CV data set was used to identify the best structure of ANN. In this paper, a multilayer perceptron network was used; optimization was performed by the Levenberg-Marquardt algorithm. The hyperbolic tangent function was used for the hidden layer with linear transfer function at the output layer. The maximum number of epoch was fixed to 1000. Trial and error method is adopted to determine the optimal number of neurons in the hidden layer, testing a number of neurons from 1 to 10. It was observed that 7 and 8 neurons give the lowest error on CV set for 5% and 95% quantiles, respectively and the following results correspond to six hidden nodes.

90% prediction interval was estimated by building two independent models of 5% and 95% quantiles of MC realizations. Figure 3 shows the hydrograph with the 90% uncertainty bounds predicted by ANN together with the MC simulation uncertainty bounds in the verification period. The correlation coefficient between (CC) for 5% quantiles of MC realizations and ANN prediction is 0.857 in verification. CC is 0.80 for 95% quantiles. ANN reproduces the MC simulations uncertainty bounds reasonably well, in spite of the low correlation of the input variables with the PIs. Inspite of some errors, the predicted uncertainty bounds follow the general trend of MC uncertainty bounds. Noticeably the model fails to capture the observed flow during one of the peak events (bottom left figure). Note however, that the results of ANN model and MC simulations are visually closer to each other than both of them to the observed data.

Detailed analysis reveals that estimated uncertainty bounds contain 77.00% of the observed runoffs, which is very close to the MC simulation result (77.24 %). The average width of prediction intervals estimated by ANN is narrower (1.93 m^3/s) compared to the value obtained with MC simulations (2.09 m^3/s). Further analysis of the results reveals that 14.74% of the observed data are below the lower uncertainty bounds whereas 8.01% of data are above the upper bounds.

The paper presents two methods using ANNs and other machine learning technique to assess parameter and residual uncertainty of a rainfall-runoff model. These methods are computationally efficient and applicable to complex models.

The first method, to analyse the residual uncertainty of the model, assumes that the model error (mismatch between the observed and modeled value) is an indication of model uncertainty. The novelty of the approach is in the following: (a) no assumptions made about the pdf of residuals; (b) the uncertainty model is specialized for particular hydrometeorological conditions identified by fuzzy clustering; and (c) use of machine learning techniques. The second method uses ML as a surrogate fast model replicating the results of Monte Carlo based simulations for parametric uncertainty analysis.

These methods were used to estimate the uncertainty of a conceptual hydrological model. The comparisons with other uncertainty estimation methods (GLUE, meta-Gaussian) show that the presented methods generate consistent and interpretable uncertainty estimates, and this is an indicator that they can be valuable tools for assessing uncertainty of various predictive models.

References

1. Kuczera, G., Parent, E.: Monte Carlo assessment of parameter uncertainty in conceptual catchment models: the Metropolis algorithm. J. of Hydrol. 211, 69–85 (1998)
2. Beven, K., Binley, A.: The future of distributed models: model calibration and uncertainty prediction. Hydrol. Processes 6, 279–298 (1992)
3. Montanari, A., Brath, A.: A stochastic approach for assessing the uncertainty of rainfall-runoff simulations. Water Resour. Res. 40, W01106 (2004) doi:10.1029/2003WR002540
4. Maskey, S., Guinot, V., Price, R.K.: Treatment of precipitation uncertainty in rainfall-runoff modelling: a fuzzy set approach. Adv. in Water Resour. 27, 889–898 (2004)
5. Shrestha, D.L., Kayastha, N., Solomatine, D.P.: A novel approach to parameter uncertainty analysis of hydrological models using neural networks. Hydrol. Earth Syst. Sci. Discuss. 6, 1677–1706 (2009)
6. Shrestha, D.L., Solomatine, D.P.: Machine learning approaches for estimation of prediction interval for the model output. Neural Networks 19(2), 225–235 (2006)
7. Bergström, S.: Development and application of a conceptual runoff model for Scandinavian catchments, SMHI Reports RH (7), Norrkping, Sweden (1976)
8. Solomatine, D.P., Dibike, Y., Kukuric, N.: Automatic calibration of groundwater models using global optimization techniques. Hydrol. Sci. J. 44(6), 879–894 (1999)
9. Quinlan, J.R.: Learning with continuous classes. In: Proc. AI92-Fifth Australian Joint Conf. on Artificial Intelligence, pp. 343–348. World Scientific, Singapore (1992)

Comparison of Adaptive Algorithms for Significant Feature Selection in Neural Network Based Solution of the Inverse Problem of Electrical Prospecting

Sergey Dolenko[1], Alexander Guzhva[1], Eugeny Obornev[2], Igor Persiantsev[1], and Mikhail Shimelevich[2]

[1] D.V. Skobeltsyn Institute of Nuclear Physics, M.V. Lomonosov Moscow State University, Leninskie Gory, Moscow, 119991 Russia
dolenko@srd.sinp.msu.ru

[2] S.Ordjonikidze Russian State Geological Prospecting University, RSGPU, 23 Miklukho-Maklai st., Moscow, 117997 Russia
eugenyo@mail.ru

Abstract. One of the important directions of research in geophysical electrical prospecting is solution of inverse problems (IP), in particular, the IP of magnetotellurics – the problem of determining the distribution of electrical conductivity in the thickness of earth by the values of electromagnetic field induced by ionosphere sources, observed on earth surface. Solution of this IP is hampered by very high dimensionality of the input data ($\sim 10^3$–10^4). Selection of the most significant features for each determined parameter makes it possible to simplify the IP and to increase the precision of its solution. This paper presents a comparison of two modifications of the developed algorithm for multi-step selection of significant features and the results of their application.

Keywords: inverse problems, feature selection, data compression, neural networks.

1 Introduction

This study is devoted to development of an algorithm for selection of significant features in neural network based solution of the inverse problem (IP) of magnetotellurics (MT) in geophysics. Solution of such problem is the process of creation of an operator mapping a data vector of the values of electromagnetic field observed on earth surface to the vector of the sought-for geophysical parameters of the section. These parameters include distribution of electrical conductivity in different points of the studied region, geometrical dimensions of separate sub-regions (geological structures) etc. Actual sections are extremely complex and require a very large number of parameters to describe them, thus leading to the known instability (incorrectness) of MT IP [1].

Neural networks (NN) are one of the instruments used to solve IP, including solving IP of MT [2]. The merits of NN used in solution of IP are high noise

C. Alippi et al. (Eds.): ICANN 2009, Part II, LNCS 5769, pp. 397–405, 2009.

immunity, stability in respect to contradictory data, possibility of training by examples etc. [3].

If an adequate analytical or computational model of the studied object is available, it can be used to produce (by solving the direct problem) a data array for NN training, with necessary representativity in the space of the determined parameters. It is clear that in this case the precision of problem solution depends on adequateness of the model. As a rule, MT models have a large number of input (observed) features and a large number of determined parameters even for two-dimensional problems, leading to a decrease in stability of the IP solution in the whole, and to a significant increase of the demands to computational resources. Thus, the dimensionality of 2D MT IP considered in this study is about $D_I = 6.5 \cdot 10^3$ at the input (the dimension of the vector of the observed values) and about $D_O = 3 \cdot 10^2$ at the output (dimensionality of the vector describing the distribution of electrical conductivity). The simplest method to reduce the output dimensionality of a problem is to divide it into D_O problems with one output each. However, this does not eliminate the difficulties connected with high input dimensionality of the problem.

Reduction of the input dimensionality can be achieved in two ways: by selection of the most significant input features for the given problem or by compression of the input data with transformation of coordinates in the initial feature space. For the latter, linear Principal Component Analysis (PCA) or non-linear PCA implemented as an auto-associative memory NN [4] can be used. This second approach is more computationally expensive, but it may provide stronger compression. However, the first approach makes it possible (along with saving computational cost) to check the sense of the extracted set of significant features from physical point of view, i.e. using the available *a priori* information on the possible influence of some or other input variables on the output variable. In this study, the reduction of the input dimensionality by selection of significant features (SSF) is considered.

2 Structure of Input Data of the Problem

The input data in the 2D MT IP considered in this study have the following structure. Registration of intensity of the variable electro-magnetic field is performed separately for modules and phases of electrical and magnetic fields; therefore, the data are naturally separated into 4 components: ρ_E, φ_E, ρ_H and φ_H. Then, registration is carried out separately at 13 frequencies. Finally, registration is performed in 126 points (pickets) on earth surface, separated from each other by a distance from 1 to 10 km. Therefore, the total dimensionality of the input data makes:

$$4(components) \times 13(frequencies) \times 126(pickets) = 6552. \quad (1)$$

It is clear that such knowledge of the input data structure enables using physical considerations to test the plausibility of the selected features. Thus, for example, electromagnetic vibrations at different frequencies penetrate earth down to different depth; the most significant features for determination of electrical conductivity in a given point should correspond to fields measured in the pickets nearest to this point. It was expected that using adaptive SSF algorithms will allow one to determine quantitative borders of the pointed effects.

3 SSF Algorithms

In this study, the following three-step algorithm for IP solution using SSF has been suggested and considered (Fig. 1, 2).

Prior to solution (step 0), the 'base' neural network (NN) is trained that solves the inverse problem without SSF (6552 inputs, 1 output). The statistics of this network are recorded for subsequent comparison with the results demonstrated by networks after SSF. In this study, SSF was performed by two alternative methods: with NN weight analysis (NNWA, [5]) only, and with the help of correlation analysis (CA).

To eliminate influence of random factors in initialization of NN weights and in random choice of the order of pattern selection, every experiment with NN included 5 runs of networks with the same parameters; the tables below display the results averaged over all such runs.

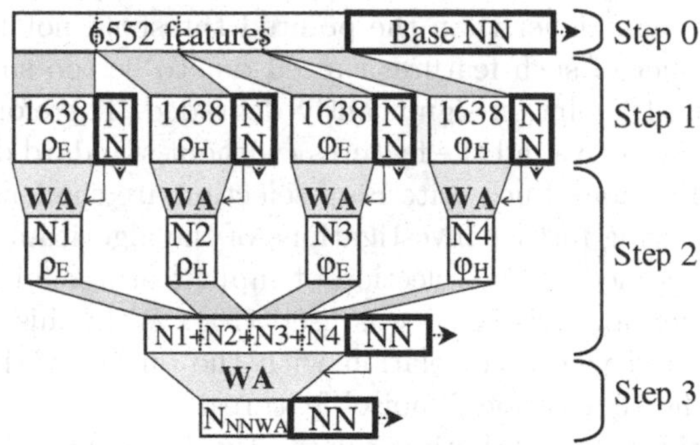

Fig. 1. General scheme of the algorithm with NNWA

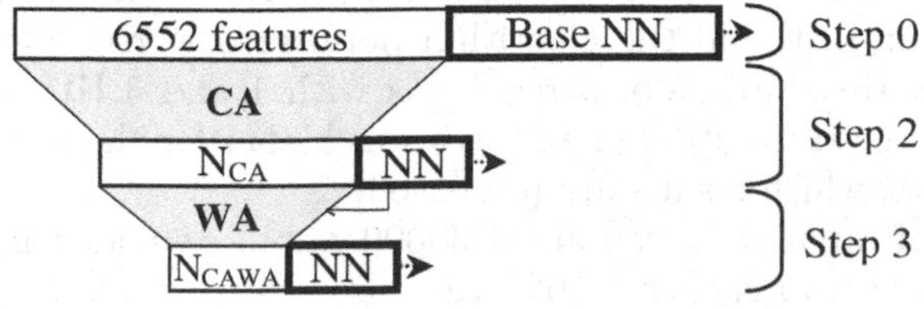

Fig. 2. General scheme of the algorithm with CA

At the first step of the first modification of the algorithm (Fig. 1), all the input features of the problem are divided into 4 groups corresponding to 4 components of the field, and for each of the groups the SSF task is performed by NNWA method. The quality of the IP solution by the obtained networks is estimated for each of the components. To determine the significance threshold for feature selection, the initial set of input features for each network was enlarged with 10 'noise' features taking on random values.

To take part in the second step, only those input features were selected that had larger weight than every of the 'noise' features in all the 5 runs of the first step. If there turned out to be too few of such features (20 or less), the features that had larger weight that all 'noise' features in 4 runs of 5, were also selected.

At the second step (Fig. 1), all the input features selected at the first step were combined together; new 10 'noise' features were added to them. This joint set was used to solve the IP again, and the most significant variables were determined by NNWA method.

It should be noted that at this step the determined values of significance of the 'noise' features turned out to be unsuitable as a threshold for further selection of features, as it was done at the preceding step. The reason of this was that the values of their significance were nearly always lower than those for all the informative (non-'noise') features. That is why at this step we used the average significance over all the features plus one standard deviation as the threshold value for selection. Only those features were selected to pass to the next stage, whose significance was higher than the pointed threshold not less than in 4 of 5 runs. If the number of such features turned out to be too small (20 or less), the precision of the IP solution significantly degraded. Therefore, in such cases the average significance over all the features (without standard deviation added) was used as the threshold. Such criteria of selection are challengeable and they may be modified during further investigations of the algorithm.

As an alternative method for selection of input features for the second step networks, correlation analysis (CA) was used (Fig. 2). In this case, only those features were selected whose correlation with the output variable was higher than that for all the 10 additional 'noise' features.

Finally, at the third stage in both variants (Fig. 1, 2) the IP was solved using only the significant variables selected at the second step.

In addition, note the following peculiarities of the algorithm implementation.

1. To perform NNWA, perceptrons with a single hidden layer (HL) were used. If a network with several HL was used for this purpose, the picture turned out to be less contrast, and the algorithm performed worse. For final solution of the IP at the third step, perceptrons with 1 and 3 HL were used. The precision of the IP solution in all cases was higher for the perceptron with 3 hidden layers, which results are presented.
2. The array of input data contained 30000 samples divided into sets in the following way: training set – 70%, test set – 20%, examination set – 10%. All the presented results were obtained on the examination set.

4 Results of the Inverse Problem Solution Using SSF

The results of the IP solution at different steps of the algorithm are presented in Tables 1, 2, 3 and in Figs. 3, 4, 5.

It should be noted that the inverse problems have been solved for the whole set of 336 output variables, corresponding to the values of electrical conductivity in 336 different blocks of the studied underground region. The presented results correspond to 170 blocks lying in the central part of the studied region; solution of the IP for all the other blocks is needed only for consequent solution of the direct problem based on the results of the IP solution in the whole region. Note that such calculations have been also conducted, and the errors for the reconstructed field values were small enough, thus confirming the validity of the approach in the whole. However, these results are not presented here, as they lie apart from NN application.

The quality of the solution depends strongly on the depth where the corresponding block is located – IP solution quality degrades with depth. This effect can be easily explained from physical point of view – but the results of the IP solution by NN allow estimating the quantitative borders and the scale of the pointed effect.

The most informative statistics describing the quality of the IP solution are the multiple determination coefficient, R squared (RSQ), and the mean squared error (MSE). Figures 3, 4 display the dependence of these statistics on depth (in km), each value being averaged over all runs of each network and over all studied underground blocks residing at the same depth, for each step of both algorithms. Fig. 5 displays the dependence of the number of selected features on depth, with the same averaging, for steps 2 and 3 of both algorithms (as the number of variables at steps 0 and 1 was constant, it is not displayed in Fig. 5). The error bars for each dependence correspond to the standard deviation for averaging over all blocks residing at the same depth (for some curves, the error bars are not visible as they lie within the size of the markers). For step 1, the results displayed are those for the best component (φ_H for depth 0.5 km and ρ_E for all the other depths).

Tables 1, 2 and 3 present the same statistics in numerical form, for several most informative depths, including the one with the best solution quality (1 km) and the one with the worst solution quality (22.5 km).

From the comparative analysis of the obtained results, the following conclusions can be drawn.

1. The suggested three-step algorithm of significant features selection turned out to be efficient. The results obtained at the third step demonstrate noticeable increase in the precision of the IP solution compared to the base network, with significant (about two orders of magnitude) reduction of the number of the input features.
2. The key issue in the efficiency of IP solution is presumably the number of input features. The optimal number of input features lies in the range from 20 to about 100, depending on the complexity of the specific problem being solved.

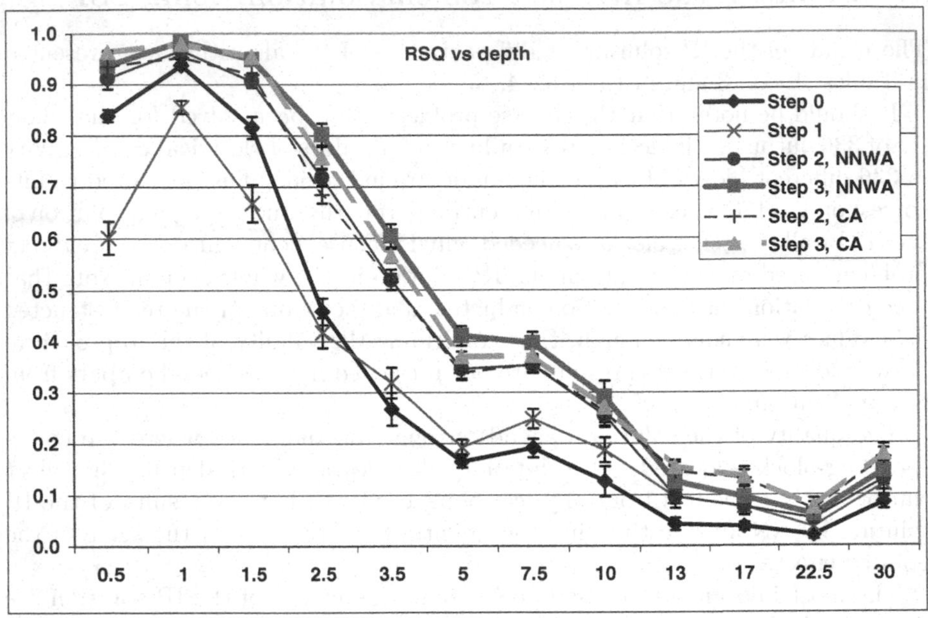

Fig. 3. Average values of RSQ vs depth in kilometers, for different steps of both algorithms

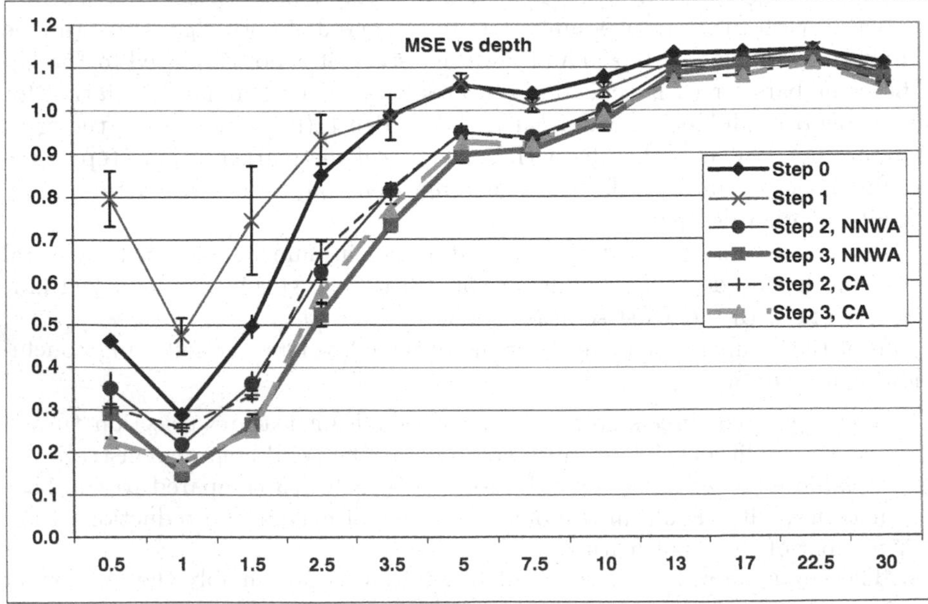

Fig. 4. Average values of mean squared error versus depth in kilometers, for different steps of both algorithms

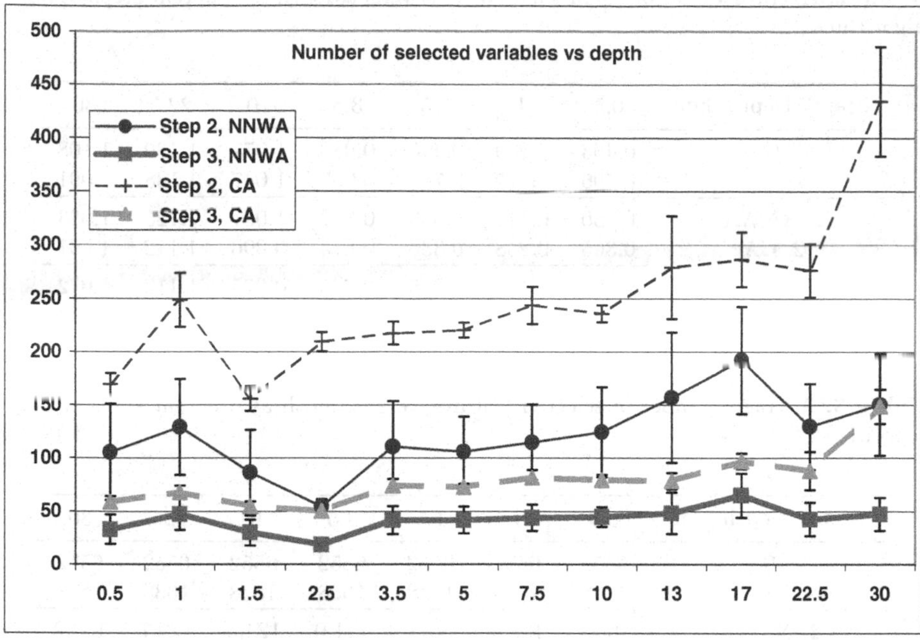

Fig. 5. Average number of selected features versus depth in kilometers, for different steps of both algorithms

Table 1. Average values of R squared versus depth at different steps of both algorithms

Step \ Depth, km	0.5	1	1.5	3.5	10	22.5	30
0	0.839	0.937	0.815	0.267	0.126	0.021	0.085
1	0.602	0.856	0.668	0.319	0.188	0.037	0.120
2, NNWA	0.912	0.967	0.906	0.519	0.255	0.053	0.130
2, CA	0.935	0.953	0.918	0.516	0.262	0.070	0.173
3, NNWA	0.937	0.984	0.949	0.605	0.291	0.061	0.145
3, CA	0.964	0.979	0.954	0.566	0.271	0.075	0.180

3. The way of selection of significant features influences the process of calculations noticeably. In general, use of much more computationally efficient CA as a method of preliminary selection at the second step (Fig. 2) brings results (Fig. 3, 4) similar with those obtained by the algorithm using NNWA at all the three stages (Fig. 1). However, great speed increase obtained by using CA at step 2 instead of training four NN at step 1, is to a significant degree compensated by much longer training of NN at steps 2 and 3, due to the fact that the number of significant features (used as input features of the

Table 2. Average values of mean squared error versus depth at different steps of both algorithms

Step \ Depth, km	0.5	1	1.5	3.5	10	22.5	30
0	0.463	0.289	0.496	0.988	1.077	1.139	1.108
1	0.796	0.475	0.745	0.983	1.047	1.138	1.091
2, NNWA	0.350	0.218	0.363	0.812	1.001	1.123	1.083
2, CA	0.305	0.258	0.336	0.815	0.996	1.112	1.061
3, NNWA	0.292	0.149	0.265	0.735	0.975	1.117	1.072
3, CA	0.226	0.173	0.252	0.768	0.987	1.109	1.050

Table 3. Average number of selected features versus depth at different steps of both algorithms

Step \ Depth, km	0.5	1	1.5	3.5	10	22.5	30
0	6552	6552	6552	6552	6552	6552	6552
1	1638	1638	1638	1638	1638	1638	1638
2, NNWA	105.7	129.4	86.3	111.0	124.6	129.7	150.2
2, CA	169.5	248.6	155.9	217.3	235.6	276.1	434.3
3, NNWA	32.9	47.2	29.8	41.6	44.6	42.5	47.2
3, CA	58.8	68.2	54.7	74.2	79.3	88.1	148.5

networks of step 2) selected by CA is much greater (nearly twice as great as that selected by the algorithm using only NNWA). In the whole, both algorithms have their merits: the one with NNWA only (Fig. 1) provides a smaller set of significant variables, while the one with CA (Fig. 2) in the whole works faster, especially if all NN calculations are performed on CPU without using GPU-based NN accelerator.

4. Analysis of the list of variables adaptively selected by the SSF algorithms shows that this list agrees well with *a priori* physical considerations on relative importance of the features. The most significant features are values of the fields measured in some area around the point at earth surface lying above the center of the block whose conductivity is being measured, in some range of the most significant frequencies. With increasing depth of the studied block, the area of the significant pickets becomes wider, and the set of the most significant frequencies is shifted towards lower frequencies. The obtained results provide quantitative estimates of these dependences.
5. With increasing depth of the studied block, the quality of IP solution rapidly degrades (Figs. 3, 4; Tables 1, 2). One can point out three depth ranges with different quality of IP solution: ranges of good IP solution (depths down to about 2.5 km), moderate quality IP solution (from 2.5 to about 10 km), and poor IP solution (deeper than 10 km). Note that the dependence of the

number of the selected significant features on the depth of the studied block is very weak (Fig. 5, Table 3).

5 Conclusion

An adaptive algorithm for selection of significant features in neural network based solution of the inverse problem of magnetotellurics, has been elaborated. The algorithm has two modifications: one is based only on neural network weight analysis, the other one also uses correlation analysis at one of the steps of the algorithm.

It has been demonstrated that use of the developed algorithm allows increasing precision of the IP solution, with simultaneous reduction of the necessary computational resources. Analysis of the set of the selected features depending on the depth of the point whose conductivity is determined, allows one to draw some conclusions that are interesting from physical point of view, and that match a priori physical considerations.

Further development of the studies will include use of other SSF methods, perfection of the selection algorithm, and further increase of the precision of the IP solution by combining classification of initial data and selection of significant features.

Acknowledgments. This study has been conducted under partial financial support of the Russian Foundation for Basic Research (RFBR), grant no. 07-07-00139-a.

References

1. Dmitriev, V.I., Berdichevsky, M.N.: Inverse problems in modern magnetotellurics. In: Spichak, V.V. (ed.) Electromagnetic Sounding of the Earth's Interior. Elsevier, Amsterdam (2007)
2. Shimelevich, M.I., Obornev, E.A., Gavryushov, S.A.: Rapid neuronet inversion of 2D magnetotelluric data for monitoring of geoelectrical section parameters. Annals of Geophysics 50(1), 105–109 (2007)
3. Dolenko, S.A., Dolenko, T.A., Persiantsev, I.G., Fadeev, V.V., Burikov, S.A.: Solution of Inverse Problems of Optical Spectroscopy with the Help of Neural Networks // Neirokompjutery: razrabotka, primenenie (Neurocomputers: Development, Application) (1-2), 89–97 (2005) (in Russian)
4. Hassoun, M.H.: Fundamentals of artificial neural networks, pp. 97–102. The MIT Press, Cambridge (1995)
5. Gevrey, M., Dimopoulos, I., Lek, S.: Review and comparison of methods to study the contribution of variables in artificial neural network models. Ecological Modelling 160, 249–264 (2003)

Efficient Optimization of the Parameters of LS-SVM for Regression versus Cross-Validation Error

Ginés Rubio[1], Héctor Pomares, Ignacio Rojas, Luis Javier Herrera, and Alberto Guillén

Department of Computer Architecture and Technology, University of Granada, C/ Periodista Daniel Saucedo sn, 18071 Granada, Spain
grubio@atc.ugr.es

Abstract. Least Squares Support Vector Machines (LS-SVM) are the state of the art in kernel methods for regression and function approximation. In the last few years, these models have been successfully applied to time series modelling and prediction. A key issue for the good performance of a LS-SVM model are the values chosen for both the kernel parameters and its hyperparameters in order to avoid overfitting the underlying system to be modelled. In this paper an efficient method for the evaluation of the cross validation error for LS-SVM is revised. The expressions for its partial derivatives are presented in order to improve the procedure for parameter optimization. Some initial guesses to set the values of both kernel parameters and the regularization factor are also presented. We finally conduct some experiments on a time series data example using a number of methods for parameter optimization for LS-SVM models. The results show that the proposed partial derivatives and heuristics can improve the performance with respect to both execution time and the optimized model obtained.

1 Introduction

Least Square Support Vector Machines (LS-SVMs) [1] have been successfully applied to regression and function approximation problems. Although they have proved to be very reliable obtaining good performances and very accurate results, they present some drawbacks:

- The selection of the kernel function could be difficult.
- The optimization of the parameters of the kernel function is computationally intensive.
- The generated models could be huge, because they include all training data inside.

In the literature, the kernel functions used are almost all variants of radial basis functions (RBF) [2] and the analysis of the problem is centered mainly in the feature selection procedure, i.e. which features must be taken into account for

C. Alippi et al. (Eds.): ICANN 2009, Part II, LNCS 5769, pp. 406–415, 2009.

the problem [3] [4]. The values of the parameters in most cases are usually set by optimizing the cross validation error of the models, although some other criteria are available such as the Bayesian Likelihood [5] [6] or some bounds on the training error [7]. In the most popular implementation of LS-SVM for Matlab, LSSVMlab[1], both the Bayesian and the cross validation error-based optimization are available. In the latter case, the software searches in the parameters' space by means of a grid search algorithm, that is very computational intensive for large grid sizes, high cross validation orders and big number of parameters. The exception to this is the special case of the Leave-One-Out cross validation, for which in LSSVMlab an efficient procedure is implemented based on the inversion of the kernel matrix for all the training data. In [8] and [9] expressions for the evaluation of the cross validation error are presented that are efficient and with a computational complexity independent of the order. This paper focus on the efficient and accurate optimization of the cross validation error of arbitrary order versus the parameters using information about the gradient.

The rest of the paper is organized as follows: Section 2 will briefly introduce the LS-SVM model for regression, and we will provide expressions for the evaluation of the cross validation error and for the partial derivatives of the cross validation error with respect to the parameters of the model; Section 3 will tackle the same objectives, but for the special case of unbiased LS-SVM models; Section 4 will present some initial guesses for the non-linear parameters of the model; Finally, in Sections 5 and 6 some experiments will be presented and final conclusions will be drawn.

2 Least Squares Support Vector Machines (LS-SVM)

Given a set of function samples of the form $\{(x_1, y_1), \ldots, (x_n, y_n)\} \subset X \times \mathbb{R}$, where n is the number of samples, the classic linear LS-SVM model for regression relates inputs X with the output Y by minimizing the objective function:

$$\begin{aligned} \min_{w \in X,\ b, e_i \in \mathbb{R}} \tau(w, b, e) &= \tfrac{1}{2}\|w\|^2 + \gamma \tfrac{1}{2} \sum_{i=1}^{n} e_i^2, \\ (y_i - (\langle w, x \rangle + b)) &= e_i, \quad \forall i = 1, \ldots, n. \end{aligned} \tag{1}$$

where w, b are the parameters of the linear approximator, $\gamma > 0$ is a regularization parameter and e_i is the error for the i-th sample ($e = \{e_1, e_2, \cdots, e_n\}$). Since this is a typical problem of optimization of a differentiable function with restrictions, it can be solved by using Lagrange multipliers (α_i), leading to:

$$\left[\begin{array}{c|c} 0 & 1_N^T \\ \hline 1_N & \Omega + I/\gamma \end{array}\right] \left[\begin{array}{c} b \\ \hline \alpha \end{array}\right] = \left[\begin{array}{c} 0 \\ \hline y \end{array}\right] \tag{2}$$

where $\Omega_{ij} = \langle x_i, x_j \rangle$ and I is the identity. Applying the well-known *Kernel Trick*, $\Omega_{ij} = k(x_i, x_j)$, the problem can be extended to non linear cases. The modelled

[1] `http://www.esat.kuleuven.ac.be/sista/lssvmlab/`

function can be written in terms of coefficients α_i, b and values of the kernel function k:

$$f(x) = \sum_{i=1}^{n} \alpha_i k(x, x_i) + b. \tag{3}$$

2.1 Evaluation of the l-fold Cross-Validation Error for LS-SVM

In order to avoid that the LS-SVM model overfits the data it learns from, it is important to measure its performance when approximating data which have not been used during the training process. One of the most common methods of obtaining a model with good generalization capabilities consist in minimizing not just the training error but the cross-validation error. The l-fold cross-validation error of a model is obtaining by dividing the available data in l sub-sets and, alternately using one of the sub-sets as test data and the rest as training data. Therefore, a total of l models are trained and cross-validated.

In [9], a reduced cost method for the evaluation of l-fold cross-validation error for LS-SVM is presented. To compute the cross-validation error of order l the following expression can be used:

$$MSE_{l\text{-fold}} = \frac{1}{n} \sum_{k=1}^{l} \sum_{j=1}^{|\beta^{(m)}|} {\beta_j^{(m)}}^2 \tag{4}$$

where $\beta = [\beta^{(1)}, \beta^{(2)}, \cdots, \beta^{(l)}]^T$ is defined as $\beta^{(m)} = C_{mm}^{-1} \alpha^{(m)}$. Please, note that $\alpha = [\alpha^{(1)}, \alpha^{(2)}, \cdots, \alpha^{(l)}]^T$ here is not the same of equation (2), but given from:

$$\alpha = Cy \tag{5}$$

$$C = \begin{bmatrix} C_{11} & C_{12} & \cdots & C_{1l} \\ C_{12}^T & C_{22} & \cdots & C_{2l} \\ \vdots & \vdots & \ddots & \vdots \\ C_{l2}^T & C_{2l}^T & \cdots & C_{ll} \end{bmatrix} = K_\gamma^{-1} + \frac{1}{d} K_\gamma^{-1} 1_n 1_n^T K_\gamma^{-1} \tag{6}$$

$$d = -1_n^T K_\gamma^{-1} 1_n \tag{7}$$

$$K_\gamma = K + I/\gamma \tag{8}$$

$$K_{ij} = k(x_i, x_j; \Theta) \tag{9}$$

where I is the identity matrix and Θ is the vector of parameters for the kernel k. The computational cost is independent from l, the fold order, being dominated by the costs of the inversion of the matrix K_γ.

2.2 Partial Derivatives of the l-Fold Cross-Validation Error for LS-SVM

In the last sub-section we have stated that our main objective when optimizing a LS-SVM model with good generalization properties is to minimize the cross-validation error. In order to accomplish the actual minimization, we need the expressions of the partial derivatives of the l-fold cross-validation error (Eq. 4) with respect to a given parameter p (that can be the regularization factor γ or any kernel parameter). After some algebra, we obtain the following expressions:

$$\frac{\partial MSE_{l\text{-fold}}}{\partial p} = \frac{2}{n} \sum_{k=1}^{l} \sum_{j=1}^{|\beta^{(m)}|} \beta_j^{(m)} \left[\frac{\partial \beta^{(m)}}{\partial p} \right]_j \tag{10}$$

$$\frac{\partial \beta^{(m)}}{\partial p} = \frac{\partial C_{mm}^{-1}}{\partial p} \alpha^{(m)} + C_{mm}^{-1} \frac{\partial \alpha^{(m)}}{\partial p} \tag{11}$$

$$\frac{\partial \alpha^{(m)}}{\partial p} = \frac{\partial C_{mm}}{\partial p} y^{(m)} \tag{12}$$

$$\frac{\partial C_{mm}^{-1}}{\partial p} = -C_{mm}^{-1} \frac{\partial C_{mm}}{\partial p} C_{mm}^{-1} \tag{13}$$

$$\frac{\partial C}{\partial p} = \begin{bmatrix} \frac{\partial C_{11}}{\partial p} & \frac{\partial C_{12}}{\partial p} & \cdots & \frac{\partial C_{1l}}{\partial p} \\ \frac{\partial C_{12}^T}{\partial p} & \frac{\partial C_{22}}{\partial p} & \cdots & \frac{\partial C_{2l}}{\partial p} \\ \vdots & \vdots & \ddots & \vdots \\ \frac{\partial C_{l2}^T}{\partial p} & \frac{\partial C_{2l}^T}{\partial p} & \cdots & \frac{\partial C_{ll}}{\partial p} \end{bmatrix} \tag{14}$$

$$\frac{\partial C}{\partial p} = \frac{\partial K_\gamma^{-1}}{\partial p} + \frac{\partial d^{-1}}{\partial p} K_\gamma^{-1} 1_n 1_n^T K_\gamma^{-1} \cdots \tag{15}$$

$$+ \frac{1}{d} \left(\frac{\partial K_\gamma^{-1}}{\partial p} 1_n 1_n^T K_\gamma^{-1} + K_\gamma^{-1} 1_n 1_n^T \frac{\partial K_\gamma^{-1}}{\partial p} \right) \tag{16}$$

$$\frac{\partial d^{-1}}{\partial p} = \frac{1}{d^2} 1_n^T \frac{\partial K_\gamma^{-1}}{\partial p} 1_n \tag{17}$$

$$\frac{\partial K_\gamma^{-1}}{\partial p} = -K_\gamma^{-1} \frac{\partial K_\gamma}{\partial p} K_\gamma^{-1} \tag{18}$$

Therefore, as expected, the partial derivative of the cross-validation error with respect to parameter p depends on $\frac{\partial K_\gamma}{\partial p}$, the partial derivative of $K_\gamma = K + I/\gamma$. For each $p = \theta \in \Theta$, the derivative depends on the specific kernel function k, but for $p = \gamma$ it is $\frac{\partial K_\gamma}{\partial p} = -I/\gamma^2$.

The evaluation of the partial derivatives implies the inversion of the matrix K_γ, that has a computational complexity of $O(N) = N^3$ with exact methods, and depends on the computation of the partial derivatives of the kernel function with respect to its parameters, that must be provided for each particular kernel.

In sum, with the use of the above equations a Conjugate Gradient (CG) scheme can be used to optimize both the regularization factor γ and the kernel parameters in a LS-SVM model. We should here point out that in [1] it is recommended to optimize in different levels the kernel parameters and the regularization factor γ). As we are not guaranteed the convergence to the global optimum, it is recommended to use several optimization stages from different starting points. In section 4 we will also provide some initial guesses for these starting points in order to provide a faster route to the global solution.

3 Unbiased LS-SVM

For function approximation, it is often preferred to use an unbiased version of LS-SVM. The elimination of the bias simplifies the expressions, considering that the functions to model have zero mean, a condition that can be imposed without loss of generality. Given a set of function samples $\{(x_1, y_1), \ldots, (x_n, y_n)\} \subset X \times \mathbb{R}$, where n is the number of samples, the unbiased LS-SVM model for regression relates inputs X with the output Y by the following optimization problem:

$$\begin{aligned} \min_{w, e_i \in \mathbb{R}} \tau(w, e) &= \tfrac{1}{2}\|w\|^2 + \gamma \tfrac{1}{2} \sum_{i=1}^{n} e_i^2, \\ (y_i - (\langle w, x \rangle)) &= e_i, \quad \forall i = 1, \ldots, n. \end{aligned} \tag{19}$$

that leads again to the use of Lagrange multipliers α and, finally, to solve the linear system:

$$[\Omega + I/\gamma]\,[\alpha] = [y] \tag{20}$$

where $\Omega_{ij} =< x_i, x_j >$, the scalar product between a pair of input vector, I is the identity matrix and γ is the regularization factor. If the scalar product operation in the *input space* is substituted by a scalar product in a *feature space* given by a kernel function $k(x, x') =< \phi(x), \phi(x') >$, where ϕ is the function that map points from input space to the feature space, then $\Omega_{ij} = k(x_i, x_j)$ and the actual model would be given by:

$$f(x) = \sum_{i=1}^{n} \alpha_i k(x, x_i). \tag{21}$$

3.1 *l*-Fold Cross-Validation Error for Unbiased LS-SVM

The reduced cost method to evaluate the l-fold cross-validation error in the special case of unbiased LS-SVM is:

$$MSE_{l\text{-fold}} = \frac{1}{n} \sum_{k=1}^{l} \sum_{j=1}^{|\beta^{(m)}|} {\beta_j^{(m)}}^2 \tag{22}$$

$$\beta^{(m)} = C_{mm}^{-1} \alpha^{(m)} \tag{23}$$

$$C = K_\gamma^{-1} \tag{24}$$

where α is the same of equation (20) and K_γ the same of equation 8.

3.2 Partial Derivatives of the l-Fold Cross-Validation Error for Unbiased LS-SVM

Again, in the special case of using unbiased LS-SVM, the partial derivatives of the l-fold cross-validation error (22) with respect to a given parameter p (that can be γ or a kernel parameter) would stay as:

$$\frac{\partial MSE_{l\text{-fold}}}{\partial p} = \frac{2}{n} \sum_{k=1}^{l} \sum_{j=1}^{|\beta^{(m)}|} \beta_j^{(m)} \left[\frac{\partial \beta^{(m)}}{\partial p} \right]_j \tag{25}$$

$$\frac{\partial \beta^{(m)}}{\partial p} = \frac{\partial C_{mm}^{-1}}{\partial p} \alpha^{(m)} + C_{mm}^{-1} \frac{\partial \alpha^{(m)}}{\partial p} \tag{26}$$

$$\frac{\partial \alpha}{\partial p} = \frac{\partial K_\gamma^{-1}}{\partial p} y \tag{27}$$

4 Guesses for the Initialization of the LS-SVM Parameters

In this section, some guesses are given to initialize the regularization factor γ from training data. Nevertheless, it is not possible to do the same for the parameters of any arbitrary kernel. Therefore, we will concentrate on the most common case, the Radial Basis Function (RBF) kernel (29) which has only one parameter, σ.

4.1 Initialization of the Regularization Parameter γ

In the paper [6] it is shown that the optimization problem of a LS-SVM model given by equation (19) can be rewritten as:

$$\begin{gathered} \min_{w, e_i \in \mathbb{R}} \tau(w, e) = \mu \tfrac{1}{2} \|w\|^2 + \xi \tfrac{1}{2} \sum_{i=1}^{n} e_i^2, \\ (y_i - (\langle w, \phi(x) \rangle)) = e_i, \quad \forall i = 1, \ldots, n. \end{gathered} \tag{28}$$

where $\gamma = \xi/\mu$, $1/\xi = \sigma_e^2$, where σ_e^2 is the variance of the noise in the data and μ is the parameter that controls the regularization of the model.

Without prior information to set the μ value, we let μ be set to 1 provided all data is normalized with mean 0 and standard deviation 1. The ξ parameter depends on the variance of the noise in the data. We can estimate this noise variance, $\hat{\sigma}_e^2$, using the non parametric estimators delta or gamma test, as recommended in [10] in which it is also pointed out that the gamma test is more reliable than the delta test for data of dimensionality higher than four. Thus, our initial guess for the regularization factor γ in a LS-SVM model will be $\gamma_h = 1/\hat{\sigma}_e^2$, where $\hat{\sigma}_e^2$ is evaluated from the data using the delta or the gamma test.

In the worst case, all data are pure noise, so $\sigma_e^2 = \sigma_y^2$ and a minimum γ value can be set as $\gamma_{min} = 1/\sigma_y^2$. A reasonable range to search for an initial value of γ

could be $[\gamma_{min}, \gamma_h]$, but if we rely on non parametric noise estimation methods we can use values very close to γ_h.

In [11] it is demonstrated that the presence of noise in the input vectors makes that the delta and gamma test, instead of approximating the variance of the noise, give a measure of the variance of the effective noise in the output. The value of the variance of the effective noise is larger than the variance of the noise itself, but a better approximation cannot be performed anyway.

Since in order to use regression methods for time series prediction the input/ouput vectors have to be generated from the time series, the input vectors will always have noise. So a reasonable range to search the γ value in a range centered on γ_h could be $[0.5 \cdot \gamma_h, 2 \cdot \gamma_h]$.

4.2 Initial Guess for the σ Parameter of the RBF Kernel

Considering the most common case of LS-SVM model for regression, i.e., the LS-SVM model with RBF kernel, whose equation is:

$$k(x, x') = \exp\left(-\frac{1}{\sigma^2}\|x - x'\|^2\right). \tag{29}$$

the σ value is related to the distance between training points and to the smooth interpolation of the resulting model from the I/O data: higher values of σ give smoother functions. It can be expected to find a good σ in $[\sigma_{min}, \sigma_{max}]$, where σ_{min} is the minimum distance (non zero) between 2 training points and σ_{max} is the maximum distance between 2 training points. But a smaller, but reasonable, parameter search space for σ could be the range $[0.5 \cdot \overline{\sigma}, 2 \cdot \overline{\sigma}]$, centered in $\overline{\sigma} = 0.5 \cdot (\sigma_{max} + \sigma_{min})$ that can be used as a good enough starting point.

In order to use this kernel in the expressions obtained in the previous sections, we need the derivative of the kernel function with respect to its parameter σ:

$$\frac{\partial k(x, x')}{\partial \sigma} = \exp\left(-\frac{1}{\sigma^2}\|x - x'\|^2\right)\frac{1}{\sigma^4}\|x - x'\|^2. \tag{30}$$

5 Experiments

To test the proposed optimization methods and the heuristics for the initialization of the parameters for LS-SVM models with RBF kernel, the Mackey-Glass time series is used. The Mackey-Glass time series [12] is a very well-known benchmark for time series modelling and prediction. It is an artificial time series without noise, that has been widely used in the literature for the comparison of neuro-fuzzy and other nonlinear models. In our experiments, 1000 data points of the series, $y_1, ..., y_{1000}$, were taken, and normalized with zero mean and unit variance. Gaussian additive noise of zero mean and controlled variance $\sigma_e^2 = \{0, 1, 10^{-1}, 10^{-2}, 10^{-3}, 10^{-4}\}$ was added to this sequence of data, $\hat{y}_i = y_i + N(0, \sigma_e^2)$, in order to test the non parametric noise estimation, and confirm that the presence of noise in the inputs can make it fail. From

Table 1. Optimizations with RBF kernel using heuristics for Mackey-Glass with noise on all series. σ_e^2: variance of the added Gaussian noise; NNE: evaluated variance of noise with non parametric noise estimation; $NRMSE_{NNE}$ best training error that can be reached evaluated from NNE; Method: Heuristic, using $\gamma_h, \overline{\sigma}$, Gridsearch using reduced cost evaluation of 10 fold cross-validation error in range $[0.5 \cdot \gamma_h, 2 \cdot \gamma_h]$, BFGS applied from $\gamma_h, \overline{\sigma}$, Iterative Local Search (of BFGS) applied from $\gamma_h, \overline{\sigma}$; T: time in seconds; f.e.: number of cross-validation error evaluations performed; 10-CVE: final value of 10 fold cross-validation error reached. Best values of T and 10-CVE in bold.

		Method	Training	Test	T	f.e.	10-CVE
		Heuristic	5.58e-03	4.95e-03	2.10e-01	0	1.38e-04
σ_e^2	0	GS	5.51e-03	4.88e-03	1.33e+01	100	1.37e-04
NNE	1.00e-06	BFGS	4.49e-03	4.03e-03	**6.00e+00**	22	1.25e-04
$NRMSE_{NNE}$	1.00e-03	ILS	4.49e-03	4.03e-03	4.47e+01	100	**1.25e-04**
		LSSVMlab	5.61e-03	4.97e-03	3.61e+02	100	1.38e-04
		Heuristic	1.19e+00	1.27e+00	1.91e-01	0	1.53e+00
σ_e^2	1	GS	1.20e+00	1.27e+00	1.34e+01	100	1.52e+00
NNE	1.38e+00	BFGS	1.20e+00	1.27e+00	**3.68e+00**	12	**1.52e+00**
$NRMSE_{NNE}$	1.17e+00	ILS	1.20e+00	1.27e+00	3.22e+01	100	1.52e+00
		LSSVMlab	1.20e+00	1.27e+00	3.55e+02	100	1.53e+00
		Heuristic	4.16e-01	4.27e-01	1.91e-01	0	2.02e-01
σ_e^2	1.00e-01	GS	4.23e-01	4.29e-01	1.34e+01	100	2.01e-01
NNE	1.86e-01	BFGS	4.21e-01	4.27e-01	**3.40e+00**	11	**2.00e-01**
$NRMSE_{NNE}$	4.31e-01	ILS	4.21e-01	4.27e-01	3.10e+01	100	2.00e-01
		LSSVMlab	4.18e-01	4.28e-01	3.56e+02	100	2.02e-01
		Heuristic	1.45e-01	1.46e-01	1.91e-01	0	2.55e-02
σ_e^2	1.00e-02	GS	1.31e-01	1.48e-01	1.34e+01	100	2.45e-02
NNE	1.63e-02	BFGS	1.33e-01	1.46e-01	**4.33e+00**	14	**2.44e-02**
$NRMSE_{NNE}$	1.28e-01	ILS	1.33e-01	1.46e-01	3.10e+01	100	2.44e-02
		LSSVMlab	1.45e-01	1.46e-01	3.61e+02	100	2.55e-02
		Heuristic	5.38e-02	5.60e-02	1.91e-01	0	3.74e-03
σ_e^2	1.00e-03	GS	4.72e-02	5.50e-02	1.34e+01	100	3.42e-03
NNE	2.44e-03	BFGS	4.79e-02	5.55e-02	**5.55e+00**	18	**3.41e-03**
$NRMSE_{NNE}$	4.94e-02	ILS	4.79e-02	5.55e-02	2.91e+01	100	3.41e-03
		LSSVMlab	5.39e-02	5.61e-02	3.62e+02	100	3.74e-03
		Heuristic	1.77e-02	2.19e-02	1.91e-01	0	5.43e-04
σ_e^2	1.00e-04	GS	1.64e-02	2.09e-02	1.35e+01	100	5.27e-04
NNE	3.18e-05	BFGS	1.61e-02	2.07e-02	**4.34e+00**	14	**5.26e-04**
$NRMSE_{NNE}$	5.64e-03	ILS	1.61e-02	2.07e-02	2.80e+01	100	5.26e-04
		LSSVMlab	1.77e-02	2.19e-02	3.62e+02	100	5.43e-04

the noisy data $\hat{y}_1, ..., \hat{y}_{1000}$, inputs and output vectors were created in the form $[\hat{y}_{t-24}, \hat{y}_{t-18}, \hat{y}_{t-12}, \hat{y}_{t-6}, \hat{y}_t]$, and half of the vectors were used for training and half for test. For the LS-SVM models, the heuristics were applied and tested using the reduced-cost evaluation procedure of the 10 fold cross-validation error (which is one of the most common orders used in literature):

1. Directly: fixing the values of the parameters $\gamma_h, \overline{\sigma}$.
2. A grid-search procedure: the same used on LSSVMlab with the range suggested in Section 4.
3. A local search procedure: using the Broyden-Fletcher-Goldfarb-Shanno method, BFGS, with initial point $(\gamma_h, \overline{\sigma})$
4. A global search procedure: Iterative Local Search using as local search BFGS with initial point $(\gamma_h, \overline{\sigma})$.

The LSSVMlab toolbox was also used. The training and test errors are given independently from translation and scale factors using the Normalized Root Mean Square Error [13] ($NRMSE = \sqrt{MSE/\sigma_y^2}$, where MSE is the Mean Square Error and σ_y^2 the variance of the function to model). The evaluated best training NRMSE error computed from the non parametric noise error will also be shown. All the software was implemented in Matlab, and the experiments were executed on a personal computer with an Intel Core 2 Quad CPU at 2.83GHz and 8 GB of RAM. Finally, the number of evaluations of the cross-validation procedure was limited to 100 in the experiments.

The results obtained are shown in Table 1. As can be seen from the table, the procedure to evaluate the cross-validation error from [9] is faster than the naive implementation for this case (see rows labelled *GS* y *LSSVMlab*). It is also shown that the evaluation of the partial derivatives (used in *BFGS* and *ILS*) adds some overhead to the evaluation of the cross-validation error but allows us to reach a local optima with less number of evaluations. The initial point for the optimization procedures has an important influence on the results. In this case the global search procedure (*ILS*) only reached a better solution than a local search procedure (*BFGS*) in the case of noise-free data. It is worth to recall that the number of evaluations required by BFGS to converge is quite smaller than the maximum allowed in most cases (exceptions for $\sigma_e^2 = 1.0e-2$ y $\sigma_e^2 = 1.0e-4$). It is remarkable the relative good results reached using only the heuristic initialization of the parameters in the models (rows labelled *Heuristic*).

6 Conclusions

In this paper we have obtained the expressions of the derivatives of the reduced cost evaluation of an arbitrary order cross-validation error for LS-SVM models. Some guesses have also been given for the initial values of the regularization factor γ of the LS-SVM model and for the RBF kernel parameter σ. All the expressions were implemented in Matlab. In the experiments we have compared the results obtained by a derivative-free procedure (gridsearch), a conjugate gradient procedure, and a global search procedure (Simulated Annealing guided by gradient). Time and error measures were taken and compared with the most popular implementation of LS-SVM for matlab, LS-SVMlab, for the well-known Mackey-Glass time series. The results confirm that the use of our reduced-cost expressions achieves performances that outperforms the naive implementation used in LS-SVMlab, and that gradient-based procedures can optimize the parameters more efficiently that other methods that don't use this information. The validity of the given heuristics has also been confirmed for this example.

Acknowledgment

This work has been supported by the Spanish CICYT Project TIN2007-60587 and Junta Andalucia Project P07-TIC-02768.

References

1. Suykens, J.A.K., Van Gestel, T., De Brabanter, J., De Moor, B., Vandewalle, J.: Least Squares Support Vector Machines. World Scientific, Singapore (2002)
2. Rojas, I., et al.: Analysis of the functional block involved in the design of radial basis function networks. Neural Processing Letters 12, 1–17 (2000)
3. Müller, K.-R., Smola, A.J., Rätsch, G., Schölkopf, B., Kohlmorgen, J., Vapnik, V.: Using support vector machines for time series prediction (2000)
4. Rubio, G., Guillen, A., Herrera, L.J., Pomares, H., Rojas, I.: Use of specific-to-problem kernel functions for time series modeling. In: ESTSP 2008: Proceedings of the European Symposium on Time Series Prediction, pp. 177–186 (2008)
5. Scholkopf, B., Smola, A.J.: Learning with Kernels: Support Vector Machines, Regularization, Optimization, and Beyond. MIT Press, Cambridge (2001)
6. Van Gestel, T., Suykens, J.A.K., Baestaens, D.-E., Lambrechts, A., Lanckriet, G., Vandaele, B., De Moor, B., Vandewalle, J.: Financial time series prediction using least squares support vector machines within the evidence framework. IEEE Transactions on Neural Networks 12(4), 809–821 (2001)
7. Lendasse, A., Ji, Y., Reyhani, N., Verleysen, M.: Ls-svm hyperparameter selection with a nonparametric noise estimator. In: ICANN (2), pp. 625–630 (2005)
8. Ying, Z., Keong, K.C.: Fast leave-one-out evaluation and improvement on inference for ls-svms. In: Proceedings of the 17th International Conference on Pattern Recognition, ICPR 2004, vol. 3, pp. 494–497 (2004)
9. An, S., Liu, W., Venkatesh, S.: Fast cross-validation algorithms for least squares support vector machine and kernel ridge regression. Pattern Recogn. 40(8), 2154–2162 (2007)
10. Liitiäinen, E., Lendasse, A., Corona, F.: Non-parametric residual variance estimation in supervised learning. In: Sandoval, F., Prieto, A.G., Cabestany, J., Graña, M. (eds.) IWANN 2007. LNCS, vol. 4507, pp. 63–71. Springer, Heidelberg (2007)
11. Jones, A.J., Evans, D., Kemp, S.E.: A note on the Gamma test analysis of noisy input/output data and noisy time series. Physica D Nonlinear Phenomena 229, 1–8 (2007)
12. Mackey, M.C., Glass, L.: Oscillation and Chaos in Physiological Control Systems. Science 197(4300), 287–289 (1977)
13. Herrera, L.J., et al.: TaSe, a Taylor Series-based fuzzy system model that combines interpretability and accuracy. Fuzzy Sets and Systems 153, 403–427 (2005)

Noiseless Independent Factor Analysis with Mixing Constraints in a Semi-supervised Framework. Application to Railway Device Fault Diagnosis

Etienne Côme[1], Latifa Oukhellou[1,2], Thierry Denœux[3], and Patrice Aknin[1]

[1] INRETS-LTN, 2 Av Malleret Joinville, 94114 Arcueil- France
[2] Université Paris 12- CERTES, 61 av du Gal de Gaulle, 94100 Créteil- France
[3] Heudiasyc, UTC - UMR CNRS 6599, B.P 20529, 60205 Compiègne - France

Abstract. In Independent Factor Analysis (IFA), latent components (or sources) are recovered from only their linear observed mixtures. Both the mixing process and the source densities (that are assumed to be generated according to mixtures of Gaussians) are learned from observed data. This paper investigates the possibility of estimating the IFA model in its noiseless setting when two kinds of prior information are incorporated: constraints on the mixing process and partial knowledge on the cluster membership of some examples. Semi-supervised or partially supervised learning frameworks can thus be handled. These two proposals have been initially motivated by a real-world application that concerns fault diagnosis of a railway device. Results from this application are provided to demonstrate the ability of our approach to enhance estimation accuracy and remove indeterminacy commonly encountered in unsupervised IFA such as source permutations.

Keywords: Independent Factor Analysis, mixing constraints, semi-supervised learning, diagnosis, railway device.

1 Introduction

The generative model involved in Independent Component Analysis (ICA) assumes that observed variables are generated by a linear mixture of independent and nongaussian latent variables. Furthermore, when the IFA model is considered, each latent variable has its own distribution, modeled semi-parametrically by a mixture of Gaussians (MOG) and the number of mixtures can differ from the number of sources. The IFA model introduced by [5] can indeed handle both square noiseless mixing and the general case where the data are noisy. These models yield reliable results provided the independence assumption is satisfied and the postulated mixing model suited to the physics of the system. Otherwise, they fail to recover the sources. Several extensions of the basic ICA model have been proposed to improve its performance. The main approaches exploit temporal correlation [6], positivity [7,3,11] or sparsity [8,9].

C. Alippi et al. (Eds.): ICANN 2009, Part II, LNCS 5769, pp. 416–425, 2009.

In this paper, we propose two extensions of the basic noiseless IFA model. The first one concerns the possibility of incorporating independence hypotheses between some latent and observed variables. Such hypotheses can be derived from physical knowledge available on the mixing process. This kind of approach has not been applied within the framework of IFA, but it has been widely considered in factor analysis [10] and, more specifically, in the structural equation modeling domain [12]. The second extension incorporates additional information on cluster membership of some samples to estimate the IFA model. In this way, the semi-supervised learning framework can be handled. Considering the graphical model of IFA shown in Figure 1, the mixing process prior consists in omitting some connections between observed (X) and latent (Z) variables. The second prior means that additional information on the discrete latent variables (Y) is taken into account.

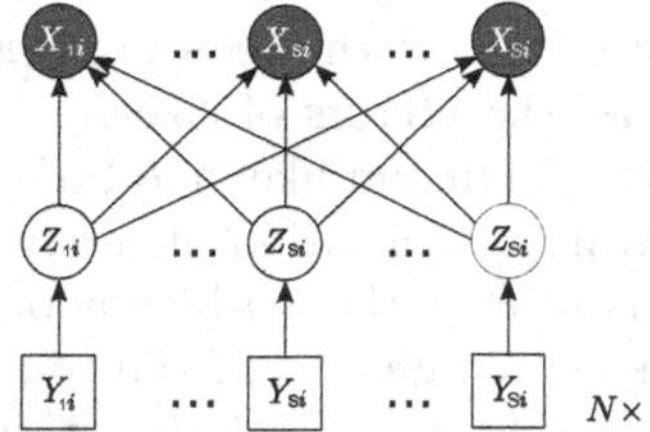

Fig. 1. Graphical model for Independent Factor Analysis

This article is organized as follows. We will first present IFA model estimation by maximum likelihood in a noiseless setting. In Section 3 and 4, the problem of learning the IFA model with prior knowledge on the mixing process and on the cluster membership of some samples will be addressed. In Section 5, the approach will be applied to diagnosis problem for which the impact of using priors will be evaluated. The paper ends with a conclusion.

2 Background on Independent Factor Analysis

ICA and IFA aim at recovering independent latent components from their observed linear mixtures. In its noiseless formulation (used throughout this paper), the ICA model can be expressed as $\mathbf{x} = A\,\mathbf{z}$, where A is a square matrix of size $S \times S$, $\mathbf{x}$ the random vector whose elements $(\mathbf{x}_1, \ldots, \mathbf{x}_S)$ are the mixtures and $\mathbf{z}$ the random vector whose elements $(\mathbf{z}_1, \ldots, \mathbf{z}_S)$ are the latent components. Thanks to the noiseless setting, a deterministic relationship between the distributions of observed and latent variables can be expressed as: $f^{\mathcal{X}}(\mathbf{x}) = \frac{1}{|\det(A)|} f^{\mathcal{Z}}(A^{-1}\mathbf{x})$. The probability density functions of the sources can be fixed using prior knowledge, or according to some indicator that allows switching between sub and super Gaussian densities [1]. An alternative solution, referred to as IFA, consists in modeling each source density as a mixture of Gaussians (MOG) so that a wide class of densities can be approximated [4,5]:

$$f^{Z_s}(z_s) = \sum_{k=1}^{K_s} \pi_k^s \varphi(z_s; \mu_k^s, \nu_k^s), \tag{1}$$

with $\varphi(.; \mu, \nu)$ the density of a Gaussian random variable of mean μ and variance ν. This model is close to ICA with a Mixture of Gaussians model for the sources. The problem consists in estimating both the mixing matrix and the MOG parameters from the observed variables alone. Considering an iid random sample of size N, the log-likelihood has the form:

$$\mathcal{L}(\psi; \mathbf{X}) = -N \log(|\det(A)|) + \sum_{i=1}^{N}\sum_{s=1}^{S} \log\left(\sum_{k=1}^{K_s} \pi_k^s \varphi\left((A^{-1}\mathbf{x}_i)_s, \mu_k^s, \nu_k^s\right)\right). \tag{2}$$

where $\boldsymbol{\psi}$ is the IFA parameter vector $\boldsymbol{\psi} = (A, \boldsymbol{\pi}^1, \ldots, \boldsymbol{\pi}^S, \boldsymbol{\mu}^1, \ldots, \boldsymbol{\mu}^S, \boldsymbol{\nu}^1, \ldots, \boldsymbol{\nu}^S)$, with A the mixing matrix, $\boldsymbol{\pi}^s$ the vector of cluster proportions of source s which sum to 1, $\boldsymbol{\mu}^s$ and $\boldsymbol{\nu}^s$ the vectors of size K_s containing the means and the variances of each cluster. Maximum likelihood of the model parameters can be achieved by an alternating optimization strategy. The gradient algorithm [14] is indeed well suited to optimize the log-likelihood function with respect to the mixing matrix A when the parameters of the source marginal densities are frozen. Conversely, with A kept fixed, an EM algorithm can be used to optimize the likelihood function with respect to the parameters of each source. These remarks have led to the development of a Generalized EM algorithm (GEM) able to simultaneously maximize the likelihood function with respect to all the model parameters [18].

3 Constraints on the Mixing Process

This section investigates the possibility of incorporating independence hypotheses concerning relationships between some latent and observed variables in the ICA model. Such hypotheses are often deduced from physical knowledge of the mixing process. The hypothesis that we consider in this section has the following form: $X_h \perp\!\!\!\perp Z_g$, which means that X_h is statistically independent from Z_g. Making this kind of hypothesis constraints the form of the mixing matrix as shown by the following proposition, which has been proven in [13]:

Proposition 1. *In the noiseless ICA model, we have :*

$$X_h \perp\!\!\!\perp Z_g \Leftrightarrow A_{hg} = 0. \tag{3}$$

The log-likelihood has to be maximized under the constraint that some of the mixing coefficients are null, and gradient ascent is only performed with respect to the non-null coefficients. In this case, the initialization and the update rule of the mixing matrix are given by $A^{(0)} = C \bullet A^{(0)}$, $A^{(q+1)} = A^{(q)} + \tau C \bullet \Delta A^{(q)}$ where $\bullet$ denotes the Hadamard product between two matrices and C a binary matrix of which the elements are $C_{hk} = 0$ if $Z_k \perp\!\!\!\perp X_h$, $C_{hk} = 1$ otherwise.

4 Semi-supervised Learning in IFA

The IFA model is often considered within an unsupervised learning framework. This section considers the learning of this model (in its noiseless setting) in a partially-supervised learning context where partial knowledge of the cluster membership of some samples is available. For that purpose, a generalized likelihood function has to be defined and an EM algorithm dedicated to its optimization has to be set up. In the general case, we will assume a learning set of the form : $\mathbf{X}^{iu} = \{(\mathbf{x}_1, m_1^{\mathcal{Y}_1}, \ldots, m_1^{\mathcal{Y}_S}), \ldots, (\mathbf{x}_N, m_N^{\mathcal{Y}_1}, \ldots, m_N^{\mathcal{Y}_S})\}$, where $m_i^{\mathcal{Y}_1}, \ldots, m_i^{\mathcal{Y}_S}$ is a set of basic belief assignments or Dempster-Shafer mass functions [15,13] encoding our knowledge on the cluster membership of sample i for each one of the S sources, $\mathcal{Y}_s = \{c_1, \ldots, c_{K_s}\}$ is the set of all possible clusters for source s. Depending on the choice of the mass functions, this formulation can therefore be seen as addressing a more general framework that encompasses unsupervised, supervised and partially-supervised learning paradigms as mentioned in Table 1. The concept of likelihood function has strong relations with that of possibility and, more generally, plausibility, as already noted by several authors [15]. Furthermore, selecting the simple hypothesis with highest plausibility given the observations $\mathbf{X}^{iu}$ is a natural decision strategy in the belief function framework. We thus propose as an estimation principle to search for the parameter value with maximal conditional plausibility given the data: $\hat{\psi} = \arg\max_{\psi} pl^{\Psi}(\psi|\mathbf{X}^{iu})$.

Table 1. Different learning paradigms and soft labels

	Mass function	*plausibility*
Unsupervised	$m_i^s(\mathcal{Y}_s) = 1,$	$pl_{ik}^s = 1, \forall k$
Supervised	$m_i^s(c_k) = 1$	$pl_{ik}^s = 1, pl_{ik'}^s = 0, \forall k' \neq k$
Partially supervised	$m_i^s(C) = 1$	$pl_{ik}^s = 1 \; if \; c_k \in C, \; pl_{ik}^s = 0 \; if \; c_k \notin C$

Parameter estimation in a mixture model with belief function-based labels was already addressed in [13]. In this context, a likelihood criterion taking into account *soft* labels has been defined and an EM algorithm dedicated to its optimization has been presented. In this article, we propose an extension of such study to the IFA model in which partial knowledge of some cluster memberships is incorporated. The following proposition, proved in [13], gives the expression of the generalized likelihood criterion for the IFA model.

Proposition 2. *If the labels are assumed to be mutually independent and independent from the samples* $\mathbf{X}$ *that are i.i.d. generated according to the the generative IFA model setting, then the logarithm of the conditional plausibility of the model parameters vector* ψ *given the learning set* $\mathbf{X}^{iu}$ *is given by:*

$$\log\left(pl^{\Psi}(\psi|\mathbf{X}^{iu})\right) = -N\log(|\det(A)|) + \sum_{i=1}^{N}\sum_{s=1}^{S}\log\left(\sum_{k=1}^{K_s} pl_{ik}^s \pi_k^s \varphi\left((A^{-1}\mathbf{x}_i)_s, \mu_k^s, \nu_k^s\right)\right) + cst. \quad (4)$$

where pl_{ik}^s is the plausibility that the sample i belong to cluster k of the latent variable s, (computed from the soft labels $m_i^{\mathcal{Y}_s}$), and cst is a constant independent of ψ.

In a semi-supervised learning context, the IFA model is built from a combination of M labeled and $N - M$ unlabeled samples. For labeled samples, the plausibilities used as labels are crisp and we have $pl_{ik}^s = l_{ik}^s \in \{0,1\}^{K_s}$, $l_{ik}^s = 1$ if sample i comes from cluster c_k of sources s and $l_{ik}^s = 0$ otherwise. For unlabeled samples, $pl_{ik}^s = 1$ for all clusters k and sources s. Consequently, the criterion can be decomposed into two parts corresponding, respectively, to the supervised and unsupervised learning examples and criterion (4) can be rewritten as:

$$\mathcal{L}(A;\mathbf{X}) = -N\log(|\det(A)|) + \sum_{i=1}^{M}\sum_{s=1}^{S}\sum_{k=1}^{K_s} l_{ik}^s \log\left(\pi_k^s \varphi\left((A^{-1}\mathbf{x}_i)_s, \mu_k^s, \nu_k^s\right)\right) + \sum_{i=M+1}^{N}\sum_{s=1}^{S}\log\left(\sum_{k=1}^{K_s}\pi_k^s\varphi\left((A^{-1}\mathbf{x}_i)_s,\mu_k^s,\nu_k^s\right)\right). \quad (5)$$

A Generalized EM algorithm (GEM), (Algorithm 1) can be designed to simultaneously maximize the likelihood function with respect to all the model parameters. This algorithm is similar to the EM algorithm used to estimate IFA parameter in an unsupervised setting, except for the E step, where the posterior probabilities t_{ik}^s are only computed for the unlabeled samples. The updating of the mixing matrix also takes into account the mixing constraints and depends not only of the latent variables, but also of the labels.

5 Fault Diagnosis in Railway Track Circuits

The application considered in this paper concerns fault diagnosis in railway track circuits. This device will first be described and the problem addressed will be exposed. An overview of the proposed diagnosis method will be presented.

5.1 Track Circuit Principle

The track circuit is an essential component of the automatic train control system [17]. Its main function is to detect the presence or absence of vehicle traffic within a specific section of railway track. On French high speed lines, the track circuit is also a fundamental component of the track/vehicle transmission system. It uses a specific carrier frequency to transmit coded data to the train, for example the maximum authorized speed on a given section on the basis of safety constraints. The railway track is divided into different sections. Each one of them has a specific track circuit consisting of the following components:

- A transmitter connected to one of the two section ends, which delivers a frequency modulated alternating current;
- The two rails that can be considered as a transmission line;

Algorithm 1. Pseudo-code for noiseless IFA with prior knowledge on labels and mixing constraints.

Input: Centered observation matrix $\mathbf{X}$, cluster belonging for the M labeled data l_{ik}^{s}, constraints matrix encoding independence hypothesis C.

 # Random initialization of parameters vector $\psi^{(0)}, \quad q = 0$

while *Convergence test* **do**

$\mathbf{Z} = \mathbf{X}\left(A^{(q)^{-1}}\right)^{t}$ *# Source update*

forall $s \in \{1,\ldots,S\}$ *and* $k \in \{1,\ldots,K_s\}$ **do**

$$t_{ik}^{s(q)} = l_{ik}^{s}, \quad \forall i \in \{1,\ldots,M\}$$

$$t_{ik}^{s(q)} = \frac{\pi_k^{s(q)} \varphi(z_{is}; \mu_k^{s(q)}, \nu_k^{s(q)})}{\sum_{k'=1}^{K_s} \pi_{k'}^{s(q)} \varphi(z_{is}; \mu_{k'}^{s(q)}, \nu_{k'}^{s(q)})}, \quad \forall i \in \{M+1,\ldots,N\}$$

forall $s \in \{1,\ldots,S\}$ *and* $k \in \{1,\ldots,K_s\}$ **do**

$$\pi_k^{s(q+1)} = \frac{1}{N}\sum_{i=1}^{N} t_{ik}^{s(q)}$$

$$\mu_k^{s(q+1)} = \frac{1}{\sum_{i=1}^{N} t_{ik}^{s(q)}} \sum_{i=1}^{N} t_{ik}^{s(q)} z_{is}$$

$$\nu_k^{s(q+1)} = \frac{1}{\sum_{i=1}^{N} t_{ik}^{s(q)}} \sum_{i=1}^{N} t_{ik}^{s(q)} (z_{is} - \mu_k^{s(q+1)})^2$$

$\mathbf{G} = \mathbf{g}^{(q+1)}(\mathbf{Z})$ *# Update of G,* $g_s(z_{is}) = \sum_{k=1}^{K_s} t_{ik}^{s(q+1)} \frac{(z_{is} - \mu_k^{s(q+1)})}{\nu_k^{s(q+1)}}$,

 # Natural gradient

$\Delta A = \left(A^{(q)^{-1}}\right)^{t} \left(\frac{1}{N}\sum_{i=1}^{N} \mathbf{g}\left(\mathbf{z}_i^{(q)}\right) \mathbf{z}_i^{(q)^{t}} - \mathbf{I}\right)$

$\tau^{*} = \text{Linearsearch}(A^{(q)}, C \bullet \Delta A)$ *# Linear Search for* τ

$A^{(q+1)} = A^{(q)} + \tau^{*}.C \bullet \Delta A$ *# mixing matrix Update*

 # source normalization to remove scale indetermination

$q \leftarrow q + 1$

- At the other end of the track section, a receiver that essentially consists of a trap circuit used to avoid the transmission of information to the neighboring section;
- Trimming capacitors connected between the two rails at constant spacing to compensate for the inductive behavior of the track. Electrical tuning is then performed to limit the attenuation of the transmitted current and improve the transmission level. The number of compensation points depends on the carrier frequency and the length of the track section.

The rails themselves are part of the track circuit, and a train is detected when its wheels and axles short-circuit the track. The presence of a train in a given section induces the loss of track circuit signal due to shorting by train wheels. The drop of the received signal below a preset threshold indicates that the section is occupied. The different parts of the system are subject to malfunctions that must be detected as soon as possible in order to maintain the system at the required safety and availability levels. In the most extreme case, this causes an unfortunate attenuation of the transmitted signal that leads to the stop of the train. The

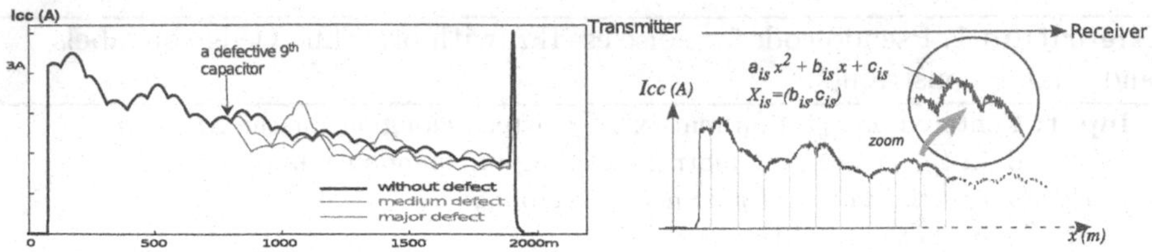

Fig. 2. Examples of inspection signals

purpose of diagnosis is to inform maintainers about track circuit failures on the basis of the analysis of a specific current, recorded by an inspection vehicle. This paper will focus on trimming capacitor faults that affect their capacitance. Figure 2 shows an example of the inspection signal when the system is fault-free while the others correspond to a defective 9th capacitor. The aim of the diagnosis system is to detect faults in the track circuit and localize the defective capacitor by analyzing the measurement signal.

5.2 Overview of the Diagnosis Method

The track circuit can be considered as a large-scale system made up of a series of spatially related subsystems that correspond to the trimming capacitors. A defect on one subsystem is represented by a continuous value of the capacitance parameter. The proposed method is based on the following two observations (see Figure 2). First, the inspection signal has a specific structure, which is a succession of so many arches as capacitors; an arch can be approximated by a quadratic polynomial $ax^2 + bx + c$. Second, each observed arch is influenced by the capacitors located upstream (on the transmitter side). The proposed method consists in extracting features from the measurement signal, and building a generative model as shown in Figure 3, where each observed variable X_{is} corresponds to the coefficients (b_{is}, c_{is}) of the local polynomial approximating the arch located between two subsystems. Only two coefficients are used because of continuity constraints between each polynomial, as their exists a linear relationship between the third coefficient and the three coefficients of the previous polynomial. The continuous latent variable Z_{is} is the capacitance of the ith capacitor and the discrete latent variable Y_{is} corresponds to the membership of the capacitor state to one of the three states: fault-free, minor defect, major defect. As there is no influence between a trimming capacitor state and the inspection signal located upstream from it, some connections between latent and observed variables are omitted. This information will be also introduced in the model estimation using constraints on the mixing matrix. We can clearly see that this model is closely linked to the IFA model represented in Figure 1. Considering the diagnosis task as a blind source separation problem, the IFA model can be used to estimate the mixing matrix A and thereby to recover the latent components (capacitances) from the observed variables alone. As already explained, a piecewise approach is adopted for the signal representation: each arch is approximated by a second

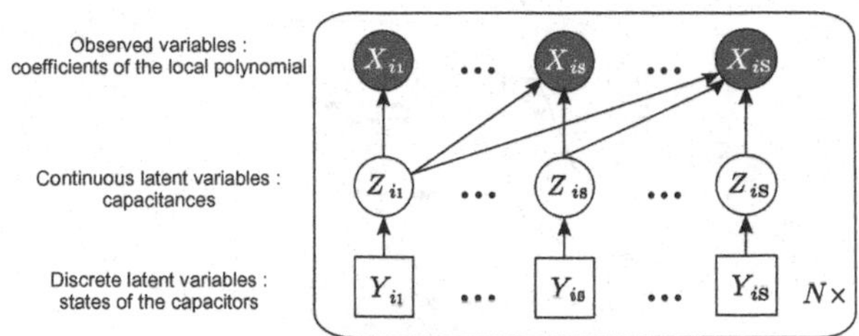

Fig. 3. Generative model for the diagnosis of track circuits represented by a graphical model including both continuous and discrete latent variables

degree polynomial of which two coefficients are used as observed variables for each node in the model of Figure 1, which results in $2 \times S$ observed variables. Given an observation matrix, the aim is to recover S latent variables from $2 \times S$ observed ones with the hope that they will be strongly correlated with the variable of interest, i.e., capacitances. As prior information on the mixing matrix is available, PCA cannot be used for preprocessing because the mixing structure would be lost. $2 \times S$ latent variables are therefore extracted: S latent variable densities corresponding to capacitances are assumed to be mixtures of three Gaussian components, one for each state of the capacitors while the S other variables are assumed to be noise variables and are thus modeled by simple Gaussian distributions. It can be noticed that with standard IFA model, S latent variables can be recovered from $2 \times S$ observed ones. But in this paper, we consider the noiseless IFA model which seems to be straightforward to incorporate prior information and to recover the sources from the data.

6 Results and Discussion

To assess the performances of the approach, we considered a track circuit of $S = 18$ subsystems (capacitors) and a database containing 2500 noised signals obtained for different values of the capacitance of each capacitor. 500 were used in the training phase, while the 2000 others were employed for the test phase. The experiments aim at illustrating the influence of both the number of labeled samples and the use of the mixing matrix constraints on the results. The model provides two levels of interpretation corresponding to discrete and continuous latent variables, but we only discuss in this paper the results for the continuous latent variables. Figure 4 shows the mean of the absolute value of the correlation between estimated latent variables and capacitances as a function of the number of labeled training samples, when the mixing matrix is constrained or not. Note that the case of unlabeled samples without constraints illustrates the performances of the traditional IFA model (without any prior), which are very poor as our criterion is sensitive to source permutation. When more labeled samples are used, the permutations of the sources are avoided and the performances

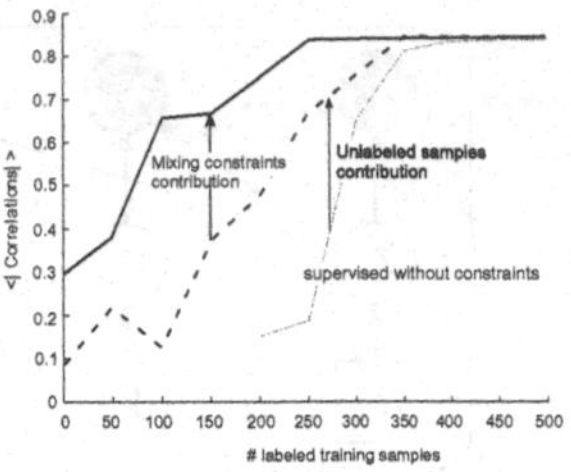

Fig. 4. Results of IFA with (–), without constraints (- -) when the number of labeled samples varies between 0 and 500 and supervised IFA without constraints (-. -.)

reach a more satisfactory level. Twenty random starting points were used for the GEM algorithm and only the best solution according to the likelihood was kept. This figure clearly highlights the benefit of using constraints when the amount of labeled samples is small. As expected, when the number of labeled data increases, the mean correlation also increases to reach a maximal value of 0.84 for the constrained IFA model with 250 labeled sampled and for the unconstrained one with 350 labeled samples. When a sufficient amount of labeled samples is provided to the model (> 350), the prior on the mixing process does not significantly improve the performances. It can also be noticed that unlabeled samples improve the performances of the approach, particularly when the size of the labeled learning data is small. Further improvement of the overall performance level would require a non-linear model.

7 Conclusion

In this paper, we have proposed a method for learning parameters of the IFA model while incorporating two kinds of prior information related to the mixing process on the one hand, and the cluster membership of some training samples on the other hand. In this context, a criterion was defined and a GEM algorithm dedicated to its optimization was described. The proposed method has been applied to fault diagnosis in railway track circuits. The diagnosis system aims at recovering the latent variables linked to the defects from their linear observed mixtures (features extracted from the inspection signal). A comparison between standard and proposed IFA models has been carried out to show that our approach is able to take advantage of prior information, thus significantly improving estimation accuracy and removing indeterminacy of the unsupervised IFA such as permutation of sources. Further studies will be carried out to incorporate nonlinearity and also to take into account imprecise and uncertain cluster memberships such as supplied by human experts.

References

1. Hyvärinen, A., Karhunen, J., Oja, E.: Independent Component Analysis. Wiley, Chichester (2001)

2. Bell, A.J., Sejnowski, T.J.: An information maximization approach to blind separation and blind deconvolution. Neural Computation 7(6), 1129–1159 (1995)
3. Jutten, C., Comon, P.: Séparation de source 2, au-delà de l'aveugle et application. Hermés (2007)
4. Moulines, E., Cardoso, J., Cassiat, E.: Maximum likelihood for blind separation and deconvolution of noisy signals using mixture models. In: Proceedings of the IEEE International Conference on Acoustics, Speech and Signal Processing, pp. 3617–3620 (1997)
5. Attias, H.: Independent factor analysis. Neural Computation 11(4), 803–851 (1999)
6. Attias, H.: Independent factor analysis with temporally structured factors. In: Proceedings of the 12th NIPS Conference, pp. 386–392. MIT Press, Cambridge (2000)
7. Moussaoui, S., Hauksdóttir, H., Schmidt, F., Jutten, C., Chanussot, J., Brie, D., Douté, D., Benediktosson, J.: On the decomposition of Mars Hyperspectral data by ICA and Bayesian positive source separation. Neurocomputing for Vision Research; Advances in Blind Signal Processing 71, 2194–2208 (2008)
8. Hyvärinen, A., Karthikesh, R.: Imposing sparsity on the mixing matrix in independent component analysis. Neurocomputing 49(1), 151–162 (2002)
9. Zhang, K., Chan, L.W.: ICA with sparse connections. In: Proceedings of Intelligent Data Engineering and Automated Learning Conference (IDEAL), pp. 530–537. Springer, Heidelberg (2006)
10. Bartholomew, D.J., Martin, K.: Latent variable models and factor analysis, 2nd edn. Kendall's library of satistics, Arnold (1999)
11. Bakir, T., Peter, A., Riley, R., Hackett, J.: Non-negative maximum likelihood ICA for blind source separation of images and signals with application to hyperspectral image subpixel demixing. In: Proceedings of the IEEE International Conference on Image Processing, pp. 3237–3240 (2006)
12. Bollen, K.A.: Structural Equations with Latent Variables. Wiley, Chichester (1989)
13. Cǒme, E.: Apprentissage de modèles génératifs pour le diagnostic de systèmes complexes avec labellisation douce et contraintes spatiales PhD thesis, Université de Technologie de Compiégne (2009)
14. Amari, S., Cichocki, A., Yang, H.H.: A New Learning Algorithm for Blind Signal Separation. In: Proceedings of the 8th Conference on Advances in Neural Information Processing Systems (NIPS). MIT Press, Cambridge (1995)
15. Shafer, G.: A mathematical theory of evidence. Princeton University Press, Princeton (1976)
16. Cǒme, E., Oukhellou, L., Denœux, T., Aknin, P.: Learning from partially supervised data using mixture models and belief functions. Pattern recognition 42, 334–348 (2009)
17. Debiolles, A., Oukhellou, L., Denoeux, T., Aknin, P.: Output coding of spatially dependent subclassifiers in evidential framework. Application to the diagnosis of railway track-vehicle transmission system. In: Proceedings of FUSION 2006, Florence, Italy (July 2006)
18. Mclachlan, G.J., Krishnan, T.: The EM algorithm and Extension. Wiley, Chichester (1996)
19. Cichocki, A., Amari, S.: Adaptive Blind Signal and Image Processing. Wiley, Chichester (2002)

Speech Hashing Algorithm Based on Short-Time Stability*

Ning Chen and Wang-Gen Wan

School of Communication and Information Engineering, Shanghai University, Shanghai 200072, China
chenning@shu.edu.cn, wanwg@staff.shu.edu.cn

Abstract. The performance of a perceptual hashing system, which is often measured by discrimination and robustness, is directly related to the features that the system extracts. In this letter, a new speech hashing scheme based on short-time stability is presented. The characteristic of natural speech that the principal components of linear prediction coefficients among neighboring frames tend to be very similar is utilized to generate the hash sequence. Experimental results demonstrate the effectiveness of the proposed scheme in terms of discrimination and robustness.

Keywords: Perceptual hashing, short-time stability, Linear Prediction Coefficients (LPCs), Principal Component Analysis (PCA).

1 Introduction

Perceptual hashing (also known as fingerprinting) scheme summarizes the multimedia content into a concise signature sequence, which can then be used to identify the original content. It provides fast and reliable means for protection, management, and indexing of multimedia contents. Promising applications of audio perceptual hashing include broadcast monitoring, connected audio, filtering for file sharing, and automatic organization of music library [1]. Unlike cryptographic hashing, which dramatically changes when a single bit of the input content changes [2], perceptual hashing is insensitive to "reasonable" degradations, such as filtering, amplitude boosting/cutting, re-quantizing, normalizing, inverting and so forth, but is sensitive to the change in content.

The most important properties that perceptual hashing scheme must satisfy are discrimination and robustness [3]:

* This work was supported in part by the National Natural Science Foundation of China (No. 60872115), in part by the Shanghai's Key Discipline Development Program (No.J50104), and in part by the International Cooperation Foundation Program of Shanghai (No. 075107035).

C. Alippi et al. (Eds.): ICANN 2009, Part II, LNCS 5769, pp. 426–434, 2009.

- Discrimination (collision free): The hash sequence should reflect the content of the data in a unique way. In other words, data with perceptually different content should yield different hash sequences;
- Robustness (invariance under perceptual similarity): The hash sequence generated from the data, which is subjected to certain non-malicious manipulations, such as low-pass filtering, amplitude boosting/cutting, re-quantizing, normalizing, inverting and so forth, is equal or similar to that generated from the original data.

Due to wide application, audio perceptual hashing scheme has been widely studied recently [4]. Most of the available audio hashing schemes are based on well known audio features, such as Fourier coefficients [5], Short Time Fourier Transform (STFT) feature [6], Mel Frequency Cepstral Coefficients (MFCC) [7], spectral flatness [8], and the derived quantities, such as derivatives, means and variances of audio features. Haitsma [1] gets the hash sequence based on extracting 32 bit sub-fingerprint, which are generated by looking at energy differences along the frequency and the time axes. Park [9] introduces alternatives to the frequency-temporal filtering combination for an extension method of [1] to achieve robustness to channel and background noise under the conditions of the real situation. Özer [10] presents audio hash function based on periodicity series of the fundamental frequency and on singular-value description of the cepstral frequencies. They achieve satisfactory discrimination and robustness. Seo utilizes the spectral sub-band centroid (due to its resilience against equalization, compression and noise addition) [11], the first-order normalized moment, the second-order normalized moment and the spectral flatness measure [12], respectively, to generate hash sequence. Jiao [13] proposes a perceptual audio hashing scheme in compressed domain based on Modified Discrete Cosine Transform (MDCT) coefficients, which are intermedial decoding results in MPEG Layer 3 (MP3) and MPEG Advanced Audio Coding (AAC) systems. In [14], a randomization scheme controlled by a random seed is introduced during both compressed domain feature extraction and hash modeling to increase the security of the scheme proposed in [13].

There is few speech perceptual hashing scheme available [15]. In [16], a speech hash scheme integrated with Mixed Excitation Linear Prediction (MELP) codec is proposed. It utilizes partial bits of the speech bitstream, the Linear Spectral Frequencies (LSFs), for hash generation, so that it is highly robust to speech parametric coding. However, it is just suited for compressed domain speech signal. In this paper, a speech hash scheme on PCM source is proposed. The characteristic of natural speech that the principal components of LPCs among neighboring frames tend to be very similar is utilized to generate the hash sequence. Experimental results demonstrate the effectiveness of the proposed hashing scheme in terms of discrimination and robustness. The remainder of this letter is organized as follows. Section 2 describes the proposed speech hashing scheme in detail. Section 3 evaluates the proposed hashing scheme. Section 4 concludes the paper.

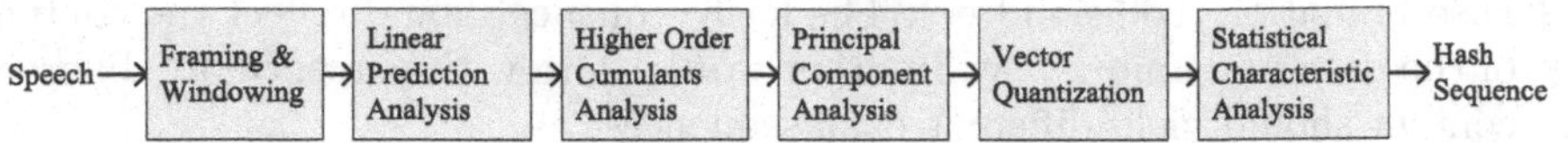

Fig. 1. Block diagram of the hash sequence generation based on short-time stability

2 Proposed Speech Hashing Scheme

2.1 Fundamental Theory

Linear prediction modeling is a widely used method for representing the frequency shaping attributes of the vocal tract. It characterizes the shape of the spectral of a short segment of speech with a small number of parameters, the Linear Prediction Coefficients (LPCs), which are widely used in speech coding, speech enhancement and speech recognition [17]. In the proposed speech hashing scheme, first, the original speech is segmented into equal frames, on which the linear prediction analysis is performed to extract corresponding LPCs. Next, considering the Gaussian signal suppression property of high-order cumulant [18], the fourth order cumulant of the LPCs of each frame is calculated to reduce Gaussian noise interference. Then, Principal Component Analysis (PCA) is performed on the obtained fourth order cumulant to get the principal components, on which Vector Quantization (VQ) is performed to get the VQ indices. Finally, the statistical characteristic of the VQ indices among neighboring frames is analyzed to generate the hash sequence. The Block diagram of the hash sequence generation based on short-time stability is shown in Fig. 1.

2.2 Hash Generation

Let $\mathbf{s} = \{s(n)|n = 1, \cdots, L\}$ be the original speech signal, the hash sequence generation procedure can be described as follows:

Step 1: Segment the original speech $\mathbf{s}$ into M equal frames, denoted as $\mathbf{f}_i = \{f_i(n)|n = 1, \cdots, l; i = 1, \cdots, M\}$. And to reduce edge effect, window each frame $\mathbf{f}_i, i = 1, \cdots, M$ with hamming window to get $\hat{\mathbf{f}}_i, i = 1, \cdots, M$.
Step 2: Perform linear prediction analysis on $\hat{\mathbf{f}}_i, i = 1, \cdots, M$ to get its pth-order LPCs, denoted as $\mathbf{x}_i = \{x_i(n)|n = 1, \cdots, p; i = 1, \cdots, M\}$, with the autocorrelation method of autoregressive modeling [19]. And for each $\mathbf{x}_i, i = 1, \cdots, M$, calculate its fourth order cumulant, denoted as $\hat{\mathbf{x}}_i = \{\hat{x}_i(n)|n = 1, \cdots, p; i = 1, \cdots, M\}$.
Step 3: First, generate matrix $\mathbf{x}$ by using $\hat{\mathbf{x}}_i, i = 1, \cdots, M$ as its column vectors as follows

$$\mathbf{x} = [\hat{\mathbf{x}}_1, \hat{\mathbf{x}}_2, \cdots, \hat{\mathbf{x}}_M] \tag{1}$$

and calculate the covariance matrix of $\mathbf{x}$, denoted as $\mathbf{c_x}$, with

$$\mathbf{c_x} = E(\mathbf{x} - \bar{\mathbf{x}})(\mathbf{x} - \bar{\mathbf{x}})^T \tag{2}$$

where $\bar{\mathbf{x}}$ is the mean vector of each sub-vectors $\hat{\mathbf{x}}_i, i = 1, \cdots, M$. Assume the eigenvalues of $\mathbf{c}_\mathbf{x}$ are $\lambda_1, \lambda_2, \cdots, \lambda_p$ ($\lambda_1 \geq \lambda_2 \geq \cdots \geq \lambda_p$), and the corresponding eigenvectors are $\mathbf{e}_1, \mathbf{e}_2, \cdots, \mathbf{e}_p$, then the matrix $\mathbf{e} = [\mathbf{e}_1, \mathbf{e}_2, \cdots, \mathbf{e}_p]$ is called the basic function of PCA. Next, de-correlate $\mathbf{x}$ using $\mathbf{e}$ as follows

$$\mathbf{y} = \mathbf{e} \times \mathbf{x} \tag{3}$$

to get matrix $\mathbf{y}$ and generate a new matrix $\hat{\mathbf{y}} = [\hat{\mathbf{y}}_1, \hat{\mathbf{y}}_2, \cdots, \hat{\mathbf{y}}_M]$ with 1 to Q rows of $\mathbf{y}$.
Step 4: Obtain the Vector Quantization (VQ) code book $\mathbf{c}$ by performing LBG algorithms [20] on the training set that is composed of $\hat{\mathbf{y}}_i, i = 1, \cdots, M$. And then perform VQ on each $\hat{\mathbf{y}}_i, i = 1, \cdots, M$ with $\mathbf{c}$ to get the corresponding VQ index, denoted as $y(i), i = 1, \cdots, M$.
Step 5: Calculate the variance of $y(i)$ and its surrounding indices $y(i-1)$ and $y(i+1)$ to get $\sigma^2(i)$ with

$$\sigma^2(i) = \frac{1}{3} \sum_{m=i-1}^{m=i+1} y^2(m) - \left[\frac{1}{3} \sum_{m=i-1}^{m=i+1} y(m) \right]^2 \tag{4}$$

And generate the hash sequence, denoted as $\mathbf{h} = \{h(i) | i = 1, \cdots, M\}$, by comparing $\sigma^2(i), i = 1, \cdots, M$ with the the median of $\sigma^2(i), i = 1, \cdots, M$ as follows

$$h(i) = \begin{cases} 1, & \text{if } \sigma^2(i) \geq \widetilde{\sigma} \\ 0, & \text{otherwise} \end{cases} \tag{5}$$

where $\widetilde{\sigma}$ is the median of $\sigma^2(i), i = 1, \cdots, M$.

2.3 Hash Matching

In the hash matching, two speech clips are declared similar if the distance between their hash sequences is below a certain threshold. The problem could be formulated as the hypothesis testing process with the following two hypothesizes [21]:

- H0: Two speech clips are from the same speech clip if the distance of their hash sequences is below the threshold α;
- H1: Two speech clips are from the different speech clip if the distance of their hash sequences is above the threshold α.

For the selection of α, there is a tradeoff between the False Accept Rate (FAR) and the False Rejection Rate (FRR). FAR is the probability that hypothesis H0 is accepted when the hypothesis H1 is true. FRR is the probability that hypothesis H1 is accepted when the hypothesis H0 is true.

In this letter, the similarity of two speech clips is measured by Hamming distance (i.e. the number of different bits of two hash sequences). Bit Error Rate (BER), which is the ratio of the number of different bits to the total number of bits, is used in our experiments.

3 Performance Evaluation

The performance of the proposed speech hashing scheme in terms of discrimination and robustness is tested. It is shown that perceptually distinct input speech clips result in different hash sequences, while perceptually similar speech clips result in similar hash sequences.

In the database we do experiments on, there are 1000 speech clips (16 bits signed, 8 kHz sampled and about 4 seconds long) extracted from speech of different content spoken by male and female talkers. And nine manipulations are performed on the original speech clip to get the manipulated ones. These manipulations include low-pass filtering, re-quantizing, amplitude boosting/cutting, normalizing, inverting and so forth, which are described in detail later in this section.

3.1 Discrimination

Since the speech clips are randomly chosen, BER value is a random variable. Then the distribution of BER is analyzed to evaluate the discrimination of the proposed hashing scheme. The hash sequences of 1000 speech clips are cross matched and 499500 comparisons between speech clips with different content have been done. The probability distribution of BER is shown in Fig. 2.

It is shown that BER has a normal distribution approximately. The expected value μ is 0.4242 and the standard deviation σ is 0.0464. And then, the FAR could be given in (6). The FAR at different BER threshold are listed in TABLE

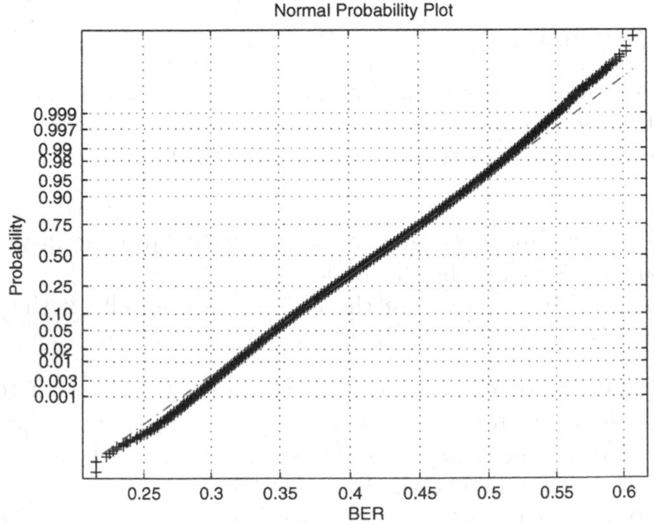

Fig. 2. Normal probability plot of the BER between each pair of hash sequences generated from speech clips with different content

Table 1. *FAR* at different *BER* threshold

α	FAR
0.10	2.2177×10^{-10}
0.15	2.2905×10^{-7}
0.20	7.4182×10^{-5}
0.25	7.5335×10^{-3}

I.

$$FAR = f(\alpha|\mu,\sigma) = \frac{1}{\sigma\sqrt{2\pi}}e^{\frac{-(\alpha-\mu)^2}{2\sigma^2}} \quad (6)$$

It is proved that the hash sequence extracted by the proposed hashing scheme is unique to different content.

3.2 Robustness

On each of the 1000 speech clips, nine kinds of signal processing manipulations are performed to generate 9000 processed ones. The nine processing manipulations include:

- Amplitude boosting: The amplitude of the speech is boosted by 2dB and 3dB, respectively;
- Amplitude cutting: The amplitude of the speech is cut by 2dB and 3dB, respectively;
- Inverting: The phase of the speech is inverted;
- Hard-limit: Max amplitude limit is -0.1 dB, input is boosted by 6 dB, look ahead time is 7 ms and release time is 100 ms;
- Normalizing: The speech is normalized to 90%;
- Re-quantizing: The 16-bit speech signal has been re-quantized up to 32 bits/sample and back to 16 bits/sample;
- Low-pass filtering: The speech is low-pass filtered using 10-tap Butterworth filter with cutoff frequency 4kHz.

To evaluate the risk of collision, the worst Intra matching is compared with the best Inter matching for each original speech clip in the database. Given a reference original speech clip, the 9000 processed speech clips are classified into Intra and Inter processed speech clips depending on whether they have been derived from the reference speech clip or not. The *BER* between the hash sequence extracted from the original speech clip and each one of the hash sequences extracted from the processed speech clips are computed. Fig. 3 presents the worst Intra *BER* and the best Inter *BER* for each speech clip in the database. It is shown that all Intra *BER* lie below 0.20 and that all Inter *BER* are larger than 0.2. So *BER* is an efficient way to compare two hash sequences and that 0.2 is a good threshold to decide whether two speech clips are perceptually similar or not.

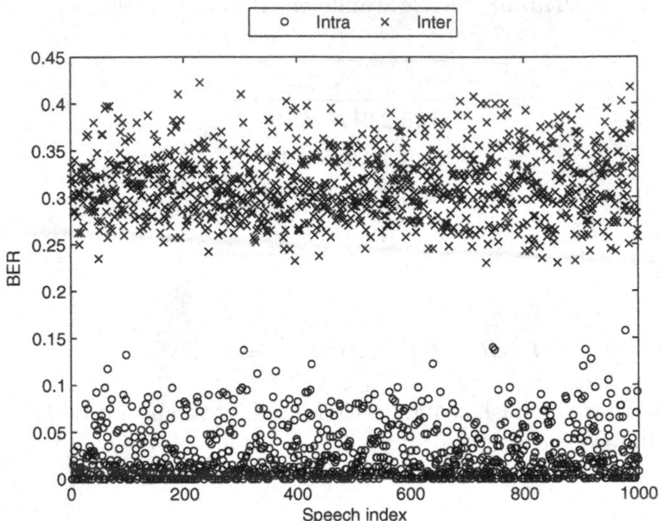

Fig. 3. BER of worst Intra matching and best Inter matching for each of 1000 speech clips

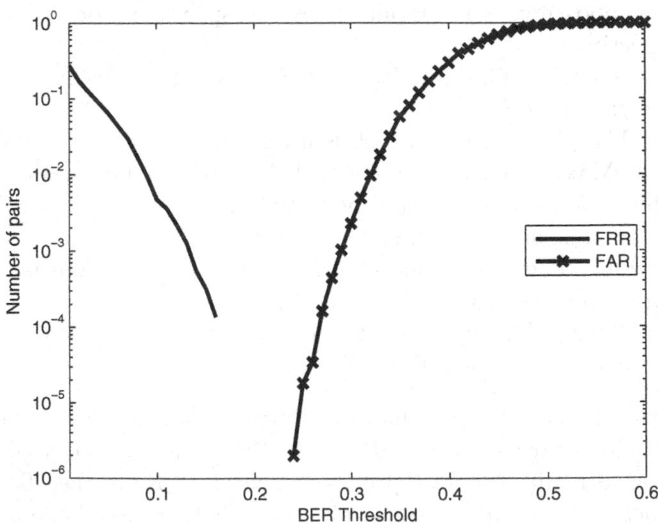

Fig. 4. $FAR - FRR$ curves

The $FAR - FRR$ curves are shown in Fig. 4, where FRR is the number of mismatched hash sequences normalized by the number of comparisons between speech clips of similar content, and FAR is the number of matched hash sequences normalized by the number of comparisons between speech clips of different content. It can be seen that there is no intersecting point on the $FAR-FRR$ curves in Figure 4 and that if the BER threshold is set to 0.20, it is reliable for us to tell whether two speech clips are perceptually similar or not.

The above experimental results and the data analysis show that the proposed hashing scheme achieves satisfactory discrimination and robustness needed for perceptual hashing system.

4 Conclusion

An efficient speech hash scheme based on short-time stability is presented. The characteristic of natural speech that the principal components of LPCs among neighboring frames tend to be very similar is utilized to generate the hash sequence. Although the VQ indices of the speech manipulated by the signal processing manipulations, such as amplitude boosting/cutting, inverting, hard-limiting, normalizing, re-quantizing, low-pass filtering and so forth, may be very different from the original one, the variance of neighboring indices does not vary too much. This is the reason why the proposed scheme achieves great robustness. Furthermore, since the short-time stability varies among different speech signal, the hash vector generated from distinct speech source is different. Thus the proposed scheme has satisfactory discrimination.

Security issue becomes important when perceptual hash is applied in speech content authentication. In order to increase unpredictability of the proposed scheme, our future work will concentrate on the design of key-based randomization scheme for the proposed hash function.

References

1. Haitsma, J., Kalker, T.: A highly robust audio fingerprinting system. In: Proc. ISMIR, vol. 2002 (2002)
2. Menezes, A.J., Van Oorschot, P.C., Vanstone, S.A.: Handbook of applied cryptography. CRC Press, Boca Raton (1997)
3. Seo, J.S., Haitsma, J., Kalker, T., Yoo, C.D.: A robust image fingerprinting system using the Radon transform. Signal Processing: Image Communication 19(4), 325–339 (2004)
4. Cano, P., Batlle, E., Kalker, T., Haitsma, J.: A review of audio fingerprinting. The Journal of VLSI Signal Processing 41(3), 271–284 (2005)
5. Fragoulis, D., Rousopoulos, G., Panagopoulos, T., Alexiou, C., Papaodysseus, C.: On the automated recognition of seriously distorted musical recordings. IEEE Transactions on Signal Processing 49(4), 898–908 (2001)
6. Ramalingam, A., Krishnan, S.: Gaussian mixture modeling of short-time Fourier transform features for audio fingerprinting. IEEE Transactions on Information Forensics and Security 1(4), 457–463 (2006)
7. Sethi, I., Kulesh, V., Petrushin, V.: Indexing and retrieval of music via gaussian mixture models. In: Proc. 3rd Int. Workshop on Content Based Multimedia Indexing, vol. 2003 (2003)

8. Allamanche, E., Herre, J., Hellmuth, O., Froba, B., Cremer, M.: AudioID: Towards content-based identification of audio material. Preprints-Audio Engineering Society (2001)
9. Park, M., Kim, H., Yang, S.H.: Frequency-temporal filtering for a robust audio fingerprinting scheme in real-noise environments. ETRI journal 28(4), 509–512 (2006)
10. Özer, H., Sankur, B., Memon, N., Anarim, E.: Perceptual audio hashing functions. EURASIP Journal on Applied Signal Processing 2005(1), 1780–1793 (2005)
11. Seo, J.S., Jin, M., Lee, S., Jang, D., Lee, S.: Audio fingerprinting based on normalized spectral subband centroids. In: IEEE International Conference on Acoustics, Speech, and Signal Processing, 2005. Proceedings(ICASSP 2005), vol. 3 (2005)
12. Seo, J.S., Jin, M., Lee, S., Jang, D., Lee, S., Yoo, C.D.: Audio fingerprinting based on normalized spectral subband moments. IEEE Signal Processing Letters 13(4), 209–212 (2006)
13. Jiao, Y., Yang, B., Li, M., Niu, X.: MDCT-Based Perceptual Hashing for Compressed Audio Content Identification. In: IEEE 9th Workshop on Multimedia Signal Processing, 2007. MMSP 2007, pp. 381–384 (2007)
14. Jiao, Y., Li, M., Li, Q., Niu, X.: Key-Dependent Compressed Domain Audio Hashing. In: Eighth International Conference on Intelligent Systems Design and Applications, ISDA 2008, vol. 3 (2008)
15. Niu, X.M., Jiao, Y.H.: An overview of perceptual hashing. Acta Electronica Sinica 36(7), 1405–1411 (2008)
16. Jiao, Y., Li, Q., Niu, X.: Compressed Domain Perceptual Hashing for MELP Coded Speech. In: International Conference on Intelligent Information Hiding and Multimedia Signal Processing, IIHMSP 2008, pp. 410–413 (2008)
17. Nichols, R.K., Lekkas, P.C.: Wireless security: models, threats, and solutions. McGraw-Hill, New York (2002)
18. Green, D.R.: The utility of higher-order statistics in gaussian noise suppression. US Government Authored or Collected Report, Naval Postgraduate School, Memory, Calif, USA (2003)
19. Jackson, L.B.: Digital filters and signal processing. Kluwer Academic Publishers, Dordrecht (1985)
20. Gersho, A., Gray, R.M.: Vector quantization and signal compression. Springer, Heidelberg (1992)
21. Swaminathan, A., Mao, Y., Wu, M.: Robust and secure image hashing. IEEE Transactions on Information Forensics and Security 1(2), 215–230 (2006)

A New Method for Complexity Reduction of Neuro-fuzzy Systems with Application to Differential Stroke Diagnosis

Krzysztof Cpałka[1,2], Olga Rebrova[3], and Leszek Rutkowski[1,2]

[1] Department of Computer Engineering,
Częstochowa University of Technology, Poland
[2] Department of Artificial Intelligence,
Academy of Humanities and Economics in Łódź, Poland
[3] Institute of Neurology,
Russian Academy of Medical Sciences, Russia
Krzysztof.Cpalka@kik.pcz.pl, olga@neurology.ru, lrutko@kik.pcz.czest.pl

Abstract. In the paper we propose a new method for designing and reduction of neuro-fuzzy systems for stroke diagnosis. The concept of the weighted parameterized triangular norms is applied and neuro-fuzzy systems based on fuzzy S-implications are derived. In subsequent stages we reduce the linguistic model. The results are implemented to solve the problem of stroke diagnosis.

Keywords: fuzzy logic, logical neuro-fuzzy systems, interpretability, reduction, merging, stroke diagnosis.

1 Introduction

In recent years various neuro-fuzzy systems have been studied in the literature (see e.g. [9]-[12], [17]-[18], [24]-[27]). They have many applications in different fields (see e.g. [20]), e.g. they allow to overcome problems of analyzing medical data characterized by qualitative character and incompleteness of the information. Recently several algorithms have been proposed to increase interpretability and accuracy and decrease complexity of fuzzy rule-based systems. For various methods of designing fuzzy rule-based systems the reader is referred to [2]-[8], [13]-[22]. In this paper we propose a new algorithm, called ABGE (best global eliminations algorithm), to design and reduce neuro-fuzzy systems. Our previous algorithm CEA (algorithm of consecutive eliminations), presented in [5] and [8], was oriented to the most possible simplification of the system structure for the price of a system accuracy. In the context of differential stroke diagnosis problem it was not desirable property. The algorithm studied in this paper allows to obtain a better accuracy and a reasonable degree of a complexity reduction.

C. Alippi et al. (Eds.): ICANN 2009, Part II, LNCS 5769, pp. 435–444, 2009.

We will consider multi-input, single-output a neuro-fuzzy system of the logical type (see e.g. [24], [26]), mapping $\mathbf{X} \to \mathbf{Y}$, where $\mathbf{X} \subset \mathbf{R}^n$ and $\mathbf{Y} \subset \mathbf{R}$. The fuzzy rule base of these systems consists of a collection of N fuzzy IF-THEN rules in the form

$$R^k : \left[\text{IF } x_1 \text{ is } A_1^k \text{ AND} \ldots \text{AND } x_n \text{ is } A_n^k \text{ THEN } y \text{ is } B^k\right], \tag{1}$$

where $\mathbf{x} = [x_1, \ldots, x_n] \in \mathbf{X}$, $y \in \mathbf{Y}$, $A_1^k, A_2^k, \ldots, A_n^k$ are fuzzy sets characterized by membership functions $\mu_{A_i^k}(x_i)$, $\mathbf{A}^k = A_1^k \times A_2^k \times \ldots \times A_n^k$, and B^k are fuzzy sets characterized by membership functions $\mu_{B^k}(y)$, respectively, $k = 1, \ldots, N$, and N is the number of rules. The defuzzification is realized by the COA (center of area) method defined by the following formula

$$\bar{y} = \sum_{r=1}^{R} \bar{y}_r^B \mu_{B'}\left(\bar{y}_r^B\right) / \sum_{r=1}^{R} \mu_{B'}\left(\bar{y}_r^B\right), \tag{2}$$

where membership function of B', obtained from the linguistic model (1), is given by

$$\mu_{B'}\left(\bar{y}_r^B\right) = \mathop{T}_{k=1}^{N} \left\{\mu_{\bar{B}^k}\left(\bar{y}_r^B\right)\right\}. \tag{3}$$

Each of N rules (1) determines a fuzzy set $\bar{B}^k \subset \mathbf{Y}$ characterized by

$$\mu_{\bar{B}^k}\left(\bar{y}_r^B\right) = \mu_{\mathbf{A}^k \to B^k}\left(\bar{\mathbf{x}}, \bar{y}_r^B\right) = I_{fuzzy}\left(\mu_{\mathbf{A}^k}\left(\bar{\mathbf{x}}\right), \mu_{B^k}\left(\bar{y}_r^B\right)\right), \tag{4}$$

where $\bar{y}_r^B$ denotes centers of the output membership functions $\mu_{B^r}(y)$, i.e. for $r = 1, \ldots, R$,

$$\mu_{B^r}\left(\bar{y}_r^B\right) = \max_{y \in \mathbf{Y}} \left\{\mu_{B^r}(y)\right\}, \tag{5}$$

and R is the number of discretization points of the integrals in the continuous version of the COA method.

This paper is organized into six sections. In Section 2 the description of a neuro-fuzzy system is given. In Section 3 the reduction of a neuro-fuzzy system is presented. In Section 4 we describe algorithm for merging of similar input and output fuzzy sets. Section 5 shows the simulation results. Conclusions are drawn in Section 6.

2 Description of Neuro-fuzzy System

Let us introduce weights $w_{i,k}^{\tau} \in [0, 1]$, $k = 1, \ldots, N$, $i = 1, \ldots, n$, describing importance of antecedents and weights $w_k^{\text{agr}} \in [0, 1]$, $k = 1, \ldots, N$, describing importance of rules. Now the linguistic model (1) is transformed to the following description

$$R^{(k)}\text{:}\left[\text{IF} x_1 \text{is} A_1^k \left(w_{1,k}^{\tau}\right) \text{AND} \ldots \text{AND} x_n \text{is} A_n^k \left(w_{n,k}^{\tau}\right) \text{THEN} y \text{is} B^k\right] \left(w_k^{\text{agr}}\right). \tag{6}$$

In order to incorporate weights into description of neuro-fuzzy systems (2) we proposed [26] the adjustable weighted t-norm

$$\begin{aligned} \overleftrightarrow{T}^{*}\left\{a_{1,k},\ldots,a_{n,k};w_{1,k}^{\tau},\ldots,w_{n,k}^{\tau},p_{k}^{\tau}\right\} = \overleftrightarrow{T}^{*}\left\{\mathbf{a}_{k};\mathbf{w}_{k}^{\tau},p_{k}^{\tau}\right\} = \\ \overleftrightarrow{T}\left\{1-w_{1,k}^{\tau}\left(1-a_{1,k}\right),\ldots,1-w_{n,k}^{\tau}\left(1-a_{n,k}\right);p_{k}^{\tau}\right\} \end{aligned} \tag{7}$$

to connect the antecedents in each rule, $k = 1, \ldots, N$, and the adjustable weighted t-norm

$$\begin{aligned} \overleftrightarrow{T}^{*}\left\{a_{1},\ldots,a_{N};w_{1}^{\mathrm{agr}},\ldots,w_{N}^{\mathrm{agr}},p^{\mathrm{agr}}\right\} = \overleftrightarrow{T}^{*}\left\{\mathbf{a};\mathbf{w}^{\mathrm{agr}},p^{\mathrm{agr}}\right\} = \\ \overleftrightarrow{T}\left\{1-w_{1}^{\mathrm{agr}}\left(1-a_{1}\right),\ldots,1-w_{N}^{\mathrm{agr}}\left(1-a_{N}\right);p^{\mathrm{agr}}\right\} \end{aligned} \tag{8}$$

to aggregate the individual rules in the logical models, respectively. In formula (7) parameters $a_{i,k}$, $i = 1, \ldots, n$, $k = 1, \ldots, N$, correspond to the values of $\mu_{A_i^k}(\bar{x}_i)$, whereas parameters a_k, $k = 1, \ldots, N$, in formula (8) correspond to the values of $\mu_{\bar{B}^k}\left(\bar{y}_r^B\right)$. It is easily seen that formula (7) can be applied to the evaluation of an importance of input linguistic values, and the weighted t-norm (8) to a selection of important rules. We use notation $\overleftrightarrow{T}^{*}\{a_1, \ldots, a_n; w_1, \ldots, w_n, p\}$ and $\overleftrightarrow{S}^{*}\{a_1, \ldots, a_n; w_1, \ldots, w_n, p\}$ for adjustable weighted triangular norms, and notation $\overleftrightarrow{T}^{*}\{a_1, \ldots, a_n; p\}$ and $\overleftrightarrow{S}^{*}\{a_1, \ldots, a_n; p\}$ for adjustable triangular norms. The hyperplanes corresponding to them can be adjusted in the process of learning of an appropriate parameter p. Fuzzy norms (7) and (8) are parameterized by parameters p_k^{τ}, $k = 1, \ldots, N$, and p^{agr}, respectively.

In the paper we propose the adjustable version of an S-implication

$$I_{fuzzy}\left(\mu_{A^k}(\bar{\mathbf{x}}), \mu_{B^k}(y)\right) = \overleftrightarrow{S}\left\{1-\mu_{A^k}(\bar{\mathbf{x}}), \mu_{B^k}(y); p_k^I\right\}. \tag{9}$$

t-conorm in formula (9) is parameterized by parameters p_k^I, $k = 1, \ldots, N$. In our study we use Dombi families [24] of adjustable triangular norms.

Neuro-fuzzy architectures developed so far in the literature are based on the assumption that number of terms R in a formula (2) is equal to the number of rules N. In view of the above assumptions formula (2) takes the form

$$\bar{y} = f(\bar{\mathbf{x}}) = \frac{\sum\limits_{r=1}^{R} \bar{y}^r \overleftrightarrow{\underset{k=1}{\overset{N}{T}}}^{*} \left\{ \begin{array}{c} \overleftrightarrow{S}\left\{ 1-\overleftrightarrow{T}^{*}\left\{ \begin{array}{c} \mu_{A_1^k}(\bar{x}_1),\ldots,\mu_{A_N^k}(\bar{x}_N); \\ w_{1,k}^{\tau},\ldots,w_{n,k}^{\tau},p_k^{\tau} \end{array} \right\}, \\ \mu_{B^k}(\bar{y}^r); p_k^I \end{array} \right\}; \\ p^{\mathrm{agr}} \end{array} \right\}}{\sum\limits_{r=1}^{R} \overleftrightarrow{\underset{k=1}{\overset{N}{T}}}^{*} \left\{ \begin{array}{c} \overleftrightarrow{S}\left\{ 1-\overleftrightarrow{T}^{*}\left\{ \begin{array}{c} \mu_{A_1^k}(\bar{x}_1),\ldots,\mu_{A_N^k}(\bar{x}_N); \\ w_{1,k}^{\tau},\ldots,w_{n,k}^{\tau},p_k^{\tau} \end{array} \right\}, \\ \mu_{B^k}(\bar{y}^r); p_k^I \end{array} \right\}; \\ p^{\mathrm{agr}} \end{array} \right\}}. \tag{10}$$

The following parameters of a system (10) for stroke diagnosis are subject to learning by using the backpropagation method: (i)Parameters $\bar{x}_{i,k}^A$ and $\sigma_{i,k}^A$ of input fuzzy sets A_i^k, $k = 1, \ldots, N$, $i = 1, \ldots, n$, and parameters $\bar{y}_k^B$ and σ_k^B of output fuzzy sets B^k, $k = 1, \ldots, N$, (ii)Weights $w_{i,k}^{\tau} \in [0,1]$, $k = 1, \ldots, N$, $i = 1, \ldots, n$, of importance of antecedents and weights $w_k^{\mathrm{agr}} \in [0,1]$, $k = 1, \ldots, N$, of importance of rules, (iii)Parameters p^{agr}, p_k^{I}, p_k^{τ}, $k = 1, \ldots, N$, of adjustable triangular norms used for aggregation of rules, connections of antecedents and consequences and aggregation of antecedents, respectively, (iv)Discretization points $\bar{y}^r$, $r = 1, \ldots, R$.

3 Reduction of a Neuro-fuzzy System (10)

Now we will develop new algorithms of reduction of a neuro-fuzzy system (10). The algorithms are based on analysis of weights in antecedents of the rules $w_{i,k}^{\tau} \in [0,1]$, $i = 1, \ldots, n$, $k = 1, \ldots, N$, and weights in aggregation of the rules $w_k^{\mathrm{agr}} \in [0,1]$, $k = 1, \ldots, N$.

The flowcharts in Fig. 2.a and Fig. 2.b comprise 4 parts. First, we determine performance of the initial system (before the reduction process); for example, in the case of the classification we determine a percentage of mistakes of the system. The weights $w_i^x \in [0,1]$, $i = 1, \ldots, n$, are calculated using

$$w_i^x = \frac{1}{N} \sum_{k=1}^{N} w_{i,k}^{\tau}. \tag{11}$$

In consecutive stages we reduce number of discretization points, number of inputs, number of rules and number of antecedents. If, e.g. reduction of the i-th input is acceptable (i.e. it does not worsen the system accuracy determined before reduction) then that input is eliminated, otherwise we do not reduce it.

The algorithm of consecutive eliminations (CEA),developed by us in [8], is oriented to the most possible simplification of the system structure. The underlying idea based on consecutive eliminations of contradicted, non-active and unimportant elements of the system starting from discretization points, and next inputs, whole rules and, finally, antecedents of rules. If a specific reduction, e.g. reduction of a specific input, is acceptable (accuracy of the system is not worse than before the reduction), then the reduction is accepted, otherwise it is cancelled. The flowchart of the algorithm is depicted in Fig. 2.a.

The algorithm proposed in this paper, called by us the best global eliminations algorithm (ABGE), is oriented to the searching, across all the parameters of the system, of an element whose reduction is the most advantageous from the point of view of the accuracy. If there is such an element, its reduction is performed and the search is repeated, if not, then the reduction algorithm is stopped. This idea takes into account the fact that if e.g. an element of the linguistic model causes the biggest mistakes in the system's performance, then the reduction should be started from that element, temporary ignoring elements with less adverse impact on the system. The flowchart of the algorithm is depicted in Fig. 2.b.

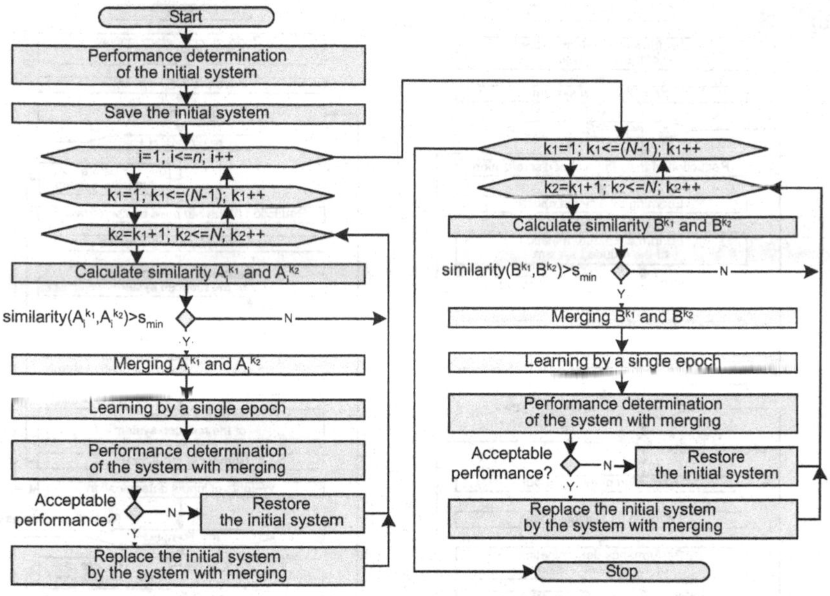

Fig. 1. Consecutive mergings algorithm (CMA)

4 Merging of Similar Input and Output Fuzzy Sets in a Neuro-fuzzy System (10)

In Section 3 we eliminated in a system (10) such elements as discretization points, input features, rules and antecedents. The elimination of these elements did not worsen the accuracy of a system (10) and simultaneously increased its transparency and decreased computational burden. Despite of the reduction, in the linguistic model still exist similar fuzzy sets. Such fuzzy sets should be automatically detected, merged and shared by different rules.

In Fig. 1 we present the algorithm of consecutive mergings algorithm (CMA). The algorithm is initialized by the performance determination (number of correctly classified samples) of a system (10) before merging. Next, we compare all combinations of input fuzzy sets corresponding to particular input features. The comparison is based on the discrete version of similarity measure [4].

$$\text{similarity}\left(A_i^{k_1}, A_i^{k_2}\right)=\frac{\sum\limits_{j=0}^{J-1}\min\left\{\begin{array}{c}\text{Gauss}\left(\bar{x}_{\min}+j\frac{(\bar{x}_{\max}-\bar{x}_{\min})}{J-1};\bar{x}_{i,k_1}^A,\sigma_{i,k_1}^A\right),\\ \text{Gauss}\left(\bar{x}_{\min}+j\frac{(\bar{x}_{\max}-\bar{x}_{\min})}{J-1};\bar{x}_{i,k_2}^A,\sigma_{i,k_2}^A\right)\end{array}\right\}}{\sum\limits_{j=0}^{J-1}\max\left\{\begin{array}{c}\text{Gauss}\left(\bar{x}_{\min}+j\frac{(\bar{x}_{\max}-\bar{x}_{\min})}{J-1};\bar{x}_{i,k_1}^A,\sigma_{i,k_1}^A\right),\\ \text{Gauss}\left(\bar{x}_{\min}+j\frac{(\bar{x}_{\max}-\bar{x}_{\min})}{J-1};\bar{x}_{i,k_2}^A,\sigma_{i,k_2}^A\right)\end{array}\right\}}, \tag{12}$$

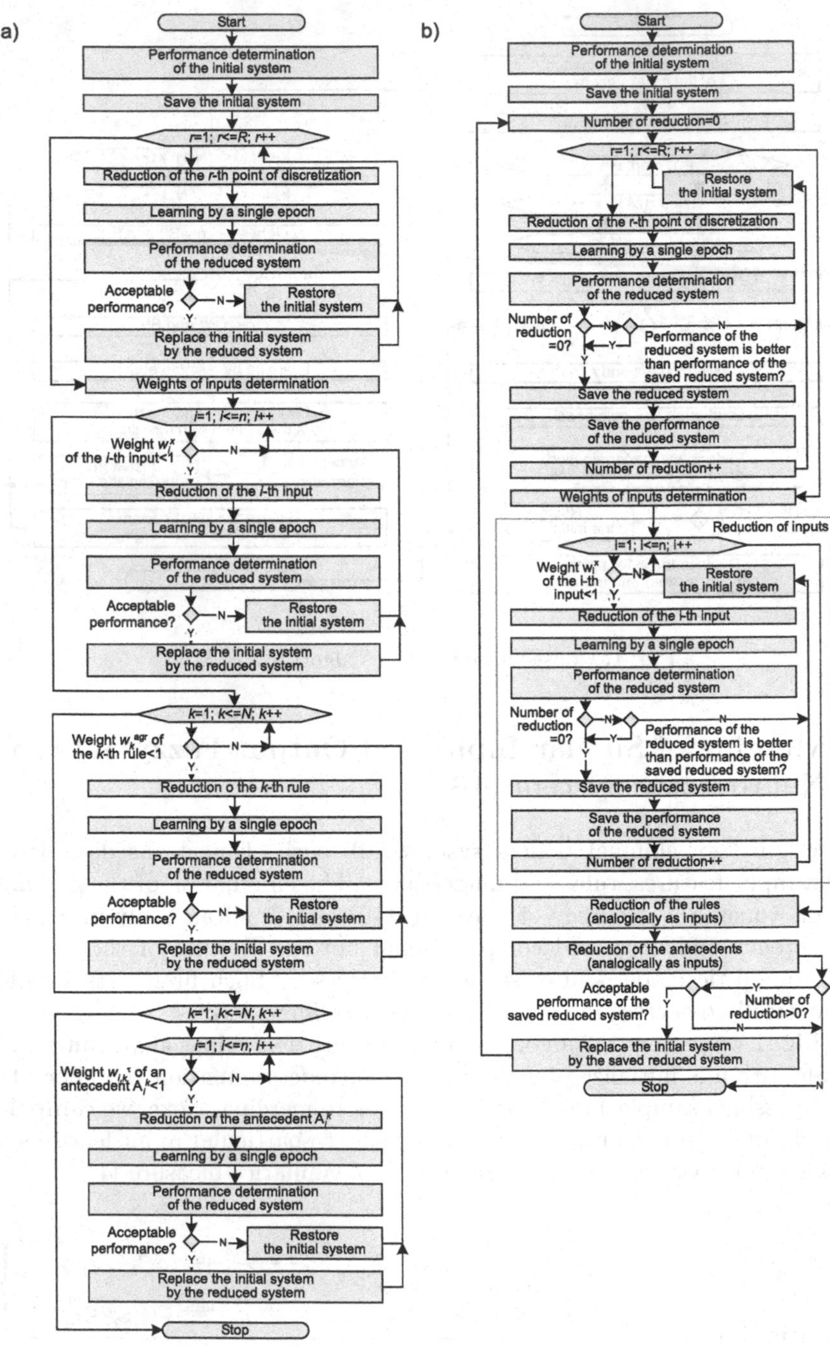

Fig. 2. Algorithms of reduction of a neuro-fuzzy system (10) for stroke diagnosis: a) consecutive eliminations algorithm (CEA), b) algorithm of the best global eliminations (ABGE)

where J is the number of discretization points, and $A_i^{k_1}$ and $A_i^{k_2}$, $i = 1, \ldots, i$, $k_1 = 1, \ldots, N$, $k_2 = 1, \ldots, N$, are fuzzy sets described by Gaussian membership functions. Their centers are located in points $\bar{x}_{i,k_1}^A$ and $\bar{x}_{i,k_2}^A$, widths are denoted by σ_{i,k_1}^A and σ_{i,k_2}^A. The discretization is defined in the interval $[\bar{x}_{\min}, \bar{x}_{\max}]$, where

$$\bar{x}_{\min}=\min\left\{\bar{x}_{\min_1}, \bar{x}_{\min_2}\right\}=\min\left\{\bar{x}_{i,k_1}^A-\sigma_{i,k_1}^A\sqrt{-\ln\left(\psi\right)}, \bar{x}_{i,k_2}^A-\sigma_{i,k_2}^A\sqrt{-\ln\left(\psi\right)}\right\} \tag{13}$$

and

$$\bar{x}_{\max}=\max\left\{\bar{x}_{\max_1}, \bar{x}_{\max_2}\right\}=\max\left\{\bar{x}_{i,k_1}^A+\sigma_{i,k_1}^A\sqrt{-\ln\left(\psi\right)}, \bar{x}_{i,k_2}^A+\sigma_{i,k_2}^A\sqrt{-\ln\left(\psi\right)}\right\}. \tag{14}$$

In our simulations $J = 100$ and $\psi = 0.01$. Values $\bar{x}_{\min_1}$, $\bar{x}_{\min_2}$, $\bar{x}_{\max_1}$, and $\bar{x}_{\max_2}$ in formulas (13) and (14) follow from the solution of equations

$$\begin{cases} \text{Gauss}\left(\bar{x}; \bar{x}_{i,k_1}^A, \sigma_{i,k_1}^A\right) = \psi \Rightarrow \left\{\bar{x}_{\min_1}, \bar{x}_{\max_1}\right\} \\ \text{Gauss}\left(\bar{x}; \bar{x}_{i,k_2}^A, \sigma_{i,k_2}^A\right) = \psi \Rightarrow \left\{\bar{x}_{\min_2}, \bar{x}_{\max_2}\right\} \end{cases} \tag{15}$$

with respect to $\bar{x}$.

For each pair of input fuzzy sets we determine the value of similarity measure given by formula (12). If that value exceeds the threshold $s_{\min}$ (in our simulations in Section 5 we assume that $s_{\min} = 0.5$) then the input fuzzy sets are merged. More precisely, Gaussian fuzzy sets $A_i^{k_1}$ and $A_i^{k_2}$ are replaced by fuzzy set A_i^k, which is also Gaussian with the center and width given by

$$\bar{x}_{i,k}^A = \frac{\bar{x}_{i,k_1}^A \cdot w_{i,k_1}^\tau + \bar{x}_{i,k_2}^A \cdot w_{i,k_2}^\tau}{w_{i,k_1}^\tau + w_{i,k_2}^\tau} \tag{16}$$

and

$$\sigma_{i,k}^A = \frac{\sigma_{i,k_1}^A \cdot w_{i,k_1}^\tau + \sigma_{i,k_2}^A \cdot w_{i,k_2}^\tau}{w_{i,k_1}^\tau + w_{i,k_2}^\tau}. \tag{17}$$

In formulas (16) and (17) we take into account the importance of merged antecedents, described by values of weights w_{i,k_1}^τ and w_{i,k_2}^τ. The importance of the antecedent A_i^k being a result of merging antecedents $A_i^{k_1}$ and $A_i^{k_2}$ is described by

$$w_{i,k}^\tau = \left(w_{i,k_1}^\tau + w_{i,k_2}^\tau\right)/2. \tag{18}$$

After each merging a simplified a neuro-fuzzy system is trained by a single epoch and then tested. Testing enables to evaluate the influence of merging on accuracy of the simplified system. If the merging does not worsen the accuracy then the simplified system replaces the previous one. Otherwise, the merging is canceled and the initial system is restored. This procedure is performed for all combinations of antecedents corresponding to all input features.

In a similar way we merge the output fuzzy sets B^k, $k = 1, \ldots, N$. The only difference is that the centers and widths of merged fuzzy sets are given by

$$\bar{y}_k^B = \frac{\bar{y}_{k_1}^B + \bar{y}_{k_2}^B}{2} \tag{19}$$

and

$$\sigma_k^B = \frac{\sigma_{k_1}^B + \sigma_{k_2}^B}{2}. \tag{20}$$

The procedure described in points 3 and 4 leads to reduced (Fig. 2) and simplified (Fig. 1) fuzzy model which is less complex and more understandable than the initial system (10).

5 Simulations Results

The stroke data, obtained from Institute of Neurology, Russian Academy of Medical Sciences, contains 298 instances and each instance is described by 30 attributes (for details see [21]). In our experiments all sets are divided into a learning sequence (268 sets) and a testing sequence (30 sets). Out of 298 data samples, 211 cases represent ischemic stroke (IS), 73 cases represent hemorrhagic stroke (HS) and 14 cases describe subarachnoid hemorrhage (SAH). In our simulations we used a neuro-fuzzy system (10) with seven rules, thirty inputs, Dombi triangular norms, Gaussian membership functions and the momentum backpropagation learning method with 100000 epochs, and learning coefficients $alpha$=0.25 and mi=0.05. The experimental results are depicted in Table 1 and Table 2. Comparing our previous algorithm CEA with the algorithm ABGE we see that algorithm studied in this paper allows to obtain a better accuracy and a reasonable degree of a reduction.

Our interest in differential stroke diagnosis is justified by the following circumstances. The death rate caused by vascular diseases in Russia is one of the highest in the world. Among all death cases caused by cerebral-vascular pathology the stroke diagnosis with the pointing of its type is fixed only in 21% of the cases, non-differential diagnosis like "poor brain blood circulation" - in 39% of cases. It is registered about 400 000 stroke annually and the rate of diseases among the population of able-bodied age is permanently increasing. Death rate because of stroke reaches 40% and 62% of the persons who survived after a stroke become invalids. The prevention has crucial importance in death reduction and invalidity reduction caused by the stroke. The essential effect gives the optimization of the patient aid system based on heterogeneity concept. That's why it is of great importance to provide comprehensive support to medical staff in making the fast and exact diagnostic decision that defines further tactics of treatment. It is important for stroke type diagnosis as well as for pathogenetic subtype of ischemic stroke diagnosis. It is also very important to make quality differential clinical diagnosis of stroke type in short time. Usually rate of mistakes in the diagnosis is in the range of 20-45%. So the problem of decision support in making a competent diagnosis is of a great interest for medical doctors.

Table 1. Simulation results

No	Sequence	Neuro-fuzzy system (10) for stroke diagnosis		
		before reduction	after reduction CEA	after reduction ABGE
1	learning	94.40[%]	94.40[%]	95.90[%]
2	testing	93.33[%]	93.33[%]	96.67[%]

Table 2. Simulation results

No	Degree of	Neuro-fuzzy system (10) for stroke diagnosis	
		after reduction CEA	after reduction ABGE
1	parameters number reduction	46.80[%]	15.30[%]
2	rules number reduction	14.28[%]	0.00[%]
3	inputs number reduction	20.00[%]	13.33[%]
4	antecedents number reduction	49.04[%]	16.19[%]
5	discretization points number reduction	14.28[%]	14.28[%]
6	degree of learning time reduction	59.33[%]	25.10[%]

6 Conclusions

In the paper we described a new methods for designing and reduction of a neuro-fuzzy system for stroke diagnosis. From simulations it follows that the reduction process of neuro-fuzzy structure based on adjustable weighted triangular norms do not worsen the performance of this structure. It was possible to quickly detect the inputs which can be eliminated. Our methods allows to decrease the number of parameters in neuro-fuzzy structure for stroke diagnosis and consequently the learning time.

References

1. Aliev, R.A., Aliev, R.R.: Soft Computing and its Applications. World Scientific Publishing, Singapore (2001)
2. Alonso, J.M., Cordon, O., Guillaume, S., Magdalena, L.: Highly Interpretable Linguistic Knowledge Bases Optimization: Genetic Tuning versus Solis-Wetts. Looking for a good interpretability-accuracy trade-off. In: Proc. of the 2007 IEEE Int. Conf. on Fuzzy Systems, pp. 1–6 (2007)
3. Amaral, T.G., Crisostomo, M.M.: An Approach to Improve the Interpretability of Neuro-Fuzzy Systems. In: Proc. of the 2006 IEEE Int. Conf. on Fuzzy Systems, pp. 1843–1850 (2006)
4. Casillas, J., Cordon, O., Herrera, F., Magdalena, L. (eds.): Interpretability Issues in Fuzzy Modeling. Springer, Heidelberg (2003)

5. Cpałka, K.: A New Method for Design and Reduction of Neuro-Fuzzy Classification Systems. IEEE Transactions on Neural Networks 20(4), 701–714 (2009)
6. Cpałka, K.: On evolutionary designing and learning of flexible neuro-fuzzy structures for nonlinear classification. Nonlinear Analysis Series A: Theory, Methods & Applications, vol. 71. Elsevier, Amsterdam (2009)
7. Cpałka, K., Rebrova, O., Gałkowski, T., Rutkowski, L.: On differential stroke diagnosis by neuro-fuzzy structures. In: Rutkowski, L., Tadeusiewicz, R., Zadeh, L.A., Zurada, J.M. (eds.) ICAISC 2008. LNCS (LNAI), vol. 5097, pp. 974–980. Springer, Heidelberg (2008)
8. Cpalka, K., Rutkowski, L.: An application of weighted triangular norms to complexity reduction of neuro-fuzzy systems. In: Rutkowski, L., Tadeusiewicz, R., Zadeh, L.A., Zurada, J.M. (eds.) ICAISC 2008. LNCS (LNAI), vol. 5097, pp. 207–216. Springer, Heidelberg (2008)
9. Czabanski, R.: Neuro-Fuzzy Modelling Based on a Deterministic Annealing Approach. Int. J. Appl. Math. Comput. Sci. 15(4), 561–576 (2005)
10. Czogała, E., Łęski, J.: Fuzzy and Neuro-Fuzzy Intelligent Systems. Physica-Verlag, New York; A Springer Company, Heidelberg (2000)
11. Fodor, C.: On fuzzy implication. Fuzzy Sets and Systems 42, 293–300 (1991)
12. Gorzałczany, M.: Computational Intelligence Systems and Applications: Neuro-Fuzzy and Fuzzy Neural Synergisms. Springer, Heidelberg (2002)
13. Guillaume, S.: Designing fuzzy inference systems from data: An interpretability-oriented review. IEEE Trans. Fuzzy Syst. 9(3), 426–443 (2001)
14. Kasabov, N.: Foundations of Neural Networks, Fuzzy Systems and Knowledge Engineering. The MIT Press, CA (1996)
15. Kecman, V.: Learning and Soft Computing. MIT, Cambridge (2001)
16. Kumar, M., Stoll, R., Stoll, N.: A robust design criterion for interpretable fuzzy models with uncertain data. IEEE Trans. Fuzzy Syst. 14(2), 314–328 (2006)
17. Łęski, J., Henzel, J.: A Neuro-Fuzzy System Based on Logical Interpretation of If-then Rules. Int. J. Appl. Math. Comput. Sci. 10(4), 703–722 (2000)
18. Łęski, J.: A Fuzzy If-Then Rule-Based Nonlinear Classifier. Int. J. Appl. Math. Comput. Sci. 13(2), 215–223 (2003)
19. Manley-Cooke, P., Razaz, M.: An efficient approach for reduction of membership functions and rules in fuzzy systems. In: Proc. of the 2007 IEEE Int. Conf. on Fuzzy Systems, pp. 1–6 (2007)
20. Mitra, S., Hayashi, Y.: Neuro-fuzzy rule generation: survey in soft computing framework. IEEE Trans. Neural Networks 11(3), 748–768 (2000)
21. Rebrova, O., Kilikowski, V., Olimpieva, S., Ishanov, O.: Expert system and neural network for stroke diagnosis. International Journal of Information Technology and Intelligent Computing 1(2), 441–453 (2006)
22. Riid, A., Rustern, E.: Interpretability of Fuzzy Systems and Its Application to Process Control. In: Proc. of the 2007 IEEE Int. Conf. on Fuzzy Systems, pp. 1–6 (2007)
23. Rutkowski, L.: Computational Intelligence. Springer, Heidelberg (2007)
24. Rutkowski, L.: Flexible Neuro-Fuzzy Systems. Kluwer Academic Publishers, Dordrecht (2004)
25. Rutkowski, L.: New Soft Computing Techniques for System Modeling. In: Pattern Classification and Image Processing. Springer, Heidelberg (2004)
26. Rutkowski, L., Cpałka, K.: Flexible neuro-fuzzy systems. IEEE Trans. Neural Networks 14(3), 554–574 (2003)
27. Yager, R.R., Filev, D.P.: Essentials of fuzzy modelling and control. John Wiley & Sons, Chichester (1994)

LS Footwear Database - Evaluating Automated Footwear Pattern Analysis

Maria Pavlou and Nigel M. Allinson

University of Sheffield
m.pavlou@shefield.ac.uk,
n.allinson@sheffield.ac.uk

Abstract. Footwear marks recovered from crime scenes are an important source of forensic intelligence or evidence - for some crime types, there is a greater probably to recover footwear marks than fingerprint ones. Currently the process of identifying a specific shoe model from the 10,000s of possibilities is a time-consuming task for expert examiners. As with many other crime marks, for example latent fingerprints, there is an increasing need for automation. The emergent research effort in this field has been hampered by the lack of a suitable dataset of footwear impressions. We present, here, a substantial and fully characterized dataset together with a proposed methodology for its use.

1 Introduction

This paper introduces the important class of forensic imagery, namely footwear impressions and describes the first openly available database that will not only be an aid in the study of footwear pattern recognition but provide a new challenge to the computer vision (CV) community. Details of the database can be found at URL: `http://eeepro.sheffield.ac.uk/footwear`

Footwear marks provide a useful source of intelligence and evidence in the application of forensics for policing and security. Similar to latent fingerprints, footwear marks are very frequently left behind on surfaces at crime scenes [1] and can be recovered to provide useful evidential clues and sometimes strong courtroom evidence. Footwear mark examination has historically been an important tool of forensics, which is now increasingly being employed by authorities. A heightened demand for computer-assisted automation will require considerable research effort as well as the development of accepted standards and experimental methodologies for evaluation. We provide our database to deal with some of the shortcomings in this research area with the intention of encouraging wider participation and elevating the quality of contributions.

1.1 Footwear Mark Etiology

The outsole is the underside of footwear and commonly describes a tread pattern which is usually distinctive to the make and model of a shoe as friction skin ridge patterns of fingers are unique to an individual. The contact of the outsole with

C. Alippi et al. (Eds.): ICANN 2009, Part II, LNCS 5769, pp. 445–454, 2009.

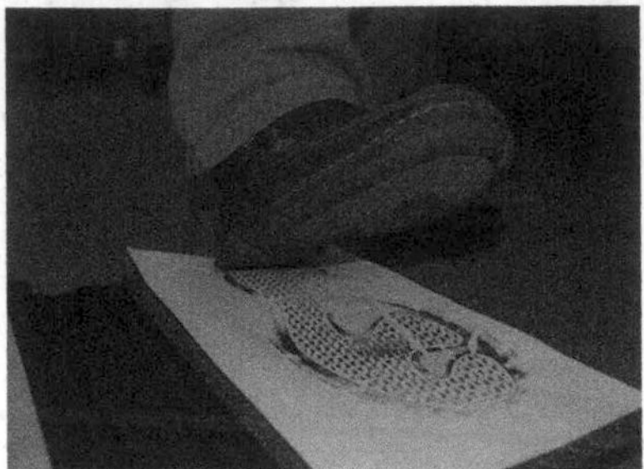

Fig. 1. Recording the outsole pattern using inkless chemical printing

various surfaces results in the formation of a (footwear) mark via the deposition or removal of material such as dust or blood. There are numerous methods by which these are detected, recorded and preserved and details can be found in [1, 2]. Briefly, these range from using specialized lighting, chemical developers and adhesive lifting techniques. Prints of the outsole can also be made directly by inking or dusting the outsole, e.g. with fine aluminum powder; and more commonly with inkless chemical printing (ICP) involving the transfer of chemical reactants onto special sensitized paper [3]. (Fig. 1). The last mentioned method is most commonly employed by Police (detention centers) to record impressions of a suspect's shoes. The resulting impression can then be used for one-to-one comparison, or can be scanned for computer-based processing.

1.2 Function and Application

Footwear marks found at crime scenes can contain sufficient characteristics to ascertain the manufacturer and model of the outsole, and can potentially be linked to the wearer. These characteristics are called ***class*** and ***individual*** characteristics. The former refers to the tread pattern and other manufacturing defects which are distinct to a particular model of footwear and the process of its production, while the latter refers to wear-and-tear artifacts making the impression unique to a specific outsole [1, 3].

In a forensic setting, comparisons of these characteristics can be made between items of recovered footwear, their reproduced marks and marks found as evidence. These comparisons can provide conclusive links and be used as court room evidence. Outside the forensic setting, ***class*** characteristics can be useful for proactive screening and intelligence gathering. The widespread collection footwear evidence from detainees and suspects at detention centers can be correlated with evidence in relation to other offenses in a local area, such as burglaries [3].

1.3 Challenges for Computer Vision and Machine Learning

The comparison of ***class*** and ***individual*** characteristics in crime scene marks (commonly called latent marks) has largely been performed manually by forensic professionals with an intimate knowledge of their etiology. Impressions are often of poor quality, confounded by details of the underlying surface and may only represent a partial impression of the entire outsole. These factors make their automatic comparison a very difficult, if not impossible, option and means that expert examiners will continue to play an active role for the foreseeable future as in the case for latent fingerprints. The automation of mark identification based on ***class*** characteristics however has been recently considered in the research literature [4–9]. The majority of these works, however, have considered essentially what are correlation-based methods and this has been due to the nature of footwear pattern data.

Typically face and other general object recognition tasks employ models constructed on *a priori* knowledge, such as their structure, which depend on these characteristics remaining fairly consistent to allow discrimination between other objects. Footwear patterns, however, cannot easily be characterized with some consistent structural form, such as possessing two eyes, a nose and mouth with a fairly consistent spatial arrangement. As a result footwear patterns have a significantly higher inter-class variation (Fig. 2) and learning a model over the general object class (i.e. outsoles) is therefore difficult. This problem is becoming more evident in the general object recognition literature [10], and is increasingly tackled using information theoretic models due to the increasing availability of image data. The CV problem is further exacerbated by large intra-class variations evident in real-world datasets of footwear marks. These variations manifest as large appearance changes due to the process of collection and capture and also importantly due to wear-and-tear of outsoles over their usable lifetime. These variations can be quite extreme sometimes changing the appearance of a mark completely or making it look similar to other marks. Therefore a solution relying on optimizing some discriminant (e.g. inter- over intra-class correlation) over the classes becomes less suitable especially in the absence of some stabilizing *a priori* structure (e.g. a physical model of how outsoles wears over use). In addition footwear pattern identification is a large scale, high multi-class task (in the order of 10^4 classes) with some classes having similar appearance and little discriminating characteristics. Combined with the operational needs for robustness, accuracy and efficiency in a forensic setting using the type of real-world data found in such a setting we see automated footwear pattern matching/retrieval as being an interesting challenge for the research community.

1.4 Proposal and Contribution

This application area is still quite immature compared to, say, face recognition, fingerprint matching and some other general object recognition tasks, which have gained mainstream acceptance and adoption. There is also an increasing demand for robust and accurate analysis and search technologies in the forensic arena, which should be developed around common protocols and widespread evaluation. Specifically for footwear pattern analysis there is a gap in accepted standards, datasets, methods and measures for evaluating systems and algorithms. These have been a key shortcoming in the past ten years of research on the topic

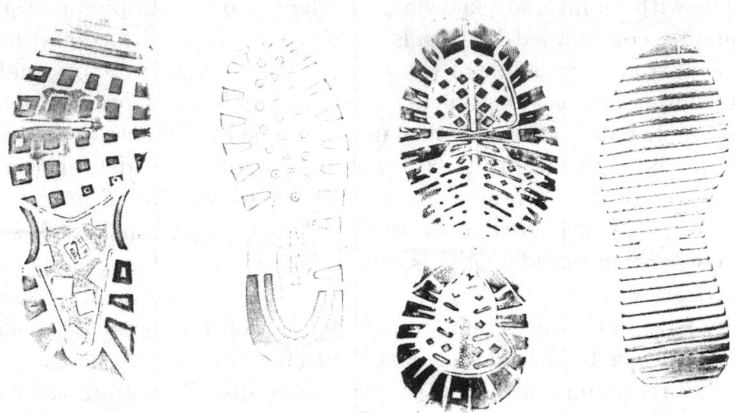

Fig. 2. Some examples form the LS footwear dataset showing (from left to right) variations in appearance due to wear, print fading, pattern complexity, scale and footedness

that can, in part, be attributed to the absence of publicly available datasets and a lack of consensus for evaluation. To encourage and support increased research in this field we provide an extensive dataset of footwear patterns and experimental protocols that draw from lessons learned in CV research on face and object recognition [11–14]. We believe this will encourage open research and lay ground for a fruitful effort towards acceptance and adoption of pattern analysis technologies for footwear and other forensic marks.

2 The Lancashire-Sheffield Dataset

We present a dataset derived from footwear mark archives held by the Lancashire Police, UK, and prepared for digital consumption at the University of Sheffield. Hence our database is named Lancashire-Sheffield (LS) Footwear dataset and is intended to address a main shortcoming of research in computer-based footwear mark analysis. The chief aspects of the database are that it is readily available and, in part, open. Importantly it is also representative of real-world footwear marks and this refers to the large range of variation in appearance evident in marks typically collected across police forces. This includes variations due to aspects described in Sect. 1.1 and 1.2 (e.g. wear-and-tear), degradation of intermediate preservation methods (e.g. print fading), and also due to the process of digital capture (e.g. capture fidelity, compression). Fig. 2 shows examples of the database images. These variational elements are important as they depict real-world noise with which to provide more robust assessment. Without this real-world noise, researchers are tempted to artificially degrade samples by adding artifacts such as Gaussian or salt-and-pepper noise which provide no meaningful assessment to the task. We view this one of the main contributions of our dataset.

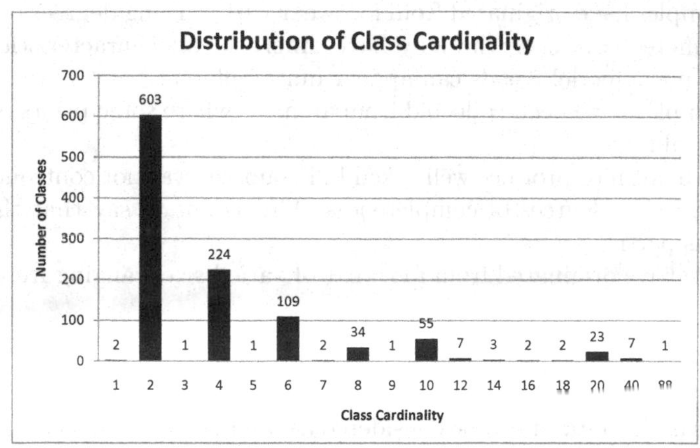

Fig. 3. Distribution of the make-model classes in the LS Dataset

Before proceeding with the details of the database, we present some summary statistics and properties.

- The database contains 4,633 images of whole impressions of footwear outsoles.
- The make and model of each impression in the image is given as ABC###, where ABC and ### correspond to the manufacturer name and model number respectively.
- The database contains images of 1,077 different make-model marks or classes. Of these the majority have a corresponding left and right imprint, and a proportion have more than one example of left and right pair from a different outsole of the same make and model. Figure 3 shows a distribution of the class numbers.
- The images are available as ~4,000 by ~2,000 pixel, 8bit grey, JPEG 2000 image files with 15:1 compression and 8 levels of compression.
- A proportion of this dataset is sequestered as suggested in the Face Recognition Grand Challenge [12], to limit bias when algorithm development is unfairly tuned to the test data. Appropriate suggestions for constructing target and query subsets from the public portion of this database are discussed in Sect. 2.2

2.1 Database Construction and Composition

The LS dataset consists of images of footwear patterns captured using the ICP method, which reliably captures the *state* of the tread pattern normally in contact with the ground. We use *state* to refer to the stage of wear of the outsole tread that will produce different patterns throughout a shoe's usable life. The ICP samples are then scanned at 600 dpi and saved as compressed JPEG 2000 images as specified above. It should be noted that the quality of prints in this dataset are highly varied for the following reasons:

- The samples have originated from footwear with varying degrees of wear and from different owners (who may have different gait characteristics). Hence similar make-model treads can appear quite different.
- The samples have been collected from archives where some prints have faded considerably.
- The care taken to produce well-inked ICP samples was not controlled, resulting in varying degrees of completeness. This is not to say that the dataset contains partial prints.
- Samples have originated from footwear of varied size, ranging from UK 3 to UK 11.
- Although all samples are upright they are not consistently aligned or centered within the ICP print and subsequently for any sample image.

All samples in the dataset can be considered as full prints, i.e. depicting the whole of the outsole tread pattern. Both left and right tread patterns are present in the dataset, although these may not be of the corresponding left or right shoe pair. As mentioned above some ICP may not be consistently developed but they are not considered as partial prints. For testing purposes partial prints can be artificially generated and this is discussed in Sect. 3.1. Samples have not been aligned or centered as they reflect real-world forensic data for which some means of normalization needs to be proposed. Providing any pre-proposed normalization would imply some agreed consensus on the structure of footwear patterns, which as suggested in Section 1.3 is not entirely possible and is left open to the research community. The high degree of variability in appearance of the dataset makes it a very difficult task for algorithm development, yet it exemplifies very closely the type of data which can be expected in the field.

We briefly summarize some basic statistics of the dataset from Fig 3 which represents the cardinality distribution of pattern classes:

- More than half of the database consists of pattern classes with only two examples or less.
- Less than 10% of patterns classes have 10 or more examples per class.

The prevalence of the pattern classes in the database loosely reflects their prevalence in real-life as collected by the Lancashire Police. However strong statements cannot be made due to the selection criteria used when scanning the database and a fixed upper limit on the maximum number of examples to retain (this was set at 40 images/class). Moreover, the types of footwear and hence pattern class will always be a function of the current fashion trends and the production life of any particular footwear model by manufacturers.

A further characteristic of the dataset is that it is populated with a number of pattern classes which have similar appearance yet have different class labels. These are not mislabeled samples or duplications but occur when manufacturers have used the same outsole pattern for a range of shoe models or where outsoles have been copied in the production of counterfeits. It is clear that these patterns will impose an upper limit on performance scores as we cannot expect automated methods to distinguish their classes. However, we have chosen not to remove these patterns as they can be useful in assessing how automated system perform with regard to pattern similarity ranking. That is, we can begin to ask, "How well does a system return similar looking patterns?"

2.2 Suggested Dataset Usage

The first task a forensic officer will undertake with any collected footwear pattern is to ascertain the make and model of the outsole. This is a pattern search or matching task where we would like to determine the similarity of patterns in a reference set to a query pattern. This is similar to the classification paradigm in face recognition in which there is a fixed gallery of test subjects for whom training images are available. The goal is to find matches of so-called probe images to members of the gallery.

As suggested by the statistics of the dataset and feedback from users there are two modes under which any system will perform. The first is where samples of a pattern class are abundant allowing the modeling of within class variations. Of concern here is whether the model learnt is robust enough to these variations while still being discriminative. The second operational mode arises where users are concerned with identifying uncommon patterns with few examples per class. In this case it is harder to model within-class pattern variations and hence the task becomes a filtering problem. Here we are concerned with how patterns can be compared despite there possibly being large within-class variations between the query and available reference patterns.

The LS dataset is divided into two subsets that are constructed to reflect the distribution of sample prevalence of the whole dataset. The first subset, "A", is for public release while subset "B" is reserved for competitive and controlled assessment of algorithms. We also propose two experimental views of these subsets, based on the two modes of operation described above, and suggest suitable guidelines to construct target/query (train/test) partitions.

Public Release Dataset - Subset A. This dataset is composed of 300 pattern classes having 2 samples each and 100 classes having 6 samples each. This dataset can be obtained by request from the URL given in Sect. 1 and will be delivered on suitable storage media along with guideline documentation and some predefined experimental partitions as described below. We encourage researchers to pay attention to these guidelines.

Sequestered Dataset - Subset B. This dataset is composed of 827 classes for which there is an overlap with subset A for 150 classes. The total number of samples amounts to 3433. We invite research teams to participate in a competitive evaluation of their work on this subset. Requests for participation can be made at the URL in Sect. 1 previously mentioned, and teams must then provide binaries of their algorithms.

Filter Experiment - View 1. This view is useful when assessing the performance of systems where few samples per class are available and for which learning a model over the data is difficult. An example of such an approach is described in [7] where footwear patterns are filtered using the Fourier Transform. For dataset A, sample target and query divisions are specified in *view1_target.txt* and *view1_query.txt*. Only one sample per class is available in the target list, which must be compared to those in the query list. These partitions do not describe a closed set over their class associations. In other words not all sample classes in the target list also appear in the query list. Other partitions of this data are possible and we ask the reader to take note of the guidelines in Sect.3.

Model Experiment - View 2. This view can be used to assess approaches using some form of model learning. For dataset A, a leave-one-out train/test partition is specified in *view2_train.txt* and *view2_test.txt* which can be applied in a classification paradigm. Again these partitions do not describe a closed set over their class associations.

3 Experimental Guidelines

Although a section of the LS dataset (subset B) is reserved for controlled and competitive assessment we anticipate that the public portion (subset A) will be used in reporting results within the research community. We hope that some basic guidelines will be adhered to as these will benefit the community in general and raise the quality of research in this field. As a first step, we recommend any performance results be reported using the suggested experimental view partitions descried above. Other partitions of subset A may be possible if it suits some specific aspects or scenarios under which a system is being tested. We only ask that some general points be adhered to, namely:

- Any target/query (train/test) partitions must be exclusive over their samples. That is to say no sample should appear in both the target and train partitions as this would be fitting to the test data and produces biased performance results [11, 15, 16]. This also applies to samples for which mirrored duplicates and "partials" have been generated (see Sect. 3.1 below).
- Images should not be "enhanced" or "degraded" with artificial noise. We feel that there is a large degree of variation already present in the data and no further modification is needed.
- Image sizes may be changed as needed however we ask that any modification is clearly reported.
- Other modifications such as scaling, translation, and rotation can be applied when testing systems on these aspects. However our first point should still be observed; that is no generated sample or its original should appear in both target and query partitions, and they should not be distributed across partitions. For example, given patterns X and Y where Y is generated from X, Y should not feature in a query partition with X as its target and visa versa.

3.1 Partial Prints and Mirrored Duplicates

We expect that some researchers would like to evaluate performance on so-called "partial" prints. These are not to be confused with partial latent prints obtained from crime scenes. In general the idea here is to ascertain if identification is still possible despite fairly significant occlusions of the footwear pattern. We recommend that researchers employ the methodology described in [7].

As mentioned previously, samples have originated from both left and right footed outsoles, and it will not always be the case that footwear will be matched in this respect between target and query. This scenario will also be typical in any real-world application, and so no attempt has been made to align target and query within the experimental views described above. It may be useful therefore, to generate mirrored versions of samples at training time or when comparing targets. This obviously will depend on the sensitivity to pattern structure of a proposed solution. For example, it is expected that correlation-based methods will be more sensitive than information theoretic approaches which might only consider content with perhaps a loose reliance on structure.

Again we stress that for both variations on the dataset described here we ask that the experimental guidelines be observed.

3.2 Performance Evaluation

There are many methods for reporting the final performance of a classification, verification or retrieval systems, including percentiles-by-rank, Receiver Operator Characteristic (ROC) and Precision-Recall (PR) curves. We ask that researchers, at a minimum, report their results using the ROC curve in the classification and verification paradigms and with PR curves in the context of information retrieval systems. Usually these quantities are calculated as a function of the threshold imposed on an similarity metric between target and query samples. Further reading on these topics can be found in [17, 18].

It will also be informative to present results using a percentiles-by-rank curve as this can give some indication of the ranking ability of proposed algorithms. This makes sense in the context of footwear pattern searching where users would like to see a list of potential matches returned by similarity strength.

4 Conclusion

We have introduced a new database, The Lancashire-Sheffield Footwear dataset to provide help in building some common consensus for evaluation and experimental procedures within the CV research community. Additionally we have indicated a means of undertaking widespread assessment by inviting competitive participation on substantial and real-world dataset. We believe the LS database is, at the time of writing, the only publicly available footwear database and one that provides a significant challenge to the CV research community. We hope this will provide stimulus and lay ground for other rigorous contributions.

Acknowledgements

We would like to thank the Lancashire Police Force who have made their archives available to the University of Sheffield - especially Kathryn Mashiter and Danyela

Kellett. Additionally we thank James Screaton at the University for putting endless hours into consistently digitizing this data.

References

1. Bodziak, W.J. (ed.): Footwear Impression Evidence. CRC Press, Boca Raton (2000)
2. Hilderbrand, D.S. (ed.): Footwear, The Missed Evidence. Staggs Publishing (1999)
3. Pavlou, M., Allinson, N.M.: Footwear Recognition. In: Encyclopedia of Biometrics, pp. 1–10. Springer, Heidelberg (2009)
4. Pavlou, M., Allinson, N.M.: Automated encoding of footwear patterns for fast indexing. Image and Vision Computing 27(4), 402–409 (2009)
5. Alexander, A., Bouridane, A., Crookes, D.: Automatic classi cation and recognition of shoeprints. In: Proc. Seventh International Conference on (Conf Image Processing and Its Applications Publ. No. 465), July 13–15, vol. 2, pp. 638–641 (1999)
6. Bouridane, A., Alexander, A., Nibouche, M., Crookes, D.: Application of fractals to the detection and classi cation of shoeprints. In: Proc. International Conference on Image Processing, September 10–13, vol. 1, pp. 474–477 (2000)
7. de Chazal, P., de Chazal, P., Flynn, J., Reilly, R.: Automated processing of shoeprint images based on the fourier transform for use in forensic science. Transactions on Pattern Analysis and Machine Intelligence 27(3), 341–350 (2005)
8. Su, H., Crookes, D., Bouridane, A., Gueham, M.: Local image features for shoeprint image retrieval. In: British Machine Vision Conference 2007 (2007)
9. Zhang, L., Allinson, N.: Automatic shoeprint retrieval system for use in forensic investigations. In: 5th Annual UK Workshop on Computational Intelligence (2005)
10. Everingham, M., Van Gool, L., Williams, C.K.I., Winn, J., Zisserman, A.: The PASCAL Visual Object Classes Challenge 2007 (VOC 2007) Results (2007), `http://www.pascal-network.org/challenges/VOC/voc2007/workshop/index.html`
11. Jain, A., Duin, R., Jianchang, M.: Statistical pattern recognition: a review. IEEE Transactions on Pattern Analysis and Machine Intelligence 22(1), 4–37 (2000)
12. Phillips, P., Flynn, P., Scruggs, T., Bowyer, K., Chang, J., Hoffman, K., Marques, J., Min, J., Worek, W.: Overview of the face recognition grand challenge. In: IEEE Computer Society Conference on Computer Vision and Pattern Recognition, CVPR 2005, June 2005, vol. 1, pp. 947–954 (2005)
13. Li, S.Z., Jain, A.K. (eds.): Handbook of Face Recognition. Springer, Heidelberg (2005)
14. Datta, R., Joshi, D., Li, J., Wang, J.Z.: Image retrieval: Ideas, influences, and trends of the new age. ACM Compututing Surveys 40(2), 1–60 (2008)
15. Kulkarni, S., Lugosi, G., Venkatesh, S.: Learning pattern classi cation-a survey. IEEE Transactions on Information Theory 44(6), 2178–2206 (1998)
16. Poggio, T., Rifkin, R., Mukherjee, S., Niyogi, P.: General conditions for predictivity in learning theory. Nature 428(6981), 419–422 (2004)
17. Fawcett, T.: An introduction to roc analysis. Pattern Recognition Letters 27(8), 861–874 (2006); ROC Analysis in Pattern Recognition
18. Makhoul, J., Kubala, F., Schwartz, R., Weischedel, R.: Performance measures for information extraction. In: Proceedings of DARPA Broadcast News Workshop, pp. 249–252 (1999)

Advanced Integration of Neural Networks for Characterizing Voids in Welded Strips

Matteo Cacciola, Salvatore Calcagno, Filippo Laganá, Giuseppe Megali, Diego Pellicanó, Mario Versaci, and Francesco Carlo Morabito

University Mediterranea of Reggio Calabria,
Via Graziella Feo di Vito, 89100 Reggio Calabria, Italy
{matteo.cacciola,calcagno,filippo.lagana,giuseppe.megali,
diego.pellicano,mario.versaci,morabito}@unirc.it

Abstract. Within the framework of aging materials inspection, one of the most important aspects regards defects detection in metal welded strips. In this context, it is important to plan a method able to distinguish the presence or absence of defects within welds as well as a robust procedure able to characterize the defect itself. In this paper an innovative solution that exploits a rotating magnetic field is presented. This approach has been carried out by a Finite Element Model. Within this framework, it is necessary to consider techniques able to offer advantages in terms of sensibility of analysis, strong reliability, speed of carrying out, low costs: its implementation can be a useful support for inspectors. To this aim, it is necessary to solve inverse problems which are mostly ill-posed: in this case, the main problems consist on both the accurate formulation of the direct problem and the correct regularization of the inverse electromagnetic problem. In the last decades, a useful and very performing way to regularize ill-posed inverse electromagnetic problems is based on the use of a Neural Network approach, the so called "learning by sample techniques".

Keywords: Neural Networks, Void Characterization, Welded strips, Rotating Magnetic Field.

1 Introduction

In many industrial and civil applications, materials and structures are subjected to various manufacturing and service conditions which make it imperative to enhance the predictive capabilities of modeling various types of defects. For instance micro-crack, micro-voids which precede possible fracture growth. A typical framework where the problem can be encountered is the welding process, i.e. the application of a joint on two or more pieces. In the welding strip, matter of discontinuity appears at micro-scale as either spherical or elliptical air bubbles. These discontinuities cause a stress concentration, modifying the constitutive

C. Alippi et al. (Eds.): ICANN 2009, Part II, LNCS 5769, pp. 455–464, 2009.

response of the material, or, in other words, building a damage within the material. The latter phenomenon represents the initial step to crack extension and, consequently, the voids' detection and control should be investigated by means of reliable devices and procedures. The quality of a welded joint depends on the product allocation. In fact, some types of welding are suitable for a particular case, the same type of welding will not be eligible in another situation. The quality is devised according to the intended use of the joint, but it takes into account all factors that may affect the welding. Scientific literature suggests a lot of different solutions for the problem of material inspection. Nowadays, the mostly used techniques are based on Non Destructive Testing and Evaluation (NDT/E), having a very important role in inspecting aging materials for industrial applications or within the framework of civil engineering. Within this framework, it is very important to plan a suitable method able to distinguish the presence or absence of defects within welds as well as a robust procedure able to characterize the defect itself: its implementation can be a useful support for inspectors. But, in order to characterize the defects within the inspected materials, it is necessary to solve an inverse problem which are mostly ill-posed. In this case, the main problems consist on both the accurate formulation of the direct problem and the correct regularization of the inverse electromagnetic problem. In the last decades, a useful and very performing way to regularize ill-posed inverse electromagnetic problems is based on the use of the so called "learning by sample techniques". They allow to heuristically solve the inverse problem, starting from the experience, and so implementing such intelligent and non-crisp algorithms as Neural Networks, Fuzzy Inference Systems and so on. In this paper, we propose a new way of determining the diameter of air bubbles within welding, exploiting heuristic approaches and starting from a well known way of inspecting welding strips [1,2]. The latter exploits a rotating magnetic field, i.e. a magnetic field generated by a three-phase system, able to rotate in the time-space domain [3]. The magnetic field induces eddy currents within the inspected specimens, which are influenced by the presence of a possible defect, e.g. voids, and in turn influences the external magnetic field. Its component normal to the upper surface of the modeled plate is measured on the specimen's surface. In the past, a lot of experimentations [4] and numerical modeling [4,5,6] have been carried out in order to understand the behavior of eddy currents if a crack occurs into the inspected materials [4,6]. In our work, we did not want to focus our attention to cracks in homogeneous materials, but we used a Finite Element Method (FEM) in order to characterize eddy currents into welded objects. We studied the case in which an air bubble is present into the welding, thus weakening the strength of the finally obtained object. A number of simulations have been carried out, involving a number of welding strips with different elliptical voids, having varying shapes, locations and orientations. All the collected data have been subsequently used in order to train and test a suitable Neural Network in order to characterize the voids. The performances are satisfying: the approach is able to recognize position and dimensions of very small cracks.

1.1 An Overview of Defectiveness in Welding Strips

Welding process could induce the following relevant defects, compromising the structural integrity of the same strips and consequently, of specific components and structures:

- **lack of fusion:** if the fusion of the basic metal is excessive, continue grooves are formed on the sides, causing a depression along the sides of the cord; sometimes the incisions can be eliminated by using a thin covering material;
- **excess of penetration on the top (dripping):** if a huge quantity of metal is caught at the top of the weld: a toe crack could be determined;
- **incomplete penetration:** it is caused by a lack of fusion at the welding apex, and seriously reduces the resistance of the joint; in heading welding made with one or more rubs, the defect can be eliminated by chiseling out and giving an additional rub;
- **gluing:** it occurs when, during the welding process, the complete fusion of metal does not take place, i.e., when the welding metal overlaps the not-yet-fused material to weld, without a mixing between the metals;
- **cracks:** the most serious kind of flaws, because they originate from phenomena of metallurgical nature. Since they depend on the cracking temperature, they are named as hot or cold cracks.

Fig. 1 shows the most typical defects in metal welded strips. Sizes can vary, greatly depending on the welding process and conditions. For instance, according to European laws UNI EN 287-2 and UNI EN 288-4, the maximum tolerable crack dimension is fixed to 0.5 (mm) of diameter for a circular defect. In many cases, it is very difficult to distinguish between the kind of defects starting from typical NDTs' measurements. In fact, at the state-of-the-art, non destructive identification systems allow to allocate a defect but without being able to determine its shape. In addition, different kinds of defects can cause similar signals. Therefore, a soft computing-based approach can be very useful for an automatic and, for instance, for real-time classification. In the following section, Rotating Magnetic Field based on Finite Element Analysis (FEA) will be described. Subsequently, techniques used for defect classification will theoretically be presented.

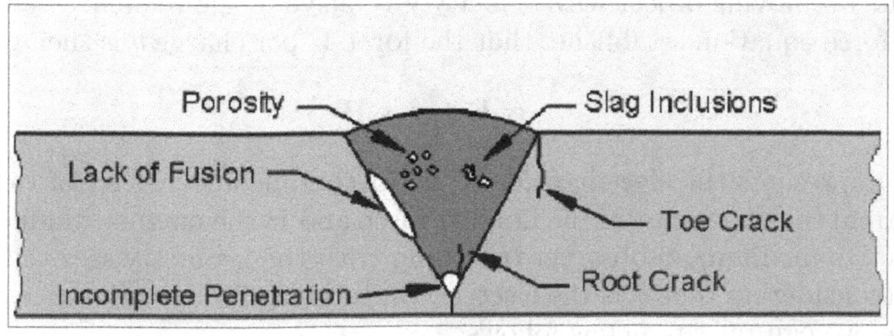

Fig. 1. Categories of discontinuity in metal welded strips

2 On the A-ψ Formulation for Building the Dataset

For our purposes, we need to implement a suitable database, useful for subsequent regularization of the ill-posed inverse problem. It can be stated as follows: evaluating the extension of voids starting from suitable data. The goal is the detection of defects in the welding and the realization of a data set in order to train and test a suitable Neural Network. Usually, data are represented by in-lab measurements or numerical simulations [7]. In our case, we exploited the latter path, modeling the direct problem through a commercial FEM software. We aim to analyze the effectiveness of a welding performed on two slabs of structural steel. The geometry of the examined case presents a "V" type welding, where a bubble shaped defect has been modeled. Geometrical dimensions of our model are listed in Table 1. For our purpose we verified the distortion of the magnetic

Table 1. General settings of numerical models

Property	Setting
Material used for the specimen	Stainless steel, not-magnetic, isotropic
Dimension of specimen	0.2 (m) x 0.01 (m)
Welding thickness	0.02 (m)
Electric conductivity	4.032*106 (S/m)
Minimum diameter of the defect	0.5 (mm)
Maximum diameter of the defect	1 (mm)

field density (T) [8,9] to detect the defect presence of a porosity bubble, formed during welding process. The simulations have been carried out by changing the size of the flaw, and exploiting the phenomenon of magnetic rotating field, since its main advantage is the insensitivity to crack's orientation. In the proposed approach, we exploit the A-ψ formulation [10]. In the case of magnetostatic and quasi-static fields, Ampère-Maxwell's equation can be written as:

$$\nabla \times \mathbf{H} = \mathbf{J} \tag{1}$$

where $\mathbf{H}$ represents the magnetic field and $\mathbf{J}$ the current density, respectively. If we consider a moving object with velocity $\mathbf{v}$ relative to the reference system, the Lorentz force equation establishes that the force $\mathbf{F}$ per charge q is then given by:

$$\frac{\mathbf{F}}{q} = \mathbf{E} + \mathbf{v} \times \mathbf{B} \tag{2}$$

where $\mathbf{E}$ represents the electric field; $\mathbf{v}$ the instantaneous velocity of the object derived from the expression of the Lorentz force and $\mathbf{B}$ the magnetic induction. In a conductive medium, an observer travelling with the geometry sees the current density (considering that σ is the electric conductivity) $\mathbf{J} = \sigma(\mathbf{E} + \mathbf{v} \times \mathbf{B}) + \mathbf{J}^{\mathbf{e}}$; therefore, we can rewrite (1) as follows:

$$\nabla \times \mathbf{H} = \sigma(\mathbf{E} + \mathbf{v} \times \mathbf{B}) + \mathbf{J}^{\mathbf{e}} \tag{3}$$

where $\mathbf{J^e}$ [A/m^2] is an externally generated current density. Considering, for a transient analysis, the definitions of magnetic vector potential $\mathbf{A}$ and electric scalar potential V:

$$\mathbf{B} = \nabla \times \mathbf{A} \tag{4}$$
$$\mathbf{E} = -\nabla V - \frac{\partial \mathbf{A}}{\partial t}$$

and the constitutive relationships:

$$\mathbf{B} = \mu_0 \mu_r \mathbf{H} \Leftrightarrow \mathbf{H} = \mu_0^{-1} \mu_r^{-1} \mathbf{B} \tag{5}$$

where μ_0 and μ_r are free space and relative magnetic permeability; we may rewrite (3), by substituting (4) and (5) in it, as:

$$\sigma \frac{\partial \mathbf{A}}{\partial t} + \nabla \times \left(\mu_0^{-1} \mu_r^{-1} \nabla \times \mathbf{A}\right) - \sigma \mathbf{v} \times (\nabla \times \mathbf{A}) + \sigma \nabla V = \mathbf{J^e} \tag{6}$$

Since we are interested in perpendicular induction current, only the z-component of $\mathbf{A}$ is non null. Therefore, the formulation of the 3D Equation (5) is simplified to:

$$\sigma \frac{\partial \mathbf{A}_z}{\partial t} + \nabla \times \left(\mu_0^{-1} \mu_r^{-1} \nabla \times \mathbf{A}_z\right) - \sigma \mathbf{v} \times (\nabla \times \mathbf{A}_z) = \sigma \frac{\Delta V}{L} + \mathbf{J^e}_z \tag{7}$$

where ΔV is the difference of electric potential and L is the thickness along the z-axis. The Partial Difference Equation (PDE) formulation of Equation (7) can be written as:

$$\sigma \frac{\partial A_z}{\partial t} + \nabla \cdot \left(\mu_0^{-1} \mu_r^{-1} \nabla A_z\right) - \sigma \mathbf{v} \cdot \nabla A_z = \sigma \frac{\Delta V}{L} + J_z^e \tag{8}$$

In this way we calculated the magnetic vector potential $\mathbf{A}$ in a generic subdomain Ω. For our aim, it is necessary to impose the boundary conditions as follows. Magnetic field ($\mathbf{n} \times \mathbf{H} = \mathbf{n} \times \mathbf{H_0}$) for boundary of air where acting the rotating magnetic field; for remaining boundaries, included the defect, the continuity is assured by the expression $\mathbf{n} \times (\mathbf{H_1} - \mathbf{H_2}) = 0$ [11,12]. The rotation effect of the magnetic field vector has been simulated by applying a uniform $\mathbf{B}$ vector, timely rotated according to the following Euler rotation formulation [13]:

$$\begin{bmatrix} \mathbf{x}(t+\tau) \\ \mathbf{y}(t+\tau) \end{bmatrix} = \begin{bmatrix} \cos(\omega t) & \sin(\omega t) \\ -\sin(\omega t) & \cos(\omega t) \end{bmatrix} \begin{bmatrix} \mathbf{x}(t) \\ \mathbf{y}(t) \end{bmatrix} \tag{9}$$

Table 2 resumes the values of set electrical parameters. The collected database is composed by 1000 numerically simulated signals, characterized by defect's presence with different frequency values and 200 numerically simulated signals showing absence of defect. We added gaussian noise with different magnitudes to the retained simulation results.

Table 2. Electrical parameters

Parameter	Dimension
Frequency	from 10 to 50 [Hz]
Pulse	$2\pi f$ [rad/sec]
Magnetic Field	$3 * 10^{-3}$ [T] in magnitude

3 A Wavelet Artificial Neural Network for the Solution of Inverse Problem

The problem of estimate the diameter of the bubble within the welding starting from experimental measurements can be solved as a typical inverse problem of pattern regression starting from computed measurements. The proposed approach, useful to detect the flaw presence, exploits Artificial Neural Network (ANN) [14,15,16,17,18] as well as an hybrid Wavelet Transform (WT)-Principal Component Analysis (PCA) based feature extraction approach. In this way, peculiar features, characterizing the defect, can be selected without increasing the computational complexity of the ANN-based classifier, in order to the requirement of flaw detection. A biorthogonal 3.1 WT has been used, with four-level multiresolution analysis (MRA). In this case, the signal is split into a_i $(i = 0, ..., 3)$ set of so called approximation coefficients, and into d_i $(i = 0, ..., 3)$ so called details coefficients. According to the Wavelet theory [19] the approximation coefficients give information about the lower frequencies, whilst the details coefficients contain information about the signal at higher frequencies. In our experimentations, on each signal, only the 47 a_3 coefficients have been subsequently selected, because they completely describe the macro-trend of each signal. This particular WT has been selected after various tests. Moreover, with an expert system, just the trends of these signals could be useful to discriminate the defect presence. That is why we used only the Wavelet Approximation Coefficients (WACs). We considered a fourth MRA level after a careful inspection of the wavelet analysis at different MRA levels in order to retain as much spatial information as possible. But, due to a still large dimension of feature space, a PCA [20] has been exploited in reducing the number of inputs by only considering the Principal Components (PCs) whose contributions to total statistical variance of the whole set of PCs are bigger than 4%. In this way the input number has been reduced from 47 to 6 elements, avoiding the so called problem of "curse of dimensionality" [21]. It describes the problem caused by the exponential increase in volume associated with adding extra dimensions to a mathematical space. Please, pay attention that a linear mapping has been applied from wavelet domain to PCA domain. Mathematically, a linear mapping is a linear function between a set (domain) of independent variables and a set (co-domain) of dependent variables. If a linear mapping is applied, all the mathematical properties and characteristics of data into the domain set are retained into the co-domain set. Thus the six inputs are not the six main important WACs, but the six PCs, participating to the total variance more than 4%, of the mapped WACs. Moreover, let us remark on how the WACs provide a redundant spatial information, whereas the six PCs cover 96% of the whole wavelet spatial information, unless those PCs have very small residuals. Thus, loosely speaking, the threshold of 4% has been set by considering a trade-off between the PCA's residuals and the number of components itself, in order to balance the retained information and the problem of "curse of dimensionality". Subsequently, collected data sets have been normalized and split in a training and test subset. In order to set the amount of training signals, we made a trade-off between the requirements of an as large as possible training subset and a significant availability of testing signals. Thus, in our experimentations, training set has been composed by 80% of collected signals. Remaining signals compose the test subset.

Proposed computational intelligence applications utilizes a ANN as heuristic pattern classifier. Its inputs are the six PCs describes above. The output of the system is represented by the defect geometrical dimension (diameter). The WT-PCA-ANN, having 13-neurons hidden layer according to the Kurková's theorem [22], uses a Back-Propagation (BP) algorithm with adaptive rate of learning during a period of 1000 epochs. The stopping criteria is based on the minimization of a Mean Squared Error (MSE) and Relative Mean Real Error (RMAE) learning performance indexes, defined in equations (10) and (11) respectively, as a consequence of application of the BP algorithm.

$$MSE = \frac{1}{nw} \sum_{w=1}^{nw} \left(d(w) - d_{w_j}(w) \right)^2 \tag{10}$$

$$RMAE = \frac{1}{nw} \sum_{w=1}^{nw} \frac{\left| d(w) - d_{w_j}(w) \right|}{\left| d(w) \right|} \tag{11}$$

where nw represents the random sample size; $d(w)$ represents the density of distribution and $d_{w_j}(w)$ represents the density of distribution of the *j-th* sample. Best WT-PCA-ANN results are obtained with log-sigmoid activation function between input and hidden layers, always considering a linear function between the latter and the output layers (see Table 3 for details). Verification of the trained network is shown in Fig. 2. Table 4 shows a comparison between actual and estimated diameters, about the worst performances for each one of the considered dimensions. The performances of WT-PCA-ANN are resumed by the square of the so called Pearson's coefficient of regression [23], i.e. the regression

Table 3. Comparison of WT-PCA-ANN training performances according to differently exploited activation functions

Activation Function	MSE [mm^2]	RMAE
Linear	0.0038	6.24
Tan-Sigmoid	0.003	4.20
Log-Sigmoid	0.0020	3.27

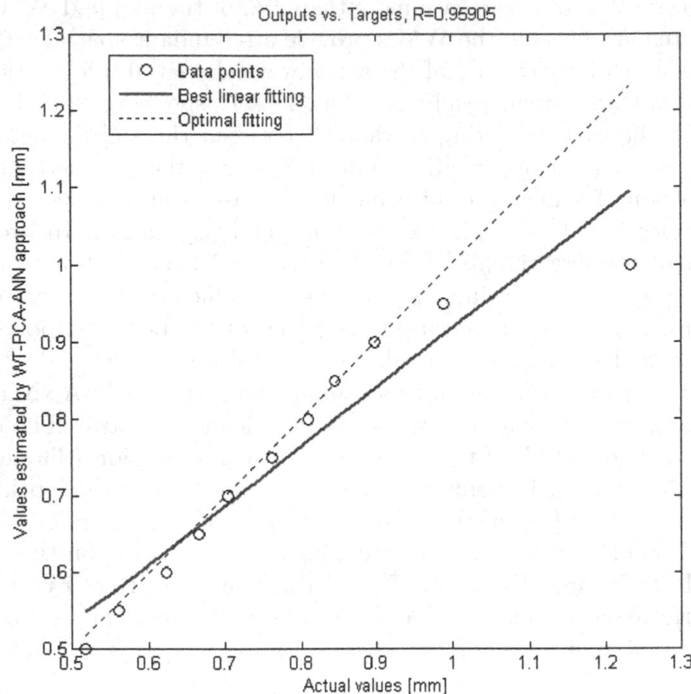

Fig. 2. Diameter evaluation: WT-PCA-ANN estimated values vs. actual values

R-value. Please, note that $R = 1$ means perfect correlation. In our case, we obtained $R = 0.959$ (see Fig. 2).

Table 4. Comparison between real diameter and reconstruction diameter

Actual diameter [mm]	WT-PCA-ANN estimated diameter [mm]
0.5	0.517
0.55	0.559
0.6	0.621
0.65	0.664
0.7	0.703
0.75	0.761
0.8	0.81
0.85	0.846
0.9	0.899
0.95	0.99
1	1.233

4 Conclusions

On the basis of the numerical method presented in this paper, the authors have developed a finite-element code for the analysis of the rotating magnetic field for metal welding strips. For our analysis we used a classical "V"-profile welding and simulated different sizes of defects according to the UNI EN law. Specifically, exploiting Rotating Magnetic Field using a self implemented FEA code, a bi-dimensional time dependent model has been studied to evaluate the distortion of the magnetic field and the magnetic field density due to the defect presence. The variation of the magnetic field **H**, induced by the variation of eddy currents, particularly, the normal component of **H**, i.e. $\mathbf{H}\perp$, is measured by suitable sensors in order to detect the presence of cracks, since it is not influenced by the exciting coils. But, if the crack's orientation is orthogonal to the longitudinal direction of the sensor, it could be insensitive to the crack's presence itself. The magnetic rotating field represents an insensitive solution to the crack's orientation, which induces variation of the eddy current density without a mechanical movement, with a remarkable economic saving. With these information, a WT-PCA-ANN approach has been exploited in order to estimate the diameter of the defect starting from signals obtained by computer simulations. Numerically obtained rotating magnetic field signals have been used, and WACs have been considered as features useful to train the ANN based regressor. The proposed method provides a good overall accuracy in reconstructing the defect's diameter, as our experimentations demonstrate. This aspect represents an useful support to the inspector, specially regarding the detection of defects with small size, improving their resolution. At the same time, the procedure should be validate for defect with different shape. The presented results can be considered as preliminary results; anyway, they are very encouraging, and suggest the possibility of increasing and generalizing the performance of the WT-PCA-ANN based classifier just refining its training step, for instance including, within the training set, rotating magnetic field signals able to describe flaws with different spatial extension and for different positions. The authors are actually engaged in this direction.

References

1. Grimberg, R., Radu, E., Mihalache, O., Savin, A.: Calculation of the induced electromagnetic field created by an arbitrary current distribution located outside a conductive cylinder. J. Phys. D: Appl. Phys. 30, 2285–2291 (1997)
2. Savin, A., Grimberg, R., Mihalache, O.: Analytical solutions describing the operation of a rotating magnetic field transducer. IEEE Transactions on Magnetics 33, 697–701 (1997)
3. Wikipedia, T.F.E.: Rotating magnetic field (2008), `http://en.wikipedia.org/wiki/Rotating_magnetic_field`
4. Takagi, T., Hashimoto, M., Fukutomi, H., Kurokawa, M., Miya, K., Tsuboi, H.: Benchmark models of eddy current testing for steam generator tube: experiment and numerical analysis. Int. J. Appl. Elect. Mater. 4, 149–162 (1994)

5. Bowler, J.: Eddy current interaction with an ideal crack: I. the forward problem. J. Appl. Phys. 75, 8128–8137 (1994)
6. Takagi, T., Huang, H., Fukutomi, H., Tani, J.: Numerical evaluation of correlation between crack size and eddy current testing signals by a very fast simulator. IEEE Transactions on Magnetics 34, 2582–2584 (1998)
7. Kohavi, R.: A study of cross-validation and bootstrap for accuracy estimation and model selection. In: Proceedings of the 14th International Joint Conference on Artificial Intelligence, pp. 1137–1143 (1995)
8. Lewis, A.M.: A theoretical model of the response of an eddy-current probe to a surface-breaking metal fatigue crack in a flat test piece. J. Appl. Phys. 25, 319–326 (1992)
9. Trevisan, F., Kettunen, L.: Geometric interpretation of discrete approaches to solving magnetostatics. IEEE Transaction on Magnetics 40, 361–365 (2004)
10. Rodger, D., Leonard, P., Lai, H.: Interfacing the general 3d a-ψ method with a thin sheet conductor model. IEEE Transactions on Magnetics 28, 1115–1117 (1992)
11. Tsuboi, H., Misaki, T.: Three dimensional analysis of eddy current distribution by the boundary element method using vector variables. IEEE Transactions on Magnetics 23, 3044–3046 (1987)
12. Weissenburger, D., Christensen, U.: A network mesh method to calculate eddy currents on conducting surfaces. IEEE Transactions on Magnetics 18, 422–425 (1982)
13. Weisstein, E.: Euler angles. mathworld, a wolfram web resource (2008), `http://mathworld.wolfram.com/EulerAngles.html`
14. Bishop, C.: Neural Networks for Pattern Recognition. Oxford University Press, Oxford (1996)
15. Gurney, K.: An Introduction to Neural Networks. Routledge, London(1997)
16. Haykin, S.: Neural Networks: A Comprehensive Foundation, 2nd edn. Prentice-Hall, London (1999)
17. Lawrence, J.: Neural Networks for Pattern Recognition. Oxford University Press, Oxford (1996)
18. Wasserman, P.: Introduction to Neural Networks: Design, Theory and Applications, 6th edn. California Scientific Software Press, San Francisco (1994)
19. Daubechies, I.: Ten Lectures on Wavelets. Society for Industrial and Applied Mathematics, Philadelphia (1992)
20. Jolliffe, I.: Principal component analysis. Springer Series in Statistics. Springer, Heidelberg (2002)
21. Powell, W.: Approximate Dynamic Programming: Solving the Curses of Dimensionality. Wiley Press, New York (2007)
22. Kurková, V.: Kolmogorov's theorem and multilayer neural networks. Neural Networks 5, 501–506 (1992)
23. Draper, N., Smith, H.: Applied Regression Analysis. Wiley-Interscience, New York (1998)

Connectionist Models for Formal Knowledge Adaptation

Ilianna Kollia, Nikolaos Simou,
Giorgos Stamou, and Andreas Stafylopatis

Department of Electrical and Computer Engineering,
National Technical University of Athens,
Zographou 15780, Greece
nsimou@image.ntua.gr

Abstract. Both symbolic knowledge representation systems and artificial neural networks play a significant role in Artificial Intelligence. A recent trend in the field aims at interweaving these techniques, in order to improve robustness and performance of classification and clustering systems. In this paper, we present a novel architecture based on the connectionist adaptation of ontological knowledge. The proposed architecture was used effectively to improve image segment classification within a multimedia application scenario.

1 Introduction

Intelligent systems based on symbolic knowledge processing, on the one hand, and artificial neural networks, on the other, differ substantially. Nevertheless, they are both standard approaches to artificial intelligence and it is very desirable to combine the robustness of neural networks with the expressivity of symbolic knowledge representation. This is the reason why the importance of the efforts to bridge the gap between the connectionist and symbolic paradigms of artificial intelligence has been widely recognised. As the amount of hybrid data containing symbolic and statistical elements, as well as noise, increases, in diverse areas, such as bioinformatics, or text and web mining, including multimodal application scenarios, neural-symbolic learning and reasoning becomes of particular practical importance. Notwithstanding the progress in this area, this is not an easy task. The merging of theory (background knowledge) and data learning (learning from examples) in neural networks has been indicated to provide learning systems that are more effective than purely symbolic and purely connectionist systems, especially when data are noisy. This has contributed decisively to the growing interest in developing neural-symbolic systems [9,5,6,4].

While significant theoretical progress has recently been made on knowledge representation and reasoning using neural networks, and on direct processing of symbolic and structured data using neural methods, the integration of neural computation and expressive logics is still in its early stages of methodological development [6].

C. Alippi et al. (Eds.): ICANN 2009, Part II, LNCS 5769, pp. 465–474, 2009.

Adaptation of symbolic ontological knowledge from raw data is an ideal use-case for further development and exploitation of neural-symbolic systems. Since the pioneering work of McCulloch and Pitts, a number of systems have been developed in the 80s and 90s, including Towell and Shavlik's KBAN, Shastri's SHRUTI, the work by Pinkas [9], Holldobler [6] and Artur S. d'Avila Garcez et al[5][4]. These systems, however, have been developed for the study of general principles, and are in general not suitable for real data or application scenarios that go beyond propositional logic. Only very recently, the theory has advanced far permitting the implementation of systems which can deal with logics beyond the propositional case [6].

This integration can be realised by an incremental workflow for knowledge adaptation. Symbolic knowledge bases can be embedded into a connectionist representation, where the knowledge can be adapted and enhanced from raw data. This knowledge may in turn be extracted into symbolic form, where it can be further used. This workflow is generally known as the neural-symbolic learning cycle, as depicted in the following diagram.

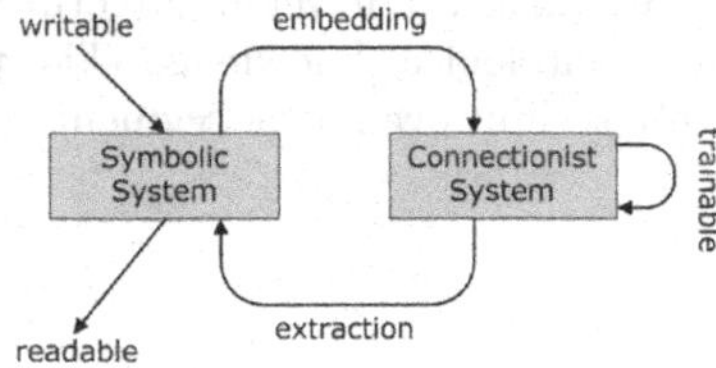

Fig. 1. The neural-symbolic learning cycle

In this paper we focus on developing connectionist adaptation of ontological knowledge, in particular of knowledge represented using expressive description logics. We then show that neural-symbolic methods can be used effectively to enhance knowledge adaptation within a multimedia application scenario. The rest of the paper is organized as follows. Section 2 presents the proposed architecture that mainly consists of the formal knowledge and the knowledge adaptation components, which are described in sections 3 and 4 respectively. Section 5 presents a multimedia analysis experimental study illustrating the theoretical developments. Conclusions and planned future activities are presented in section 6.

2 The Proposed Architecture

Capitalizing these experiences our system is designated as a learning, evolving and adapting cognitive model. Starting with basic knowledge about the nature of the problem and by using powerful reasoning mechanisms the proposed system gradually attempts to evolve its knowledge. In that way it incorporates its observations along with its own or the user's evaluation.

Figure 2 summarizes the proposed system architecture, consisting of two main components: the *Formal Knowledge* and the *Knowledge Adaptation*. The Formal Knowledge stores, the terminology and assertions, constraints that describe the problem under analysis in the appropriate knowledge representation formalism. More specifically, the *Ontologies module* formally represents the general knowledge about the problem.

It is actually a formal ontological description representing the concepts and relationships of the field, providing formal definitions and axioms that hold in every similar environment. This forms the system's knowledge which generated during the Development Phase by knowledge engineers and experts.

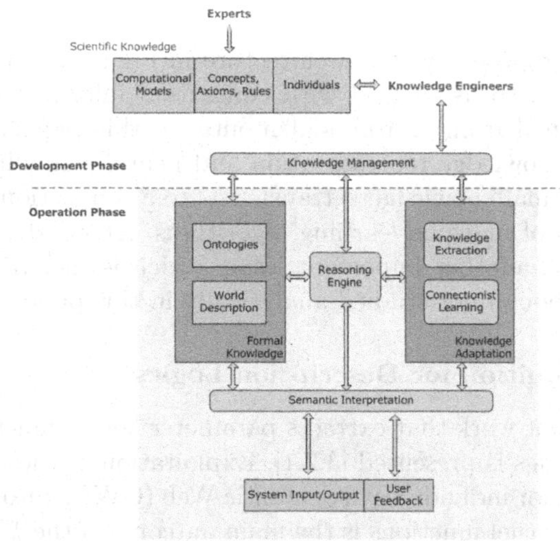

Fig. 2. The semantic adaptation architecture

Moreover, the *Formal Knowledge* contains the *World Description* that is actually a representation of all objects and individuals of the world, as well as their properties and relationships in terms of the Ontology.It is evident that most of the above data involve different types of uncertain information and, thus, they can be represented as formal (fuzzy) description logic assertions connecting the objects and individuals of the world with the concepts and relationships of the Ontology. This operation is performed by the *Semantic Interpretation* module.

In real environments however, this is a rather optimistic claim. Unfortunately, there may be lot of reasons that cause inconsistencies in the *Formal Knowledge*. For example, it is impossible to model all specific environments and thus, in some cases, conflicting assertions can arise. As a more abstract example (and more difficult to handle), the personality and expressivity of a specific user makes some of the axioms and constraints of the Ontology non-applicable or even wrong, according to logical entailments or user feedback. These inconsistencies make the formal use of knowledge that the *Reasoner* provides rather problematic. In such cases, the *Knowledge Adaptation* component of the system tries to resolve the inconsistency through a recursive learning process.

The knowledge adaptation improves the knowledge of the system by changing the world description and to some degree the axioms of the terminology of the system. The new information as represented in a connectionist model and, with the aid of learning algorithms, is adapted and then re-inserted in the knowledge base through the *Knowledge Extraction* and the *Semantic Interpretation* module for adaptation purposes.

3 The Formal Knowledge Component

3.1 Formal (Ontological) Knowledge and Connectionist Models

The focus of the proposed system architecture in Figure 2 is the adaptation of the knowledge base, so as to deal with contextual information and raw data peculiarities obtained from multimodal inputs. In this paper we adopt recent results in formal knowledge representation and neural-symbolic integration. In this framework, formal knowledge is transferred to a connectionist system and is adapted by means of machine learning algorithms. Knowledge extraction from trained networks is another important issue, which is included in the neural-symbolic loop, although not studied analytically in this paper.

3.2 Kernel Definition for Description Logics

In this section recent work that extracts parameter kernel functions for individuals within ontologies is presented [3,2,1]. Exploitation of these kernels permits inducing classifiers for individuals in Semantic Web (OWL) ontologies. In this paper, extraction of kernel functions is the main outcome of the *Formal Knowledge* component - assisted by the reasoning engine - for feeding the connectionist-based *Knowledge Adaptation* module.

The basis for developing these functions in the framework of the formal knowledge is the encoding of similarity between individuals, as they are presented to the knowledge base of the system, by exploiting semantic aspects of the reference representations.

The family of kernel functions $k_p^F : Ind(A) \times Ind(A) \rightarrow [0, 1]$, for a knowledge base $K = \langle T, A \rangle$ consisting of the TBox T (set of terminological axioms of concept descriptions-Ontology) and the ABox A (assertions on the world state-World Description); $Ind(A)$ indicates the set of individuals appearing in A), and $F = \{F_1, F_2, \ldots, F_m\}$ is a set of concept descriptions. These functions are defined as the L_p mean of the, say m, simple concept kernel functions κ_i , $i = 1, \ldots, m$, where, for every two individuals a,b, and $p > 0$,

$$\kappa_i(a,b) = \begin{cases} 1 & (F_i(a) \in A \wedge F_i(b) \in A) \vee (\neg F_i(a) \in A \wedge \neg F_i(b) \in A) \\ 0 & (F_i(a) \in A \wedge \neg F_i(b) \in A) \vee (\neg F_i(a) \in A \wedge F_i(b) \in A) \\ \frac{1}{2} & \text{otherwise} \end{cases} \tag{1}$$

$$\forall a, b \in Ind(A) \qquad k_p^F(a,b) := \left[\sum_{i=1}^{m} \left| \frac{\kappa_i(a,b)^p}{m} \right| \right]^{1/p} \tag{2}$$

The rationale of these kernels is that similarity between individuals is determined by their similarity with respect to each concept F_i , i.e, if they both are instances of the concept or of its negation. Because of the Open World Assumption for the underlying semantics, a possible uncertainty in concept membership is represented by an intermediate value of the kernel. A value of $p = 1$ has generally been used for implementing (2) in [3]. In our case, we have used the mean value of the above kernel, which is computed through high level feature relations and a normalized linear kernel which is computed through low level feature values.

3.3 The Reasoning Engine

It should be stressed that the reasoning engine, included in Figure 2, is of major importance for the whole procedure, because it assists the operation of all knowledge related components. First, during the knowledge development phase, it is responsible for enriching manual generation of concepts and relations, so that computation of the kernels in (1), (2) includes the fewest ambiguities possible, and any inconsistencies are removed from the knowledge representation. In fact (1), (2) are computed, by relating every two individuals w.r.t each concept in the knowledge base, by using the reasoning engine. In the operation phase, it interacts with the semantic interpretation layer and the connectionist system for achieving knowledge adaptation to real life environments. Both crisp and fuzzy reasoners can form this engine. In our case, we have been using the FIRE engine [12].

The FIRE system is based on Description Logic f-$\mathcal{SHIN}$ [11] that is a fuzzy extension of the DL $\mathcal{SHIN}$ [7] and it similarly consists of an alphabet of distinct concept names ($\mathbf{C}$), role names ($\mathbf{R}$) and individual names ($\mathbf{I}$). The main difference of the fuzzy extended Description Logics (DL) is their assertional component. Hence, in fuzzy DLs *ABox* is a finite set of fuzzy assertions of the form $\langle a : C \bowtie n \rangle$, $\langle (a,b) : R \bowtie n \rangle$, where $\bowtie$ stands for $\geq, >, \leq, <$, for $a, b \in \mathbf{I}$. Fuzzy representation enriches expressiveness, so a fuzzy assertion of the form $\langle a : C \geq n \rangle$ means that a participates in the concept C with a membership degree that is at least equal to n. In this case a contradiction is formed when an individual participates in a concept with a membership degree at least equal to n and at the same time with a membership degree at-most equal to l, with $l < n$.

The main reasoning services supported by crisp reasoners are *Abox consistency, entailment* and *subsumption.* These services are also available by FiRE together with greatest lower bound queries which incorporate the fuzzy element. Since a fuzzy *ABox* might contain many positive assertions for the same individual, without forming a contradiction, it is of interest to compute what is the best lower and upper truth-value bounds of a fuzzy assertion. For that purpose the term of *greatest lower bound* (GLB) of a fuzzy assertion with respect to a knowledge base is defined.

The reason why we use fuzzy reasoning is that fuzzy assertional component permits more detailed descriptions of a domain. In order to compute (1), (2) the GLB reasoning service of FiRE is used, but the resulting greatest lower bound is treated crisply. In other words, if GLB for $F_i(a) > 0$, then $F_i(a) \in A$, while if GLB for $F_i(a) = 0$, then $\neg F_i(a) \in A$. As a future extension, we intend to incorporate the fuzzy element in the estimation of kernel functions using fuzzy operations like fuzzy aggregation and fuzzy weighted norms for the evaluation of the individuals.

4 The Knowledge Adaptation Mechanism

4.1 The System Operation Phase

In the proposed architecture of Figure 2, let us assume that the set of individuals (with their corresponding features and kernel functions), that have been used to generate the formal knowledge representation in the development phase, is provided, by the *Semantic Interpretation Layer*, to the *Knowledge Adaptation* component.

Support Vector Machines constitute a well known method which can be based on kernel functions to efficiently induce classifiers that work by mapping the instances into an embedding feature space, where they can be discriminated by means of a linear classifier. As such, they can be used for effectively exploiting the knowledge-driven kernel functions in (1), (2), and be trained to classify the available individuals in different concept categories included in the formal knowledge. In [3] it is shown that SVMs can exploit such kernels, so that they can classify the (same) individuals - used for extracting the kernels - accurately; this is validated by several test cases. A Kernel Perceptron is another connectionist method that can be trained using the set of individuals and applied to this linearly separable classification problem.

Let us assume that the system is in its - real life - operation phase. Then, the system deals with new individuals, with their corresponding - multimodal - input data and low level features being captured by the system and being provided through the semantic interpretation layer to the connectionist subsystem for classification to a specific concept. It is well known that due to local or user oriented characteristics, these data can be quite different from those of the individuals used in the training phase; thus they may be not well represented by the existing formal knowledge. In the following we discuss adaptation phase of the system to this local information, taking place through the connectionist architecture.

4.2 Adaptation of the Connectionist Architecture

Whenever a new individual is presented to the system, it should be related, through the kernel function to each individual of the knowledge base w.r.t a specific concept - category; the input data domain is, thus, transformed to another domain - taking into account the semantics that have been inserted to the kernel function.

There are some issues that should be solved in this procedure. The first is that the number of individuals can be quite large, so that transporting them in different user environments is quite difficult. A Principal Component Analysis (PCA), or a clustering procedure can reduce the number of individuals so as to be capable of effectively performing approximate reasoning. Consequently, it is assumed that through clustering, individuals become the centers of clusters, to which a new individual will be related through (1), (2).

The second issue is that the kernel function in (1), (2) is not continuous w.r.t individuals. Consequently, the values of the kernel functions when relating a new individual to any existing one should be computed. To cope with this problem, it is assumed that the semantic relations, that are expressed through the above kernel functions, also hold for the syntactic relations of the individuals, as expressed by their corresponding low level features, estimated and presented at the system input. Under this assumption, a feature based matching criterion using a k-means algorithm, is used to relate the new individual to each one of the cluster centers w.r.t the low level feature vector. Various techniques can be adopted for defining the value of the kernel functions at the resulting instances. A vector quantization type of approach, where each new individual is replaced by its closest neighbor, when computing the kernel value, is a straightforward choice. To extend the approach to a fuzzy framework, weighted averages and Gaussian functions around the cluster centers are used to compute the new instances' kernel values.

In cases that classification - of the new individual - in the specific (local) environment and the specific individual characteristics or behaviour, remains linearly separable, the SVM or Kernel Perceptron are retrained - including the new individuals in the training data set, while getting the corresponding desired responses by the *User* or by the *Semantic Interpretation Layer* - thus, adapting its architecture / knowledge to the specific context and use.

In case the problem doesn't remain linearly separable, we propose to use an hierarchical, multilayer kernel perceptron, the input layer of which is identical to the trained kernel perceptron, and which is - constructively - created, by adding hidden neurons, and learning the resulting additional weights through a tractable adaptation procedure [10]. The latter is achieved through linearization of the added neurons' activation function, while taking into account both the new input/desired output data, as well as the previous knowledge and individuals. To stress, however, the importance of current training data, a constraint that the actual network outputs are equal to the desired ones, for the new individuals, is used. As a result of this network adaptation, the system will be able to operate satisfactorily within the user's environment

The problem will, in parallel, be reported back to formal knowledge and reasoning mechanism, for updating system's knowledge for the specific context, and then (off-line) providing again the connectionist module of the user with a new, knowledge-updated, version of the system. This case is discussed in the following subsection.

4.3 Adaptation of the Knowledge Base

Knowledge extraction from trained neural networks, e.g. perceptrons, or neuro-fuzzy systems, has been a topic of extensive research [8]. Such methods can be used to transfer locally extracted knowledge to the central knowledge base. Nevertheless, the - most characteristic - new individuals obtained in the local environment, together with the corresponding desired outputs - concepts of the knowledge base, can be transferred to the knowledge development module of the main system (in Figure 2), so that with the assistance of the reasoning engine, the system's formal knowledge, i.e., both the TBox and the ABox, can be updated, w.r.t the specific context or user.

More specifically the new individuals obtained in the local environment form an ABox $A^{'}$. In order to adapt a knowledge base $K = \langle T, A \rangle$ for a defined concept F_i using atomic concepts denoted as C, we check all related concepts denoted as $R_{F_i}C_1 \ldots R_{F_i}C_n$ under the specific context, i.e. in $A^{'}$. Let $|R_{F_i}C_n|$ denote the occurrences of $R_{F_i}C_n \in A$, t denote a threshold defined according to the data size and $Axiom(F_i)$ denote the axiom defined for the concept F_i in the knowledge base (i.e. $Axiom(F_i) \in T$). Furthermore, we write $R_{F_i}C_n \in Axiom(F_i)$ when the concept $R_{F_i}C_n$ is used in $Axiom(F_i)$ and $R_{F_i}C_n \notin Axiom(F_i)$ when it is not used. Knowledge adaptation is made according to the following criteria:

$$|R_{F_i}C_n| = \begin{cases} 0 - t/4 & \text{If } R_{F_i}C_n \in Axiom(F_i) \rightarrow \text{Remove } R_{F_i}C_n \text{ from } Axiom(F_i) \\ t/4 - t & \text{No adaptation in } K \\ > t & \text{If } R_{F_i}C_n \notin Axiom(F_i) \rightarrow Axiom(F_i) \cup R_{F_i}C_n \end{cases} \quad (3)$$

Equation (3) implies that the related concepts with the most occurrences in $A^{'}$ are selected for the adaptation of the terminology, while those that are not significant are removed. Currently, the DL constructor that is used for the incorporation of the related concept, in order to adapt the knowledge base, is specified by the domain expert. Future work includes a semi-automatic selection of constructors, that will be based on the inconsistencies formed by the use of specific DL constructors for the update of the knowledge base.

5 A Multimedia Analysis Experimental Study

The proposed architecture was evaluated in solving segment classification in images and video frames from the summer holiday domain. Such images typically include persons swimming or playing sports in the beach and therefore we selected as concepts of interest for this domain the following: Natural-Person, Sand, Building, Pavement, Sea, Sky, Wave, Dried-Plant, Grass, Tree, Trunk and Ground.

Following a region-based segmentation procedure, we let each individual correspond to an image segment. The low level features used as input to the system for each individual are the MPEG-7 Color Structure Descriptor, Scalable Color and Homogeneous Texture together with the dominant color of each segment. The colours used in this case are White, Blue, Green, Red, Yellow, Brown, Grey and Black.

We used equations (1)-(2) to compute the kernel functions and transferred them to the connectionist subsystem. In that way we trained threshold (and multilayer) perceptrons to classify more than 3000 individuals (i.e., regions extracted from 500 images), regions to the above-mentioned concepts. We tested the classification performance with new segments, with results reported in Table 1. The next step was to use the improved performance of the connectioninst model which forms a new ABox, in order to adapt the knowledge base. The roles used in our knowledge base are $above-of$, $below-of$, $left-of$ and $right-of$ that indicate the neighboring segments, and are extracted by a segmentation algorithm, included in the semantic interpretation layer. The new axioms referred to concepts Sea, $Sand$, Sky, $Tree$ and $Building$ using a neighbor criterion, that is the related concept in the specific context. For example, the concept Sea was defined as $Sea \equiv Blue \sqcap \exists below-of.Blue$. Assuming Sea as F_1, then the concepts formed by the combination of spatial relations with the other concepts i.e. $\exists below-of.Blue, \exists below-of.Brown, \ldots, \exists above-of.Green$, form the set of the related concepts $R_{F_1}C_1 \ldots R_{F_1}C_n$.

Using the technique described in section 4.3, the relative concepts that play a significant role, according to the Abox that is formed by the connectionist model, were defined. An adapted axiom was
$Sea \equiv Blue \sqcap (\exists below-of.Blue \sqcup \exists above-of.Brown \sqcup \exists above-of.White \sqcup \exists right-of.White \sqcup \exists left-of.White \sqcup \exists left-of.Blue \sqcup \exists right-of.Blue \sqcup \exists above-of.Blue \sqcup \exists below-of.Blue)$.

The adapted knowledge was again transferred , through (1) and (2) to the connectionist system, which was then able to improve its classification performance, w.r.t the five concepts, as shown in third column of Table 1. It is important to

Table 1. Performance after the adaptation of the knowledge base

Label	Regions	NN Performance Precision	NN Performance Recall	Adapted KB Precision	Adapted KB Recall
Person	76	56.25%	47.3	56.25%	47.3%
Sand	116	**75%**	**51.7%**	**83.1%**	**72.1%**
Building	108	**58.8%**	**37**	**72.7%**	**53.6%**
Pavement	64	25%	18%	25%	18%
Sea	80	**68.1%**	**75%**	**88%**	**79.2%**
Sky	88	**64.7%**	**50%**	**75.3%**	**64%**
Wave	36	33.3%	66.6%	33.3%	66.6%
Dried Plant	64	50%	37.5%	50%	37.5%
Grass	80	52.3%	55%	52.3%	55%
Tree	92	**63.1%**	**52.1%**	**71.2%**	**63.1%**
Trunk	72	57.1%	22.2%	57.1%	22.2%
Ground	112	24.5%	53.5%	24.5%	53.5%

note that the performance obtained is similar to that provided by adaptation of the (kernel) multilayer perceptron presented in 4.2.

6 Conclusion

In this paper we presented a novel architecture based on connectionist adaptation of ontological knowledge. The proposed architecture was evaluated using a multimedia analysis experimental study presenting very promising results. Future work, includes the incorporation of fuzzy set theory in the kernel evaluation. Additionally, we intend to further examine the adaptation of a knowledge base using the connectionist architecture, mainly focusing on the selection of the appropriate DL constructors and on inconsistency handling.

References

1. Bloehdorn, S., Sure, Y.: Kernel methods for mining instance data in ontologies. In: Aberer, K., Choi, K.-S., Noy, N., Allemang, D., Lee, K.-I., Nixon, L.J.B., Golbeck, J., Mika, P., Maynard, D., Mizoguchi, R., Schreiber, G., Cudré-Mauroux, P. (eds.) ASWC 2007 and ISWC 2007. LNCS, vol. 4825, pp. 58–71. Springer, Heidelberg (2007)
2. Fanizzi, N., d Amato, C., Esposito, F.: Randomised metric induction and evolutionary conceptual clustering for semantic knowledge bases. In: CIKM 2007 (2007)
3. Fanizzi, N., d Amato, C., Esposito, F.: Statistical learning for inductive query answering on owl ontologies. In: Proceedings of the 7th International Semantic Web Conference (ISWC), pp. 195–212 (2008)
4. Avila Garcez, A.S., Broda, K., Gabbay, D.: Symbolic knowledge extraction from trained neural networks: A sound approach. Artificial Intelligence 125, 155–207 (2001)
5. Avila Garcez, A.S., Broda, K., Gabbay, D.: The connectionist inductive learning and logic programming system. Applied Intelligence, Special Issue on Neural networks and Structured Knowledge 11, 59–77 (1999)
6. Hitzler, P., Holldobler, S., Seda, A.: Logic programs and connectionist networks. Journal of Applied Logic, 245–272 (2004)
7. Horrocks, I., Sattler, U., Tobies, S.: Reasoning with Individuals for the Description Logic $\mathcal{SHIQ}$. In: McAllester, D. (ed.) CADE 2000. LNCS (LNAI), vol. 1831, pp. 482–496. Springer, Heidelberg (2000)
8. Kolman, E., Margaliot, M.: Are artificial neural networks white boxes? IEEE Trans. on Neural Networks 16(4), 844–852 (2005)
9. Pinkas, G.: Propositional non-monotonic reasoning and inconsistency in symmetric neural networks. In: Proceedings of the 12th International Joint Conference on Artificial Intelligence, pp. 525–530 (1991)
10. Simou, N., Athanasiadis, T., Kollias, S., Stamou, G., Stafylopatis., A.: Semantic adaptation of neural network classifiers in image segmentation. In: 18th International Conference on Artificial Neural Networks, pp. 907–916 (2008)
11. Stoilos, G., Stamou, G., Pan, J.Z., Tzouvaras, V., Horrocks, I.: Reasoning with very expressive fuzzy description logics. Journal of Artificial Intelligence Research 30(8), 273–320 (2007)
12. Stoilos, G., Simou, N., Stamou, G., Kollias, S.: Uncertainty and the semantic web. IEEE Intelligent Systems 21(5), 84–87 (2006)

Modeling Human Operator Controlling Process in Different Environments

Darko Kovacevic[1], Nikica Pribacic[1], Mate Jovic[1], Radovan Antonic[1], and Asja Kovacevic[2]

[1] Faculty of Maritime Studies SPLIT, Zrinsko-Frankopanska 38, 21000 Split, Croatia
[2] Clinical Hospital SPLIT, Spinciceva 1, 21000 Split, Croatia
dkovac@pfst.hr

Abstract. An algorithm representing distribution and sell of foreign newspaper in Croatia controlled by a human operator is designed and performed in different modelling media including a hardware environment. The operator's behavior modeled in hardware can be taken as a base for developing a new kind of controller.

Keywords: operator's algorithm, modeling newspaper sell control.

1 Introduction

Informally, an operator's algorithm is well-defined computational procedure that takes some value (remnant), or set of values, as input and produces some value (order), or set of values, as output. An algorithm is thus a sequence of evolving computational steps that transform the input into the output [1]. We can also see the algorithm as a tool for solving a similar well-specified computational problem. The statement of the problem specifies in general terms the desired input/output relationship. The algorithm describes a specific evolving computational procedure for achieving that input/output relationship. As input/output relationship is in many cases represented by transfer function, it looks natural to correlate algorithms and transfer functions especially in control engineering. In this paper we will try to explain our efforts in seeking for the algorithm(s) that describe human (agent) behavior in very complex environment as socio-economic environments are by definition, but in the same time we are looking for new inspirations (evolving hardware) that could be used in technical environments as control tools. The guiding principle for developing one algorithm in a software and the other one (the same or almost the same) in two hardware environments (SO-HA-triplet) is to compare output results and behaviors (stability) of algorithms (models). One must know that the model(s) stability is out of the scope of this paper but it will be shown implicitly.

2 A Word about Newspaper Distribution and Selling Process

In manufacturing and other commercial settings, it is often important to allocate resources in the most beneficial way. An oil company may wish to know where to

C. Alippi et al. (Eds.): ICANN 2009, Part II, LNCS 5769, pp. 475–484, 2009.

place its wells in order to maximize its expected profit. A newspaper distribution and selling agency always wants to know when, where and how to distribute newspapers (different titles) to satisfy very variable tourist demands (see Fig. 1. a)) and to keep remnant (the number of unsold newspapers) in a certain limits (see Fig. 1. b)).

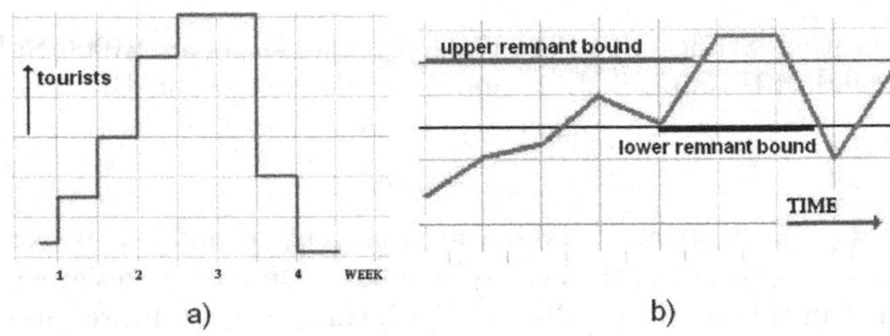

Fig. 1. a) An illustration of a possible normalized time distribution of Italian tourists in Dubrovnik during August. b) Remnant of the Austrian newspaper "Kronen Zeitung".

A newspaper distribution and selling process (NDSP) is controlled by dedicated human operator. As the process become more complex (number of different tourists grows, as well as number of newspaper titles and number of kiosks), greater are the demands on the human operator who controls the process.

However, the human operator when performing a control task in a socio-economic system (NDSP) often shows remarkable versatility and adaptability in handling stable and unstable systems. Philosophically, in control engineering terms, such "an element" human operator would be considered a time varying, variable gain, nonlinear element. In NDSP, the human operator is performing a single axis discrete tracking task (one track for each newspaper title) based on a certain algorithm and is seen to be an integral part of the closed-loop control system (see Fig. 2).

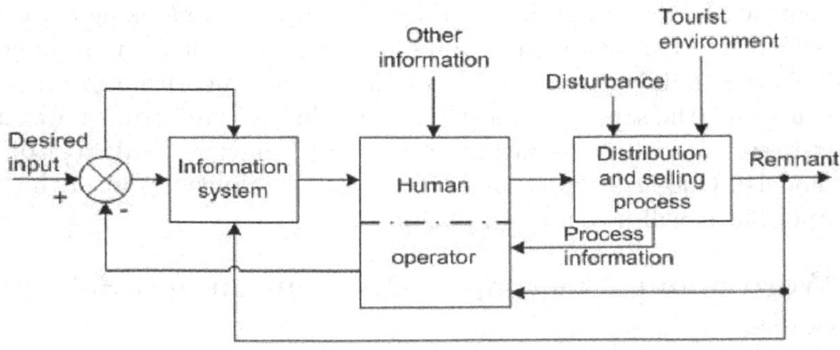

Fig. 2. Basic model of human controlled newspaper distribution and selling process

Tracking task is done on two levels; one level is dedicated to a certain sell point (kiosk), and the other one is correlated with tracking the remnant of each newspaper title on the agency level. In principle every sell point has the different tracking algorithm for a certain newspaper title. On the other hand, for every newspaper title on the agency level, human operator has to establish a distinguished overall algorithm. For an agency with at least one hundred sell points and with daily input of fifty foreign newspaper titles, five thousand plus one tracking task must be performed daily if the NDSP is to be controlled efficiently. In principle, the tracking task means making the decision for a new (or confirming the old one) order of a certain newspaper title for a certain selling point. In practice the decision is mainly based on remnant data, going one to three days in the past, that are available from agency information system. For an experienced human operator, even a quick look on those data takes about thirty seconds to create the new (old) ordering. A field investigation done by authors showed that the human operator in the agency needs (in average) about 16 hours for completing all his tracking tasks in a proper manner. A term "a proper manner" in terms of control theory means that the operator, working in a compensatory or a pursuit mode, is trying to drive the current remnants of all selling points to a given remnant channel. The channel boundaries are set up (given) by publishers and they are fuzzy; for German "Bild" the bound is about 30 percent of a daily order and for Austrian "Kronen" the remnant channel is in the range from about 30 percent to round 50 percent of a daily order.

When undertaking a compensatory discrete tracking task, the operator is presented with the error between the system input signal and the output [2].

Thus the human operator can effectively be considered as the controller element in a servomechanism. An information system can be considered as a display that only provides relative information. Hence, when the operator is tracking a desired optimal ordering he cannot be certain if the displayed remnant error is a result of his or agency performance, or the change in tourist environment (or any combination of both). The notion of "the change in tourist environment" is extremely wide and will not be elaborated in this paper.

In pursuit tracking, the operator perceives information about possible number of buyers (consumers) and about the related remnant.

In practice, both tracking modes are highly affected by information delays introduced in the NDSP. Field investigation has showed that poor management and human resources, bad system organization and lack of adequate equipment are main causes of many delays in the agency information system.

3 Algorithm

The algorithm defining NDSP, and an operator as an integral part of the process, will be specified as geometric problem to be more illustrative. That approach leads us to the notion and implementation of computational geometry as a possible problem solving tool.

Computational geometry is the branch of computer science that studies algorithm for solving geometric problems [1]. In modern engineering and mathematics, computational geometry has applications in, among other fields, computer graphics, robotics, VLSI design etc. The input to a computational-geometry problem is typically a description of asset of geometric objects, such as a set of points, a set of line segments, or the vertices of a polygon in counterclockwise order. The output is often a response to a query about the objects, such as whether any of lines intersect, or perhaps a new geometric object [1].

The transfer characteristic of an operator can be regarded as geometric object.

In a general way, transfer characteristic(s) of an operator (TCO) has reactive character expressing a negative feedback sense of operator role in a NDSP:

- When the daily remnant of a certain title (newspaper) is high (over upper bound - u.b.), the control action goes to decrease of next order. New order is processed.

- When the daily remnant of a certain title (newspaper) is low (under lower bound - l.b.), the control action goes to increase of next order. New order is processed.

- When the daily remnant of a certain title (newspaper) is within permitted remnant levels the control action will be ceased. There is no new order.

These three simple rules define a simple TCO, i.e., a simple rule base. In a real life, the geometry (shape) of TCO emerges as result of interactive mutual interactions between an operator and his environment. Therefore, new rules must be added in the rule base if linguistic model of an algorithm concerning TCO must be close to the NDSP-reality as much as it is possible. Seven IF THEN rules are added to the rule base:

- If a remnant is zero, or close to zero level then limit the response to a certain value (highest order level prescribed by a publisher)

- If a remnant is small and number of tourists rises then increase a next order more then usually.

- If a remnant is small and number of tourists rises then increase a next order more then usually taking into consideration limits posed by publisher(s).

- If remnant is to be controlled on immediate demand (publisher policy), then make lower and upper bounds controllable via outside information.

- If a remnant is very high and a number of tourist decreases, then decrease a next order more then usually.

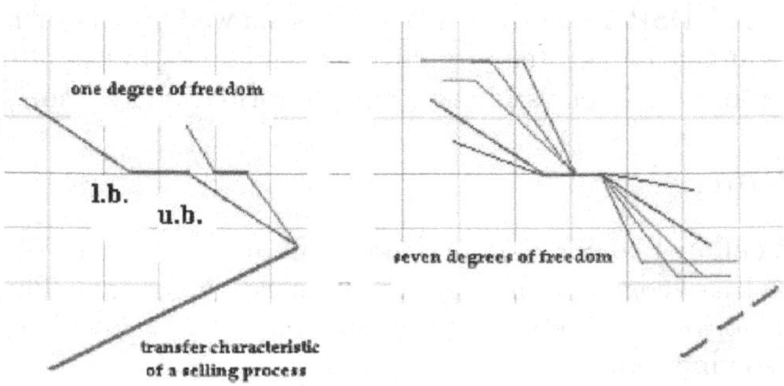

Fig. 3. The geometry of TCO related to a degree of freedom (constraints)

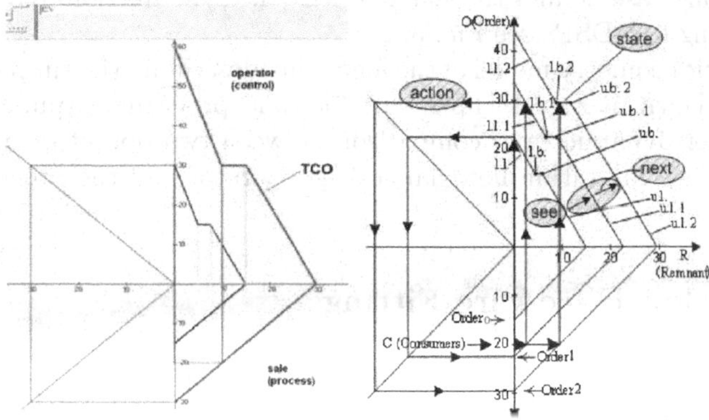

Fig. 4. Responses of two different operators for the same inputs (remnant, consumers)

```
Enter l.b. and u.b.!
l.b.= 0.33
u.b.= 0.5
Enter Order and Consumers!
O=10
C=60

Upper line:
-2R+20
Lower line:
-2R+16.6
Lower bound=3.3
Upper bound=5
R=0

O1=17
Upper line1:
-2R+33.2
Lower line1:
-2R+27.556
Lower bound1=5.478
Upper bound1=8.3
R=0

O2=28
Upper line2:
-2R+55.112
Lower line2:
-2R+45.743
Lower bound2=9.09348
Upper bound2=13.778
R=0

O3=46
Upper line3:
-2R+91.4859
Lower line3:
-2R+75.9333
Lower bound3=15.0952
Upper bound3=22.8715
R=0

O4=76
Upper line4:
-2R+151.867
Lower line4:
-2R+126.049
Lower bound4=25.058
Upper bound4=37.9667
R=16

O5=94
Upper line5:
-2R+188.365
Lower line5:
-2R+156.343
Lower bound5=31.0803
Upper bound5=47.0913
R=34

End! O5=94
```

The response of an operater with one degree of freedom.

```
Enter l.b. and u.b.!
l.b.= 0.33
u.b.= 0.5
Enter Order and Consumers!
O=10
C=60
Enter ki=3.5
Enter kd=3
Enter T=1

Upper line:
-6R+40
Lower line:
-7R+33.1
Lower bound=3.3
Upper bound=5
R=0

O1=33
Upper line1:
-6R+132.4
Lower line1:
-7R+109.561
Lower bound1=10.923
Upper bound1=16.55
R=0

O2=110
Upper line2:
-6R+438.244
Lower line2:
-7R+362.647
Lower bound2=36.1551
Upper bound2=54.7805
R=50

End! O2=110
```

The response of an operater with seven degrees of freedom.

Fig. 5. The way of comparing responses (orders) of two different operators for the same inputs

- If a remnant is very high and a number of tourist decreases, then decrease a next order more then usually taking into consideration demands posed by publisher(s).

- If a remnant is 100 percent or close to 100 percent then a next order must be limited to a minimum distribution level (some publishers want to be present on a market regardless to an existing remnant).

The new rule base defines the shape of TCO i.e., the freedom (constraints) of operator acting in NDSP (see Fig. 3.).

The analytic geometry model of the algorithm describing the simplified version of NDSP is shown in Fig. 4. Input variables and program output representing a certain order dynamics and comparison between two operators are shown in Fig. 5. The C++ algorithm flowchart of the operator and the process is shown in Fig. 6.

4 Algorithm Hardware Fitting

Algorithm hardware fitting, in general, is the translation of a mathematical or empirical (linguistic) relationship (between a dependent variable, i.e. output and one or more independent variables, i.e. input (s)), from one medium,such as a table, a mathematical formula, a string of numbers, a set of curves or a collection of logical (linguistic) control rules, to another medium, usually a physical-realizable device or system having an output and one or more inputs [3].

An algorithm may be fit by an "exact" relationship, or it may be somehow approximated.

There are three basic steps in the algorithm hardware fitting. The first step is the establishment of a close-enough approximation in terms of ideal building blocks, that is, a conceptual model. The second step is the hardware embodiment of the algorithm specifics such as statements concerning "IF THEN - condition(s)" and "FOR/NEXT- loop(s)", distributive ALU-elements, etc. originated in the conceptual model.

On that stage input/output communications must be considered, as well as taking hardware-body information sensitivity into account if hardware model must act as an hardware agent in a certain environment. The third step is successful employment of actual (new) circuit devices to embody the function within an acceptable set of constrains, such is range, scale factor, drift, response time, complexity, cost, etc.

5 D-Operators

TCO-synthesis will be based on hardware circuits called D-operators that use ideal diodes. For switching purposes, the "ideal diode" is a one way switch that is open when the imposed voltage is of one polarity and closed when the polarity is opposite. The ideal diode operator is a voltage to voltage circuit that would have the same response as a circuit that used an ideal diode as switching element: the output voltage is zero for one polarity; it increases linearly with input when the polarity changes (see Fig. 7.).

The ideal diode operator can also be considered as a "zero-bound" circuit [3]. In a special cases, when $\mid V_o \mid = \mid V_i \mid$ these circuits can be considered as D - operator circuits because they can be used in synthesis of different transfer characteristics. The ideal diode operators are useful in precision dead zone,

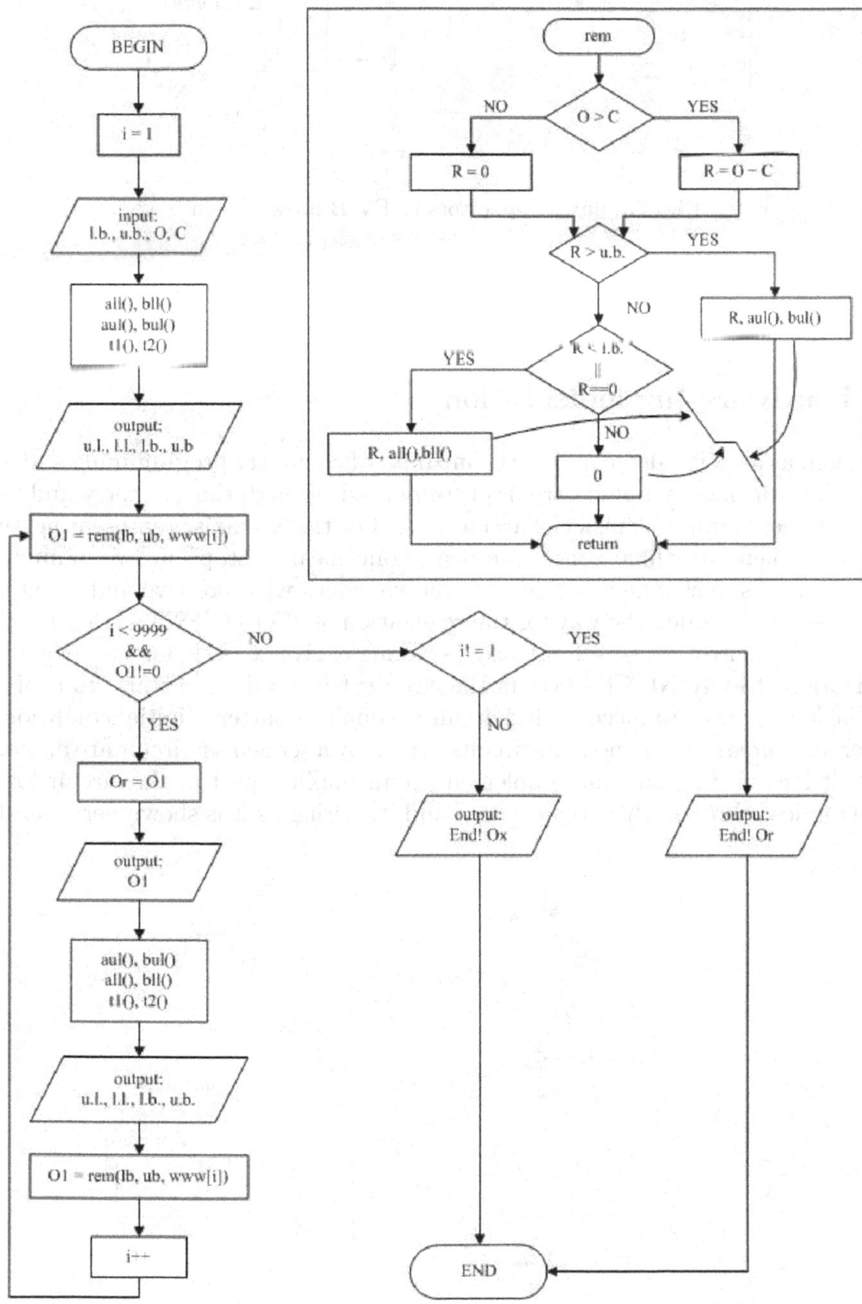

Fig. 6. Developing the algorithm in C++ environment

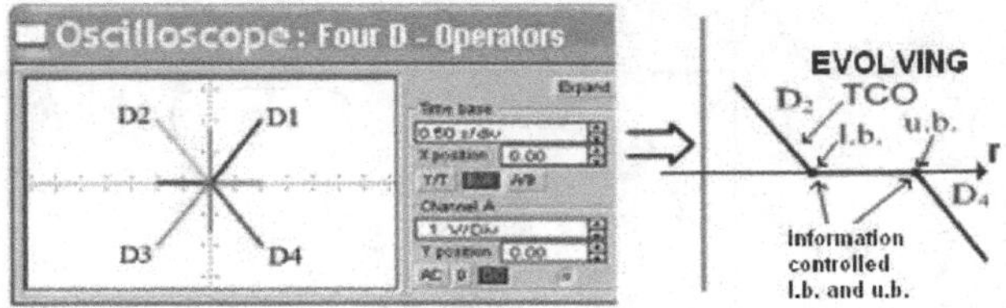

Fig. 7. Four D operators in EWB presentation

bounds, and absolute-value circuits and in function fitting with piecewise-linear approximations (different fuzzy membership functions).

6 Hardware Implementation

Decision, as a notion dedicated to the medium of computer programming and/or control, in the medium of electronics is presented through the geometry and the scale of the operator transfer characteristic. On the x - axis, representing the remnant, there are three zones; the dead zone means "stop" in programming sense (order is unchanged), while two reactive zones with positive and negative gain respectively open the way for the realization of "FOR/NEXT - statement". Wiring (in control sense - feedback) is done to give a path for the effective realization of FOR/NEXT - loop in the circuit; triggered TCO starts to evolve.

The logic unit introduced in this circuit is simple; a battery (initial condition), adder and three sample and hold circuits driven by a sequencer circuit are making explicit logic in the hardware implementation. Implicit part of the circuit logic is introduced through the circuit design and its wiring as it is shown very clearly

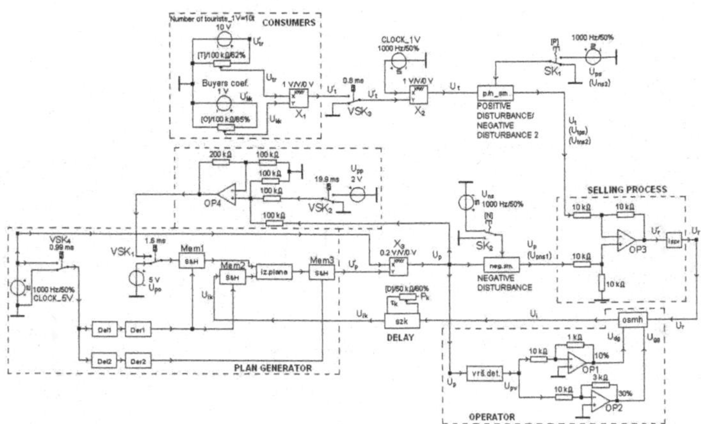

Fig. 8. Hardware model of the algorithm includes the process, operator and environment

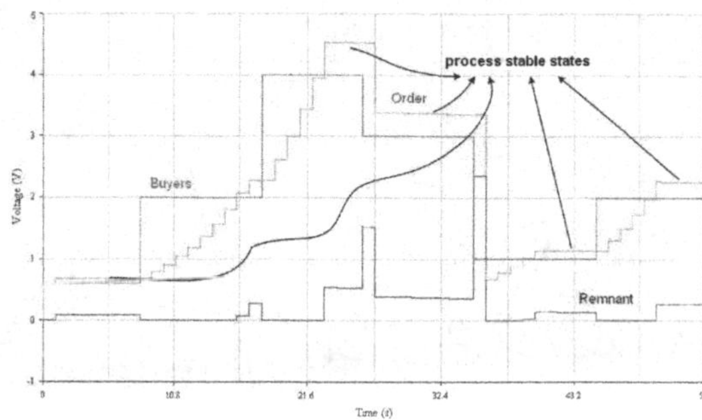

Fig. 9. The response of AI operator to tourist dynamics

in Fig. 8. Existing arithmetic is distributed through this circuit; operations of addition, division, subtraction and multiplication are done around the circuit (see Fig. 8.).

Graph shown in Fig. 9. presents states in the hardware model of the process, including AI-operator in his process work.

If we want to identify our hardware model of the operator as an agent in AI (DAI) sense then we must consider standard AI (DAI) - agent attributes and architectures.

7 Discussion

In this paper an algorithm derived from human behavior (operator) in a socio-economic process is presented. The algorithm has been modeled in two (three) different media; as computer program and as a hardware circuit.

SO-HA-triplet is of crucial importance when human behavior is to be modeled in an artificial environment and when humans will be involved in a process of verification and validation of the model(s) and its outputs (results). SO-HA-triplet generate virtual reference environment as state and responses of the system can be compared.

Hardware modeling of the operator is based on a family of IC circuits called D - operators. D - operators can be used in synthesis of different transfer characteristics that can relate electronic and human environments.

The shape of those characteristics can be efficiently controlled by applied voltage (information).

The implementation of the algorithm in a hardware environment was chosen deliberately as the first step (premise) of a possible application of the algorithm for control tasks in a technical environment.

The future work can be multidimensional: from comparing models' behavior(s) and stability and fitting models in possible applications in a real NDSP to analyzing the operator as an agent (see Fig. 10).

Operator can be easily seen as an agent (controller) that maintain state. It has internal data structure which is used to record information about the

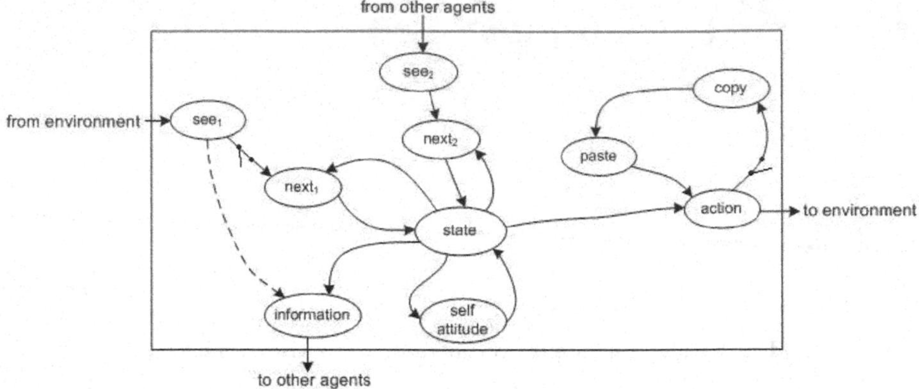

Fig. 10. Abstract architecture of an operator that maintain state

environment state and history. The agent starts in some initial internal state i_0. It then observes ($see_1(k)$ function) its process environment state $s_1 = r(k)$ (remnant in k moment). The internal state of the agent is then updated via $next_1$ function, becoming set to $next_1(i(k), r(k))$. The action selected by the agent is then $action(next_1(i(k), r(k)))$ (see Fig. 4). This action (order) is performed, and agent enters another cycle, perceiving the world via see_1 and see_2 (if it exists) updating its current initial state via $next_2$ (if see_2 exists) and current process state $next_1$. Note that current initial internal states $i_0(k) = next_2(i_0(k), see_2(s_2))$ representing information about current remnant channel) of the agent depends on percepts ($see_2(s_2)$) from business environment (for example publisher).

References

[1] Cormen, T.H., Leiserson, C.E., Rivest, R.L., Stein, C.: Introduction to algorithms. The MIT Press, Cambridge (2001)

[2] Sutton, R.: Modeling Human Operators. Research Studies Press LTD, Taunton (1990)

[3] Sheingold, D.H.: Nonlinear Circuits Hanbook. Analog Devices, Inc., Norwood (1976)

Discriminating between V and N Beats from ECGs Introducing an Integrated Reduced Representation along with a Neural Network Classifier

Vaclav Chudacek[1], George Georgoulas[2], Michal Huptych[1], Chrysostomos Stylios[3], and Lenka Lhotska[1]

[1] Dept. of Cybernetics - Czech Technical University in Prague, Prague, Czech Republic
{chudacv,huptycm,lhotska}@fel.cvut.cz

[2] Dept of Computer Applications in Finance and Management, TEI of Ionian Islands, Lefkas, Greece
georgoul@teiion.gr

[3] Dept. of Informatics and Telecommunications Technology TEI of Epirus, Arta, Greece
stylios@teiep.gr

Abstract. The main objective of this paper is to investigate and propose a new approach to distinguish between two classes of beats from the ECG holter recordings - the premature ventricular beats (V) and the normal ones (N). The integrated methodology consists of a specific sequence: R-peak detection, feature extraction, Principal Component Analysis dimensionality reduction and classification with a neural classifier. ECG beats of holter recordings are described using means as simple as possible resulting in a description of the QRS complex by features derived mathematically from the signal using only R-peak detection. For this research work, normal (N) and ventricular (V) beats from the well known MIT-BIH database were used to test the proposed methodology. The results are promising paving the way for the more demanding multiclass classification problem.

Keywords: Holter monitoring, PCA, Neural Networks, MLP classification.

1 Introduction

Many different methods have been proposed to solve the crucial problem of long-term holter recordings evaluation, which could be transformed into the classification problem of discriminating between normal 'N' and a variety of other beats, mainly premature ventricular 'V' beats and supraventricular beats (S).

A lot of research effort has been put to investigate and propose methods to examine and classify the holter recordings based on beat-shape description

C. Alippi et al. (Eds.): ICANN 2009, Part II, LNCS 5769, pp. 485–494, 2009.

parameters [1], shape descriptive parameters transformed with the Karhunen-Loeve method [2], and hermite polynomials [3]. Some other research proposals use time-frequency features [4] and features obtained from heartbeat interval measurements [5,6] in order to identify cardiac arrhythmia.

For any classification problem, in order to compare different approaches, the setup of the experiments, where the type of the training and the selection of testing sets are defined, is of major importance. In most problems, there are two main setups to be considered: training based on a local learning set [7] and on a global learning set [8].

In holter monitoring, the main reason for using local learning is the fact that beats within one patient tend to look alike - but tend to differ widely among different patients - therefore using locally trained classifiers usually leads to better overall results in a personalized medical approach where the patient himself is its own control. In the case of global training the records used for training the classifier are distinct from those used for testing - warranting therefore better, if the data are representative, generality of the classifier.

In this paper we investigate a new configuration to deal solely with distinguishing between normal and ventricular beats [9] where global training fashion is considered; for our testing we use a reduced representation of the original beats coming from the MIT-BIH database [10] and a neural network classifier. The results of the proposed approach are then compared with other published methods using the metrics of sensitivity and specificity.

The rest of the paper is organized as follows: Section 2 presents the proposed methodology to handle the data and extract the specific set of features. Section 3 briefly describes the classification methods involved. In Section 4 the selection of the training and testing sets is described and the results are presented. Section 5 concludes the paper giving some directions for future research.

2 Handling of Data and Feature Extraction

2.1 Preprocessing

For this research work in order to prepare the Holter records, all data records were re-sampled to 500Hz from the original 360Hz. No filtering was performed on any of the signals.

The detection and localization of the R-peak is of paramount importance in the subsequent analysis. For the detection of R-peaks of our data set, the method proposed firstly by Christov [11] was applied. The feature set that is used here is based solely on the R-peak findings - we do not use any other measurement of beat's characteristic points. More specifically the maximum of the major R-peak is found and a window of 128 samples with R peak centered on position 64 is selected for further computations. Fig. 1 presents the result of the applied method for the two different classes under investigation. As it can be seen the "mean" N and V beats have quite a distinct morphology (even though some beats can deviate quite a lot from these shapes, constituting what is widely known as "outliers"). The extracted features are described in the next section.

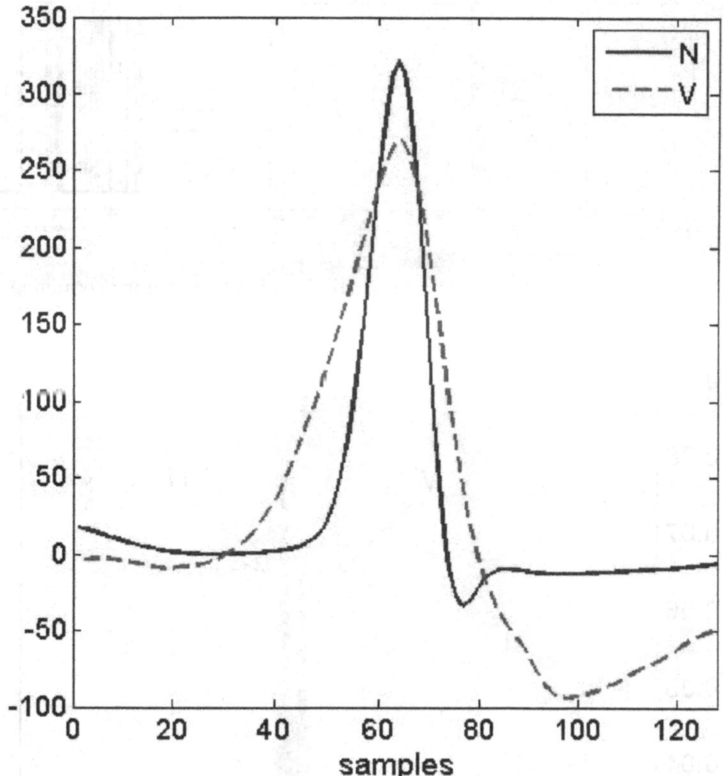

Fig. 1. The "mean" N and V beat waveforms as they were calculated using the MIT database

2.2 Feature Extraction and Selection

Usually feature sets for beat characterization use time intervals, amplitudes and their ratios based on the important points measured from the signal [9].

Here, we propose the use of a feature set consisting of features that are computed solely on the 128 samples around the R-peak. We propose to use features that were selected after visually inspecting the waveforms coming from different classes. These features as proven by the classification results can quantify the difference between the 2 classes.

More specifically the feature set consists of nine measures. Three of them are directly calculated from the truncated signal; namely, the minimum value of the second half of the signal (i.e. from sample 64 till the end of the signal) along with the location of the minimum, and the standard deviation (using all 128 samples).

The rest 6 features are calculated by processing the binary sequences that are constructed after thresholding the first order difference of the original signal and

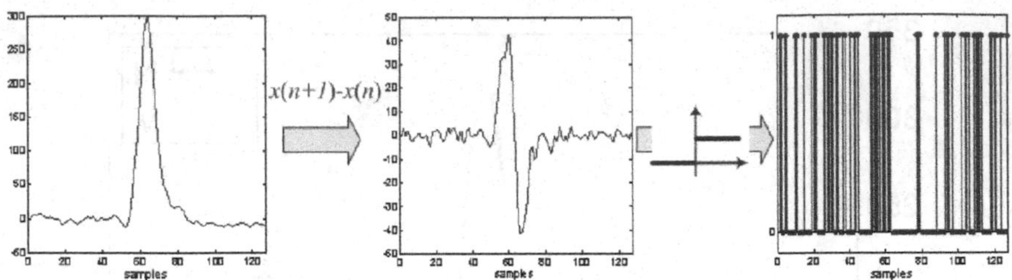

Fig. 2. Transformation of the original signal into a binary sequence

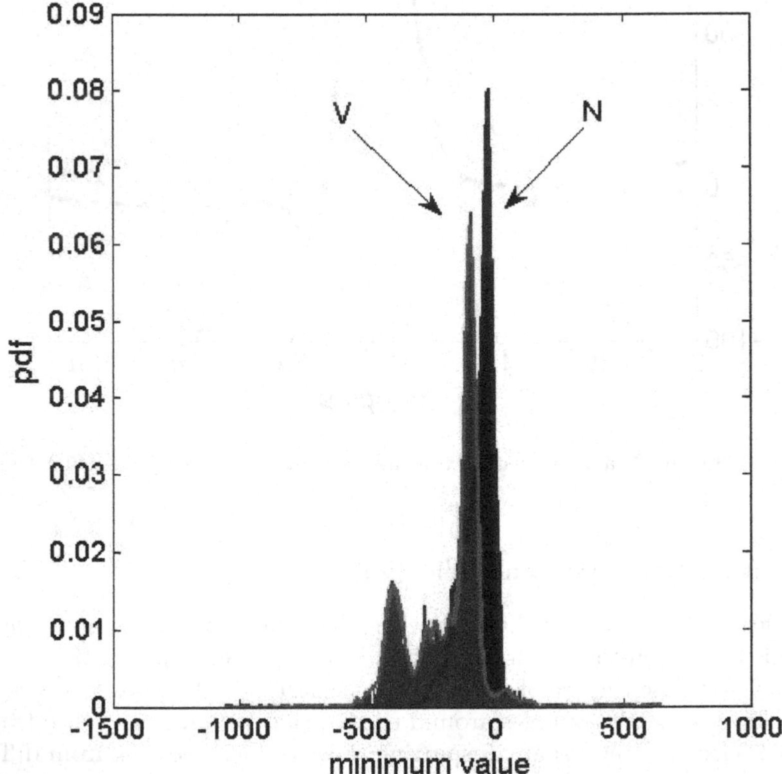

Fig. 3. Experimental "pdfs" of the minimum value for the N (blue) and V (red) class

the second order difference of the original signal. Fig. 2 depicts the transformation of the original signal into a binary sequence as described above.

For each one of the two aforementioned sequences the Shannon entropy is calculated (eq. 1). The two binary sequences are then combined creating a 4-level sequence (a two digit binary word can be described by one digit of an "alphabet" with base 4). The Shannon entropy of this sequence constitutes the third feature and the probabilities (ratios of occurrences) of the three out of the four levels (including the fourth would be redundant) completes the feature set.

$$Entropy = -\sum_{n=1}^{K} p_n \ln(p_n) \tag{1}$$

Fig. 3 depicts the "pdfs" (histograms with a superimposed spline curve for illustration purposes) of the distribution of the minimum value of the waveform for the two classes. As it can be seen this feature captures the variation between the 2 classes. The rest of the features (not shown here) also have distributions that show the potential benefit of being used for this particular discrimination task.

The feature set is mainly based on the ability to find correctly the maxima of the major R-peak. Then all the features are computed based on the "truncated" signal itself without the need of any other measurements. Therefore this feature set could be a very useful model for classifying the data obtained by telemedicine application devices. In the proposed approach, we have no information from the depolarization phase of the beat cycle - since the behavior of the T-wave varies wildly in terms of shape and length and therefore it would be necessary to employ additional measurement of the end of T-wave.

It is also apparent that some of those features might be correlated. But, it is well known that when we use neural networks classifiers, it is beneficial to feed them with uncorrelated features and also to get rid of redundant information. Thus, in the proposed methodology a dimensionality reduction stage was included before the neural network stage.

2.3 Dimensionality Reduction

It is well known that in pattern recognition tasks, usually potential improvement (better generalization) can be achieved by using fewer features than those available [12]. Actually, literature proposes during the development of a classifier to extract several features, which may convey redundant information about the pattern-class of interest. Therefore, in the proposed approach we included a Principal Component Analysis (PCA) stage so that to un-correlate the originally extracted features using a linear transformation [12].

PCA, or Karhunen-Loeve transformation, is an approach to perform dimensionality reduction by linear combination of the original features in such a way that preserves as much of the relevant information as possible [12,13]. This method computes eigenvalues of the correlation matrix of the input data vector and then projects the data orthogonally onto the subspace spanned by the eigenvectors (principal components) corresponding to the dominant eigenvalues. Even if the whole set of the eigenvectors is retained, this may also lead to an improvement of the classification performance, because the new set has features that are uncorrelated and this, in general, improves the classification capabilities of a classifier.

3 Neural Network Classification

Artificial Neural Networks (ANNs) are increasingly and successfully used in classification problems. They are structures composed of many simple processing elements, that operate in parallel and whose main function is determined by the network's structure, the strength of their connection and the processing carried out by the processing elements (artificial neurons). They are capable of finding commonalities in a set of seemingly unrelated data and for this reason are used in a growing number of classification tasks.

Among the numerous ANN paradigms encountered in the literature [12], the Multi-layer Perceptron (MLP) is the most widely used in the field of pattern recognition [12,13,14]. Training of an MLP is often formulated as the minimization of an error function, such as the total mean square error between the actual output and the desired output summed over all available data. While the sum-of-squares error function is appropriate for regression, for classification problems it is often advantageous [14,15] to optimize the network using the cross entropy error function (eq. 2), i.e. optimizing the network to represent the posterior probabilities of each class [12,13].

$$E = -\sum_{n=1}^{N}\sum_{k=1}^{c}\{t_k^n \ln y_k^n + (1 - t_k^n)\ln(1 - y_k^n)\} \tag{2}$$

where N is the number of training samples and c the number of classes, $t_k^n \in \{0,1\}$ is a binary class label, (k=1,...,c) of the nth data sample and y_k^n is the actual output of the kth neuron of the ANN, when the nth data sample is presented at its input.

For this case, we use the logistic activation function for the hidden layer units and the softmax (eq. 3) activation function for the output nodes [13,14].

$$y_j = \frac{\exp(a_j)}{\sum_i \exp(a_i)} \tag{3}$$

where a_i is the intermediate linear output of an artificial neuron.

The above configuration has proven to be more appropriate for classification purposes with many successful implementations [13,14]. Therefore, in this research work the above formulation has been adopted.

4 Experimental Results

For evaluation of the proposed approach, we used the commonly used MIT-BIH database [10]. There are two ways of training the classifiers with this database.

The first one is to use local training - using vertical division of the database. That means that, usually, the beginnings of each of the recordings from the database are used for training and remaining parts of each of the recordings are used as a testing set. Although this type of training brings usually results close to absolute sensitivity and specificity as it is often encountered in the literature e.g.

[1,7], it is very controversial from the point of view that any practical application would require additional annotation of at least a short part of each patient's recordings. On the other hand global training implies that the records used for training the classifier are distinct from those used for testing. This means that no additional annotation is needed and the classifier can be directly used on any new patient.

After considering the above mentioned advantages and disadvantages the results reported in the next section are based on the global classification approach using 44 of the MIT recordings and employing the leave one out technique. In other words each time 43 recordings were used for training the classifier and 1 for testing.

Since one of the classes is heavily underrepresented in the given dataset (this is not a flaw of the data, it is "just the way things are"), this makes training of the MLP problematic. This means that we are running the risk to build a classifier heavily biased to classify everything as N class. Different approaches have been proposed in order to alleviate this problem. In our case we downsampled the N class (only during the creation of the train set) taking one every 14 samples. By doing so we have "pushed" the MLP to better learn the V class since in the problem at hand having a high sensitivity is a bit more significant than having a very high specificity (a very high specificity is achieved in almost all similar studies as reported in the conclusion section).

As mentioned in Section 2, after the feature extraction stage, we have included a dimensionality reduction stage based on PCA. In PCA, selecting the number of the retained Principal Components constitutes another design parameter and more than one "criteria" can be found in order to guide the selection process [15]. However, usually the selection is based on a trial and error approach. In our case through an initial experimentation phase using a simple classifier we found out that five to seven Principal Components yield similar results. As a result we selected to retain six of them. Among the different configurations of the MLP (10, 15, 20 and 25 neurons in the hidden layer were tested using a small subset of the dataset in a few preliminary runs without however a thorough search into the parameter search) the one with 20 neurons yields slightly better results.

The overall procedure is depicted in Fig. 4. In total the classifier (6-20-2) managed to classify correctly 58651 out of the 67264 N beats and 5535 out of the 5997 V beats resulting in sensitivity equal to 92.30% and specificity equal to 87.20%. The results are summarized in Table 1 for each one of the 44 recordings.

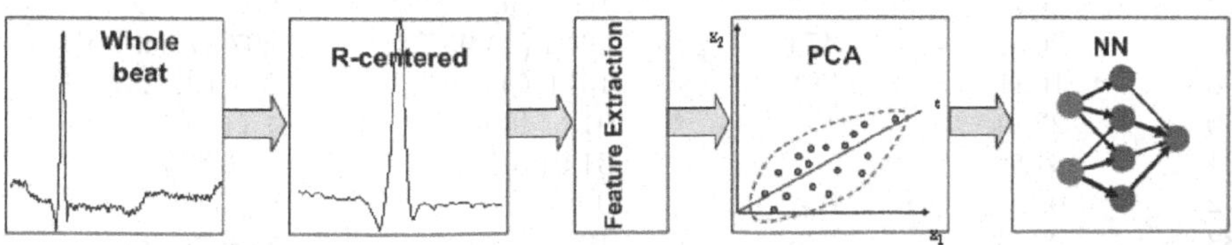

Fig. 4. Overall procedure

Table 1. Classification results for all 44 recordings

Record Number	# N beats	# V beats	Correctly classified N beats	Correctly classified V beats
100	2234	1	2234 (100%)	1 (100%)
101	1855	0	1841 (99.25%)	0 (-)
103	2077	0	2070 (99.66%)	0 (-)
105	2521	41	2190 (86.87%)	38 (92.68%)
106	1504	518	1501 (99.80%)	441 (85.14%)
108	1735	14	860 (49.57%)	6 (42.86%)
109	0	38	0 (-)	37 (97.37%)
111	0	1	0 (-)	0 (0%)
112	2532	0	2388 (94.31%)	0 (-)
113	1784	0	1781 (99.83%)	0 (-)
114	1815	43	1644 (90.58%)	39 (90.70%)
115	1948	0	1935 (99.33%)	0 (-)
116	763	41	148 (19.40%)	41 (100%)
117	1529	0	1224 (80.05%)	0 (-)
118	0	2	0 (-)	1 (50%)
119	1539	443	1538 (99.94%)	440 (99.32%)
121	1856	1	1849 (99.62%)	1 (100%)
122	2471	0	2462 (99.64%)	0 (-)
123	1510	3	1462 (96.82%)	3 (100%)
124	0	47	0 (-)	42 (89.36%)
200	1479	724	1294 (87.49%)	701 (96.82%)
201	588	4	585 (99.49%)	0
202	2056	19	2024 (98.44%)	7 (36.84%)
203	2519	396	1093 (43.39%)	328 (82.83%)
205	448	10	448 (100%)	10 (100%)
207	0	104	0 (-)	80 (76.92%)
208	1584	990	1556 (98.23%)	943 (95.25%)
209	2616	1	2582 (98.70%)	1 (100%)
210	319	24	297 (93.10%)	14 (58.33%)
212	920	0	895 (97.28%)	0 (-)
213	2636	220	2078 (78.83%)	220 (100%)
214	0	18	0	18 (100%)
215	3191	164	2005 (62.83%)	143 (87.20%)
219	2010	64	1677 (83.43%)	58 (90.63%)
220	1949	0	1670 (85.69%)	0
221	2026	396	2015 (99.46%)	388 (97.98%)
222	2057	0	1962 (95.38%)	0
223	2024	473	1942 (95.95%)	367 (77.59%)
228	1684	361	1571 (93.29%)	351 (97.23%)
230	2250	1	912 (40.53%)	1 (100%)
231	314	2	314 (100%)	1 (50%)
232	0	0	0 (-)	0 (-)
233	2226	830	1915 (86.02%)	813 (97.95%)
234	2695	3	2689 (99.78%)	1 (33.33%)

5 Conclusions

It is essential to compare the proposed integrated methodology with the work of other researchers but it is important to bear in mind two distinguishing points where this work is unique. First of all, there are, at least according to the best of our knowledge, no recent works dealing with **global** classification of ECG signals. And most important in this work, we introduced and used only **mathematically obtained features** derived from the ECG signal utilizing only the detected R-peak.

However, there exist research works dealing with global classifiers, using the MIT-BIH database to distinguish 'N' and 'V' beats, usually with slight modifications in the way of obtaining the global training/testing for each one of them. Hu and his coworkers [8] achieved global accuracy of 62.2% for distinguishing 'N' and 'V' beats. The sensitivity and specificity achieved in [7] is about 80%. Jekova et al [17] reports sensitivity 78.79% and specificity of 80.61% on the global training set when distinguishing also right and left bundle branch blocks. Lower numbers but on a more difficult task are reported in [1] with 86.7% specificity and 67.3% sensitivity for V beats when classifying holter beats into five classes on the MIT database.

There are also works trying to distinguish between N and V beats using simple features derived from one-lead signal where only R-peaks were computed. In [18] Tsipouras et al. have used HRV for classification obtaining sensitivity 87.27% and specificity of 94.77% on the MIT database - but they did not use global training. In [19] four descriptive parameters were used for beat classification but the experiments were performed on the selected signals only, with unspecified training routine.

To sum up our results are at all times at least as good as and in some occasions better than those reported in the literature. The prime novelty of this work is the proposal of a new combination of features for the discrimination of "V" and "N" beats. A neural network classifier has been employed using the cross entropy error function which usually performs better for classification problems. The results are very promising and in the next phase of our research we will test the usefulness of our approach on the more demanding problem of distinguishing between five beat categories. Towards this path we will also experiment with more advanced methods for the construction of our classifier (i.e. an incremental building of the hidden layer) since the discrimination of five classes increases the need for a more customized classifier. Moreover more elaborated techniques for handling imbalanced data sets might be needed (i.e. Synthetic Minority Oversampling Technique (SMOTE) [20]). Finally, we will test our method using the AHA database which will allow for the generality of our approach to be examined.

References

1. Chazal, P., O'Dwyer, O., Reilly, R.B.: Automatic Classification of Heartbeats Using ECG Morp. Heartbeat Interval Features. IEEE Trans. Biom. Eng. 51(7), 1196–1206 (2004)

2. Cuesta-Frau, D., Perez-Cortes, J.C., Andreu-Garcia, G.: Clustering of electrocardiograph signals in computer-aided Holter analysis. Computer methods and programs in Biomedicine 72, 179–196 (2003)
3. Moody, G., Mark, R.: QRS morphology representation and noise estimation using the Karhunen-Loeve transform. Comput. Cardiol. 16, 269–272 (1989)
4. Lagerholm, M., Peterson, C., Braccini, G., Edenbrandt, L., Sornmo, L.: Clustering ECG complexes using hermite functions and Self-organizing maps. IEEE Trans. Biomed. Eng. 47, 838–848 (2000)
5. Christov, I., Herrero, G.G., Krasteva, V., Jekova, I., Gotchev, A., Egiazarian, K.: Comparative study of morphological and time-frequency ECG descriptors for heartbeat classification. Medical Engineering & Physics 28, 876–887 (2006)
6. Tsipouras, M.G., Voglis, C., Lagaris, I.E., Fotiadis, D.I.: Cardiag Arrhytmia Classification Using Support Vector Machines. In: 3rd European Medical and Biological Engineering Conference (2005)
7. Bortolan, G., Jekova, I., Christov, I.: Comparison of Four Methods for Premature Ventricular Contraction and Normal Beat Clustering. Computers in Cardiology 32, 921–924 (2005)
8. Hu, Y.H., Palreddy, S., Tompkins, W.J.: A patient-adaptable ECG beat classifier using a mixture of experts approach. IEEE Trans. Biomed. Eng. 44, 891–900 (1997)
9. Chudáček, V., Lhotská, L., Stylios, C., Georgoulas, G.: Comparison of Methods for Premature Ventricular Beat Detection. In: ITAB Conference. IEEE, Piscataway (2006)
10. Goldberger, L., Amaral, L., Glass, L., Hausdorf, J.M., Ivanov, P.C., Mark, R.G., Mietus, J.E., Moody, G.B., Peng, C.K., Stanley, H.E.: PhysioBank, PhysioToolkit, and PhysioNet: Components of a New Research Resource for Complex Physiologic Signals. Circulation 101(23), e215–e220
11. Christov, I.: Real time electrocardiogram QRS detection using combined adaptive threshold. Biomed. Eng., `http://www.biomedical-engineering-online.com/content/3/1/28`
12. Haykin, S.: Neural Networks: A Comprehensive Foundation., 2nd edn. Prentice Hall, Englewood Cliffs (1999)
13. Bishop, C.M.: Neural Networks for Pattern Recognition. Oxford University Press, New York (1995)
14. Dunne, R.A.: A statistical Approach to Neural Networks for Pattern Recognition. John Wiley & Sons, New Jersey (2006)
15. Berthold, M., Hand, D.J.: Intelligent Data Analysis. Springer, Heidelberg (2003)
16. Simard, P.Y., Steinkraus, D., Platt, J.C.: Best practices for convolutional neural networks applied to visual document analysis. In: International Conference on Document Analysis and Recogntion (ICDAR), pp. 958–962 (2003)
17. Jekova, I., Bortolan, G., Christov, I.: Assessment and comparison of different methods for heartbeat classification. Medical Engineering & Physics 30(2), 248–257 (2008)
18. Tsipouras, M.G., Fotiadis, D.I., Sideris, D.: An Arrhythmia Classification System Based on the RR-Interval Signal. J. Artificial Intelligence in Medicine 33, 237–250 (2004)
19. Augustyniak, P.: The Use of Shape Factors for Heart Beats Classification in Holter Recordings. In: Conference on Computers in Medicine, pp. 47–52 (1997)
20. Chawla, N.V., Hall, L., Kegelmeyer, W.: SMOTE: Synthetic Minority Over-Sampling Technique. J. Artif. Intell. Res. 16, 3241–3357 (2002)

Mental Tasks Classification for a Noninvasive BCI Application

Alexandre Ormiga G. Barbosa[1], David Ronald A. Diaz[1], Marley Maria B.R. Vellasco[2], Marco Antonio Meggiolaro[1], and Ricardo Tanscheit[2]

[1] Department of Mechanical Engineering
Pontifical Catholic University of Rio de Janeiro - PUC-Rio
[2] Department of Electrical Engineering
Pontifical Catholic University of Rio de Janeiro - PUC-Rio
aormiga@yahoo.com.br, david251@aluno.puc-rio.br, marley@ele.puc-rio.br, meggi@puc-rio.br, ricardo@ele.puc-rio.br

Abstract. Mapping brain activity patterns in external actions has been studied in recent decades and is the base of a brain-computer interface. This type of interface is extremely useful for people with disabilities, where one can control robotic systems that assist, or even replace, non functional body members. Part of the studies in this area focuses on noninvasive interfaces, in order to broaden the interface usage to a larger number of users without surgical risks. Thus, the purpose of this study is to assess the performance of different pattern recognition methods on the classification of mental activities present in electroencephalograph signals. Three different approaches were evaluated: Multi Layer Perceptron neural networks; an ensemble of adaptive neuro-fuzzy inference systems; and a hierarchical hybrid neuro-fuzzy model.

Keywords: Brain Computer Interface, artificial neural network, neuro-fuzzy, hierarchical network.

1 Introduction

The development of interfaces between humans and machines has been an expanding field in the last decades, including several interfaces using voice, vision, haptics, electromyography (EMG) signals, electroencephalography (EEG) signals, as well as and combinations of these, as communication support [1].

Recent studies [2-3] have shown the possibility of online brainwaves analyses to derive information about the subject's mental state, which can then be mapped onto some external action such as selecting a letter from a virtual keyboard or moving robotics devices. Systems that utilize these brainwaves are called Brain Computer Interface (BCI) [4].

People who have severe motor disabilities, that are partially or totally paralyzed, can use BCI as an alternative communication and control channel that does not depend on the brain's normal output pathway - peripheral nerves and muscles. Hence, BCI enhances these persons' quality of life [5].

C. Alippi et al. (Eds.): ICANN 2009, Part II, LNCS 5769, pp. 495–504, 2009.

BCIs can be noninvasive or invasive. The latter faces substantial technical difficulties and entails significant clinical risks: they require that recording electrodes are implanted in the cortex and are functional for long periods. Most importantly, however, is the risk of infections and other damages to the brain [6]. On the other hand, non-invasive BCIs are based on the EEG analysis associated with various brain function aspects [7], offering, therefore, a more secure and accessible interface.

Pattern classification of brain activity is one of the important aspects in BCI systems [8]. Artificial Neural Networks have already been applied to the classification of brain activities, attaining better performance than traditional methods [9].

In [10], a Probabilistic Neural Network (PNN) [11-12] and a Multi-Layer Perceptron (MLP) neural network [12-13] have been used as classifiers to recognize five different mental activities in noninvasive BCIs. These models presented promising results, but due to the inherent complexity of the problem, more complex models are necessary to achieve more accurate classification rates. As in [10], no user-independence is evaluated in this paper.

Therefore, this paper presents the study of more efficient classifiers in order to increase the hit rate in pattern classification of mental activities. Three different models were developed: a MLP neural network, for comparison purposes, and two hybrid approaches, consisting of an ensemble of ANFIS (Adaptive Neuro-Fuzzy Inference Systems) [14-15] models and a hierarchical neuro-fuzzy classifier.

This paper is organized in four additional sections. Section 2 presents the real database used in this study. Section 3 describes in details the proposed classification models. Section 4 presents the results obtained with all three models and, finally, section 5 discusses the conclusions of this work.

2 Mental Activities Database

The mental activity database was obtained from [10], where an electroencephalograph, composed of ten electrodes placed on the user's scalp (according to International System 10-20 [16]), was implemented (Fig. 1). Usual classification of the main EEG rhythms is based on five frequency ranges [17], called: delta (0 to 4 Hz), theta (4 to 8 Hz), alpha (8 to 13 Hz), beta (13 to 30 Hz), and gamma (higher than 30 Hz). After an independent analysis of each frequency range, better results were obtained in delta band [10]. Therefore, in order to reduce the number of inputs for the neural network and hybrid models, as well as to allow a direct comparison between studies, only delta band was considered in this study. Using the knowledge of the brain activity specialization and electrodes positions (see Fig. 1), it is possible to discard six electrodes readings, reducing the relevant signals to four (C3, C4, P3 and P4) [10].

In order to capture useful information in the time and frequency domain, wavelet transform [18] was used to preprocess the EEG signals. The mean of the wavelet coefficients of the four relevant signals in the delta band were selected as inputs of the neural network and hybrid models.

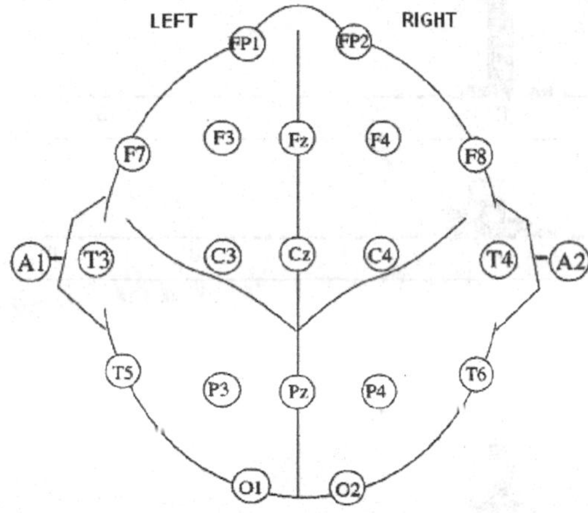

Fig. 1. Electrodes positions from the International System 10-20

The database was created by asking the user to carry out 100 trials for each of the five chosen mental activities: motor imagery of the forefinger movement to the right side (RM), motor imagery of the forefinger movement to the left side (LM), 3D rotation of a cube (CR), arithmetic operation of subtraction (AS), and the mental state of relax (MR) [10]. The produced data was divided into 70% for training, 15% for validation, and 15% for testing. Figure 2 presents the histogram analysis of wavelet coefficients averages obtained from the four selected electrodes. As can be noticed from Figure 2, the EEG signals contain outlier values that can be visually detected as the ones located far from the value with samples concentration.

To reduce outliers, two distinct data pre-processing were evaluated. In the first approach (see Fig. 3), outliers were replaced by neutral values (mean of the other values). In the second approach, signal values were also linear normalized, in addition to outliers' replacement (see Fig. 4).

3 Brain Activities Classification Models

3.1 MLP Neural Network

The first proposed classification model is a single MLP Neural Network, which was developed in Matlab Neural Network Toolbox for comparison reasons. The MLP is composed of four inputs, one hidden layer and five neurons in the output layer, one for each of the five chosen mental activities. The best MLP configuration in terms of number of neurons in the hidden layer and number of training epochs was obtained by evaluating the best performance in the validation set. Different neural networks were trained for maximum 5000 epochs, changing the number of neurons on the hidden layer from 2 to 15 and the learning rate from

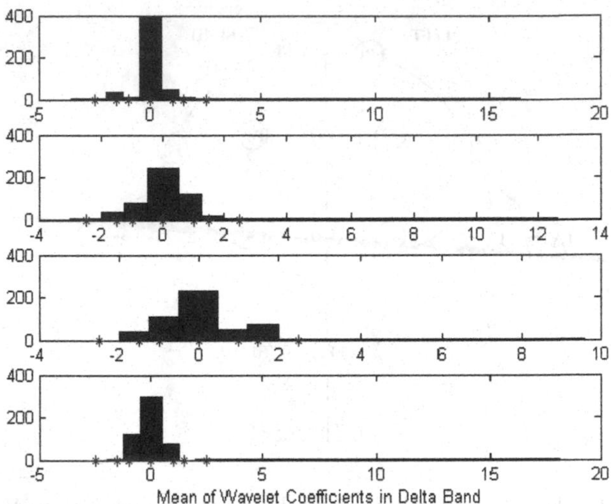

Fig. 2. Histogram of C3, C4, P3 and P4 respectively

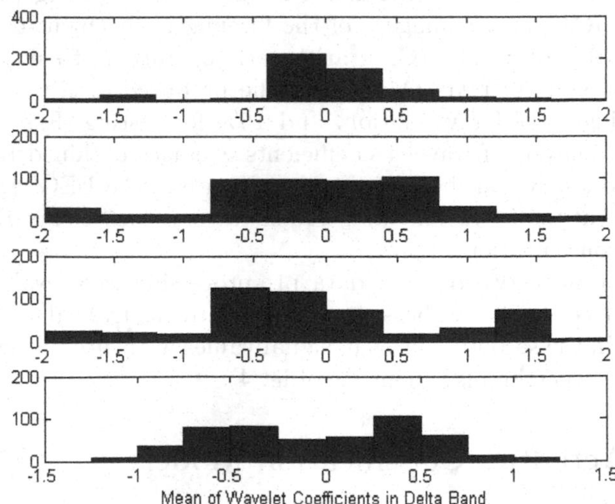

Fig. 3. Histogram with outliers' replacement of C3, C4, P3 and P4 respectively

0.2 to 0.8. The number of neurons in the hidden layer was chosen as the one that presented the lowest mean between minimum validation errors for all learning rates used. Similarly, the learning rate was chosen to present the lowest validation error for the selected number of neurons on the hidden layer.

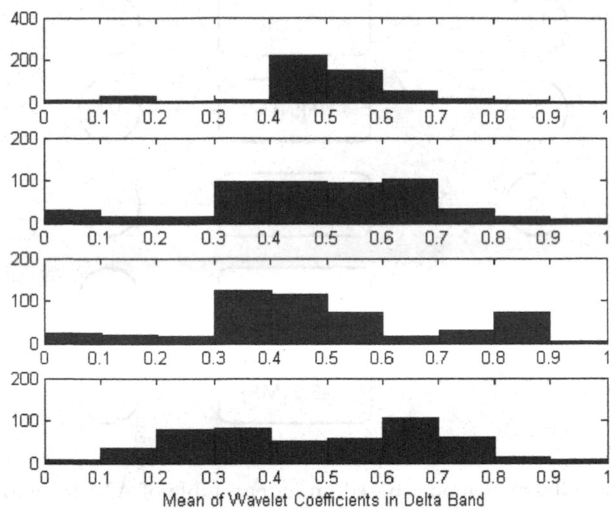

Fig. 4. Histogram with outliers' replacement and normalization of C3, C4, P3 and P4 respectively

Table 1. MLP Neural Network parameters

	Database		
	No Pre-Processing	Outlier Replacement	Outlier Replacement and Normalized
Neurons on Hidden Layer	10	5	8
Learning Rate	0.72	0.59	0.43
Momentum	0.9	0.9	0.9

This methodology was applied to define the neural network topology for each pre-processing applied to the database, resulting in different topologies. Table 1 presents the obtained topologies for each of the three different configurations.

3.2 Ensemble of ANFIS Models

Adaptive Neuro-Fuzzy Inference Systems (ANFIS) [14-15] were already proposed as good classifiers to pattern recognition for brain-computer interfaces [19]. The main advantage of this hybrid neuro-fuzzy system is the ability to provide linguistic rules that indicate the relation of the input variables and the output classification variable. Although ANFIS models are of Takagi-Sugeno type [20] (rules' consequents are singletons or a linear combination of the input variables), they are more interpretable than artificial neural networks.

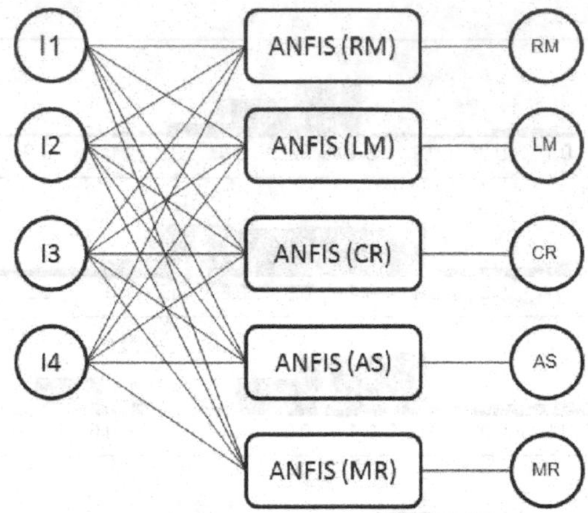

Fig. 5. Classifier model based on an ensemble of ANFIS models

In this study, an ensemble of five ANFIS classifiers is proposed (Fig. 5): each one specialized in one of the five possible mental tasks classification. The final ensemble classification is accomplished by applying the MAX operator among subsystems output value, that is, the final classification of the mental activity is indicated by the ANFIS model with the highest output value.

Each subsystem was trained with backpropagation algorithm in the Matlab Fuzzy Toolbox, with maximum training epochs specified by the validation set (early stopping process) to avoid overfitting.

Each ANFIS subsystem was trained with two and three fuzzy sets per input signal. The best generalization performance was obtained with two fuzzy sets, resulting in 16 fuzzy rules. Different shapes were also evaluated for the fuzzy sets (triangular and bell function), with the best performance attained with bell shape.

3.3 Hierarchical Hybrid Model

The third classification model was proposed after evaluating the classification performance of the ANFIS ensemble. By analyzing the resulting confusion matrix of the ANFIS ensemble (see results presented in Table 3 for the database with outlier replacement), it is possible to verify that the majority of missed classification is bettween "LM" and "CR" patterns.

Therefore, a hierarchical hybrid structure was modeled (Fig. 6), composed of four ANFIS classifiers, trained to recognize "RM", "LM *or* "CR", "AS" and "MR", and one MLP neural network to identify between "LM" and "CR" patterns when "LM *or* CR" has been pre-classified by its respective ANFIS subsystem.

The same four input signals are applied to all classifiers, and the final system response depends on the ANFIS classifiers. If the ANFIS subsystem trained to

Table 2. ANFIS classifier confusion matrix

	RM	LM	CR	AS	MR
RM	9	3	0	3	0
LM	0	7	8	0	0
CR	0	0	15	0	0
AS	0	2	0	13	0
MR	0	0	0	0	15

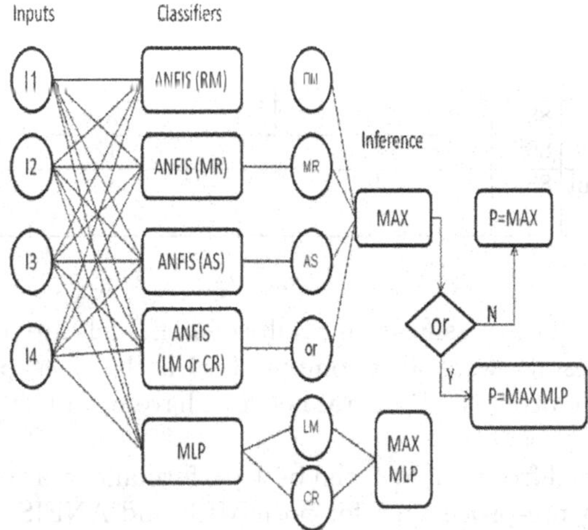

Fig. 6. Hierarchical model

identify LM or CR provides the highest output level among all subsystems, the hierarchical hybrid system response is given by the MLP network classification (MAX between LM and CR outputs). Otherwise, the final response is provided by ANFIS with the highest output value.

The methodology described in Section 3.1 used to define the MLP neural network topology was also applied to train the MLP subsystem in the hierarchical hybrid model. The ANFIS topology used in the hierarchical hybrid model is also the same applied in the ANFIS ensemble.

Table 3 presents the new confusion matrix obtained with the hierarchical hybrid model, using the same dataset provided in Table 2. As can be observed, the discrimination between "LM" and "CR" classes has improved considerably, maintaining the accuracy in the other classes.

4 Results

The three classification models described in the previous sections were evaluated in the testing datasets described in Section 2, that is, the dataset with no pre-processing, dataset with outlier replacement and dataset with normalization and

Table 3. Confusion matrix of hierarchical model

	RM	LM	CR	AS	MR
RM	9	1	3	2	0
LM	0	13	2	0	0
CR	0	0	15	0	0
AS	0	0	0	15	0
MR	0	0	0	0	15

Table 4. Classifiers hit rates

	MLP	ANFIS Ensemble	Hierarchical Hybrid Model	PNN[1]	MLP1
No Pre-Processing	83%	72%	X	83%	63%
Outlier Replacement	86%	78%	89%	X	X
Outlier Replacement and Normalization	85%	76%	X	X	X

outlier replacement. The classification results of all models are presented in Table 4. The obtained results were also compared with the ones presented in [10], where a PNN and another MLP neural network have been tested with the same database.

As can be observed from Table 4, the best performance was obtained with the outlier replacement pre-processing, for both MLP and ANFIS ensemble. Therefore, the hierarchical hybrid model was only evaluated with this dataset, improving accuracy to almost 90%. By using data pre-processing and the hierarchical structure, better results than the ones presented in [10] were obtained.

5 Conclusions

This paper presented the evaluation of three different classification models for the discrimination of mental activities for a noninvasive BCI (Brain Computer Interface) application.

The models presented in this paper were proposed to improve the classification performance presented in previous work [10], where a Probabilistic Neural Network and a Multi-Layer Perceptron were used as classifiers. By analyzing the difficulty in separating some of the brain activities, a hierarchical hybrid model was proposed, which led to a better overall classification accuracy, as well as better classification per class.

The ANFIS ensemble classifier proposed did not provide good results when compared to a simple MLP neural network. Better results can be obtained if neuro-fuzzy systems specifically developed for classification problems are used in the ensemble formation, such as the Inverted BSP System [21-22]

[1] Results obtained in [10].

According to other studies [23-24], alpha and beta bands contain the relevant information for mental activity analyses in a BCI application. So a new database including these bands will be created for future work to test the classification models proposed. Closed-loop feedback learning will be also implemented in order to improve signals quality, making pattern recognition easier. Considering a better scenario, future works should present better results.

References

1. Garcia, G.N.: Direct brain-computer communication through scalp recorded EEG signals. Doctor's thesis, Dept. of Electricity, Ecole Polytechnique Fédérale de Lausanne (2004)
2. Galán, F., Nuttin, M., Lew, E., Ferrez, P.W., Vanacker, G., Philips, J., del Millán, J.R.: A Brain-Actuated Wheelchair: Asynchronous and Non-Invasive Brain-Computer Interfaces for Continuous Control of Robots. Clinical Neurophysiology 119, 119–2159 (2008)
3. Babiloni, F., Cincotti, F., Lazzarini, L., Millan, J., Mourino, J., Varsta, M., Heikkonen, J., Bianchi, L., Marciani, M.G.: Linear Classification of Low-Resolution EEG Patterns Produced by Imagined Hand Movements. IEEE Transactions Rehabilitation Engineering 8(2), 186–188 (2000)
4. del Millán, J.R.: Handbook of Brain Theory and Neural Networks, 2nd edn. (2002)
5. Wolpaw, J.R., McFarland, D.J., Vaughan, T.M.: Brain Computer Interface Research at the Wadsworth Center. IEEE Transactions on Neural Systems and Rehabilitation Engineering 8, 222–226 (2000)
6. Wolpaw, J.R., McFarland, D.J.: Control of a Two-dimensional Movement Signal by a Noninvasive Brain-Computer Interface in Humans. Proc. Nat'l Academy of Sciences 101(51), 17849–17854 (2004)
7. del Millán, J.R., Mouriño, J., Franzé, M., Cincotti, F., Varsta, M., Heikkonen, J., Babiloni, F.: A local neural classifier for the recognition of EEG patterns associated to mental tasks. In: IEEE Trans. on Neural Networks, vol. 11, pp. 678–686 (2002)
8. Phothisonothai, M., Nakagawa, M.: EEG-Based Classification of New Imagery Tasks Using Three-Layer Feedforward Neural Network Classifier for Brain–Computer Interface. J. Phys. Soc. Jpn. 75 (2006)
9. Phothisonothai, M., Nakagawa, M.: EEG-Based Classification of Motor Imagery Tasks Using Fractal Dimension and Neural Network for Brain-Computer Interface. IEICE Transactions on Information and Systems E91-D(1), 44–53 (2008)
10. Diaz, D.R.A.: Activation of a Mobile Robot Through a Brain Computer Interface Based on Electroencephalographic Signal Processing. M.Sc. Thesis, Pontifical Catholic University of Rio de Janeiro (2009)
11. Wasserman, P.D.: Advanced Methods in Neural Computing. Van Nostrand Reinhold (1993)
12. Haykin, S.: Neural Networks - A Comprehensive Foundation. McMillan College Publishing Co. (1999)
13. Reed, R.D., Mark II, R.J.: Neural Smithing – Supervised Learning in Feedforward Artificial Neural Networks. Bradford Book, The MIT Press (1999)
14. Jang, J.-S.R., Sun, C.-T., Mizutani, E.: Neuro-Fuzzy and Soft Computing: A Computational Approach to Learning and Machine Intelligence. Prentice-Hall, Englewood Cliffs (1997)

15. Jang, J.-S.R.: ANFIS: Adaptive-Network-Based Fuzzy Inference Systems. IEEE Trans. On Systems, Man and Cybernetics 23(3), 665–685 (1993)
16. Harner, P.F., Sannit, T.: A review of the International Ten-Twenty system of electrode placement. Grass Instruments Company (1974)
17. Garcia, G.N., Hoffmann, U., Ebrahimi, T.: Direct Brain-Computer Communication through EEG Signals. IEEE EMBS Book Series on Neural Engineering (2004)
18. Ogden Tood, R.: Essential Wavelets for Statistical and Data Analysis (1997)
19. Ben Dayan Rubin, D.D., et al.: An adaptative neuro-fuzzy method (ANFIS) for estimating single-trial movement-related potentials (2004)
20. Takagi, T., Sugeno, S.: Fuzzy Identification of Systems and its Application to Modelling and Control. IEEE Trans. on Systems, Man and Cybernetics 15, 116–132 (1985)
21. Gonçalves, L., Vellasco, M., Pacheco, M., Souza, F.: Inverted Hierarchical Neuro-Fuzzy BSP System: A Novel Neuro-Fuzzy Model for Pattern Classification and Rule Extraction in Databases. IEEE Trans. on S.M.C., Part C 36(2), 236–248 (2006)
22. Vellasco, M., Pacheco, M., Figueiredo, K., Souza, F.: Hierarchical Neuro-Fuzzy Systems - Part I. Encyclopedia of Artificial Intelligence. In: Rabuñal, J., Dorado, J., Pazos, A. (eds.) Information Science Reference (2008)
23. Millán, J.d.R., Renkens, F., Mouriño, J., Gerstner, W.: Noninvasive: Brain-Actuated Control of a Mobile Robot by Human EEG. IEEE Transactions on Biomedical Engineering 51(6), 1026–1033 (2004)
24. Millán, J.d.R., Mouriño, J.: Asynchronous BCI and Local Neural Classifiers: An Overview of the Adaptive Brain Interface Project. IEEE Transactions on Neural Systems and Rehabilitation Engineering 11(2), 159–161 (2003)

Municipal Creditworthiness Modelling by Radial Basis Function Neural Networks and Sensitive Analysis of Their Input Parameters

Vladimir Olej and Petr Hajek

Institute of System Engineering and Informatics
Faculty of Economics and Administration
University of Pardubice
Studentska 84, 53210 Pardubice
Czech Rcpublic
vladimir.olej@upce.cz, petr.hajek@upce.cz

Abstract. The paper presents concept of vector parameters characterizing creditworthiness of municipalities and its modelling possibilities. Based on designed model and structures of radial basic functions neural networks, the modelling is realized with the aim to classify municipalities into classes. Further, the article includes sensitivity analysis of individual parameter vector components. Sensitivity analysis represents exploring contributions of individual vector components to classification quality.

Keywords: Municipal creditworthiness, radial basis functions neural network, classification, multinomial regression, sensitive analysis.

1 Introduction

Municipal creditworthiness is an independent expert evaluation based on complex analysis of all known municipal creditworthiness parameters. Municipal creditworthiness evaluation is currently being realized by methods combining mathematical-statistical methods and expert opinion [1]. However, they are considered to be rather subjective and inaccurate [1]. Besides these methods for municipality creditworthiness evaluation, there were also models based on computational intelligence designed. For example, hierarchical structures of fuzzy inference systems [2], unsupervised (supervised) methods [3], [4] and neuro-fuzzy systems [5] were designed for municipal creditworthiness evaluation. The use of the mentioned methods has proved to be problematic in some respects. Expert knowledge is required for rule base design in hierarchical structures of fuzzy inference systems [2]. Similarly, clusters have to be labelled by expert and, moreover, generalization ability of unsupervised methods is limited [3]. Large rule base has been obtained by neuro-fuzzy systems leading to difficult interpretation of the models. On the other hand, low number of rules implies low classification accuracy in neuro-fuzzy systems [5]. The output of thc methods is represented by an assignment of the i-th object $o_i \in O$, $O=\{o_1,o_2, \ldots, o_i, \ldots, o_n\}$ to the j-th class

C. Alippi et al. (Eds.): ICANN 2009, Part II, LNCS 5769, pp. 505–514, 2009.

$\omega_{i,j} \in \Omega$, $\Omega = \{\omega_{1,j}, \omega_{2,j}, \ldots, \omega_{i,j}, \ldots, \omega_{n,j}\}$ [2], [3], [4] and [5]. Based on the facts mentioned we can state, that the methods capable of processing and learning the expert knowledge, enabling their user to generalize and properly interpret, have proved to be most suitable for municipal creditworthiness modelling. The use of the outputs of unsupervised methods along with the classification capabilities of Radial basis function (RBF) neural networks [6] seems to realize the presented needs.

Radial basis function neural networks were independently proposed by many researchers [7], [8], [9], [10], [11], and are a popular alternative to the feed-forward neural networks (FFNNs) [10]. They are also good at modelling nonlinear data and can be trained in one stage rather than using an iterative process as in FFNNs, and also learn the given application quickly. The way in which the RBF neural networks are used for data modelling is different when realizing classification process and approximating time series. In the first case, the inputs of the RBF neural networks are represented by feature vectors, while each output corresponds to a class. In the second case, RBF neural network inputs are represented by data samples and certain time-lags, while the RBF neural network has only one output representing a signal value.

In the paper there is a concept of parameters vector $\mathbf{x} = (x_1, x_2, \ldots, x_k, \ldots, x_m)$ for municipal creditworthiness evaluation [2], [3], [4] and [5]. Further, based on data analysis, data representation method via data matrix $\mathbf{P}$ is designed and formalized. Then it is possible to design a model for modelling municipal creditworthiness evaluation with designed RBF structure. That consists of data preprocessing, clustering by Kohonen's self-organizing feature maps (KSOFMs) and class $\omega_{i,j} \in \Omega$ labelling based on expert opinion and municipal creditworthiness classification $o_i \in O$ to classes $\omega_{i,j} \in \Omega$. By means of sensitivity analysis, we survey contribution of individual parameters vector $\mathbf{x} = (x_1, x_2, \ldots, x_k, \ldots, x_m)$ components to classification quality of the i-th object (municipalities) $o_i \in O$ to the j-th class $\omega_{i,j} \in \Omega$. The final part of the paper includes the analysis of the results and comparison to other classification methods.

2 Problem of Municipal Creditworthiness Evaluation

In [2], [3], [4] and [5] common categories of parameters there are mentioned (economic, debt, financial and administrative categories). The economic, debt and financial parameters are pivotal. Economic parameters (x_1, x_2, x_3, x_4) affect long-term credit risk. The municipalities with more diversified economy and more favourable social and economic conditions are better prepared for the economic recession. Debt parameters (x_5, x_6, x_7) include the size and structure of the debt. Financial parameters $(x_8, x_9, x_{10}, x_{11}, x_{12})$ inform about the budget implementation. Their values are extracted from the municipality budget. The design of parameters vector $\mathbf{x} = (x_1, x_2, \ldots, x_k, \ldots, x_m)$, m=12, based on previous correlation analysis and recommendations of notable experts, can be realized as presented in Table 1. The parameters x_3 and x_4 are defined in the r-th year and parameters x_5 to x_{12} as the average value of the r-th and (r-1)th years.

Table 1. Municipal creditworthiness parameters design

	Parameters
Economic	$x_1 = PO_r$, PO_r is population in the r-th year. Higher value of x_1 entails especially higher municipal tax revenues.
	$x_2 = PO_r/PO_{r-s}$, PO_{r-s} is population in the year r-s, and s is the selected time period. Economic growth of the municipality leads to the growing number of its inhabitants.
	$x_3 = U$, U is the unemployment rate in a municipality.
	$x_4 = \sum_{i=1}^{k} (EPO_i/EIN)^2$, EPO_i is the employed population of the municipality in the i-th economic sector, i=1,2, …,k, EIN is the total number of employed inhabitants, k is the number of the economic sector.
Debt	$x_5 = DS/PR$, $x_5 \in <0,1>$, DS is debt service, PR are periodical revenues. It measures the ability of the municipality to pay off the DS from regular budget revenues.
	$x_6 = TD/PO$, TD is a total debt.
	$x_7 = SD/TD$, $x_7 \in <0,1>$, SD is short-term debt.
Financial	$x_8 = PR/CE$, $x_8 \in R^+$, CE are current expenditures. If it is constantly greater than 1, the municipality implements the budget well.
	$x_9 = OR/TR$, $x_9 \in <0,1>$, OR are own revenues, TR are total revenues.
	$x_{10} = CAE/TE$, $x_{10} \in <0,1>$, CAE are capital expenditures, TE are total expenditures. It indicates capital activity of the municipality.
	$x_{11} = CAR/TR$, $x_{11} \in <0,1>$, CAR are capital revenues.
	$x_{12} = LA/PO$, [Czech Crowns], LA is the size of the municipal liquid assets. Municipal assets are often used as bank's credit collateral.

Several Czech municipalities $o_i \in O$ have the class $\omega_{i,j} \in \Omega$ assigned by specialized agencies. Moreover, the municipalities in micro-region Pardubice, the Czech Republic, have no class $\omega_{i,j} \in \Omega$ assigned. However, the descriptions of classes $\omega_{i,j} \in \Omega$ can be designed based on expert opinion (Table 2). Then the municipalities $o_i \in O$ can be labelled with classes $\omega_{i,j} \in \Omega$ following this description.

Based on the presented facts, the data matrix **P** can be designed as follows

$$\mathbf{P} = \begin{array}{c|cccccc|c} & x_1 & \dots & x_k & \dots & x_m & \\ \hline o_1 & x_{1,1} & \dots & x_{1,k} & \dots & x_{1,m} & \omega_{1,j} \\ \dots & \dots & \dots & \dots & \dots & \dots & \dots \\ o_i & x_{i,1} & \dots & x_{i,k} & \dots & x_{i,m} & \omega_{i,j} \\ \dots & \dots & \dots & \dots & \dots & \dots & \dots \\ o_n & x_{n,1} & \dots & x_{n,k} & \dots & x_{n,m} & \omega_{n,j} \end{array},$$

where $o_i \in O$, $O=\{o_1,o_2, \dots ,o_i, \dots ,o_n\}$ are objects (municipalities), x_k is the k-th parameter, $x_{i,k}$ is the value of the parameter x_k for the i-th object $o_i \in O$, $\omega_{i,j}$ is the j-th class assigned to the i-th object $o_i \in O$, $\mathbf{p}_i=(x_{i,1},x_{i,2}, \dots ,x_{i,k}, \dots ,x_{i,m})$ is the i-th pattern, and $\mathbf{x}=(x_1,x_2, \dots ,x_k, \dots ,x_m)$ is the parameters vector.

Table 2. Descriptions of classes $\omega_{i,j} \in \Omega$

	Description
$\omega_{i,1}$	High ability of a municipality to meet its financial obligation. Very favourable economic conditions, low debt and excellent budget implementation.
$\omega_{i,2}$	Very good ability of a municipality to meet its financial obligation.
$\omega_{i,3}$	Good ability of a municipality to meet its financial obligation.
$\omega_{i,4}$	A municipality with stable economy, medium debt and good budget implementation.
$\omega_{i,5}$	Municipality meets its financial obligation only under favourable economic conditions.
$\omega_{i,6}$	A municipality meets its financial obligations with difficulty, the municipality is highly indebted.
$\omega_{i,7}$	Inability of a municipality to meet its financial obligation.

3 Basic Notions of RBF Neural Networks

The term RBF neural network [6] means any kind of FFNN that uses RBF as an activation function. Using RBF neural network for classification is suitable, since in most cases a specific group of input vectors $\mathbf{p}_i$ belongs to one of classes $\omega_{i,j} \in \Omega$, which are sought by RBF neural network. It is, therefore, possible to pick a group representative and consider its surroundings as the set within output of required class $\omega_{i,j} \in \Omega$. Moreover, RBF neural networks defined in this fashion are, in term of approximation natural, because approximation is realized by functions, which influence the final function only in the surroundings' center c_i of the RBF neuron and not in the whole function range. The j-th output $f(\mathbf{x}, H, \mathbf{w})$ of RBF neural network can be defined this way

$$f(\mathbf{x}, H, \mathbf{w}) = \sum_{i=1}^{q} w_{j,i} \times h_i(\mathbf{x}), \tag{1}$$

where $H=\{h_1(\mathbf{x}),h_2(\mathbf{x}), \ldots ,h_i(\mathbf{x}), \ldots ,h_q(\mathbf{x})\}$ is a set of activation functions of RBF neurons (RBF functions) in hidden layer and $w_{j,i}$ are synapse weights. Each of m components of vector $\mathbf{x}=(x_1,x_2, \ldots ,x_k, \ldots ,x_m)$ is an input value for q activation functions $h_i(\mathbf{x})$ of RBF neurons. The j-th output $f(\mathbf{x}, H, \mathbf{w})$ of RBF neural network represents linear combination of outputs from q RBF neurons and corresponding synapse weights $w_{j,i}$.

Input layer of RBF neural network provides loading of individual input samples $\mathbf{p}_i=(x_{i,1},x_{i,2}, \ldots ,x_{i,k}, \ldots ,x_{i,m})$. Synapse weights $\mathbf{W}(m,q)$ between input and hidden layer are not used by RBF neural networks. This neural network includes exactly one hidden layer. Reason to hidden layer number limitation is the fact, that each from m input vector values $\mathbf{x}=(x_1,x_2, \ldots ,x_k, \ldots ,x_m)$ is used as an activation function parameter $H=\{h_1(\mathbf{x}),h_2(\mathbf{x}), \ldots ,h_i(\mathbf{x}), \ldots ,h_q(\mathbf{x})\}$ of RBF neurons, where q is a number of neurons in the hidden layer. Activation function $h_i(\mathbf{x})$ of RBF neurons in the hidden layer is a special class of mathematical functions, whose main characteristics is monotonous rising or falling with increasing distance from center c_i of activation function $h_i(\mathbf{x})$ of RBF neuron. Neurons of hidden layer can use as activation function $h_i(\mathbf{x})$ of RBF neurons for example Gaussian and rotary Gaussian activation function (one- and two-dimensional RBF), multisquare and inverse multisquare activation function, Cauchy's, etc. For classification problem, Gaussian activation function $h_i(\mathbf{x})$ is preferred. It is possible to present it like this

$$f(\mathbf{x}, C, R) = \sum_{i=1}^{q} \exp(-\frac{\|\mathbf{x} - c_i\|^2}{r_i}), \tag{2}$$

where $\mathbf{x}=(x_1,x_2, \ldots ,x_k, \ldots , x_m)$ represents input vector, $C=(c_1,c_2, \ldots ,c_i, \ldots ,c_q)$ are centres of activation functions $h_i(\mathbf{x})$, and $R=(r_1,r_2, \ldots ,r_i, \ldots ,r_q)$ are radiuses of activation functions $h_i(\mathbf{x})$.

Neurons of output layer represent only weighted sum of all inputs coming from the hidden layer. Activation function of neurons in the output layer can be linear, eventually unit jump in order to convert the output to binary form. In RBF neural network learning process [12], [13], [14] it is required to set a number of centres c_i of activation function $h_i(\mathbf{x})$ of RBF neurons and to find the most suitable positions for RBF centres c_i. Other parameters are radiuses r_i of centres c_i, gradient of activation functions $h_i(\mathbf{x})$ of RBFs and synapse weights $\mathbf{W}(q,n)$ setup between hidden and output layer. Design of appropriate number of RBF neurons in hidden layer is presented in [8] and [9]. Possibilities of centres c_i recognition are mentioned in [12] as a random choice. This easiest method uses fixed gradient of activation functions $h_i(\mathbf{x})$ of RBF neurons. Their position is chosen randomly from a set of training data. This approach presumes that randomly picked centres c_i will sufficiently represent data entering the RBF neural network. This method is suitable only for small sets of input data. If used on larger sets, it often means quick and needless increase in RBF neuron numbers in hidden layer and therefore unjustified complexicity of neural network. The second approach to locating centres c_i of activation functions $h_i(\mathbf{x})$ of RBF neurons can be realized by K-means algorithm [15].

4 Modelling and Analysis of the Results

Municipal creditworthiness modelling represents a classification problem. It is possible to be modelled by supervised methods (if classes $\omega_{i,j} \in \Omega$ of the objects are known) or by unsupervised methods (if classes $\omega_{i,j} \in \Omega$ are not known). Data pre-processing is carried out by means of data standardization. Thereby, the dependency on units is eliminated. The KSOFMs assign municipalities to clusters [4], [5]. Subsequently, the clusters are labeled with classes $\omega_{i,j} \in \Omega$ based on expert opinion. The outputs from the KSOFM are used as the inputs of the RBF neural networks, or other neural network structures and statistic methods which realize advantages of supervised methods. Finally, sensitivity analysis for individual vector $\mathbf{x}=(x_1,x_2, \ldots ,x_k, \ldots ,x_m)$ parameters is carried out, which represents evaluation of contributions of individual components of vector parameters to classification quality. Based on presented facts, the model is designed

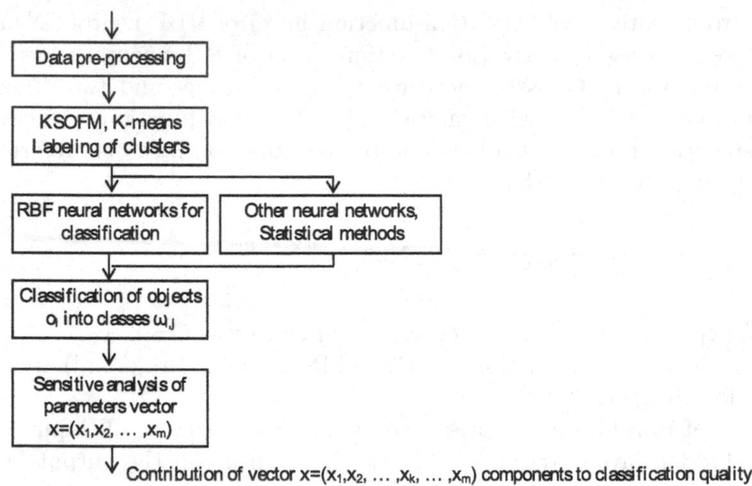

Fig. 1. Model for classification of municipalities $o_i \in O$ into classes $\omega_{i,j} \in \Omega$

for the classification of municipalities $o_i \in O$ into classes $\omega_{i,j} \in \Omega$, Fig. 1. The frequencies f of municipalities $o_i \in O$ in classes $\omega_{i,j} \in \Omega$ for the RBF neural network are presented in Fig. 2. From Table 2 and Fig. 2 results that in term of classes $\omega_{i,j} \in \Omega$ best objects (municipalities) are placed to class $\omega_{i,1}$, worst to class $\omega_{i,7}$. As the data matrix **P** includes $o_i \in O$, $O=\{o_1,o_2, \ldots ,o_i, \ldots ,o_n\}$, n=452, 10-fold cross-validation was employed for testing the model. The model realizes an assignment of the i-th object $o_i \in O$ to the j-th class $\omega_{i,j} \in \Omega$, $\Omega=\{\omega_{1,j},\omega_{2,j}, \ldots ,\omega_{i,j}, \ldots ,\omega_{n,j}\}$ so that classes $\omega_{i,2},\omega_{i,3}$, and $\omega_{i,4}$ have the highest percent occurrence. This means that objects (municipalities), which are average in terms of economic, debt and financial parameters, prevail. Further, experiments with different RBF neural networks' structures show, that, with increasing number of q neurons in the hidden layer of RBF neural network the classification accuracy value ξ [%] rises to value q=72, then classification accuracy ξ [%] decreases. Centers c_i of activation functions of RBF neurons are found by K-means algorithm, where radius value r_i of activation function $h_i(\mathbf{x})$ is r_i=1.6. Results of experiments are shown in Fig. 3 and Fig. 4. Classification accuracy ξ [%] increases with rising number of cycles up to value pc=350, then it stays without change.

In Table 3, there is a comparison of the classification accuracy ξ [%] on the testing set to other designed and analyzed structures of neural networks and representatives of statistical models. Concretely, we used a RBF neural network, Learning Vector Quantization (LVQ) neural networks [16], FFNN [10], an Adaptive Resonance Theory and Mapfield (ARTMAP) [17], a Linear neural network (LNN) [10], a Probabilistic neural network (PNN) [18], Support Vector Machines (SVM) [19], K-Nearest Neighbour (KNN) [20], and Multinomial Logistic Regression Model (MLRM) [20]. The RBF neural network represents excellent results with the maximum classification accuracy ξ_{max}=94.69[%], the average classification accuracy ξ_a=89.93[%], and the standard deviation SD=2.88[%].

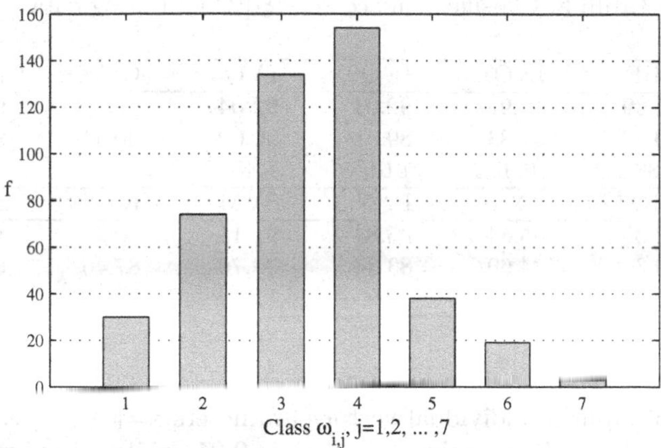

Fig. 2. Frequencies f of municipalities in classes $\omega_{i,j} \in \Omega$ for the RBF neural network

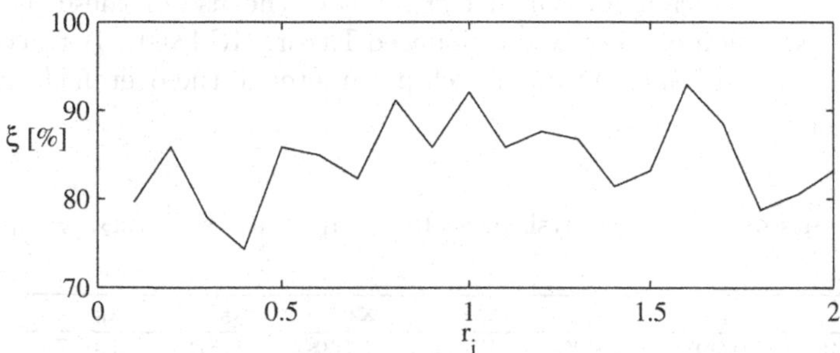

Fig. 3. Dependency of classification accuracy ξ [%] on r_i values

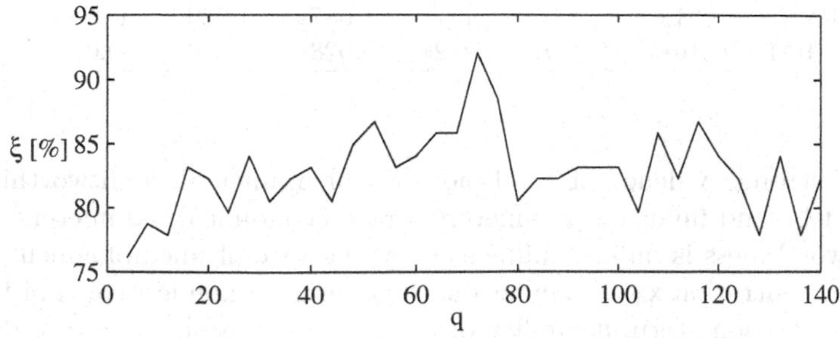

Fig. 4. Dependency of classification accuracy ξ [%] on the number of q

The maximum classification accuracy ξ_{max}=94.69[%] was reached based on a number of designed RBF neural network structures.

Table 3. Classification accuracy ξ [%] on testing data

	RBF	LVQ1	LVQ2	LVQ3	OLVQ1	FFNN
ξ_{max}[%]	94.69	92.92	92.04	92.04	92.92	92.04
ξ_a[%]	89.93	91.33	89.91	90.09	90.44	90.56
SD[%]	2.88	0.97	1.61	1.45	1.45	1.33
	ARTMAP	LNN	PNN	SVM	KNN	MLRM
ξ_{max}[%]	93.36	85.84	85.84	91.15	90.27	86.73
ξ_a[%]	90.34	84.60	83.34	89.76	87.46	81.42
SD[%]	3.81	0.79	1.89	1.83	3.38	5.31

The effect of input of individual vector parameters $\mathbf{x}=(x_1,x_2, \ldots ,x_k, \ldots ,x_m)$ is tested for the MLRM on significance level p=0.05, while using Error/Baseline (E/B) for neural networks. The results of the modelling are presented in Table 4. Statistically significant parameters of the vector $\mathbf{x}=(x_1,x_2, \ldots ,x_k, \ldots ,x_m)$ are marked with an asterisk. For some structures from Table 3, those input parameters are significant for which E/B>1 (i.e. the use of these parameters leads to the reduction of Root Mean Squared Error, (RMSE)). Beta coefficient evaluates the relative contribution of each parameter to the overall classification of the class $\omega_{i,j}\in\Omega$.

Table 4. Results of sensitive analysis of vector components $\mathbf{x}=(x_1,x_2, \ldots ,x_k, \ldots ,x_m)$

	x_1	x_2	x_3	x_4	x_5	x_6
RBF	0.997	1.000	1.033*	1.038*	1.572*	1.077*
FFNN	0.990	1.030*	1.146*	1.162*	1.582*	1.388*
MLRM	-0.016	0.046	0.147*	0.155*	0.350*	0.417*
	x_7	x_8	x_9	x_{10}	x_{11}	x_{12}
RBF	1.030*	1.120*	1.134*	1.161*	1.022*	1.005*
FFNN	1.021*	1.543*	2.181*	1.687*	1.029*	1.078*
MLRM	0.210*	-0.389*	0.272*	0.028	0.010	-0.018

There is strong evidence of a relationship of municipal creditworthiness to economic, debt and financial parameters. From economical parameters, municipal creditworthiness is mainly influenced by the rate of unemployment x_3 and economy concentration x_4 as they evaluate general economic wealth of the municipality, and a long-term flexibility of the municipal economy. From debt parameters, it is generally debt service indicator x_5 measuring the ability of the municipality to pay off the debt service DS from regular budget revenues. Financial parameters reporting on the quality of the budget implementation x_8, fiscal autonomy x_9 and capital activity of the municipality x_{10} have also been shown as key parameters to measuring creditworthiness of municipalities.

5 Conclusion

The paper presents the problem of municipal creditworthiness evaluation. The model design realizes municipal creditworthiness evaluation. The previous analysis of unsupervised methods [3] (KSOFM, ART-type neural networks, cluster analysis and fuzzy cluster analysis) showed that the KSOFM is the most suitable one for municipal creditworthiness modelling. The classes $\omega_{i,j} \in \Omega$ obtained based on the labelling the outputs of the KSOFM are used as the inputs of the RBF neural networks. The RBF neural networks structures were designed and studied for the classification of municipalities $o_i \in O$ into classes $\omega_{i,j} \in \Omega$ due to its high maximum classification accuracy ξ_{max} [%] and average classification accuracy ξ_u[%] with a low standard deviation SD[%]. In context of classification accuracy ξ[%], dependencies of classification accuracy ξ[%] on number of neurons q and radius value r_i of activation function $h_i(\mathbf{x})$ are studied. The results of the designed model for classification of municipalities $o_i \in O$ into classes $\omega_{i,j} \in \Omega$ show the possibility of evaluating municipal creditworthiness of the given municipalities in years to come. Based on sensitivity analysis of individual vector $\mathbf{x}=(x_1,x_2,\ldots,x_k,\ldots,x_m)$ components, contributions to classification quality of i-th object (municipalities) $o_i \in O$ to the j-th class $\omega_{i,j} \in \Omega$ are studied in terms of economic, debt and financial parameters. Further, the model presents an easier conception of the municipal creditworthiness for the public administration managers. Classification by the RBF neural networks was carried out in program environment SPSS Clementine 10.1.

Acknowledgements

The work is in part supported by the National Science Foundation of the Czech Republic, Grant No. 402/09/P090 with title Modelling of Municipal Finance by Computational Intelligence Methods and Grant No. 402/08/0849 with title Model of Sustainable Regional Development Management.

References

1. Loviscek, L.A., Crowley, F.D.: Municipal Bond Ratings and Municipal Debt Management. Marcel Dekker, New York (2003)
2. Olej, V., Hajek, P.: Hierarchical Structure of Fuzzy Inference Systems Design for Municipal Creditworthiness Modelling. WSEAS Transactions on Systems and Control 2, 162–169 (2007)
3. Olej, V., Hajek, P.: Modelling of Municipal Rating by Unsupervised Methods. WSEAS Transactions on Systems 7, 1679–1686 (2006)
4. Hájek, P., Olej, V.: Municipal creditworthiness modelling by kohonen's self-organizing feature maps and LVQ neural networks. In: Rutkowski, L., Tadeusiewicz, R., Zadeh, L.A., Zurada, J.M. (eds.) ICAISC 2008. LNCS (LNAI), vol. 5097, pp. 52–61. Springer, Heidelberg (2008)

5. Hajek, P., Olej, V.: Municipal creditworthiness modelling by kohonen's self-organizing feature maps and fuzzy logic neural networks. In: Kůrková, V., Neruda, R., Koutník, J. (eds.) ICANN 2008, Part I. LNCS, vol. 5163, pp. 533–542. Springer, Heidelberg (2008)
6. Broomhead, D.S., Lowe, D.: Multivariate Functional Interpolation and Adaptive Networks. Complex Systems 2, 321–355 (1988)
7. Moody, J., Darken, C.J.: Fast Learning in Networks of Locally Tuned Processing Units. Neural Computing 1, 281–294 (1989)
8. Poggio, T., Girosi, F.: Regularization Algorithms for Learning that are Equivalent to Multilayer Networks. Science 247, 978–982 (1990)
9. Niyogi, P., Girosi, F.: On the Relationship between Generalization Error, Hypothesis Complexity, and Sample Complexity for Radial Basis Functions. Massachusetts Institute of Technology Artificial Intelligence Laboratory, Massachusetts (1994)
10. Haykin, S.: Neural Networks: A Comprehensive Foundation. Prentice-Hall Inc., Englewood Cliffs (1999)
11. Poggio, T., Girosi, F.: Networks for Approximation and Learning. Proc. of the IEEE 78, 1481–1497 (1990)
12. Park, J., Sandberg, I.W.: Universal Approximation using Radial Basis Function Network. Neural Computing 3, 246–257 (1991)
13. Wettschereck, D., Dietterich, T.: Improving the Performance of Radial Basis Function Networks by Learning Center Locations. In: Advances in Neural Information Processing Systems, vol. 4, pp. 1133–1140. Morgan Kaufman Publishers, San Francisco (1992)
14. Orr, M.J.: Regularisation in the Selection of Radial Basis Function Center. Neural Computing 7, 606–623 (1995)
15. Tou, J.T., Gonzales, R.C.: Pattern Recognition Principles. Addition-Walley Publishing Comp., Massachusetts (1974)
16. Kohonen, T.: Self-organizing Maps. Springer, New York (2001)
17. Carpenter, G.A., Grossberg, S., Reynolds, J.H.: ARTMAP: Supervised Real-time Learning and Classification of Non-stationary Data by a Self-organizing Neural Network. Neural Networks 5, 565–588 (1991)
18. Speckt, D.F.: Probabilistic Neural Networks. Neural Networks 1, 109–118 (1990)
19. Cristianini, N., Shawe-Taylor, J.: Introduction to Support Vector Machines and other Kernel-based Learning Methods. Cambridge University Press, Cambridge (2000)
20. Bishop, C.M.: Pattern Recognition and Machine Learning. Springer, Cambridge (2006)

A Comparison of Three Methods with Implicit Features for Automatic Identification of P300s in a BCI

Luigi Sportiello, Bernardo Dal Seno, and Matteo Matteucci

Politecnico di Milano, Department of Electronics and Information, IIT-Unit,
Piazza Leonardo da Vinci 32, 20133 Milano, Italy

Abstract. When using a pattern recognition technique to classify signals, a common practice is to define a set of features to be extracted and, possibly after feature selection/projection, to use a learning machine in the resulting feature space for classification. However it is not always easy to devise the "right" set of features for a given problem, and the resulting classifier might turn out to be suboptimal because of this, especially in presence of noise or incomplete knowledge of the phenomenon. In this paper we present an off-line comparison of three methods (genetic algorithm, time-delay neural network, support vector machines) that leverage different ideas to handle features; we apply them to the recognition of the P300 potential in an EEG-based brain-computer interface. They all performed good, with the genetic algorithm being slightly better.

Keywords: Brain-Computer Interface, P300 Potential, Time-Delay Neural Networks, Genetic Algorithms, Support Vector Machines.

1 Introduction

A brain-computer interface (BCI) is an interface that does not entail muscle movements, but it bypasses any muscle or nerve mediation to connect a computer directly with the brain through the signals generated by the brain activity. Among the different kinds of brain activity that can be used in a BCI, the *P300* phenomenon has been known [1] and investigated for many years. It is an event-related potential (ERP), visible in an EEG recording as a positive peak at approximately 300 ms from the event. It follows unexpected, rare, or particularly informative stimuli, and it is stronger in the parietal area. The shape of the P300 depends on the characteristics of the stimuli and their presentation.

The P300 has been widely used for BCIs, with many variations, but in all cases the paradigm is the same: the BCI system presents the user with some choices, one at a time; when it detects a P300 potential the associated choice is selected. The user is normally asked to count the number of times the choice of interest is presented, so as to remain concentrated on the task (although the counting is not mandatory). As the P300 is an innate response, it does not require training on part of the user, but algorithms must adapt to the particular shape of each user's P300 in order to detect it. Detecting a P300 in a single trial is very difficult

C. Alippi et al. (Eds.): ICANN 2009, Part II, LNCS 5769, pp. 515–524, 2009.

and, therefore, repeated stimuli are normally used to facilitate the detection of the correct stimulus. The number of repetitions can be predetermined for each user to get the best trade-off between speed and accuracy.

In [2], Donchin and colleagues presented the first P300-based BCI, called also P300 speller, which permits to spell words. A 6×6 grid of letters and symbols is presented to the user, and entire columns or rows are flashed one after the other in random order (see Fig. 1 for an example). When the column/row containing the desired letter is flashed, a P300 is elicited. Each one of the 6 rows and 6 columns is flashed exactly once in the first 12 stimulations; then another round of 12 stimulations is repeated, with flashing of rows and columns done in a new random order, and this procedure is repeated for a predefined number of times for each letter. In [2], epochs 1.1 s long are extracted around each stimulation, and classification is made through Stepwise Discriminant Analysis (SWDA) applied to averages of samples from epochs relative to the same stimulation (same row or same column).

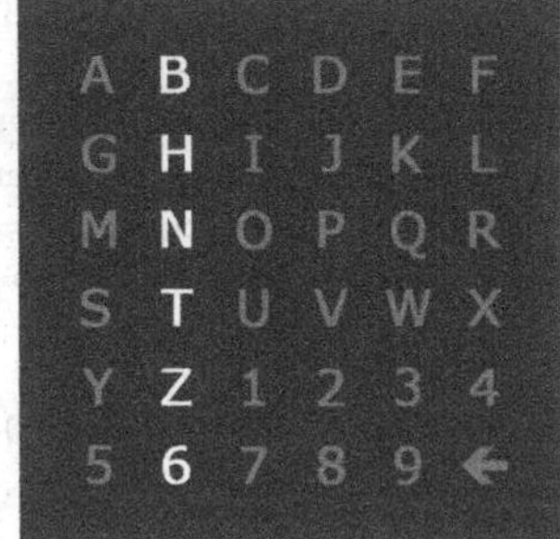

Fig. 1. The visual stimulator for a P300-based BCI speller. Lines and rows are flashed one at a time in random order, and the intersection between the row and the column that elicit a P300 indicates the selected letter.

P300 has been used in other BCI experiments, with different classification techniques. In [3], a virtual-reality system is presented where subjects operate objects selected through the P300. Classification is made by comparing the correlation of single responses with the averages of all target and non-target responses. In [4] a wheelchair driven by a BCI is described, and P300 is recognized through a genetic algorithm. In [5], the subjects (healthy and impaired ones) control a cursor by choosing among four commands (up, down, left, right) via the P300. Single-sweep detection is performed; independent component analysis (ICA) is used to decompose the EEG signals, a fuzzy classifier selects one of the components extracted by ICA, and a neural network classifies it as target or non-target. The system is more effective with healthy subjects, though no exact reason could be pinpointed. In [6], an initial attempt at using a BCI in a home environment is reported: A person with ALS uses a P300 speller on a daily basis.

Often in the literature (and in the works cited in the previous review) when a pattern recognition technique is used to recognize the P300 signal, a two-stage process is employed: feature extraction, followed by classification in this new feature space. This approach starts with the definition of a suitable set of features to be extracted from the signal and, possibly after feature selection/projection, it applies a learning machine on those features in order to generate a classifier. Especially when the signal to noise ratio is very low and the a-priori information/knowledge about the observed phenomenon is only partial, it is not always easy to define the "right" set of features for a given signal or problem, and the resulting classifier might turn out to be suboptimal due to the suboptimal choice of the feature space. In these situations, the use of methods able to automatically find the best features for classification is of uttermost importance.

In this paper we present a comparison of three methods for implicit-feature classification applied to the recognition of the P300 event-related potential in an EEG based brain-computer interface. These methods either use the raw signals for classification or implement an implicit, data driven, feature extraction that is not based on any knowledge about the phenomenon, but it is automatically optimized for the classification task. These methods are genetic algorithms, time-delay neural networks and support vector machines, and they are described in details in the next section. In Sect. 3, we present and briefly discuss experimental results on real data coming from an international BCI competition and from recordings at the AIRLab of Politecnico di Milano.

2 Models and Methods

In this section the three methods for implicit-feature classification of the P300 are presented. Two methods, one based on a genetic algorithm and one based on time-delay neural networks, have been developed by our group; the third, based on support vector machines, has been replicated from the literature. Our method to recognize P300s with a time-delay neural networks is presented here for the first time, while the other methods have been already published. All methods can be applied in real time on recent processors.

2.1 Genetic Algorithm

We applied the genetic algorithm described in [7] to data recorded from people using a P300 speller in an offline fashion. In this section, only a very brief description of the algorithm is given; details are given for the fitness function, as it differs from the one used in the cited work.

Genetic algorithms are a class of optimization algorithms that mimic the way natural evolution works. In a genetic algorithm, a set of possible solutions to an optimization problem are coded in strings called *chromosomes*; solutions are evaluated, and the best ones (those with highest *fitness*) are selected and combined together to form new possible solutions. After enough repetitions of this procedure, good solutions emerge.

In the genetic algorithm used in this work, each individual (chromosome) represents a set of possible features for discriminating the presence of a P300 in epochs of EEG recordings. Each gene encodes a feature and an EEG channel from which to extract it; a feature is obtained by multiplying the EEG channel by a weight function (see Fig. 2 for two examples), whose exact shape is encoded by parameters in the genes. Genetic operators are variants of 1-point crossover and uniform mutation, and tournament selection with elitism is used.

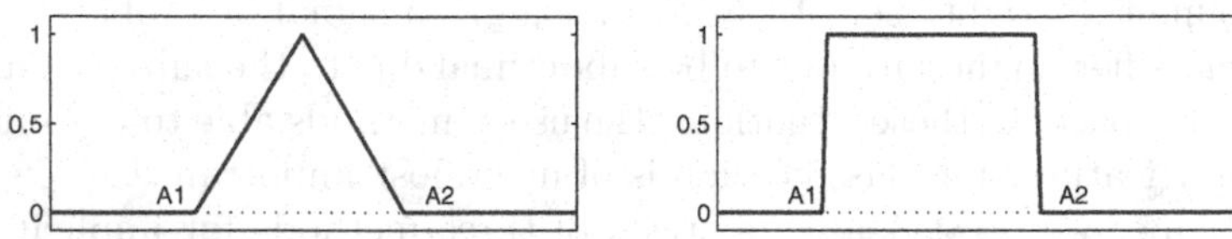

Fig. 2. Weight functions used for feature extraction

The fitness of a chromosome is determined by measuring the performance of a logistic classifier on the features it encodes. To have a fair estimate of the performance, a 4-fold cross-validation scheme on the training set is used, and the mean performance on the 4 folds is used as the fitness.

The performance is measured as the number of correctly predicted letters, with a little bonus for letters that can be correctly predicted with less than the maximum number of repetitions, i.e., the number of times the whole grid is flashed for each letter. Let us call l the number of correctly predicted letters out of a total of n, N the number of repetitions in the data set, and r_i, $i = 1 \dots n$, the number of repetitions needed for the prediction of the letter i. The fitness f, used by the genetic algorithm, is then given by

$$f = \frac{1}{n}\left(l + \frac{1}{l}\sum_{i \in I} \frac{N - r_i}{N + 1}\right), \qquad (1)$$

where I is the set of correctly predicted letters. The second term in the parentheses computes an index, averaged over the l correct letters, that grows with the decreasing of r_i; this index is always strictly less than 1, and therefore it contributes to the fitness less than a single correctly predicted letter. In this way, a higher number of correct letters is always preferred to a lower number of repetitions needed for correct prediction. Repetitions are taken in their occurring order, and r_i is computed in way such that if a letter is correctly predicted by using the first r_i repetitions, then it must be correctly predicted also by using the first $r_i + 1, \dots, N$ repetitions. In other words, if a letter were predicted correctly after 3 repetitions, wrongly after 4, and again correctly when using 5 repetitions or more, then r_i would be 5, and not 3.

Typically a chromosome contains about 100–150 different unique features at the end of a run of the genetic algorithm; an analysis of the combination of the encoded features and the classifier trained on the training set allows to compute weights assigned to individual EEG samples (see Fig. 3).

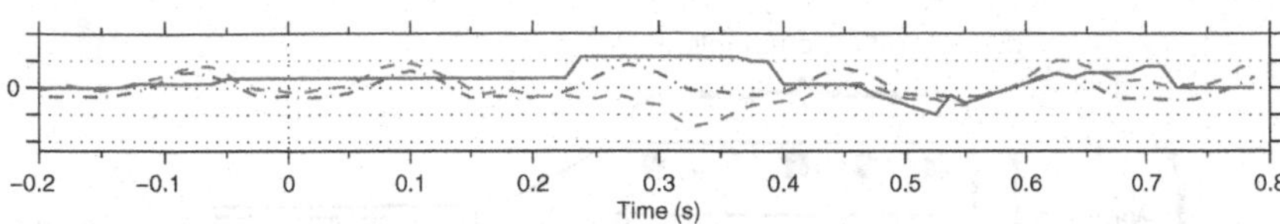

Fig. 3. The green solid line is an example of a template found by the GA for one EEG channel. For comparison, also the average of target (dashed red line) and non-target (dot-dashed blue line) responses are shown.

2.2 Time-Delay Neural Networks

Neural networks can be applied for the detection of patterns in signals. In case the signals to be processed are affected by translations and distorsions, the position of the patterns in the input may be subject to variation. In this case a neural network of relevant size may be required for the recognition task, with overfitting problem that may occur if the training data is scarce.

Time-delay neural networks (TDNNs) are multilayer backpropagation neural networks that present a modular architecture with replicated weights [8]. TDNNs use sliding windows to process input signals, with windows represented by sets of units in the input layer. The windows are partially overlapped and are connected to a set of neurons in the next layer, which use the windows as local receptive fields to look for local features in the input. All the windows share the same weights for the connections to the relative neurons, enabling the network to process several parts of the input in the same way. This replicated architecture allow to be invariant to shifts and distortions of the input, as local feature detectors scan the input. The structure is replicated in the following layers combining the extracted features of previous layers, with the network trained through a backpropagation algorithm. The reduced number of weights due to replication reduces the capacity of the network, improves its generalization ability, and tends to result in a reduced amount of data required for the training.

The idea is to feed a TDNN with raw EEG data, with the first layers that act as feature extractors, while the remaining part of the network performs the classification. As all the network parameters are set through the learning procedure, it would not be necessary to explicitly design a feature extractor.

In our application a TDNN has been designed to discern between presence or absence of a P300 in an EEG. For this classification task the window length in the input layer has been set to 400 ms, a time interval sufficient to contain a single elicited P300. Considering that a P300 peak is elicited roughly 300 ms after a stimulation, we use 1 s epochs derived from the EEG recordings in correspondence of each stimulus, starting from 200 ms before the stimulation. The derived patterns are classified by a 3-layer TDNN with a $[S \times C]$ unit matrix as input layer, where C represents the number of EEG channels, and S the number of EEG samples contained in the length of an epoch (see Fig. 4). According to the 400 ms time interval set to process the input, windows of size $[0.4S \times C]$ are applied in the input layer, with windows sliding one sample each time. Each

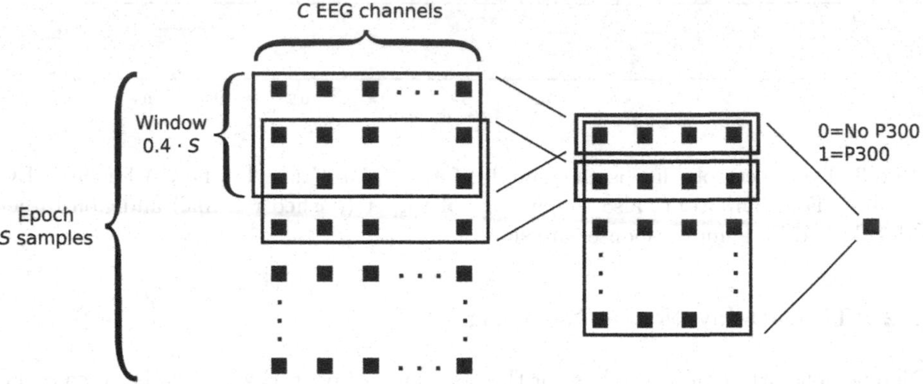

Fig. 4. The TDNN used to process EEG patterns

window is fully linked to 4 neurons in the hidden layer forming a $[(0.6S+1)\times 4]$ matrix. The hidden layer and the single output unit adopt the logistic function and are fully interconnected to perform the final classification.

As most of the examples in the data sets are relative to non-target stimuli, a balancing of the training and validation sets is necessary to avoid undesired bias towards non-target patterns. We have balanced the sets by discarding some non-target recordings so that the number of non-targets is the same as targets. For the definition and evaluation of the designed TDNNs, the Stuttgart neural network simulator (SNNS) tool and its Java counterpart, the JavaNNS [14], were adopted.

2.3 Support Vector Machines

We replicated also the method used by the winners of the BCI Competition 2003 [9,10], data set *IIb*, Kaper and colleagues from the University of Bielefeld, Germany [11]. Their method relies more on the power of the classifier employed, an SVM, than on signal processing, and the feature extraction is done implicitly by the SVM; in other words, this is a blind algorithm, which does not rely on specific knowledge about the P300.

A *support vector machine* (SVM) [12,13] is a supervised learning method used for classification and regression developed by Vladimir Vapnik in the late 1970s. In the simplest case, an SVM is a hyperplane in the space X of the samples $\boldsymbol{x}$. This hyperplane separates the space in two regions, one for each of the possible labels. Samples are assigned labels depending on which side of the hyperplane they lie (see Fig. 5). In formulas:

$$f(\boldsymbol{x}) = \operatorname{sign}(f^*(\boldsymbol{x})) \tag{2}$$

$$f^*(\boldsymbol{x}) = \langle \boldsymbol{w}, \boldsymbol{x} \rangle + b\,. \tag{3}$$

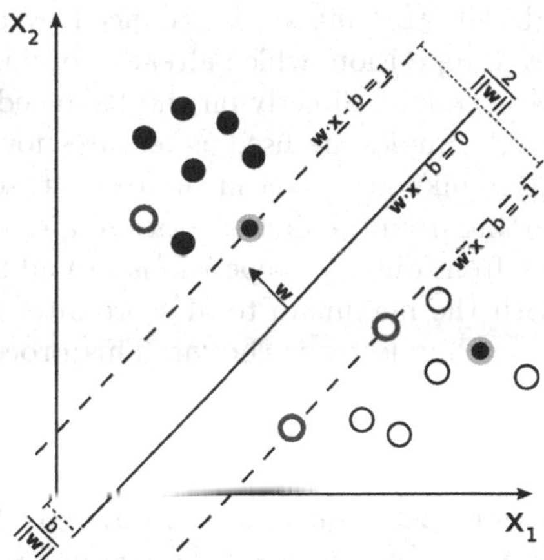

Fig. 5. SVM: the maximum-margin (optimal) hyperplane in the non-separable case. Support vectors are surrounded by a circle.

The hyperplane is found through an optimization algorithm in such a way to maximize the distance of the plane from the nearest samples (*margin* in SVM terminology) and minimize the number of misclassified samples. In mathematical terms, this goal can be written by introducing *slack variables* ξ_i, $i = 1 \dots N$ so that the SVM minimizes

$$\|\boldsymbol{w}\|^2 + C \sum_i \xi_i \tag{4}$$

subject to

$$y_i \left(\langle \boldsymbol{w}, \boldsymbol{x} \rangle + b \right) \geq 1 - \xi_i \tag{5}$$

$$\xi_i \geq 0 \tag{6}$$

The parameter C can be varied to shift the trade-off between margin and errors.

It is possible to extend the idea to non-linear SVMs by mapping samples $\boldsymbol{x}$ in a higher-dimensional space $\mathcal{H}$ by means of non-linear function $\boldsymbol{\Phi} : X \to \mathcal{H}$. The separating hyperplane is now to be found in $\mathcal{H}$. By using a $\boldsymbol{\Phi}$ such that $\mathcal{H}$ and $\boldsymbol{\Phi}$ satisfy Mercer's condition [13], it is possible to find a *kernel function* K such that $\langle \boldsymbol{\Phi}(\boldsymbol{x}_i), \boldsymbol{\Phi}(\boldsymbol{x}) \rangle = K(\boldsymbol{x}_i, \boldsymbol{x})$, where $K(\cdot, \cdot)$ is much easier to compute than the inner product in $\mathcal{H}$. The discriminating function becomes

$$f^*(\boldsymbol{x}) = \sum_i \alpha_i y_i K(\boldsymbol{x}_i, \boldsymbol{x}) + b \,, \tag{7}$$

where $\boldsymbol{x}_i$ are support vectors, i.e., the training samples closest to separating hyperplane and all the misclassified training samples, and α_i are coefficients found by the training process.

In Kaper's method, an epoch begins at the time of stimulus, and ends 600 ms after. Epochs are bandpass filtered between 0.5 and 30 Hz, and then normalized to the $[-1, +1]$ interval. The training set is balanced by taking only two non-target examples from each repetition, which already contains exactly two target examples, and an SVM is trained directly on the balanced training set. In this case, the normalized EEG samples are used as features for the classification.

In order to find out the unknown letter in the test set, several repetitions are combined together. Starting from the first repetition, a score is assigned to each row and column. Scores from different repetitions are added together, and the row and the column with the maximum total score after the last repetition is selected and the corresponding letter is chosen. This procedure is repeated for each letter in the test set.

The score mentioned above is the SVM discriminant function $f^*(\cdot)$ in equation (7). In this case, the kernel function $K(\cdot)$ is Gaussian. The parameter σ of the kernel and the penalization coefficient C in the objective function of the SVM are found with a cross-validation scheme on the training set.

3 Experiments and Results

We applied the methods described above to two different data sets: the data set *IIb* from the BCI Competition 2003 [10], and recordings made at AIRLab, our laboratory. Both data sets consist in EEG recordings of subjects using a P300 speller like the one described in [2], made in three sessions; in our classification experiments, the first two sessions were used for training, and the last one was used for testing. For the competition data set, EEG was recorded from one subject with 64 electrodes in all positions of the 10-10 system and a sampling frequency of 240 Hz; stimuli were repeated 15 times for each letter. We used only channels Fz, Cz, Pz, Oz, C3, C4, P3, P4, Po7, and Po8, as they gave good results and they are the same selected by the competition winners [11]. Also, we recorded data from five subjects in our laboratory using four channels (Fz, Cz, Pz, Oz) at 512 Hz; stimuli were repeated only 5 times per letter. The two data sets differ also for the timing: a stimulus was given every 175 ms in the competition data set, and every 200 ms in AIRLab recordings. Care has been taken to avoid the presence strong 50 Hz noise in the data; spectral analyses confirmed this.

For GA and TDNN epochs started from 200 ms before each stimuli and were 1 s long, while for SVMs they started from the stimulus time and were 600 ms long. A decimation factor of 4 was used always for AIRLab data, resulting in a 128 Hz frequency, while data from the competition were decimated with different factors because of memory constraints; a factor of 3 (resulting in 80 Hz) was used for GA, no decimation was applied for SVMs, and a factor of 2 (down to 120 Hz) was used for TDNNs.

For all three classification methods, the test results are computed in terms of correctly classified letters in the test set, the same criterion used in the competition. Letters are selected by considering all repetitions for each letter: Classification results related to the same stimulus (i.e., same column or row) are added together, and the row and the column with the highest score are selected. Table 1 shows the test results of the three methods applied to data sets described

Table 1. Test results of the three methods applied to different data from P300 speller tasks

Data set / Subj.	GA	TDNN	SVM
Competition	31/31 (100%)	30/31 (97%)	31/31 (100%)
AIRLab S1	121–126/143 (85–88%)	118/143 (83%)	119/143 (83%)
AIRLab S2	89–95/165 (54–58%)	92/165 (56%)	89/165 (54%)
AIRLab S3	91–94/144 (63–65%)	65/144 (45%)	64/144 (44%)
AIRLab S4	80–85/199 (40–43%)	64/199 (32%)	92/199 (46%)
AIRLab S5	46-52/135 (34–39%)	41/135 (30%)	51/135 (38%)

above as the ratio (and the equivalent percentage) of correct letters over the total number of letters in the test sets. For the GA, a range of values is shown instead of a single number, as we have decided to show the entire range of values obtained for all chromosomes with a fitness that is at least 99% of the top fitness. In fact, a single GA run returns many different chromosomes that reaches the top fitness and all these chromosomes are closely related, due to the mixing of the genetic material. For this reason, it is not possible to summarize their performance in one number, as by averaging, for example.

The results of the three methods are similar, though their relative strength vary a little depending on the data set. Overall, the GA obtains the best results, while the results of SVMs are slightly better than those of TDNNs. The performances on some subjects are very low for all methods, but that is expected, because for the AIRLab data set only 5 repetitions and 4 EEG channels are used; such a "hard" data set makes it easier to compare the various methods.

The three methods implement the idea of implicit or automatically optimized features in different ways. The genetic algorithm uses a fixed classifier family and features are in a way created so as give the optimal classification results. Although features are involved, they are not chosen in advance, and when combined with the logistic classifier they naturally lead to a featureless template. In the TDNN model, features can be thought of as if learned implicitly in the weights of the neurons. In the case of an SVM, features cannot be pinpointed in any particular aspect of the model.

Beside the architecture presented in Sect. 2.2, we examined other network topologies for TDNNs, varying the number and the sizes of the hidden layers, increasing or decreasing the epoch length and the window size, and using neurons in the hidden layer connected to input-layer neurons relative to different numbers of channels. The results were always worse than for the proposed TDNN, or at most comparable with it.

The fact that the GA performed better than TDNNs and SVMs may seem strange, as the discriminant function found by the GA is linear, while the other two methods can find a nonlinear separation. There are two possible explanations for this: Either the training of TDNNs and SVMs makes a bad use of the available degrees of freedom because it needs a better tuning, or the highly noisy nature of the problem calls for simpler classifier. Possibly, further experiments could tell which is the best explanation.

References

1. Sutton, S., Braren, M., Zubin, J., John, E.R.: Evoked-potential correlates of stimulus uncertainty. Science 1187, 1187–1188 (1965)
2. Donchin, E., Spencer, K.M., Wijesinghe, R.: The mental prosthesis: Assessing the speed of a P300-based brain-computer interface. IEEE Trans. Rehabil. Eng. 8(2), 174–179 (2000)
3. Bayliss, J.D., Inverso, S.A., Tentler, A.: Changing the P300 brain computer interface. Cyberpsychol. & Behav. 7(6), 694–704 (2004)
4. Blatt, R., Ceriani, S., Dal Seno, B., Fontana, G., Matteucci, M., Migliore, D.: Brain control of a smart wheelchair. In: Burgard, W., Dillmann, R., Plagemann, C., Vahrenkamp, N. (eds.) Intelligent Autonomous Systems 10 – IAS-10, Baden, Germany, pp. 221–228. IOS Press, Amsterdam (2008)
5. Piccione, F., Giorgi, F., Tonin, P., Priftis, K., Giove, S., Silvoni, S., Palmas, G., Beverina, F.: P300-based brain computer interface: reliability and performance in healthy and paralysed participants. Clin. Neurophysiol. 117(3), 531–537 (2006)
6. Vaughan, T.M., Mcfarland, D.J., Schalk, G., Sarnacki, W.A., Krusienski, D.J., Sellers, E.W., Wolpaw, J.R.: The Wadsworth BCI research and development program:At home with BCI. IEEE Trans. Neural Syst. Rehabil. Eng. 14(2), 229–233 (2006)
7. Dal Seno, B., Matteucci, M., Mainardi, L.: A genetic algorithm for automatic feature extraction in P300 detection. In: 2008 International Joint Conference on Neural Networks (IJCNN), Hong Kong, China, June 2008, pp. 3145–3152 (2008)
8. Waibel, A., Hanazawa, T., Hinton, G., Shikano, K., Lang, K.J.: Phoneme recognition using time-delay neural networks. IEEE Trans. Acoust., Speech, Signal Process. 37(3), 328–339 (1989)
9. Blankertz, B.: BCI competition 2003 (2003), `http://ida.first.fraunhofer.de/projects/bci/competition_ii/`
10. Blankertz, B., Müller, K.R., Curio, G., Vaughan, T.M., Schalk, G., Wolpaw, J.R., Schlögl, A., Neuper, C., Pfurtscheller, G., Hinterberger, T., Schröder, M., Birbaumer, N.: The BCI competition 2003: Progress and perspectives in detection and discrimination of EEG single trials. IEEE Trans. Biomed. Eng. 51(6), 1044–1051 (2004)
11. Kaper, M., Meinicke, P., Grossekathoefer, U., Lingner, T., Ritter, H.: BCI competition 2003 data set IIb: support vector machines for the P300 speller paradigm. IEEE Trans. Biomed. Eng. 51(6), 1073–1076 (2004)
12. Vapnik, V.N.: An overview of statistical learning theory. IEEE Trans. Neural Netw. 10(5), 988–999 (1999)
13. Vapnik, V.N.: Statistical learning theory. Wiley, Chichester (1998)
14. Zell, A.: Stuttgart neural network simulator (SNNS) and Java neural network simulator (JavaNNS) (2009), `http://www.ra.cs.uni-tuebingen.de/SNNS/`

Computing with Probabilistic Cellular Automata*

Martin Schüle[1], Thomas Ott[2], and Ruedi Stoop[1]

[1] Institute of Neuroinformatics, ETH and University of Zurich, 8057 Zurich, Switzerland
[2] Institute of Applied Simulation, ZHAW Zurich, 8820 Wädenswil, Switzerland

Abstract. We investigate the computational capabilities of probabilistic cellular automata by means of the density classification problem. We find that a specific probabilistic cellular automata rule is able to solve the density classification problem, i.e. classifies binary input strings according to the number of 1's and 0's in the string, and show that its computational abilities are related to critical behaviour at a phase transition.

1 Preliminaries

Cellular automata (CA) models have been widely studied and applied in physics, biology and computer science. They are among the simplest mathematical systems which exhibit self-organisation, complex patterning and capability of universal (Turing) computation [1,2,3]. Various authors have suggested CA as the generic model for parallel, biologically inspired computing [1,4,5]. As such they are closely related to neural networks [5]. In this contribution we investigate claims regarding the computational abilities of elementary probabilistic cellular automata.

1.1 Deterministic Cellular Automata

A *deterministic* cellular automaton (DCA) is specified by a d-dimensional regular discrete lattice L with given boundary conditions, a finite set Σ of states x_i assigned to each node or cell i of the lattice and a local rule f acting on the states in the range k of the neighbourhood N_k^i of each cell i in discrete time steps. Given some initial configuration of states, the local rule completely determines the dynamics of the cellular automaton. In this paper we deal with *finite* DCA, that is DCA with a finite number N of cells, and *elementary* DCA, that is DCA with $d = 1$, $\Sigma = \{0, 1\}$ and nearest neighbourhood $k = 3$. In this case, there are 256 different possible local rules ${x_i}^{t+1} = f({x_{i-1}}^t, {x_i}^t, {x_{i+1}}^t)$. The N cells are subject to periodic boundary conditions and their states x_i are updated synchronously by the local rule. Local rules are given by a rule table.

* This work was supported by ETH Research Grant TH-04 07-2.

C. Alippi et al. (Eds.): ICANN 2009, Part II, LNCS 5769, pp. 525–533, 2009.

Example 1 (Rule 232). The rule table of the so-called *majority* DCA rule 232 is $(f(111) = 1, f(110) = 1, f(101) = 1, f(100) = 0, f(011) = 1, f(010) = 0, f(001) = 0, f(000) = 0)$.
It is customary to assign a decimal number to such rule tables. One speaks of *rule 232* as the binary expansion of the decimal number 232 ($232 = 11101000$) which encodes the rule table when read from left to right.

A configuration or *global state* $\mathbf{x}^t$ of a CA is the string of the states of the N cells at the time t, that is $\mathbf{x}^t = (x_0^t, x_1^t, ..., x_{N-1}^t)$. Starting from an initial configuration $\mathbf{x}^0$, the global function or map F then maps configuration $\mathbf{x}^t$ to $\mathbf{x}^{t+1} = F(\mathbf{x}^t)$, thereby generating a space-time pattern. The global map F is only indirectly given through the local rule f. A *quiescent* or *stationary* global state $\mathbf{x}^*$ is defined as $\mathbf{x}^* = F(\mathbf{x}^*)$.

Any DCA rule can be represented as a *Boolean function*, which is expressible as a *disjunctive normal form* (DNF) [1]. The DNF is a disjunction of clauses, where a clause is a conjunction of Boolean variables.

Example 2 (DNF of rule 232). DCA rule 232 written as a DNF is $(X_{i-1} \wedge X_i \wedge X_{i+1}) \vee (X_{i-1} \wedge X_i \wedge \neg X_{i+1}) \vee (X_{i-1} \wedge \neg X_i \wedge X_{i+1}) \vee (\neg X_{i-1} \wedge X_i \wedge X_{i+1})$ with the Boolean variables X_i, the disjunction denoted by $\vee$, the conjunction by $\wedge$ and the negation by $\neg$.

As outlined in [6] the DNF of CA rules can then be rewritten as algebraic expressions which represents CA dynamics in a concise form.

Example 3 (Algebraic expression of rule 232). The algebraic expression for DCA rule 232 is

$$x_i{}^{t+1} = x_{i-1}{}^t \cdot x_i{}^t + x_i{}^t \cdot x_{i+1}{}^t + x_{i-1}{}^t \cdot x_{i+1}{}^t - 2x_{i-1}{}^t \cdot x_i{}^t \cdot x_{i+1}{}^t \quad (1)$$

We now turn to a stochastic generalisation of deterministic CA, that is probabilistic CA.

1.2 Probabilistic Cellular Automata

Probabilistic cellular automata (PCA) are generalized DCA in the regard that the states x_i are stochastically updated, that is by some local probability transition function. In the case of elementary PCA this means that the probability of having cell i the value $\tilde{x}_i = 1$ at the time $t+1$ is given by $p[\tilde{x}_i^{t+1}|(x_i^t)]$, where (x_i^t) are the states of the cells in the next-nearest-neighbourhood of cell i, i.e. $(x_i^t) = (x_{i-1}^t, x_i^t, x_{i+1}^t)$. The local probability transition function is subject to the normalisation condition $\sum_{\tilde{x}_i^{t+1}=\{0,1\}} p[\tilde{x}_i^{t+1}|(x_i^t)] = 1$. The *probabilistic majority rule* we will work with exemplifies the notion of elementary PCA.

Example 4 (Majority PCA rule). The rule table of the majority PCA rule is $(p(111) = 1, p(110) = \epsilon, p(101) = \epsilon, p(100) = 1-\epsilon, p(011) = \epsilon, p(010) = 1-\epsilon, p(001) = 1-\epsilon, p(000) = 0)$.

That is the probability for $\tilde{x}_i^{t+1} = 1$ given $(x_{i-1}^t, x_i^t, x_{i+1}^t)$ is $(1, \epsilon, \epsilon, 1-\epsilon, \epsilon, 1-\epsilon, 1-\epsilon, 0)$.

The properties of PCA from a statistical mechanics viewpoint have been widely discussed [7,8]. The general dynamics of PCA is given by a *master equation* [7,9] which reads to

$$P[\tilde{\mathbf{x}}_i^{t+1}] = \sum_{(\mathbf{X}_i)} P[(\mathbf{x}_i^t)] \prod_i p[\tilde{x}_i^{t+1}|(x_i^t)] \qquad (2)$$

where $\sum_{(\mathbf{X}_i)} P[(\mathbf{x}_i^t)]$ is the sum (over all global states $(\mathbf{x}_i^t)$) of the probabilities to find the PCA in some state $\mathbf{x}_i$ at the time t and $p[\tilde{x}_i^{t+1}|(x_i^t)]$ is the local probability transition function.

Following the algebraic approach outlined above, the local probability transition function of the majority PCA rule can be written as

$$\begin{aligned} p[\tilde{x}_i^{t+1}|(x_i^t)] = {} & {x_{i-1}}^t + {x_i}^t + {x_{i+1}}^t - 2{x_{i-1}}^t{x_i}^t - 2{x_{i-1}}^t{x_{i+1}}^t - 2{x_i}^t{x_{i+1}}^t + 4{x_{i-1}}^t{x_i}^t{x_{i+1}}^t \\ & -\epsilon({x_{i-1}}^t + {x_i}^t + {x_{i+1}}^t - 3{x_{i-1}}^t{x_i}^t - 3{x_{i-1}}^t{x_{i+1}}^t - 3{x_i}^t{x_{i+1}}^t + 6{x_{i-1}}^t{x_i}^t{x_{i+1}}^t). \end{aligned}$$

Accordingly, the dynamics of the majority PCA rule can be written as

$$\begin{aligned} \tilde{x}_i^{t+1} = {} & x_{i-1} + x_i + x_{i+1} - 2x_{i-1}x_i - 2x_{i-1}x_{i+1} - 2x_ix_{i+1} + 4x_{i-1}x_ix_{i+1} \\ & -Y_i(x_{i-1} + x_i + x_{i+1} - 3x_{i-1}x_i - 3x_{i-1}x_{i+1} - 3x_ix_{i+1} + 6x_{i-1}x_ix_{i+1}) \quad (3) \end{aligned}$$

where $\{Y_i\}_{i=0}^{N-1}$ is a set of iid random variables with probability distribution $p[Y_i = 1] = \epsilon, p[Y_i = 0] = 1 - \epsilon$ (and x_i^t shortened to x_i).

Solving the master equation for large N is usually not feasible. In order to simplify the treatment one works in a *mean field approximation* (MF). The MF approximation assumes that the values x_i^t are independent of each other and the probability of having cell i in state x_i^t is therefore given through the *global density* $\rho^t = \frac{1}{N}\sum_i x_i^t$. In the MF, the dynamics of the global density variable becomes (with ρ^t shortened to ρ)

$$\rho^{t+1} = 3\rho - 6\rho^2 + 4\rho^3 - \epsilon(3\rho - 9\rho^2 + 6\rho^3) \qquad (4)$$

PCA are *finite Markov chains* [14]. In the case of the majority PCA rule defined above, we have an *absorbing* finite Markov chain. Again, as an array of N cells yields 2^N different configurations, i.e. global states, most techniques developed in the field of finite Markov chains are not practicable for large N. Theorems regarding general properties of absorbing finite Markov chains can however be of use.

2 Computing with Probabilistic Cellular Automata

The computational abilities of deterministic cellular automata (DCA) have been early recognized and discussed in Wolfram's contributions and by ensuing papers [10]. The basic idea is that some input string $\mathbf{x}^0$ at time $t = 0$ is processed to some output string $\mathbf{x}^T$ at time T through the DCA's time evolution. We therefore define *computation*, in this context, as the global map $G(\mathbf{x}^0) = \mathbf{x}^\mathbf{T}$ with $G = F^T(\mathbf{x}^0)$, that is the global map F is iterated T times. The *computation time* T is the number of the discrete time steps in the DCA evolution from the input to the output string. The output string $\mathbf{x}^T$ is usually, but not always, some quiescent, i.e. stationary state.

Wolfram and other authors have described computing by DCA within formal language theory. We follow here a different approach by, generally, adhering to a dynamical system viewpoint on CA and, specifically, by discussing a particular exemplary problem, the so-called *density classification problem.* In fact, it can be questioned whether any global map G should be called "computational" irrespective of a specific, well-defined *computational problem* at hand.

The density classification problem is the computational task to classify input strings according to their densities of 1's and 0's. Usually this means that the CA should asymptotically evolve to either the 0- or 1-quiescent state, that is the global state with all states equal 0, or 1 respectively, depending on the initial densities. In the Markov chain liteature, this final, stationary global state is called the *absorbing* state. The density classification problem is an obvious, well-defined computational task which any basic computing device should be able to carry out. Land and Belew have however shown that there exists no elementary DCA able to solve the problem [11]. Later Fuks has demonstrated that a specific PCA rule can solve the problem in a stochastic sense [12]. We investigate a different PCA rule, e.g. the majority PCA rule introduced above, and discuss its computational capabilities in broader terms.

2.1 PCA and DCA

As we will see, the majority PCA rule solves the density classification problem in a stochastic sense. The basic reason for this is, that the PCA, unlike DCA, will not get stuck in certain non-intended periodic patterns, that is certain quiescent or periodic states. As pointed out before, a PCA rule is a stochastic combination of DCA rules [8]. In the case of the majority PCA rule this means that we have with a certain probability ϵ DCA rule 232 (the deterministic majority rule) and with probability $1-\epsilon$ DCA rule 150.

Example 5 (The majority PCA rule as a combination of DCA rules). With probability ϵ the majority PCA rule is equal to DCA rule 232

$$x_i^{t+1} = {x_{i-1}}^t {x_i}^t + {x_{i-1}}^t {x_{i+1}}^t + {x_i}^t {x_{i+1}}^t - 2{x_{i-1}}^t {x_i}^t {x_{i+1}}^t$$

and with probability $1-\epsilon$ DCA to rule 150

$$x_i^{t+1} = {x_{i-1}}^t + {x_i}^t + {x_{i+1}}^t - 2{x_{i-1}}^t {x_i}^t - 2{x_{i-1}}^t {x_{i+1}}^t - 2{x_i}^t {x_{i+1}}^t + 4{x_{i-1}}^t {x_i}^t {x_{i+1}}^t.$$

It has been shown [13] that a pair of elementary DCA, namely rules 184 and 232, can solve the density classification problem exactly. In view of the above considerations, this comes as no surprise as a pair of DCA rules is, in a certain sense, equivalent to a single PCA rule. In future work, we intend to study further which combinations of DCA are equivalent to which PCA and why.

2.2 Density Classification and Phase Transitions

We now discuss simulation results for the majority PCA and compare it with the MF predictions. As mentioned before, the majority PCA is a finite, elementary probabilistic cellular automata with periodic boundary conditions and synchronous updating. Because the majority PCA is a finite absorbing Markov chain, every individual majority PCA will end up in an absorbing state after a finite number of time steps [14], that is every input string will eventually be classified. An ensemble of equivalent majority PCA will however show a distinct behaviour which can be approximated by the MF approach. In the MF

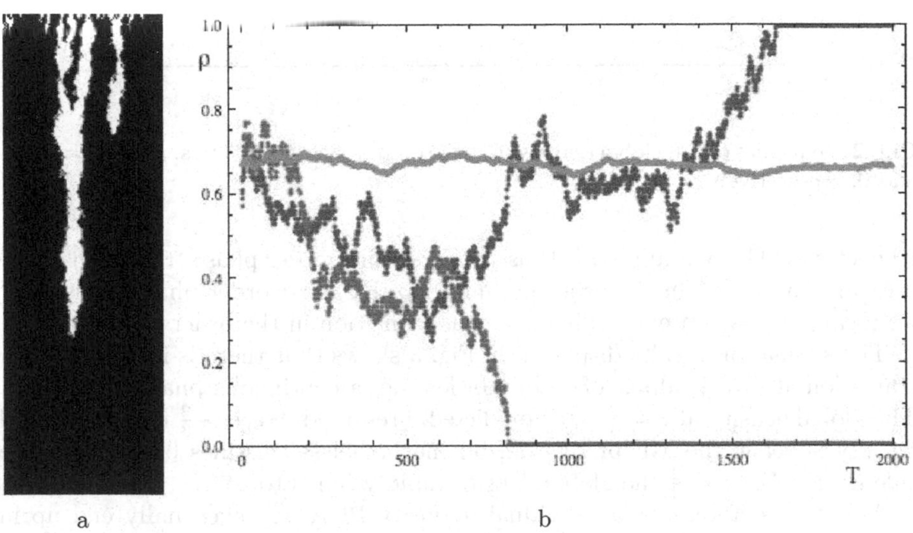

Fig. 1. a Space-time pattern of a single majority PCA with $\rho^0 = \frac{2}{3}$, $\epsilon = \frac{2}{3}$ and $N = 100$. Time axis is from top down. **b** Dynamics of the global density for two single majority PCA with $\rho^0 = \frac{2}{3}$, $\epsilon = \frac{2}{3}$ and $N = 100$ and of the mean global density of an ensemble of 100 majority PCA with $\rho^0 = \frac{2}{3}$, $\epsilon = \frac{2}{3}$ and $N = 100$. The mean global density fluctuates around $\rho^0 = \frac{2}{3}$.

approach the global density for $\epsilon = \frac{2}{3}$ is $\rho^{t+1} = \rho^t$, that is the global density for an ensemble of equivalent majority PCA is preserved. For $\epsilon = \frac{2}{3}$ this result holds also when using the exact local probability transition function, which can be seen by taking expectation values of both sides of equation (3) which yields $E[\rho^{t+1}] = \frac{1}{N}\sum_i E[\tilde{x}_i^{t+1}] = \frac{1}{N}\sum_i E[x_i^t] = E[\rho^t]$. For $\epsilon = \frac{2}{3}$ the majority PCA thus solves the density classification problem in the sense that input strings will be classified correctly with a probability equal to the initial density of the input string. For $\epsilon > \frac{2}{3}$ the MF approach predicts an unstable fixed point at $\rho = \frac{1}{2}$ and stable fixed points at $\epsilon = 0$, $\epsilon = 1$ respectively. For $\epsilon < \frac{2}{3}$ the MF approach predicts a stable fixed point at $\rho = \frac{1}{2}$ and unstable fixed points at $\epsilon = 0$, $\epsilon = 1$

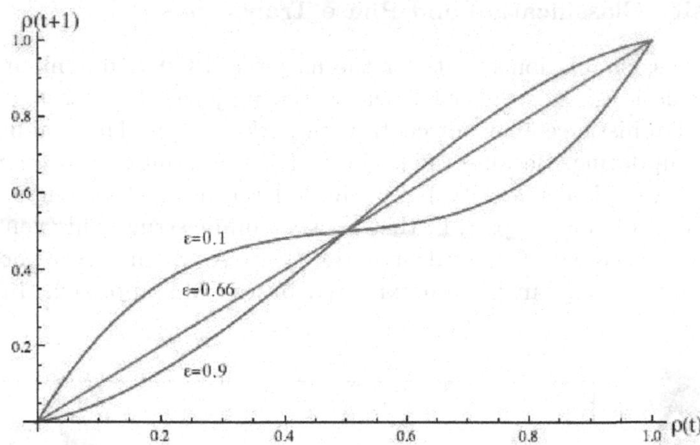

Fig. 2. Dynamics of the global density in the MF approach, i.e. ρ^{t+1} vs. ρ^t, for $\epsilon = 0.1$, $\frac{2}{3}$ and 0.9 respectively

respectively. The MF approach thus predicts a first-order phase transition at the critical point $\epsilon = \frac{2}{3}$ in the order parameter ρ. By "first-order phase transition" we mean, in this context, a discontinuous transition in the order parameter.

The simulation results displayed in Fig. 3 shows that there is indeed a phase transition at $\epsilon = \frac{2}{3}$, albeit of second order (i.e. a continuous phase transition). The global density at $\epsilon = \frac{2}{3}$ is, as predicted, preserved. For $\epsilon > \frac{2}{3}$ the final global density is below the MF prediction, but nevertheless classifies better than the rule in [12]. For $\epsilon < \frac{2}{3}$ the global density rapidly tends to $\rho^T = \frac{1}{2}$.

As stated before, every individual majority PCA will eventually end up in the absorbing state, that is in either the 0- or 1-quiescent state. We define the *accurancy* of the computation as the fraction of correct solutions by the majority PCA to the density classification problem, that is the final global density. The *efficiency* of the computation is then the ratio of time to absorption T to the accurancy. As can be inferred from Fig. 3 and 4 the computational efficiency is highest around the critical point $\epsilon = \frac{2}{3}$ for initial densities $\rho^0 = 0.5, \frac{1}{3}$, and $\frac{2}{3}$ and at around $\epsilon = 0.9$ for initial densities $\rho^0 = 0.1$ and 0.9. In ongoing work we study the possibilities to derive analytical expressions formalising these observations.

There has been much speculation in recent years about computation at the "edge of chaos" or near critical points of phase transitions, albeit no formal theory, as far as we know, has elaborated on these speculations. From the majority PCA, we first see that we have a phase transition at $\epsilon = \frac{2}{3}$ and an enhanced computational capacity in the sense that there is a transition from solving one problem, the density classification problem, to another which could be termed the "reshuffling problem" as an input string with an arbitrary initial density will be effectively reshuffled to a state with $\rho = \frac{1}{2}$. Secondly, we observe that the "optimal" computational capability of the majority PCA, that is its efficiency in the sense defined above, is in the vicinity of the critical point $\epsilon = \frac{2}{3}$.

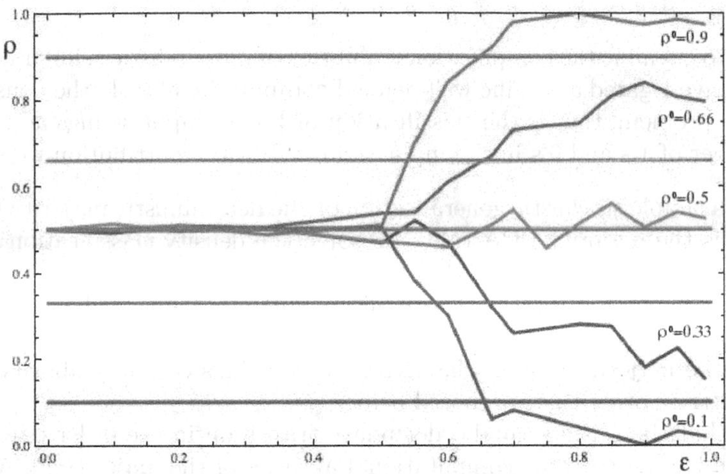

Fig. 3. The global density ρ at time $T = 10000$ for an ensemble of 100 identical majority PCA with $N = 100$ cells in dependence of the parameter ϵ for different initial global densities $\rho^0 = 0.9, \frac{2}{3}, 0.5, \frac{1}{3}, 0.1$

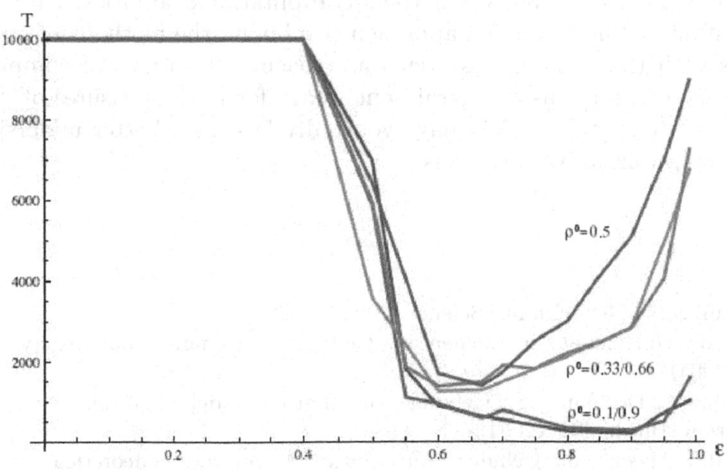

Fig. 4. The absorption time T for an ensemble of 100 identical majority PCA with $N = 100$ in dependence of parameter ϵ for different initial global densities ρ^0. The curves show the *minimum* average absorption time as the running time of the simulation was bounded by $T = 10000$.

3 Summary

In order to examine the computational abilities of probabilistic cellular automata we have investigated a specific well-defined computational task, the density classification problem, that is the classification of binary input strings according to the number of 1's and 0's in the input strings. In this contribution we showed:

1. that a simple stochastic generalisation of the deterministic majority rule 232, that is the majority PCA rule, can solve the density classification problem (which is not solvable by DCA) in a stochastic sense,
2. that, compared with the deterministic rule, the enhanced computational capability is due to a stochastic combination of deterministic CA rules,
3. that the majority PCA is classifying more strings correctly above a critical point than other rules proposed before
4. and that there is a second-order phase transition in the order parameter ρ which is related to the computational abilities of the majority PCA.

This preliminary contribution focuses on illustrating the computational abilities of probabilistic cellular automata within a specific computational task. Probabilistic cellular automata are simple computational systems which can closely model biological of physical systems. We believe that further investigation into probabilistic cellular automata will shed light on the connection between actual physical or biological systems and their computational abilities. Probabilistic cellular automata offer a unified approach combining the methods of statistical mechanics with the dynamical systems approach and notions of computation. As such, they can serve as a natural benchmark for novel measures of "natural computation" (e.g. [15]), which may eventually lead to a better understanding of what "computation by nature" is.

References

1. Wolfram, S.: A New Kind of Science. B&T (2002)
2. Cook, M.: Universality in Elementary Cellular Automata. Complex Systems 15, 1–40 (2004)
3. Deutsch, A., Dormann, S.: Cellular Automaton Modeling of Biological Pattern Formation, Birkhäuser (2004)
4. Kari, J.: Theory of Cellular Automata: A Survey. Theoretical Computer Science 334, 3–33 (2005)
5. Garzon, M.: Models of Massive Parallelism: Analysis of Cellular Automata and Neural Networks. Springer, Heidelberg (1995)
6. Schüle, M., Ott, T., Stoop, R.: Global dynamics of finite cellular automata. In: Kůrková, V., Neruda, R., Koutník, J. (eds.) ICANN 2008, Part I. LNCS, vol. 5163, pp. 71–78. Springer, Heidelberg (2008)
7. Grinstein, G., Jayaprakash, C.: Statistical Mechanics of Probabilistic Cellular Automata. Phys. Rev. Lett. 55, 2527–2530 (1985)
8. Lebowitz, J.L., Maes, C., Speer, E.R.: Statistical Mechanics of Probabilistic Cellular Automata. Journal of Statistical Physics 59, 117–170 (1990)

9. Van Kampen, N.G.: Stochastic Processes in Physics and Chemistry. North Holland Publishing Company, Amsterdam (1981)
10. Wolfram, S.: Computation Theory of Cellular Automata. Communications in Mathematical Physics 96, 15–57 (1985)
11. Land, M., Belew, R.K.: No Perfect Two-State Cellular Automata for Density Classification Exists. Phys. Rev. Lett. 74, 5148–5150 (1984)
12. Fuks, H.: Nondeterministic density classification with diffusive probabilistic cellular automata. Phys. Rev. E 66, 066106 (2002)
13. Fuks, H.: Solution of the density classification problem with two cellular automata rules. Phys. Rev. E 55, R2081 - R2084 (1997)
14. Kemeny, J.G., Snell, J.L.: Finite Markov Chains. D. Van Nostrand Co. (1960)
15. Stoop, R., Stoop, N.: Natural computation measured as a reduction of complexity. Chaos 14, 675–679 (2004)

Delay-Induced Hopf Bifurcation and Periodic Solution in a BAM Network with Two Delays

Jian Xu[1,⋆], Kwok Wai Chung[2], Ju Hong Ge[1], and Yu Huang[1]

[1] School of Aerospace Engineering and Applied Mechanics, Tongji University, Shanghai 200092, P.R. China
Fax: +86 21 5504 1398
xujian@tongji.edu.cn

[2] Department of Mathematics, City University of Hong Kong, Kowloon, Hong Kong

Abstract. The time delays in neural networks come from the transformation of the information processing between neurons. To consider delay-induced dynamics in neural networks, the four neurons are coupled to model a called bidirectional associative memory neural network. If the transformation time is distinct in the two direction, then two delays occurs in the model. a simple but efficient method is first introduced and then the delay-induced Hopf bifurcation is investigated and the periodic approximate solution derived from the Hopf bifurcation is obtained analytically. It can be seen that theoretical prediction is in good agreement with the result from the numerical simulation, which shows that the provided method is valid. The results show also that the method has higher accuracy than the center manifold reduction (CMR) with norm form theory.

Keywords: Artificial neural network, bidirectional associative memory, delayed differential equation, Hopf bifurcation, periodic solution.

1 Introduction

To clarify the mechanism of the information processing in the brain of living organisms, and investigate information coding of a neural network, a reasonable mathematical model of the network is needed. In such mathematical model, one has to consider the transformation delay in the information processing between the neurons, such as Hopfield model [1] from which the interest in investigating the dynamics of neural networks has been steadily increasing. Starting from Kosko's works [2], a called bidirectional associative memory neural network (BAMNN) has kept many scientists attention for a few years. Due to bidirectional structure, the BAMNNs have practical applications in storing paired patterns or memories and possess the ability of searching the desired patterns via both directions: forward and backward.

⋆ Corresponding author.

C. Alippi et al. (Eds.): ICANN 2009, Part II, LNCS 5769, pp. 534–543, 2009.

Until now, most studies was focused on the local and global stability analysis for the BAMNNs with constant or varying delays, or subjected to impulses by using the Lyapunov method or its extensive techniques [3,4,5,6,7]. It is well known that studies on neural dynamical systems involve not only a discussion of stability properties, but also many dynamic behaviors such as periodic oscillation, bifurcation and chaos. In many application, the properties of periodic solution are of great interest. Therefore, it is quite interesting to obtain the periodic solution in a closed form from a Hopf bifurcation. Very recently, there have been extensive literatures on bifurcation analysis of some special BAMNNs [8,9,10]. Song et. al [8] was proposed a delay-differential equation to model a bidirectional associative memory (BAM) neural network with three neurons. The center manifold reduction (CMR) with the normal form technique is used to investigate the stability and direction of the Hopf bifurcation due to delays. Cao et al. [9] extended results of [8] to the model with four coupled neurons in terms of the same approach. For the model considered in [8], Yan [10] employed also the CMR to obtain the local codimension-two bifurcation. These researches merely treated the delay-induced bifurcation qualitatively but not quantitatively due to the restriction of using the CMR.

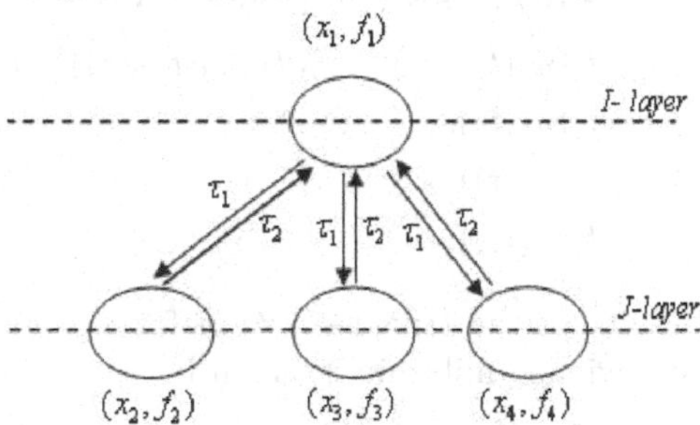

Fig. 1. The graph of architecture for model (1)

Motivated by such problem, we consider here a BAMNN with four neurons and two delays. In particular, when there is only one neuron with the activation function f_1 on the I-layer and there are three neurons with respective activation functions f_2, f_3 and f_4 on the J-layer, the time delay of the transformation of the information processing from the I-layer to J-layer is τ_1 and that from J-layer back to the I-layer is τ_2, as shown in Fig. 1. The network schemed is modeled as [9]

$$\begin{aligned}
\dot{x}_1(t) &= -\mu_1 x_1(t) + c_{21} f_1(x_2(t-\tau_2)) + c_{31} f_1(x_3(t-\tau_2)) + c_{41} f_1(x_4(t-\tau_2)),\\
\dot{x}_2(t) &= -\mu_2 x_2(t) + c_{12} f_2(x_1(t-\tau_1)),\\
\dot{x}_3(t) &= -\mu_3 x_3(t) + c_{13} f_3(x_1(t-\tau_1)),\\
\dot{x}_4(t) &= -\mu_4 x_4(t) + c_{14} f_4(x_1(t-\tau_1)),
\end{aligned} \tag{1}$$

where x_k $(k = 1, 2, 3, 4)$ denote the state of the kth neuron, $\mu_k > 0$ $(k = 1, 2, 3, 4)$ describe the stability of internal neuron processing between the I-layer and

J-layer respectively, the real constants c_{k1} and c_{1k} $(k = 2, 3, 4)$ are the connected weights between the I-layer and the J-layer neurons. Throughout this paper, we assume that the activation functions $f_k : \mathbf{R} \to \mathbf{R}$ $(k = 2, 3, 4)$, the connected weights c_{k1} and $c_{1k}(k = 2, 3, 4)$ in (1) satisfy the hypotheses given by

(H_1) $f_k(x) \in C^2(\mathbf{R}, \mathbf{R})$, $f_k'(x) \geq 0$ and there exists a constant $L > 0$ such that $|f_k(x)| \leq L$ $(k = 1, 2, 3, 4)$ for $x \in \mathbf{R}$.

(H_2) $xf_k(x) > 0$, $xf_k''(x) < 0$ for $x \neq 0$ $(k = 1, 2, 3, 4)$.

In general, if the activation function is taken as $\tanh(x)$ or $(1-e^{-x})/(1+e^{-x})$ for $x \in \mathbf{R}$ to imitate the switch function, then Hypotheses (H_1) and (H_2) always hold.

Our goal in this paper is to obtain the periodic solution from a Hopf bifurcation in equation (1) quantitatively by introducing a simple but efficient method [11], which has higher accuracy than the CMR with normal form.

2 Stability Analysis

It follows from (H_2) that $f_k(0) = 0$, which implies that $(0, 0, 0, 0)$ is always an equilibrium of (1). Letting $\tau = \tau_1 + \tau_2$, $u_1(t) = x_1(t - \tau_1)$, $u_2(t) = x_2(t)$, $u_3(t) = x_3(t)$ and $u_4(t) = x_4(t)$ yields that equation (1) is rewritten as

$$\begin{aligned}
\dot{u}_1(t) &= -\mu_1 u_1(t) + c_{21} f_1(u_2(t-\tau)) + c_{31} f_1(u_3(t-\tau)) + c_{41} f_1(u_4(t-\tau)),\\
\dot{u}_2(t) &= -\mu_2 u_2(t) + c_{12} f_2(u_1(t)),\\
\dot{u}_3(t) &= -\mu_3 u_3(t) + c_{13} f_3(u_1(t)),\\
\dot{u}_4(t) &= -\mu_4 u_4(t) + c_{14} f_4(u_1(t)). \qquad (2)
\end{aligned}$$

To determine the stability of the trivial solution for $\tau \neq 0$, the characteristic equation of (2) at the trivial equilibrium is given by

$$\lambda^4 + d_3\lambda^3 + d_2\lambda^2 + d_1\lambda + d_0 + (e_2\lambda^2 + e_1\lambda + e_0)e^{-\lambda\tau} = 0, \qquad (3)$$

where $d_0 = \mu_1\mu_2\mu_3\mu_4 > 0$, $d_1 = \mu_1\mu_2\mu_3 + \mu_1\mu_2\mu_4 + \mu_1\mu_3\mu_4 + \mu_2\mu_3\mu_4 > 0$, $d_2 = \mu_1\mu_2 + \mu_1\mu_3 + \mu_1\mu_4 + \mu_2\mu_3 + \mu_2\mu_4 + \mu_3\mu_4 > 0$, $d_3 = \mu_1 + \mu_2 + \mu_3 + \mu_4 > 0$, $e_0 = -(\alpha_{21}\alpha_{12}\mu_3\mu_4 + \alpha_{31}\alpha_{13}\mu_2\mu_4 + \alpha_{41}\alpha_{14}\mu_2\mu_3)$, $e_1 = -[(\alpha_{21}\alpha_{12}(\mu_3 + \mu_4) + \alpha_{31}\alpha_{13}(\mu_2 + \mu_4) + \alpha_{41}\alpha_{14}(\mu_2 + \mu_3)]$, $e_2 = -(\alpha_{21}\alpha_{12} + \alpha_{31}\alpha_{13} + \alpha_{41}\alpha_{14})$. When $\tau > 0$, one of roots in (3) can be represented in $i\omega$ $(\omega > 0)$ if and only if ω^2 satisfies with $h(z) = 0$, where

$$h(z) = z^4 + az^3 + bz^2 + cz + d, \qquad (4)$$

and $a = d_3^2 - 2d_2 > 0$, $b = d_2^2 + 2d_0 - 2d_1d_3 - e_2^2$, $c = d_1^2 - 2d_0d_2 + 2e_0e_2 - e_1^2$, $d = d_0^2 - e_0^2, z = \omega^2$. To consider the delay-induced periodic solution, we give two hypothesis as follows

(H_3) $d_0 + e_0 > 0$, $d_1 + e_1 > 0$, $d_3(d_2 + e_2)(d_1 + e_1) > (d_1 + e_1)^2 + d_3^2(d_0 + e_0)$,

(H_4) $z_1^* = -\dfrac{a}{4} + \sqrt[3]{-\dfrac{q}{2} + \sqrt{\Delta}} + \sqrt[3]{-\dfrac{q}{2} - \sqrt{\Delta}} > 0$, $h(z_1^*) < 0$ for $\Delta > 0$,

where $p = (8b - 3a^2)/16$, $q = (a^3 - 4ab + 8c)/32$, $\Delta = q^2/4 + p^3/27$.

It can be seen from (H_3) that the Routh-Hurwitz criterion holds and all roots of equation (3) have negative real parts if $\tau = 0$, which implies the trivial equilibrium of (1) is stable for $\tau = 0$. One can conclude that $h(z) = 0$ has at least one positive root z_0, i.e. $\omega_0 = \sqrt{z_0}$ if Hypothesis (H_4) holds. Correspondingly, values of τ can be solved in

$$\tau_j = \frac{1}{\omega_0}[\arccos(\frac{P}{Q}) + 2j\pi], \tag{5}$$

where $P = e_2\omega_0^6 + (d_3e_1 - d_2e_2 - e_0)\omega_0^4 + (d_0e_2 + d_2e_0 - d_1e_1)\omega_0^2 - d_0e_0$, $Q = e_2^2\omega_0^4 + (e_1^2 - 2e_0e_2)\omega_0^2 + e_0^2$. Thus, equation (3) has a pair of simple purely imaginary roots $\pm i\omega_0$ at $\tau = \tau_j$. The obtained results is in good agreement with that by Cao [9].

Now we illustrate $Re(\frac{d\lambda(\tau_j)}{d\tau}) > 0$. According to [9], we know that the sign of $Re(\frac{d\lambda(\tau_j)}{d\tau})$ of root $\lambda(\tau)$ of (3) is the same as that of $h'(\omega_0^2)$. If $\Delta > 0$ and $z_1^* > 0$, then $h'(\omega_0^2) > 0$. In fact, If $\Delta > 0$, equation $h'(z) = 0$ has only one real root z_1^*. Noticing that $\lim_{z\to\pm\infty} h(z) = +\infty$, we know that z_1^* is an unique minimum value point of h(z) on $\mathbf{R}$. Therefore, $z_1^* < \omega_0^2$. In addition, when $z_1^* > 0$ and $h(z_1^*) < 0$ hold, $h'(z) > 0$ for $z > z_1^*$. So we have the following theorem.

Theorem 1. *For equation (2) with Hypotheses (H_1) and(H_2), three conclusions are obtained as follows.*

(a) *If (H_4) holds and $d_0 - e_0 < 0$ or if (H_3) and (H_4) hold but $d_0 - e_0 \geq 0$, then the trivial equilibrium is asymptotically stable for $\tau \in [0, \tau_0)$ and unstable for $\tau > \tau_0$. Equation (2) undergoes Hopf bifurcation at $\tau = \tau_0$, where τ_0 is given by (5).*
(b) *The trivial equilibrium is unstable for any $\tau \geq 0$ if $d_0 + e_0 < 0$.*
(c) *Equation (2) undergoes a pitchfork bifurcation for $d_0 + e_0 = 0$.* □

Now, we apply Theorem 1 seeking for the Hopf bifurcation point in (2). Taking

$$c_{21} = c_{31} = c_{41} = 1,\ c_{12} = c_{13} = -2, c_{14} = -1,$$

$$\mu_1 = \mu_2 = \mu_3 = \mu_4 = 2,\ f_i(x) = \tanh(x) \tag{6}$$

in equation (2), one has that $a = 16$, $b = 71$, $c = 56$ and $d = -144$ in (4) since $f'(0) = 1$, which yields that (H_3) and (H_4) hold, and $d_0 - e_0 \geq 0$. It is easily to see that $h(z) = 0$ has only a positive root $z_0 = 1$ such that $\omega_0 = 1$ and $\tau_j = \arccos(-3/5) + 2j\pi$ $(j = 0, 1, 2, ...)$. Theorem 1 tell us that a Hopf bifurcation occurs in (2) at $\tau = \tau_0$. The numerical method is employed to verify the analytical prediction. Fig. 2(a) shows that all eigenvalues of (2) have strictly negative real parts when $\tau_1 + \tau_2 < \tau_0 = 2.2143$ for $\tau_1 = 1.2$, $\tau_2 = 0.8$. However, a pair of conjugate eigenvalues have positive real parts but all the others have negative real parts when $\tau_1 + \tau_2 > \tau_0 = 2.2143$ with $\tau_1 = 1.2$, $\tau_2 = 1.3$, as shown in Fig. 2(b). It follows from Fig. 2 that the trivial solution loses its stability when τpasses through τ_0 and Hopf bifurcation occurs in (4). Thus, the delay induces the Hopf bifurcation. This is in agreement with the analytical prediction mentioned above. In the next section, we will give a simple but efficient method in order to quantitatively obtain the periodic solution derived from the Hopf bifurcation, in which one does not require the tedious CMR and normal form.

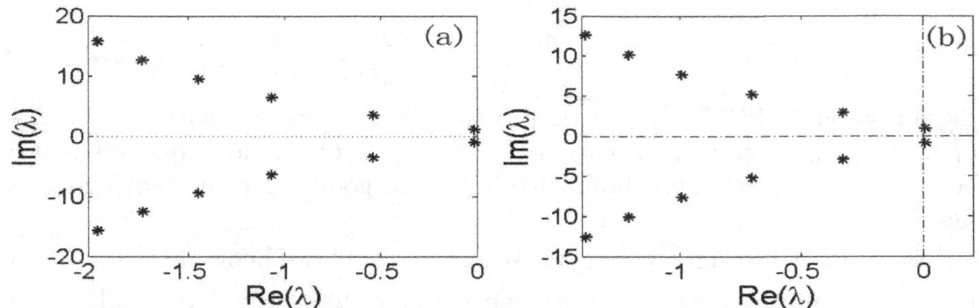

Fig. 2. Distribution of eigenvalues in (3) for (a) $\tau_1 = 1.2$, $\tau_2 = 0.8$, $\tau_1+\tau_2 < \tau_0 = 2.2143$ and (b) $\tau_1 = 1.2$, $\tau_2 = 1.3$, $\tau_1 + \tau_2 > \tau_0 = 2.2143$, where the eigenvalues with positive real part are $0.00996238 \pm 0.91491i$

3 Hopf Bifurcation and Periodic Solution

In our previous work [11], a called perturbation-incremental scheme (PIS) is proposed to obtain the periodic solution in a type of delayed differential equations. The PIS is a simple but efficient method and can investigate the periodic solution derived from Hopf bifurcation due to time delays in a system of first-order delayed differential equations both qualitatively and quantitatively. It posses of advantages of both the CMR and method of multiple scale (MMS) [12], from which the tedious calculation with normal form is avoided. The scheme is described in two steps, namely, the perturbation step (noted as step one) for bifurcation parameters close to the bifurcation point and the incremental step (noted as step two) for those far away from the bifurcation point. In this paper, we only use the first step to obtain a periodic solution near a Hopf bifurcation point in (2). Therefore, we only introduce here the first step of the PIS for readers' convenience.

A class of delay models can be written as

$$\dot{\mathbf{Z}}(t) = \mathbf{C}\mathbf{Z}(t) + \mathbf{D}\mathbf{Z}(t-\tau) + \epsilon\mathbf{F}(\mathbf{Z}(t), \mathbf{Z}(t-\tau)) \tag{7}$$

where $\mathrm{Z}(t) = (z_1(t), z_2(t), ..., z_n(t))^T \in \mathbf{R}^n$, $\mathbf{C}$ and $\mathbf{D}$ are $n \times n$ real constant matrices, $\mathbf{F}(\cdot)$ is a nonlinear function in its variables with $\mathbf{F}(\mathbf{0},\mathbf{0}) = \mathbf{0}$, ϵ is a parameter representing the strength of nonlinear coupling, τ is the time delay, and n a positive integer. We assume τ_0 to be the critical value of a simple Hopf bifurcation. Furthermore, the crossing speed of the root is assumed to be nonzero, and all eigenvalues are neither zero nor purely imaginary pairs. Thus, the system at $\tau = \tau_0$ has a simple Hopf bifurcation.

A perturbation to τ, i. e. $\tau = \tau_0 + \epsilon\tau_\epsilon$, in (7) yields

$$\dot{\mathbf{Z}}(t) = \mathbf{C}\mathbf{Z}(t) + \mathbf{D}\mathbf{Z}(t-\tau_0) + \tilde{\mathbf{F}}(\mathbf{Z}(t), \mathbf{Z}(t-\tau_0), \mathbf{Z}(t-\tau_0-\epsilon\tau_\epsilon), \epsilon), \quad (8)$$

where

$$\tilde{\mathbf{F}}(\mathbf{Z}(t), \mathbf{Z}(t-\tau_0), \mathbf{Z}(t-\tau_0-\epsilon\tau_\epsilon), \epsilon) = \mathbf{D}[\mathbf{Z}(t-\tau_0-\epsilon\tau_\epsilon) - \mathbf{Z}(t-\tau_0)] + \epsilon\mathbf{F}(\mathbf{Z}(t), \mathbf{Z}(t-\tau_0-\epsilon\tau_\epsilon)), \quad (9)$$

if equation (8) undergoes a Hopf bifurcation at $\tau = \tau_0$. We assume that $\mathbf{Z}(t)$ has a periodic solution in period $2\pi/\omega$ at $\tau \approx \tau_0$, given by

$$\mathbf{Z}(t) = \mathbf{a}\cos(\varphi) + \mathbf{b}\sin(\varphi), \quad (10)$$

where $\varphi = \omega t$, $\mathbf{a} = (a_1, a_2, ..., a_n)^T$, $\mathbf{b} = (b_1, b_2, ..., b_n)^T$. If a_1 and b_1 are independent, a_i and b_i $(i = 2, ..., n)$ are functions of a_1 and b_1, given by

$$\mathbf{M}\mathbf{b} = \mathbf{N}\mathbf{a}, \quad -\mathbf{M}\mathbf{a} = \mathbf{N}\mathbf{b}, \quad (11)$$

where $\mathbf{M} = \omega\mathbf{I} + \mathbf{D}\sin(\omega\tau_0)$ and $\mathbf{N} = \mathbf{C} + \mathbf{D}\cos(\omega\tau_0)$. Equation (10) in a polar coordinate can be expressed as

$$Z_i(t) = r_i\cos(\varphi + \theta_i), \quad (12)$$

where $\mathbf{Z}(t) = (z_1(t), z_2(t), ..., z_n(t))^T$, $\mathbf{r} = (r_1, r_2, ..., r_n)^T$. Correspondingly, r_i are functions of r_1 and θ_i are functions of θ_1 $(i = 2, \ldots, n)$.

Based on the expression in (12), we consider the solution of (8) for a small $\epsilon\tau_\epsilon$, which means the harmonic solution of (8) can be considered as a perturbation to that of (12) , given by

$$\mathbf{Z}(t) = \mathbf{r}(\epsilon\tau_\epsilon)\cos\phi, \quad (13)$$

where $\dfrac{d\phi}{dt} = \omega + \sigma(\epsilon\tau_\epsilon)$, $\mathbf{r}(0) = \mathbf{r}$, $\sigma(0) = 0$, $r_i(\epsilon\tau_\epsilon) = r_i(r_1(\epsilon\tau_\epsilon))$ $(i = 2, ..., n)$. Symbol $\bullet(\epsilon\tau_\epsilon)$ is denoted as $\bullet(\epsilon)$ for convenience, such as, $r_i(\epsilon\tau_\epsilon)$ are denoted as $r_i(\epsilon)$ and so on. A key problem arises from equation (13), that is, how to express $r_1(\epsilon)$ and $\sigma(\epsilon)$ in (13) as an analytical form. The following theorem provides a method to determine them.

Theorem 2. *If* $\mathbf{W}(t)$ *is a periodic solution of the equation*

$$\dot{\mathbf{W}}(t) = -\mathbf{C}^T\mathbf{W}(t) - \mathbf{D}^T\mathbf{W}(t+\tau_0) \quad (14)$$

and $\mathbf{W}(t) = \mathbf{W}(t + \frac{2\pi}{\omega})$, *then*

$$\int_{-\tau_0}^{0}[\mathbf{D}^T(\mathbf{W}(t+\tau_0) - \mathbf{W}(t+\tau_0+\frac{2\pi}{\omega+\sigma})]^T\mathbf{Z}(t)dt - [\mathbf{W}^T(\frac{2\pi}{\omega+\sigma}) - \mathbf{W}^T(0)]\mathbf{Z}(0) + \int_0^{\frac{2\pi}{\omega+\sigma}}\mathbf{W}^T(t)\tilde{\mathbf{F}}(\mathbf{Z}(t), \mathbf{Z}(t-\tau_0), \mathbf{Z}(t-\tau_0-\epsilon\tau_\epsilon), \epsilon)dt = 0. \quad (15)$$

□

The proof of Theorem 2 is here neglected due to the limit pages. One may also see Ref. [11].

To apply Theorem 2 for solving $r_1(\epsilon)$ and $\sigma(\epsilon)$, one must obtain the expression of $\mathbf{W}(t)$ in (14). The periodic solution of (14) can be written as

$$\mathbf{W}(t) = \mathbf{p}\cos(\varphi) + \mathbf{q}\sin(\varphi) \tag{16}$$

where $\varphi = \omega t$, $\mathbf{p} = (p_1, p_2, ..., p_n)^T$, $\mathbf{q} = (q_1, q_2, ..., q_n)^T$. Substituting (16) into (14) and using the harmonic balance, one may obtain that

$$\mathbf{M}^T\mathbf{p} = \mathbf{N}^T\mathbf{q}, \quad \mathbf{M}^T\mathbf{q} = -\mathbf{N}^T\mathbf{p}. \tag{17}$$

If p_1 and q_1 are chosen to be independent, then p_i and q_i $(i = 2, ..., n)$ can be determined by (17) in terms of p_1 and q_1.

Now, we return to equation (2) with (6) and hope to obtain analytically the periodic solution derived from the Hopf bifurcation by applying Theorem 2 mentioned above. To this end, the equation is rewritten as the form in (8), where

$$\mathbf{C} = \begin{pmatrix} -2 & 0 & 0 & 0 \\ -2 & -2 & 0 & 0 \\ -2 & 0 & -2 & 0 \\ -1 & 0 & 0 & -2 \end{pmatrix}, \quad \mathbf{D} = \begin{pmatrix} 0 & 1 & 1 & 1 \\ 0 & 0 & 0 & 0 \\ 0 & 0 & 0 & 0 \\ 0 & 0 & 0 & 0 \end{pmatrix}, \mathbf{F}(\mathbf{Z}(t), \ \mathbf{Z}(t-\tau)) = \begin{pmatrix} F_1(t) \\ F_2(t) \\ F_3(t) \\ F_4(t) \end{pmatrix},$$

$$\widetilde{\mathbf{F}} = \mathbf{D}[\mathbf{Z}(t - \tau_0 - \epsilon^2\tau_\epsilon) - \mathbf{Z}(t - \tau_0)] + \epsilon\mathbf{F}(\mathbf{Z}(t), \mathbf{Z}(t - \tau_0 - \epsilon^2\tau_\epsilon)),$$

and

$$F_1 = -\frac{1}{3}\epsilon x_2^3(t - \tau_0 - \epsilon^2\tau_\epsilon) - \frac{1}{3}\epsilon x_3^3(t - \tau_0 - \epsilon^2\tau_\epsilon) - \frac{1}{3}\epsilon x_4^3(t - \tau_0 - \epsilon^2\tau_\epsilon) + o(\epsilon^3),$$

$$F_2 = \frac{2}{3}\epsilon x_1^3(t) + o(\epsilon^3), \quad F_3 = \frac{2}{3}\epsilon x_1^3(t) + o(\epsilon^3), \quad F_4 = \frac{1}{3}\epsilon x_1^3(t) + o(\epsilon^3).$$

If $\mathbf{W}(t)$ is assumed to be a periodic solution of (14), then it can be represented in the form of

$$\mathbf{W}(t) = \begin{pmatrix} p_1\cos(t) + q_1\sin(t) \\ p_2\cos(t) + q_2\sin(t) \\ p_3\cos(t) + q_3\sin(t) \\ p_4\cos(t) + q_4\sin(t) \end{pmatrix}. \tag{18}$$

Substituting (18) into (17) yields that $p_1 = -2p_2 - q_2$, $q_1 = p_2 - 2q_2$, $p_3 = p_2$, $q_3 = q_2$, $p_4 = p_2$, $q_4 = q_2$. Consequently, the periodic solution of (2) with (6) can be expressed as

$$\mathbf{Z}(t) = \begin{pmatrix} -\frac{1}{2}\epsilon r_1(2\cos((1+\epsilon^2\sigma_2)t+\theta) - \sin((1+\epsilon^2\sigma_2)t+\theta)) \\ \epsilon r_1\cos((1+\epsilon^2\sigma_2)t+\theta) \\ \epsilon r_1\cos((1+\epsilon^2\sigma_2)t+\theta) \\ \frac{1}{2}\epsilon r_1\cos((1+\epsilon^2\sigma_2)t+\theta) \end{pmatrix}. \tag{19}$$

Substituting (18) and (19) into (15), one can obtain an approximation (19) at $O(\epsilon^3)$, and r_1, σ_2 satisfy the following equations given by

$$\frac{21}{8}\pi(\epsilon r_1)^3 - \frac{5}{2}\pi\arccos(-\frac{3}{5})(\epsilon^2\sigma_2)(\epsilon r_1) - \frac{5}{2}\pi(\epsilon^2\tau_\epsilon)(\epsilon r_1) = 0,$$
$$\frac{21}{16}\pi(\epsilon r_1)^3 + 5\pi\left(1 + \arccos(-\frac{3}{5})\right)(\epsilon^2\sigma_2)(\epsilon r_1) + 5\pi(\epsilon^2\tau_\epsilon)(\epsilon r_1) = 0, \quad (20)$$

where $\epsilon^2\tau_\epsilon = \tau - \tau_0$ and $\tau_0 = 2.2143$. Solving (20) yields that

$$\epsilon r_1 = 4\sqrt{\frac{5(\tau-\tau_0)}{21(4+5\arccos(-\frac{3}{5}))}}, \quad \epsilon^2\sigma_2 = -\frac{5(\tau-\tau_0)}{4+5\arccos(-\frac{3}{5})}. \quad (21)$$

To consider the accuracy expressed in (19) with (21), we compare the present result with that from the CMR and the numerical simulation, respectively, as

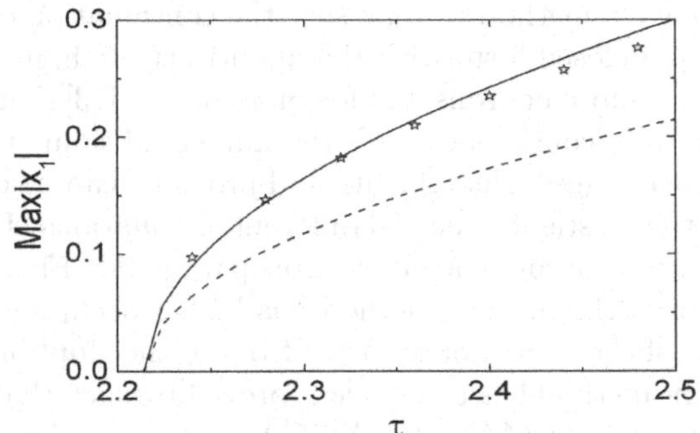

Fig. 3. Hopf bifurcation curves from the trivial solution in (2) with (6), where solid line denotes the present solution (19) with (21), dashing line represents solution from the CMR from [9] and star that from numerical simulation for (2) with (6)

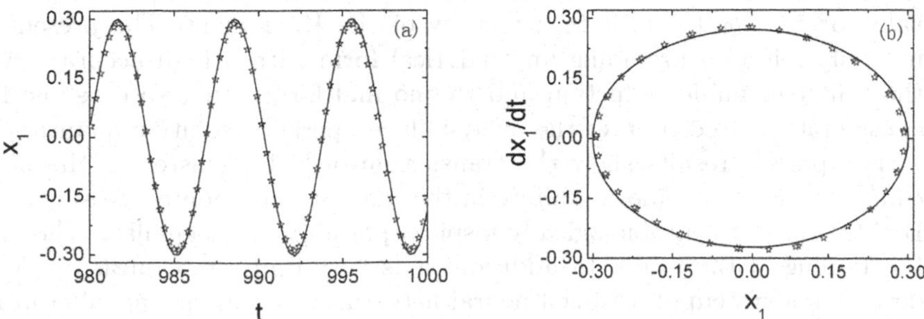

Fig. 4. Comparison in (a) time history and (b) phase plane, between the present solution (19) with (21) (solid line) and numerical simulation (symbol stars) from (2) with (6), where $\tau_0 = 2.2143$, $\tau = 2.5$ or $\epsilon\tau_\epsilon = 0.2857$

shown in Fig. 3 where the result from the present method is marked in solid, that from the CMR in dashing and the numerical simulation in symbol stars. It can be seen that the solution from the present method is much closer to the numerical solution than that from the CMR, especially for values far away from the Hopf bifurcation point at τ_0. For example, the approximation error of the accuracy related to the numerical solution at $\tau = 2.48$ is 25% for the CMR but 5.07% for the PIS. Figs. 4(a) and 4(b) display the corresponding time history and phase plane when $\tau = 2.5$, It can be seen from Fig. 4 that the approximate solution (19) with (21) is in good agreement with that from the numerical simulation. This implies the present method has very high accuracy in quantity.

4 Conclusion

Based on Cao's work [9] and our previous result [11], a simple but efficient method is extended to investigate the periodic solution derived from the Hopf bifurcation due to time delay in a BAMNN with four neurons and two delays. This paper contributes a method to represent the continuation of the bifurcated periodic solutions in a closed form with the quantitatively high accuracy. It can be easily extended for more neurons and for more delays [13]. The validity of the results is shown by their consistency with the numerical simulation. The results obtained in this paper suggest that the method provided can be considered as an effective approach to investigate delayed differential equations (DDEs) when the time delay is taken into account a bifurcation parameter. Firstly, the periodic solution obtained from the present method has higher accuracy than that from the CMR for values of the delay not far away from to the Hopf bifurcation point. Secondly, the present method has a very clear procedure such that some symbolic algebraic packages, such as MATHEMATICA, can easily be programmed to compute the solution. Thirdly, using the present method can avoid the tedious computation often encountered in the CMR. resulting in being extended to the study of high-codimension DDEs.

It should be emphasized that the parameter ϵ is not required to be small in the present method. In fact, a suitable choice of τ_ϵ can always ensure that the value of $\tau = \tau_0 + \epsilon^2\tau_\epsilon$ is in a vicinity of the Hopf point. The present paper not only relies on providing an analytical form with a high accuracy without the center manifold reduction and the normal form, but also on some further considerations to computer the delay-induced periodic solution quantitatively.

The present results show the transformation delay existed in the networks can cause storing memories failure in the brain information processing since the BAMNNs are a neurobiologically inspired paradigm that emulates the memory functioning of the brain. In addition, it is very difficult to measure delays in designing a system of artificial neural networks to mimic phenomena in nature or engineering. One has to guess these delays by the output signals of the system in designing a system of artificial neural networks to mimic phenomena in nature or engineering. Thus, an analytical form of the solution, even an approximation, becomes of important significance since it may considered as a basis for estimating delays existed. This is reason that we are interested in seeking for an analytical form of the delay-induced periodic solution in artificial neural networks.

Acknowledgments. This work is supported by Hong Kong Research Grants Council under CERG Grant Cityu 1005/07E and Key Project of National Nature and Science Foundation of China under grant No. 10532050. The senior author also thanks for the support of the National Outstanding Young Funds of China under Grant No. 10625211 and the Program of Shanghai Subject Chief Scientist under Grant No. 08XD14044.

References

1. Hopfield, J.J.: Neural Networks and Physical Systems with Emergent Collective Computational Abilities. In: Proceedings of the National Academy of sciences of the United States of America-Biological Sciences, vol. 79, pp. 2554–2558 (1982)
2. Kosko, B.: Adaptive Bidirectional Associative Memories. Applied Optics 26, 4947–4960 (1987)
3. Cao, J.D., Wang, L.: Exponential Stability and Periodic Oscillatory Solution in BAM Networks with Delays. IEEE Transactions on Neural Networks 13, 457–463 (2002)
4. Li, Y.K.: Existence and Stability of Periodic Solution for BAM Neural Networks with Distributed Delays. Mathematical and Computer modelling 40, 757–770 (2004)
5. Zhou, T.J., Chen, A.P., Zhou, Y.Y.: Existence and Global Exponential Stability of Periodic Solution to BAM Neural Networks with Periodic Coefficients and Continuously Distributed Delays. Physics Letters A 343, 336–350 (2005)
6. Ho, D.W.C., Liang, J.L., Lam, J.: Global Exponential Stability of Impulsive High-order BAM Neural Networks with Time-varying Delays. Neural Networks 19, 1581–1590 (2006)
7. Mohamad, S., Gopalsamy, K.: A Unified Treatment for Stability Preservation in Computer Simulations of Impulsive BAM Networks. Computers & Mathematics with Applications 55, 2043–2063 (2008)
8. Song, Y.L., Han, M.A., Wei, J.J.: Stability and Hopf Bifurcation Analysis on a Simplified BAM Neural Network with Delays. Physica D 200, 185–204 (2005)
9. Cao, J.D., Xiao, M.: Stability and Hopf Bifurcation in a Simplified BAM Neural Network with Two Time Delays. IEEE Transactions on Neural Networks 18, 416–430 (2007)
10. Yan, X.P.: Bifurcation Analysis in a Simplified Trineuron BAM Network Model with Multiple Delays. Nonlinear Ananlysis 9, 963–976 (2008)
11. Xu, J., Chung, K.W., Chan, C.L.: An Efficient Method for Studying Weak Resonant Double Hopf Bifurcation in Nonlinear Systems with Delayed Feedbacks. SIAM J. Applied Dynamical Systems 6, 29–60 (2007)
12. Xu, J., Chung, K.W.: Effects of Time Delayed Position Feedback on a van der PolCDuffing Oscillator. Physica D 180, 17–39 (2003)
13. Xu, J., Chung, K.W.: A perturbation-incremental scheme for studying Hopf bifurcation in delayed differential systems. Science in China Series E: Technological Sciences 52, 698–708 (2009)

Response Properties to Inputs of Memory Pattern Fragments in Three Types of Chaotic Neural Network Models

Hamada Toshiyuki, Jousuke Kuroiwa, Hisakazu Ogura, Tomohiro Odaka, Haruhiko Shirai, and Yuko Kato

Graduate School of Engineering, University of Fukui, 3-9-1, Bunkyo, Fukui, Fukui, 910-8507, Japan
{hamada,jou,ogura,odaka,shirai}@i.his.fukui-u.ac.jp

Abstract. In this paper, we investigate response properties to inputs of memory pattern fragments in chaotic wandering states among three types of chaotic neural network (CNN) models, related with the instability of their orbits. From the computer experiments, Aihara model shows the highest success ratio and the shortest steps for all the memory pattern fragments. On the other hand, Nara & Davis model and Kuroiwa & Nara model show quite higher success ratio and shorter averaged steps than random search. Thus, choas in the three model is practical in the memory pattern search.

Keywords: Instability of orbit, CNN, memory search, Memory pattern fragments, Chaotic wandering state.

1 Introduction

Recently, chaotic behaivors have been invesigated and observed in the various fieldincluding the brain nervous system[1,2]. From the EEG experiments and computational research on the olfactory bulb, Skarda and Freeman have shown the following two features[2].

1. During the waiting states for unknown inputs, the dynamical response falls into highly developed choatic attractor.
2. The response of system to memorized input shows the dynamical behaivors of the weak chaotic attractor or limit cycles.

Thus, choatic behaviors would play the important roles in a learning process and a recall process. In the learning process, chaos could provide driving activity essential for memorizing novel inputs. In the recall process, chaos could ensure rapid and unbiased access to proviously trained patterns. From the theoretical viewpoints, Nara and Davis presented the interesting results in complex memory search of neuron network model with multi-cyclic memory patterns[3,4].Hence, in order to access the rapid to memory patterns, it is important to control the chaotic wandering states.

C. Alippi et al. (Eds.): ICANN 2009, Part II, LNCS 5769, pp. 544–551, 2009.

In this paper, we investigate response properties in memory pattern fragments in chaotic wandering states among three types of chaotic neural network model. We focus on the following three types of chaotic neural network models:

1. Aihara model: The existence of relative refractoriness and the continuity of the output function introduce chaotic wandering state in system even if the system is an isolated neuron element[5].
2. Nara & Davis model: By reducing the connectivity of synaptic connections, chaotic wandering state is induced[3].
3. Kuroiwa & Nara model: By partly inverting synaptic connections, chaotic wandering state apperars[6].

We hae shown that these three types of chaotic neural network model reveals similar chaotic wandering dynamics in the space of memory patterns. Furthermore, they have the similar functional potentiality of memory searching and synthesis[3,7,8].However, it has been still unknown whether the different potentiality does underly in them, or not, In this paper, therefore, we investigate the relationships betwwon the response properties to memory pattern fragments and the instability of their orbits.

2 Chaotic Neural Network Model

2.1 Recurent Neural Network Model with Assocciatiev Memory

Let us explain a reccurent neural network model, briefly. The updating rule of the reccurent neural network model is writen as follows,

$$u_i(t+1) = \sum_{j=1}^{N} w_{ij} z_j(t), \tag{1}$$

where $u_i(t+1)$ represents an internal state of the ith element at discreet time t, $z_i(t)$ is its output, w_{ij} is a synaptic connection between the ith element and the jth element, and N is the total number of elements in the reccurent neural network model.

In this paper, we employ continuous output function in order to evaluate Lyapunov dimension,

$$z_i(t+1) = \tanh(\beta u_i(t+1)), \tag{2}$$

where β controls steepness of the output function. In computer experiments, we apply $\beta = 100.0$ that the function corresponds to a sign function approximately.

In this paper, the synaptic connection is defined as,

$$w_{ij} = \sum_{a=1}^{P} \sum_{\mu=1}^{L} v_i^{a\ \mu+1} (v_j^{a\ \mu})^{\dagger}, \tag{3}$$

where $\boldsymbol{v}^{a\ \mu}$ denotes a memory pattern vector, L and P are the number of cycles and the number of patterns per cycle, respectively, and $\boldsymbol{v}^{a\ P+1} = \boldsymbol{v}^{a\ 1}$. The dagger vector of $(\boldsymbol{v}^{a\ \mu})^{\dagger}$ is given by

$$(\boldsymbol{v}^{a\ \mu})^{\dagger} = \sum_{b=1}^{L}\sum_{\nu=1}^{P}(\boldsymbol{o}^{-1})_{a\mu b\nu}\boldsymbol{v}^{b\ \nu}, \tag{4}$$

where $\boldsymbol{o}^{-1}$ is the inverse of the $\boldsymbol{o}$ defined by

$$(\boldsymbol{o}_{a\mu b\nu}) = \sum_{k=1}^{N} v_k^{a\ \mu} v_k^{b\ \nu}, \tag{5}$$

Thus, according to Eq(3), P cycles and L patterns per cycle are embedded in the recurrent neural netwrok model.

2.2 Aihara Model

In Aihara model[5], the updating rule is given by

$$u_i(t+1) = ku_i(t) + \sum_{j=1}^{N} w_{ij}z_j(t) - \alpha z_i(t) + A_i, \tag{6}$$

where k control a decay effect, $\alpha z_i(t)$ denotes a relative refractoriness of the ith element, and A_i represent a constant bias input. The output of $z_i(t)$ is given by Eq.(2). The parameters are k, α and A which control dynamics.

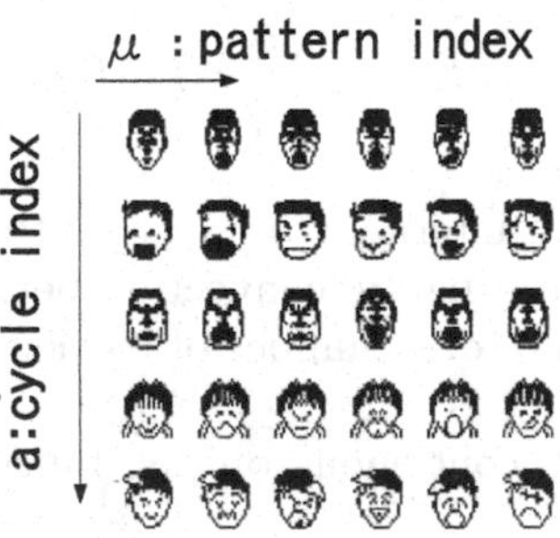

Fig. 1. Memory patterns[3]. The number of cycles, $L = 5$, the number of patterns per cycle, $P = 6$, and each pattern consists of 20×20 pixels which take ± 1. In this paper, we identify the one face pattern with two notations, ath cyde and μth pattern, or kth face pattern($k = a \times \mu$, $k = 1, 2, \ldots, 30$).

2.3 Nara and Davis Model

In Nara & Davis model,the updating tule is represented by

$$u_i(t+1) = \sum_{j=1}^{N} w_{ij}\epsilon_{ij}(d)z_j(t), \tag{7}$$

where $\epsilon_{ij}(d)$ denotes a matrix of binary activity values, that is,

$$\epsilon_{ij}(d) = \begin{cases} 0 & (j \in F_i(d)) \\ 1 & (\text{otherwise}) \end{cases} \tag{8}$$

where $F_i(d)$ represents a configuration of elements at which the in-coming synaptic connection is inhibited, and d is the connectivity which represents of remaining synaptic connection, that is, $\sum_{ij} \epsilon_{ij}(d) = d$. In thist model, the connectivity d is a system parameter which control dynamics.

2.4 Kuroiwa and Nara Model

In Kuroiwa & Nara model,the updating rule is written by

$$u_i(t+1) = \begin{cases} \sum_{j=1}^{N} w_{ij}\epsilon_{ij}z_j(t) & (t = \text{even}) \\ \sum_{j=1}^{N} w_{ij}z_j(t) & (t = \text{odd}), \end{cases} \tag{9}$$

where

$$\epsilon_{ij}(r) = \begin{cases} -1 & (j \in F_i(r)) \\ 1 & (\text{otherwise}), \end{cases} \tag{10}$$

and $F_i(r)$ defines a configuration of elements at which a value of in-coming synaptic connections is partly inverted with the number of r. In this model, the inverted number of r is a system parameter which control dynamics.

3 Response properties in Memory Pattern Fragments

3.1 The Purpose and the Method of Experiment

we investigate the diffference of the response properties to memory pattern fragments in chaotic wandering states among three types of chaotic neural network model. In the computer experiments, we add a certain fragment of a memory pattern to a chaotic wandering state, and then we focus on the folloing two problem.

1. Success ratio : How many times does it reach the target basin within 30 iteration steps starting from different points of chaotic wandering state? In other words, how is the searching procedure by means of chaotic wandering effective?

2. Averaged steps : How many steps does it take to reach the target basin related with a memory fragment? In other words, how short is the "access" time for the memory basin corresponding to the external input of a memory fragment?

In adding a memory pattern fragment, the updating rule change, respectively, as follows :

$$u_i(t+1) = \sum_{j=1}^{N} w_{ij} z_j(t) + k u_i(t) - \alpha z_i(t) + A_i + c\rho I_i \quad \text{if } i \in F \tag{11}$$

$$u_i(t+1) = \sum_{j=1}^{N} w_{ij} \epsilon_{ij} z_j(t) + c\rho I_i \quad \text{if} \quad i \in F \tag{12}$$

$$u_i(t+1) = \begin{cases} \sum_{j=1}^{N} w_{ij} \epsilon_{ij} z_j(t) + c\rho I_i \quad \text{if} \quad i \in F & (t = \text{even}) \\ \sum_{j=1}^{N} w_{ij} z_j(t) + c\rho I_i \quad \text{if} \quad i \in F & (t = \text{odd}) \end{cases} \tag{13}$$

where I_i represents a memory pattern fragment with 40 pixels as shown in the upper part of Fig.2, F denotes the configuration set of theirs, p denotes the strength of the input, and c denotes the normalized parameter which corresponds to the standard deviation of $u_i(t)$. In the experiments, we employ a chaotic wandering state for each model. We set parameter values as follows:

- For Aihara model, $k = 0.9$, $\alpha = 1.2$, $A = 0.1$ for all elements and the standard deviation $c = 0.86$
- For Nara & Davis model, $d = 390$, and $c = 0.04$
- For Kuroiwa & Nara model, $r = 214$ and $c = 0.34$

The system parameter values are chosen in giving optimal results base on the preliminary experiments. If the system converge the pattern related with memory pattern fragment within 30 steps of the updating rule in adding a memory

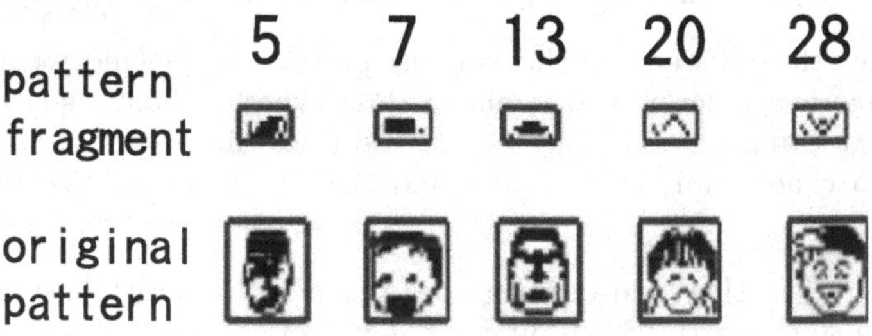

Fig. 2. Memory pattern fragments

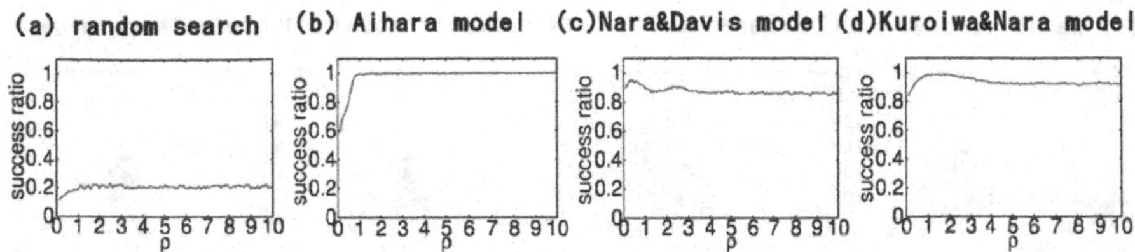

Fig. 3. Success ration in accessing to the target pattern. (a) Random search. (b)Aihara model. (c) Nara & Davis model. (d) Kuroiwa & Nara model.

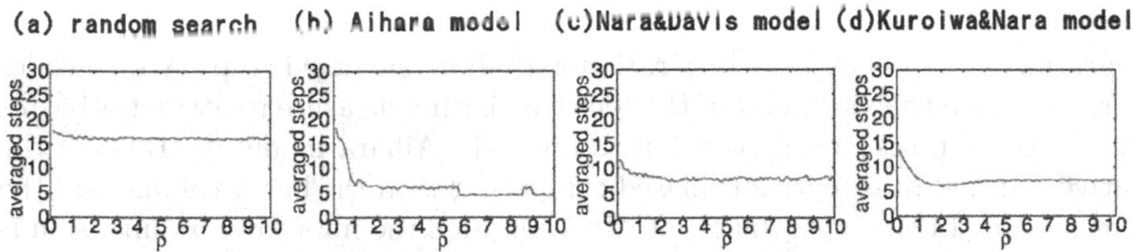

Fig. 4. Averaged steps in accessing to the target pattern. (a) Random search. (b)Aihara model. (c) Nara & Davis model. (d) Kuroiwa & Nara model.

pattern fragment, we identify the search of memory pattern as success. On the other hand, if the chaotic wandering state doesn't converge the pattern related with memory pattern fragment within 30 steps , we regard that it take 30 steps. In addition, we evaluate the response feature of memory pattern fragments with 1000 steps in chaotic wandering state. Comparing with the response feature between the random pattern and chaotic wandering state, we employ the random search. In random search, we employ the random patterns instead of the output of chaotic neural network model and perform the same process.

3.2 Results

Results of success ration and averaged steps are given in Fig.3 and 4, respectively. From the results, comparing the random search, the others show high success ratio and short steps. Especially, Aihara model reveals a 99 % probability of success ratio over the region of $\rho > 1$. Thus, the system accesses to the target memory pattern with the 10 percent of memory pattern. In addition, Aihara model shows that the averaged steps is quite small, almost around 5 steps. On the other hand, Nara & Davis model and Kuroiwa & Nara model take lower success ratio than Aihara model. Therefore, the average aceess time are 7 steps and 6 steps. However, comparing with the random search, Nara & Davis model and Kuroiwa & Nara model quickly acceess to the target pattern with high success ratio.

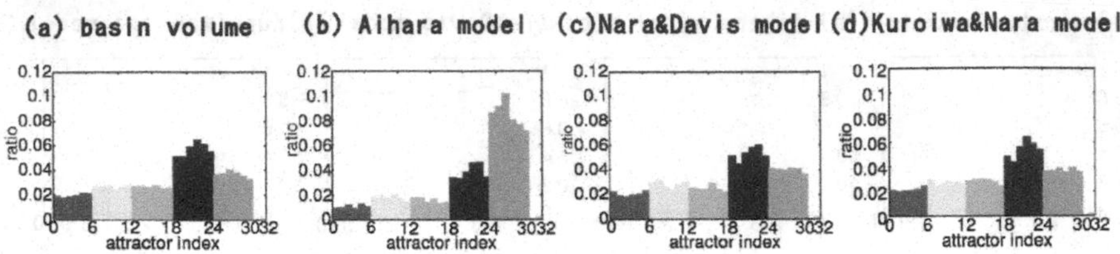

Fig. 5. The relative visiting measure to attractor basins of memory patterns

4 Discussions

Aihara model shows high success ration and short averaged steps. We consider the reason from the viewpoint of the relative visiting measure to attractor basins of memory patterns. Result is given in Fig.5. In Aihara model,the ratio of the relative visiting measures show biased comparing with the basin volume and the other models. In these parameter of Aihara model, the information dimension is 1.31. If we employ the other parameters where Aihara model reveals quite strong chaotic behaviors with the information dimension of 2.97 and the equal ratio of the relative visiting measures, the success ratio decreases and the averaged steps become long. In intuitive, the equal ratio means that one can realize easier access to each attractor of memory patterns than the case of biased ratio. However, our results are opposite. In the case of the biased ratio with weak chaotic behaviors, we succeeded to realize easy access. We consider that this result would relate with edge of chaos.

5 Conclusions

In this paper, we investigate the response properties to memory pattern fragments in chaotic wandering states. The three models, which are in chaotic wandering states, show the higher success ratio and the shorter averaged steps to reach the target memory pattern than random search. In other words, chaotic wandering states are quite sensitive to the memory pattern fragments. The sensitivity could play important role in realizing in memory pattern search.

References

1. Aihara, K., Matsumoto, G.: Chaotic oscillations and bifurcations in squid giant axons. In: Holden, A.V. (ed.) Chaos, pp. 257–269. Manchester University and Princeton University Press, Princeton (1987)
2. Freeman, W.J., Skarda, C.: How brains make chaos in order to make sense of the world. Behavioral and Sciences 10, 161–195 (1987)
3. Nara, S., Davis, P., Kawachi, M., Totsuji, H.: Chaotic Memory Dynamics in a Recurent Neural Network with Cycle Memories Embeded by Pseudoinverse Method. Int. J. Bifurcation & Chaos 5, 1205–1212 (1995)

4. Nara, S.: Can potentially useful dynamics to solve complex problems emerge from constrained chaos and / or chaotic itinerancy? Chaos 13, 1110–1121 (2003)
5. Aihara, K., Takabe, T., Toyoda, M.: Chaostic Neural Networks. Phys. Lett. A 144, 333–339 (1990)
6. Nakayama, S., Kuroiwa, J., Nara, S.: Partly Inverted Synaptic Connections and Complex Dynamics in a Symmetric Recurrent Neural Network Model. In: Proceedings of international joint conference on neural network ICONIP 2000, Taejon, Korea, vol. 2, pp. 1274–1279 (2000)
7. Kuroiwa, J., Masutani, S., Nara, S., Aihara, K.: Sensitive responses of chaotic wandering to memory pattern fragment inputs in a chaotic neural network mode. Int. J. Bifurcation & Chaos 14, 1413–1421 (2004)
8. Ishii. T., Kuroiwa. J., Ushijima. N., Takahashi. I., Shirai. H., Odaka. T., Ogura. T.: Sensitivity to memory fragment in chaotic wandering by partly inverted synaptic connection method. IEICE Technical Report, NLP2005-128 (2006) (in Jananese)

Partial Differential Equations Numerical Modeling Using Dynamic Neural Networks

Rita Fuentes[1], Alexander Poznyak[1], Isaac Chairez[2], and Tatyana Poznyak[3]

[1] CINVESTAV-IPN, Automatic Control Department, México, D.F.
rfuentes@ctrl.cinvestav.mx
[2] UPIBI-IPN, Bioelectronics Department, México, D.F.
[3] ESIQIE-IPN, Postgraduete Division, México, D.F.

Abstract. In this paper a strategy based on differential neural networks (DNN) for the identification of the parameters in a mathematical model described by partial differential equations is proposed. The identification problem is reduced to finding an exact expression for the weights dynamics using the DNNs properties. The adaptive laws for weights ensure the convergence of the DNN trajectories to the PDE states. To investigate the qualitative behavior of the suggested methodology, here the non parametric modeling problem for a distributed parameter plant is analyzed: the anaerobic digestion system

Keywords: Neural Networks, Adaptive identification, Distributed Parameter Systems, Partial Differential Equations and Practical Stability.

1 Introduction

Many problems in science and engineering are reduced to a set of partial differential equations (DES) through a process of mathematical modeling. For instance, linear second order parabolic partial differential equations (PDEs) appear in time dependent diffusion problems, such as the transient flow of heat conduction . These equations define a state in both space and time. It is not easy to obtain their exact solutions, so numerical methods must be resorted to. There are a lot of techniques available such as the finite difference method (FDM) [1] and the finite element method (FEM) [2]. These numerical techniques, require large number of iterations in calculation and process the data in series. Besides, all those methods are well defined if the PDE structure is perfectly known. Actually, the most of suitable numerical solutions could be achieved just when the PDE is linear. Nevertheless, there are not so many methods to solve or approximate the PDE solution when its structure (even in linear case) is uncertain.

It is well known that Radial Basis Function Neural Networks (RBFNN) and Multi-Layer Perceptrons (MLP) are universal approximators [3]: any continuous function defined on a compact set can be approximated to arbitrary accuracy by such neural networks [4]. Since the solutions to the PDE of interest are known to be uniformly continuous and the viable sets that arise in safety problems are often compact, neural networks seem like ideal candidates for approximating

C. Alippi et al. (Eds.): ICANN 2009, Part II, LNCS 5769, pp. 552–562, 2009.

viability problems. There are, however, relatively few works that exploit neural networks to solve PDE. [5] proposed a method for solving PDE defined in orthogonal boxes, that relies upon an approximation composed by a MLP network added to the boundary condition. The method is illustrated by solving a variety of model problems and comparing the result with the exact solution. Convergence is a difficult issue in this case. The results of [6] show that their feed-forward neural network (FFNN) method for solving an elliptic PDE in 2D required 1000 iterations for convergence. Another method for solving PDE is presented in [7], based on a Multi- Quadric RBFNN. The proposed procedure showed high accuracy of the solution. It is important to note, however, that the accuracy of the RBFNN solution is heavily dependent on parameters such as the "width" of basis functions for which there is no systematic method for determining their values. These approaches have to approximate PDE solutions by neural networks have been also been applied to some problems in control theory. In [8] a method like this is used to solve a class of first-order partial differential equations that arise in input-to-state linearizable control systems. The solution of the PDE, together with its Lie derivatives, yields a change of coordinates required for feedback linearization. In [9] a method similar to the method of [5], has been successfully applied to steady-state heat transfer problems.

Within the NN framework, differential neural network (DNN) methodology avoids many problems related to global extremum searching [10]. If mathematical continuous model of the considered process is incomplete or partially known, the DNN methodology provides an effective instrument to analyze a wide range of problems in control theory such as identification, state estimation, trajectories tracking and etc. [11]. Most of real systems are really difficult to be controlled because of the lack of information on its internal structure or-and their current states trajectories. The paper is organized as follows: in Section II, we introduce a model given by a partial differential equation with unknown structure and formulate the problem. In Section III the DNN identifier is proposed. Some simulation results are presented in Section IV to show its performance. Section V finishes the paper with some particular conclusions.

2 Distributed Parameters Plant and NN Approximation

Let us consider the partial differential equation

$$u_t(x,t) = f\left(u(x,t), u_x(x,t), u_{xx}(x,t)\right) \tag{1}$$

for $x \in (0,1)$, $t > 0$, with boundary conditions:

$$u(0,t) = u_0, \quad u(x,0) = c, \quad u_x(0,t) = 0 \tag{2}$$

Let $f(x,t)$ be a piecewise continuous in t. Suppose that the uncertain nonlinear function $f(x,t)$ satisfies the Lipschitz condition $\|f(t,x) - f(t,y)\| \le L\|x-y\|$, $\forall\ x,y \in B := \{x \in \Re^n \mid \|x - x_0\| \le r\}$, $\forall\ t \in [t_0, t_1]$, where L is constant and $\|f\|^2 = (f,f)$ just to ensure that there exists some $\delta > 0$ such that the state equation $\dot{x} = f(x,t)$ with $x(t_0) = x_0$ has a unique solution over $[t_0, t_0 + \delta]$ [12]. The norm defined above in (8) is given in a Sobolev space.

Definition. Sobolev space [13], $H^{m,p}(\Omega)$: Let Ω be an open set in $\mathbb{R}^n$ and let $u \in C^m(\Omega)$. Define a norm on u by

$$\|u\|_{m,p} := \sum_{0\leq|\alpha|\leq m} \left(\int_{\Omega} |D^\alpha u(x)|^p \, dx \right)^{1/p}, \ 1 \leq p < \infty$$

This is the Sobolev norm in which the integration is in the Lebesgue sense. The completion of $u \in C^m(\Omega)$: $\|u\|_{m,p} < \infty$ with respect to $\|\cdot\|_{m,p}$ is the Sobolev space $H^{m,p}(\Omega)$. For $p = 2$, the Sobolev space is a Hilbert space.

Now lets consider a function $h_0(\cdot)$ in $H^{m,2}(\Omega)$. By [14], $h_0(\cdot)$ can be rewritten as

$$h_0(x) = \sum_i \sum_j \theta_{ij} \Psi_{ij} x, \ \theta_{ij} = \int_{-\infty}^{+\infty} h_0(x) \Psi_{ij}(x) \, dx, \ \forall i,j \in \mathbb{Z}$$

Last expression is called a function series expansion of $h_0(x)$. Based on this series expansion, a neural network takes the following mathematical structure

$$\hat{h}_0(x,\theta) := \sum_{i=M_1}^{M_2} \sum_{j=N_1}^{N_2} \theta_{ij} \Psi_{ij}(x) = \Theta^\intercal W(x)$$

that can be used to approximate any nonlinear function $h_0(x) \in S$ with the adequate selection of integers M_1, M_2, N_1, $N_2 \in \mathbb{Z}$ where

$$\begin{aligned} \Theta &= [\theta_{M_1 N_1}, \ldots, \theta_{M_1 N_2}, \ldots \theta_{M_2 N_1}, \ldots \theta_{M_2 N_2}]^\intercal \\ W(x) &= [\Psi_{M_1 N_1}, \ldots, \Psi_{M_1 N_2}, \ldots \Psi_{M_2 N_1}, \ldots \Psi_{M_2 N_2}]^\intercal \end{aligned}$$

Following the Stone Wiestrass Theorem [15], if $\epsilon(M_1, M_2, N_1, N_2) = h_0(x) - \hat{h}_0(x,\theta)$ is the NN approximation error, then for any arbitrary positive constant ε there are some constants $M_1, M_2, N_1, N_2 \in \mathbb{Z}$ such that

$$\|\epsilon(M_1, M_2, N_1, N_2)\|_2 \leq \varepsilon \tag{3}$$

for all $x \in X \subset \Re$. In the case when $x \in X^n \subset \Re^n$ ($x := [x_1, x_2, \ldots, x_n]^\intercal$), the Ψ_{ij} argument (x) should be modify to $(x,c) = c^\intercal x = \sum_{i=1}^n x_i c_i$ with $c \in X^n$ as a weighting constant vector.

Remark 1. Appropriate selection of functions $\Psi_{ij}(\cdot)$ is an important task to construct an adequate approximation of nonlinear functions. Many functions have been reported in literature [16] that have remarkable results to approximate nonlinear unknown functions. Which one is the most suitable basis in practical application depends on each particular design specifications.

Remark 2. M_1, M_2, N_1, N_2 parameters in neural network design are closely related to the quality approximation $\epsilon(M_1, M_2, N_1, N_2)$. Although, the NN has been demonstrated to be effective to reproduce uncertain nonlinear functions which satisfies the Lipschitz condition.

Following the methodology of differential neural networks, we assume that there exists a set of parameters $W_1^{i,*} \in \Re^{s_1}$, $W_2^{i,*} \in \Re^{s_2}$, $W_3^{i,*} \in \Re^{s_3}$ such that

$$u_i - \int_0^t \left(A^i u_i + \left[W_1^{i,*}\right]^\mathsf{T} \sigma(x_i) + \left[W_2^{i,*}\right]^\mathsf{T} \varphi(x_i) u_{i-1} - \left[W_3^{i,*}\right]^\mathsf{T} \gamma(x_i) u_{i-2} - \tilde{f}^i \right) d\tau = 0$$

where the functions $\sigma(u_i) \in \Re^{s_1}$, $\varphi(u_i) \in \Re^{s_2}$, $\gamma(u_i) \in \Re^{s_3}$ obey the following sector conditions:

$$\|\sigma(v_i) - \sigma(\tilde{v}_i)\|^2 \le L_\sigma \|v_i - \tilde{v}_i\|^2, \ \|\varphi(v_i) - \varphi(\tilde{v}_i)\|^2 \le L_\varphi \|v_i - \tilde{v}_i\|^2$$
$$\|\gamma(v_i) - \gamma(\tilde{v}_i)\|^2 \le L_\gamma \|v_i - \tilde{v}_i\|^2$$

which are bounded in U, i.e., $\|\sigma(\cdot)\|^2 \le \sigma^+$, $\|\varphi(\cdot)\|^2 \le \varphi^+$, $\|\gamma(\cdot)\|^2 \le \gamma^+$. Note that, since one requires $\partial u(x,t)/\partial t$ in (1), the NN weights are selected to be time varying. However, here $\sigma(x_i)$, $\varphi(x_i)$, $\gamma(x_i)$ are NN activation vectors, not a set of eigen-functions. That is, the NN approximation property significantly simplifies the specification of $\sigma(\cdot)$, $\varphi(\cdot)$, $\gamma(\cdot)$.

The terms $\tilde{f}^i$, called *modeling errors* of each NN applied for the approach of the PDE, that is,

$$\tilde{f}^i := f^i - \left(A^i u_i + \left[W_1^{i,*}\right]^\mathsf{T} \sigma(x_i) + \left[W_2^{i,*}\right]^\mathsf{T} \varphi(x_i) u_{i-1} + \left[W_3^{i,*}\right]^\mathsf{T} \gamma(x_i) u_{i-2} \right)$$

- *Assumption 1.* The modeling error $\tilde{f}^i$ satisfy the following group of inequalities:

$$\left\|\tilde{f}^i\right\|^2 \le f_0^i \|u_i\|^2 + f_1^i \|u_{i-1}\|^2 + f_2^i \|u_{i-2}\|^2 + f_3^i \tag{4}$$

  and $\left\|\bar{f}^i\right\|^2 \le F_0^i \|u_i\|^2 + F_1^i \|u_{i-1}\|^2 + F_2^i \|u_{i-2}\|^2 + F_3^i \left\|\Delta^i(t,x)\right\|^2 + F_4^i$ where $\|\bar{f}^i\|^2 := \left\|\tilde{f}^i + A^i \Delta^i(t,x)\right\|^2$

- *Assumption 2.* The error modeling gradient is bounded as follows : $\left\|\nabla_x \tilde{f}^i\right\|^2 \le f_4^i \|u_i\|^2 + f_5^i \|u_{i-1}\|^2 + f_6^i \|u_{i-2}\|^2 + f_7^i$ yielding to

$$\left\|\nabla_x \bar{f}^i\right\|^2 := \left\|\nabla_x \tilde{f}^i + A^i \Delta_x^i(t,x)\right\|^2 \le$$
$$F_5^i \|u_i\|^2 + F_6^i \|u_{i-1}\|^2 + F_7^i \|u_{i-2}\|^2 + F_8^i \left\|\Delta_x^i(t,x)\right\|^2 + F_9^i$$

where $\Delta^i(t,x) := \hat{u}_i(x,t) - u_i(x,t)$ and f_j^i, F_k^i ($j = \overline{0,7}$, $k = \overline{0,9}$) are constants.

3 DNN Neural Identifier for Distributed Systems

Consider the following structure of adaptive identifier

$$\begin{aligned}\frac{d}{dt}\hat{u}_i(x,t) = & A^i\hat{u}_i(x,t) + \left[W_{1,t}^i\right]^\intercal \sigma(\hat{u}_i) + \\ & \left[W_{2,t}^i\right]^\intercal \varphi(\hat{u}_i)\hat{u}_{i-1}(x,t) + \left[W_{3,t}^i\right]^\intercal \gamma(\hat{u}_i)\hat{u}_{i-2}(x,t)\end{aligned} \tag{5}$$

$\forall i \in [3, N]$ where $A^i \in \Re^-$ and where the variant in time matrices $W_{1,t}^i \in \Re^{s_1}$, $W_{2,t}^i \in \Re^{s_2}$, $W_{3,t}^i \in \Re^{s_3}$ and $\hat{u}_i(x,t)$ is the estimate of $u_i(x,t)$. Satisfy the matrix differential equations

$$\begin{aligned}
\dot{W}_1^i(t) = & -\frac{2}{K_1}\sum_{i=1}^{N}\hat{u}_i^\intercal(t,x)\,T^i\sigma(\hat{u}_i) - \frac{2}{K_1}\sum_{i=1}^{N}\left(\Delta^i(t,x)\right)^\intercal P^i\sigma(\hat{u}_i) \\
& -\alpha_m^i\tilde{W}_1^i(t) - \frac{2}{K_1}\sum_{i=1}^{N}\left(\Delta_x^i(t,x)\right)^\intercal S^i\nabla_x\sigma(\hat{u}_i) \\
\dot{W}_2^i(t) = & -\frac{2}{K_2}\sum_{i=1}^{N}\hat{u}_i^\intercal(t,x)\,T^i\varphi(\hat{u}_i)\hat{u}_{i-1}(t,x) - \alpha_m^i\tilde{W}_2^i(t) \\
& -\frac{2}{K_2}\sum_{i=1}^{N}\left(\Delta^i(t,x)\right)^\intercal P^i\varphi(\hat{u}_i)\hat{u}_{i-1}(t,x) - \frac{2}{K_2}\sum_{i=1}^{N}\Delta_x^i(t,x)^\intercal S^i\nabla_x\varphi(\hat{u}_i)\hat{u}_{i-1}(t,x) \\
\dot{W}_3^i(t) = & -\frac{2}{K_3}\sum_{i=1}^{N}\hat{u}_i^\intercal(t,x)\,T^i\gamma(\hat{u}_i)\hat{u}_{i-2}(t,x) - \alpha_m^i\tilde{W}_3^i(t) \\
& -\frac{2}{K_3}\sum_{i=1}^{N}\left(\Delta_x^i(t,x)\right)^\intercal S^i\nabla_x\gamma(\hat{u}_i)\hat{u}_{i-2}(t,x) - \frac{2}{K_3}\sum_{i=1}^{N}\left(\Delta^i(t,x)\right)^\intercal P^i\gamma(\hat{u}_i)\hat{u}_{i-2}(t,x)
\end{aligned} \tag{6}$$

with P^i, S^i and T^i $(i = \overline{3, N})$ are positive definite solutions ($P^i > 0$, $S^i > 0$ and $T^i > 0$) of the algebraic Riccati equations given by

$$\begin{aligned}
& P^iA^i + \left[A^i\right]^\intercal P^i + P^i\Lambda_\alpha^iP^i + \lambda_{\max}\left(\left[\Lambda_P^i\right]^{-1}\right)F_3^iI_{n\times n} + Q_P^i = 0 \\
& S^iA^i + \left[A^i\right]^\intercal S^i + S^i\Lambda_S^iS^i + \lambda_{\max}\left(\left[\Lambda_S^i\right]^{-1}F_8^iI_{n\times n}\right) + Q_S^i = 0 \\
& T^iA^i + \left[A^i\right]^\intercal T^i + T^i\Lambda_T^iT^i + Q_T^i + \lambda_{\max}\left(\left[\Lambda_P^i\right]^{-1}\right)F_0^iI_{n\times n} + \\
& \quad \left(\lambda_{\max}\left(\left[\Lambda_S^i\right]^{-1}F_5^i\right) + \lambda_{\max}\left(\left[\Lambda_T^i\right]^{-1}\right)f_0^i\right)I_{n\times n} = 0
\end{aligned} \tag{7}$$

Special class of Riccati equation $PA + A^\intercal P + PRP + Q = 0$ has positive solution if and only if [10] the following four conditions given below are fulfilled.

- Matrix A is stable,
- Pair $\left(A, R^{1/2}\right)$ is controllable,
- Pair $\left(Q^{1/2}, A\right)$ is observable,
- Matrices $(A,\ Q,\ R)$ should be selected in such a way to satisfy the following inequality $\frac{1}{4}\left(A^\intercal R^{-1}\text{-}R^{-1}A\right)R\left(A^\intercal R^{-1}\text{-}R^{-1}A\right)^\intercal + Q \le A^\intercal R^{-1}A$

Last condition restricts the largest eigenvalue of R avoiding the inexistence of Riccati equation positive solution. State estimation problem for uncertain nonlinear systems analyzed in this study, could be now stated as follows:

Problem Statement. Under the nonlinear system with an adequate selection of matrices A^i and with the neural network identifier structure supplied with the adjustment law (6) (including the selection of W_i^*, $i = 1, 2, 3$), the upper bound for the estimation error β defined as

$$\beta := \overline{\lim_{t\to\infty}} \left\| \hat{u}\left(t,x\right) - u\left(t,x\right) \right\|_P^2 \tag{8}$$

$P > 0$, $P = P^{\intercal} \in \mathbb{R}^{n\times n}$ *must be obtained,* and if it is possible, to reduce to its less achievable value, using any of the free parameters participating into the NN structure.

The following definition and proposition are needed for the main results of the paper. Consider the following ODE nonlinear system

$$\dot{z}_t = g(z_t, v_t) + \varpi_t \tag{9}$$

with $z_t \in \Re^n$, $v_t \in \Re^m$ and ϖ_t an external perturbation or uncertainty such that $\left\|\varpi_t\right\|^2 \leq \varpi^+$

Definition 1 (Practical Stability). *Assume that a time interval T and a fixed function $v_t^* \in \Re^m$ over T are given. Given $\varepsilon > 0$, the nonlinear system (9) is said to be ε-practically stable over T under the presence of ϖ_t if there exists a $\delta > 0$ (δ depends on and the interval T) such that $z_t \epsilon B[0,\varepsilon], \forall\ t \epsilon T$, whenever $z_{t_0} \epsilon B[0,\delta]$.*

Similarly to the Lyapunov stability theory for nonlinear systems, it was applied the aforementioned direct method for the ε-practical stability of nonlinear systems using-practical Lyapunov-like functions under the presence of external perturbations and model uncertainties. Note that these functions have properties differing significantly from the usual Lyapunov functions in classic stability theory.

The following proposition requires the following Lemma.

Lemma 1. *Let a nonnegative function $V(t)$ sastisfying the following differential inequality $\dot{V}(t) \leq -\alpha V(t) + \beta$ where $\alpha > 0$ and $\beta \geq 0$. Then $\left[1 - \mu\left(\sqrt{V(t)}\right)^{-1}\right]_+ \to 0$ with $\mu = \sqrt{\beta/\alpha}$ and the function $[\cdot]_+$ defined as*

$$[z]_+ := \begin{cases} z \; if \; z \geq 0 \\ 0 \; if \; z < 0 \end{cases}$$

Proof. The proof of this Lemma can be found in [17].

Proposition 1. *Given a time interval T and a function $v(\cdot)$ over a continuously differentiable real-valued function $V(z,t)$ satisfying $V(0,t) = 0, \forall\ t \epsilon T$ is said to be $\varepsilon-$practical Lyapunov-like function over T under v if there exist a constant $\alpha > 0$ such that $\dot{V}(z,t) \leq -\alpha V(z,t) + H(\varpi^+)$ with H a bounded non-negative nonlinear function with upper bound H^+. Moreover, the trajectories of z_t belongs to the zone $\varepsilon := \dfrac{H^+}{\alpha}$ when $t \to \infty$. In this proposition $\dot{V}(z_t,t)$ denotes the derivative of $V(z,t)$ along z_t, i.e., $\dot{V}(z,t) = V_z(z,t) \cdot (g(z_t,v_t) + \varpi_t) + V_t(x,t)$.*

Proof. The proof follows directly from Lemma 1.

Definition 2. *Given a time interval T and a function $v(\cdot)$ over T, nonlinear system (9) is ε-practically stable, T under v if there exists an $\varepsilon-$practical Lyapunov-like function $V(x,t)$ over T under v.*

Theorem. Let be the non linear system described on PDE's unknown and perturbed on the state and the output (1) with the conditions at the border of Dirichlet and Neumman type defined on (2). Moreover, suppose the structure of non-parametric adaptive identifier (5) whose parameters are adjusted as the adaptable law given in (6). If there exists matrices Q_P^i, Q_S^i and Q_T^i positive defined such that the Riccati equations (7) have positive definite solutions P^i, S^i and T^i $(i = \overline{3,N})$, then the following upper bound $\lim_{t\to\infty} \|\hat{u}_i(t,x) - u_i(t,x)\|_P \leq \rho$ is ensured for the state nonparametric identification process where

$$\rho := \sqrt{\min_i (\alpha_m^i)^{-1} N \max_i \left(\lambda_{\max}\left(\left[\Lambda_P^i\right]^{-1}\right) F_4^i + \lambda_{\max}\left(\left[\Lambda_S^i\right]^{-1}\right) F_9^i\right)} + \sqrt{\min_i (\alpha_m^i)^{-1} N \max_i \left(\lambda_{\max}\left(\left[\Lambda_T^i\right]^{-1}\right) f_3^i\right)}$$

Moreover, the weights $W_{1,t}$, $W_{2,t}$ and $W_{3,t}$ are bounded with the following bounds $\|W_{1,t}\| \leq K_1\rho$, $\|W_{2,t}\| \leq K_2\rho$, $\|W_{3,t}\| \leq K_3\rho$.

Proof. The detailed proof is given in the appendix.

4 Simulation Results

For the purpose of illustrating the main theoretical results derived in previous sections, here it is considered an anaerobic degradation system, which is realized in a fixed bed reactor with a recirculation tank. The dynamics of the state variables in this process are described by the following energy and mass balance PDE:

$$\begin{gathered}
\frac{\partial X_1}{\partial t} = (\mu_1 - \varepsilon D) X_1, \quad \frac{\partial X_2}{\partial t} = (\mu_2 - \varepsilon D) X_2 \\
\frac{\partial S_1}{\partial t} = \frac{E_z}{H^2}\frac{\partial^2 S_1}{\partial x^2} - D\frac{\partial S_1}{\partial x} - k_1\mu_1 X_1 \\
\frac{\partial S_2}{\partial t} = \frac{E_z}{H^2}\frac{\partial^2 S_2}{\partial x^2} - D\frac{\partial S_2}{\partial x} + k_1\mu_1 X_1 - k_2\mu_2 X_2 \\
\mu_1 = \frac{\mu_{1,\max} S_1}{S_1 + K_{S_1}}, \quad \mu_2 = \frac{\mu_{2,\max} S_2}{S_2 + K_{S_2} + \frac{S_2^2}{K_{I_2}}}
\end{gathered} \tag{10}$$

In this equations: $x(\cdot) \in [0,1]$, $t\,[d]$ is the evolution time of the digestion process, $E_z\,[m^2d^{-1}]$ is the dispersion axial coefficient, $D\,[d^{-1}]$ is the dilution factor, $H\,[m]$ is the length of the reactor, $X_1\,[gL^{-1}]$ is the acid-genic biomass, $X_2\,[gL^{-1}]$ is the metano-genic biomass, $S_1\,[gL^{-1}]$ is the oxygen chemical demand, $S_2\,[gL^{-1}]$ is the volatile acid concentration and ε is the fraction of bactery

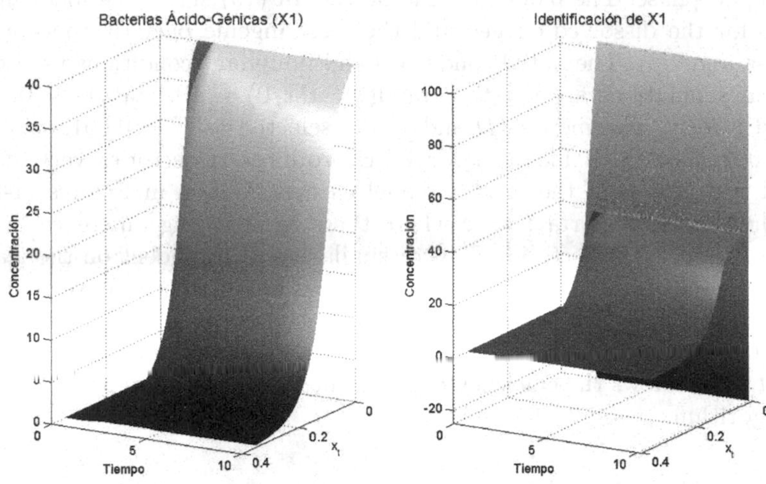

Fig. 1. Chemical Oxygen Demand dynamics. Numerical trajectory produced by the mathematical model during 10 hours (a) and estimated trajectories produced by the DNN based identifier (b). Both trajectories are really close one each other except during the first 1 hour.

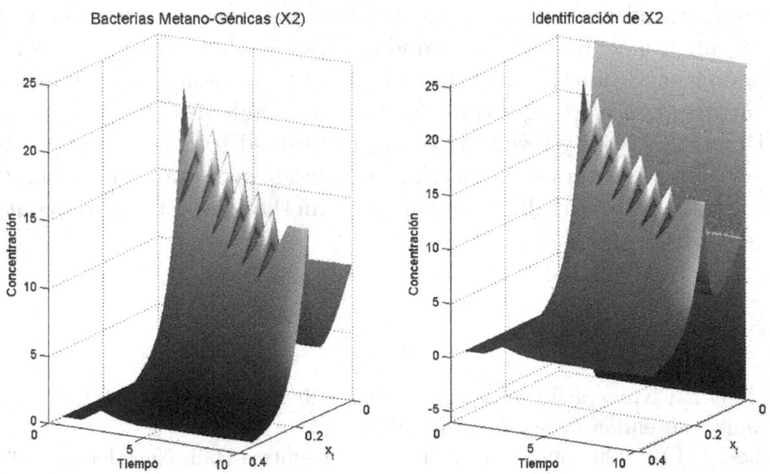

Fig. 2. Methangenic bacteria dynamics. Numerical trajectory produced by the mathematical model during 10 hours (a) and estimated trajectories produced by the DNN based identifier (b). Both trajectories are really close one each other except during the first 1 hour.

in the liquid phase. The biological reactor system (10) shows the following trajectories for the dissolved oxygen and the methangenic bacteria concentration (Figs. 1-a and 2-a). The initial conditions and boundary conditions used in this numerical simulations are $u(0,t) = \text{rand}(1)$, $u(x,0) = 10$, $u_x(0,t) = 0$. Diffusion and velocity parameters (D and v) are selected as $D = 0.001$, $v = 0.01$, $c = 0.0001$. The DNN identifier for PDE produces trajectories very close to the real trajectories of the reactor model as can be seen in Figures (1-b and 2-b). There is an important zone where there exists a big difference between real and estimated trajectories. This dissimilarity is dependent on the learning period required to adjust the DNN identifier. The difference between the real PDE trajectories and the estimated state produced by the DNN identifier is just perceptible during first seconds. This error is close to zero almost during all x and all t. This shows the efficiency of the identification process provided by the DNN algorithm.

5 Conclusions

In this paper a new methodology to identify a class of nonlinear distributed parameter systems has been introduced. The suggested approach solves the problem of non parametric identification of uncertain nonlinear described by partial differential equations. Asymptotic convergence for the identification error has been demonstrated applying a Lyapunov-like analysis using a special class of Lyapunov functional. Besides, the same analysis leads to the generation of the corresponding conditions for the upper bound of the weights involved in the identifier structure. Learning algorithms for the adjustable weights have been introduced. Practical stability results have been obtained to generate upper bounds for the weights trajectories. Numerical example showing an anaerobic dynamics demonstrates the workability of this new methodology based on continuous neural networks.

References

1. Smith, G.D.: Numerical solution of partial differential equations: finite difference methods. Clarendon Press, Oxford (1978)
2. Hughes, T.J.R.: The finite element method. Prentice Hall, New Jersey (1987)
3. Haykin, S.: Neural Networks. A comprehensive Foundation. IEEE Press, New York (1994)
4. Cybenko, G.: Approximation by superposition of a sigmoidal function. Mathematical Control Signals Systems 2, 303–314 (1989)
5. Lagaris, I.E., Likas, A., Fotiadis, D.I.: Artificial neural networks for solving ordinary and partial differential equations. IEEE Transactions on Neural Networks 9, 987–1000 (1998)
6. Dissanayake, M.W.M.G., Phan-Thien, N.: Neural-network based approximations for solving partial differential equations. Communications in Numerical Methods in Engineering 10, 195–201 (2000)

7. Mai-Duy, N., Tran-Cong, T.: Numerical solution of differential equations unsing multiquadric radial basis function networks. Neural Networks 14, 185–199 (2001)
8. He, S., Reif, K., Unbehauen, R.: Multilayer neural networks for solving a class of partial differential equations. Neural Networks 13, 385–396 (2000)
9. Jaime, E.: Approximation Analytique de la solution dequations differentielles partielles par reseau de neurons artificiels: Application a la Simulation Thermique dans les Micro-Systemes. PhD thesis, Institut National des Scienes Appliquees de Toulouse (2004)
10. Poznyak, A., Sanchez, E., Yu, W.: Differential Neural Networks for Robust Nonlinear Control (Identification, state Estimation an trajectory Tracking). World Scientific, Singapore (2001)
11. Lewis, F.L., Yesildirek, A., Liu, K.: Multilayer neural-net robot controller with guaranteed tracking performance. IEEE Trans. Neural Netw. 7, 1–11 (1996)
12. Khalil, H.K.: Nonlinear systems. Prentice-Hall, Upper Saddle River (2002)
13. Adams, R., Fournier, J.: Sobolev spaces., 2nd edn. Academic Press, New York (2003)
14. Delyon, B., Juditsky, A., Benveniste, A.: Accuracy analysis for wavelet approximations. IEEE Trans. Neural Network 6, 332–348 (1995)
15. Cotter, N.E.: The stone-weierstrass theorem and its application to neural networks. IEEE Transactions on Neural Networks 1, 290–295 (1990)
16. Daubechies, I.: Ten lectures on Wavelets. SIAM, Philadelphia (1992)
17. Poznyak, A.: Deterministic Output Noise Effects in Sliding Mode Observation. In: Sliding Modes: From Principles to Implementation, ch. 3, pp. 123–146. IEEE Press, Los Alamitos (2001)

Appendix

Consider the Lyapunov-like functional

$$\begin{array}{c} V(t) := \sum_{i=1}^{N} \bar{V}_i(t,x) + \sum_{r=1}^{3} \operatorname{tr}\left\{\left[\tilde{W}_r^i(t)\right]^{\intercal} K_r \tilde{W}_r^i(t)\right\} \\ \bar{V}_i(t,x) := \left\|\Delta^i(t,x)\right\|_{P^i}^2 + \left\|\Delta_x^i(t,x)\right\|_{S^i}^2 + \left\|\hat{u}_i(t,x)\right\|_{T^i}^2 \end{array} \tag{11}$$

Following the procedure for the second Lyapunov method, the time derivative of V_t is

$$\begin{array}{r} \dot{V}(t) = 2\sum_{i=1}^{N}\left(\left(\Delta^i(t,x)\right)^{\intercal}(t) P^i \frac{d}{dt}\Delta^i(t,x) + \sum_{r=1}^{3} \operatorname{tr}\left\{\left[\tilde{W}_r^i(t)\right]^{\intercal} K_r \dot{W}_r^i(t)\right\}\right) \\ +2\sum_{i=1}^{N}\left(\left[\Delta_x^i(t,x)\right]^{\intercal} S^i \frac{d}{dt}\Delta_x^i(t,x) + \hat{u}_i^{\intercal}(t,x) T^i \frac{d}{dt}\hat{u}_i(t,x)\right) \end{array} \tag{12}$$

Using last results and the following matrix inequality $XY^{\intercal} + YX^{\intercal} \leq X\Lambda X^{\intercal} + Y\Lambda^{-1}Y^{\intercal}$ valid for any $X, Y \in R^{r\times s}$ and any $0 < \Lambda = \Lambda^{\intercal} \in R^{s\times s}$ then by the Riccati equations defined in (7) and in view of the adjust equations of the weights (6), the previous equality is simplified to

$$\dot{V}(t) \leq -\alpha_m^i V(t) + \sum_{i=1}^{N}\left(\lambda_{\max}\left(\left[\Lambda_S^i\right]^{-1} F_9^i + \left[\Lambda_T^i\right]^{-1} f_3^i + \left[\Lambda_P^i\right]^{-1} F_4^i\right)\right)$$

Taking the maximum value over i, we obtain

$$\dot{V}(t) \leq -\min_i \left(\alpha_m^i\right) V(t) + N \max_i \left(\lambda_{\max}\left(\left[\Lambda_S^i\right]^{-1} F_9^i + \left[\Lambda_T^i\right]^{-1} f_3^i + \left[\Lambda_P^i\right]^{-1} F_4^i\right)\right)$$

Applying the Lemma 1, one has $\left[1-\mu\left(\sqrt{V(t)}\right)^{-1}\right]_+ \to 0$ that completes the proof.

The Lin-Kernighan Algorithm Driven by Chaotic Neurodynamics for Large Scale Traveling Salesman Problems

Shun Motohashi[1], Takafumi Matsuura[1],
Tohru Ikeguchi[1,3], and Kazuyuki Aihara[2,3]

[1]Graduate school of Science and Engineering, Saitama University,
255 Shimo-Ohkubo Saitama 338–8570, Japan
{motohashi,takafumi,tohru}@nls.ics.saitama-u.ac.jp
[2]Graduate school of Information Science and Technology, The University of Tokyo,
4-6-1 Komaba, Meguro-ku, Tokyo, 153-8505, Japan
aihara@sat.t.u-tokyo.ac.jp
[3]Aihara Complexity Modelling Project, ERATO, JST,
4-6-1 Komaba, Meguro-ku, Tokyo, 153-8505, Japan

Abstract. The traveling salesman problem (TSP) is one of the typical $\mathcal{NP}$-hard problems. Then, it is inevitable to develop an effective approximate algorithm. We have already proposed an effective algorithm which uses chaotic neurodynamics. The algorithm drives a local search method, such as the 2-opt algorithm and the adaptive k-opt algorithm, to escape from undesirable local minima. In this paper, we propose a new chaotic search method using the Lin-Kernighan algorithm. The Lin-Kernighan algorithm is one of the most effective algorithms for solving TSP. Moreover, to diversify searching states, we introduce the double bridge algorithm. As a result, the proposed method exhibits higher performance than the conventional algorithms. We validate the applicability of the proposed method for very large scale instances, such as 10^5 order TSPs.

1 Introduction

In our daily life, we are often confronted with difficulties of realizing optimization: for example, scheduling, delivery planning, circuit designing and drilling, computer wiring, and so on. These problems are classified into discrete optimization. In case of solving these problems, we often try to find a solution intuitively. However, such an intuition often leads to worse situations, then it is important to design effective algorithms systematically. To develop effective algorithms for solving these problems, the traveling salesman problem (TSP), which is one of the typical combinatorial optimization problems, is widely studied. The TSP is described as follows: given a set of N cities and each distance d_{ij} between cities i and j, find an optimal solution, or a shortest-length tour. Namely, the goal of

C. Alippi et al. (Eds.): ICANN 2009, Part II, LNCS 5769, pp. 563–572, 2009.

the TSP is to find a permutation σ of the cities that minimizes the following quantity:

$$\sum_{k=1}^{N-1} d_{\sigma(k)\sigma(k+1)} + d_{\sigma(N)\sigma(1)}. \tag{1}$$

If $d_{ij} = d_{ji}$ for all i and j, the TSP is symmetric, otherwise, asymmetric. In this paper, we deal with the symmetric TSP.

For an N-city TSP, the total number of different tours is $(N-1)!/2$. Then, if the number of cities increases, the number of all possible tours exponentially diverges, because $N! \simeq N^N$. The TSP generally belongs to a class of $\mathcal{NP}$-hard. Therefore, it is commonly believed that no polynomial algorithm exists. Thus, it is required to develop an effective approximate algorithm for finding near optimal solutions or approximate solutions in a reasonable time.

As the approximate algorithm, several local search algorithms have already been proposed. Among them, the 2-opt algorithm, the 3-opt algorithm and the Lin-Kernighan algorithm[1] are famous to find near optimal solutions of the TSP. The basic mechanism of these algorithms is to explore better solutions from neighborhoods of the current state. Then, the algorithms search a new state in a local space to obtain a shorter tour. For example, the 2-opt algorithm which is one of the simplest local search methods is described as follows: first, two links are deleted from a current tour. Second, other two links are added in such a way that the length of a new tour is shorter than the current tour (Fig.1). Such an exchange continues until no further improvements can be obtained.

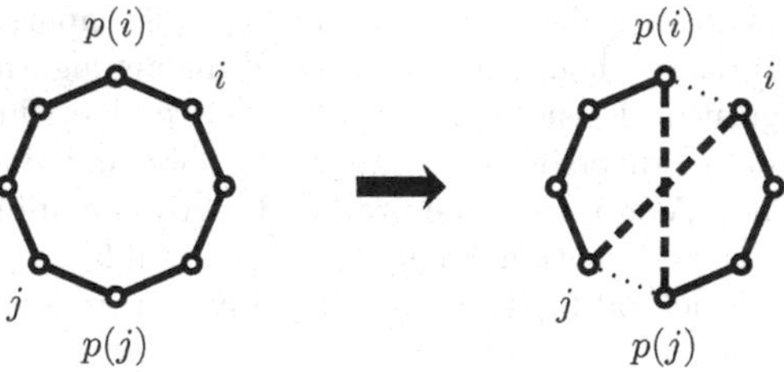

Fig. 1. An example of the 2-opt algorithm. $p(i)$ is the previous city of the city i.

However, it is hard to find an optimal solution by the local search methods, because the local search methods usually get stuck at local minima. To avoid such local minima, a wide variety of strategies have been proposed, such as a tabu search method[2], a simulated annealing[3], a genetic algorithm[4], and so on.

We have already proposed effective algorithms for solving TSP[5, 6, 7, 8]. In the algorithms, to avoid local minima, local search methods are driven by chaotic neurodynamics. To realize such a chaotic search method, we introduce a chaotic neuron model proposed by Aihara, Takabe and Toyoda[9]. The chaotic neuron model can reproduce one of important properties which real nerve cells have; refractoriness. Using the refractoriness implemented in the chaotic neuron model, the chaotic search methods can explore a searching space by avoiding local minima.

First, a chaotic search method which drives the simplest local search method, the 2-opt algorithm, was proposed[5, 6]. Then, this method shows higher performance than the tabu search method[2] which has almost the same strategy of searching solutions as the chaotic search methods.

Next, to improve the performance of the chaotic search method, a chaotic search method using a more powerful local search method, the adaptive k-opt algorithm (Fig. 2), has been proposed[7]. The adaptive k-opt algorithm is one of the variable depth neighborhood search methods and changes the number of exchanged links adaptively. As a result, this chaotic search method shows higher performance than the chaotic search method using the 2-opt algorithm.

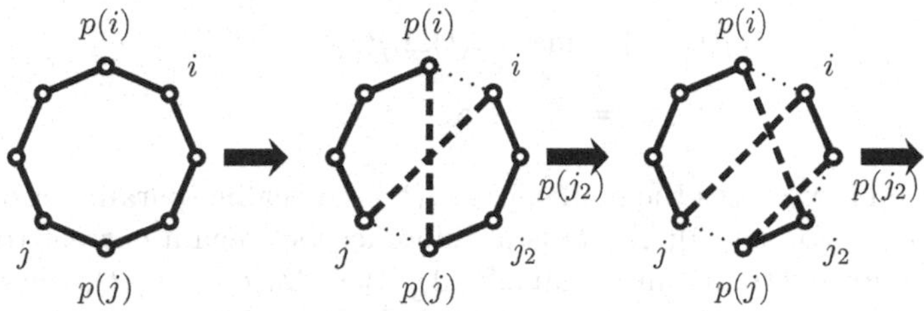

Fig. 2. An example of the adaptive k-opt algorithm. $p(i)$ is the previous city of the city i.

The Lin-Kernighan algorithm[1] is generally considered to be one of the most effective local search methods for symmetric TSPs. The Lin-Kernighan algorithm is also a variable depth neighborhood search method and improves a tour by changing the number of exchanged links dynamically. The Lin-Kernighan algorithm can search better solutions than the adaptive k-opt algorithm, because the Lin-Kernighan algorithm can search more deeply than the adaptive k-opt algorithm. We have already proposed a basic algorithm for avoiding the local-minimum problem in the Lin-Kernighan algorithm by chaotic neurodynamics[8]. Although the proposed method in Ref.[8] exhibits better performance than the conventional chaotic search methods[5, 6, 7], it is still hard to apply the proposed method to very large scale TSP instances, because it requires much computational time to decide values of parameters[8]. In the method[8], the parameters are decided by the standard deviation of the link length. Thus, the computational costs for calculating the standard deviation become large for very large scale TSP instances, such as 10^5 order. To resolve this problem, in this paper, we introduce a new parameter tuning method. This method tunes the scaling parameter adaptively.

To evaluate the performance of the proposed method, we use instances with the order of $10^3 \sim 10^5$. As a result, the proposed method shows better solutions than the previous chaotic search methods[5, 6, 7, 8].

2 Proposed Method

2.1 The Lin-Kernighan Algorithm Driven by Chaotic Neurodynamics[8]

In the proposed method, we use chaotic neurodynamics[9] to drive the Lin-Kernighan algorithm[1]. To realize the chaotic neurodynamics, we use a chaotic neural network constructed by chaotic neurons[9]. The number of chaotic neurons is the same as the number of cities. Each chaotic neuron is assigned to each city, then an execution of the Lin-Kernighan algorithm is controlled by firing of the corresponding chaotic neuron.

The chaotic neuron has a gain effect and a refractory effect. The first effect, or the gain effect, is defined by the following equations[8]:

$$\xi_i(t+1) = \max_j\{\beta(t)\Delta_{ij}(t) + \zeta_j(t)\}, \tag{2}$$

$$\beta(t+1) = \beta(t) + \gamma, \tag{3}$$

where $\xi_i(t+1)$ expresses the gain effect; $\beta(t)$ is a scaling parameter of the gain effect at time t ($\beta(t) > 0$); $\Delta_{ij}(t)$ is a difference between a current tour length and a new tour length at time t, namely $\Delta_{ij}(t) = D_0(t) - D_{ij}(t)$, where $D_0(t)$ is the length of the current tour at time t, and $D_{ij}(t)$ is the new tour length offered by the Lin-Kernighan algorithm which links the cities i and j at time t. If the new tour length is shorter than the current tour length, the value of the gain effect becomes positive because $\Delta_{ij}(t) > 0$. Then, the gain effect encourages the chaotic neuron to fire.

In Eq. (3), γ is a scaling parameter of the annealing effect. The scaling parameter $\beta(t)$ increases with time t. By increasing the value of $\beta(t)$ gradually, a searching space is increasingly limited as the simulated annealing[3].

The second effect, or the refractory effect, is defined by the following equation:

$$\zeta_i(t+1) = -\alpha\sum_{d=0}^{s-1} k_r^d x_i(t-d) + \theta, \tag{4}$$

where $\zeta_i(t+1)$ expresses the refractory effect. α is a scaling parameter of the refractory effect after a neuron firing ($\alpha > 0$); k_r is a decay parameter of the refractory effect ($0 < k_r < 1$); s is a temporal period for memorizing past outputs; $x_i(t)$ is an output of the ith neuron at time t; θ is a threshold value. If the neuron fires in the past s steps, the right hand side of Eq. (4) becomes negative. Namely, the refractory effect inhibits the neuron from firing for a while. In Eq. (4), if $s-1=t$, it means that the neuron memorizes its all history from $t=0$. If we use Eq. (4) directly, it needs much amount of memory to memorize its all history. However, Eq. (4) can be transformed into the following simple one-dimensional difference equation:

$$\zeta_i(t+1) = k_r\zeta_i(t) - \alpha x_i(t) + (1-k_r)\theta. \tag{5}$$

Then, the output of the ith neuron is defined by the following equation:

$$x_i(t+1) = f\left(\xi_i(t+1) + \zeta_i(t+1)\right), \tag{6}$$

where $f(y) = 1/(1+e^{-y/\epsilon})$. If $x_i(t+1) \geq 1/2$, the ith neuron fires at time $t+1$ and the Lin-Kernighan algorithm which links the cities i and j is executed. In the proposed method, each neuron is updated asynchronously.

To solve an N-city TSP, the procedure of a single iteration in the proposed method is shown below.

1. Let $i = 1$.
2. Choose the city j which maximizes the value of the gain effect of the ith neuron. In this step, the Lin-Kernighan algorithm which links the cities i and j is temporarily executed to obtain the value of $\Delta_{ij}(t)$.
3. Calculate the output of the ith neuron $x_i(t+1)$.
4. If $x_i(t+1) \geq 1/2$, the ith neuron fires, then the Lin-Kernighan algorithm which links cities i and j is executed.
5. If $i < N$, let $i = i + 1$ and go to Step 2. Otherwise finish this iteration.

Next, the procedure of the Lin-Kernighan algorithm used in the Step 2 is shown below. Note that i and j have already been selected. The Lin-Kernighan algorithm searches better solutions by repeating a choice of a deleted link x and an added link y.

1. Let T be initial tour (Fig. 3(a)).
2. Let $G^* = 0$, $m = 1$, t_1 is $p(i)$, t_2 is i, t_3 is j, x_1 is the link (t_1, t_2), and y_1 is the link (t_2, t_3) (Fig.3(b)). Here, G^* is a value of the best improvement in the previous searches, m is the number of pairs of a deletion and an addition of links, and $p(i)$ is the previous city of the city i.
3. Let m increase by one. Delete $x_m(t_{2m-1}, t_{2m})$ and add $y_m(t_{2m}, t_{2m+1})$ by the following steps (a)–(d). If such x_m and y_m cannot be found, go to Step 4.
   (a) Delete x_m until the following conditions is satisfied:
      i. x_m is not previously added.
      ii. If t_{2m} is connected to t_1, the resulting configuration is a feasible tour.
   (b) If $f(T) - f(T') > G^*$, set $G^* = f(T) - f(T')$ and $k = m$, where T is an initial tour, T' is a tour constructed by connecting t_{2m} to t_1 (Fig.3(c)), $f(T)$ is a length of T and k is the number of exchanged links to achieve G^*.
   (c) Add y_m until the following conditions is satisfied (Fig.3(d)):
      i. y_m is not previously deleted.
      ii. $G_m > 0$, where $G_m = \sum_{j=1}^{m}(|x_j| - |y_j|)$.
      iii. In case that y_m is added, a next link x_{m+1} exists.
      iv. $|x_{m+1}| - |y_m|$ is maximum for all candidates of y_m.
   (d) If $G_m > G^*$, go to Step 3.

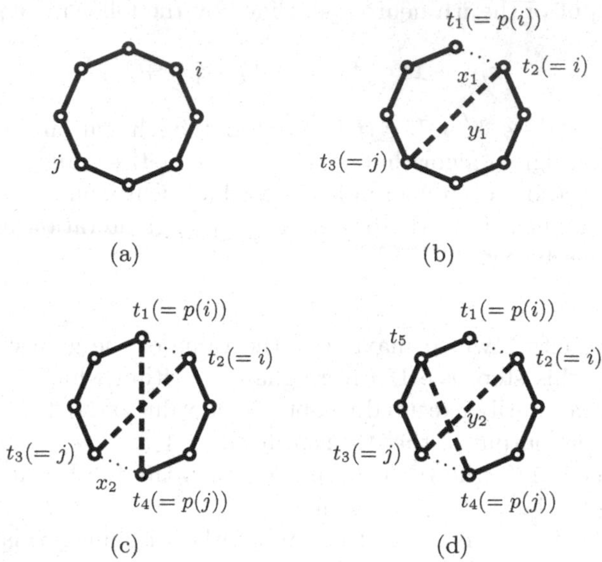

Fig. 3. An example of the Lin-Kernighan algorithm. (a) shows an initial tour, (b) shows a case in which depth of search is 1, (c) shows a tour constructed by connecting t_{2m} to t_1 at $m = 2$, and (d) shows a case in which depth of search is 2. $p(i)$ is the previous city of the city i, x_1 and x_2 are the deleted links, and y_1 and y_2 are the added links.

4. If $G^* = 0$ and $m = 2$, set $\Delta_{ij}(t)$ to $f(T) - f(T')$. Otherwise set $\Delta_{ij}(t)$ to G^*.
5. The procedure terminates.

In this algorithm, we also introduce a double bridge algorithm. The double bridge algorithm is a special 4-opt local search method as shown in Fig. 4. The double bridge algorithm is decomposed into two 2-opt improvements. The Lin-Kernighan algorithm cannot realize such an improvement. In the proposed method, the double bridge algorithm is executed to diversify obtained solutions if a better solution could not be found for more than 5 iterations.

2.2 A Parameter Tuning Method

In Eq. (3), a scaling parameter $\beta(t)$ increases linearly based on γ, which is decided by the standard deviation of the link length for each instance because the range of Δ_{ij} depends on each instance[8]. However, the standard deviation of the link length is not enough to adjust $\beta(t)$, because the value of $\Delta_{ij}(t)$ temporally depends on solution states. Moreover, it is hard to apply this method to large scale TSP instances because the calculation costs of the standard deviation is too large to calculate in a reasonable time frame for such a case.

Then we introduce a new parameter tuning method. In this method, to obtain the same range of $\xi_i(t)$ for all instances, the scaling parameter $\beta(t)$ is adjusted by

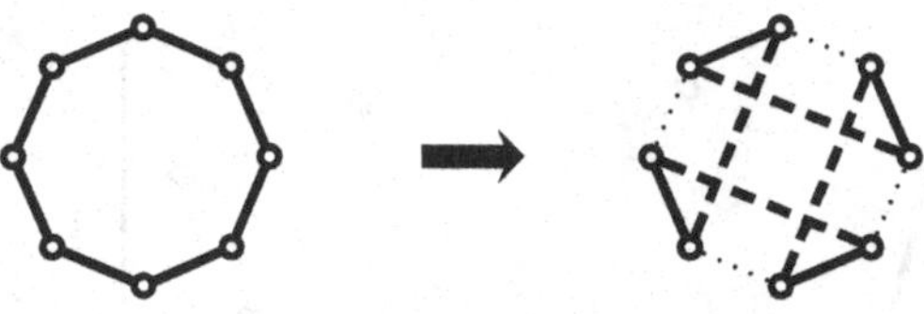

Fig. 4. An example of the double bridge algorithm

Δ_{ij}, because the range of Δ_{ij} reflects a structural property of problem instances. In addition, because the value of Δ_{ij} temporally depends on solution states, an average value of Δ_{ij} at time t is used to adjust $\beta(t+1)$. Then, we transform Eq. (3) into the following equations:

$$\beta(t+1) = \beta(t) + \frac{q}{\overline{\Delta}(t)}, \tag{7}$$

$$\overline{\Delta}(t) = \frac{1}{N}\sum_{i=1}^{N} |\Delta_{ij}(t)|, \tag{8}$$

where q is a scaling parameter of the annealing effect.

3 Results

To evaluate the performance of the proposed method, we used two benchmark problems: TSPLIB[10] and classes E and C in "8th DIMACS Implementation Challenge"[11]. The difference between classes E and C is how cities are distributed: class E has a uniform distribution in a square and class C is clustered in a square. We conducted numerical simulations using the gcc compiler on a Mac mini of 1.5GHz Intel Core Solo processor with 2GB memory running on MAC OS X 10. 4. 11.

Initial solutions are constructed by the nearest neighbor method. The parameters of the proposed method $\beta(0), q, \alpha, k_r, \theta$ and ϵ are fixed for all instances: $\beta(0) = 0, q = 0.045, \alpha = 1.0, k_r = 0.5, \theta = 1.0$ and $\epsilon = 0.002$. The proposed method is applied for 200 iterations. Then, to reduce computational costs, we use two different candidate lists: nearest neighbors and quadrant neighbors. In the nearest neighbors, the nearest cities are added to the candidate list for each city (Fig. 5(a)). In the quadrant neighbors, first, considering each city as an origin, the Euclidian plane is divided into four quadrants. Next, for the four quadrants, the nearest cities are added to the candidate list. For example, if the number of the candidates is four, for each quadrant, one nearest city is added to the candidate list (Fig. 5(b)). If no city exists or all cities have already been selected as candidates, no cities are added to the candidate list.

The candidate lists are constructed from 10 nearest neighbors (10NN) and 8 quadrant neighbors (8QN: 2 nearest neighbors are selected in each quadrant of the Euclidian plane). The candidate list is used, when we decide the city j in

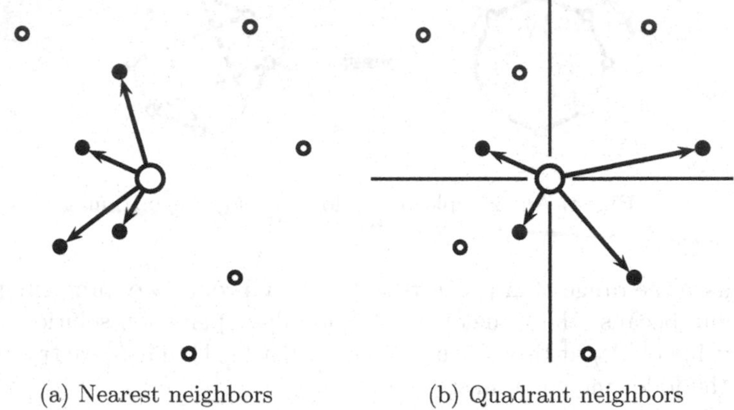

Fig. 5. Difference between the nearest neighbors and the quadrant neighbors. Each point shows cities. The city distribution is the same for both cases. The number of the candidates is four. Black points show candidate cities for the center city(larger circles).

Step 2 of the procedure of a single iteration in the proposed method (Eq. (2)) and we decide added link y_m in Step 3 of the procedure of the Lin-Kernighan algorithm. We investigate which candidate list is more effective for the proposed method. Then, the best obtained tour is improved by the Lin-Kernighan algorithm and the double bridge algorithm until no further improvements are possible.

Table 1 shows results for the chaotic search method using Lin-Kernighan algorithm without the double bridge algorithm (CS+LK) and that with the double bridge algorithm (CS+LK+DB). In Table 1, the results for TSPLIB instances are expressed by percentages of average gaps between obtained solutions and the optimal solutions. However, the optimal solution of DIMACS instances are currently unknown. Thus, the results for DIMACS instances of Table 1 are expressed by percentages of average gaps between obtained solutions and the Held-Karp lower bound[12, 13].

From Table 1, the performance of CS+LK+DB is higher than that of CS+LK. These results indicate that the double bridge algorithm is effective. However, the running time is almost same. Then, the quadrant neighbors obtain better solutions than the nearest neighbors for TSPLIB instances except for pcb1173 and DIMACS instances of class C. On the other hand, for the DIMACS instances of class E, the nearest neighbors obtain better solutions. The reason is that in the DIMACS instances of class E, the nearest neighbors and the quadrant neighbors produce almost the same candidate lists, because the cities are uniformly distributed in a square. Thus, the nearest neighbors obtain better solutions than the quadratic neighbors, because the number of candidates is larger than the quadratic neighbors.

Table 1. Results for the chaotic search method using Lin-Kernighan algorithm without the double bridge algorithm (CS+LK) and that with the double bridge algorithm (CS+LK+DB). Average gaps between obtained solutions and the best known solutions are shown in percentage. Bold faces are the best results. Running times are shown in second.

Instance	Gaps[%]				Running times[s]			
	CS+LK		CS+LK+DB		CS+LK		CS+LK+DB	
	Candidate lists							
	10NN	8QN	10NN	8QN	10NN	8QN	10NN	8QN
pcb1173	0.759	0.797	**0.530**	0.647	16	13	10	13
rl5915	1.011	0.906	0.886	**0.650**	92	119	95	127
rl11849	0.890	0.807	0.694	**0.594**	302	372	313	381
pla33810	1.175	0.789	1.065	**0.680**	1637	1715	1669	1778
pla85900	0.973	0.638	0.904	**0.562**	7952	10834	8283	11063
E10k.0	1.502	1.588	**1.325**	1.367	337	337	345	340
E31k.0	1.449	1.564	**1.264**	1.346	1957	2031	2010	2088
E100k.0	1.463	1.571	**1.291**	1.347	12636	13833	13272	14311
E316k.0	1.496	1.573	**1.329**	1.381	101352	115714	105696	123157
C10k.0	7.533	2.111	7.352	**1.976**	239	205	254	216
C31k.0	7.398	2.210	7.202	**2.000**	831	829	918	893
C100k.0	7.433	2.255	7.258	**2.137**	3787	4376	4559	4886
C316k.0	7.211	2.378	7.048	**2.103**	35821	39212	43814	41045

4 Conclusions

In this paper, we proposed a new chaotic search method using the Lin-Kernighan algorithm for solving TSP. From the computational experiments, the proposed method shows higher performance for large scale instances. In the future work, to improve performance of the proposed method, we should modify the chaotic search method, for example we will develop an effective parameter tuning method. The research of T.I. is partially supported by Grant-in-Aid for Scientific Research (B) (No.20300085) from the JSPS.

References

[1] Lin, S., Kernighan, B.: An effective heuristic algorithm for the traveling-salesman problem. Operations Research 21, 498–516 (1973)
[2] Glover, F.: Tabu search–part I. ORSA J. Computing 1, 190–206 (1989)
[3] Kirkpatrick, S., Gelatt, C.D., Vecchi, P.M.: Optimization by simulated annealing. Science 220, 671–680 (1983)

[4] Holland, J.: Adaptation in Natural and Artificial Systems: An Introductory Analysis with Applications to Biology, Control, and Artificial Intelligence. The University of Michigan Press (1975)

[5] Hasegawa, M., Ikeguchi, T., Aihara, K.: Combination of chaotic neurodynamics with the 2-opt algorithm to solve traveling salesman problems. Physical Review Letters 79, 2344–2347 (1997)

[6] Hasegawa, M., Ikeguchi, T., Aihara, K.: Solving large scale traveling salesman problems by chaotic neurodynamics. Neural Networks 15, 271–283 (2002)

[7] Hasegawa, M., Ikeguchi, T., Aihara, K.: On the effects of the k-opt method with chaotic neurodynamics. IEICE Technical Report 101, 25–32 (2001)

[8] Motohashi, S., Matsuura, T., Ikeguchi, T.: Chaotic search method using the Lin-Kernighan algorithm for traveling salesman problems. In: Proceedings of International Symposium on Nonlinear Theory and its Applications (NOLTA), pp. 144–147 (2008)

[9] Aihara, K., Takabe, T., Toyoda, M.: Chaotic neural networks. Physics Letters A 144, 333–340 (1990)

[10] TSPLIB, `http://www.iwr.uni-heidelberg.de/groups/comopt/software/TSPLIB95/`

[11] Johnson, D.S., McGeoch, L.A., Glover, F., Rego, C.: 8th DIMACS implementation challenge: The traveling salesman problem (2000), `http://www.research.att.com/~dsj/chtsp/`

[12] Held, M., Karp, R.M.: The traveling-salesman problem and minimum spanning trees. Operations Research 18, 1138–1162 (1970)

[13] Held, M., Karp, R.M.: The traveling-salesman problem and minimum spanning trees: Part II. Mathematical Programming 1, 6–25 (1971)

Quadratic Assignment Problems for Chaotic Neural Networks with Dynamical Noise

Takayuki Suzuki[1,2], Shun Motohashi[1], Takafumi Matsuura[1], Tohru Ikeguchi[1,3], and Kazuyuki Aihara[3,4]

[1] Graduate school of Science and Engineering, Saitama University, 255 Shimo-Ohkubo Saitama 338–8570, Japan

[2] suzuki@nls.ics.saitama-u.ac.jp

[3] Aihara Complexity Modelling Project, ERATO, JST, 4-6-1 Komaba, Meguro-ku, Tokyo, 153–8505, Japan

[4] Graduate School of Information Science and Technology, The University of Tokyo, 4-6-1 Komaba, Meguro-ku Tokyo, Japan

Abstract. The quadratic assignment problem (QAP) is one of the combinatorial optimization problems which belong to a class of NP-hard. To solve QAP, various algorithms for finding near optimal solutions have already been proposed. Among them, the Hopfield-Tank neural network approach is very attractive from a viewpoint of an application of neural dynamics to combinatorial optimization, this approach is not so effective because of local minimum problem. To overcome this problem, a method which uses chaotic dynamics has already been proposed. On the other hand, to avoid undesirable local minima, dynamical noise is often used. In this paper, we combine these two approaches–chaotic dynamics and dynamical noise–to realize an effective approach for solving combinatorial optimization problems: we add dynamical noise to chaotic neural network for solving QAP. The results show that when the small amount of dynamical noise is added, the solving performance is much improved. We also analyze the influence of dynamical noise to the chaotic dynamics, and show that dynamical noise diversifies the searching states to explore much better solutions.

1 Introduction

Many optimization problems exist in real world, for example, VLSI design, scheduling problem, routing problem, facility layout problem, and so on. In these problems, it is important to find the optimal solution, because it leads to the reductions of operation costs. These problems are often formulated as a quadratic assignment problem (QAP). QAP is one of the typical combinatorial optimization problems, and it is widely acknowledged that QAP is one of the most difficult NP-hard problems. Thus, it is almost impossible to find an optimal solution in realistic time. Then, it is required to develop approximate algorithms for finding near optimal solutions in reasonable time.

C. Alippi et al. (Eds.): ICANN 2009, Part II, LNCS 5769, pp. 573–582, 2009.

As one of the approximate algorithms, a method which uses the Hopfield-Tank neural network (HNN), or the mutual connection neural network, has been proposed[1]. In this method, a firing pattern of HNN represents a solution. The synaptic weights of HNN are decided so that an optimal solution is embeded in a stable equilibrium point of HNN. Thus, by providing a good initial solution with HNN, dynamics of HNN offers a firing pattern corresponding to the optimal solution. However, this method gets trapped into local minima. As an approach for avoiding such local minimum problems, a method which uses chaotic neural network (CNN)[2] has already been proposed. It is shown that complex dynamics of CNN is effective to avoid the local minima[3–5]. As another approach for solving the local minimum problems, a method which uses additive dynamical noise to HNN has been proposed[6]. By using fluctuation of chaotic noise[7, 8], the searching state can escape from the local minima. This method shows high performance.

In this paper, we proposed a new algorithm for solving QAP by combining chaotic dynamics[2] and dynamical noise[6]. Namely, we add dynamical noise to nonlinear dynamics of CNN. Combination of chaotic dynamics and dynamical noise could lead to better performance. We first investigate how solving performance of CNN with dynamical noise depends on the amplitude of the noise. Next, we analyze the effect of dynamical noise to the chaotic dynamics and show that the dynamical noise diversifies the searching state to explore much better solutions.

2 Solving Quadratic Assignment Problem with Chaotic Neural Networks

2.1 Quadratic Assignment Problem

The quadratic assignment problem (QAP) is described as follows: given two $N \times N$ matrices, a distance matrix D, which gives mutual distances between locations, and a flow matrix C, which gives mutual relationships between units, find a permutation $\boldsymbol{p}$ which minimizes the value of the objective function $F(\boldsymbol{p})$ shown below:

$$F(\boldsymbol{p}) = \sum_{i=1}^{N} \sum_{j=1}^{N} d_{ij} c_{p(i)p(j)}, \qquad (1)$$

where d_{ij} is the (i, j)th element of D, $p(i)$ is the ith element of $\boldsymbol{p}$, $c_{p(i)p(j)}$ is the $(p(i), p(j))$th element of C.

2.2 Chaotic Neural Network

The chaotic neural network model is proposed by Aihara, Takabe and Toyoda[2]. This neural network model can reproduce a chaotic dynamics observed in real neural membrane. The dynamics of the ith chaotic neuron in the chaotic neural network is defined as follows:

$$x_i(t+1) = f\{\sum_{j=0}^{M} v_{ij} \sum_{d=0}^{t} k_e^d A_j(t-d) + \sum_{j=0}^{N} w_{ij} \sum_{d=0}^{t} k_f^d x_j(t-d) - \alpha \sum_{d=0}^{t} k_r^d x_i(t-d) - \theta_i\}, \quad (2)$$

where $x_i(t+1)$ is an output of the ith neuron at time $t+1$; k_e, k_f and k_r are decay parameters for external inputs, feedback inputs and a refractoriness, respectively; $A_j(t)$ is the amplitude of the jth external input at the time t; M is the number of the external inputs; v_{ij} is a connection weight from the jth externally applied input to the ith neuron; w_{ij} is a connection weight from the jth neuron to the ith neuron; N is the number of neurons; α is a strength parameter of the refractory effect; θ_i is a threshold of the ith chaotic neuron; f is a continuous output function.

In the case that all three decay parameters (k_e, k_m and k_r) are equal to k, the ith neural dynamics is reduced to the following simple forms:

$$y_i(t+1) = ky_i(t) + \sum_{j=1}^{N} w_{ij} f(y_j(t)) - \alpha f(y_i(t)) + a_i, \quad (3)$$

$$x_i(t+1) = f(y_i(t+1)). \quad (4)$$

where $y_i(t)$ is an internal state of the ith neuron at time t, and $a_i = -\theta_i(1-k)$.

2.3 An Application of Chaotic Neural Network to QAP

To solve QAP of the size N by the Hopfield-Tank neural network (HNN), $N \times N$ neurons are prepared, and they are arranged on an $N \times N$ grid. The outputs of neurons X represent a solution $\boldsymbol{p}$ of QAP. Namely, The (i, m)th neuron x_{im} controls whether the ith unit is assigned to the mth location. If the ith unit is assigned to the mth location, $x_{im} = 1$, otherwise, $x_{im} = 0$. Then, using X, Eq.(1) can be written as follows:

$$F(X) = \sum_{i=1}^{N} \sum_{j=1}^{N} \sum_{m=1}^{N} \sum_{n=1}^{N} d_{ij} c_{mn} x_{im} x_{jn}. \quad (5)$$

To produce a feasible solution for QAP, X must satisfy the constraints that one unit is assigned to one location and one location is assigned to one unit. These constraints are described by the following equations:

$$\sum_{m=1}^{N} x_{im} = 1, \tag{6}$$

$$\sum_{i=1}^{N} x_{im} = 1. \tag{7}$$

Using Eqs.(5), (6) and (7), the objective function is calculated as follows:

$$\begin{aligned} F(X) = A \sum_{i=1}^{N} (\sum_{m=1}^{N} x_{im} - 1)^2 \\ + B \sum_{m=1}^{N} (\sum_{i=1}^{N} x_{im} - 1)^2 \\ + \sum_{i=1}^{N} \sum_{j=1}^{N} \sum_{m=1}^{N} \sum_{n=1}^{N} d_{ij} c_{mn} x_{im} x_{jn}, \end{aligned} \tag{8}$$

where A and B are control parameter for the constraints of Eqs.(6) and (7).

The energy function E of such an $N \times N$ neural network is defined as follows:

$$E(X) = -\frac{1}{2} \sum_{i=1}^{N} \sum_{j=1}^{N} \sum_{m=1}^{N} \sum_{n=1}^{N} w_{im;jn} x_{im} x_{jn} + \sum_{i=1}^{N} \sum_{m=1}^{N} \theta_{im} x_{im}, \tag{9}$$

where $w_{im;jn}$ is the synaptic weight from the (j, n)th neuron toj the (i, m)th neuron and θ_{im} are the threshold of the (i, m)th neuron. From Eqs.(5) and (9), we determine the synaptic weight and the threshold as follows:

$$w_{im;jn} = -2 \left(A(1 - \delta_{mn})\delta_{ij} + B\delta_{mn}(1 - \delta_{ij}) + \frac{d_{ij} c_{mn}}{q} \right), \tag{10}$$

$$\theta_{im} = -(A + B), \tag{11}$$

where δ is Kronecker's delta and q is a normalization parameter for the product of d_{ij} and c_{mn}. Then, an internal state of the (i, m)th neuron is defined as follows:

$$y_{im}(t+1) = k y_{im}(t) + \sum_{j=1}^{N} \sum_{n=1}^{N} w_{im;jn} f(y_{jn}(t)) - \alpha f(y_{im}(t)) + \theta_{im}(1 - k), \tag{12}$$

where f is a sigmoidal function

$$f(y) = \frac{1}{1 + \exp(-\frac{y}{\epsilon})}, \tag{13}$$

where ϵ is a gradient parameter of a sigmoid function.

Each neuron is updated asynchronously. However, it is not so easy to generate feasible solutions, because an internal state of the chaotic neuron takes an analog value. Thus, we use the firing decision method[3] which can always generate a feasible solution for QAP. The procedure is described as follows:

1. Choose an index (i, m) whose internal state y_{im} takes the maximum value among all the neurons. Then set the (i, m)th neuron as a firing state by letting $x_{im} = 1$(Fig.1(a)).
2. Next, other neurons in the ith row and the mth column are set to a resting state. Namely, let $x_{ik} = 0 (k \neq m)$ and $x_{ml} = 0 (l \neq i)$. Then exclude the neurons which have already been selected at Steps 1 and 2(Fig.1(b))
3. Repeat Steps 1 and 2 until all the states of neurons are decided(Figs.1(c),(d) and (e)).

3 Proposed Method

A method for solving QAP by using Hopfiled-Tank neural network(HNN) has the local minimum problem. To overcome this problem, some approaches have already been proposed, for example, additive chaotic dynamical noise to HNN [7],[8], chaotic neural network(CNN)[3],[4], and so on. It was shown that the method of using CNN can obtain better solutions than the method of adding chaotic dynamical noise to HNN[9]. However, we expected that it would be more effective to use both CNN and dynamical noise. Then, in the proposed method, we add dynamical noise to CNN for solving QAP. The noise is added each neuron every iteration. The internal state of the (i, m)th neuron with noise is defined as follows:

$$y_{im}(t+1) = ky_{im}(t) + \sum_{j=1}^{N}\sum_{n=1}^{N} w_{im;jn} f(y_{jn}(t)) - \alpha f(y_{im}(t)) + \theta_{im}(1-k) + \gamma\beta_{im}(t), \tag{14}$$

where γ is a weight of dynamical noise and $\beta_{im}(t)$ is the dynamical noise which is added to the internal state of the (i, m)th neuron at time t.

4 Results

4.1 Performance with Respect to Weight of Noise

To evaluate the performance of the proposed method, we used Nug15, Nug20, Nug25, and Nug30 from QAPLIB[10]. Parameters of chaotic neural networks (in Eq.(14))are as follows: $A = 0.29$, $B = 0.29$, $k = 0.83$, $\alpha = 1.01$, $\epsilon = 0.017$, $\theta = 0.58$, $q = 350$ for Nug15, $q = 540$ for Nug20, $q = 820$ for Nug25, $q = 880$ for Nug30. We used white Gaussian noise whose average is zero and variance is unity. The proposed method is applied for 2000 iteration for each instance.

$$\begin{array}{c|ccccc} & y_{i1} & y_{i2} & y_{i3} & y_{i4} & y_{i5} \\ \hline y_{1j} & 0.1 & 0.4 & 0.2 & 0.3 & 0.7 \\ y_{2j} & 0.5 & \boxed{0.9} & 0.7 & 0.4 & 0.3 \\ y_{3j} & 0.1 & 0.3 & 0.4 & 0.4 & 0.3 \\ y_{4j} & 0.3 & 0.6 & 0.6 & 0.2 & 0.5 \\ y_{5j} & 0.8 & 0.2 & 0.3 & 0.5 & 0.4 \end{array} \Rightarrow \begin{array}{c|ccccc} & x_{i1} & x_{i2} & x_{i3} & x_{i4} & x_{i5} \\ \hline x_{1j} & ? & ? & ? & ? & ? \\ x_{2j} & ? & 1 & ? & ? & ? \\ x_{3j} & ? & ? & ? & ? & ? \\ x_{4j} & ? & ? & ? & ? & ? \\ x_{5j} & ? & ? & ? & ? & ? \end{array}$$

(a) The $(2, 2)$th neuron is selected, because y_{22} is the largest. Then let $x_{22} = 1$.

$$\begin{array}{c|ccccc} & y_{i1} & y_{i2} & y_{i3} & y_{i4} & y_{i5} \\ \hline y_{1j} & 0.1 & 0.4 & 0.2 & 0.3 & 0.7 \\ y_{2j} & 0.5 & \boxed{0.9} & 0.7 & 0.4 & 0.3 \\ y_{3j} & 0.1 & 0.3 & 0.4 & 0.4 & 0.3 \\ y_{4j} & 0.3 & 0.6 & 0.6 & 0.2 & 0.5 \\ y_{5j} & 0.8 & 0.2 & 0.3 & 0.5 & 0.4 \end{array} \Rightarrow \begin{array}{c|ccccc} & x_{i1} & x_{i2} & x_{i3} & x_{i4} & x_{i5} \\ \hline x_{1j} & ? & 0 & ? & ? & ? \\ x_{2j} & 0 & 1 & 0 & 0 & 0 \\ x_{3j} & ? & 0 & ? & ? & ? \\ x_{4j} & ? & 0 & ? & ? & ? \\ x_{5j} & ? & 0 & ? & ? & ? \end{array}$$

(b) Let $x_{2k} = 0 (k \neq 2)$ and $x_{l2} = 0 (l \neq 2)$. Then exclude neurons in the 2nd row and in the 2nd column.

$$\begin{array}{c|ccccc} & y_{i1} & y_{i2} & y_{i3} & y_{i4} & y_{i5} \\ \hline y_{1j} & 0.1 & 0.4 & 0.2 & 0.3 & 0.7 \\ y_{2j} & 0.5 & \boxed{0.9} & 0.7 & 0.4 & 0.3 \\ y_{3j} & 0.1 & 0.3 & 0.4 & 0.4 & 0.3 \\ y_{4j} & 0.3 & 0.6 & 0.6 & 0.2 & 0.5 \\ y_{5j} & \boxed{0.8} & 0.2 & 0.3 & 0.5 & 0.4 \end{array} \Rightarrow \begin{array}{c|ccccc} & x_{i1} & x_{i2} & x_{i3} & x_{i4} & x_{i5} \\ \hline x_{1j} & ? & 0 & ? & ? & ? \\ x_{2j} & 0 & 1 & 0 & 0 & 0 \\ x_{3j} & ? & 0 & ? & ? & ? \\ x_{4j} & ? & 0 & ? & ? & ? \\ x_{5j} & 1 & 0 & ? & ? & ? \end{array}$$

(c) The $(5, 1)$th neuron is selected, because y_{51} is the largest. Then let $x_{51} = 1$.

$$\begin{array}{c|ccccc} & y_{i1} & y_{i2} & y_{i3} & y_{i4} & y_{i5} \\ \hline y_{1j} & 0.1 & 0.4 & 0.2 & 0.3 & 0.7 \\ y_{2j} & 0.5 & \boxed{0.9} & 0.7 & 0.4 & 0.3 \\ y_{3j} & 0.1 & 0.3 & 0.4 & 0.4 & 0.3 \\ y_{4j} & 0.3 & 0.6 & 0.6 & 0.2 & 0.5 \\ y_{5j} & \boxed{0.8} & 0.2 & 0.3 & 0.5 & 0.4 \end{array} \Rightarrow \begin{array}{c|ccccc} & x_{i1} & x_{i2} & x_{i3} & x_{i4} & x_{i5} \\ \hline x_{1j} & 0 & 0 & ? & ? & ? \\ x_{2j} & 0 & 1 & 0 & 0 & 0 \\ x_{3j} & 0 & 0 & ? & ? & ? \\ x_{4j} & 0 & 0 & ? & ? & ? \\ x_{5j} & 1 & 0 & 0 & 0 & 0 \end{array}$$

(d) Let $x_{5k} = 0 (k \neq 1)$ and $x_{l1} = 0 (l \neq 5)$. Then exclude neurons in the 5th row and in the 1st column.

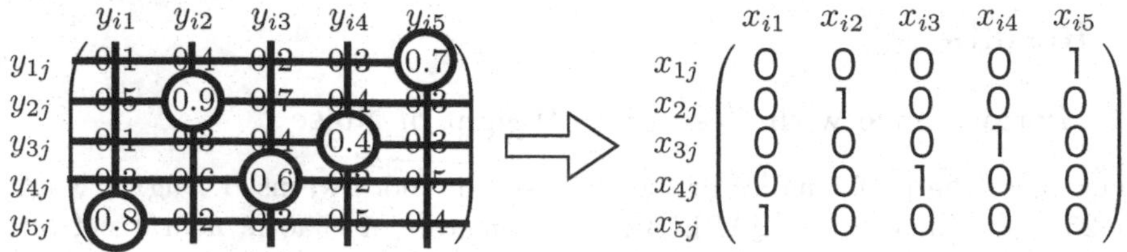

(e) Repeat the above procedures to select and exclude neurons.

Fig. 1. How to obtain a feasible solution. The left matrix is the internal state of neurons, and the right matrix is the outputs of neurons.

Figure 2 shows results for the proposed method, when we change γ from 0 to 0.005 at intervals of 0.0001. In Fig.2, the results are expressed by the average gap of 100 trial.

$$\text{Gap} = \frac{\text{Best Obtained Solution} - \text{Optimal Solution}}{\text{Optimal Solution}} \times 100[\%]. \tag{15}$$

From Fig.2, to add small amount of noise, the performance of the proposed method is improved for all problems. However, to add larger amplitude of noise, the performance of the proposed method becomes worse gradually. The small amount of noise leads to more effective search, because the searching space is extended appropriately. On the other hand, the larger amount of noise disturbs a search, because it leads to a random search.

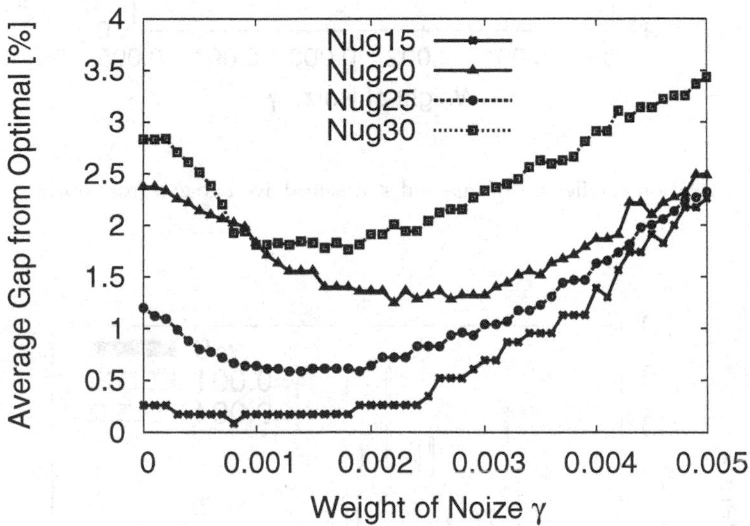

Fig. 2. Average gaps from the optimal solution when noise weight γ is changed for Nug15, Nug20, Nug25 and Nug30

4.2 Correlation Coefficient of Internal States and Fire Patterns

Next, to analyze an influence of noise, we calculate an average correlation coefficient $\overline{r}$ of the internal states of neurons at every iteration, defined as follows:

$$\overline{r} = \frac{1}{T-1} \sum_{t=1}^{T-1} \left(\frac{\sum_{i=1}^{N} \sum_{m=1}^{N} (y_{im}(t) - \overline{y}(t))(y_{im}(t+1) - \overline{y}(t+1))}{\sqrt{\sum_{i=1}^{N} \sum_{m=1}^{N} (y_{im}(t) - \overline{y}(t))^2} \sqrt{\sum_{i=1}^{N} \sum_{m=1}^{N} (y_{im}(t+1) - \overline{y}(t+1))^2}} \right), \tag{16}$$

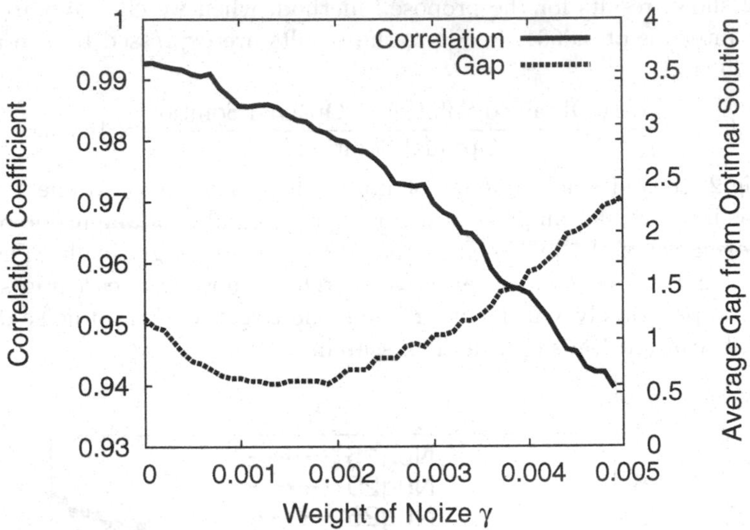

Fig. 3. Correlation coefficients of internal states and average gap from optimal solution for Nug25

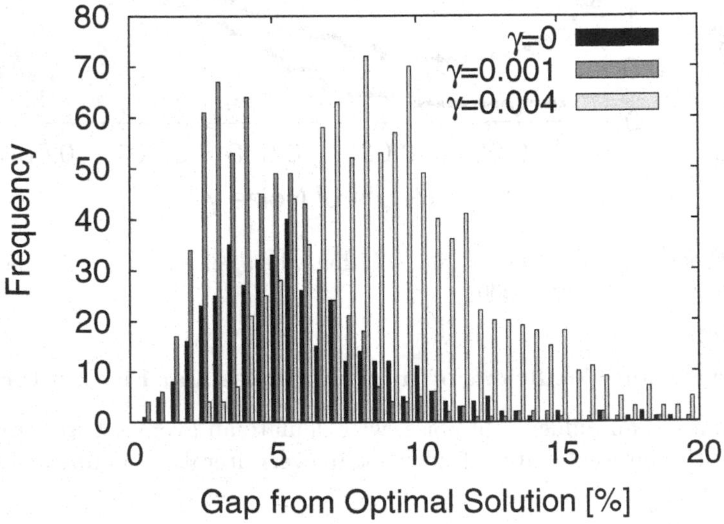

Fig. 4. Frequency distributions of instantaneous gap of each iteration for Nug25

where T is the number of iterations, $\overline{y}(t)$ is an average of the internal state of the neurons at time t, and is defined as follows:

$$\overline{y}(t) = \frac{1}{N^2} \sum_{i=1}^{N} \sum_{m=1}^{N} y_{im}(t). \tag{17}$$

Figure 3 shows the results for average correlation coefficient of internal states of the neurons and average gaps from optimal solution. From Fig.3, when the noise is not added, the correlation coefficients take large value. Namely, internal states of neurons change slowly. On the other hand, the correlation coefficients decrease when the amount of noise increased. The reason is that due to dynamical noise, the dynamic range of the internal states becomes large, then the solution search has been improved.

Next, we examine how additive dynamical noise effects quality of solutions offered by CNN. Figure 4 shows frequency distributions of instantaneous gap of each iteration. In Fig.4 we plot the result just after the 200th iteration. From Fig.4, if the small amount of dynamical noise is added to the dynamics of CNN, the frequency of good solutions ($0 \sim 5\%$) increases. If the amount of dynamical noise is too large, the frequency of good solutions reduces and that of worse solutions ($5 \sim 15\%$) increases. Namely, the solution search is promoted by adding the small amount of noise. However, the large amount of noise reduces original searching ability, then good solutions cannot be obtained.

5 Conclusions

We propose a new method for solving QAP which uses chaotic dynamics with dynamical noise. As a result, in case of adding the small amount of dynamical noise, the proposed method shows higher performance than the conventional method. To analyse the property of CNN with respect to noise, we investigate correlation coefficients of internal states of the neurons and frequency distribution of solutions. From the results, when we add dynamical noise to the chaotic dynamics, variation of internal states of the chaotic neuron becomes large and much better solutions are obtained. The research of T.I. is partially supported by Grant-in-Aid for Scientific Research (B) (No.20300085) from the JSPS.

References

1. Hopfield, J.J.: Neural networks and physical systems with emergent collective computational abilities. Proceedings of the National Academy of Sciences of the United States of America, 79 (1982)
2. Aihara, K., Takabe, T., Toyoda, M.: Chaotic neural networks. Physics Letters A 144 (1990)
3. Ikeguchi, T., Sato, K., Hasegawa, M., Aihara, K.: Chaotic optimization for quadratic assignment problems. In: Proceedings of International Symposium on Circuits and Systems, vol. 3 (2002)

[4] Horio, Y., Ikeguchi, T., Aihara, K.: A mixed analog/digital chaotic neuro-computer system for quadratic assignment problems. Neural Networks 18(5-6), 505–513 (2005)
[5] Horio, Y., Aihara, K.: Analog computation through high-dimensional physical chaotic neuro-dynamics. Physica D 237, 1215–1225 (2008)
[6] Kirkpatrick, S., Gelatt Jr, C.D., Vecchi, M.P.: Optimization by simulated annealing. Science 220 (1983)
[7] Hayakawa, Y., Sawada, Y.: Effects of the chaotic noise on the performance of a neural network model for optimization problems. Physical Review E 51, 2693–2696 (1995)
[8] Uwate, Y., Nishio, Y., Ueta, T., Kawabe, T., Ikeguchi, T.: Performance of chaos and burst noises injected to the hopfield nn for quadratic assignment problems. IEICE Trans. on Fund E87–A(4) (2004)
[9] Sato, K., Ikeguchi, T., Hasegawa, M., Aihara, K.: A combinatorial optimization method by chaotic neurodynamics. IEICE Technical Report 102, 31–34 (2003)
[10] QAPLIB, `http://www.opt.math.tu-graz.ac.at/qaplib`

Global Exponential Stability of Recurrent Neural Networks with Time-Dependent Switching Dynamics

Zhigang Zeng[1], Jun Wang[2], and Tingwen Huang[3]

[1] Department of Control Science and Engineering, Huazhong University of Science and Technology, Wuhan, Hubei, 430074, China
zgzeng@mail.hust.edu.cn

[2] Department of Mechanical and Automation Engineering, The Chinese University of Hong Kong, Shatin, New Territories, Hong Kong
jwang@mae.cuhk.edu.hk

[3] Texas A&M University at Qatar, Doha, P.O. Box 5825, Qatar
tingwen.huang@qatar.tamu.edu

Abstract. In this paper, the switching dynamics of recurrent neural networks are studied. Sufficient conditions on global exponential stability with an arbitrary switching law or a dwell time switching law and the estimates of Lyapunov exponent are obtained. The obtained results can be used to analyze and synthesize a family of continuous-time configurations with the switching between the configurations. Specially, the obtained results are new and efficacious for the switching between the stable and unstable configurations. Finally, simulation results are discussed to illustrate the theoretical results.

1 Introduction

In designs and applications of dynamic systems, it is not uncommon for their dynamics to switch between two or more configurations. A switching dynamic system is composed of a family of configurations and a switching rule between the configurations. Recently, there has been increasing interest in the stability analysis and switching control design of dynamic systems (see, for example, [1]-[11]). The motivation for studying switching systems comes from the fact that many practical systems are inherently multimodal in the sense that several dynamic systems are required to describe their behaviors, depending on various environmental factors [2], and from the fact that the methods of intelligent control design are based on the idea of switching between different controllers [3].

For a family of linear time-invariant configurations, it was shown in [4] that, when all configuration matrices are Hurwitz stable, the entire system is exponentially stable for any switching law if the time between consecutive switchings (called the 'dwell time') is sufficiently large. Hespanha and Morse extended the

C. Alippi et al. (Eds.): ICANN 2009, Part II, LNCS 5769, pp. 583–592, 2009.

'dwell time' concept to an 'average dwell time' concept [5], which means that the average time interval between consecutive switchings is no less than a specified constant. And it is proved that, if such a constant is sufficiently large, then the switching system is exponentially stable. The idea of specifying the total activation time period ratio of Hurwitz stable configurations to unstable configurations is motivated by the work in [6], where all configuration matrices are assumed to be pairwise commutative.

Hu and Michel [3] analyzed a dwell time scheme for local asymptotic stability of nonlinear switching systems with the activation time being used as the dwell time. In [7], some stability properties of switching systems composed of a family of linear time-invariant configurations are studied. In addition, the obtained results are applied to the perturbed switching systems where nonlinear norm-bounded perturbations exist in the linear time-invariant configurations.

In [8], a class of switching Hopfield neural networks with time-varying delay is investigated. In [9], robust stability of switching Cohen-Grossberg neural networks with mixed time-varying delays is considered. In [10] and [11], new stability results for recurrent neural networks with Markovian switching are presented.

In this paper, we will analyze the stability of switching neural networks. The remainder of this paper is organized as follows. Useful notations and definitions are given in Section 2. In Section 3, the global exponential stability results with arbitrary switching law and the estimates of Lyapunov exponent are discussed. In Section 4, the global exponential stability results with dwell time switching law and the estimates of the Lyapunov exponent are derived. Illustrative examples are provided in Section 5. Finally, concluding remarks are given in Section 6.

2 Preliminaries

2.1 Model

Consider the recurrent neural network model with time-varying delays and switching law $\sigma(t)$:

$$\dot{x}(t) = -D_\sigma x(t) + A_\sigma f(x(t)) + B_\sigma g(x(t-\tau(t))) + J_\sigma. \tag{1}$$

This system can be regarded as the result of the following N configurations

$$\dot{x}^{(\ell)}(t) = -D_\ell x^{(\ell)}(t) + A_\ell f(x^{(\ell)}(t)) + B_\ell g(x^{(\ell)}(t-\tau(t))) + J_\ell \tag{2}$$

switching from one to the others according to a switching law:

$$\sigma : [t_0, +\infty) \to \{1, 2, \cdots, N\}. \tag{3}$$

Thus, A_σ is a piecewise constant function:

$$A_\sigma : [t_0, +\infty) \to \{A_1, A_2, \cdots, A_N\}. \tag{4}$$

Here, $\{A_\ell : 1 \le \ell \le N\}$ is a family of constant matrices describing the configurations, and the integer $N > 1$ denotes the number of configurations. Similarly, D_σ, B_σ, J_σ can be obtained.

In recurrent neural network model (1), $x(t) = (x_1(t), x_2(t), \cdots, x_n(t))^T$ is a state vector of neurons. In recurrent neural network (2), $x^{(\ell)}(t) = (x_1^{(\ell)}(t), x_2^{(\ell)}(t), \cdots, x_n^{(\ell)}(t))^T$ is a state vector of neurons of configuration, $D_\ell = \mathrm{diag}(d_1^{(\ell)}, d_2^{(\ell)}, \cdots, d_n^{(\ell)})$ is a self-feedback connection weight matrix, $A_\ell = (a_{ij}^{(\ell)})_{n\times n}$ and $B_\ell = (b_{ij}^{(\ell)})_{n\times n}$ are connection weight matrices without delays and with delays, respectively, $J_\ell = (J_1^{(\ell)}, J_2^{(\ell)}, \cdots, J_n^{(\ell)})^T$ is an external input (bias) to the network, $f(x(\cdot)) = (f_1(x_1(\cdot)), f_2(x_2(\cdot)), \cdots, f_n(x_n(\cdot)))^T$ and $g(x(\cdot)) = (g_1(x_1(\cdot)), g_2(x_2(\cdot)), \cdots, g_n(x_n(\cdot)))^T$ are activation functions which satisfy

A$_1$: $\forall j \in \{1, 2, \cdots, n\}, \forall r_1, r_2, r_3, r_4 \in \Re$, there exist real numbers γ_j and μ_j such that

$$|f_j(r_1) - f_j(r_2)| \le \gamma_j |r_1 - r_2|, \quad |g_j(r_3) - g_j(r_4)| \le \mu_j |r_3 - r_4|.$$

Hence, (2) can be rewritten as the following form:

$$\dot{x}_i^{(\ell)}(t) = -d_i^{(\ell)} x^{(\ell)}(t) + \sum_{j=1}^{n} a_{ij}^{(\ell)} f_j(x_j^{(\ell)}(t)) + \sum_{j=1}^{n} b_{ij}^{(\ell)} g_j(x_j^{(\ell)}(t - \tau_j(t))) + J_i^{(\ell)} \quad (5)$$

where $0 \le \tau_j(t) \le \tau$. Throughout of this paper, we denote $|u|$ as the absolute-value vector; i.e., $|u| = (|u_1|, |u_2|, \cdots, |u_n|)^T$, $|A|$ as the absolute-value matrix; i.e., $|A| = [|a_{ij}|]$. Denote the vector $u > 0$ as $u_i > 0$, $\forall i \in \{1, 2, \cdots, n\}$. Denote $I_{n\times n}$ as the $n \times n$ identity matrix. Denote $\|u\|_p$ as the vector p-norm of the vector u with p satisfies $1 \le p < \infty$. $\|u\|_\infty = \max_{i=1,2,\cdots,n} |u_i|$ is the vector infinity norm. Denote $\|A\|_p$ as the p-norm of the matrix A induced by the vector p-norm. Denote C^0 as the set of continuous functions.

2.2 Definitions

Definition 1. If $\forall \ell \in \{1, 2, \cdots, N\}$, $J_\ell = 0$, and an arbitrary solution $x(t)$ of the neural network (1) satisfies

$$|x_i(t)| \le (\sum_{j=1}^{n} \sup_{t_0 - \tau \le \zeta \le t_0} |x_j(\zeta)|)\beta_i \exp\{-\alpha_i(t - t_0)\},$$

where $t \ge t_0 > 0$, β_i and α_i are positive constants, then the equilibrium point of the neural networks (1) is said to be globally exponentially stable.

Definition 2. Let Ω be solution set of the neural network (1),

$$\lambda = \sup_{x(t)\in\Omega} \overline{\lim}_{t\to+\infty} (\ln \|x(t)\|_\infty / (t - t_0))$$

is called as Lyapunov exponent.

From Definition 2, there exist positive constants $\bar{\beta}_i$, $(i = 1, 2, \cdots, n)$ such that the state vector $x(t)$ of the neural network (1) satisfies

$$|x_i(t)| \le \bar{\beta}_i \exp\{\lambda(t - t_0)\},$$

where $t \geq t_0 > 0$. Hence, the neural network (1) is exponentially stable when its Lyapunov exponent is negative.

2.3 Dwell Time

Given a positive constant τ_d, let $\mathcal{S}[\tau_d]$ denote the set of all switching signals with interval between consecutive discontinuities no smaller than τ_d. The constant τ_d is called the dwell time.

Morse A.S. showed that, if τ_d is sufficiently large, then the switching system

$$\dot{x}(t) = A_\sigma x(t) \tag{6}$$

is exponentially stable for any switching law $\sigma \in \mathcal{S}[\tau_d]$ [4].

Definition 3. Given two positive constants $\underline{\tau_d}, \bar{\tau_d}$, let $\mathcal{BS}[\underline{\tau_d}, \bar{\tau_d}]$ denote the set of all switching signals with interval between consecutive discontinuities at least $\underline{\tau_d}$ and at most $\bar{\tau_d}$. The constant $\bar{\tau_d}$ is called the dwell time upper bound and $\underline{\tau_d}$ is called the dwell time lower bound.

3 Arbitrary Switching Laws

In this section, the global exponential stability results with arbitrary switching law and the estimates of Lyapunov exponent are discussed.

Theorem 1. If $\forall \ell \in \{1, 2, \cdots, N\}$, $J_\ell = 0$, and there exist positive numbers $\alpha_1, \cdots, \alpha_n$ such that $\forall i \in \{1, 2, \cdots, n\}$

$$d_i^{(\ell)} \alpha_i > \sum_{j=1}^{n} (|a_{ij}^{(\ell)}|\gamma_j + |b_{ij}^{(\ell)}|\mu_j)\alpha_j, \tag{7}$$

then the switching neural network (1) under any switching laws is globally exponentially stable with its estimated Lyapunov exponent $\hat{\lambda}$, where

$$\hat{\lambda} = \max_{1 \leq \ell \leq N} \Big\{ \lambda | \ (d_i^{(\ell)} - \lambda)\alpha_i - \sum_{j=1}^{n} (|a_{ij}^{(\ell)}|\gamma_j + |b_{ij}^{(\ell)}|\mu_j \exp\{\lambda\tau\})\alpha_j \geq 0, \ \ i = 1, \cdots, n \Big\}. \tag{8}$$

Proof. Let $y(t) = (y_1(t), y_2(t), \cdots, y_n(t))^T = (|x_1(t)|/\alpha_1, \cdots, |x_n(t)|/\alpha_n)^T$, $y^{(\ell)}(t) = (y_1^{(\ell)}(t), \cdots, y_n^{(\ell)}(t))^T = (|x_1^{(\ell)}(t)|/\alpha_1, |x_2^{(\ell)}(t)|/\alpha_2, \cdots, |x_n^{(\ell)}(t)|/\alpha_n)^T$, where $x(t) = (x_1(t), \cdots, x_n(t))^T$ is a state of (1), $x^{(\ell)}(t) = (x_1^{(\ell)}(t), \cdots, x_n^{(\ell)}(t))^T$ is a state of (2). Then from (5),

$$D^+ y_i^{(\ell)}(t) \mid_{(5)} \leq -d_i^{(\ell)} y_i^{(\ell)}(t) + \sum_{j=1}^{n} \left| a_{ij}^{(\ell)} \right| \gamma_j \frac{\alpha_j}{\alpha_i} y_j^{(\ell)}(t) + \sum_{j=1}^{n} \left| b_{ij}^{(\ell)} \right| \mu_j \frac{\alpha_j}{\alpha_i} y_j^{(\ell)}(t - \tau_{ij}(t)), \tag{9}$$

where D^+ denotes upper-right Dini-derivative operator.

Without loss of generality, let

$$\sigma(t) = \begin{cases} 1, & t \in (t_0, t_1], \\ 2, & t \in (t_1, t_2], \\ \vdots, & \\ N, & t \in (t_{N-1}, t_N]. \end{cases} \tag{10}$$

Let $\bar{y}(t_0) = \max_{1 \le i \le n} \sup_{t_0 - \tau < \zeta \le t_0} |x_i(\zeta)/\alpha_i|$, $z_i(t) = y_i(t) - \bar{y}(t_0)\exp\{-\hat{\lambda}(t - t_0)\}$, $z_i^{(\ell)}(t) = y_i^{(\ell)}(t) - \bar{y}(t_0)\exp\{-\hat{\lambda}(t - t_0)\}$, then $\forall t \ge t_0$, $z_i(t) \le 0$. Thus $y_i(t) \le \bar{y}(t_0)\exp\{-\hat{\lambda}(t - t_0)\}$, for all $i \in \{1, 2, \cdots, n\}, t \ge t_0$. □

Corollary 1. If $\forall \ell \in \{1, 2, \cdots, N\}$, $J_\ell = 0$, and $\forall i \in \{1, 2, \cdots, n\}$

$$d_i^{(\ell)} > \sum_{j=1}^{n} (|a_{ij}^{(\ell)}|\gamma_j + |b_{ij}^{(\ell)}|\mu_j), \tag{11}$$

then the switching neural network (1) is globally exponentially stable with estimated Lyapunov exponent $\hat{\lambda}$ for any switching laws, where

$$\hat{\lambda} = \min_{1 \le \ell \le N} \left\{ \lambda | (d_i^{\ell)} - \lambda - \sum_{j=1}^{n} (|a_{ij}^{(\ell)}|\gamma_j + |b_{ij}^{(\ell)}|\mu_j \exp\{\lambda\tau\}) \ge 0, i = 1, \cdots, n \right\} \tag{12}$$

Proof. Choose $\alpha_i = 1$ $(i = 1, 2, \cdots, n)$ in (15). According to Theorem 1, Corollary 1 holds. □

Remark 1. In [8] and [9], the robust stability of switching neural networks with similar state characteristic configurations are considered. The method in proof of Theorem 1 can also be utilized to study the robust stability of switching neural networks with similar state characteristic configurations.

Remark 2. When $N = 1$, the results of Theorem 1 is the same as the so called M-matrix criterion in [12].

4 Switching Laws with Dwell Time

In practical applications, (11) does not always hold. Without loss of generality, we assume that $\forall \ell \in \{1, 2, \cdots, N_1\}$,

$$d_i^{(\ell)} > \sum_{j=1}^{n} (|a_{ij}^{(\ell)}|\gamma_j + |b_{ij}^{(\ell)}|\mu_j), \quad i = 1, 2, \cdots, n; \tag{13}$$

and $\forall \ell \in \{N_1 + 1, N_1 + 2, \cdots, N\}$, there exist $i \in \{1, 2, \cdots, n\}$, such that

$$d_i^{(\ell)} \le \sum_{j=1}^{n} (|a_{ij}^{(\ell)}|\gamma_j + |b_{ij}^{(\ell)}|\mu_j). \tag{14}$$

For $\ell \in \{N_1+1, N_1+2, \cdots, N\}$, it is possible that the configuration has multiple equilibrium points. Hence, the switching recurrent neural network with configurations satisfying (13) and (14) has dissimilar stability property.
Theorem 2. If $\forall \ell \in \{1, 2, \cdots, N\}$, $J_\ell = 0$, and

$$\underline{\tau_d}\lambda_1 > 2\bar{\tau_d}\lambda_2, \tag{15}$$

and for any interval $[\bar{t}, \bar{t}+\bar{\tau_d}]$, there exist $\bar{\ell} \in \{1, 2, \cdots, N_1\}$ and $t^* \in [\bar{t}, \bar{t}+\bar{\tau_d}]$ such that $\sigma(t^*) = \bar{\ell}$, then the switching neural network (1) under the switching laws $\sigma(t) \in \mathcal{BS}[\underline{\tau_d}, \bar{\tau_d}]$ is globally exponentially stable with its estimated Lyapunov exponent $\hat{\lambda}$, where

$$\hat{\lambda} = \underline{\tau_d}\lambda_1 - 2\bar{\tau_d}\lambda_2, \tag{16}$$

$$\lambda_1 = \max_{1 \le \ell \le N_1, 1 \le i \le n} \Big\{\lambda | \ d_i^{(\ell)} - \lambda - \sum_{j=1}^{n}(|a_{ij}^{(\ell)}|\gamma_j + |b_{ij}^{(\ell)}|\mu_j \exp\{\lambda\tau\}) \ge 0\Big\}, \tag{17}$$

$$\lambda_2 = \max_{N_1+1 \le \ell \le N, 1 \le i \le n} \Big\{ - d_i^{(\ell)} + \sum_{j=1}^{n}(|a_{ij}^{(\ell)}|\gamma_j + |b_{ij}^{(\ell)}|\mu_j)\Big\}. \tag{18}$$

Proof. From (5),

$$\begin{aligned} D^+ x_i^{(\ell)}(t) \mid_{(5)} \le -d_i^{(\ell)} x_i^{(\ell)}(t) + \sum_{j=1}^{n} \Big(\left|a_{ij}^{(\ell)}\right| \gamma_j \left|x_j^{(\ell)}(t)\right| \\ + \left|b_{ij}^{(\ell)}\right| \mu_j \left|x_j^{(\ell)}(t - \tau_{ij}(t))\right| \Big), \end{aligned} \tag{19}$$

where D^+ denotes upper-right Dini-derivative operator. Let

$$\bar{x}(t_0) = \max_{1 \le i \le n} \sup_{t_0 - \tau < \zeta \le t_0} |x_i(\zeta)|.$$

We will prove that $\forall t \ge t_0$, $\forall i \in \{1, 2, \cdots, n\}$,

$$x_i(t) \le \bar{x}(t_0) \exp\{\hat{\lambda}(t - t_0)\} \tag{20}$$

via three steps.

The first step, when $t \in [t_0, t_0 + \bar{\tau_d}]$, let

$$z_i(t) = x_i(t) - \bar{x}(t_0) \exp\{\lambda_2(t - t_0)\},$$
$$z_i^{(\ell)}(t) = x_i^{(\ell)}(t) - \bar{x}(t_0) \exp\{\lambda_2(t - t_0)\}.$$

Similar to the proof of Theorem 1, we can prove that $\forall t \in [t_0, t_0 + \bar{\tau_d}]$, $\forall \ell \in \{1, \cdots, N\}$, $z_i^{(\ell)}(t) \le 0$.

The second step, when $t \in [t_0, t_0 + \underline{\tau_d} + \bar{\tau_d}]$, let

$$z_i(t) = x_i(t) - \bar{x}(t_0) \exp\{\lambda_2 \bar{\tau_d}\} \exp\{-\lambda_1(t - t_0 - \bar{\tau_d})\}, \tag{21}$$

$$z_i^{(\ell)}(t) = x_i^{(\ell)}(t) - \bar{x}(t_0) \exp\{\lambda_2 \bar{\tau_d}\} \exp\{-\lambda_1(t - t_0 - \underline{\tau_d})\}. \tag{22}$$

Then $\forall t \in [t_0, t_0 + \underline{\tau_d} + \bar{\tau_d}]$,

$$z_i(t) \leq 0. \tag{23}$$

The third step, similarly, it can be proved that for all $i \in \{1, 2, \cdots, n\}$, $t \in [t_0, t_0 + \underline{\tau_d} + 2\bar{\tau_d}]$,

$$x_i(t) \leq \bar{x}(t_0)\exp\{\lambda_2\bar{\tau_d}\}\exp\{-\lambda_1\underline{\tau_d}\}\exp\{\lambda_2(t - t_0)\}; \tag{24}$$

for all $i \in \{1, 2, \cdots, n\}$, $t \in [t_0, t_0 + 2\underline{\tau_d} + 2\bar{\tau_d}]$,

$$\begin{aligned} x_i(t) &\leq \bar{x}(t_0)\exp\{2\lambda_2\bar{\tau_d}\}\exp\{-\lambda_1\underline{\tau_d}\}\exp\{-\lambda_1(t - t_0 - 2\underline{\tau_d})\} \\ &\leq \bar{x}(t_0)\exp\{-\hat{\lambda} - \lambda_1(t - t_0 - 2\underline{\tau_d})\}. \end{aligned} \tag{25}$$

Hence, for $k > 1$, $i \in \{1, 2, \cdots, n\}$, $t \in [t_0, t_0 + 2k\underline{\tau_d} + 2k\bar{\tau_d}]$,

$$x_i(t) \leq \bar{x}(t_0)\exp\{-k\hat{\lambda} - \lambda_1(t - t_0 - 2k\underline{\tau_d})\}. \tag{26}$$

Thus, the conclusion of Theorem 2 holds. □

Remark 3. When $\bar{\tau_d}$ is very large, if $\lambda_2 \leq 0$, (15) still holds.

5 Illustrative Examples

In this section, we give two numerical examples to illustrate the new results.

Example 1: Consider two configurations:

$$\begin{pmatrix} \dot{x}_1(t) \\ \dot{x}_2(t) \end{pmatrix} = -\begin{pmatrix} 2 & 0 \\ 0 & 4 \end{pmatrix}\begin{pmatrix} x_1(t) \\ x_2(t) \end{pmatrix} + \begin{pmatrix} 0.8 & 1 \\ 1 & -1 \end{pmatrix}\begin{pmatrix} f(x_1(t)) \\ f(x_2(t)) \end{pmatrix}; \tag{27}$$

$$\begin{pmatrix} \dot{x}_1(t) \\ \dot{x}_2(t) \end{pmatrix} = -\begin{pmatrix} 5 & 0 \\ 0 & 2 \end{pmatrix}\begin{pmatrix} x_1(t) \\ x_2(t) \end{pmatrix} + \begin{pmatrix} 1 & -1 \\ -1 & 0.9 \end{pmatrix}\begin{pmatrix} g(x_1(t)) \\ g(x_2(t)) \end{pmatrix}, \tag{28}$$

where $f(x(t)) = (\exp\{x(t)\} - \exp\{-x(t)\})/(\exp\{x(t)\} + \exp\{-x(t)\}); g(x(t)) = (|x(t) + 1| - |x(t) - 1|)/2$.

It is easy to prove that (27) and (28) are globally exponentially stable. In addition, (11) holds. According to Corollary 1, a switching neural network between configurations (27) and (28) is globally exponentially stable for any switching laws.

The transient states with some random initial conditions of the switching system with a switching law

$$\sigma(t) = \begin{cases} 1, & t \in (2k, 2k+1], k = 0, 1, 2, \cdots \\ 2, & t \in (2k+1, 2k+2], k = 0, 1, 2, \cdots \end{cases} \tag{29}$$

are depicted in Figure 1.

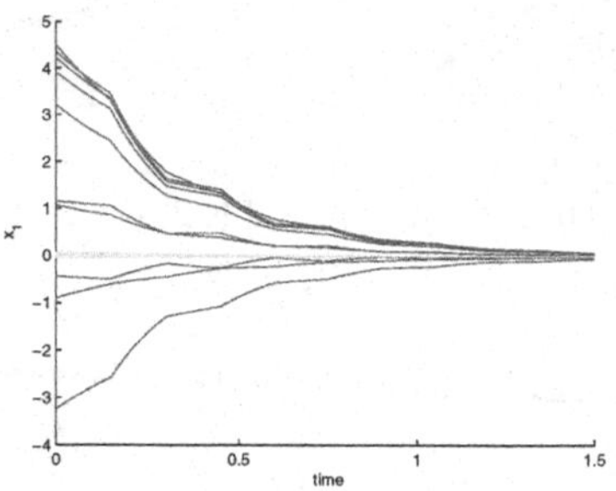

(a) Transient behaviors of state $x_1(t)$.

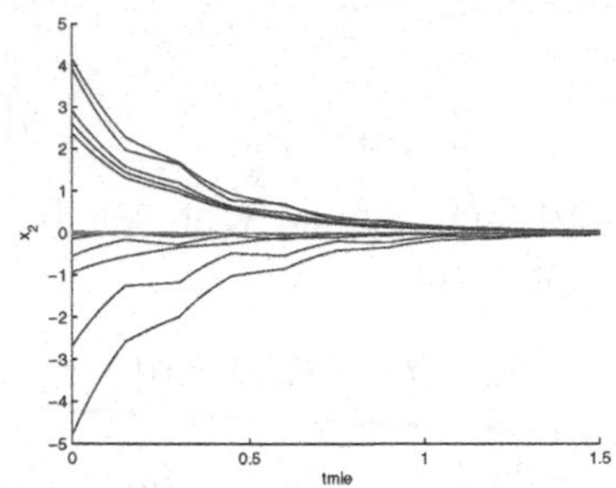

(b) Transient behaviors of state $x_2(t)$.

Fig. 1. The states with random initial conditions and the switching law (29) in Example 1

Example 2: Consider two configurations:

$$\begin{pmatrix} \dot{x}_1(t) \\ \dot{x}_2(t) \end{pmatrix} = -\begin{pmatrix} 6 & 0 \\ 0 & 6 \end{pmatrix} \begin{pmatrix} x_1(t) \\ x_2(t) \end{pmatrix} + \begin{pmatrix} -1 & 1 \\ 1 & 1 \end{pmatrix} \begin{pmatrix} f(x_1(t)) \\ f(x_2(t)) \end{pmatrix}; \tag{30}$$

$$\begin{cases} \dot{x}_1(t) = -0.1x_1(t) + 0.9f(x_1(t)) + 0.2f(x_2(t)), \\ \dot{x}_2(t) = -0.1x_2(t) + 0.2f(x_1(t)) + 0.9f(x_2(t)), \end{cases} \tag{31}$$

where $f(x(t)) = (\exp(x(t)) - \exp(-x(t)))/(\exp(x(t)) + \exp(-x(t)))$. It is easy to prove that (30) is globally exponentially stable. In contrast, (31) has 8 equilibrium points where only 4 equilibrium points are locally stable. The transient states with some random initial conditions of (30) and (31) are depicted in Figure 2.

Hence, it is impossible that these exists a common Lyapunov function for these two configurations (30) and (31). Thus, it is necessary to find a proper switching law such that the switching neural network is globally exponentially stable. The switching laws with dwell time are worth studying. In addition, it is necessary to find new methods in stability analysis of the nonlinear switching systems. According to Theorem 2, we can always find a proper switching law such that the switching recurrent neural network is globally exponentially stable. In fact, from (30) and (31), $\lambda_1 = 3, \lambda_2 = 1$. If we choose the switching law such that $\underline{\tau_d} = \bar{\tau_d} = 1$, then the conditions of Theorem 2 are all satisfied. Hence, the switching neural network with this switching law is globally exponentially stable. The states with some random initial conditions of the switching system with a switching law

$$\sigma(t) = \begin{cases} 1, & t \in (2k, 2k+1], k = 0, 1, 2, \cdots \\ 2, & t \in (2k+1, 2k+2], k = 0, 1, 2, \cdots \end{cases} \tag{32}$$

are depicted in Figure 3.

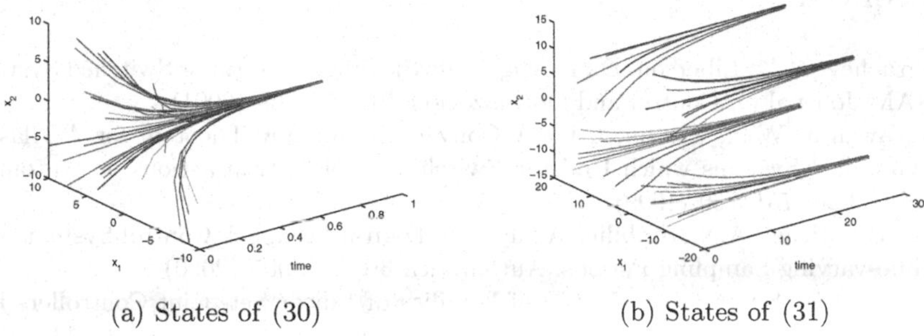

(a) States of (30) (b) States of (31)

Fig. 2. Transient behaviors of states of (30) and (31) with random initial conditions in Example 2

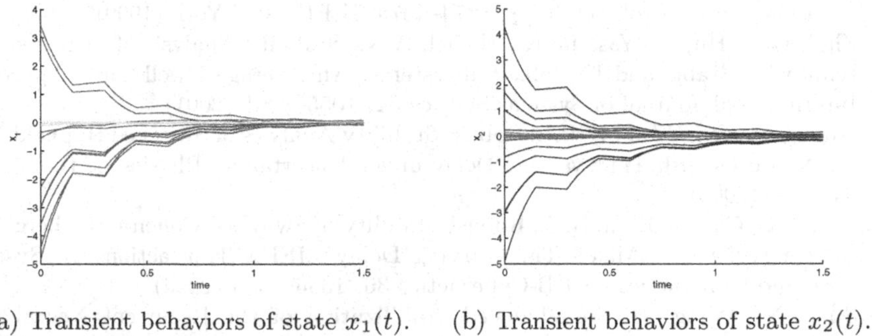

(a) Transient behaviors of state $x_1(t)$. (b) Transient behaviors of state $x_2(t)$.

Fig. 3. Transient behaviors of states with random initial conditions and the switching law (32) in Example 2

6 Concluding Remarks

In this paper, it is shown that the convergence behaviors of switching neural networks with the difference switching laws on dissimilar configurations could be significantly different. The global exponential stability results with an arbitrary switching law or a dwell time switching law and the estimates of Lyapunov exponent are obtained. The success of the switching law with dwell time encouraged us for further studies.

Acknowledgement. This work was supported by the Hong Kong Research Grants Council under Grant CUHK4176/08E, the Natural Science Foundation of China under Grant 60774051, Program for New Century Excellent Talents in Universities of China under Grant NCET-06-0658 and Fok Ying Tung Education Foundation under Grant 111068.

References

1. Agrachev, A.A., Liberzon, D.: Lie-algebraic Stability Criteria or Switched Systems. SIAM Journal on Control and Optimization 40, 253–269 (2001)
2. Dayawansa, W.P., Martin, C.F.: A Converse Lyapunov Theorem for A Class of Dynamical Systems which Undergo Switching. IEEE Transactions on Automatic Control 44, 751–760 (1999)
3. Hu, B., Michel, A.N.: Stability Analysis of Digital Feedback Control Systems with Time-varying Sampling Periods. Automatica 36, 897–905 (2000)
4. Morse, A.S.: Supervisory Control of Families of Linear Set-point Controllers Part 1: Exact Matching. IEEE Transactions on Automatic Control 41, 1413–1431 (1996)
5. Hespanha, J.P., Morse, A.S.: Stability of Switched Systems with Average Dwell-time. In: Proceedings of the 38th IEEE Conference on Decision and Control, vol. 3, pp. 2655–2660. IEEE, New York (1999)
6. Hu, B., Xu, X., Michel, A.N., Antsaklis, P.J.: Stability Analysis for A Class of Nonlinear Switched Systems. In: Proceedings of the 38th IEEE Conference on Decision and Control, vol. 3, pp. 4374–4379. IEEE, New York (1999)
7. Zhai, G.S., Hu, B., Yasuda, K., Michel, A.N.: Stability Analysis of Switched Systems with Stable and Unstable Subsystems: An Average Dwell Time Approach. International Journal of Systems Science 32, 1055–1061 (2001)
8. Huang, H., Qu, Y.Z., Li, H.X.: Robust Stability Analysis of Switched Hopfield Neural Networks with Time-varying Delay under Uncertainty. Physics Letters A 345, 345–354 (2005)
9. Yuan, K., Cao, J.D., Li, H.X.: Robust Stability of Switched Cohen-Grossberg Neural Networks with Mixed Time-varying Delays. IEEE Transactions on Systems Man and Cybernetics Part B-Cybernetics 36, 1356–1363 (2006)
10. Shen, Y., Wang, J.: Noise-induced Stabilization of the Recurrent Neural Networks with Mixed Time-varying Delays and Markovian-switching Parameters. IEEE Transactions on Neural Networks 18, 1857–1862 (2007)
11. Shen, Y., Wang, J.: Almost Sure Exponential Stability of Recurrent Neural Networks with Markovian Switching. IEEE Transactions on Neural Networks 20, 840-855 (2009)
12. Zeng, Z.G., Wang, J., Liao, X.X.: Global Exponential Stability of A General Class of Recurrent Neural Networks with Time-varying Delays. IEEE Trans. Circ. Syst. I 50, 1353–1358 (2003)

Approximation Capability of Continuous Time Recurrent Neural Networks for Non-autonomous Dynamical Systems

Yuichi Nakamura and Masahiro Nakagawa

Nagaoka University of Technology,
1603-1, Kamitomioka-machi, Nagaoka-shi, Niigata-ken, 940-2188, Japan
{nyuichi,masanaka}@vos.nagaokaut.ac.jp

Abstract. The main goal of this study is to elucidate the theoretical capability of the continuous time recurrent neural network. In this paper, we show that the approximation capability of the continuous time recurrent network can be extended to non-autonomous dynamical systems with external inputs. Moreover, if the dynamical system has an asymptotically stable periodic solution for a periodic external input, it is shown that the approximation can be extended to the global time interval.

Keywords: Continuous Time Recurrent Neural Network, Dynamical system, Non-autonomous, Approximation, Capability.

1 Introduction

There are two types of neural networks. The one is the feed-forward network with only feed-forward connections, and the other is the recurrent network with arbitrary feedback connections. The feed-forward network defines a mapping as the input-output relation of the network. The recurrent network defines a dynamical system as the dynamics of the network. Clarifying the theoretical capability of these neural networks gives important information to the learning algorithm and its applications [1,12].

As the theoretic capability of the feed-forward neural network, the approximation possibility to several mappings has been considered. The input-output mapping of neural networks can approximately realize any continuous mapping with an arbitrary accuracy [2,3]. Moreover, if a given mapping is smooth, the derivative of the mapping can be also approximated by an appropriate feed-forward network[7]. As the theoretic capability of the recurrent neural network, the approximation possibility to several dynamical systems has been considered. A given trajectory of a dynamical system can be approximately realized by an appropriate continuous time recurrent network [4]. The similar result of the discrete time recurrent network for discrete systems has been shown[8,11]. However, there are restrictions that the approximation only on some finite time interval, and the capability of continuous time recurrent networks only to autonomous dynamical systems.

C. Alippi et al. (Eds.): ICANN 2009, Part II, LNCS 5769, pp. 593–602, 2009.

In this paper, the capability of the continuous time recurrent network is further discussed. It is shown that the approximation capability is extended to non-autonomous dynamical systems. Moreover, if the dynamical system has an asymptotically stable periodic solution for a periodic input, the approximation can be extended to the global time interval.

2 Continuous Time Recurrent Neural Networks

This paper deals with the continuous time recurrent neural network (CTRNN) defined by the following differential equations[1]

$$\frac{dx_i}{dt} = -\frac{1}{\tau_i}x_i + \sum_{j=1}^{N} w_{ij}\sigma(x_j) + B_i + I_i(t), \quad (i = 1, \ldots, N), \tag{1}$$

where N is the number of neurons in the network. x_i, τ_i, B_i, and $I_i(t)$ are the internal state, the time constant, the internal bias, and the external input of the i-th neuron, respectively. w_{ij} is the coupling coefficient from the j-th neuron to the i-th neuron. σ is the activation function and thus $y_i \equiv \sigma(x_i)$ is the output state of the i-th neuron. In this paper, as the activation function we use the sigmoid function (bounded, increasing, and smooth). For example, $1/(1+e^{-x})$,$\tanh x$, and $\tan^{-1} x$ satisfy the conditions.

The vectors of the internal states, the internal biases, and the external inputs are denoted by

$$x(t) = \begin{bmatrix} x_1(t) \\ \vdots \\ x_N(t) \end{bmatrix}, \quad B = \begin{bmatrix} B_1 \\ \vdots \\ B_N \end{bmatrix}, \quad \text{and} \quad I(t) = \begin{bmatrix} I_1(t) \\ \vdots \\ I_N(t) \end{bmatrix}, \tag{2}$$

respectively. The matrix of the coupling coefficients is denoted by

$$W = \begin{bmatrix} w_{11} & \cdots & w_{1N} \\ \vdots & \ddots & \vdots \\ w_{N1} & \cdots & w_{NN} \end{bmatrix}. \tag{3}$$

The activation mapping $\sigma : \mathbb{R}^N \to \mathbb{R}^N$ operates the sigmoid function to each component. When all the time constants equal τ, then (1) is represented by

$$\frac{dx}{dt} = -\frac{1}{\tau}x + W\sigma(x) + B + I(t). \tag{4}$$

There are two types of neurons in the recurrent network. The one is the output neuron that the state is observed as the output of the network. The other is the hidden neuron. We assume that the number of output neurons N^O, the number

[1] The recurrent higher-order neural network (RHON) is considered the enhance model of this network. Therefore the RHON inherits the approximation ability.

of hidden neurons N^H, and the number of all neurons $N = N^O + N^H$. The states and parameters in (2) and (3) are denoted by

$$x(t)=\begin{bmatrix}x^O(t)\\x^H(t)\end{bmatrix},\ B=\begin{bmatrix}B^O\\B^H\end{bmatrix},\ I(t)=\begin{bmatrix}I^O(t)\\I^H(t)\end{bmatrix},\ \text{and } W=\begin{bmatrix}W^{OO}&W^{OH}\\W^{HO}&W^{HH}\end{bmatrix}, \quad (5)$$

respectively. x^O, B^O, and I^O are $\mathbb{R}^{N^O}$ dimensional vectors of the internal states, the internal biases, and the external inputs of the output neurons, respectively. x^H, B^H, and I^H are $\mathbb{R}^{N^H}$ dimensional vectors of the internal states, the internal biases, and the external inputs of the hidden neurons, respectively. W^{OO} is the $N^O \times N^O$ coupling matrix from the output neurons to themselves, W^{OH} is the $N^O \times N^H$ coupling matrix from the hidden neurons to the output neurons, W^{HO} is the $N^H \times N^O$ coupling matrix from the output neurons to the hidden neurons, and W^{HH} is the $N^H \times N^H$ coupling matrix from the hidden neurons to themselves.

Therefore, (4) is represented by

$$\frac{dx^O}{dt}=-\frac{1}{\tau}x^O+W^{OO}\sigma(x^O)+W^{OH}\sigma(x^H)+B^O+I^O(t) \quad (6)$$

$$\frac{dx^H}{dt}=-\frac{1}{\tau}x^H+W^{HO}\sigma(x^O)+W^{HH}\sigma(x^H)+B^H+I^H(t) \quad (7)$$

separating the output neurons and the hidden neurons, where σ is the activation mapping on $\mathbb{R}^{N^O}$ or on $\mathbb{R}^{N^H}$.

3 Feed-Forward Neural Networks

The approximation capability of CTRNN to dynamical systems is proved by using of the mapping approximation theorem of the three layer feed-forward neural network (TLFNN). Here, we consider the TLFNN composed of sigmoid middle-layer neurons, linear input-layer neurons and linear output-layer neurons. Let N^I be the number of input-layer neurons, N^M be the number of middle-layer neurons, and N^O be the number of output-layer neurons. Then, the TLFNN defines a mapping $g : \mathbb{R}^{N^I} \to \mathbb{R}^{N^O}$

$$g_i(x_1,\ldots,x_{N^I}) = \sum_{j=1}^{N^M} w_{ij}^{\mathrm{OM}}\sigma\left(\sum_{k=1}^{N^I} w_{jk}^{\mathrm{MI}}x_k + b_j^{\mathrm{M}}\right) + b_i^{\mathrm{O}} \quad (8)$$

as the input-output relation of the network. w_{jk}^{MI} and w_{ij}^{OM} are the coupling coefficients from the input-layer to the middle-layer and from the middle-layer to the output-layer, respectively. b_j^M and b_i^O are the internal biases of the middle-layer and the output-layer, respectively.

The vector expression of (8) is derived by

$$g(x) = W^{OM}\sigma\left(W^{MI}x + B^M\right) + B^O \quad (9)$$

where $W^{MI} = [w^{MI}_{jk}]$ is an $N^M \times N^I$ matrix, and $W^{OM} = [w^{OM}_{ij}]$ is an $N^O \times N^M$ matrix. $B^M = [b^M_j] \in \mathbb{R}^{N^M}$ and $B^O = [b^O_i] \in \mathbb{R}^{N^O}$ are vectors. $\sigma : \mathbb{R}^{N^M} \to \mathbb{R}^{N^M}$ is the activation mapping operating the sigmoid function to each component.

It has been shown that the TLFNN has the approximation capability for any continuous mapping [2,3,7].

Foundation Theorem of TLFNN. *Let $U \subset \mathbb{R}^m$ be an open set, $K \subset U$ be a compact set, and $f : U \longrightarrow \mathbb{R}^n$ be a continuous mapping. Then, for an arbitrary $\varepsilon > 0$, there is a TLFNN (9) such that $|f(x) - g(x)| < \varepsilon$ holds for all $x \in K$, where $N^I = m$ and $N^O = n$.*

We use the following corollary that is a restated form of the foundation theorem.

Corollary 1. *Let $K_u \subset \mathbb{R}^{N_u}$ and $K_x \subset \mathbb{R}^{N_x}$ be compact sets, $U \subset \mathbb{R}^{N_u} \times \mathbb{R}^{N_x}$ be an open set including $K_u \times K_x$, and $f : U \longrightarrow \mathbb{R}^n$ be a continuous mapping. Then, for an arbitrary $\varepsilon > 0$, there is a TLFNN such that $|f(u,x) - g(u,x)| < \varepsilon$ holds for all $(u,x) \in K_u \times K_x$, where $N^I = N_u + N_x$, $N^O = n$, and g is the input-output mapping of the TLFNN.*

Proof. In the foundation theorem, let $m = N_u + N_x$. The first N_u components of $x \in \mathbb{R}^m$ are denoted by u, and the rest N_x components are denoted by x. Then, (9) is represented by the form

$$g(u,x) = W^{OM}\sigma\left(W^{MIu}u + W^{MIx}x + B^M\right) + B^O \tag{10}$$

where W^{MIu} is an $N^M \times N_u$ matrix, and W^{MIx} is an $N^M \times N_x$ matrix. Q.E.D.

4 Pseudo Neural Systems

In [5], the relation of CTRNN and TLFNN is demonstrated by introducing the pseudo neural system (PNS)[2]. The PNS is defined by

$$\frac{dx}{dt} = -\frac{1}{\tau}x + W_1\sigma(W_2x + B_1) + B_2. \tag{11}$$

We extend the PNS to adapt non-autonomous dynamical systems. The Extended PNS (EPNS) is defined by

$$\frac{dx}{dt} = -\frac{1}{\tau}x + W^{OO}\sigma(x) + W^{OM}\sigma(W^{MIu}u + W^{MIx}x + B^M) + B^O. \tag{12}$$

The third and fourth term of the right side of (12) correspond to the input-output mapping (10) of the TLFNN. Therefore, the approximation capability of EPNS to non-autonomous dynamical systems is shown as the following.

[2] The affine neural dynamical system (A-NDS) was proposed as the generalized form of the PNS, and the relation of CTRNN and A-NDS was considered[9].

Theorem 1. *Let $E \subseteqq \mathbb{R}^n$ be a normed vector space, $W \subset \mathbb{R}^{N_u} \times E$ be an open set, $u : \mathbb{R} \to \mathbb{R}^{N_u}$ be a C^1-mapping, $f : W \to E$ be a C^1-mapping, and suppose that $x' = f(u(t), x(t))$ defines a non-autonomous dynamical system. Let $K \subset E$ be a compact set and we consider solutions of the system on a finite time interval $J = [0, T]$. Then, for an arbitrary $\varepsilon > 0$, there is an EPNS such that $|x(t) - z(t)| < \varepsilon$ holds for all $x(0) \in K$ and $t \in J$, where $z(t)$ is the solution of the EPNS with the initial condition $z(0) = x(0)$.*

The vector field $f(u(t), x(t))$ in the previous theorem can be restated by $F(t, x(t))$. Thus, we can use the following lemma.

Lemma 1 (Theorem3, Chap.15§1, [6]). *Let $W \subset \mathbb{R} \times E$ be open and $f, g : W \to E$ continuous. Suppose that for all $(t, x) \in W$, $|f(t, x) - g(t, x)| < \varepsilon$. Let L be a Lipschitz constant in x for $f(t, x)$. If $x(t)$, $y(t)$ are solutions to $x' = f(t, x)$, $y' = g(t, y)$, respectively, on some interval J, and $x(t_0) = y(t_0)$, then*

$$|x(t) - y(t)| \leq \frac{\varepsilon}{L}\left(e^{L|t-t_0|} - 1\right) \tag{13}$$

for any $t \in J$.

This lemma shows that if there are two dynamical systems with sufficiently close vector fields, the trajectories with the same initial condition of each system are close on a finite time interval.

Proof (Theorem 1). Because K, J are compact and u, x are continuous,

$$K_J = \{(u(t), x(t)); t \in J, x(0) \in K\} \tag{14}$$

is a compact subset of W. Thus,

$$K_\varepsilon = \{(u, x); \exists(u, y) \in K_J, |(u, x) - (u, y)| \leq \varepsilon\} \tag{15}$$

is a compact set including K_J. Then, $\varepsilon > 0$ is reselected such that $K_\varepsilon \subset W$ holds for the given ε. Because f is C^1, the restriction mapping $f|K_\varepsilon$ satisfies Lipschitz condition. Let L be a Lipschitz constant in x for $f|K_\varepsilon$. We set a time constant $\tau > 0$ and an $n \times n$ matrix W^{OO}. For the given mapping f,

$$h(u, x) = f(u, x) + \frac{1}{\tau}x - W^{OO}\sigma(x) \tag{16}$$

is considered. By Corollary 1, there is a TLFNN (10),

$$|h(u, x) - g(u, x)| < \eta < \frac{\varepsilon L}{e^{LT} - 1} \tag{17}$$

holds for $(u, x) \in K_\varepsilon$. Then, the mapping

$$G(u, x) = -\frac{1}{\tau}x + W^{OO}\sigma(x) + g(u, x) \tag{18}$$

satisfies

$$|f(u,x) - G(u,x)| < \eta < \frac{\varepsilon L}{e^{LT} - 1} \tag{19}$$

for $(u, x) \in K_\varepsilon$. The non-autonomous dynamical system $z' = G(u, z)$ is an EPNS.

By Lemma 1, if a solution of the dynamical system $x(t)$ and a solution of the EPNS $z(t)$ have the same initial condition $x(0) = z(0) \in K$,

$$|x(t) - z(t)| \leq \frac{\eta}{L}\left(e^{Lt} - 1\right) \leq \frac{\eta}{L}\left(e^{LT} - 1\right) < \varepsilon \tag{20}$$

holds for all $t \in J$. Q.E.D.

5 Approximation Capability of CTRNN

The following theorem is one of the main results in this paper.

Theorem 2. *Let $E \subseteqq \mathbb{R}^n$ be a normed vector space, $W \subset \mathbb{R}^{N_u} \times E$ be an open set, $u : \mathbb{R} \to \mathbb{R}^{N_u}$ be a C^1 mapping, $f : W \to E$ be a C^1 mapping, and suppose that $x' = f(u(t), x(t))$ defines a non-autonomous dynamical system. Let $K \subset E$ be a compact set and we consider solutions of the system on a finite time interval $J = [0, T]$. Then, for an arbitrary $\varepsilon > 0$, there is a CTRNN such that $\left|x(t) - z^O(t)\right| < \varepsilon$ holds for all $x(0) \in K$ and $t \in J$, where $z^O(t)$ is the internal state vector of the output neurons of the CTRNN with an appropriate initial condition.*

It has been shown that a given trajectory of a non-autonomous dynamical system is approximated by an EPNS in Theorem 1. Therefore, we only show that the EPNS can be represented by a CTRNN.

Proposition 1. *For an arbitrary EPNS, there is a CTRNN that the state of the output neurons is equal to the solution of the EPNS.*

Proof. For the n dimensional EPNS(12), we consider a CTRNN with $N^O(= n)$ output neurons and $N^H(= N^M)$ hidden neurons. Let z^O be an N^O dimensional state vector, and z^H be an N^H dimensional state vector such that

$$z^O = x \tag{21}$$

$$z^H = W^{MIu}u + W^{MIx}z^O + B^M. \tag{22}$$

Then, (12) is denoted by

$$\frac{dz^O}{dt} = -\frac{1}{\tau}z^O + W^{OO}\sigma(z^O) + W^{OM}\sigma(z^H) + B^O. \tag{23}$$

By differentiating (22) in t,

$$\frac{dz^H}{dt} = W^{MIu}\frac{du}{dt} + W^{MIx}\frac{dz^O}{dt}. \tag{24}$$

(23) is substituted for the right side of (24),

$$\begin{aligned}\frac{dz^H}{dt} = &-\frac{1}{\tau}z^H + W^{MIx}W^{OO}\sigma(z^O) + W^{MIx}W^{OM}\sigma(z^H) \\ &+\frac{1}{\tau}B^M + W^{MIx}B^O + W^{MIu}\left(\frac{1}{\tau}u + u'\right)\end{aligned} \tag{25}$$

holds. Let

$$z(t) = \begin{bmatrix} z^O \\ z^H \end{bmatrix}, \quad B = \begin{bmatrix} B^O \\ \frac{1}{\tau}B^M + W^{MIx}B^O \end{bmatrix}, \quad I(t) = \begin{bmatrix} 0 \\ W^{MIu}\left(\frac{1}{\tau}u + u'\right) \end{bmatrix}, \tag{26}$$

and

$$W = \begin{bmatrix} W^{OO} & W^{OM} \\ W^{MIx}W^{OO} & W^{MIx}W^{OM} \end{bmatrix}. \tag{27}$$

Then, (23) and (25) are represented by (4). This system is a CTRNN. For an initial condition $x(0)$ of the EPNS we set the initial condition of the CTRNN as follows:

$$z^O(0) = x(0)\ , \quad \text{and} \quad z^H(0) = W^{MIu}u(0) + W^{MIx}x(0) + B^M. \tag{28}$$

By the uniqueness of solution [3], $z^O(t) = x(t)$ holds for $t \in \mathbb{R}$. Q.E.D.

Therefore, Theorem 2 has been proved.

6 Approximate Extended Condition

In Theorem 2, the approximation capability of CTRNN is restricted to a finite time interval. However, we can expect that the restriction is removed, if the non-autonomous system with a periodic external input has an asymptotically stable periodic solution.

We describe some definitions necessary for the following discussion. Let $E \subseteq \mathbb{R}^n$, $W \subset \mathbb{R}^{N_u} \times E$, and $f : W \to E$ be the same defined in Theorem 2. Moreover, let $u : \mathbb{R} \to \mathbb{R}^{N_u}$ be a C^1 external input with a period $T > 0$ such that

$$u(t + T) = u(t). \tag{29}$$

Then the flow $\phi_t : S \to S$ of the dynamical system $x' = f(u(t), x(t))$ on the phase space $S \subseteq E$ satisfies the following;ϕ_0 is the identical mapping on S and $\phi_t \circ \phi_s = \phi_{t+s}$. By using the flow, the solution with an initial value $x_0 \in S$ of the dynamical system is represented by $\phi_t(x_0)$. Moreover, in case of considering an initial phase $u(s)$ of the external input, we use the form $\phi_t(x_0; u(s))$.

[3] Theorem 1,Chap.15§1,[6].

The periodic solution with the period T satisfies

$$\phi_{t+T}(x_0; u(s)) = \phi_t(x_0; u(s)) \tag{30}$$

for any $t \in \mathbb{R}$. Then,

$$\gamma = \{[u(t+s), \phi_t(x_0; u(s))]; t \in \mathbb{R}\} \subset W \tag{31}$$

is called the closed orbit. We define the asymptotically stability of a periodic solution by the following:

Definition 1. *The periodic solution is called asymptotically stable when the following is satisfied: for any neighborhood $U_1 \in W$ of the closed orbit γ, there is a neighborhood U_2, $\gamma \subset U_2 \subset U_1$, such that*

$$[u(t+s), \phi_t(x_0; u(s))] \in U_1, \tag{32}$$

for all $[u(s), x_0] \in U_2$, $t \geq 0$ and

$$\lim_{t \to \infty} d\left([u(t+s), \phi_t(x; u(s))], \gamma\right) = 0, \tag{33}$$

where $d\left([u(t+s), \phi_t(x; u(s))], \gamma\right)$ is the distance of $\phi_t(x; u(s))$ and the point in γ at the same phase $u(t+s)$.

The following theorem is another main result of this paper.

Theorem 3. *Let $x' = f(u(t), x)$ be a non-autonomous dynamical system with an external input $u(t)$ of a period T. Let ϕ_t be the flow of the system, $\phi_t(x_p; u(0))$ be an asymptotically stable solution with the period T, and $\gamma \subset W$ be the closed orbit of the solution. Then, for an arbitrary $\varepsilon > 0$, there is a neighborhood $V \subset W$ of γ and a CTRNN such that $\left|\phi_t(x_0; u(s)) - z^O(t)\right| < \varepsilon$ holds for $[u(s), x_0] \in V$ and $t \geq 0$, were $z^O(t)$ is the internal state of the output neurons of the CTRNN with an appropriate initial condition.*

Proof. Because $u(t)$ has the period T,

$$\{u(t); t \in [0, T]\} \times E = \{u(t); t \in [mT, (m+1)T]\} \times E \tag{34}$$

holds for $m \in \mathbb{N}$. It is considered that the spaces every the time T are the same. Let the r-neighborhood of γ be defined by

$$V_r(\gamma) = \{[u(s), x]; |x - \phi_s(x_p; u(0))| \leq r, s \in [0, T]\} \subset W \tag{35}$$

for $r > 0$. For the given ε, we reselect a smaller $\varepsilon > 0$ that satisfies $V_\varepsilon(\gamma) \subset W$. By the definition of asymptotically stable, there are $0 < R < \varepsilon/2$ and $V_R(\gamma)$ that satisfy $[u(t+s), \phi_t(x_0; u(s))] \in V_{\varepsilon/2}(\gamma)$, and $\lim_{t \to \infty} d\left([u(t+s), \phi_t(x_0; u(s))], \gamma\right) = 0$ for $[u(s), x_0] \in V_R(\gamma)$ and $t \geq 0$. Then, there is $m \in \mathbb{N}$ such that

$$[u(t+s), \phi_t(x_0; u(s))] \in V_{R/2}(\gamma) \tag{36}$$

holds for $t \geq T_C \equiv mT$.

Let D be a compact set including $V_\varepsilon(\gamma)$. By Theorem1, for D and T_C, there is an EPNS such that $|\phi_t(x_0; u(s)) - z(t)| < R/2$ holds for all $t \in [0, T_C]$, $[u(s), x_0] = [u(s), z(0)] \in D$. Here $z(t)$ is the solution of the EPNS. Let $V = V_R(\gamma)$ be the neighborhood of γ. Then, for $[u(s), x_0] = [u(s), z(0)] \in V$ and $t \in [0, T_C]$, $|\phi_t(x_0; u(s)) - z(t)| < R/2 < \varepsilon$ is satisfied. Moreover,

$$[u(T_C + s), \phi_{T_C}(x_0; u(s))] = [u(s), \phi_{T_C}(x_0; u(s))] \in V_{R/2}(\gamma) \tag{37}$$

is satisfied by (36). Then,

$$\begin{aligned}
&|z(T_C) - \phi_{T_C+s}(x_p; u(0))| \\
&\quad \le |z(T_C) - \phi_{T_C}(x_0; u(s))| + |\phi_{T_C}(x_0; u(s)) - \phi_{T_C+s}(x_p; u(0))| < R
\end{aligned} \tag{38}$$

and thus $[u(T_C + s), z(T_C)] = [u(s), Z(T_C)] \in V$ holds.

Next, we consider the case of $t \ge T_C$. From (36), $[u(t+s), \phi_t(x_0)] \in V_{R/2}(\gamma)$ is satisfied. Let $\phi_t(y_0; u(s))$ be a solution of the dynamical system with the initial condition $[u(s), y_0] = [u(s), z(T_C)] \in V$. Then, $|\phi_t(y_0; u(s)) - z(t+T_C)| < R/2$ and $[u(t+s), \phi_t(y_0; u(s))] \in V_{\varepsilon/2}(\gamma)$ hold for $t \in [0, T_C]$. Thus, for $t \in [T_C, 2T_C]$,

$$\begin{aligned}
&|\phi_t(x_0; u(s)) - z(t)| \\
&\quad \le |\phi_t(x_0; u(s)) - \phi_{t+s}(x_p; u(0))| + |\phi_{t+s}(x_p; u(0)) - \phi_{t-T_C}(y_0; u(s))| \\
&\qquad + |\phi_{t-T_C}(y_0; u(s)) - z(t)| \\
&\quad < \frac{R}{2} + \frac{\varepsilon}{2} + \frac{R}{2} < \varepsilon.
\end{aligned} \tag{39}$$

Moreover, $[u(s), z(2T_C)] \in V$ is satisfied again by $[u(T_C + s), \phi_{T_C}(y_0; u(s))] = [u(s), \phi_{T_C}(y_0; u(s))] \in V_{R/2}(\gamma)$. Therefore, by the similar discussion, the approximation can be extended, such that $|\phi_t(x_0) - z(t)| < \varepsilon$ holds for $t \ge 0$. By Proposition 1, there is a CTRNN, and $z(t)$ can be represented by the state of the output neurons of the CTRNN. This theorem has been proved. Q.E.D.

In Theorem 3, we treated the initial condition in the appropriate neighborhood V of γ. However, the initial condition is extended to the region in the basin of γ.

Corollary 2. *In Theorem 3, let $K \subset W$ be a compact set in the basin of γ. Then, for an arbitrary $\varepsilon > 0$, there is a CTRNN such that $|\phi_t(x_0) - z^O(t)| < \varepsilon$ holds for $[u(s), x_0] \in K$ and $t \ge 0$, where $z^O(t)$ is the state vector of the output neurons of the CTRNN with an appropriate initial condition.*

Proof. Because K is in the basin of γ and ϕ_t is continuous, there is $k \in \mathbb{N}$ such that $[u(t+s), \phi_t(x_0; u(s))] \in V_{R/2}(\gamma)$ holds for $t \ge T_K \equiv kT$ and $[u(s), x_0] \in K$. For T_C and D in the proof of Theorem 3, $T_M = \max(T_C, T_K)$ and a connected compact set $K_D \supset D \cap \{(u(t+s), \phi_t(x_0)); [u(s), x_0] \in K, t \in [0, T_K]\}$ are considered. Then, for T_M and K_D, there is an appropriate CTRNN that holds this corollary condition by the similar discussion of Theorem3. Q.E.D.

7 Conclusions

We show that the approximation capability of the continuous time recurrent network can be extended to non-autonomous dynamical systems. Moreover, if the dynamical system has an asymptotically stable periodic solution for a periodic external input, it is shown that the approximation to trajectories which converge to the periodic solution can be extended to the global time interval. In this paper, although the sigmoid function is assumed as the activation function, the similar approximation capability with a more extensive activation function is also derived by the result of [10]. The considerations about the approximation capability to a dynamics system with bifurcation parameters, and about the relation between the learning algorithm and the network structure based on the approximation capability are interesting subjects.

References

1. Alanis, A.Y., Sanchez, E.N., Loukianov, A.G.: Discrete-Time Adaptive Backstepping Nonlinear Control via High-Order Neural Networks. IEEE Transactions on Nural Networks 18(4), 1185–1195 (2007)
2. Cybenko, C.: Approximation by Superpositions of Sigmodial Function. Mathematics of Control, Signals and Systems 2, 303–314 (1989)
3. Funahashi, K.: On the approximate realization of continuous mapping by neural networks. Neural Networks 2(3), 183–191 (1989)
4. Funahashi, K., Nakamura, Y.: Approximation of dynamical systems by continuous time recurrent neural networks. Neural Networks 6(6), 801–806 (1993)
5. Funahashi, K.: On the Approximation of Hyperbolic Dynamical Systems by Recurrent Neural Networks I. Technical Report of IEICE (The Institute of Electronics, Information and Communication Engineers), NC2001-54, 39–45 (2001) (in Japanese)
6. Hirsch, M., Smale, S.: Differential Equations, Dynamical Systems, and Linear Algebra. Academic Press, London (1974)
7. Hornik, K.: Approximation Capabilities of Multilayer Feed-forward Networks. Neural Networks 4, 251–257 (1991)
8. Jin, L., Nikiforuk, P.N., Gupta, M.M.: Approximation of Discrete-Time State-Space Trajectories Using Dynamic Recurrent Neural Networks. IEEE Transactions on Automatic Control 40(7), 1266–1270 (1995)
9. Kimura, M., Nakano, R.: Learning dynamical systems by recurrent neural networks from orbits. Neural Networks 11, 1589–1599 (1998)
10. Leshno, M., Lin, V.Y., Pinkus, A., Schocken, S.: Multilayer Feedforward Networks With a Nonpolynomial Activation Function Can Approximate Any Function. Neural Networks 6, 861–867 (1993)
11. Patan, K.: Approximation of state-space trajectories by locally recurrent globally feed-forward neural networks. Neural Networks 21, 59–64 (2008)
12. Sato, M.: A Learning Algorithm to Teach Spatiotemporal Patterns to Recurrent Neural Networks. Biological Cybernetics 1, 256–263 (1990)

Spectra of the Spike Flow Graphs of Recurrent Neural Networks

Filip Piekniewski

Faculty of Mathematics and Computer Science,
Nicolaus Copernicus University, 87-100 Torun, Poland
philip@mat.umk.pl
http://www.mat.umk.pl/~philip

Abstract. Recently the notion of power law networks in the context of neural networks has gathered considerable attention. Some empirical results show that functional correlation networks in human subjects solving certain tasks form power law graphs with exponent approaching ≈ 2. The mechanisms leading to such a connectivity are still obscure, nevertheless there are sizable efforts to provide theoretical models that would include neural specific properties. One such model is the so called *spike flow model* in which every unit may contain arbitrary amount of charge, which can later be exchanged under stochastic dynamics. It has been shown that under certain natural assumptions about the Hamiltonian the large-scale behavior of the *spike flow model* admits an accurate description in terms of a *winner-take-all* type dynamics. This can be used to show that the resulting graph of charge transfers, referred to as the *spike flow graph* in the sequel, has scale-free properties with power law exponent $\gamma = 2$. In this paper we analyze the spectra of the spike flow graphs with respect to previous theoretical results based on the simplified *winner-take-all* model. We have found numerical support for certain theoretical predictions and also discovered other spectral properties which require further theoretical investigation.

Keywords: power law network; spike flow model; graphs spectrum.

1 Introduction

Power law networks (often referred to as scale-free networks, which sometimes causes confusion [1]) are now an established field of study in random graph theory. Diverse empirical evidence have shown that power law connectivity emerges spontaneously in miscellaneous systems ranging from the World Wide Web [2], science collaboration networks [3], citation networks [4], ecological networks [5], linguistic networks [6], cellular metabolic networks [7,8] to telephone call network [9,10] and many others. In many cases networks featuring power law degree distributions also include certain structural properties which enhance tolerance against attacks or bandwidth (the correspondence between power law degree distribution and structural properties is not straightforward and has been discussed in [1]). It is quite natural to ask whether neural systems could benefit

C. Alippi et al. (Eds.): ICANN 2009, Part II, LNCS 5769, pp. 603–612, 2009.

from such an architecture, and if so, whether there are any mechanisms inherent to neural activity that might lead to a power law connectivity. Early studies of C.elegans worm nervous system showed exponential decay of degree distribution [11,12], however the network of C. elegans is very small (the whole organism has only about 1000 cells, and the total number of neurons is just above 300) whereas the mechanisms of self-organization leading to a power law structure might emerge in larger populations of neurons with significant feedback wirings. One strong empirical evidence that this might be the case are the results of [13,14], which show that the network composed of centers of activity observed in human brain by FMRI, connected whenever their activity is correlated[1] above a certain threshold is scale-free with power law exponent ≈ 2.

It is worth noting that power law graphs are usually sparse (in the sense that the number of edges depends linearly on the number of vertices) and yet well connected (power law graphs are more likely to form a giant component than corresponding – in terms of edge density – Erdős- Rényi random graphs - see chapter 6 in [15] for related study). These features seem to be advantageous for recurrent neural networks, and indeed some studies [16,17] have proved that power law architectures are useful for artificial NN. In that case however, the connectivity was not a result of neural activity but was rather imposed as a background for already existing models.

Many of the existing models describing the development of power law networks are stemming from the model of Barabási and Albert [18] based on growth and preferential attachment. This model however does not describe well the situation considered in [13] since growth in this case is very limited. Another reason why Barabási-Albert model is inadequate to the situation is that in its most natural setup it leads[2] to power law exponent $\gamma = 3$ while empirical studies of [13] strongly suggest $\gamma = 2$. In our attempt to provide a more adequate theoretical description we have developed the so called *spike flow model* [19] which essentially resembles a typical Boltzmann machine but has more capacitive space of states and a bit tweaked Hamiltonian (the details are supplied in the next section) . Quite unexpectedly the *spike flow model* turned out to be mathematically tractable (at low enough temperatures), which allowed to establish explicit results [20] on the asymptotic properties of the dynamics and the emergence of a power law charge transfer graph (referred to as the *spike flow graph* in the sequel). Further theoretical research allowed to characterize the spectra of the *spike flow graph* in the asymptotic regime [21]. The study of spectral properties is particularly important to determine, whether the *spike flow model* is an adequate description of the mechanisms leading to power law connectivity in nervous system. The results from [21] impose that a certain kind of power law-like dependence should also be present in the distribution of graph eigenvalues (in section 4 below there is a brief discussion concerning the details). In this paper we provide numerical simulations which support claims of [21] which can be

[1] The patient was asked to perform certain simple tasks during the measurement.

[2] There are ways of reaching exponent 2 with variants of Barabási-Albert model, but they are even less suitable for the phenomena discussed.

regarded as a theoretical foundations of the results presented here. The theory in [21] however, is based on the simplified asymptotic version of the model (described below as well) whereas the presented material is based on the full-blown version of the *spike flow model.* Nevertheless the results show the existence of a spectral regime in which the predicted dependency is present. There are also other features of the spectra which require further theoretical study.

The rest of the paper is organized as follows. In section 2 we briefly describe the spike flow model, its basic properties and theoretical results (subsection 2.1) and motivations for studying spectral characteristics (subsection 2.2). In section 3 we describe the numerical setup of the simulation. In further two sections we provide results of the simulation and conclusions.

2 The Spike Flow Model

2.1 Basic Properties

The model consists of nodes σ_i, $i = 1 \ldots N$. Each node's state is described by a natural number from some fixed interval $[0, M_i]$. In the scope of this paper we assume $M_i = \infty$, that is the state space is unbounded (when $M_i = 1$ on the other hand the model much resembles Hopfield network). The network is built on a complete graph in that there is a connection between each pair of neurons σ_i, σ_j, $i \neq j$, carrying a real-valued weight $w_{ij} \in \mathbb{R}$ satisfying the usual symmetry condition $w_{ij} = w_{ji}$, moreover $w_{ii} := 0$. The values of w_{ij} are drawn independently from the standard Gaussian distribution $\mathcal{N}(0, 1)$ and are assumed to remain fixed in the course of the network dynamics. The model is equipped with the Hamiltonian of the form:

$$\mathcal{H}(\bar{\sigma}) := \frac{1}{2} \sum_{i \neq j} w_{ij} |\sigma_i - \sigma_j| \tag{1}$$

if $0 \leq \sigma_i \leq M_i$, $i = 1, \ldots, N$, and $\mathcal{H}(\bar{\sigma}) = +\infty$ in the other case. Here $\bar{\sigma}$ denotes of the state of the whole system. The dynamics of the network is defined as follows: at each step we randomly choose a pair of neurons (units) (σ_i, σ_j), $i \neq j$, and denote by $\bar{\sigma}^*$ the network configuration resulting from the original configuration $\bar{\sigma}$ by decreasing σ_i by one and increasing σ_j by one, that is to say by *letting a unit charge transfer from* σ_i *to* σ_j, whenever $\sigma_i > 0$ and $\sigma_j < M_j$. Next, if $\mathcal{H}(\bar{\sigma}^*) \leq \mathcal{H}(\bar{\sigma})$ we accept $\bar{\sigma}^*$ as the new configuration of the network whereas if $\mathcal{H}(\bar{\sigma}*) > \mathcal{H}(\bar{\sigma})$ we accept the new configuration $\bar{\sigma}^*$ with probability $\exp(-\beta[\mathcal{H}(\bar{\sigma}^*) - \mathcal{H}(\bar{\sigma})])$, $\beta > 0$, and reject it keeping the original configuration $\bar{\sigma}$ otherwise, with $\beta > 0$ standing for an extra parameter of the dynamics, in the sequel referred to as the inverse temperature conforming to the usual language of statistical mechanics. In the present paper we will assume β fixed and large, that is the system is in low temperature regime and so such "stochastic" jumps are rare.

Note that in this setup positive weights $w_{i,j}$ favor agreement of states σ_i and σ_j, while negative weight favor disagreement. Whenever a unit of charge is exchanged between two nodes that fact is recorded by increasing the counter associated with a corresponding edge. The edges (and nodes) being frequently visited by units of charge are in the focus of our interest. We refer to the resulting weighted[3] graph as to the *spike flow graph.*

In [20] a number of results related to the *spike flow model* have been established:

- In contrast to a seemingly complex dynamics, with high probability there is a unique ground state of the system, in which all the charge is gathered in a unit that maximizes
$$S_i := -\sum_{j \neq i} w_{ij}. \tag{2}$$
referred to as *support* in the sequel. The proof goes by a mixture of rigorous and semi rigorous calculations and has a rather asymptotic character, but is in full agreement with numerical simulations for systems containing between a couple hundreds to a couple of thousands of nodes.
- The system's behavior eventually admits a particularly simple approximation in terms of a kind of winner-take-all dynamics: almost all transfers converge to units of higher support (referred to as the *elite*, while the others referred to as the *bulk*), which then compete in draining charge from each other. That is to say, whenever a pair of units is chosen, the transfer occurs from the unit of lower support to the unit of higher support. Ultimately the unit of maximal support gathers all of the charge and the system freezes in a ground state. This approximation was used in [21] to establish explicit theoretical results on the properties of the spectra of the *spike flow graph.*
- The node degree distribution (where by degree we mean the sum of counters of edges adjacent to a given node[4]) obeys a power law with exponent $\gamma = 2$. The proof is based on the elite/bulk approximation and properties of ordering sequences. Again there is a strong agreement with numerical results

2.2 Spectral Properties

The graph's spectrum (in this paper by *graph's spectrum* we mean the set of eigenvalues of the adjacency matrix, not the eigenvalues of combinatorial laplacian which are also studied in the literature, see [22] for comprehensive introduction) is among the most basic characteristic features, yet it provides insights into various properties which are usually faint in the typical analysis. In [21] some basic properties of the spectra of simplified *spike flow model* were investigated. By the simplified version of the model here we mean a model equipped with the asymptotic winner-take-all version of the dynamics, that is each charge transfer occurs according the direction of increasing support (however for the validity of spectral analysis edge directions were dropped). It is worth noting,

[3] Weighted by edge counters that are not directly related to $w_{i,j}$ which remain fixed as a background to the energy function.

[4] Since charge transfers are directed, we distinguish in and out degrees, but asymptotically these two are equal in terms of distributions.

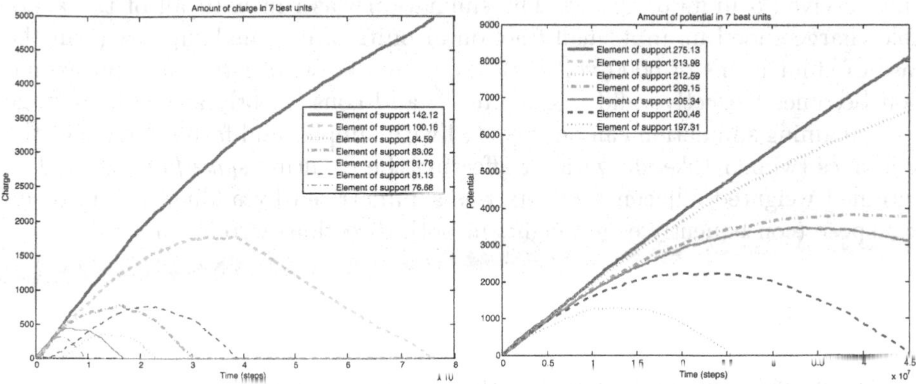

Fig. 1. Typical evolution of the amount of charge in seven units of highest support in the spike flow simulation. The left figure shows the system of 1000 vertices, while the right one of 5000 vertices. At the late stage of simulation when only a few units of highest support contain any charge the winner-take-all dynamics is certainly valid. At this stage the simulation can be simplified according to the restricted dynamics for efficiency. It is not obvious however, at which stage the winner-take-all approximation becomes acceptable. The study of spectral properties might shed some light into such issues.

that even though such a dynamics becomes reasonably valid in large instances of the *spike flow model* after some number of steps, it is not valid in the early stages of the simulation when there is still a lot of charge in the bulk units. Correspondingly, the resulting *spike flow graph* may be noisy and contain various distortions. Nevertheless the shape of the spectrum depends on global features and should to some extent exhibit the predicted properties. The results of [21] imply that when sorted descending, the k-th eigenvalue behaves like $\frac{C}{k^2}$ for some constant C (note, this paper [21] provides a theoretical background of the results presented here). This result, established by investigating the spectrum of appropriate Hilbert-Schmidt type operators associated to the random evolution in the asymptotic regime, is valid for the simplified *spike flow graph* truncated at both ends by some δ_1 and δ_2 i.e. the nodes of degree less than δ_1 and more than δ_2 are removed from the graph. Recall that removing single vertices from the graph is not easily expressible in terms of graph's eigenvalues and in particular it does not correspond to removal of any particular eigenvalues from the spectrum[5].

3 Numerical Setup

The simulations were carried out with the spike flow model consisting of 5000 units in the low temperature regime ($\beta = 100$ which results in extremely rare transitions against the energy factor). At the beginning of the simulation each

[5] The resulting graph has less vertices and consequently there are less eigenvalues in the spectrum, but these remaining eigenvalues correspond to a different adjacency matrix.

unit received 5 units of charge. The simulation was run until all of the available charge ended up in a small fraction of units of maximal support (5 in the case of simulations of 5000 vertices). By then, the winner-take-all approximation becomes perfectly valid (see figure 1), and consequently at the final stage the remaining simulation can be executed with simpler and faster version of the dynamics (winner-take-all) without affecting the resulting *spike flow graph*. The obtained weighted adjacency matrix was symmetrized by adding matrix to its transpose (consequently edge weights in both directions were summed).

4 Results

In the present study we simulated the setup described in subsection 2.2 by truncating empirical *spike flow graph* at various levels, having in mind however, that the winner-take-all approximation is itself valid for the upper part of the range of vertex degrees. That is, we expect that significant cutoff of low degree vertices should not alter (interesting part of) the spectrum significantly whereas even minor cutoff of high degree vertices might have a devastating effects on the shape of the spectrum[6] (at least at the part which is in the focus of our interest, that is the set of largest eigenvalues). .

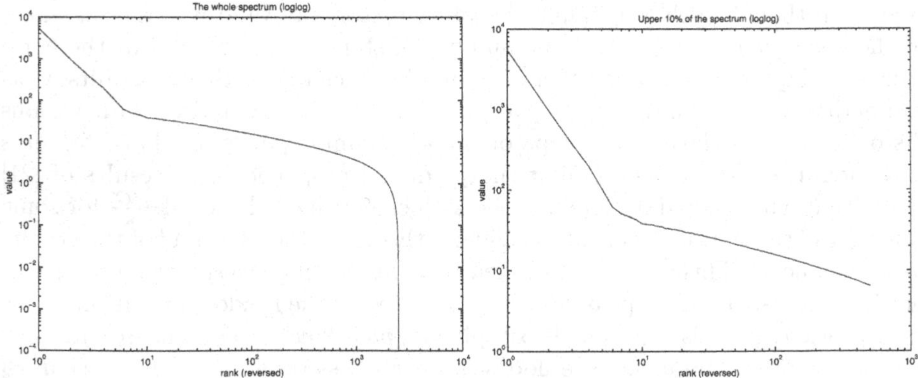

Fig. 2. An example spectrum of a graph consisting of 5000 vertices. The eigenvalues were sorted descending and plotted on a log-log plot. The left figure shows the whole spectrum, the right one only 10% largest eigenvalues. Clearly, the left linear part of the plot behaves like $y = -2x$ which implies that k-th eigenvalue is proportional to k^{-2}. This relation is only visible for the top ten eigenvalues.

The above considerations proved to be true for the investigated model. To make things clearly visible we plotted the spectrum in the descending order on a log-log plots (figures 2,3,4,5) with various cutoffs. We expected that the initial

[6] Consequently, finding out at which point the predicted spectral properties give up provides an insight into how accurate the winner-take-all approximation actually is.

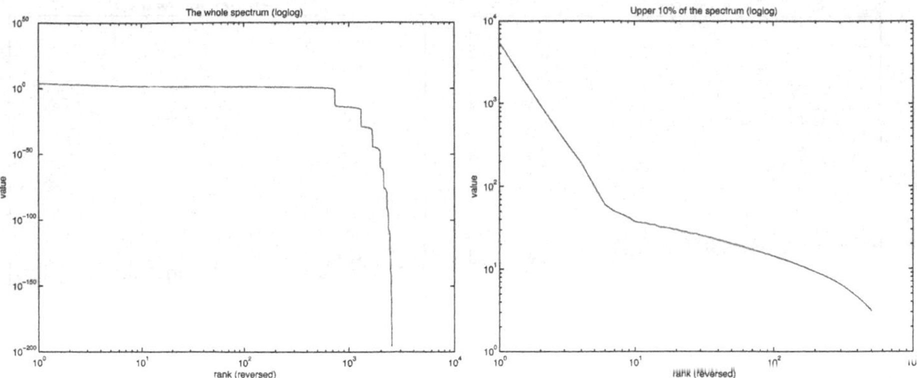

Fig. 3. An example spectrum of a graph consisting of 5000 vertices. The eigenvalues were sorted descending and plotted on a log-log plot. The left figure shows the whole spectrum, the right one only 10% largest eigenvalues. In this case however, 70% of the nodes of lowest degrees were cut off. Nevertheless the left linear part of the plot still behaves like $y = -2x$. The rightmost part of the spectrum (which corresponds to small eigenvalues) exhibits interesting "stair regime", which requires further investigation.

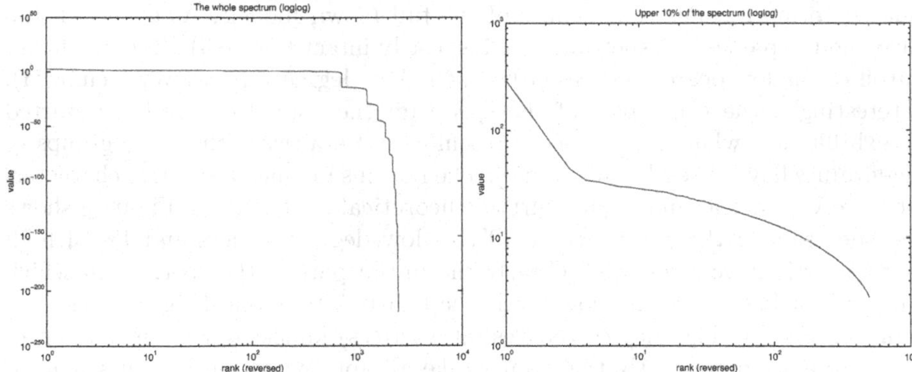

Fig. 4. An example spectrum of a graph consisting of 5000 vertices. The eigenvalues were sorted descending and plotted on a log-log plot. The left figure shows the whole spectrum, the right one only 10% largest eigenvalues. In this case however, 70% of the nodes of lowest degrees and 1% (50) of the nodes of highest degrees were cut off. A couple of top eigenvalues behave like $y = -2x$. The "stair regime" is clearly visible at right of the spectrum.

(leftmost) part of the spectrum would form a straight line on the log-log plot with slope of ≈ -2. Clearly such a straight line is visible on figure 2 where the spetrum of the full (not truncated) graph is presented. This regime is valid for about 10 largest eigenvalues (it might not seem that significant, but note that these 10 largest eigenvalues contain notable part of the total mass of the spectrum), for smaller eigenvalues the approximation breaks down due to distortions

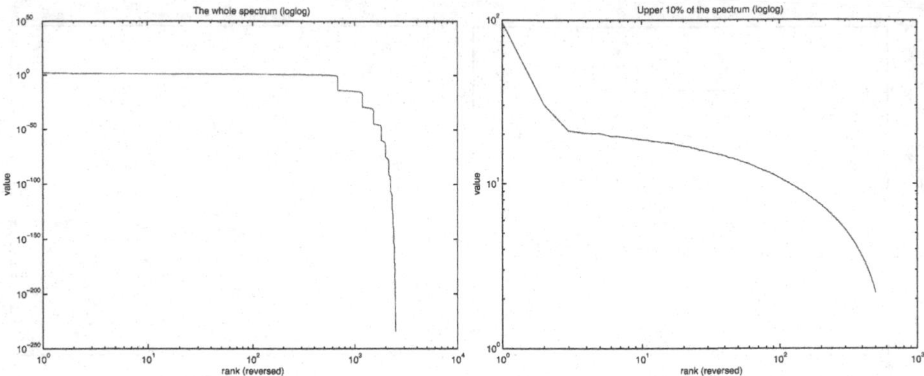

Fig. 5. An example spectrum of a graph consisting of 5000 vertices. The eigenvalues were sorted descending and plotted on a log-log plot. The left figure shows the whole spectrum, the right one only 10% largest eigenvalues. In this case however, 70% of the nodes of lowest degrees and 3% (150) of the nodes of highest degrees were cut off. Here the $y = -2x$ becomes irrelevant, which suggests that the property is related to the elite part of the graph which has been fairly cut off.

related to more complex dynamics of the full blown model[7]. As expected, the investigated part of the spetrum remains nearly intact (figure 3) after significant cutoff of the low degree vertices (70% of the low degree vertices were removed). Interestingly, the other part of the spectrum (i.e., small eigenvalues) started to exhibit somewhat discrete decay resembling a stairway (there are groups of eigenvalues having nearly same value). The reasons for such a spectral characteristic are yet unclear and require further theoretical explaination. Figure 4 shows the spectrum of the graph, whose 70% of low degree vertices and 1% of high degree vertices were removed. Clearly the initial part of the spetrum in which the straight line approximation is valid had shrunk to about 4 eigenvalues. The removal of 1% of high degree vertices (50 vertices) knocks down fair amount of the elite and consequently the winner-take-all approximation becomes inaccurate. This effect is even more visible in figure 5 where 3% (150) of high degree vertices were removed and the straight line regime is nearly absent, although still the first two eigenvalues seem to follow the expected relation. As mentioned earlier, the largest eigenvalues are not directly related to largest degree vertices and the above result should rather be interpreted in terms of validity of the winner-take-all approximation of the resulting truncated graph.

5 Conclusions

It seems that the theoretical predictions of [21] are to some extent observable in the full blown *spike flow model* equipped with significantly more complex dynamics

[7] This is not very surprising since the winner-take-all dynamics is certainly not valid for low degree vertices.

that the winner-take-all simplification. The results of [21] are themselves of rather asymptotic character, and consequently it was not obvious whether any of the predicted properties would be visible in the simulation of 5000 vertices, where the winner-take-all approximation is only valid for some fraction of the units of high support. Empirical data give insight into more complex spectral properties of the *spike flow model*, notably the *stairway regime* which might be related to the fact, that the truncated version of the graph can become disconnected, and consequently the particular flat regions in the spectrum could be attributed to various connected components, but this requires further investigation.

Acknowledgments. The author gratefully acknowledges the support from the Polish Minister of Scientific Research and Higher Education grant N N201 385234 (2008-2010). The author also appreciates fruitful collaboration with Dr Tomasz Schreiber.

References

1. Li, L., Alderson, D., Tanaka, R., Doyle, J.C., Willinger, W.: Towards a theory of scale-free graphs: Definition, properties, and implications (extended version) (2005)
2. Albert, R., Jeong, H., Barabási, A.L.: Diameter of the world-wide web. Science 401, 130–131 (1999)
3. Barabási, A.L., Jeong, H., Néda, Z., Ravasz, E., Schubert, A., Vicsek, T.: Evolution of the social network of scientific collaborations. Physica A 311(4), 590–614 (2002)
4. Redner, S.: How popular is your paper? an empirical study of the citation distribution. European Physical Journal B 4(2), 131–134 (1998)
5. Montoya, J.M., Solé, R.V.: Small world patterns in food webs. Journal of Theoretical Biology 214(3), 405–412 (2002)
6. i Cancho, R.F., Solé, R.V.: The small-world of human language. Proceedings of the Royal Society of London B 268(1482), 2261–2265 (2001)
7. Bhalla, U.S., Iyengar, R.: Emergent properties of networks of biological signaling pathways. Science 283(5400), 381–387 (1999)
8. Jeong, H., Tombor, B., Albert, R., Oltvai, Z.N., Barabási, A.L.: The large-scale organization of metabolic networks. Nature 407(6804), 651–653 (2000)
9. Abello, J., Buchsbaum, A., Westbrook, J.: A functional approach to external graph algorithms. In: Bilardi, G., Pietracaprina, A., Italiano, G.F., Pucci, G. (eds.) ESA 1998. LNCS, vol. 1461, pp. 332–343. Springer, Heidelberg (1998)
10. Aiello, W., Chung, F., Lu, L.: A random graph model for massive graphs. In: STOC 2000: Proceedings of the thirty-second annual ACM symposium on Theory of computing, pp. 171–180. ACM, New York (2000)
11. Amaral, L.A., Scala, A., Barthelemy, M., Stanley, H.E.: Classes of small-world networks. Proc. Natl. Acad. Sci. U.S.A. 97(21), 11149–11152 (2000)
12. Koch, C., Laurent, G.: Complexity and the Nervous System. Science 284(5411), 96–98 (1999)
13. Eguíluz, V.M., Chialvo, D.R., Cecchi, G.A., Baliki, M., Apkarian, A.V.: Scale-free brain functional networks. Phys. Rev. Lett. 94(1) (2005)
14. Sporns, O., Chialvo, D.R., Kaiser, M., Hilgetag, C.C.: Organization, development and function of complex brain networks. Trends. Cogn. Sci. 8(9), 418–425 (2004)

15. Chung, F., Lu, L.: Complex Graphs and Networks (Cbms Regional Conference Series in Mathematics). American Mathematical Society, Boston (2006)
16. Perotti, J.I., Tamarit, F.A., Cannas, S.A.: A scale-free neural network for modelling neurogenesis. Physica A Statistical Mechanics and its Applications 371, 71–75 (2006)
17. Stauffer, D., Aharony, A., da Fontoura Costa, L., Adler, J.: Efficient hopfield pattern recognition on a scale-free neural network. The european physical journal B (32), 395–399 (2003)
18. Barabási, A.L., Albert, R.: Emergence of scaling in random networks. Science (286), 509–512 (1999)
19. Piękniewski, F., Schreiber, T.: Emergence of scale-free spike flow graphs in recurrent neural networks. In: Proc. IEEE Symposium Series in Computational Intelligence - Foundations of Computational Intelligence, Honolulu, Hawai USA, pp. 357–362 (2007)
20. Piękniewski, F., Schreiber, T.: Spontaneous scale-free structure of spike flow graphs in recurrent neural networks. Neural Networks 21(10), 1530–1536 (2008)
21. Schreiber, T.: Spectra of winner-take-all stochastic neural networks (2008). In review, available at arXiv, `http://arxiv.org/abs/0810.3193`
22. Chung, F.R.K.: Spectral Graph Theory. CBMS Regional Conference Series in Mathematics, vol. 92. American Mathematical Society (1997)

Activation Dynamics in Excitable Maps: Limits to Communication Can Facilitate the Spread of Activity

Andreas Loengarov and Valery Tereshko

School of Computing, University of the West of Scotland, Paisley, PA1 2BE, UK
{valery.tereshko,andreas.loengarov}@uws.ac.uk

Abstract. We explore the map $x \mapsto (\alpha x + \gamma x^3)e^{-\beta x^2}$, which depending on the parameters displays a variety of different behaviours, both as a single map and when arranged in a coupled map lattice. The map can take on excitable and various oscillatory guises. For parameter values that result in excitable behaviour, we obtain a system that is roughly similar to a network of neurons that build up activation until they exceed the threshold and then fire some activation to their neighbours, depleting themselves in the process. We found that a higher communication rate and a lower threshold do not necessarily, and do not linearly result in a faster or more pervasive spread of activation. In fact, limits to communication can help the spread of activity.

Keywords: Coupled Maps; Excitability; Activation Dynamics.

1 Map

A main characteristic of neural tissue is its excitability. Biologists define an excitable cell as a cell that can generate an action potential at its membrane in response to depolarization and may transmit an impulse along the membrane [1]. To model a neuron, it can be viewed as a dynamical system unit. A dynamical system with a stable equilibrium is excitable if there is a large-amplitude piece of trajectory that starts in a small neighborhood of the equilibrium, leaves the neighborhood, and then returns to the equilibrium [2]. This behaviour can be modelled by various dynamical systems, such as differential equations, maps, and cellular automata [3].

In this work, we study a map proposed as an "excitable unit" by De Monte *et al.* [4,5]:

$$F : x \mapsto f(x) = (\alpha x + \gamma x^3)e^{-\beta x^2}. \tag{1}$$

De Monte *et al.* discuss just one set of parameter values (α=0.4, β=1, and γ=8) and the corresponding behaviour of f, which in that case is indeed excitable. But the map can display a greater variety of types of behaviours, not all of which indicate excitability, when different parameter values are used. In this paper

C. Alippi et al. (Eds.): ICANN 2009, Part II, LNCS 5769, pp. 613–622, 2009.

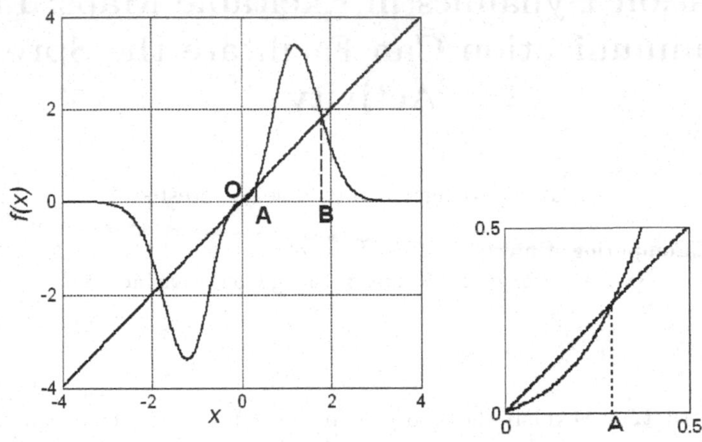

Fig. 1. $f(x) = (0.4x + 8x^3)e^{-x^2}$. Also shown are the fixed points in the positive quadrant (O, A and B) and the diagonal $f(x) = x$. The positive region near the origin is shown at a higher magnification.

we explore this variety. We describe the variants for of a single, individual map first; subsequently we explore some of the global phenomena obtained when such maps are coupled into a coupled map lattice (CML).

We can get some first insight in the nature of f by showing a plot (Fig. 1) of $f(x)$ with respect to x for some particular parameter values, for instance the values which are used in [4]. Since it passes through the origin (O), the map has a fixed point (a value of x for which $x(t+1) = x(t)$) there. De Monte et al. claim that the origin is the unique global attractor of this map. This means that given infinite iterations of the map, x will always approach the origin no matter which $x(0)$ one starts out with. They also indicate that f has another fixed point that is unstable: x-values close to it will, through iteration of the map, move away from this point rather than approach it. If the map is "excited" beyond that unstable fixed point there will be a chaotic but eventually downward transition towards the global attractor, zero. If we concentrate on the positive quadrant, this point corresponds to A in Fig. 1. There is a small region between the fixed points O and A where $f(x) < x$. Therefore, when x is in this region, iterating the map will make x smaller and approach zero.

But the assessment in [4] is only accurate for the particular parameter values used. When f is excited beyond A, the next fixed point B comes into play. There may indeed be a chaotic transition back to the $[O\ A]$ region and then an approach to O. But depending on the parameters, B can also acts as a second attractor for the map, preventing a return to the origin. Moreover, again depending on the parameters there can be no fixed point A and only a fixed point B. Thus, we must investigate these different possibilities.

2 Analysis

The geometrical insight in the map's variety was obtained by varying each of the three parameters of f while keeping the other two constant [5]. α determines whether the map has a fixed point of type A (as in Fig. 1) or not: when f always stays above the diagonal before it starts sloping down, there is no point A at all. The width and height of the "bell" of f, and the fixed point B are relatively if not entirely unaffected by the choice of α, and therefore we will use the latter to vary the existence of A while keeping the behaviour around B roughly constant. By varying γ we can change the height of the bell of f without affecting its width. We will use γ mainly to vary the behaviour around B. But these fairly independent influences of α and γ hold only when γ is relatively large. As can be gleaned from Fig. 2 and Fig. 3 there is a more complex interplay between the two parameters when γ is small: then γ also affects whether there is a point A at all. Finally, β does not affect the nature of A while it affects the whole "magnitude" of the bell. In order to evade the extra complexity of these magnitude effects, we have chosen to keep $\beta = 1$ in the remainder of the current paper; thus we

Table 1. Iteration of map (2) with example values illustrating various types of behaviour of the map. Runs for 350 iterations, where $x(0) = 10^{-5}$ and external pulses of magnitude 0.5 and 3.5 are given at $t = 100$ and $t = 200$ respectively. α determines whether the map leaves the origin spontaneously; γ whether it returns (or approaches) it spontaneously.

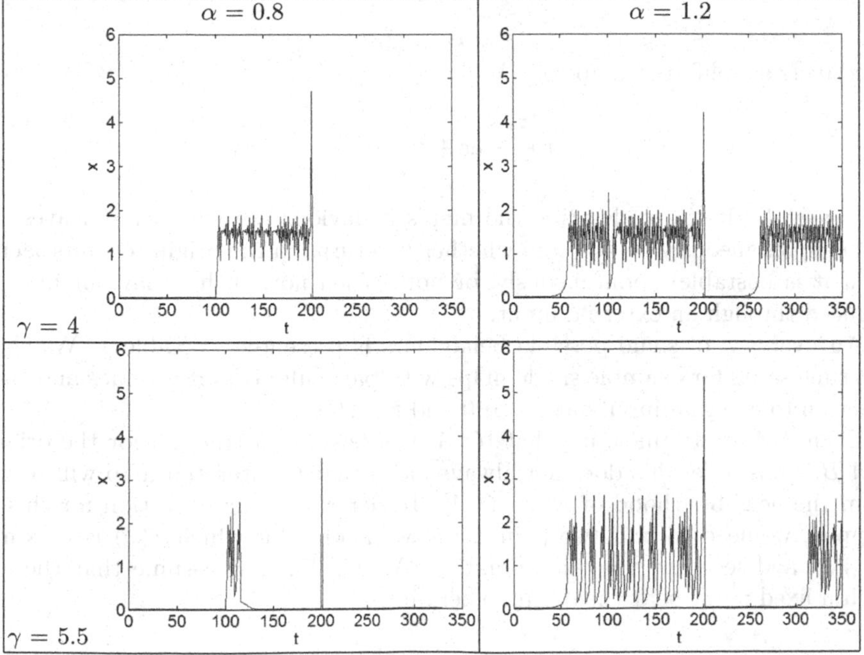

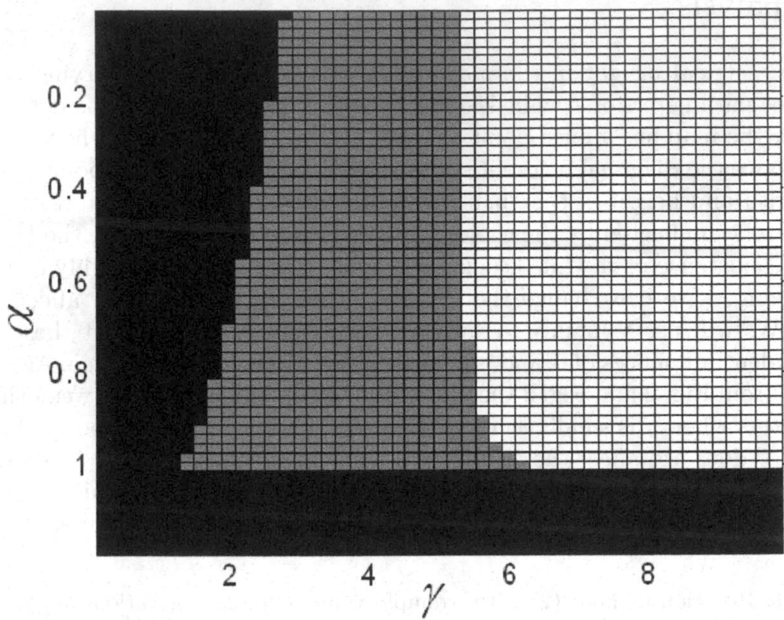

Fig. 2. Nature of the fixed point A. Black: there is no A. Grey: A exists and f does not return below it from the neighbourhood of B. White: A exists and f returns below it spontaneously from the neighbourhood of B.

effectively simplify the map to:

$$x \mapsto (\alpha x + \gamma x^3)e^{-x^2}. \quad (2)$$

Two useful criteria to describe the map's behaviour are whether it leaves the origin spontaneously or not, and whether it returns to the origin (or approaches it, if it is unstable) spontaneously, or not. When not, such behaviour may be induced through an external input.

An overview of some possible behaviours is presented in Table 1. We show the time series for example single maps with particular α and γ values and their reaction to external inputs at $t = 100$ and $t = 200$.

Even if A exists (meaning that $[O\ A]$ is a basin of attraction for the origin) and B is unstable, this does not always mean that the iterated map will return from the neighbourhood of B to $[O\ A]$. To determine the condition for this to happen, we need to find two points: the value of x for which $f(x)$ is maximal (x_{max}), and secondly the fixed point A. We will for now assume that there is such a fixed point near the origin. Then, if

$$f\big(f(x_{max})\big) < A \quad (3)$$

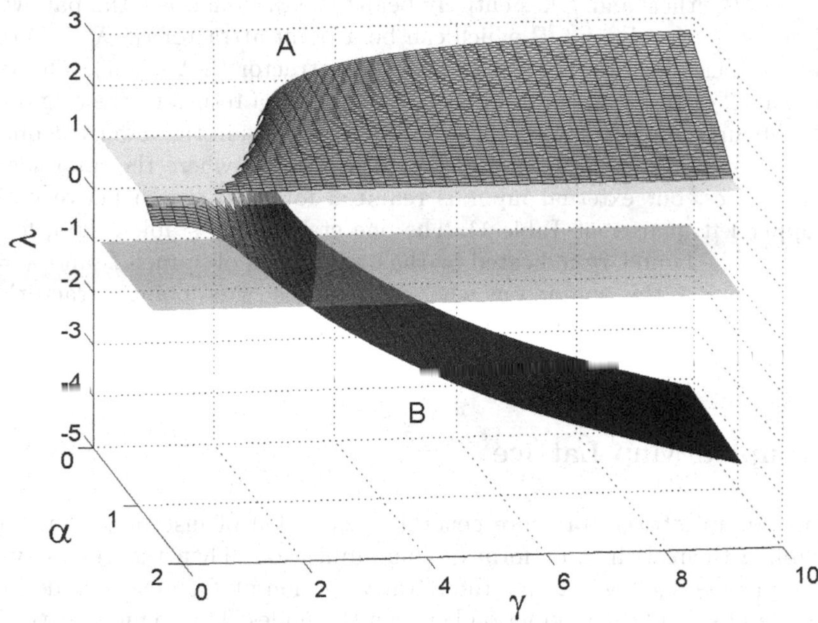

Fig. 3. $\lambda = df/dx$ for fixed points A and B, depending on α and γ. The surfaces $\lambda = 1$ and $\lambda = -1$ are for reference: between them a fixed point is stable, outside them unstable.

it means that this particular value of x_{max} results in the map returning to the origin. Since x_{max} depends on both α and particularly γ we want to express inequality (3) in terms of those parameters.

Requiring $\frac{df(x)}{dx} = 0$ we find that

$$x_{max} = \frac{1}{2}\sqrt{3 - \frac{2\alpha - \sqrt{4\alpha^2 - 4\alpha\gamma + 9\gamma^2}}{\gamma}}. \tag{4}$$

The fixed points are found by requiring that $f(x) - x = 0$. Solving this, we find

$$A, B = \sqrt{-\frac{\alpha}{\gamma} - W\left(\frac{-e^{-\frac{\alpha}{\gamma}}}{\gamma}\right)}, \tag{5}$$

where W is the Lambert W function, with A given using the principal branch of W and B using the other real branch.

Using the above, Fig. 2 shows whether there exists a fixed point A, and if so, whether f returns below it from B as per condition (3). Fig. 3 shows $\lambda = \frac{df}{dx}$ for A and B (where these exist): wherever $|\lambda| < 1$ the fixed point is stable.

Let us summarize Fig. 2 and Fig. 3. The black area in Fig. 2 is where there is no (unstable, by definition) fixed point A. For the part of it where $\alpha < 1$

there is no B either, and f lies entirely below the diagonal. For the part where $\alpha > 1$ there is a fixed point B which can be a point attractor (if $|\lambda| < 1$) or an oscillatory (periodic, intermittent or chaotic) attractor (if $\lambda < -1$). The white area in Fig. 2 displays two unstable fixed points and a return to the origin from the neighbourhood of B: this is the "excitable" case as in the lower left quarter of Table 1. Finally, the grey area in Fig. 2 is the case where the map has two different states, but external input is required for it to switch between these two (upper left quarter of Table 1). The one state is the stable origin; for the other state, the boundary indicated by the intersection of branch B and $\lambda = -1$ plane that divides this area in one where B is stable (fixed point attractor) and one where B is unstable but f nevertheless does not escape B's neighbourhood (periodic, intermittent or chaotic behaviour).

3 Coupled Map Lattice

We can set up lattices (or more generally networks) of instances of the map F, coupling them with some form of communication. Then the global system has two processes going on: one, the iterative action of f on every node in the network, and second the interaction between the nodes. The communication can take many forms, such as classical diffusion as in reaction-diffusion systems. We work with a *threshold* variant of this:

$$x_{i,j} \mapsto f(x_{i,j})+D\Big(-nH\big(f(x_{i,j})-\theta\big)f(x_{i,j})+\sum_{k,l\in R_{i,j}} H\big(f(x_{k,l})-\theta\big)f(x_{k,l})\Big), \quad (6)$$

where D is the communication rate, H is the Heaviside step function, θ is a communication threshold, and $R_{i,j}$ is the set of n nearest neighbours for vertex i, j. Equation (6) specifies that the map F is first applied to every node in the network; subsequently those vertices that then exceed θ distribute some of their activation to their neighbours. If this is combined with parameter values for F that result in excitable behaviour, we obtain a system that is roughly similar to a network of neurons that build up activation until they exceed the threshold and then "fire" some activation to their neighbours, depleting themselves in the process.

To observe a wave-like activation, consider the lattice of excitable maps (Fig. 4). The values of D and θ are chosen so that when a node fires, its own activation drops near zero, since with four neighbours 4×0.54 will be subtracted from it, a value slightly below θ. This allows for activation to be *passed* on rather than just diffused. We can see the effect of this in the following plots, showing the lattice at different times. The spread of activation in patterns similar to circular waves is clearly visible. The sixth plot for $t = 151$ shows the interference which occurs once the boundary of the lattice is reached and the activity bounces back into the lattice. These two phases (circular waves and interference) can also be recognized quite clearly from Fig. 5: the rim of the lattice is reached around $t \approx 75$ and after that the more chaotic activation spreads through the grid, after

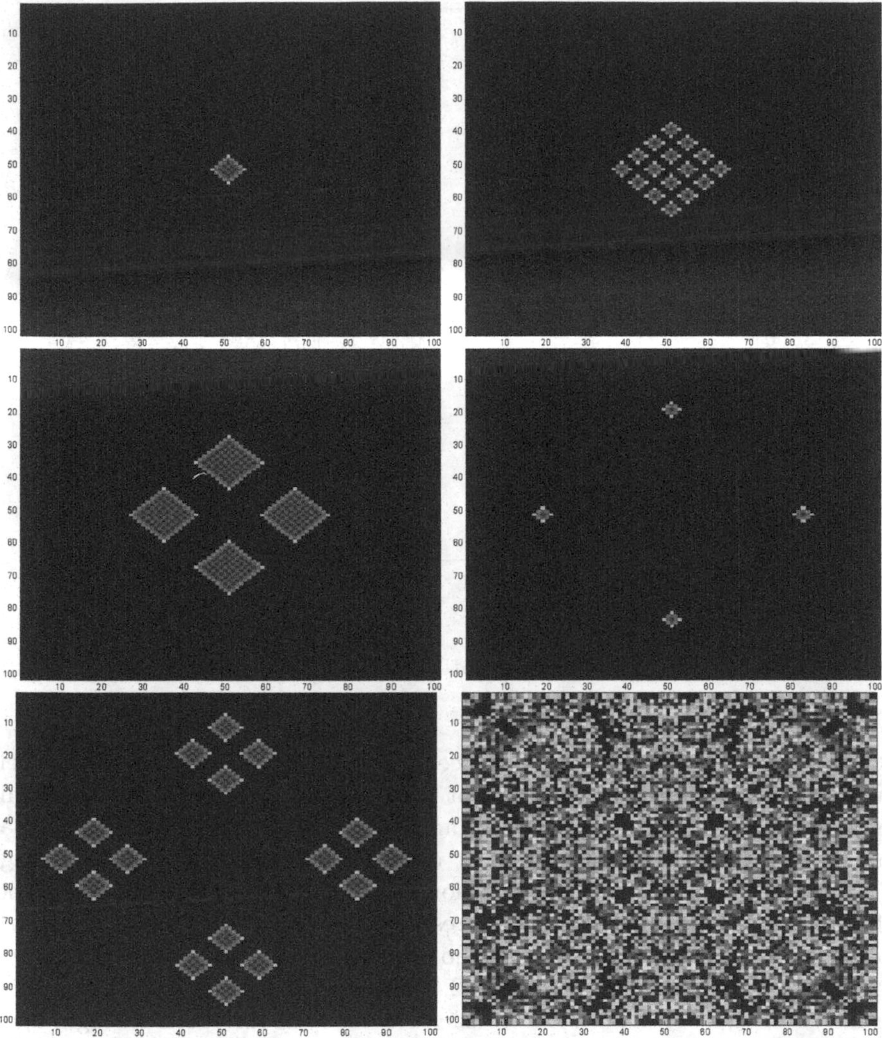

Fig. 4. Activation spreading in a 101 × 101 excitable map lattice at t = [11, 21, 31, 41, 51, 151]. $\alpha = 0.8$ and $\gamma = 5.5$; the initial value central vertex $x_{51,51}(0) = 0.5$ with all other initial values zero; $D = 0.54$; $\theta = 2.19$.

which the overall activation level fluctuates just below $\sum x = 3000$ (or average $x = 0.2941$).

To investigate the effect of diffusion threshold θ and diffusion rate D on total activation, consider an 11 × 11 grid of excitable maps, with activation spreading from one central node. All units start off deactivated except for the central unit, which starts at $x(0) = 0.3$. In Fig. 6, we tend to get either a fully activated or a zero activated grid, with a steep regime change in between. More intermediate

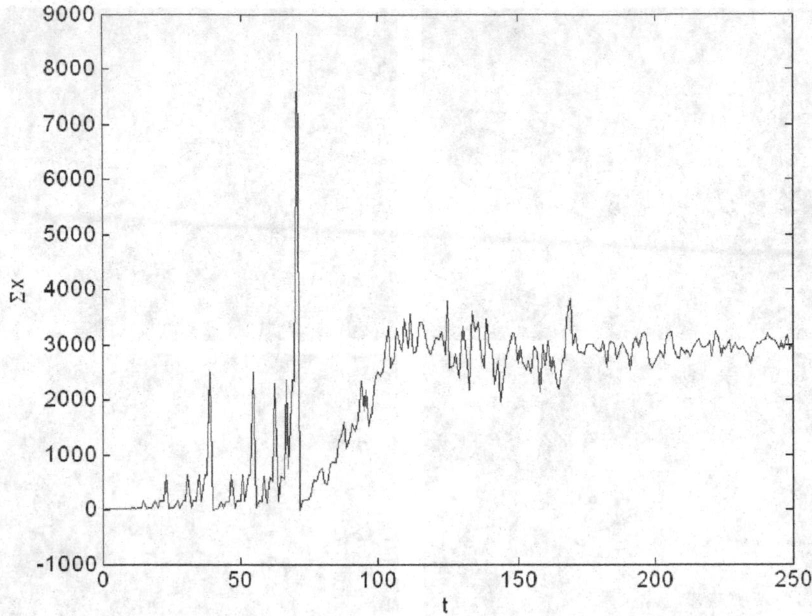

Fig. 5. Dynamics of total activation for the lattice shown in Fig. 4

values are rarely observed. We can see from the upper case, with relatively low initial activation, that there is not necessarily a monotonous relationship between our independent parameters and $\sum x$ (the summed activation of all nodes at $t = 100$). In that particular case, for very low diffusion rates high $\sum x$ is only obtained when the diffusion threshold is low; whereas for higher diffusion rates a very low diffusion threshold precisely results in no (or nearly no) activation. Note that we also get no activation when the threshold value is simply higher than the map can go, ~ 2.4 in this case. We can understand these results as follows. When both the diffusion rate and the threshold are high, maps will build up activation and then spread it liberally, resulting in high $\sum x$ (after 100 time steps). Fig. 4 gives an example of this. When both are low (but not *too* low in the case of the diffusion rate), the spread of activation will be slow but steady, and $\sum x$ will be high as well. With a low diffusion rate and a high threshold however, the spread of activation will be slow, and $\sum x$ is low. Finally with a high diffusion rate and a low threshold, diffusion can spread what little activation there is so thinly that the then only slightly activated neighbour maps subsequently "fall asleep" towards their stable origin. This loss of activation from the system results in low $\sum x$. With a high diffusion rate it is therefore best (if activity is desired) to have a higher threshold, so that when diffusion occurs, it is substantial and allows the recipients of the diffused activity to stay awake for a while and diffuse in turn. This particular nonmonotonous relationship is not observed for higher initial activation (lower part of Fig. 6), which suggests a less

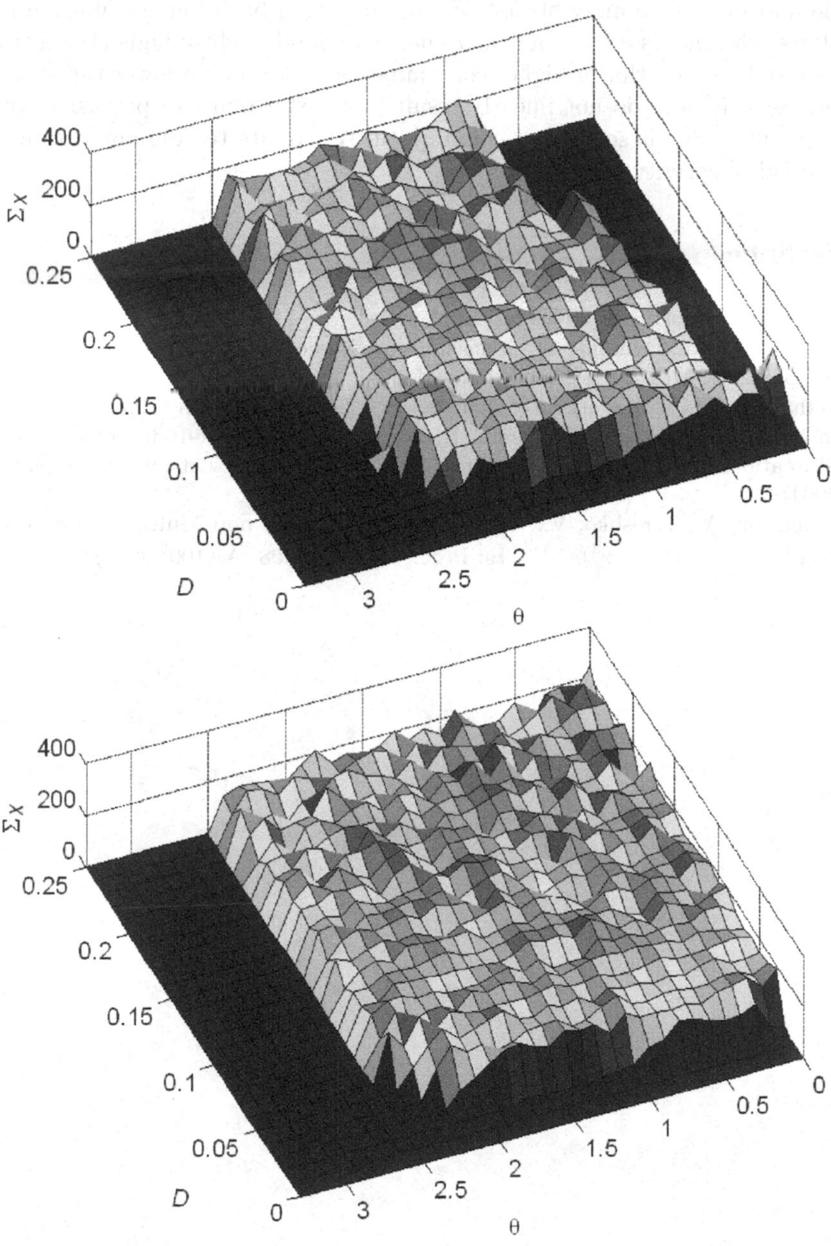

Fig. 6. Pattern of total activation (summed activation of all nodes, at $t = 100$) in 11×11 excitable map lattice. $\alpha = 0.8$ and $\gamma = 5.5$; the initial value central vertex $x_{6,6}(0) = 0.3$ (top) and $x_{6,6}(0) = 1.2$ (bottom) with all other initial values zero.

subtle and more seemingly obvious picture in which both higher diffusion rates and lower thresholds lead to a better chance of (and slightly higher) activation.

It must be noted that a higher communication rate and a lower threshold do not necessarily, and do not linearly result in a faster or more pervasive spread of activation. This is somewhat counterintuitive: limits to communication help the spread of activity.

References

1. Dayan, P., Abbott, L.F.: Theoretical Neuroscience. MIT press, Cambridge (2001)
2. Izhikevich, E.M.: Dynamical Systems in Neuroscience: The Geometry of Excitability and Bursting. MIT Press, Cambridge (2006)
3. Mikhailov, A.S.: Foundations of Synergetics I. Springer, Berlin (1990)
4. De Monte, S., d'Ovidio, F., Chate, H., Mosekilde, E.: Noise-Induced Macroscopic Bifurcations in Globally Coupled Chaotic Units. Phys. Rev. Lett. 92, 254101–254104 (2004)
5. Loengarov, A., Tereshko, V.: Excitability, Oscillations and Multiple Attractors in the Map $x \mapsto (\alpha x + \gamma x^3)e^{-\beta x^2}$. In: Proc. CHAOS 2008, AA100 (2008)

Learning Features by Contrasting Natural Images with Noise

Michael Gutmann[1] and Aapo Hyvärinen[1,2]

[1] Dept. of Computer Science and HIIT, University of Helsinki, P.O. Box 68, FIN-00014 University of Helsinki, Finland
[2] Dept. of Mathematics and Statistics, University of Helsinki
{michael.gutmann,aapo.hyvarinen}@helsinki.fi

Abstract. Modeling the statistical structure of natural images is interesting for reasons related to neuroscience as well as engineering. Currently, this modeling relies heavily on generative probabilistic models. The estimation of such models is, however, difficult, especially when they consist of multiple layers. If the goal lies only in estimating the features, i.e. in pinpointing structure in natural images, one could also estimate instead a discriminative probabilistic model where multiple layers are more easily handled. For that purpose, we propose to estimate a classifier that can tell natural images apart from reference data which has been constructed to contain some known structure of natural images. The features of the classifier then reveal the interesting structure. Here, we use a classifier with one layer of features and reference data which contains the covariance-structure of natural images. We show that the features of the classifier are similar to those which are obtained from generative probabilistic models. Furthermore, we investigate the optimal shape of the nonlinearity that is used within the classifier.

Keywords: Natural image statistics, learning, features, classifier.

1 Introduction

Natural scenes are built up from several objects of various scales. As a consequence, pictures that are taken in such an environment, i.e. "natural images", are endowed with structure. There is interest in modeling structure of natural images for reasons that go from engineering considerations to sensory neuroscience, see e.g. [1,2].

A prominent approach to model natural images is to specify a generative probabilistic model. In this approach, the probabilistic model consists of a parameterized family of probability distributions. In non-overcomplete ICA, for example, where each realization of the natural images $\mathbf{x} \in \mathbb{R}^N$ can be written as unique superposition of some basic features $\mathbf{a}_i$,

$$\mathbf{x} = \sum_{i=1}^{N} \mathbf{a}_i s_i, \tag{1}$$

C. Alippi et al. (Eds.): ICANN 2009, Part II, LNCS 5769, pp. 623–632, 2009.

the parameters in the statistical model are given by the $\mathbf{a}_i$, see e.g. [2]. In overcomplete models, either more than N latent variables s_i are introduced or the parameterization of the probability distribution is changed such that the parameters are some feature vectors $\mathbf{w}_i$ onto which the natural image is projected (non-normalized models, see e.g. [3,4]). Estimation of the features, i.e the $\mathbf{a}_i$ or $\mathbf{w}_i$, yields then an estimate of the probability distribution of the natural images.

The interest in this approach is threefold: (1) The statistical model for natural images can be used as prior in work that involves Bayesian inference. (2) It can be used to artificially generate images that emulate natural images. (3) The features visualize structure in natural images.

However, the estimation of latent variables, or non-normalized models, pose great computational challenges [2]. If the main goal in the modeling is to find features, i.e. structure in natural images, an approach that circumvents this difficult estimation can be used: For the learning of distinguishing features in natural images, we propose to estimate instead a discriminative probabilistic model. In other words, we propose to estimate a classifier (a neural network) that can tell natural images apart from certain reference data. The trick is to choose the reference data such that it incorporates some known structure of natural images. Then, the classifier teases out structure that is not contained in the reference data, and makes in that way interesting structure of natural images visible. We call this approach contrastive feature learning.

This paper is structured as follows: In Section 2, we present the three parts of contrastive feature learning: the discriminative model, the estimation of the model, as well as the reference data. The discriminative model has one layer of features. It further relies on some nonlinear function $g(u)$. In Section 3, we first discuss some properties which a suitable nonlinearity should have. Then, we propose some candidates and go on with presenting learning rules to optimize the nonlinearity. Section 4 presents simulation results, and Section 5 concludes the paper.

2 Contrastive Feature Learning

2.1 The Model

Since we want to discriminate between natural images and reference data, we need a classifier $h(.)$ that maps the data $\mathbf{x}$ onto two classes: $C = 1$ if $\mathbf{x}$ is a natural image and $C = 0$ if it is reference data.

We choose a classification approach where we first estimate the regression function $r(\mathbf{x}) = E(C|\mathbf{x})$, which is here equal to the conditional probability $P(C = 1|\mathbf{x})$. Then, we classify the data based on Bayes classification rule, i.e. $h(\mathbf{x}) = 1$ if $r(\mathbf{x}) > 1/2$ and $h(\mathbf{x}) = 0$ if $r(\mathbf{x}) \leq 1/2$.

Our model for $r(\mathbf{x})$ is a nonlinear logistic regression function:

$$r(\mathbf{x}) = \frac{1}{1 + \exp(-y(\mathbf{x}))}, \tag{2}$$

where

$$y(\mathbf{x}) = \sum_{m=1}^{M} g(\mathbf{w}_m^T \mathbf{x} + b_m) + \gamma \tag{3}$$

for a suitable nonlinearity $g(u)$ (see Section 3), and where M is not necessarily related to the dimension N of the data $\mathbf{x}$.

In a neural network interpretation, $g(\mathbf{w}_m^T \mathbf{x} + b_m)$ is the output of node m in the first layer. The weights of the second layer are all fixed to one. The second layer pools thus the outputs of the first layer together and adds an offset γ, the result of which is $y(\mathbf{x})$. The network has only one output node which computes the probability $r(\mathbf{x})$ that $\mathbf{x}$ is a natural image.

Parameters in our model for $r(\mathbf{x})$ are the features $\mathbf{w}_m$, the bias terms b_m and the offset γ. The features $\mathbf{w}_m$ are the quantities of interest in this paper since they visualize structure that can be used to tell natural images apart from the reference data.

2.2 Cost Function to Estimate the Model

Given the data $\{\mathbf{x}_t, C_t\}_{t=1}^T$, where $C_t = 1$ if the t-th input data point $\mathbf{x}_t$ is a natural image and $C_t = 0$ if it is reference data, we estimate the parameters by maximization of the conditional likelihood $L(\mathbf{w}_m, b_m, \gamma)$. Given $\mathbf{x}_t$, class C_t is Bernoulli distributed so that

$$L(\mathbf{w}_m, b_m, \gamma) = \prod_{t=1}^{T} P(C_t = 1|\mathbf{x}_t)^{C_t} P(C_t = 0|\mathbf{x}_t)^{1-C_t} \tag{4}$$

$$= \prod_{t=1}^{T} r_t^{C_t} (1 - r_t)^{1-C_t}, \tag{5}$$

where we have used the shorthand notation r_t for $r(\mathbf{x}_t; \mathbf{w}_m, b_m, \gamma)$. Maximization of $L(\mathbf{w}_m, b_m, \gamma)$ is done by minimization of the cost function $J = -1/T \log L$,

$$J(\mathbf{w}_m, b_m, \gamma) = \frac{1}{T} \sum_{t=1}^{T} \left(-C_t \log r_t - (1 - C_t) \log(1 - r_t)\right). \tag{6}$$

This cost function J is the same as the cross-entropy error function [5]. Furthermore, minimizing the cost function J is equivalent to minimization of the Kullback-Leibler distance between $P(C|\mathbf{x})$ and an assumed true conditional probability $P_{\text{true}}(C|\mathbf{x}) = C$.

2.3 Reference Data

In contrastive feature learning, we construct the reference data set such that it contains the structure of natural images which we are familiar with so that the features of the classifier can reveal novel structure.

A simple way to characterize a data set is to calculate its covariance matrix. For natural images, the covariance-structure has been intensively studied, see e.g. [2] (keyword: approximate $1/f^2$ behavior of the power spectrum). Hence, we take data which has the same covariance matrix as natural images as reference data. Or equivalently, we take white noise as reference data and contrast it with whitened natural images.

3 Choice of the Nonlinearity

3.1 Ambiguities for Linear and Quadratic Functions

We show here that the function $g(u)$ in Equation (3) should not be linear or quadratic.

Plugging a linear $g(u) = u$ into the formula for $y(\mathbf{x})$ in Equation (3) leads to

$$y(\mathbf{x}) = \sum_{m=1}^{M} (\mathbf{w}_m^T \mathbf{x} + b_m) + \gamma = \left(\sum_{m=1}^{M} \mathbf{w}_m^T \right) \mathbf{x} + \left(\sum_{m=1}^{M} b_m \right) + \gamma, \tag{7}$$

so that instead of learning M features $\mathbf{w}_m$ one could also learn only a single one, namely $\sum_m \mathbf{w}_m$ with bias $\sum_m b_m$. In other words, having more than a single feature introduces into the cost function J an ambiguity regarding the values of the parameters $\mathbf{w}_m$ and b_m.

There are also ambiguities in the cost function if $g(u) = u^2$. In that case, $y(\mathbf{x})$ equals

$$y(\mathbf{x}) = ||\tilde{\mathbf{W}}^T \tilde{\mathbf{x}}||^2 + \gamma, \tag{8}$$

where $\tilde{\mathbf{x}} = [\mathbf{x}; 1]$ and the $m - th$ column of $\tilde{\mathbf{W}}$ is $[\mathbf{w}_m; b_m]$. As

$$||\tilde{\mathbf{W}}^T \tilde{\mathbf{x}}||^2 = ||\mathbf{Q}\tilde{\mathbf{W}}^T \tilde{\mathbf{x}}||^2 \tag{9}$$

for any orthogonal matrix $\mathbf{Q}$, choosing a quadratic nonlinearity leads to a rotational ambiguity in the cost function. Again, many different sets of features will give exactly the same classifier.

While the arguments just given show that the features are ambiguous for linear and quadratic $g(u)$ for any data set, there is another reason why they are not suitable for the particular data sets used in this paper. In this paper, the natural image data and the reference data have, by construction, exactly the same mean and covariance structure. Thus, any linear or quadratic function has, on the average, the same values for both data sets. Therefore, any linear or quadratic classifier is likely to perform very poorly on our data. Note that such poor performance is not logically implied by the ambiguity in the features discussed above.

3.2 Candidates for the Nonlinearity

In the neural network literature, two classical choices for $g(u)$ in Equation (3) are the tanh and the logistic function $\sigma(u)$,

$$\sigma(u) = \frac{1}{1 + \exp(-u)}. \tag{10}$$

For zero-mean natural images, it seems reasonable to assume that if $\mathbf{x}$ is a natural image then also $-\mathbf{x}$. Thus, the regression function should verify $r(\mathbf{x}) \approx r(-\mathbf{x})$ if $\mathbf{x}$ is a natural image. This holds naturally if we omit the bias terms b_m and choose $g(u)$ even-symmetric. A symmetric version of the logistic function is

$$g(u) = \sigma(u - u_0) + \sigma(-u - u_0), \tag{11}$$

where $2u_0$ is the length of the "thresholding zone" where $g(u) \approx 0$. Other simple symmetric functions are obtained when we add a thresholding zone to the linear and quadratic function, i.e.

$$g(u) = [\max(0, u - u_0)]^{1+\epsilon} + [\max(0, -u - u_0)]^{1+\epsilon} \tag{12}$$

where we added $\epsilon \ll 1$ in the exponent to avoid jumps in the derivative $g'(u)$, and

$$g(u) = [\max(0, u - u_0)]^2 + [\max(0, -u - u_0)]^2. \tag{13}$$

In the following, we call this two functions linear-thresholding nonlinearity and squared-thresholding nonlinearity, respectively.

3.3 Optimizing the Nonlinearity

Instead of using a fixed nonlinearity, one can also learn it from the data. For instance, $g(u)$ can be written as weighted superposition of some parameterized functions $g_i(u; \theta)$,

$$g(u) = \sum_{i=1}^{I} \alpha_i g_i(u; \theta). \tag{14}$$

Then, we can optimize the conditional likelihood, or in practice the cost function J of Equation (6), also with respect to α_i and the parameters θ.

We consider here the special case where

$$g(u) = \alpha_1[\max(0, u - \beta_1)]^{\eta_1} + \alpha_2[\max(0, -(u - \beta_2))]^{\eta_2}, \tag{15}$$

for $\alpha_i \in \mathbb{R}$, $\beta_i \in \mathbb{R}$, and $\eta_i \in (1, 4]$. We optimize thus the type of nonlinearity of Equation (12) and (13) with respect to the size of the thresholding zone and the power-exponent. Furthermore, the signs of the α_i control whether the two power functions in (15) are each concave-up or concave-down.

4 Simulations

4.1 Settings

We estimate the classifier with a steepest descent algorithm on the cost function J of Equation (6), where we sped up the convergence by using the rprop algorithm [6].[1] Preliminary simulations with a fixed stepsize yielded similar results. The classifier was estimated several times starting from different random

[1] The multiplicative factors were $\eta_+ = 1.2$ and $\eta_- = 0.5$, maximal allowed change was 2, and minimal change 10^{-4}.

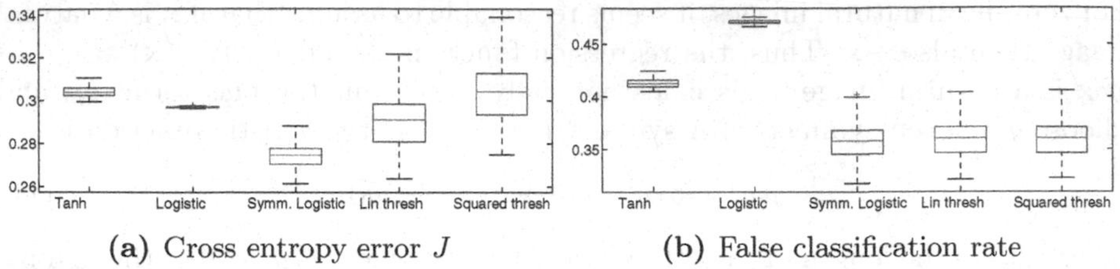

(a) Cross entropy error J

(b) False classification rate

Fig. 1. Distributions of the cross entropy error J and the false classification rate for the nonlinearities of Section 3.2. The distributions were obtained from the validation sets and are shown as box plots. The central red line is the median, the edges of the box are the 25th and 75th percentiles, and the whiskers extend to the most extreme data points. Outliers are marked with a cross. The settings were as follows: Bias terms b_m were only included for the tanh and the logistic function. The shift amount u_0 was $u_0 = 8$ for the symmetric logistic function (Symm. Logistic, see Equation (11)). For the linear-thresholding nonlinearity (Lin thresh, see Equation (12)) and the squared-thresholding nonlinearity (Squared thresh, see Equation (13)), we used $u_0 = 2$. On the training set, the minimal cross entropies J and false classification rates ("*er*") for each nonlinearity were $J = 0.282$, $er = 0.372$ for Tanh; $J = 0.294$, $er = 0.457$ for Logistic; $J = 0.231$, $er = 0.223$ for Symm. Logistic; $J = 0.202$, $er = 0.223$ for Lin tresh; $J = 0.199$, $er = 0.220$ for Squared thresh.

initializations. For computational reasons, we used only 5 initializations for the simulations of Section 4.2 and 20 for those of Section 4.3. We stopped optimization when the average change in the parameters was smaller than 10^{-3}. The classifier that had the smallest cost was selected for validation. The number of features M was set to 100.

Each training sample $\mathbf{x}_t$ was normalized to have an average value (DC component) of zero and norm one to reduce the sensitivity to outliers. The training set consisted of 80000 patches of natural images (size: 14× 14 pixels), and an equal number of reference data. For validation, we used 50 data sets of the same size as the training set. We also reduced the dimensions from $14 \times 14 = 196$ to 49, i.e. we retained only 25% of the dimensions.

4.2 Results for Fixed Nonlinearities

Performance. First, we validated our reasoning of Section 3.1 that a linear or quadratic $g(u)$ is not suitable to discriminate between whitened natural images and white Gaussian noise. Indeed, the false classification rate for the validation sets were distributed around chance level for the linear function (mean 0.5) and above chance level for the quadratic nonlinearity (mean 0.52).

Then, we performed simulations for the nonlinearities discussed in Section 3.2. The generalization performance as measured by the cross entropy error function J for validation sets is summarized in Figure 1a. The classifier with the symmetric logistic function has the best generalization performance. The squared-thresholding nonlinearity, which attained the minimal value of J for the training

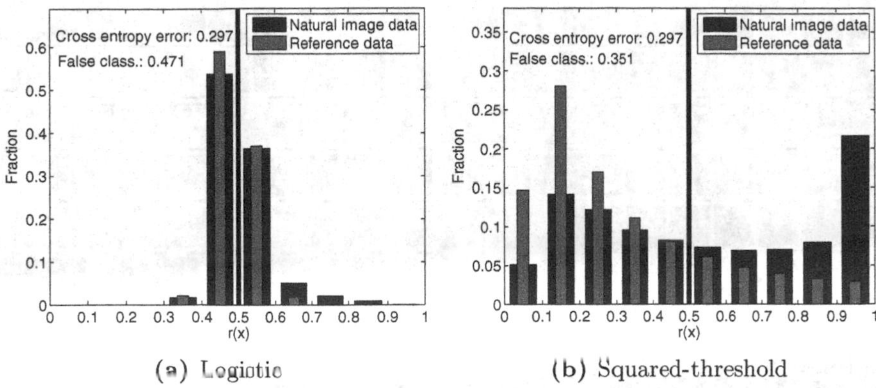

(a) Logistic

(b) Squared-threshold

Fig. 2. Conditional probability distributions $r(\mathbf{x}) = P(C = 1|\mathbf{x})$ when the input is natural image data (blue) or reference data (red). For natural image input, $r(\mathbf{x})$ should be 1. For reference data, $r(\mathbf{x})$ should be 0. The data set was chosen from the validation sets such that the cross entropy error J was approximately the same for the logistic nonlinearity and the squared-threshold nonlinearity. It is intuitively clear that the distribution in (b) is better for classification, although the cross-entropies are equal in the two cases. This seems to be because the cross-entropy gives a lot of weight to values near 0 or 1 due to the logarithmic function.

set (see caption of Figure 1), leads to the distribution with the highest median and a large dispersion, which seems to indicate some overlearning.

Figure 1b shows the false classification rates for the validation data. The symmetric nonlinearities, i.e. the symmetric logistic function and the nonlinearities with a thresholding zone, perform all equally well. Furthermore, they outperform the tanh and the logistic function. The performance as measured by the false classification rate and the cross entropy lead thus to different rankings.

Figure 2 gives a possible explanation for the discrepancy between the cross-entropies and false classification rates. The figure shows that two distributions of the conditional probability $r(\mathbf{x})$ can be rather different but, nevertheless, attain the same cross entropy error J. For the logistic nonlinearity in Figure 2a, $r(\mathbf{x})$ is clustered around chance level 0.5. The false classification rate is therefore also close to 0.5. On the other hand, for the same cross entropy error, the squared-thresholding nonlinearity leads to a false classification rate of 0.35. The reason behind its high cross entropy is that natural images (reference data) which are wrongly assigned a too low (high) conditional probability $r(\mathbf{x})$ enter logarithmically weighted into the calculation of the cross entropy.

Features. The estimated features $\mathbf{w}_m$ when the nonlinearity $g(u)$ is the symmetric logistic function are shown in Figure 3a. They are localized, oriented, and indicate bright-dark transitions. They are thus "gabor-like" features. For the linear-thresholding and squared-thresholding function, the features were similar. For the tanh and the logistic function, however, they did not have any clear structure.

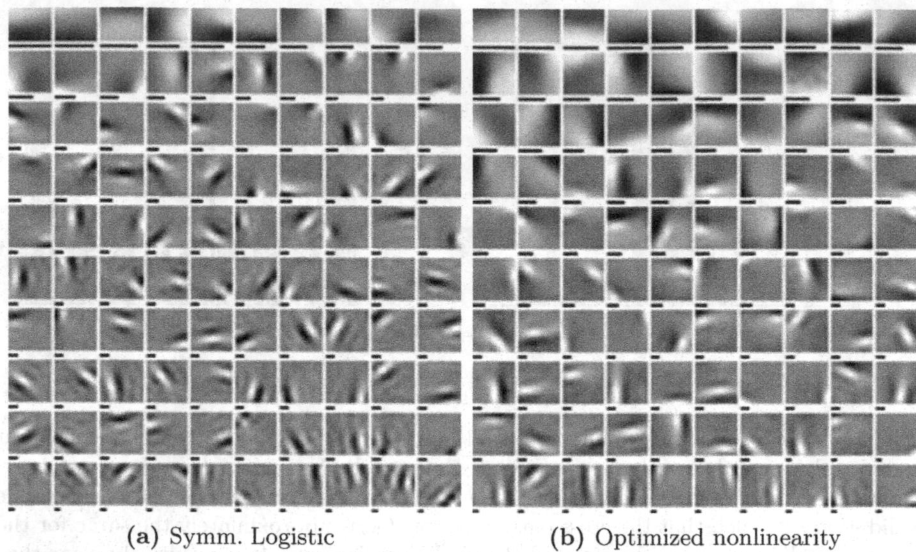

(a) Symm. Logistic (b) Optimized nonlinearity

Fig. 3. Features $\mathbf{w}_m$, $m = 1 \ldots M = 100$. The features are shown in the original image space, i.e. after multiplication with the dewhitening matrix. For visualization purposes, each feature was normalized to use the full range of the color-map. The black bar under each image panel indicates the euclidean norm $||\mathbf{w}_m||$ of the feature.

For the symmetric logistic function the learned offset γ in Equation (3) is -3.54. For the linear- and squared-thresholding functions, we also have $\gamma < 0$. For natural image input, $y(\mathbf{x})$ in Equation (3) must be as large as possible so that $r(\mathbf{x}) \to 1$. For reference data, $y(\mathbf{x})$ should be as negative as possible. Since the nonlinearities $g(u)$ attain only positive values, $y(\mathbf{x}) < 0$ is only possible when $\mathbf{w}_m^T\mathbf{x}$ falls into the thresholding zone of the nonlinearity. The negative γ leads then to a negative $y(\mathbf{x})$. Hence, the classifiers work by thresholding the outputs of gabor-like features. Large outputs are indicators for natural image input while small outputs indicate the presence of reference data.

4.3 Results for Optimized Nonlinearity

Optimization of the nonlinearity in Equation (15) leads to a classifier with a better performance than the fixed nonlinearities, see Figure 4.

Figure 5a shows the optimal nonlinearity where the offset per feature has been added, i.e. $g_{\text{eff}}(u) = g(u) + \gamma/M$ is shown. If $g_{\text{eff}}(\mathbf{w}_m^T\mathbf{x}) > 0$, feature m signals the presence of natural image data. Negative outputs indicate the presence of reference data. The outputs are negative when, approximately, $\mathbf{w}_m^T\mathbf{x} < 0$. This is in contrast to the fixed nonlinearities where for reference data $\mathbf{w}_m^T\mathbf{x}$ had to be in the thresholding zone.

The features for the optimal nonlinearity are shown in Figure 3b. They are also "gabor-like". Visual inspection, as well as a histogram of the normalized

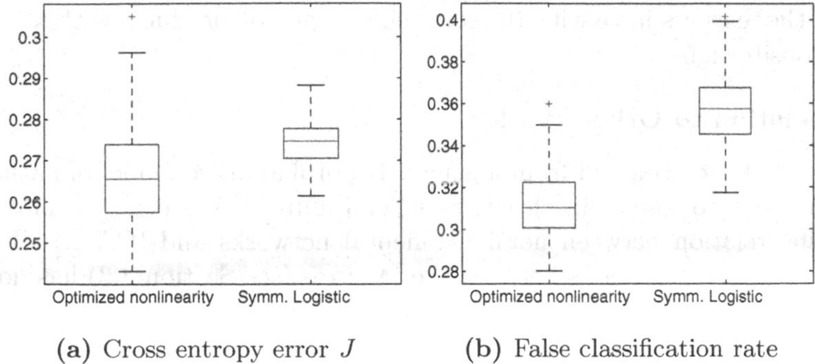

(a) Cross entropy error J (b) False classification rate

Fig. 4. Distribution of the cross entropy error J and the false classification rate for the learned nonlinearity and, for reference, the symmetric logistic function. The optimized nonlinearity achieves better performance both in terms of the cross entropy error J and the false classification rate. We tested if the distributions give enough evidence to conclude that the mean cross entropy error and the mean false classification rate are different for the two nonlinearities. For the cross entropy error, the p-value was 0.0014. For the false classification rate, the p-value was 0 (below machine precision). Hence, there is statistically significant evidence that their means are different, i.e. that, on average, the classifier with the optimized nonlinearity performs better than the symmetric logistic function. On the training set, the cross entropy error J was 0.186, and the false classification rate 0.203, cf. Figure 1.

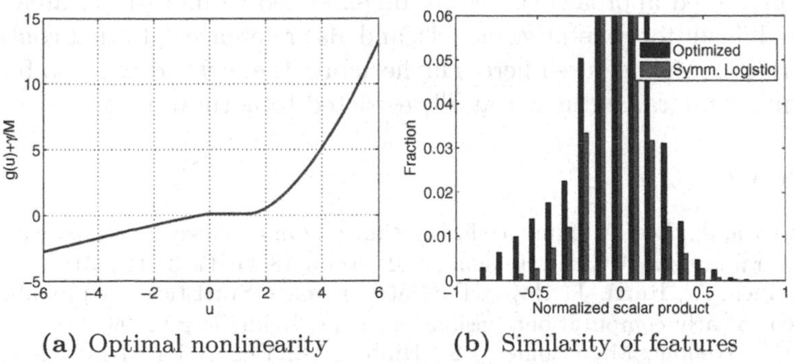

(a) Optimal nonlinearity (b) Similarity of features

Fig. 5. (a) The optimal nonlinearity of Equation (15) has the parameters $\alpha_1 = 1.00$, $\alpha_2 = -0.40$, $\beta_1 = 1.40$, $\beta_2 = -0.01$, $\eta_1 = 1.69$, $\eta_2 = 1.10$, and $\gamma = 11.76$. The negative α_2 makes the nonlinearity highly asymmetric. Note that $\eta_2 = 1.10$ is the smallest exponent which was allowed in the optimization. (b) In the calculation of the scalar product between the features, we first normalized them to unit norm.

dot-products between the features in Figure 5b, shows, however, that the features are more similar to each other than the features of the symmetric logistic function. Together with the shape of the optimal nonlinearity, this suggests that the classifier is using a different strategy than with the fixed (symmetric) nonlinearities. The negative part of the nonlinearity can be interpreted as leading to an interaction between features. An input is likely to be a natural image if some of the features have with the same sign large dot-products with $\mathbf{x}$, and not with opposite signs.

4.4 Relation to Other Work

Features that are obtained from a generative probabilistic model of natural images lead also to gabor-like features as in Figure 3, see e.g. [2]. This might reflect the relation between nonlinear neural networks and ICA [7]. However, sign-dependent interactions between the features (see Section 4.3) has not been found so far in generative models of natural images. Other work where learning a discriminative model led to gabor-like features is [8] where the features emerged from learning shape-from-shadings.

5 Conclusions

We presented an alternative to generative probabilistic modeling for the learning of features in natural images. The features are learned by contrasting natural image data with reference data that contains some known structure of natural images. Here, we used a classifier with only one layer of features and reference data with the same covariance-structure as natural images to validate the concept. The learned features were similar to those of generative models. When we optimized the nonlinearity in the classifier, we obtained a function which seems to facilitate interaction between the features.

The presented approach can easily be extended to multi-layer architectures, which is difficult for generative models, and also reference data that contain more structure than the one used here. Furthermore, the method can also be used on other kinds of data, and is not at all restricted to natural images.

References

1. Srivastava, A., Lee, A., Simoncelli, E., Zhu, S.: On advances in statistical modeling of natural images. J. Math. Imaging and Vision 18(1), 17–33 (2003)
2. Hyvärinen, A., Hurri, J., Hoyer, P.: Natural Image Statistics - A probabilistic approach to early computational vision. Springer, Heidelberg (2009)
3. Teh, Y., Welling, M., Osindero, S., Hinton, G.: Energy-based models for sparse overcomplete representations. J. Mach. Learn. Res. 4(7-8), 1235–1260 (2004)
4. Hyvärinen, A.: Estimation of non-normalized statistical models using score matching. J. Mach. Learn Res. 6, 695–709 (2005)
5. Bishop, C.: Neural networks for pattern recognition. Oxford University Press, Oxford (1995)
6. Riedmiller, M., Braun, H.: A direct adapatative method for faster backpropagation learning: The rprop algorithm. In: Ruspini, H. (ed.) Proc. IEEE Int Conference on Neural Networks (ICNN), pp. 586–591 (1993)
7. Hyvärinen, A., Bingham, E.: Connection between multilayer perceptrons and regression using independent component analysis. Neurocomp. 50(C), 211–222 (2003)
8. Lehky, S., Sejnowski, T.: Network model of shape-from-shading: neural function arises from both receptive and projective fields. Nature 333, 452–454 (1988)

Feature Selection for Neural-Network Based No-Reference Video Quality Assessment*

Dubravko Ćulibrk, Dragan Kukolj, Petar Vasiljević, Maja Pokrić, and Vladimir Zlokolica

Faculty of Technical Sciences, University of Novi Sad, Trg Dositeja Obradovića 6, 21000 Novi Sad, Serbia
culibrk@iis.ns.ac.yu, dragan.kukolj@rt-rk.com, petarv@uns.ac.rs, {maja.pokric,vladimir.zlokolica}@rt-rk.com
http://www.ftn.uns.ac.rs

Abstract. Design of algorithms that are able to estimate video quality as perceived by human observers is of interest for a number of applications. Depending on the video content, the artifacts introduced by the coding process can be more or less pronounced and diversely affect the quality of videos, as estimated by humans. In this paper we propose a new scheme for quality assessment of coded video streams, based on suitably chosen set of objective quality measures driven by human perception. Specifically, the relation of large number of objective measure features related to video coding artifacts is examined. Standardized procedure has been used to calculate the Mean Opinion Score (MOS), based on experiments conducted with a group of non-expert observers viewing SD sequences. MOS measurements were taken for nine different standard definition (SD) sequences, coded using MPEG-2 at five different bit-rates. Eighteen different published approaches for measuring the amount of coding artifacts objectively were implemented. The results obtained were used to design a novel no-reference MOS estimation algorithm using a multi-layer perceptron neural-network.

Keywords: Video quality assessment, no-reference approach, perceptual quality, neural-networks, multi-layer perceptron.

1 Introduction

There is an increased need to measure and assess the quality of video sequences, as it is perceived by the multimedia content consumers. The quality greatly depends on the video codec, bit-rates required and the content of video material. User oriented video quality assessment (VQA) research is aimed at providing means to monitor the perceptual service quality.

It is well understood that the overall degradation in the quality of the sequence, due to encoder/decoder implementations as part of transport stream at

* This work was supported in part by Ministry of Science and Technological Development of Republic of Serbia, under Grant 161003.

C. Alippi et al. (Eds.): ICANN 2009, Part II, LNCS 5769, pp. 633–642, 2009.

various bit rates, is a compound effect of different coding artifacts. Three types of artifacts are typically considered, pertinent to pertinent to DCT block (JPEG and MPEG) coded data: blocking, ringing and blurring. Blocking appears in all block-based compression techniques due to coarse quantization of frequency components [1][2]. It can be observed as surface discontinuity (edge) at block boundaries. These edges are perceived as abnormal high frequency components in the spectrum. Ringing is observed as periodic pseudo edges around original edges [3]. It is due to improper truncation of high frequency components. This artifact is also known as the Gibbs phenomenon or Gibbs effect. In the worst case, the edges can be shifted far away from the original edge locations. This effect is observed as false edge. Blurring, which appears as edge smoothness or texture blur, is due to the loss of high frequency components when compared with the original image. Blurring causes the received image to be smoother than the original one [4].

There is a myriad of published papers that propose different measures of prominent artifacts which appear in coded images and video sequences [1]-[2]. The goal of each no-reference approach is to create an estimator based on the proposed features that would predict the Mean Opinion Score (MOS)[5] of human observes, without using the original (not-degraded) image or sequence data.

In the paper, the applicability of a large set of published features to the problem of MPEG coded video quality assessment is evaluated. An approach to the selection of the optimal set of measures is proposed, where a non-linear estimator is trained to predict MOS. The selection of a smaller subset of objective measures is performed by means of statistical analysis, resulting in a final set of five basic measures. Based on the selected features, a Multi-Layer Perceptron (MLP)[6] as a nonlinear estimator was trained to predict the MOS.

Section 2 provides an overview of the relevant published work. The methodology used is described in Section 3. Section 4 presents the experiments conducted to evaluate the proposed approach and results obtained. Conclusions and some directions for future work can be found in Section 5.

2 Background and Related Work

The work presented falls within the scope of no-reference methodologies [7]. No information regarding the original (not-coded) video is used to estimate video quality, as perceived by human observers. A subjective quality measure typically used is the mean opinion score (MOS), which is obtained by averaging scores from a number of human observers[8][1]. The correct procedure for conducting such experiments was derived from ITU-R BT.500-10 recommendations[5].

In the research presented here, 18 different measures of image and video quality have been evaluated. Since the goal of the research is to create a VQA approach able to achieve real-time processing, the measures have been selected both for their reported results and simplicity.

Most perceived blockiness measures are based on the notion that the block-edge-related effects can be masked by high spatial activity in the image itself, and that the blockiness cannot be observed in very bright and very dark regions. Wang *et al.*[1] proposed a no-reference approach to quality assessment in JPEG coded images. His final measure is derived as a non-linear combination of a blockiness, local activity and a so-called zero-crossing measure. The combination is supposed to provide information regarding both blockiness and blurring in JPEG coded images.

Recently, Babu *et al.* [8] proposed a blockiness measure for use in VQA, which takes effects along each edge of the block into account separately. Thus, they derive a measure surpassing the Wang *et al.* approach.

Kusuma and Zepernick [7] describe three additional measures focusing on image activity and contrast. They propose using two different image activity measures edge and gradient activity, as a way to detect and measure ringing and lost blocks.

Spatial activity of the images and video frames in general has a profound effect on the quality of video coding. Within the work presented here, additional measures related to texture have been used to ensure a better description of the spatial activity within the frames of the sequence. These are based on the work Idrissi *et al.* [9].

Kim and Davis [10] proposed a noise and blur measure, aimed at evaluating the quality of video within the framework of automatic surveillance. They show their local-variance-based measure, dubbed fine-structure, able to describe video degradation well, in terms of noise and blur. In order to arrive at a single measure for the quality of a video sequence, based on the values of their proposed measure obtained for the inspected frames of the sequence, they used median as a statistic robust to outliers.

Kirenko [3] proposed simple measures for ringing effects detection, allowing for efficient real-time implementation.

In addition to spatial activity, the coded video quality depends on the temporal dynamics of the sequence. In order to be able to capture the characteristics of video material two motion intensity measures have been devised to describe the average magnitude of motion in a frame: (i) global motion intensity, calculated from the global motion field, and (ii) object motion intensity, calculated by subtracting the global motion from the MPEG motion vectors [2].

In 2005, Babu and Perkis proposed using their proposed quality measures to train a MLP estimator of MOS [11], when JPEG coded images are concerned. MLP has not, to the best of our knowledge, been used for VQA.

3 The Proposed Method for Video Quality Assesment

An set of 18 different features has been evaluated based on the VQEG sequences [12]. The features,with their respective references, are listed in Table 1. To make for an efficient VQA approach the set of features has been reduced to 5 features deemed to describe the quality best. These five features have subsequently been

Table 1. List of measures evaluated with pertinent references

#	Feature	Reference
1	Two field difference	[13]
2	Variance ratio	[10]
3	Blockiness	[8]
4	Ringing	[3]
5	Ringing 2	[3]
6	Global motion vector intensity	[2]
7	Activity	[1]
8	Blocking effect	[1]
9	Zero-crossing rate	[1]
10	Z score	[1]
11	Gradient activity	[7]
12	Edge activity	[7]
13	Contrast	[7]
14	Correlation	[9]
15	Energy	[9]
16	Homogeneity	[9]
17	Variance	[9]
18	Contrast	[9]

used to train a multi-layer perceptron neural-network, as an estimator for the MOS of new sequences.

3.1 Creating the Training Set

The training set used is based on nine SD sequences made available by Video Quality Experts Group (VQEG) for purposes of testing the quality of video codecs. Each sequence has been encoded using five different bit-rate settings (0.5Mb, 1Mb, 2Mb, 3Mb, 4Mb). Values of the features have been calculated for 110 frames of the sequences, i.e. half of the frames of the sequence, distributed uniformly. The mean opinion score (MOS), which is a subjective quality measure obtained by averaging scores from a number of human observers, is derived from tests created according to ITU-R BT.500-10 [5] recommendations. Double Stimulus Continuous Quality Scale (DSCQS) method was used, where pairs of sequences were presented to the viewer. The first one being an original sequence and the other the processed impaired sequence. The final test video was formed by pairing original and degraded video sequence and the observers were asked to evaluate the quality of overall impaired sequences using a five-point grading scale, from 1 to 5, according to perceived quality. Number of viewers had to be at least 20 for each test run to be able to obtain statistically meaningful results, and the test run was kept to maximum of 30 minutes in order to maintain viewer attention. The final MOS value for a sequence is the average score over for all observers for the sequence at a specific bit rate. The MOS values obtained for the sequences are shown in Table 2.

Table 2. MOS for the training sequences

Test sequence	Bit rate [Mb/s] 0.5	1	2	3	4
"Parade"	1.800	1.200	2.900	3.850	4.300
"Harp"	1.150	2.100	2.850	4.200	4.450
"Ant"	1.077	2.038	3.269	3.538	4.500
"Kayak"	1.100	1.850	3.300	3.950	4.700
"Formula"	1.885	2.385	3.308	4.192	4.231
"Food court"	1.150	2.150	3.550	4.400	4.800
"Scrolling titles"	1.450	2.800	3.650	3.950	4.400
"Football"	1.200	1.800	3.150	3.800	4.700
"Train"	1.962	1.615	3.231	4.154	4.654

3.2 Feature Ranking and Selection

To evaluate the predictive capability of each feature (measure), when MOS estimation is concerned, a wrapper methodology for attribute selection has been used [14]. Each feature was evaluated separately by providing it as input of a simple MLP, whose output was the MOS prediction. A simple Multi-layer perceptron (MLP) neural-network estimator has been trained based on a single measure. The estimators contained 3 nodes in a single hidden layer and were trained using 50% of our data, 25% was used for validation and another 25% for testing. A set of statistics was collected for the performance of each estimator, including: root mean square error (RMSE), Pearson correlation, Spearman correlation, maximum absolute prediction error (MAPE) and outlier ratio (OR). The features were than ranked according to the performance of the estimators. The ranking of measures determined through evaluation conducted on the VQEG sequences is shown in Table 3. The values of the statistics are listed along with the feature number corresponding to numbers in Tables 1 and 4. Table 4 provides the descriptions of the top-ranking features.

As the tables show, the highest ranking feature is the combined measure of Wang *et al.* (the Z-score). However, since the two out of three constituents of this measure ranked high (blocking effect and zero-crossing rate), the Z-score was not selected for the final set of features. The rationale for this being the fact that the MLP should be able to combine the constituents in a more informed way and achieve better performance. Thus, the final set of features selected includes: the blockiness measure of Babu *et al.*, the blocking effect measure and the zero-crossing rate of Wang *et al.*, the edge activity measure by Kusuma and Zepernick and the second ringing metric proposed by Kirenko. These are indicated in bold print in Table 4.

Forward selection has been explored as an alternative to the proposed approach, where features have been added to the selected set, using progressively more complex MLP estimators to rank the growing feature sets. Selecting the best feature set after each iteration, yielded exactly the same ranking as the independent analysis.

Table 3. Feature ranking

RMSE	#	MAPE	#	Spearman	#	Pearson	#	OR	#
1.0264	10	0.2853	10	0.4443	10	0.5344	10	0.0576	10
1.1320	3	0.3198	3	0.3192	8	0.3560	9	0.0365	8
1.1349	9	0.3330	8	0.2964	9	0.3534	8	0.0239	3
1.1357	8	0.3331	5	0.2635	3	0.3506	3	0.0179	6
1.1528	5	0.3377	9	0.2156	6	0.2947	5	0.0135	12
1.1684	12	0.3424	18	0.2048	5	0.2535	4	0.0100	16
1.1714	11	0.3433	12	0.1962	11	0.2495	16	0.0095	5
1.1746	4	0.3445	6	0.1833	12	0.2466	12	0.0077	4
1.1748	16	0.3446	11	0.1773	16	0.2464	7	0.0069	15
1.1755	15	0.3454	15	0.1657	7	0.2460	11	0.0064	18
1.1768	7	0.3454	4	0.1539	1	0.2460	6	0.0063	9
1.1769	6	0.3456	16	0.1362	13	0.2440	15	0.0056	11
1.1790	18	0.3471	7	0.1351	15	0.2392	18	0.0048	13
1.1945	13	0.3507	1	0.1318	18	0.1767	13	0.0043	1
1.1976	1	0.3508	13	0.0989	4	0.1638	1	0.0021	7
1.2059	17	0.3529	17	0.0661	2	0.1170	17	0.0020	17
1.2102	14	0.3532	14	0.0405	17	0.0836	14	0.0014	14
1.2154	2	0.3547	2	0.0345	14	0.0451	2	0.0002	2

Table 4. Description of top ranking features with pertinent references

#	Feature	Reference
3	**Blockiness**	[8]
5	**Ringing 2**	[3]
8	**Blocking effect**	[1]
9	**Zero-crossing rate**	[1]
12	**Edge activity measure**	[7]
10	Z score	[1]

3.3 VQA Estimator

A block diagram of the proposed video-quality estimator is shown in Fig. 1. Based on the selected set of features a MLP neural network is trained. The network contains 5 input nodes, 7 nodes in the hidden layer and a single output node corresponding to the MOS. No significant gain in prediction performance has been observed when increasing the number of nodes in the hidden layer.

The video quality assessment is conducted by calculating the five selected features for half of the frames of the sequence, uniformly distributed (i.e. the frame rate was halved to make the approach more efficient). The features obtained for each evaluated frame were fed into the neural network and the measure of the quality for that frame obtained.

Table 5. Cross-validation results

Test sequence	RMSE train	RMSE test	RMSE test stddev
"Parade"	0.6631	0.5142	0.0747
"Harp"	0.7424	0.9697	0.0825
"Ant"	1.1410	1.1460	0.1494
"Kayak"	0.7741	0.9475	0.0428
"Formula"	0.8178	0.6938	0.1487
"Food court"	0.8263	0.9351	0.0793
"Scrolling titles"	0.7852	0.6113	0.1204
"Football"	0.6941	0.7898	0.0218
"Train"	0.8127	0.7052	0.1223

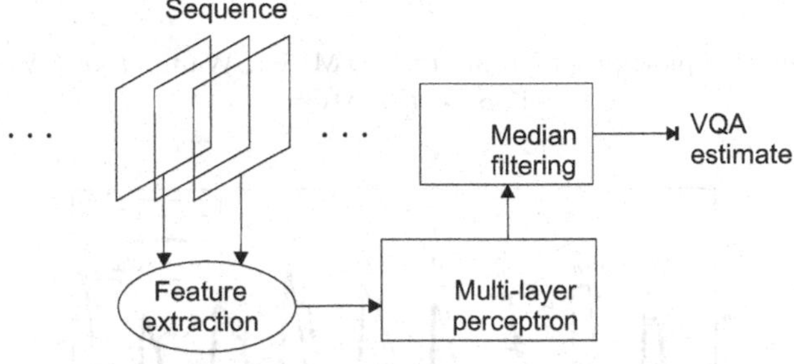

Fig. 1. Block diagram of the proposed approach

Since the standard deviation of the estimator RMSE (RMSE test stddev) over the frames of a single sequence is relatively high, robust statistics should be used to arrive at the final single measure of sequence video quality. Kim and Davis [10] suggest using the median of the quality values across the frames to achieve this. We followed their recommendation and adopted the median of values across the evaluated frames of the sequence as the final measure of sequence quality. Median is known to be a measure robust to the outliers, which commonly occurred in the experiments performed.

4 Results and Discussion

Two different approaches to the testing of the proposed approach were taken: using a part of the data as a separate test set and cross-validation.

Based on a test set comprised of 25% of data available, the proposed estimator achieved the RMSE value of 0.6364 averaged over 20 trial runs, with a standard deviation of 0.0241. The best published quality measure evaluated (Z-score of Wang *et al.*) achieved significantly higher RMSE (1.0264), suggesting that the proposed approach benefited from additional features introduced. The plots of of the test set results achieved per test case (sequence coded at a specific bit rate)

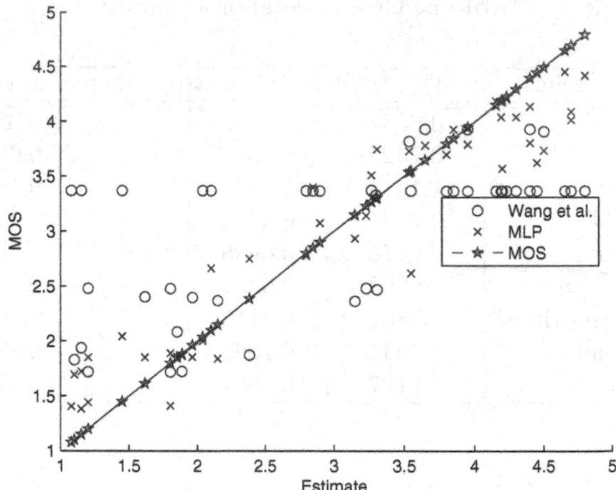

(a) Estimate scatter plot: proposed approach (MLP), Wang *et al.* (Wang) and true MOS values (MOS)

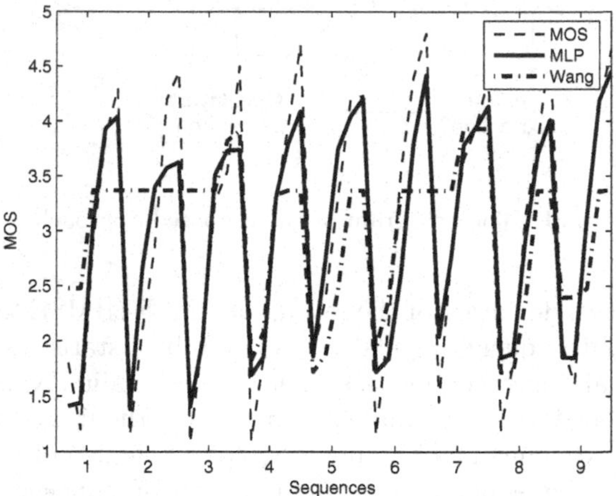

(b) Estimate over the sequences: proposed approach (MLP), Wang *et al.* (Wang) and true MOS values (MOS)

Fig. 2. Test results for the test set containing 25% of data

are shown in Fig. 2, for both the proposed approach (MLP) and Wang *et al.*. Estimate for a specific test case is the median value of quality estimates across all the evaluated frames of the sequence. As the figure shows, the proposed approach is able to achieve significantly better prediction than that of Wang *(*et al.) approach.

The error on the training set comprised of 50% of the data (another 25% was used for validation), was 0.6268, indicating that there was no over-fitting.

To evaluate the applicability of the proposed approach to a more realistic scenario, where quality evaluation is to be done for new sequences the likes of which may not be present in the training set, cross-validation was performed. This was done in a supervised way, by excluding all data pertinent to a single sequence. The results of this nine-fold cross-validation are shown in Table 5. While the estimator maintained a good RMSE, the results indicate that the training set is not diverse enough to allow for balanced performance when whole sequences are excluded. This suggests that the training set should be extended, and possibly that specialized estimators should be constructed based on the sequence characteristics and/or content.

5 Conclusion

A large number of features designed to detect and measure the coding artifacts introduced by DCT block coding algorithms, has been evaluated in terms of applicability to video-quality assessment of MPEG2 coded video sequences. An approach to determining the correct reduced set of features, based on the training data available has been described. A multi-layer perceptron based estimator of MOS has been trained using the five selected features. The proposed estimator achieved results superior to those of the single features evaluated, in terms of RMSE, when compared on frame-by-frame basis. The results of the experiments conducted suggest that a larger set of sequences should be used for MLP training in order to improve performance in a general case. In addition, the sequences could be separated into similar groups and specialized estimators constructed for each cluster, in order to improve the performance even further.

References

1. Wang, Z., Sheikh, H.R., Bovik, A.C.: No-reference perceptual quality assessment of jpeg compressed images, pp. 477–480 (2002)
2. Warwick, G., Thong, N.: Classification of Video Sequences in MPEG Domain. In: Signal Processing for Telecommunications and Multimedia, ch. 6, Springer, Heidelberg (2004)
3. Kirenko, I.: Reduction of coding artifacts using chrominance and luminance spatial analysis. In: International Conference on Consumer Electronics, ICCE 2006, Digest of Technical Papers, pp. 209–210 (2006)
4. Ferzli, R., Karam, L.: A no-reference objective image sharpness metric based on just-noticeable blur and probability summation. In: IEEE International Conference on Image Processing, 2007. ICIP 2007, vol. 3, III -445–III -448 (2007)
5. BT.500, I.R.: Methodology for the Subjective Assessment of the Quality of Television Pictures (2002)
6. Haykin, S.: Neural Networks: A Comprehensive Foundation. Macmillan, New York (1994)
7. Kusuma, T., Caldera, M., Zepernick, H.: Utilising objective perceptual image quality metrics for implicit link adaptation, IV: 2319–2322 (2004)

8. Venkatesh Babu, R., Perkis, A., Hillestad, O.I.: Evaluation and monitoring of video quality for uma enabled video streaming systems. Multimedia Tools Appl. 37(2), 211–231 (2008)
9. Idrissi, N., Martinez, J., Aboutajdine, D.: Selecting a discriminant subset of co-occurrence matrix features for texture-based image retrieval, pp. 696–703 (2005)
10. Kim, K., Davis, L.: A fine-structure image/video quality measure using local statistics, V: 3535–3538 (2004)
11. Babu, R., Perkis, A.: An hvs-based no-reference perceptual quality assessment of jpeg coded images using neural networks (2005)
12. `ftp://ftp.crc.ca/crc/vqeg/TestSequences/Reference/`
13. Wolf, S., Pinson, M.: Ntia report 02-392: Video quality measurement techniques. Technical report (Institute for Telecommunication Sciences)
14. Witten, I.H., Frank, E.: Data Mining: Practical machine learning tools and techniques, 2nd edn. Morgan Kaufmann, San Francisco (2005)

Learning from Examples to Generalize over Pose and Illumination

Marco K. Müller and Rolf P. Würtz

Institute für Neural Computation, Ruhr-University, 44780 Bochum, Germany
{marco.mueller,rolf.wuertz}@neuroinformatik.rub.de

Abstract. We present a neural system that recognizes faces under strong variations in pose and illumination. The generalization is learnt completely on the basis of examples of a subset of persons (the model database) in frontal and rotated view and under different illuminations. Similarities in identical pose/illumination are calculated by bunch graph matching, identity is coded by similarity rank lists. A neural network based on spike timing decodes these rank lists. We show that identity decisions can be made on the basis of few spikes. Recognition results on a large database of Chinese faces show that the transformations were successfully learnt.

Keywords: rank order coding, face recognition, pose invariance, illumination invariance, learning from examples, controlled generalization.

1 Introduction

Invariant recognition of objects is one of the most important features of the visual system and a classical classification task for artificial neural networks. However, invariance is not a natural generalization performed by known network architectures.

Invariances can, to a limited degree, be learnt from real-world data based on the assumption that temporally continuous sequences leave the object identity unchanged [2,6,1,13].

Nevertheless, successful recognition systems have the desired invariances built in by hand. This includes elastic graph matching [7,12], where the graph dynamics explicitly have to probe all possible variations in order to compare an input image with the stored models. Neural architectures that perform this matching include [14,8,15], with the more recent ones being massively parallel and can account for invariant recognition with processing times comparable to that of the visual system. These methods work fine for the recognition of identity under changes in translation, scale, and small deformations. The latter includes small changes in three-dimensional pose.

Invariances for which explicit modeling is difficult, like large pose differences or illumination changes, can be handled by elastic bunch graph matching only if bunch graphs are supplied for a coarsely sampled set of variants, e.g., 10 different head poses. This is problematic from a technical point of view [10], because for a

C. Alippi et al. (Eds.): ICANN 2009, Part II, LNCS 5769, pp. 643–652, 2009.

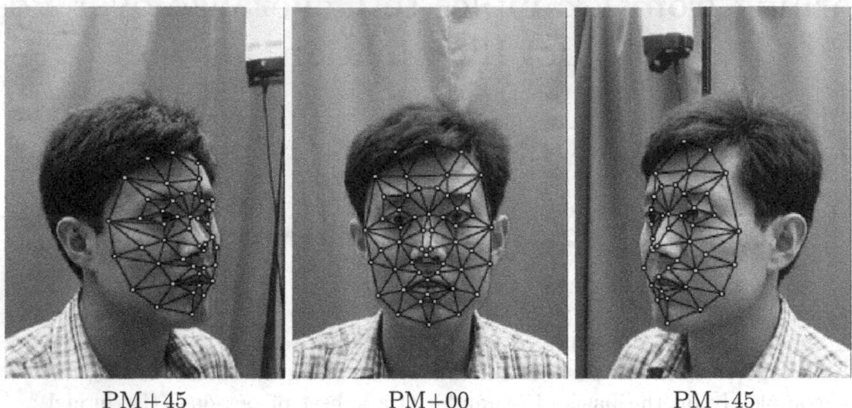

Fig. 1. Bunch graphs for different poses in the CAS-PEAL database. Images in different poses are not directly comparable because of different node numbers and strongly distorted features.

recognition system for many persons it is infeasible to store and match all persons in all possible poses or illuminations. It is also improbable that the brain would employ such a strategy because of the same waste of memory resources.

We here present a system that can learn invariances in a supervised way from a set of examples of individual objects in several instances of variations. For lack of a better term, we refer to each coarsely sampled constant illumination or pose angle as one *situation*. Invariant recognition generalizes to other objects that are known only in one situation.

We have recently reported that such a recognition scheme can achieve pose-invariance on the basis of *similarity rank lists* [9]. Here we extend this technique by a neuronal network that implements these similarity rank lists by relative spike timing [11]. This implementation on the one hand gives a plausible neural network for recognition under learnt invariances. On the other hand, it suggests a similarity function, which is different from the one used in [9]. We show that this yields better recognition results for pose-invariant face recognition. In this paper, we also tested the performance on illumination invariance.

2 Recognition by Similarity Rank Lists

Recognition by graph matching [7,12] compares a given *probe* image P with *gallery images* G_g of all known persons. It first estimates the correspondences between image points on the basis of N local features (Gabor jets) in a process called landmark finding. Then, it calculates a similarity between persons by adding (or averaging) local similarities $S_J(P, G_g, n)$ of *corresponding* features (n being a local feature index). The local similarity function is usually different from the one used for landmark finding. The recognized person is then the G_g

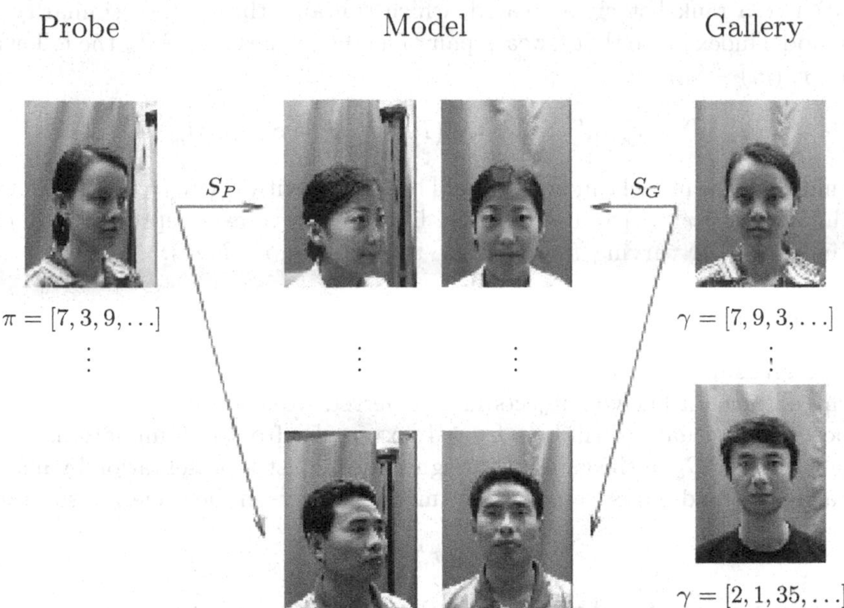

Fig. 2. Situation-independent recognition is mediated by a model database of some persons in all situations. Probe and gallery images are coded into rank lists π and γ by their similarities to the models. These rank lists are comparable, while the similarities are not (feature indices have been dropped for clarity).

with

$$g = \arg\max_g \frac{1}{N} \sum_n S_J(P, G_g, n) . \tag{1}$$

This cannot work between different situations, because the features are heavily distorted by pose and illumination changes, for large pose differences some feature points even disappear, leaving no visual features to compare with. In order to overcome this problem we construct a system that can look up the variations in a set of faces which are known in all situations. A number of N_V situations are coded into a *model database* with N_M subjects. The respective graphs are denoted by M_m^v, where m is an index of personal identity and v one of situation. Graphs with the same value of m are derived from images of the same person, the ones with the same value of v show the same situation. On the basis of these examples the variations are learnt.

Each situation requires its own similarity S_v, because the correspondence between features in different situations can not be assumed. Especially, the graphs in different poses contain different numbers of features N^v (see figure 1).

Personal identity is coded by a similarity rank list to the models of the same situation. The rank list for a test subject T is created as follows. First, all local similarities S_v to all model images M_m^v are calculated. For each index n and situation v a rank list r_n^v is created, which contains the rank of similarity for each model index m, so that for each pair of model images $M_m^v, M_{m'}^v$ the following holds ($r_n^v(m) \in \mathbb{N}_0$):

$$r_n^v(m) < r_n^v(m') \quad \Rightarrow \quad S^v(T, M_m^v, n) \geq S^v(T, M_{m'}^v, n) \,. \tag{2}$$

The most similar model candidate would be the one with $r_n^v(m) = 0$, the follower-up the one with $r_n^v(m) = 1$, etc. These lists now serve as a representation of a test image T. For varying T we will use the notation $r_n^v(T, m)$.

2.1 Invariant Recognition

For the recognition of an arbitrary subject a large *gallery* database is created, which contains all known subjects in a preferred situation $v = 0$. For practical purposes, this situation will be a frontal pose under frontal illumination.

Each subject G_g in the gallery is assigned a rank list representation by matching each of its landmarks to those of the model subjects in the preferred situation:

$$\gamma_{g,n}(\cdot) = r_n^0(G_g, \cdot) \,. \tag{3}$$

For recognition we assume that a *probe* P^v image appears in the known situation v. This probe is also represented as a similarity rank list for each landmark of all models in situation v:

$$\pi_n^v(\cdot) = r_n^v(P^v, \cdot) \,. \tag{4}$$

The requirement to know the situation beforehand will be removed in section 2.4.

Now the identity of the probe image is coded into the lists π_n^v, and the gallery images into $\gamma_{g,n}$. Each entry in a rank list is the rank of similarity of that model image to the probe or gallery image.

As the model database contains the same persons in different situations the rank lists should be similar for the same person. This is basically a continuity assumption on the transformations between situations: People that are similar in one situation are also similar in any other situations.

What is required now is a similarity function between rank lists. In contrast to the function chosen in [9] we here construct one on the basis of a neural network, which recognizes patterns on the basis of spike arrival times.

This similarity function enables the comparison of images under pose and illumination variation. For identification tasks it is now sufficient to store a single image of a person in a neutral view. Images taken in different situations can be compared to this gallery image using the rank list similarity.

2.2 Neuronal Rank List Comparison

Thorpe et al [11] have proposed a neural network that can evaluate rank codes. A set of feature detectors responds to an input pattern such that the most similar detector fires first. The order in which the spikes arrive can then be decoded by a circuit depicted in the left half of figure 3.

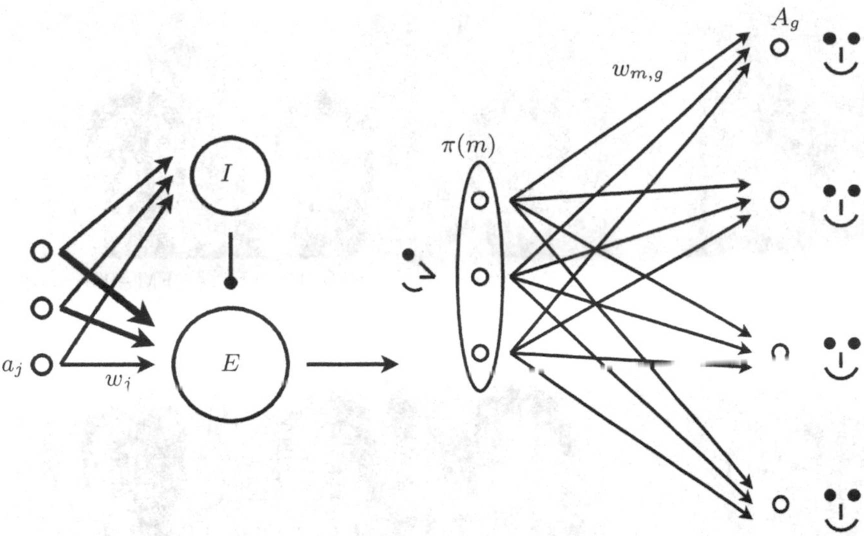

Fig. 3. Left: A neural circuit sensitive to the order of firing neurons, the preferred order is stored in the weights w_j (after [11]). Right; The same circuit is repeated for each gallery image. The probe image is represented as a rank list π according according to similarities with model images in the same situation. The similarities of the gallery to the model images in neutral situation are coded in the weights $w_{m,g}$.

We assume a neuronal module that calculates the similarity of stored model images to the actual probe image. Each gallery subject has one representing neuron. The similarity influences the time a neuron corresponding to this subject sends a spike. The higher the similarity the earlier the spike.

The activation in response to a spike train a_j is calculated as

$$A = \sum_{j=1}^{K} \exp\left(\frac{\text{order}(a_j)}{\lambda}\right) w_j\,, \tag{5}$$

with λ determining the activity decrease per spike. This parameter has to be optimized, it varies with the size of the rank list. If b_j is the sequence to elicit the largest activation the weights must be

$$w_j = \frac{1}{K} \exp\left(\frac{\text{order}(b_j)}{\lambda}\right)\,. \tag{6}$$

For our purposes, such a decoding circuit is required for each gallery image G_g. π is the rank list or the firing order of a number of N_M model neurons firing according to their similarity of each model image with index m to the probe image. The rank list γ_g of gallery image G_g is coded in the synaptic weights $w_{m,g}$ as follows:

$$w_{m,g} = \frac{1}{N_M} \exp\left(\frac{\gamma_g(m)}{\lambda}\right)\,. \tag{7}$$

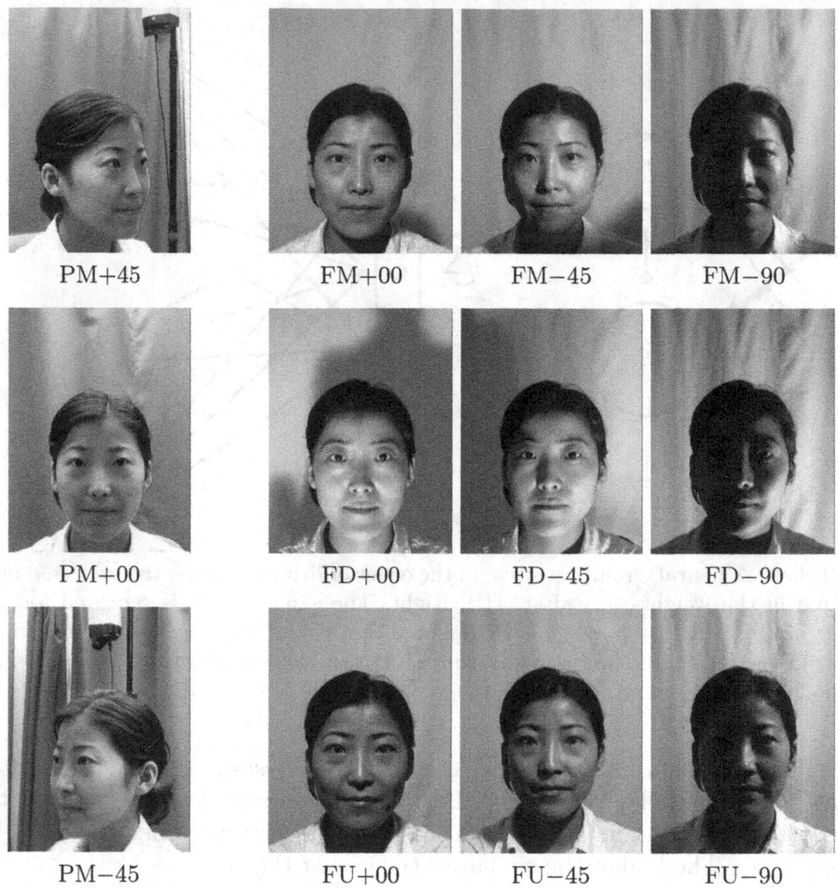

Fig. 4. Examples for pose variation (left column) and illumination variation in frontal pose handled by the system

The activity A_g then becomes

$$A_g = \sum_m \exp\left(\frac{\pi(m)}{\lambda}\right) w_{m,g} \,, \tag{8}$$

$$= \frac{1}{N_M} \sum_m \exp\left(\frac{\pi(m) + \gamma_g(m)}{\lambda}\right) \,, \tag{9}$$

and is interpreted as a similarity function between the rank lists π and γ_g.

$$S_{\mathrm{rank}}(\pi, \gamma_g) = \frac{1}{N_M} \sum_m^{N_M} \exp\left(\frac{\pi(m) + \gamma_g(m)}{\lambda}\right) \,. \tag{10}$$

Besides the neural interpretation, this similarity function has yielded better recognition results than the one used in [9].

2.3 Recognition

So far, the feature index n has been omitted from the rank list derivations. Clearly, the above circuit can be repeated for each feature, and the resulting similarities are averaged over all features for a similarity between the persons.

$$S_{\text{recog}}(g) = \frac{1}{N^v} \sum_{n=1}^{N^v} S_{\text{rank}}(\pi_n^v, \gamma_{gn}) . \tag{11}$$

As usual, the recognized person is the one with the index g that maximizes this similarity.

2.4 Automatic Estimation of Situation

In a realistic setting, the situation of the probe image is, of course, unknown. It can be estimated by matching with bunch graphs of all situations, and assigning the situation with the highest similarity:

$$v_{\text{est}} = \arg\max_v \frac{1}{N^v} \frac{1}{N_M} \sum_{n=1}^{N^v} \sum_{m=1}^{N_M} S^v(T, M_m^v, n) . \tag{12}$$

In case of v situations, bunch graph matching leads to v graphs for a given test image T. For each situation, the average similarity of that graph to all corresponding graphs of the model is calculated. The highest similarity indicates the estimated situation v_{est}, which is used instead of the known situation in the above procedure.

3 Experimental Setup

The network was tested on the CAS-PEAL face database [4]. The landmarks are found by elastic bunch graph matching, starting from very few images, that were labeled by hand. 24 subjects have been set aside for manual labeling. From these, the basic bunch graphs have been built (12 for pose, 8 for illumination).

The remaining 1015 subjects have been split up into model sets and testing sets (500 model and 515 testing for the pose case, and 100 model and 91 testing for illumination).

From the basic bunch graphs the landmarks on the model set database have been determined by incremental bunch graph building [9,5]. After EBGM was performed on one situation of the model set, good matches have been added to the bunch graph to achieve also a good match on previously poor matches. Each situation creates a separate bunch graph. After landmarks for all model images have been found and each bunch graph has grown to a convenient size (15 model graphs have been added in 3 iterations), gallery registration could begin. For registration of a gallery image, a single match has to be performed with the bunch graph of the corresponding situation. After that, similarities to the model images are calculated and the rank lists are created.

Table 1. Recognition rates (all in %) with known situation are only slightly impaired when the situation is estimated

	Pose	**Illumination**
Recognition rate with given situation	99.02	89.01
Rate of correct situation estimation	99.89 ± 0.09	91.96 ± 0.89
Recognition rate with automatically determined situation	97.75 ± 0.50	89.97 ± 1.36
Best recognition rate reported in [3]	71	51

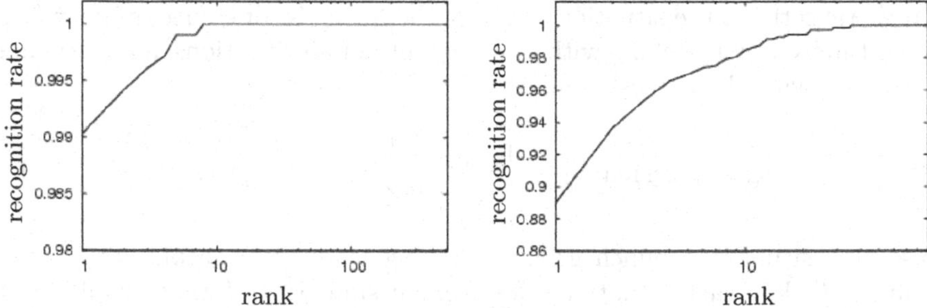

Fig. 5. Cumulative match score with known situation for pose (left) and illumination variation (right). A recognition rate of 100% is reached at rank 8 out of 515 (for pose) and 36 of 91 (for illumination). Rank-1 recognition rates are 99% and 89%, respectively.

Identifying a probe image works as follows. A single match with the bunch graph of the appropriate situation has to be done for landmark finding. A comparison with each model subject is done to calculate the rank lists. Then the rank lists can be compared to the ones in the gallery in a cross run.

4 Results

Figure 5 shows the cumulative match scores for recognition under pose and illumination variations. 100% recognition rate has been achieved at rank 8 for pose and 36 for illumination. To estimate the uncertainty in the recognition rate, the available subjects have been assigned to model or test in 100 randomly chosen partitions. The resulting recognition rates with error bars are shown in table 1.

In a final experiment, the decision was made on the basis of subsets of the k most similar model candidates. This means, a decision was already made when the first k spikes had reached the gallery neurons. The resulting recognition rates are shown in figure 6. This shows that recognition rates are not impaired if only the 10 most similar model candidates are used.

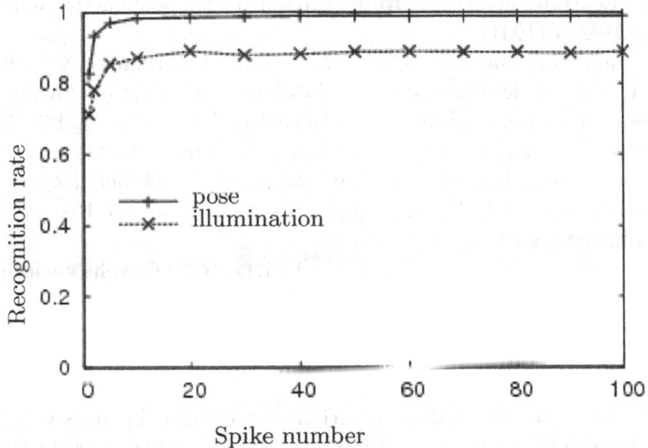

Fig. 6. This curve shows the recognition rates when a recognition decision is made before the spikes from all gallery representations are in. It can be seen that the first 10 spikes suffice to make the correct decision and even the first one is usually a good guess.

5 Discussion

We have presented a neural network based on spike timing, which is capable of learning the variations caused by pose and illumination changes on the basis of examples. Decisions are made from spike timing with the most similar template firing first. The model database holding the variations for a limited number of persons allows the generalization of identities known only in a single situation. The high recognition rates in comparison with previously published recognition results on the CAS-PEAL database demonstrate that a usable model of the variations due to pose and illumination changes has been learnt from examples. The recognition decision can be made using early stopping, which makes the system very fast in a parallel architecture.

Acknowledgments

We gratefully acknowledge funding from the German Research Foundation (WU 314/2-2 and WU 314/5-2). Portions of the research in this paper use the CAS-PEAL face database collected under the sponsorship of the Chinese National Hi-Tech Program and ISVISION Tech. Co. Ltd. [4,3].

References

1. Bartlett, M.S., Sejnowski, T.J.: Learning viewpoint-invariant face representations from visual experience in an attractor network. Network – Computation in Neural Systems 9(3), 399–417 (1998)

2. Földiák, P.: Learning invariance from transformation sequences. Neural Computation 3(2), 194–200 (1991)
3. Gao, W., Cao, B., Shan, S., Chen, X., Zhou, D., Zhang, X., Zhao, D.: The CAS-PEAL large-scale Chinese face database and baseline evaluations. IEEE Transactions on Systems, Man, and Cybernetics Part A 38(1), 149–161 (2008)
4. Gao, W., Cao, B., Shan, S., Zhou, D., Zhang, X., Zhao, D.: The CAS-PEAL large-scale Chinese face database and baseline evaluations. Technical Report JDL-TR-04-FR-001, Joint Research & Development Laboratory for Face Recognition, Chinese Academy of Sciences (2004)
5. Heinrichs, A., Müller, M.K., Tewes, A.H., Würtz, R.P.: Graphs with principal components of Gabor wavelet features for improved face recognition. In: Cristóbal, G., Javidi, B., Vallmitjana, S. (eds.) Information Optics: 5th International Workshop on Information Optics; WIO 2006, pp. 243–252. American Institute of Physics (2006)
6. Hinton, G.: Learning translation invariant recognition in massively parallel networks. In: Goos, G., Hartmanis, J. (eds.) PARLE Parallel Architectures and Languages Europe. LNCS, vol. 258, pp. 1–13. Springer, Heidelberg (1987)
7. Lades, M., Vorbrüggen, J.C., Buhmann, J., Lange, J., von der Malsburg, C., Würtz, R.P., Konen, W.: Distortion invariant object recognition in the dynamic link architecture. IEEE Transactions on Computers 42(3), 300–311 (1993)
8. Lücke, J., Keck, C., von der Malsburg, C.: Rapid convergence to feature layer correspondences. Neural Computation 20(10), 2441–2463 (2008)
9. Müller, M.K., Heinrichs, A., Tewes, A.H.J., Schäfer, A., Würtz, R.P.: Similarity rank correlation for face recognition under unenrolled pose. In: Lee, S.-W., Li, S.Z. (eds.) ICB 2007. LNCS, vol. 4642, pp. 67–76. Springer, Heidelberg (2007)
10. Murphy-Chutorian, E., Trivedi, M.M.: Head pose estimation in computer vision: A survey. IEEE Transactions on Pattern Analysis and Machine Intelligence 31(4), 607–626 (2009)
11. Thorpe, S., Delorme, A., Van Rullen, R.: Spike-based strategies for rapid processing. Neural Networks 14(6-7), 715–725 (2001)
12. Wiskott, L., Fellous, J.-M., Krüger, N., von der Malsburg, C.: Face recognition by elastic bunch graph matching. IEEE Transactions on Pattern Analysis and Machine Intelligence 19(7), 775–779 (1997)
13. Wiskott, L., Sejnowski, T.J.: Slow feature analysis: Unsupervised learning of invariances. Neural Computation 14(4), 715–770 (2002)
14. Wiskott, L., von der Malsburg, C.: Recognizing faces by dynamic link matching. Neuroimage 4(3), S14–S18 (1996)
15. Wolfrum, P., Wolff, C., Lücke, J., von der Malsburg, C.: A recurrent dynamic model for correspondence-based face recognition. Journal of Vision 8(7) (2008)

Semi–supervised Learning with Constraints for Multi–view Object Recognition

Stefano Melacci, Marco Maggini, and Marco Gori

Department of Information Engineering,
University of Siena,
Via Roma 56, 53100 Siena, Italy
{mela,maggini,marco}@dii.unisi.it

Abstract. In this paper we present a novel approach to multi–view object recognition based on kernel methods with constraints. Differently from many previous approaches, we describe a system that is able to exploit a set of views of an input object to recognize it. Views are acquired by cameras located around the object and each view is modeled by a specific classifier. The relationships among different views are formulated as constraints that are exploited by a sort of collaborative learning process. The proposed approach applies the constraints on unlabeled data in a semi–supervised framework. The results collected on the COIL benchmark show that constraint based learning can improve the quality of the recognition system and of each single classifier, both on the original and noisy data, and it can increase the invariance with respect to object orientation.

Keywords: semi–supervised learning, constraints, multi–view object recognition, kernel methods.

1 Introduction

Object recognition from static images is a wide and challenging research topic in the fields of computer vision and pattern recognition. In the last few years several systems and techniques have been proposed for this task[1–12]. Some of them are *single–view*, in the sense that they process a single viewpoint of an object. Objects are captured in different conditions of illumination, with occlusions or in presence of noise [1]. In those contexts the focus is posed in finding a compact, discriminative and robust representation of the objects in the feature space [1, 2].

When multiple viewpoints are introduced, object recognition usually performs more accurately [3–12]. In this scenario, referred to as *multi–view* object recognition, a single object is represented by a set of views captured at different angles. Some existing approaches use local feature representations to exploit the correspondences among the available views [4]. The generation of 3D models from local image features for viewpoint invariant object recognition has been studied in [5]. Other authors jointly modeled object appearance and viewpoint or extended single–view techniques, such as the Implicit Shape Model (ISM) [13], to

C. Alippi et al. (Eds.): ICANN 2009, Part II, LNCS 5769, pp. 653–662, 2009.

the multi–view scenario [6]. However, many of these approaches assume that a single image is available at test time [8–12].

In this paper we investigate the problem of object recognition from multiple views. In this case, a set of views of an object is fed as input to the system at test time. In a real scenario this model corresponds to the situation in which a set of cameras acquire images of a given object from different viewpoints. The recognition system must be able to exploit the availability of multiple views to enhance its discriminative power.

In our approach, we adopt kernel machines [14] to model each view and then we reinforce the classifiers by combining the single decisions in a constraint based framework, requiring coherence in the decision among different views. In particular, unlabeled data is exploited in a semi–supervised fashion to force the fulfillment of coherence constraints. In a wider context, our method could be applied also with other kind of classifiers and in every situation when there is a relationship among corresponding decisions on different representations of the same object.

This paper is organized as follows. In Section 2 the multi–view object recognition scenario is formalized. Section 3 describes constraint based learning in the semi–supervised framework. Experimental results are collected in Section 4 and concluding remarks are presented in Section 5.

2 Multi–view Object Recognition

In multi–view object recognition, each object is represented by a set of images acquired from different viewpoints. Given a collection of known objects, the goal is to correctly classify the input element into one of the known object categories. The information contained in multiple views is more informative than the one in a single image and it can increase the accuracy of the classifier but it can also contain redundant data due to, for example, the overlapping regions among different images.

In details, given a set D of objects, we consider k cameras c_i, $i = 1, \ldots, k$ that simultaneously acquire k pictures of the same object $\boldsymbol{x} \in D$ from k different points of view. Each camera produces a bidimensional representation of $\boldsymbol{x}$, indicated with $\boldsymbol{x}_i$. Such process can be modeled by an unknown function $g_i : D \rightarrow I\!R^d$, where d is the number of pixels of each acquired image, and $g_i(\boldsymbol{x}) = \boldsymbol{x}_i$. The functions g_i describe a complex relationship that maps the object $\boldsymbol{x}$ in the three dimensional object space to a planar image belonging to $I\!R^d$. A collection of k views is referred to as *viewset* and it is indicated with $X = \{\boldsymbol{x}_1, \ldots, \boldsymbol{x}_k\}$. Viewsets belong to the cartesian product of k sets in $I\!R^d$, $V = I\!R^d \times I\!R^d \times \cdots \times I\!R^d$. In particular, we can define a distribution $\mathcal{P}$ on V of the viewsets representing objects from D. The distribution $\mathcal{P}$ expresses the correlation between different views of the same object, and regions with zero probability correspond to unknown objects.

Given a collection of q viewsets representing the objects in D, acquired in different conditions of illumination or with slight orientation/position changes, we define the set of labeled instances as $L = \{(X_{j,h}, t_j) \mid X_{j,h} \in V;\ j = 1,\ldots,n;\ h = 1,\ldots,v_j\}$, where t_j is the actual label of the j–th object described by the viewset $X_{j,h}$, and v_j is the number of viewsets available for that object (note that $q = \sum_{j=1}^{n} v_j$).

We model the system using n binary multi–view classifiers, in a one–against–all strategy [15]. Moreover, we indicate with the function $o_j : V \to [0,1]$ the output of each classifier.

First, as baseline approach, we use a single discriminating function $f_j : \mathbb{R}^d \to [0,1]$ as base of the j–th classifier, that makes no distinctions among the views of an object, since it does not include any information on viewpoints. The output of such classifier for a generic input X is then

$$o_j(X) = \frac{1}{k}\sum_{i=1}^{k} f_j(\boldsymbol{x}_i), \tag{1}$$

where the k outputs are averaged to obtain a single combined output given the k input images.

Secondly, we separately model the data $\boldsymbol{x}_i$ acquired by the camera c_i with a specific function $f_{j,i} : \mathbb{R}^d \to [0,1]$. The output function becomes

$$o_j(X) = \frac{1}{k}\sum_{i=1}^{k} f_{j,i}(\boldsymbol{x}_i). \tag{2}$$

In both cases, the output of each binary classifier is compared with a reject threshold $\tau_j \in (0,1]$. If all $o_j(X)$, $j = 1,\ldots,n$, are less than their corresponding thresholds, the object is classified as not belonging to the set D. Otherwise, the predicted class label $c(X)$ corresponds to the index of the binary classifier with the highest confidence, as formalized in

$$c(X) = \begin{cases} \arg\max_j \frac{o_j(X)-\tau_j}{1-\tau_j} & \textit{if } \exists j \ (o_j(X) \geq \tau_j) \\ \text{unknown} & \textit{otherwise.} \end{cases} \tag{3}$$

We exploit kernel machines [14] to model the functions f_j and $f_{j,i}$. Focusing on the second approach, given a positive definite Kernel function $K_j : \mathbb{R}^d \times \mathbb{R}^d \to \mathbb{R}$, we indicate with $\mathcal{H}$ the Reproducing Kernel Hilbert Space (RKHS) corresponding to it, and with $\|\cdot\|_{\mathcal{H}}$ the norm of $\mathcal{H}$. From Tikhonov regularization in a RKHS, when the loss function $\mathcal{L}$ is the classic squared loss, the problem becomes an instance of ridge regression [15]. In details, for each of the k functions of the j–th classifier we have $\mathcal{L}_{j,i} = \sum_{r=1}^{q}(y_r - f_{j,i}(\boldsymbol{x}_i^r))^2$, where $\boldsymbol{x}_i^r$ indicates the r–th instance of the i–th view and $y_r \in \{0,1\}$ is the corresponding label. The k functions $f_{j,i} \in \mathcal{H}$ are chosen such that

$$\min_{f_{j,i}\in\mathcal{H}} \sum_{i=1}^{k}\sum_{r=1}^{q}(y_r - f_{j,i}(\boldsymbol{x}_i^r))^2 + \lambda_j \sum_{i=1}^{k}\|f_{j,i}\|_{\mathcal{H}}^2, \tag{4}$$

where λ_j is the weight of the regularization term.

From the Representer Theorem [14] the form of functions $f_{j,i}$, solution to the Tikhonov minimization problem, is given by

$$f_{j,i}(\boldsymbol{x}_i^{\cdot}) = \sum_{r=1}^{q} w_{j,i}^r K_j(\boldsymbol{x}_i^{\cdot}, \boldsymbol{x}_i^r), \tag{5}$$

where $w_{j,i}^r$ are the function weights and $\boldsymbol{x}_i^{\cdot}$ is a generic input. Using this representation when minimizing Eq. 4 with respect to the function $f_{j,i}$, is equivalent to solving a linear system of equations in the weights $w_{j,i}^r$, $r = 1, \ldots, q$ [15]. In matrix notation, $\boldsymbol{w}_{j,i} \in \mathbb{R}^q$ is the weight vector that collects the q weights $w_{j,i}^r$, $G_{j,i} \in \mathbb{R}^{p,p}$ is the Gram matrix associated to the selected kernel function, $\boldsymbol{y}_j \in \{0,1\}^q$ is the vector that collects the q labels y_r and $I \in \mathbb{R}^{p,p}$ is the identity matrix. Finally,

$$\boldsymbol{w}_{j,i} = (\lambda_j I + G_{j,i})^{-1} \boldsymbol{y}_j. \tag{6}$$

The solution for the baseline approach (Eq. 1) is straightforward, since it is a just simplified case of the described one. Note that the number of parameters for the j–th classifier in both the approaches is exactly the same. In particular each of the k functions $f_{j,i}$ is composed by q weights for a total of $k \cdot q$, that is equivalent to the number of weights of f_j since its representation includes all the $k \cdot q$ training views.

3 Semi–supervised Learning with Constraints

Each input viewset X belongs to the space V, and in particular to regions of V where the distribution $\mathcal{P}$ is non–zero. The classification approach described by Eq. 2 models different views with independent functions, that share only the selected kernel function and regularization weight. The set L of labeled training instances implicitly includes the information on the data distribution, since views of the same object are marked with the same label. If the classifier accurately approximates training data, it is assured to model the distribution $\mathcal{P}$ but only in regions of V that correspond to such data.

When unlabeled data is available, the correlation among the k views expressed by $\mathcal{P}$ can be exploited as prior knowledge to improve the discriminative power of the classifier. In particular, it introduces a dependency among the functions $f_{j,i}$ that can be modeled by constraining the learning process. Each function can benefit by taking into account the shape of the others in different, but corresponding, regions of the space.

Ideally the functions should produce exactly the same output for the k views of a given viewset X, since they belong to the same object. More formally, we require the fulfillment of the following constraints

$$\begin{cases} f_{j,1}(\boldsymbol{x}_1) = f_{j,2}(\boldsymbol{x}_2) \\ f_{j,2}(\boldsymbol{x}_2) = f_{j,3}(\boldsymbol{x}_3) \\ \quad \cdots \\ f_{j,k-1}(\boldsymbol{x}_{k-1}) = f_{j,k}(\boldsymbol{x}_k). \end{cases} \tag{7}$$

Given a collection of m unlabeled viewsets $U = \{X_u \in V \mid u = 1, \ldots, m\}$, a penalty term is added to the cost function of Eq. 4 to bias the learning process by the described constraints, leading to the following new cost

$$\sum_{i=1}^{k} \sum_{\boldsymbol{x}_i^r \in L} (y_r - f_{j,i}(\boldsymbol{x}_i^r))^2 + \lambda_j \sum_{i=1}^{k} \|f_{j,i}\|_{\mathcal{H}}^2 + \mu \sum_{i=1}^{k-1} \sum_{\boldsymbol{x}_i^u \in U} (f_{j,i}(\boldsymbol{x}_i^u) - f_{j,i+1}(\boldsymbol{x}_{i+1}^u))^2. \quad (8)$$

The parameter μ is the weight associated to the penalty term and it determines how strictly the system is forced to fulfill the given constraints. The accurate selection of the value of μ is crucial for the system performances. In fact, high values of μ could result in a worse fitting of the labeled data, and the overall accuracy could degenerate, moving the system towards a trivial solution where all the functions assume values close to zero.

We solved the minimization problem of Eq. 8 by gradient descent. Since labeled data already fulfill the constraints, training the unconstrained classifiers by solving the linear system of Eq. 6 will lead to a solution that is probably close to the constrained one. Exploiting this consideration, the solution of Eq. 6 is a promising starting point for the gradient descent, in order to reduce the number of iteration required to achieve convergence.

4 Experimental Results

The COIL-100 database [16] is one the most used benchmarks for object recognition algorithms. It consists of a collection of multiple views of 100 objects. Each object was placed on a turntable and every 5° an image was acquired, generating a total of 72 views for object. The database is composed by the collection of 7200 color images at the resolution of 128x128 pixels (Fig. 1).

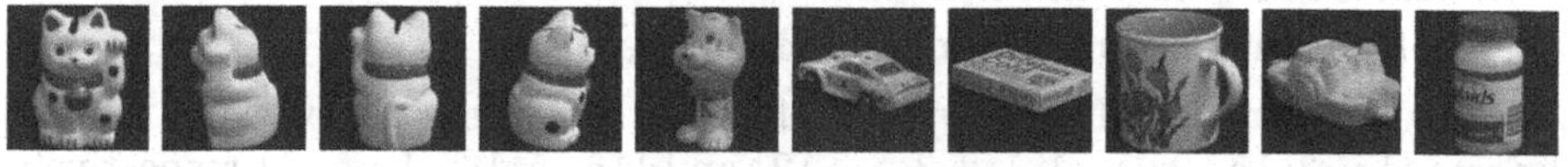

Fig. 1. Sample images from the COIL-100 database

In the last decade, a large number of experiments have been performed on this collection [7–12]. As in many previous approaches [8–10] we rescaled each image to 32x32 gray scale pixels in the interval $[0, 1]$, since it has been shown that the information coming from color is highly discriminative among objects and it makes the learning task quite trivial [9, 11].

In a multi–view scenario we consider four cameras c_i, $i = 1, \ldots, 4$, equally spaced around the object, that simultaneously acquire four images at $90° \cdot (i-1)$ considering the reference angles provided in the COIL-100 database. Each viewset $X = \{\boldsymbol{x}_1, \boldsymbol{x}_2, \boldsymbol{x}_3, \boldsymbol{x}_4\}$ is identified by the degree of rotation of the image acquired by the first camera, c_1, that falls in the range $[-45°, 45°]$.

Differently from the experiments available in the literature, we decided to make the recognition task more challenging by considering only a relatively small amount of views of a sub selection of objects to train the recognizer. We defined a set K of *known objects*, composed by the first 50 ones, and a set U of the remaining 50 *unknown objects*. For each element in K we selected only 3 viewsets (12 images) to train the system, each separated from the previous one by 30°, starting at −30°. Similarly other 3 viewsets where selected to cross–validate the system parameters, alternatively starting at −15° or −45° for each object[1]. The other viewsets were used to test the recognition accuracy in two different scenarios, *test K* and *test KU*. In the former, only the remaining 12 viewsets (48 images) of the known objects K are considered, whereas in the latter, also the 18 ones (72 images) that are available for each unknown object in U are added. In other words we do not only require the ability to recognize and discriminate known objects but also to correctly reject the unknown ones. Table 1 summaries the details of the described experimental framework.

Table 1. The selected experimental setup. The left portion of the table details the list of objects and total number of images in each set, whereas the right one collects information on viewsets for "each" object of the list ($j = 0, \dots,$Viewsets−1).

Set	Objects	Images
Training	$1, \dots, 50$	600
Validation	$2, \dots, 50$ (even only)	300
	$1, \dots, 49$ (odd only)	300
Test K	$1, \dots, 50$	2400
Test KU	$1, \dots, 50$	2400
	$51, \dots, 100$	3600

Set	Viewsets	Positions
Training	3	$-30^\circ + (30 \cdot j)^\circ$
Validation	3	$-15^\circ + (30 \cdot j)^\circ$
	3	$-45^\circ + (30 \cdot j)^\circ$
Test K	12	The remaining ones
Test KU	12	The remaining ones
	18	All

We trained 50 binary classifiers in a one–against–all strategy and we selected as kernel a Gaussian function of the form $K_j(x, y) = \exp \frac{-\|x-y\|}{2 \cdot \sigma_j^2}$. For every classifier the optimal values of σ_j and of λ_j are determined by varying them in the sets $\{1e{-}3, 1e{-}2, 1e{-}1, 1, 2, 3, \dots, 12\}$ and $\{1e{-}5, 1e{-}4, \dots, 1\}$ respectively, in order to maximize the sum of accuracies on training and validation data. The optimal rejection threshold τ_j^* is determined with the same criterion.

We approached the problem using three different methods, in order to show how the new constraints can improve the performances. First, the baseline approach of Eq. 1, where we discarded the information about the four cameras and their positions, modeling each classifier with a single function. In the second approach the output of every classifier is composed by the contribution of 4 functions, one for each image of the viewset, as described in Eq. 2. Finally, we constrained the 4 functions to be coherent in a semi–supervised framework, by minimizing the cost function of Eq. 8.

[1] The views located at $45^\circ \cdot (i - 1)$, with $i = 1, \dots, 4$, were alternatively considered as acquired by camera c_i or by the following one.

We smoothly increased the value of the penalty weight μ, ranging in $[1e{-}2, 25]$. Constraints were forced on validation data, then the thresholds τ_j^* and, in particular, the optimal value of μ were determined. We selected the value of μ that yields the best performances on both training and validation data first, and, secondly, the value that causes a better accuracy in approximating the given constraints. In Table 2 the resulting macro accuracies of the three described approaches are reported. They are referred as *single* (classifiers with a single function), *multi* (classifiers with four functions), and *constrained* (classifiers with four functions and constraints) respectively. In Fig. 2(a) the accuracy of the complete constraint based learner with respect to the value of μ is shown, and the selected optimal value μ^* is indicated with a vertical line. Similarly, in Fig. 2(b) the average penalty value on the 50 classifiers is reported. The violation of the constraints on the validation data decreases as the value of μ grows but the opposite behavior can be observed on training data, since the contribution of the approximation error becomes less important that the constraint penalty. The optimal value μ^* can be selected in correspondence of a roughly equivalent violation of constraints on the two data sets, as a trade–off between an appropriate labeled data fitting and a good fulfillment of the given constraints.

Table 2. Recognition (macro) accuracies of the three proposed approaches (in percentage). The better results on test data are reported in bold.

Technique	Training Data	Validation Data	Test K Data	Test KU Data
Single	100	100	99.67	90.07
Multi	100	100	99.67	92.53
Constrained	100	100	**99.83**	**94.67**

The recognition accuracy of the multiple function approach is equivalent to the single one for known objects, but when unknown objects are introduced the multiple function technique is more robust. This is mainly due to the specific training of each function on a specific view that allows them to achieve a more tight fitting around the positive training instances. The introduction of constraints offers another significant increment of accuracy on such data and a slight increment on the discrimination capability of the system. It can be clearly seen that increasing the weight of the constraints increases the accuracy on the test data. Moreover, beyond a certain value, the contribution of the squared loss on labeled data becomes less significant in the cost function, and performances decrease or become really unstable.

We tested the performances of the constraint based learner also in other different tasks: robustness with respect to object orientation, to noise and to missing cumulative information.

Assuming that an input object is given to the system but its actual orientation is unknown, we checked if the model is still able to correctly recognize it. As a consequence, if the object is rotated by 90° four times and four viewsets are acquired, one of such sets must be oriented consistently with the training data.

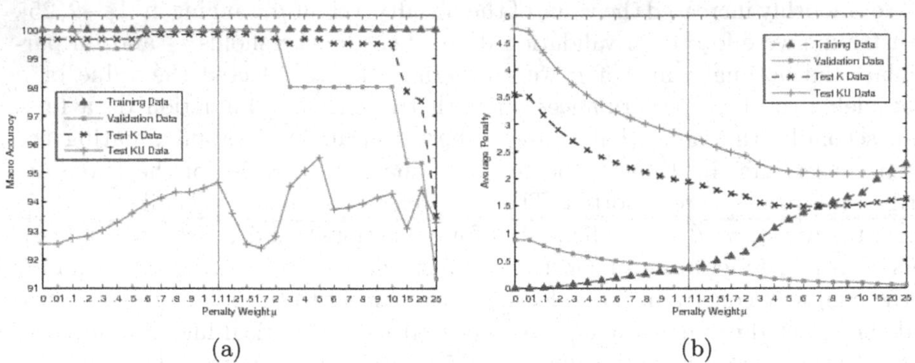

Fig. 2. Recognition (macro) accuracy (a) and average penalty value (b) on training, validation and test data in function of the penalty weight μ. The vertical line represent the selected value of μ accordingly to the described validation criterion.

If the object is highly asymmetric and differs among the four views, then the system should have more confidence only on the viewset aligned with respect to the training data. Following this idea we generated the required four viewsets for each data set in Table 1 and we fed them to the system, selecting, for each classifier, the prediction with the highest confidence on the four "rotated" inputs. The recognition accuracies are reported in Table 3.

Table 3. Recognition (macro) accuracies of the three proposed approaches (in percentage) discarding information on the right viewset orientation. The better results on test data are reported in bold.

Technique	Training Data	Validation Data	Test K Data	Test KU Data
Single	100	100	99.67	90.07
Multi	100	99.33	99.67	91.87
Constrained	100	99.33	**99.83**	**93.53**

The results for the single function case are obviously the same of Table 2, since we are not differently modeling the four views. The other techniques achieve the same results on test objects with or without the information on viewset position but when unknown objects are introduced, performances are slightly reduced. This indicates that a small portion of unknown objects, under some viewset orientations are wrongly recognized as known ones. The constraint based learner keeps showing better accuracy than the other approaches on test data and, in particular, it is still the most accurate recognizer when unknown objects are introduced.

Another test scenario involves the introduction of noise into the acquired images. In a real scenario this could be due to low quality or damaged cameras or to a noisy transmission channel from cameras to the recognizing software. We artificially introduced pseudo–random noisy values drawn from a normal

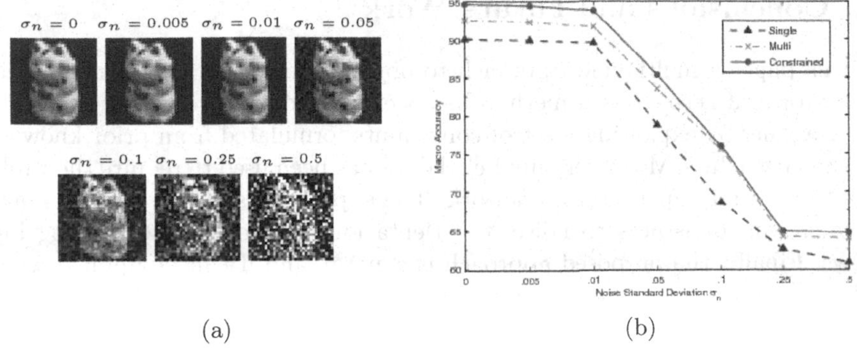

Fig. 3. (a) An object from COIL-100 with increasing noise ratios – (b) Recognition (macro) accuracy on test data *KU* in presence of noise

distribution, with zero mean and incremental values of the standard deviation σ_n, to each pixel of the images (Fig. 3(a)).

The recognition accuracies are reported in Fig. 3(b). As expected, while the noise standard deviation increases, the performances of the three techniques degrades gracefully. The constraint based classifier keep showing more robustness to noisy images.

Finally, we investigate how the recognition performances of the functions that model each view are changed after applying the constraints to the four function classifier. We "turned off" three of the four cameras and we tried to recognize the object by a single image. In Table 4 the resulting accuracies are reported.

Table 4. Recognition (macro) accuracies based on only one of the four functions that compose the multi function system, with $(+C)$ and without constraints. The better results on test data between each pair of functions are reported in bold.

Data	$f_{j,1}$	$f_{j,1}$ + C	$f_{j,2}$	$f_{j,2}$ + C	$f_{j,3}$	$f_{j,3}$ + C	$f_{j,4}$	$f_{j,4}$ + C
Training	100	100	100	100	100	100	100	100
Validation	85.33	91.33	74.67	75.33	95.33	95.33	62.67	71.33
Test K	94.5	**97.83**	85.5	**92.5**	98.83	**99**	83.5	**89.17**
Test KU	85.87	**87.07**	**87.07**	86.87	86.87	**90.2**	87.07	**90.2**

Interestingly, the role of the constraints appears determinant for the increments of accuracy of the single functions. The improvement of the functions that model each view from the constrained classifier with respect to the ones from the unconstrained system is evident. These results show that the interaction among functions due to the constraints can enhance the cumulative decision of the classifier but also the single power of each $f_{j,i}$. Moreover, the lower performances of the pair of functions $f_{j,2}$ and $f_{j,4}$ with respect to $f_{j,1}$ and $f_{j,3}$ indicates how the frontal and backward views, associated to the former pair, are more discriminative that the side views for the object set of COIL-100.

5 Conclusions and Future Work

In this paper a multi–view approach to object recognition has been presented. The proposed kernel based method has been proved to increase the accuracy of the classifier by exploiting a set of constraints formulated from prior knowledge on the viewpoints. Moreover, unlabeled data has been used to require their fulfillment in a semi–supervised framework. The experiments on the COIL database have shown robustness to noise, to orientation changes and to missing input views. Finally, the proposed approach is general, and it can be applied when a coherent decision on different representations of the same input is required.

References

1. Lowe, D.G.: Object recognition from local scale-invariant features. In: Proc. of the Int. Conf. on Computer Vision, vol. 2, p. 1150 (1999)
2. Belongie, S., Malik, J., Puzicha, J.: Shape matching and object recognition using shape contexts. IEEE Trans. PAMI 24(4), 509–522 (2002)
3. Mokhtarian, F., Abbasi, S.: Automatic selection of optimal views in multi-view object recognition. In: Proc. of the British Machine Vision Conf., pp. 272–281 (2000)
4. Torralba, A., Murphy, K.P.: Sharing visual features for multiclass and multiview object detection. IEEE Trans. PAMI 29(5), 854–869 (2007)
5. Rothganger, F., Lazebnik, S., Schmid, C., Ponce, J.: 3d object modeling and recognition using local affine-invariant image descriptors and multi-view spatial constraints. Int. J. Comput. Vision 66(3), 231–259 (2006)
6. Thomas, A., Ferrari, V., Leibe, B., Tuytelaars, T., Schiele, B., Van Gool, L.: Towards multi-view object class detection. In: Proc. of CVPR, pp. 1589–1596 (2006)
7. Christoudias, C., Urtasun, R., Darrell, T.: Unsupervised feature selection via distributed coding for multi-view object recognition. In: Proc. of CVPR, pp. 1–8 (2008)
8. Pontil, M., Verri, A.: Support vector machines for 3D object recognition. IEEE Transactions on Pattern Analysis and Machine Intelligence 20(6), 637–646 (1998)
9. Roobaert, D., Van Hulle, M.: View-based 3D object recognition with support vector machines. Neural Networks for Signal Processing, 77–84 (1999)
10. Wallraven, C., Caputo, B., Graf, A.: Recognition with local features: the kernel recipe. In: Proc. of Int. Conf. on Computer Vision, vol. 1, pp. 257–264 (2003)
11. Caputo, B., Dorko, G.: How to Combine Color and Shape Information for 3D Object Recognition: Kernels do the Trick. Advances in NIPS, 1399–1406 (2003)
12. Lyu, S.: Mercer Kernels for Object Recognition with Local Features. In: Proc. of Int. Conf. on CVPR, vol. 2, pp. 223–229 (2005)
13. Leibe, B., Schiele, B.: Scale-invariant object categorization using a scale-adaptive mean-shift search. In: Rasmussen, C.E., Bülthoff, H.H., Schölkopf, B., Giese, M.A. (eds.) DAGM 2004. LNCS, vol. 3175, pp. 145–153. Springer, Heidelberg (2004)
14. Shawe-Taylor, J., Cristianini, N.: Kernel Methods for Pattern Analysis. Cambridge University Press, New York (2004)
15. Rifkin, R., Klautau, A.: In defense of one-vs-all classification. Journal of Machine Learning Research 5, 101–141 (2004)
16. Nene, S., Nayar, S., Murase, H.: Columbia Object Image Library (COIL-100). Techn. Rep. No. CUCS-006-96, Dept. Comp. Science, Columbia University (1996)

Large-Scale Real-Time Object Identification Based on Analytic Features

Stephan Hasler, Heiko Wersing, Stephan Kirstein, and Edgar Körner

Honda Research Institute Europe GmbH
D-63073 Offenbach/Germany
stephan.hasler@honda-ri.de

Abstract. Inspired by biological findings, we present a system that is able to robustly identify a large number of pre-trained objects in real-time. In contrast to related work, we do not restrict the objects' pose to characteristic views but rotate them freely in hand in front of a cluttered background. We describe the essential system's ingredients, like prototype-based figure-ground segmentation, extraction of brain-like analytic features, and a simple classifier on top. Finally we analyze the performance of the system using databases of varying difficulty.

1 Introduction

The recognition of objects under real-world conditions is a difficult problem. Because of this, most approaches limit the complexity by using only few objects, restricting the pose to canonical views, or by providing controlled background conditions. In contrast to this, we freely rotated the objects in hand in front of a cluttered background. For this unconstrained setting, we describe a system that can robustly identify a large number of objects in real-time.

In general, the recognition task and the given setting define the generalization capabilities the system requires. These have to be achieved by the interplay of the system components, but most strongly by the chosen type of object representation. On the one hand, the representation must be specific, i.e contain enough details to distinguish the objects. On the other hand, it must be general to yield invariance to the expected variations.

A main distinction with regard to representations can be made between holistic and parts-based approaches. Both types differ in the way they handle spatial information. Holistic approaches look at the whole image and represent global patterns in fixed relation to the image frame. All features are bound to a certain image location. Such representations are very specific and break down if the constellation of features changes strongly as it is the case for occlusion and 3D rotation. A simple holistic method might use the images directly as templates, or learn simple global features [1]. A more advanced processing is described in [2]. Here a hierarchical processing related to the ventral visual pathway is used, where stages of local spatial pooling soften the rigid coding of patterns. We will use this approach to provide a baseline for our results.

C. Alippi et al. (Eds.): ICANN 2009, Part II, LNCS 5769, pp. 663–672, 2009.

In contrast to holistic processing, parts-based methods have in common that they detect the presence of features or parts independent of their position in the image. The relative position between the parts of an object can be handled differently. Some approaches store the constellation of parts on a reduced resolution [3] or by explicitly modeling there position by means of a Gaussian distribution. If multiple objects are in an image, this information is necessary to bind features to the corresponding object models. The handling of spatial information is less specific than for holistic coding but still leads to problems when the constellation undergoes strong changes as it is the case for 3D rotation. Additionally, these approaches often extract features at so-called keypoints only. Keypoints are determined by saliency detectors that favor parts whose position is not ambiguous (like vertices or highly textured regions, but not parallel lines or shadings). This is a limitation since meaningful information might be neglected.

Other parts-based approaches, like the one we use here, leave out spatial information by determining only the maximum response of an alphabet of features to an image [4,5]. The use of such an alphabet is motivated by biological findings. The experiments in [6] revealed that columns in inferotemporal cortex represent a large set of complex features that can be recognized invariant to position and other transformations. Combinations of activated columns then code for the presence of an object [7]. Keeping only the maximum activation per feature can be interpreted by means of neural latency coding where the highest activations provoke the fastest response and non-optimal local responses are delayed and usually do not contribute to further feed-forward processing.

By leaving out spatial information, these approaches implicitly assume that only a single object is in view so that no binding is necessary. To balance this more general type of representation the parts themselves have to be more specific and meaningful. This can be achieved in different ways. The work in [5] uses a similar hierarchical processing like the holistic framework in [2]. But on the highest feature layer a maximum step is performed using an alphabet with millions of local features that were randomly selected. Finally a support vector machine (SVM) is trained to separate the classes in this high-dimensional space. Here the final SVM learns which parts of the large set are meaningful. In contrast to this, we use a much smaller alphabet of so-called analytic features, which are optimized using the supervised selection method described in [4], which will be explained later. Because of this smaller subset our system runs in real-time.

Besides feature selection and handling of spatial information, also the coding of the parts is important. For our analytic approach we describe the parts by means of SIFT descriptors [3]. A SIFT descriptor is made up of a grid of local gradient histograms. Thus it shows similarity to the response properties of neurons in primary visual cortex. Gray-scale gradients are a very simple form of edge detectors found in the so-called simple cells, while building of local histograms is comparable to spatial pooling which is attributed to so-called complex cells. Simple cells of a higher visual area respond to activation patterns of these complex cells. Such patterns can be interpreted as a grid.

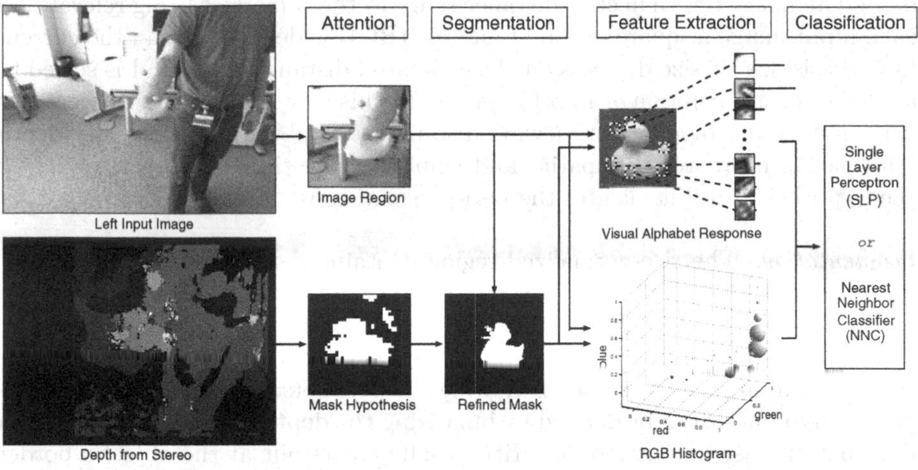

Fig. 1. Basic system architecture. Using a depth criterion a region of interest is cropped from the input image. After computing an improved mask, the response to an alphabet of parts is calculated together with a histogram in RGB color-space. The resulting activations are presented to the final classifier.

Previous experiments in [4] revealed that SIFT descriptors outperform the use of gray-scale patches, which are too specific, and also patches from the output of the hierarchy in [2], which are too general. Please note, that instead of using the whole framework usually associated with SIFT, we use the simple maximum step as outlined before. Additionally, we omit the use of keypoint detectors to avoid restrictions on the parts that can be learned.

With regard to the recognition task and the system architecture, the work of [8] is quite similar to ours. But they use a rather large alphabet of features which is trained in an unsupervised fashion and they represent spatial relations. This more complex and slower processing is not reflected in a gain in performance as we report a similar performance for an even higher number of objects.

We describe the building blocks of our system in Sect. 2 with a special focus on the learning and use of the analytic features. Later we investigate and discuss its performance in Sect. 3 and present our conclusions in Sect. 4.

2 System

In this section we describe the essential building blocks of the system, whose overall architecture is shown in Fig. 1.

Attention. When performing recognition tasks in unconstrained environments (presence of background clutter and variation in object position), the system has to decide which part of the input image should be processed. Here we use the concept of peri-personal space [9] to generate such a hypothesis. This concept defines an image region in close distance range to the camera as being relevant. In each input image a square region of interest (ROI) is defined around the current hypothesis, whose size depends on the estimated distance. This ROI is scaled to a fixed output resolution of 144x144 pixels. In this way, we normalize object size variation caused by different viewing distances. We obtain the necessary depth information from stereo disparity and employ a pan-tilt unit to actively track the hypothesis until it violates the peri-personal constraints.

Segmentation. The size-normalized region contains the object, but also a substantial amount of background clutter. Since we do not represent spatial information, features detected on the background would be wrongly associated to the object. As this would harden the task of the classifier, we need to segment the object from its background. Following the peri-personal concept, a first foreground hypothesis can be derived by binarizing the depth image. Since the depth information based on stereo disparity usually wears out at the object's border and cannot be estimated for non-textured regions, we apply the segmentation method proposed by [10]. As pre-processing, this method removes all skin-colored pixels from the foreground hypothesis, because otherwise the hand holding the object would have a systematic influence on the result. Second, based on color and position information a prototype-based model for foreground (i.e. everything activated in the initial hypothesis) is learned and a model for background correspondingly. Finally, these models are used to classify each pixel as being figure or ground, where the learned prototype-specific distance metrics leads to a good generalization performance at the border of an object. In the following, features are extracted only at locations marked as foreground.

Feature extraction. The feature extraction is the most important part of the system. In this work we extract features for texture and color. Texture is represented by means of analytic features as proposed in [4] which are a preselected alphabet of SIFT-descriptors. These descriptors are widely used for coding local texture with invariance to lighting and planar rotation [3]. For a given input image i the response of a feature $\mathbf{w}_m$ is determined by $r_{mi} = \max_n (\mathbf{w}_m \cdot \mathbf{p}_{in})$, where the $\mathbf{p}_{in}$ are SIFT-descriptors from all image locations n, and $\cdot$ denotes the dot product. Keeping only the maximum response of each analytic feature over the image, we measure the pure presence of a certain object part and do not represent their spatial constellation. This yields invariance to translation of parts together with a strong reduction of dimensionality. In contrast to this, the standard SIFT framework calculates descriptors at interesting keypoints and their constellation is then matched to those of the training images. In Sect. 3 we show that this has shortcomings for several reasons.

The alphabet of analytic features is optimized for the scenario at hand using the selection method proposed in [4]. Starting from a large set of candidate SIFT-descriptors, this method first evaluates how well each element m can separate views from a single class. This is done by assigning scores s_{mi} for each combination of feature and image as shown in Fig. 2. After that, out of the candidates a subset M is selected that can separate most of the views among the training

Feature $\mathbf{w}_m$	Image i									
	Response r_{mi}	0.43	0.45	0.48	0.49	0.54	0.56	0.60	0.85	0.90
	Score s_{mi}	0	0	0	0	0	0	0	1	1

Threshold t_m

Fig. 2. Score table of single feature. For visualization images are sorted on their response r_{mi}. The threshold t_m is chosen to separate a maximal number of views of one class (smiley cup) from all other images. To these views a score $s_{mi} = 1$ is assigned.

images. This subset has a predefined cardinality (usually several hundreds) and maximizes $\sum_i f\left(\sum_{m \in M} s_{mi}\right)$ with $f(z) = \frac{1}{1+e^{-kz}}$ and $k = 3$. $f(z)$ saturates quickly, thus forcing the selected features to distribute their scores s_{mi} over all images. Because trying all possible subsets M is intractable, a greedy iterative selection is used instead. The described method is dynamic in the way that it selects more features for objects with strong variation in appearance.

To represent color we calculate a histogram in RGB color-space ($6x6x6 = 216$ bins) and normalize it by dividing by the highest entry. Histograms combine robustness against view and scale changes with computational efficiency [11]. Before the calculation, we apply the color constancy method proposed by [12].

The activation of the RGB histogram bins are combined with the responses of the analytic features to form the final feature vector.

Classification. To associate an object label to the current input image we use simple classifiers as a nearest neighbor classifier (NNC) or a single layer perceptron (SLP). The NNC stores the feature vectors of the training images as representatives and determines the object label for a test image based on closest Euclidean distance. The SLP has a neuron for each object. Using the training data, the weights of one neuron are adapted to produce a strong response for the corresponding object and a low response for views of other objects. The object label of a test image is determined by the highest activated neuron. For the real-time system we use the SLP because it consumes drastically less memory and CPU time, and also has a slightly higher performance for the combined use of analytic and color features.

As outlined before, the usual platform is a stereo camera head mounted on a pan-tilt unit. When using our humanoid robot ASIMO instead, its degrees of freedom are used to track but also follow the current peri-personal hypothesis [13]. The proposed system runs in real-time with a frame-rate of 6Hz. The limiting factor is the calculation of the analytic feature response for each possible location in the size-normalized region of interest.

3 Results

In this section we first present results for an object database which has been acquired to train and optimize the final real-time recognition system. Using a

Fig. 3. HRI126 database. Database contains 126 objects with 1200 views each. Objects were rotated in hand in front of a cluttered background.

simpler database, we later distinguish the analytic feature approach from the standard SIFT framework.

The HRI126 database which corresponds to the scenario for the real-time system is shown in Fig. 3. It contains 126 objects with 1200 views each. The objects were freely rotated in hand in front of a cluttered background. Because of this unconstrained setting the database is very difficult compared to ones used in related work. In the following we evaluate the recognition performance and scalability of the proposed approach and test the necessity of the segmentation step. All training was done on the first 1000 views per object while the offline performance was evaluated on the remaining 200 views.

In the first training step we selected 441 analytic features (see Fig. 4a) using the algorithm proposed in [4]. Combined with the 216 RGB histogram bins, this yields a 657 dimensional feature vector for each view. Although in the final system an SLP is used, Fig. 4b gives the result of some NNC experiments using different representations and varying the number of representatives. Here especially the result of two holistic approaches GRAY and C2 is important to judge the difficulty of the database. GRAY simply uses the holistic gray-scale images as representatives while the so-called C2-activation is the output of the biologically inspired, edge-based, feed-forward hierarchy proposed in [2]. Both holistic methods show a very weak performance. With many training views C2 outperforms GRAY but still does not generalize as well as the analytic approach using few training views. One reason for this is that the coding of spatial information is too rigid. Additionally, the local features underlying C2 are too coarse to separate certain objects in the database, e.g. individual mobile phones. In contrast to this ANALYTIC uses very specific features while neglecting spatial information completely. ANALYTIC also outperforms the color histograms while the concatenation of both complementary feature types yields the best result.

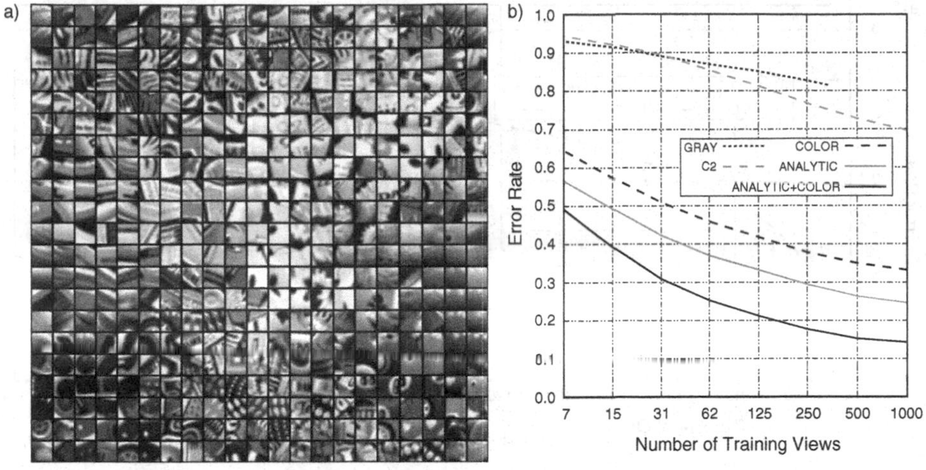

Fig. 4. a) Selected 441 analytic features. Most of the features are quite object specific. E.g., different versions of keypads and wheels are necessary to distinguish individual mobile phones and cars respectively. The set contains features that would not have passed usual keypoint criteria (e.g. several versions of parallel lines). For visualization the features are arranged using a self organizing map. b) Results of NNC experiments. Error rate over number of training views. For GRAY not all views could be used because of the required memory.

Similar conclusions can be drawn from the SLP experiments shown in Fig. 5a. Here the error rates of recognition are given depending on the number of used objects. This should help to predict the scalability of the approach towards larger number of objects. The value for 126 objects is directly the performance of the SLP on the test images. The performance for less objects was determined by choosing a random subset of objects and removing their test-views and SLP neurons from the experiment. The SLP was not retrained on the remaining objects. Interestingly, this yields better results because the SLP profits from a high number of negative training examples. The curves show the average of 100 runs.

In general, the order from the NNC experiment is preserved. Only for ANALYTIC+COLOR the SLP is better than the NNC, because it finds a better weighting between both feature types than the simple concatenation used for the NNC experiment. The selective use of ANALYTIC and especially COLOR prevents the SLP from finding a good separation because of the low input dimensionality. In contrast to this we observed some over-fitting for C2.

For a given error rate much more objects can be distinguished by means of analytic features than by C2. The combination ANALYTIC+COLOR again provides the best result with an error rate of only 10.35% for 126 objects. Taken the difficulty of the database into account this is a very high performance and a big step towards invariant 3D object recognition. In the real-time system we accumulate the classification results over 10 successive frames and only output

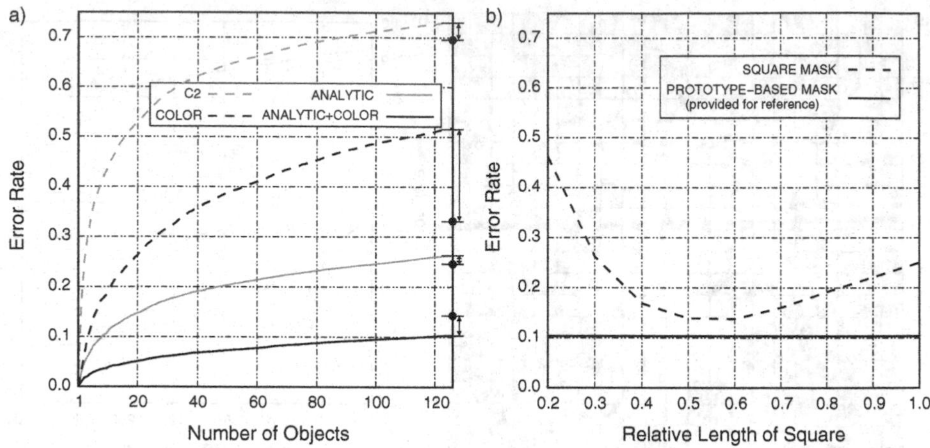

Fig. 5. a) Error rate of SLP over the number of objects. The symbol • denotes the performance of the corresponding NNC experiments. b) Error rates for differently sized square masks. For 1.0 the whole region is used for feature extraction (no masking) while for decreasing values only a smaller square inner part contributes.

the most voted object label. This removes outliers and thus leads to further improvement and stability.

After having investigated the contribution of the feature extraction, Fig. 5b sheds light on the importance of the segmentation step. The horizontal line is the reference performance for 126 objects when using the prototype-based segmentation proposed in [10], while the other curve gives the results when simply placing differently sized, square masks in the center of the region. Even in the best case such a simple mask has a 4% higher error rate. Using no mask gives a 15% higher error rate. For too small masks there is an even higher loss in performance. These results show that a good segmentation helps our position invariant object representation to counteract the binding problem.

In Sect. 2 we shortly compared the basics of the analytic feature approach with the standard SIFT framework. The effect of the differences become clear in the results in Fig. 6. For this experiment we used the simple COIL100 database [14]. Because of the non-cluttered background we did not use a mask and we also abandoned the color features to get a fair comparison. Fig. 6b shows the result of different nearest neighbor classifications where we varied the number of stored representatives (out of 72 available ones per object) for different approaches. For the analytic approach we used the same set of 441 features that was selected for the HRI126 database. This was done because of the low number of available training views in the COIL100 database and the strong similarity in the types of objects in both databases. For the SIFT framework we applied the visual pattern recognition system (ViPR) by Evolution Robotics (see [15], www.evolution.com) which is claimed to be the "gold standard" implementation of the SIFT approach. We optimized the parameters of this software for

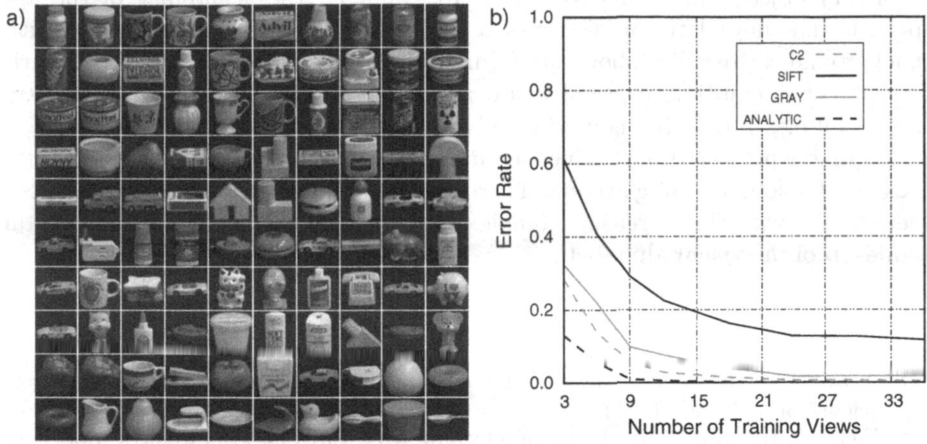

Fig. 6. a) COIL100 database. b) Error rates of NNC over number of training views. For 9 representatives the analytic approach has an error rate of 1% while the standard SIFT approach has 28%. The objects in a) are sorted in ascending order on their individual probability of being misclassified using the SIFT approach with 9 representatives.

best performance. As a baseline we again provide results for the use of holistic gray-scale images and for C2.

For few training views, the analytic approach generalizes well while the SIFT framework shows a very weak performance. A reason for this is the different handling of spatial information. SIFT tries to re-detect the rigid constellation of parts that was present in the training images, which usually changes strongly under rotation in depth. Additionally, there are several objects for which the SIFT framework completely fails, as shown by the bad convergence towards larger numbers of training views. A reason for this is the dependency on the (re-)detection of interesting keypoints. This breaks down for objects with little texture, which is underlined by the order of objects in Fig. 6a. Both holistic approaches show good convergence and an intermediate capability to generalize from few training views, as they also use a rigid spatial representation. This rigidness is a little softened by the hierarchical processing underlying C2.

To also compare our approach to that in [5], we trained a set of 300 analytic features for an animal vs. non-animal separation task. On the test data we reached an error rate of 20% compared to 18% reported in [5]. This small difference makes it questionable if the hierarchical processing and the use of millions of local features do provide a gain over our simpler and much faster method.

4 Conclusion

On the basis of a biologically motivated, parts-based representation, we developed a real-time system capable of robustly recognizing a large number of arbitrary objects under 3D rotation. We evaluated the scalability of the approach and showed the necessity of a good object segmentation to deal with background clutter. The shown performance marks a major step towards invariant object recognition, especially in comparison to existing work where mostly more complex processing is used to solve easier tasks.

Using the presented pre-trained architecture as a starting point, we target at a flexible, life-long learning system. Therefore we investigate in hierarchical classifiers to deal with the increasing complexity of the scenario and in an incremental build-up of the visual alphabet.

References

1. Turk, M., Pentland, A.: Eigenfaces for Recognition. Journal of Cognitive Neuroscience 3(1), 71–86 (1991)
2. Wersing, H., Körner, E.: Learning Optimized Features for Hierarchical Models of Invariant Object Recognition. Neural Computation 15(7), 1559–1588 (2003)
3. Lowe, D.G.: Distinctive Image Features from Scale-Invariant Keypoints. International Journal of Computer Vision 60(2), 91–110 (2004)
4. Hasler, S., Wersing, H., Körner, E.: A comparison of features in parts-based object recognition hierarchies. In: Artificial Neural Networks – ICANN, pp. 210–219 (2007)
5. Serre, T., Oliva, A., Poggio, T.: A feedforward architecture accounts for rapid categorization. Proc. of the National Academy of Science, 6424–6429 (2007)
6. Tanaka, K.: Inferotemporal Cortex And Object Vision. Annual Review of Neuroscience 19, 109–139 (1996)
7. Tsunoda, K., Yamane, Y., Nishizaki, M., Tanifuji, M.: Complex objects are represented in inferotemporal cortex by the combination of feature columns. Nature Neuroscience 4(8), 832–838 (2001)
8. Kim, H., Chutorian, E.M., Triesch, J.: Semi-autonomous learning of objects. In: IEEE CVPR Workshop: Vision for Human-Computer Interaction, vol.145 (2006)
9. Goerick, C., Mikhailova, I., Wersing, H., Kirstein, S.: Biologically motivated visual behaviours for humanoids: Learning to interact and learning in interaction. In: Proc. IEEE/RSJ Int. Conf. on Humanoid Robots, Tsukuba, Japan (2006)
10. Denecke, A., Wersing, H., Steil, J.J., Körner, E.: Online figure-ground segmentation with adaptive metrics in generalized LVQ. Neurocomputing 72(7-9), 1470–1482 (2009)
11. Swain, M.J., Ballard, D.H.: Color indexing. International Journal of Computer Vision 7(1), 11–32 (1991)
12. Pomierski, T., Gross, H.M.: Biological neural architecture for chromatic adaptation resulting in constant color sensations. In: IEEE International Conference on Neural Networks, pp. 734–739 (1996)
13. Bolder, B., Dunn, M., Gienger, M., Janssen, H., Sugiura, H., Goerick, C.: Visually guided whole body interaction. In: IEEE International Conference on Robotics and Automation (2007)
14. Nayar, S.K., Nene, S.A., Murase, H.: Real-time 100 object recognition system. In: Proc. IEEE Conference on Robotics and Automation, vol. 3, pp. 2321–2325 (1996)
15. Munich, M.E., Pirjanian, P., Bernardo, E.D., Goncalves, L., Karlsson, N., Lowe, D.G.: SIFT-ing through features with ViPR: Application of visual pattern recognition to robotics and automation. IEEE Robotics and Autom. Mag., 72–77 (2006)

Estimation Method of Motion Fields from Images by Model Inclusive Learning of Neural Networks

Yasuaki Kuroe[1] and Hajimu Kawakami[2]

[1] Department of Information Science, Kyoto Institute of Technology
Matsugasaki, Sakyo-ku, Kyoto 606-8585, Japan
kuroe@kit.ac.jp

[2] Department of Electronics and Informatics, Ryukoku University
1-5, Yokotani, Ohe-cho, Seta, Ohtsu 520-2194, Japan
kawakami@rins.ryukoku.ac.jp

Abstract. The problem of estimating motion fields from image sequences is essential for robot vision and so on. This paper discusses a method for estimating an entire continuous motion-vector field from a given set of image-sequence data. One promising method to realize accurate and efficient estimations is to fuse different estimation methods. We propose a neural network-based method to estimate motion-vector fields. The proposed method fuses two conventional methods, the correlation method and the differential method by model inclusive learning, which enables approximation results to possess inherent property of vector fields. It is shown through experiments that the proposed method makes it possible to estimate motion fields more accurately.

Keywords: estimation of motion field, neural network, model inclusive learning, fusion, optical flow, correlation method, differential method.

1 Introduction

Estimating motion fields from video images is an essential problem for robot vision and so on. In particular, it is important to develop a method for estimating an entire continuous motion-vector field from a given set of image-sequence data. This problem is a nonlinear function approximation problem.

In recent years, there have been increasing research interests in artificial neural networks and many efforts have been made on applications of neural networks to various fields. The most significant feature of artificial neural networks is the extreme flexibility due to the learning ability and their capability of nonlinear function approximation.

In this paper we propose a neural-network based method for estimating an entire continuous motion-vector field from a given set of image-sequence data. The features of the proposed methods are the model inclusive learning of neural networks and the fusion of two different methods of estimating motion fields.

C. Alippi et al. (Eds.): ICANN 2009, Part II, LNCS 5769, pp. 673–683, 2009.

The model inclusive learning could make approximation results possess inherent property of vector fields and reasonable approximation accuracy. The inherent property of vector fields, "any vector field is composed of the sum of two vector fields: an irrotational vector field and a solenoidal vector field" is embedded into the approximation results.

For the problem of estimating motion fields from video images, several methods have already been proposed. Typical representatives of them are the correlation method and the differential method [1]. In this paper we propose a method to fuse these conventional methods by the model inclusive learning of neural networks. The fusion of the two methods could compensate their defects each other and bring better approximation results. In order to check the performance of the proposed method experiments have been done. It is shown that the proposed method makes it possible to estimate vector fields more accurately.

2 Model Inclusive Learning for Estimating Vector Fields

We consider a two dimensional motion-vector field and discuss a method for approximating its entire field from a set of sample data. Let

$$\boldsymbol{F}(\boldsymbol{x}) := \begin{bmatrix} F_x(\boldsymbol{x}) \\ F_y(\boldsymbol{x}) \end{bmatrix}$$

be a motion-vector field considered in this paper. Here $\boldsymbol{x} = [x, y]^T$ denotes a position in the two dimensional plane $\boldsymbol{R}^2$. The problem is to estimate the vector field $\boldsymbol{F}(\boldsymbol{x}) \in \boldsymbol{R}^2$ from a given set of sample data $\{\boldsymbol{F}(\boldsymbol{x}_p)\}$ where $\boldsymbol{x}_p$ denotes a sample point.

A usual method to solve this problem by neural networks is as follows. Considering that $\boldsymbol{F}(\boldsymbol{x})$ is a nonlinear mapping with two inputs and two outputs, we prepare a two-input and two-output neural network. The network is trained such that, for a given set of example data $\{\boldsymbol{F}(\boldsymbol{x}_p)\}$, $\boldsymbol{x}_p$ is fed to the neural network as its input and its output comes close to the data $\boldsymbol{F}(\boldsymbol{x}_p)$. Figure 1 shows a block diagram of this learning method. The backpropagation is usually utilized in order to adjust the weights of the neural network so as to minimize the square error:

$$\sum_p \{(O_1 - F_x(\boldsymbol{x_p}))^2 + (O_2 - F_y(\boldsymbol{x_p}))^2\}$$

where O_1 and O_2 are the outputs of the neural network.

Note that $\boldsymbol{F}(\boldsymbol{x})$ is not merely a nonlinear function but a vector field. If a priori knowledge on inherent properties of vector fields can be embedded into the learning problem, the approximation result will possess the inherent properties, and approximation accuracy and learning efficiency will be improved. Mussa-Ivaldi proposed a method for approximating a vector field [2]. They define basis fields by using derivatives of the Green functions and reconstruct a vector field by superposition of the basis fields.

We have already proposed a learning method of neural networks for reconstructing vector fields which we call model inclusive learning [3]. In the method

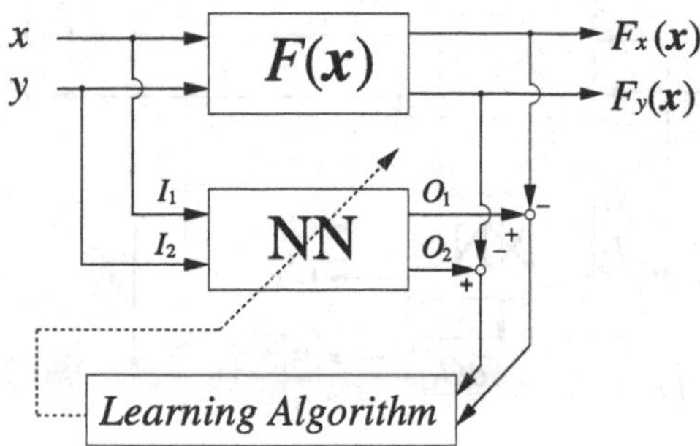

Fig. 1. Conventional learning method of vector fields

we formulate the learning problem in such a way that the knowledge on the inherent property of vector fields is included in the learning loop of neural networks. It is known that any vector field $\boldsymbol{F}(\boldsymbol{x}) \in \boldsymbol{R}^2$ is composed of the sum of two vector fields as follows.

$$\boldsymbol{F}(\boldsymbol{x}) = \boldsymbol{C}(\boldsymbol{x}) + \boldsymbol{S}(\boldsymbol{x}), \quad \boldsymbol{x} = [x, y]^T \tag{1}$$

where $\boldsymbol{C}(\boldsymbol{x})$ and $\boldsymbol{S}(\boldsymbol{x})$ are an irrotational vector field and a solenoidal vector field, respectively, and satisfy the following relations.

$$\mathrm{curl}(\boldsymbol{C}) = \nabla \times \boldsymbol{C}(\boldsymbol{x}) = 0, \qquad \mathrm{div}(\boldsymbol{S}) = \nabla \cdot \boldsymbol{S}(\boldsymbol{x}) = 0 \tag{2}$$

Here we introduce scalar functions $U_1(\boldsymbol{x})$ and $U_2(\boldsymbol{x})$, and express the vector fields $\boldsymbol{C}(\boldsymbol{x})$ and $\boldsymbol{S}(\boldsymbol{x})$ as follows.

$$\boldsymbol{C}(\boldsymbol{x}) = \alpha \nabla U_1(\boldsymbol{x}), \;\; \boldsymbol{S}(\boldsymbol{x}) = \begin{pmatrix} 0 & \beta \\ -\beta & 0 \end{pmatrix} \nabla U_2(\boldsymbol{x}) \tag{3}$$

where α and β are scalars. Since the above equations (3) satisfy eqs.(2), the vector field $\boldsymbol{F}(\boldsymbol{x})$ can be expressed by using the scalar functions $U_1(\boldsymbol{x})$ and $U_1(\boldsymbol{x})$ as follows.

$$\boldsymbol{F}(\boldsymbol{x}) = \boldsymbol{C}(\boldsymbol{x}) + \boldsymbol{S}(\boldsymbol{x}) = \begin{bmatrix} \alpha \dfrac{\partial U_1(\boldsymbol{x})}{\partial x} + \beta \dfrac{\partial U_2(\boldsymbol{x})}{\partial y} \\ \alpha \dfrac{\partial U_1(\boldsymbol{x})}{\partial y} - \beta \dfrac{\partial U_2(\boldsymbol{x})}{\partial x} \end{bmatrix} =: \begin{bmatrix} F_x(\boldsymbol{x}) \\ F_y(\boldsymbol{x}) \end{bmatrix} \tag{4}$$

Considering the relation (4), we now give a new learning formulation for reconstructing vector fields by neural networks. We formulate the learning problem in such a way that a neural network reconstructs the vector field $\boldsymbol{F}(\boldsymbol{x})$ with the relation (4) being satisfied. For this purpose we train the neural network such that the vector field $\boldsymbol{F}(\boldsymbol{x})$ itself is not realized on the neural network as is in the conventional method (Fig. 1), but the scalar functions $U_1(\boldsymbol{x})$ and $U_2(\boldsymbol{x})$ satisfying the relation (4) are realized on the network. Figure 2 shows a block diagram of the proposed model-inclusive learning method.

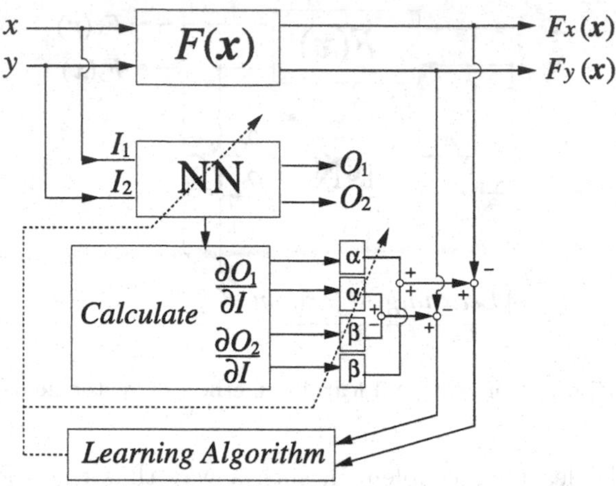

Fig. 2. Model inclusive learning method of vector fields

We prepare a neural network with two inputs denoted by $\boldsymbol{I} = [I_1, I_2]^T$ and two outputs denoted by $\boldsymbol{O} = [O_1, O_2]^T$. For a set of sample data $\{\boldsymbol{F}(\boldsymbol{x}_p)\} = \{[F_x(\boldsymbol{x}_p), F_y(\boldsymbol{x}_p)]^T\}$, the network is given the position data $\boldsymbol{x}_p = [x_p, y_p]^T$ as its input data ($\boldsymbol{I} = \boldsymbol{x}_p$) and is trained such that the outputs O_1 and O_2 come close to $U_1(\boldsymbol{x}_p)$ and $U_2(\boldsymbol{x}_p)$ satisfying the relation (4), respectively. This can be realized in the following manner. We input the position data $\boldsymbol{x}_p = [x_p, y_p]^T$ to the neural network and calculate the derivatives of the output of the neural network with respect to the input,

$$\left.\frac{\partial O_k}{\partial \boldsymbol{I}}\right|_{\boldsymbol{I}=\boldsymbol{x}_p} = \left[\left.\frac{\partial O_k}{\partial I_1}\right|_{\boldsymbol{I}=\boldsymbol{x}_p}, \left.\frac{\partial O_k}{\partial I_2}\right|_{\boldsymbol{I}=\boldsymbol{x}_p}\right]^T, \quad (k = 1, 2).$$

According to eq.(4), their linear combination

$$\left(\alpha \left.\frac{\partial O_1}{\partial I_1}\right|_{\boldsymbol{I}=\boldsymbol{x}_p} + \beta \left.\frac{\partial O_2}{\partial I_2}\right|_{\boldsymbol{I}=\boldsymbol{x}_p}, \alpha \left.\frac{\partial O_1}{\partial I_2}\right|_{\boldsymbol{I}=\boldsymbol{x}_p} - \beta \left.\frac{\partial O_2}{\partial I_1}\right|_{\boldsymbol{I}=\boldsymbol{x}_p}\right)^T \tag{5}$$

is calculated and the result is compared with the given vector data $F(\boldsymbol{x}_p) = [F_x(\boldsymbol{x}_p), F_y(\boldsymbol{x}_p)]^T$. The neural network is trained so as to minimize the error between them. Define the performance index by

$$J = \frac{1}{2}\sum_p \left\{\left\{\left(\alpha \left.\frac{\partial O_1}{\partial I_1}\right|_{\boldsymbol{I}=\boldsymbol{x}_p} + \beta \left.\frac{\partial O_2}{\partial I_2}\right|_{\boldsymbol{I}=\boldsymbol{x}_p}\right) - F_x(\boldsymbol{x}_p)\right\}^2 + \left\{\left(\alpha \left.\frac{\partial O_1}{\partial I_2}\right|_{\boldsymbol{I}=\boldsymbol{x}_p} - \beta \left.\frac{\partial O_2}{\partial I_1}\right|_{\boldsymbol{I}=\boldsymbol{x}_p}\right) - F_y(\boldsymbol{x}_p)\right\}^2\right\}. \tag{6}$$

where $\sum_p$ stands for the summation with respect to the set of data. By minimizing the performance index thus introduced, the scalar functions $U_1(\boldsymbol{x})$ and $U_2(\boldsymbol{x})$ satisfying eq.(4) are realized as input-output relation of the neural network. As the result an approximation of the vector field $\boldsymbol{F}(\boldsymbol{x})$ is obtained. Thus the problem is reduced to finding the parameters of the neural network which minimize J. Note that, since the parameters α and β in eq.(4) are not known a prior, we choose not only weights of connections w_{ij} but also α and β as the learning parameters. In order to search the values of the learning parameters which minimize J, gradient based methods can be used, in which several useful algorithms are available: the steepest decent algorithm, the conjugate gradient algorithm, the quasi-Newton algorithm and so on. In gradient based algorithms, it becomes a key how to compute the gradients of performance index, $\partial J/\partial w_{ij}$, $\partial J/\partial \alpha$, and $\partial J/\partial \beta$, efficiently and accurately. Efficient algorithms for computing them can be derived in a systematic manner by introducing the adjoint model of neural networks [3,4].

3 Estimation Method of Motion-Vector Fields

In this section we propose an estimation method of motion-vector fields from video images. Typical methods proposed so far are the correlation method and the differential method [1], which possess their own good and week points. It is desired to develop an estimation method which makes the most use of their good points and also compensates their weaknesses. We propose a new estimation method which fuses these two methods by the model inclusive learning of neural networks.

3.1 Estimation of Motion Fields Based on Correlation Method

In the correlation method, firstly, motion vectors at some feature positions in scenes of video images are calculated by using the corresponding position data obtained from the subsequent image frames. Secondly, the entire vector field over the whole image is estimated by using the calculated motion vectors as sample data. Suppose that a feature of a scene of video images moves from a position (x_p, y_p) to a position (x'_p, y'_p), the motion vector (u, v) at the feature position is calculated as the displacement vector:

$$(u, v) = (x'_p - x_p, y'_p - y_p). \tag{7}$$

Let $\boldsymbol{x}_{p_c} = [x_{p_c}, y_{p_c}]^T$ be a feature position and (u_{p_c}, v_{p_c}) be the corresponding motion vector thus obtained. The correlation method estimates the entire vector field over the image from a set of the obtained motion vector data $\{(u_{p_c}, v_{p_c})\}$ at a given set of feature positions $\{\boldsymbol{x}_{p_c}\}$. In order to realize the method by neural networks, we utilize the method of vector field approximation by the model inclusive learning discussed in the previous section. In the proposed correlation method, the performance index (6) is replaced by

$$J_c = \frac{1}{2}\sum_{p_c}\left\{\left\{\left(\alpha \left.\frac{\partial O_1}{\partial I_1}\right|_{\boldsymbol{I}=\boldsymbol{x}_{p_c}} + \beta \left.\frac{\partial O_2}{\partial I_2}\right|_{\boldsymbol{I}=\boldsymbol{x}_{p_c}}\right) - u_{p_c}\right\}^2 + \left\{\left(\alpha \left.\frac{\partial O_1}{\partial I_2}\right|_{\boldsymbol{I}=\boldsymbol{x}_{p_c}} - \beta \left.\frac{\partial O_2}{\partial I_1}\right|_{\boldsymbol{I}=\boldsymbol{x}_{p_c}}\right) - v_{p_c}\right\}^2\right\}. \tag{8}$$

The problem is reduced to finding the parameters of the neural network w_{ij}, and α and β which minimize J_c.

3.2 Estimation of Motion Fields Based on Differential Method

In this subsection we propose an estimation method of motion-vector fields based on the differential method. Let $B(x, y, t)$ be the brightness of a position (x, y) in an image at time t. Suppose that the brightness at each position does not vary in a short time, the following equation holds.

$$B_x u + B_y v + B_t = 0 \tag{9}$$

where $B_x = \partial B/\partial x$, $B_y = \partial B/\partial y$, $B_t = \partial B/\partial t$, $u = dx/dt$ and $v = dy/dt$. This equation is called the optical flow constraint. The method of estimating motion vectors (u, v) over the entire vector field so as to satisfy the condition (9) is called the differential method. We propose a new differential method by using the method of vector field approximation by the model inclusive learning of neural networks discussed in the previous section. Let $B_x(\boldsymbol{x}_{p_d})$, $B_y(\boldsymbol{x}_{p_d})$ and $B_t(\boldsymbol{x}_{p_d})$ be the derivatives of the brightness $B(x, y, t)$ at a point $\boldsymbol{x}_{p_d} = [x_{p_d}, y_{p_d}]^T$ and time t, respectively. Given a set of data $\{B_x(\boldsymbol{x}_{p_d}), B_y(\boldsymbol{x}_{p_d}), B_t(\boldsymbol{x}_{p_d})\}$, we train a neural network such that approximations (5) of motion vectors obtained by the the model inclusive learning of neural networks satisfy the condition (9):

$$B_x(\boldsymbol{x}_{p_d})\left(\alpha \left.\frac{\partial O_1}{\partial I_1}\right|_{\boldsymbol{I}=\boldsymbol{x}_{p_d}} + \beta \left.\frac{\partial O_2}{\partial I_2}\right|_{\boldsymbol{I}=\boldsymbol{x}_{p_d}}\right) + B_y(\boldsymbol{x}_{p_d})\left(\alpha \left.\frac{\partial O_1}{\partial I_2}\right|_{\boldsymbol{I}=\boldsymbol{x}_{p_d}} - \beta \left.\frac{\partial O_2}{\partial I_1}\right|_{\boldsymbol{I}=\boldsymbol{x}_{p_d}}\right) + B_t(\boldsymbol{x}_{p_d}) = 0. \tag{10}$$

Define the performance index by

$$J_d = \frac{1}{2}\sum_{p_d}\left\{B_x(\boldsymbol{x}_{p_d})\left(\alpha \left.\frac{\partial O_1}{\partial I_1}\right|_{\boldsymbol{I}=\boldsymbol{x}_{p_d}} + \beta \left.\frac{\partial O_2}{\partial I_2}\right|_{\boldsymbol{I}=\boldsymbol{x}_{p_d}}\right) + B_y(\boldsymbol{x}_{p_d})\left(\alpha \left.\frac{\partial O_1}{\partial I_2}\right|_{\boldsymbol{I}=\boldsymbol{x}_{p_d}} - \beta \left.\frac{\partial O_2}{\partial I_1}\right|_{\boldsymbol{I}=\boldsymbol{x}_{p_d}}\right) + B_t(\boldsymbol{x}_{p_d})\right\}^2 \tag{11}$$

The problem is now reduced to finding the parameters of the neural network w_{ij}, and α and β which minimize J_d.

3.3 Fusion of Correlation and Differential Methods

As state above the correlation method and the differential method possess their own good points and week points. The correlation method is very simple, but the accuracy of the estimated motion vector (u_{p_c}, v_{p_c}) sometimes becomes a problem because of the accuracy of displacement vector estimation. Furthermore the number of the feature positions specified in a real image is usually very small, that is, only sparse motion vector data can be obtained. On the other hand, the optical flow constraint (9) holds at every position over entire area of images and the derivative data $B_x(\boldsymbol{x}_{p_d})$, $B_y(\boldsymbol{x}_{p_d})$ and $B_t(\boldsymbol{x}_{p_d})$ are obtained over entire area of images. However in the area where the derivatives are small or almost zero, the accuracy of the those data sometimes deteriorate significantly or it is impossible to obtain those data. In order to make the most use of good points and compensates weaknesses of the correlation and differential methods, we propose a method to fuse those two methods.

Define a new performance index by the weighted sum of the performance indexes J_c and J_d:

$$J_{cd} = aJ_c + bJ_d \tag{12}$$

where $a \geq 0$ and $b \geq 0$ are weighting coefficients. The problem is now to reduce to finding the parameters of the neural network w_{ij}, and α and β which minimize J_{cd}, which could bring better estimation results because of the fusion of the two methods. Note that in order to search the values of the learning parameters which minimize J_{cd}, gradient based methods can also be used and efficient algorithms for computing gradients of J_{cd} can be derived by introducing the adjoint model of neural networks. Details of the derivation are omitted here.

4 Experiments

In this section we show some results of the experiments in order to demonstrate the performance of the proposed method. In the following examples a four-layer feedforward neural network with 5 hidden units is used. We utilize the Davidon-Fletcher-Powell algorithm [5] as a gradient based method.

4.1 Example 1

Firstly, in order to evaluate accuracy of the proposed method, we use an artificially generated motion field obtained by moving an image. Figure 3 shows the image which is used in order to generate a motion field artificially. The size of the image is 256 × 256. We turn the image clockwise by 0.5 degree round the center position (63,63) and then translate it by (+1,+1). The motion field thus obtained is shown in Fig.4 which is the target field we will estimate. In the images before and after this movement, five corresponding points are selected as feature positions and the vectors at those positions as shown in Fig. 5 are utilized in the correlation method. In the differential method, we use brightness data at 28 × 28 lattice points which are shown in Fig. 4 as starting points of the

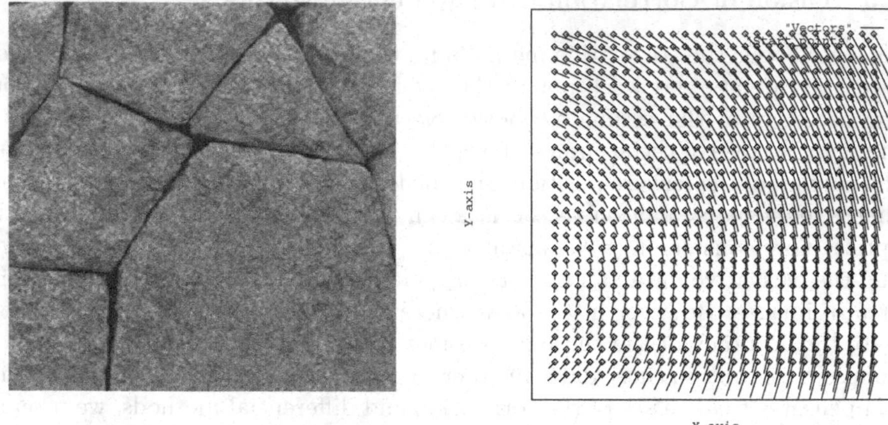

Fig. 3. Image used in order to generate a motion field artificially

Fig. 4. Artificially generated motion field

Table 1. Accuracy of Estimated Motion Field

	errors in direction (rad)	errors in magnitude
proposed fusion method	1.38787×10^{-1}	2.30084×10^{-1}
correlation method	1.50335×10^{-1}	2.76285×10^{-1}
differential method	3.52294×10^{-1}	1.00388

motion vectors. Under the above conditions, we perform experiments of estimating motion field by using three methods, that is, by using only the correlation method, only the differential method and the proposed fusion method. Figure 6 shows the obtained motion field by using the proposed fusion method. Note that the target is the artificially generated motion field and we can evaluate accuracy quantitatively by calculating the average errors between the target vectors and estimated ones. Table 1 shows the results thus calculated. It is observed that the result obtained by the proposed fusion method is more accurate than those by the other individual methods.

4.2 Example 2

In the second experiment we estimate real motion fields from image sequence data by using the proposed method. Figure 7 shows subsequent two images taken by a digital video camera which is set in a car moving straight ahead. The time difference between the left and right images is 1/30 second. In the images ten corresponding points are selected as feature positions and the vectors at those positions shown in Fig. 8 are utilized in the correlation method. In the differential method, we use brightness data at 28 $\times$ 28 lattice points. Figures 9, 10 and 11

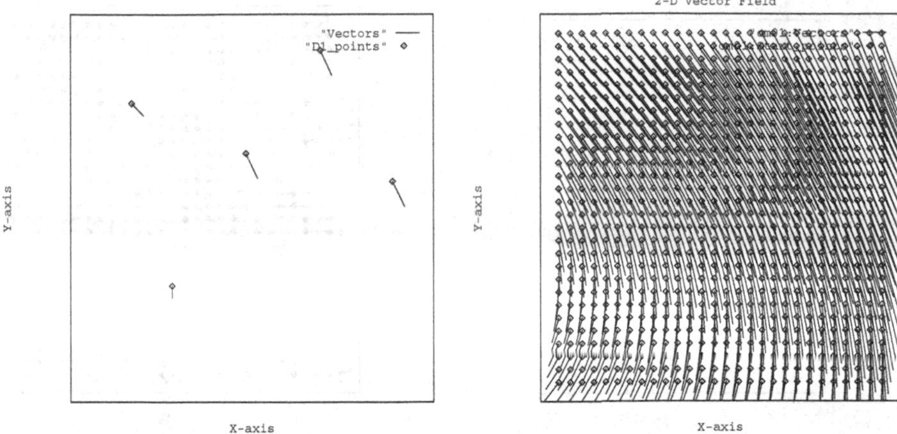

Fig. 5. Vectors at feature positions utilized in the correlation method

Fig. 6. Estimated motion field by the proposed fusion method

Fig. 7. Sequence of images taken by a digital video camera

show the result obtained by only the correlation method, only the differential method and the proposed fusion method, respectively. The feature of the target motion vector field is as follows. The images in Fig. 7 are taken with a video camera which is set in a car moving straight ahead and the preceding car in the images is turning to the left. From this, the generated motion vector field should be:

- in the area of the stationary scene the motion vectors are radiating in all directions from a point and the length of each vector is in inverse proportion to the distance between its position and the camera position,

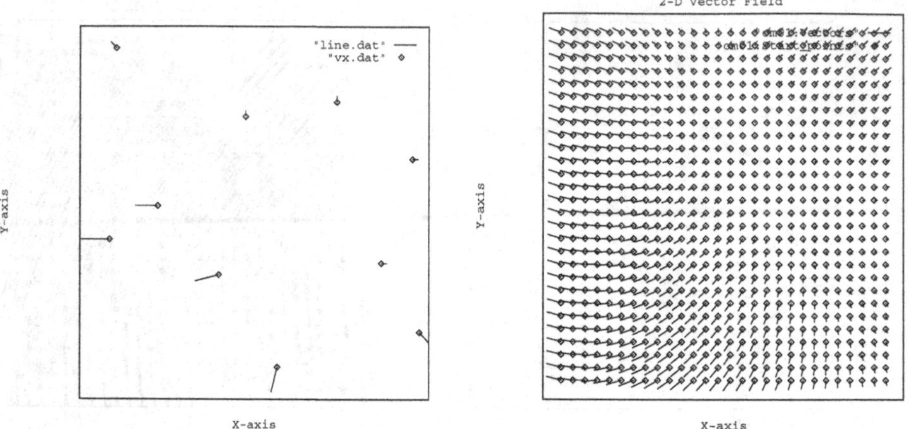

Fig. 8. Vectors at feature positions utilized in the correlation method

Fig. 9. Estimated motion field by only the correlation method

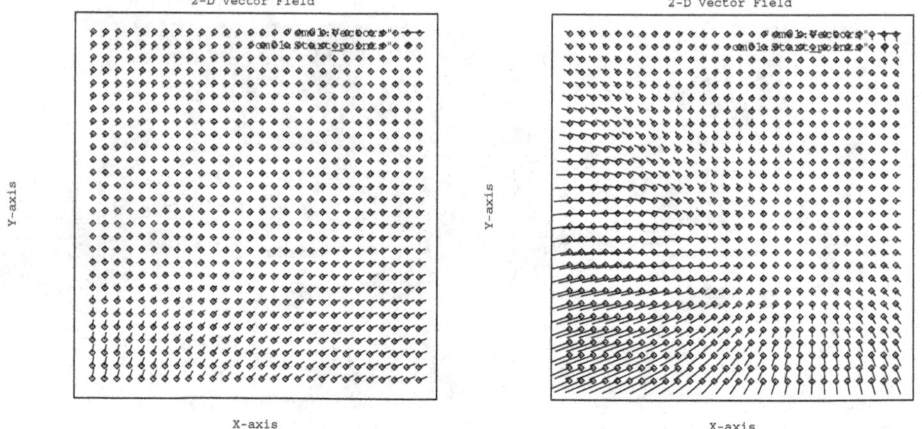

Fig. 10. Estimated motion field by only the differential method

Fig. 11. Estimated motion field by the proposed fusion method

– in the area of the the preceding car, the motion vectors gain further in left directional velocity, because the car is turning to the left.

It is observed from Figs. 9, 10 and 11 that the proposed fusion method catches the feature most accurately.

5 Conclusion

In this paper we have proposed a neural-network based method for estimating an entire continuous motion-vector field from a given set of image-sequence data.

The features of the proposed method are the model inclusive learning and the fusion of two conventional methods of estimating motion fields, the correlation method and the differential method. The model inclusive learning enables approximation results to possess inherent property of vector fields with reasonable approximation accuracy. The fusion can compensate their defects each other and bring better approximation results. It was shown through experiments that the proposed fusion method makes it possible to estimate motion fields more accurately.

References

1. Jahne, B.: Digital Image Processing, 6th edn. Springer, Heidelberg (2005)
2. Mussa-Ivaldi, F.A.: From basis functions to basis fields: vector field approximation from sparse data. Biological Cybernetics 67, 479–489 (1992)
3. Kuroe, Y., Kawakami, H.: Vector field approximation by model inclusive learning of neural networks. In: de Sá, J.M., Alexandre, L.A., Duch, W., Mandic, D.P. (eds.) ICANN 2007. LNCS, vol. 4668, pp. 717–726. Springer, Heidelberg (2007)
4. Kuroe, Y., Nakai, Y., Mori, T.: A Learning Method of Nonlinear Mappings by Neural Networks with Considering Their Derivatives. In: Proc. IJCNN, Nagoya, Japan, pp. 528–531 (1993)
5. Luenberger, D.G.: Introduction to Linear and Nonlinear Programming, pp. 194–197. Addison-Wesley, Reading (1973)

Hybrid Neural Systems for Reduced-Reference Image Quality Assessment

Judith Redi, Paolo Gastaldo, and Rodolfo Zunino

Dept. of Biophysical and Electronic Engineering (DIBE), Genoa University
Via Opera Pia 11a, 16145 Genoa, Italy
{judith.redi,paolo.gastaldo,rodolfo.zunino}@unige.it

Abstract. Reduced-reference paradigms are suitable for supporting real-time modeling of perceived quality, since they make use of salient features both from the target image and its original, undistorted version, without requiring the full original information. In this paper a reduced-reference system is proposed, based on a feature-based description of images which encodes relevant information on the changes in luminance distribution brought about by distortions. Such a numerical description feeds a double-layer hybrid neural system: first, the kind of distortion affecting the image is identified by a classifier relying on Support Vector Machines (SVMs); in a second step, the actual quality level of the distorted image is assessed by a dedicated predictor based on Circular Back Propagation (CBP) neural networks, specifically trained to assess image quality for a given artifact. The general validity of the approach is supported by experimental validations based on subjective quality data.

Keywords: Image quality assessment, neural networks, svm.

1 Introduction

Modern electronic imaging applications are required to guarantee the high quality of the displayed signal, and therefore should cope with the various artifacts, due to the transmission and/or the displaying processes, that can deteriorate original images. A post-processing system, which can both estimate the perceived quality of displayed images and enhance them when needed, can be beneficial for consumer electronics. Hence, understanding the mechanisms underlying the association of a particular quality level to a visual stimulus not only is of great interest from a scientific perspective, but is also crucial from an applicative viewpoint. An accurate measurement of perceived quality can be attained by a panel of human assessors in a subjective experiment session [1]. In fact, when aiming at building an automated real-time system that optimizes the quality of any input signal, an objective system for quality assessment is needed [2,3], which bypasses any human evaluation. Objective methods rely on extracting salient features from images and correlating them to the perceived quality. While *full reference* (FR) paradigms [2,3] require full knowledge of the original image to

C. Alippi et al. (Eds.): ICANN 2009, Part II, LNCS 5769, pp. 684–693, 2009.

assess the quality of the incoming signal, *reduced reference* (RR) methods only involve a limited amount of numerical features characterizing the original signal [2,4]. Thus, the RR paradigm can provide a successful approach for supporting real-time modeling of perceived quality. The research presented in this paper exploits the modeling power of computational intelligence to support a reduced-reference, objective system for image quality assessment. A double-layer system based on connectionist paradigms is designed to map a numerical description of the image into quality scores. To handle different distortions, first a classification algorithm determines the type of artifact affecting the image; then a specific regression machine (targeted to the specific type of distortion) attains a reliable estimation of the loss in quality of the present sample with respect to its original version. A major advantage of the use of computational intelligent methods is that, by training the neural network to mimic perceived image quality, one bypasses the design of an explicit model of the human visual system. Luminance information supports image representation; by analyzing the luminance distribution, the method gathers a feature-based description of a picture. The artifacts brought about by digital processing affect the original luminance content of the image, and the peculiar effects often depend on the specific artifact. The rationale of the present approach is that by comparing the statistics of the original and distorted image one can identify both the kind and the extent of distortion. Previous works [5,6] showed that second order statistics can apply successfully toward that end, hence this research adopts a set of features derived from the correlogram-based description [7,8], yielding a second-order histogram for luminance. The LIVE database [9] provides the performance-evaluation testbed, including three types of distortion: White noise, Gaussian Blur and JPEG compression. Empirical results confirm the validity of the connectionist paradigm and the effectiveness of luminance statistics for predicting image quality.

2 Objective Quality Assessment: A Reduced Reference Approach

In reduced-reference approaches to objective quality assessment, few parameters are extracted from the original, undistorted image and are used as a reference for predicting the quality level of the target, distorted image. Let $I^{(n)}$ be the reference image, and $\bar{I}^{(n,r)}$ the image resulting from the insertion of some artifact to $I^{(n)}$, being r the distortion level. Let $\mathbf{x}^{(n)}$ and $\mathbf{x}^{(n,r)}$ denote the numerical representations of $I^{(n)}$ and $\bar{I}^{(n,r)}$, respectively. Finally, let $q^{(n)}$ and $q^{(n,r)}$ be the subjective quality ratings for $I^{(n)}$ and $\bar{I}^{(n,r)}$, respectively. A reduced-reference paradigm compares the numerical descriptors, $\{\mathbf{x}^{(n)}, \mathbf{x}^{(n,r)}\}$, and estimates the discrepancy, $d_S(q^{(n)}, q^{(n,r)})$, between the subjective scores associated with the images. At runtime, the objective information related to the original image can be worked out before transmission, so that the resulting descriptor, $\mathbf{x}^{(n)}$, can be broadcast as metadata. Vector $\mathbf{x}^{(n,r)}$ is computed from the signal at the receiver end, and the two vectors are eventually processed to obtain quality estimates. The assessment process includes two steps. First, the distortion affecting the

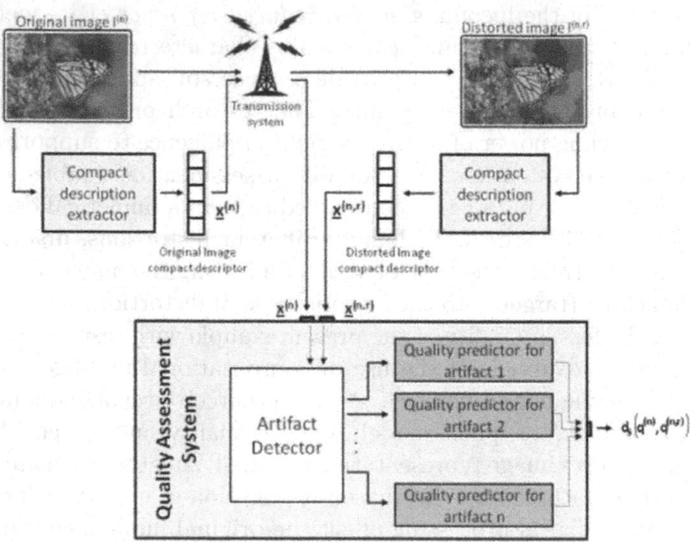

Fig. 1. Overall scheme of the proposed system for reduced-reference objective quality assessment

incoming video signal is identified; then, a specific estimation module quantifies the gain/loss in quality brought about by the detected distortion. The rationale behind such an approach is that every kind of distortion modifies the image in a different way, thus requiring a specific objective metric. So, as soon as the set $A = \{a_1, a_2, ..., a_n\}$ of artifacts of interest is established, implementing a dedicated quality predictor for each of them can be a successful strategy. As no a-priori knowledge can be assumed on the nature of the distortion affecting the input signal, an artifact detector is needed to identify the distortion ($a_i \in A$) applied to the image $I^{(n)}$. This module then forwards the image descriptors to the quality predictor $\Omega(a_i)$, which finally provides an estimation for the difference in quality $d_S(q^{(n)}, q^{(n,r)})$ between the reference and incoming signal, distorted by a_i. Both operations are based on the descriptors $\mathbf{x}^{(n)}$ and $\mathbf{x}^{(n,r)}$. Figure 1 shows the complete framework for image quality estimation.

3 Luminance Information Extraction for Quality Evaluation

To balance effectiveness and performance, image descriptors should be small-sized to satisfy constraints in transmission bandwidth, yet still informative to allow both artifact identification and quality assessment. The approach presented here retrieves such information by analyzing the image luminance distribution, for two main reasons: first, the relevance of luminance in quality assessment has been widely proved [3]; secondly, incoming signals are usually encoded in the YC_bC_r colorspace, hence the luminance channel is immediately available for computation. The analysis of luminance distribution is based on the correlogram tool [7]. The correlogram belongs to the family of the second order histograms, just like the co-occurrence matrix, which has already been proved effective for objective image quality estimation [5,6]. Aiming at maintaining both local and global information in the final description, the correlogram is computed on equally sized sub-regions of the image, and a feature-based representation is worked out for each sub-region. Global information is obtained by using a statistical approach, finally assembling local information into a single descriptive vector.

3.1 Local Information Extraction

A correlogram H_b^k of a sub-region b (including $W_b \times H_b$ pixels) for a predefined distance k, is a three-dimensional matrix that describes the spatial distribution of luminance, color or any other image property that can be represented by quantized values. More precisely, the correlogram describes how the spatial correlation of pairs of pixels changes with distance. Formally, the entry $H_b^k(i,j)$ of the correlogram matrix is defined as:

$$H_b^k(i,j) = \left| \left\{ \begin{array}{c} (m,n), m < W_b, n < H_b s.t. \\ b[m,n] = Y_i; b[p,q] = Y_j; dist(b[m,n], b[p,q]) = k \end{array} \right\} \right| \quad (1)$$

In this research, the luminance distribution is targeted. Thus, each matrix element $H_b^k(i,j)$ specifies the probability of finding a pixel with luminance Y_j at a distance k from a pixel with luminance Y_i. The $dist()$ operator denotes the measure of distance between pixels selected to calculate the correlogram. In this study, the $dist()$ operator embeds the L_1-norm, so, in practice, only those pairs of pixels with a distance k in horizontal and vertical directions are considered in the computation. The proposed system preserves local information by computing the color correlogram on sub-regions of the image. To this end, images are split into non-overlapping square regions, each holding $N_b \times N_b$ pixels. For each block, the set of features $\Phi = \{f_0, f_1, ..., f_5\}$, as defined in table I, is extracted from the correlogram.

Table 1. Features extracted from the color correlogram for images description

Feature name	Definition	Feature name	Definition
Energy	$f_0 = \sum_{i,j}[H_b^k(i,j)]^2$	*Contrast*	$f_3 = \sum_z z^2 P(z)$
Diagonal Energy	$f_1 = \sum_i [H_b^k(i,i)]^2$	*Homogeneity*	$f_4 = \sum_{i,j} H_b^k/[1+(i-j)]^2$
Entropy	$f_2 = -\sum_{i,j} H_b^k(i,j) \log_2 H_b^k(i,j)$	*Energy Ratio*	$f_5 = f_0/f_1$
$P(z) = \sum_{\|i-j\|=z} H_b^k(i,j)$			

3.2 Global-Level Numerical Representation

The local information extraction phase outputs as many values for each feature of Φ as the number of blocks. In fact, the eventual amount of data is too large to fit a descriptor that should be sent in real time through a communication channel. Moreover, subjective scores usually express the quality of a whole image, and not of a set of blocks. So, from a modeling perspective, the feature-based image description should consist of a single vector, to be associated with the single quality score. Toward this end, block-based information is combined using statistical descriptors, namely percentiles, to represent the distribution of a feature f_u over the image. Recalling the notations used in the previous paragraphs, the following pseudo-code can be applied to both the reference and the distorted image for constructing the global descriptors.

Algorithm 1. Numerical representation extraction

1: **(Block-Level feature extraction)**
- Split $\bar{I}$ into N_b non-overlapping square blocks, and obtain the set $B = \{b_m; m = 1, ..., N_b\}$
- For each block $b_m \in B$ compute the associate correlogram: $H^k_{b_m}(i, j)$
- For each matrix $H^k_{b_m}$: compute the value $x_{u,m}$ of each feature and obtain the sets: $X_u = \{x_{u,m}; m = 1, ..., N_b\}$ with $u \in \Phi$

2: **(Global level numerical representation)**
- Compute a percentile-based description of X_u; let p_α be the αth percentile: $\psi_{\alpha,u} = p_\alpha(X_u)$
- Assemble the objective descriptor vector, $\bar{\mathbf{x}}_u$, for the feature f_u on the image $\bar{I}$

$$\bar{\mathbf{x}}_u = \{\psi_{\alpha,u}; \alpha \in [0, 100]\} \tag{2}$$

Eventually, the extraction process results in a descriptor of an image, $\bar{I}$, being a global pattern, $\bar{\mathbf{x}}_u$ (2). This output of the extraction process is forwarded to the double-layer assessment system for quality estimation.

4 Machine Learning Methods for Objective Quality Assessment

The system described in section 2 consists of two steps. First, the artifact affecting the image has to be identified; secondly, the numerical representation of the image has to be mapped into a quality score by a dedicated predictor, which is specifically trained to assess image quality for a given artifact. The first layer is required to solve a classification problem. When aiming to detect the artifact a_i affecting the sample $\bar{I}^{(n)}$, given m artifacts of interest $a_1, a_2, ..., a_m$, the system is required to relate the input vector $\bar{\mathbf{x}}_u$ to a discrete value, representing a_i. On the other hand, the second layer aims at mapping the numerical descriptor $\bar{\mathbf{x}}_u$ into a quality score, which cannot be expressed by discrete values to achieve acceptable accuracy. Therefore, this module can be designed to solve a regression problem. In both cases, the use of computational intelligence methods is appealing. The machine learning world provides excellent tools able to handle both classification and regression supervised problems. Moreover, from a modeling point of view, such methods appear particularly suitable to model a highly non-linear context such as perception. Hence, in this work, two well known tools are involved to implement the proposed system: Support Vector Machines (SVM) for the classification task and a feed forward neural network, namely the Circular Back-Propagation (CBP) network for the regression task.

4.1 Support Vector Machines

Support vector machines are a powerful and effective tool for binary classification problems. Given a set of np patterns $Z = \{(x_l, y_l); l = 1, .., n_p; y_l \in \{-1, +1\}\}$, a SVM relies on the solution of the following Quadratic Programming problem to find the optimal hyper-plane $\mathbf{w}$ separating the two classes:

$$min_\alpha \frac{1}{2} \sum_{l,m=1}^{n_p} \alpha_l \alpha_m y_l y_m K(\mathbf{x}_l, \mathbf{x}_m) - \sum_{l=1}^{n_p} \alpha_l \quad (3)$$

subject to $0 \leq \alpha_l \leq C \; \forall l$ and $\sum_{l=1}^{n_p} y_l \alpha_l = 0$.
In (3), α_i are the SVM parameters setting the class-separating surface and C is a fixed regularization term that rules the trade-off between accuracy and complexity. The kernel function $K()$ allows inner products of patterns in a higher dimensional, transformed space, yet disregarding the specific mapping of each single pattern; let $\Phi(\mathbf{x}_1)$ and $\Phi(\mathbf{x}_2)$ be the points in the feature space that are associated with $\mathbf{x}_1$ and $\mathbf{x}_2$, respectively, then their dot product can be written as $< \Phi(\mathbf{x}_1), \Phi(\mathbf{x}_2) >= K(\mathbf{x}_1, \mathbf{x}_2)$. In this paper, the conventional Radial Basis Function (RBF) kernel has been adopted. Problem setting (3) has the crucial advantage of involving a quadratic-optimization problem with linear constraints, ensuring that the solution is unique. Actually, the specific choice for the kernel parameters $\{C, \sigma\}$ has an impact on the eventual generalization performance of the SVM. Both theoretical [10] and empirical [11] approaches can be adopted to determine the generalization limits. The present research follows an empirical approach involving k-fold cross validation [11].

4.2 Circular Back-Propagation Neural Networks

In the proposed framework a Circular Back Propagation (CBP) network [12] maps feature-based image descriptions into the associated estimates of perceived quality, which, in the present formulation, are scalar values. The CBP network augments the conventional MultiLayer Perceptron (MLP) architecture with an additional input, being the sum of the squared values of all the network inputs. The CBP architecture can be formally described as follows. The j-th hidden neuron performs a non-linear transformation on the weighted combination of the input values (i.e, the image numerical descriptor (2)) with coefficients

$w_{j,i}$ ($j = 1, ..., n_h$; $i = 1, ..., n_i$), as well as the output layer, which eventually provides the actual network responses, y_k, ($k = 1, ..., n_o$):

$$y_k = sigm(w_{k,0} + \sum_{j=1}^{n_h} w_{k,j} a_j) \;\; a_j = sigm(w_{j,0} + \sum_{i=1}^{n_i} w_{j,i} x_i + w_{j,n_i+1} \sum_{i=1}^{n_i} x_i^2) \quad (4)$$

where $sigm(z) = (1+\exp(-z))^{-1}$, and a_j is the neuron activation. The quadratic term allows the CBP network either to adopt the standard sigmoidal behavior, or a bell-shaped radial function, depending on the data. As a major result, the CBP network does not need any a-priori assumption to formulate the model without affecting the fruitful properties of an MLP structure. The structural CBP enhancement still allows adopting conventional back-propagation algorithms [13] for weight adjustment, yielding an effective training:

$$E = \frac{1}{n_O n_p} \sum_{l=1}^{n_p} \sum_{k=1}^{n_O} (t_k^{(l)} - y_k^{(l)})^2 \quad (5)$$

where n_p is the number of training patterns, and t_k is the desired reference output. In the present application, $k = 1$ and the expected output is given by the quality score obtained from subjective panel tests.

5 Effective Implementation

In this section a possible implementation of the double-layer system is proposed. The artifact identification problem is tackled by an SVM-based module, while a CBP-based architecture, maps feature-based description of images into quality scores.

5.1 Images Feature-Based Description

Image descriptors $\bar{\mathbf{x}}_u^{(n)}$ and $\bar{\mathbf{x}}_u^{(n,r)}$, corresponding to the reference image $I^{(n)}$ and the target image $\bar{I}^{(n,r)}$ respectively, are computed according to the procedure reported in Section 3. The input image is divided into squared sub-regions of 32x32 pixels and the color correlogram is computed on the luminance component (Y-layer) of the blocks. The set of features Φ defined in table I is then extracted from the correlogram corresponding to each block. Finally, the global descriptor is assembled by computing 6 percentiles of the distribution of each feature f_u, and combining them in the vector $\bar{\mathbf{x}}_u^{(n,r)} = \{\psi_{\alpha,u}^{(n,r)}; \alpha = 0, 20, 40, 60, 80, 100; u = 0, 1, ..., 5\}$ For each feature, the objective representation of the stimulus is obtained simply by combining the descriptors of the original and the distorted image:

$$\mathbf{z}_u^{(n,r)} = [\mathbf{x}_u^{(n)}, \mathbf{x}_u^{(n,r)}] \quad (6)$$

The resulting 12-dimensional vectors $\mathbf{z}_u^{(n,r)}$, $u \in \Phi$ feed the double-layer quality estimator.

5.2 Neural-Based Quality Loss/Gain Quantification

The role of the artifact identifier module in the system is to solve a multiclass problem, associating each $\mathbf{z}_u$ to the appropriate artifact $a_i \in A$. To implement this multiclass classifier, several binary predictors are connected in series. In this study three possible artifacts are considered: White noise, Gaussian Blur and JPEG compression. The classification module is implemented using a first SVM to select images distorted by White Noise, and a second SVM to separate blurred images from JPEG compressed images. Both SVMs receive as input a vector based on a single feature, namely Entropy (see f_2 in table I). This feature proved sufficiently effective for the distortion classification task. The module assigned to quality level prediction (the second layer of the system) is composed of as many subsystems as the number of considered distortions; in this case three. The three prediction modules exploit an ensemble strategy [14]. To this end, each single predictor involved in the ensemble is fed with a single, distinct feature, thus satisfying the required hypothesis of disjoint input sub-spaces. Since the three modules are independent from one another, each artifact-dedicated predictor makes use of a particular pair of features, designed to be as much informative as possible for the specific artifact. In the proposed research to quantify the effect of white noise on quality, the pair Entropy, Contrast is used; the pair Contrast, Homogeneity decribes blurred images and the couple Entropy, Energy Ratio is used to characterize JPEG stimuli.

6 Experimental Results

The second release of the LIVE database [9] was used as a testbed for the performance evaluation of our proposed model, being a recognized benchmark in the image quality assessment field. In particular, the three datasets including samples distorted with White noise, Gaussian Blur and JPEG compression were considered. A k-fold-like strategy was adopted to prove the system to be able to generalize independently of the specific image content used for the training. 5 folds of images were created for every dataset, each containing all the distorted versions of a few original images, in such a way that none of the 29 image contents of the LIVE belonged to more than one group. Both layers were trained performing 5 runs, in each of which alternatively 4 of the 5 folds were used as training data and the remaining one was used as test data. For all experiments, the correlogram was computed for a distance $k = 1$ using the L_1 norm. The two SVMs of the first layer were trained independently, and for both, to tune the kernel parameters the k-fold cross-validation technique was applied. The first SVM was trained on a dataset resulting from the merge of the three LIVE sets, to recognize noisy images. For this SVM, a RBF kernel hyper-parameter was set to 1, and the parameter C was finally set to 10^4. The second SVM was trained on a subset of the previous dataset, including only blurred and compressed images. Based on the cross-validation output, the parameter C was set to 10^5 and $\sigma = 1$. Table II reports the classification errors for each run and for both SVM classifiers. It clearly illustrates that the first SVM has an almost-perfect

Table 2. Performance of each SVM machine for artifact identification in terms of % of misclassified patterns

Performance	Run#1	Run#2	Run#3	Run#4	Run#5	Avg
Noise detection	0.00%	1.07%	1.09%	0.00%	0.00%	0.43%
Blur VS Jpeg Detection	1.49%	1.58%	4.84%	19.35%	12.76%	7.56%

Table 3. Performance of quality estimators in terms of percentage absolute error and RMS between predicted and subjective quality scores

	White Noise		*JPEG Compression*		*Gaussian Blur*	
	$\nu_{\lvert err \rvert}$	ρ	$\nu_{\lvert err \rvert}$	ρ	$\nu_{\lvert err \rvert}$	ρ
Run #1	0.062	0.98	0.107	0.94	0.107	0.95
Run #2	0.104	0.94	0.148	0.85	0.247	0.67
Run #3	0.054	0.99	0.139	0.92	0.077	0.96
Run #4	0.062	0.98	0.092	0.94	0.129	0.91
Run #5	0.093	0.98	0.171	0.88	0.209	0.89
Average	0.075	0.97	0.131	0.91	0.154	0.88

performance, since only an insignificant percentage of patterns is misclassified. The performance of the second SVM is worse, due to the intrinsically more complex task. Indeed, the problem of distinguishing blur from compression artifacts is intuitively more difficult than the previous one, since the visual effects produced by the two distortions often overlap. Nonetheless, on average the percentage of misclassified images is less than 8%. The three ensembles of two CBP neural networks implementing the second layer were each trained on a different dataset. The datasets for White Noise and Blurred images contained 145 patterns; the remaining testbed included 159 JPEG compressed images. For each image a quality score was provided in the LIVE database. The Difference Mean Opinion Score (DMOS), originally ranging between [0,100] was remapped for computational reasons into the range [-1, +1]. Due to the flexibility of the system, it was possible to design the network architecture on purpose for each task. For coherence, all the 6 neural networks shared the same input layer with the dimensionality of the global descriptor $\mathbf{z}_u$. The networks for the quality estimation of noisy images were equipped with 3 hidden neurons, those used for blurred images were equipped with a 5-neurons hidden layer, while, to maximize the generalization ability, the quality estimator for the JPEG compressed images counted 7 hidden neurons. For performance evaluation, we define the accuracy as the discrepancy between the objectively predicted quality change $\hat{d}_S(q^{(n)}, q^{(n,r)})$ and the quality score provided by the LIVE database $d_S(q^{(n)}, q^{(n,r)})$. Two different indicators are used to evaluate the second layer subsystems accuracy: the mean value of the absolute prediction error between d_S and $\hat{d}_S$, $\nu_{|err|}$, and Pearsons correlation coefficient ρ [15]. Table III shows how the second layer in our system is able to produce satisfactory quality predictions. Assessing the quality gain/loss of noisy images, the systems mean absolute prediction error, averaged over the five folds, is less than 0.08 on a two points range. This implies that the confidence in the system estimates is, on average, higher than 96%, improving previous results [6]. Dealing with compressed images, the mean absolute prediction error is 0.131, which is comparable with the results obtained in [6]. Finally, the quality estimator for the blurred images achieves an accuracy of 92.3%, which represents an improvement of about 1.00% with respect to previous results [6].

References

1. ITU: Methodology for the subjective assessment of the quality of television pictures. International Telecommunication Union (1995)
2. Wang, Z., Sheikh, H., Bovik, A.: The Handbook of Video Databases: Design and Applications. CRC Press, Boca Raton (2003)
3. Sheikh, H., Sabir, M., Bovik, A.: A statistical evaluation of recent full reference image quality assessment algorithm. IEEE Trans. Image Processing (2006)
4. Wang, Z., Simoncelli, E.P.: Reduced-reference image quality assessment using a wavelet-domain natural image statistic model. Proc. SPIE Human Vision and Electronic Imaging (2005)
5. Gastaldo, P., Zunino, Z.: Neural networks for the no-reference assessment of perceived quality. Journal of Electronic Imaging (2005)
6. Redi, J., Gastaldo, P., Zunino, R., Heynderickx, I.: Co-occurrence matrixes for the quality assessment of coded images. In: Kůrková, V., Neruda, R., Koutník, J. (eds.) ICANN 2008, Part I. LNCS, vol. 5163, pp. 897–906. Springer, Heidelberg (2008)
7. Huang, J., Kumar, S.R., Mitra, M., Zhu, W., Zabih, R.: Image indexing using color correlograms. In: Proc. IEEE CVPR 1997, pp. 762–768 (1997)
8. Haralick, R., Shanmugam, K., Dinstein, I.: Textural features for image classification. IEEE Trans. on Systems, Man and Cybernetics SMC-3 (1973)
9. Sheikh, H., Wang, Z., Cormack, L., Bovik, A.: Image quality assessment database, `http://live.ece.utexas.edu/research/quality`
10. Vapnik, V.: Statistical Learning Theory. Wiley, Chichester (1998)
11. Bartlett, P., Boucheron, S., Lugosi, G.: Model selection and error estimation. Machine Learning, 85–113 (2002)
12. Ridella, S., Rovetta, S., Zunino, R.: Circular back-propagation networks for classification. IEEE Trans. on Neural Networks, 84–97 (1997)
13. Rumelhart, D., McClelland, J.: Parallel distributed processing (1986)
14. Kittler, J., Hatef, M., Duin, R., Matas, J.: On combining classifiers. IEEE Trans. Pattern Analysis and Machine Intelligence (1998)
15. Kittler, J., Hatef, M., Duin, R., Matas, J.: Final report from the video quality experts group on the validation of objective models of video quality assessment (2000), `http://www.vqeg.org/`

Representing Images with χ^2 Distance Based Histograms of SIFT Descriptors

Ville Viitaniemi and Jorma Laaksonen

Department of Information and Computer Science, Helsinki University of Technology, P.O.Box 5400, FI-02015 TKK, Finland
{ville.viitaniemi,jorma.laaksonen}@tkk.fi

Abstract. Histograms of local descriptors such as SIFT have proven to be powerful representations of image content. Often the histograms are formed using a clustering algorithm that compares the SIFT descriptors with the Euclidean distance. In this paper we experimentally investigate the usefulness of basing the comparisons of the SIFT descriptors on the χ^2 distance measure instead. The modified approach results in improved image category detection performance when it is incorporated into a Bag-of-Visual-Words type category detection system.

1 Introduction

In this paper we consider the problem of automatic supervised recognition of image content. The recognised content can then be used as a basis for classification, indexing or other further content-based processing of the image. Histograms of local features have proven to be powerful representations for image classification and object detection. Consequently, their use has become commonplace in image content analysis tasks (e.g. [1,2]). This paradigm is also known by the name Bag of Visual Words (BoV) in analogy with the successful Bag of Words paradigm in text retrieval. In this analogue, images correspond to documents and quantised local feature values to words.

Use of local image feature histograms for supervised image classification and characterisation can be divided into several stages:

1. Selecting image locations of interest.
2. Describing each location with suitable visual descriptors (e.g. Scale-Invariant Feature Transform (SIFT) [3]).
3. Characterising the distribution of the descriptors within each image with a histogram, each histogram bin corresponding to a visual word.
4. Using the histograms as feature vectors representing the images in a supervised vector space algorithm, such as the Support Vector Machine (SVM).

All parts of the BoV pipeline are subject of continuous study. However, in this paper we concentrate on stage 3 and regard the other stages as given. In particular, we take the local image neigbourhoods to be described with the SIFT descriptor.

C. Alippi et al. (Eds.): ICANN 2009, Part II, LNCS 5769, pp. 694–703, 2009.

In order to characterise the descriptor distribution of an image with a histogram, one needs to define the bins of the histogram in the descriptor space. Often this is accomplished by employing a clustering algorithm such as K-means to the combined set of descriptors of all the considered images (e.g. [1,4]). This general approach is followed also in this paper. The modification we propose is that of basing the K-means clustering and histogram bin assignment on the χ^2 distance measure instead of the (squared) Euclidean distance that is traditionally used for comparing SIFT descriptors. The modification is motivated by the observation that one can improve the performance of stage 4 of the BoV pipeline if the χ^2 distance is used to compare histograms of SIFT descriptors instead of the Euclidean distance (e.g. [2]). However, also SIFT descriptors themselves (and several other similar descriptors) resemble normalised histograms, namely histograms of local edge directions. This leads to the idea of comparing also the SIFT descriptors with the χ^2 distance. This only motivates the modification, no theoretical reason guarantees the χ^2 distance to be better suited for matching SIFT descriptors than the Euclidean distance. The actual justification needs to be obtained empirically.

We consider the main contribution of this paper to be the experimental investigation whether the proposed distance measure modification is useful in an image category detection task defined in the PASCAL NoE VOC 2007 object detection benchmark [5]. More generally, we investigate whether it is useful to compare SIFT descriptors with the χ^2 distance instead of the Euclidean distance.

The rest of this paper is organised as follows. In Section 2 we review the clustering problem and describe how the K-means clustering algorithm can be modified to use the χ^2 distance. In Sect. 3 we describe our implementation of the BoV pipeline. Section 4 details the image category detection task and experimental procedures that are used in the experiments of Sect. 5 to compare the clustering and histogram forming based on either Euclidean or χ^2 distance measures. Finally, in Sect. 6 we draw our conclusions from the experiments.

2 Clustering

Given a set of N D-dimensional data points $\{\mathbf{x}_i\}_{i=1}^N$, the goal of unsupervised clustering is to find K D-dimensional cluster centers $\{\mathbf{m}_i\}_{i=1}^K$ and variables $\{c_i\}_{i=1}^N$, $c_i \in \{1, 2, \ldots, K\}$ that assign the N data points among the K clusters so that a criterion C measuring the incompatibility of sets $\{\mathbf{x}_i\}$, $\{\mathbf{m}_i\}$ and $\{c_i\}$ is minimised. As the criterion, one often uses the mean point-wise quantisation error of the form

$$C = \frac{1}{N}\sum_{i=1}^{N} E(\mathbf{x}_i, \mathbf{m}_{c_i}) = \frac{1}{N}\sum_{k=1}^{K}\sum_{i|c_i=k} E(\mathbf{x}_i, \mathbf{m}_k). \quad (1)$$

Here $E(\cdot)$ is a function measuring the dissimilarity of two D-dimensional vectors.

Clustering can be employed in the BoV paradigm by choosing the visual words to be the cluster centers $\{\mathbf{m}_i\}$ that result from clustering of all the local descriptors of interest points in all the considered images. Each single image is then represented by a K-dimensional histogram of the cluster assignment variables $\{c_i\}$ corresponding to the interest points within the image in question.

2.1 Two Distance Measures

Let us denote $\mathbf{x}_i = [x_{i1}\, x_{i2} \ldots x_{iD}]^T$. The square of the Euclidean distance metric between two vectors is given by

$$d_{\mathrm{E}}(\mathbf{x}_i, \mathbf{x}_j)^2 = \sum_{d=1}^{D} (x_{id} - x_{jd})^2. \tag{2}$$

The Euclidean distance is a natural choice for a distance measure since it usefully reflects the closeness of points in the three-dimensional physical world when the components of the vectors are the points' Cartesian coordinates.

Another distance measure is the χ^2 distance

$$d_{\chi^2}(\mathbf{x}_i, \mathbf{x}_j) = \sum_{d=1}^{D} \frac{(x_{id} - x_{jd})^2}{x_{id} + x_{jd}}. \tag{3}$$

This measure can be used only when all the vector components are non-negative. The measure can be interpreted as a weighted squared Euclidean distance where the weighting compresses the variation in the components' values by assigning less weight to components with large values. In practice the χ^2 distance has proven to be useful for comparing histograms, e.g. in BoV systems [2,6].

2.2 K-means Clustering Algorithm

K-means [7] is a widely-used and simple clustering algorithm. In its basic form it defines an iteration that finds a local optimum of the clustering criterion

$$C_{\mathrm{MSE}} = \frac{1}{N} \sum_{i=1}^{N} d_{\mathrm{E}}(\mathbf{x}_i, \mathbf{m}_{c_i})^2 \tag{4}$$

obtained by setting the dissimilarity function

$$E(\mathbf{x}_i, \mathbf{x}_j) = d_{\mathrm{E}}(\mathbf{x}_i, \mathbf{x}_j)^2 \tag{5}$$

in (1). Each step of the iteration is guaranteed not to increase the value of the criterion.

The iteration is started by choosing initial values for $\{\mathbf{m}_i\}$. Then the following two steps are repeated until convergence:

1. Updating the cluster assignments $\{c_i\}$ according to

$$c_i = \arg\min_k d_{\mathrm{E}}(\mathbf{x}_i, \mathbf{m}_k). \tag{6}$$

2. Updating the cluster centers $\{\mathbf{m}_i\}$ according to

$$\mathbf{m}_k = \frac{\sum_{i|c_i=k} \mathbf{x}_i}{\sum_{i|c_i=k} 1}. \tag{7}$$

For later use we, note that step 2 can be connected to the squared Euclidean distance measure by noticing it to be the solution of

$$\mathbf{m}_k = \arg\min_{\mathbf{m}} \sum_{i|c_i=k} d_{\mathrm{E}}(\mathbf{x}_i, \mathbf{m})^2. \tag{8}$$

2.3 K-means Based on χ^2 Distance

If all the components of the data points are non-negative, the squared Euclidean distance can be replaced in the K-means algorithm by the χ^2 distance by taking the iteration steps to be

1. Updating the cluster assignments $\{c_i\}$ according to

$$c_i = \arg\min_k d_{\chi^2}(\mathbf{x}_i, \mathbf{m}_k). \tag{9}$$

2. Updating the cluster centers $\{\mathbf{m}_i\}$ according to

$$\mathbf{m}_k = \arg\min_{\mathbf{m}>0} \sum_{i|c_i=k} d_{\chi^2}(\mathbf{x}_i, \mathbf{m}). \tag{10}$$

These iteration steps are guaranteed not to increase the criterion

$$C_{\chi^2} = \frac{1}{N} \sum_{i=1}^{N} d_{\chi^2}(\mathbf{x}_i, \mathbf{m}_{c_i}). \tag{11}$$

obtained from (1) with

$$E(\mathbf{x}_i, \mathbf{x}_j) = d_{\chi^2}(\mathbf{x}_i, \mathbf{x}_j). \tag{12}$$

Thus a local minimum will be reached when the iteration is repeated until convergence.

In the case of the χ^2 distance the step 2 of the iteration is not as straightforward as in the squared Euclidean case. In fact, the optimisation must be performed numerically. Fortunately, because of independence of the vector components in the χ^2 distance measure, we may optimise each of the components of $\mathbf{m}_k$ separately. In other words, the optimisation problem of step 2 can be recast to D one-dimensional optimisation problems

$$m_{kd} = \arg\min_{m>0} \sum_{i|c_i=k} \frac{(x_{id} - m)^2}{x_{id} + m}, \quad d \in \{1, \ldots, D\}. \tag{13}$$

Figure 1 illustrates the one-dimensional optimisation problem. The crosses on the x-axis denote sample data points. The y-coordinates of the solid and dashed

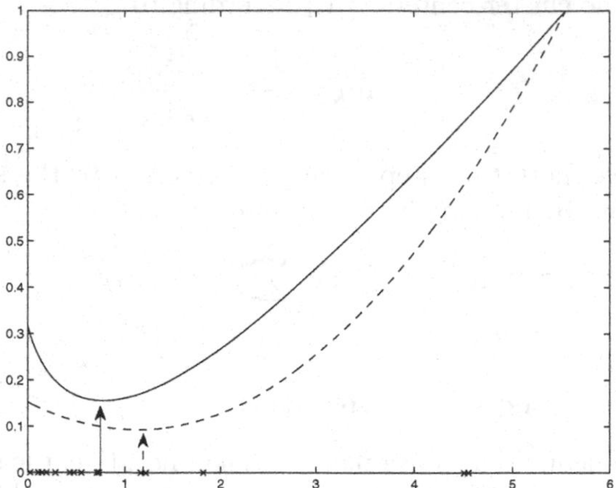

Fig. 1. Sum of χ^2 (solid line) and squared Euclidean (dashed line) distances in one dimension

lines correspond to the sum of χ^2 distances and squared Euclidean distances, correspondingly, if the cluster center m would receive the value on x-axis. We observe that the minima (indicated with arrows) result from somewhat different choices of m for the two distance measures. The minimisation of sum of χ^2 distances thus cannot be replaced by the more straightforward minimisation of the sum of squared Euclidean distances if one wants to attain the convergence to a minimum of C_{χ^2}.

The optimisation can be performed by first converting the constrained optimisation problem (13) into an unconstrained one

$$z_* = \arg\min_z \sum_{i|c_i=k} \frac{(x_{id} - z^2)^2}{x_{id} + z^2} \tag{14}$$

and after solution doing the back-substitution $m_{kd} = z_*^2$. This idea corresponds to the Newton-Raphson (NR) iteration

$$z_{t+1} = z_t - \frac{\partial f}{\partial z} / \frac{\partial}{\partial z}\left(\frac{\partial f}{\partial z}\right) = z_t - \frac{\partial f}{\partial m}\frac{\partial m}{\partial z} / \frac{\partial}{\partial z}\left(\frac{\partial f}{\partial m}\frac{\partial m}{\partial z}\right) \tag{15}$$

with

$$f = \sum_{i|c_i=k} \frac{(x_{id} - m)^2}{x_{id} + m} = \sum_{i|c_i=k} \frac{(x_{id} - z^2)^2}{x_{id} + z^2}. \tag{16}$$

Further evaluation of the derivatives is omitted here due to space reasons. In practice just a few rounds of NR iteration is sufficient per each round of the K-means iteration. The computational cost of the χ^2 K-means is approximately two times the cost of the Euclidean K-means, resulting mainly from more costly evaluation of the distance measure in step 1 of the iteration.

3 Implementation of the BoV System

In this section we describe our implementation of the Bag of Visual Words pipeline outlined in Sect. 1. In the first stage, a number of interest points are identified in each image. For these experiments, the interest points are detected with a combined Harris-Laplace detector [8] that outputs around 1200 interest points on average per image for the images used in this study. In stage 2 the image area around each interest point is individually described with a 128-dimensional SIFT descriptor [3], a widely-used and rather well-performing descriptor that is based on local edge statistics.

In stage 3 each image is described by forming a histogram of the SIFT descriptors. In the experiments we investigate several alternative methods for generating the histograms. The methods are detailed in Sect. 5 in conjunction with the experiments.

In the final fourth stage the histogram descriptors of both training and test images are fed into supervised probabilistic classifiers, separately for each of the 20 object classes. As classifiers we use weighted C-SVC variants of the SVM algorithm, implemented in the version 2.84 of the software package LIBSVM [9]. As the kernel function g we employ the exponential χ^2-kernel

$$g_{\chi^2}(\mathbf{x}, \mathbf{x}') = \exp\left(-\gamma \sum_{i=1}^{d} \frac{(x_i - x_i')^2}{x_i + x_i'}\right) . \tag{17}$$

The χ^2-kernel provides histogram comparison performance superior to e.g. RBF kernel (e.g. [2]) and is thus often used in BoV systems.

The free parameters of the C-SVC cost function and the kernel function are chosen on basis of a search procedure that aims at maximising the six-fold cross validated area under the receiver operating characteristic curve (AUC) measure in the training set. To limit the computational cost of the classifiers, we perform random sampling of the training set. Some more details of the SVM classification stage can be found in [10].

4 Image Category Detection Task and Experimental Procedures

In the current experiments we consider the supervised image category detection problem. Specifically, we measure the performance of several algorithmic variants for the task using images and categories defined in the PASCAL NoE Visual Object Classes (VOC) Challenge 2007 collection [5]. In the collection there are altogether 9963 photographic images of natural scenes. In the experiments we use the half of them (5011 images) denoted "trainval" by the challenge organisers.

Each of the images contains at least one occurrence of the defined 20 object classes, including e.g. several types of vehicles (bus, car, bicycle etc.), animals and furniture. The presences of these objects in the images were manually annotated by the organisers. In many images there are objects of several classes present. In the experiments (and in the "classification task" of VOC Challenge) each object class is taken to define an image category.

In the experiments the 5011 images are partitioned approximately equally into training and test sets. Every experiment was performed separately for each of the 20 object classes. The category detection accuracy is measured in terms of non-interpolated average precision (AP) [11]. The AP values are averaged over the 20 object classes, resulting in mean average precision (MAP) values. In the experiment section of this paper, we report the average MAP values over the six different train/test partitionings, along with their 95% confidence intervals. The confidence intervals are based on the usual assumptions of normal distributions and are not to be taken literally. For the purpose of comparing various techniques, the confidence intervals arguably underestimate the reliability of the results because of systematic differences between the six trials.

5 Experiments and Results

In the experiments we kept other parts of our BoV implementation constant but varied the histogram generation stage. Three different types of histogram codebooks were generated by applying three different clustering algorithms to the same random sample of all the SIFT descriptors in all the images, containing 20 interest points from each image. As the baseline clustering we selected the codebook vectors randomly among the data points of the sample. Against this baseline we compared the K-means clustering based on both the squared Euclidean distance and the χ^2 distance. Given the set of codebook vectors, descriptors of all interest points were assigned to the histogram bin with the nearest codebook vector. In the case of random and Euclidean K-means codebooks, we employed the Euclidean distance as the basis for nearest bin selection. The χ^2 distance was used with codebooks generated by the χ^2 K-means. The experiments were repeated for several codebook sizes, ranging from 256 to 4096.

5.1 Quantisation Error

To see how well the different clustering algorithms solve the clustering problem they were devised for—defined in terms of quantisation error—we evaluated the mean quantisation errors defined in Sect. 2, i.e. C_{MSE} and C_{χ^2}. The errors were evaluated for the same sample of SIFT descriptors that was used to form the codebooks.

The quantisation errors are shown seen in Table 1. We note that in terms of C_{MSE}, Euclidean K-means is consistently somewhat better than K-means based on the χ^2 distance, and vice versa for C_{χ^2}. In terms of both errors, the K-means variants are clearly better than the random baseline. However, we also notice that in this kind of SIFT descriptor data set, the quantisation error differences between the K-means variants are not large. It thus seems that the squared Euclidean distance does not order SIFT descriptors very differently from the χ^2 distance. It remains to be seen in the next section, whether the orderings are different enough to result in significantly different performance in image category detection.

Table 1. Quantisation errors by the different clustering algorithms

	256 bins		512 bins		1024 bins		2048 bins		4096 bins	
	C_{MSE}	C_{χ^2}	C_{MSE}	C_{χ^2}	C_{MSE}	C_{χ^2}	C_{MSE}	C_{χ^2}	C_{MSE}	C_{χ^2}
random selection	0.362	2.28	0.331	2.12	0.306	1.98	0.283	1.85	0.259	1.71
Euclidean K-means	0.238	1.66	0.222	1.56	0.207	1.45	0.193	1.38	0.178	1.29
χ^2 K-means	0.248	1.58	0.231	1.49	0.222	1.43	0.206	1.34	0.191	1.25

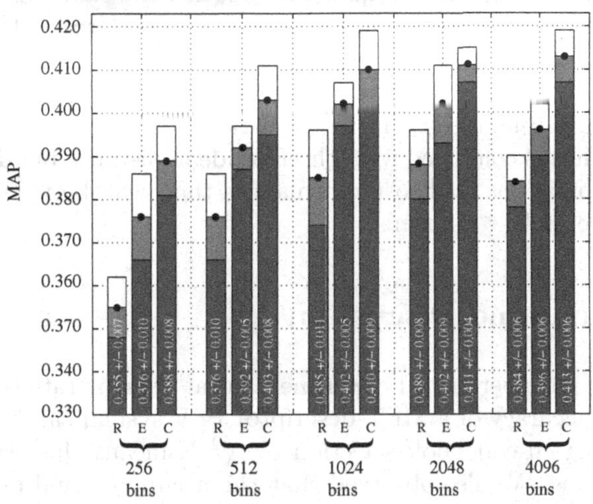

Fig. 2. Mean average precisions (MAP) in image category detection and their 95% confidence intervals. Bars corresponding to different clustering algorithms are indicated by letters R (random codebook selection), E (Euclidean K-means) and C (χ^2 K-means).

5.2 Category Detection Performance

The quantisation errors investigated in the previous section confirm that the clustering algorithms solve the problem they were devised for. However, in BoV systems clustering is used for a different purpose. There one hopes that optimising the quantisation error leads to codebooks that would also lead to good category detection performance. This sounds rather reasonable as these two different objectives can easily be believed to be rather similar. In our earlier experiments [6], however, we have seen that this is not always the case. Clusterings having smaller quantisation errors sometimes lead to worse category detection performance. Thus the performance of different methods for obtaining codebooks must be experimentally verified to see if they are useful for the purpose of BoV.

Figure 2 shows the image category detection performance in the category detection task of Sect. 4 when the three different histogram generation methods are inserted into the BoV category detection system of Sect. 3. In the SVM detector stage of the system, all the three types of histograms are regarded as feature vectors in the same way, i.e. the χ^2 kernel is used for comparing the histograms in all the cases. From the figure we see that both the K-means variants lead to better category detection performance than the random baseline. χ^2 distance based K-means performs still markedly better than K-means based on the squared Euclidean distance. This result is consistently seen with codebooks of all sizes. The development of performance with increasing codebook size is similar to what we have observed in our earlier experiments with several clustering algorithms. For this image categorisation task, the performance peaks with codebooks of a couple of thousand bins [6].

For the results included in the figure we applied the same distance measure for clustering and nearest histogram bin selection when forming the histograms, i.e. the Euclidean distance for Euclidean K-means and the χ^2 distance for χ^2 K-means. We also performed experiments where we mixed the distance measures: Euclidean K-means combined with the χ^2 distance based nearest bin determination, and χ^2 K-means combined with the Euclidean distance bin determination. Both these combinations lead to performances that were between those shown for the Euclidean and χ^2 K-means.

6 Conclusions and Discussion

In this paper we have performed experiments that demonstrate that image category detection accuracy of a SIFT-descriptor BoV system can be improved by basing the histogram codebook selection on χ^2 K-means clustering instead of Euclidean K-means. We also observed that the accuracy could be increased by selecting the nearest histogram bin based on the χ^2 distance instead of the Euclidean distance even if the codebook selection was not originally based on the χ^2 distance measure. We consider the likely explanation to be that the χ^2 distances between SIFT descriptors separate the 20 image categories in the considered VOC 2007 task better than the Euclidean distances do. We think it is plausible to hypothesise that this result would generalise also to other similar image category and object detection tasks. However, there is no reason to believe that even the χ^2 distance would be the most suitable distance measure. A possible future direction could be to continue along the same lines in the search for the optimal distance measure.

Selecting histogram codebooks with the Euclidean distance based K-means results in a statistically significant improvement in category detection MAP when compared with random codebook selection. The use of χ^2 K-means results in almost 80% larger improvement. However, in absolute terms the MAP improvements may be argued to be rather small. It naturally depends on the application whether or not the improvements in MAP are considered significant enough to justify the computational cost of the codebook selection algorithms.

In this paper we have evaluated the χ^2 distance based techniques in the case of SIFT descriptors of image neighbourhoods. The same kind of improvements could be expected for many other local descriptors that share the property of resembling histograms of local edge distribution [12,13], such as variants of the ColorSIFT descriptor [14]. This is left as a possible subject of further studies.

One could also experiment with the incorporation of the χ^2 distance in other clustering algorithms in addition to K-means.

Acknowledgments. Supported by the Academy of Finland in the *Finnish Centre of Excellence in Adaptive Informatics Research* project.

References

1. Sivic, J., Zisserman, A.: Video Google: A text retrieval approach to object matching in videos. In: Proc. of ICCV 2003, October 2003, vol. 2, pp. 1470–1477 (2003)
2. Zhang, J., Marszałek, M., Lazebnik, S., Schmid, C.: Local features and kernels for classification of texture and object categories: a comprehensive study. International Journal of Computer Vision 73(2), 213–238 (2007)
3. Lowe, D.G.: Distinctive image features from scale-invariant keypoints. International Journal of Computer Vision 60(2), 91–110 (2004)
4. Yang, J., Jiang, Y.-G., Hauptmann, A.G., Ngo, C.-W.: Evaluating bag-of-visual-words representations in scene classification. In: Proc. of MIR 2007, pp. 197–206 (2007)
5. Everingham, M., Van Gool, L., Williams, C.K.I., Winn, J., Zisserman, A.: The PASCAL Visual Object Classes Challenge (VOC2007) Results (2007), `http://www.pascal-network.org/challenges/VOC/voc2007/workshop/index.html`
6. Viitaniemi, V., Laaksonen, J.: Experiments on selection of codebooks for local image feature histograms. In: Sebillo, M., Vitiello, G., Schaefer, G. (eds.) VISUAL 2008. LNCS, vol. 5188, pp. 126–137. Springer, Heidelberg (2008)
7. Schalkoff, R.J.: Pattern Recognition: Statistical, Structural and Neural Approaches. John Wiley & Sons, Ltd., Chichester (1992)
8. Mikolajcyk, K., Schmid, C.: Scale and affine point invariant interest point detectors. International Journal of Computer Vision 60(1), 68–86 (2004)
9. Chang, C.-C., Lin, C.-J.: LIBSVM: a library for support vector machines (2001), `http://www.csie.ntu.edu.tw/~cjlin/libsvm`
10. Viitaniemi, V., Laaksonen, J.: Improving the accuracy of global feature fusion based image categorisation. In: Falcidieno, B., Spagnuolo, M., Avrithis, Y., Kompatsiaris, I., Buitelaar, P. (eds.) SAMT 2007. LNCS, vol. 4816, pp. 1–14. Springer, Heidelberg (2007)
11. TREC 2006 common evaluation measures. In: Proceedings of 15th Text Retrieval Conference (TREC 2006) (2006), `http://trec.nist.gov/`
12. Mikolajczyk, K., Schmid, C.: A performance evaluation of local descriptors. IEEE Transactions on Pattern Analysis and Machine Intelligence 27(10), 1615–1630 (2005)
13. Winder, S.A.J., Brown, M.: Learning local image descriptors. In: Proc. of IEEE CVPR 2007, pp. 1–8 (2007)
14. van de Sande, K.E.A., Gevers, T., Snoek, C.G.M.: Evaluation of color descriptors for object and scene recognition. In: Proc. of IEEE CVPR 2008, Anchorage, Alaska, USA (June 2008)

Modelling Image Complexity by Independent Component Analysis, with Application to Content-Based Image Retrieval

Jukka Perkiö[1,2] and Aapo Hyvärinen[1,2,3]

[1] Helsinki Institute for Information Technology
[2] Department of Computer Science
[3] Department of Mathematics and Statistics
P.O. Box 68, FI-00014, University of Helsinki, Finland
{jperkio,ahyvarin}@cs.helsinki.fi
http://www.hiit.fi

Abstract. Estimating the degree of similarity between images is a challenging task as the similarity always depends on the context. Because of this context dependency, it seems quite impossible to create a universal metric for the task. The number of low-level features on which the judgement of similarity is based may be rather low, however. One approach to quantifying the similarity of images is to estimate the (joint) complexity of images based on these features. We present a novel method to estimate the complexity of images, based on ICA. We further use this to model joint complexity of images, which gives distances that can be used in content-based retrieval. We compare this new method to two other methods, namely estimating mutual information of images using marginal Kullback-Leibler divergence and approximating the Kolmogorov complexity of images using Normalized Compression Distance.

Keywords: Image complexity; ICA; NCD; Kolmogorov complexity.

1 Introduction

Measuring image similarity is not a simple task. Similarity is always defined at two levels: The semantics and the syntax of an image. Two images containing cars may be judged similar based on the fact that there are cars in both of the images but on the other hand they may be judged dissimilar based on the make of the car. This is an example of the semantic level. Similarly two versions of the same image may be judged similar or dissimilar based on – for example – different colorspaces, which is an example of the syntactic level.

The semantics of an image are dependent on the context. When one decides whether the images containing cars are similar, it is the context that defines whether similarity is dependent on the bare fact that there are cars in the image or whether the make of the cars is also important. The less context-dependently

C. Alippi et al. (Eds.): ICANN 2009, Part II, LNCS 5769, pp. 704–714, 2009.

one defines similarity, the simpler the interpretation of semantics is and the more general the similarity measure is.

There certainly exist features which give a lot of information on the similarity of images. The problem is that sometimes one simply does not know what the discriminating features are and sometimes there are no clear dominating features. In general, manually selecting one or a few simple low level features works only for specific tasks, whereas using a large number of low level features raises the complexity of estimation process to impractical level.

The complexity of images is a universal property which is related to similarity. Intuitively it may be easy to decide between two images which one is more complex, but one can also imagine situation when semantically completely different images may appear equally complex. This is not a desirable result, hence complexity alone may not be very good measure of similarity or distance between images. If one is mostly interested in pair-wise distances, one can try remedy this by looking at the joint complexity of images versus the complexity of images separately [7]. The difference between complexity of a single image and the joint complexity of two images is more descriptive than arbitrary complexity values of arbitrary images alone. Of course – depending on the method used – these values have to be normalized appropriately.

Whether the difference between joint complexity and complexity of single image is good enough measure of similarity depends on the task in hand. As in all data-analysis, results depend a lot on the preprocessing and especially feature extraction. For example, measuring general image similarity may not require any specific feature extraction (pixel level intensity and color are the lowest level features and directly available) but if one wants to perform more specific tasks, the importance of features used grows. For specific tasks, there may be well established working methods and complexity-based measures of similarity may not be very attractive. On the other hand, the attractiveness of using complexity-based similarities is based on its universality, and the fact that in principle one can do this completely model-free—although the results will depend on the complexity measure chosen.

Two options for estimating the complexity of images are Shannon's classical information theory and algorithmic information theory. Although fundamentally different in some basic concepts, the two theories are connected [3]. Classical information theory have been utilized extensively in data analysis for clustering, feature selection, blind signal separation, etc. These methods maximize or minimize certain information theoretic measures. Kolmogorov complexity based similarity measures have been studied and used for different data [7,2]. In those papers the authors develop and use data compression based techniques to approximate the Kolmogorov complexity. They call the distance measure *normalized compression distance* [7].

Complexity-based methods have been applied to image analysis. In [1] these methods are applied to earth observation imagery and in [8] approximation of Kolmogorov complexity is applied to image classification. Both of the above papers use normalized compression distance as the measure of difference, hence they belong to the methods based on algorithmic information theory.

In this paper we present a new method based on a model that approximates the complexity of the data. The model that we use is *independent component analysis* (ICA) [5]. We first build the ICA model and then estimate the image complexity from the properties of the model. Our method can be justified from the information-theoretic framework, and it incorporates the sparsity of data in the complexity measure. Sparsity is a prominent statistical property of images which may not be well-captured by other methods.

The rest of this paper is organized as follows: In Section 2 we present our method and discuss it in the context of other complexity measures, namely measuring complexity by marginal Kullback-Leibler divergence and approximating Kolmogorov complexity. In Section 3 we present experiments using natural images and in Section 4 we present our conclusions.

2 Estimating Image Complexity

Given a general complexity measure $\mathcal{C}(x)$ for an image x one can try to estimate similarities between images. A naive assumption would be that the difference $|\mathcal{C}(x_0) - \mathcal{C}(x_1)|$ tells the similarity between images x_0 and x_1. Unfortunately such a general complexity measure does not exist. The closest thing that exists is the Kolmogorov complexity or algorithmic entropy $K(x)$ of the image (or any string) x. Kolmogorov complexity is not computable, however.

Even if the complexity measure $\mathcal{C}(x)$ existed or Kolmogorov complexity were computable, their value as measures of similarity would be questionable. Intuitively, the similarity between images does not always equal to the difference in complexity. This is because the context plays an important role even at the syntactic level, although not as much as in the semantic level.

An obvious way of introducing the context in the picture is to estimate the joint complexity of images. This is still at a very low level but estimating the complexity in the context of other image versus the complexity of single image is more informative than arbitrary complexity values alone. Hence we are interested in the distance that is defined as

$$D(x_0, x_1) = \mathcal{C}(x_0|x_1) - \min\{\mathcal{C}(x_0), \mathcal{C}(x_1)\}, \tag{1}$$

assuming that the joint complexity is symmetric, i.e. $\mathcal{C}(x_0|x_1) = \mathcal{C}(x_1|x_0)$. Also one wants to ensure that the distance is normalized appropriately.

As it was noted above the ideal complexity measure does not exist and Kolmogorov complexity is not computable. One can approximate the ideal complexity measure in different manners, however. Shannon's information theory introduced the concept of entropy, which is easily estimated from data. Entropy can be seen also as a statistical measure of complexity. Even though Kolmogorov complexity is not computable it can be approximated using compression based methods. Complexity can also be estimated from a model that approximates the log-pdf of data as we do in this paper.

2.1 Relative Entropy as Distance Measure

Given a discrete probability distribution P Shannon's entropy $H(x)$ is defined as

$$H(x) = -\sum_{x} P(x) \log P(x). \tag{2}$$

Entropy is a natural measure of complexity, since it estimates the degree of uncertainty with random variables. Intuitively it is appealing: The more uncertain we are about an outcome of an event, the more complex the phenomenon (data, image, etc.) is.

Given another distribution Q, the Kullback-Leibler divergence is defined as

$$KL(P||Q) = \sum_{x} P(x) \log \frac{P(x)}{Q(x)}. \tag{3}$$

KL-divergence is also called relative entropy and it can be interpreted as the amount of extra bits that is needed to code samples from P using code from Q. If the distributions are the same, the need for extra information is zero and the divergence is zero as well. KL-divergence is nonnegative but not symmetric and as such it can not be used directly as a measure of distance or dissimilarity between distributions. The symmetry is easy to obtain, however, just by calculating and summing the KL-divergence from Q to P and from P to Q, hence the symmetric[1] version is simply

$$KLS(P, Q) = KL(P||Q) + KL(Q||P). \tag{4}$$

This is not a true metric but it can be used directly as measure of distance or dissimilarity between distributions.

Using the symmetric version of KL-divergence (Eq. 4) as the pair-wise distance between two images is straight forward. It is not quite the ideal distance measure in Eq. 1, but it captures the idea of estimating the complexity in the context of another image.

2.2 Algorithmic Complexity

Kolmogorov complexity $K(x)$ of string x is the length of shortest program p using given description language L on a universal Turing machine U that produces the string x.

$$K(x) = \min_{p}\{|p| : U(p) = x\}, \tag{5}$$

[1] Actually this is the original formulation that Kullback and Leibler give [6].

where $|p|$ denotes the length of the program p. Kolmogorov complexity is not computable.

Conditional Kolmogorov complexity $K(x_0|x_1)$ of string x_0 given string x_1 is the length of shortest program that produces output x_0 from input x_1

$$K(x_0|x_1) = \min_p\{|p| : U(p|x_1) = x_0\}. \tag{6}$$

Normalized information distance [7] is based on the Kolmogorov complexity and is defined as

$$NID(x_0, x_1) = \frac{\max\{K(x_0|x_1), K(x_1|x_0)\}}{\max\{K(x_0), K(x_1)\}}. \tag{7}$$

As Kolmogorov complexity is not computable, NID neither is computable. It can be approximated, however, using the normalized compression distance (NCD) [7]. NCD approximates NID by using a real world compressor C and it is defined as

$$NCD(x_0, x_1) = \frac{C(x_0, x_1) - \min\{C(x_0), C(x_1)\}}{\max\{C(x_0), C(x_1)\}}. \tag{8}$$

To use the NCD for measuring pair-wise distances between images one just compresses images separately and concatenated and observes the difference between the compression results.

2.3 Using ICA as an Approximation for Entropy

A practical approximation of entropy can be attained by fixing some model which approximates the log-pdf. We propose here to use this approach, in connection with the model of independent component analysis (ICA), or equivalently sparse coding [4]. These models are widely used in statistical image modelling. In ICA, the pdf is approximated as

$$\log p(\mathbf{x}; \mathbf{W}) = \sum_i G(\mathbf{w}_i^T \mathbf{x}) + \log|\det \mathbf{W}| \tag{9}$$

where n is the dimension of the space, the $\mathbf{w}_i$ are linear features, collected together in the matrix $\mathbf{W}$. The function G is a non-quadratic function which measures the sparsity of the features; typically $G(u) = -|u|$ or $G(u) = -\log\cosh(u)$ are used. The latter can be considered as a smooth approximation of the former, which improves the convergence of the algorithm. A number of algorithms have been developed for estimation of the ICA model, in particular the matrix of features $\mathbf{W}$ [5].

After the model has been estimated, we can then approximate the complexity of $\mathbf{x}$ as

$$-E\{\log p(\mathbf{x}; \mathbf{W})\} = E\{-\sum_i G(\mathbf{w}_i^T \mathbf{x}) - \log|\det \mathbf{W}|\} \tag{10}$$

where the expectation is taken, in practice, over the sample.

An intuitive interpretation of the ensuing complexity measure is also possible. First, note that in ICA, the variance of the $\mathbf{w}_i^T\mathbf{x}$ is fixed to one. The first term on the right-hand-side in (10) can thus be considered as a measure of sparsity. In other words, it measures the non-Gaussian aspect of the components, completely neglecting the variance-covariance structure of the data. In fact, this term is minimized by sparse components. What is interesting is that the second term does measure the covariance structure. In fact, we have in ICA the well-known identity

$$2|\det \mathbf{W}| = |\det \mathbf{W}\mathbf{W}^T| = |\det C(\mathbf{x})|^{-1} \quad (11)$$

where $C(\mathbf{x})$ is the covariance matrix of the data. This formula shows that the second term in (10) is a simple function of the data covariance matrix. In fact, $\log|\det \mathbf{W}|$ is maximum if the data covariance has a minimum determinant. A minimum determinant for a covariance matrix is obtained if the variances are small in general, or, what is more interesting for our purposes, if some of the projections of the data have a very small variances. Since in ICA, we constrain the variances of the components to be equal to one, only the latter case is possible. Thus, our entropy measure becomes small if the data is concentrated in a subspace of a limited dimension.

Thus, this measure of entropy (complexity) is small if the components are very sparse, or if the data is concentrated in a subspace of limited dimension, both of which are in line with our intuition of structure of multivariate data.

Practicalities. Remembering the ideal complexity distance in Eq. 1 we present some remarks about the use of ICA model.

- Assuming that we want to estimate the distance between two images, we estimate the ICA model from both images separately and combined.
- The complexity value that we get using Eq. 10 is normalized in similar manner as the NCD in Eq. 8.
- In practice the ICA model for images is estimated from data that contains a large number of randomly sampled image patches.

3 Experiments

We wanted to evaluate how our method relates to other complexity based methods. For that we performed experiments using a subset of images in the University of Washington content-based image retrieval database[2].

We estimated the pair-wise distances between the subset of images using ICA, marginal KL-divergence and NCD. All the images were in RGB colorspace. The experiments were conducted as follows:

[2] http://www.cs.washington.edu/research/imagedatabase/groundtruth/

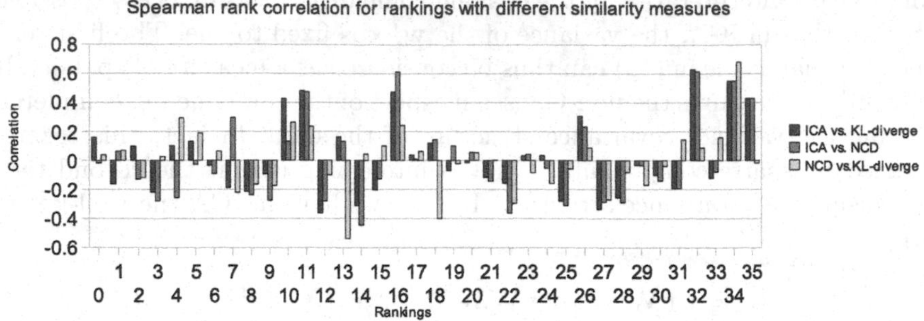

Fig. 1. The Spearman rank correlation between the different methods is showed when the test images are ranked relative to every other image shown. Within each experiment and ranking, the significance level $\alpha = 0.05$ is attained by an absolute value 0.26 or higher of correlation.

- The ICA models were estimated from data that contained 10,000 16×16 randomly sampled patches for each image. The data was normalized to be of zero-mean and of unit variance as is customary.
- Marginal KL-divergences were estimated from RGB intensity histograms.
- NCDs were estimated from RGB image matrices using zlib[3], which uses the DEFLATE algorithm for compression.

All the experiments were implemented in Python[4]. KL-divergence and NCD experiments were done for comparison. At this point we are not interested in image classification or clustering: We want to inspect the results visually and using some quantitative measure.

For the quantitative evaluation we turned the distances into rankings. This was done relative to every image in the data set. Rankings capture quite nicely the essential differences between the methods. For the rankings we calculated the Spearman rank correlation in order to understand the differences. Figure 1 shows for each image the rank correlation between all the methods we tried.

First, we observe that the correlations between rankings differ significantly depending on the image the ranking is relative to. This is actually somewhat surprising. Second, we notice that for most statistically significant correlations our method agrees more with both the KL-divergence- and the NCD-based methods, whereas the KL-divergence and NCD rankings are less correlated. This may suggest that our method captures more general features than the other two. Whether this works in real world applications is not sure though. Lastly we also observe surprisingly many negative correlations and the average correlation is rather low. This is different though if we only observe the absolute values of

[3] http://www.zlib.net/

[4] http://www.python.org

Fig. 2. Two-dimensional Sammon mapping calculated from the pair-wise distances between images, when the distances were estimated using ICA as an approximation for entropy. Even though the Sammon mapping is used to preserve the distances in the two dimensional visualization as well as possible, the individual rankings are not directly comparable to the mapping.

the correlation, which is justifiable, since correlation – negative or positive – is interesting, whereas non-correlated data does not tell us much.

Images 2 and 3 show two-dimensional Sammon mappings estimated from the pair-wise distances between images using ICA, KL-divergence and NCD respectively. Image 4 show example rankings for one reference image using all the methods.

Fig. 3. Two-dimensional Sammon mapping calculated from the pair-wise distances between images, when the distances were estimated using the KL-divergence (left) and compression-based approximation for Kolmogorov complexity, NCD (right). Even though the Sammon mapping is used to preserve the distances in the two dimensional visualization as well as possible, the individual rankings are not directly comparable to the mapping.

Visually inspecting it is clear that all the methods produce different results. It is harder to judge one better than the other, however.

It seems that the ICA method (Fig. 2, Fig. 4 left) is affected mostly by the texture of the images. It is able to nicely group different kinds of trees according to their appearance. The method do not seem to be very specific with regards to the grass appearing in the images.

For the marginal KL-divergence visual experiment (Fig. 3 left, Fig. 4 middle) the first impression is that it seem to be mostly affected by the different intensity in the lighting in the images. That is actually quite natural since the distances were estimated from RGB-intensity histograms. Nevertheless it also produces reasonable results.

The results for NCD visual experiment (Fig. 3 right, Fig. 4 right) are quite intuitive also but it is quite hard to find a common factor on which the grouping is based. NCD seems to be mostly affected by the complexity of rather low level features.

Finally one have to note that at their current state none of the methods presented can compete with more specialized application specific image similarity measures. The similarity that the methods measure is rather generic low level similarity. On the other hand that is exactly what one expects from complexity based similarity measures.

Fig. 4. An example of rankings produced by the three methods. The four rows below the reference image show two most similar and two least similar images to the reference image. The columns are from left to right ICA, KL-divergence and NCD.

4 Conclusions

We have presented a novel method to estimate image complexity in order to derive a pair-wise similarity measure for natural images. Our method is based on using ICA model to estimate the entropy of images separately and combined.

The similarity is derived from the normalized difference between the single image complexity and the pair-wise complexity. This method is comparable but not similar to other complexity based measures such as normalized compression distance and other information theoretic entropy based methods.

Based on quantitative analysis our method seem to be somewhere in between NCD and KL-divergence based distance measures. Visually all the methods tried, produce reasonable results, the ICA method being more responsive to textures.

For future work one has to consider applications of the method for clustering and classification, if not for other reasons than to get more decisive quantitative results than those obtained from the present analysis.

Acknowledgements

This work was supported in part by the IST Programme of the European Community under the PASCAL Network of Excellence, and in part by the Academy of Finland through the Finnish Center of Excellence for Algorithmic Data Analysis. The authors would also like to express their gratitude to Dr. Teemu Roos for providing the Sammon mapping code used in the visualizations.

References

1. Cerra, D., Mallet, A., Gueguen, L., Datcu, M.: Complexity based analysis of earth observation imagery: an assessment. In: ESA-EUSC 2008: Image Information Mining: pursuing automation of geospatial intelligence for environment and security (March 2008)
2. Cilibrasi, R., Vitányi, P.: Clustering by compression. IEEE Transactions in Information Theory 51, 1523–1545 (2005)
3. Grunwald, P., Vitanyi, P.: Shannon information and kolmogorov complexity (October 2004), `http://arxiv.org/abs/cs.IT/0410002`
4. Hyvärinen, A., Hurri, J., Hoyer, P.O.: Natural Image Statistics. Springer, Heidelberg (2009)
5. Hyvärinen, A., Karhunen, J., Oja, E.: Independent Component Analysis. Wiley Interscience, Hoboken (2001)
6. Kullback, S., Leibler, R.A.: On information and sufficiency. The Annals of Mathematical Statistics 22(1), 79–86 (1951)
7. Li, M., Chen, X., Li, X., Ma, B., Vitányi, P.: The similarity metric. IEEE Transactions in Information Theory 50, 3250–3264 (2004)
8. Li, M., Zhu, Y.: Image classification via lz78 based string kernel: A comparative study. In: PAKDD, pp. 704–712 (2006)

Adaptable Neural Networks for Objects' Tracking Re-initialization

Anastasios Doulamis

Technical University of Crete Decision Support Lab.,
Kounoupidiana, Chania, Greece
Tel.: (+30) 28210 37430
adoulam@cs.ntua.gr

Abstract. In this paper, we propose an automatic tracking recovery tool which improves the performance of any tracking algorithm each time the results are not acceptable. For the recovery, we include an object identification task, implemented through an adaptable neural network structure, which classifies image regions as objects. The neural network structure is automatically modified whenever environmental changes occur to improve object classification in very complex visual environments like the examined one. The architecture is enhanced by a decision mechanism which permits verification of the time instances in which track-ing recovery should take place. Experimental results on a set of different video sequences that present complex visual phenomena reveal the efficiency of the proposed scheme in proving tracking in very difficult visual content conditions. *abstract* environment.

Keywords: computer vision, object tracking.

1 Introduction

In today's world, security of citizens in public areas continually gains research interest. Computer vision tools, able to automatically detect and then recognize actions (simple or more complex) and behaviors by examining the raw video data, can play an important role. It is usual nowadays the most frequently used public areas and infrastructures to be monitored by multiple cameras and surveyed by specialized employees who are responsible for setting an alert in case of an emergency. However, it is very difficult for a human to continuously monitor different video sequences in which many humans act and behave, especially when the overwhelming majority of the content of such video files are of no important abnormal action. In addition, there is subjectivity as far as humans' perception is concerned. Different humans or even the same under different circumstances may interpret the same visual content differently affecting the security of an area. Consequently, it would be very useful if one can embed intelligent tools and mechanisms in surveillance systems which can (or at least assist to) detect abnormal actions and behaviors in critical public infrastructures and areas in which security of the visitors is of prime importance.

C. Alippi et al. (Eds.): ICANN 2009, Part II, LNCS 5769, pp. 715–724, 2009.

Probably the most important research aspect for detecting behaviors and actions in video surveillance systems is motion segmentation and tracking [1] [2]. Despite the fact, however, that background subtraction methods and temporal differencing techniques have been applied in the literature for motion segmentation [3], [4], in real-life surveillance applications there exists complex visual phenomena which makes those simple methods to be not efficiently applicable. For this reason, other complicated moving object detection/tracking algorithms have been proposed in the computer vision society for object tracking in complex visual environments, which can be discriminated into three main categories; motion models, search methods and appearance-based techniques [5]. In motion models, motion information is exploited to predict the new location of an object using either linear or non-linear approaches [6] [7]. The main drawback of these approaches, however, is that their accuracy is dropped in the existence of agile motion, distraction and occlusions. The search techniques exploit the assumption that objects' appearances do change from time to time and thus it presents similar properties within adjacent frames of a video sequence. Approaches towards this direction are the methods of [8] and [9]. These techniques, however, are also sensitive to background distractors, clutter, and occlusions issues.

Other approaches use stochastic methodologies, such as the Kalman filter [10] and the particle filter techniques [11]. The performance of a particle filter algorithm, however, actually depends on appearance models and the similarity measures used for object matching. For implementing accurate models for objects appearance visual descriptors are used such as color histograms [12], contours [13] and texture [14]. In real-life environments, however, the appearance models can change over time due to illumination variations, complex objects' motion, occlusions, image distortion phenomena, etc [5]. As stochastic approaches, one can include the Gaussian mixture models, [15], kernel density methods [16] and Hidden Markov Models. The performance of these approaches, however, still remains unacceptable when partial/full occlusions occur.

Generally, trackers are not involved re-initialization strategies. Though particle filter trackers support some naive re-initialization schemes, their performance deteriorates in case of sudden changes and occlusions. Objects characteristics vary from frame to frame. New objects may enter in the scene or others can disappear. Objects can be partially or fully occluded while their color/texture properties vary from time to time. Thus, it was very useful if we could be able to include intelligent mechanisms in any tracking process which would be able to re-initialize the tracker each time its performance is unacceptable. Such approaches, only very recently has been proposed in the literature combining tracking methods with object detection techniques [17] [18]. In particular the work of [17] uses a neural network structure to reclassify image regions as objects blocks. However, the technique is applicable only for stereoscopic video sequences or two-dimensional sequences presenting video conferences scenarios. Similarly, in [18] tracking of multiple objects is accomplished using a coupled optimization problem which combines a Minimum Description Length hypothesis framework which allows our system to recover from mismatches and temporarily lost tracks.

Despite its ability, the method of [18] suffers from reconfiguration which permits automatic tracker initialization necessary in broad domain application scenarios.

In this paper, we address these drawbacks by proposing a novel adaptable neural network architecture which is able to automatically recover the results of a tracking algorithm whenever its performance is not acceptable. The network labels in a non-linear fashion image regions as objects by exploiting visual descriptors appropriately. The network is designed so that, apart from approximating the non-linear functions used for object labeling it also handles the problem that the non-linear function to be modeled should be time varying due to the environmental changes. For this reason, an adaptable neural network structured is adopted as in [19]. In particular, the proposed adaptable strategy is implemented in a way that a) the non-linear model trusts as much as possible the current conditions, and (b) a minimal degradation of the already obtained knowledge is achieved. The proposed methodology is framed by a decision mechanism which defines the time instances in which a new tracking activation is required.

2 Overview of the Proposed Architecture

The adopted architecture improves the performance of any object tracking algorithm in complex visual conditions by incorporating an automatic tracking recovery mechanism. A block diagram of the architecture is shown in Figure 1.

The first component of the architecture is the tracking algorithm which identifies either correctly or erroneously the objects in a scene by taking into account an initial estimate of their position. Simultaneous to object tracking acts another component which labels image regions of a scene as objects taking into account their color/texture properties. Object identification is activated whenever the Decision Mechanism ascertains that the tracking performance is not acceptable with the use of a neural network structure. Since, however, the objects characteristics change form time to time, an adaptable learning strategy is implemented for the identification process to improve its performance in highly dynamic environments, like the examined one. The adaptable strategy should take into account information about the current visual content so as to update the models of the object labeler performance to the current conditions.

Finally, a decision mechanism is included in the proposed architecture which defines those time instances (frames) in which the tracking is not acceptable and thus recovery should take place. The mechanism exploits the probabilistic nature of the tracker as well as the evolution of the tracker through time.

3 Non-linear Object Identification

In this section, we propose a novel neural network structure able to automatically improve the performance of any tracking algorithm whenever a severe deterioration takes place. To identify objects, the neural network structure takes as input visual descriptors and then it classifies image regions as objects with respect to these descriptors. Since, significant variations in visual descriptors are expected

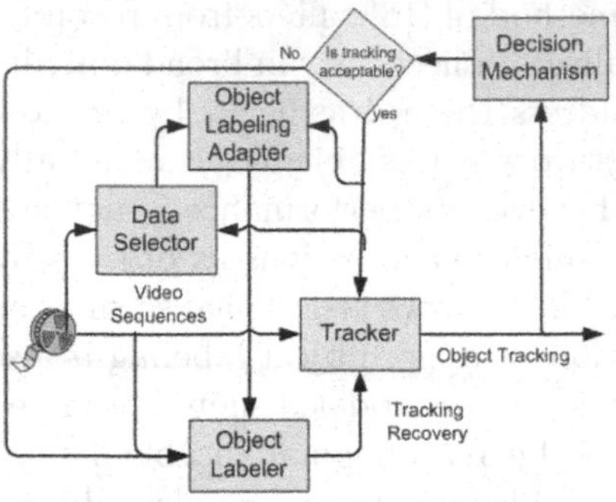

Fig. 1. proposed tracking recovery architecture

the proposed object identification scheme should vary form time to time to fit the dynamic changes of the environment.

Let us denote as $\mathbf{f}_i(n) \in R^M$ a feature vector of M visual descriptors for the nth frame of a sequence. Index ith corresponds to the ith image region of frame nth, for instance to the ith image block. Let us denote as $P_j^{(n)}(\mathbf{f}_i(n))$, $j = 1, 2, , L$ the probability of the i-th image region at n-th frame to be assigned to the j-th tracked object. Thus, we assume that L objects are available. The function $P_j^{(n)}(\mathbf{f}_i(n))$ is unknown and is modeled in our case by a feedforward neural network structure. That is,

$$P^{(n)}(\mathbf{f}_i(n)) \approx \mathbf{v}^T(n) \cdot \mathbf{\Phi}(\mathbf{W}(n) \cdot \mathbf{f}_i(n)) \tag{1}$$

where we have omitted the object index jth for simplicity purposes. In equation (1), we have assumed a one hidden layer neural network with one output neuron since we focus on classification of an object. For the case of identifying multiple objects, the scalar $P^{(n)}(\mathbf{f}_i(n))$ is transformed to a vector. The $\mathbf{v}(n)$ are the weights that connect the hidden neurons with the output neuron. Similarly, $\mathbf{W}(n)$ is matrix the columns of which are the weights that connect the input vector ele-ments with one hidden neuron. The $\mathbf{\Phi}(\cdot)$ is a vector-valued function which returns the activation functions of the hidden neurons. In the modeling of (1), we have assumed a linear relationship for the output neuron to yield any degree of relevance of an image region to an object. The unknown components of equation (1), i.e., the elements of vector $\mathbf{v}(n)$ and $\mathbf{W}(n)$ are estimated through a training process, which uses a reduced Levenberg-Marquardt (LM).

4 The Adaptation Strategy

The parameters of the neural network model should vary from time to time. Assuming that a slight modification of the non-linear function is adequate for modeling the following stage since the environmental conditions cannot be changed rapidly we can relate the model parameters as follows.

$$\mathbf{v}(n+1) = \mathbf{v}(n) + d\mathbf{v} \quad \mathbf{W}(n+1) = \mathbf{W}(n) + d\mathbf{W} \tag{2}$$

where $d\mathbf{v}$ and $d\mathbf{W}$ are small perturbations of parameters $\mathbf{v}$ and $\mathbf{W}$. Let us also assume that at the nth frame, a reliable mask for all the L available objects is derived through the tracking algorithm. Then, the labels for all the L tracked objects and the background can be considered as known. Thus,

$$P^{(n+1)}(\mathbf{f}_i(n)) = D_i(n) \tag{3}$$

where $D_i(n)$ are the labels (IDs) for the i-th image region at the n-th frame ranged in $[1L]$ since we have assumed that L objects are available. In (3), the superscript $(n+1)$ means that the output of the identification module is calculated using the new model parameters, i.e., the $\mathbf{v}(n+1), \mathbf{W}(n+1)$.

Exploiting equation (3), we can linearize equation (1) using a first order Taylor series expansion. Based on [19], the differences between the labels of an image regions before and after the adaptation, i.e., using the weights $\mathbf{v}(n+1)$ and $\mathbf{W}(n+1)$ and $\mathbf{v}(n)$, $\mathbf{W}(n)$ is linearly related with the small perturbations at time n while the parameters of the linear model only depend on the previous coefficients $\mathbf{v}(n)$, $\mathbf{W}(n)$. That is, it can be proved in [19] that

$$P^{(n+1)}(\mathbf{f}_i(n)) - P^{(n)}(\mathbf{f}_i(n)) = \mathbf{v}^T(n) \cdot \mathbf{H} \cdot d\mathbf{W} \cdot \mathbf{f}_i(n) + d\mathbf{v}^T \cdot \mathbf{\Phi}(\mathbf{W}(n) \cdot \mathbf{f}_i(n)) \tag{4}$$

which can be written as

$$P^{(n+1)}(\mathbf{f}_i(n)) = P^{(n)}(\mathbf{f}_i(n)) + \mathbf{J}d\mathbf{w} \tag{5}$$

where J is a matrix including elements coming from the current coefficients at nth iteration and the small increments. More information can be found in the appendix of [19].

Taking into account the effect of all image regions R for all the available objects L (including the background), we can form a vector, say $\mathbf{d}$, that contains all differences of equation (5). That is,

$$d\mathbf{w} = \mathbf{J}^{-1} \cdot \mathbf{d} \tag{6}$$

The number of unknown parameters $d\mathbf{w}$ depends on i) the number of visual descriptors used for modeling the content of an image region, and ii) the complexity of the hidden layer. As a result, three difference cases can be obtained. The first is the one that the number of network weights is greater than the number of linear equations of (6). Instead, for a small number of descriptors and network size it is probable that the unknowns of (6) to be smaller than the number of linear equations. Finally, when the number of unknowns equals the number of linear equations then, the small increments can be straightforwardly estimated by solving the linear system of (6).

The first two cases can be handled by the pseudo-inverse of matrix J.

5 New Data Selection and Decision Mechanism

In this section, we describe how can be obtain an estimation of (3). The labels $D_i(n)$ can be supervisedly (manually) provided but it is more efficient to be provided under an automatic framework. The proposed data selection algorithm exploits the tracking performance. In particular, initially, we detect all image regions that have been assigned to an object by the tracker. Then we estimate the region that is closest to the center of gravity of the tracking output. We assume that the most confident regions are within a 67% confidence in a Gaussian framework.

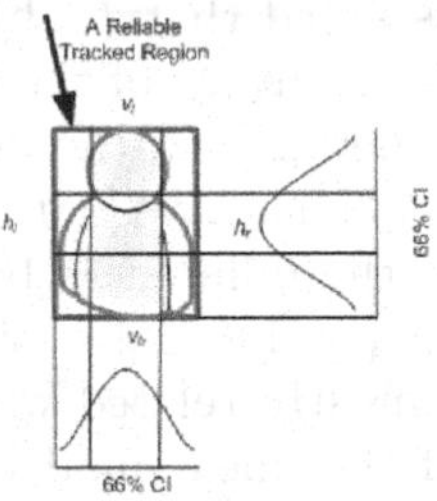

Fig. 2. A graphical representation of the proposed optimal data selection algorithm

Figure 2 presents a graphical representation of the proposed method adopted for optimal data selection. In this case, a reliable tracked mask has been detected and the most left, right, bottom and top lines of the region have been detected. Then, the center of gravity of the region is calculated and the standard deviation to achieve a very high confidence interval (i.e., 99.99%). Then, we select as data the ones lying within a 66% confidence interval.

The goal of the decision mechanism is to automatically detect those time instance (frames) which tracking recovery should take place since the performance of the tracking algorithm cannot be considered as acceptable. Our implementation includes i) an indicator about the performance of the tracking algorithm and ii) con-sistency between the tracking of successive frames.

If both criteria are active, meaning that a significant visual change of the environment takes place with a simultaneous low confidence of the tracke, the decision mechanism should be undoubtedly activated. In the vague case that one criterions is active while the other inactive, we include additional frames to verify the performance and with respect to their values we activate or not tracking re-initialization.

6 Experimental Results and Comparisons with/without Re-initialization

In the following, we evaluate the performance of the proposed tracking recovery algorithm in a set of different video sequences, which present complex phenomena such as occlusions, illumination changes, presence of multiple objects, etc. the proposed methodology is generic and it can be applied for any tracking algorithm. Thus, we need not to compare our approach with some other trackers but with its performance with or without initialization. Some sequences are publicly available, such as the PETS one, so as to compare our results under a common framework, while some others have been recorded under the framework of European Union funded research projects (such as POLYMNIA [20] and SCOVIS [21]).

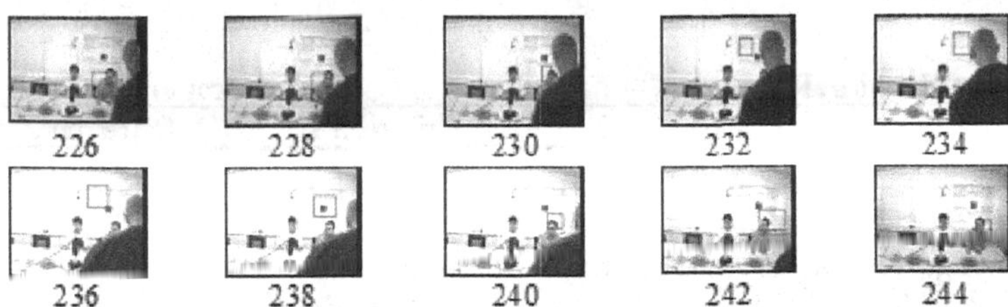

Fig. 3. Tracking results for a characteristic shot of PETS sequence using without the proposed recovery strategy

Figure 3 shows the results of a tracking algorithm (in our case using a particle filter) for a shot of PETS sequence. The specific shot depicts 19 frames in which a full occlusion is encountered. As we observe, the tracker performance deteriorates in the occluded regions since it is difficult in this case to monitor the correct trajectory of the objects. We also notice that tracking is deteriorated after the occlusion since the algorithm cannot initialize correct the samples at the previous video frames. The results after the proposed tracking recovery scheme are shown in Figure 4. We observe a significant improvement of the tracking performance, robust to the full occlusion.

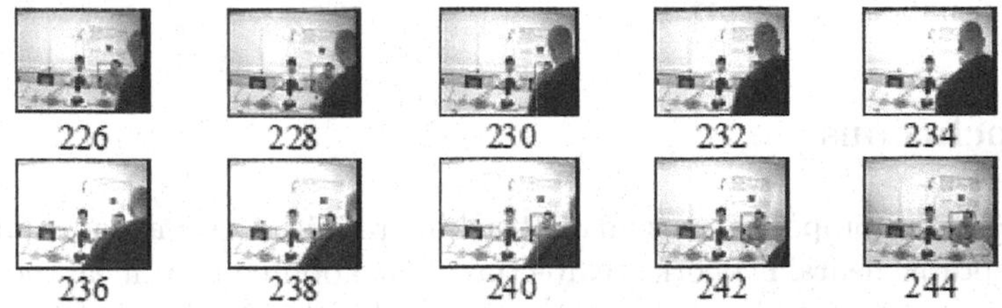

Fig. 4. Tracking results for a characteristic shot of PETS sequence using with the use of the proposed recovery strategy

Figure 5 shows the results of the adaptable object identification at frames before, during and after the full occlusion in which the tracking performance deteriorates. In all cases, blocks 8x8 have been detected as image regions, while the DC along with some of the 9 zig-zag scanned AC coefficients of each block are used as appropriate visual descriptors. We notice that correct labeling is accomplished even for this complex visual content case.

Fig. 5. The results of the adaptable object labeling module before, during and after the occlusion

Without Recovery		With Recovery	
Criterion 1	Criterion 2	Criterion 1	Criterion 2
46.78%	73.53%	48.67%	75.44%

Fig. 6. Average Values of both Objective Criteria over several different video sequences

Apart from the previous subjective evaluation, we have compared all these sequences using two objective criteria. The first expresses how close the tracked mask is with the reference one. This criterion is not adequate since it is possible large parts of the reference actual object to be located outside the tracked mask even though when C takes values close to one. This is for example the case when the tracked mask coincides with a part (even small) of the object. For this reason, we need another one, which presents the percentage of the reference object that is located within the tracked mask. In case of light changes the model will be robust as we select visual feature for object modeling that are also robust to light changes. Otherwise, the models will be modified and the automatic selection of the new training set will be inefficient.

Figure 6 shows the average performance for both criteria in case of 15,000 different frames of these sequences. It is clear that the proposed tracking recovery scheme improves the performance but this improvement is more evident in complex visual environments.

7 Conclusions

In this paper, we propose an automatic tracking re-initialization algorithm based on an adaptable neural network architecture. The adopted non-linear models are time-varying since the visual characteristics of the objects change from time to time. The architecture is enhanced with a decision mechanism able to verify the time instances in which tracking recovery from take place. The efficiency and robustness of the proposed scheme has been tested on a set of real-life video sequences in which complex motions (full and partial occlusions), illumination changes and presence of multiple objects in the scene are encountered. The evaluation has been performed subjectively by comparing the results among effective tracking methods (like the particle filter one) with the proposed recovery methodology. Additionally, two criteria are presented to objectively assess the tracking recovery performance and compare it with other approaches presented in the literature.

References

1. Jeyakar, J., Venkatesh Babu, R., Ramakrishnan, K.R.: Robust Object Tracking with Back-ground-weighted Local Kernels. Computer Vision and Image Understanding 112, 296–309 (2008)
2. Nascimento, J.C., Marques, J.S.: Performance Evaluation of Object Detection Algorithms for Video Surveillance. IEEE Trans. on Multimedia 8, 761–774 (2006)
3. Jodoin, P.M., Mignotte, M., Konrad, J.: Statistical Background Subtraction Using Spatial Cues. IEEE Trans. on Circuits and Systems for Video Technology 17, 1758–1763 (2007)
4. Tsai, D.-M., Lai, S.-C.: Independent Component Analysis-Based Background Subtraction for Indoor Surveillance. IEEE Trans. on Image Processing 18, 158–167 (2008)
5. Chen, D., Yang, J.: Robust Object Tracking via Online Dynamic Spatial Bias Appearance Models. IEEE Trans. on Pattern Analysis and Machine Intelligence 29, 2157–2169 (2007)
6. Lucas, B., Kanade, T.: An Iterative Image Registration Technique with an Application to Stereo Vision. In: DARPA Image Understanding Workshop, pp. 121–130 (1981)
7. Davatzikos, C., Prince, J., Bryan, R.: Image Registration based on Boundary Mapping. IEEE Trans. Medical Imaging 15, 112–115 (1996)
8. Shi, J., Tomasi, C.: Good Features to Track. In: Inter. Conf. Computer Vision and Pattern Recognition, pp. 593–600. IEEE Press, Washington (1994)
9. Comanicu, D., Ramesh, V., Meer, P.: Real-Time Tracking of Non-Rigid Objects Using Mean Shift. In: Int. Conf. Computer Vision and Pattern Recognition, pp. 142–149. IEEE Press, South Carolina (2000)
10. Medeiros, H., Park, J., Kak, A.: Distributed Object Tracking Using a Cluster-Based Kalman Filter in Wireless Camera Networks. IEEE Journal of Selected Topics in Signal Processing 2, 448–463 (2008)
11. Isard, M., Blake, A.: Condensation c Conditional Density Propagation for Visual Tracking. Int'l J. Computer Vision 29, 5–28 (1998)
12. Heisele, B., Kressel, J., Ritter, W.: Tracking Non-Rigid, Moving Objects Based on Color Cluster Flow. In: Int'l Conf. Computer Vision and Pattern Recognition, pp. 253–257. IEEE Press, Puerto Rico (1997)
13. Kass, M., Witkin, A., Terzopoulos, D.: Snakes: Active Contour Models. Int'l J. Computer Vision 1, 321–331 (1988)
14. Shahrokni, A., Drummond, T., Fua, P.: Texture Boundary Detection for Real-Time Tracking. In: European Conf. Computer Vision, Prague, pp. 566–577 (2004)
15. Wang, H., Suter, D., Schindler, K., Shen, C.: Adaptive Object Tracking Based on an Effective Appearance Filter. IEEE Trans. on Pattern Analysis and Machine Intelligence 29, 1661–1667 (2007)
16. Leichter, I., Lindenbaum, M., Rivlin, E.: Tracking by Affine Kernel Transformations Using Color and Boundary Cues. IEEE Trans. on Pattern Analysis and Machine Intelligence 31, 164–171 (2009)
17. Doulamis, A., Doulamis, N., Ntalianis, K., Kollias, S.: An efficient fully unsupervised video object segmentation scheme using an adaptive neural-network classifier architecture. IEEE Transactions on Neural Networks 14, 616–630 (2003)
18. Leibe, B., Schindler, K., Cornelis, N., Van Gool, L.: Coupled Object Detection and Tracking from Static Cameras and Moving Vehicles. IEEE Trans. on Pattern Analysis and Machine Intelligence 30, 1683–1698 (2008)

19. Doulamis, A., Doulamis, N., Kollias, S.: On-line Retrainable Neural Networks: Improving the Performance of Neural Networks in Image Analysis Problems. IEEE Trans. On Neural Networks 11, 137–155 (2000)
20. Doulamis, A., Kosmopoulos, D., Christogiannis, C., Varvarigou, T.: Polymnia: Personalised Leisure And Entertainment Over Cross Media Intelligent Platforms. In: European Workshop on Integration of Knowledge, Semantics and Digital Media Technology, London, vol. 25 (2004)
21. Doulamis, A., Kosmopoulos, D., Sardis, E., Varvarigou, T.: An Architecture for a Self Configurable Video Supervision. In: ACM Workshop on Analysis and Retrieval of Events, Actions and Workflows in Video Streams (2008)

Lattice Independent Component Analysis for fMRI Analysis*

Manuel Graña, Maite García-Sebastián, and Carmen Hernández

Grupo de Inteligencia computacional,
University of the Basque Country
www.ehu.es/ccwintco

Abstract. Pursuing an analogy to the Independent Component Analysis (ICA) we propose a Lattice Independent Component Analysis (LICA), where ICA signal sources correspond to the so-called endmembers and the mixing matrix corresponds to the abundance images. We introduce an approach to fMRI analysis based on a Lattice Computing based algorithm that induces endmembers from the data. The endmembers obtained this way are used to compute the linear unmixing of each voxel's time series independently. The resulting mixing coefficients roughly correspond to the General Linear Model (GLM) estimated regression parameters, while the set of endmembers corresponds to the GLM design matrix. The proposed approach is model free in the sense that the design matrix is not fixed *a priori* but induced from the data. Our approach does not impose any assumption on the probability distribution of the data. We show on a well known case study that this unsupervised approach discovered activation patterns are similar to the ones detected by an Independent Component Analysis (ICA).

1 Introduction

Human brain mapping is a rapidly expanding discipline, and in recent years the interest in novel methods for imaging human brain functionality has grown. Noninvasive techniques can measure cerebral physiologic responses during neural activation. One of the relevant techniques is functional Magnetic Resonance Imaging (fMRI) [13], which uses the blood oxygenation level dependent (BOLD) contrast to detect physiological alterations, such as neuronal activation resulting in changes of blood flow and blood oxygenation. Since these methods are completely noninvasive, using no contrast agent or ionizing radiation, repeated single-subject studies are becoming feasible [12].

The fMRI experiment consists of a functional template or protocol (e.g., alternating activation and rest for a certain time) that induces a functional response in the brain.The aim of an fMRI experiment is to detect the response to this time varying stimulus, through the examination of the signal resulting from the BOLD effect, in a defined volume element (voxel). The functional information of a voxel has to be extracted from its time series. One fMRI volume is recorded at each sampling time instant during the experiment. The frequency of the time sampling being determined by the

* The Spanish Ministerio de Educacion y Ciencia supports this work through grant DPI2006-15346-C03-03.

C. Alippi et al. (Eds.): ICANN 2009, Part II, LNCS 5769, pp. 725–734, 2009.

resolution of the fMRI imaging pulse sequence. The complete four-dimensional dataset (three spatial dimensions plus one time dimension) consists of subsequently recorded three-dimensional (3-D) volumes. The acquisition of the complete series of functional volumes runs over periods lasting up to several minutes.

The most extended analysis approach for fMRI signals is the Statistical Parametric Maps (SPM) [5,6] which has evolved into a free open source software package. This method consists in the separate voxel estimation of the regression parameters of General Linear Model (GLM), whose design matrix has been built corresponding to the experimental design. A contrast is then defined on the estimated regression parameters, which can take the form of a t-test or an F-test. The theory of Random Fields is then applied to correct the test thresholds, taking into account the spatial correlation of the independent test results.

Approaches to fMRI analysis based on the Independent Component Analysis (ICA) [4] assume that the time series observations are linear mixtures of independent sources which can not be observed. ICA assumes that the source signals are non-Gaussian and that the linear mixing process is unknown. The solutions to the ICA problem obtain both the independent sources and the linear unmixing matrix. These approaches are unsupervised because no *a priori* information about the sources or the mixing process is included, hence the alternative name of Blind Deconvolution.

In the present work we propose the use of an heuristic algorithm, called Endmember Induction Heuristic Algorithm (EIHA) described in detail in [7] to attack the fMRI analysis problem. The basic assumption in this approach is that the data is generated by a hidden process as a convex combination of a set of endmembers which are the vertices of a convex polytope covering the data observations. This assumption is similar to the linear mixture assumed by the ICA approach, however EIHA does not impose any probabilistic assumption on the data. This EIHA algorithm falls more properly in the field of Lattice Computing algorithms [8]. The endmembers discovered by the EIHA are equivalent to the GLM design matrix columns, and the unmixing process is identical to the conventional least squares estimator. Therefore, our approach is a kind of unsupervised GLM whose regressor functions are discovered in the data. When we establish an analogy with the ICA, the endmembers correspond to the unknown ICA sources and the mixing is solved by least squares estimation.

We call Lattice Independent Component Analysis (LICA) the overall process of applying EIHA and computing the unmixing process that gives the abundance matrices (in remote sensing terminology). The EIHA relies on the conjecture that Strong Lattice Independent sets of vectors are Affine Independent, and, therefore, the vertices of the convex polytope that explains (contains) the data. The algorithm searches for these Strong Lattice Independent vectors by using the properties of Lattice Autoassociative Memories (LAM). The main advantages that LICA can produce respect to ICA for data analysis are the lack of strong probabilistic assumptions (independence, non-Gaussianity) that may fail in many realistic situations. Besides the EIHA is a computationally light algorithm that works on one pass over the data and does not need optimization steps.

The outline of the paper is as follows: Section 2 introduces the linear mixing model so that the proposed approach can be understood. Section 3 presents an sketch of the theoretical relation between Lattice Independence and Linear (Affine) Independence through the LAM theory. Section 4 recalls the definition of our Endmember Induction Heuristic Algorithm (EIHA). Section 5 gives a brief recall of ICA. Section 6 presents results of the proposed approach on a case study. Section 7 provides some conclusions.

2 The Linear Mixing Model

The linear mixing model can be expressed as follows: $\mathbf{x} = \sum_{i=1}^{M} a_i \mathbf{s}_i + \mathbf{w} = \mathbf{S}\mathbf{a} + \mathbf{w}$, where $\mathbf{x}$ is the d-dimension pattern vector corresponding to the fMRI voxel time series vector, $\mathbf{S}$ is the $d \times M$ matrix whose columns are the d-dimension vertices of the convex region covering the data corresponding to the so called endmembers $\mathbf{s}_i$, $i = 1, .., M$, $\mathbf{a}$ is the M dimension fractional abundance vector, and $\mathbf{w}$ is the d-dimension additive observation noise vector. The heuristic algorithm EIHA described in [7] provides the estimation of the endmembers from the data. We can not review EIHA here due to the lack of space. The linear mixing model is subjected to two constraints on the abundance coefficients. First, to be physically meaningful, all abundance coefficients must be non-negative $a_i \geq 0, i = 1, .., M$. Second, to account for the entire composition, they must be fully additive $\sum_{i=1}^{M} a_i = 1$. That means that we expect the vectors in $\mathbf{S}$ to be affinely independent and that the convex region defined by them includes *all* the data points.

Once the convex region vertices have been determined the unmixing process is the computation of the matrix inversion that gives the coordinates of the point relative to the convex region vertices. The simplest approach is the unconstrained least squared error (LSE) estimation given by: $\widehat{\mathbf{a}} = \left(\mathbf{S}^T\mathbf{S}\right)^{-1} \mathbf{S}^T\mathbf{x}$. The coefficients that result from this equation do not necessarily fulfill the non-negativity and full additivity conditions. Moreover, the EIHA [7] always produces convex regions that lie inside the data cloud, so that enforcing the non-negative and additivity to one conditions would be impossible for some data points. Negative values are considered as zero values and the additivity to one condition is not important as long as we are looking for the maximum abundances to assign meaning to the resulting spatial distribution of the coefficients. These coefficients are interpreted as the regressor coefficients corresponding to the decomposition of the fMRI voxel time series into the set of endmembers. That is, high positive values are interpreted as high positive correlation with the time pattern of the corresponding endmember. The interpretation of the endmember time series pattern is rather straightforward in some cases (i.e. the background noise), but it is difficult in general. Therefore, the unmixing process aims to to find regions of related behavior, as it done in ICA-based studies [4].

3 Lattice Independence and Lattice Autoassociative Memories

The work on Lattice Associative Memories (LAM) stems from the consideration of the algebraic lattice structure $(\mathbb{R}, \vee, \wedge, +)$ as the alternative to the algebraic framework given by the mathematical field $(\mathbb{R}, +, \cdot)$ for the definition of Neural Networks computation. The operators $\vee$ and $\wedge$ denote, respectively, the discrete max and min operators (resp. sup and inf in a continuous setting). Given a set of input/output pairs of pattern $(X,Y) = \{(\mathbf{x}^\xi, \mathbf{y}^\xi); \xi = 1,..,k\}$, a linear heteroassociative neural network based on the pattern's cross correlation is built up as $W = \sum_\xi \mathbf{y}^\xi \cdot (\mathbf{x}^\xi)'$. Mimicking this constructive procedure [14,15] propose the following constructions of Lattice Memories (LM): $W_{XY} = \bigwedge_{\xi=1}^k \left[\mathbf{y}^\xi \times (-\mathbf{x}^\xi)'\right]$ and $M_{XY} = \bigvee_{\xi=1}^k \left[\mathbf{y}^\xi \times (-\mathbf{x}^\xi)'\right]$, where $\times$ is any of the $\boxtimes$ or $\boxtimes$ operators. Here $\boxtimes$ and $\boxtimes$ denote the max and min matrix product [14,15]. respectively defined as follows: $C = A \boxtimes B = [c_{ij}] \Leftrightarrow c_{ij} = \bigvee_{k=1,...,n} \{a_{ik} + b_{kj}\}$, and $C = A \boxtimes B = [c_{ij}] \Leftrightarrow c_{ij} = \bigwedge_{k=1,...,n} \{a_{ik} + b_{kj}\}$.

Definition 1. *Given a set of vectors* $\{\mathbf{x}^1, ..., \mathbf{x}^k\} \subset \mathbb{R}^n$ *a* linear minimax combination *of vectors from this set is any vector* $\mathbf{x} \in \mathbb{R}^n_{\pm\infty}$ *which is a* linear minimax sum *of these vectors:* $x = \mathcal{L}(\mathbf{x}^1, ..., \mathbf{x}^k) = \bigvee_{j\in J} \bigwedge_{\xi=1}^k (a_{\xi j} + \mathbf{x}^\xi)$, *where J is a finite set of indices and* $a_{\xi j} \in \mathbb{R}_{\pm\infty}\ \forall j \in J$ *and* $\forall \xi = 1, ..., k$.

Definition 2. *The* linear minimax span *of vectors* $\{\mathbf{x}^1, ..., \mathbf{x}^k\} = X \subset \mathbb{R}^n$ *is the set of all linear minimax sums of subsets of X, denoted* $LMS(\mathbf{x}^1, ..., \mathbf{x}^k)$.

Definition 3. *Given a set of vectors* $X = \{\mathbf{x}^1, ..., \mathbf{x}^k\} \subset \mathbb{R}^n$, *a vector* $\mathbf{x} \in \mathbb{R}^n_{\pm\infty}$ *is* lattice dependent *if and only if* $x \in LMS(\mathbf{x}^1, ..., \mathbf{x}^k)$. *The vector* $\mathbf{x}$ *is* lattice independent *if and only if it is not lattice dependent on X. The set X is said to be* lattice independent *if and only if* $\forall \lambda \in \{1, ..., k\}$, $\mathbf{x}^\lambda$ *is lattice independent of* $X \backslash \{\mathbf{x}^\lambda\} = \{\mathbf{x}^\xi \in X : \xi \neq \lambda\}$.

Definition 4. *A set of vectors* $X = \{\mathbf{x}^1, ..., \mathbf{x}^k\} \subset \mathbb{R}^n$ *is said to be* max dominant *if and only if for every* $\lambda \in \{1, ..., k\}$ *there exists and index* $j_\lambda \in \{1, ..., n\}$ *such that*

$$x_{j_\lambda}^\lambda - x_i^\lambda = \bigvee_{\xi=1}^k \left(x_{j_\lambda}^\xi - x_i^\xi\right) \forall i \in \{1, ..., n\}.$$

Similarly, X is said to be min dominant *if and only if for every* $\lambda \in \{1, ..., k\}$ *there exists and index* $j_\lambda \in \{1, ..., n\}$ *such that*

$$x_{j_\lambda}^\lambda - x_i^\lambda = \bigwedge_{\xi=1}^k \left(x_{j_\lambda}^\xi - x_i^\xi\right) \forall i \in \{1, ..., n\}.$$

Definition 5. *A set of lattice independent vectors* $\{\mathbf{x}^1, ..., \mathbf{x}^k\} \subset \mathbb{R}^n$ *is said to be* strongly lattice independent *(SLI) if and only if X is max dominant or min dominant or both.*

Conjecture 1. [17] If $X = \{\mathbf{x}^1, ..., \mathbf{x}^k\} \subset \mathbb{R}^n$ is strongly lattice independent then X is affinely independent.

4 Endmember Induction Heuristic Algorithm (EIHA)

Let us denote $\{\mathbf{f}(i) \in \mathbb{R}^d; i = 1,..,n\}$ the time series in fMRI voxels, $\overrightarrow{\mu}$ and $\overrightarrow{\sigma}$ are, respectively, the mean vector and the vector of standard deviations computed componentwise over the voxels, α the noise correction factor and E the set of already discovered vertices. The gain parameter α controls the amount of flexibility in the discovering

1. Shift the data sample to zero mean $\{\mathbf{f}^c(i) = \mathbf{f}(i) - \overrightarrow{\mu}; i = 1, .., n\}$.
2. Initialize the set of vertices $E = \{\mathbf{e}_1\}$ with a randomly picked sample. Initialize the set of lattice independent binary signatures $X = \{\mathbf{x}_1\} = \left\{\left(e_k^1 > 0; k = 1, .., d\right)\right\}$
3. Construct the LAM's based on the lattice independent binary signatures: M_{XX} and W_{XX}.
4. For each pixel $\mathbf{f}^c(i)$
   (a) compute the noise corrections sign vectors$\mathbf{f}^+(i) = (\mathbf{f}^c(i) + \alpha\overrightarrow{\sigma} > \mathbf{0})$ and $\mathbf{f}^-(i) = (\mathbf{f}^c(i) - \alpha\overrightarrow{\sigma} > \mathbf{0})$
   (b) compute $y^+ = M_{XX} \boxtimes \mathbf{f}^+(i)$
   (c) compute $y^- = W_{XX} \boxtimes \mathbf{f}^-(i)$
   (d) if $y^+ \notin X$ or $y^- \notin X$ then $\mathbf{f}^c(i)$ is a new vertex to be added to E, execute once 3 with the new E and resume the exploration of the data sample.
   (e) if $y^+ \in X$ and $\mathbf{f}^c(i) > \mathbf{e}_{y^+}$ the pixel spectral signature is more extreme than the stored vertex, then substitute $\mathbf{e}_{y^+}$ with $\mathbf{f}^c(i)$.
   (f) if $y^- \in X$ and $\mathbf{f}^c(i) < \mathbf{e}_{y^-}$ the new data point is more extreme than the stored vertex, then substitute $\mathbf{e}_{y^-}$ with $\mathbf{f}^c(i)$.
5. The final set of endmembers is the set of original data vectors $\mathbf{f}(i)$ corresponding to the sign vectors selected as members of E.

Algorithm 1. Endmember Induction Heuristic Algorithm (EIHA)

of new endmembers. The detailed description of the steps in the heuristic algorithm is presented as Algorithm 1. The starting endmember set consists of a randomly picked pixel. However, this selection is not definitive, because the algorithm may later change this endmember for another, more extreme, one. The noise correction parameter α has a great impact on the number of endmembers found. Low values imply large number of endmembers. It determines if a vector is interpreted as a random perturbation of an already selected endmember. This algorithm does not need a priori information about the nature of the data points that we want to detect. It runs once over the image and finds the most salient data samples on the fly. For this reason we say that it is an on-line algorithm.

5 Independent Component Analysis (ICA)

The Independent Component Analysis (ICA) [11] assumes that the data is a linear combination of non Gaussian, mutually independent latent variables with an unknown mixing matrix. The ICA reveals the hidden independent sources and the mixing matrix. That is, given a set of observations represented by a d dimensional vector $\mathbf{x}$, ICA assumes a generative model $\mathbf{x} = \mathbf{As}$, where $\mathbf{s}$ is the M dimensional vector of independent sources and $\mathbf{A}$ is the $d \times M$ unknown basis matrix. The ICA searches for the linear transformation of the data $\mathbf{W}$, such that the projected variables $\mathbf{Wx} = \mathbf{s}$ are as independent as possible. It has been shown that the model is completely identifiable if the sources are statistically independent and at least $M - 1$ of them are non Gaussian. If the sources are gaussian the ICA transformation could be estimated up to an orthogonal transformation. Estimation of mixing and unmixing matrices can be done maximizing diverse objective functions, among them the non gaussianity of the sources and the likelihood of the sample. We have used the FastICA [10] algorithm available at http://www.cis.hut./projects/ica/fastica

Application of ICA to fMRI has been reviewed by [4]. Reports on the research application of ICA to fMRI signals include the identification of signal types (task related and physiology related) and the analysis of multisubject fMRI data. The most common approach is the spatial ICA that looks for spatial disjoint regions corresponding to the identified signal types. It has been claimed that ICA has identified several physiological noise sources as well as other noise sources (motion, thermodynamics) identifying task related signals. Diverse ICA algorithms have been tested in the literature with inconclusive results. Among them, fastICA, the one that we will apply in the case study, did identify the task related signals consistently. Among the clinical applications, ICA has been used to study the brain activation due to pain in healthy individuals versus those with chronic pain [1], the discrimination of Alzheimer's patients from healthy controls [9], the classification of schizophrenia [2] and studies about the patterns of brain activation under alcohol intoxication [3].

6 A Case Study

The experimental data corresponds to auditory stimulation test data of single person[1]. These data are the result of the preprocessing pipeline that removes many noise sources. These whole brain BOLD/EPI images were acquired on a modified 2T Siemens MAGNETOM Vision system. Each acquisition consisted of 64 contiguous slices. Each slice being a 2D image of one head volume cut. There are 64x64x64 voxels of size 3mm x 3mm x 3mm. The data acquisition took 6.05s, with the scan-to-scan repeat time (RT) set arbitrarily to 7s., 96 acquisitions were made (RT=7s) in blocks of 6, i.e., 16 blocks of 42s duration. The condition for successive blocks alternated between rest and auditory stimulation, starting with rest. Auditory stimulation was bi-syllabic words presented binaurally at a rate of 60 per minute. Due to T1 effects it is advisable to discard the first few scans (there were no "dummy" lead-in scans). We have discarded the first 10 scans.

Voxel time series are normalized susbtracting the mean value of each voxel time series independently, so that the plots are collapsed around the origin. This mean substraction corresponds to an scale normalization in the Lattice Computing sense. It removes scale effects that hinder the detection of meaningful lattice independent vectors. In the context of the GLM this normalization corresponds to the estimation of the voxel linear model offset.

The application of the EIHA algorithm with $\alpha = 20$ to the lattice normalized time series of the whole 3D volume produces the collection of eleven endmembers shown in figure 1. Attending to the intensity scale it can be assumed that the first endmember (top left plot) corresponds to the background (thermodynamical) noise, while the remaining endmembers correspond to some kind of hemodynamic response pattern or noise source. To identify the endmember which is closer to modeling the task, we compute the correlation of the endmembers with the square wave represent the task. We

[1] The dataset is freely available from ftp://ftp.fil.ion.ucl.ac.uk/spm/data, the file name is snrfM00223.zip. The functional data starts at acquisition 4, image snrfMOO223-004.

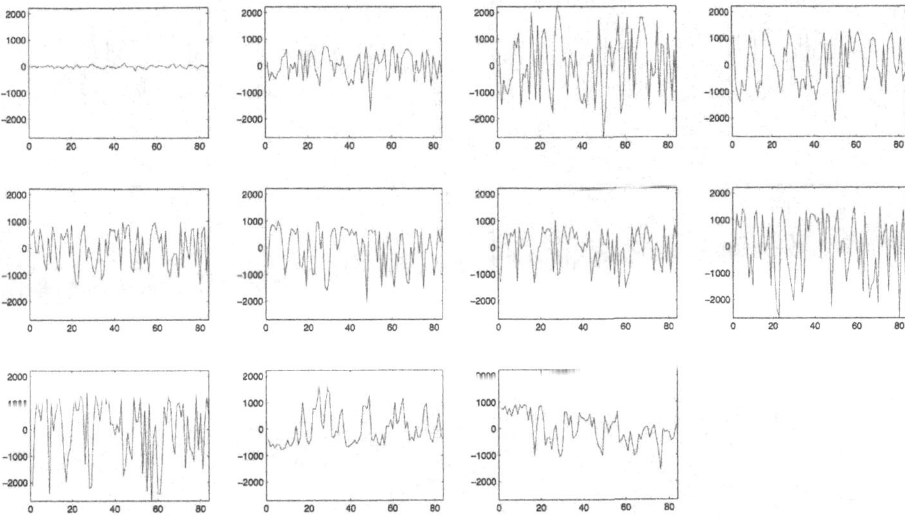

Fig. 1. Eleven endmbers detected by EIHA over the lattice normalized time series of the whole 3D volume

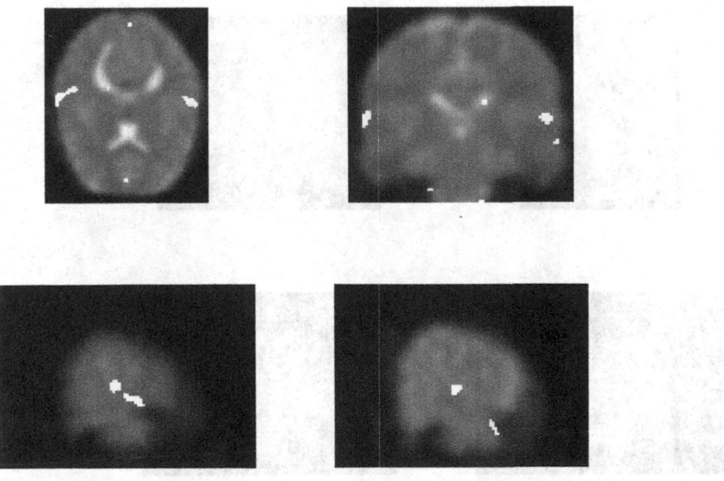

Fig. 2. Detected task related activations for endmember #9 from figure 1. White voxels correspond to abundance values above the 99% percentile of the distribution of the abundances for this endmember on the whole volume.

find that endmember #9 (counting row-wise in figure 1) has the maximum such correlation. We present in figure 2 the activations corresponding to it, where the slices shown roughly try to show the region of the auditory cortex were the task-related activations are expected. Top row is the axial and coronal cut, and the bottom row shows two

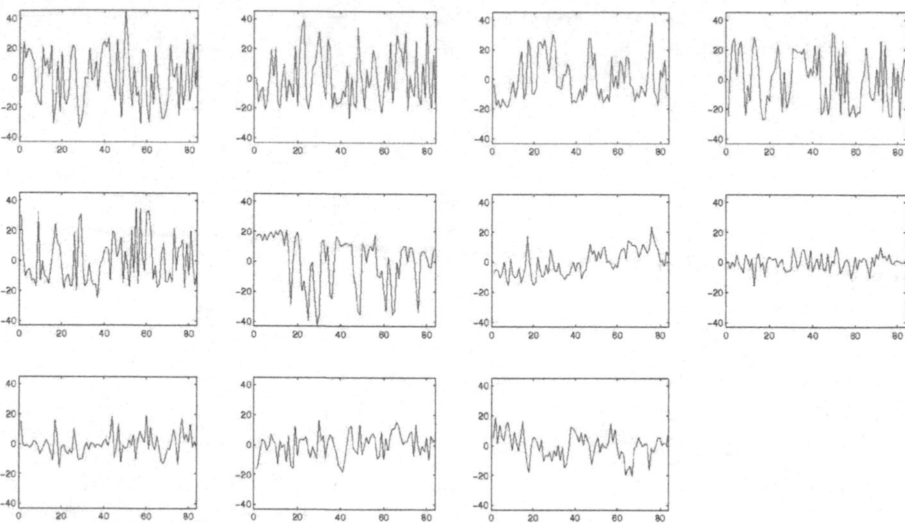

Fig. 3. Eleven time series sources detected by fastICA over the lattice normalized time series of the whole 3D volume

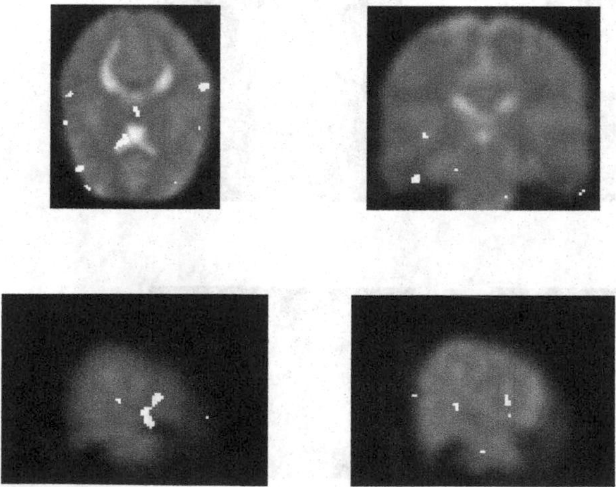

Fig. 4. Detected task related activations for source #6 from figure 3. White voxels correspond to mixing values above the 99% percentile of the distribution of the mixing coefficients for this source on the whole volume.

sagital cuts at both sides of the brain. White voxels in this figure correspond to voxels with abundance value above the 99% percentile of the distribution of this endmember abundance over the whole volume. It can be appreciated that most of the activations fall in the auditory cortex region.

The application of the fastICA algorithm with the number of sources set to 11, to match the number of endmembers found by the EIHA, to the lattice normalized time series of the whole 3D volume produces the collection of eleven endmembers shown in figure 3. Counting row-wise, source #8 may correspond to the background noise, while source #6 is the one most correlated with the task. Figure 4 shows the axial, coronal and sagital cuts corresponding to the auditory cortex, organized like in figure 2, with activation clusters, computed as the 99% percentile of the distribution of the spatial mixing values for source #6 over all the volume, superimposed. The task-related activations are localized in the auditory cortex, but they are less coherent than the EIHA found ones.

7 Conclusions and Discussion

We have proposed and applied the endmember induction algorithm EIHA discussed in [7] to the model-free (unsupervised) analysis of fMRI. We have discussed the similarities of our approach to the ICA application to fMRI activation detection [4,18]. In our approach the temporal sources correspond to endmembers detected by the EIHA algorithm and the spatial mixing coefficients correpond to the abundance volumes obtained by unmixing the voxel time series on the basis of the found endmembers.

The first obstacle that we find in this endeavor is that the distribution of the fMIR voxel time series is not well aspected for the detection of Lattice Independence as a meaningful characteristic. In fact the voxel's fMRI time series show a dense distribution of intensity displacements from the origin, so almost all of them are lattice dependent and our proposed algorithm only identifies two endmembers on the raw data. To overcome this problem we apply a Lattice Normalization which corresponds to a scale normalization in the sense of Lattice Computing. We substract the mean of its time series to each voxel time series. The resulting lattice normalized data set shows a much more rich structure in terms of Lattice Independence. Our computational experiment with a well known fMRI data set, provided with the distribution of the SPM software, show some promising results in the sense that we can at least identify a task related endmember, and that other effects, such as thermodynamical noise, are also clearly identified. We have found also a strong agreement of the spatial localizations of a task related source found by the fastICA on the same dataset, with the ones provided by our approach. Future works must address works on clinical research, and the extension of the approach to groups and multiple groups.

References

1. Buffington, A.L., Hanlon, C.A., McKeown, M.J.: Acute and persistent pain modulation of attention-related anterior cingulate fMRI Activations. Pain 113(12), 172–184 (2005)
2. Calhoun, V.D., Kiehl, K.A., Liddle, P.F., Pearlson, G.D.: Aberrant localization of synchronous hemodynamic activity in auditory cortex reliably characterizes schizophrenia. Biol. Psych. 55(8), 842–849 (2004)
3. Calhoun, V.D., Pekar, J.J., Pearlson, G.D.: Alcohol intoxication effects on simulated driving: Exploring alcohol-dose effects on brain activation using functional MRI. Neuropsychopharmacology 29(11), 2097–2107 (2004)

4. Calhoun, V.D., Adali, T.: Unmixing fMRI with independent component analysis. Engineering in Medicine and Biology Magazine, IEEE 25(2), 79–90 (2006)
5. Friston, K.J., Holmes, A.P., Worsley, K.J., Poline, J.P., Frith, C.D., Frackowiak, R.S.J.: Statistical parametric maps in functional imaging: A general linear approach. Hum. Brain Map 2(4), 189–210 (1995)
6. Friston, K.J., Ashburner, J.T., Kiebel, S.J., Nichols, T.E., Penny, W.D. (eds.): Statistical Parametric Mapping, the analysis of functional brain images. Academic Press, London (2007)
7. Graña, M., Villaverde, I., Maldonado, J.O., Hernandez, C.: Two Lattice Computing approaches for the unsupervised segmentation of Hyperspectral Images. Neurocomputing (2008) DOI 10.1016/j.neucom.2008.06.026
8. Graña, M.: A Brief Review of Lattice Computing. In: Proc. WCCI 2008, pp. 1777–1781 (2008)
9. Greicius, M.D., Srivastava, G., Reiss, A.L., Menon, V.: Default-mode network activity distinguishes alzheimers disease from healthy aging: Evidence from func- tional MRI. Proc. Nat. Acad. Sci. U.S.A. 101(13), 4637–4642 (2004)
10. Hyvarynën, A., Oja, E.: A fast fixed-point algorithm for independent component analysis. Neural Comp. 9, 1483–1492 (1999)
11. Hyvarynën, A., Karhunen, J., Oja, E.: Independent Component Analysis. John Wiley & Sons, New York (2001)
12. Muller, H.-P., Kraft, E., Ludolph, A., Erne, S.N.: New methods in fMRI analysis. Engineering in Medicine and Biology Magazine, IEEE 21(5), 134–142 (2002)
13. Pekar, J.J.: A brief introduction to functional MRI. Engineering in Medicine and Biology Magazine, IEEE 25(2), 24–26 (2006)
14. Ritter, G.X., Sussner, P., Diaz-de-Leon, J.L.: Morphological associative memories. IEEE Trans. on Neural Networks 9(2), 281–292 (1998)
15. Ritter, G.X., Diaz-de-Leon, J.L., Sussner, P.: Morphological bidirectional associative memories. Neural Networks 12, 851–867 (1999)
16. Ritter, G.X., Gader, P.: Fixed points of lattice transforms and lattice associative memories. In: Hawkes, P. (ed.) Advances in Imaging and Electron Physics, vol. 144, pp. 165–242. Elsevier, San Diego (2006)
17. Ritter, G.X., Urcid, G., Schmalz, M.S.: Autonomous single-pass endmember approximation using lattice auto-associative memories Neurocomputing (2008) (in press)
18. Sarty, G.E.: Computing Brain Activation Maps from fMRI Time-Series Images. Cambridge University Press, Cambridge (2007)

Adaptive Feature Transformation for Image Data from Non-stationary Processes

Erik Schaffernicht[1], Volker Stephan[2], and Horst-Michael Gross[1]

[1] Ilmenau University of Technology
Neuroinformatics and Cognitive Robotics Lab
98693 Ilmenau, Germany
[2] Powitec Intelligent Technologies GmbH
45219 Essen-Kettwig, Germany
Erik.Schaffernicht@Tu-Ilmenau.de

Abstract. This paper introduces the application of the feature transformation approach proposed by Torkkola [1] to the domain of image processing. Thereto, we extended the approach and identifed its advantages and limitations.

We compare the results with more common transformation methods like Principal Component Analysis and Linear Discriminant Analysis for a function approximation task from the challenging domain of video-based combustion optimization. It is demonstrated that the proposed method generates superior results in very low dimensional subspaces.

Further, we investigate the usefulness of an adaptive variant of the introduced method in comparison to basic subspace transformations and discuss the results.

1 Introduction

Optimizing the combustion of coal in power plants is an important task, since increasing efficiency equals a reduction of carbon oxides (CO and CO2), nitrogen oxides (NOx) and other greenhouse gases in the flue gas. But all data normally measured at a plant is insufficient to build meaningful models and controllers. Therefore, our approach includes cameras to actively observe the flame itself. On one hand, with this additional information about the combustion process improved controllers can be built automatically. On the other hand, relying on image data introduces additional challenges.

The use of the original pixel space for learning algorithms that operate on image data is a rare occurrence. The high dimensionality of this space is a major obstacle in this respect, because this leads to a high complexity of the learning problem and a high number of free parameters to be estimated for an approximation or classification task. Furthermore, the feasibility of this approach is restricted by the computational effort required to handle the data.

Hence, preprocessing is applied to extract useful information from the original images. One way to achieve this is the use of designed features like certain geometric shapes, intensity values or certain texture patterns. This implicitly

C. Alippi et al. (Eds.): ICANN 2009, Part II, LNCS 5769, pp. 735–744, 2009.

requires at least a bit expert knowledge by the system designer to decide which methods are meaningful for the given problem.

Another way to cope with the problem are feature transformation algorithms which attempt to find an image subspace that contains much useful information. Typically these methods are guided by a statistical criterion to achieve this goal. Perhaps the best known representatives are *Principal Component Analysis* (PCA) [2], *Independent Component Analysis* (ICA) [3], *Nonnegative Matrix Factorization* (NMF) [4] and *Linear Discriminant Analysis* (LDA) [2]. The basic forms of these algorithms produce linear transformations only, but there are several non-linear (e.g. kernel-based) extensions for all methods, but PCA and ICA specifically attracted a lot of attention in this respect.

The PCA transforms data into a subspace based on the eigenvectors of the data covariance matrix, hence this produces axes along the most variant parts of the data. High eigenvalues mark high variant directions. The resulting subspaces are often named according to the task, like *eigenfaces* or, for combustion optimization, *eigenflames*. This technique, as well as ICA and NMF, are purely data driven. They only consider the data intrinsic relations, but not the recognition or approximation task to be solved. ICA tries to find subspaces that represent independent data parts. A contrast function like *Negentropy* or *Mutual Information* is used to measure the independence of the resulting subspace dimensions. The NMF transformation's unique selling point is that all subspace dimensions and resulting data points are in fact non negative, which is a constraint for certain application areas.

Unlike the aforementioned methods, algorithms like the LDA take the target of the learning problem into account to find a suitable subspace representation. It derives itself from the Fisher criterion [5] and aims at a subspace transformation that allows a good approximation with linear learning machines.

The *Maximal Mutual Information* (MMI) transformation introduced by Torkkola [1] is similar in this respect. It takes the target values into account, but unlike the LDA it does not make any assumptions about a specific learning machine. Instead, it tries to maximize the information content about the target in the new subspace. The basic ideas and mechanisms of this approach are recapped in Sect. 2.

The application of this approach to image data is straightforward, but requires the consideration of its limitations for this high dimensional domain. Additionally, we propose an supplemental step in the algorithm to capture image specific traits. A comparison to PCA approaches on a flame image prediction task completes Sect. 3.

Since our intended application area, the intelligent control of combustion processes in power plants, is non-stationary, the feature extraction's requirements include a certain adaptivity. A comparison of different initializations, PCA and LDA is given, and we will discuss the use of the MMI transformation as adaptive system and the pitfalls associated in Sect. 4.

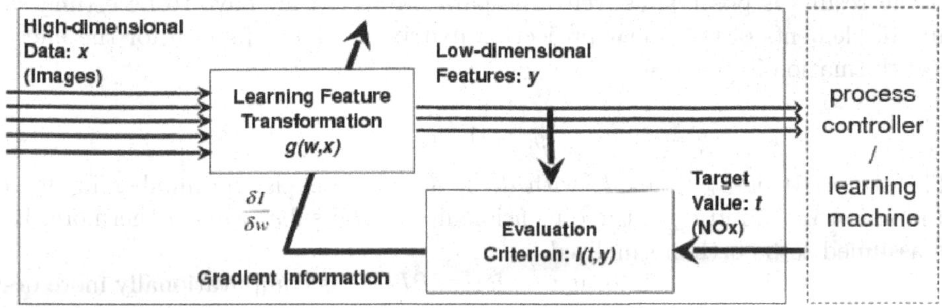

Fig. 1. The original image data x is transformed by some function g into a lower-dimensional space. An evaluation criterion, the Quadratic Mutual Information, measures the correspondence to the desired target value t, e.g. the nitro oxides to the reduced images y. From this criterion, a gradient information $\delta I/\delta w$ is derived and used to adapt the transformation parameters w.

2 Feature Extraction Using Mutual Information Maximization

The Maximal Mutual Information approach of Torkkola [1] is built upon the *Information-Theoretic Learning* (ITL) framework introduced by Principe [6]. The idea is to find a transformation that maximizes the *Mutual Information* (MI) between the transformed data in a certain subspace and the desired target values. A number of "forces" is computed to be used as the direction in a gradient ascent to maximize the MI.

The basic adaption loop for the optimization process is shown in Fig. 1. The original input data sample x_i is transformed by some transformation g with the parameters w into a lower dimensional space. The transformed data is denoted by y_i. The goal is to find those transformation parameters w that confer the most information into the lower dimensional space with respect to the target.

The update rule for the parameters of the transformation is given by the following equation, where α denotes the learning rate

$$w_{t+1} = w_t + \alpha \frac{\partial I}{\partial w} = w_t + \alpha \sum_{i=1}^{N} \frac{\partial I}{\partial y_i} \frac{\partial y_i}{\partial w}. \tag{1}$$

Finding the gradient $\partial I/\partial w$ can be split into the sample wise computation of the information forces $\partial I/\partial y_i$ and the adaption of the parameters $\partial y_i/\partial w$.

The second part is the simple one, since there exists a number of suitable transformations g, e.g. linear transformations or neural networks like Radial Basis Function Networks [1] or Multi Layer Perceptrons [7]. The only requirement is that they have to use the gradient information $\partial y_i/\partial w$ to adapt their parameters. All following examinations are limited to the linear transformation case, because this allows easy comparison with PCA or LDA and a visual inspection of the results is possible as well. The parameters w that have to be estimated are all elements of the linear projection matrix W. The equation for the linear transformation is given by

$$y_i = W^T x_i. \tag{2}$$

The size of W is d_x times d_y with $d_x > d_y$ where d_x is the number input of dimension in X and d_y is the dimensionality of the subspace. Furthermore, W is assumed to be orthonormalized.

The calculation of the information forces $\partial I / \partial y_i$ is computationally more demanding. The straightforward approach would be to use the well known *Mutual Information*

$$I(Y,T) = \int_y \int_t P(y,t) \log \frac{P(y,t)}{P(y)P(t)} dtdy \tag{3}$$

to evaluate the correspondence between the transformed data and the target values. But due to the associated problems of estimating this criterion in high dimensional spaces, Torkkola proposes a non-parametric estimation based on *Quadratic Mutual Information* I^2

$$I^2(Y,T) = \int_y \int_t (p(y,t) - p(y)p(t))^2 dtdy \tag{4}$$

and kernel density estimation with Parzen windows. Application of the binomial formula splits equation 4 in three parts which are interpreted as information potentials and the derivatives as information forces.

$$\frac{\partial I^2}{\partial y_i} = \frac{\partial V_{IN}}{\partial y_i} + \frac{\partial V_{ALL}}{\partial y_i} - 2\frac{\partial V_{BTW}}{\partial y_i} \tag{5}$$

V_{IN} represents the "attractive potential" of all samples with the same/similar target value, V_{ALL} is the same but for all samples, and V_{BTW} is the "repulsive potential" (negative sign) between samples of different target values. The derivatives show the direction each sample has to move to maximize the objective function. The actual computation of these terms is reduced to interactions between all pairwise samples using Gaussian kernel density estimates. The reader is referred to [1] for the details that are omitted here.

In Algorithm 1 the procedure for one adaption step is given. These steps are repeated until convergence of the parameters w.

3 Image Data Processing

According to Torkkola [1], the previously described system is suitable for small input dimensions, but higher dimensions can be problematic. On one hand, image data is intrinsically high dimensional, because each pixel position is considered an input. On the other hand, treating each pixel as an independent input channel neglects the fact that neighbor pixels from the camera are dependent on each

Algorithm 1. Maximal Mutual Information Adaption Step

Input: current transformation W_t, the input data X and the target values T
Output: new transformation W_{t+1}

$Y = g(W, X) = W^T X$ // computation of the transformed data
$\frac{\partial I^2}{\partial y_i} = \frac{\partial V_{IN}}{\partial y_i} + \frac{\partial V_{ALL}}{\partial y_i} - 2\frac{\partial V_{BTW}}{\partial y_i}$ //estimation of the different forces
$\frac{\partial Y}{\partial W} = X^T$ //The gradient of the linear transformation matrix W
$W'_{t+1} = W_t + \alpha \frac{\partial I}{\partial W} = w_t + \alpha \sum_{i=1}^{N} \frac{\partial I}{\partial y_i} \frac{\partial y_i}{\partial W}$ //Adaptation step
$W'_{t+1} =$ GaussianFilter(W'_{t+1}) //Supplemental step for images, see Sec. 3
$W_{t+1} =$ GramSchmidt(W'_{t+1}) //Orthonormalization step to ensure $W^T W = I$

other. We assume that informative parts of the image are not defined at pixel level, but by a more general, arbitrary shaped region, that is approximated on the pixel level. Thus, it is very unlikely that neighboring pixel have a rank different information content.

To cope with this problem and forcing the filter to consider these neighborhood dependencies, we introduced an additional step into the algorithm. After computing the new filter according to the gradient information and before the orthonormalization step, we perform a smoothing with a Gaussian filter in the two dimensional image space on the filter mask. This does not only distribute information between neighbor input dimensions, but increases stability and convergence speed, because the algorithm finds smooth solutions. An additional benefit is the obvious reduction of measurement noise in the observations.

This approach is not suitable for images only, but every continuous domain that is sampled and approximated at certain points and has a clear neighborhood definition.

We used 1.440 small images of the size 40x32 pixels which equals 1.280 input dimensions for each sample. All images are flame pictures taken from a coal burning power plant. An example image with a higher resolution is shown in Fig. 2. The respective targets are measurements like the nitrogen oxides (NOx), carbon monoxides (CO) or excess air (O2) produced by the combustion. We used Multi Layer Perceptrons as function approximators and evaluated the performance of different instances of the ITL framework with different parameters like the dimensionality of transformed data, and compared them with results from PCA based transformations.

The first tests were made with a visual examination of the resulting filter masks, similar to the well known eigenfaces. Images with obvious structures are considered "stable" solutions, while "unstable" results are characterized by high frequent noise and no structures in the filter masks. See Fig. 3 for examples on a higher resolution (134x100 pixels). Interestingly, if stable solutions were found, they tend to be similar to each other, besides differences in the sign of the filter masks, which relates to the same axis but the opposite direction. More discussions on this topic are following in Sec. 4.

Different initializations for the transformation parameters w result in the clear preference for PCA or LDA, since randomly initialized filters tend to produce

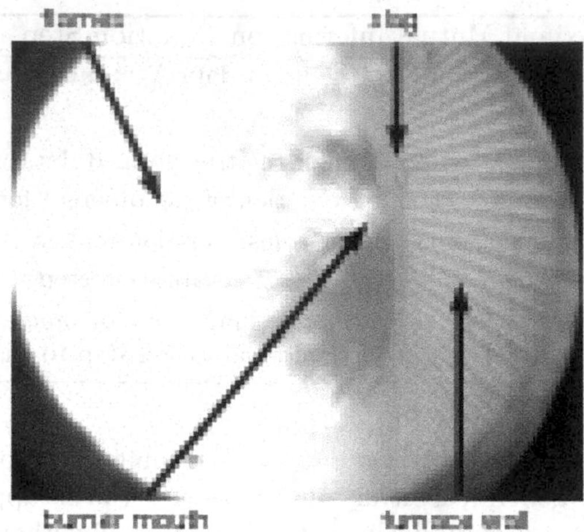

Fig. 2. An example of a black and white image taken from the furnace of a coal fired power plant. Clearly visible is the furnace located wall on the right. Roughly in the middle of the images is the burner mouth were coal dust is inserted to the furnace and ignites. Around this area on the wall, slag (molten ash) is visible.

unstable results. Hence the MMI method is more of an objective driven refinement for these plausible starting guesses.

The possible dimensionality of the reduced feature space d_y is greatly dependent on the number of available samples. This makes sense with respect to the curse of high dimensionality, because the higher the dimensionality, the more difficult it is to estimate the required probability distributions. For the presented setup of data we noticed two things: First, the bigger d_y, the more it deviates from a PCA initialization. Second we observed that the breaking point, where it switches from stable subspace transformations to unstable results, is between $d_y = 4$ and $d_y = 5$. By doubling the number of samples to 3.600, we get stable results in the five dimensional subspace, but $d_y = 6$ and higher remain unstable.

Further experiments where conducted with images subscaled to an even smaller size of 10 x 8 pixels per image. The reductions of the input dimensionality d_x does not improve the results considerably. On the other hand, using images with 160x120 pixels decreased stable results to 3 dimensions. This is due to the linear connection between d_x and the number of parameters w compared to the exponential influence of d_y as discussed above.

The next experiments are conducted to test whether the MMI subspace transformation yields any improvements compared to PCA-based *eigenflames*. Taking the previous results into account, the target dimensionality is limited to $d_y <= 3$ and the MMI subspace search started with an PCA initialization.

The results clearly demonstrate the benefits of the MMI approach. The approximation errors are smaller or at least in the same magnitude of the PCA-based approach. By adding more channels, the PCA can achieve similar results to the MMI transformation, but there is always the need of additional dimensions

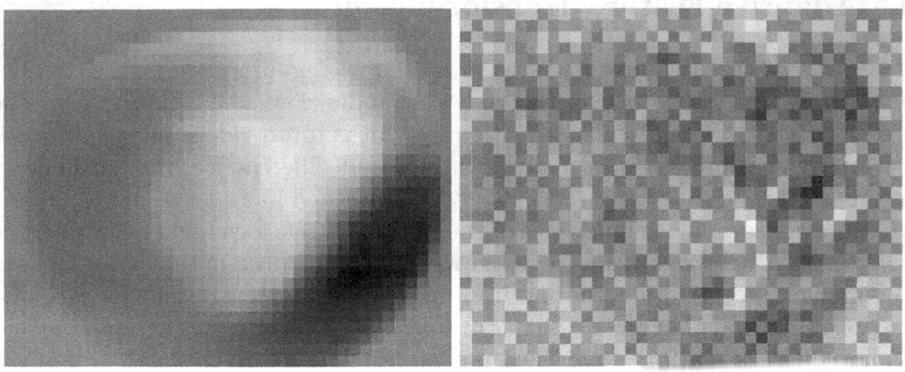

Fig. 3. (Left) A stable filter mask. **(Right)** An unstable one. Both subspace transformations are the results of the optimization procedure described in this section and depict the first dimension of the new subspace. The white areas are coding positive values, the black regions negative ones, while the gray areas are near zero and thus unimportant, like the round margin in the left image. One important fact to remember is, that the sign does not tell anything about the importance of this pixel, while the absolute value does. This kind of visualization is comparable to Eigenflames produced by PCA, besides in this case it doesn't depict the variances in the data, but the information.

Table 1. Comparison of the same MLP trained with PCA subspace features or MMI subspace features respectively for three different targets. All prediction errors are the MSE from an independent test set. The high level of noise present in the data leads sometimes to the effect of increasing errors when providing additional input features.

	Prediction Error for CO		Prediction Error for O2		Prediction Error for NOx	
d_y	PCA	MMI	PCA	MMI	PCA	MMI
1	3.11	3.07	0.90	0.24	28.88	25.99
2	3.33	2.43	0.25	0.29	35.50	25.00
3	4.07	2.66	0.22	0.28	27.65	30.26

to represent the information. Hence, we conjecture that the MMI method is able to compress the relevant information better than the PCA *eigenflames*.

One negative aspect concerning the MMI approach are the computational costs associated with the density estimation and gradient computation. While PCA is fast to compute, MMI takes a lot of time (which is mainly dependent on the number of input images used). For several thousand images it can easily take one or two hours to obtain the filter masks. Hence, the MMI methods can be applied only if there are no hard time constraints.

To conclude this section, the experiments show that it is beneficial to use the MMI system to improve PCA based subspace transformation for image data.

4 Adaptive Feature Transformation

It is assumed that the presented system is used as a preprocessor for a controller or function approximator which is able to handle slow adaptations itself. The goal for the adaptive feature extraction system is to provide similar features if the underlying process is in a similar state, and different features in different process states.

There are several configurations of the subspace transformation parameters w possible that achieve a maximal value with respect to the optimization goal of maximizing the Quadratic Mutual Information even for the optimistic case of a single, global maximum. For example, a scaling of the matrix W with a non-zero scalar will not change the information content of the results. All but two of the possible solutions are eliminated during the orthonormalization step. This step restrains all configurations in the parameter space to the hyper unit sphere. The two remaining valid solutions are w^* and $-w^*$. These transformations obviously contain the same information, since the only discriminating feature between data transformed by the two filter is the inverted sign. This behavior is not desired since the same state can yield two different subspace transformations that produce opposite transformed data, which are completely different to the system using the transformed features.

If the process is stationary, this problem can be overcome by the use of a suitable similarity measure to compare the old filter configuration w_{old} to w_{new} and $-w_{new}$ accepting the better match. But it is quite hard to define good similarity measures and thresholds if the process is transient. The most obvious work around is a different starting initialization. Instead of starting from the PCA subspace, it is possible to use the previous MMI subspace as initialization point. Assuming that the transient process changes are slow, compared to the adaptive updates of the filter, these changes will yield slow changes of the relevant feature areas. Thus, the subspace transformations will be similar to each other, which justifies the use of the previous solution as a starting point.

The actual adaptivity can be achieved on different time scales. One possibility is to adapt the current filter into the new one after a few measurements, using the techniques described in [1](Appendix A) where not the whole available data is used for a adaption step but only a small subset. The extreme case is the use of two samples. Torkkola draws them randomly, while for an online system these samples are the last measurements. For those samples one adaption step is applied (see Algorithm 1).

For applications with very noisy measurements, this may introduce the problem that the systems tries to adapt to the noise, rather than the underlying process changes. Hence, slower timescales change the procedure to collecting a certain amount of data before performing a batch update of the filter.

For the online application of the system of the power plant, we are interested mainly in very slow changes induced by wear and tear of the furnace or coal type changes. There are other changes on a much faster timescale, but they are even harder to detect, due to the presence of a high measurement noise. For the experiments presented here, a daily batch update was used. We used the

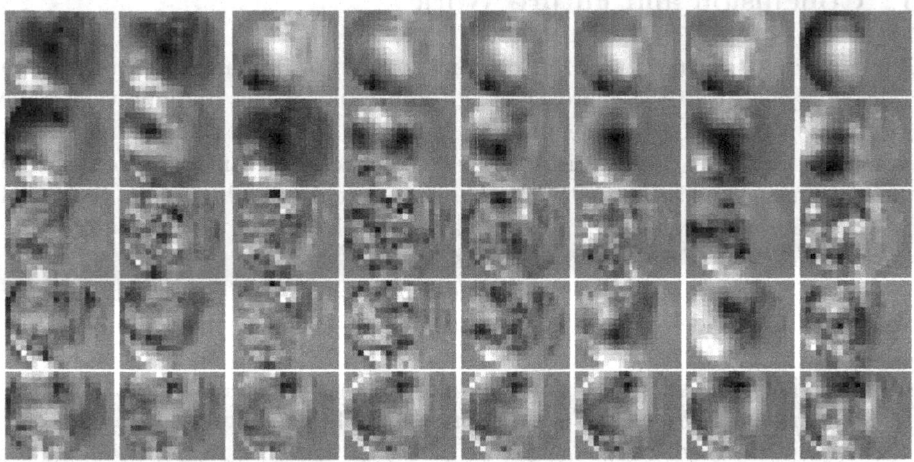

Fig. 4. Each of the columns represents the results of the MMI adaption process for the flame images and always shows the first component of the new subspace. The target values used for the training process of the LDA and MMI are the corresponding nitrogen oxides (NOx) measurements. The differences in each row are the used subspace transformations. **First row:** PCA. **Second row:** LDA. **Third row:** MMI initialized with the current PCA result. **Fourth row:** MMI initialized with a global PCA result. **Fifth row:** MMI initialized with the previous MMI result. The most interesting observations are the changes over time (from column to column), since smaller changes are desired.

collected data of five elapsed days for training purposes, and the most recent day as test set. We used this data to form PCA, LDA and MMI subspaces for eight consecutive days. For the MMI method we employed three different initialization points. First, we used the result of the PCA on that time frame for this purpose. Second, a fixed *eigenflames* subspace transformation calculated over the complete data was used, and third, the previous MMI result was used for the initialization.

Some results of these experiments are shown in Fig. 4. The PCA results (first row) are the most stable ones over time, the variance in the data over time is similar. But here again is the possible pitfall of the sign inversion problem between column 2 and 3. The LDA results (second row) identify big connected regions, but the shapes are completely different each day. Independent of the initialization, all MMI results share the tendency to produce less homogenous regions. The MMI results based on the the PCA initializations (third and fourth row) behave similar to the LDA subspace transformations, they are different each day. Using the previous MMI subspaces as starting points (last row) yield very useful but adapting filter masks.

These experiments show that the initialization with the previous MMI results is the most promising way to handle the adaptation task in a changing environment.

5 Conclusion and Future Work

Our experiments using the MMI feature subspace transformations for image data processing show that the approach is indeed useful, but has its limitations. The information extracted is either more informative for a classifier than a PCA-based subspace, or at least it is possible to compress the same information into a lower dimensional subspace than PCA. But to achieve stable results the use of PCA as a initialization is required anyway, so the MMI is in practice a objective driven refinement of the results obtained by PCA or LDA.

These positive results only hold true for rather low dimensional subspace constructs. If the desired transformation projects into a still high dimensional space, the MMI approach will get stuck at a local minimum very soon or venture into directions were stable solutions are hard to find by gradient descent. In these cases the use of LDA or PCA is superior to the MMI method.

The stepwise gradient estimation of the MMI subspace is an advantage for an adaptive online system. It allows the use of previous solutions to estimate a similar subspace which captures at least some changes of the underlying process without a complete redefinition of the channels in the new subspace.

Possible directions for future work include the investigation of the extension to nonlinear transformations like neural networks in the image domain. The adaptive changes of the subspace transformations focus on finding that subspace which is most important to the tasks at hand are engineered from the practitioners point of view. Hence investigating the connection of our proposed system to biological inspired, attention-based systems would be an interesting venue, too.

References

1. Torkkola, K.: Feature Extraction by Non Parametric Mutual Information Maximization. Journal of Machine Learning Research 3, 1415–1438 (2003)
2. Duda, R., Hart, P., Stork, D.: Pattern Classification, 2nd edn. Wiley, New York (2001)
3. Hyvärinen, A., Karhunen, J., Oja, E.: Independent Component Analysis. Wiley, New York (2001)
4. Lee, D.D., Seung, H.S.: Algorithms for Non-negative Matrix Factorization. Advances in Neural Information Processing Systems, vol. 13, pp. 556–562. MIT Press, Cambridge (2001)
5. Fisher, R.A.: The use of multiple measurements in taxonomic problems. Annals of Eugenics 7, 179–188 (1936)
6. Principe, J., Xu, D., Fisher, J.: Information theoretic learning. In: Haykin, S. (ed.) Unsupervised Adaptive Filtering, pp. 265–319. Wiley, Chichester (2000)
7. Torkkola, K.: Nonlinear feature transforms using maximum mutual information. In: Proc. Int. Conf. Neural Networks (IJCNN), pp. 2756–2761 (2001)

Bio-inspired Connectionist Architecture for Visual Detection and Refinement of Shapes

Pedro L. Sánchez Orellana and Claudio Castellanos Sánchez

Laboratory of Information Technology of Centre for Research and Advanced Studies
LTI Cinvestav - Tamaulipas, Ciudad Victoria, Tamaulipas, México
{psanchez,castellanos}@tamps.cinvestav.mx

Abstract. In this paper we propose a bio-inspired architecture for the visual reconstruction of silhouettes of moving objects, based on the behaviour of simple cells, complex cells and the Long-Range interactions of these neurons present in the primary visual cortex of the primates. This architecture was tested with real sequences of images acquired in natural environments. The results combined with our previous results show the flexibility of our propose since it allows not only to reconstruct the silhouettes of objects in general, but also, allows to distinguish between different types of objects in motion. This distinction is necessary since our future objective is the identification of people by their gait.

Keywords: Bio-inspired; visual cortex; silhouettes reconstruction.

1 Introduction

Nowadays, the use of video sequences to obtain information about objects has become one of the most challenging issues in artificial vision area. Special attention has been focused on the task of detecting and recognising human figures in motion, the main reason is that through this task many applications can be derived, like video surveillance for security [1–3]. Basically and according to literature, the main approaches used for these applications can be classified in three main categories.

The first one is called the *3D model-based approaches*. Some of the approaches use Bayesian information to construct a 3D model from 2D views [4–6]. Although, the majority requires the use of multiple-views of the human body in order to construct an accurate model of the body [7]. A lot of research has been done under this approach with very good results. So the problem here is the integration of the selected interest points, from multiple perspectives, for the construction of the model, it would turn into a very complex problem as shape turns into a more complex figure.

Another alternative to this problem is the second approach called *2D silhouette-based models*. This alternative consist of identifing the subject by matching the hypotheses of pose structure from the observation and choosing the most similar hypothesis to the DB [8, 9]. Even though the main problem in the results shown (over 90% of success on their DB's) that it is obvious since the

C. Alippi et al. (Eds.): ICANN 2009, Part II, LNCS 5769, pp. 745–754, 2009.

accuracy depends on the amount and quality of the sample data. So changing from a motion type to another (like from walking/running to dancing) is not as transparent as it seems since new sample data must be provided in order to achieve the recognition of the new actions [10–12].

Also there is an important detail about the last two approaches, that is, they use databases composed of only one object in the scene. So searching for the human shape figure, in most of the cases can be done by simply subtracting the object from environment. But what can be done if we don't even know the shape of the object in motion? one possible solution is by trying to locate the areas where motion exists, this is the main goal of the third approach, the *models based on motion*, which are based on energy measurements in a spatial-temporal way [13–15]. Besides, this approach has the capability of being environment invariant, so changes in environment configuration can be easily overcome. However, the main problem with this approach is that it uses extra information, such as learning human actions, to identify the interest points (joints or extremities) in order to make the recognition.

Even though this detection and recognition task requires a huge computational effort, it seems that it is a simple task for the primates. They have the capability to generalise the detection of object shapes in environments, no matter the presence of background motion, neither the uniformity of the illumination conditions. Several experiments with functional MRI [16, 17] have shown clues about the allocation and functioning of key areas involved in the localisation and recognition of figures in motion. In the case of the visual cortex, the areas are divided into two major pathways [18, 19], the ventral one (processing the form) and the dorsal one (which process the spatial localisation).

Although there are very few computational models that successfully explain or describe the functioning of this area, some good results can be found [20–22]. To achieve this goal we will explore some work that describe the capabilities of the brain to recognise and isolate objects from background (with uncontrolled conditions). In the following sections we will mention the biological motivations and foundations of our work, our proposed architecture, also analyse and describe our results, and finally we'll mention the conclusions we have achieved so far.

2 Biological Bases

In this section we describe the biological foundations of our methodology. Recent research on computational neuroscience has provided an improved understanding of human brain functionality and bio-inspired models have been proposed to mimic the computational capabilities of the brain for both motion and shape perception and understanding. So the main question for us is: how does brain subtracts figures in motion from the environment? First of all, early visual processing comprises of the magnocellular and parvocellular pathways. Broadly speaking, the magnocellular pathway carries low-spatial-frequency information, it feeds primarily into the dorsal stream and is concerned with the spatial information (also known as "where" pathway). The parvocellular

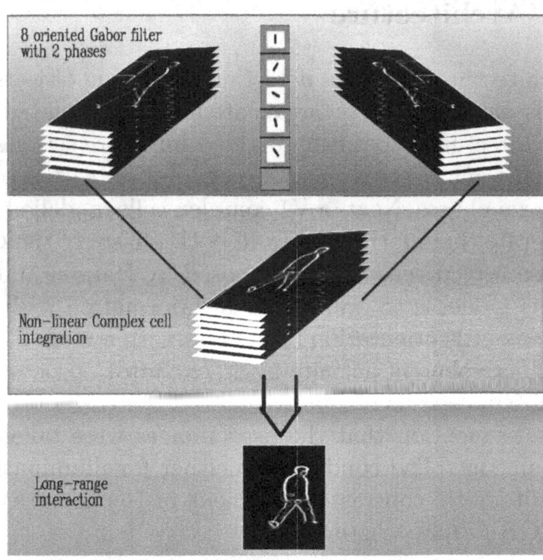

Fig. 1. Proposed architecture. It is composed of a bank of 2 phases and 8 oriented simple cells, a non-liner integration complex cell model, and a Long-range mechanism for orientations integration.

pathway carries higher-spatial-frequency information and is thought to contribute to fine form vision in the ventral stream (also known as "what" pathway).

A cornerstone of our investigation are the functionality of a more specialised type of cells in V1 area, which are presumably in charge of computation of object's contour, refinement and subtraction from background [23]. These tasks are achieved by an anatomical pyramidal connectivity in V1. According to literature, it is clear that V1 primary and complex cells both react according to its inner orientation and direction organisation. This functionality has inspired some work that considered the functionality of (V1) with a strong neural cooperative-competitive interactions that converge to a local, distributed and oriented auto-organisation [24–26].

However, one good question here is how information from oriented neurons in V1 can be integrated to isolate the contour of objects in motion?. Thus, at this point the problem can be separated into two tasks, the first one focused on the extraction of objects in motion, since neurons in V1 are contrast sensitive motion can be obtained by modelling the V1 neuron's function in a temporal base [27]. The second problem involves the refinement of the contour, this task is presumably located in area V1/V2, which can be explained by long-range connections and lateral feedback between the layers of oriented cells in V1 [28–30].

Based on these clues we constructed our architecture for both human silhouette detection and refinement, which is described in the following section.

3 Proposed Architecture

The proposed architecture for visual detection of moving objects is an extension from our previous work to detect articulated/non-articulated objects in motion [31]. This architecture is basically divided into three stages (see figure 1). The first stage is the spatial treatment, a convolution with our Gabor-like oriented filters with two phases. Next, a V1 complex cells modelled with a temporal treatment to integrate the information from both phases of the Gabor-like filter. And finally, we use a recurrent model proposed by Hansen and Neumann [32], the purpose of this model is to evaluate local information with certain orientation (due to complex cells orientation responses) with a more global context and to selectively enhance coherent activity by an excitatory process that modulates the feedback.

It is important to mention that the used images were taken in outdoor environments, with uncontrolled conditions (neither for illumination, background motion nor the automatic contrast adjustment of the camera), and they were converted from RGB format to grey scale.

3.1 First Stage (A)

Our architecture applies the Gabor-like oriented filters that modelled the responses of the simple cells in V1. This filtering ensures the capability to detect the local motion in a simple and local way, defined as follows.

Let $I(x, y, t)$ be an image sequence representing the shape of intensity in the time-varying image, assuming that every point has an invariant brightness. By applying an oriented Gabor filter, $G_{\theta,\phi}(x, y)$, we obtain :

$$
\begin{aligned}
D_\theta(t) &= \int\int_{t=0} \frac{dI(x,y,t)}{dt} * G_\theta(\hat{x},\hat{y})\,dx\,dy \\
&= \frac{d\int\int_{t=0} I(x,y,t) * G_\theta(\hat{x},\hat{y})\,dx\,dy}{dt}
\end{aligned}
\tag{1}
$$

where D is the result of the convolution between the Gabor functions and the image, $*$ is the convolution function, $\hat{x}$, $\hat{y}$ the rotational values, ϕ is the phase ($\phi = \pi, -\frac{\pi}{2}$) and $G_\theta(\hat{x}, \hat{y})$ is computed in a standard way:

$$
G_{\lambda,\theta,\phi}(\hat{x},\hat{y}) = \frac{1}{2\pi\sigma_x\sigma_y} e^{\left(-\frac{\hat{x}^2}{2\sigma_x^2} - \frac{\gamma^2\hat{y}^2}{2\sigma_y^2}\right)} e^{\left(2\pi_i \frac{\hat{x}}{\lambda} + \phi\right)} \tag{2}
$$

This is our Gabor-like filter model the simple cells in V1 where γ is the eccentricity of the receptive field, $\sigma_x\sigma_y$ its dimensions, λ the wavelength and ϕ the phase. For simplicity $0 \leq \theta = n\pi/4 < 2 \cdot \pi$ and the other parameters in the filter were set by experimentation and considering the parameters described by Castellanos [33].

The result of this stage is a set of oriented responses $D_\theta(t)$ which contain the preferred responses of the simple cells in V1, for both orientations and phases.

Generating a filter bank, for a given image, resulting into a 16-dimensional feature space for each point in the image.

The symmetrical and anti-symmetrical responses from the filter bank are then combined to simulate the action of V1 complex cells as it is shown in the following.

3.2 Second Stage (B)

There are biological clues ([34–36]) about the integration of several phases of simple cells in V1 by complex cells in the same area. These complex cells have the capability to respond selectively to lines or edges at particular orientations. This property can be modelled by the so called Gabor energy function, which is related to the behaviour of complex cells in V1 [37] and is defined in the following way:

$$C_{\lambda,\theta}(x,y) = \sqrt{G^2_{\lambda,\theta,0}(x,y) + G^2_{\lambda,\theta,(-\frac{1}{2})}(x,y)} \tag{3}$$

where $G_{\lambda,\theta,0}(x,y)$ and $G^2_{\lambda,\theta,(-\frac{1}{2})}(x,y)$ are the responses of the linear symmetric and antisymmetric Gabor filters. The result is a new nonlinear filter bank of 8 channels, where each channel represents an orientation θ.

3.3 Third Stage (C)

In order to enhance the contribution of every oriented complex cell we use the stage of "recurrent long-range interaction" model proposed by Hansen and Neumann, at this stage, the contextual influences from complex cell responses are modelled. Orientation-specific, anisotropic long-range connections provide the excitatory input. The inhibitory input is given by isotropic interactions in both the spatial and orientation domain. The spatial weighting function of the long-range filter is narrowly tuned to the preferred orientation, reflecting the highly significant anisotropies of long-range fibres in visual cortex. The equation for the shunting for the long-range stage reads:

$$\partial_t W_\theta = -\alpha_W W_\theta + \beta_W C_\theta (1 + \eta^+ net^+_\theta) - \eta^- W_\theta net^-_\theta \tag{4}$$

where the activity of the long-range stage results from interactions between the excitatory long-range input net^+_θ and the inhibitory input net^-_θ. The excitatory long-range input net^+_θ is gated by the activity $C_{\lambda,\theta}$ to implement a modulating rather than generating effect of lateral interaction on complex cell activities. Similarly, the inhibition from net^+_θ is divisive. A difference between our architecture and the one used by Hanse and Neumann is that we are using the Gabor filters for the simulations of simple cells in V1, since it has been proven that its behaviour is more similar to this type of cells. Although its computational expensiveness, due to the Gabor filter, our architecture shows higher resolution of the silhouettes in the scene as it can be seen in the figure 2.

Fig. 2. Comparison between responses computed with the model of Hanse and Neumann (right column), and the responses computed by our architecture (left column)

Table 1. Results of the integration stage. On the left the input image, in the middle a bank of 8-oriented complex cells from V1, and in the right the integration by the long range mechanisms.

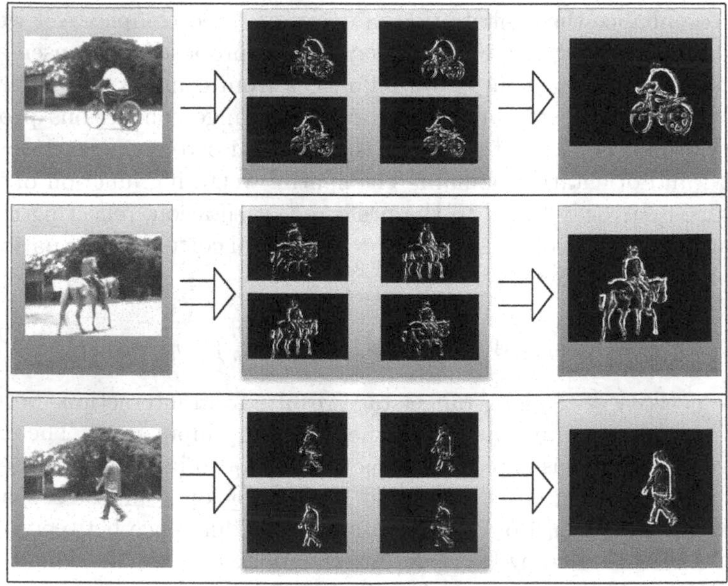

Table 2. Activation patterns of neurons after Long-Range interaction process. These graphs describes the activation patterns over the time. Note that here we remark the correlations between local minimum values in the graph and the occlusion of a leg.

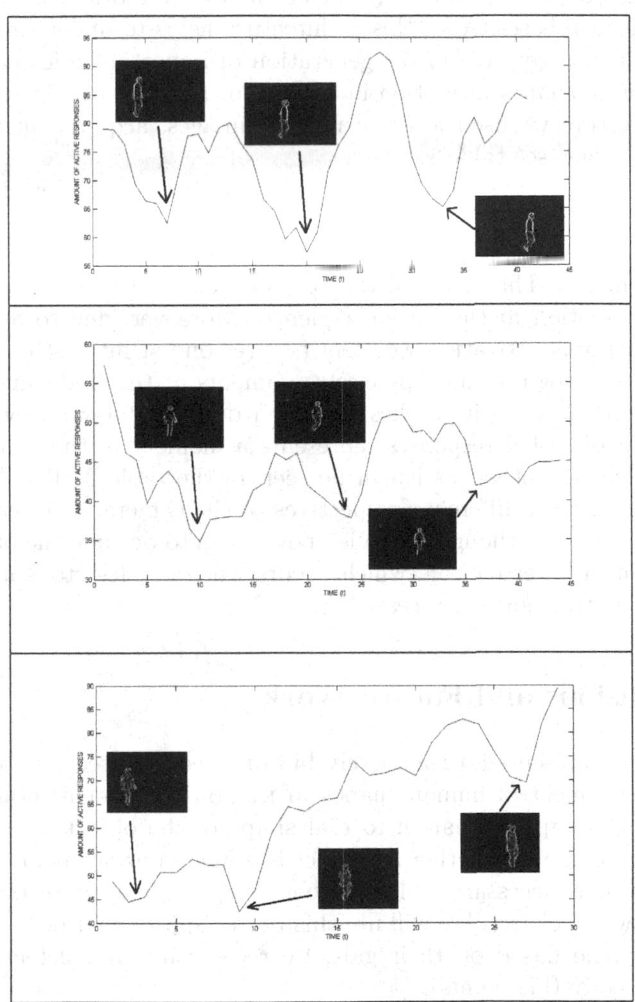

3.4 Human from Neurons Responses

Since we are searching for different patterns of activations of neurons that depend on the type of motion presented in the sequence. Our theory is that due to the way an articulated object moves (might be a human) it is possible to discriminate it from other type of moving objects that are not articulated (vehicles). For this we counted the responses and estimated the number of global activation of the cells over the time. According to our previous results [31] this process allowed us to distinguish between objects with an articulated (Humans) type of motion from those that have no articulations (cars or bicycles).

4 Results

One of the main contributions of this work is the capacity of our architecture to achieve the process of completely extract figures in motion from background by using biological bases. Also, this architecture helps to understand the mechanisms that have a key role in the generation of important information for the reconstruction of shapes in motion in our brain. In order to test the behaviour of the architecture we used a set of natural images, acquired in uncontrolled outdoor conditions, see table 1.

Besides, by combining our previous technique for human detection and the refinement of the shapes in video sequences we can isolate (by reconstructing from multiple orientations from V1 complex cells) an object through all the sequence of images. The result of the combination still allows the recognition of humans in motion in the image sequence. Moreover, due to a more stable patterns of neurons activation we identified regions of interest on the graphs, which allow to recognise more precisely moments of the body in motion. For example, the parts when it reaches a sudden decrease followed by an increase of the amount of global responses represents a moment in the scene when the one leg occludes the other, as it can be seen in the table 2. This behaviour is repeated in even with different perspectives of the camera. The reason for this behaviour is that even though if one leg never gets to occlude the other there is always a point of lower motion, which occurs when one leg stops moving while the other starts the motion process.

5 Conclusions and Future Work

In this work we presented a completely bio-inspired architecture for the recognition and refinement of human shapes in motion. The architecture allows us to achieve a good approximation to real shape of the object in motion. Even though it can be known whether an object is a human or not before the process of refinement, it is necessary to have a better approximation of the parts that compose the whole object. We will use this better approximation to a real shape to identify people based on their gait, by recognising and detecting relevant points in the body (like joints).

However, it is important to mention that an important part of the behaviour of the architecture is linked to the acquisition process. One of the most significant troubles in general in the Artificial Vision area is about the uniformity of the existing databases, it means that there is not a methodology for measuring nor the conditions of the environment, nor the objects in motion that appear on it. This problem requires a deeper study of the more relevant characteristics and a methodology that allow to extract a measurement of the videos and its conditions in order to establish a better reference point to evaluate the behaviour of the exciting models.

Acknowledgement. This research was partially funded by project number 51623 from "Fondo Mixto Conacyt-Gobierno del Estado de Tamaulipas" and the project number 78885 from "Ciencia Básica 2007" of Conacyt.

References

1. Cunado, D., Nixon, M., Carter, J.: Automatic extraction and description of human gait models for recognition purposes. Comput. Vis. Image Underst. 90, 1–41 (2003)
2. Moldenhauer, J., Boesnach, I., Stein, T.: Composition of complex motion models from elementary human motions. Articulated Motion and Deformable Objects 12, 68–77 (2006)
3. Xu, H., Jiwei, L., Lei, L., Zhiliang, W.: Gait recognition considering directions of walking. IEEE Cybernetics and Intelligent Systems, 1–5 (2006)
4. Mahmoudi, S., Daoudi, M.: A probabilistic approach for 3d shape retrieval by characteristic views. Pattern Recogn. Lett. 28(13), 1705–1718 (2007)
5. Yang, H.D., Lee, S.W.: Reconstruction of 3d human body pose from stereo image sequences based on top-down learning. Pattern Recogn. 40(11), 3120–3131 (2007)
6. Chen, H., Bhanu, B.: 3d free-form object recognition in range images using local surface patches. Pattern Recogn. Lett. 28(10), 1252–1262 (2007)
7. Yan, P., Bowyer, K.W.: A fast algorithm for icp-based 3d shape biometrics. Comput. Vis. Image Underst. 107(3), 195–202 (2007)
8. Lin, W.S., Fang, C.H.: Synthesized affine invariant function for 2d shape recognition. Pattern Recogn. 40(7), 1921–1928 (2007)
9. Mokhber, A., Achard, C., Milgram, M.: Recognition of human behavior by space-time silhouette characterization. Pattern Recogn. Lett. 29(1), 81–89 (2008)
10. Rogez, G., Orrite-Uruñuela, C., Martínez-del Rincón, J.: A spatio-temporal 2d-models framework for human pose recovery in monocular sequences. Pattern Recognition 41(9), 2926–2944 (2008)
11. Schlei, B.R.: A new computational framework for 2d shape-enclosing contours. CoRR cs.CV/0405029 (2004)
12. Shen, C., Lin, X., Shi, Y.: Human pose estimation from corrupted silhouettes using a sub-manifold voting strategy in latent variable space. Pattern Recogn. Lett. 30(4), 421–431 (2009)
13. Ahmad, M., Lee, S.W.: Human action recognition using shape and clg-motion flow from multi-view image sequences. Pattern Recogn. 41(7), 2237–2252 (2008)
14. Denman, S., Chandran, V., Sridharan, S.: An adaptive optical flow technique for person tracking systems. Pattern Recogn. Lett. 28(10), 1232–1239 (2007)
15. Laptev, I., Lindeberg, T.: Velocity-adapted spatio-temporal receptive fields for direct recognition of activities. IVC 22, 61–66 (2002)
16. Sereno, M.I., Dale, A.M., Belliveau, J.W., Brady, T.J., Rosen, B.R., Tootell, R.B.H.: Borders of multiple visual areas in humans revealed by functional magnetic resonance imaging. Science 268, 889–893 (1995)
17. Lerner, Y., Epshtein, B., Ullman, S., Malach, R.: Class information predicts activation by object fragments in human object areas. J. Cognitive Neuroscience 20(7), 1189–1206 (2008)
18. Van Essen, D.C., Anderson, C.H.: Information processing strategies and pathways in the primate retina and visual cortex, pp. 43–72. Academic Press Professional, Inc., San Diego (1990)

19. Kouh, M., Poggio, T.: A canonical neural circuit for cortical nonlinear operations. Neural Comput. 20(6), 1427–1451 (2008)
20. Giese, M., Poggio, T.: Neural mechanisms for the recognition of biological movements and acti on. Nature Reviews Neuroscience 4, 179–192 (2003)
21. Laxmi, V., Carter, J.N., Damper, R.I.: Biologically-inspired human motion detection. In: 10th European Symposium on Artificial Neural Networks, pp. 95–100 (2002)
22. Lange, J.,, M.L.: The role of spatial and temporal information in biological motion perception. In: Advances in Cognitive Psychology, vol. 3, pp. 419–428. University of Muenster, Germany (2007)
23. Livingstone, M.S.,, D.: Connections between layer 4b of area 17 and the thick cytochrome oxidase stripes of area 18 in the squirrel monkey. J. Neurosci. 7, 3371–3377 (1987)
24. Fellez, W.A., Taylor, J.G.: Establishing retinotopy by lateral-inhibition type homogeneous neural fields. Neurocomputing 48, 313–322 (2002)
25. Latham, P.E., Nirenberg, S.: Computing and stability in cortical networks. Neural Computation, 1385–1412 (2004)
26. Moga, S.: Apprendre par imitation: une nouvelle voie d'apprentissage pour les robots autonomes. PhD thesis, Université de Cergy-Pontoise, Cergy-Pontoise, France (September 2000)
27. Collins, K.: 13. In the Primate Visual System, pp. 311–337. CRC Press, Boca Raton (2004)
28. Sirosh, J., Miikkulainen, R.: Topographic receptive fields and patterned lateral interaction in a self-organizing model of the primary visual cortex. Neural Computation 9(3), 577–594 (1997)
29. Bair, W., Cavanaugh, J.R., Movshon, J.A.: Time course and time-distance relationships for surround suppression in macaque v1 neurons. J. Neurosci. 23(20), 7690–7701 (2003)
30. Schwabe, L., Obermayer, K., Angelucci, A., Bressloff, P.C.: The role of feedback in shaping the extra-classical receptive field of cortical neurons: A recurrent network model. J. Neurosci. 26(36), 9117–9129 (2006)
31. Sánchez Orellana, P.L., Castellanos Sánchez, C.: A bio-inspired connectionist architecture for visual classification of moving objects. In: ICANN (1), pp. 982–990 (2008)
32. Hansen, T., Neumann, H.: A recurrent model of contour integration in primary visual cortex. J. Vis. 8(8), 1–25 (2008)
33. Castellanos Sánchez, C.: Neuromimetic indicators for visual perception of motion. In: 2nd International Symposium on Brain, Vision and Artificial Intelligence, vol. 103, pp. 134–143 (2007)
34. Kjaer, T.W., Gawne, T.J., Hertz, J.A., Richmond, B.J.: Insensitivity of V1 Complex Cell Responses to Small Shifts in the Retinal Image of Complex Patterns. J. Neurophysiol. 78(6), 3187–3197 (1997)
35. Lauritzen, T.Z., Miller, K.D.: Different Roles for Simple-Cell and Complex-Cell Inhibition in V1. J. Neurosci. 23(6), 10201–10213 (2003)
36. Ersoy, B., Kagan, I., Rucci, M., Snodderly, M.: Modeling the responses of v1 complex cells to natural temporal inputs. J. Vis. 4(8), 278–278 (2004)
37. Petkov, N., Kruizinga, P.: Computational models of visual neurons specialised in the detection of periodic and aperiodic oriented visual stimuli: bar and grating cells. Biological Cybernetics 76(2), 83–96 (1997)

Evolving Memory Cell Structures for Sequence Learning

Justin Bayer, Daan Wierstra, Julian Togelius, and Jürgen Schmidhuber

IDSIA, Galleria 2, 6928 Manno-Lugano, Switzerland
{justin,daan,julian,juergen}@idsia.ch

Abstract. Long Short-Term Memory (LSTM) is one of the best recent supervised sequence learning methods. Using gradient descent, it trains *memory cells* represented as differentiable computational graph structures. Interestingly, LSTM's cell structure seems somewhat arbitrary. In this paper we optimize its computational structure using a multi-objective evolutionary algorithm. The fitness function reflects the structure's usefulness for learning various formal languages. The evolved cells help to understand crucial features that aid sequence learning.

1 Introduction

The problem of sequence learning is to learn the underlying function of a dynamic system, so as to be able to either produce the next step in a sequence produced by the system (sequence prediction), or to correctly classify a sequence (sequence classification). Sequence learning is tremendously important in various applications, e.g. stock market prediction and speech and handwriting recognition.

Neural networks are among the best tools available for general sequence learning. Most often, a *sliding time window* approach is used, where a finite subsequence is presented to a feedforward neural network. This approach is ultimately limited by the size of the time window. In the last decade, sequence prediction using *recurrent* neural networks has attracted some attention because of their simplicity and potential power. Here, the whole sequence is presented to the network, which is then trained by backpropagation through time (BPTT) [16]. However, there are some serious practical limitations to most types of RNNs due to their inability to capture long-term time dependencies. They suffer from the problem of *vanishing gradient* [8], the fact that the gradient signal vanishes as the error signal is propagated back through time. Because of this, events more than 10 time steps apart can typically not be related.

1.1 Dealing with Vanishing Gradient: LSTM

One method purposely designed to avoid this problem is Long Short-Term Memory (LSTM [9]), which is a special RNN architecture capable of capturing long term time dependencies. The defining feature of this architecture is that it consists of a number of *memory cells*, which can be used to store activations during

C. Alippi et al. (Eds.): ICANN 2009, Part II, LNCS 5769, pp. 755–764, 2009.

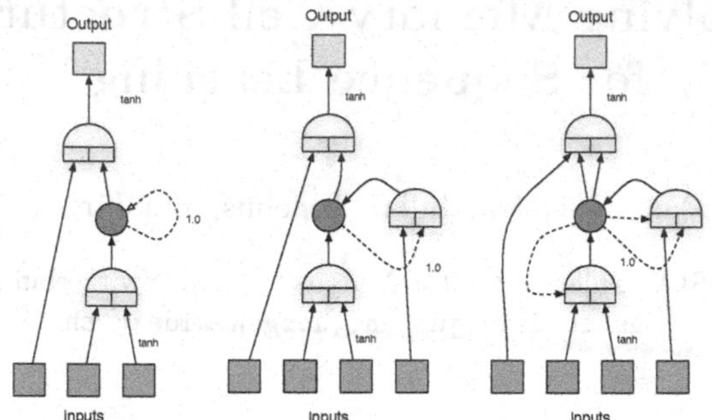

Fig. 1. The incremental development of the LSTM cell. Input units are in teal, output units in yellow. The gate units are shown as a half circle with as the output part and two different squares as inputs. The time delayed connection is dashed, and the red circle is the state unit. The second version of the LSTM cell adds a forget gate, and the third version adds peepholes.

arbitrarily long time spans. Access to the memory cell is *gated* by units that learn to open or close depending on the context. The memory cell's internal structure consists of a number of computational units, including the sigmoid function, the tanh function, and the gating function, which are connected in a graph structure. The fact that these units are differentiable ensures the memory cell as a whole can be used in conjunction with BPTT, using the chain rule as a connecting principle.

LSTM, unlike conventional RNNs, has been shown to be able to capture long-term time dependencies, learn precise timing, and generalize well on examples of both context-free and context-sensitive languages such as a^nb^n and $a^nb^nc^n$, respectively, whereas normal RNNs completely failed to capture the underlying structure of the problem [12]. LSTM networks have been shown to outperform other RNNs on numerous time series requiring the use of deep memory [13].

Interestingly, the development of LSTM was incremental (see figure 1). First, the concept of an internal *state* was introduced, guarded by input and output gates [9]. A time delay connection from the state to itself with weight one ensured that the state retained its value, unless the input gate was opened. Then, the concept of a *forget gate* was introduced, which modulates the state's self-connection and enables precise timing abilities [4]. Finally, *peepholes* were devised, which are direct connections from the state to all gates [5]. This final step enabled LSTM to learn the underlying structure of the context-sensitive language $a^nb^nc^n$ up to hundreds of time steps using just 10 sample sequences for training [3]. LSTM has recently been shown to perform excellently on many tasks, including speech processing and handwriting recognition (e.g. see [11]).

The incremental design evolution of the LSTM cell outlined above, taken together with its somewhat arbitrary structure, suggests that the development of LSTM could be retraced with artificial evolution, and that LSTM's design could even be bettered using the same means. In particular, we propose to use techniques introduced to evolve neural network topologies to evolve the internal structure of LSTM-like memory cells, using the sequence learning capability of networks of such cells as fitness functions.

1.2 Evolving Neural Topologies

A large body of work exists where evolutionary algorithms are used to create and optimize topologies of neural networks. Topologies have been evolved for a number of different purposes, including direct function approximation (without subsequent learning), reinforcement learning, and the capacity to be trained by gradient descent methods.

A core distinction can be made between *indirect* or *generative* approaches to topology evolution, and *direct* approaches. The former try to replicate nature's ability to encode complex phenotypes (e.g. human brains) with vastly simpler genotypes (e.g. human DNA), using graph rewriting systems or models of biological processes [10,7]. Apparently, the promise of scalability motivating these approaches has so far not been realized. The latter category, which includes the empirically successful NEAT algorithm [15], instead encodes the structure directly into the genome. A central concept of NEAT is complexification; a network starts out small, but the mutation operators can add new connections as well as split existing connections to insert new neurons. The algorithm used in this paper has similarities to NEAT, but lacks the recombination operator for simplicity.

Usually, the weights of the neural connections are evolved at the same time as the topology. However, Whiteson [17] evolved network topologies without weights, with a fitness function based on their ability to be used as function approximators for TD-learning. Similarly, in this paper we do not evolve connection weights, but use fitness functions based on capacity for sequence learning.

1.3 This Paper: Evolving Cell Structures

The purpose of our work is to investigate the space of architectural alternatives to LSTM and to understand the structural features promoting successful sequence learning through evolving structures of memory cells so as to optimize their sequence learning capability. We view each memory cell as a miniature neural network, consisting of a graph of connected computational units such as the sigmoid, the tanh and the gating unit. For every run, the structure of the cell is replicated a number of times to form a complete recurrent neural network. We then use a NEAT-inspired direct topology evolution algorithm to evolve this structure.

The fitness functions for structures are based on how well networks of memory cells can learn different sequences using gradient descent. (Note that connection weights are reset between fitness evaluations; evolution is thus *not* "Lamarckian"). As it is crucial that all cell structures can be trained by gradient descent, we constrain the structures to be directed acyclic graphs (*DAGs*) of differentiable

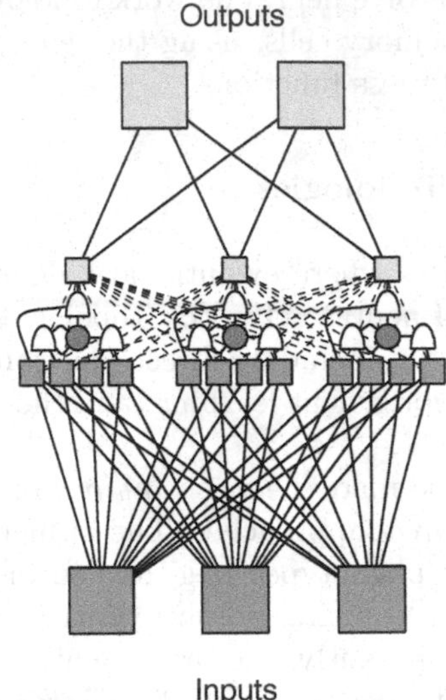

Fig. 2. A network constructed with a hidden layer of three LSTM cells. The recurrent connections from the hidden layer to itself, necessary for the cells to communicate with each other, are shown as dashed.

units, plus time delay connections: time delay connections which may break the DAG property but only propagate activations between time steps.

We start with evolving cells capable of learning simple versions of the problems; once these problems can be learnt satisfactorily, we increase the complexity of the problem, a practice known as incremental evolution [6]. So as not to overspecialize and develop cell structures only capable of learning solutions to one type of problem, we test each cell on two problems. Using the learning capability on each problem as a separate fitness measure means that we pose cell structure evolution as a *multiobjective* optimization problem, requiring the use of a multiobjective evolutionary algorithm (*MOEA*) in our case the *NSGA-II* [2].

2 Methods

2.1 Memory Cell Representation

A memory cell structure is a set of computational units and a graph connecting them to each other. Connections between units possess a flag indicating whether the connection is time delayed and another flag indicating whether the connection is parameterized (i.e. has a trainable *weight*) or has a fixed weight of 1.0. The former case is called a *linear connection* while the latter is called an *identity connection.* There are several types of computational nodes: linear, sigmoid, the hyperbolic tangent and the 'gate' unit, each having its own transfer function.

- The linear node takes input x and produces output $id(x) = x$.
- The sigmoid node takes input x, and produces output $\sigma(x) = 1/(1 + e^{-x})$.
- The tanh node is the hyperbolic tangent $\tau(x) = tanh(x)$.
- The gating transfer has two inputs x_1 and x_2 and produces $g(x_1, x_2) = \sigma(x_1)x_2$.

The most interesting type of node used in this paper is the *gating* unit that was first introduced in the LSTM cell. Its structure can be thought of as a continuous version of the `if ... then ...` statement, and has two inputs: one condition and one signal. It is this unit type that enables LSTM's internal state to open and close to incoming signals, depending on the context.

All units have two additional flags: one indicating whether a unit is an input unit to the cell, i.e. receives input from outside the cell, and one indicating whether the unit is an output unit, connecting to other cells and network outputs.

2.2 Evolutionary Algorithm

We used the *NSGA-II* multiobjective evolutionary algorithm (MOEA), as it is one of the most widely used MOEAs and known for robust performance under diverse conditions [2]. A population size of 100 was used. For simplicity, no recombination was used; mutation was the only variation operator.

A cell structure is mutated by applying mutations from the list below, a geometrically distributed number of times. The expected amount of mutations is given by $E_M = \sum_{m \in M} \frac{1}{1-\pi[m]}$, where $\pi[m]$ is the probability of each mutation type. The probabilities used in our experiments are given in parentheses in the following list of available mutations; these probabilities were chosen carefully in order to prevent bloating of the structure. If any mutation breaks the DAG property by making the structure cyclic, that mutation is simply rolled back.

- *Add unit.* A random connection is split into two parts with a new linear unit in between. ($\pi[\cdot] = 0.1$)
- *Add gate unit.* A unit is added as in *Add unit* but also assigned the gate transfer function. Its second input is connected to a random unit. ($\pi[\cdot] = 0.2$)
- *Add connection.* Two units are randomly chosen and connected by an identity connection which is not time delayed. ($\pi[\cdot] = 0.15$)
- *Add time delay connection.* Two units are connected by an identity connection which is time delayed. This connection is allowed to break the DAG property. ($\pi[\cdot] = 0.15$)
- *Change transfer function.* The transfer function of a randomly chosen unit is set to another transfer function. In the case of the gate transfer function, a new connection to the second input of the unit is made. ($\pi[\cdot] = 0.3$)
- *Change connection.* The type of a randomly chosen connection is switched from identity to linear or vice versa. ($\pi[\cdot] = 0.25$)
- *Flip time delay.* The time delay flag of a connection is flipped. ($\pi[\cdot] = 0.25$)
- *Flip input.* The input flag of a random unit is flipped. (($\pi[\cdot] = 0.15$)
- *Flip output.* The output flag of a random unit is flipped. ($\pi[\cdot] = 0.15$)
- *Tidy up.* If a random unit is not reachable from the input, or the output is not reachable from that unit, it is removed. ($\pi[\cdot] = 0.5$)

2.3 Fitness Function

At every fitness evaluation, a cell structure was used to create a recurrent network with 5 hidden memory cells connected to all inputs and all outputs. (Similar to the LSTM network in figure 2, except for the nature and number of the cells.) To calculate the fitness of the structure, three separate BPTT training runs were performed using different weight initializations. (Since each unit is differentiable, we can apply standard BPTT to learn the parameters of the network.) The negative of the highest mean squared error was taken to be the actual fitness value. Weights were initialized between -0.1 and 0.1, and learning rate 0.001 with momentum 0.99 was used. Training time was set to 2000 epochs.

Formal languages. Determining whether a string of symbols belongs to a particular formal language often requires remembering some symbols in the string seen so far, which rules out the use of non-recurrent architectures. In order to evolve memory cells, we chose the context-free language $a^n b^n$ [19] (yielding strings ST, $SabT$, $SaabbT$, $SaaabbbT$, etc.) and the context-sensitive language $a^n b^n c^n$ (which yields ST, $SabcT$, $SaabbccT$, $SaaabbbcccT$, etc.), which require memory of up to n and $2n$ time steps, respectively. Symbol strings were presented sequentially to the network, with each symbol's corresponding input unit set to `1`, and the other set to `-1`. At each time step, the network must predict the possible symbols that could come next in a legal string. The $a^n b^n c^n$ is too hard for regular RNNs but LSTM achieves decent to superb performance on this task [3]. To ensure that the evolved cells were not limited to being able to learn a single language, we used the related but significantly different language $a^n b^{\mathrm{m}} a^n$ as an additional objective. See [3] for a more complete explanation.

3 Results

A typical evolutionary run required roughly one hour per objective per generation on a 3 Ghz processor. Cell structures capable of learning the desired languages were typically found within 10 generations. An overview of their performance on the selected languages is given in figure 3.

In one configuration, the context-free language $a^n b^n c^n$ was used as one objective and the context-free language $a^n b^n$ as the other. n was increased incrementally as learning capacity increased; when structures had evolved that could learn to recognize string of lengths 1-5, maximum length was increased to 10. During runs with this configuration, the cells shown in figure 4 were evolved.

In a second configuration, evolution started out with a context-free language ($a^n b^n, n \in [1, 5]$) and moved on to a multiobjective setting with one context-free and one context-sensitive language ($a^n b^m c^n$ and $a^t b^t c^t, (m, n) \in [1, 4] \times [1, 4], t \in [1, 5]$). In most runs with this configuration, a cell capable of learning both languages was found.

	Benchmark	
Cell	$a^nb^n, n_t = 0..5$	$a^nb^n, n_t = 0..10$
Ana	8.6	19.5
Cathy	0	8.5
Charlotte	8.2	19.5
Mary	1.675	3.325
LSTM	9.6	27.5
	$a^nb^nc^n, n_t = 0..5$	$a^nb^nc^n, n_t = 0..10$
Ana	18.55	47.0
Cathy	6.05	7.7
Charlotte	18.95	44.9
Mary	5.3	1.1
LSTM	15.05	44.85
	$a^nb^mc^n, n_t = m_t = 0..4$	
Ana	8.0	
Cathy	4.37	
Charlotte	8.0	
Mary	2.72	
LSTM	8.0	

Fig. 3. Results of four evolved cells, named `Ana`, `Cathy`, `Charlotte` and `Mary`, on grammar benchmarks compared to LSTM. The table reports the biggest parameter to which a network constructed out of the indicated cells could generalize after training, averaged over twenty runs. n_t and m_t give the ranges of the training sets.

3.1 Genealogical Analysis

Figure 5 depicts the evolution of a cell capable of learning the $a^nb^nc^n$ language in about 20% of the training runs. It is interesting to note that the very first step is just a simple recurrent network, which cannot even learn the a^nb^n language to more than a rudimentary level. The third stage added a new node, with a time delay connection in and a linear connection back to the input, essentially creating three types of recurrence to the input node. The final mutation turned the linear connection back from the new unit into a time delay connection, and added a new recurrent connection on the output. This suddenly enabled several steps of recurrence, which seems to be necessary to handle more complex languages. On the other hand, the cells `Ana` and `Charlotte`, which outperform `Mary` significantly, feature only a single recurrent internal connection themselves and are mostly constructed out of identity connections and gate units – this makes them similar to LSTMs.

3.2 Validation: Long-Term Dependency T-Maze

In order to validate the cells found, we performed validation tests on the deep memory T-Maze task as described in Bakker's work [1]. The T-Maze task was

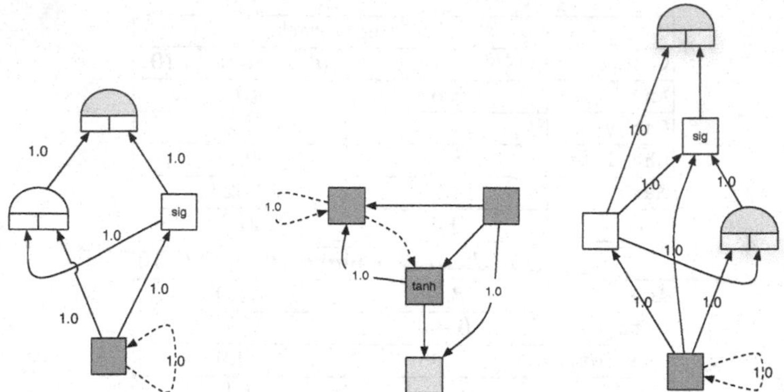

Fig. 4. An evolved cell, named `Charlotte` that can reliably learn the $a^nb^nc^n$ grammar (left), and two others (`Cathy` and `Ana`) that can learn the a^nb^n grammar (right and middle). Standard RNNs cannot learn these languages. Note the absence of any substantial similarity in their structure.

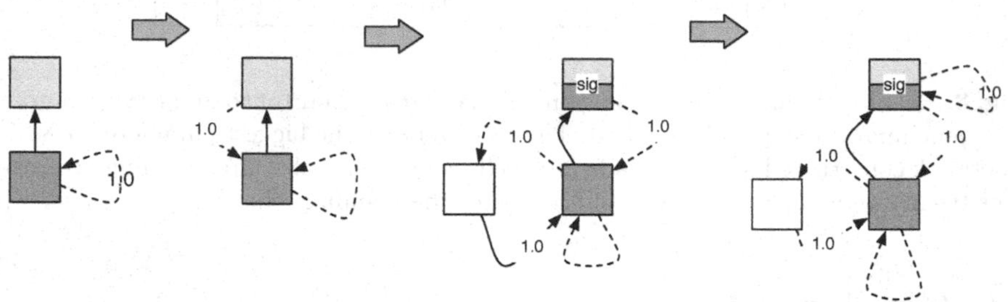

Fig. 5. The evolution of cell `Mary`. Although happening over the course of nine generations, only four mutations were needed in order to evolve a cell which is casually able to learn the underlying structures of the context-sensitive language $a^nb^nc^n$ and the context-free language $a^nb^ma^n$.

Cell	Success ratio	Average reward
`Ana`	0.45	-3.895
LSTM	0.35	-7.545
`Charlotte`	0.25	-8.915
`Cathy`	0.0	-54.485
`Mary`	0.0	-57.74

Fig. 6. The cells tested on the T-Maze task. Each cell was evaluated 20 times.

specifically designed to test a reinforcement learning algorithm's capability to relate events far apart in history. It involves having to remember a single observation at the beginning of the task until the very last time step. Applying the recurrent policy gradient algorithm [18], a learning rate of 0.01 and a momentum of 0.99 was used in conjunction with a batch size of 100 and a discount factor of 0.99. The corridor length was set to 15.

We found that the cell structure `Ana` outperforms LSTM (see figure 6). Note that this is a reinforcement learning tasks instead of a supervised training task. This is significant, since although we evolved the cell structure to perform well on sequence prediction, it actually peforms well on an unrelated reinforcement learning task. This suggests the evolved cell structures might be quite general and capable of performing substantially different tasks.

4 Conclusion and Discussion

Using an algorithm similar to neural network topology evolution algorithms, we evolved structures for memory cells capable of learning context-sensitive formal languages through gradient descent. The fitness functions were based on the learning capacity of networks of such cells. The evolved memory cells were in many ways comparable in performance to LSTM, the current state-of-the-art in gradient-based sequence learning.

Analysis of the (very diverse) evolved cell structures and their genealogies provided interesting insights into what features contribute to the power of LSTM. The essential ingredients of LSTM's success seem to be (1) linear units with fixed self-connections and (2) gate units while the precise connection structure seems less important. It is important to note that the cells with gates significantly outperform those without. An open question is how big the tradeoff between performance and generality of a specific cell is. Since LSTM is used in a wide range of applications, we believe that evolving general cells is actually quite possible. In order to evaluate the generality of our approach, it is crucial to try our methods on more benchmark problems from other domains, combining unrelated objectives in one single run. These could include learning to predict continuous functions (e.g. superimposed sines), real-world sequence learning problems (e.g. speech processing), and even reinforcement learning problems. It could also mean using non-gradient-based training algorithms, such as evolutionary algorithms, for some objectives. Cells developed using this method could also be incorporated into hybrid algorithms such as *Evolino* [14].

References

1. Bakker, B., Linker, F., Schmidhuber, J.: Reinforcement learning in partially observable mobile robot domains using unsupervised event extraction. In: Proc. IROS 2002, pp. 938–943 (2002)
2. Deb, K., Pratap, A., Agarwal, S., Meyarivan, T.: A fast and elitist multiobjective genetic algorithm: Nsga-ii. IEEE Transactions on Evolutionary Computation 6, 182–197 (2002)
3. Gers, F.A., Schmidhuber, J.: LSTM recurrent networks learn simple context free and context sensitive languages. IEEE Transactions on Neural Networks 12, 1333–1340 (2001)

4. Gers, F.A., Schmidhuber, J., Cummins, F.: Learning to forget: Continual prediction with LSTM. Neural Computation 12, 2451–2471 (2000)
5. Gers, F.A., Schraudolph, N.: Learning precise timing with LSTM recurrent networks. Journal of Machine Learning Research 3, 2002 (2002)
6. Gomez, F., Miikkulainen, R.: Incremental evolution of complex general behavior. Adaptive Behavior 5, 317–342 (1997)
7. Gruau, F.: Genetic synthesis of modular neural networks. In: Proceedings of the Fifth International Conference on Genetic Algorithms, pp. 318–325. Morgan Kaufmann, San Francisco (1993)
8. Hochreiter, S., Bengio, Y., Frasconi, P., Schmidhuber, J.: Gradient flow in recurrent nets: the difficulty of learning long-term dependencies. In: Kremer, S.C., Kolen, J.F. (eds.) A Field Guide to Dynamical Recurrent Neural Networks. IEEE Press, Los Alamitos (2001)
9. Hochreiter, S., Schmidhuber, J.: Long short-term memory. Neural Computation 9(8), 1735–1780 (1997)
10. Kitano, H.: Designing neural networks using genetic algorithms with graph generation system. Complex Systems 4, 461–476 (1990)
11. Liwicki, M., Graves, A., Bunke, H., Schmidhuber, J.: A novel approach to on-line handwriting recognition based on bidirectional long short-term memory networks. In: Proc. 9th Int. Conf. on Document Analysis and Recognition, vol. 1, pp. 367–371 (2007)
12. Rodriguez, P., Wiles, J.: Recurrent neural networks can learn to implement symbol-sensitive counting. In: NIPS 1997: Proceedings of the 1997 conference on Advances in neural information processing systems, vol. 10, pp. 87–93. MIT Press, Cambridge (1998)
13. Schmidhuber, J.: RNN overview (2004), `http://www.idsia.ch/~juergen/rnn.html`
14. Schmidhuber, J., Wierstra, D., Gagliolo, M., Gomez, F.: Training recurrent networks by evolino. Neural Computation 19(3), 757–779 (2007)
15. Stanley, K.O., Miikkulainen, R.: Evolving neural networks through augmenting topologies. Evolutionary Computation 10(2), 99–127 (2002)
16. Werbos, P.: Backpropagation through time: What it does and how to do it. Proceedings of the IEEE 78, 1550–1560 (1990)
17. Whiteson, S., Taylor, M.E., Stone, P.: Empirical studies in action selection with reinforcement learning. Adaptive Behavior 15, 33–50 (2007)
18. Wierstra, D., Foerster, A., Peters, J., Schmidhuber, J.: Solving deep memory pOMDPs with recurrent policy gradients. In: de Sá, J.M., Alexandre, L.A., Duch, W., Mandic, D.P. (eds.) ICANN 2007. LNCS, vol. 4668, pp. 697–706. Springer, Heidelberg (2007)
19. Wiles, J., Elman, J.: Learning to count without a counter: A case study of dynamics and activation landscapes in recurrent networks. In: Proceedings of the Seventeenth Annual Conference of the Cognitive Science Society, pp. 482–487 (1995)

Measuring and Optimizing Behavioral Complexity for Evolutionary Reinforcement Learning

Faustino J. Gomez, Julian Togelius, and Juergen Schmidhuber

IDSIA, Galleria 2
6928 Manno-Lugano
Switzerland

Abstract. Model complexity is key concern to any artificial learning system due its critical impact on generalization. However, EC research has only focused phenotype structural complexity for static problems. For sequential decision tasks, phenotypes that are very similar in structure, can produce radically different behaviors, and the trade-off between fitness and complexity in this context is not clear. In this paper, behavioral complexity is measured explicitly using compression, and used as a separate objective to be optimized (not as an additional regularization term in a scalar fitness), in order to study this trade-off directly.

1 Introduction

A guiding principle in inductive inference is the concept of parsimony: given a set of competing models that equally explain the data, one should prefer the simplest according to some reasonable measure of complexity. A simpler model is less likely to overfit the data, and will therefore generalize better to new data arising from the same source. In EC, this principle has been applied to encourage minimal phenotypic structure (e.g. GP programs, neural network topologies) by penalizing the fitness of overly complex individuals so that selection drives the search toward simpler solutions [13, 6, 15, 12].

The advantage of incorporating this *parsimony pressure* has been demonstrated convincingly in supervised learning tasks, producing solutions that are significantly more general. However, for dynamic tasks involving sequential decisions (e.g. reinforcement learning), a phenotype's structural complexity may not be a good predictor of its *behavioral complexity* [5] (i.e. the complexity of the observation-action sequences generated by the evolving policies). Phenotypes that are very similar in structure, can produce radically different behaviors, and the trade-off between fitness and complexity in this context is not clear. In this paper, behavioral complexity is measured explicitly using compression, and used as a separate objective to be optimized (not as an additional regularization term in a scalar fitness), in order to study this trade-off directly.

Multi-Objective approaches have been used previously to control structural complexity (or promote diversity [10]), but only in a supervised learning context [3, 2], and always to promote parsimonious solutions. The goal here is to look at complexity more generally, and analyze how encouraging both low *and*

C. Alippi et al. (Eds.): ICANN 2009, Part II, LNCS 5769, pp. 765–774, 2009.

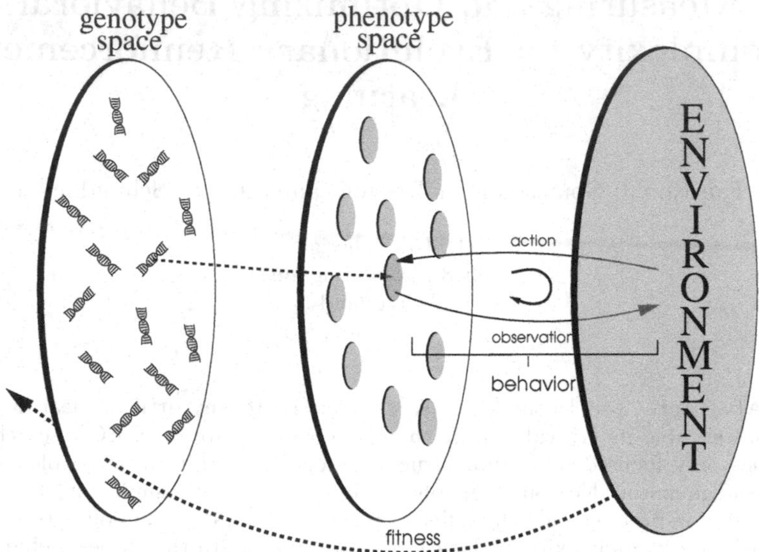

Fig. 1. Genotype-Phenotype map. The complexity of evolving candidate solutions can be computed at different levels. In sequential decision tasks, measuring the structural (model) complexity in phenotype space may not give a reliable indication of the relative complexity of the phenotype behavior (shown as the cycling of actions and observations of the two highlighted phenotypes).

high behavioral complexity relates to and can affect performance (fitness) in reinforcement learning tasks.

The next section describes the general idea of complexity within the context of EC. Section 3, presents our experiments in evolving neural network controllers using a multi-objective evolutionary algorithm for two different reinforcement learning tasks: the Tartarus problem, and the Simplerace car driving task. Section 4 provides some analysis of our results and direction for future research.

2 Measuring Complexity

In evolutionary algorithms, the complexity of an individual can be measured in the genotype space where the solutions are encoded as strings, or in the phenotype space where the solutions are manifest.

For some problem classes and genetic representations, measuring complexity in one space is equivalent to applying it in the other: the genotype→phenotype mapping, G, preserves the relative complexity between individuals. When this is not the case, it is more informative to measure the complexity phenotypes (figure 1), after all what we are truly interested in is the complexity of solutions, not there encodings.

For sequential decision tasks (e.g. reinforcement learning), G maps each individual x to some form of policy, π, that implements a probability distribution over a set of possible actions, conditioned on the observation from the environment. More generally, the choice of action at time t can be conditioned on the entire *history* of previous observations, $o \in O$, and actions, $a \in A$: $a_t \leftarrow \pi(o_{t-1}, a_{t-1}, \ldots, o_0, a_0)$, where O is the set of possible observations, and A is the set of possible actions. In this case, structural complexity can be misleading as policies that are structurally similar with respect to a chosen metric may be very different in terms of behavior when they interact with the environment. We define the *behavior* of individual x to be a set of one or more histories resulting from one or more evaluations in the environment. A behavior is therefore an approximation of the *true* behavior of the individual that can only be sampled by interaction with the environment.

Measuring behavior complexity requires computing a function over the space of possible behaviors for a given $\{A, O\}$. A general framework, rooted in algorithmic information theory [7], that can be used to quantify complexity is the Minimum Description Length Principle [8], which states that any regularity in the data can be used to compress it, i.e. recoding it such that it can be represented using fewer symbols. For a given compressor, and two objects (e.g. bit-strings) of equal length, the object with the shortest compressed representation can be considered less complex as it contains more identifiable regularity [1]. In the experiments that follow, this idea is applied to assess the complexity of evolved neural network behaviors, using an real-world compressor.

MDL inspired complexity measures have been used in conjunction with evolutionary algorithms before to address *bloat* in Genetic Programming [6] and to evolve minimal neural networks [13, 15, 12], i.e. to control phenotype structural complexity. In the next section, data compressibility is used to measure the complexity of phenotype behaviors, and is used as additional objective to be optimized in order study the interaction between fitness and complexity at the behavioral level.

3 Experiments

To ensure a degree of generality, our experiments were conducted in two substantially different reinforcement learning benchmark domains: Tartarus and Simplerace. The three following objectives were used in various combinations:

1. $\boldsymbol{P}$: the standard performance measure or *fitness* for the task.
2. $\boldsymbol{C}$: the length of the behavior after applying the Lempel-Ziv [16] based `gzip` compressor to it. Behaviors with low C are considered less complex as they contain more regularity for the compressor to exploit.
3. $\boldsymbol{H}$: the Shannon entropy of the behavior: $-\sum p(x_i)log(p(x_i))$, where each x_i is one of the possible symbols representing an action or observation in the behavior. The entropy computes the lower bound on the average number of bits per symbol required to represent the behavior.

Four sets of multi-objective experiments were conducted using the well-known *NSGA-II* algorithm [4]. Each set used a different pairing of objectives: $\boldsymbol{M_{PH}}$,$\boldsymbol{M_{P-H}}$,$\boldsymbol{M_{PC}}$,$\boldsymbol{M_{P-C}}$, where the first subscript is the first objective which is always maximized (P in all cases), and the second subscript is the second objective which is maximized, unless it is preceded by a minus sign, in which cased it is minimized. At each generation, the scores on all three objectives, $\{P, C, H\}$ were recorded for two individuals in the Pareto front: the one with highest P, and the one with the best score on the chosen complexity-related objective.

In all of experiments, the controllers were represented by recurrent neural networks (figure 3, details below), and the population size was 100. Each run lasted for 4000 generations for the Tartarus problem, and 200 generations for the Simplerace problem. No recombination was used; the only variation operator was mutation, consisting in adding real numbers drawn from a Gaussian distribution with mean 0 and standard deviation 0.1 to all weights in the network.

For both tasks it was not necessary to include observations in the behaviors because the environments are deterministic and the initial states were fixed for all individuals in a single run, so that each sequence of actions only has one corresponding sequence of observations.

3.1 The Tartarus Problem

Figure 2a describes the Tartarus problem [9], used in the experiments. Although the grid-world is quite small, the task is challenging because the bulldozer can only see the adjacent grid cells, so that many observations that require different actions look the same, i.e. perceptual aliasing. In order to perform the task successfully, the bulldozer must remember previous observations such that it can compute its location relative to the walls and record the locations of observed blocks for the purpose quickly acquiring them later. In short, the agent is quite blind which means that evolutionary search can quickly discover simple, mechanical behaviors that produce better than random performance but do not exhibit the underlying memory capability to perform well on the task.

The Tartarus controllers were represented by fully recurrent neural networks with five sigmoidal neurons (figure 3a). Each controller was evaluated on 100 random board configurations. To reduce evaluation noise, the set of 100 initial boards was chosen at random for each simulation, but remained fixed for the duration of the simulation. That is, in a given run all networks were evaluated on the same 100 initial boards. The behaviors consisted of sequences of 80 {Left=1, Right=2, Forward=3} actions executed in each of the 100 trials.

3.2 Simulated Race Car Driving

The *simplerace* problem involves driving a car in a simple racing simulation in order to reach as many randomly placed waypoints as possible in a limited amount of time (figure 2b). There are plenty of good controllers to compare our results with, as the game has previously been used as a benchmark problem in several papers, and in two competitions associated with recent conferences[1].

The Simplerace controllers were represented by simple recurrent networks (SRN; figure 3b) with six inputs, eight hidden sigmoidal units, and two outputs.

[1] A more complete description of the problem is available in [11], and source code can be downloaded from http://julian.togelius.com/cec2007competition

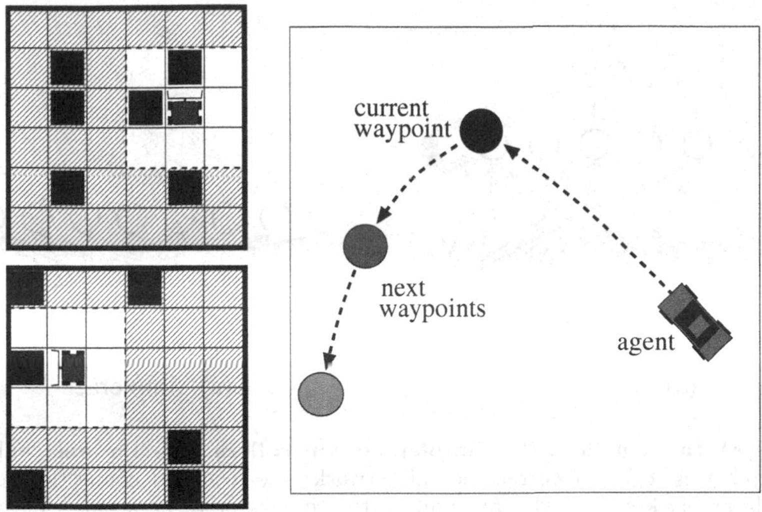

Fig. 2. The Tartarus (left) and Simplerace (right) tasks. The upper Tartarus board shows a possible initial state with the six blocks and the bulldozer placed at random squares away from the walls; the orientation the bulldozer is also random. The bulldozer must select an action (either turn left, turn right, or go forward) at each time-step based on the situation within its visual field (shown in white), and its internal state (memory). The bulldozer can only move forward if its path is unobstructed or the block in its way has no block behind it, otherwise it will remain its current position. The lower board is a possible final state after the alloted 80 moves. The score for this configuration is 7: two blocks receive a score of two for being in the corner, plus one point for the other three blocks that are against a wall. The object is the drive the car (both accelerator and steering) through as many randomly place waypoints in an alloted amount of time.

The inputs consisted of: (1) the speed of the car, (2) the angle and (3) distance to the current waypoint, the (4) angle and (5) distance to the next way point, and (6) a bias term. The two output units encode nine actions using the following scheme: the first unit steers the car, an activation of < -0.3 means "turn left", between -0.3 and 0.3 means "go straight", and > 0.3, means "turn right". The second unit controls the forward-backward motion, < -0.3 means "go forward", between -0.3 and 0.3 means "put the car in neutral", and > 0.3, means "brake" (if the car is no longer moving forward this action puts the car in reverse). Each network was evaluated using the same set of 10 cases (i.e. waypoint locations) chosen at random at the beginning of each simulation, and each lasting 1000 time-steps (actions).

3.3 Results

Figures 4a and 4b show the performance, P, for the four configurations. The "high complexity" configurations, M_{PH} and M_{PC}, performed significantly better than the "low complexity", M_{P-H} and M_{P-C}, on both tasks, but the effect was

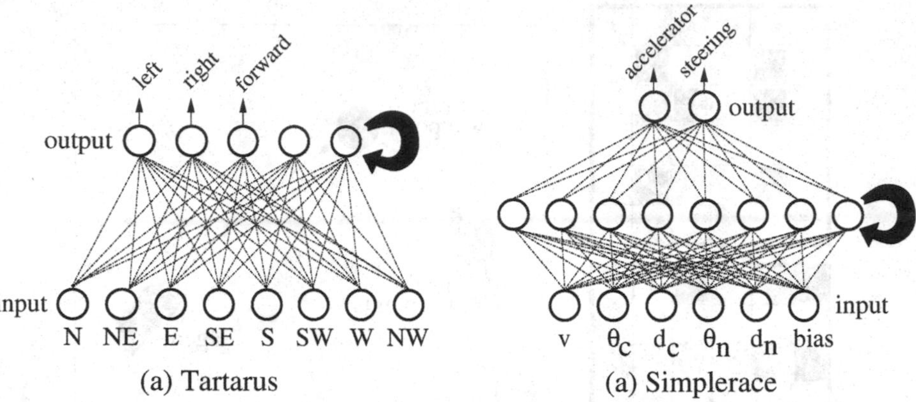

Fig. 3. (a) Tartarus and (b) Simplerace controllers. The Tartarus "bulldozer" is controlled by a fully recurrent neural network (the recurrent connections denoted by the large black arrow) with five units. At each time step the network outputs the action corresponding to action unit (left, right, forward) with the highest activation based on the state of the eight surrounding grid cells. The Simplerace car is controlled by a simple recurrent network with eight hidden units; v, speed of the car, θ_c, and d_c, the angle and distance to the current waypoint, θ_n, and d_n, the angle and distance to the next waypoint.

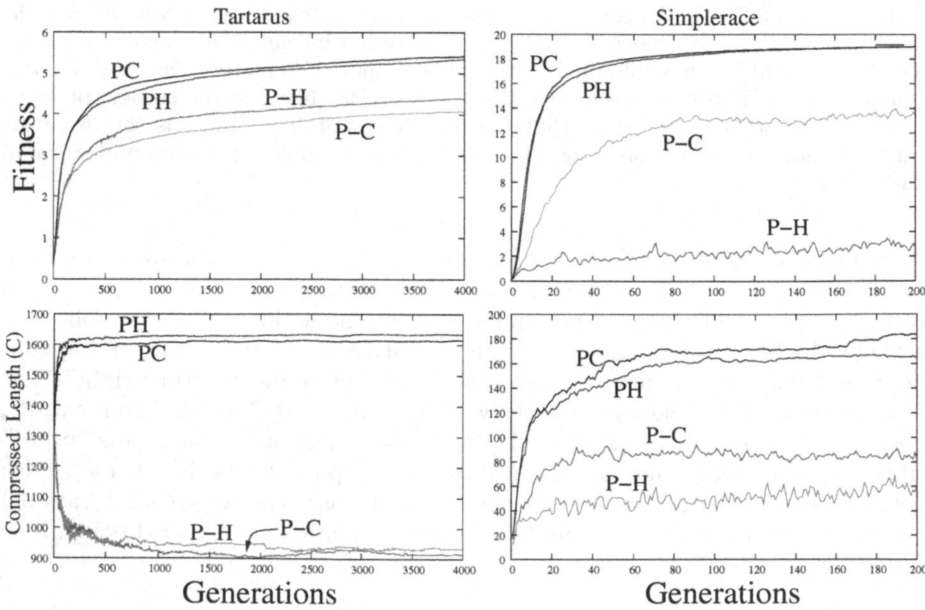

Fig. 4. Performance on Tartarus and Simplerace. Each curve denotes the fitness of the best individual in each generation for each of the four multi-objective configurations. Each curve shows, for the each configuration, the compressed length of the behavior of the most fit individual from each generation. Average of 50 runs.

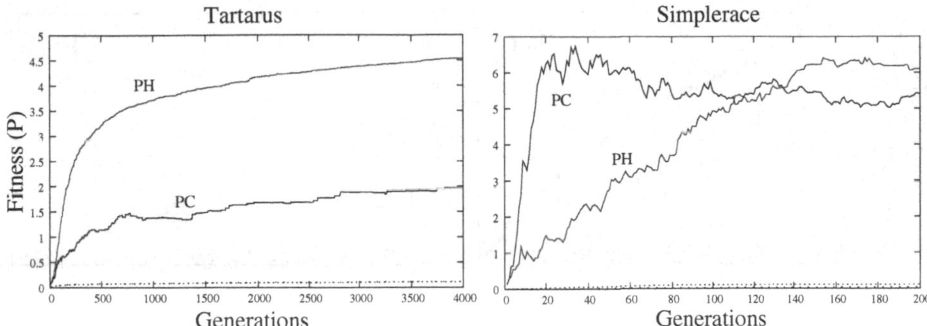

Fig. 5. Fitness of Most/Least complex individual. Each curve shows, for the each configuration, the fitness score of the individual in each generation with the best complexity (highest or lowest, depending on whether it is being maximized or minimized), in terms of either C or H, depending on which is being used as the second objective. The minimization runs are at the bottom of graph and are indistinguishable. Average of 50 runs.

more pronounced for the Simplerace task where minimizing H interferes strongly with fitness. The problem with selecting for low entropy solutions in Simplerace may be that, because the task has nine actions (compared with 3 for Tartarus), entropy can be reduced greatly by restricting the number of actions used, whereas compression can work by forcing the sequence of actions into regular patterns that still utilize all actions.

The difference between M_{PC} and M_{PH} on both tasks was not statistically significant. Pushing complexity, either by maximizing C or H, promotes policies that make more full use of their action repertoire. As there are many more high complexity sequences, of a given length, than low complexity sequences (i.e. low complexity sequences tend to be more similar), diversity in the population is better maintained allowing evolutionary search to discover more fit solutions.

Figures 4c and 4d show the compressed length (C) of the most fit individual in the population for the four configurations. Here, again, there is a clear distinction between the maximization and minimization runs, as should be expected, but the two tasks have very different regimes. In both, maximizing complexity (C *or* H) increases the compressed length of the most fit individual. For Tartarus the C of the most fit individual starts at an intermediate value of around 1300, and then rises or drops sharply until reaching a steady value for rest of the run. In contrast, the Simplerace runs always start with very compressible behaviors, which gradually become more complex, even for the minimization configurations. The reason for this is very likely due, at least in part, to the output representation used in the Simplerace network. Because each output unit can select one of three actions (as opposed the one-action-per-unit scheme for Tartarus), the initial random networks will tend to have units that saturate at 0 or 1 such that the behaviors will have very low complexity.

Figure 5 shows the fitness of the most complex individual (either in terms of C or H, depending on the measure being optimized). For Tartarus, the most complex individual in M_{PC} is less correlated with fitness compared to M_{PH},

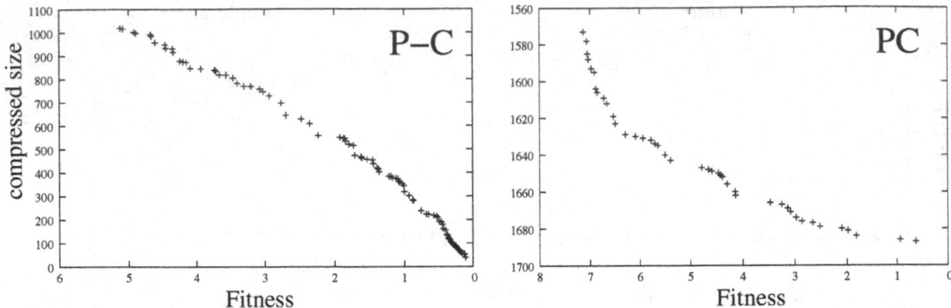

Fig. 6. Pareto Front: fitness/complexity trade-off. The plot on the left shows a typical final Pareto front for M_{P-C}. When behavioral complexity as measured by compressed length is minimized, high fitness ($P > 6$) is not achieved, and the front is most densely sampled near zero fitness. When behavioral complexity is maximized, M_{PC}, the complexity of the entire front is much higher, but the most fit individuals are those with relatively low complexity.

suggesting that is P are C conflict more than P and H. For Simplerace, the fitness of the least compressible behavior increases rapidly and then gradually trends downward, whereas the fitness of the behavior with the highest entropy rises steadily throughout the run.

The overall result is that the two measures of complexity encourage similar performance in both tasks. This is quite different from [5]...

Figure 6 shows the Pareto fronts of the final generation of a typical M_{PC} and M_{P-C} run for Tartarus (Simplerace produces very similar results, also for the PH runs). For M_{P-C} the most fit non-dominated individuals are also the most complex, with most solutions concentrated around the lowest complexity. For M_{PC}, the complexity is in a much higher range (note the y-axis is inverted w.r.t. M_{P-C}), and, in contrast with M_{P-C}, the most fit solutions are the *least* complex. So while parsimony is favorable for given level of fitness, suppressing complexity from the outset, as in M_{P-C}, works against acquiring high fitness (compare the max fitness in figure 4).

Figure 7 examines the relationship between complexity and generalization in the Tartarus task (similar results were obtained for Simplerace). Controllers were collected throughout the entire course of a run and grouped according to *training fitness* (i.e. the fitness awarded on the 100-case training set used during evolution) into fitness classes, each class spanning one fitness point. Each data point in the graph denotes the correlation between the C of the controller's behavior, as measured on the training set, and the fitness on a set of 100 test cases, for a given fitness class. For controllers with low fitness, there is a positive correlation. That is, for a low fitness class, those controllers with high behavioral complexity generalize better within that class. As the training fitness increases there is a clear trend (indicated by the regression line), toward a negative correlation between C and P: the lower the behavioral complexity within a given class the better the generalization. Therefore, as performance improves it is better to behave in a simpler (more compressible) manner in order to cope with new cases.

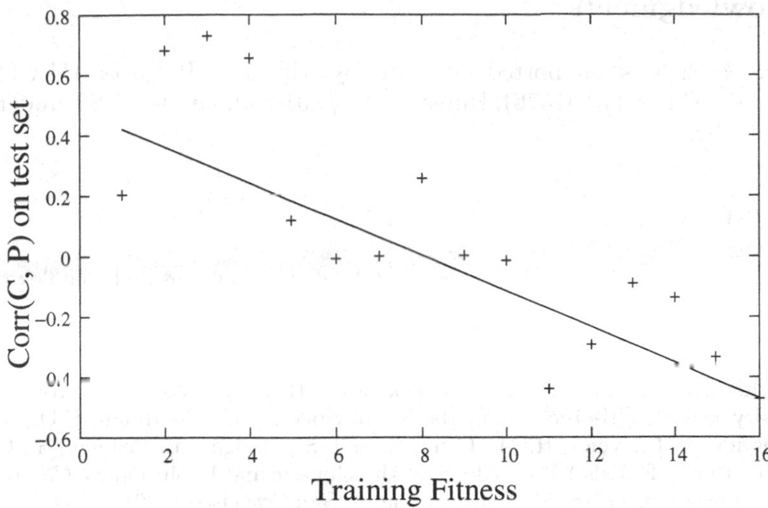

Fig. 7. Generalization trend. Each data point denotes the correlation between C (as measure on the training set behaviors) and the fitness on a set of 100 new test cases, for each training fitness class.

4 Discussion and Conclusions

To our knowledge this paper represents the first attempt at using an explicit measure of *behavioral complexity* (as opposed to model complexity) in the context of evolutionary reinforcement learning.

Although we have barely scratched the surface, the results of these preliminary experiments are interesting and consistent with those of heuristic shaping techniques used in supervised learning (illustrated in figures 6 and 7), where model complexity is given a lower priority in the early stages of learning so that the learner acquires more degrees of freedom with which to reduce error. Once the error reaches a set threshold, the complexity of the model is penalized to reduce the number of free parameters and in order to improve generalization [14].

The overall effect on performance of both entropy and Lempel-Ziv (e.g. `gzip`) was very similar, even though entropy is only concerned with the expected occurrence of each symbol in the behavior, not the ordering or structure of the behavior; the compressor also relies on entropy to encode the behavior, though only after analyzing the structure of the symbol sequence. The behaviors themselves should be analyzed to see if qualitatively different policies arise when complexity of driven in terms of entropy, gzip, or other compressors (e.g. PPM, bzip2) that exploit different algorithmic regularities.

For sequential decision tasks, behavior seems to be the right level at which to compare individuals [5], but, of course, model complexity is critical in determining the range of possible behaviors available to the agent. Future work will also look at combining structural and behavioral complexity criteria for evolutionary methods that search, e.g. both neural network topology and weight space.

Acknowledgments

This research was supported in part by the EU Projects IM-CLEVER (#231711), STIFF (#231576), Humanobs (#231453), and the NSF under grant EIA-0303609.

References

1. Baronchelli, A., Caglioti, E., Loreto, V.: Artificial sequences and complexity measures. Journal of Statistical Mechanics (2005)
2. De Jong, E.D., Pollack, J.B.: Multi-objective methods for tree size control. Genetic Programming and Evolvable Machines 4(3), 211–233 (2003)
3. De Jong, E.D., Watson, R.A., Pollack, J.B.: Reducing bloat and promoting diversity using multi-objective methods. In: Spector, L., Goodman, E.D., Wu, A., Langdon, W.B., Voigt, H.-M., Gen, M., Sen, S., Dorigo, M., Pezeshk, S., Garzon, M.H., Burke, E. (eds.) Proceedings of the Genetic and Evolutionary Computation Conference, pp. 11–18. Morgan Kaufmann, San Francisco (2001)
4. Deb, K., Pratap, A., Agarwal, S., Meyarivan, T.: A fast and elitist multiobjective genetic algorithm: NSGA-II. IEEE Transaction on Evolutionary Computation 6, 182–197 (2002)
5. Gomez, F.: Sustaining diversity using behavioral information distance. In: Proceedings of the Genetic and Evolutionary Computation Conference (to appear, 2009)
6. Iba, H., Garis, H.D., Sato, T.: Genetic programming using a minimum description length principle. In: Advances in Genetic Programming, pp. 265–284. MIT Press, Cambridge (1994)
7. Li, M., Vitányi, P.M.B.: An introduction to Kolmogorov complexity and its applications. In: van Leeuwen, J. (ed.) Handbook of Theoretical Computer Science, pp. 188–254. Elsevier Science Publishers B.V., Amsterdam (1990)
8. Rissanen, J.: Modeling by shortest data description. Automatica, 465–471 (1978)
9. Teller, A.: Advances in Genetic Programming, ch. 9. MIT Press, Cambridge (1994)
10. Toffolo, A., Benini, E.: Genetic diversity as an objective in multi-objective evolutionary algorithms. Evolutionary Computation 11(2), 151–167 (2003)
11. Togelius, J.: Optimization, Imitation and Innovation: Computational Intelligence and Games. PhD thesis, Department of Computing and Electronic Systems, University of Essex, Colchester, UK (2007)
12. Zhang, B.-T., Muhlenbein, H.: Evolving optimal neural networks using genetic algorithms with occam's razor. Complex Systems 7, 199–220 (1993)
13. Zhang, B.-T., Muhlenbein, H.: Balancing accuracy and parsimony in genetic programming. Evolutionary Computation 3, 17–38 (1995)
14. Zhang, B.-T., Mühlenbein, H.: MDL-based fitness functions for learning parsimonious programs. In: Siegel, E.V., Koza, J.R. (eds.) Working Notes for the AAAI Symposium on Genetic Programming, November 10–12, pp. 122–126. MIT, Cambridge (1995) AAAI
15. Zhang, B.-T., Ohm, P., Mühlenbein, H.: Evolutionary induction of sparse neural trees. Evolutionary Computation 5(2), 213–236 (1997)
16. Ziv, J., Lempel, A.: Compression of individual sequences via variable-rate coding. IEEE Transactions on Information Theory (September 1978)

Combining Multiple Inputs in HyperNEAT Mobile Agent Controller

Jan Drchal[1], Ondrej Kapral[1], Jan Koutník[2], and Miroslav Šnorek[1]

[1] Computational Intelligence Group,
Department of Computer Science and Engineering,
Faculty of Electrical Engineering,
Czech Technical University in Prague
Tel.: +420 224 357 470, Fax: +420 224 923 325
{drchaj1,kaprao1,snorek}@fel.cvut.cz
[2] IDSIA, Galleria 2, 6928 Manno-Lugano, Switzerland
Tel.: +41 58 666 6669, Fax: +41 58 666 6661
hkou@idsia.ch

Abstract. In this paper we present neuro-evolution of neural network controllers for mobile agents in a simulated environment. The controller is obtained through evolution of hypercube encoded weights of recurrent neural networks (HyperNEAT). The simulated agent's goal is to find a target in a shortest time interval. The generated neural network processes three different inputs - surface quality, obstacles and distance to the target. A behavior emerged in agents features ability of driving on roads, obstacle avoidance and provides an efficient way of the target search.

1 Introduction

Exhaustive preprocessing techniques are usually used in design of controllers for artificial agents (robots). Environment sensors such as cameras, radars etc. with possibly high resolution in space and time domain are used and their outputs are utilized to perform the desired task.

Our goal is to generate robotic controllers based on recurrent artificial neural networks trained with evolutionary algorithm. Recurrent neural networks [1] are capable of effective temporal information processing because feedback connections form a short term memory within the networks. Such controllers can express more complex behavior.

There are many options how to transform preprocessed sensory input to actions that the robot performs in order to fulfill goals. Artificial neural networks can play the role of a such controlling system. In artificial neural networks the dimensionality of the sensory input was the obstacle that blocked direct processing of e.g. camera images. To overcome this limitation, we use a hypercube encoding of neural network weights [2], which allows to increase input vector as well as amount of artificial neurons in the networks.

Hypercube encoding allows the large-scale neural networks to be effectively encoded into population of individuals. A single genome size does not grow with

C. Alippi et al. (Eds.): ICANN 2009, Part II, LNCS 5769, pp. 775–783, 2009.

the number of neurons in the network. Similarly, a resolution of the network inputs can be extended without growth of the network genome. This is the property of HyperNEAT algorithm used.

HyperNEAT algorithm was introduced in [2] and [3]. It is an evolutionary algorithm able to evolve large-scale networks utilizing so called generative encoding. HyperNEAT evolves neural networks in a two step process: the NEAT (see below) is used to create networks combining a set of transfer functions into a special function. The transfer functions allow to encode symmetry, imperfect symmetry and repetition with variation. The composed functions are called the Compositional Pattern Producing Networks (CPPNs). In the second step, planned neurons are given spatial coordinates. The previously evolved CPPN is then used to determine synaptic weights between all pairs (or subset of pairs) of neurons. The coordinates of both neurons are fed into the CPPNs inputs, the CPPN then outputs their connection weight. The weight is not expressed if its absolute value is below a given threshold. Such connectivity pattern created by CPPN is called the substrate. The important feature of HyperNEAT substrate is that it can be scaled to higher resolutions approximately preserving its inner structure and function.

NEAT (NeuroEvolution of Augmenting Topologies) [4] is an algorithm originally developed for evolution of both parameters (weights) and topology of artificial neural networks. It was extended to produce the CPPNs in the HyperNEAT algorithm instead of producing the neural networks directly. It works with genomes of variable size. NEAT introduced a concept of historical markings, which are gene labels allowing effective genome alignment in order to facilitate crossover-like operations. Moreover, historical markings are used for computation of a genotypical distance of two individuals. The distance measure is needed by niching evolutionary algorithm, which is a core of the NEAT. Because NEAT evolves networks of different complexity (sizes) niching was found to be necessary for protection of new topology innovations. The important NEAT property is the complexification – it starts with simple networks and gradually adds new neurons and connections. For evolving CPPNs, NEAT was extended to evolve heterogeneous computational units (nodes).

1.1 Related Work

Evolution of artificial neural networks is a robust technique for development of neural systems. Many techniques were developed for evolution of either weights or even a structure of neural networks like e.g. Analog Genetic Encoding [5,6,7], Continual Evolution Algorithm [8], GNARL for recurrent neural networks [9], Evolino [10] and NeuroEvolution of Augmenting Topologies (NEAT) [4]. The NEAT algorithm became a part of HyperNEAT algorithm as a tool for evolution of CPPNs.

HyperNEAT algorithm was already applied to control artificial robots in a food gathering problem [2]. A robot with a set of range-finder sensors is controlled to approach the food. It was shown that HyperNEAT is able to evolve very large neural networks with more than eight million connections. A very interesting property of the HyperNEAT is the ability to change a resolution of the substrate. For example 11×11 grid was resized to 55×55 while preserving the underlying neural network function. In the food gathering experiment the inputs indicating whether the food is in a particular direction were arranged parallel or concentric with the robot body. Each sensor was geometrically linked with an effector, which drives the robot.

Our approach differs in the organization of the input sensors, which are arranged in polar rays having particular angular and distance resolution. The sensors are sensitive to color of the surface and in fact represent a camera with arbitrary pixel resolution.

In [11] HyperNEAT algorithm was applied in a very efficient way so that each agent shares a portion of the substrate and neural network. The neural network splits to local areas in the substrate geometrically but all agents share a single substrate. This can be exploited in agents' cooperative behavior.

In [12] it is shown that robots can complete common goals with a minimum information coming from sensors. The robots are controlled by evolved feed-forward neural networks.

In [13] we shown that the HyperNEAT is capable to generate neural networks that can keep the agents stay and drive on roads. Further more, we replaced the NEAT in HyperNEAT with Genetic Programming [14] with comparable results.

In our approach, we reduced an effort typically required to build hardware robotic platforms such as described in [12]. We moved directly to a simulation to concentrate on development of the robot's control algorithms. First, we created a simulation environment described in Section 2.1. The environment allows a rapid development and an experimentation with simulated robots.

This paper is organized as follows. In the next section a simulation environment and a robot setup is described. Section 3 describes the experimental results. A final section concludes the paper.

2 Experimental Setup

2.1 Simulation Environment

Experiments were performed in a simulation environment called ViVAE (Visual Vector Agent Environment) featuring easy design of simulation scenarios in a SVG vector format [13]. There are two types of surfaces in the simulation (a road and a grass) with different frictions. The grass has a friction 5 times higher than the road. Additionally, solid unmovable and movable objects can be placed into the simulation environment. In the current experiments, we used the fixed objects only.

ViVAE supports number of different agents equipped with various sensors for surfaces and other objects in the scenario.

ViVAE allows easy snapshoting of the whole simulation into a sequence of SVG frames. All agents can be tracked and their tracks recorded as a SVG path displayed as a simulation result, see Figure 2.

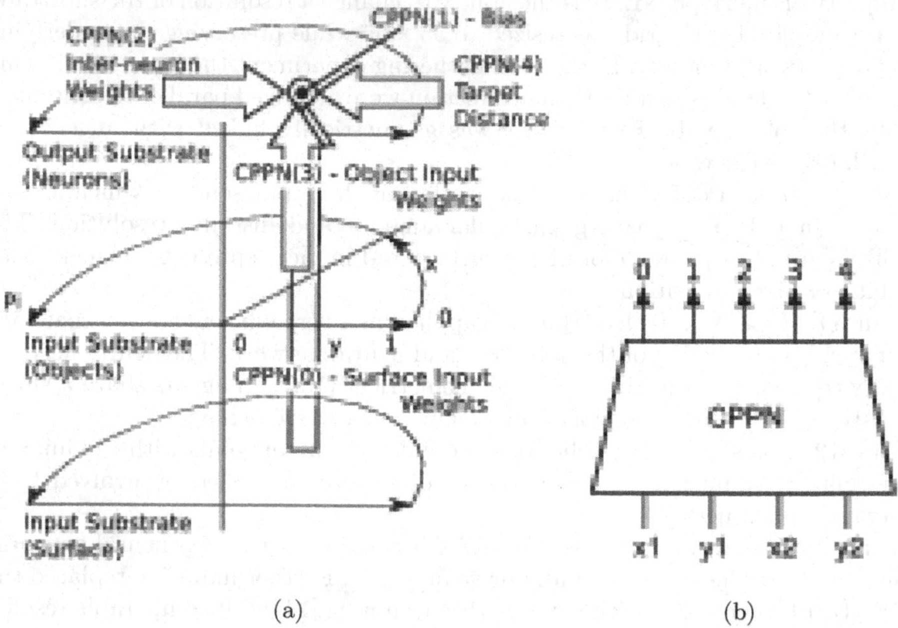

Fig. 1. Organization of the HyperNEAT substrate. There are three distinct substrates used (a) and the CPPN has 5 outputs (b). CPPN(0) output is a weight between input surface substrate and a neuron in the upper substrate. Second, CPPN(1) output is used as bias for neurons in the upper substrate. For a bias calculation the third and the fourth CPPN inputs are set to 0. Third, CPPN(2) output represents connection weights among neurons in the upper substrate. Fourth, CPPN(3) output represents connection weights between input object substrate and neurons in the upper substrate. Last, CPPN(4) output is used for weights between distance to target input variable and neurons in the upper substrate. In this case, first two CPPN inputs are set to 0.

2.2 Agent Setup

The agent is driven by two simulated wheels and is equipped with sensors of three different types. The controlling neural network is organized in a single layer of possibly fully interconnected perceptron (global) type neurons (neurons compute biased scalar product, which is transformed by a bipolar logistic sigmoidal function). Steering angle is proportional to an inverse actual speed of the robot.

The sensors as well as the neural network are spread in a substrate. Neurons and sensors are addressed with polar coordinates, see Figure 1. Two of the neurons in the output substrate are dedicated to control acceleration of the wheels.

During a simulation, an agent is controlled by a neural network controller constructed using HyperNEAT. The neural network neurons and connections are mapped into three substrates. The CPPN has 5 outputs. Three outputs are used for obtaining weights among the neurons and between neurons and inputs from the input substrates (CPPN outputs 0, 2 and 3). One CPPN output is used to set up neurons biases (CPPN output 1). The last CPPN output (4) determines a weight of a connection between distance to target input and particular neuron in the neurons substrate.

The substrate resolution was chosen to be 5 polar rays of 3 sensors in both input layers and 3 × 3 neurons in the layer of neurons.

Table 1. CPPN node functions

Name	Equation
Bipolar Sigmoid	$\frac{2}{1+e^{-4.9\,x}} - 1$
Linear	x
Gaussian	$e^{-2.5\,x^2}$
Absolute value	$\lvert x \rvert$
Sine	$sin(x)$
Cosine	$cos(x)$

Table 2. HyperNEAT parameters

Parameter	Value
population size	100
CPPN weights amplitude	3.0
CPPN output amplitude	1.0
controller network weights amplitude	3.0
distance threshold	15.0
distance C_1	2.0
distance C_2	2.0
distance C_3	0.5
distance C_{ACT}	1.0
mating probability	0.75
add link mutation probability	0.3
add node mutation probability	0.1
elitism per species	5%

2.3 HyperNEAT Setup

We have used our own implementation of the HyperNEAT algorithm. The NEAT part resembles Stanley's original implementation. The HyperNEAT extension is inspired mainly by the David D'Ambrosio's HyperSharpNEAT[1]. Table 1 shows CPPN node functions.

The parameter settings are summarized in Table 2. Note, that we have extended the original set of constants which determine the genotype distance between two individuals (C_1, C_2 and C_3) by the new constant C_{ACT}. The constant C_{ACT} was added due to the fact that, unlike in classic NEAT, we evolve networks (CPPNs) with heterogeneous nodes. C_{ACT} multiplies the number of not matching output nodes of aligned link genes. The CPPN output nodes were limited to bipolar sigmoidal functions in order to constrain the output.

3 Experimental Results

Experimental results described in Section 3.1 were intended to learn the agents to drive on roads instead of grass surface, which has 5 times greater friction than a road.

3.1 On Road Driving

In this experiment the agent controller used three substrates (surface input, neurons and biases) only. The scenario contained no obstacles. And there were 5 agents in the simulation performing concurrently. The agents had no particular target to find. Instead, the agents were trained to gain a maximum average speed in the simulation, according to the following fitness function:

$$f_1 = \frac{distanceTraveled}{simulationSteps + 1} \tag{1}$$

Fitness function f_1 is an average speed of the simulated robots. The 1 is added to prevent a division by zero. The speed is a meaningless variable (number of pixels per a simulation step) but can be computed in a straightforward way and is suitably proportional.

Figure 2 shows a final solution which was found in a generation 324. The trajectories are smooth. Moreover, robots learned to drive on a one side of the road to avoid mutual collisions. The complete experiment is described in [13].

3.2 Obstacle Avoidance

In this experiment the agents were equipped with an additional substrate for connections generated by a CPPN output number 3. The last CPPN output

[1] Both Stanley's original NEAT implementation and D'Ambrosio's HyperSharpNEAT can be found on `http://www.cs.ucf.edu/~kstanley`.

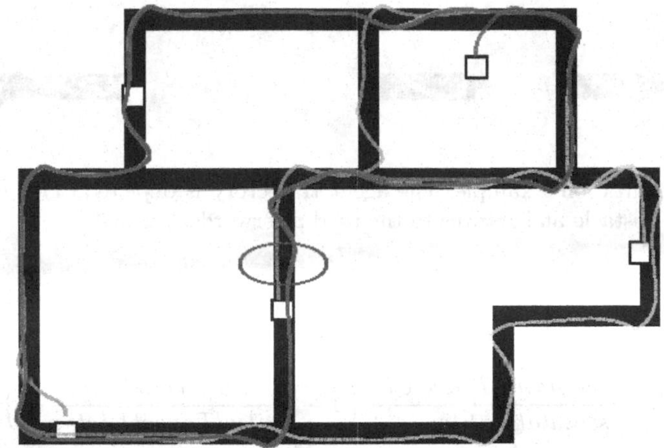

Fig. 2. Trajectories of the robots controlled by a neural network found in a generation 324. The trajectories are smooth. The robots learned how to drive on a one side of the road (as emphasized by the red ellipse).

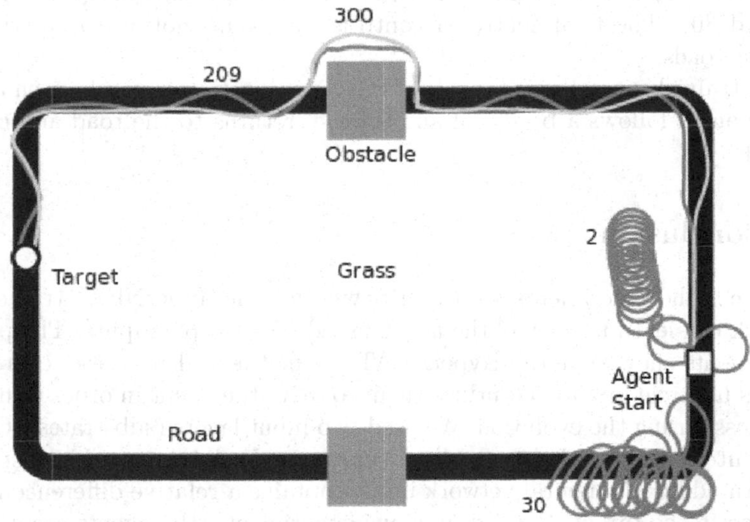

Fig. 3. Obstacle avoidance. This figure shows how an obstacle avoidance emerges during an evolution. There are trajectories of an agent controlled by the best controller found in a particular generation. We can see that after 300 generations of the evolution run, the agent can successfully drive around the obstacle and return to the road to reach the target. The path improved in comparison to generation 209 in which the controller moves the agent periodically from one side of the road to the other one.

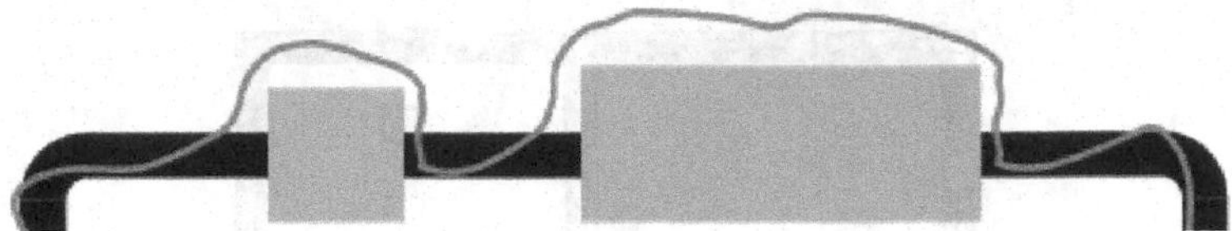

Fig. 4. Generalization Example. The agent trajectory is depicted. The agent follows border of an obstacle and returns to the road afterwards.

controls weights of connections to input containing sign of the actual target distance difference. The fitness function is the following one:

$$f_1 = \frac{distanceTraveled}{simulationSteps + 1}\left(1 - \frac{targetDistance}{initialTargetDistance}\right) \tag{2}$$

The fitness from the previous experiment is multiplied by a relative distance to the target. Agents that find the target faster are preferred to those, which drive on road but do not approach the target.

The controller performances obtained in the evolution are depicted in Figure 3. Agent trajectories evolved in generations 2, 30, 209 and 300 are depicted in a single scenario. We can see how the path was precised between generations 200 and 300. The final controller controls the agent motion to be straight on straight roads.

The trained controller is capable of the generalization as can be seen in Figure 4. The agent follows a border of an obstacle, returns to the road and continues the ride.

4 Conclusion

The aim of the experiments was to verify whether the HyperNEAT trained neural network can learn to control the agent based on multiple inputs. The presented experiments show that the HyperNEAT trained neural network controller can process multiple inputs and utilize them to drive the agent in order to maximize its fitness during the evolution. We used two input layers (substrates). One layer represented surfaces, the second layer represented solid unmovable obstacle sensors. An additional neural network input contains a relative difference in actual distance to the target. Beyond previous experiments, the agents are capable to bypass solid obstacles and drive in a direction to the target. The fitness was the agent average speed multiplied by a relative distance to the target reached. The agents learned to follow the roads in a direction to the target. The agents trails were improved during the evolution to be more smooth and straight on straight roads.

Further experiments should discover dependencies of the controller capabilities on a density of the input substrates as well as a coevolution among agents in a population, possibly in a different more complex task than approaching a single target.

Acknowledgement

This work has been supported by the research program "Transdisciplinary Research in the Area of Biomedical Engineering II" (MSM6840770012) sponsored by the Ministry of Education, Youth and Sports of the Czech Republic and partially by Humanobs EU Project (#231453).

References

1. Elman, J.L.: Finding structure in time. Cognitive Science 14(2), 179–211 (1990)
2. D'Ambrosio, D.B., Stanley, K.O.: A novel generative encoding for exploiting neural network sensor and output geometry. In: GECCO 2007: Proceedings of the 9th annual conference on Genetic and evolutionary computation, pp. 974–981. ACM, New York (2007)
3. Gauci, J., Stanley, K.: Generating large-scale neural networks through discovering geometric regularities. In: GECCO 2007: Proceedings of the 9th annual conference on Genetic and evolutionary computation, pp. 997–1004. ACM, New York (2007)
4. Stanley, K.O., Miikkulainen, R.: Evolving neural networks through augmenting topologies. Evolutionary Computation 10, 99–127 (2002)
5. Mattiussi, C.: Evolutionary synthesis of analog networks. PhD thesis, EPFL, Lausanne (2005)
6. Dürr, P., Mattiussi, C., Floreano, D.: Neuroevolution with analog genetic encoding. In: Runarsson, T.P., Beyer, H.-G., Burke, E.K., Merelo-Guervós, J.J., Whitley, L.D., Yao, X. (eds.) PPSN 2006. LNCS, vol. 4193, pp. 671–680. Springer, Heidelberg (2006)
7. Dürr, P., Mattiussi, C., Soltoggio, A., Floreano, D.: Evolvability of Neuromodulated Learning for Robots. In: The 2008 ECSIS Symposium on Learning and Adaptive Behavior in Robotic Systems, pp. 41–46. IEEE Computer Society, Los Alamitos (2008)
8. Buk, Z., Šnorek, M.: Hybrid evolution of heterogeneous neural networks. In: Kůrková, V., Neruda, R., Koutník, J. (eds.) ICANN 2008, Part I. LNCS, vol. 5163, pp. 426–434. Springer, Heidelberg (2008)
9. Angeline, P.J., Saunders, G.M., Pollack, J.B.: An evolutionary algorithm that constructs recurrent neural networks. IEEE Transactions on Neural Networks 5, 54–65 (1993)
10. Schmidhuber, J., Wierstra, D., Gagliolo, M., Gomez, F.: Training recurrent networks by evolino. Neural computation 19(3), 757–779 (2007)
11. D'Ambrosio, D.B., Stanley, K.O.: Generative encoding for multiagent learning. In: GECCO 2008: Proceedings of the 10th annual conference on Genetic and evolutionary computation, pp. 819–826. ACM, New York (2008)
12. Waibel, M.: Evolution of Cooperation in Artificial Ants. PhD thesis, EPFL (2007)
13. Drchal, J., Koutník, J., Šnorek, M.: HyperNEAT controlled robots learn to drive on roads in simulated environment. In: Accepted to IEEE Congress on Evolutionary Computation (CEC 2009) (2009)
14. Buk, Z., Koutník, J., Šnorek, M.: NEAT in HyperNEAT substituted with genetic programming. In: Accepted to International Conference on Adaptive and Natural Computing Algorithms (ICANNGA 2009) (2009)

Evolving Spiking Neural Parameters for Behavioral Sequences

Thomas M. Poulsen and Roger K. Moore

Department of Computer Science, University of Sheffield
Regent Court, 211 Portobello, Sheffield S1 4DP, United Kingdom
{t.poulsen,r.k.moore}@dcs.shef.ac.uk

Abstract. Sequential behavior has been the subject of numerous studies that involve agent simulations. In such research, investigators often develop and examine neural networks that attempt to produce a sequence of outputs. Results have provided important insights into neural network designs but they offer a limited understanding of the underlying neural mechanisms. It is therefore still unclear how relevant neural parameters can advantageously be employed to alter motor output throughout a sequence of behavior. Here we implement a biologically based spiking neural network for different sequential tasks and investigate some of the neural mechanisms involved. It is demonstrated how a genetic algorithm can be employed to successfully evolve a range of neural parameters for different sequential tasks.

1 Introduction

The ability to produce a sequence of movement is a central issue of study in agent simulations [1,2,3]. In particular, sequences where decisions are dependent on previous actions have been the subject of attention as they underlie many types of complex behavior [4]. Sequence generation of this form requires short-term memory (STM) [4] and biologically inspired solutions have employed recurrent neural networks for STM with considerable success [5,6]. Such approaches employ rate based neural models which do not take temporal effects of neurons into account, yet the temporal domain plays a fundamental role in many behavioral sequences [5]. There is also accumulating evidence that timing is central to the underlying neural circuitry in biological systems (see for example Bothe [7] for a survey of the biological evidence, and Rieke et al.[8] for a statistical analysis). This has prompted investigators to employ spiking networks (which take temporal behavior into account) for agent simulations that involve tasks such as sound localization [9,10] and navigation [11].

Such approaches typically involve altering synaptic weights using spike-timing dependent plasticity (STDP) or by employing a genetic algorithm (GA). Spiking models however, employ additional parameters such as time delays and refractory constants that have significant effects on network functionality. These parameters cannot be adjusted by employing a formalized approach such as Hebbian

C. Alippi et al. (Eds.): ICANN 2009, Part II, LNCS 5769, pp. 784–793, 2009.

learning, but are typically either estimated or determined through trial an error. Yet for sequences where the temporal domain is fundamental, altering such parameters could have considerable effects on results.

An important first step for investigating the role of spiking neural mechanisms in sequence generation is therefore to demonstrate how relevant parameters can be advantageously altered. If such an approach can be shown, it would present a methodology for investigating neural mechanisms and developing spiking network models for different behavioral sequences, in particular where timing of movement is essential.

The current investigation describes an important proof of concept that demonstrates how different parameters in a spiking neural network can play an important role in sequence generation, and how such parameters can be adjusted using a genetic algorithm. This involves a population of agents that each utilize a recurrent spiking network to perform sequential tasks that require temporal integration across events. An analysis is then made of the best performing agents to assess the role of the most significant parameters and mechanisms in agent neural networks.

2 Experimental Set-Up

Based on Fuster [4], we define a sequence where each subsequent event is dependent on previous occurrences as:

$$f(t_1) \rightarrow f(t_2|f(t_1)) \rightarrow f(t_3|f(t_1), f(t_2)) \rightarrow f(t_n|f(t_1)...f(t_n-1)) \qquad (1)$$

where *f(t)* represents an event in a sequence at time t and an event corresponds to an action in a behavioral sequence such as vocalization, head movement, or walking. In animals, such an action is propagated by motor neurons that adjust the relevant muscles (see for example Squire et al. [12]). Computational neural network models often adopt a similar approach for sequence generation, and for example employ two groups of neurons to adjust left and right movement of a robot [13], or to alter the pressure and tension for songbird vocalization [14].

The agent task we define in this work is based on the same principle of adjusting two parameters to generate a sequence. However, as we aim for a general proof of concept, we simplify the model and avoid any biophysical implementation such as a vocalization tract. This also avoids assumptions about the biophysical model selection as well as what constitutes a successful sequence (for example what is considered correct vocalization). Therefore, an abstract representation, in which agents must adjust two parameters that are dependent on previous events (as defined in Equation 1) is adopted instead. As it is helpful to provide a visualization of this abstract representation, in particular with respect to neural dynamics, we provide agents with the ability to execute simple movement by employing two "wheels" (i.e. each wheel represents a variable controlled by neural output). Further specification of this movement and the sequential task, is made after a description of the spiking neuron model and agent neural network implementation has been presented.

2.1 Spike Response Model

Neural firing was modeled using the spike response model (SRM) which has demonstrated that it can successfully capture many of the dynamic behaviors of biological neurons [15]. Simulations were made with two SRMs: one with spike frequency adaptation (SFA) (Equation 2), and one without (Equation 3):

$$u_i(t) = \sum_{t_i^{(f)} \in F_i} \eta(t - t_i^{(f)}) + \sum_j w_{ij} \sum_{t_j^{(f)} \in F} \varepsilon_{ij}(t - t_j^{(f)}) \quad (2)$$

$$u_i(t) = \eta(t - t_i^{'}) + \sum_j w_{ij} \sum_{t_j^{(f)} \in F} \varepsilon_{ij}(t - t_j^{(f)}) \quad (3)$$

where presynaptic neuron j connects to neuron i, η and ε are kernels (we use the same kernels as Floreano & Mattiusi [13]), w denotes the synaptic efficacy, t is the current time, t' is the last time a neuron fired and t^f denotes a neuron's previous firing times (indexed by f). A neuron fires when u exceeds a threshold value given by u_{thresh}. Thereafter, the neuron is reset to its resting potential u_{rest}. The SFA in Equation 2 results from negative feedback in a neuron due to firing (also referred to as output-driven adaptation) and it is included by summating η over previous firing times [15]. The feedback described by η depicts the afterhyperpolarization (AHP):

$$\eta(s) = -(u_{thresh} - u_{rest})e^{-\frac{s}{\tau_{ref}}} \quad (4)$$

where $s = t - t'$ and τ_{ref} sets the time it takes for a neuron to return to its resting potential after firing. The effect of incoming spikes declines over time such that more recent spikes yield greater influence on the membrane potential. This effect is dependent on the properties of the neural membrane and the synaptic connection (time constants τ_m and τ_s respectively), expressed by:

$$\varepsilon(s) = e^{-\frac{s-\Delta_{abs}}{\tau_m}}(1 - e^{-\frac{s-\Delta_{abs}}{\tau_s}})\,\Theta(s - \Delta_{abs}) \quad (5)$$

Spikes do not arrive immediately, but after a time delay of Δ_{abs}, enforced by the Heaviside step function Θ: $\Theta(\mathrm{x}) = 1$, $\mathrm{x} > 0$ otherwise $\Theta(\mathrm{x}) = 0$. Input neurons did not follow any equation but were set to fire ten times within the first 100ms (as a response to an imaginary stimulus that signaled agents to begin).

2.2 Agent Neural Network

Each agent's network consisted of an input, a hidden and an output layer with 18 neurons in total. This number was determined through experimentation to be the most suitable in terms of computation time and likelihood of producing a solution. Agents employed a 3-layered feed forward network with four different types of neurons: input, hidden, left output, and right output. Neuron types, time constants, and synaptic efficacies were defined in an agent's genome such that network structures and functionality changed throughout the application of the genetic algorithm (see Figure 1).

Time constants were experimentally derived within biologically plausible ranges (which were employed to help find suitable limits): τ_{ref} = 0.1ms to 50.0ms [16], τ_m = 2.0ms to 60.0ms [17], τ_s = 0.1ms to 60.0ms [18], Δ_{abs} = 0.1ms to 60.0ms [19]. The neural threshold and resting potential (u_{thresh} = 0.5 and u_{rest} = 0) were set with respect to synaptic values (-1.0 to +1.0). Constraints were applied to the neural network such that input neurons did not receive connections from other neurons, and only connected to hidden neurons (Figure 2). Hidden neurons adjoined the input and output layer but also connected to one another.

Agent neural networks employed recurrent connections which have been shown to provide short-term memory (STM) [5]. In a recurrent network structure, neural output is dependent on previous activity in the network, and this allows temporal integration across previous events in a sequence of events. This provided agents with memory of previous movement such that the next step in the sequence could be made.

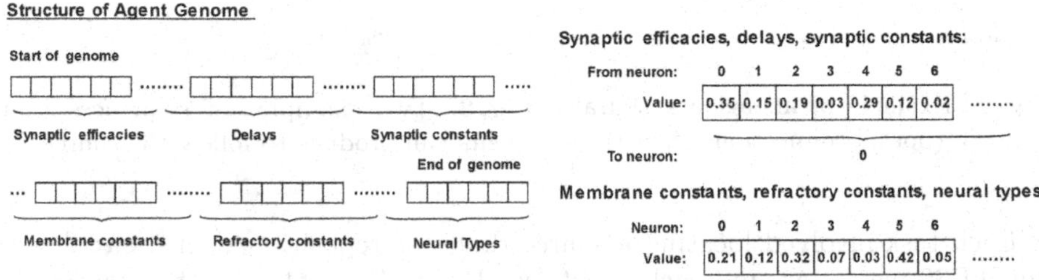

Fig. 1. Each agent network is defined in an agent's genome and consists of an input, a hidden and an output layer. For the sake of clarity, not all connections shown.

Agents wheels provided simple motor function that allowed left and right turns. The wheels could only turn forwards and the turning force of each wheel was calculated according to the number of spikes received from output neurons (denoted by n_l and n_r for the left and right wheel respectively). This allowed the agents to turn at an angle a:

$$n_l > n_r : \ \ a = 2\pi\,(\frac{n_r}{n_l} - 1)\ ,\ \ n_r > n_l : \ \ a\ = 2\pi\,(1 - \frac{n_l}{n_r}) \tag{6}$$

Previous research has shown that directional movement can be derived from the firing rate of neurons [20] and Equation 6 was developed on this basis. Also, Equation 6 was designed such that altering a requires non-linear changes in n_l and n_r which entails that agents cannot simply scale neural output across events to produce a correct sequence.

2.3 Agent Task

Initially, experiments were performed where agents had to perform a sequence that will be termed Task 1. Following an analysis of agent neural network mechanisms and performance in Task 1, three additional sets of experiments (Tasks 2-4) were performed with new sequences to ascertain results. All tasks were designed such that they required changes in the neural dynamics across timeframes, could be executed using different sequences and neural network solutions, and required different types of neural dynamics in terms of how quickly and substantially changes in neural output needed to occur across timeframes.

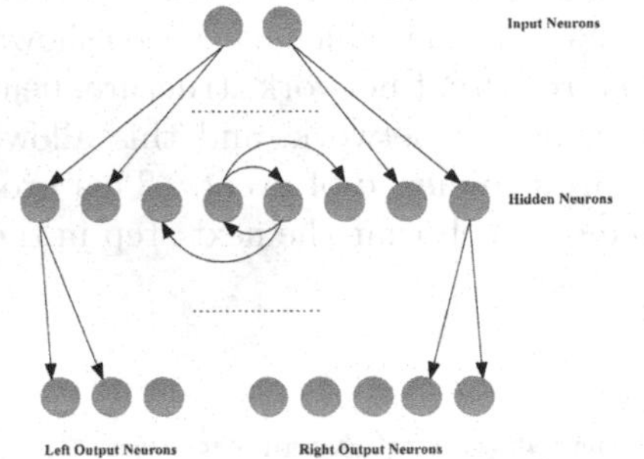

Fig. 2. Example of an agent's neural network (not all connections shown)

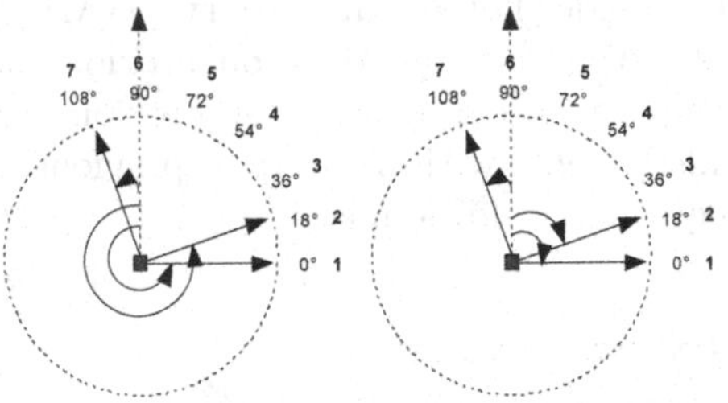

Fig. 3. Two examples of sequences that agents can produce to follow the source

Each task involved locating a source that was repositioned in a circular radius of 30 meters around each agent (see Figure 3). In Task 4, the source was positioned at the same positions as in Task 1, but for different time durations at each position (see Table 1). In Tasks 1-3, the source was positioned at the angles shown in Table 1. Agents had to turn towards the source every 100ms (except in Task 4 where this timeframe was variable) using Equation 6, but agents only turned if both n_l and n_r were greater than zero.

After each turn, agents were repositioned to their original direction (set to 90° as shown in Figure 3) and the spike counts n_l and n_r in Equation 6 were reset

Table 1. Locations (for Tasks 1-3) and timeframes (for Task 4) of the source

Position	Task 1	Task 2	Task 3	Task 4	Position	Task 1	Task 2	Task 3	Task 4
1	0°	0°	114°	80ms	11	180°	135°	294°	120ms
2	18°	22.5°	126°	100ms	12	198°	157.5°	306°	80ms
3	36°	45°	150°	80ms	13	216°	180°	330°	100ms
4	54°	67.5°	162°	80ms	14	234°	202.5°	342°	80ms
5	72°	45°	186°	120ms	15	252°	180°	6°	120ms
6	90°	67.5°	198°	80ms	16	270°	202.5°	18°	100ms
7	108°	90°	222°	100ms	17	288°	225°	42°	80ms
8	126°	112.5°	234°	160ms	18	306°	247.5°	54°	80ms
9	144°	135°	258°	80ms	19	324°	270°	78°	140ms
10	162°	112.5°	270°	140ms	20	342°	247.5°	90°	80ms

back to zero. Since agents were turned back to their original position after each turn, they were not able to follow the source simply by making fixed increments in their turns. Instead, they were required to retain memory of at least their last turn and continue the sequence accordingly (in agreement with Equation 1). Figure 3 illustrates two possible solutions that an agent could employ to follow the source. In the first solution the agent turns the same direction each time, while in the second solution both left and right turns are made. The first solution might appear relatively simple, but similar to other solutions, it requires a non-linear increase in firing rates of output neurons (as a result of Equation 6) and agents therefore did not appear to favor this solution.

2.4 Genetic Algorithm and Agent Scoring

Agent fitness values were calculated by using the distance between the source and agent (d_{agent}), and the distance between the source and a point situated one meter directly in front of the agent (d_{point}):

$$d_{agent} = \sqrt{(x_{agent} - x_{source})^2 + (y_{agent} - y_{source})^2} \tag{7}$$

$$d_{point} = \sqrt{(x_{point} - x_{source})^2 + (y_{point} - y_{source})^2} \tag{8}$$

$$d_{diff} = (d_{agent} - d_{point})^{28} \tag{9}$$

where x and y refer to the coordinates of the agent, the point and the source (the coordinate system was set with the agent in the center such that x_{agent}=0 and y_{agent}=0). Each agent was scored with a value v (initially set to zero) that was adjusted after each timeframe (when an agent was supposed to have turned) according to: if $d_{point} < d_{agent}$: $v = v + d_{diff}$, if $d_{point} > d_{agent}$: $v = v - d_{diff}$, otherwise if agent did not turn: $v = v - 1$.

At the end of each generation, any negative values of v were set to zero and the fitness value was calculated as $f = v/n_{angles}$ (with n_{angles} set to 20 - the number of different source positions). Intuitively, it would seem better to calculate the fitness values using the average angle error for an agent's turns, but this did not yield as good results.

The BLX-α crossover algorithm [21] was employed for the entire genome; α was set to 0.5 and the probability of a crossover was set to 0.8. The mutation operator was applied twice (with a probability of 0.20 in both cases): in the first instance to end of the genome containing the neuron types, and the second time to the rest of the genome. This approach was adopted because changing neuron types had a substantial effect on the neural network. Without this targeted mutation operator, neuron types were not affected often enough due to the length of agent genomes. The probability of mutation was for similar reasons set high. Elitism was employed (with two elites) once an agent in the population reached an average error angle of 14°. The population size was set to 100 and simulations were run for 3000 generations.

3 Results

3.1 Task 1

In Task 1, agents with average error scores of less than 7° were analyzed from 250 different simulations. Approximately 10% of simulations evolved agents with error scores less than 7°. All of these simulations utilized SFA (Equation 2); simulations without SFA (Equation 3) are discussed in Section 3.2.

The best agents were able to perform Task 1 with an average angle error of about 2° (four simulations evolved agents with such scores). One agent for example, had an average angle error of 1.97° with a standard deviation of 1.48°. We only discuss this agent's neural network mechanisms in the following as most other network solutions demonstrated similar functionality.

The agent's neural network consisted of three input neurons, ten hidden neurons, and five output neurons (Figure 4 and 5). Figure 6 displays the total spikes produced by the left and right output neurons respectively during the 100ms time frame for each source angle. Network output to the left and right wheel increased up until a source angle of 36°, at which point it reached a maximum for the right wheel and started to decline for the left wheel.

An analysis of neural dynamics found that the change at 36° was a result of time delays (in particular for reccurent connections between hidden neurons) and SFA. An example of the how time delays functioned can been seen by looking at Figure 5 which depicts connectivity between the most active hidden neurons. There is a strong excitatory connection from neuron 16 to 3 and from neuron 3 to 17, but a strong inhibitory connection from neuron 17 to 16. The total time delay across this series of connections is 54ms (neglecting delays caused by processing time in individual neurons). Therefore, if neuron 16 increases its output, it will likely receive additional inhibition later in the sequence. This delayed recurrent activity, along with SFA, caused a decrease in neuron 16's firing rate at source angle 36°, prompting the firing decrease of output neurons at 36° (Figure 6).

An example SFA functionality can be seen by first looking at the output neurons in Figure 6 which shows that the left wheel received input almost exclusively from neurons 5 and 10. To measure the effects of AHP throughout the sequence, two different summations of AHPs were made across each 100ms movement time frame. One summation was made for iterations where neuron 5 did not produce a spike (we term this "suppressing AHP") and another summation was made across all iterations ("Total AHP"). Comparing these two AHP totals throughout the time course of a sequence provided a measure of the AHP's effectiveness with respect to SFA. Looking at the curves in Figure 7, it can be seen that the slope of the suppressing AHP follows the direction of the slope of the total AHP, except between 18° and 36°, and between 54° and 72°. The result between 18° and 36° is particularly interesting as there is a significant decrease in the total AHP but a large increase in the suppressing AHP. This is a result of SFA, which can be seen in the drop in neural output between 18° and 36° in Figure 6, and it causes the total AHP to decrease as well. However, the large increase in the suppressing AHP shows that the AHP played a significant role in decreasing

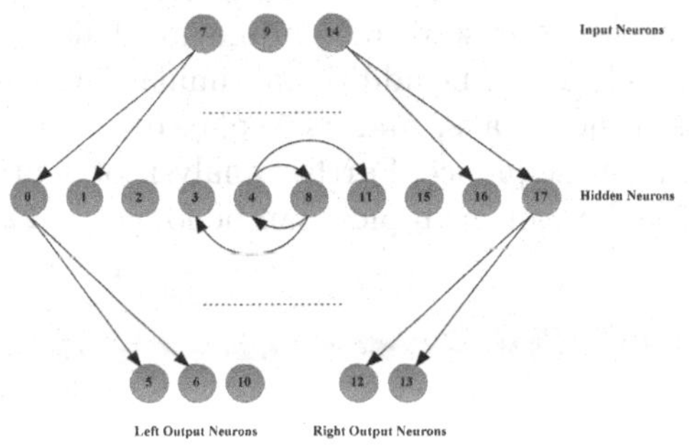

Fig. 4. Neural network of agent selected for analysis (not all connections shown)

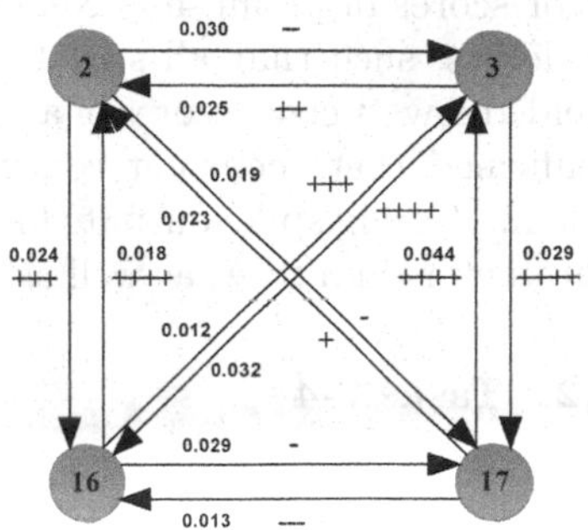

Fig. 5. Inhibitory and excitatory connections of hidden neurons that fired the most. Synaptic efficacies shown by minus and plus symbols (one = weakest, four = strongest).

spike output between these two angles. A similar occurrence, albeit to a lesser degree, occurs between 54° and 72° (Figures 6 and 7). In general throughout a sequence, the degree of this SFA can be set by τ_{ref} (see Equations 2 and 4) which was therefore a central parameter in this functionality.

Additional simulations were performed to confirm the significance of the neural mechanisms observed and 100 simulations without spike frequency adaptation were made. Agent neural networks without SFA yielded lowest error scores of 12° to 13°, considerably worse than the best results of 2° where SFA was employed.

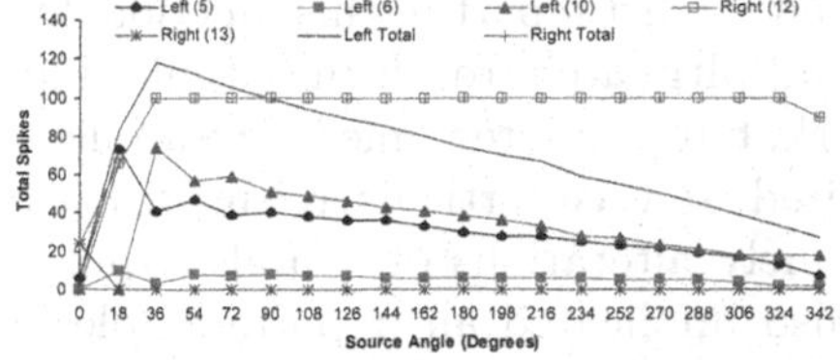

Fig. 6. The total number of spikes produced by left (neurons 5, 6, and 10) and right (neurons 12 and 13). Neuron 12 was the only right output neuron that fired for most of the sequence and it overlaps with the 'Right Total'.

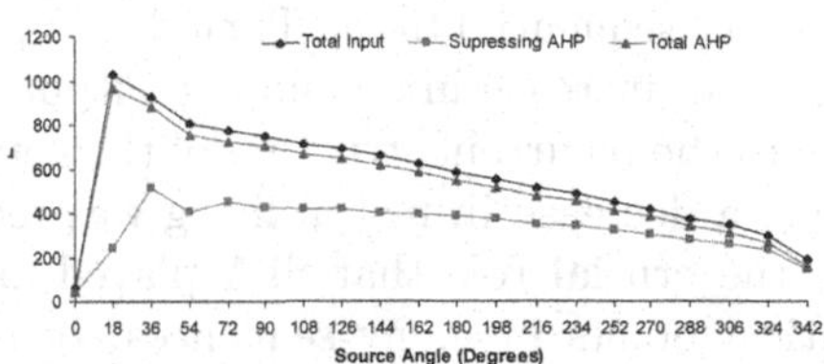

Fig. 7. The total input received, AHP summation across iterations where no spikes were produced ("Suppressing AHP"), and summation across all iterations ("Total AHP") observed for neuron 5. AHP summations are shown as absolute values.

To verify the importance of agent recurrent neural networks, 100 simulations were performed where neural networks did not utilize recurrent or inhibitory connections. In both cases, agents performed very poorly and achieved lowest error scores of about 40°. Neural networks were also evolved without any form of delays, such that all signals between neurons were instantaneous. This also yielded lowest error scores of approximately 40°. The additional simulations thus confirmed that recurrent connections, time delays, and SFA played a central role in altering spike output throughout the sequence. Further analysis of neural parameters and SFA, as well as additional agent examples, can be found in [22].

3.2 Tasks 2-4

Tasks 2-4 focused on the role of SFA that was observed in Task 1 as this was considered the most significant and surprising finding. To further assess the significance of SFA, two sets of 30 simulations were performed for each task. In the first set of simulations, agent neural networks utilized SFA, but in the second set of simulations no SFA was implemented. The results of Tasks 2-4 were similar to those observed in Task 1 and achieved the following best error scores:

$$\text{Task 2(SFA)} = 5.46°, \text{Task 2(No SFA)} = 15.10°$$
$$\text{Task 3(SFA)} = 2.05°, \text{Task 3(No SFA)} = 18.34°$$
$$\text{Task 4(SFA)} = 4.12°, \text{Task 4(No SFA)} = 17.01°$$

The neural networks that evolved in Tasks 2-4 revealed many of the same characteristics observed in Task 1. While our analysis of these networks focused on the spike frequency adaptation, it was also observed that time delays and recurrent connections appeared to play a significant role as described in Task 1.

4 Concluding Remarks

Agents were able to successfully evolve neural network solutions for a range of different sequential tasks. Time delays played a central part in this functionality, where an increase in a neuron's output could ultimately result in self-inhibition due to the recurrent structure of the network, but at a later time in the sequence when a decrease in neural firing was required. It was particularly surprising to see the crucial role that SFA played in timely alterations of neural output at critical points in many sequences - this also implicated an important role for refractory time constants (see Equations 2 and 4).

Our results have shown that it is important to consider different neural parameters for sequences that require temporal integration and we have demonstrated an approach that shows how such parameters can be determined. The study described here encompasses a broader research program that involves the development of neural networks for sequential behavior. Follow-up research is applying the approach presented here to more physiologically detailed scenarios to investigate vocalization sequences. This will involve further assessment of parameters such as refractory constants and time delays and how they shape neural dynamics. We anticipate that this line of research will provide insights into the neural mechanisms that are involved in behavioral sequences and help further development of neural network models where timing is important.

References

1. Yamauchi, B.M., Beer, R.D.: Sequential behavior and learning in evolved dynamical neural networks. Adap. Behav. 2, 219–246 (1994)
2. Dauce, E., Quoy, M., Doyon, B.: Resonant spatiotemporal learning in large random recurrent networks. Bio. Cyb. 87, 185–198 (2002)
3. Capi, G., Doya, K.: Evolution of neural architecture fitting environmental dynamics. Adap. Behav. 13, 53–66 (2005)
4. Fuster, J.M.: The prefrontal cortex - an update: time is of essence. Neuron. 30, 319–333 (2001)
5. Zipser, D., Kehoe, B., Littlewort, G., Fuster, J.: A spiking network model of short-term active memory. J. Neurosci. 13, 3406–3420 (1993)
6. Xing, J., Andersen, R.A.: Memory activity of lip neurons for sequential eye movements simulated with neural networks. J. Neurophysiol. 84, 651–665 (2000)
7. Bohte, S.M.: The evidence for neural information processing with precise spike-times: a survey. Natural Computing 2, 195–206 (2000)
8. Rieke, F., Warland, D., Steveninck, R.R., Bialek, W.: Spikes Exploring the Neural Code. MIT Press, Cambridge (1997)
9. Poulsen, T.M., Moore, R.K.: Sound localization through evolutionary learning applied to spiking neural networks. In: IEEE Symposium on Foundations of Computational Intelligence (FOCI 2007), pp. 350–356 (2007)
10. Gonzlez-Nalda, P., Cases, B.: Spiking neural networks for temporal pattern recognition in complex real sounds. Neurocomputing 71, 721–732 (2008)
11. Saggie-Wexler, K., Keinan, A., Ruppin, E.: Neural processing of counting in evolved spiking and mcculloch-pitts agents. Artificial Life 12, 1–16 (2006)
12. Squire, L.R., Bloom, F., Spitzer, N.: Fundamental Neuroscience, 3rd edn. Academic Press, London (2008)
13. Floreano, D., Mattiussi, S.: Evolution of spiking neural controllers for autonomous vision-based robots. In: Evolutionary Robotics IV, pp. 38–61. Springer, Heidelberg (2001)
14. Abarbanel, D.I., Gibb, L., Mindlin, G.B., Talathi, S.: Mapping neural architectures onto acoustic features of birdsong. J. Neurophysiol. 92, 96–110 (2004)
15. Kistler, W.M., Gerstner, W., Hemmen, J.: Reduction of Hodgkin-Huxley equations to a single-variable threshold model. Neural Comput. 9, 1015–1045 (1997)
16. Storm, J.F.: An after-hypolarization of medium duration in rat hippocampal pyramidal cells. J. Physiol. 409, 171–190 (1989)
17. Koch, C., Rapp, M., Segev, I.: A brief history of time (constants). Cerebral Cortex 6, 93–101 (1996)
18. Kinney, G.A., Peterson, B.W., Slater, N.T.: The synaptic activation of n-methyl-d-aspartate receptors in the rat medial vestibular nucleus. J. Neurophysiol. 72, 1588–1595 (1994)
19. Kistler, W.M., De Zeeuw, C.I.: Dynamical working memory and timed responses: the role of reverberating loops in the olivo-cerebellar system. Neural Comput. 14, 2597–2626 (2002)
20. Georgopoulos, A., Caminiti, R., Kalaska, J., Massey, J.: Spatial coding of movement: a hypothesis concerning the coding of movement direction by motor control populations. Exp. Br. Res. 7, 327–336 (1983)
21. Eshelman, L.J., Schaffer, J.D.: Real-coded genetic algorithms and interval-schemata. In: Foundations of Genetic Algorithms, vol. 2. Morgan Kaufman, San Francisco (1993)
22. Poulsen, T.M.: The use of computational simulations to investigate neural models of acoustic signalling and hearing. PhD thesis (in preperation)

Robospike Sensory Processing for a Mobile Robot Using Spiking Neural Networks

Michael F. Mc Bride[1], T.M. McGinnity[2], and Liam P. Maguire[3]

Intelligent Systems Research Centre, University of Ulster (Magee),
Northland Road, Derry,
Northern Ireland,U.K.
mcbride-m@email.ulster.ac.uk, {tm.mcginnity,lp.maguire}@ulster.ac.uk

Abstract. Current research in intelligent systems investigates their deployment in dynamic and complex environments. Such systems require the capability to be aware of their operating environment and to process effectively sensory information from multiple sensory sources. The abilities observed in the animal kingdom to process sensory information in varying conditions, from many different sensory sources, is an inspiration for intelligent systems research. Sensory processing in the mammalian brain involves thousands of neurons in cortical columns, with extensive interconnect. However it is known that interconnections between neurons and thus the source of spiking activity within these biological columns is locally based. Cortical columns are also stimulated by connections from related areas within the brain which are dedicated to the processing of alternative sensory stimuli. This paper reports on an approach to emulate biological sensory fusion, based on Spiking Neural Networks (SNN) and Liquid State Machines (LSM), and is assessed in experiments involving the control of a mobile robot in a reactive manner. The results show that the sensory processing provided by the Liquid State Machine enables the reactive control of the robot within its environment.

Keywords: Spiking Neural Network, Sensory Processing, Liquid State Machines, Cortical Columns.

1 Introduction

The task of a robot exploring an environment and interacting with it, is enhanced by access to real-time information concerning the environment's structure, layout, obstacles and configuration. This means that the robot should have continuous access to the data of different sensory sources, the validity and confidence with which the data can be regarded and a method of integrating the information in a robust and reliable manner. Clearly the fusion of data from multiple sensors is crucial, particularly in the case of conflicting, incomplete or uncertain data. The development of artificial intelligence techniques to solve this problem has traditionally been directed towards rule oriented solutions [1,2]. In these approaches, which involve the use of many complex logical tests and rules, a sensor S_1 is treated as separate and discrete from other available sensors $S_2, .., S_n$ and

C. Alippi et al. (Eds.): ICANN 2009, Part II, LNCS 5769, pp. 794–803, 2009.

does not reflect the possibility that there may be further information available from the other senses $S_2, .., S_n$ which could potentially be of use in interpreting S_1. This is in contrast to biological systems [3,4] which consistently integrate sensory data to optimize decision and responses. Approaches to emulate the sensory fusion evident in biology include variations of Kalman filters [5] and Dempster-Schafer methods [6]. These methods attempt to remove uncertainties between different sensor streams $S_1, .., S_n$. The result can be a value from the sensors qualified by their reliability. This makes the artificial system more robust; as one sensor's output degrades, the fusion process may allow other sources of sensory information to mitigate the data which is lost.

Biologically inspired approaches for sensory processing include the work of Burgcnsteiner, [7] who investigated the control of a mobile robot using a Liquid State Machine (LSM), and achieved results comparable to a Braitenberg controller [8]. The work is limited to infra-red sensory information and obstacle avoidance, and the author omits a path plot for a sample run from the results [7] instead reporting on the change of wheel speeds. Control of a robotic arm has also been researched using LSMs [9]; this provided a method of controlling the robotic arm and represented feedback from the robotic arm's state; it also enabled the LSM itself to predict the feedback expected.

A Liquid state Machine (LSM) is a computational construct designed to capture the dynamics of spiking neurons. In this work the LSM as described within [10] is created using the toolbox presented within [11]. It is comprised of inputs, a column of neurons and a separate readout. It is possible to have full connectivity from an input neuron to all neurons within the LSM column. The number of connections is normally controlled by using the weight, length and connection chance variables for the creation of connections. The neurons within the column are also connected using the same method. The state of the neurons within the column is changed by the values from the inputs. The changes are passed through the column of the LSM via the interconnecting synapses within the column. The changes cause ripples of activity to pass through the column of the LSM. The state of the liquid, which is changing constantly based upon the inputs to the LSM column, holds a record of the past inputs to the liquid which has been measured to last as long as 80 ms. The state of the liquid is classified with a readout function. The readout's function is to classify the state information which is available within the liquid. The readouts which are suggested within [10,11] include Artificial Neural Networks (ANN). ANN's are not biologically plausible but are used within the LSM paradigm to classify the output of the liquid which is responsible for establishing the state dependent on the inputs and recent history.

Cortical columns are both the inspiration for the LSM [10] and can be considered as computational units within the primary sensory processing cortices. The columns within the somatosensory cortex provide similar processing of sensory stimulus for different areas of the body. Due to this locality of processing cortical columns have a high degree of local connectivity but are also connected to other brain areas which are involved in sensory processing [12].

This paper investigates the use of Spiking Neural Networks (SNN) and LSMs for sensory fusion of information from multiple instances of a single sense represented by infrared sensors on a mobile robotic platform. In section 2 the approach for using the LSM for sensory processing is discussed. Section 3 explains the experimental setup and the specific LSM architecture utilised. In section 4 the experimental results are presented. Finally in section 5, a discussion of the experiments is presented and future plans for the research summarised.

2 SNN for Sensory Processing

SNNs are believed to be biologically compatible processing structures. They are modeled on the neurons within the brain [13] and provide a progression in biological accuracy from second generation Artificial Neural Networks (ANN). To date SNNs have achieved success at performing tasks such as edge detection or motion tracking [14]. It has been shown that evolving robotic controllers is possible using SNN's [15].

LSM uses biological inspiration of the spiking neuron and combines this approach with established techniques from ANNs to create a hybrid method in which the states of the liquid are interpreted using a ANN readout function. The LSM was developed [10,11] to allow more biologically inspired SNN states within the computations of the state of its liquid. The LSM uses recurrent connections which are built into the liquid to maintain a memory of the previous stimulus to the LSM [16].

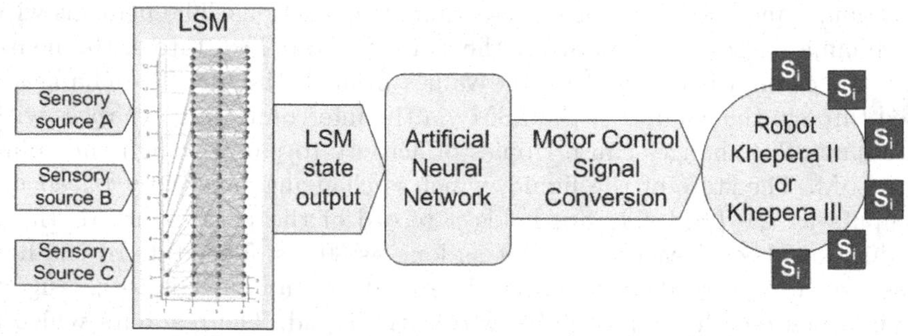

Fig. 1. Structures of the LSM in use in this research

In this work, the Liquid State Machine (LSM), as a special case of the Spiking Neural Network (SNN), is used as a sensory combination element of the sensory processing system. Figure 1 shows the steps involved within the sensory processing approach described in this paper. The sensors indicated on the right of the figure 1 as S_i are processed using the methods described in section 3.2 and passed into the LSM, as indicated on the left of figure 1. The hypothesis is that the LSM integrates the various sensory data to form a complete, time-dependent picture of the environment that stimulated the sensors. This approach means it is unnecessary to quantify the differences between sensory sources as in other fusion systems, utilizing for example the Dempster-Schafer approach. In this work an ANN is trained to interpret the complex states which are contained within the LSM's internal state.

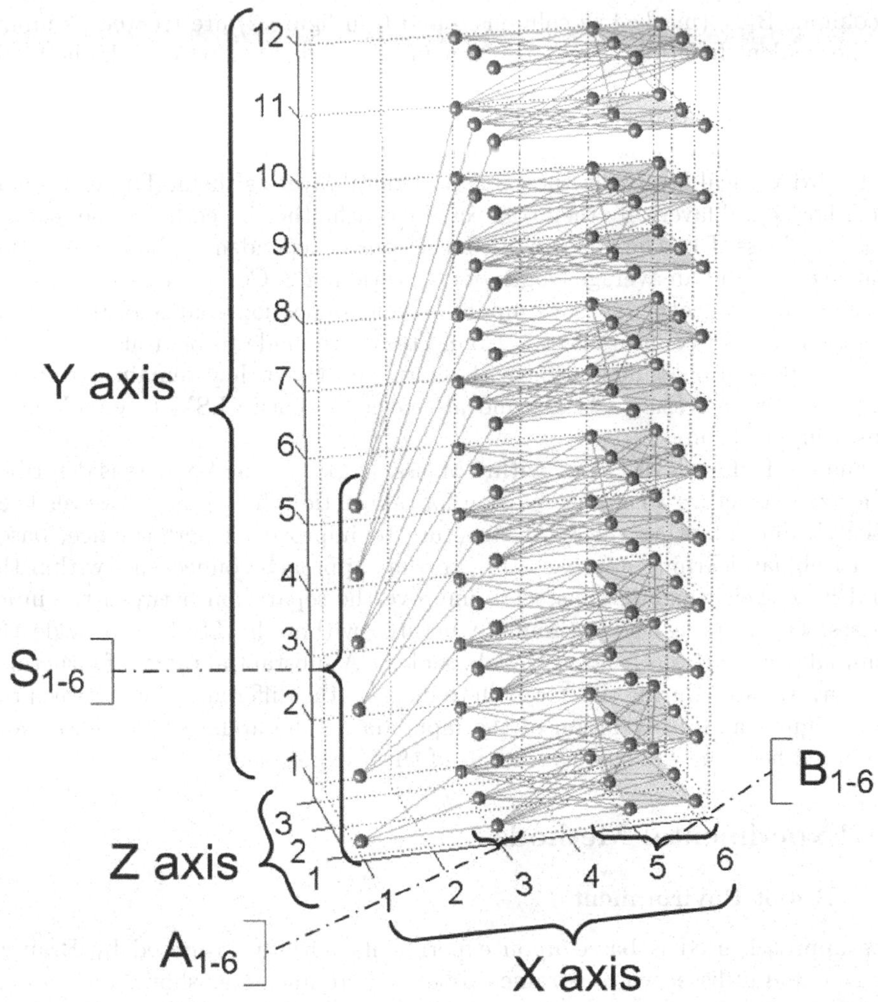

Fig. 2. Structure of the LSM in use in this research

Figure 2 illustrates the LSM structure used in this work. The sensor data is input to the LSM via the neurons labeled as S_{1-6} in figure 2. The figure shows the entire LSM structure in use for the experiments (with all six input neurons S_{1-6}). Each section of the LSM deals with a specific locale of stimulus on the robot and is highly locally interconnected. The structure of the LSM allows the processing of the individual sensory inputs without distorting the processing with data relevant to the other sensors. The LSM has an input for each of the sensors putting data into the column of the LSM. The inputs labeled as S_{1-6} are connected only to neurons in the first vertical layer of the LSM, labeled as A_{1-6}. These are the neurons marked 3 on the X axis of the figure 2. In the complete structure there are 36 neurons in this section of the LSM column. The remaining 2 columns B_{1-6} (marked as columns 5 and 6 in figure 2) are treated as inputs to the readout function. There are 72 neurons within that part of the network. There are 2 horizontal layers created for each input $S^{6}_{i=1}$. Each input neuron (S_i in figure 2) is connected to 3 of the neurons in the first layer of the LSM column. These 3 neurons are then connected to the 6 neurons from the second and third vertical layers, on the same horizontal layer as them. The neurons in the 2 horizontal layers are then connected to each other based upon the method used by Maass [10]. The number of connections is dependent upon a connection chance of 0.9 and an average length of connection of 2. Connections to the areas of the column for other sensory inputs have a connection chance of 0.9 and an average length of connection of 5. Connections are made to neurons within the single sensory processing section with short lengths so they remain within the single sensor processing section, and also over the entire LSM to enable cross sensor input fusion.

The LSM which is discussed within Maass' [10,11] original papers is described as a fixed structure with no provision for adaptation. There has however been research into adapting the LSM structure to improve its performance, based on a Hebbian learning approach [17]. In that approach connections within the LSM have their weights adjusted to improve the separation between the input classes. Determining the optimum structures within the LSM to provide the required computations is not a trivial problem. A substantial range of structures were investigated during this research to increase the differentiation between the input signals and allow fusion of the input data. The architecture chosen and presented here has emerged as a result of these experiments.

3 Experimental Methods

3.1 Robot Environment

The approach used is based upon experiments initially proposed by Braitenberg [8] and utilises two constrained robots environments as shown in figure 3a and figure 3b. Two Khepera robots were used, a Khephera II for the enclosure of figure 3a and a Khephera III for the enclosure of figure 3b. The environment in which the Khepera II robot is deployed is lined with white to increase the reflectivity and thus the performance of the IR sensors on the robot. The environment in Figure 3a is a rectangular environment with a floor area of approximately $0.3m^2$. while that of Figure 3b is set in an environment with a floor area of approximately $0.9m^2$. In each experimental run, the robot is set to start from the same position in the environment. The experiment is then performed, allowing the robot to run through the environment reacting to avoid obstacles and the path taken by the robot is recorded. Infra red sensors are used in both experiments. The Khepera II provides 8 Infra Red (IR) sensors of which the 6 forward and sideward facing sensors were used; for Khepera III only the 6 forward and sidewards facing sensors were used. The sensor positions on the robots are marked on the 'robot' depicted in Figure 1. The IR sensors on the Khepera II have a detection range from 0 to 1024 (sensor units) which varies depending on the reflected IR values which are detected by the sensor. The Khepera III uses a more recent sensor which has a larger response range of between 0 and 4096 (sensor units); again this is dependent upon reflected IR light being returned. Different robots in two separate environments are used to demonstrate the generalisability of the solution for use on different robots in alternate simple environments.

3.2 Processing

$$inputfrequency = \frac{1}{\frac{S_i}{\frac{MR}{x}} - 1} \tag{1}$$

The input values to the LSM are calculated from the sensor inputs S_i into a linear frequency dependent spike train. The sensory inputs are taken from the robot sensors marked S_i upon the right hand side of figure 1. They correspond to the sensor inputs on the left of the figure 1. The input frequency is measured in seconds and must be within the range of 0 to 1. The input frequency is calculated using equation (1), dependent upon the maximum range MR of the sensor on the robot. This range is adjusted by a scaling value x to ensure an appropriate range of spiking activity in the LSM. This frequency is then converted into a series of spike times which acts as the input to the LSM's input neurons indicated as S_i in figure 2.

The inputs were presented to the LSM and the observed set of spiking activity from the liquid recorded. Output values from the LSM are converted from spike time arrays into an array of averaged activation of the corresponding neuron for the spiking activity, which has occurred over the last few milliseconds, using equation (2). These values are calculated every 25ms of the LSM's simulation. Equation (2) converts the state of the LSM into a format compatible with the ANN. This is identified in figure 1 as the LSM state output.

$$z(l,k) = \left[{}^{p}_{l=0} \left[{}^{m}_{k=0} \sum_{j=0}^{n} e^{(t_s(l,j)-t_i(k))/\tau_1} \right] \right]^{n<t_i(k)} \tag{2}$$

In equation (2) where t_s represents the spike times from the LSM recorder; t_i represents the sample times for the conversion operation; n represents the number of spike times $t_s(l)$ recorded for the current synapse. The number of sample time points within t_i is represented by m. The number of synapses for which records exist is represented by p. l corresponds to the number of neurons recorded by the recorder. The sample time periods which the output of the LSM are being recorded for corresponds to k. The time steps here are 0.025ms in size and cover the full period of the simulation. J corresponds to the spike time value within the record of spiking activity. The stimulus value returned from this connection l for the k time segment is z.

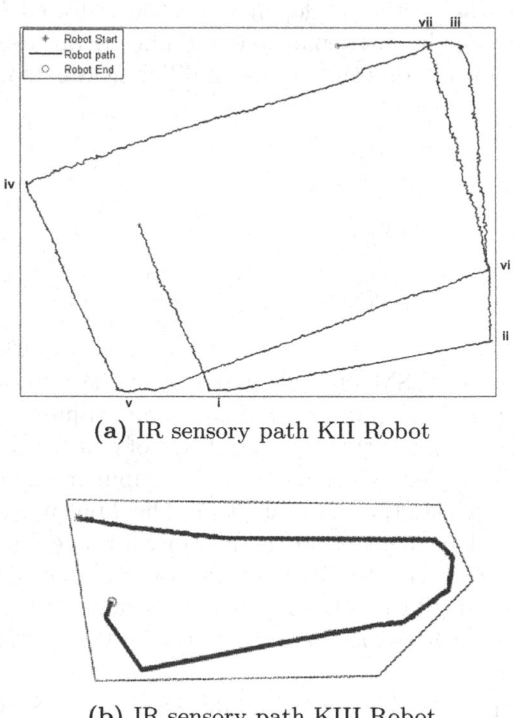

(a) IR sensory path KII Robot

(b) IR sensory path KIII Robot

Fig. 3. Robot paths for infrared obstacle avoidance

3.3 Readout

Figure 4 provides an illustration of the spiking information returned from the LSM in figure 4a those spikes are then converted as described in Equation (2). From these values the representative samples of the data are passed to the readout for classification. 8 samples are taken within a 1 second window, (see figure 4b) from the LSM recorder to act as inputs to the readout network. The data is presented to the readout network, provided by a ANN. The training method used in the ANN is error back propagation based on the Scaled Conjugate Gradient method from Matlab's Neural Network tool. This training method was chosen

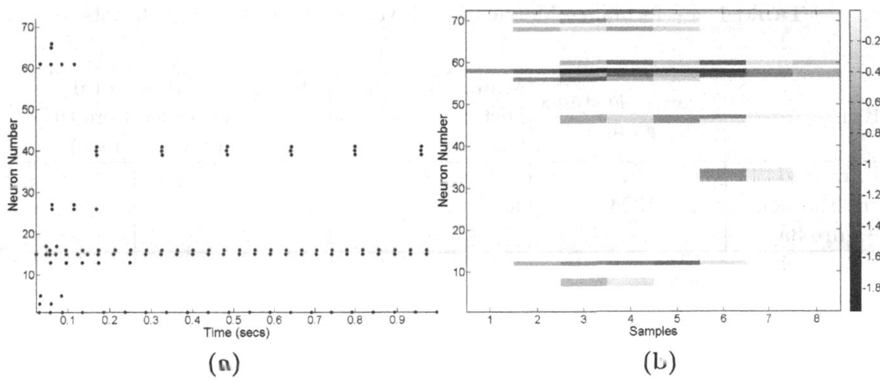

Fig. 4. Spiking output from the LSM and values returned from the output conversion (2)

as it scales well over large training sets. The learning rate is 0.1. The structure of the ANN used of the readout is a hidden layer of 4 log sigmoid neurons and an output layer of 3 log sigmoid neurons. The output of the readout network is a signal to turn right or left or to move forward.

4 Results

The behaviours observed showed that the edges of the environment were avoided, and the environments of Figure 3a and Figure 3b were explored. When an edge was encountered the behaviour exhibited was to turn from the obstacles, veering off at an angle and not following the wall of the environment. A short range was used for the sensory processing during the experiment in figure 3a to avoid the interference from which IR distance sensors suffer. The activation at a much greater range range than the Khepera II, was used with the Khepera III experimental setup. The behaviour of the Khepera III in the environment in figure 3b shows that this was effective in increasing the robots distance from the walls.

Table 1 attempts to quantify the performance of the approach. The values are based upon the metrics proposed by Ceballos et al. [18]. The experimental values are reported with the value recorded x together with the maximum possible value y in the format x/y. This representation of gives a way of interpreting the results which are not reported as distance values due to the accuracy of the sensors. The first metric is represented in IR sensor units, it shows the Mean Distance (MD) between the robot and any obstacles throughout the experiment. The second metric is the Minimum Mean Distance (MMD) which is the minimum value recorded for any of the IR sensors, for every time step of the experiment, between the robot and the nearest wall. The Absolute Minimum Distance (AMD) is the closest that the robot gets to the wall during the experiment on any of the sensors. The encounters with the walls of the environment for the Khepera II

Table 1. Performance Metrics for LSM Sensory Processing Robots

Run	Mean Distance (sensor unit)	Minimum Mean Distance (sensor Units)	Absolute Minimum Distance (sensor units)	Total Length (mm)
Infra red 1st environment Figure 3a	806 / 1024	484/ 1024	4/ 1024	2301
Infra red 2nd environment Figure 3b	1782 / 4096	628/ 4096	33/ 4096	3238.7

are seen in figure 3a they are marked i to vii. The encounters with the edges of the environment is reflected in the metric values shown in the table 1. The value which represents the fact that the distance between the robot and the wall was at times small is the AMD. The Total Length of the path gives an indication of how much of the environment the robot encountered. The greater distance traveled which is shown for the second experiment is caused by the size of the environment

5 Conclusions

This work demonstrates it is possible to control a mobile robot from multiple sensors using an LSM and ANN solution. The experiment were conducted in a simple environment. Future work will expand the sensory inputs to include sensors of different modalities such as ultrasound sensors which are available on the Khepera III. It will also examine more complex experimental environments for the mobile robot.

These experiments demonstrate an approach for combining multiple instances of a single sensory modality into a bio-inspired architecture of spiking neurons, and successfully processing the relevant data. The paper describes an effective means of combining this information and enabling basic robotic control. It is planned to continue this research by combining multiple different sensory modalities into an extended LSM structure based upon the structure reported in this paper,this provides a method of multi-modal sensory fusion inspired by the somatosensory cortex of the brain.

References

1. Russell, S.J., Norvig, P.: Artificial Intelligence: a modern approach. Prentice-Hall, Upper Saddle River (1995)
2. Pfeifer, R., Scheier, C.: Understanding Intelligence. MIT Press, Cambridge (1999)

3. Knudsen, E.I., Esterly, S.D., du Lac, S.: Stretched and upside-down maps of auditory space in the optic tectum of blind-reared owls; acoustic basis and behavioral correlates. The Journal of neuroscience: the official journal of the Society for Neuroscience 11(6), 1727–1747 (1991); LR: 20041117; PUBM: Print; JID: 8102140; ppublish
4. Witten, I.B., Knudsen, E.I.: Why seeing is believing: Merging auditory and visual worlds. Neuron 48(3), 489–496 (2005)
5. Ge, S.S., Lewis, F.L.: Autonomous Mobile Robots: Sensing, Control, Decision Making and Applications, vol. 22. CRC Press, Taylor & Francis Group, Boca Raton (2006)
6. Murphy, R.R.: Dempster- shafer theory for sensor fusion in autonomous mobile robots. IEEE Transactions on Robotics and Automation 14(2), 197–206 (1998)
7. Burgsteiner, H.: Training networks of biological realistic spiking neurons for real-time robot control. In: Proceedings of the 9th International Conference on Engineering Applications of Neural Networks, pp. 4620–4626
8. Braitenberg, V.: Vehicles, experiments in Synthetic Psychology. Bradford Book (1984)
9. Joshi, P., Maass, W.: Movement generation with circuits of spiking neurons. Neural Computation 17(8), 1715–1738 (2005)
10. Maass, W., Natschlager, T., Markram, H.: Real-time computing without stable states: A new framework for neural computation based on perturbations. Neural computation 14(11), 2531–2560 (2002)
11. Natschlager, T., Maass, W., Markram, H.: The "liquid computer": A novel strategy for real-time computing on time series. Special Issue on Foundations of Information Processing of TELEMATIK 8(1), 39–43 (2002)
12. Fregnac, Y., Ren, A., Durand, J.B., Trotter, Y.: Brain encoding and representation of 3d-space using different senses, in different species. Journal of Physiology-Paris 98, 1–18 (2004)
13. Gerstner, W., Kistler, W.M.: Spiking Neuron Models: Single Neurons, Populations, Plasticity. Cambridge University Press, Cambridge (2002)
14. Wu, Q., McGinnity, T., Maguire, L., Belatreche, A., Glackin, B.: Processing visual stimuli using hierarchical spiking neural networks. Neurocomputing (2008)
15. Hagras, H., Pounds-Cornish, A., Colley, M., Callaghan, V., Clarke, G.: Evolving spiking neural network controllers for autonomous robots, vol. 5, pp. 4620–4626 (2004); ID: 54; M1: Conference Proceedings
16. Vreeken, J.: On real-world temporal pattern recognition using liquid state machines (2004)
17. Norton, D., Ventura, D.: Preparing more effective liquid state machines using hebbian learning. In: Proceedings of the International Joint Conference on Neural Networks, pp. 4243–4248 (2006)
18. Ceballos, N.D.M., Valencia, J.A., Ospina, N.L.: Performance metrics for robot navigation. In: Electronics, Robotics and Automotive Mechanics Conference. CERMA 2007, pp. 518–523 (2007)

Basis Decomposition of Motion Trajectories Using Spatio-temporal NMF

Sven Hellbach[1], Julian P. Eggert[2], Edgar Körner[2], and Horst-Michael Gross[1]

[1] Ilmenau University of Technology, Neuroinformatics and Cognitive Robotics Labs, POB 10 05 65, 98684 Ilmenau, Germany
sven.hellbach@tu-ilmenau.de

[2] Honda Research Institute Europe GmbH, Carl-Legien-Strasse 30, 63073 Offenbach/Main, Germany
julian.eggert@honda-ri.de

Abstract. This paper's intention is to present a new approach for decomposing motion trajectories. The proposed algorithm is based on non-negative matrix factorization, which is applied to a grid like representation of the trajectories. From a set of training samples a number of basis primitives is generated. These basis primitives are applied to reconstruct an observed trajectory, and the reconstruction information can be used afterwards for classification. An extension of the reconstruction approach furthermore enables to predict the observed movement further into the future. The proposed algorithm goes beyond the standard methods for tracking, since it doesn't use an explicit motion model but is able to adapt to the observed situation. In experiments we used real movement data to evaluate several aspects of the proposed approach.

Keywords: Non-negative Matrix Factorization, Prediction, Movement Data, Robot, Motion Trajectories.

1 Introduction

The understanding and interpretation of movement trajectories is a crucial component in dynamic visual scenes with multiple moving items. Nevertheless, this problem has been approached very sparsely by the research community. Most approaches for describing motion patterns, like [1], rely on a kinematic model for the observed human motion. This causes the drawback that the approaches are difficult to adapt to other objects. Here, we aim at a generic, model-independent framework for decomposition, classification and prediction. In this paper, we focus on the decomposition and prediction problem, while the classification is not yet further investigated.

Consider the simple task for a robot of grasping an object which is handed over by the human interaction partner. To avoid a purely reactive behavior, which might lead to 'mechanical' movements of the robots, it is necessary to predict the further movement of the human's hand.

In [2] an interesting concept for a decomposition task is presented. Like playing a piano a basis alphabet – the different notes – are superimposed to reconstruct

C. Alippi et al. (Eds.): ICANN 2009, Part II, LNCS 5769, pp. 804–814, 2009.

the observation (the piece of music). The much less dimensional description of when each basis primitive is used, can be exploited for further processing. While the so-called piano model relies on a set of given basis primitives, our approach is able to learn these primitives from the training data.

Beside the standard source separation approaches, like PCA and ICA, another promising algorithm exists. It is called non-negative matrix factorization (NMF) [3]. The system of basis vectors which is generated by the NMF is not orthogonal. This is very useful for motion trajectories, since one basis primitive is allowed to share a common part of its trajectory with other primitives and to specialize later.

The next section introduces the standard non-negative matrix factorization approach and two extensions that can be found in the literature. In section 3 the new approach for decomposing motion trajectories is presented. The experiments with their conditions and results are presented in section 4, while the paper concludes in section 5.

2 Non-negative Matrix Factorization

Like other approaches, e. g. PCA and ICA, non-negative matrix factorization (NMF) [3] is meant to solve the source separation problem. Hence, a set of training data is decomposed into basis primitives:

$$\mathbf{V} \approx \mathbf{W} \cdot \mathbf{H} \tag{1}$$

Each training data sample is represented as a column vector $\mathbf{V}_i$ within the matrix $\mathbf{V}$. Each column of the matrix $\mathbf{W}$ stands for one of the basis primitives. In matrix $\mathbf{H}$ the element H_i^j determines how the basis primitive $\mathbf{W}_j$ is activated to reconstruct training sample $\mathbf{V}_i$. Since NMF is an iterative method, the training data $\mathbf{V}$ can only be approximated by the product of $\mathbf{W}$ and $\mathbf{H}$. This product will be referred to as reconstruction $\mathbf{R} = \mathbf{W} \cdot \mathbf{H}$ later.

Unlike PCA or ICA, NMF aims to a decomposition, which only consists of non-negative elements. This means that the basis primitives can only be accumulated. There exists no primitive which is able to erase a 'wrong' superposition of other primitives. This leads to a more specific set of basis primitives, which is an advantage for certain applications, like face recognition [4].

For generating the decomposition, optimization-based methods are used. Hence, an energy function E has to be defined:

$$E(\mathbf{W}, \mathbf{H}) = \frac{1}{2} \left\| \mathbf{V} - \mathbf{W} \cdot \mathbf{H} \right\|^2 \tag{2}$$

By minimizing the energy equation, it is now possible to achieve a reconstruction using the matrices $\mathbf{W}$ and $\mathbf{H}$. This reconstruction is aimed to be as close as possible to the training data $\mathbf{V}$. No further constraints are given in the standard formulation of the NMF. As it can be seen in equation 2, the energy function depends on the two unknown matrices $\mathbf{W}$ and $\mathbf{H}$.

Since both matrices usually have a large number of elements, the optimization problem seems to be an extensive task. Fortunately, the training samples can be presented to the algorithm one after the other:

$$\mathbf{V_i} \approx \sum_j H_i^j \cdot \mathbf{W_j} \tag{3}$$

Furthermore, both matrices are adapted in an alternating fashion. This helps to reduce the number of dimensions for the optimization process and allows a training regime with fewer examples. The algorithm is formulated in the following description in vector-wise notation:

1. Calculate the reconstruction

$$\mathbf{R}_i = \sum_j H_i^j \mathbf{W}_j \tag{4}$$

2. Update the activities

$$H_i^j \leftarrow H_i^j \odot \frac{\mathbf{V}_i^T \mathbf{W}_j}{\mathbf{R}_i^T \mathbf{W}_j} \tag{5}$$

3. Calculate the reconstruction with the new activities

$$\mathbf{R}_i = \sum_j H_i^j \mathbf{W}_j \tag{6}$$

4. Update the basis vectors

$$\mathbf{W}_j \leftarrow \mathbf{W}_j \odot \frac{\sum_i H_i^j \mathbf{V}_i}{\sum_i H_i^j \mathbf{R}_i} \tag{7}$$

Where the operation $\odot$ denotes a component-wise multiplication. Steps 1 to 4 are iterated until a defined convergence criterion is reached, e. g. a threshold for the energy or the change of the energy. Details about convergence properties are discussed in [3].

2.1 Sparse Coding

As it could be seen in equation 2 the energy function is formulated in a very simple way. This results in a decomposition, which is quite arbitrary with no further characteristics. This can lead, for example, to redundant information. Especially, if the number of basis primitives is chosen higher than needed to decompose the given training data. To compensate this drawback, it is useful to introduce a constraint which demands a sparse activation matrix, like it was introduced in [5]. This avoids the fact that several basis primitives are activated at the same time, and hence are being superimposed.

$$E(\mathbf{W}, \mathbf{H}) = \frac{1}{2} \|\mathbf{V} - \mathbf{W} \cdot \mathbf{H}\|^2 + \lambda \sum_{i,j} H_i^j \tag{8}$$

The influence of the sparsity constraint can be controlled using the parameter λ. In this paper, we only discuss a special case for the sparsity term. A more detailed discussion can be found in [5]. The algorithmic description is similar to the one of the standard NMF. The only thing that has to be considered is that the basis primitives need to be normalized.

2.2 Transformation Invariance

Beside the sparsity constraint another extension to NMF has been published in [6]. The concept of transformation invariance allows moving, rotating, and scaling the basis primitives for reconstructing the input. In this way, we do not have to handle each possible transformation using separate basis vectors. This is achieved by adding a transformation matrix $\mathbf{T}$ to the decomposition formulation:

$$\mathbf{V} \approx \mathbf{T} \cdot \mathbf{W} \cdot \mathbf{H} \tag{9}$$

However the activation matrix $\mathbf{H}$ has to be adapted in a way that each possible transformation carries its own activation. This can be regarded as a set of activity matrices, with each single matrix being indexed as $\mathbf{H}^{\mathbf{m}}$, while $\mathbf{m}$ is an index vector describing the transformation parameters (rotation, scaling and translation).

$$\mathbf{V}_\mathbf{i} \approx \sum_j \sum_\mathbf{m} H_i^{j,\mathbf{m}} \cdot \mathbf{T}^\mathbf{m} \cdot \mathbf{W}_\mathbf{j} \tag{10}$$

For each allowed transformation the corresponding activity has to be trained individually.

3 Decomposing Motion Trajectories

For being able to decompose and to predict the trajectories of the surrounding dynamic objects, it is necessary to identify them and to follow their movements. For simplification, a tracker is assumed, which is able to provide such trajectories in real-time. A possible tracker to be used is presented in [7]. The given trajectory of the motion is now interpreted as a time series $\mathcal{T}$ with values $\mathbf{s}_i = (x_i, y_i, z_i)$ for time steps $i = 0, 1, \ldots, n-1$: $\mathcal{T} = (\mathbf{s}_0, \mathbf{s}_1, \ldots, \mathbf{s}_{n-1})$.

It is now possible to present the vector $\mathcal{T}$ directly to the NMF approach. But this could result in an unwanted behavior, while trying to reconstruct the motion by use of the basis primitives. Imagine two basis primitives, one representing a left turn and another representing a right turn. A superposition of those basis primitives would result in a straight movement.

The goal is to have a set of basis primitives, which can be concatenated one after the other. Furthermore, it is necessary for a prediction task to be able to formulate multiple hypotheses. For achieving these goals, the x-t-trajectory is transferred into a grid representation, as it is shown in figure 1. Then, each grid cell (x_i, t_j) represents a certain state (spatial coordinate) x_i at a certain time t_j.

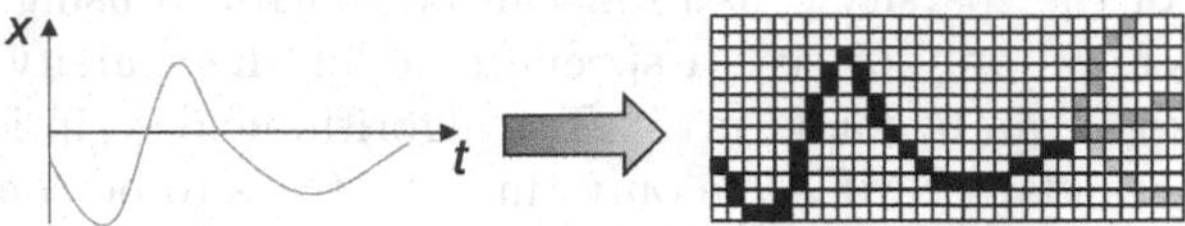

Fig. 1. Motion Trajectories are transferred into a grid representation. A grid cell is set to 1 if it is in the path of the trajectory and set to zero otherwise. Each dimension has to be regarded separately. During the prediction phase multiple hypotheses can be gained by superimposing several basis primitives. This is indicated with the gray trajectories on the right side of the grid.

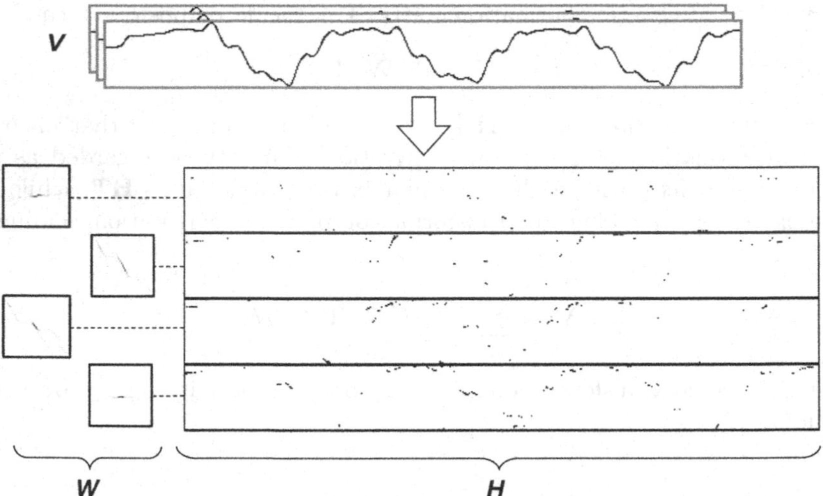

Fig. 2. Training with Spatio-Temporal NMF. Given is a set of training samples in matrix **V**. The described algorithm computes the weights **W** and the corresponding activities **H**. Only the weights are used as basis primitives for further processing.

Since most of the state-of-the-art navigation techniques rely on grid maps, the prediction can be integrated easily. Grid Maps were first introduced in [8]. This 2D-grid is now presented as image-like input to the NMF algorithm using the sparsity constraint as well as transformation invariance (See section 2.1 and 2.2 respectively). Using the grid representation of the trajectory also supports the non-negative character of the basis components and their activities.

It has to be mentioned, that the transformation to the grid representation is done for each of the dimensions individually. Hence, the spatio-temporal NMF has to be processed on each of these grids. Regarding each of the dimensions separately is often used to reduce the complexity of the analysis of trajectories (compare [9]). Theoretically, the algorithm could also handle multi-dimensional grid representation.

While applying an algorithm for basis decomposition to motion trajectories it seems to be clear that the motion primitives can undergo certain transformations to be combined to the whole trajectory. For example, the same basis primitive standing for a straight move can be concatenated with another one standing for a left turn. Hence, the turning left primitive has to be moved to the end of the straight line, and transformation invariance is needed while decomposing motion data. For our purposes, we concentrate on translation. This makes it possible to reduce the complexity of the calculations and to achieve real time performance.

The sparse coding constraint helps to avoid trivial solutions. Since the input can be compared with a binary image, one possible solution would be a basis component with only a single grid cell filled. These can then be concatenated one directly after another. So, the trajectory is simply copied into the activities.

3.1 Training Phase

The goal of the training phase is to gain a set of basis primitives which allow to decompose an observed and yet unknown trajectory (see Fig. 2). As it is discussed in section 3, the training samples are transferred into a grid representation. These grid representations are taken as input for the NMF approach and are therefore represented in matrix $\mathbf{V}$. On this matrix $\mathbf{V}$ the standard NMF approach, extended by the sparsity constraint and by translation invariance, is applied. The algorithm is summarized in Fig. 3.

Beside the computed basis primitives, the NMF algorithm also provides the information of how each of the training samples can be decomposed by these basis primitives.

3.2 Application Phase

As it is indicated in Fig. 4, from the training phase a set of motion primitives is extracted. During the application phase, we assume that the motion of a dynamic object (e. g. a person) is tracked continuously. For getting the input for the NMF algorithm, a sliding window approach is taken. A certain frame in time is transferred into the already discussed grid like representation. For this grid the activation of the basis primitives is determined by trying to reconstruct the input. For the computation the algorithm is identical to the one sketched in Fig. 3 besides, that steps 4 and 5 can be skipped.

The standard approach to NMF implies that each new observation at the next time step demands a new random initialization for the optimization problem. Since an increasing column number in the grid representation stands for an increase in time, the trajectory is shifted to the left while moving further in time. For identical initialization, the same shift is then reflected in the activities after the next convergence. To reduce the number of iterations until convergence, the shifted activities from the previous time step are used as initialization for the current one.

To fulfill the main goal discussed in this paper – the prediction of the observed trajectory into the future – the proposed algorithm had to be extended. Since

1. Normalize the basis vectors according to

$$\overline{\mathbf{W}}_j = \frac{\mathbf{W}_j}{\|\mathbf{W}_j\|} \tag{11}$$

2. Calculate the reconstruction

$$\mathbf{R}_i = \sum_j \sum_{\mathbf{m}} H_i^{j,\mathbf{m}} \mathbf{T}^{\mathbf{m}} \overline{\mathbf{W}}_j \tag{12}$$

3. Update the activities

$$H_i^{j,\mathbf{m}} \leftarrow H_i^{j,\mathbf{m}} \odot \frac{\mathbf{V}_i^T \mathbf{T}^{\mathbf{m}} \overline{\mathbf{W}}_j}{\mathbf{R}_i^T \mathbf{T}^{\mathbf{m}} \overline{\mathbf{W}}_j} \tag{13}$$

4. Calculate the reconstruction with the new activities

$$\mathbf{R}_i = \sum_j \sum_{\mathbf{m}} H_i^{j,\mathbf{m}} \mathbf{T}^{\mathbf{m}} \overline{\mathbf{W}}_j \tag{14}$$

5. Update the basis vectors

$$\mathbf{W}_j \leftarrow \mathbf{W}_j \odot \frac{\sum_i \sum_{\mathbf{m}} H_i^{j,\mathbf{m}} \mathbf{V}_i^T \mathbf{T}^{\mathbf{m}} + \overline{\mathbf{W}}_j \overline{\mathbf{W}}_j^T \sum_i \sum_{\mathbf{m}} H_i^{j,\mathbf{m}} \mathbf{R}_i^T \mathbf{T}^{\mathbf{m}}}{\sum_i \sum_{\mathbf{m}} H_i^{j,\mathbf{m}} \mathbf{R}_i^T \mathbf{T}^{\mathbf{m}} + \overline{\mathbf{W}}_j \overline{\mathbf{W}}_j^T \sum_i \sum_{\mathbf{m}} H_i^{j,\mathbf{m}} \mathbf{V}_i^T \mathbf{T}^{\mathbf{m}}} \tag{15}$$

Fig. 3. Algorithmic description of the Spatio-temporal NMF

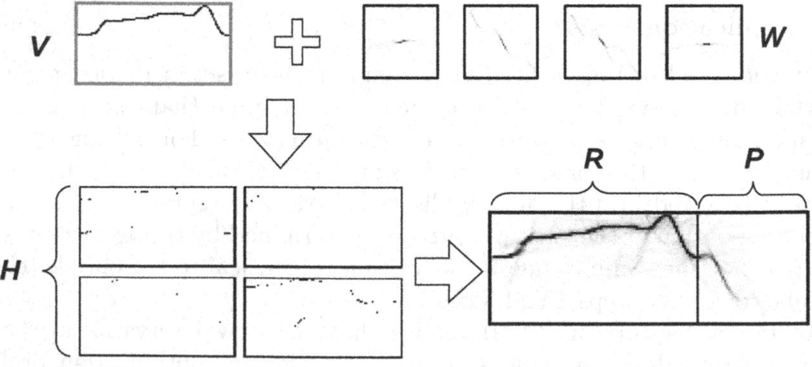

Fig. 4. The basis primitives **W**, which were computed during the training, are used to reconstruct (matrix **R**) the observed trajectory **V**. This results in a set of sparse activities – one for each basis primitive – which describe on which position in space and time a certain primitive is used. Beside the reconstruction of the observed trajectory (shown in Fig. 4), it is furthermore possible to predict a number of time steps into the future. Hence, the matrix **R** is extended by the prediction horizon **P**.

the algorithm contains the transformation invariance constraint, the computed basis primitives can be translated to an arbitrary position on the grid. This means that they can also be moved in a way that they exceed the borders of the grid. Up to now, the size of reconstruction was chosen to be the same size as the input grid. Hence, using the standard approach means that the overlapping information has to be clipped. To be able to solve the prediction task, we simply extend the reconstruction grid to the right – or into the future (see Fig. 4). So, the previously clipped information is available for prediction.

4 Evaluation

Taking a closer look at the example scenario from introductory section 1 reveals that a robust identification and tracking of the single body parts is needed. To be comparable and to avoid errors from the tracking system influencing the test results, movement data from the Perception Action Cognition Lab at the University of Glasgow [10] is used. The data contains trajectories from 30 persons is recorded performing different actions in different moods. The movement data has a resolution of 60 time steps per second, so that an average prediction of about 50 steps means a prediction of 0.83 seconds into the future. Since most trackers work with a lower resolution, a prediction further into the future is still possible.

In the next subsections, two aspects of the proposed algorithm are investigated in detail. First, it is shown that activity shifting brings a great benefit towards real time performance. Afterwards the focus is set to the quality of the prediction part.

For the experiments, the size of the basis primitives was chosen to be 50×50 grid cells (for an example see Fig. 5). The input grid size was set to 500×50 during the training phase and to 100×50 during application phase for each of the trajectories.

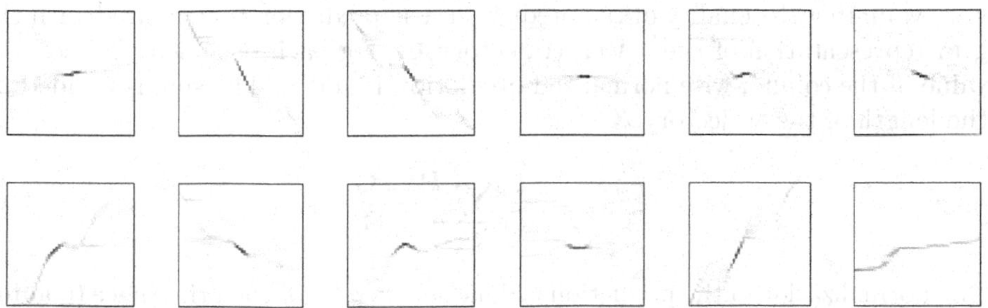

Fig. 5. Basis primitives gained by Spatio-Temporal NMF. The value for each grid cell is coded in gray scale from white (low) to black (high). A certain value stands for the influence of this grid cell, in the sense that light gray parts can be superimposed well, while dark gray to black parts indicate unambiguous trajectory segments.

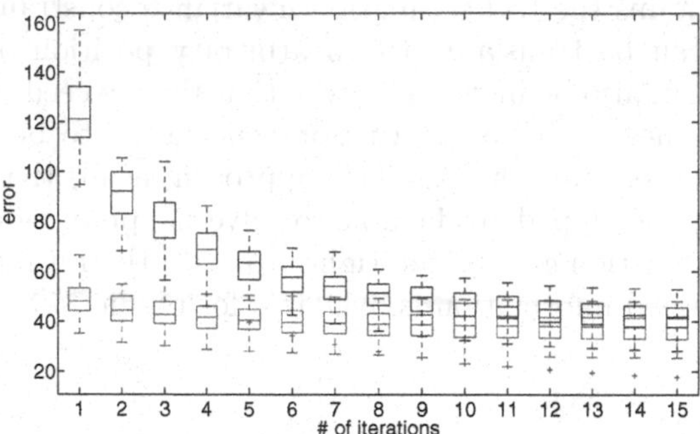

Fig. 6. Box whiskers plot showing the convergence characteristics of the energy function (see eqn. 2) for 15 iteration steps. For the upper (blue) plot the activities are initialized randomly after each shift of the input data. For the lower (red) curve the activities from the previous computations are shifted and used as initialization.

4.1 Activity Shifting

In section 3.2 it has been mentioned that the information from the previous time step can be used as initialization for the current one. Figure 6 shows the energy function, which is defined in equation 2, for both possibilities of initialization. It is plotted only for a low number of iteration steps (up to 15), since already there the effect can be observed. A single iteration step takes a time of 92 ms with our current implementation on an Intel T2050 CPU with 1.6 GHz. For the upper (blue) plot a random initialization of the activities was used. For the lower (red) curve the activities from the previous computations are shifted and used as initialization. It can clearly be seen that the convergence is faster by a number of about 10 steps in average.

4.2 Prediction

For evaluating the quality of the prediction, the prediction is compared with the grid representation of the actual trajectory $\mathbf{G}$. For each occupied grid cell the value of the column-wise normalized prediction is added. The sum is divided by the length of the trajectory:

$$S_{GT} = \frac{1}{|T|} \sum_{t \in T} \frac{\mathbf{P}_t^T \cdot \mathbf{G}_t}{\sum_i \mathbf{G}_t^i} \tag{16}$$

The normalization of the prediction is done separately for each time slice (column in the grid).

The basis primitives can at most be shifted by their width out of the reconstruction grid $\mathbf{R}$. So the maximal size of the prediction horizon equals the width of the basis primitives. Practically this maximum can not be reached, because the basis primitives need a reliable basis in the part where the input is known.

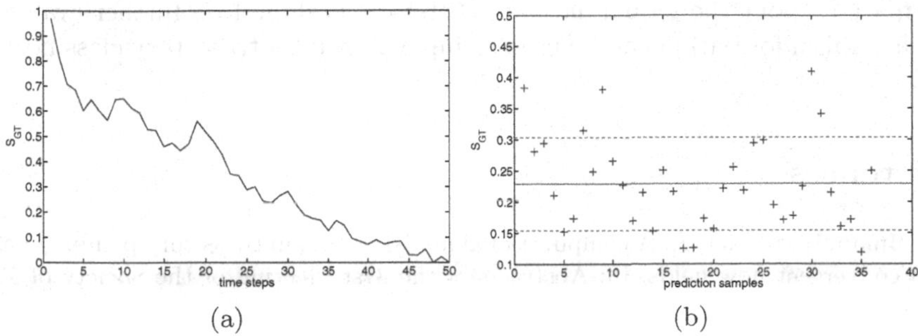

Fig. 7. (a) The mean correlation S_{GT} (see eqn. 16) between the ground truth trajectory and the prediction is plotted for each time step of the prediction horizon. A fit value of 1.0 stands for a perfect prediction over the whole prediction horizon. As it is expected the accuracy of the prediction decreases for a longer prediction period. (b) The plot shows the prediction accuracy for predictions along a sample trajectory. The 36 predictions were performed at each tenth time step of the chosen trajectory. A fit value of 1.0 stands for a perfect prediction over the whole prediction horizon. The constant and dotted lines (red) indicate mean and variance respectively.

Nevertheless, we have chosen to use the theoretical maximum as basis for the evaluation.

The results are depicted in Fig. 7. The first plot (Fig. 7(a)) shows the expected decrease of the average prediction quality over the prediction horizon. Nevertheless, the decrease is smooth and no sudden collapses can be observed. For Fig. 7(b) an example trajectory has been selected for the reasons of clearness. The plot is intended to show how the algorithm behaves in practical applications. The predictions were performed at each tenth time step of the chosen trajectory. A fit value of 1.0 stands for a perfect prediction over the maximum prediction horizon, with only a single hypothesis for the prediction. The value decreases significantly with multiple hypotheses being present.

5 Conclusion and Outlook

This paper presented a new approach for decomposing motion trajectories using non-negative matrix factorization. To solve this problem, sparsity constraint and transformation invariance have been combined. The trajectories were then decomposed using a grid-based representation. It could be demonstrated that the concept of activity shifting clearly decreases the number of iterations needed until convergence. Furthermore it was shown that the proposed algorithm is able to predict the motion into the future. The prediction occurs by a superposition of possible trajectory alternatives, yielding a quasi-probabilistic description. At this point, the information about the sparse activation of the basis primitives was used only for reconstruction purposes, even though it contains significant information about the global motion. Therefore, it should be further evaluated whether this information can be used as input to solve a trajectory classification task.

References

1. Hoffman, H., Schaal, S.: A computational model of human trajectory planning based on convergent flow fields. In: Abstracts of the 37st Meeting of the Society of Neuroscience (2007)
2. Cemgil, A., Kappen, B., Barber, D.: A generative model for music transcription. IEEE Transactions on Speech and Audio Processing 14, 679–694 (2006)
3. Lee, D.D., Seung, H.S.: Algorithms for non-negative matrix factorization. Advances in Neural Information Processing 13, 556–562 (2001)
4. Rajapakse, M., Wyse, L.: NMF vs ICA for face recognition. In: Guo, M. (ed.) ISPA 2003. LNCS, vol. 2745, pp. 605–610. Springer, Heidelberg (2003)
5. Eggert, J., Körner, E.: Sparse Coding and NMF. In: IJCNN, pp. 2529–2533 (2004)
6. Eggert, J., Wersing, H., Körner, E.: Transformation-invariant represenatation and NMF. In: IJCNN, pp. 2535–2539 (2004)
7. Otero, N., Knoop, S., Nehaniv, C., Syrdal, D., Dautenhahn, K., Dillmann, R.: Distribution and Recognition of Gestures in Human-Robot Interaction. In: The 15th IEEE International Symposium on Robot and Human Interactive Communication, ROMAN 2006, pp. 103–110 (2006)
8. Elfes, A.: Using Occupancy Grids for Mobile Robot Perception and Navigation. Computer 12(6), 46–57 (1989)
9. Naftel, A., Khalid, S.: Classifying spatiotemporal object trajectories using unsupervised learning in the coefficient feature space. MM Syst. 12(3), 227–238 (2006)
10. http://paco.psy.gla.ac.uk/data_ptd.php

An Adaptive NN Controller with Second Order SMC-Based NN Weight Update Law for Asymptotic Tracking

Haris Psillakis

Department of Electronic & Computer Engineering
Technical University of Crete, Chania, 73100, Greece
psilakish@hotmail.com

Abstract. The asymptotic tracking control problem of a class of single-input single-output (SISO) uncertain nonlinear systems is addressed in this paper. A single-hidden layer neural network is used as a controller with a novel online weight training algorithm. The proposed NN weight update law mimics standard second order sliding mode control (2-SMC) approaches to ensure semi-global asymptotic convergence of the tracking error to the origin with continuous control effort. A simulation study verifies the effectiveness of the NN controller with 2-SMC-based online training.

1 Introduction

Neural network-based adaptive control is a well-established methodology to control several classes of uncertain nonlinear systems [1]-[6]. In most of these approaches NNs are considered for their ability to approximate general unknown nonlinear functions [7],[8]. Using Lyapunov stability arguments semiglobal uniform ultimate boundedness of the tracking error within some region of the origin can be proved [1]-[6]. Asymptotic tracking results can be obtained if one augments standard adaptive NN controllers with sliding mode control terms. A well known drawback of such an approach is the fact that the resulting control law will suffer from chattering [9].

Recently, a novel NN-based control scheme was proposed in [10] that achieves asymptotic tracking results without using a discontinuous sliding mode controller. The proposed controller is based on the robust integral sign error (RISE) approach of [11] which is further developed in [12]. In [10], the NN is used to account (approximate) for a time varying nonlinearity and is augmented by the RISE terms to guarantee asymptotic tracking.

In the spirit of [10], we attempt to answer the following important question in this paper: *Can we achieve semiglobal asymptotic tracking results using only a NN as a (continuous) controller ?* The answer to this question is affirmative and this is accomplished by the proper selection of the NN's weight vector update rule.

The majority of existing NN control schemes employ some odd power of the filtered tracking error (FTE) in their weight update laws. For example, the first

C. Alippi et al. (Eds.): ICANN 2009, Part II, LNCS 5769, pp. 815–822, 2009.

power is used in [1]-[6] while the third power is considered in stochastic control [13],[14],[15] and sampled-data control [16]. In this paper, we propose a sign-like NN weight update law (see (11)). An initial design is considered to transform the unknown uncertain nonlinear system augmented model (filtered tracking error and its derivative) into a standard form ((2) of [20]). In this way existing results from 2-SMC theory [17]-[20] can be invoked to prove the desired asymptotic tracking property. Finally, a simulation study is carried out to verify the effectiveness of the proposed scheme and investigate the role of the selected parameters (NN input vector, number of nodes) on the tracking performance.

The following notation will be used throughout the paper: $||\cdot||$ denotes the standard Euclidean norm and A^T is the transpose of some matrix A.

2 Plant Description and Control Problem Formulation

Let the following class of SISO systems in normal form

$$\begin{aligned} \dot{x}_i &= x_{i+1}\,, \quad i = 1, 2, \ldots, n-1 \\ \dot{x}_n &= a(x) + b(x)u\,, \quad y = x_1 \end{aligned} \tag{1}$$

where $x := [x_1, x_2, \cdots, x_n]^T$ is the state vector and $u \in \mathbf{R}$, $y \in \mathbf{R}$ the control input and the system output, respectively. The functions $a : \mathbf{R}^n \to \mathbf{R}$, $b : \mathbf{R}^n \to \mathbf{R}$ are the system's unknown smooth nonlinearities. The control objective is the output $y(t)$ to asymptotically track some desired reference signal $y_d(t)$.

Assumption 1. The sign of $b(x)$ is known and there exists some constant $b_0 > 0$ such that $b_0 \leq |b(x)|$, $\forall x \in \mathbf{R}^n$.

Remark 1. Assumption 1 represents a controllability condition on system (1) ([5],[6]) since it implies that the smooth function $b(x)$ is either strictly positive or strictly negative. From now on, we assume without loss of generality that $b(x) > 0$, $\forall x \in \mathbf{R}^n$.

Let us also define the vectors $x_d := \left[y_d, \dot{y}_d, \ldots, y_d^{(n-1)}\right]^T$, $e := x - x_d = [e_1, e_2, \cdots, e_n]^T$ and the filtered tracking error $s := (d/dt + \lambda)^{n-1} e_1 = [\Lambda^T\, 1]e := [\bar{\Lambda}^T\ (n-1)\lambda\ 1]e$ with $\lambda > 0$ and $\Lambda := \left[\lambda^{n-1},\ (n-1)\lambda^{n-2},\ \cdots,\ (n-1)\lambda\right]^T$. Then, the time derivative of s can be written as

$$\dot{s} = a(x) + b(x)u - v \tag{2}$$

with $v := y_d^{(n)} - \left[0\ \Lambda^T\right]e$ a known time-varying signal.

Assumption 2. The reference signal y_d and its derivatives $\dot{y}_d, \ddot{y}_d, \ldots, y_d^{(n)}$ are all known, smooth and bounded i.e. there exist known $r_i \geq 0$ such that $|y_d^{(i)}| \leq r_i$, $i = 0, 1, \ldots, n$.

Lemma 1. *For the filtered tracking error s it holds true that: i) if $e(0) = 0$, $|s(t)| \leq C$, $\forall t \geq 0$ with $C \geq 0$, then $e(t) \in \Omega_c$ with $\Omega_c := \{e : |e_i| \leq 2^{i-1}\lambda^{i-n}C,\ i = 1, 2, \ldots, n\}$ and ii) if $e(0) \neq 0$ and $|s(t)| \leq C$ then $e(t)$ will converge to Ω_c within a time constant $(n-1)/\lambda$ (see [9]).*

Lemma 1 allows us to study the convergence of s to zero in order to ensure tracking of the output y to y_d.

3 Adaptive NN Control Design

3.1 NN as a Controller

In most adaptive NN control schemes, a single-hidden-layer neural network $W^T S(z)$ is used directly in the control law and is augmented by a linear control term [1]-[6] or a sign error term [21] or even an integral of the sign error term [10]. The NN input vector z usually takes the form of $z_1 := [x^T, s, v]^T \subset \mathbf{R}^{n+2}$ but we shall also consider the vectors $z_2 := [x^T, s]^T \subset \mathbf{R}^{n+1}$ and $z_3 := x \subset \mathbf{R}^n$. In this paper, we address the following problem: If the NN control law is employed

$$u_i = W^T S(z_i) \ , \qquad (i = 1, 2, 3) \tag{3}$$

can we design an NN update algorithm such that the system output y asymptotically tracks the desired reference signal y_d? The answer is affirmative and a solution can be obtained by using standard results from second-order sliding mode control theory [17]-[20].

Remark 2. The results of this paper hold true for any type of neural networks with a globally supported basis vector such as high-order neural-networks [7], functional link neural networks [8] etc.

3.2 NN Weight Update Law Selection Based on 2-SMC

For the control law (3) the dynamics of the filtered tracking error s given by (2) take the form

$$\dot{s} = a(x) - v + b(x) S^T(z_i) W. \tag{4}$$

If we now define the variables $r_i := \dot{s}$ $(i = 1, 2, 3)$ and consider a differentiation of (4) we obtain

$$\begin{aligned} \dot{s} &= r_i \ , \quad (i = 1, 2, 3) \\ \dot{r}_i &= f_i(x, W, t) + b(x) S^T(z_i) \dot{W} \end{aligned} \tag{5}$$

where

$$\begin{aligned} f_i(x, W, t) := & \sum_{j=1}^{n-1} \Big(\frac{\partial a}{\partial x_j} + \frac{\partial b}{\partial x_j} S^T(z_i) W \Big) x_{j+1} + \left[a(x) + b(x) W^T S(z_i) \right] \\ & \times \left[\frac{\partial a}{\partial x_n} + \frac{\partial b}{\partial x_n} S^T(z_i) W + (n-1)\lambda \right] - y_d^{(n+1)} - (n-1)\lambda y_d^{(n)} \\ & + [0\, 0\, \bar{\Lambda}^T] e + b(x) W^T \sum_j \frac{\partial S}{\partial z_{ij}} \dot{z}_{ij} \end{aligned}$$

and z_{ij} are the elements of the vector z_i. Selecting then the NN weight update algorithm

$$\dot{W} = \eta \frac{S(z_i)}{\|S(z_i)\|^2}, \quad (i = 1, 2, 3) \tag{6}$$

(5) yields

$$\begin{aligned} \dot{s} &= r_i, \quad (i = 1, 2, 3) \\ \dot{r}_i &= f_i(x, W, t) + b(x)\eta \end{aligned} \tag{7}$$

where η is some signal to be designed. Standard second-order sliding mode control results [17]-[20] can now be employed to design $\eta(t)$. Herein, we will use the approach of [20] where a suitable methodology robust to measurement noise is presented. To this end, the following Assumption is needed.

Assumption 3. The drift term $f_i(x, W, t)$ and the gain $b(x)$ satisfy the following inequalities

$$|f_i(x, W, t)| \leq \Phi_i, \quad 0 < b_0 \leq b(x) \leq b_1 \, \forall [x^T, W^T]^T \in \Omega \quad (i = 1, 2, 3) \tag{8}$$

for some positive constants b_0, b_1, Φ_i $(i = 1, 2, 3)$ with Ω some compact set wherein the state and NN weight trajectories evolve.

The scheme proposed in [20] considers a discrete-event system that switches among 4 different states S_M^+, S_M^-, S_m^+, S_m^- with the following selection for η:

$$\eta = \begin{cases} \Gamma & \text{when in } S_m^+ \text{ or } S_m^- \\ -\Gamma & \text{when in } S_M^+ \text{ or } S_M^- \end{cases} \tag{9}$$

where $\Gamma > 0$ is the NN weight adaptation gain. Initially, if $s(0) \geq 0$ the state is set to S_M^+, otherwise the state is set to S_m^-. We define also the variables

$$\begin{aligned} s_m(t) &= s_M(t) = s(t), \quad t \in [0, \tau) \\ s_m(t) &= \min\{s(t), s_m(t - \tau)\}, \quad t \geq \tau \\ s_M(t) &= \max\{s(t), s_M(t - \tau)\}, \quad t \geq \tau \end{aligned} \tag{10}$$

for some small $\tau > 0$ introduced to hold the maximum and minimum values of the filtered tracking error subject to resetting at state switching.

The switching state strategy is described below (for further details see [20]):

- When in S_M^+: if $s(t) \leq \beta s_M(t)$ with $\beta \in (0.75, 1) \cap (b_1/(b_0 + b_1), 1)$ the state switches to S_m^+.
- When in S_m^+: if $s < 0$ switch to S_m^- or if $s \geq s_m + (s_M - s_m)/N$ then switch to S_M^+ and reset s_M to s with $N > 1/(1 - \beta)$.
- When in S_m^-: if $s(t) \geq \beta s_m(t)$ switch to S_M^-.
- When in S_M^-: if $s > 0$ switch to S_M^+ or if $s \leq s_M - (s_M - s_m)/N$ then switch to S_m^- and reset s_m to s.

From (6),(9) the NN weight update algorithm takes the form

$$\dot{W} = \begin{cases} \Gamma \frac{S(z_i)}{\|S(z_i)\|^2} & \text{when in } S_m^+ \text{ or } S_m^- \\ -\Gamma \frac{S(z_i)}{\|S(z_i)\|^2} & \text{when in } S_M^+ \text{ or } S_M^- \end{cases} \tag{11}$$

Then the following theorem can be proved.

Theorem 1. *Consider system* (1) *with the NN controller* (3) *and the NN weight update law* (11). *If the resulting second-order sliding variable dynamics* (7) *satisfies conditions* (8) *of Assumption 3, then, the selection of the weight adaptation gain*

$$\Gamma = \gamma \Phi_i \quad \textit{with}\ \gamma \geq \frac{1}{2[\beta b_0 - (1-\beta) b_1]}$$

ensures that the system output y *asymptotically tracks the reference signal* y_d.

Proof. The proof follows directly from the analysis carried out in Theorem 1 and Remark 4 of [20].

Remark 3. Practically, the above Theorem ensures asymptotic tracking if a sufficiently large value of the adaptation gain Γ is selected. We note that there are no theoretical results on how the selection of the NN input vector or the NN node number affect the output tracking performance. This is in fact an open research problem. In the following section, wherein a simulation study is performed, we investigate this dependence in more detail.

4 Simulation Study

To evaluate the effectiveness of the proposed approach, consider now a pendulum plant with variable length described in [16] with dynamics given by (1), $n = 2$,

$$a(x) = \frac{0.5 \sin x_1 (1 + 0.5 \cos x_1) x_2^2 - 10 \sin x_1 (1 + \cos x_1)}{0.25(2 + \cos x_1)^2}$$

$$b(x) = \frac{1}{0.25(2 + \cos x_1)^2}$$

initial conditions $x(0) = [0.5 \quad 0]^T$ and reference signal $y_d(t) = (\pi/6)[\sin(t) + \sin(\sqrt{2}t)]$. The adaptive NN control law of Theorem 1 is implemented with $\Gamma = 90$, $\beta = 0.95$, $\lambda = 5$, $\tau = 10^{-4}$, $N = 100$ for the three cases ($i = 1, 2, 3$) of NN input vectors z_1, z_2, z_3. A second order NN is used in all cases with 4,3,2 inputs and 15, 10, 6 nodes respectively. Simulation results (Figs. 1-5) illustrate that all controllers provide effective tracking of the desired reference signal with bounded NN weights and continuous control signals. From Fig. 1 one can observe that the NN input vector z_1 yields the best transient performance. Another simulation test was carried out to evaluate the effect of the node number on the controllers' tracking ability. Particularly, for $i = 2$ we examine the case of a second, a first and a zero-th order NN with 10, 4 and 1 nodes respectively. We note that the last case is identical to the controller of Theorem 1 of [20]. Fig. 6 shows that both 1st and second order controllers perform efficiently while the zero-th order controller fails to stabilize the system.

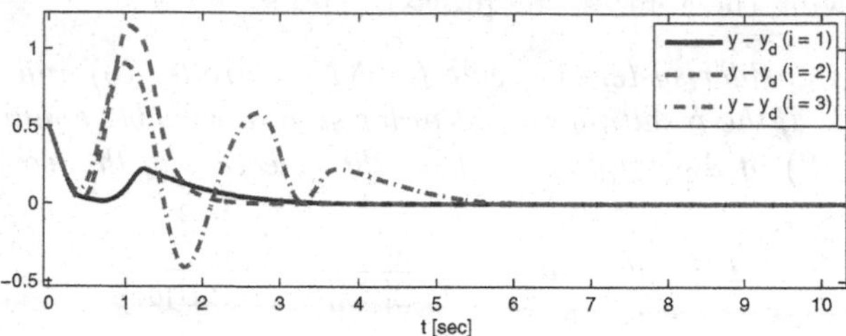

Fig. 1. The tracking error for $i = 1, 2, 3$

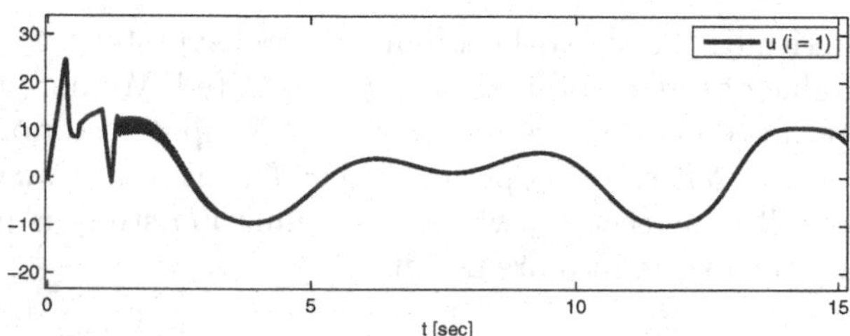

Fig. 2. The control input for $i = 1$

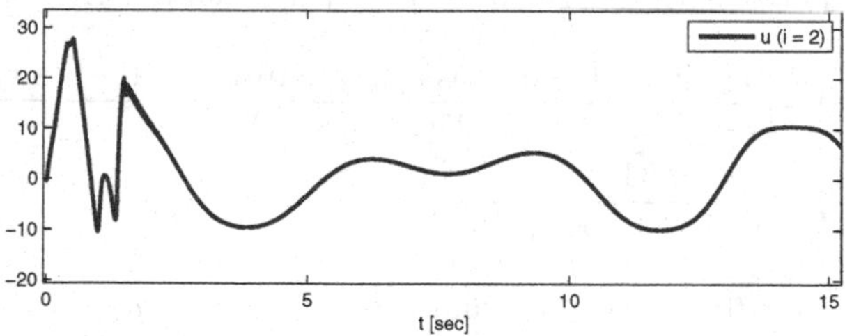

Fig. 3. The control input for $i = 2$

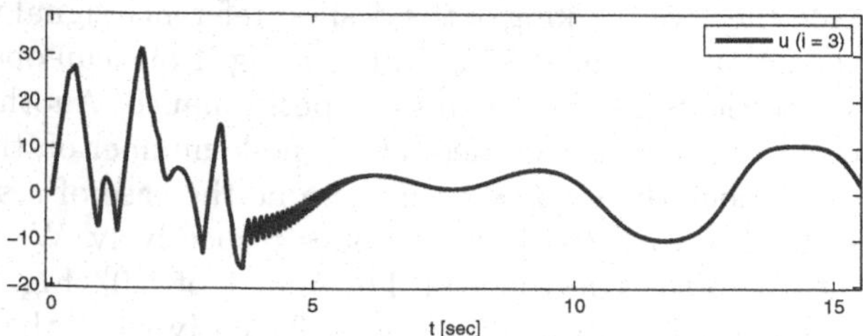

Fig. 4. The control input for $i = 3$

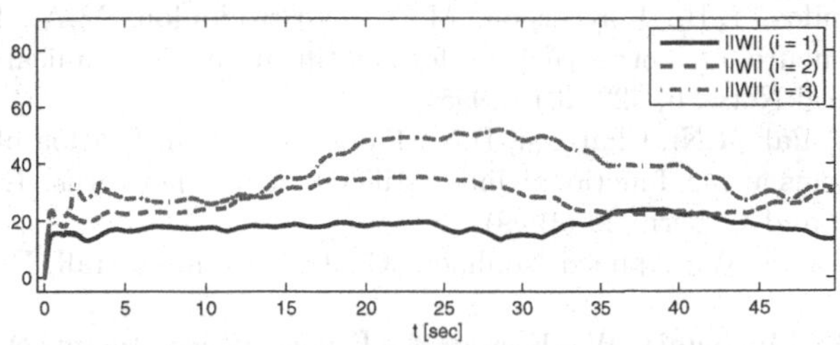

Fig. 5. The NN weight vector norm $\|W\|$ for $i = 1, 2, 3$

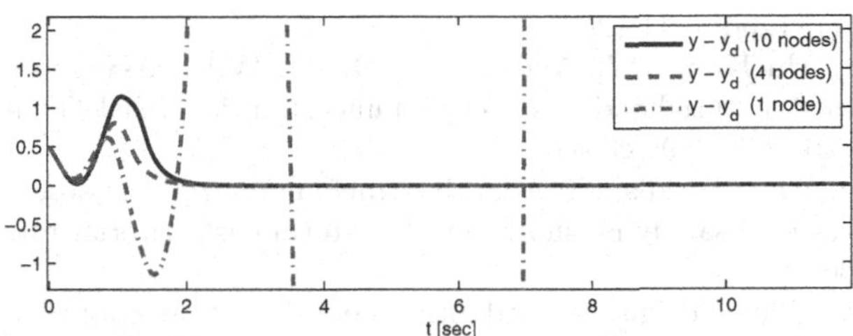

Fig. 6. The tracking error for various numbers of neurons

5 Conclusion

A novel NN weight training algorithm is presented in this paper to ensure asymptotic output tracking when a simple single-hidden-layer NN is used as a controller. The simulation study reveals that a relatively small number of nodes suffices to obtain an excellent tracking performance.

References

1. Ge, S.S., Lee, T.H., Harris, C.: Adaptive Neural Network Control of Robotic Manipulators. World Scientific, London (1998)
2. Lewis, F.L., Jagannathan, S., Yesildirek, A.: Neural Network Control of Robot Manipulators and Nonlinear Systems. Taylor and Francis, London (1999)
3. Rovithakis, G.A., Christodoulou, M.A.: Adaptive Control with Recurrent High-Order Neural Networks. Springer, London (2000)
4. Spooner, J.T., Maggiore, M., Ordonez, R., Passino, K.M.: Stable Adaptive Control and Estimation for Nonlinear Systems- Neural and Fuzzy Approximator Techniques. Wiley, New York (2002)
5. Ge, S.S., Hang, C.C., Lee, T.H., Zhang, T.: Stable Adaptive Neural Network Control. Kluwer, London (2002)
6. Farrell, J.A., Polycarpou, M.M.: Adaptive Approximation Based Control: Unifying, Neural, Fuzzy and Traditional Adaptive Approximation Approaches. Wiley, New York (2006)

7. Kosmatopoulos, E.B., Polycarpou, M.M., Christodoulou, M.A., Ioannou, P.A.: High-order neural network structures for identification of dynamical systems. IEEE Trans. Neural Netw. 6, 422–431 (1995)
8. Patra, J.C., Pal, R.N., Chatterji, B.N., Panda, G.: Identification of nonlinear dynamic systems using functional link artificial neural networks. IEEE Trans. on Syst., Man, and Cybern. 29 (1999)
9. Slotine, J.E., Li, W.: Applied Nonlinear Control. Prentice Hall, Englewood Cliffs (1991)
10. Patre, P.M., MacKunis, W., Kaiser, K., Dixon, W.E.: Asymptotic tracking for uncertain dynamic systems via a multilayer neural network feedforward and RISE feedback control structure. IEEE Trans. Automat. Contr. 53(9), 2180–2185 (2008)
11. Xian, B., Dawson, D.M., de Queiroz, M.S., Chen, J.: A continuous asymptotic tracking control strategy for uncertain nonlinear systems. IEEE Trans. Automat. Control 49(7), 1206–1211 (2004)
12. Patre, P.M., MacKunis, W., Makkar, C., Dixon, W.E.: Asymptotic tracking for systems with structured and unstructured uncertainties. IEEE Trans. Contr. Syst. Technol. 16(2), 373–379 (2008)
13. Psillakis, H., Alexandridis, A.: Adaptive tracking control for stochastic uncertain non-linear systems satisfying short- and long-term cost criteria. Int. J. Control 79, 107–118 (2006)
14. Psillakis, H., Alexandridis, A.: Adaptive neural motion control of n-link robot manipulators subject to unknown disturbances and stochastic perturbations. IEE Proc.-D: Contr. Th. & Appl. 153(2), 127–138 (2006)
15. Psillakis, H., Alexandridis, A.: NN-based adaptive tracking control of uncertain nonlinear systems disturbed by unknown covariance noise. IEEE Trans. Neural Netw. 18(6), 1830–1835 (2007)
16. Psillakis, H.: Sampled-data NN tracking control of uncertain nonlinear systems. IEEE Trans. Neural Netw. 20(2), 336–355 (2009)
17. Bartolini, G., Pydynowski, P.: Asymptotic linearization of uncertain nonlinear systems by means of continuous control. Int. J. Robust Nonlin. Contr. 3, 87–103 (1993)
18. Bartolini, G., Pydynowski, P.: An improved chattering free VSC scheme for uncertain dynamical systems. IEEE Trans. Automat. Contr. 41, 1220–1226 (1996)
19. Bartolini, G., Ferrara, A., Usai, E.: Chattering avoidance by second order sliding mode control. IEEE Trans. Automat. Contr. 43, 241–246 (1998)
20. Bartolini, G., Pisano, A., Usai, E.: An improved second-order sliding-mode control scheme robust against the measurement noise. IEEE Trans. Automat. Contr. 49(10), 1731–1736 (2004)
21. Ge, S.S., Lee, T.H.: Robust adaptive neural network control for a class of non-linear systems. Proc. Inst. Mech. Engr. Part I 211, 171–181 (1997)

Optimizing Control by Robustly Feasible Model Predictive Control and Application to Drinking Water Distribution Systems

Vu Nam Tran[1] and Mietek A. Brdys[1,2]

[1] Department of Electronic, Electrical and Computer Engineering, College of Engineering and Physical Sciences. University of Birmingham, Birmingham, B15 2TT, U.K.
m.brdys@bham.ac.uk, tvn_th@yahoo.com

[2] Department of Control Systems Engineering, Faculty of Electrical and Control Engineering, Gdansk University of Technology, ul. Narutowicza 11/12, 80 - 233 Gdansk, Poland

Abstract. The paper consider optimizing Model Predictive Control (MPC) for nonlinear plants with output constraints under uncertainties. Although the MPC technology can handle the constraint in the model by solving constraint model based optimization task, satisfying the plant output constraints still remains a challenge. The paper proposes Robustly Feasible MPC (RFMPC), which achieves feasibility of the outputs in the controlled plant. The RFMPC is applied to control quantity which is illustrated by application to a Drinking Water Distribution Systems (DWDS) example.

Keywords: predictive control, robust feasibility, genetic algorithms, robust output prediction, optimization, relaxation algorithm, drinking water distribution systems.

1 Introduction

Model Predictive Control has been an advanced technology and widely used in process control industry due to its ability to control multivariable systems with the presence of constraints. MPC actually belongs to a class of model based controller design concepts. The basic idea of the MPC algorithm remains unchanged regardless whatever kind of plant models are considered. It determines the optimal control actions by minimizing the user-defined objective function, or performance index. The current control actions are determined on-line, at each control step, by solving a finite-horizon open loop optimization problem, using the current state of the plant process as the initial state. However, only the first part of the optimized control input sequence is applied to the plant in the next time step. At the next control step, the prediction horizon moves forward and the same procedure repeat [2] [1]. Due to its operation on a receding horizon, MPC is also referred as receding control horizon or moving horizon optimal control. There are two significant factors that determine how effectively an MPC is. The

C. Alippi et al. (Eds.): ICANN 2009, Part II, LNCS 5769, pp. 823–834, 2009.

first factor is the accuracy of the plant model since it is explicitly used to predict the plant outputs. The second factor is how effective optimization solvers are. Although with the best plant models, MPC technology is still challenged by the uncertainty existing in the system such as model structure error, state estimation error, and disturbances. Fulfilling constraints is essential in many process plants for reason of safety, productivity, and environment protection. The controller outputs, which are based on the plant model, may not meet the plant output constraint due to the model-reality mismatch. The mismatch is often caused by the difference between predicted disturbance and actual disturbance. Feasible control input may become infeasible when they are applied to the plant if there is no robustly feasible controller. The robustness meeting of the output constraints or state constraints under system uncertainties is the main objective of the robustly feasible MPC. In this paper the optimizing RFMPC is considered. The robust feasibility will be assessed by robust output prediction over reduced horizon. Safety zones are employed to tighten the output constraints in order to achieve robustly feasible control input. The control method is applied to control quantity in DWDS.

2 Presentation of RFMPC

The structure of the RFMPC [3] consists of several units as illustrated in Fig. 1. The MPC optimizer solves the MPC optimization task to produce control inputs. In this task, the plant outputs are predicted based on nominal model of the plant. In the nominal model the disturbance inputs are represented by their predictions, while the internal model uncertainties are represented by a selected scenario. Before the control input is applied to the plant its robust feasibility is assessed by the "Constraint Violation Checking" unit. The feasibility assessment is based on the robust output prediction that is generated by the "Robust Output Prediction" unit. Given the control input the corresponding robust output predictions over the prediction horizon is a region in the output space in which

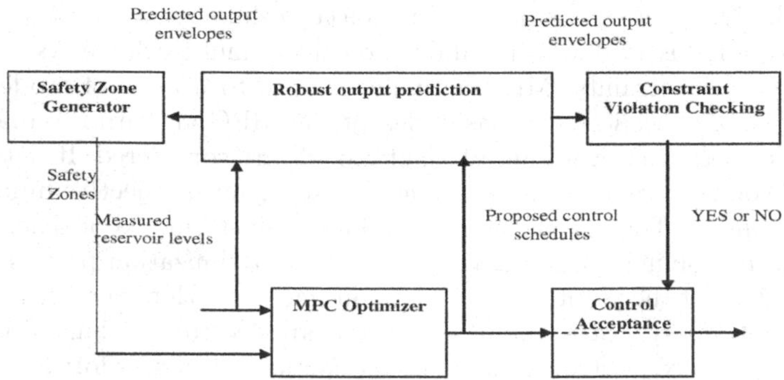

Fig. 1. Structure of Robustly Feasible MPC

all the plant outputs generated by the control input and all possible scenarios of the disturbance inputs are contained. The input robust feasibility is checked by confronting the output constraints with the robust output prediction. If the control feasibility passed its assessment, then the proposed control input is applied to the plant. Otherwise, robust output prediction is fed into the "Safety Zone Generator" unit. The safety zones as such are used to tighten the output constraints. The control actions produced by the MPC optimizer under modified (tighten) output constraints are expected to produce the real plant outputs that satisfy the plant constraints although they still may violate the modified constraints. Such control actions and the corresponding safety zones are called robustly feasible.

3 Robust Output Prediction (ROP)

Vector of control input and output over the prediction horizon are respectively defined as:

$$\hat{U} = \left[u(t \mid t)...u(t + H_m - 1 \mid t) \underbrace{u(t + H_m \mid t)...u(t + H_p - 1 \mid t)}_{from(t+H_m)to(t+H_p)} \right] \tag{1}$$

$$\hat{Y} = [y(t + 1 \mid t)...y(t + H_p \mid t)]^T \tag{2}$$

where $u(t + i \mid t), y(t + i \mid t)$are the control input and model output at time instant t. H_m and H_p are the *input horizon* and prediction horizon respectively. The robust output prediction (ROP) is an envelope over the prediction horizon

$$Y_p^l = \left[y_p^l(t + 1 \mid t).....y_p^l(t + H_p \mid t)\right]^T \tag{3}$$

$$Y_p^u = \left[y_p^u(t + 1 \mid t).....y_p^u(t + H_p \mid t)\right]^T \tag{4}$$

where $y_p^l(t + k \mid t)$ and $y_p^u(t + k \mid t)$ are the upper and lower limits that bound the plant output robustly at prediction time step :

$$y_p^l(t + k \mid t) \leq y(t) \mid_{t=t+k} \leq y_p^u(t + k \mid t) \tag{5}$$

The least conservative bounding envelopes $y_p^l(t + k \mid t)$ and $y_p^u(t + k \mid t)$ can be determined as:

$$y_p^l(t + k \mid t) = \min_{z(1),z(2),...,z(k)} y(t + k \mid t) \tag{6}$$

$$y_p^u(t + k \mid t) = \max_{z(1),z(2),...,z(k)} y(t + k \mid t) \tag{7}$$

where uncertainty at time step i-th is $z(i) \in \left[z^{min}, z^{max}\right], \forall i \in \overline{1 : H_p}$ Since the robust output prediction is calculated over the horizon H_p, there are H_p optimization problems to be solved to find H_p values of $y_p^l(t+k \mid t)$ and $y_p^u(t+k \mid t)$. As k increases from 1 to H_p, the optimization also increases the number of variables from 1 to H_p.Indeed, when k $=H_p$, (6) and (7) have H_p variables z(1),z(2),...,z(H_p). The more variables the optimization has, the more computing time the solvers require. As these computations are carried out online, it is desired to reduce the time computing as much as possible.

3.1 Stepwise Robust Output Prediction (SWROP)

In previous section, solving optimization problems (6) (7) give a least conservative solution of robust output prediction (ROP). This approach is so called exact optimization method. In contrast to the exact optimization method, we propose in this section an approximated optimization method where its advantage is to reduce the optimization process computing time.

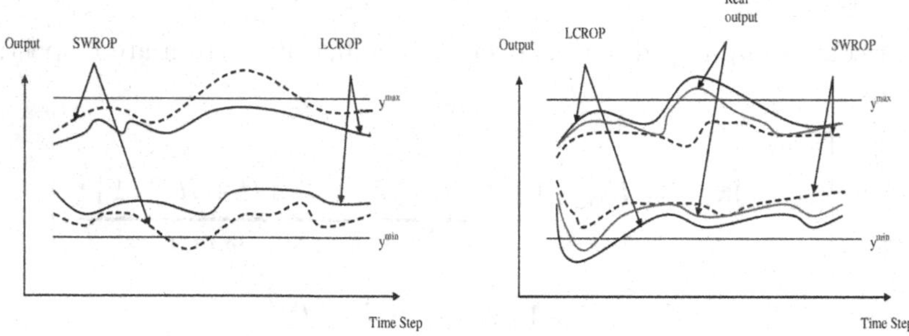

(a) SWROP stays outside LCROP (b) SWROP lies entirely inside LCROP

Fig. 2.

Instead of solving the optimization task with respect to k variables , one could approximate least conservative robust output prediction (LCROP) by solving the optimization tasks (6) and (7) with respect to only one variable where z(1),z(2),..,z(k-1) are obtained from the optimization in the previous time steps. In other words, in stead of simultaneous optimization with respect to all disturbance inputs, a step by step optimization is applied with respect to one disturbance input at the time.

$$y_p^l(t+k \mid t) = \min_{z(k)} y(t+k \mid t) \mid_{z(1)=z^{min}(1),\ldots,z(k-1)=z^{min}(k-1)} \tag{8}$$

$$y_p^u(t+k \mid t) = \max_{z(k)} y(t+k \mid t) \mid_{z(1)=z^{max}(1),\ldots,z(k-1)=z^{max}(k-1)} \tag{9}$$

where $\overline{z(i)}$ can be obtained by solving:

$$z^{min}(i) = arg \min_{z(i)} y(t+i \mid t) \mid_{z(1)=z^{min}(1),\ldots,z(k-1)=z^{min}(i-1)} \forall i \in \overline{1:k} \tag{10}$$

$$z^{max}(i) = arg \max_{z(i)} y(t+i \mid t) \mid_{z(1)=z^{max}(1),\ldots,z(k-1)=z^{max}(i-1)} \forall i \in \overline{1:k} \tag{11}$$

The resulting bounding envelopes are more conservative but the computing time is vastly reduced. Unfortunately, the expressions (8) and (9) generate the ROP only for some class of systems. The paper objective is to apply RFMPC to DWDS and there such class has clear interpretation, hence can clearly be identified. In order to assess the robust feasibility by SWROP, one should ensure that the LCROP entirely remains inside the SWROP as described in Fig. 2a. Otherwise the real output may possibly violate the upper or lower constraint even though the SWROP does not as described in Fig. 2b. In practice, there are some classes of system that have the characteristic as depicted in Fig. 2a while some will have the characteristic of Fig. 2b. Hence, in order to avoid the situation of having robustly infeasible control input, designers in practice should take that into consideration of choosing the appropriate method to calculate the robust output prediction.

3.2 Reduced Robust Feasibility Horizon

So far the ROP has been considered over the whole output prediction horizon H_p set up for the RFMPC. This has been done in order to secure existence of the robustly feasible safety zones at any control time step. However as computing of ROP over H_p is computationally very demanding and this may not meet the time constraints set up by on-line computing requirements. We should consider reducing this demand by shortening the ROP horizon. Clearly the cost to be paid

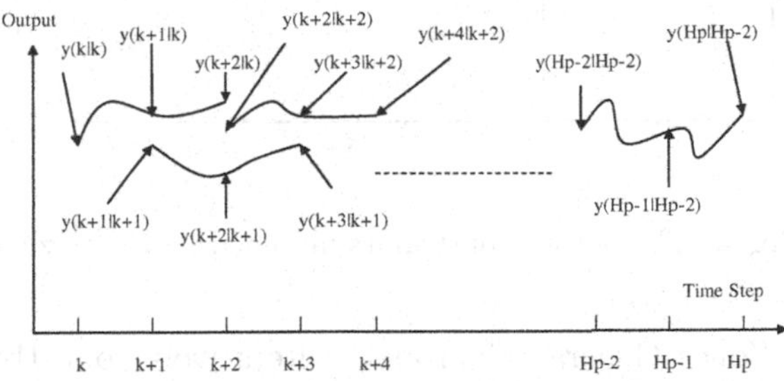

Fig. 3. Example of reduced robust feasibility horizon to two time steps Hr=2

is an increased risk of non existence of robustly feasible safety zones at certain control time steps. As only the first control action out of a whole sequence determined by the RFMPC is applied to the plant, we must secure the robust feasibility over the first time step. This is how far we can go with reduction of the ROP horizon from H_p to H_r. An attractive outcome of the ROP horizon reduction is that the very attractive computing SWROP method may become applicable over the reduced horizon while may not be applicable over the entire horizon. (see Fig. 3).

4 Safety Zone Generator

Using safety zones is not a new idea to meet system constraint under unknown factors. It is widely used in engineering area, such as conservative design in many electrical devices. When the input from the nominal model based MPC controller is applied to the plant, due to the uncertainties of the system, the output constraints may not be fulfilled and their violations may be unacceptable at certain time instants. If the violation occurs, it is important to correct or modify the constraints that apply to the nominal MPC. Safety zones generator is the unit that modify the output constraints via iterative scheme.

Consider over the prediction horizon, the vectors of the lower and upper limits on the plant output $Y^{min} = [y^{min}...y^{min}], Y^{max} = [y^{max}...y^{max}]$ and the vectors of the safety zones $\sigma^l = \left[\sigma_1^l, ..., \sigma_{H_p}^l\right]^T$, $\sigma^u = \left[\sigma_1^u, ..., \sigma_{H_p}^u\right]^T$ for the lower and upper output constraints, respectively where σ_i^u and σ_i^l are non negative real numbers. The vectors $Y_s^{min} = Y^{min} + \sigma^l$ and $Y_s^{max} = Y^{max} - \sigma^u$ are composed of the lower and upper bounds of the modified output constraints over H_p, respectively.

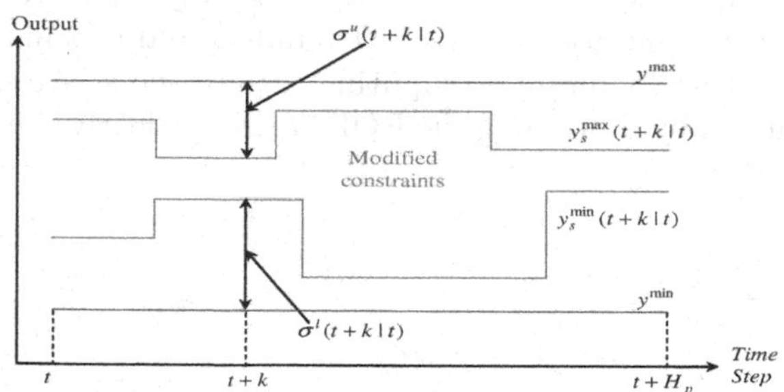

Fig. 4. The output constraints modified by safety zones

The "Safety Zones Generator" produces iteratively robustly feasible safety zones by using the following relaxation algorithm [3] :

(i) Set $x = [\sigma^l \quad \sigma^u] = 0$;

(ii) Solve MPC optimization task with modified output constraints and assess robust feasibility of the solution over H_p.

(iii) A vector V composed of the output constraint violations over the prediction horizon H_p is calculated as:

$$V = \left[V_1...V_{2H_p}\right]^T \triangleq \left[(Y^{min} - Y_p^l) \quad (Y_p^u - Y^{max})\right]^T$$

Define: $f(V_i) \triangleq max(0, V_i)$ and $C(\sigma^l, \sigma^u) \triangleq \left[f(V_1), ...f(V_{2H_p})\right]^T$

If

$$C(\sigma^l, \sigma^u) = 0 \quad (12)$$

is satisfied then go to step (vi),

Else

go to step(iv);

(iv) Calculate the safety zone corrections by using $\delta^{(k)} = -\nu C(x^{(k)})$ where $\nu = max([diag\,[\bigtriangledown C(0)]]^{-1})$ is called the relaxation gain

(iv) $x^{(k+1)} = x^{(k)} + \delta^{(k)}$, go to step (ii);

(v) The robustly feasible safety zones have now been found and the control input $u(t \mid t)$ is applied to the plant.

5 Optimizing Control of DWDS by RFMPC

5.1 Formulation of the Optimizing Control Problem

The main goal of DWDS is to supply water to customers and satisfy their quantity and quality demand. There are two major aspects in the control of DWDS: quantity and quality. The quality control deals with water quality parameters. Having disallowed concentration of chemical parameter, for instance chlorine, cause serious heath dangers. Maintaining concentrations of the water quality parameters within the prescribed limits throughout the network is a major objective. When the quantity control is considered, the objective is to minimize the electrical energy cost of pumping, while satisfying consumer water demand and physical constraints such as pressure at nodes or reservoir levels, by producing optimized control input such as optimized pump speeds and valve control schedules [4]. The uncertainty is in the demand and structure and parameters of DWDS model. In this paper, only the quantity control aspect is considered by applying RFMPC technique. The quality issues are addressed in [5] [6] for example.

Objective function- pumping cost control: It is a very common control objective to achieve the least pumping cost while satisfying constraints. Moreover, in order to achieve a sustainable operation day after day, it is expected that tank levels can come back to their original states after a certain period. For the DWDS example, the network is operated daily and the prediction horizon is $H_p = 24h$. It is desired that after 24 hours, the tank level could have similar level.
Hence, the overall objective function at $t = \bar{t}$ reads:

$$J = \sum_{t=\bar{t}}^{\bar{t}+H_p-1} \gamma(t)\Delta t \sum_{j=1}^{G} \sum_{i=1}^{U_j} \frac{\xi q_{j,i}(t)\Delta h_j(t)}{\eta_{j,i}(t)} + \rho \sum_{s=1}^{S} \mid r_s(\bar{t}+H_p) - r_s(\bar{t}) \mid$$

where H_p is the prediction time horizon,ρ is a weighting factors,$\gamma(t)$ is a power unit charge in /kWh for the (t+1) time stage,r_s is the s-th reservoir/tank level, s = 1,...,S, ξ is a unit conversion factor for electrical power relating water quantities to electrical energies, and η_i is the pump efficiency of the i-th pump in the j-th pump group, $i = 1, ..., U_j$, and j=1,...,G.

Optimization constraints are composed of:

- Nodal flow continuity equations: $\sum_{j\in J_i^-} q_j - \sum_{j\in J_i^+} q_j - d_i = \begin{cases} 0 & i \in M \\ lq_i & i \in M_l \end{cases}$
- Water elements head-flow equations: $h_{N_j^+} - h_{N_j^-} = \Delta h_j(q_j, u_j)$
  where q_j is the flow at arc *j-th*(liter/sec) ; h_i is the head at node *i-th*(m); d_i is the demand flow at node *i-th*(liter/sec); lq_i is the leakage flow at node *i-th* (liter/sec); u_j is the control variable representing the state of valve or pump at arc *j-th*; $\Delta h_j$is the head-flow characteristic function at arc *j-th*; $M(M_l)$ is the set of non-leaky (leaky) nodes; $J_i^+(J_i^-)$ is the set of arcs whose start (end) node are *i-th*; and $N_j^+(N_j^-)$ is the start (end) node of arc *j-th*.
- Volume mass balance equations of tanks/reservoirs.
- Output constraints. They are in the form of lower and upper bounds on certain flows, junction heads, an on all tank heads in order to meet the tank capacity constraints.
- Control input constraints. It can be the sequence of pump speed schedule or the ON/OFF state of pumps and need to satisfy the physical constraints: $u^{min} \leq u(t+k \mid t) \leq u^{max}$

5.2 Application of RFMPC to Example Case Study DWDS

Computer implementation is based on Matlab-Epanet simulation environment. The optimization problems are solved by standard Genetic Algorithm (GA) [10] [7] which can be called through Matlab toolbox. Epanet is used as the water network simulator generating "real-life" data [9], which are fed back to update the initial state of the predictive controllers for each time steps. For complex DWDS, the standard GA needs to be enhanced in order to exploit specific features of the optimization task and achieve required computing efficiency [8].

The DWDS, which is depicted in Fig. 5 includes 1 source reservoir and 1 storage tank. Water is pumped from the reservoir source by the pump station to the consumption nodes 2,3,4,5, and 6. The prediction horizon is $H_p = 24$. The interested control input is pump speed sequence over 24 hours period. RFMPC

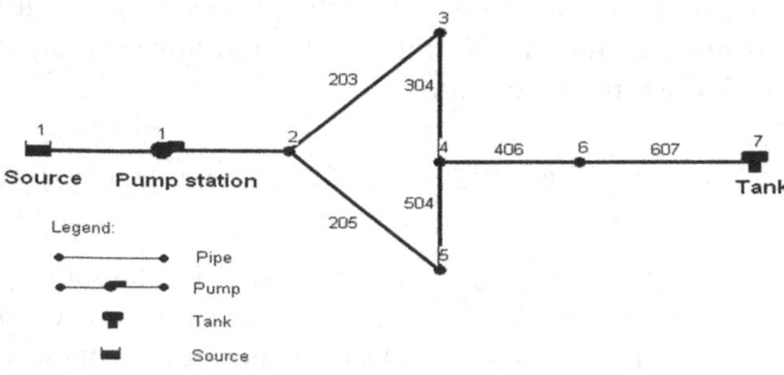

Fig. 5. Diagram of an DWDS example

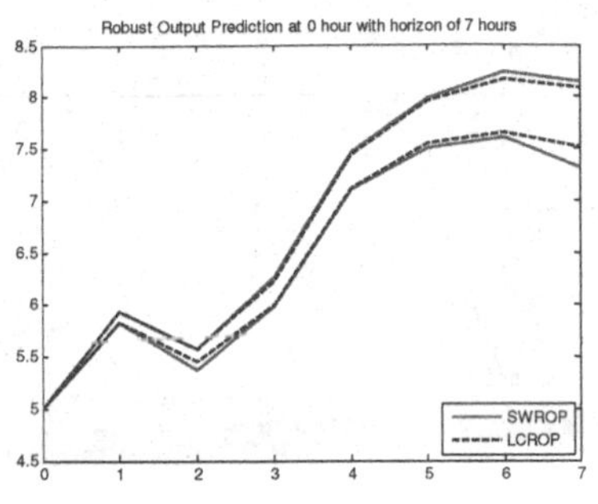

(a) Robust output prediction at t=0;Hr=7

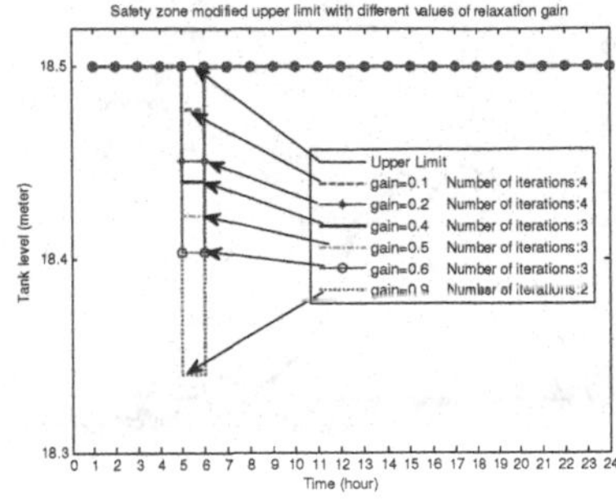

(b) Robustly feasible safety zones and the corresponding modified tank upper limit for different relaxation gain values

Fig. 6.

is applied to produce control input sequence. The tank level limits are: $r_s^{min} \leq r_s \leq r_s^{max}$, and these are the output constraints. The demands are predicted with the error of 10% at each consumption node.

Designing RFMPC. The MPC task is solved by GA solver with the optimization search in the reduced space. In this search, the GA is coupled to the Epanet simulator solving the DWDS equality constraints. A method for generating ROP is chosen by observing the simulation results shown in Fig. 6a. The SWROP and the LCROP are applied at t = 0 over 7 time steps. It can be seen in Fig. 6a that the SWROP method generates envelopes that are outside the region determined by the LCROP method. Hence, the SWROP is applicable to our example DWDS. Moreover, the envelopes calculated by the two methods are very close over the first 6 steps. The ROP horizon therefore is further reduced to 2 steps and the SWROP method is to be applied.

Also the relaxation gain in the algorithm for determining the robustly safety zones (RFSZ) is selected by simulation where several gain values are tried and the results are illustrated in Fig. 6b. The equality (12) in the step (iii) of the RFSZ relaxation algorithm has more than one solution. Clearly, the smaller safety zones are, the less conservative control actions are, and consequently better controller performance is achieved. From Fig. 6b this is obtained for small gain values. On the other hand, the computing time is essential; hence the number of iterations needed to reach the RFSZ should be minimized. This is obtained for high gain values as described in Fig. 6b. Therefore, gain is chosen in order to trade between the two aspects.

Simulation results. First the RFMPC is applied to the example DWDS at t = 0. Robust feasibility at the obtained control sequence is checked over the horizon $H_r = 2$ and the first two control inputs are assessed as robustly feasible. Hence, there is no need to activate the "Safety Zone Generator". In Fig. 7a two

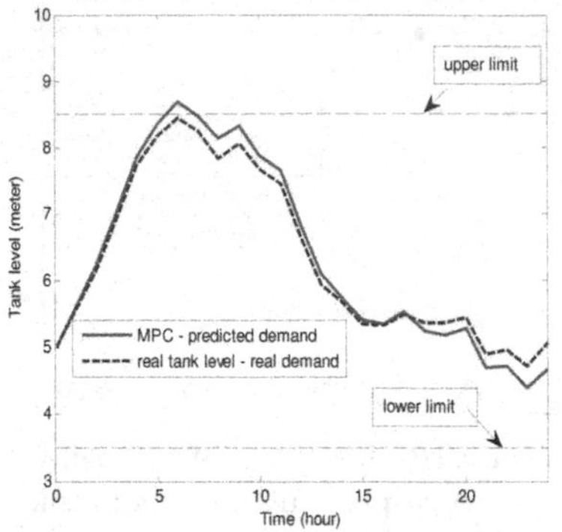

(a) Predicted tank level trajectory by RFMPC over the horizon at time instant t=0 and t=2

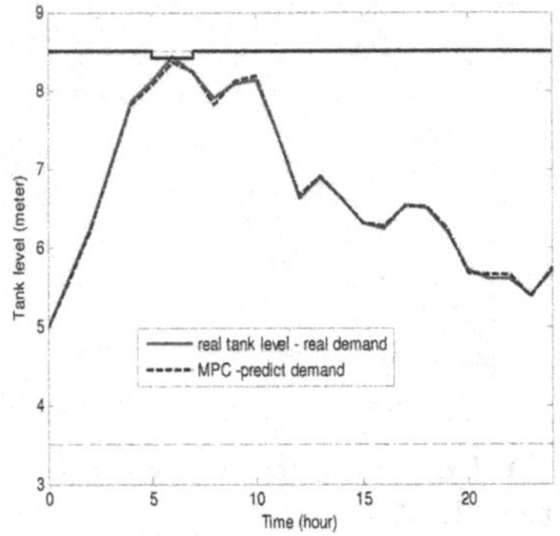

(b) Tank trajectory over the 24 hours

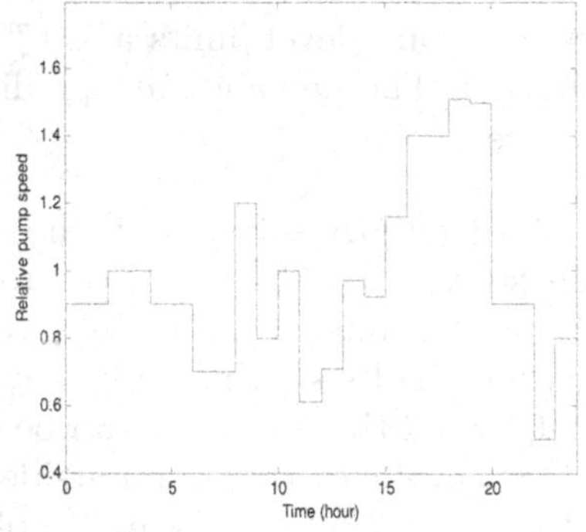

(c) Control actions - relative pump speed

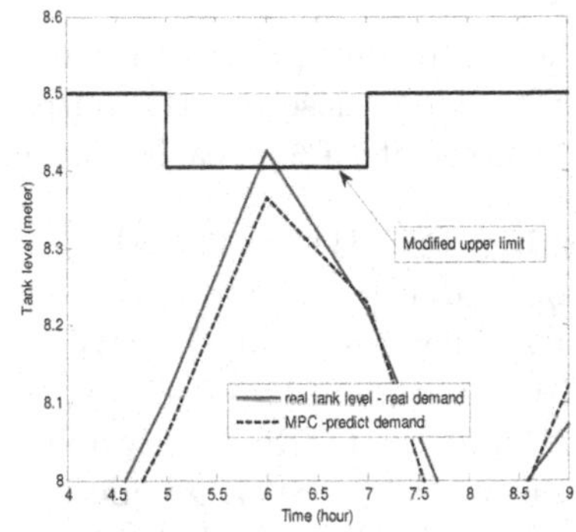

(d) Zoom-in of Fig. 7b during 4-9 hours

Fig. 7.

tank trajectories are illustrated: one in dash line is obtained by applying the control sequence to the model with the demand prediction while the second one in solid is the tank trajectory seen in the real system where the demand may differ from the predicted one up to 10%. It can be seen in Fig. 7a that the upper limit tank constraint is violated during 5 hour to 7 hour time period. Clearly, we are not aware about this violation at t = 0. However, a lesson to be learnt is that applying a whole control sequence obtained at t = 0 to the network is not recommended not only in this case but in general. Therefore the RFMPC is kept applying to produce the control actions on-line by employing feedback and all its mechanism described in this paper.

The results are illustrated in Fig. 7b,7c,7d. It can be seen in Fig. 7b that the upper tank level constraint had to be modified by robustly feedback safety zones over 5, 6, and 7 time steps in order to achieve robust feedback of the control action over these time steps. Although the modification does not tighten the constraints excessively its conservatism would be improved by extending the robust prediction horizon. The details of the situation over 5, 6, and 7 time steps are illustrated in Fig. 7d.

o assess the RFMPC feedback strength, the control actions generated on-line are also applied to the DWDS model. The resulting tank trajectory and the control input are shown in Fig. 7b and Fig. 7c, respectively. The two trajectories are much closer in Fig. 7b than in Fig. 7a. Hence, possible impact of the feedback in compensating the demand error impact is noticeable.

Lastly, as shown in Fig. 7d, the modified constraints are satisfied in the model but not in reality. However, the actual constraint is met in reality showing the effectiveness of the RFSZ mechanism.

6 Conclusions

This paper has further developed Robustly Feasible Model Predictive Control Method for on line optimizing control of nonlinear plants with output constraints under uncertainty. The RFMPC has been applied to quantity control in Drinking Water Distribution Systems. It has been illustrated by simulation based on an example DWDS. The effects of robust output prediction, shortening the robust output prediction horizon, robustly feasible safety zones, and the feedback strength of RFMPC have been shown.

References

1. Bemporad, A., Morari, M.: Robust model predictive control: A Survey. LNCIS, vol. 245, pp. 207–226. Springer, Heidelberg (1999), `http://control.ethz.ch/`
2. Maciejowski, J.M.: Predictive Control with Constraints. Prentice Hall, Englewood Cliffs (2000)
3. Brdys, M.A., Chang, T.: Robust model predictive control under output constraints. In: Proc. of the 15th IFAC World Congress, Barcelona, July 21-22 (2002)
4. Brdys, M.A., Ulanicki, B.: Operational Control of Water Systems: Structures, Algoritms and Applications. Prentice Hall International Ltd., Englewood Cliffs (1994)
5. Brdys, M., Duzinkiewicz, K., Chang, T., Polycarpou, M.M., Wang, Z., Uber, J.G., Propato, M.: Hierarchical control of integrated quality and quantity in drinking water distribution systems. In: Proc. of I International Conference on Technology, Automation and Control of Wastewater and Drinking Water Systems - TiASWiK 2002, Sobieszewo -Gdansk, Poland, June 19-21 (2002)
6. Chang, T., Brdys, M.A.: Performance comparison of chlorine controllers based on input-output and state-space models. In: Proc. of the 1st Annual Environmental & Water Resources Systems Analysis (EWRSA) Symposium, A.S.C.E. Environmental & Water Resources Institute (EWRI) Annual Conference, Roanoke, Virginia (2002)
7. Deb, K.: An efficient constraint handling method for genetic algorithms. Computer Methods in Applied Mechanics and Engineering 186, 311–338 (2000)

8. Kurek, W., Brdys, M.A.: Genetic solver of optimization task of MPC for optimizing control of integrated quantity and quality in drinking water distribution systems. In: Proc. of the 11th IFAC/IFORS/IMACS/IFIP Symposium on Large Scale Systems: Theory and Applications, Gdan'sk, Poland (2007)
9. US EPA. Epanet 2.0 (2001), `http://www.epa.gov/ORD/NRMRL/wswrd/epanet.html`
10. MATLAB Gentic Algorithm Toolbox manual, `http://www.mathworks.com/access/helpdesk/help/toolbox/gads/index.html?/access/helpdesk/help/toolbox/gads/`

Distributed Control over Networks Using Smoothing Techniques

Ion Necoara

Automatic Control and Systems Engineering Department
Politehnica University of Bucharest
060042 Bucharest, Romania

Abstract. In this paper we propose two dual decomposition methods based on smoothing techniques, called here the proximal center method and the interior-point Lagrangian method, to solve distributively separable convex problems. We show that some relevant centralized control problems can be recast as a separable convex problem for which our dual methods can be applied. The new dual optimization methods are suitable for application to distributed control since they are highly parallelizable, each subsystem uses local information and the coordination between the local controllers is performed via the Lagrange multipliers corresponding to the coupled dynamics or constraints.

1 Introduction

For the control problem of large-scale networked systems, centralized control is considered impractical, inflexible and unsuitable due to information requirement and computational aspects. The subsystems in the network may have different authorities that prevent sending all necessary information to one processing center. Moreover, the optimization problem yielded by centralized control is too big for online computation. Distributed control is proposed for control of such large-scale systems, by decomposing the overall system into small subsystems with distinct controllers for each subsystem that collaborate to achieve global decisions. In order to derive the local controllers we decompose the centralized problem into a set of subproblems each solved by an individual agent. The coordination of the subproblems is achieved by local communication among the agents.

Solving the problem of how distributing effectively the computations among the subsystems, has challenged many researchers in the last decades. Several contributions on this subject appeared for general control problems (see e.g. [13,11,12]) and in the model predictive control framework (see e.g. [2,3,9,10]).

Most of the computational methods for the above distributed control problems are based on Jacobi or subgradient type algorithm which are well-known to have a very slow convergence rate (see e.g. [8]). However, a control scheme is practical as long as its computational complexity is very low since the solution has to be computed fast. In this paper we show that using smoothing techniques we can derive distributed control algorithms with a great improvement of the

C. Alippi et al. (Eds.): ICANN 2009, Part II, LNCS 5769, pp. 835–844, 2009.

convergence rate compared to the previous methods. The algorithms, called here the proximal center algorithm and the interior-point Lagrangian algorithm (see [6,7] for more details), involve every subsystem optimizing an objective function that is the sum of his own objective function and a smoothing term while the coordination between the subsystems is performed via the Lagrange multipliers. We show that the solution of our distributed algorithms converges to the solution of the centralized control problem and we also provide estimates for the rate of convergence, which improve the estimates of the existing methods with at least one order of magnitude.

The paper is organized as follows. In Section 2 we formulate a general separable convex problem followed by a brief description of two dual-based decomposition algorithms (the proximal center and the interior-point Lagrangian algorithm) developed recently in [6,7] for solving this type of problems. In Section 3 we show that many relevant control problems can be recast as particular instances of a separable convex problem and we provide alternatives to solve them based on these two distributed algorithms but which exploit the specific control problem structure.

2 Application of Smoothing Techniques to Separable Convex Problems

In this section we devise two distributed algorithms for solving the *separable convex* optimization problem:

$$\begin{aligned} f^* = & \min_{x^1 \in X_1 \cdots x^M \in X_M} \sum_{i=1}^{M} f_i(x^i) \\ & \text{s.t.} \ \sum_{i=1}^{M} F_i x^i = a, \ \ G_i x^i = a^i \ \forall i = 1 \cdots M, \end{aligned} \tag{1}$$

where $f_i : \mathbb{R}^n \to \mathbb{R}$ are convex functions, X_i are closed convex sets, $G_i \in \mathbb{R}^{m \times n}$, $F_i \in \mathbb{R}^{p \times n}$, $a^i \in \mathbb{R}^m$ and $a \in \mathbb{R}^p$. For simplicity of the exposition we define $x = [(x^1)^T \cdots (x^M)^T]^T$, $f(x) = \sum_{i=1}^{M} f_i(x^i)$, $X = \prod_{i=1}^{M} X_i$, and $F = [F_1 \cdots F_M]$. The following assumptions for optimization problem (1) will be valid:

Assumption 1

(i) Each function f_i is convex quadratic and X_i are compact convex sets
(ii) The matrix $[\text{diag}(G_1^T \cdots G_M^T) \ F^T]$ has full column rank and $\{x \in \text{int}(X) : G_i x^i = a^i \ \forall i, \ Fx = a\} \neq \emptyset$.

Let $\langle \cdot, \cdot \rangle / \| \cdot \|$ denote the Euclidian inner product/norm on $\mathbb{R}^n$. By forming the Lagrangian corresponding to the coupling constraints (with Lagrange multipliers $\lambda \in \mathbb{R}^p$) we obtain in general a nonsmooth dual function and thus for maximizing it we have to use involved nonsmooth optimization techniques such as subgradient algorithm with slow convergence rate.

In order to obtain a smooth dual function we need to use smoothing techniques applied to the ordinary Lagrangian L_0 (see e.g. [1]). In [6,7] we proposed two dual decomposition methods for (1) in which we add to the standard Lagrangian a smoothing term $\mu \sum_{i=1}^{M} \phi_{X_i}$, where each function ϕ_{X_i} associated to the set X_i (usually called *prox function*) must have certain properties explained below. The two algorithms differ in the choice of the prox functions ϕ_{X_i}. In this case we define the *augmented Lagrangian*:

$$L_\mu(x, \lambda) = \sum_{i=1}^{M} [f_i(x^i) + \mu \phi_{X_i}(x^i)] + \langle \lambda, Fx - a \rangle, \qquad (2)$$

We also define the corresponding *augmented dual function*:

$$d(\mu, \lambda) = \min_{x^i \in X_i, G_i x^i = a^i} L_\mu(x, \lambda). \qquad (3)$$

Denote by $x^i(\mu, \lambda)$ the optimal solution of minimization problem in x^i:

$$x^i(\mu, \lambda) = \arg \min_{x^i \in X_i, G_i x^i = a^i} [f_i(x^i) + \mu \phi_{X_i}(x^i) + \langle \lambda, F_i x^i \rangle].$$

We are interested in the properties of the family of augmented dual functions $\{d(\mu, \cdot)\}_{\mu > 0}$. It is obvious that $\lim_{\mu \to 0} d(\mu, \lambda) = d_0(\lambda) \ \ \forall \lambda$.

2.1 Proximal Center Method

In the sequel we briefly describe the proximal center decomposition method whose efficiency estimates improves with one order of magnitude the bounds on the number of iterations of the classical dual subgradient method (see [6] for more details). In the proximal center method, the functions ϕ_{X_i} are chosen to be continuous, nonnegative and strongly convex on X_i with strong parameter σ_i. Since X_i are compact, we can choose $D_{X_i} \geq 0$ such that

$$D_{X_i} \geq \max_{x^i \in X_i} \phi_{X_i}(x^i) \ \forall i.$$

Lemma 1. *[6] If Assumption 1 holds and the functions ϕ_{X_i} are continuous, nonnegative and strongly convex on X_i, then the family of dual functions $\{d(\mu, \cdot)\}_{\mu > 0}$ is concave and differentiable at any λ. Moreover, the gradient $\nabla d(\mu, \lambda) = \sum_{i=1}^{M} F_i x^i(\mu, \lambda) - a$ is Lipschitz continuous with Lipschitz constant $D_\mu = \sum_{i=1}^{M} \frac{\|F_i\|^2}{\mu \sigma_i}$. The following inequalities also hold: $d(\mu, \lambda) \geq d_0(\lambda) \geq d(\mu, \lambda) - \mu \sum_{i=1}^{M} D_{X_i} \ \ \forall \lambda$.*

We now describe a distributed optimization method for (1), called the *proximal center algorithm*:

Algorithm PCM

0. *input*: λ_0 and $p = 0$
1. given λ_p compute in *parallel* for all i

$$x^i_{p+1} = \arg \min_{x^i \in X_i, G_i x^i = a_i} f_i(x^i) + \mu \phi_{X_i}(x^i) + \langle \lambda_p, F_i x^i \rangle$$

2. compute $\nabla d(\mu, \lambda_p) = \sum_{i=1}^M F_i x^i_{p+1} - a$
3. find $u_p = \arg\max_\lambda \langle \nabla d(\mu, \lambda_p), \lambda - \lambda_p \rangle - \frac{D_\mu}{2} \|\lambda - \lambda_p\|^2$
4. find $v_p = \arg\max_\lambda -\frac{D_\mu}{2}\|\lambda\|^2 + \sum_{l=0}^p \frac{l+1}{2} \langle \nabla d(\mu, \lambda_l), \lambda - \lambda_l \rangle$
5. set $\lambda_{p+1} = \frac{p+1}{p+3} u_p + \frac{2}{p+3} v_p$.

Note that the maximization problems in Steps 3 and 4 of Algorithm PCM can be solved explicitly and thus computationally very efficient. The main computational effort is done in Step 1. However, in some applications, e.g. distributed control, Step 1 can be performed also very efficiently (see Section 3), making the MPC algorithm suitable for online implementation. After p iterations of Algorithm PCM we define:

$$\hat{x}^i = \sum_{l=0}^p \frac{2(l+1)}{(p+1)(p+2)} x^i_{l+1} \text{ and } \hat{\lambda} = \lambda_p.$$

In the next theorem we provide estimates for the rate of convergence:

Theorem 1. *[6] Under the hypothesis of Lemma 1 and taking* $\mu = \epsilon / \sum_i D_{X_i}$ *and* $p + 1 = 2\sqrt{(\sum_i \|F_i\|^2/\sigma_i)(\sum_i D_{X_i})}\ \frac{1}{\epsilon}$, *then after* p *iterations* $-\|\lambda^*\| \|\sum_i F_i \hat{x}^i - a\| \leq f(\hat{x}) - f^* \leq \epsilon$ *and the constraints satisfy* $\|\sum_i F_i \hat{x}^i - a\| \leq \epsilon \big(\|\lambda^*\| + \sqrt{\|\lambda^*\|^2 + 2} \big)$, *where* λ^* *is the minimum norm optimal multiplier.*

Therefore, the efficiency estimates of Algorithm PCM is of order $\mathcal{O}(\frac{1}{\epsilon})$ and thus improves with one order of magnitude the complexity of the subgradient algorithm, whose efficiency estimates is $\mathcal{O}(\frac{1}{\epsilon^2})$.

2.2 Interior-Point Lagrangian Method

In this section we describe the second decomposition method, called the interior-point Lagrangian algorithm. In this decomposition method we add to the standard Lagrangian a smoothing term $\mu \sum_{i=1}^M \phi_{X_i}$, where each function ϕ_{X_i} is a N_i-self-concordant barrier associated to the convex set X_i.

Lemma 2. *[7] If Assumption 1 holds and* ϕ_{X_i}*'s are* N_i*-self-concordant barriers associated to* X_i, *then the family of dual functions* $\{-d(\mu, \cdot)\}_{\mu > 0}$ *is self-concordant. Moreover, the Hessian of* $-d(\mu, \cdot)$ *is positive definite and given by:*

$$\nabla^2 d(\mu, \lambda) = \sum_{i=1}^M F_i \big[H_i^{-1}(\mu, \lambda) G_i^T \big(G_i H_i^{-1}(\mu, \lambda) G_i^T \big)^{-1} G_i H_i^{-1}(\mu, \lambda) - H_i^{-1}(\mu, \lambda) \big] F_i^T,$$

where $H_i(\mu, \lambda) = \nabla^2 f_i(x^i(\mu, \lambda)) + \mu \nabla^2 \phi(x^i(\mu, \lambda))$.

In conclusion dual functions $\{-d(\mu,\cdot)\}_{\mu>0}$ are self-concordant. This opens the possibility of deriving a dual interior-point based method for (1) using Newton directions for updating the multipliers to speed up the convergence rate. Denote the Newton direction associated to function $d(\mu,\cdot)$ at λ as follows:

$$\Delta\lambda(\mu,\lambda) = -\left(\nabla^2 d(\mu,\lambda)\right)^{-1}\nabla d(\mu,\lambda).$$

For every $\mu > 0$, we also define the Newton decrement:

$$\delta(\mu,\lambda) = \sqrt{-1/\mu \nabla d(\mu,\lambda)^T \left(\nabla^2 d(\mu,\lambda)\right)^{-1}\nabla d(\mu,\lambda)}.$$

Algorithm IPLM

0. *input*: (μ_0, λ_0) satisfying $\delta(\mu_0,\lambda_0) \leq \epsilon_V$, $p = 0$, $0 < \tau < 1$ and $\epsilon > 0$
1. if $\mu_p \leq \epsilon$, then stop
2. (outer iteration) let $\mu_{p+1} = \tau\mu_p$ and go to inner iteration (step 3)
3. (inner iteration) initialize $\lambda = \lambda_p$, $\mu = \mu_{p+1}$ and $\delta = \delta(\mu_{p+1}, \lambda^p)$
   while $\delta > \epsilon_V$ do
   3.1 determine a step size α and compute $x^i(\mu,\lambda)$, $\lambda^+ = \lambda + \alpha\Delta\lambda(\mu,\lambda)$
   3.2 compute $\delta^+ = \delta(\mu,\lambda^+)$ and update $\lambda = \lambda^+$ and $\delta = \delta^+$
4. $x^i_{p+1} = x^i(\mu,\lambda)$, $\lambda_{p+1} = \lambda$, replace p by $p+1$ and go to step 1

Theorem 2. *[7] Under the hypothesis of Lemma 2 the following convergence rate holds for the Algorithm IPLM:* $0 \leq f(x_p) - f^* \leq N_\phi \mu_p$, *where* $x_p = \arg\min_x L_{\mu_p}(x,\lambda_p)$ *and* $N_\phi = \sum_i N_i$. □

Therefore, the complexity of the interior-point Lagrangian method (Algorithm IPLM) is of order $\mathcal{O}(\ln(\frac{\mu_0}{\epsilon}))$.

3 Distributed Control

In the rest of the paper we explore the potential of the previous two distributed algorithms, the proximal center and interior-point Lagrangian method, in distributed control problems. We show that many relevant centralized control schemes can be recast as separable convex problems for which our two algorithms can be applied but exploiting the specific problem structure.

3.1 Distributed Control for Coupled Dynamics

The application that we will discuss in this section is distributed control of large-scale networked systems with interacting subsystem dynamics, which can be found in a broad spectrum of applications ranging from robotics to regulator systems. We assume that the overall system model can be decomposed into M appropriate subsystem models:

$$x^i(k+1) = \sum_{j\in\mathcal{N}(i)} A_{ij}x^j(k) + B_{ij}u^j(k)\ \forall i = 1\cdots M, \tag{4}$$

where $\mathcal{N}(i)$ denotes the set of subsystems that interact with the ith subsystem, including itself. The control and state sequence must satisfy local constraints:

$$x^i(k) \in \Omega_i,\ u^i(k) \in U_i\ \ \forall i = 1 \cdots M \text{ and } \forall k \geq 0,$$

where the constraint sets $\Omega_i \subseteq \mathbb{R}^{n_{x_i}}$ and $U_i \subseteq \mathbb{R}^{n_{u_i}}$ are usually convex, compact, with the origin in their interior.

Performance of the system is expressed via a stage cost, which is composed of individual separate costs assumed to have the following form: $\ell(x,u) = \sum_{i=1}^{M} \ell_i(x^i, u^i)$, where usually $\ell_i(x^i, u^i)$ is a convex quadratic function, not necessarily strict.

The centralized control problem for this application is formulated as follows:

$$\begin{aligned}
&\min_{x_l^i, u_l^i} \sum_{l=0}^{N-1} \sum_{i=1}^{M} \ell_i(x_l^i, u_l^i) + \sum_{i=1}^{M} \ell_i^f(x_N^i) \qquad (5)\\
&\text{s.t.}:\ x_0^i = x^i,\ x_{l+1}^i = \sum_{j \in \mathcal{N}(i)} A_{ij} x_l^j + B_{ij} u_l^j\\
&\quad x_N^i \in \Omega_i,\ x_l^i \in \Omega_i,\ u_l^i \in U_i\ \forall l = 0 \cdots N-1,\ \forall i = 1 \cdots M,
\end{aligned}$$

where N denotes the prediction horizon and $\ell_i^f(x_N^i)$ denotes some terminal cost introduced for stability reasons.

Theorem 3. *The centralized optimization problem* (5) *can be written as a separable convex problem* (1).

Proof: Let us introduce the following notation:

$$\mathbf{x}^i = (x_1^i \cdots x_N^i\ u_0^i \cdots u_{N-1}^i),\ \ X_i = \Omega_i^N \times U_i^N,$$

$$f_i(\mathbf{x}^i) = \sum_{l=0}^{N-1} \ell_i(x_l^i, u_l^i) + \ell_i^f(x_N^i),$$

where f_i's are convex quadratic functions as in Assumption 1, but not necessarily strictly convex. Then, (5) can be recast as a separable convex program:

$$\min_{\mathbf{x}^i \in X_i} \Big\{ \sum_{i=1}^{M} f_i(\mathbf{x}^i) : \sum_{i=1}^{M} F_i \mathbf{x}^i = a \Big\}, \qquad (6)$$

where the matrices F_i and a are defined accordingly. □

3.2 Distributed Control for Consensus Constraints

Another important problem that arises in such large-scale networked systems is related to consensus or rendezvous seeking which has received increasing attention in the recent literature, see e.g. [3,4], etc. In this problem each subsystem is uncoupled of each other but the coupling is determined by the consensus or rendezvous point which may be either fixed a priori or considered as an additional variable in the control problem formulation.

We consider that the whole network can be decomposed into M subsystems having decoupled dynamics given by the following discrete-time state equations:

$$x^i(k+1) = A_i x^i(k) + B_i u^i(k) \; \forall i = 1 \cdots M.$$

Similarly as in the previous section, we assume that the control and state sequence must satisfy local constraints:

$$x^i(k) \in \Omega_i, \; u^i(k) \in U_i \;\; \forall i = 1 \cdots M \text{ and } \forall k \geq 0.$$

The system reaches a consensus at some time N if each subsystem is at an equilibrium at time N and the consensus or rendezvous point is also attained (see also [4]). We assume that these two requirements can be written mathematically as follows: there exists equilibrium points $(x^i(N), u^i(N))$, i.e. satisfying the local linear equalities

$$x^i(N) = A_i x^i(N) + B_i u^i(N)$$

and the consensus or rendezvous condition is expressed by the following coupling linear (in)equalities:

$$\sum_{i=1}^{M} \bar{F}_i x^i(N) = \bar{a}. \tag{7}$$

A typical example of consensus constraints is $C_i x^i - C_j x^j = d_{ij}$, which for example might describe relative distance between vehicles.

The optimal control problem for the consensus or rendezvous settings can be described as follows:

$$\begin{aligned} \min_{x_l^i, u_l^i} \; & \sum_{l=0}^{N-1} \sum_{i=1}^{M} \ell_i(x_l^i, u_l^i) + \sum_{i=1}^{M} \ell_i^f(x_N^i, u_N^i) \\ \text{s.t.}: \; & x_0^i = x^i, \; x_{l+1}^i = A_i x_l^i + B_i u_l^i \\ & x_l^i \in \Omega_i, \; u_l^i \in U_i \; \forall l = 0 \cdots N, \; \forall i = 1 \cdots M, \\ & x_N^i = A_i x_N^i + B_i u_N^i, \; \sum_{i=1}^{M} \bar{F}_i x_N^i = \bar{a}. \end{aligned} \tag{8}$$

Our goal is to design a distributed control algorithm for solving the consensus or rendezvous problem (8) where each agent performs individual planing of its trajectory and then negotiate with each other locally in order to achieve the consensus.

In the next theorem we show that the optimal control problem (8) can be written as a separable convex problem:

Theorem 4. *The centralized optimization problem* (8) *can be written as a separable convex problem* (1).

Proof: Similarly as in Theorem 3.

In the next section we show that using the distributed algorithms based on smoothing techniques that we have described in Section 2 the efficiency estimates can be improved.

3.3 Practical Implementation

In this section we describe the practical implementation of the Algorithm PCM and IPLM for the two control problems (5) and (8). Our algorithms can be an alternative to the classical methods (e.g. Jacobi algorithm, (incremental) subgradient algorithm, etc), leading to new methods of solution in a much faster way making them suitable for online implementation. We will show that due to the special structure of the control problems (5) and (8) our two algorithms leads to decomposition in both "space" and "time", i.e. not only over M but also over N.

Remark 1. Note that the proximal center method can be applied to separable convex problems with general convex functions f_i, not necessarily quadratic.

For simplicity of the exposition we assume that the sets are normalized Euclidian balls:

$$\Omega_i = \{x^i \in \mathbb{R}^{n_{x_i}} : \|x^i\| \leq 1\},\ U_i = \{u^i \in \mathbb{R}^{n_{u_i}} : \|u^i\| \leq 1\}.$$

In the proximal center algorithm we need to properly choose the function ϕ_{X_i} according to the structure of X_i. Given the specific structure of the set X_i in the control formulations (5) and (8), i.e. $X_i = \Omega_i^N \times U_i^N$, we choose

$$\phi_{X_i}(\mathbf{x}^i) = \|\mathbf{x}^i\|^2 = \sum_{l=1}^{N} \|x_l^i\|^2 + \sum_{l=0}^{N-1} \|u_l^i\|^2.$$

In the interior-point Lagrangian algorithm the prox function ϕ_{X_i} must be chosen as self-concordant barrier functions for the set X_i. Let $b(x)$ be the self-concordant barrier function for the Euclidian set $\{x : \|x\| \leq 1\}$. Since in the control formulations that we have just described previously the set $X_i = \Omega_i^N \times U_i^N$, we choose

$$\phi_{X_i}(\mathbf{x}^i) = \sum_{l=1}^{N} b(x_l^i) + \sum_{l=0}^{N-1} b(u_l^i).$$

We further assume that the stage costs have the following quadratic structure:

$$\ell_i(x_l^i, u_l^i) = x_l^{iT} Q_i x_l^i + u_l^{iT} R_i u_l^i,$$

where the matrices Q_i and R_i are positive semidefinite.
Constructing the variables $\mathbf{x}_i$, the sets X_i and the matrices F_i, G_i as described in Section 3, the centralized control problems (5) and (8) can be rewritten as:

$$\min_{\mathbf{x}^i \in X_i} \{\sum_{i=1}^{M} \mathbf{x}^{iT} \mathbf{Q}_i \mathbf{x}_i : \sum_{i=1}^{M} F_i \mathbf{x}^i = a,\ G_i \mathbf{x}^i = a^i\ \forall i\}, \tag{9}$$

where $\mathbf{Q}_i = \text{diag}(Q_i, \cdots, Q_i, R_i, \cdots, R_i)$ with N terms of Q_i and N terms of R_i. Note that $X_i \subseteq \{\mathbf{x} : \|\mathbf{x}\| \leq \sqrt{2N}\}$. Furthermore, in the control formulation (5) local equalities $G_i\mathbf{x}^i = a^i$ are not present.
In the Algorithm PCM the most expensive computations are done in Step 1, where we must solve for all $i = 1 \cdots M$, the following minimization problems:

$$\min_{\mathbf{x}^i \in X_i} \mathbf{x}^{iT}\mathbf{Q}_i\mathbf{x}^i + \langle \lambda^p, F_i\mathbf{x}^i \rangle + \mu\|\mathbf{x}^i\|^2,$$

with the Lagrange multipliers λ^p computed at previous iteration. In fact, since $\mathbf{Q}_i$ has a diagonal structure we can further decompose each minimization problem into $2N$ quadratic cost quadratic constraints problems with a particular structure:

$$\min_{\|x\| \leq 1} x^T Q x + \langle q, x \rangle, \tag{10}$$

where Q is a positive definite diagonal matrix (for our example $Q = Q_i + \mu I_{n_{x_i}}$ for the state variables x_l^i or $Q = R_i + \mu I_{n_{x_i}}$ for the input variables u_l^i. Here x represents the state variable x_l^i or control variable u_l^i at step l. Note that in some particular cases (e.g. $Q = \beta I$), the solution of (10) can be computed analytically.

In the Algorithm IPLM we must solve for all $i = 1 \cdots M$ the following minimization problems:

$$\min_{\mathbf{x}^i \in X_i} \mathbf{x}^{iT}\mathbf{Q}_i\mathbf{x}^i + \langle \lambda^p, F_i\mathbf{x}^i \rangle + \mu[\sum_{l=1}^{N} b(x_l^i) + \sum_{l=0}^{N-1} b(u_l^i)].$$

which as before it can be further decomposed over the prediction horizon N in $2N$ problems of the form

$$\min_{\|x\| \leq 1} x^T Q x + \langle q, x \rangle + b(x),$$

where again x represents the state x_l^i or the input u_l^i at step l. Such problem can be solved with standard interior-point solvers very easy since the dimension of the problem is very small: n_{x_i} or n_{u_i}. Note that compared to Theorem 3, in Theorem 4 we can eliminate the variables x_l^i from the optimization problem. But then the decomposition over the prediction horizon N will not be possible.
In summary, the special structure of the control problems (5) and (8) shows that our two algorithms lead to decomposition in both "space" and "time", i.e. the centralized control problem can be decomposed into small subproblems corresponding to the spatial structure of the system (M subsystems) but also to the prediction horizon (N the length of the prediction). Note that this is *not* the case with Jacobi or primal (incremental) subgradient type algorithms. Due to space limitations we opted to include simulations elsewhere (see also [6,7]).

4 Conclusions

The proximal center and the interior-point Lagrangian algorithm are applied for solving distributed control problems for either dynamically coupled systems or uncoupled systems but with consensus constraints. We show that the corresponding centralized control problems can be recast as separable convex problems for which our two algorithms can be applied. We proved that the solution generated by our distributed algorithms converges to the solution of the centralized control problem and we provided also estimates for the rate of convergence which greatly improves the convergence rates of the existing distributed algorithms, e.g. Jacobi or subgradient algorithms. It was also proved that the main steps of the two algorithms can be computed efficiently for control problems by making use of the specific structure of the underlying control problem and thus making these methods suitable for online implementation of the corresponding distributed control scheme.

References

1. Bertsekas, D.P., Tsitsiklis, J.N.: Parallel and distributed computation: Numerical Methods. Prentice-Hall, Englewood Cliffs (1989)
2. Camponogara, E., Jia, D., Krogh, B., Talukdar, S.: Distributed model predictive control. IEEE Control Systems Magazine 22(1), 44–52 (2002)
3. Dunbar, W., Murray, R.: Distributed receding horizon control for multi-vehicle formation stabilization. Automatica 42, 549–558 (2006)
4. Keviczky, T., Johansson, K.: A study on distributed model predictive consensus. In: 17th IFAC World Congress, Seoul, South Korea (2008)
5. Maciejowski, J.M.: Predictive Control with Constraints. Prentice Hall, Harlow (2002)
6. Necoara, I., Suykens, J.: Application of a smoothing technique to decomposition in convex optimization. IEEE Transactions on Automatic Control 53(11), 2674–2679 (2008)
7. Necoara, I., Suykens, J.: An interior-point Lagrangian decomposition method for separable convex optimization. Journal of Optimization Theory and Applications (in press, 2009)
8. Nesterov, Y.: Introductory Lectures on Convex Optimization: A Basic Course. Kluwer, Boston (2004)
9. Richards, A., How, J.: Robust distributed model predictive control. International Journal of Control 80(9), 1517–1531 (2007)
10. Venkat, A., Hiskens, I., Rawlings, J., Wright, S.: Distributed MPC strategies with application to power system automatic generation control. IEEE Transactions on Control Sys. Techn. 16(6), 1192–1206 (2006)
11. Shamma, J.S.: Cooperative Control of Distributed Multiagent Systems. Wiley, Chichester (2008)
12. Rantzer, A.: On Prize Mechanism in Linear Quadratic Team Theory. In: Proc. 46th IEEE Conf. Decision Contr. (2007)
13. Kalvins, E., Murray, R.M.: Distributed algorithms for cooperative control. IEEE Pervasive Computing 3(1), 56–65 (2004)

Trajectory Tracking of a Nonholonomic Mobile Robot Considering the Actuator Dynamics: Design of a Neural Dynamic Controller Based on Sliding Mode Theory

Nardênio A. Martins, Douglas W. Bertol, and Edson R. De Pieri

Federal Unversity of Santa Catarina, Florianpolis, Brasil
{nardenio,dwbertol,edson}@das.ufsc.br
http://www.das.ufsc.br

Abstract. In this paper, a trajectory tracking control for a nonholonomic mobile robot by the integration of a kinematic controller and neural dynamic controller is investigated, where the wheel actuator dynamics is integrated with mobile robot dynamics and kinematics so that the actuator input voltages are the control inputs. The proposed neural dynamic controller (PNDC), based on the sliding mode theory, is constituted by a neural voltage controller (NVC) and a neural robust compensator (NRC), which has as objective compensates the uncertainties and disturbances in the dynamics. Stability analysis and numerical simulation are provided to show the effectiveness of the PNDC.

Keywords: mobile robot, trajectory tracking, neural dynamic control, actuator dynamics, sliding mode theory, Lyapunov method.

1 Introduction

In this paper, the wheel actuator (e.g., dc motor) dynamics is integrated with mobile robot dynamics and kinematics so that the actuator input voltages are the control inputs. Differently from other investigations using neural networks in the dynamic control of mobile robots [1]-[9], the contributions are: the implementation of the proposed neural dynamic controller (PNDC) based on the partitioning of the RBFNN into several smaller subnets in order to obtain more efficient computation; the modelling by RBFNNs of the centripetal and Coriolis matrix through of the inertia matrix of the mobile robot dynamics. As a result, the obtained neural voltage controller (NVC) is modeled with static RBFNNs only, which makes possible the reduction of the size of the RBFNNs, of the computational load and the implementation in real time; an neural sliding mode controller as neural robust compensator (NRC) is used as the replacement of the discontinuous parts of the classical sliding mode controller to avoid the chattering as well as to suppress the neural network modeling errors, bounded unknown disturbances, and influence of payload; the PNDC neither require the knowledge of the mobile robot dynamics nor the time-consuming training process; the stability analysis and convergence of the mobile robot control system,

C. Alippi et al. (Eds.): ICANN 2009, Part II, LNCS 5769, pp. 845–854, 2009.

and the learning algorithms for weights are proved by using Lyapunov theory, considering the presence of bounded unknown disturbances.

2 Kinematics and Dynamics of Mobile Robots

Disregarding surface friction $F(\dot{q})$ and gravitational torques $G(q)$, the dynamic equations under nonholonomic constraints can be described by Euler-Lagrange formulation as:

$$H(q)\ddot{q} + C(q,\dot{q})\dot{q} = B(q)\tau + A^T(q)\lambda - \tau_d, \tag{1}$$

where the properties are maintained, as well as matrices, vectors, and variables are defined as in [1]. The right side of (1) can be rewritten in the Lagrange-Euler formulation as:

$$\begin{gathered} \textstyle\sum_{j=1}^{n} h_{kj}(q)\ddot{q}_j + \sum_{j=1}^{n} c_{kj}\dot{q}_j = B(q)\tau + A^T(q)\lambda - \tau_d, \\ c_{kj} = \textstyle\sum_{i=1}^{n} c_{ijk}\dot{q}_i = \sum_{i=1}^{n} \frac{1}{2}\left(\frac{\partial h_{kj}}{\partial q_i} + \frac{\partial h_{ki}}{\partial q_j} - \frac{\partial h_{ij}}{\partial q_k}\right)\dot{q}_i, \end{gathered} \tag{2}$$

with the coefficient c_{ijk} is known as Christoffel symbols. The dynamic equations of the nonholonomic mobile robot for control purposes are:

$$\begin{gathered} \dot{q} = S(q)v, \;\; \bar{H}(q)\dot{v} + \bar{C}(q,\dot{q})v = \bar{\tau} - \bar{\tau}_d, \\ \bar{H}(q) = S^T H S, \;\; \bar{C}(q,\dot{q}) = S^T(H\dot{S} + CS), \;\; \bar{\tau} = \bar{B}\tau = S^T B\tau, \;\; \bar{\tau}_d = S^T\tau_d, \end{gathered} \tag{3}$$

being $S(q)$ the Jacobian matrix, and v the actual velocity of the mobile robot [1].

3 Actuator Dynamics

Neglecting motor inductance in the electrical part of the actuator [10], the equations governing the actuator motor can be written as:

$$\tau_m = K_T i, \;\; u = R_a i + K_b\dot{\phi}_m, \tag{4}$$

where τ_m is the torque generated by the motor, K_T is the motor torque constant, i is the current, u is the actuator input voltage, R_a is the resistance, K_b is the counter electromotive force coefficient, and $\dot{\phi}_m$ is the velocity of the actuator motor. The angular velocity of the actuator motor, $\dot{\phi}_m$, and the corresponding wheel angular velocity $\dot{\varphi}$ are related by gear ratio N as:

$$\dot{\varphi} = \frac{\dot{\phi}_m}{N}, \tag{5}$$

and the motor torque τ_m is related to the wheel torque τ as:

$$\tau = N\tau_m. \tag{6}$$

The relationship between the angular wheel velocities $\dot{\varphi}$ and the velocity vector v is given by:

$$\begin{bmatrix} \dot{\varphi}_r \\ \dot{\varphi}_l \end{bmatrix} = \begin{bmatrix} \frac{1}{r} & \frac{R}{r} \\ \frac{1}{r} & -\frac{R}{r} \end{bmatrix} \begin{bmatrix} v_L \\ \varpi_A \end{bmatrix} = Xv. \tag{7}$$

Using (3)-(7), the mobile robot dynamics equation (including actuator dynamics) can be written as:

$$\bar{H}(q)\dot{v} + \bar{C}(q,\dot{q})v + \bar{\tau}_d = \frac{NK_T}{R_a}\bar{B}u - \frac{N^2 K_T K_b}{R_a}\bar{B}Xv = \bar{\tau}. \tag{8}$$

4 Neural Networks Modeling by RBFNNs

Based on (1) and (3), it can be verified that $H(q)$ is function of q only, thus, static neural networks are enough to model them. As a consequence, the size of the network can be much smaller compared with its dynamic counterparts. The stability of the neural networks can be analyzed, where Ge-Lee (GL) matrix [11], defined by $\{.\}$, and its product operator '$\bullet$' are used. The ordinary matrix and vector are denoted by $[.]$. It is well known that $h_{kj}(q)$ of (2) is infinite differentiable. Thus, for $h_{kj}(q)$ (Figure 1) has:

$$h_{kj}(q) = \sum_l W_{H_{kjl}} \xi_{H_{kjl}}(q) + \varepsilon_{H_{kj}}(q) = W_{H_{kj}}^T \xi_{H_{kj}}(q) + \varepsilon_{H_{kj}}(q), \tag{9}$$

where l denotes the number of hidden neurons, $W_{H_{kjl}}$ is the weight, and:

$$\xi_{H_{kjl}}(q, m_H, \sigma_H) = \exp\left(\frac{-\|q-m_H\|^2}{\sigma_H^2}\right) = \exp\left(\frac{-(q-m_H)^T(q-m_H)}{\sigma_H^2}\right). \tag{10}$$

Since

$$\frac{\partial h_{ij}}{\partial q_k} = -2\frac{1}{\sigma_H^2} W_{H_{ij}}^T \xi_{H_{ij}} (q_k - m_k) + \phi_{H_{ijk}}, \tag{11}$$

with $\phi_{H_{ijk}} = \frac{\partial \varepsilon_{h_{ij}}}{\partial q_k}$. It is assumed that $\left|\phi_{H_{ijk}}\right|$ has an upper limit given by $\varepsilon_{h_{ijk}}$.

Based on the definition of Christoffel symbols in (2), it has:

$$\begin{aligned} c_{ijk} &= -\tfrac{1}{\sigma_H^2}\big(W_{H_{kj}}^T \xi_{H_{kj}} (q_i - m_{H_i}) + W_{H_{ki}}^T \xi_{H_{ki}} (q_j - m_{H_j}) - \\ &\quad - W_{H_{ij}}^T \xi_{H_{ij}} (q_k - m_{H_k})\big) + \varepsilon_{C_{ijk}}, \\ \varepsilon_{C_{ijk}} &= -\phi_{H_{kji}} - \phi_{H_{kij}} + \phi_{H_{ijk}}, \end{aligned} \tag{12}$$

which leads to:

$$\begin{aligned} c_{kj}(q,\dot{q}_i) &= \sum_{i=1}^{n} c_{ijk}\dot{q}_i \\ &= -\tfrac{1}{\sigma_H^2} \sum_{i=1}^{n} \big(W_{H_{kj}}^T \xi_{H_{kj}} (q_i - m_{H_i}) + \\ &\quad + W_{H_{ki}}^T \xi_{H_{ki}} (q_j - m_{H_j}) - W_{H_{ij}}^T \xi_{H_{ij}} (q_k - m_{H_k})\big)\dot{q}_i + \varepsilon_{C_{kj}}, \end{aligned} \tag{13}$$

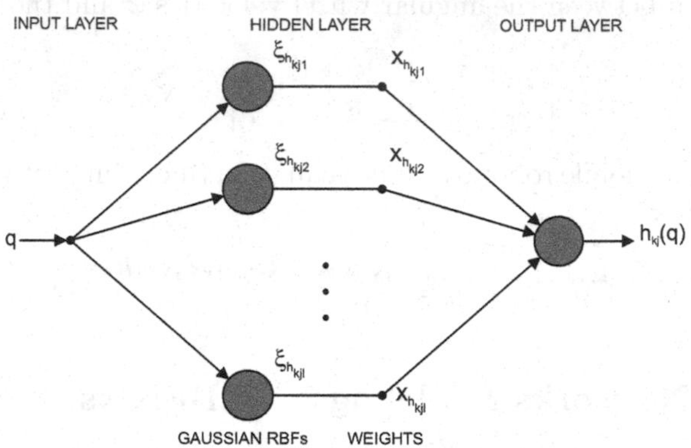

Fig. 1. Implementation of the static neural network of $h_{kj}(q)$

where $\varepsilon_{C_{kj}} = \sum_{i=1}^{n} \varepsilon_{C_{ijk}} \dot{q}_i$. It can easily be seen that the dynamics (3) of mobile robots can be constructed by using the subnets for $H(q)$, because $C(q,\dot{q})$ can be constructed based on the parameters of $H(q)$. Note that since the $H(q)$ is function of q only, the subnets are static instead of dynamic, the size of the network is much smaller by introducing deterministic factors into the neural network model. Thus, the matrix $C(q,\dot{q})$ is a function of $H(q)$ (Figure 2), i. e.,

$$\begin{aligned} C(q,\dot{q}) = &-\tfrac{1}{\sigma_H^2}\left[\{W_H\}^T \bullet \{\xi_H(q)\}\right](q-m_H)^T\dot{q} - \\ &-\tfrac{1}{\sigma_H^2}\left[\{W_H\}^T \bullet \{\xi_H(q)\}\right]\dot{q}\,(q-m_H)^T + \\ &+\tfrac{1}{\sigma_H^2}(q-m_H)\,\dot{q}^T\left[\{W_H\}^T \bullet \{\xi_H(q)\}\right] + E_C(q,\dot{q}). \end{aligned} \tag{14}$$

In summary, the dynamics (3) and (8) results in:

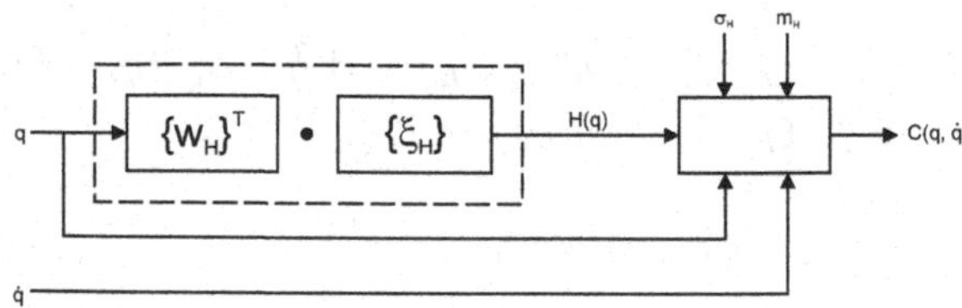

Fig. 2. Implementation of the $C(q,\dot{q})$ through $H(q)$

$$\bar{\tau} = S^T\left[\{W_H\}^T \bullet \{\xi_H(q)\}\right] z + S^T \tau_d + \bar{E}(q,\dot{q},v,\dot{v}), \tag{15}$$

with:

$$z = S\dot{v} + \dot{S}v - \underbrace{\left(-\frac{1}{\sigma_H^2}(q - m_H)^T \dot{q} - \frac{1}{\sigma_H^2}\dot{q}(q - m_H)^T\right)}_{x} Sv - \underbrace{-\left(\frac{1}{\sigma_H^2}(q - m_H)\dot{q}^T\right)}_{y} Sv, \tag{16}$$

$$\bar{E}(.) = S^T(q)E_H(q)S(q)\dot{v} + S^T(q)E_H(q)\dot{S}(q)v + S^T(q)E_C(q,\dot{q})S(q)v.$$

5 Kinematic Control

Let velocity and position of a reference robot be given as:

$$q_r = \left[x_d \; y_d \; \theta_d\right]^T, \; v_{ref} = \left[v_d \; \omega_d\right]^T, \\ \dot{x}_r = v_d \cos(\theta_d), \; \dot{y}_r = v_d \sin(\theta_d), \; \dot{\theta}_r = \omega_d, \tag{17}$$

where $v_d > 0$ for all t is the reference linear velocity and ω_d is the reference angular velocity. Thus, the position tracking error vector is expressed in the basis of a frame linked to the mobile robot platform as:

$$e_q = \begin{bmatrix} e_1 \\ e_2 \\ e_3 \end{bmatrix} = \begin{bmatrix} \cos(\theta) & \sin(\theta) & 0 \\ -\sin(\theta) & \cos(\theta) & 0 \\ 0 & 0 & 1 \end{bmatrix} \begin{bmatrix} x_d - x \\ y_d - y \\ \theta_d - \theta \end{bmatrix}. \tag{18}$$

The position error dynamics can be obtained from the time derivative of (18) as:

$$\dot{e}_q = \begin{bmatrix} \dot{e}_1 \\ \dot{e}_2 \\ \dot{e}_3 \end{bmatrix} = \begin{bmatrix} \omega e_2 - v_1 + v_d \cos(e_3) \\ -\omega e_1 + v_d \sin(e_3) \\ \omega_d - \omega \end{bmatrix}. \tag{19}$$

An auxiliary velocity control input v_c [1] that achieves tracking for (3) is given by:

$$v_c = \begin{bmatrix} v_d \cos(e_3) + k_1 e_1 \\ \omega_d + k_2 v_d e_2 + k_3 v_d \sin(e_3) \end{bmatrix}, \tag{20}$$

where k_1, k_2, and k_3 are positive parameters. To design the actuator voltage input and generate the desired velocities v_c, the auxiliary velocity tracking error is defined as:

$$e_c = v_c - v = \begin{bmatrix} v_{c_1} - v_1 \\ v_{c_2} - \omega \end{bmatrix} = \begin{bmatrix} e_4 \\ e_5 \end{bmatrix}. \tag{21}$$

Stability and convergence analysis of this controller will be described later, through of the choice of a Lyapunov candidate function V_1, but further details in [1].

6 Neural Dynamic Controller (PNDC)

Let Λ be a symmetric diagonal positive definite matrix, one defines:

$$
\begin{aligned}
v_r &= v_c + \Lambda \int_0^t e_c dt, \;\; \dot{v}_r = \dot{v}_c + \Lambda e_c, \\
r &= v_r - v = e_c + \Lambda \int_0^t e_c dt, \;\; \dot{r} = \dot{v}_r - \dot{v} = \dot{e}_c + \Lambda e_c
\end{aligned}
\tag{22}
$$

where r is a filtered tracking error term, $\int_0^t e_c dt$ is an auxiliary position tracking error, which does not reflect the position tracking error e_q directly, besides not having physical meaning. One defines the control input to be of the form:

$$
\begin{aligned}
u &= \tfrac{R_a}{NK_T}\bar{B}^{-1}\left(\underbrace{S^T\left[\left\{ \hat{W}_H \right\}^T \bullet \{\xi_H(.)\} \right] z(.)}_{NVC} + \rho - \gamma \right), \\
\rho &= \tfrac{N^2 K_T K_b}{R_a}\bar{B} X v_r,
\end{aligned}
\tag{23}
$$

where $\left\{ \hat{W}_H \right\}$ represent estimate of true parameters of matrix $\{W_H\}$ of (15), (16), and γ is the constant plus proportional rate reaching law with the aim of compensating the bounded unknown disturbances, which is defined as:

$$
\gamma = -G sgn(r) - (Q + K + I_n) r, \tag{24}
$$

with $G^T = G > 0$, $(Q + K + I_n)^T = (Q + K + I_n) > 0$, and I_n is identity matrix. In (24) appears the chattering phenomenon, and for his elimination or minimization, in this control design is proposed a RBFNN, as continuous approximation of $Gsgn(r)$ in γ, (24), which is referred as NRC. Then,

$$
\gamma = -\hat{P} - (Q + I_n) r = -\underbrace{\left[\left\{ \hat{W}_P \right\}^T \bullet \{\xi_P(r)\} \right]}_{NRC} - (Q + K + I_n) r, \tag{25}
$$

with $\hat{P}(r)$ being an $n \times 1$ vector in which $\hat{p}_k$ is the output of the *i-th* RBFNN. Let us consider Lyapunov candidate function:

$$
\begin{aligned}
V_2 = \tfrac{1}{2}\Bigg(& r^T \bar{H}(q) r + \sum_{k=1}^{n} \tilde{W}_{H_k}^T \Gamma_{H_k}^{-1} \tilde{W}_{H_k} + \\
& + \sum_{k=1}^{n} \tilde{W}_{P_k}^T \Gamma_{P_k}^{-1} \tilde{W}_{P_k} \Bigg) + \left(\int_0^t e_c dt \right)^T \Lambda \int_0^t e_c dt,
\end{aligned}
\tag{26}
$$

being $\Gamma_{.k}$ the dimensional compatible symmetric positive definite matrices, and $\left\{ \tilde{W}_{.k} \right\} = \{W_{.k}\} - \left\{ \hat{W}_{.k} \right\}$ the parameter error. Clearly, $V_2 \geq 0$, and $V_2 = 0$

if only if $\int_o^t e_c dt = 0$, $r = 0$, and $\left\{\tilde{W}_.\right\} = 0$. The parameter learning laws of RBFNNs are chosen as:

$$\begin{aligned}\dot{\hat{W}}_{H_k} = \Gamma_{H_k} \bullet \{\xi_{H_k}\} \Big(\Big(S\dot{v}_r + \dot{S}v_r - xSv_r\Big) r_{r_k} - \\ -r_r \left(\tfrac{1}{\sigma_H^2}(q - m_H)\right) Sv_r \dot{q}_k \Big) - K_{H_k} \Gamma_{H_k} \|r\| \hat{W}_{H_k},\end{aligned} \quad (27)$$

$$\dot{\hat{W}}_{P_k} = \Gamma_{P_k} \bullet \{\xi_{P_k}(r)\} r_k - K_{P_k} \Gamma_{P_k} \|r\| \hat{W}_{P_k},$$

where $K_{.k} = K_. > 0$ are positive constants. After the necessary mathematical manipulations and assumptions (similar to [12]), $\dot{V}_2$ stays:

$$\begin{aligned}\dot{V}_2 \leq & -e_c^T e_c - \left(\int_0^t e_c dt\right)^T \Lambda^T \Lambda \int_0^t e_c dt - \tfrac{N^2 K_T K_b}{R_a} r^T \bar{B} X r - \\ & - \textstyle\sum_{k=1}^n (Q_k - \beta_k) |r_k|^2 - \|r\| (K_{\min} \|r\| \\ & -K_{\bar{H}}(\left\|\tilde{W}_{\bar{H}}\right\|_F - \tfrac{W_{\bar{H}\max}}{2})^2 + K_{\bar{H}} \tfrac{W^2_{\bar{H}\max}}{4} \\ & -K_{\bar{C}}(\left\|\tilde{W}_{\bar{C}}\right\|_F - \tfrac{W_{\bar{C}\max}}{2})^2 + K_{\bar{C}} \tfrac{W^2_{\bar{C}\max}}{4} - \\ & -K_P(\left\|\tilde{W}_P\right\|_F - \tfrac{W_{P\max}}{2})^2 + K_P \tfrac{W^2_{P\max}}{4}),\end{aligned} \quad (28)$$

with K_{min} being the minimum singular value of K, and $W_{.max}$ the maximum value of $W_.$. Thus, $\dot{V}_2$ is negative as long as $Q_k > \beta_k$ and the term in parentheses in (28) is positive. To ensure that the global system is stable, the Lyapunov candidate function is given as:

$$V = V_1 + V_2, \; V_1 = k_1(e_1^2 + e_2^2) + 2\frac{k_1}{k_2}(1 - \cos(e_3)), \quad (29)$$

where V_2 refers to (26). Moreover, $V \geq 0$, and $V = 0$ if only if $e_q = 0$, $e_c = 0$, $\int_o^t e_c dt = 0$, $r = 0$, and $\left\{\tilde{W}_.\right\} = 0$. Since $\dot{V}_1$ and $\dot{V}_2$ are negative, then $\dot{V}$ is also negative [12]. According to a standard Lyapunov theory and LaSalle lemma, all signals $\|e_q\|$, $\|e_c\|$, $\left\|\int_0^t e_c dt\right\|$, $\|r\|$, and $\left\{\tilde{W}_.\right\}$ are uniformly ultimately bounded.

7 Numerical Simulations

In the realization of the simulations, the kinematic and the dynamic (including actuator dynamics) models described in [1] are used. The model parameters of the prototype wheeled mobile robot estimated in [13] are: $m = 11.0$ kg, $I = 1.057$ kgm^2, $R = 0.265$ m, $r = 0.125$ m, $N = 21$, $K_T = \left[0.057\; 0.051\right]^T$ Vs, $K_b = \left[0.057\; 0.051\right]^T$ Vs, and $R_a = \left[0.476\; 0.233\right]^T$ Ω. The reference trajectory is an elliptical trajectory given by $x_r = \cos(t/3)$, $y_r = 0.8\sin(t/3)$, $\theta_r = a\tan 2(\dot{y}_r, \dot{x}_r)$. The reference linear and angular velocity is given by $v_r = \sqrt{\dot{x}_r^2 + \dot{y}_r^2}$, $\omega_r = \frac{\ddot{y}_r \dot{x}_r - \ddot{x}_r \dot{y}_r}{\dot{x}_r^2 + \dot{y}_r^2}$. The trajectory starts from $[x_r(0), y_r(0), \theta_r(0)] = [1, 0, \pi/2]$ and the

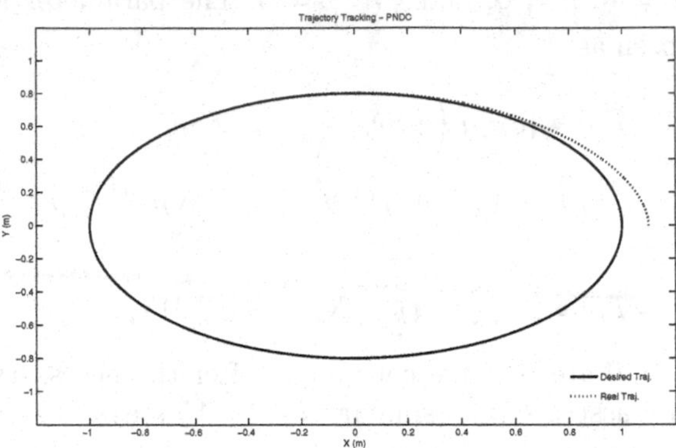

Fig. 3. Trajectory tracking using PNDC

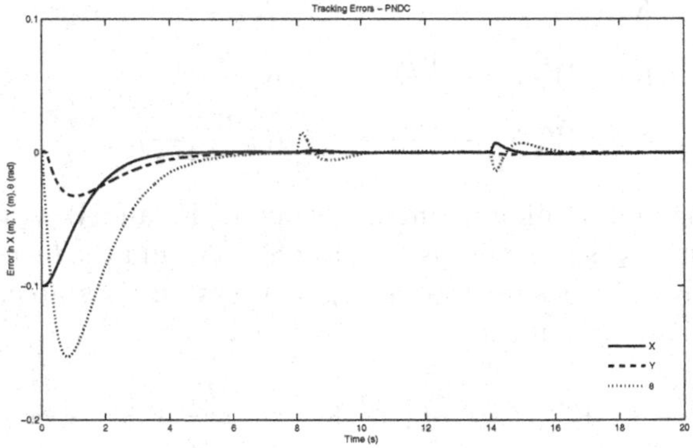

Fig. 4. Tracking errors using PNDC

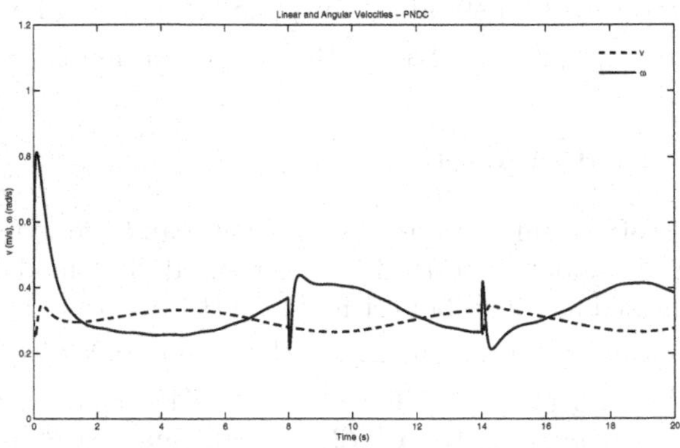

Fig. 5. Mobile robot velocities using PNDC

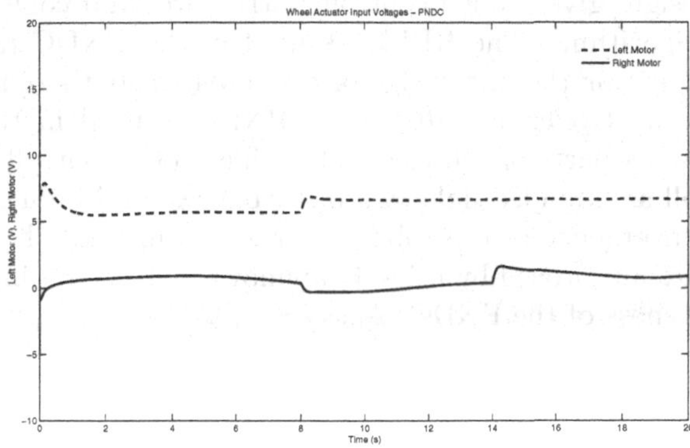

Fig. 6. Wheel actuator input voltages using PNDC

robot initial posture is taken as $[x_c(0), y_c(0), \theta(0)] = [1.1, 0, \pi/2]$. The parameters of the KC are chosen as $k_1 = 10$, $k_2 = 20$, $k_3 = 10$; and the gains of the PNDC as: NVC - $\Lambda = diag[2]$, $\Gamma_{Hk} = 0.01$, $\sigma_H^2 = 9$, $K_H = 0.01$, $Q = diag[1]$, and NRC - $\Lambda = diag[2]$, $\Gamma_{Pk} = 5$, $\sigma_P^2 = 4$, $K_P = 0.01$. The centers $m_.$ of the localised Gaussian RBFs are evenly distributed to span the input space. A Coulomb friction and a bounded periodic disturbance term are added to the robot system as:

$$\bar{\tau}_d = \begin{bmatrix} (f_1 + f_1(t))sgn(v_1) + 0.1\sin(2\pi t) \\ (f_2 + f_2(t))sgn(\omega) + 0.1\cos(2\pi t) \end{bmatrix}, \tag{30}$$

where $f_1 = 20.0$, $f_2 = 10.0$. Function $f(t)$ is nonlinear, defined as: $\left[f_1(t)\ f_2(t) \right] = \left[0.0\ 0.0 \right]^T$ if $t < 8$; $\left[f_1(t)\ f_2(t) \right] = \left[0.0\ 10.0 \right]^T$ if $8 \leq t < 14$; $\left[f_1(t)\ f_2(t) \right] = \left[40.0\ 0.0 \right]^T$ if $t \geq 14$, respectively. Thus, $\bar{\tau}_d$ is subject to a sudden change at time goes to 8s and 14s. Moreover, in 14s, the mobile robot suddenly dropped of an object of 2.75 kg, that is, a quarter of its original mass. The tracking performance of the PNDC can be observed in the: Figure 3, since the mobile robot naturally describes a smooth path tracking over the elliptical trajectory; Figure 4 shows that the tracking errors tend to zero; Figure 5 shows that the robot velocities tend to desired values; Figure 6 shows that the behavior of the wheel actuator input voltages, where is important to emphasize that the PNDC eliminate entirely the chattering.

8 Conclusions

A neural control algorithm (PNDC), considering uncertainties and disturbances in the dynamics, as an alternative trajectory tracking problem applied to non-holonomic mobile robot was proposed in this work. The implementation of the PNDC is based on the partitioning of neural networks into several smaller neural subnets, in order to obtain more efficient computation. This implementation simplifies the design; gives added controller structure; and contribute to faster weight tuning algorithms. The RBFNN's used in the PNDC neither requires an off-line learning nor the knowledge of the mobile robot's dynamics. A neural sliding mode controller as NRC of the PNDC is used in the replacement of the discontinuous parts of the classical sliding mode controller to avoid the chattering as well as bounded unknown disturbances, and influence of payload. Stability and convergence of the robot control system, and the learning algorithms for weights are proved by using Lyapunov theory. The simulation results show the effectiveness of the PNDC.

References

1. Fierro, R., Lewis, F.L.: Control of a nonholonomic mobile robot using neural networks. IEEE Trans. on Neural Networks 9(4), 589–600 (1998)
2. de Oliveira, V.M., De Pieri, E.R., Lages, W.F.: Feedforward control of a mobile robot using a neural network. In: Proc. IEEE Int. Conf. on Systems, Man, and Cybernetics, pp. 3342–3347 (2000)
3. Hu, T., Yang, S.X.: An efficient neural controller for a nonholonomic mobile robot. In: Proc. IEEE Int. Symp. Comput. Intel. in Robotics and Automation, pp. 461–466 (2001)
4. de Sousa Jr., C., Hemerly, E.M., Galvao, R.K.H.: Adaptive control for mobile robot using wavelet networks. IEEE Trans. on Systems, Man, and Cybernetics, Part B 32(4), 493–504 (2002)
5. Oh, C., Kim, M.-S., Lee, J.-Y., Lee, J.-J.: Control of mobile robots using RBF network. In: Proc. IEEE/RSJ Int. Conf. Intel. Robots and Systems, vol. 4, pp. 3528–3533 (2003)
6. Oh, C., Kim, M.-S., Lee, J.-J.: Control of a nonholonomic mobile robot using an RBF network. Journal Artificial Life and Robotics 8(1), 14–19 (2004)
7. Liu, S., Yu, Q., Lin, W., Yang, S.X.: Tracking control of mobile robots based on improved RBF neural networks. In: Proc. IEEE/RSJ Int. Conf. on Intel. Robots and Systems, pp. 1879–1884 (2006)
8. Purwar, S., Gupta, N., Kar, I.N.: Trajectory tracking control of mobile robots using wavelet networks. In: Proc. IEEE Int. Symp. on Intel. Control, pp. 550–555 (2007)
9. Peng, J., Wang, Y., Yu, H.: Neural network based robust tracking control for nonholonomic mobile robot. In: Liu, D., Fei, S., Hou, Z.-G., Zhang, H., Sun, C. (eds.) ISNN 2007. LNCS, vol. 4491, pp. 804–812. Springer, Heidelberg (2007)
10. Mills, J.K.: Hybrid actuator for robot manipulators: design, control and performance. In: Proc. IEEE Int. Conf. in Robotics and Automation, pp. 1872–1878 (1990)
11. Ge, S.S.: Robust adaptive NN feedback linearization control of nonlinear systems. Int. J. Systems Science, 1327–1338 (1996)
12. Martins, N.A., Bertol, D.W., Lombardi, W.C., de Pieri, E.R., Castelan, E.B.: Trajectory tracking of a nonholonomic mobile robot with parametric and nonparametric uncertainties: a proposed neural control. International Journal of Factory Automation, Robotics and Soft Computing 2, 103–110 (2008)
13. Huang, C.-L.: A Motion Platform of Autonomous Mobile Robot, Master thesis, Department of Electrical Engineering, National Taiwan University, pp. 27–30 (2003)

Tracking with Multiple Prediction Models

Chen Zhang[1] and Julian Eggert[2]

[1] Darmstadt University of Technology, Institute of Automatic Control, Control Theory and Robotics Lab, Darmstadt D-64283, Germany
[2] Honda Research Institute Europe GmbH, Offenbach D-63073, Germany

Abstract. In Bayesian-based tracking systems, prediction is an essential part of the framework. It models object motion and links the internal estimated motion parameters with sensory measurement of the object from the outside world. In this paper a Bayesian-based tracking system with multiple prediction models is introduced. The benefit of multiple model prediction is that each of the models has individual strengths suited for different situations. For example, extreme situations like a rebound can be better coped with a rebound prediction model than with a linear one. That leads to an overall increase of prediction quality. However, it is still an open question of research how to organize the prediction models. To address this topic, in this paper, several quality measures are proposed as switching criteria for prediction models. In a final evaluation by means of two real-world scenarios, the performance of the tracking system with two models (a linear one and a rebound one) is compared concerning different switching criteria for the prediction models.

1 Introduction

Visually tracking an object means to locate a moving object in space over time by estimating the state of its dynamics. The state estimation process happens by a fusion of state prediction for the next time slot according to a motion model on the one hand side and a measurement of its position by means of visual sensory input data on the other hand side. The sensory measurement has the function to confirm or reject the state prediction ([1]).

Tracking *arbitrary* objects in arbitrary environments is a sophisticated task, since several challenges have to be overcome. One challenge is to cope with the temporarily changing environment conditions, which let the object's features get temporarily unselective and so the measurement unreliable. Another challenge is the change of object's appearance, which makes the comparison with the original template difficult. All these possibly cause a measurement failure which may lead to a temporarily loss of the object for several frames. For coping with these measurement challenges, several works exist concerning multi-cue approach to overcome temporarily failures in some features (see e.g. [2], [3]) or concerning template adaptation to overcome appearance changes (see e.g. [4], [5]). However, the best measurement is of no help, if the state prediction is unreliable, since sensory measurement is only an additional information for confirmation

C. Alippi et al. (Eds.): ICANN 2009, Part II, LNCS 5769, pp. 855–864, 2009.

or rejection of the state prediction. State prediction requires a model of object motion which is used to predict the object's state in the next time slot. Since for arbitrary objects, there is usually no knowledge about specific prediction models available, tracking frameworks (see e.g. [6]) have to rely on rather generic prediction models which cope well with a large variety of situations. Therefore, a linear motion model based on a constant acceleration or even a constant velocity assumption is often a choice. But a real object can also undergo a sudden transposition maneuver, rebound, or other heavily accelerated motions. In these cases a linear prediction model is not always appropriate.

The key idea of this paper is that a reliable prediction system should contain multiple prediction models, where each model has individual advantages for a special situation. So, the overall prediction system benefits from individual strengths of each of the single models. However, having multiple prediction models poses the question of how to manage them. Several approaches were proposed concerning probabilistic model management for multiple-model estimations (see e.g. [7], [8]). Here, we analyze the advantages of having *multiple structurally different prediction models* for visual object tracking and propose concrete *quality measures* as methods for deterministic switching between the models. This paper is structured as follows. We first introduce a simple Bayesian tracking framework. Then we extend it by multiple prediction models, and introduce methods to switch between them. Finally, we evaluate the performance of our tracking system with multiple prediction models on test sequences.

2 Tracking Framework

The system we used to test the multiple prediction models is a correlation-based, particle-filter tracker for locating an arbitrary object in a sequence of 2-D images. It estimates the object's state $\mathbf{x} = (x, y, v_x, v_y, a_x, a_y)$ in a recursive Bayesian way ([1]) by incorporating measurement results gained from multiple cues.

Let $\mathbf{x}_k$ be the state and $\mathbf{z}_k$ the measurement at the k-th frame. Starting from the propagation and measurement equations with additive noises ζ_{k-1} and η_k

$$\mathbf{x}_k = f(\mathbf{x}_{k-1}) + \zeta_{k-1} \tag{1}$$

$$\mathbf{z}_k = g(\mathbf{x}_k) + \eta_k \tag{2}$$

and its probabilistic notation via the Bayesian state tracking formulation [1], the belief probability density function (pdf) about the object state (posterior)

$$p(\mathbf{x}_k|\mathbf{z}_{1:k}) = \frac{p(\mathbf{z}_k|\mathbf{x}_k)p(\mathbf{x}_k|\mathbf{z}_{1:k-1})}{p(\mathbf{z}_k|\mathbf{z}_{1:k-1})} \tag{3}$$

is constructed as a fusion of $p(\mathbf{z}_k|\mathbf{x}_k)$ as the measurement expectation (likelihood) and $p(\mathbf{x}_k|\mathbf{z}_{1:k-1})$ as the predicted state pdf (prior) which evolves from the posterior pdf of the last time step by applying a transformation using a given prediction model for state transition $p(\mathbf{x}_k|\mathbf{x}_{k-1})$ according to

$$p(\mathbf{x}_k|\mathbf{z}_{1:k-1}) = \int p(\mathbf{x}_k|\mathbf{x}_{k-1})p(\mathbf{x}_{k-1}|\mathbf{z}_{1:k-1})d\mathbf{x}_{k-1}. \tag{4}$$

Here $p(\mathbf{z}_k|\mathbf{z}_{1:k-1})$ is a normalization constant with

$$p(\mathbf{z}_k|\mathbf{z}_{1:k-1}) = \int p(\mathbf{z}_k|\mathbf{x}_k)p(\mathbf{x}_k|\mathbf{z}_{1:k-1})d\mathbf{x}_k. \tag{5}$$

In our tracking framework, the likelihood $\mathbf{L} := p(\mathbf{z}_k|\mathbf{x}_k)$ is obtained by comparing the measurement result of the target object with the expected measurement result as stated in (2). From an input image $\mathbf{I}$ a set of cues $\mathbf{C}_i$ with $i = 1, \ldots, N$ is extracted, including e.g. RGB color, DoG edges, structure tensors. On the other hand template cues containing the tracked object inside are stored in $\mathbf{T}_i$ with $i = 1, \ldots, N$. In addition, a window $\mathbf{W}$ for weighting the target object in the templates cues $\mathbf{T}_i$ is given. The measurement $\mathbf{M}_i$ for the target object position is gained by correlation of $\mathbf{C}_i$ and $\mathbf{T}_i$ with window $\mathbf{W}$ by

$$\mathbf{M}_i = \mathbf{Corr2D}\left\{\mathbf{C}_i, \mathbf{T}_i, \mathbf{W}\right\}. \tag{6}$$

The object's expected measurement $\mathbf{S}_i$ is calculated by auto-correlating the template cues $\mathbf{T}_i$ according to

$$\mathbf{S}_i = \mathbf{Corr2D}\left\{\mathbf{T}_i, \mathbf{T}_i, \mathbf{W}\right\}. \tag{7}$$

The operations in (6) and (7) are accelerated by multiplication of $\mathbf{C}_i$ resp. $\mathbf{T}_i$ and $\mathbf{T}_i$ in the Fourier domain, weighted by $\mathbf{W}$. With the measurement $\mathbf{M}_i$ and the expected measurement $\mathbf{S}_i$, likelihood $\mathbf{L}_i$ is gained (assuming a normal distribution of measurement noise η_k with a variance of σ_η^2) by

$$\mathbf{L}_i(x,y) \sim \exp\left(-\frac{1}{2\sigma_{\eta_i}^2}\left\|\left(\mathbf{M}_i - \mathbf{A}_{x,y}\left(\mathbf{S}_i\right)\right) \odot \mathbf{A}_{x,y}\left(\mathbf{W}\right)\right\|^2\right), \tag{8}$$

with $\mathbf{A}_{x,y}$ as a translatory transformation operator to shift a block by (x, y) and $\odot$ as a pixel-wise multiplication of two blocks. Fusion of the likelihoods of all cues delivers an overall likelihood $\mathbf{L} = \mathbf{F}\left\{\mathbf{L}_1, \ldots, \mathbf{L}_N\right\}$.

The likelihood $\mathbf{L}$ is used to weight the prior pdf $p(\mathbf{x}_k|\mathbf{z}_{1:k-1})$, which is obtained according to formula (4), in the resampling phase of particle filtering. The estimation process of the posterior $p(\mathbf{x}_k|\mathbf{z}_{1:k})$ is evolved by a Sample Importance Resampling (SIR) Particle Filter ([1], [9]) where prior and posterior pdfs are approximately represented by 5000 particles in the six dimensional state space $\mathbf{x}$.

3 Multiple Prediction Models

In a Bayesian tracking framework like presented here, measurement is a supplementary information for correcting the guess coming from the motion prediction model. In the case of an inappropriate motion prediction model even a good likelihood coming from the measurement can not prevent a loss of the object. Since a single motion prediction model can never cope with all situations, it is beneficial to have multiple few-parameterized prediction models specialized for

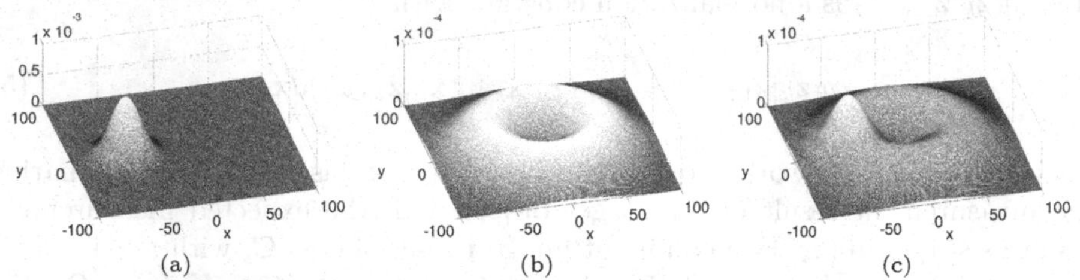

Fig. 1. Visualization of three different prediction models, projected to the x, y-plane. (a) visualizes the prior distribution of a linear prediction model. One can see the uni-directional motion from the origin and normal distribution due to noise. (b) Elastic rebound prediction model. It shows the omnidirectional characteristic of a rebound with no knowledge about the rebound direction and uncertainty of the rebound reflection factor. (c) visualizes a rebound prediction model with a preferred reflection direction.

different kinds of motion. In this case, each of them plays its strengths on current situations where others are unreliable. In this way the models complement one another.

In order to show the limitation of a single prediction model, we tested our tracking system in combination of a linear prediction model of the form

$$\begin{bmatrix} x_k \\ y_k \\ v_{x,k} \\ v_{y,k} \\ a_{x,k} \\ a_{y,k} \end{bmatrix} = \begin{bmatrix} 1 & 0 & \Delta T & 0 & 0 & 0 \\ 0 & 1 & 0 & \Delta T & 0 & 0 \\ 0 & 0 & 1 & 0 & \Delta T & 0 \\ 0 & 0 & 0 & 1 & 0 & \Delta T \\ 0 & 0 & 0 & 0 & 1 & 0 \\ 0 & 0 & 0 & 0 & 0 & 1 \end{bmatrix} \begin{bmatrix} x_{k-1} \\ y_{k-1} \\ v_{x,k-1} \\ v_{y,k-1} \\ a_{x,k-1} \\ a_{y,k-1} \end{bmatrix} + \begin{bmatrix} \zeta_{x_{k-1}} \\ \zeta_{y_{k-1}} \\ \zeta_{v_{x,k-1}} \\ \zeta_{v_{y,k-1}} \\ \zeta_{a_{x,k-1}} \\ \zeta_{a_{y,k-1}} \end{bmatrix} \tag{9}$$

with $\zeta_{\ldots,k-1} \sim N(0, \sigma^2_{\zeta_{\ldots}})$ as model noise (an illustration of the linear prediction model can be seen in figure 1(a)) using a sequence of a falling ball which rebounds on a can, as illustrated in figure 2(a). The tracking result plotted in figure 3 shows that the tracker loses the object after the rebound.

Since a linear prediction model has problems at the rebound, we used a second, non-linear prediction model

$$\begin{bmatrix} x_k \\ y_k \\ v_{x,k} \\ v_{y,k} \\ a_{x,k} \\ a_{y,k} \end{bmatrix} = \begin{bmatrix} v_{x,k-1} \cdot \Delta T + x_{k-1} + \zeta_{x,k-1} \\ v_{y,k-1} \cdot \Delta T + y_{k-1} + \zeta_{y,k-1} \\ \left(\sqrt{v^2_{x,k-1} + v^2_{y,k-1}} + \zeta_{r,k-1}\right) \cdot \cos(\xi_\varphi) \\ \left(\sqrt{v^2_{x,k-1} + v^2_{y,k-1}} + \zeta_{r,k-1}\right) \cdot \sin(\xi_\varphi) \\ a_{x,k-1} + \zeta_{a_{x,k-1}} \\ a_{y,k-1} + \zeta_{a_{x,k-1}} \end{bmatrix} \tag{10}$$

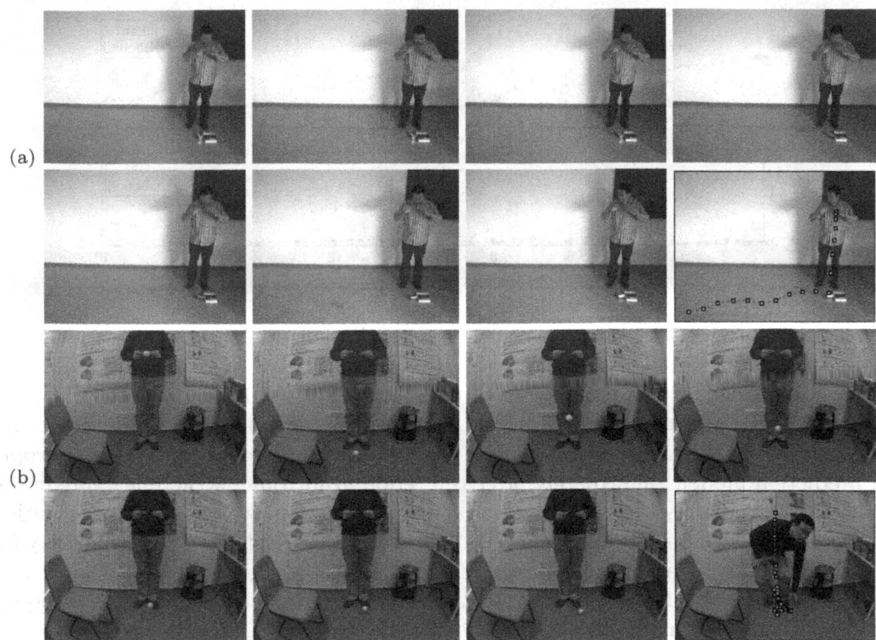

Fig. 2. This figure shows two real-world scenarios containing 18 and 39 frames with 400×300 pixel resolution, respectively. In the first scenario (a) a ball is falling on a can and rebounds to the left. A selection of the 18 frames is shown here to illustrate the rebound. The lower right image illustrates the complete trajectory of the ball. In the second scenario (b) a tennis ball is falling down to the floor and rebounds several times up and down. A selection of the 39 frames is shown in these figures. The lower right one contains the complete trajectory of the tennis ball.

with $\zeta_{\ldots,k-1} \sim N(0, \sigma^2_{\zeta_{\ldots}})$ and ξ_φ equally distributed in $[0, 2\pi[$. This is a noisy rebound prediction model (see figure 1(b)), that assumes that the object changes its direction arbitrarily while keeping its velocity approximately constant. Figure 3 shows the tracking result of our framework using a rebound model with a preferred direction (see in figure 1(c)) as a single prediction model, i.e. a mixture between (9) and (10). The reason for using a rebound model with a preferred direction is that a pure rebound model is obviously not suited for describing the linear phases of the motion with sufficient accuracy. Here, the object is tracked throughout the sequence, but the confidence is not as high as in the case of linear prediction before rebound, since the rebound model is more unselective.

At this point it seems straightforward to assume that a switching between both models, which corresponds to the confirmation-rejection-concept of tracking, is a good solution to overcome the rebound in the scenario and still to have high confidence for the posterior. The question is how to automatically find out when to switch between prediction models. For this purpose, in the following several quality measures for a prediction model are taken into consideration.

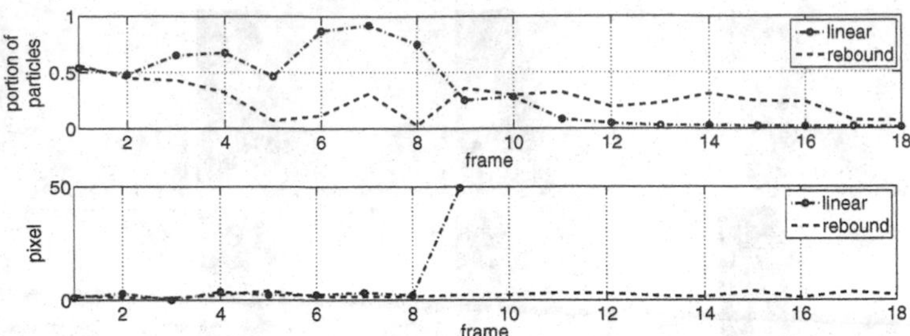

Fig. 3. Tracking results using a single linear prediction model vs. using a single rebound prediction model with a preferred direction, without switching between both prediction models, for the scenario shown in figure 2(a). The first plot shows the value of the highest posterior peak, the second one the distance of the peak to the ground truth position of the target object. Before the rebound the linear model is an appropriate prediction model. Immediately after the rebound in frame 9 the linear prediction model further predicts the object motion in same direction, whereas the target object rebounds on the can and turns to the left. So, the target object gets lost. Using only the rebound model with a preferred direction the target object is tracked over all frames (with a distance of 2.19px to ground truth in average), but the standard deviation of posterior is quite high (63.44px in average) which indicates a high uncertainty.

Highest posterior peak. The first quality measure for selecting prediction model is the value of the highest peak of the posterior. So, the prediction model $\hat{\imath}$ with the highest overall value of its posterior is chosen as the operative prediction model:

$$\hat{\imath} = \arg\max_i \hat{p}_i \qquad \text{with} \qquad \hat{p}_i = \max_{\mathbf{x}_k} p_i(\mathbf{x}_k|\mathbf{z}_{1:k}). \tag{11}$$

Looking at the posterior value of the highest peak plot in figure 3 it can be seen that the highest posterior peak value of the linear model decreases during rebound (frame 9), whereas the highest posterior peak value of the rebound model surpasses that of the linear model. Taking this as a switching criterion, the object can be tracked successfully over the entire sequence resulting in an overall higher posterior peak value as compared to the single prediction models. In figure 4, we show the respective contributions of the two prediction models (linear and pure rebound) and the posterior result gained by selection of the best prediction model at each time step.

Quotient of standard deviations of prior to posterior. A second quality measure is the ratio between the standard deviations of prior and posterior. A strong decrease from the standard deviation of prior to the standard deviation of posterior is an indication for a reliable likelihood that is consistent with the prediction. So, the model $\hat{\imath}$ with the highest quotient of standard deviation of prior to posterior is taken as the operative model:

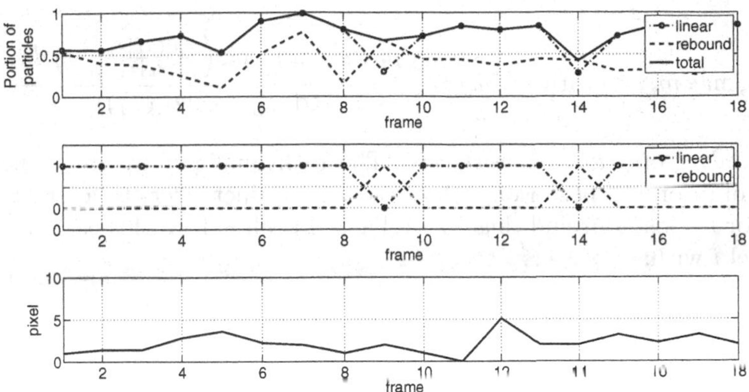

Fig. 4. Switching behavior between two prediction models using the value of the highest posterior peak as switching criterion, for scenario in figure 2(a). In the first plot the values of the highest posterior peaks of both participating models (linear and pure rebound) and that of the currently selected model are shown. In the second plot it is shown which prediction model was active (the one which has the greater value at the highest posterior peak). In the third plot the distance to the ground truth position is shown. An average distance to the ground truth position of 2.13px indicates that the object is never lost over the frames. With an average standard deviation of posterior of only 16.47px the confidence is quite high.

$$\hat{\imath} = \arg\max_i \hat{q}_i \qquad \text{with} \qquad \hat{q}_i = \frac{\text{stdev}(p_i(\mathbf{x}_k|\mathbf{z}_{1:k-1}))}{\text{stdev}(p_i(\mathbf{x}_k|\mathbf{z}_{1:k}))}. \tag{12}$$

Kullback-Leibler-divergence. The next quality measure is the Kullback-Leibler-divergence ([10]), which quantifies the change of entropy of two pdfs. A higher K-L value refers to a stronger decrease of entropy of prior to that of posterior due to a reliable likelihood which is consistent with the prediction. So, the model $\hat{\imath}$ with the highest K-L-divergence then becomes the operative model:

$$\hat{\imath} = \arg\max_i \hat{k}_i \qquad \text{with} \qquad \hat{k}_i = \int p_i(\mathbf{x}_k|\mathbf{z}_{1:k}) \cdot \log\left(\frac{p_i(\mathbf{x}_k|\mathbf{z}_{1:k})}{p_i(\mathbf{x}_k|\mathbf{z}_{1:k-1})}\right) d\mathbf{x}_k. \tag{13}$$

Modified Kullback-Leibler-divergence. A property of the K-L-divergence is that it only takes the change of prior to posterior into account, but not the fact that, on a reliable likelihood and a consistent prediction, it is easier for a prior with a higher standard deviation to get a larger change towards posterior. That means, under this circumstance, a model with a widely spread prior, e.g. a rebound model, gets a higher K-L-divergence more easily than a model with a more selective prior, e.g. a linear model. So a modified K-L-divergence weighted by the standard deviation of prior is taken as the next quality measure, in order to compensate this bias effect:

$$\hat{i} = \arg\max_i \hat{m}_i \qquad \text{with} \qquad \hat{m}_i = \frac{\int p_i(\mathbf{x}_k|\mathbf{z}_{1:k}) \cdot \log\left(\frac{p_i(\mathbf{x}_k|\mathbf{z}_{1:k})}{p_i(\mathbf{x}_k|\mathbf{z}_{1:k-1})}\right) d\mathbf{x}_k}{\mathrm{stdev}(p_i(\mathbf{x}_k|\mathbf{z}_{1:k-1}))}. \tag{14}$$

Scalar product of prior and posterior. The fifth quality measure is the scalar product of prior and posterior. A lower scalar product refers to a larger change from prior to posterior and thus to a reliable likelihood. In this case, we choose the model $\hat{i}$ with:

$$\hat{i} = \arg\max_i \hat{s}_i \qquad \text{with} \qquad \hat{s}_i = \frac{p_i(\mathbf{x}_k|\mathbf{z}_{1:k}) \cdot p_i(\mathbf{x}_k|\mathbf{z}_{1:k-1})}{||p_i(\mathbf{x}_k|\mathbf{z}_{1:k})|| \cdot ||p_i(\mathbf{x}_k|\mathbf{z}_{1:k-1})||}. \tag{15}$$

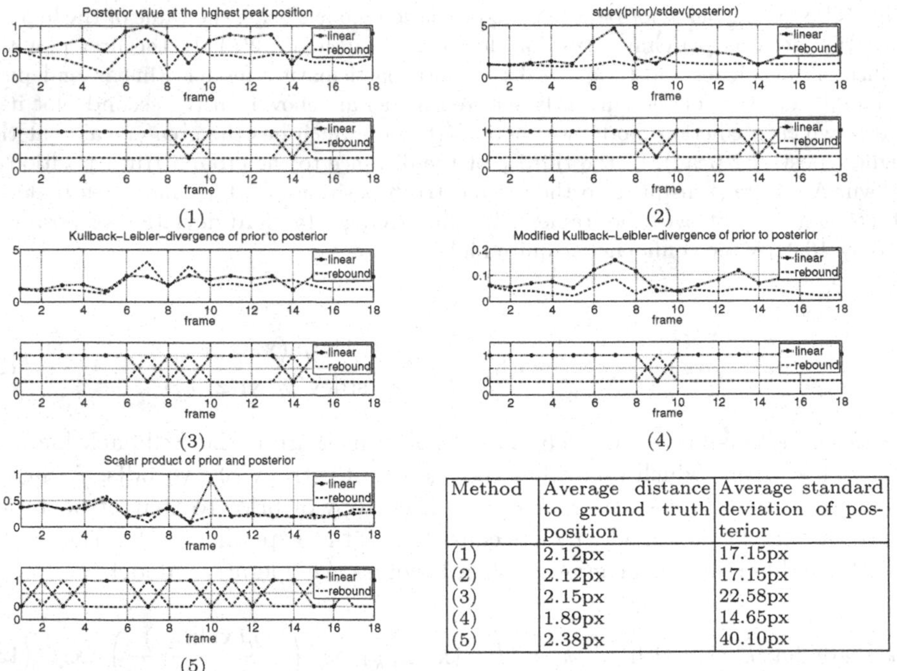

Method	Average distance to ground truth position	Average standard deviation of posterior
(1)	2.12px	17.15px
(2)	2.12px	17.15px
(3)	2.15px	22.58px
(4)	1.89px	14.65px
(5)	2.38px	40.10px

Fig. 5. Tracker evaluation results for scenario 1 (figure 2(a)) using different switching criteria for prediction models. For each of the five switching criteria its specific quality measures are shown for both models in the first plot and its switching behavior in the second plot. In the table, the average distance to ground truth position and the average standard deviation of the posterior of the methods are shown. This table reveals that the object is tracked successfully throughout the entire sequence. Methods 1, 2 and 4 exhibit the lowest standard deviation of posterior and appropriate points in time for switching (a big rebound occurs at frame 9 and a small rebound at frame 14).

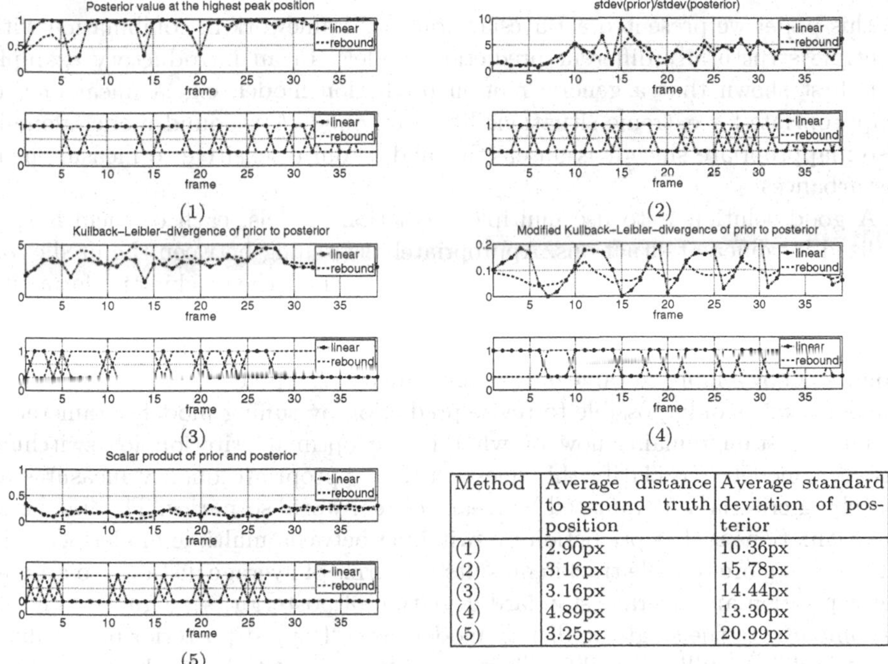

Method	Average distance to ground truth position	Average standard deviation of posterior
(1)	2.90px	10.36px
(2)	3.16px	15.78px
(3)	3.16px	14.44px
(4)	4.89px	13.30px
(5)	3.25px	20.99px

Fig. 6. This figure shows the tracker evaluation results for the scenario 2 (figure 2(b)) using different switching criteria for prediction models. For each of the five switching criteria its specific quality measures are shown for both models in the first plot and its switching behavior in the second plot. In the table the average distance to ground truth position and the average standard deviation of posterior of the methods are shown. This table reveals that the object is tracked successfully throughout the entire sequence. Methods 1 and 4 exhibit the lowest standard deviation of posterior and appropriate points in time for switching (big rebounds occur at frames 7, 14 and 20 and small rebounds at frame 24, 27 and 29).

4 Evaluation

We have evaluated the five methods for switching between prediction models by means of two scenarios. One is the scenario with one big rebound shown in figure 2(a). Another one with a series of rebounds is shown in figure 2(b). The results of the comparative evaluations can be seen in figures 5 and 6. In no case in the evaluations, the tracker loses the object. From all the five switching methods the "highest posterior peack value" and "modified Kullback-Leibler-divergence" turn out to be the best ones, since they switch at the most appropriate points in time and provide the lowest standard deviation of posterior.

5 Conclusion

In this paper we presented a Bayesian tracking framework in combination with multiple structurally different prediction models. In an introductory example it is first shown that a generic motion prediction model, e.g. a linear one, is inappropriate for extreme situations like a rebound. A rebound model alone is also inappropriate since it is unselective and so quite sensitive to measurement disturbances.

A good solution is to use multiple prediction models, each of them is specialized for different situations. Appropriately switching between the prediction models increases the overall predictive capability which the tracking performance benefits from. An essential gain of this concept consists in a further possibility for measurement to revise prediction by completely replacing an unsuitable prediction model by a more suitable one, whereas on a single prediction model tracking framework it is only possible to revise prediction by tuning model parameters.

The question remains how or what is the optimal criterion for switching between models. To clarify this question five appropriate quality measures as switching criteria are evaluated by means of real-world scenarios. The finding of the evaluations is that prediction by switching between multiple models leads in all cases to more reliable tracking results (in terms of average distance to ground truth position and average standard deviation of posterior, see figures 5 and 6) as compared to the single prediction model case. "Highest posterior peak value" and "modified Kullback-Leibler-divergence" turned out to be the best switching criteria.

References

1. Arulampalam, S., Maskell, S., Gordon, N.: A Tutorial on Particle Filters for Online Nonlinear/Non-Gaussian Bayesian Tracking. IEEE Transactions on Signal Processing 50, 174–188 (2002)
2. Triesch, J., v.d. Malsburg, C.: Democratic Integration: Self-Organized Integration of Adaptive Cues. Neural Computation 13(9), 2049–2074 (2001)
3. Spengler, M., Schiele, B.: Towards Robust Multi-Cue Integration for Visual Tracking. Machine Vision and Applications 14(1), 50–58 (2003)
4. Zhong, Y., Jain, A.K.: Object Tracking using Deformable Templates. IEEE Transactions on Pattern Analysis and Machine Intelligence 22, 544–549 (2000)
5. Comaniciu, D., Ramesh, V., Meer, P.: Kernel-Based Object Tracking. IEEE Transactions on Pattern Analysis and Machine Intelligence 25, 564–577 (2003)
6. Isard, M., Blake, A.: CONDENSATION - Conditional Density Propagation for Visual Tracking. International Journal of Computer Vision 29, 5–28 (1998)
7. Li, X.R., Jilkov, V.P., Ru, J.: Multiple-Model Estimation with Variable Structure - Part VI: Expected-Mode Augmentation. IEEE Transactions on Aerospace and Electronic Systems 41(3), 853–867 (2005)
8. Bar-Shalom, Y.: Multitarget-Multisensor Tracking: Applications and Advances, vol. III. Artech House, Norwood (2000)
9. Doucet, A., Godsill, S., Andrieu, C.: On Sequential Monte Carlo Methods for Bayesian Filtering. Statistics and Computing 10(3), 197–208 (2000)
10. Kullback, S., Leibler, R.A.: On Information and Sufficiency. Annals of Mathematical Statistics 22, 79–86 (1951)

Sliding Mode Control for Trajectory Tracking Problem - Performance Evaluation*

Razvan Solea and Daniela Cernega

Control Systems and Industrial Informatics Department,
Computer Science Faculty, "Dunarea de Jos" University of Galati,
Domneasca 111, 80021, Galati, Romania
{razvan.solea,daniela.cernega}@ugal.ro
http://www.fsc.ugal.ro

Abstract. The trajectory tracking control problem for the wheeled mobile robot is solved using the sliding mode control. The wheeled mobile robot is a nonlinear system. In this paper four control laws are modeled and the system performances are investigated. The sliding mode control laws for the trajectory tracking problem are simulated and then implemented on the PatrolBot Robot. The performances are analyzed in order to establish some rules. The analysis conclusions are based on the simulation results and on the real time implementation of the control laws on the PatrolBot Robot.

Keywords: Mobile Robots, Nonlinear Control, Sliding Mode Control, Trajectory Tracking.

1 Introduction

In this paper trajectory tracking problem is solved using four Sliding Mode Control (SMC) laws. Sliding Mode control was chose because is known to possess merits such as the invariance to parametric uncertainties as well as the capacity to reject disturbances. However, this type of control suffers from the chattering phenomenon which is due to high frequency switching over discontinuity of the control signal ([1]).

Sliding Mode is also known to posses merits such as the invariance to parametric uncertainties. Dynamic characteristics of the reaching mode are very important, and this type of control suffers from the chattering phenomenon which is due to high frequency switching over discontinuity of the control signal. This aspect of the sliding mode control is further investigated in this paper using four different control laws.

In order to handle the chattering problem two approaches are widely referred in literature. The first one is called the continuation method because the discontinuous relay type actuator is replaced by a high - gain device with saturation ([6]). Although this method eliminates the high frequency chattering it

* This work was supported by the Romanian High Education Scientific Research National Council, under project IDEI-506.

C. Alippi et al. (Eds.): ICANN 2009, Part II, LNCS 5769, pp. 865–874, 2009.

also destroys the sliding mode. In addition, the resulting physical system, often exhibits low frequency oscillations due to unmodeled dynamics. The second approach deals directly with the reaching process since chattering is caused by the nonideal reaching at the end of the reaching phase. This approach establishes the reaching mode characteristics by the use of a reaching model. The resulting method is called the reaching law method ([3]).

In this paper four different reaching laws and reaching modes are investigated. The performances of the four laws in controlling the amplitude of the chattering are compared in order to establish an on-line parameters adjusting procedure.

This paper is organized as follows: Section II is dedicated to the presentation of the reaching mode and switching functions for nonlinear systems. The three stages considered for the sliding mode used for trajectory tracking problem are presented. The four reaching laws models are also listed. In Section III the control problem for the wheeled mobile robots is presented using the kinematic model. The trajectory-tracking problem is treated using the sliding mode control, and the sliding manifolds equations for the four reaching laws in Section II are presented. The commands for each controller are also obtained. Section IV is dedicated to experimental results of the implementation of the four reaching laws presented in Section II. Section V presents the conclusions of the implementation on PatrolBot Robot and future work directions.

2 Reaching Mode and Control Law Design

[3] proposed a reaching law which directly specifies the dynamics of the switching surface by the differential equation

$$\dot{s} = -Q \cdot sgn(s) - P \cdot h(s) \tag{1}$$

where

$$Q = diag\left[q_1, q_2, ..., q_n\right], \quad P = diag\left[p_1, p_2, ..., p_n\right], q_i, p_i > 0, \; i = 1, 2, ..., n$$

$$sgn(s) = \left[sgn(s_1), sgn(s_2), ..., sgn(s_n)\right]^T, \quad h(s) = \left[h_1(s_1), h_2(s_2), ..., h_n(s_n)\right]^T$$

$$s_i \cdot h_i(s) > 0, \; h_i(0) = 0.$$

2.1 Reaching Laws

The four practical cases of the equation (1) used in this paper are given below.

A. Constant rate reaching ([3])

$$\dot{s} = -Q \cdot sgn(s) \tag{2}$$

This law forces the switching variable $s(x)$ to reach the switching manifold S at a constant rate $|\dot{s}_i| = -q_i$. The merit of this reaching law is its simplicity. But, as we know, if q_i is too small, the reaching time will be too long. On the other hand, a too large q_i will cause severe chattering.

B. Constant plus proportional rate reaching ([3])

$$\dot{s} = -Q \cdot sgn(s) - P \cdot s \tag{3}$$

Clearly, by adding the proportional rate term $-P \cdot s$, the state is forced to approach the switching manifolds faster when s is large. It can be shown that the reaching time for x to move from an initial state x_0 to the switching manifold s_i is finite, and is given by:

$$T_i = \frac{1}{p_i} \cdot ln \frac{p_i \cdot |s_i| + q_i}{q_i} \tag{4}$$

C. Power rate reaching ([3])

$$\dot{s}_i = -p_i \cdot |s_i|^{\alpha} \cdot sgn(s_i), \ 0 < \alpha < 1, \ i = 1, ..., m \tag{5}$$

This reaching law increases the reaching speed when the state is far away from the switching manifold, but reduces the rate when the state is near the manifold. The result is a fast reaching and low chattering reaching mode. Integrating (5) from $s_i = s_{i0}$ to $s_i = 0$ yields

$$T_i = \frac{|s_i(0)|^{1-\alpha}}{(1-\alpha) \cdot p_i} \tag{6}$$

showing that the reaching time T_i, is finite. Thus power rate reaching law gives a finite reaching time. In addition, because of the absence of the $-Q \cdot sgn(s)$ term on the right-hand side of (5), this reaching law eliminates the chattering.

D. Speed control rate reaching ([7])

$$\begin{aligned} &\dot{s}_i = -p_i \cdot exp\left(\alpha \cdot |s_i|\right) \cdot sgn(s_i), \\ &p_i > 0, \ \alpha > 0, \ i = 1, ..., m \end{aligned} \tag{7}$$

and the reaching time T_i becomes:

$$T_i = \frac{1}{\alpha \cdot p_i} \cdot \left(1 - exp\left(-\alpha \cdot |s_i(0)|\right)\right) \tag{8}$$

The four reaching laws presented above are used in the implementation on the PatrolBot Robot system in order to analyze their performances.

3 Control of Wheeled Mobile Robots

The application of SMC strategies in nonlinear systems has received considerable attention in recent years ([8], [9], [10], [11]). A well-studied example of a non-holonomic system is a WMR that is subject to the *rolling without slipping* constraint.

In trajectory tracking is an objective to control the non-holonomic WMR to follow a desired trajectory, with a given orientation relatively to the path tangent,

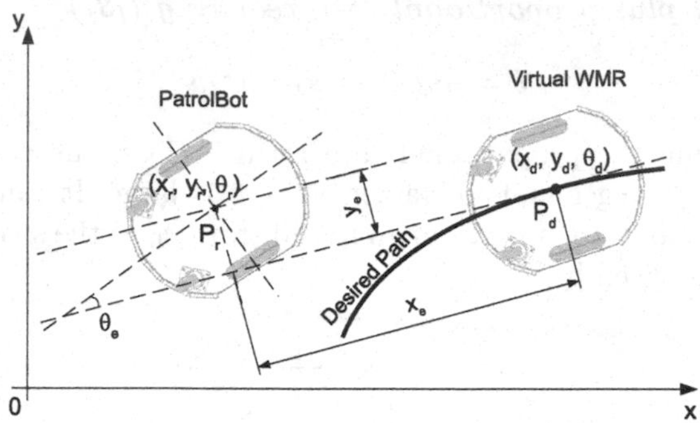

Fig. 1. Lateral, longitudinal and orientation errors (trajectory-tracking)

even when disturbances exist. In the case of trajectory-tracking the path is to be followed under time constraints. The path has an associated velocity profile, with each point of the trajectory embedding spatiotemporal information that is to be satisfied by the WMR along the path. Trajectory tracking is formulated as having the WMR following a virtual target WMR which is assumed to move exactly along the path with specified velocity profile.

3.1 Kinematic Model of a WMR

Figure 1 presents a WMR with two diametrically opposed drive wheels (radius R) and free-wheeling castors (not considered in the kinematic models). P_r is the origin of the robot coordinates system. $2L$ is the length of the axis between the drive wheels. ω_R and ω_L are the angular velocities of the right and left wheels. Let the pose of the mobile robot be defined by the vector $q_r = [x_r, y_r, \theta_r]^T$, where $[x_r, y_r]^T$ denotes the robot position on the plane and θ_r the heading angle with respect to the x-axis. In addition, v_r denotes the linear velocity of the robot, and ω_r the angular velocity around the vertical axis. For a unicycle WMR rolling on a horizontal plane without slipping, the kinematic model can be expressed by:

$$\begin{bmatrix} \dot{x}_r \\ \dot{y}_r \\ \dot{\theta}_r \end{bmatrix} = \begin{bmatrix} cos\theta_r & 0 \\ sin\theta_r & 0 \\ 0 & 1 \end{bmatrix} \cdot \begin{bmatrix} v_r \\ \omega_r \end{bmatrix} \tag{9}$$

which represents a nonlinear system.

Controllability of the system (9) is easily checked using the Lie algebra rank condition for nonlinear systems. However, the Taylor linearization of the system about the origin is not controllable, thus excluding the application of classical linear design approaches.

3.2 Trajectory-Tracking

The first case to be considered is the trajectory-tracking control. Without loss of generality, it can be assumed that the desired traj. $q_d(t) = [x_d(t), y_d(t), \theta_d(t)]^T$

is generated by a virtual unicycle mobile robot (see Fig. 1). The kinematic relationship between the virtual configuration $q_d(t)$ and the corresponding desired velocity inputs $[v_d(t), \omega_d(t)]^T$ is analog with (9):

$$\begin{bmatrix} \dot{x}_d \\ \dot{y}_d \\ \dot{\theta}_d \end{bmatrix} = \begin{bmatrix} cos\theta_d & 0 \\ sin\theta_d & 0 \\ 0 & 1 \end{bmatrix} \cdot \begin{bmatrix} v_d \\ \omega_d \end{bmatrix} \tag{10}$$

When a real robot is controlled to move on a desired path it exhibits some tracking error. This tracking error, expressed in terms of the robot coordinate system, as shown in Fig. 1, is given by

$$\begin{bmatrix} x_e \\ y_e \\ \theta_e \end{bmatrix} = \begin{bmatrix} cos\theta_d & sin\theta_d & 0 \\ -sin\theta_d & cos\theta_d & 0 \\ 0 & 0 & 1 \end{bmatrix} \cdot \begin{bmatrix} x_r - x_d \\ y_r - y_d \\ \theta_r - \theta_d \end{bmatrix} \tag{11}$$

Consequently one gets the error dynamics for trajectory tracking as

$$\begin{cases} \dot{x}_e = -v_d + v_r \cdot cos\theta_e + \omega_d \cdot y_e \\ \dot{y}_e = v_r \cdot sin\theta_e - \omega_d \cdot x_e \\ \dot{\theta}_e = \omega_r - \omega_d \end{cases} \tag{12}$$

3.3 Sliding-Mode Trajectory-Tracking Control

Uncertainties which exist in real mobile robot applications degrade the control performance significantly, and accordingly, need to be compensated. In this section, is proposed a SM-TT controller, in Cartesian space, where trajectory tracking is achieved even in the presence of large initial pose errors and disturbances.

Let us define the sliding surface $s = [s_1\ s_2]^T$ as

$$\begin{aligned} s_1 &= \dot{x}_e + k_1 \cdot x_e, \\ s_2 &= \dot{y}_e + k_2 \cdot y_e + k_0 \cdot sgn(y_e) \cdot \theta_e. \end{aligned} \tag{13}$$

where k_0, k_1, k_2 are positive constant parameters, x_e, y_e and θ_e are the trajectory tracking errors defined in (11).

If s_1 converges to zero, trivially x_e converges to zero. If s_2 converges to zero, in steady-state it becomes $\dot{y}_e = -k_2 \cdot y_e - k_0 \cdot sgn(y_e) \cdot \theta_e$. For $y_e < 0 \Rightarrow \dot{y}_e > 0$ if only if $k_0 < k_2 \cdot |y_e| \,/\, |\theta_e|$. For $y_e > 0 \Rightarrow \dot{y}_e < 0$ if only if $k_0 < k_2 \cdot |y_e| \,/\, |\theta_e|$. Finally, it can be known from s_2 that convergence of y_e and $\dot{y}_e$ leads to convergence of θ_e to zero.

From the time derivative of (13) and using the reaching laws defined in (2), (3), (5) and (7) yields:

$$\begin{aligned} \dot{s}_{1A} &= \ddot{x}_e + k_1 \cdot \dot{x}_e = -q_1 \cdot sgn(s_1) \\ \dot{s}_{2A} &= \ddot{y}_e + k_2 \cdot \dot{y}_e + k_0 \cdot sgn(y_e) \cdot \dot{\theta}_e = -q_2 \cdot sgn(s_2) \end{aligned} \tag{14}$$

$$\begin{aligned} \dot{s}_{1B} &= \ddot{x}_e + k_1 \cdot \dot{x}_e = -q_1 \cdot sgn(s_1) - p_1 \cdot s_1 \\ \dot{s}_{2B} &= \ddot{y}_e + k_2 \cdot \dot{y}_e + k_0 \cdot sgn(y_e) \cdot \dot{\theta}_e = -q_2 \cdot sgn(s_2) - p_2 \cdot s_2 \end{aligned} \tag{15}$$

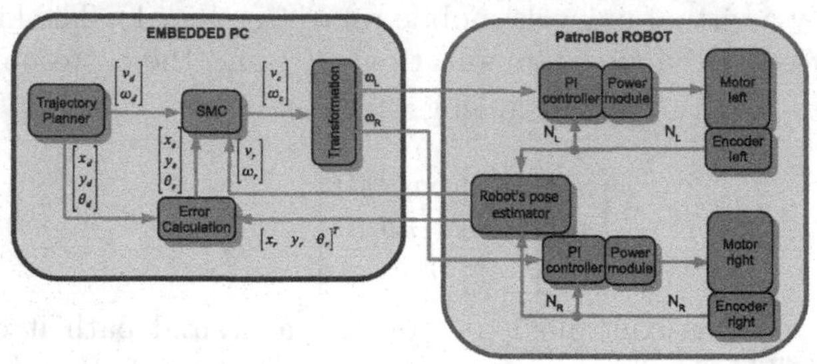

Fig. 2. Sliding-Mode Trajectory-Tracking control architecture

$$\begin{aligned} \dot{s}_{1C} &= \ddot{x}_e + k_1 \cdot \dot{x}_e = -p_1 \cdot |s_1|^{\alpha} \cdot sgn(s_1) \\ \dot{s}_{2C} &= \ddot{y}_e + k_2 \cdot \dot{y}_e + k_0 \cdot sgn(y_e) \cdot \dot{\theta}_e = -p_2 \cdot |s_2|^{\alpha} \cdot sgn(s_2) \end{aligned} \tag{16}$$

$$\begin{aligned} \dot{s}_{1D} &= \ddot{x}_e + k_1 \cdot \dot{x}_e = -p_1 \cdot exp(\alpha \cdot |s_1|) \cdot sgn(s_1) \\ \dot{s}_{2D} &= \ddot{y}_e + k_2 \cdot \dot{y}_e + k_0 \cdot sgn(y_e) \cdot \dot{\theta}_e = -p_2 \cdot exp(\alpha \cdot |s_2|) \cdot sgn(s_2) \end{aligned} \tag{17}$$

From (11), (12) and (14)-(17), and after some mathematical manipulation, we get the output commands of the sliding-mode trajectory-tracking controller:

$$\begin{aligned} \dot{v}_{cA} &= \frac{(-q_1 \cdot sgn(s_1) - k_1 \cdot \dot{x}_e - y_e \cdot \dot{\omega}_d - \dot{y}_e \cdot \omega_d + v_r \cdot \dot{\theta}_e \cdot sin\theta_e + \dot{v}_d)}{cos\theta_e} \\ \omega_{cA} &= \frac{(-k_2 \cdot \dot{y}_e - q_2 \cdot sgn(s_2) - \dot{v}_r \cdot sin\theta_e + x_e \cdot \dot{\omega}_d + \dot{x}_e \cdot \omega_d)}{v_r \cdot cos\theta_e + k_0 \cdot sgn(y_e)} + \omega_d \end{aligned} \tag{18}$$

$$\begin{aligned} \dot{v}_{cB} &= \frac{(-p_1 \cdot s_1 - q_1 \cdot sgn(s_1) - k_1 \cdot \dot{x}_e - y_e \cdot \dot{\omega}_d - \dot{y}_e \cdot \omega_d + v_r \cdot \dot{\theta}_e \cdot sin\theta_e + \dot{v}_d)}{cos\theta_e} \\ \omega_{cB} &= \frac{(-p_2 \cdot s_2 - k_2 \cdot \dot{y}_e - q_2 \cdot sgn(s_2) - \dot{v}_r \cdot sin\theta_e + x_e \cdot \dot{\omega}_d + \dot{x}_e \cdot \omega_d)}{v_r \cdot cos\theta_e + k_0 \cdot sgn(y_e)} + \omega_d \end{aligned} \tag{19}$$

$$\begin{aligned} \dot{v}_{cC} &= \frac{(-p_1 \cdot |s_1|^{\alpha} \cdot sgn(s_1) - k_1 \cdot \dot{x}_e - y_e \cdot \dot{\omega}_d - \dot{y}_e \cdot \omega_d + v_r \cdot \dot{\theta}_e \cdot sin\theta_e + \dot{v}_d)}{cos\theta_e} \\ \omega_{cC} &= \frac{(x_e \cdot \dot{\omega}_d + \dot{x}_e \cdot \omega_d - p_2 \cdot |s_2|^{\alpha} \cdot sgn(s_2) - k_2 \cdot \dot{y}_e - \dot{v}_r \cdot sin\theta_e)}{v_r \cdot cos\theta_e + k_0 \cdot sgn(y_e)} + \omega_d \end{aligned} \tag{20}$$

$$\begin{aligned} \dot{v}_{cD} &= \frac{(-p_1 \cdot exp(\alpha \cdot |s_1|) \cdot sgn(s_1) - k_1 \cdot \dot{x}_e - y_e \cdot \dot{\omega}_d - \dot{y}_e \cdot \omega_d + v_r \cdot \dot{\theta}_e \cdot sin\theta_e + \dot{v}_d)}{cos\theta_e} \\ \omega_{cD} &= \frac{(x_e \cdot \dot{\omega}_d + \dot{x}_e \cdot \omega_d - p_2 \cdot exp(\alpha \cdot |s_2|) \cdot sgn(s_2) - k_2 \cdot \dot{y}_e - \dot{v}_r \cdot sin\theta_e)}{v_r \cdot cos\theta_e + k_0 \cdot sgn(y_e)} + \omega_d \end{aligned} \tag{21}$$

Let us define $V = \frac{1}{2} \cdot s^T \cdot s$ as a Lyapunov function candidate, therefore its time derivative is $\dot{V} = s_1 \cdot \dot{s}_1 + s_2 \cdot \dot{s}_2 = s_1 \cdot (-p_1 \cdot s_1 - q_1 \cdot sgn(s_1)) + s_2 \cdot (-p_2 \cdot s_2 - q_2 \cdot sgn(s_2)) = -s^T \cdot p \cdot s - q_1 \cdot |s_1| - q_2 \cdot |s_2|$

For $\dot{V}$ to be negative semi-definite, it is sufficient to choose q_i and p_i such that $q_i, p_i > 0$.

The *signum* functions in the control laws were replaced by *saturation* functions, to reduce the chattering phenomenon ([2], [12]).

4 Experimental Results

In this section, experimental results of the proposed method are presented.

To show the effectiveness of the proposed sliding mode control law numerically, real experiments were carried out on the trajectory-tracking problem of a nonholonomic wheeled mobile robot. The parameters of sliding modes were held constant during the experiments: $k_1 = 0.75$, $k_2 = 3.75$, and $k_0 = 2.5$; and the desired trajectory is given by $v_d = 0.4$ $[m/s]$, $\omega_d = 0$ $[rad/s]$.

The robot has two-level control architecture (see Fig. 2). High-level control algorithms (including desired motion generation) are written in C++ and run with a sampling time of Ts = 100 ms on a embedded PC, which also provides a user interface with real-time visualization and a simulation environment. Wheel velocity commands, $\omega_R = \frac{v_c + L \cdot \omega_c}{R}$, $\omega_L = \frac{v_c - L \cdot \omega_c}{R}$ are sent to the PI controllers, and encoder measures N_R and N_L are received in the robots pose estimator for odometric computations.

The real-time experiments are carried out on PatrolBot, a general purpose mobile robot acquired from MobileRobots Inc (see Fig. 3).

4.1 Mobile Robot Setup

PatrolBot is a programmable autonomous general purpose Service robot rover built by MobileRobots Inc.

Technical Specifications PatrolBot has a 59cm x 48cm x 38cm, CNC aluminum body. Its 19 *cm* diameter tires handle nearly any indoor surface. The two motor shafts hold 1000-tick encoders. This differential drive platform is holonomic so it can turn in place. Moving wheels on one side only, it forms a circle of 29 *cm* radius. The robot is equipped with 1.6 GHz Intel Pentium processor and 500 MB of RAM.

Software Specifications A small proprietary μARCS transfers sonar readings, motor encoder information and other I/O via packets from the micro controller

Fig. 3. The experimental mobile robot - PatrolBot

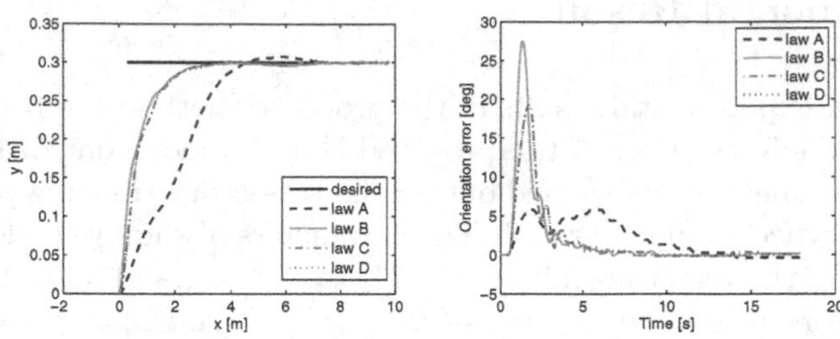

Fig. 4. Experimental SM-TT control starting from an initial error state ($x_e(0) = -0.3$, $y_e(0) = -0.3$, $\theta_e(0) = 0$)

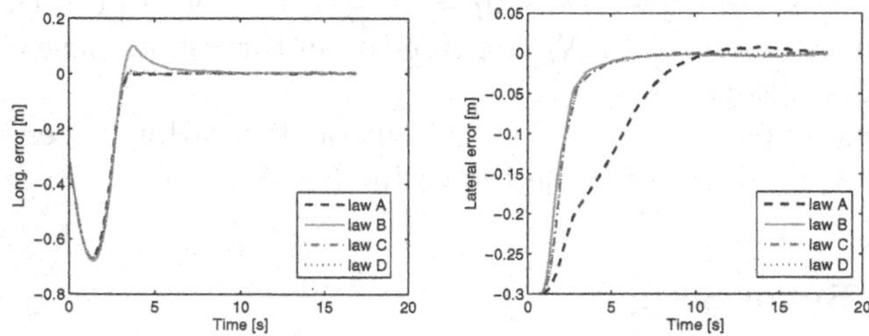

Fig. 5. Longitudinal and lateral errors for experimental SM-TT control

server to the PC client and returns control commands. PatrolBot can be operated from the client or users can design their own programs under Linux or under WIN32 using C/C++ compiler. ARIA and ARNL software supply library functions to handle navigation, path planning, obstacle avoidance and many other robotic tasks.

4.2 Real-Time Experiments Results

The real-time experiments were made for all types of reaching laws presented in (2), (3), (5) and (7).

In Table 1 are represented 36 experiments using sliding-mode trajectory-tracking controller for PatrolBot robot. Three experimental trials were executed for each parameters of reaching low. The table shows the maximum (Max) and root mean square (RMS) of errors (longitudinal - x_e, lateral - y_e and orientation - θ_e). Root mean square error is an old, proven measure of control and quality. RMS can be expressed as $RMS = \left[\frac{1}{N}\sum x^2(i)\right]^{\frac{1}{2}}$.

In order to compare all the four reaching laws there was analyzed the real-time implementation on the PatrolBot Robot. All the 36 real time experiments were realized on a single mobile robot with the same initial error ($x_e = -0.3$ m, $y_e = -0.3$ m, $\theta_e = 0$).

Table 1. Experimental Results

Q, P, α	var.	x_e Max $[m]$	x_e RMS	y_e Max $[m]$	y_e RMS	θ_e Max $[deg]$	θ_e RMS
Reaching law A.							
	$q_1 = 0.05$	0.6810	0.3762	0.3000	0.1137	6.8520	0.0520
$q_2 = 0.5$	$q_1 = 0.50$	0.6670	0.1823	0.3000	0.1115	5.7978	0.0454
	$q_1 = 0.95$	0.6670	0.3081	0.3000	0.1126	17.3056	0.0827
	$q_2 = 0.05$	0.6670	0.1820	0.3000	0.2495	1.0542	0.0124
$q_1 = 0.5$	$q_2 = 1.50$	0.6790	0.1880	0.3000	0.0792	27.4962	0.0837
	$q_2 = 2.50$	0.6730	0.1866	0.3000	0.0826	25.4754	0.1063
Reaching law B.							
$q_2 = 0.5,$	$q_1 = 0.01$	0.7190	0.2031	0.3000	0.0816	27.4962	0.0836
$p_1 = 0.75, p_2 = 1.75$	$q_1 = 0.05$	0.6790	0.1880	0.3000	0.0792	27.4962	0.0837
	$q_1 = 0.40$	0.6790	0.2966	0.3000	0.0783	27.7598	0.1037
$q_1 = 0.05,$	$q_2 = 0.05$	0.6780	0.1875	0.3000	0.0798	25.8272	0.0808
$p_1 = 0.75, p_2 = 1.75$	$q_2 = 0.50$	0.6790	0.1880	0.3000	0.0792	27.4962	0.0837
	$q_2 = 2.00$	0.6860	0.1944	0.3000	0.0854	34.8754	0.1942
$q_1 = 0.05,$	$p_1 = 0.05$	0.6790	0.3133	0.3000	0.0819	27.0568	0.0877
$q_2 = 0.5, p_2 = 1.75$	$p_1 = 0.75$	0.6790	0.1880	0.3000	0.0792	27.4962	0.0837
	$p_1 = 1.00$	0.6790	0.2020	0.3000	0.0790	27.8475	0.0841
$q_1 = 0.05,$	$p_2 = 0.05$	0.6670	0.1828	0.3000	0.1047	6.8520	0.0457
$q_2 = 0.5, p_1 = 0.75$	$p_2 = 1.75$	0.6790	0.1880	0.3000	0.0792	27.4962	0.0837
	$p_2 = 2.50$	0.6860	0.1930	0.3000	0.0796	34.8754	0.1310
Reaching law C.							
	$p_1 = 0.05$	0.6840	0.3761	0.3000	0.0880	19.6777	0.0794
$p_2 = 1.75, \alpha = 0.75$	$p_1 = 0.50$	0.6690	0.1851	0.3000	0.0829	19.3264	0.0678
	$p_1 = 0.90$	0.6690	0.2961	0.3000	0.0822	20.8196	0.0859
	$p_2 = 0.05$	0.6670	0.1820	0.3000	0.2500	1.0542	0.0122
$p_1 = 0.5, \alpha = 0.75$	$p_2 = 1.75$	0.6690	0.1851	0.3000	0.0829	19.3264	0.0678
	$p_2 = 3.00$	0.6750	0.1882	0.3000	0.0846	28.8140	0.1405
	$\alpha = 0.05$	0.6690	0.1850	0.3000	0.0829	19.3264	0.0678
$p_1 = 0.5, p_2 = 1.75$	$\alpha = 0.50$	0.6690	0.1850	0.3000	0.0829	19.3264	0.0679
	$\alpha = 0.95$	0.6690	0.1851	0.3000	0.0829	19.3264	0.0678
Reaching law D.							
	$p_1 = 0.05$	0.6840	0.3759	0.3000	0.0880	19.6777	0.0795
$p_2 = 1.75, \alpha = 0.75$	$p_1 = 0.50$	0.6690	0.1850	0.3000	0.0829	19.3264	0.0678
	$p_1 = 0.90$	0.6690	0.2961	0.3000	0.0822	20.8196	0.0860
	$p_2 = 0.05$	0.6670	0.1820	0.3000	0.2498	1.0542	0.0120
$p_1 = 0.5, \alpha = 0.75$	$p_2 = 1.75$	0.6690	0.1850	0.3000	0.0829	19.3264	0.0678
	$p_2 = 3.00$	0.6750	0.1882	0.3000	0.0852	28.8140	0.1423
	$\alpha = 0.05$	0.6690	0.1851	0.3000	0.0829	19.3264	0.0678
$p_1 = 0.5, p_2 = 1.75$	$\alpha = 0.50$	0.6690	0.1851	0.3000	0.0829	19.3264	0.0679
	$\alpha = 0.95$	0.6690	0.1851	0.3000	0.0829	19.3264	0.0681

In Figures 4 and 5 the experimental results for the most favorable cases are presented. These results offer the opportunity to distinguish the performances of the four analyzed reaching laws.

It is easy to observe in figure that the most unfavorable case is the reaching law A. The other laws, cases B, C, and D have similar characteristics with small differences.

In Table 1 one can observe the performances for laws C and D are equal. The differences between law B and laws C and D (which are identical) can be observed in Figs. 4 and 5. The same observation can be also extracted from Table 1, where one can observe differences between RMS of x_e and θ_e (in case of reaching law B these values are smaller than for cases C and D).

5 Conclusions

The paper was focused on the performances analysis of the four laws presented in Section II. This performance analysis is based on real-time implementation on PatrolBot Robot. All the experimental results have been presented in a table where the position errors and their mean root square were considered.

Analyzing the performances of the four laws it is easy to see that the most adequate laws for the trajectory tracking problem of the robot are laws C and D (equations (5) and (7)). The most unfavorable of the four laws is law A having the longest reaching time.

References

1. Perruquetti, W., Barbot, J.P.: Sliding Mode Control in Engineering. Marcel Dekker, New York (2002)
2. Slotine, J.J.E., Li, W.P.: Applied Nonliner Control. Prentice Hall, New Jersey (1991)
3. Gao, W., Hung, J.C.: Variable Structure Control of Nonlinear Systems: A New Approach. IEEE Transactions on Industrial Electronics 40, 45–55 (1993)
4. Utkin, V.I., Young, K.K.G.: Methods for Constructing Discontinuity Planes in Multidimensional Variable Structure Systems. Automat. Remote Control 39, 1466–1470 (1978)
5. Dorling, C.W., Zinober, A.Z.: Two Approaches to Hyperplane Design in Multivariable Variable Structure Control Systems. Int Journal Control 44, 65–82 (1986)
6. Burton, A.Z., Zinober, A.Z.: Continuous Approximation of Variable Structure Control. International Journal of Systems and Science 17, 875–885 (1989)
7. Loh, A.M., Yeung, L.F.: Chattering Reduction in Sliding Mode Control: An Improvement for Nonlinear Systems. WSEAS Transaction on Circuits and Systems 10, 2090–2098 (2004)
8. Chwa, D.: Sliding-mode Tracking Control of Nonholonomic Wheeled Mobile Robots in Polar Coordinates. IEEE Transactions on Control Systems Technology 12, 637–644 (2004)
9. Yang, J.M., Kim, J.H.: Sliding Mode Control for Trajectory Tracking of Nonholonomic Wheeled Mobile Robots. IEEE Transactions on Robotics and Automation 15, 578–587 (1999)
10. Chwa, D., Hong, S.K., Song, B.: Robust Posture Stabilization of Wheeled Mobile Robots in Polar Coordinates. In: The 17th International Symposium on Mathematical Theory of Networks and Systems, vol. 39, pp. 343–348 (2006)
11. Floquet, T., Barbot, J.P., Perruquetti, W.: Higher-order Sliding Mode Stabilization for a Class of Nonholonomic Perturbed Systems. Automatica 39, 1077–1083 (2003)
12. Slotine, J.J., Sastry, S.S.: Tracking Control of Non-linear Systems Using Sliding Surfaces, with Application to Robot Manipulators, pp. 1–54. Massachusetts Institute of Technology, Cambridge (1982)

Bilinear Adaptive Parameter Estimation in Fuzzy Cognitive Networks

Thodoris Kottas[1], Yiannis Boutalis[1,2], and Manolis Christodoulou[3]

[1] Democritus University of Thrace, 67100 Xanthi, Greece
ybout@ee.duth.gr
[2] Chair of Automatic Control, University of Erlangen-Nuremberg, 91058 Germany
[3] Technical University of Crete, 73100 Chania, Greece
manolis@ece.tuc.gr

Abstract. Fuzzy Cognitive Networks (FCN) have been introduced by the authors recently as an extension of Fuzzy Cognitive Maps (FCM). One important issue of their operation is the conditions under which they reach a certain equilibrium point after an initial perturbation. This is equivalent to studying the existence and uniqueness of solutions for their concept values. In this paper, we study the existence of solutions of FCNs equipped with continuous differentiable sigmoid functions. This is done by using an appropriately defined contraction mapping theorem. It is proved that when the weight interconnections and the chosen sigmoid function fulfill certain conditions the concept values will converge to a unique solution regardless the exact values of the initial concept values perturbations. Otherwise the existence or the uniqueness of equilibrium can not be assured. Assuming that these conditions are met, an adaptive bilinear weight estimation algorithm is proposed.

1 Introduction

Fuzzy Cognitive Networks (FCNs) were proposed in [1] as an extension of Fuzzy Cognitive Maps to support the close interaction with the system they describe and consequently become appropriate for control and system identification applications [2]. Fuzzy Cognitive Maps (FCM) have been initially introduced by Kosko [3] based on Axelrod's work on cognitive maps [4]. They are inference networks using cyclic directed graphs that represent the causal relationships between concepts and in the recent years have been used in various applications [5], [6]. In order to illustrate different aspects in the behavior of the system, a fuzzy cognitive map consists of nodes where each one represents a system characteristic feature. The node interactions represent system dynamics. Different methodologies to develop FCM and extract knowledge from experts have been proposed in [7].

An issue that is very important, both in FCN and FCM is the conditions under which they reach an equilibrium point. This is equivalent to studying the existence and uniqueness of solutions for their concept values. A first study on this subject has already been proposed by the authors in [8]. In this paper, we

C. Alippi et al. (Eds.): ICANN 2009, Part II, LNCS 5769, pp. 875–884, 2009.

extend the study of the existence of solutions of FCNs equipped with continuous differentiable sigmoid functions. This is done by using an appropriately defined contraction mapping theorem. It is proved that when the weight interconnections and the chosen sigmoid function fulfill certain conditions the concept values will converge to a unique solution regardless the exact values of the initial concept values perturbations. Otherwise the existence or the uniqueness of equilibrium can not be assured. Assuming that these conditions are met, an adaptive bilinear weight estimation algorithm is proposed, which updates both FCN weights and the inclination of the sigmoid functions used, based on systems' operation data. It is proved that the algorithm guarantees the error converges to zero exponentially fast.

The paper is organized as follows. Section 2 describes the representation and mathematical formulation of Fuzzy Cognitive Networks. Section 3 provides the proof of the existence solution of the concept values of a Fuzzy Cognitive Network. Section 4 presents the bilinear adaptive weight estimation algorithm with proven stability and parameter convergence, while Section 5 provides illustrative numerical examples. Finally, Section 6 concludes the work providing also hints for future extensions.

2 Fuzzy Cognitive Networks

A graphical representation of FCNs is depicted in Fig. 1. Each concept represents a characteristic of the system; in general it represents events, actions, goals, values and trends of the system . Each concept is characterized by a number A_i that represents its value and it results from the transformation of the real value of the systems variable, represented by this concept, either in the interval [0,1] or in the interval [-1,1]. All concept values form Vector A are expressed as:

$$A = \left[A_1 \ A_2 \ .... \ A_n \right]^T$$

with n being the number of the nodes (in Fig. 1 $n = 8$). Causality between concepts allows degrees of causality and not the usual binary logic, so the weights of the interconnections can range in the interval [-1,1].

The existing knowledge on the behavior of the system is stored in the structure of nodes and interconnections of the map. The value of w_{ij} indicates how strongly concept C_j influences concept C_i. The sign of w_{ij} indicates whether the relationship between concepts C_j and C_i is direct or inverse.

For the FCN of Fig. 1 matrix W is equal to

$$W = \begin{bmatrix} d_{11} & w_{12} & w_{13} & w_{14} & 0 & w_{16} & 0 & 0 \\ w_{21} & d_{22} & 0 & 0 & 0 & w_{26} & 0 & w_{28} \\ w_{31} & 0 & d_{33} & w_{34} & w_{35} & 0 & w_{37} & 0 \\ w_{41} & 0 & w_{43} & d_{44} & 0 & 0 & 0 & 0 \\ 0 & 0 & 0 & 0 & d_{55} & 0 & w_{57} & 0 \\ 0 & 0 & w_{63} & w_{64} & w_{65} & d_{66} & 0 & w_{68} \\ 0 & 0 & w_{73} & 0 & 0 & 0 & d_{77} & w_{78} \\ 0 & w_{82} & 0 & 0 & 0 & w_{86} & w_{87} & d_{88} \end{bmatrix}$$

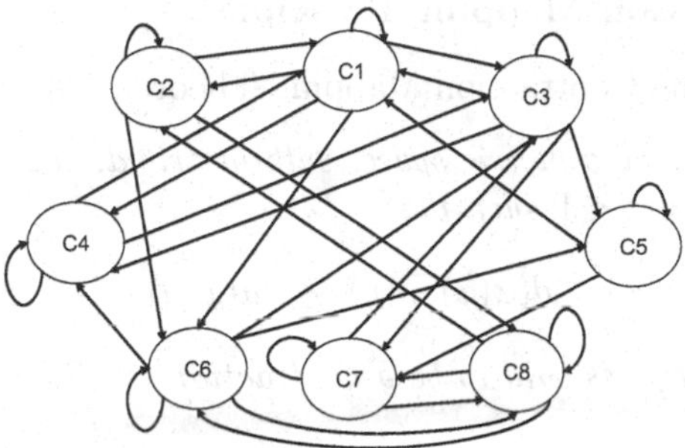

Fig. 1. An FCN with 8 nodes

The equation that calculates the values of concepts of Fuzzy Cognitive Networks, may or may not include self-feedback. In its general form it can be written as:

$$A_i(k) = f(\sum_{\substack{j=1 \\ j \neq i}}^{n} w_{ij} A_j(k-1) + d_{ii} A_i(k-1)) \tag{1}$$

Where $A_i(k)$ is the value of concept C_i at discrete time k, $A_i(k-1)$ the value of concept C_i at discrete time $k-1$ and $A_j(k-1)$ is the value of concept C_j at discrete time $k-1$. w_{ij} is the weight of the interconnection from concept C_j to concept C_i and d_{ii} is a variable that takes on values in the interval $[0,1]$, depending upon the existence of "strong" or "weak" self-feedback to node i.

f is a sigmoid function commonly used in the Fuzzy Cognitive Maps, which squashes the result in the interval [0,1] and is expressed as, $f = \frac{1}{1+e^{-c_l x}}$, where $c_l > 0$ is used to adjust its inclination.

Equation (1) can be rewritten as:

$$A(k) = f(WA(k-1)) \tag{2}$$

In the next Section we are deriving conditions, which determine the existence of a unique solution of (2), when continuous differentiable transfer functions f are used.

3 Existence and Uniqueness of Solutions in Fuzzy Cognitive Networks

In this Section we check the existence of solutions in equation (2), when a continuous and differentiable transfer function is used, such as sigmoid functions are. We know that the allowable values of the elements of FCN vectors A lie in the closed interval $[0,1]$. This is a subset of $\Re$ and is a complete metric space with the usual L_2 metric. We will define the regions where the FCN has a unique solution, which does not depend on the initial condition since it is the unique equilibrium point.

3.1 The Contraction Mapping Principle

We now introduce the Contraction Mapping Theorem [9].

Definition 1. *Let X be a metric space, with metric d. If φ maps X into X and there is a number $0 < c < 1$ such that*

$$d(\varphi(x), \varphi(y)) \leq cd(x, y) \tag{3}$$

for all $x, y \in X$, then φ is said to be a contraction of X into X.

Theorem 1. *[9] If X is a complete metric space, and if φ is a contraction of X into X, then there exists one and only one $x \in X$ such that $\varphi(x) = x$.*

In other words, φ has a unique fixed point. The uniqueness follows from the fact that if $\varphi(x) = x$ and $\varphi(y) = y$, then (3) gives $d(x, y) \leq cd(x, y)$, which can only happen when $d(x, y) = 0$ (See [9]).
Equation (2) can be written as:

$$A(k) = G(A(k-1)) \tag{4}$$

where $G(A(k-1))$ is equal to $f(WA(k-1))$.
In FCN's $A \in [0,1]^n$ and it is also clear according to (2) that $G(A(k-1)) \in [0,1]^n$.
If the following inequality is true:

$$d(G(A), G(A')) \leq cd(A, A')$$

where A and A' are different vectors of concept values and G is defined in (4), then G has a unique fixed point A such that $G(A) = A$. Before presenting the main theorem we need to explore the role of f as a contraction function.

Theorem 2. *The scalar sigmoid function f, ($f = \frac{1}{1+e^{-c_l x}}$) is a contraction of the metric space X into X, were $X = [a, b], a, b$, finite, according to Definition 1, where:*

$$d(f(x), f(y)) \leq cd(x, y) \tag{5}$$

if the above inequality is true :

$$\frac{c_l}{e^{c_l x}} f^2 < 1 \tag{6}$$

Proof. Here f is the sigmoid function, x,$y \in X$, X is as defined above and c is a real number such that $0 < c < 1$

The inclination l of a sigmoid function f is equal to:

$$l = \frac{\partial f}{\partial x} = \frac{c_l e^{-c_l x}}{(1+e^{-c_l x})^2} = \frac{c_l}{e^{c_l x}} \left(\frac{1}{1+e^{-c_l x}} \right)^2 = \frac{c_l}{e^{c_l x}} f^2 \tag{7}$$

for $x \in X$, for l also holds that:

$$\frac{d\,(f(x), f(y))}{d\,(x, y)} \leq l \tag{8}$$

From (7) and (8) we get:

$$\frac{d\,(f(x), f(y))}{d\,(x,y)} \leq l \leq c < 1 \tag{9}$$

if and only if the above inequality is true: $\frac{c_l}{e^{c_l x}} f^2 < 1$

Theorem 3. *There is one and only one solution for any concept value A_i of any FCN where the sigmoid function $f = \frac{1}{1+e^{-x}}$ is used, if:*

$$\left(\sum_{i=1}^{n} \left(c_{l_i} l_i \left\| w_i \right\| \right)^2 \right)^{1/2} < 1 \tag{10}$$

where w_i is the i_{th} row of matrix W, $\|w_i\|$ is the L_2 norm of w_i, l_i is the inclination of function f equal to $l_i = \frac{c_{l_i}}{e^{c_{l_i} w_i \cdot A}} f^2(c_{l_i} w_i \cdot A)$, and c_{l_i} is the c_l factor of function f corresponding to A_i concept.

Proof. Let X be the complete metric space $[a,b]^n$ and $G : X \rightarrow X$ be a map such that:

$$d(G(A), G(A')) \leq cd(A, A') \tag{11}$$

for some $0 < c < 1$.

Vector G is equal to:

$$G = \left[f(c_{l_1}(w_1 \cdot A)) \; f(c_{l_2}(w_2 \cdot A)) \cdots f(c_{l_n}(w_n \cdot A)) \right]^T \tag{12}$$

where n is the number of concepts of the FCN, f is the sigmoid function $f = \frac{1}{1+e^{-x}}$, w_i is the i_{th} row of matrix W of the FCN, where $i = 1, 2, ..., n$, and by $\cdot$ we denote the inner product between two equidimensional vectors which both belong in $\Re^n$.

Assume A and A' are two different concept values for the FCN. Then, we want to prove the following inequality:

$$\|G(A) - G(A')\| \leq c \|A - A'\| \tag{13}$$

But $\|G(A) - G(A')\|$ according to (12) equals to:

$$\|G(A) - G(A')\| = \left(\sum_{i=1}^{n} \left(f(c_{l_i}(w_i \cdot A)) - f(c_{l_i}(w_i \cdot A')) \right)^2 \right)^{1/2}$$

According to Theorem 2 for the scalar argument of $f(.)$, which is $c_{l_i}(w_i \cdot A)$ in the bounded and closed interval $[a,b]$ with a and b being finite numbers, it is true that:

$$|f(c_{l_i}(w_i \cdot A)) - f(c_{l_i}(w_i \cdot A'))| \leq l_i \left| (c_{l_i}(w_i \cdot A)) - (c_{l_i}(w_i \cdot A')) \right|$$

for every $i = 1, ..., n$, where $l_i = max(\frac{c_{l_i}}{e^{c_{l_i} w_i \cdot A}} f^2(c_{l_i} w_i \cdot A), \frac{c_{l_i}}{e^{c_{l_i} w_i \cdot A'}} f^2(c_{l_i} w_i \cdot A'))$.

By using the Cauchy-Schwartz inequality we get:

$$l_i|(c_{l_i}(w_i \cdot A)) - (c_{l_i}(w_i \cdot A'))| = l_i|c_{l_i}(w_i \cdot (A - A'))| \leq l_i c_{l_i} \|w_i\| \|A - A'\|$$

Subsequently, we get:

$$\|G(A) - G(A')\| = \left(\sum_{i=1}^{n} (f(c_{l_i}(w_i \cdot A)) - f(c_{l_i}(w_i \cdot A')))^2 \right)^{1/2}$$

Finally: $\|G(A) - G(A')\| \leq \|A - A'\| \left(\sum_{i=1}^{n} (c_{l_i} l_i \|w_i\|)^2 \right)^{1/2}$

A necessary condition for the above to be a contraction is:

$$\left(\sum_{i=1}^{n} (c_{l_i} l_i \|w_i\|)^2 \right)^{1/2} < 1 \tag{14}$$

4 Bilinear on Line Parameter Estimation of Fuzzy Cognitive Networks

Based on the results and observations of Section 3 we are now proposing a method of finding appropriate weight sets related to a desired equilibrium point of the FCN. Choosing a desired state A^{des} for the FCN this is equivalent to solving the equation

$$f_i^{-1}(A_i^{des}) = c_{l_i}^* w_i^* \cdot A^{des} \tag{15}$$

with w_i^* being the ith row of W^* and c_{l_i} is the c_l factor of function f corresponding to A_i concept.

In order to solve the above equation and since both $c_{l_i}^*$ and w_i^* have to be estimated, we use a bilinear adaptive estimation algorithm [10]. Taking into account that $f_i(x_i)$ is the sigmoid function, weight updating laws are given as follows:

The error $\varepsilon_i(k)$ of the parametric discrete-time adaptive law is of the form:

$$\varepsilon_i(k) = \frac{f_i^{-1}(A_i^{des}) - c_{l_i}(k-1) w_i(k-1) A^{des}}{c + c_{l_i}(0)(A^{des})^T A^{des} + \gamma(w_i(k-1)A^{des})^2} \tag{16}$$

while the updating algorithm is given by:

$$c_{l_i}(k) = c_{l_i}(k-1) + \gamma \varepsilon_i(k)(w_i(k-1)A^{des}) \tag{17}$$

$$w_i(k) = w_i(k-1) + \alpha sgn(c_{l_i}^*(k)) \varepsilon_i(k) A^{des} \tag{18}$$

where $0 < \gamma < 1$, $a, c > 0$ and $c_{l_i}(0) > 0$ is an upper bound for $\left|c_{l_i}^*(k)\right|$. $w_i(k)$ is the ith row of $W(k)$, which is the estimator of $W^*(k)$. $c_{l_i}(k)$ is the c_l factor of function f corresponiding to concept A_i^{des} and is the estimator of $c_{l_i}^*(k)$. A^{des} is constant vector and $f_i^{-1}(A_i^{des})$ is also constant and scalar. $\alpha > 0$, $c > 0$ and $\gamma > 0$ are design parameters.

By using the above updating algorithms we can now prove that the estimators converges to W^* and $c_{l_i}^*$ respectively. In Section 3 we proved that if inequality

(10) is true then W^* and $c_{l_i}^*$ corresponding to the designed FCN provide a unique solution and satisfies (15).

Proof. From (16), (17), (18) and $\tilde{w}_i(k) = w_i(k) - w_i^*$, $\tilde{c}_{l_i}(k) = c_{l_i}(k) - c_{l_i}^*$ we obtain the error equation

$$\varepsilon_i(k) = -\frac{\tilde{c}_{l_i}(k-1)w_i(k-1)A^{des} + c_{l_i}^*(k)\tilde{w}_i(k-1)A^{des}}{c + c_{l_i}(0)(A^{des})^T A^{des} + \gamma(w_i(k-1)A^{des})^2} \tag{19}$$

The updating algorithms of (17) and (18) can also be written as:

$$\tilde{c}_{l_i}(k) = \tilde{c}_{l_i}(k-1) + \gamma\varepsilon_i(k)(w_i(k-1)A^{des}) \tag{20}$$

$$\tilde{w}_i(k) = \tilde{w}_i(k-1) + \alpha sgn(c_{l_i}^*(k))\varepsilon_i(k)A^{des} \tag{21}$$

For each node i consider the function

$$V_i(k) = \frac{\tilde{c}_{l_i}^2(k)}{2\gamma} + \frac{\left|c_{l_i}^*(k)\right|}{2a}\tilde{w}_i(k)\tilde{w}_i^T(k) \tag{22}$$

Then

$$\Delta V_i(k) = \left(\frac{\tilde{c}_{l_i}^2(k)}{2\gamma} - \frac{\tilde{c}_{l_i}^2(k-1)}{2\gamma}\right) + \frac{\left|c_{l_i}^*(k)\right|}{2a}\left(\tilde{w}_i(k)\tilde{w}_i^T(k) - \tilde{w}_i(k-1)\tilde{w}_i^T(k-1)\right)$$

Using (20) and (21) $\Delta V_i(k)$ is:

$$\Delta V_i(k) = \left(\frac{\gamma\varepsilon_i^2(k)\left(w_i(k-1)A^{des}\right)^2}{2} + \tilde{c}_{l_i}(k-1)\varepsilon_i(k)\left(w_i(k-1)A^{des}\right)\right) + \\ \frac{\left|c_{l_i}^*(k)\right|}{2}\text{sgn}\left(c_{l_i}^*(k)\right)\varepsilon_i(k)\left(\tilde{w}_i(k-1)A^{des} + \left(A^{des}\right)^T\tilde{w}_i^T(k-1)\right) + \\ \frac{\left|c_{l_i}^*(k)\right|}{2}a\left(\varepsilon_i(k)\right)^2\left(A^{des}\right)^T A^{des}$$

Taking into account that $c_{l_i}^*(k) = \left|c_{l_i}^*(k)\right|\text{sgn}\left(c_{l_i}^*(k)\right)$ and $\tilde{w}_i(k-1)A^{des} = \left(A^{des}\right)^T\tilde{w}_i^T(k-1)$, $\Delta V_i(k)$ is now equal to:

$$\Delta V_i(k) = \tfrac{1}{2}\gamma\varepsilon_i^2(k)\left(w_i(k-1)A^{des}\right)^2 + \tilde{c}_{l_i}(k-1)\varepsilon_i(k)\left(w_i(k-1)A^{des}\right) + \\ c_{l_i}^*(k)\varepsilon_i(k)\left(\tilde{w}_i(k-1)A^{des}\right) + \frac{\left|c_{l_i}^*(k)\right|}{2}a\left(\varepsilon_i(k)\right)^2\left(A^{des}\right)^T A^{des}$$

Using (19), $\Delta V_i(k)$ is:

$$\Delta V_i(k) = \tfrac{1}{2}\gamma\varepsilon_i^2(k)\left(w_i(k-1)A^{des}\right)^2 - \\ \varepsilon_i(k)\varepsilon_i(k)\left(c + c_{l_i}(0)\left(A^{des}\right)^T A^{des} + \gamma(w_i(k-1)A^{des})^2\right) + \\ \frac{\left|c_{l_i}^*(k)\right|}{2}a\left(\varepsilon_i(k)\right)^2\left(A^{des}\right)^T A^{des}$$

Finally:

$$\Delta V_i(k) = -\varepsilon_i^2(k)m^2(k)\left[1 - \frac{\gamma\xi^2(k) + \left|c_{l_i}^*(k)\right|a\left(A^{des}\right)^T A^{des}}{2m^2(k)}\right]$$

where $m^2(k) = c + c_{l_i}(0)\left(A^{des}\right)^T A^{des} + \gamma\xi^2(k)$ and $\xi^2(k) = \left(w_i(k-1)A^{des}\right)^2$

It is obvious that if $0 < \gamma < 1$ and $a, c > 0$, then

$$\frac{\gamma\xi^2(k) + \left|c^*_{l_i}(k)\right| a \left(A^{des}\right)^T A^{des}}{2m^2(k)} < 1$$

which implies that:

$$\Delta V_i(k) \leq -c_0 \varepsilon_i^2(k) m^2(k) \leq 0 \tag{23}$$

for some constant $c_0 > 0$.

From (22), (23) we have that $V_i(k)$ and therefore $w_i(k) \in L_\infty$, $c_{l_i}(k) \in L_\infty$ and $V_i(k)$ has a limit, i.e., $\lim_{k\to\infty} V_i(k) = V_\infty$. Consequently, using (23) we obtain

$$c_0 \sum_{k=1}^{\infty} \left(\varepsilon_i^2(k) m_i^2(k)\right) \leq V_i(0) - V_\infty < \infty$$

which implies $\varepsilon_i(k) m_i(k) \in L_2$ and

$\varepsilon_i(k) m_i(k) \to 0$ as $k \to \infty$. Since

$m_i(k) = \sqrt{(c + c_{l_i}(0)(A^{des})^T A^{des} + \gamma\xi^2(k))} \geq c > 0$, we also have that $\varepsilon_i(k) \in L_2$ and $\varepsilon_i(k) \to 0$ as $k \to \infty$. We have

$\varepsilon_i(k) A^{des} = \varepsilon_i(k) m_i(k) \frac{A^{des}}{m_i(k)}$. Since $\frac{A^{des}}{m_i(k)}$ is bounded and

$\varepsilon_i(k) m_i(k) \in L_2$, we have that $\varepsilon_i(k) A^{des} \in L_2$ and $\left\|\varepsilon_i(k) A^{des}\right\| \to 0$ as $k \to \infty$. This implies (using (18)) that $\|w_i(k) - w_i(k-1)\| \in L_2$ and $\|w_i(k) - w_i(k-1)\| \to 0$ as $k \to \infty$. Now

$$\begin{gathered} w_i(k) - w_i(k-N) = \\ w_i(k) - w_i(k-1) + w_i(k-1) - w_i(k-2) + \ldots + w_i(k-N+1) - w_i(k-N) \end{gathered}$$

for any finite N. Using the Schwartz inequality, we have

$$\begin{gathered} \|w_i(k) - w_i(k-N)\|^2 \leq \|w_i(k) - w_i(k-1)\|^2 + \|w_i(k-1) - w_i(k-2)\|^2 + \\ \ldots + \|w_i(k-N+1) - w_i(k-N)\|^2 \end{gathered}$$

Since each term on the right-hand side of the inequality is in L_2 and goes to zero with $k \to \infty$, it follows that $\|w_i(k) - w_i(k-N)\| \in L_2$ and $\|w_i(k) - w_i(k-N)\| \to 0$ as $k \to \infty$. Since $\varepsilon_i(k) A^{des} \in L_2$ and $w_i(k-1) \in L_\infty$ then $w_i(k-1)\varepsilon_i(k) A^{des} \in L_2 \cap L_\infty$ and $\left|\varepsilon_i(k) w_i(k-1) A^{des}\right| \to 0$ as $k \to \infty$. This implies (using(17)) that $|c_{l_i}(k) - c_{l_i}(k-1)| \in L_2$ and $|c_{l_i}(k) - c_{l_i}(k-1)| \to 0$ as $k \to \infty$. Now

$$\begin{gathered} c_{l_i}(k) - c_{l_i}(k-N) = \\ c_{l_i}(k) - c_{l_i}(k-1) + c_{l_i}(k-1) - c_{l_i}(k-2) + \ldots + c_{l_i}(k-N+1) - c_{l_i}(k-N) \end{gathered}$$

for any finite N. We also have that,

$$\begin{gathered} |c_{l_i}(k) - c_{l_i}(k-N)|^2 \leq |c_{l_i}(k) - c_{l_i}(k-1)|^2 + |c_{l_i}(k-1) - c_{l_i}(k-2)|^2 + \ldots + \\ |c_{l_i}(k-N+1) - c_{l_i}(k-N)|^2 \end{gathered}$$

Since each term of the right-hand side of the inequality is in L_2 and goes to zero with $k \rightarrow \infty$, it follows that $|c_{l_i}(k) - c_{l_i}(k-N)| \in L_2$ and $|c_{l_i}(k) - c_{l_i}(k-N)| \rightarrow 0$ as $k \rightarrow \infty$.

5 Numerical Example

For the numerical example the FCN of Fig. 1 is used. The initial W and C_l matrices of FCN is equal to:

$$W = \begin{bmatrix} 1 & 0.4 & -0.5 & 0.7 & 0 & 0.1 & 0 & 0 \\ 0.1 & 1 & 0 & 0 & 0 & 0.5 & 0 & 0.2 \\ 0.2 & 0 & 1 & 0.9 & 0.7 & 0 & 0.7 & 0 \\ 0.5 & 0 & 0.8 & 1 & 0 & 0 & 0 & 0 \\ 0 & 0 & 0 & 0 & 1 & 0 & 0.4 & 0 \\ 0 & 0 & 0.5 & 0.4 & 0.9 & 1 & 0 & 0.2 \\ 0 & 0 & 0.7 & 0 & 0 & 0 & 1 & 0.1 \\ 0 & 0.4 & 0 & 0 & 0 & 0.7 & -0.1 & 1 \end{bmatrix}, C_l = \begin{bmatrix} 1 \, 1 \, 1 \, 1 \, 1 \, 1 \, 1 \, 1 \end{bmatrix}$$

where C_l is a vector containing all the individual c_{l_i} of each sigmoid function of each node. Except from the diagonal elements, which have the value 1 and the zero elements all other weights were randomly selected. With these matrices the FCN reaches an equilibrium point given by the following vector A.

$$A = \begin{bmatrix} 0.8981 \, 0.8147 \, 0.9037 \, 0.9387 \, 0.9111 \, 0.8704 \, 0.8551 \, 0.7666 \end{bmatrix}$$

This FCN equilibrium point fulfills eq. (10) because

$$\left(\sum_{i=1}^{8} l_i^2 c_{l_i}^2 \left\| w_i \right\|^2 \right)^{1/2} = 0.4585 < 1$$

Suppose that for the FCN of Fig. 1 the desired state is equal to:
$A^{des} = \begin{bmatrix} 0.76 \, 0.71 \, 0.75 \, 0.67 \, 0.69 \, 0.58 \, 0.67 \, 0.78 \end{bmatrix}$

Applying (16), (18) and (17) the W matrix concludes to W_1 matrix and c_l factors of f fuctions corresponding to A_i^{des} concept cocludes to C_{l_1} matrix, both given below.

$$W_1 = \begin{bmatrix} 0.9872 & 0.3880 & -0.5127 & 0.6887 & 0 & 0.0902 & 0 & 0 \\ -0.0204 & 0.8875 & 0 & 0 & 0 & 0.4081 & 0 & 0.0764 \\ -0.1766 & 0 & 0.6283 & 0.5680 & 0.3581 & 0 & 0.3680 & 0 \\ 0.0744 & 0 & 0.3800 & 0.6248 & 0 & 0 & 0 & 0 \\ 0 & 0 & 0 & 0 & 0.8907 & 0 & 0.2939 & 0 \\ 0 & 0 & -0.0082 & -0.0540 & 0.4325 & 0.6070 & 0 & -0.3285 \\ 0 & 0 & 0.4500 & 0 & 0 & 0 & 0.7766 & -0.1600 \\ 0 & 0.3529 & 0 & 0 & 0 & 0.6615 & -0.1445 & 0.9482 \end{bmatrix}$$

$$C_{l_1} = \begin{bmatrix} 0.9980 \, 0.9829 \, 0.9071 \, 0.9315 \, 0.9860 \, 0.9170 \, 0.9662 \, 0.9911 \end{bmatrix}$$

For the matrices W_1 and C_{l_1} eq. (10) is true:

$$\left(\sum_{i=1}^{8} l_i^2 c_{l_i}^2 \left\| w_i \right\|^2 \right)^{1/2} = 0.5359 < 1$$

6 Conclusions

In this paper the existence and uniqueness of the equilibrium values of the concepts of FCNs was studied. This study concerns FCNs equipped with continuous differentiable sigmoid functions having contractive properties and is performed using an appropriately defined contraction mapping theorem. It was proved that when the weight interconnections fulfill certain conditions, related to the size of the FCN and the inclination of the sigmoids used, the concept values will converge to a unique solution regardless their initial values. Assuming that these conditions are met, a bilinear adaptive estimation algorithm is proposed, which estimates on-line both weights and the inclinations of the sigmoids of the FCN. Future work will include the modification of the proposed algorithm so that it takes into account the conditions derived and the development of reliable control schemes which will use the concept of FCN and employ the theoretical results presented here.

References

1. Kottas, T.L., Boutalis, Y.S., Christodoulou, M.A.: Fuzzy Cognitive Networks: A General Framework. Inteligent Desicion Technologies 1(4), 183–196 (2007)
2. Kottas, T.L., Boutalis, Y.S., Karlis, A.D.: A New Maximum Power Point Tracker for PV Arrays Using Fuzzy Controller in Close Cooperation with Fuzzy Cognitive Networks. IEEE Transactions on Energy Conversion 21(3), 793–803 (2006)
3. Kosko, B.: Fuzzy Cognitive Maps. International Journal of Man-Machine Studies, 65–75 (1986)
4. Axelrod, R.: Structure of Decision. In: The Cognitive Maps of Political Elites. Princeton University Press, New Jersey (1976)
5. Kottas, T., Boutalis, Y., Devedzic, G., Mertzios, B.: A new method for reaching equilibrium points in Fuzzy Cognitive Maps. In: Proceedings of 2nd International IEEE Conference of Intelligent Systems, Varna Burgaria, pp. 53–60 (2004)
6. Zhang, W., Chen, S., Bezdek, J.: Pool2: A Generic System for Cognitive Map Development and Decision Analysis. IEEE Transactions on Systems, Man, and Cybernetics 19(1), 31–39 (1989)
7. Schneider, M., Shnaider, E., Kandel, A., Chew, G.: Constructing fuzzy cognitive maps. In: International Joint Conference of the Fourth IEEE International Conference on Fuzzy Systems and The Second International Fuzzy Engineering Symposium, vol. 4(1), pp. 2281–2288 (1995)
8. Boutalis, Y., Kottas, T., Christodoulou, M.: On the Existence and Uniqueness of Solutions for the Concept Values in Fuzzy Cognitive Maps. In: Proceedings of 47th IEEE Conference on Decision and Control - CDC 2008, Cancun, Mexico, December 9-11, pp. 98–104 (2008)
9. Rudin, W.: Principles of Mathematical Analysis, pp. 220–221. McGraw-Hill Inc., New York (1964)
10. Ioannou, P., Fidan, B.: Adaptive Control Tutorial. Society for Industrial and Applied Mathematics (SIAM), Philadelphia (2006)

AM-FM Texture Image Analysis of the Intima and Media Layers of the Carotid Artery

Christos P. Loizou*, Victor Murray, Marios S. Pattichis, Christodoulos S. Christodoulou, Marios Pantziaris, Andrew Nicolaides, and Constantinos S. Pattichis

Department of Computer Science, School of Sciences, Intercollege, P.O. Box 51604, CY-3507 Limassol-Cyprus
loizou.c@lim.intercollege.ac.cy, vmurray@ieee.org, pattichis@ece.unm.edu, cschr2@ucy.ac.cy, pantzari@cing.ac.cy, andisnicolai@gmail.com, pattichi@ucy.ac.cy

Abstract. The purpose of this paper is to propose the use of amplitude modulation-frequency modulation (AM-FM) features for describing atherosclerotic plaque features that are associated with clinical factors such as intima media thickness and a patient's age. AM-FM analysis reveals the instantaneous amplitude (IA) of the media layer decreases with age. This decrease in IA maybe attributed to the reduction in calcified, stable plaque components and an increase in stroke risk with age. On the other hand, an increase in the median instantaneous frequency (IF) of the media layer suggests the fragmentation of solid, large plaque components, which also lead to an increase in the risk of stroke. The findings suggest that AM-FM features can be used to assess the risk of stroke over a wide range of patient populations. Future work will incorporate a new texture image retrieval system that uses AM-FM features to retrieve intima and intima media layer images that could be associated with the same level of the risk of stroke.

1 Introduction

Atherosclerosis causes enlargement of the arteries and thickening of the artery walls. Thus clinically the intima-media thickness (IMT) is used as a validated measure for the assessment of atherosclerosis [1,2] (see Fig. 1). It was proposed but not thoroughly investigated [3], that not only the IMT but rather the media-layer (ML), its thickness [4,5] its textural characteristics [3], and amplitude modulation-frequency modulation (AM-FM) [6] characteristics may be used for evaluating the risk of a patient to develop a stroke and account in general the risk of the cardiovascular disease (CVD) by differentiating between patients at high and low risk for stroke. The objective of this study is to investigate the application of amplitude-modulation frequency-modulation (AM-FM) analysis of intima media complex (IMC), media layer (ML), and intima layer (IL) of the common carotid artery

* Corresponding author.

C. Alippi et al. (Eds.): ICANN 2009, Part II, LNCS 5769, pp. 885–894, 2009.

(CCA). Only one study [7] investigated AM-FM representations of the atherosclerotic carotid plague of the CCA but not for the IMC, IL, and ML.

As shown in Fig. 1, the IL is a thin layer, the thickness of which increases with age, from a single cell layer at birth to $250\mu m$ at the age of 40 for non-diseased individuals [8]. In ultrasound images the media layer (ML) is characterized by an echolucent region, predominantly composed of smooth muscle cells, enclosed by the intima and adventitia layers (see Fig. 1, band Z6) [2,9]. Earlier research [10], showed that the media layer thickness (MLT) does not change significantly with age ($125\mu m <$ MLT $< 350\mu m$). In recent studies by our group the median (IQR) of IMT, MLT and intima layer thickness (ILT), were computed from 100 ultrasound images from 42 female and 58 male asymptomatic subjects aged between 26 and 95 years old, with a mean age of 54 years to be as follows $0.66mm$ (0.18), $0.23mm$ (0.18), $0.43mm$ (0.12) respectively [3,4,5].

In [11] a method has been presented for quantifying the reflectivity of the ML of the distal CCA. It was shown that the GSM of the intima media layer is the earliest change representing atherosclerotic disease in the arterial wall that can currently be imaged in vivo. This may be the first marker of atherosclerosis and may precede the development of a significant increase in IMT. This would enable earlier identification of high-risk individuals based on the analysis of the CCA artery wall textural and AM-FM characteristics. In [12] the authors reported on the properties of the GSM of the IMC from a random sample of 1016 subjects aged exactly 70. They found that the GSM of the IMC of the CCA is closely related to the echogenecity in overt carotid plaques.

There are several studies reported earlier suggesting that the instability of the carotid atheromatous plaque can be characterized from B-mode ultrasound images [9,13]. In [9,13] the echogenecity in atherosclerotic carotid plaques was evaluated through the GSM, where as in [3] the IMC the ML, and IL were characterized based on texture feature analysis. It is evident from the visual inspections of the IMC in the CCA that a great variation in echogenecity does exist. However, the usefulness of this information has not yet been studied.

We propose to study changes in AM-FM characteristics that can be associated with disease progression for different age groups and different gender. Here, we note that for fully developed plaques in the CCA, texture features derived from statistical, model based, and Fourier based methods, have been used to characterize and classify carotid atheromatous plaques from B-mode ultrasound images [13].

To the best of our knowledge no other study carried out ML and IL ultrasound AM-FM measurements for investigating their relationship with the increase of age and gender, and the risk of stroke based on their AM-FM characteristics. We do note that the best known (related) results were presented in [14,15] where it was shown that IMT increases linearly with age.

The objective of our study is to investigate whether AM-FM characteristics extracted from the IMC, the ML, and the IL of the CCA, segmented manually by an expert and automatically by a snakes segmentation system [5,16] can be associated with the increase of IMT, MLT or ILT and how these are affected by

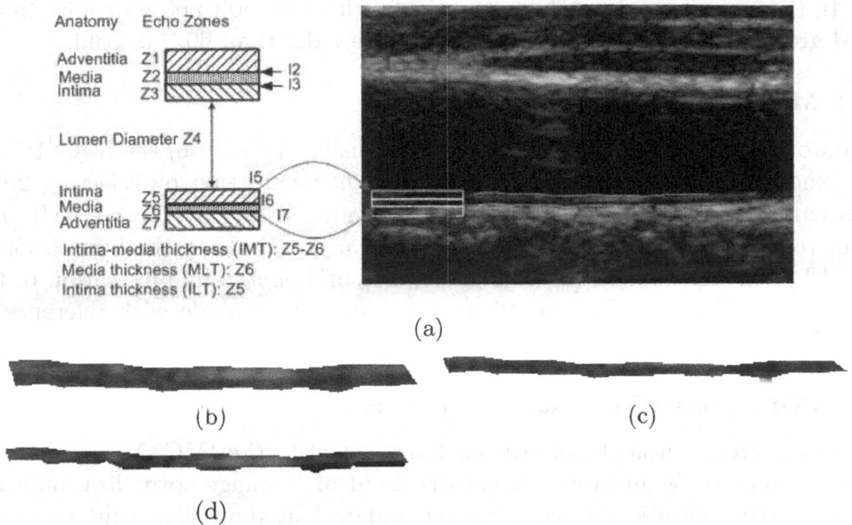

(a)

(b) (c)

(d)

Fig. 1. (a) Illustration of the intima-media-complex (IMC) of the far wall of the (b) common carotid artery and the automatic IMC segmentation [4,16]. The media layer (ML) is defined as the layer (band) between the intima-media and the media-adventitia interface (band Z6), (c) extracted automated IMC, (d) extracted automated media layer (ML) and e) extracted automated intima layer (IL).

age and gender. Ultimately, AM-FM characteristics that vary with age, gender, or IMT, MLT or ILT might be used to assess the risk of stroke.

The paper is organized as follows. In Section 2, we provide materials and methods for the current study. Results are given in Section 3. We provide discussion in Section 4, and give concluding remarks in Section 5.

2 Materials and Methods

2.1 Recording of Ultrasound Images

A total of 100 B-mode longitudinal ultrasound images of the CCA were recorded using the ATL HDI-3000 ultrasound scanner (Advanced Technology Laboratories, Seattle, USA). For the recordings, a linear probe (L74) at a recording frequency of 7 MHz was used. Assuming a sound velocity of $1550m/s$ and 1 cycle per pulse, we thus have an effective spatial pulse width of $0.22mm$ with an axial system resolution of $0.11mm$ [16]. We use bicubic spline interpolation to resize all images to a standard pixel density of $16.66pixels/mm$ (with a resulting pixel width of $0.06mm$). Furthermore, the images were normalized as described in [17]. The grayscale-normalized image was obtained through algebraic (linear) scaling of the image by linearly adjusting the image so that the median gray level value of the blood was 0-5, and the median gray level of the adventitia (artery wall) was 180-190. The images were partitioned into three different age groups. In the first group, we included 27 images from patients who were younger than 50 years old. In the second group, we had 36 patients who were 50 to 60 years old. In the third group, we included 37 patients who were older than 60 years old.

2.2 Manual Measurements

A neurovascular expert manually segmented (using the mouse) the IMC [16] the ML, and IL [4,5] on each image after image normalization by selecting 20-40 consecutive points for the adventitia, media and intima at the far wall. The measurements were performed between $1 - 2cm$ proximal to the bifurcation of the CCA, on the far wall [2], over a distance of $1.5cm$. The bifurcation of the CCA was used as a guide and all measurements were made with reference to that region.

2.3 IMC, ML and IL Snake Segmentations

All images were automatically segmented to identify the IMC, ML, and IL regions. Automatic segmentation was carried out after image normalization using the snakes segmentation system proposed and evaluated on ultrasound images of the CCA in [4,5,16]. The segmentation system is based on the Williams & Shah method [18]. Using the definitions given in Fig. 1, we first segment the IMC [16] by extracting the I5 (lumen-intima interface) and I7 boundaries (media-adventitia interface). The upper side of the ML (see Fig. 1, Z6) was then estimated by deforming the lumen-intima interface (boundary I5) by $0.36mm$ (6 pixels) downwards and then deformed by the snakes segmentation algorithm proposed in [16] in order to fit to the media boundary.

2.4 Amplitude-Modulation Frequency-Modulation (AM-FM) Methods

Two AM-FM estimates were computed from the automated IMC, ML, and the IL segmented regions of interest as follows: a) the instantaneous amplitude (IA) and b) the instantaneous frequency (IF).

The IA models average intensity variations and the IF provide us with information at a pixel level related with orientation variations, or structures in an image region. We use the IF in terms of both its amplitude and its angle. Thus, for each input image we estimate the information about: (i) the IA, (ii) the IF, and (iii) the instantaneous frequency angle. Then, for each of the three AM-FM parameters, we compute the histograms over the IMC, ML, and IL segmented regions.

We consider a multi-scale AM-FM representation of digital non-stationary images given by [19,20]:

$$I(k_1, k_2) \approx \sum_{n=1}^{M} a_n(k_1, k_2) \cos \varphi_n(k_1, k_2), \tag{1}$$

where $n = 1, 2, \ldots, M$ denote different scales, $a_n(k_1, k_2)$ denotes slowly-varying instantaneous amplitude (IA) function and $\varphi_n(k_1, k_2)$ denoted the instantaneous phase (IP). The basic idea is to let the frequency-modulated components

$\cos\varphi_n(k_1,k_2)$ capture fast-changing spatial variability in the image intensity. The IF $\nabla\varphi_n(k_1,k_2)$ is defined in terms of the gradient of the IP: $\nabla\varphi_n(k_1,k_2) = (\partial\varphi_n/\partial k_1(k_1,k_2), \partial\varphi_n/\partial k_2(k_1,k_2))$.

For a single-scale AM-FM representation ($M = 1$ in (1)), the IA and the IP are estimated using [21]:

$$\hat{a}(k_1,k_2) = |\hat{I}_{AS}(k_1,k_2)| \text{ and} \tag{2}$$

$$\hat{\varphi}(k_1,k_2) = \arctan\left(\frac{\text{imag}(\hat{I}_{AS}(k_1,k_2))}{\text{real}(\hat{I}_{AS}(k_1,k_2))}\right), \tag{3}$$

respectively, where $\hat{I}_{AS}(k_1,k_2)$ is an extended version of the one-dimensional analytic signal computed with $\hat{I}_{AS}(k_1,k_2) = I(k_1,k_2) + j\mathcal{H}_{2d}[I(k_1,k_2)]$, where $\mathcal{H}_{2d}$ denotes a two-dimensional extension of the one-dimensional Hilbert transform operator.

The IF is computed using a variable spacing, local quadratic phase (VS-LQP) method as described in [19,20]:

$$\frac{\partial\varphi(k_1,k_2)}{\partial k_1} \cong \frac{1}{n_1}\arccos\left(\frac{\bar{I}_{AS}(k_1+n_1,k_2)+\bar{I}_{AS}(k_1-n_1,k_2)}{2\bar{I}_{AS}(k_1,k_2)}\right), \tag{4}$$

and similarly for $\frac{\partial\varphi(k_1,k_2)}{\partial k_2}$. In (4) $\bar{I}_{AS}(k_1,k_2) = \hat{I}_{AS}(k_1,k_2)/|\hat{I}_{AS}(k_1,k_2)|$, and n_1 is a variable displacement from 1 to 4.

We generate a 96-bin feature vector using the histograms of each of the three AM-FM estimates described (IA, IF magnitude (|IF|), and IF angle, 32-bin each) on the ROI of IMC, ML, and IL segmentations. Additionally, the normal histogram of the ROI of the IMC, ML, and IL segmented regions, was computed for 32 equals width bins used and was used as another feature set for comparison purposes.

2.5 Statistical Analysis

The Mann-Whitney rank sum test (for independent samples of different sizes) was used in order to identify if there are significant differences (SD) or not (NS) between the extracted AM-FM features. For significant differences, we require $p < 0.05$, and compare between age groups. Similarly, for comparing independent samples from equal populations, we use the Wilcoxon rank sum test. We use the Wilcoxon rank sum test to detect AM-FM feature differences between the IL, ML, and IMC, for the automated segmentations. We use regression analysis to investigate the relationship between the IMT, MLT, and ILT and medium IF (MIF) and medium IA (MIA) and age.

3 Results

Fig. 1 illustrates an original normalized ultrasound image of the CCA with the automated segmentation of the IMC in (b) and the extracted automated IMC, ML and IL in (b), (c) and (d), respectively.

Table 1. Comparison between the High, Medium and Low AM-FM features extracted from the IMC, ML, and IL for the automated segmentation measurements based on the Mann-Whitney rank sum test for the three different age groups, Below 50 (< 50), Between 50 AND 60 (50-60) AND Above 60 (> 60) years old

			Automated segmentation measurements for the IA and IF																	
			Instantaneous Amplitude (IA)									Instantaneous Frequency (IF)								
			IMC			ML			IL			IMC			ML			IL		
			<50	50-60	>60	<50	50-60	>60	<50	50-60	>60	<50	50-60	>60	<50	50-60	>60	<50	50-60	>60
Automated Segmentation measurements	High	<50		NS (0.65)	NS (0.65)		NS (0.56)	NS (0.21)		NS (0.77)	NS (0.82)		NS (0.4)	NS (0.69)		NS (0.4)	NS (0.62)		NS (0.64)	NS (0.67)
		50-60			NS (0.15)			NS (0.56)			NS (0.19)			S **(0.001)**			S **(0.044)**			NS (0.79)
		>60																		
	Medium	<50		NS (0.15)	S **(0.014)**		NS (0.18)	NS (0.8)		NS (0.16)	NS (0.89)		NS (0.19)	NS (0.67)		NS (0.34)	NS (0.32)		NS (0.13)	NS (0.34)
		50-60			NS (0.32)			NS (0.69)			NS (0.11)			NS (0.68)			NS (0.79)			NS (0.65)
		>60																		
	Low	<50		NS (0.07)	NS (0.34)		S **(0.033)**	NS (0.16)		NS (0.08)	NS (0.82)		NS (0.76)	NS (0.59)		NS (0.22)	NS (0.46)		NS (0.75)	NS (0.67)
		50-60			S **(0.03)**			NS (0.31)			S **(0.0264)**			NS (0.97)			NS (0.24)			NS (0.35)
		>60																		

IMC: Intima-media complex, *ML*: Media layer, *IL*: Intima layer. The p value is shown in parentheses (*S*=Significant difference at p<0.05, *NS*=Non significant difference at p>0.05).

The measurements were extracted using the automated IMC/ML/IL segmentations.

Regression was also carried out for the media IA and media IF of the ML in order to investigate their relationship with age. It was found that the IA of the ML linearly decreases with age (IAml=0.825-0.00373*age, $p = 0.005$), while the IF of the ML linearly increases with age (IFml=-0.046+0.00178*age, $p = 0.001$).

Table 1 presents a comparison among the high, medium and low AM-FM features extracted from the IMC, ML, and IL for the automated segmentation measurements based on the Mann-Whitney rank sum test for the three different age groups, namely below 50 (< 50), between 50 and 60 (50-60) and above 60 (> 60) years old. It is shown that it is possible to differentiate between the three different structures (IMC, IM, IL) using AM-FM features. The AM-FM features were computed at different frequency scales, considering only horizontal oriented filters, of the three-scale filter bank used: (i) Low frequencies (11.3 to 32 pixels wavelengths), (ii) Medium frequencies (5.7 to 16 pixels wavelengths) and (iii) High frequencies (2.8 to 8 pixels wavelengths). It is shown from Table 1, that there is no single feature differentiating between ML and IL, and between the age groups.

More specifically the following observations are made using the information from Table 1:

1. It is possible to differentiate IMC:
   (a) For the ages < 50 and > 60 years old using medium IA.
   (b) For the ages 50 to 60 and > 60 years old using low IA or high IF.
2. It is possible to differentiate ML:
   (a) For the ages < 50 and 50 to 60 years old using medium IA.
   (b) 15BFor the ages 50 to 60 and > 60 years old using high IF.
3. It is possible to differentiate IL:
   (a) For the ages < 50 and > 60 years old using medium IA.
   (b) For the ages 50 to 60 and > 60 years old using low IA.
4. There is no single feature differentiating between ML and IL, and between the age groups.

4 Discussion

In this study AM-FM of the IMC, ML, and IL of 100 longitudinal ultrasound images of the CCA of asymptomatic subjects were investigated. AM-FM analysis reveals the IA of the ML decreases with age. This decrease in IA maybe attributed to the reduction in calcified, stable plaque components and an increase in stroke risk with age. On the other hand, an increase in the median IF of the ML suggests the fragmentation of solid, large plaque components, which also lead to an increase in the risk of stroke. Our study also showed that the IA high, medium, and low components for the IMC, ML, and IL show an increasing trend from high to low, while the IF high, medium, and low components for the IMC, ML, and IL show a decreasing trend from high to low. The findings suggest that AM-FM features can be used to assess the risk of stroke over a wide range of patient populations.

Our study also showed that the IA high, medium, and low components, for the IMC, ML, and IL show a decreasing trend from high to low, while the IF high, medium, and low components, for the IMC, ML, and IL show a decreasing trend from high to low. It was also shown that the AM-FM features performed slightly better than the traditional texture features and gave better results than simple histogram. It is also shown that almost all AM-FM (expect for the ML) features increase with increasing age.

Texture features analysis was also carried out on the same dataset in another study [3]. It is very important to note that texture features provided complementary information in the discrimination between age groups when compared to the AM-FM features extracted in this study. More specifically, for the ML when comparing the age groups < 50 and 50-60, there is significant difference for the AM-FM IA low component (see Table 1), whereas for the texture features GSM and SS-texture energy laws are significantly different [3]. Also, when comparing the age groups < 50 and > 60 only texture features (GSM, contrast, complexity, coarseness) are significantly different.

It has also been observed that there is an increase in the granularity in association with atherosclerotic disease [22]. A granular IMC indicates more advanced atherosclerosis, which may precede the development of significant IMT thickening. In [11] a method has been presented for quantifying the reflectivity of the IM layer of the distal CCA. It was shown that the GSM of the IM layer is the earliest change representing atherosclerotic disease in the arterial wall that can currently be imaged in vivo. This may be the first marker of atherosclerosis and may precede the development of significant increase in IMT. This would enable earlier identification of high-risk individuals based on the analysis of the CCA artery wall textural characteristics.

In [7] the use of AM-FM representations for the characterization of carotid plaques ultrasound images for the identification of individuals with asymptomatic carotid stenosis at risk of stroke was investigated. To characterize the plaques using AM-FM features, the authors computed (i) the instantaneous amplitude, (ii) the instantaneous frequency magnitude and (iii) the instantaneous frequency angle in order to capture directional information. For each AM-FM feature, they compute the histograms over the plaque regions. The study showed that the AM-FM features performed slightly better than the traditional texture features and gave better results than simple histogram. In previous work [13] on the same problem a large number of features were extracted for the classification of carotid plaques including the traditional texture features, statistical features, and shape.

5 Concluding Remarks

The AM-FM analysis presented in this study was performed on the IMC, ML, and IL, on 100 ultrasound images of the CCA of asymptomatic subjects. It was shown that the (IA) of the media layer decreases with age and that the median instantaneous frequency (IF) of the media layer increases with age. AM-FM analysis reveals the instantaneous amplitude (IA) of the media layer decreases with age. This decrease in IA maybe attributed to the reduction in calcified, stable plaque components and an increase in stroke risk with age. On the other hand, an increase in the median instantaneous frequency (IF) of the media layer suggests the fragmentation of solid, large plaque components, which also lead to an increase in the risk of stroke. The findings suggest that AM-FM features can be used to assess the risk of stroke over a wide range of patient populations. It may also be possible to identify and differentiate those individuals into high and low risk groups according to their cardiovascular risk before the development of plaques. The proposed methodology may also be applied to a group of people, which already developed plaques in order to study the contribution of the ML texture features to cardiovascular risk. Both groups of patients may be benefited by prognosing and managing future cardiovascular events. The use of AM-FM representations will also be utilised in order to provide new feature sets, which can be used successfully for the classification of the IMC, ML and IL structures in normal and abnormal. Future work will incorporate a new texture image retrieval system that uses AM-FM features to retrieve intima and intima media layer images that could be associated with the same level of the risk of stroke.

References

1. American Heart Association: Heart disease and stroke statistics-2007, update, Dallas, Texas (2007)
2. Pignoli, P., Tremoli, E., Poli, A., Oreste, P., Paoletti, R.: Intimal plus medial thickness of the arterial wall: a direct measurement with ultrasound imaging. Circulation 74(6), 1399–1406 (1986)
3. Loizou, C.P., Pantziaris, M., Pattichis, M.S., Kyriacou, E., Pattichis, C.S.: Ultrasound image texture analysis of the intima and media layers of the common carotid artery and its correlation with age and gender. Computerized Medical Imaging and Graphics 33(4), 317–324 (2009)
4. Loizou, C., Pattichis, C., Pantziaris, M., Nicolaides, A., Georgiou, N., Kyriakou, E.: Media thickness measurement of the common carotid artery. In: Annual International Conference of the IEEE Engineering in Medicine and Biology Society, pp. 2171–2174 (2007)
5. Loizou, C.P., Pattichis, C.S., Nicolaides, A.N., Pantziaris, M.: Manual and automated media and intima thickness measurements of the common carotid artery. IEEE Transactions on Ultrasonics, Ferroelectrics and Frequency Control 56(5), 983–994 (2009)
6. Murray, V., Murillo, S., Pattichis, M., Loizou, C., Pattichis, C., Kyriacou, E., Nicolaides, A.: An AM-FM model for motion estimation in atherosclerotic plaque videos. In: 41st IEEE Asilomar Conference on Signals, Systems and Computers, pp. 746–750 (2007)

7. Christodoulou, C.I., Pattichis, C.S., Murray, V., Pattichis, M.S., Nicolaides, A.: AM-FM representations for the characterization of carotid plaque ultrasound images. In: 4th European Conference of the International Federation for Medical and Biological Engineering, Antwerp, Belgium, pp. 546–549 (2008)
8. Mario, C., Gorge, G., Peters, R., Pinto, F., Hausmann, D., von Birgelen, C., Colombo, A., Murda, H., Roelandt, J., Erbel, R.: Clinical application and image interpretation in coronary ultrasound. study group of intra-coronary imaging of the working group of coronary circulation and of the subgroup of intravascular ultrasound of the working group of echocardiography of the european society of cardiology. European Heart Journal 19, 201–229 (1998)
9. Gronholdt, M., Nordestgaard, B.G., Schroeder, T.V., Vorstrup, S., Sillesen, H.: Ultrasonic echolucent carotid plaques predict future strokes. Circulation 104(1), 68–73 (2001)
10. Gussenhoven, E., Frietman, P., van Suylen, R., van Egmond, F., Lancee, C.: Assessment of medial thinning in atherosclerosis by intravascular ultrasound. American Journal of Cardiology 68, 1625–1632 (1991)
11. Ellis, S., Sidhu, P.: Granularity of the carotid artery intima-medial layer: reproducibility of quantification by a computer-based program. The British Journal of Radiology 73(870), 595–600 (2000)
12. Lind, L., Andersson, J., Rönn, M., Gustavsson, T.: The echogenecity of the intima-media complex in the common carotid artery is closely related to the echogenecity in plaques. Atherosclerosis 195(2), 411–444 (2007)
13. Christodoulou, C., Pattichis, C., Pantziaris, M., Nicolaides, A.: Texture-based classification of atherosclerotic carotid plaques. IEEE Transactions on Medical Imaging 22(7), 902–912 (2003)
14. In: Prevention of disampling and fatal strokes by successful carotid endarterectomy in patients without recent neurological symptoms: randomized control trial, vol. 363, pp. 1491–1502. The Lancer (2004)
15. Graf, S., Gariery, J., Massonneau, M., Armentano, R., Mansour, S., Barra, J., Simon, A., Levenson, J.: Experimental and clinical validation of arterial diameter waveform and intimal media thickness obtained from B-mode ultrasound image processing - fundamental principles and description of a computerized system. Ultrasound in Medicine & Biology 25(9), 1353–1363 (1999)
16. Loizou, C., Pattichis, C., Pantziaris, M., Tyllis, T., Nicolaides, A.: Snakes based segmentation of the common carotid artery intima media. Medical and Biological Engineering and Computing 45(1), 35–49 (2007)
17. Tegos, T.J., Sabetai, M.M., Nicolaides, A.N., Elatrozy, T.S., Dhanjil, S., Stevens, J.M.: Patterns of brain computed tomography infarction and carotid plaque echogenicity. Journal of Vascular Surgery 33(2), 334–339 (2001)
18. Williams, D.J., Shah, M.: A fast algorithm for active contours and curvature estimation. CVGIP: Image Underst 55(1), 14–26 (1992)
19. Victor Manuel Murray Herrera: AM-FM Methods for Image and Video Processing. PhD thesis, University of New Mexico (2008)
20. Murray, V., Rodriguez, V.P., Pattichis, M.S.: Multi-scale image AM-FM demodulation methods. In revision at IEEE Transactions on Image Processing
21. Havlicek, J.P.: AM-FM Image Models. PhD thesis, The University of Texas at Austin (1996)
22. Belcaro, G., Barsotti, A., Nicolaides, A.: "Ultrasonic Biopsy"–a non-invasive screening technique to evaluate the cardiovascular risk and to follow up the progression and the regression of arteriosclerosis. Vasa 20, 40–50 (1991)

Unsupervised Clustering of Clickthrough Data for Automatic Annotation of Multimedia Content

Klimis Ntalianis[1], Anastasios Doulamis[2], Nicolas Tsapatsoulis[3], and Nikolaos Doulamis[1]

[1] Electrical and Computer Engineering Department, National Technical University of Athens
9, Iroon Polytechniou str., Zografou 15773, Athens, Greece
kntal@image.ntua.gr, ndoulam@cs.ntua.gr

[2] Department of Production Engineering and Management, Technical University of Crete 73100, Chania, Greece
adoulam@ergasya.tuc.gr

[3] Cyprus University of Technology, Arch. Kyprianos, P.O. Box 50329, 3603, Limmasol, Cyprus
nicolas.tsapatsoulis@cut.ac.cy

Abstract. Current low-level feature-based CBIR methods do not provide meaningful results on non-annotated content. On the other hand manual annotation is both time/money consuming and user-dependent. To address these problems in this paper we present an automatic annotation approach by clustering, in an unsupervised way, clickthrough data of search engines. In particular the query-log and the log of links the users clicked on are analyzed in order to extract and assign keywords to selected content. Content annotation is also accelerated by a carousel-like methodology. The proposed approach is feasible even for large sets of queries and features and theoretical results are verified in a controlled experiment, which shows that the method can effectively annotate multimedia files.

Keywords: image retrieval, automatic annotation of multimedia, clickthrough data.

1 Introduction

The number of Web multimedia files grows in an incredibly fast way and an urging need is related to the efficient search and retrieval of content. Considering that the majority of multimedia files are not annotated and manual annotation is time/money consuming, intelligent systems for automatic annotation should be implemented.

Automatic annotation is an extremely difficult problem. Researchers try to con-front it under several constraints. In particular concept detection through supervised training on simple concepts such as city, landscape, and sunset, is

C. Alippi et al. (Eds.): ICANN 2009, Part II, LNCS 5769, pp. 895–904, 2009.

proposed in [15]. Image annotation using both a structure-composition model and a WordNet-based word saliency measure has been proposed in [7]. In [8] ALIP is proposed, which uses a 2D multiresolution HMMs-based approach to capture inter - and intrascale spatial dependencies of image features of given semantic categories. The real-time image annotation system ALIPR has been recently proposed [9]. ALIPR inherits its high-level learning architecture from ALIP, using however a simpler modeling approach, which supports real-time computations of statistical likelihoods. An approach to soft annotation, using Bayes point machines to give images a confidence level for each trained semantic label, is explored in [6]. Multiple instance learning based approaches have been proposed for semantic categorization of images [5] and to learn the correspondence between image regions and keywords [16]. Other interesting works include [13] focusing on the detection of simple concepts such as indoor/outdoor, while significant research has been directed toward detecting more challenging concepts in the context of the TREC video benchmark [12]. Additionally large sets of various concepts have been addressed in recent work, such as [3] and [4]. Popular approaches in concept classification mainly rely on SVMs [1], [10] or boosting approaches [14]. However, in case of multiple-word queries, concept classifiers are more difficult to apply since the independent training of each concept classifier requires the definition of fusion rules [4], [1].

All aforementioned techniques depend on low-level visual features. However neither a single low-level feature nor a combination has explicit semantic meaning. To overcome these problems in this paper we incorporate user clickthroughs. Having in mind that most users are unwilling to give explicit feedback, our method is based on implicit interaction, arguing that sufficient information is already hidden in the logfiles of WWW search engines. The whole framework is based on associating keywords of user queries to selected multimedia files. In particular searching is performed using keywords. The search mechanism retrieves multimedia files, some of which are selected by users. Then the proposed method properly associates query keywords to selected files according to an unsupervised clustering mechanism.

In this paper we focus on images, however the proposed scheme can be applied to any other application domain. Finally robustness, scalability and flexibility of the proposed system are evaluated in real-life settings.

This paper is organized as follows: in Section 2 the methodology of clickthrough data clustering and keyword assignment are described. Section 3 exhibits the advantages of the proposed system through analytical experiments. Finally, Section 4 concludes this paper.

2 Unsupervised Clustering of Clickthrough Data and Keyword Weighting

Clickthrough data in search engines can be thought of as triplets (q, r, c) consisting of the submitted query q (consisting of some keywords), the ranking r, and the set c of images the user clicked on. Figure 1 illustrates this with an example from Google image search, where the user asked for "amazing beach", receiving the ranking shown in Figure 1 (21 first images are presented), and then clicking on links ranked 2nd, 3rd, 7th, 15th, 16th and 18th. Clearly, users do not click on links at random, but make a somewhat informed choice. While clickthrough data is typically noisy and particular clicks are not perfect objective judgments, millions of clicks in an iterative and converging procedure are likely to convey important information. The key question is: how can we extract this information? Before analyzing clickthrough data, we first overview the recording procedure.

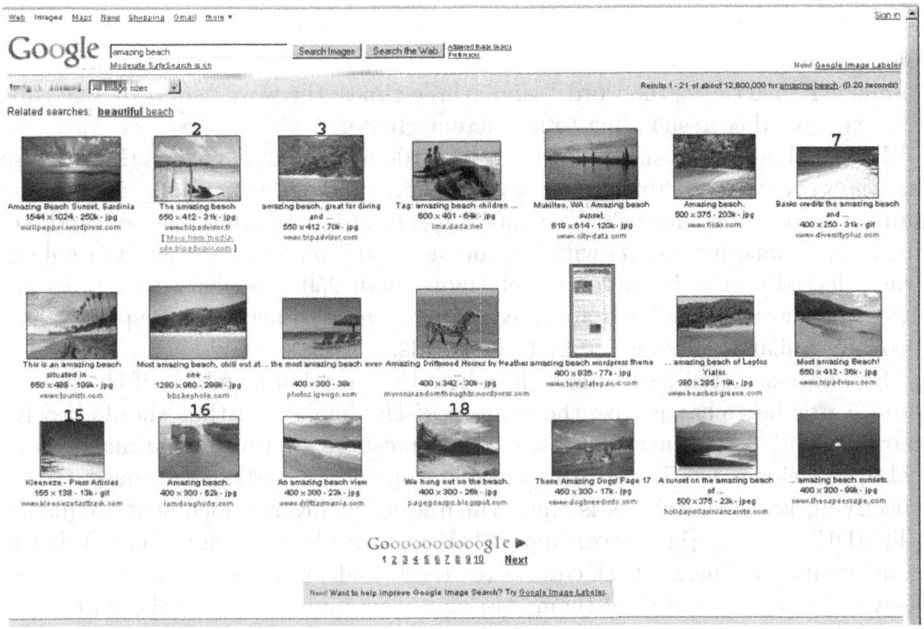

Fig. 1. An example of a search for "amazing beach" in Google and the respective user selections

2.1 Recording of Clickthrough Data

Clickthrough data recording adds little overhead without compromising the functionality of a search engine. The query q and the returned ranking r can easily be recorded whenever results are displayed to users. On the other hand clicks can be recorded by a simple proxy system that keeps a logfile. In this paper each query is assigned a unique ID which is stored in the query-log along with the query words and the presented ranking. The links on the results-page presented to the user do not lead directly to the suggested images, but point to a proxy server. These links en-code the query-ID and the URL of the suggested image. When the user clicks on the link, the proxy-server records the URL and the query-ID in the click-log. The proxy then uses the HTTP Location command to forward the user to the target URL. This process can be made transparent to the user and does not influence system performance.

2.2 Clustering of Clickthrough Data and Keyword Weights

By organizing the clickthrough recording procedure in the aforementioned way, we are able to associate keywords to images based on user selections. For example in case of the query "amazing beach" both words will be associated to every retrieved image the user selects. If one or both of the words have been already associated to an image by another user in a previous search session, then word rank increases in a similar manner to a voting scheme. By this way, for example, an image may have been associated the word "beach" ten times, the word "amazing" five times, the word "sand" three times, the word "sunbed" one time etc, each word corresponding to a different cluster.

Here it should be mentioned that the top 100 most common words (known also as stop words) of the Project Gutenberg (http://www.gutenberg.org) are ignored during association, since their semantic meaning is negligible (the, of, and, to, in, I, that, was, his, he, it, with, is, for, as, had, you, etc). Project Gutenberg was selected due to its vast test set (more than 25000 books with more than 1.8 billion words have been analyzed). Furthermore other word frequency lists provide similar results for the first 100 words.

Having associated keywords to files now the question is: which of the associated words best characterize the content of the image and thus should receive more weight? This is actually a rhetorical question. However some mathematical approaches exist. Toward this direction one of the best-known measures for specifying keyword weights is the term frequency/inverse document frequency (TF-IDF) measure [11]. According to this measure let us assume that N is the total number of images that can be retrieved and presented to users and that keyword k_i appears in n_i of them. Moreover, assume that $f_{i,j}$ is the number of times keyword k_i is associated to image I_j, according to the cluster's population. Then, $TF_{i,j}$, the term frequency (or normalized frequency) of keyword k_i in image I_j, is defined as

$$TF_{i,j} = \frac{f_{i,j}}{max_z f_{z,j}} \tag{1}$$

where the maximum is computed over the frequencies $f_{z,j}$ of all keywords k_z that are associated to image I_j. However, keywords that are associated to many images are not useful in distinguishing between a relevant image and a non-relevant one. Therefore, the measure of inverse document frequency (IDF_i) is often used in combination with simple term frequency ($TF_{i,j}$). The inverse document frequency for keyword k_i is usually defined as:

$$IDF_i = log(\frac{N}{n_i}) \tag{2}$$

Then, the TF-IDF weight for keyword k_i in image I_j is defined as:

$$w_{i,j} = TF_{i,j} \times IDF_i \tag{3}$$

and the verbal content of image I_j is defined as:

$$VCont(I_j) = (w_{1,j}, ..., w_{k,j}) \tag{4}$$

Based on the previous methodology, a high keyword weight is reached by a high term frequency (in the given image) and a low frequency of the term in the whole collection of images; this approach hence tends to filter out common terms, by as-signing higher weights to more distinctive keywords.

3 Experimental Results

In this section we test the proposed method with a general-purpose image database including 1000 images. These images where selected from flickr.com and belong to ten different general categories (landscape, monument, food, space, painting, animal, sports, music, plant, people). In order to evaluate the performance of our method as an automatic annotation scheme we have asked 50 users to interact with the system and for each user we have recorded 300 interaction sessions. By this way 15000 sessions have been recorded and analyzed in total. The first 7500 sessions have been used as initialization sessions (first phase), since in the beginning of the experiments images were not annotated at all. According to this methodology images were retrieved at random, irrespectively of the search keywords a user used. Furthermore keyword weights were also set after completing half of the experiment (when the first 7500 sessions have been completed). Of course, especially for larger sets, other automatic methods [15], [13] could also be incorporated in order to get better and more meaningful initialization results.

On completing the initialization step (first phase), we observed that users selected 2.6 images per session on average and used 1612 different keywords during searching (excluding the 100 most common words as stated in Section 2). At the end of this first phase 440 images were annotated with 3.66 keywords on average. Three of these images are presented in Figure 2 together with the associated keywords. Keyword order of appearance corresponds to the weights in decreasing order. Next a second phase of our experiment was carried. Aim of this phase was to pro-mote the rest of the images that have not been annotated at all and to validate the correctness of the initial annotations.

In order to accelerate the annotation procedure, this phase started with the estimation of a weight for each keyword, associated to each one of the 440 images. Next for each of the best three keywords of each image, clusters have been produced. For example the cluster of images that were associated to keyword “beach” is presented in Figure 3. Aim of such a clustering is to reduce the number of retrieved annotated images during the second phase, by returning

Fig. 2. Three of the 440 images that were annotated during the first phase of the experiment. The associated keywords in weight order are: (a) food, egg, apple, rice, (b) guitar, flute, resonator, and (c) boat, sea, sport, sail, people.

only a small part of them. This part is returned to users for validation purposes. For example, during the second phase, when a user submits the query "beach", only a small number (4 images) of the cluster of Figure 3 is retrieved, while the rest of the retrieved images are totally non-annotated. This small number of images changes in every query, based on a Carousel algorithm [2], so that all images belonging to a cluster have the chance to be validated in a subsequent cycle of the experiment.

The same methodology was followed for all examined keywords, leading finally to the annotation of 673 images. Of course more images would have been annotated if the experiment continued. Here it should be mentioned that the proposed method does not guarantee that all images will be finally annotated. However a large subset of them has been annotated (about 67%) and the evolution of the procedure shows that further increase in the number of annotated images can be achieved in oncom-ing sessions (Figure 4).

Now regarding accuracy of the proposed system, it is difficult to estimate the precision and recall, since images are initially considered not to be annotated. Of course users of flickr.com have explicitly assigned some keywords. For example in case of the first image of Figure 2 the tags (keywords) were "garden, ingredients, food market, olive oil, seed, meal, brown rice, oats, yeast, recipes, harvest, fall, autumn, seasons". As it can be observed some of the words refer to the real content of the image, some others are irrelevant to the actual content and the rest are general words.

In this case the only common words between our approach and the predetermined keywords of the flickr.com site are "food" and "rice". Based on such a comparison, in Figure 5 the number of images that are assigned the same keywords (by flickr.com and by the users of the proposed system) is presented. As it can be observed most frequently 1 or 2 words are common, while about 40 images do not share any common words. Of course it should be mentioned that it is very difficult to compare the perception of users, something which is also

Fig. 3. 34 out of 440 images annotated during the first phase were associated to keyword "beach" as one of the first three keywords

out of the scope of this paper. This is why in this work we let the annotation to the average user's preferences, selections and understanding. All users contribute to this task (by submitting different queries and selecting different content) and the associated keywords correspond to the average perception.

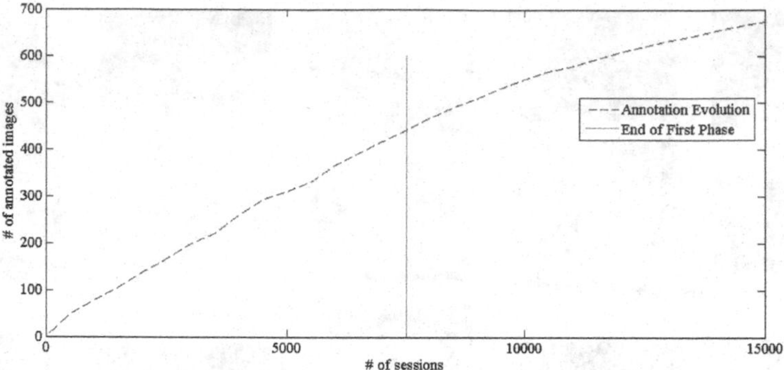

Fig. 4. Same annotation keywords for flickr.com and users of the proposed system. 1 and 2 words are encountered in 601 images, while 5 and 6 words just in 4 images.

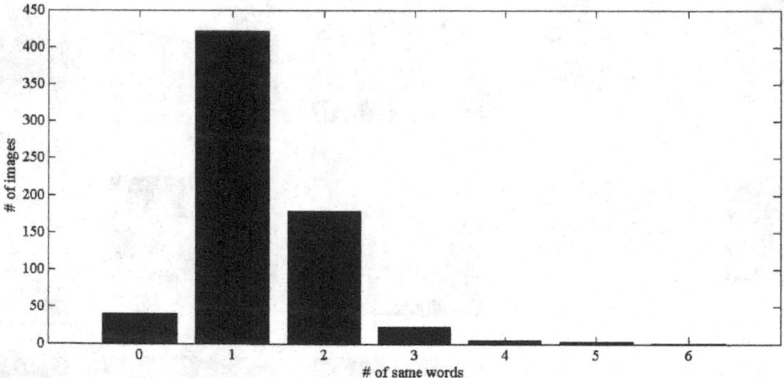

Fig. 5. Same annotation keywords for flickr.com and users of the proposed system. 1 and 2 words are encountered in 601 images, while 5 and 6 words just in 4 images.

4 Conclusion and Further Work

The Internet is currently overwhelmed with images and other multimedia files. Most of them are not annotated, while some of the annotated ones are assigned several keywords that do not reflect the perception of the average user. To ensure easy sharing and effective searching over a huge and fast growing number of online images, automatic keyword annotation is an imperative but highly challenging task. In this paper we have proposed and tested our automatic keyword annotation system, which is based on implicit user interaction. The main framework depends on recording and analyzing click through data produced during user search sessions. In particular each time a user submits a query and selects some of the retrieved images, each image is assigned the query keywords according to a voting scheme, where weights are properly set. By this way user perception is implicitly incorporated for automatically annotating images.

Additionally in order to accelerate the annotation procedure, image clusters are produced, with already annotated images. By this way in subsequent search cycles only a small number of annotated content is retrieved for validation reasons. The procedure is based on a Carousel algorithm.

Experimental results show the promising performance of the proposed system. In the performed experiments the total number of annotated images was near 67%, presenting an increasing tendency, which however is not a proof that finally all images will be annotated. However, even if a 70% is achieved this means that the proposed system can significantly decrease the manual work. Here it should be men-tioned that users participating in the experiments were informed of the content cate-gories. Furthermore they also tried to avoid selecting images in a random way, and thus avoid introducing any significant noise.

Future work can take many directions. First of all the initial annotation can be per-formed by incorporating other systems that possibly recognize simple concepts such as indoor/outdoor, landscape/cityscape, sea, sky etc. Secondly image analysis techniques can be incorporated, instead of Carousel-like methodologies, during validation. Third we should further model the way of how many opportunities should a non-annotated file receive (when retrieved and presented to users) before being excluded from the class of a specific keyword. Finally it would be interesting to test the system in larger dimensions (larger set of images, image categories and users) and under the existence of noise. In conclusion the synergy of implicit user interaction and automatic annotation is a novel and very interesting way that will probably open new horizons to the multi-media annotation scientific area.

Acknowledgments. This work was undertaken in the framework of the Commandaria project (The History of Commandaria: Digital Journeys Back to Time) project funded by the Cyprus Research Promotion Foundation (CRPF) under the contract ANTHRO/0308(BIE)/04.

References

1. Amir, A., Iyengar, G., Argillander, J., Campbell, M., Haubold, A., Ebadollahi, S., Kang, F., Naphade, M.R., Natsev, A., Smith, J.R., Tesic, J., Volkmer, T.: IBM Research TRECVID-2005 Video Retrieval System. In: Proceedings of TREC Video Workshop, Gaithersburg, MD (November 2005)
2. Bartholdi, J.J., Platzman, L.K.: Retrieval Strategies for a Carousel Conveyor. IIE Transactions 18(2), 166–173 (1986)
3. Carneiro, G., Vasconcelos, N.: Formulating Semantic Image Annotation as a Supervised Learning Problem. In: Proceedings of IEEE International Conference on Computer Vision and Pattern Recognition (CVPR 2005), vol. 2, pp. 163–168 (2005)
4. Chang, S.-F., Hsu, W., Jiang, W., Kennedy, L., Xu, D., Yanagawa, A., Zavesky, E.: Trecvid-2006 Video Search and High-Level Feature Extraction. In: Proc. TREC Video Workshop, Gaithersburg, MD (November 2006)
5. Chen, Y., Wang, J.Z.: Image categorization by learning and reasoning with regions. Journal of Machine Learning Research (5), 913–939 (2004)

6. Chang, E.Y., Goh, K., Sychay, G., Wu, G.: CBSA: Content-based soft annotation for multimodal image retrieval using Bayes point machines. IEEE Transactions Circuits Systems Video Technology 13(1), 26–38 (2003)
7. Datta, R., Ge, W., Li, J., Wang, J.Z.: Toward bridging the annotation-retrieval gap in image search. IEEE Multimedia 14(3), 24–35 (2007)
8. Li, J., Wang, J.Z.: Automatic linguistic indexing of pictures by a statistical modeling approach. IEEE Transactions on Pattern Analysis and Machine Intelligence 25(9), 1075–1088 (2003)
9. Li, J., Wang, J.Z.: Real-time computerized annotation of pictures. IEEE Transactions on Pattern Analysis and Machine Intelligence 30(9), 985–1002 (2008)
10. Naphade, M.R.: On Supervision and Statistical Learning for Semantic Multimedia Analysis. Journal on Visual Communication and Image Representation 15(3), 348–369 (2004)
11. Salton, G.: Automatic Text Processing. Addison-Wesley, Reading (1989)
12. Smeaton, A.F., Over, P., Kraaij, W.: Evaluation Campaigns and TRECVid. In: Proceedings of the 8th ACM Workshop Multimedia Information Retrieval, Santa Barbara, California, USA, pp. 321–330 (2006)
13. Szummer, M., Picard, R.W.: Indoor-Outdoor Image Classification. In: Proceedings of the IEEE International Workshop Content-Based Access of Image and Video Databases, pp. 42–51 (1998)
14. Tieu, K., Viola, P.: Boosting Image Retrieval. International Journal on Computer Vision 56(1), 17–36 (2004)
15. Vailaya, A., Figueiredo, M.A.T., Jain, A.K., Zhang, H.-J.: Image classification for content-based indexing. IEEE Transactions on Image Processing 10(1), 117–130 (2001)
16. Yang, C., Dong, M., Fotouhi, F.: Region based image annotation through multiple-instance learning. In: Proceedings of the 13th ACM International Conference on Multimedia, pp. 435–438 (2005)

Object Classification Using the MPEG-7 Visual Descriptors: An Experimental Evaluation Using State of the Art Data Classifiers

Nicolas Tsapatsoulis and Zenonas Theodosiou

Cyprus University of Technology, 31 Arch.Kyprianos,
P.O. Box 50329, 3603, Limassol, Cyprus
{nicolas.tsapatsoulis,zenonas.theodosiou}@cut.ac.cy

Abstract. MPEG-7 visual descriptors include the color, texture and shape descriptor and were introduced, after a long period of evaluation, for efficient content-based image retrieval. A total of 22 different kind of features are included, nine for color, eight for texture and five for shape. Encoded values of these features vary significantly and their combination, as a means for better retrieval, is neither straightforward nor efficient. Despite their extensive usage MPEG-7 visual descriptors have never compared concerning their retrieval performance; thus the question which descriptor to use for a particular image retrieval scenario stills unanswered. In this paper we report the results of an extended experimental study on the efficiency of the various MPEG-7 visual features with the aid of the Weka tool and a variety of well-known data classifiers. Our data consist of 1952 images from the athletics domain, containing 7686 manually annotated objects corresponding to eight different classes. The results indicate that combination of selected MPEG-7 visual features may lead to increased retrieval performance compared to single descriptors but this is not a general fact. Furthermore, although the models created using alternative training schemes have similar performance libSVM is by far more effective in model creation in terms of training time and robustness to parameter variation.

Keywords: image retrieval, multimedia annotation, MPEG-7 visual descriptors, supervised classification.

1 Introduction

Automatic image annotation has gained great attention in the research community because it deals with a real world problem which is laborious to be handled with human intervention exclusively: Searching in image repositories of thousands of images which they have not got explicit metadata assigned to them by humans. In the MPEG-7 framework there is a special foresight for this problem through the definition of the MPEG-7 visual descriptors [1]. These descriptors are low-level image features proposed after an extended evaluation procedure [2].

C. Alippi et al. (Eds.): ICANN 2009, Part II, LNCS 5769, pp. 905–912, 2009.

No doubt that much of the attention paid recently to automatic image annotation and CBIR systems is due to the MPEG-7 visual content description interface, which provides a unified framework for experimentation. Furthermore, the MPEG-7 experimentation model [3] provides practical ways for the computation of the MPEG-7 descriptors.

The performance of the MPEG-7 visual descriptors in terms of image retrieval, however, was not examined in detail. Although inclusion of these particular descriptors in the MPEG-7 protocol stack was based on experimental evaluation, the results were not published and the experiments cannot be recreated. Investigation of the performance of color and texture descriptors was reported in [2] but the main discussion there was devoted to the introduction of these descriptors to the research community rather than to experimental evaluation. The same holds for the work of Bober [4], which deals with the shape descriptors. A very interesting study on the MPEG-7 visual descriptors was conducted by Eidenberger in [5]. The descriptors are evaluated using statistics obtained by three different datasets including the one used during the MPEG-7 tests. One of the aims of the current study is to investigate experimentally whether the conclusions made by Eidenberger are valid in a different dataset and by using a variety of classifiers. Spyrou in [6] investigates a variety of methods for fusing the MPEG-7 visual descriptors for image classification. The idea is interesting but the dataset used is small and the experiments cannot be recreated based on the description given in the corresponding paper.

In this paper we deal with the experimental evaluation of the performance of the MPEG-7 descriptors [1] in terms of object classification. None of the works reported in the previous paragraph deals with object classification. This is quite logical since the MPEG-7 visual descriptors were defined primarily for image classification and not for object detection and classification. Furthermore, manual annotation of image objects through definition of the blob area is much harder than image annotation. In our study we get advantage of the availability of a large dataset of manually annotated objects created during the FP6 BOEMIE project [7] to perform extended experiments. We have used publicly available tools for the computation the MPEG-7 descriptors [3] and the object model creation (the Weka tool [8] and the libSVM [9] library integrated with Weka) and we provide both the training and test files along with the created Weka models [10] so as to allow experiments recreation and benchmark tests.

The paper is organized as follows: In Section 2 the MPEG-7 visual descriptors used in this study are presented. The dataset used, the annotation process and the object modeling method we have followed are explained in Section 3. Extended experimental results are reported in Section 4. Finally, conclusions are drawn and further works hints are given in Section 5.

2 MPEG-7 Visual Descriptors

MPEG-7 visual descriptors include the color, texture and shape descriptor. A total of 22 different kind of features are included, nine for color, eight for texture and five for shape. The various feature types are shown in Table 1. In the third column of this Table is indicated whether or not the corresponding feature type is used in holistic image and/or object description. The number of features shown in the fourth column in most cases is not fixed and depends on user choice; we indicate there the settings in our implementation. The dominant color features include color value, percentage and variance and require especially designed metrics for similarity matching. Furthermore, their length is not known a priori since they are image dependent (for example an image may be composed from a single color whereas others vary in color distribution). The previously mentioned difficulties cannot be easily handled in machine learning schemes, therefore we decided to exclude these features for the current experimentation. The texture browsing features (regularity, direction, scale) have not been included in the description vectors (for image and image segments) because in the current implementation of the MPEG-7 experimentation model [3] the corresponding descriptor cannot be reliably computed (it is a known bug of the implementation software). The shape descriptor features are computed only on specific image regions (they are not used in the holistic image description). The number of Peaks values of the contour shape descriptor vary depending on the form of an input object. Furthermore, they require a specifically designed metric for similarity matching because they are computed based on the HighestPeak value. For these reason they have been excluded also from the segment description vector at this stage.

Table 1. MPEG-7 visual descriptors used in the proposed classification scheme

Descriptor	*Type*	*# of features*	*Usage level*	*Comments*
Color	DC coefficient of DCT (Y channel)	1	Both	Part of the Color Layout descriptor
	DC coefficient of DCT (Cb channel)	1	Both	Part of the Color Layout descriptor
	DC coefficient of DCT (Cr channel)	1	Both	Part of the Color Layout descriptor
	AC coefficients of DCT (Y channel)	5	Both	Part of the Color Layout descriptor
	AC coefficients of DCT (Cb channel)	2	Both	Part of the Color Layout descriptor
	AC coefficients of DCT (Cr channel)	2	Both	Part of the Color Layout descriptor
	Dominant colors	Varies	Both	Includes color value, percentage and variance
	Scalable color	16	Both	
	Structure	32	Both	They used in both holistic image and image segment description
Texture	Intensity average	1	Both	Part of the Homogeneous Texture descriptor
	Intensity standard deviation	1	Both	Part of the Homogeneous Texture descriptor
	Energy distribution	30	Both	Part of the Homogeneous Texture descriptor
	Deviation of energy's distribution	30	Both	Part of the Homogeneous Texture descriptor
	Regularity	1	Both	Part of the Texture Browsing descriptor
	Direction	1 or 2	Both	Part of the Texture Browsing descriptor
	Scale	1 or 2	Both	Part of the Texture Browsing descriptor
	Edge histogram	80	Both	Includes the spatial distribution of five types of edges
Shape	Region shape	35	Segment	A set of angular radial transform coefficients
	Global curvature	2	Both	Part of the Contour Shape descriptor
	Prototype curvature	2	Both	Part of the Contour Shape descriptor
	Highest peak	1	Both	Part of the Contour Shape descriptor
	Curvature peaks	Varies	Both	Describes curvature peaks in term of amplitude and distance from highest peak

3 Dataset Creation and Object Modeling

For dataset creation 1952 images from the athletics domain were used. These images were collected in the framework of the FP6 BOEMIE project [7] and objects, corresponding to humans and athletic instruments, were manually marked by humans creating blobs. Example of such blobs overlayed on the original images are shown in Figure 1. A total of 7686 manually annotated object instances corresponding to eight different class objects were used in our experiments. The eight object classes are: Person Body, Person Face, Horizontal Bar, Pole, Pillar, Discus, Hammer and Javelin. The training set contains 2597 instances while the remaining 5089 were used for test. The distribution of the various object instances in the training and test sets are presented in Table 2.

Fig. 1. Images from the athletic domain showing the detected objects (a)Person Body, (b) Person Face, (c)Horizontal Bar, (d)Pillar, (e)Pole, (f)Hammer, (g)Discus, (h)Javelin

Object models were created using Weka tool [8]. Among a variety of possible classifiers we decided to use (1)libSVM [9], (2) Sequential Minimal Optimization (SMO) [11], [12] and (3) Radial Basis Function networks [13]. The latter is a reasonable choice when dealing with multidimensional and multiclustered data while libSVM and SMO are state of the art implementations of Support Vector Machines. These algorithms have been reported in several publications as the best performing machine learning algorithms for a variety of classification tasks.

During training some parameters were optimized via experimentation in order to obtain the best performing model for each descriptor. Cost, Gamma, and Epsilon were optimally selected for the libSVM models. For SMO models we have experimented on the complexity constant C and then based on the chosen kernel type, we try to get the optimum values the exponent of the polynomial kernel or the Gamma for the RBF kernel respectively. Finally, for RBF model, the number of clusters and ridge were tuned for each one of the MPEG-7 descriptors.

In addition to the construction of individual models for each MPEG-7 descriptor we also trained models for several descriptor combinations using feature fusion. The parameter optimization followed was the same as the one described earlier. All trained models as well as the Weka training files can be found at [10] for evaluation and further experimentation.

Table 2. Dataset

Objects	*Number of instances*	*Training Set*	*Test Set*
Person Body	3180	1062	2118
Person Face	3209	1044	2165
Horizontal Bar	493	164	329
Pole	229	94	135
Pillar	138	51	87
Discus	132	49	83
Hammer	142	56	86
Javelin	163	77	86
Total	**7686**	**2597**	**5089**

4 Experimental Results

We used the dataset and object modeling process described in the previous section to examine the classification performance, of the eight object classes, in terms of precision and recall values. Table 3 summarizes the results for the models of the individual MPEG-7 descriptors while Table 4 shows the corresponding figures obtained using descriptor combinations. The results shown in these tables can be examined under two perspectives: First, in terms of the efficiency of the various descriptors as far as the object classification task is concerned. Second, in terms of the ability of the machine learning algorithms to create efficient object class models for classification.

Concerning the classification efficiency of the individual MPEG-7 descriptors it is evident from Table 3 that the most reliable descriptor is Edge Histogram. Not only has the ability to discriminate the whole range of the eight classes used but the precision and recall values obtained using this descriptor are quite good irrespectively of the training algorithm used. This result is in full agreement with the conclusion drawn by Eidenberger [5] who examines the efficiency of the MPEG-7 descriptors using statistical analysis on different datasets. The second most reliable descriptor for object classification is Color Structure. Although the precision and recall values obtained for the classes with few training examples (that is, all classes but Person Body and Person Face) are rather low this descriptor has the potential to discriminate multiple classes irrespectively of the training algorithm used. The Contour Shape descriptor is effective for classification of objects having a well defined shape such as Horizontal Bar, Pole and Pillar. In contrary, it cannot be used for the classification of Discus and Hammer. These two classes although in principle they must have a circular shape their inaccurate segmentation, as created by the human annotators, make them appearing extremely variable in shape. Furthermore, they have been easily confused with Person Face as far as the shape (and probably their size) is concerned. The most disappointing classification performance is achieved by the Region Shape descriptor. Although it contains much more features than the Contour Shape descriptor, it is only able to discriminate Person Body and Person Face. These two classes have a high population of training samples and are easily discriminated by all descriptors (with some variance mainly in the precision values).

Table 3. Object classification results using the MPEG-7 visual descriptors and various data classifiers

Classifier	*Descriptor*	*Measure*	*Object Class*							
			Person Body	*Person Face*	*Horizontal Bar*	*Pole*	*Pillar*	*Discus*	*Hammer*	*Javelin*
libSVM	Color Layout (CL)	Recall	0.818	0.876	0.152	0.252	0.241	0.060	0.151	0.163
		Precision	0.772	0.796	0.370	0.301	0.236	0.208	0.342	0.219
	Color Structure (CSt)	Recall	0.990	0.819	0.176	0.185	0.287	0.241	0.233	0.279
		Precision	0.750	0.921	0.527	0.439	0.556	0.465	0.588	0.308
	Scalable Color (SC)	Recall	0.817	0.847	0.313	0.400	0.333	0.145	0.314	0.070
		Precision	0.813	0.895	0.256	0.214	0.240	0.333	0.375	0.222
	Contour Shape (CS)	Recall	0.901	0.899	0.565	0.311	0.379	0.000	0.081	0.349
		Precision	0.875	0.848	0.699	0.609	0.317	0.000	0.636	0.201
	Region Shape (RS)	Recall	0.516	0.541	0.334	0.000	0.264	0.000	0.000	0.000
		Precision	0.475	0.513	0.298	0.000	0.200	0.000	0.000	0.000
	Edge Histogram (EH)	Recall	0.986	0.870	0.818	0.615	0.529	0.349	0.209	0.651
		Precision	0.864	0.931	0.906	0.669	0.767	0.744	0.720	0.549
	Homogenous Texture (HT)	Recall	0.968	0.762	0.252	0.104	0.460	0.325	0.291	0.093
		Precision	0.783	0.824	0.653	0.304	0.444	0.297	0.379	0.170
SMO	Color Layout (CL)	Recall	0.906	0.866	0.195	0.200	0.184	0.012	0.093	0.198
		Precision	0.758	0.830	0.547	0.391	0.333	0.200	0.500	0.370
	Color Structure (CSt)	Recall	0.992	0.763	0.179	0.111	0.184	0.133	0.349	0.221
		Precision	0.720	0.931	0.476	0.283	0.410	0.500	0.612	0.171
	Scalable Color (SC)	Recall	0.996	0.306	0.000	0.000	0.000	0.000	0.012	0.000
		Precision	0.482	0.934	0.000	0.000	0.000	0.000	1.000	0.000
	Contour Shape (CS)	Recall	0.903	0.892	0.714	0.289	0.506	0.000	0.000	0.081
		Precision	0.868	0.840	0.685	0.639	0.270	0.000	0.000	0.467
	Region Shape (RS)	Recall	0.973	0.274	0.000	0.000	0.000	0.000	0.000	0.000
		Precision	0.482	0.730	0.000	0.000	0.000	0.000	0.000	0.000
	Edge Histogram (EH)	Recall	0.981	0.876	0.828	0.556	0.586	0.325	0.244	0.698
		Precision	0.874	0.928	0.906	0.688	0.680	0.692	0.636	0.546
	Homogenous Texture (HT)	Recall	0.943	0.635	0.256	0.091	0.325	0.102	0.232	0.000
		Precision	0.738	0.741	0.606	0.320	0.331	0.215	0.220	0.000
RBF Network	Color Layout (CL)	Recall	0.830	0.869	0.228	0.296	0.230	0.121	0.140	0.256
		Precision	0.797	0.825	0.346	0.342	0.198	0.227	0.200	0.339
	Color Structure (CSt)	Recall	0.949	0.818	0.374	0.311	0.322	0.133	0.326	0.326
		Precision	0.842	0.915	0.547	0.269	0.235	0.220	0.467	0.181
	Scalable Color (SC)	Recall	0.282	0.881	0.167	0.052	0.172	0.000	0.000	0.000
		Precision	0.615	0.527	0.200	0.119	0.119	0.000	0.000	0.000
	Contour Shape (CS)	Recall	0.914	0.883	0.559	0.311	0.379	0.000	0.058	0.302
		Precision	0.870	0.854	0.669	0.618	0.260	0.000	0.556	0.193
	Region Shape (RS)	Recall	0.515	0.407	0.207	0.007	0.149	0.000	0.000	0.000
		Precision	0.458	0.464	0.192	0.004	0.086	0.000	0.000	0.000
	Edge Histogram (EH)	Recall	0.985	0.785	0.520	0.637	0.482	0.361	0.670	0.581
		Precision	0.798	0.909	0.945	0.601	0.646	0.411	0.697	0.471
	Homogenous Texture (HT)	Recall	0.953	0.695	0.204	0.074	0.402	0.157	0.244	0.000
		Precision	0.743	0.765	0.632	0.312	0.321	0.245	0.212	0.000

Combinations of MPEG-7 descriptors are shown in Table 4. There, it can be seen that classification performance is increased through the use of feature based fusion for the majority of descriptor combinations. However, improvement in recall and precision values is not as significant as one might expect. This can be attributed to the variance of the feature values among different descriptors.

The efficiency of the training algorithms is examined through the effectiveness of the created models, the time required to train the models and the robustness to the variation of learning parameters. The libSVM algorithm requires by far the lower time and effort to create an effective model. This is true, however, if an RBF or a polynomial kernel is used. In such a case learning takes no more than a few seconds for the majority of the descriptor models. Furthermore, the fluctuation in classification performance during parameters' tuning is significantly lower than that of the other two training algorithms. The models created using libSVM are the ones that are able to discriminate between multiple classes for all individual descriptors used. A characteristic example is the model created for the Scalable Color descriptor. The libSVM model for this descriptor can be used for the discrimination between the seven of the eight object classes (Javelin class is an exception) while the corresponding SMO and RBF network models are only able to discriminate between three classes at most.

Table 4. Object classification results using selected combinations of the MPEG-7 visual descriptors and various data classifiers

Classifier	*Descriptors Combination*	*Measure*	*Object Class*							
			Person Body	*Person Face*	*Horizontal Bar*	*Pole*	*Pillar*	*Discus*	*Hammer*	*Javelin*
libSVM	SC and CS	Recall	0.910	0.901	0.580	0.421	0.382	0.150	0.336	0.352
		Precision	0.881	0.899	0.706	0.622	0.325	0.342	0.640	0.301
	SC and EH	Recall	0.991	0.882	0.825	0.631	0.542	0.361	0.329	0.662
		Precision	0.872	0.945	0.916	0.681	0.786	0.766	0.736	0.561
	CS and EH	Recall	0.995	0.990	0.831	0.634	0.536	0.359	0.230	0.669
		Precision	0.892	0.940	0.912	0.683	0.771	0.740	0.731	0.553
	SC and CS and EH	Recall	0.997	0.994	0.841	0.642	0.550	0.401	0.346	0.672
		Precision	0.895	0.951	0.922	0.689	0.801	0.770	0.742	0.571
SMO	SC and CS	Recall	0.998	0.895	0.725	0.291	0.520	0.000	0.020	0.092
		Precision	0.872	0.941	0.691	0.649	0.281	0.000	1.000	0.475
	SC and EH	Recall	0.999	0.881	0.835	0.568	0.589	0.331	0.251	0.703
		Precision	0.875	0.942	0.910	0.691	0.689	0.699	1.000	0.559
	CS and EH	Recall	0.982	0.899	0.832	0.560	0.591	0.335	0.251	0.702
		Precision	0.880	0.939	0.912	0.695	0.692	0.701	0.642	0.560
	SC and CS and EH	Recall	0.999	0.901	0.840	0.571	0.601	0.341	0.259	0.712
		Precision	0.882	0.945	0.915	0.680	0.682	0.701	1.000	0.565
RBF Network	SC and CS	Recall	0.915	0.888	0.669	0.325	0.388	0.000	0.062	0.306
		Precision	0.872	0.862	0.660	0.617	0.271	0.000	0.550	0.192
	SC and EH	Recall	0.988	0.895	0.529	0.652	0.488	0.370	0.679	0.592
		Precision	0.802	0.909	0.952	0.601	0.652	0.412	0.709	0.469
	CS and EH	Recall	0.985	0.890	0.572	0.642	0.492	0.360	0.682	0.592
		Precision	0.872	0.892	0.950	0.629	0.654	0.421	0.701	0.479
	SC and CS and EH	Recall	0.901	0.899	0.662	0.661	0.495	0.371	0.685	0.598
		Precision	0.879	0.912	0.960	0.631	0.659	0.431	0.712	0.481

5 Conclusion and Further Work

An extended experimental study on the efficiency of the MPEG-7 visual descriptors for object classification was presented in this paper. The use of three different training algorithm ensures that obtained results are not biased on the training scheme selected. The main conclusions of this work are:

1. There is a significant variation on the efficiency of the various descriptors with the Edge Histogram descriptor having the highest performance among all.
2. In contrary, to what one might expect the shape descriptors are not very efficient for object classification. Especially the performance of the Region Shape descriptor is disappointing.
3. Combination of descriptors increase the classification performance for the majority of object classes but in most cases the improvement is negligible.
4. The use of different training schemes leads, to some extent, to models with varying performance. However, libSVM is by far the scheme requiring the least time to train the models and presents the highest robustness with

respect to learning parameters' variation. The RBF Network on the other hand creates in several cases the most compact model in terms of file size.

Further work includes the examination of additional training algorithms as well as other classifications schemes (decision trees, fuzzy rules, etc). In addition comparison of the MPEG-7 descriptor combinations using score-based and decision-based fusion will be investigated.

References

1. ISO/IEC 15938-3:2001 Information Technology - Multimedia Content Description Interface - Part 3: Visual, Ver. 1
2. Manjunath, B.S., Ohm, J.-R., Vasudevan, V.V., Yamada, A.: Color and Texture Descriptors. IEEE Transactions on Circuits and Systems for Video Technology 11(6), 394–410 (2001)
3. MPEG-7 Visual Experimentation Model (XM), Version 10.0, ISO/IEC/JTC1/SC29/WG11, Doc. N4063 (2001)
4. Bober, M.: MPEG-7 Visual shape descriptors. Special Issue on MPEG-7. IEEE Transactions on Circuits and Systems for Video Technology 11(6), 716–719 (2001)
5. Eidenberger, H.: How good are the visual MPEG-7 features? In: Proc. SPIE, vol. 5150 (2003)
6. Spyrou, E., Le Borgne, H., Mailis, T., Cooke, E., Avrithis, Y., O'Connor, N.E.: Fusing MPEG-7 visual descriptors for image classification. In: Duch, W., Kacprzyk, J., Oja, E., Zadrożny, S. (eds.) ICANN 2005. LNCS, vol. 3697, pp. 847–852. Springer, Heidelberg (2005)
7. BOEMIE - Bootstrapping Ontology Evolution with Multimedia Information Extraction, `http://www.boemie.org`
8. Witten, I.H., Frank, E.: Data Mining: Practical machine learning tools and techniques, 2nd edn. Morgan Kaufmann, San Francisco (2005)
9. Fan, R.-E., Chen, P.-H., Lin, C.-J.: Working set selection using the second order information for training SVM. Jour. of Mach. Learn. Res. 6, 1889–1918 (2005)
10. `http://www.cs.ucy.ac.cy/~nicolast/MPEG7/data.rar`
11. Platt, J.: Fast Training of Support Vector Machines using Sequential Minimal Optimization. In: Schoelkopf, B., Burges, C., Smola, A. (eds.) Advances in Kernel Methods-Support Vector Learning. MIT Press, Cambridge (1998)
12. Keerthi, S.S., Shevade, S.K., Bhattacharyya, C., Murthy, K.R.K.: Improvements to Platt's SMO Algorithm for SVM Classifier Design. Neural Computation 13(3), 637–649 (2001)
13. Bishop, C.M.: Pattern Recognition and Machine Learning. Springer, Heidelberg (2006)

MuLVAT: A Video Annotation Tool Based on XML-Dictionaries and Shot Clustering

Zenonas Theodosiou[1], Anastasis Kounoudes[2],
Nicolas Tsapatsoulis[1], and Marios Milis[2]

[1] Cyprus University of Technology, 31 Arch.Kyprianos,
P.O. Box 50329, 3603, Limassol, Cyprus
{nicolas.tsapatsoulis}@cut.ac.cy
{z.theodosiou}@gmail.com
[2] SignalGeneriX Ltd, Arch.Leontiou A' Maximos Court B',
3rd floor, P.O. Box 51341, 3504, Limassol, Cyprus
{tasos,milis}@signalgenerix.com

Abstract. Recent advances in digital video technology have resulted in an explosion of digital video data which are available through the Web or in private repositories. Efficient searching in these repositories created the need of semantic labeling of video data at various levels of granularity, i.e., movie, scene, shot, keyframe, video object, etc. Through multilevel labeling video content is appropriately indexed, allowing access from various modalities and for a variety of applications. However, despite the huge efforts for automatic video annotation human intervention is the only way for reliable semantic video annotation. Manual video annotation is an extremely laborious process and efficient tools developed for this purpose can make, in many cases, the true difference. In this paper we present a video annotation tool, which uses structured knowledge, in the form of XML dictionaries, combined with a hierarchical classification scheme to attach semantic labels to video segments at various level of granularity. Video segmentation is supported through the use of an efficient shot detection algorithm; while shots are combined into scenes through clustering with the aid of a Genetic Algorithm scheme. Finally, XML dictionary creation and editing tools are available during annotation allowing the user to always use the semantic label she/he wishes instead of the automatically created ones.

Keywords: video annotation, hierarchical classification, XML dictionaries.

1 Introduction

Video annotation is a powerful tool for adding useful comments and explanations to video which can serve as the first step of different data access modalities. The use of an integrated system which can provide video annotation can greatly simplify the process.

C. Alippi et al. (Eds.): ICANN 2009, Part II, LNCS 5769, pp. 913–922, 2009.

However some annotation tools are available publicly, few of them are annotate in multiple levels. VideoAnnEx MPEG-7 annotation tool is implemented by IBM [1] for collaborative multimedia annotation task in distributed environment. MovieTool is developed by Ricoh for creating video content descriptions conforming to MPEG-7 syntax interactively [2]. The IBM Multimedia Mining Project released a Multimodal Annotation Tool, which is derived from an earlier version of VideoAnnEx with special features with audio signal graphs and manual audio segmentation functions [3]. Some other media annotation systems, including collaborative annotations, have been developed for various purposes. Bargeron et. al. developed an Microsoft Research Annotation System (MRAS), which is a web-based system for annotating multimedia web content [4]. Annotations include comments and audio in the distance learning scenario. Steves et. al. developed a Synchronous Multimedia and Annotation Tool (SMAT) [12]. SMAT is used to annotate images. There is no granularity for video annotations nor controlled-term labels. Nack and Putz developed a semi-automated annotation tool for audio-visual media in news [11]. The European Cultural heritage Online (ECHO) is developing a multimedia annotation tools which allows people to work collaboratively on a resource and to add comments to it [5].

In this paper we present a Video Annotation Tool based on MPEG-7 standard. The tool provides various features as annotation in multiple levels based on XML dictionaries, creation of XML dictionaries, and image watermarking. It consists of three panels and is supported by friendly graphical user interface, making it very practical and simple to use. The paper is organized as follows: Section 2 gives a brief description of the Annotation Tool. In Section 3 we present the algorithm that is used for shot clustering and scene construction. Finally, conclusions are drawn and further work hints are given in Section 4.

2 Video Annotation Tool

We have developed an annotation tool based on MPEG-7 standard using MATLAB. The tool boasts a user-friendly Graphical User Interface allowing the management of multimedia content (images and videos), video segmentation, video and image annotation, image watermarking and creation of XML dictionaries. The supported image and video file types loading by the annotation tool are presented in Table 1. For the tool application, any PC can be used as far as hardware is concerned, although as expected, the more power it has, the better performance level it will reach. In particular, a large amount of RAM memory will help to improve the performance.

The GUI consists of three major panels. The *Video Panel* provides the video segmentation and annotation while the *Image Panel* provides the image annotation and watermarking. The dictionaries used for annotation are created via the third panel named *Dictionary Panel.* A brief description of each panel is follows.

Table 1. Supported file types

Visual Content Type	File type Extension
images	jpg, bmp, gif, tiff, png
videos	mpg, avi, wav

2.1 Video Panel

An example screen of *Video Panel* is shown in Fig.1. The Panel consists of three tabs. The first and second tab provide the manual and automated segmentation respectively, while the third one accords the video annotation.

Video Segmentation. The video segmentation is performed to cut up the video sequence into smaller video units. As is shown in Fig.1, the user is able to select one of two available modes for video segmentation: manual and automated. During the manual segmentation, the opened video sequence displayed in the window on the upper left-hand corner. The user can explore it and set the shots boundaries by specified the first and last frame of each shot. Then he choose and set the representative frames as key-frames. The key-frame is a representative image of the video shot segment, and thus offer an instantaneous recap of the whole video shot. The shot frame boundaries and key-frames of each shot are saved in an XML-file.

For the automated shot detection we used the Color Histogram Differences algorithm [10]. The algorithm is one of the most trustworthy variants of histogram-based detection algorithms and is based on the idea that the color content rapidly changes across shots. So, hard cuts and other short-lasting transitions can be detected as single peaks in the time series of the differences between color histograms of contiguous frames or of frames a certain distance k apart.

Let $p_i(r,g,b)$ be the number of pixels of color (r,g,b) in frame I_i of N pixels. Each color component is discretized to 2^B different values, resulting in $r,g,b \in [0, 2^B-1]$. Usually B is set to 2 or 3 in order to reduce sensitivity to noise and slight light, object as well as view changes. Then the color histogram difference CHD_i between two color frame I_{i-1} and I_i is given by

$$CHD_i = \frac{1}{N} \cdot \sum_{r=0}^{2^B-1} \sum_{g=0}^{2^B-1} \sum_{b=0}^{2^B-1} \mid p_i(r,g,b) - p_{i-1}(r,g,b) \mid \qquad (1)$$

If within a local environment of radius I_c of frame I_i only CHD_i exceeds a certain threshold, then a hard cut is detected. As presented in [10], for particular type of hard cut which consists of one transitional frame, in a pre-processing stage double peaks (i.e groups of $s_c = 2$) contiguous CHD_i exceeding threshold were modified into single peaks at the higher CHD_i.

Video Annotation. A shot video clip can simply annotated by describing its content in its entirety. However, when the video is longer, annotation of its content can benefit from segmenting the video into smaller units. Given the shot boundaries, the annotations are assigned for each video shot by using the Video Annotation Tab. The tool uses a specific type of dictionaries based on MPEG-7 descriptions made via the *Dictionary Panel*. To be more precise the video annotation is performed through the following three steps.

First, the annotation dictionary (XML) that will be used to annotate the key-frames of each shot is loaded. Annotation dictionaries can be created using the *Dictionary Panel* as will be explained in Section 2.3. The three categories of the dictionary are shown in the list-boxes on under left-hand, as illustrated in Fig.2.

Second, the segmented video resulted from the video segmentation procedure is loaded and its shots are shown in shot axes at the upper right-hand corner. After choosing a shot, its key-frames are shown in the four axes below the shot axes. The user can choose a key-frame in order to annotate it. The chosen key-frame can be seen in the axes at the left-hand corner.

Third, the key-frame annotation can be implemented using the dictionary categories presented in list-boxes on under left-hand corner. The user ticks the boxes of the most representative annotations and adds if needed free text and key-words using the corresponding edit boxes. Annotations are shown in the list-box on the right-hand and are saved into an XML file in the video directory.

2.2 Image Panel

The *Video Annotation* tab provides the capability of annotating and embedding information into an image via the *Image Panel* A screen shot of *Image Panel* is

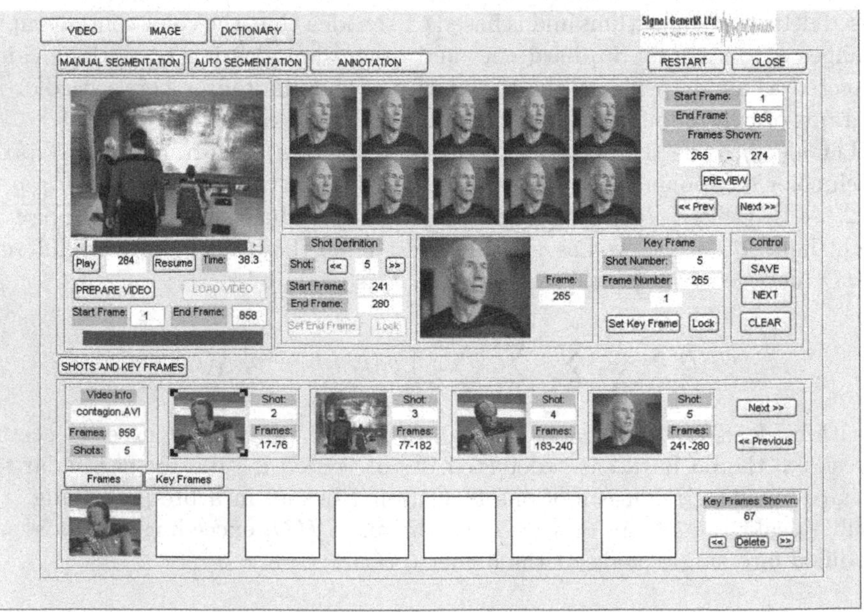

Fig. 1. Video Segmentation Tab

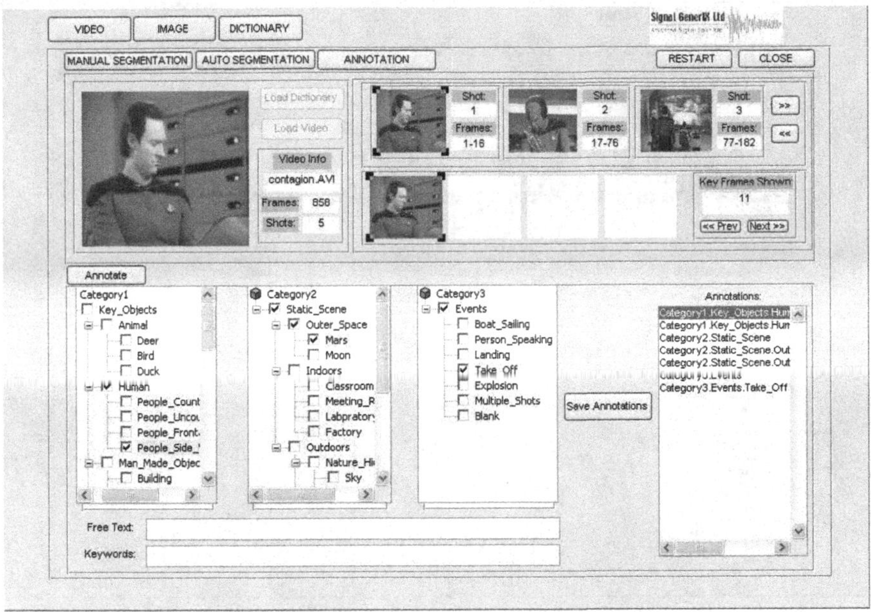

Fig. 2. Video Annotation Tab

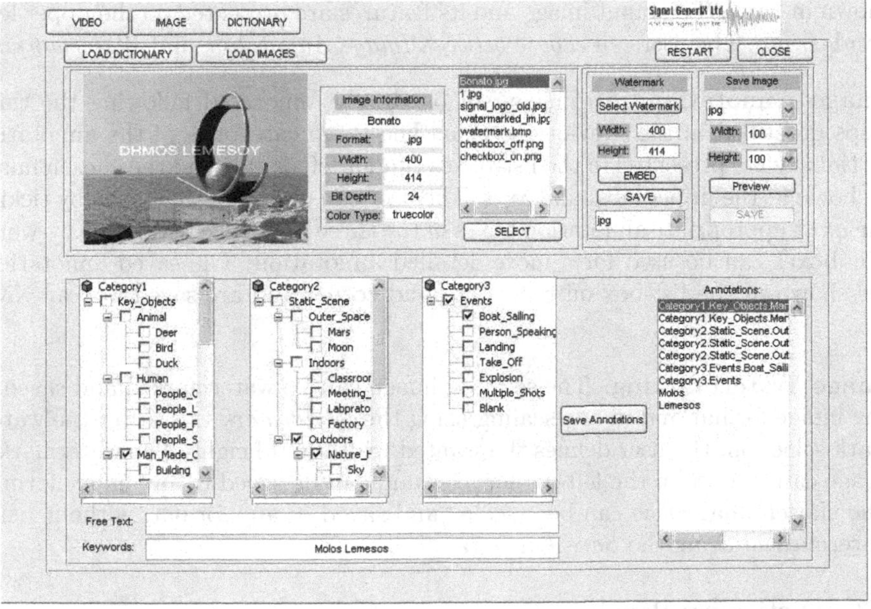

Fig. 3. Image Panel

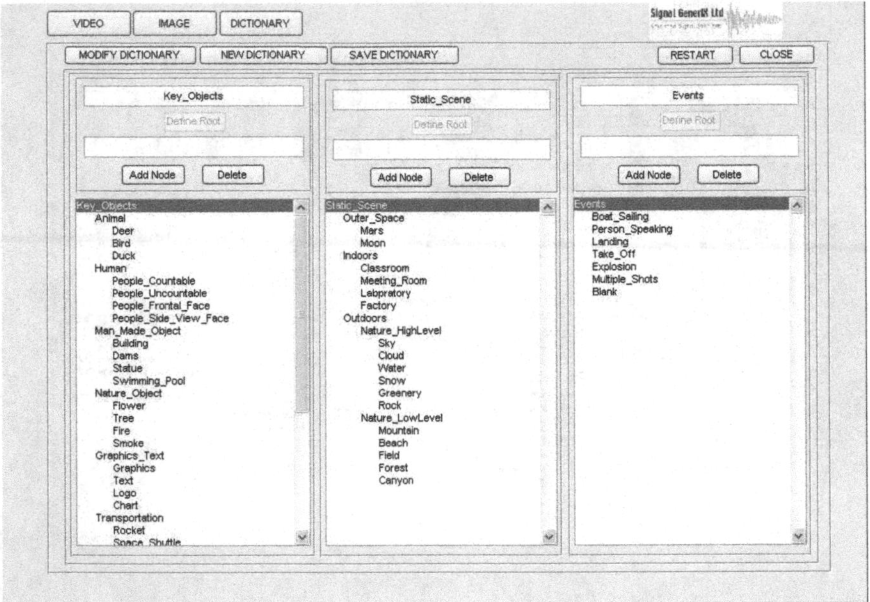

Fig. 4. Dictionary Panel

shown in Fig.3. The input image and its features are presented on the upper left-hand corner. The user can *choose between Image Annotation* and *Watermarking.*

Image Annotation. An image can be simply annotated following the three steps described in the previous section. The three categories of the annotation dictionary are presented in the list-boxes (under left-hand) and the chosen image is shown at the upper left-hand axes. The user can annotate the image by ticking the most appropriate annotation boxes of the three lists. Free text and Key-words edit boxes can be used for a more detailed annotation. The saved annotations are shown in the list-box on the right-hand corner and are saved into an XML file.

Image Watermarking. The selected image can be watermarked and saved in any image format and in any scaling using the *Watermark* tab. After the watermark selection, the user defines the wanted width and height. The watermarked image can be seen in the left-corner axes and can be saved in any image format. The chosen image also can be rescaled and saved in any format, without being watermarked using the *Save Image* tab.

2.3 Dictionary Panel

Dictionaries used for annotating via the *Video Annotation Tool* are created at *Dictionary Panel.* Each dictionary consists of three different categories as presented in Fig.4. Each category consists of a root, a root consists of nodes, and nodes consists of subnodes. The creation of a new dictionary is comprised by three simple steps. First, the user defines the root for each category and then can add nodes and subnodes in the dictionary categories. Finally, the new dictionary is saved in an XML file.

3 Shot Clustering and Scene Construction

In this section we present a genetic algorithm which is used for scene construction through clustering of consecutive shots that have similar keyframes. The algorithm assumes that the number of scenes, say N_s, to which a video is decomposed is given through the MuLVAT tool. We assume also that the input video is already cut into N_k shots with the aid of the algorithm presented in the previous section.

We define the set $\mathbf{K} = \{K_1, K_2, ..., K_{N_k}\}$ of keyframes with keyframe K_i corresponding to shot C_i. We did not devise a particular algorithm for keyframe selection; keyframes are selected and annotated by a human user through the MuLVAT tool. We consider also that only one keyframe is selected per shot.

Let us define a vector of integer values in increasing order:

$$Idx^{(i)} = \{Idx_1^{(i)}, Idx_2^{(i)}, ..., Idx_{N_s-1}^{(i)}\}, \tag{2}$$

with $1 < Idx_j^{(i)} < N_k, \quad j = 1, ..., N_s - 1$. Each vector $Idx^{(i)}$ defines a partition $P^{(i)} = \{P_1^{(i)}, P_2^{(i)}, ..., P_{N_s}^{(i)}\}$ of set $\mathbf{K}$ with $P_j^{(i)}$ corresponding to a set of consecutive keyframes $\{K_{Idx_{j-1}^{(i)}}, ..., K_{Idx_j^{(i)}-1}\}$, while $P_1^{(i)} = \{K_1^{(i)}, ..., K_{Idx_1^{(i)}-1}\}$ and $P_{N_s}^{(i)} = \{K_{Idx_{N_s-1}^{(i)}}, ..., K_{N_k}\}$. Given that each keyframe corresponds to a shot, the partition $P^{(i)}$ defines a possible decomposition of input video into N_s. The task of the genetic algorithm described next is to find a partition $P^{(\xi)}$ which creates the optimum decomposition of input video into N_s scenes given a properly defined metric.

3.1 A Genetic Algorithm for Shot Clustering

Genetic Algorithms are adaptive optimization methods that resemble the evolution mechanisms of biological species [6]. Feature selection is one of the areas that GAs present excellent performance. The main advantages of GAs are:

- they do not require the continuity of parameter space and,
- they are able to efficiently search over a wide range of parameters /parameter sets.

In a GA, the search begins from a population of P_N possible solutions (in our case strings corresponding to integer vectors $Idx^{(i)}, \ i = 1, ..., P_N$ of length $N_s - 1$, with integer values limited to the interval $[1 \quad N_k]$), and not just one possible solution. Solution refers to a partition $P^{(i)}$ as explained in the previous section. A population of solutions guarantees that search will not be trapped in a local optimum, especially if significant diversity exists among the various solutions. The population of solutions tends to evolve toward increasingly better regions of the search space through the use of certain randomized processes, called genetic operators. Typical genetic operators are the *selection*, *mutation* and *recombination*. The *selection* process chooses strings with better objective function value and reproduces them more often than their counterparts with worse objective function value. Thus, a new population is formed consisting of the strings that perform better in their environment. The *recombination* (crossover) operator allows for the mixing of parental information, which is then passed to their descendants. The initial population is randomly acquired; this means that the first and major degree of diversity is introduced in this stage of the GA. The second and lesser degree of diversity is introduced when the *mutation* operator acts upon each string of the population. The whole evolution process stops after a predefined maximum number of iterations (generations) is reached or the variation among population of solutions is too small.

Once the initial population has been created the process of creating new generations starts and consists, typically, of three stages:

1. A fitness value (measure of optimality) of each string in the random population is calculated.
2. Genetic operators, corresponding to mathematical models of simple laws of nature, like reproduction, crossover and mutation are applied to the population and result in the creation of a new population.
3. The new population replaces the old one.

In our case the fitness function F is a metric of similarity between keyframes corresponding the same shot cluster divided by the similarity of keyframes corresponding to different shot clusters. Equation (3) gives the mathematical notation of the fitness function F_i corresponding to the string $Idx^{(i)}$ ($||\cdot||$ refers to the second norm of a multidimensional matrix):

$$F_i = \sum_{j=1}^{N_s} \frac{\sum_{K_l, K_m \in P_j^{(i)},\ l \neq m} ||K_l - K_m||}{\sum_{K_l \in P_j^{(i)}} \sum_{K_p \in P_k^{(i)},\ k \neq j} ||K_l - K_p||} \tag{3}$$

The objective is to find the string that maximizes the fitness function F. The realization of the genetic operators (reproduction, mutation and crossover) is as follows:

Reproduction. The fitness function F is used in the classical "roulette" wheel reproduction operator that gives higher probability of reproduction to the strings with better fitness according to the following procedure:

1. An order number, say q, is assigned to the population strings. That is q ranges from 1 to P_N, where P_N is the size of population.
2. The sum of fitness values (F_{sum}) of all strings in the population is calculated.

3. The interval $[0\ F_{sum}]$ is divided into P_N sub-intervals each of one being $[SF_{q-1}\ SF_q]$ where

$$SF_{q-1} = \sum_{i=1}^{q-1} F_i, \quad q > 1 \tag{4}$$

($SF_{q-1} = 0$ for $q = 0$ and $q = 1$)

$$SF_q = \sum_{i=1}^{q} F_i, \quad \forall q \tag{5}$$

F_i is the value of fitness function for the $i-th$ string (see equation 3).
4. A random real number R_0 lying in the interval $[0\ F_{sum}]$ is selected.
5. The string having the same order number as the subinterval of R_0 is selected
6. Steps (4) and (5) are repeated P_N times in order to produce the intermediate population to which the other genetic operators will be applied.

Crossover. Given two strings (parents) of length $N_s - 1$ an integer number $1 < r < N_s - 1$ is randomly selected. The two strings retain their gene values up to gene r and interchange the values of the remaining genes creating two new strings (offspring). Obviously the integer numbers in offspring must be reordered so as to correspond to vectors of integer values in increasing order.

Mutation. This operator is applied to each gene of a string and it alters its content, with a small probability. The mutation operator is actually a random number that is selected and depending on whether it exceeds a predefined limit it changes the value of a gene. If gene r is to be mutated the allowable values $Idx_r^{(i)}$ for it are those in the interval $(Idx_{r-1}^{(i)}\ Idx_{r+1}^{(i)})$.

4 Conclusion

In this paper we presented a multi-level Video Annotation tool based on XML dictionaries. It consists of three different panels and provides a friendly user interface that seems to be powerful for various user profiles. It allows for semantic labeling using structured knowledge in the form of XML dictionaries and provides the user with a powerful algorithm for shot detection algorithm which minimizes the human intervention for video segmentation at lowest level. XML dictionaries, used for semantic labeling, can be derived on request and during the annotation process maximizing the flexibility of the user. Our future work includes ontology support, incorporation of keyframe selection methodologies and the automatic creation of a list of semantic labels which will be proposed to the user, for the annotation of keyframes and shots, based on machine learning processes.

References

1. Lin, C.Y., Tseng, L., Smith, R.: Video Collaboration Annotation Forum: Establishing Ground-Truth Labels on Large Multimedia Datasets. In: Proc. of NIST Text Retrieval Conference (TREC) (November 2003)
2. Ricoh Movie Tool website, http://www.ricoh.co.jp/src/multimedia/MovieTool
3. Adams, W.H., Lin, C.Y., Iyengar, B., Tseng, B.L., Smith, J.R.: IBM Multimedia Annotation Tool. IBM Alphaworks (August 2002)
4. Bargeron, D., Gupta, A., Grudin, J., Sanocki, E.: Annotations for Streaming Video on the Web:System Design and usage Studies. In: Proc. ACM 8th Conference on World Wide Web, Torondo, Canada (1999)
5. European Cultural Heritage Online (ECHO), http://www.mpi.nl/echo/
6. Goldberg, D.: Genetic Algorithms in Search, Optimization, and Machine Learning. Addison-Wesley, Reading (1989)
7. ISO/IEC 15938-3:2001 Information Technology - Multimedia Content Description Interface - Part 3: Visual, Version 1
8. ISO/IEC 15938-4:2001 Information Technology - Multimedia Content Description Interface - Part 4: Audio, Version 1
9. ISO/IEC 15938-5:2003 Information Technology - Multimedia Content Description Interface - Part 5: Multimedia Description Schemes, First edn.
10. Lienhart, R.: Comparison of Automatic Shot Boundary Detection Algorithms. In: Proc. of SPIE, Storage and Retrieval for Image and Video Databases VII, San Jose, CA, USA, vol. 3656, pp. 290–301 (1999)
11. Nack, F., Putz, W.: Semi-automated Annotation of Audio-Visual Media in News. GMD Report 121 (2000)
12. Steves, M.P., Ranganathan, M., Morse, E.L.: SMAT:Synchronous Multimedia and Annotation Tool. In: Proc. of 34th Hawaii International Conference on Systems Sciences (2001)

Multimodal Sparse Features for Object Detection

Martin Haker, Thomas Martinetz, and Erhardt Barth

Institute for Neuro- and Bioinformatics, University of Lübeck,
Ratzeburger Allee 160, 23538 Lübeck, Germany
{haker,martinetz,barth}@inb.uni-luebeck.de
http://www.inb.uni-luebeck.de

Abstract. In this paper the sparse coding principle is employed for the representation of multimodal image data, i.e. image intensity and range. We estimate an image basis for frontal face images taken with a Time-of-Flight (TOF) camera to obtain a sparse representation of facial features, such as the nose. These features are then evaluated in an object detection scenario where we estimate the position of the nose by template matching and a subsequent application of appropriate thresholds that are estimated from a labeled training set. The main contribution of this work is to show that the templates can be learned simultaneously on both intensity and range data based on the sparse coding principle, and that these multimodal templates significantly outperform templates generated by averaging over a set of aligned image patches containing the facial feature of interest as well as multimodal templates computed via Principal Component Analysis (PCA). The system achieves a detection rate of 96.4% on average with a false positive rate of 3.7%.

1 Introduction

In recent years there has been a lot of interest in learning sparse codes for data representation, and favorable properties of sparse codes with respect to noise resistance have been investigated [1]. Olshausen and Field [2] applied sparse coding to natural images and showed that the resulting features resemble receptive fields of simple cells in V1. Thus, it stands to reason that the basis functions computed by sparse coding can be used effectively in pattern recognition tasks in the fashion introduced by Serre et al. [3], who model a recognition system that uses cortex-like mechanisms.

Sparse coding has also been successfully applied to the recognition of handwritten digits [4]. The authors learn basis functions for representing patches of handwritten digits and use these to extract local features for classification.

In this work, we aim to learn a sparse code for multimodal image data, i.e. we simultaneously learn basis functions for representing corresponding intensity and range image patches. As a result, we obtain aligned pairs of basis functions that encode prominent features that co-occur consistently in both types of data. Thus, a corresponding pair of basis functions can be used to consistently extract

C. Alippi et al. (Eds.): ICANN 2009, Part II, LNCS 5769, pp. 923–932, 2009.

features from intensity and range data. To our knowledge, sparse representations have not yet been learned for multimodal signals.

The considered image data was obtained by a Time-of-Flight (TOF) camera [5] which provides a range map that is perfectly registered with an intensity image (often referred to as an *amplitude* image in TOF nomenclature). Although TOF cameras emerged on the market only recently, they have been used in a number of image processing applications, such as shape from shading [6], people tracking [7], gesture recognition [8], and stereo vision [9]. A review of publications related to TOF cameras can be found in [10].

It has already been shown that using both intensity and range data of a TOF camera in an object detection task can significantly improve performance in comparison to using either data alone [11]. The fact, that a sparse code learned simultaneously on both intensity and range data yields perfectly aligned basis functions, allows us to extract relevant features from both types of data.

Here, we aim to learn a set of basis functions that encode structural information of frontal face images in a component-based fashion. As a result, the basis functions estimated by sparse coding can be regarded as templates for facial features, such as the nose. We evaluate the resulting templates on a database of TOF images and use simple template matching to identify the presence and position of the nose in frontal face images. The importance of the nose as a facial feature for problems such as head tracking was already mentioned in [12,13].

Section 2 will discuss the computation of a set of basis functions under the constraint of the sparse coding principle. In Section 3 we discuss the procedure of determining the basis function that yields the optimal equal error rate (EER) in the nose detection task. Section 4 presents the results and shows that templates generated via sparse coding yield significantly better detection rates than templates obtained by PCA or by averaging over a set of aligned image patches.

2 Sparse Features

The investigated database of frontal face images [14] was obtained using an SR3000 TOF camera [15]. The subjects were seated at a distance of about 60 cm from the camera and were facing the camera with a maximum horizontal and/or vertical head rotation of approximately 10 degrees. As a result, the facial feature of interest, i.e. the nose, appears at a size of roughly 10×10 pixels in the image. A number of sample images are given in Fig. 1.

As a TOF camera provides a range map that is perfectly registered with an intensity image, we aim to learn an image basis for intensity and range simultaneously. To this end, the input data for the sparse coding algorithm are vectors whose first half is composed of intensity data and the second half of range data, i.e. in case we consider image patches of size 13×13, each patch is represented by a 338-dimensional vector ($d = 338 = 2 \cdot 13 \cdot 13$) where the first 169 dimensions encode intensity and the remaining 169 dimensions encode range.

In order to speed up the training process, we only considered training data that originated from an area of 40×40 pixels centered around the position

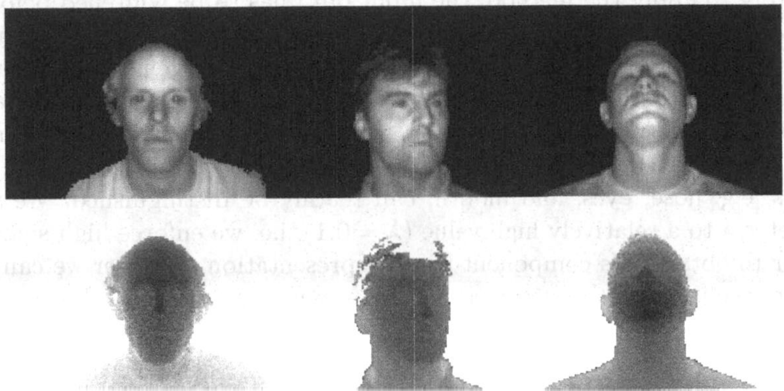

Fig. 1. Three sample images of frontal face images taken by an SR3000 TOF camera. The top row shows the intensity and the bottom row the range data.

of the nose. The position of the nose was annotated manually beforehand. By this procedure we prevent the basis functions from being attracted by irrelevant image features, and a number of 72 basis functions proved to be sufficient to represent the dominant facial features, such as the nose or the eyes.

A common difficulty with TOF images is that the range data is relatively noisy and that both intensity and range can contain large outliers due to reflections of the active illumination (e.g. if subjects wear glasses). These outliers violate the assumed level of Gaussian additive noise in the data and can lead the sparse coding algorithm astray. To compensate for this effect, we applied a 5×5 median filter to both types of data. To ensure the conservation of detailed image information while effectively removing only outliers, pixel values in the original image I_o were only substituted by values of the median filtered image I_f if the absolute difference between the values exceeded a certain threshold:

$$I_o(i,j) = \begin{cases} I_f(i,j) & \text{if } |I_o(i,j) - I_f(i,j)| \geq \theta \\ I_o(i,j) & \text{otherwise} \end{cases} .$$

There exist a number of different sparse coding approaches, see for example [2,16,17]. We employed the Sparsenet algorithm [2] for learning the sparse code. The basic principle aims at finding a basis W for representing vectors $\boldsymbol{x}$ as a linear combination of the basis vectors using coefficients $\boldsymbol{a}$ under the assumption of Gaussian additive noise: $\boldsymbol{x} = W\boldsymbol{a} + \boldsymbol{\epsilon}$. To enforce sparseness, i.e. the property that the majority of coefficients a_i are zero, the Sparsenet algorithm solves the following optimization problem:

$$\min_W E\left(\min_{\boldsymbol{a}} \left(\|\boldsymbol{x} - W\boldsymbol{a}\| + \lambda S(\boldsymbol{a})\right)\right) . \tag{1}$$

Here, E denotes the expectation and $S(\boldsymbol{a})$ is an additive regularization term that favors model parameters W that lead to sparse coefficients $\boldsymbol{a}$. The parameter λ balances the reconstruction error $\boldsymbol{\epsilon}$ against the sparseness of the coefficients.

In order to apply the method, the input data has to be whitened beforehand as indicated in [2]. We applied the whitening to both types of data individually. Only after this preprocessing step, the training data was generated by selecting random image patches of the template size, i.e. for a patch in a given image the corresponding intensity and range data were assembled in a single vector.

The resulting features for 19 × 19 image patches are given in Fig. 2. Facial features, e.g. nose, eyes, and mouth, can readily be distinguished. We set the parameter λ to a relatively high value ($\lambda = 0.1$), i.e. we enforce high sparseness, in order to obtain this component-based representation, however we can report that the results are not particularly sensitive to minor changes of this parameter.

3 Nose Detection

Since the basis functions computed by sparse coding in Section 2 represent facial features, it stands to reason that they can be used for object detection via template matching. At this point two questions arise: (i) Which basis function represents the best template, and (ii) what is the actual position of the facial feature with respect to the center of the image patch corresponding to this basis function. A straightforward solution would be to select the most promising feature by visual inspection and to annotate the position of the facial feature within the image patch manually. Obviously though, this procedure is not generally applicable and is likely to yield suboptimal results.

Thus, we decided to follow a computationally more intensive procedure that, in contrast, is fully automatic and operates in a purely data-driven fashion. For each of the 72 basis functions we trained and evaluated a nose detector for every possible position of the nose in a certain neighborhood around the center of the image patch. In the case of 13 × 13 image patches we chose this neighborhood to be 11 × 11. As a result, a total of $8712 = 72 \cdot 11 \cdot 11$ detectors were trained. The final detector uses the basis function and the position of the nose out of the 8712 configurations that produced the best EER on the training set.

The thresholds of each detector were simply determined by taking the minimum and the maximum of the filter responses at the annotated positions of the nose on a set of training images, i.e. upper and lower bounds for the filter responses that identify a nose were determined for both intensity and range data. In order to identify a nose in a new image, both intensity and range were filtered with the corresponding template images and each pixel whose filter responses complied with the identified bounds was classified as a nose pixel. To obtain an EER, these bounds were relaxed or tightened.

The procedure was evaluated on a data set of 120 TOF images of frontal faces taken from 15 different subjects. To double the amount of data, mirrored versions of the images were also added to the data set. From the total of 240 images one half was chosen at random as a training set to determine the bounds of each classifier. These bounds were then adjusted to yield an EER on the training set. Finally, the optimal classifier, i.e. the one out of the 8712 candidates yielding

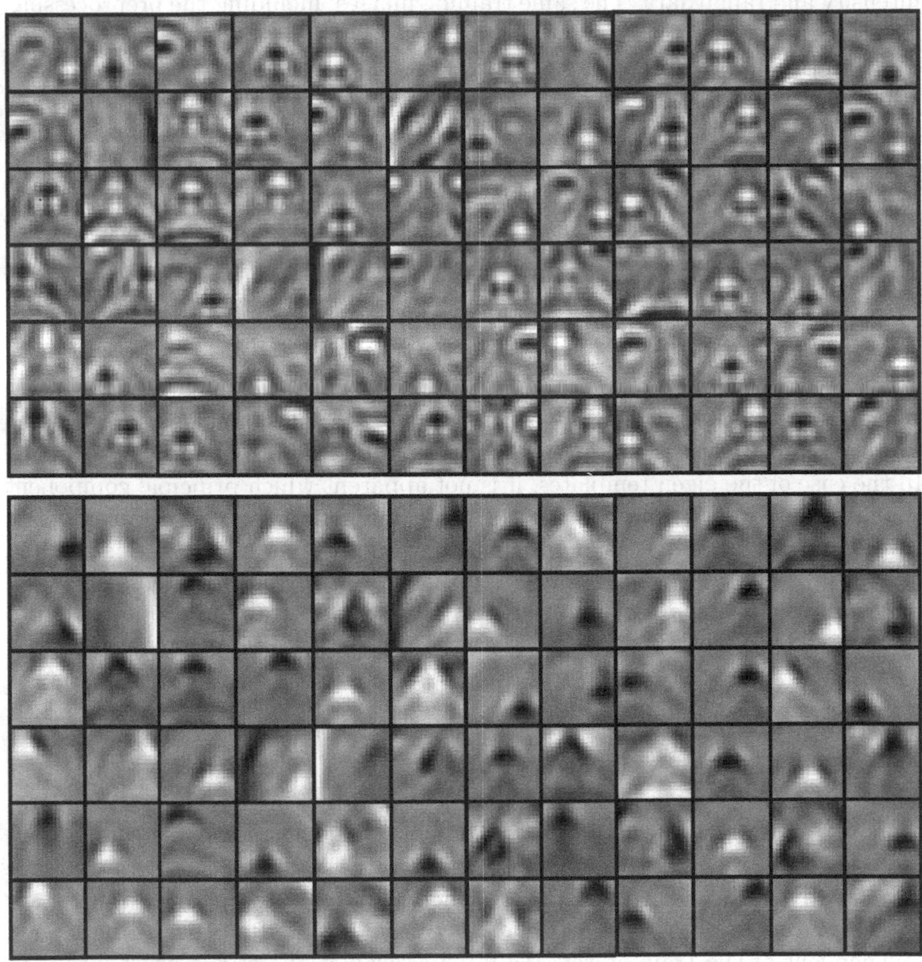

Fig. 2. Basis functions learned for frontal face images via the Sparsenet algorithm. The upper and lower part of the figure show the basis functions representing intensity data and range data, respectively. The basis functions for both types of data were learned simultaneously and correspond pairwise, i.e. the top-left intensity feature is perfectly aligned with the top-left range feature.

the best EER, was evaluated on the remaining 120 images that were not used during the training process.

In order to assess the performance of the learned templates, we also evaluated two other types of templates – "average" and "eigen" templates. The former were generated by averaging over a set of aligned image patches containing a nose. The latter were obtained as the principal components of these aligned image patches. Again, we generated corresponding pairs of templates for both intensity and range data. The same training images, including the preprocessing, were used as in Section 2 for the Sparsenet algorithm.

A fundamental difference between these two approaches to generating the average and eigen templates and the sparse coding method is, that the former only yield templates in which the nose is centered in the image patch whereas the latter also produces translated versions of the nose (see Fig. 2). To guarantee a fair comparison between the different templates we applied the following procedure: since the optimal position of the nose within the template is not known a priori, we generated a total of 121 13×13 templates centered at all possible positions in a 11×11 neighborhood around the true position of the nose, i.e. the templates were shifted so that the nose was not positioned in the center of the template. In correspondence to the procedure described above for the sparse-coding templates, each shifted template was then evaluated for every possible position of the nose in a 11×11 neighborhood around the center of the image patch. For the average templates the resulting number of possible detectors amounts to 14641. In the case of the eigen templates, it is not apparent which principal component should be used as a template. To constrain the computational complexity, we considered only the first three principal components. Nevertheless, this resulted in 43923 possible detectors. Again, the optimal average and eigen templates were determined as the ones yielding the best EER on the training set according to the procedure described above.

4 Results

The results of the training for the nose detection task using 13×13 templates are given in Fig. 3. The EER on the training set using the sparse-coding templates is 3.9%. The eigen templates achieve an EER of 6.6%, and the average templates yield an EER of 22.5%, i.e. the EERs for these two procedures are higher by roughly 50% and 500%, respectively. The EERs prove to be largely independent of the training set. We ran 100 evaluations of the procedure with random configurations of the training set and recomputed both the templates and the classifiers in each run. The standard deviations for the three EERs over the 100 evaluations were $\sigma = 0.9\%$, $\sigma = 1.6\%$, and $\sigma = 2.3\%$, respectively.

Fig. 3 also shows the ROC curves for detectors that use the sparse-coding templates computed on either intensity or range data of the TOF images alone. Note that the EERs are dramatically higher in comparison to the detector that uses both types of data together. This confirms results reported in [11], where the combination of intensity and range data also yielded markedly better results in the detection of the nose based on geometrical features.

The error rates on the test set are only slightly higher than the EERs on the training set, which shows that the method generalizes well to new data. The false positive rates (FPR) amount to 5.3%, 9.3%, and 24.4% for the sparse-coding, eigen, and average templates.

The evaluation above considered only templates of a fixed size of 13×13 pixels. However, varying the template size reveals some interesting properties of

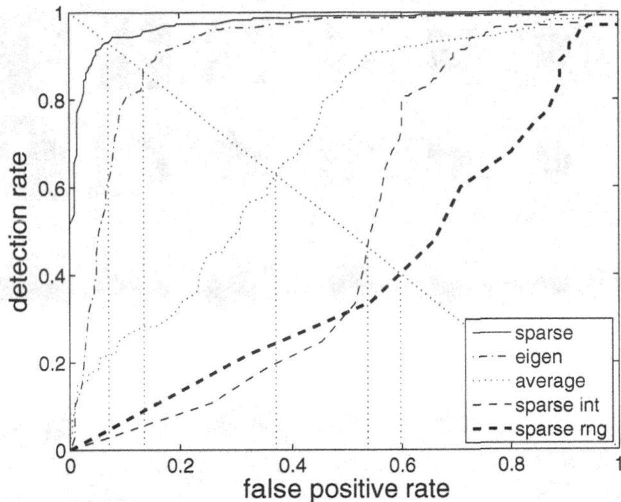

Fig. 3. ROC curves of detection rate vs. false positive rate. The curves were generated using the **sparse**-coding templates, the **eigen** templates, the **average** templates, and the sparse-coding templates using only the intensity data (**sparse int**) or only the range data (**sparse rng**). The detection rate gives the percentage of images in which the nose was identified correctly, whereas the false positive rate denotes the percentage of images where at least one non-nose pixel was misclassified. Thus, strictly speaking, the curves do not represent ROC curves in the standard format, but they convey exactly the information one is interested in for this application, that is, the accuracy with which the detector gives the correct response per image.

the different approaches. To this end, we computed templates of size $n \times n$, where $n = 1, 3, \ldots, 19$, for each approach and estimated the optimal detector according to the same procedure outlined above. To reduce the number of possible detectors to evaluate, the neighborhood sizes for positioning the nose and shifting the template were reduced to 7×7 pixels for templates with size n smaller than 13. Again, we considered 100 random configurations of training and test set.

Fig. 4 shows the configurations of template and position of the nose within the template that yielded the best EERs on the training set for each approach with respect to the different template sizes.

Note that the sparse-coding templates (first two rows) exhibit a much higher contrast in comparison to the average templates (rows three and four), especially for larger sizes of the template. This explains the bad performance of the average templates, because an increase in size does not add more information to the template. This effect becomes clearly visible in Fig. 5. The plot shows the FPRs on the test set for the different approaches over varying template sizes. One can observe that the FPR of the average template starts to increase for templates of size five. In comparison, the FPR of the sparse-coding templates continues to decrease up to a template size of 19.

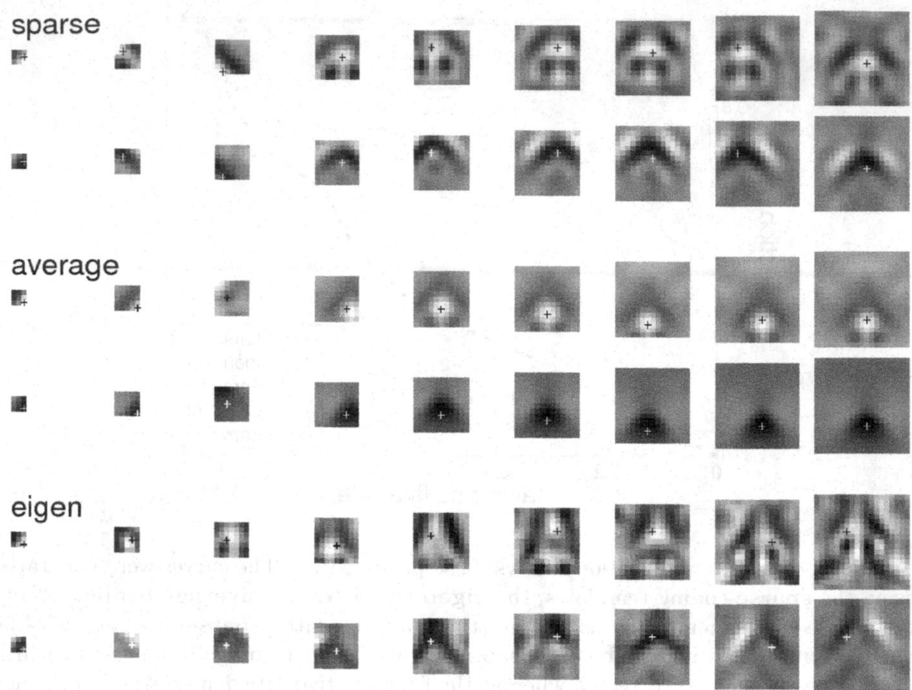

Fig. 4. Optimal templates for different template sizes, where each column shows templates of odd pixel sizes ranging from 3 to 19. The first row shows the sparse-coding templates for intensity data and the second row shows the corresponding features for range data. Rows three and four give the average templates and rows five and six show eigen templates. The crosses mark the estimated position of the nose within the templates.

A decrease of the FPR can also be observed for the eigen templates up to size 11 of the template. For larger template sizes the FPR also starts to increase, whereas the sparse-coding templates continue to achieve low FPRs, as already mentioned above. It seems that sparse coding can exploit further reaching dependencies.

A comparison of the FPRs with respect to the optimal template size for each method reveals that the average template achieves the worst overall performance with an FPR of 9.6% ($\sigma = 3.5$) for a 7×7 template. The best results for the eigen templates were obtained with templates of size 11 yielding an FPR of 7.9% ($\sigma = 3.2$). The sparse coding templates of size 19 had the best overall performance (FPR 3.7%, $\sigma = 2.3$), and the FPR improved roughly by a factor of 2.5 in comparison to the best eigen template.

Note that the false negative rate for the different approaches lies well within the error bars of the FPR in Fig. 5, as one would expect, since the classifier was set to achieve an EER during training.

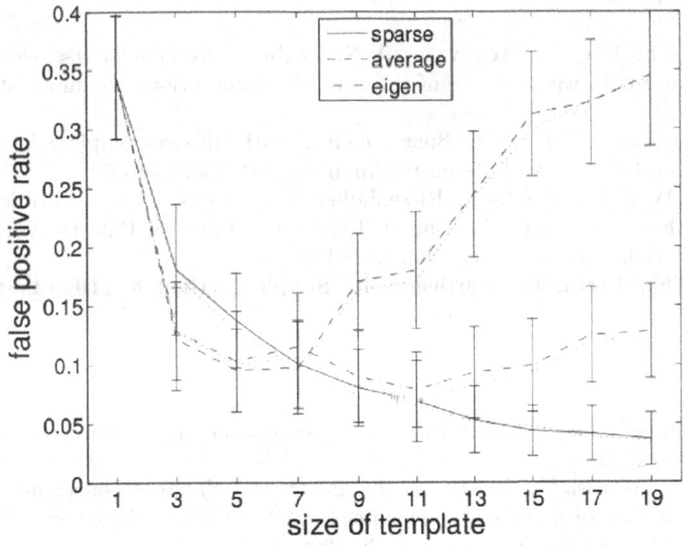

Fig. 5. The graph shows the FPRs and standard deviations on the different test sets for the different templates at different template sizes. The dotted lines show the corresponding false negative rates.

5 Discussion

We have demonstrated how a sparse code can be learned for multimodal image data. The resulting basis functions can be used effectively for template matching in detecting the nose in frontal face images. The sparse-coding templates yield significantly improved results in comparison to templates obtained by averaging over a number of aligned sample images of noses. Templates resembling the principal components of these aligned sample images were also outperformed, especially for large sizes of the template.

The sparse-coding templates were learned on intensity and range data of a TOF camera simultaneously, which yields templates that are perfectly registered for the two different input modalities. The combination of intensity and range data yields a greatly improved detector compared to either type of data alone.

Acknowledgment

This work was developed within the ARTTS project (www.artts.eu), which is funded by the European Commission (contract no. IST-34107) within the Information Society Technologies (IST) priority of the 6th Framework Programme. This publication reflects the views only of the authors, and the Commission cannot be held responsible for any use which may be made of the information contained therein.

References

1. Donoho, D.L., Elad, M., Temlyakov, V.N.: Stable recovery of sparse overcomplete representations in the presence of noise. IEEE Transactions on Information Theory 52(1), 6–18 (2006)
2. Olshausen, B.A., Field, D.J.: Sparse coding with an overcomplete basis set: A strategy employed by V1? Vision Research 37, 3311–3325 (1997)
3. Serre, T., Wolf, L., Bileschi, S., Riesenhuber, M., Poggio, T.: Robust object recognition with cortex-like mechanisms. IEEE Transactions on Pattern Analysis and Machine Intelligence 29(3), 411–426 (2007)
4. Labusch, K., Barth, E., Martinetz, T.: Simple Method for High-Performance Digit Recognition Based on Sparse Coding. IEEE Transactions on Neural Networks 19(11), 1985–1989 (2008)
5. Oggier, T., Büttgen, B., Lustenberger, F., Becker, G., Rüegg, B., Hodac, A.: SwissRanger^{TM} SR3000 and first experiences based on miniaturized 3D-TOF cameras. In: Ingensand, K. (ed.) Proc. 1st Range Imaging Research Day, Zurich, pp. 97–108 (2005)
6. Böhme, M., Haker, M., Martinetz, T., Barth, E.: Shading constraint improves accuracy of time-of-flight measurements. In: CVPR 2008 Workshop on Time-of-Flight-based Computer Vision, TOF-CV (2008)
7. Hansen, D.W., Hansen, M., Kirschmeyer, M., Larsen, R., Silvestre, D.: Cluster tracking with time-of-flight cameras. In: CVPR 2008 Workshop on Time-of-Flight-based Computer Vision, TOF-CV (2008)
8. Kollorz, E., Penne, J., Hornegger, J., Barke, A.: Gesture recognition with a Time-Of-Flight camera. International Journal of Intelligent Systems Technologies and Applications 5(3/4), 334–343 (2008)
9. Gudmundsson, S.A., Aanaes, H., Larsen, R.: Fusion of Stereo Vision and Time-of-Flight Imaging for Improved 3D Estimation. In: Dynamic 3D Imaging – Workshop in Conjunction with DAGM (2007) (in print)
10. Kolb, A., Barth, E., Koch, R., Larsen, R.: Time-of-Flight Sensors in Computer Graphics. Eurographics State of the Art Reports, 119–134 (2009)
11. Haker, M., Böhme, M., Martinetz, T., Barth, E.: Geometric invariants for facial feature tracking with 3D TOF cameras. In: Proceedings of the IEEE International Symposium on Signals, Circuits & Systems (ISSCS), Iasi, Romania, vol. 1, pp. 109–112 (2007)
12. Yin, L., Basu, A.: Nose shape estimation and tracking for model-based coding. In: Proceedings of the IEEE International Conference on Acoustics, Speech, and Signal Processing (ICASSP 2001), May 2001, vol. 3, pp. 1477–1480 (2001)
13. Gorodnichy, D.O.: On importance of nose for face tracking. In: Proceedings of the IEEE International Conference on Automatic Face and Gesture Recognition (FG 2002), Washington, D.C, May 2002, pp. 188–196 (2002)
14. ARTTS: 3D TOF Database, `http://www.artts.eu/publications/3d_tof_db`
15. Oggier, T., Büttgen, B., Lustenberger, F., Becker, G., Rüegg, B., Hodac, A.: SwissRanger^{TM} SR3000 and first experiences based on miniaturized 3D-TOF cameras. In: Proceedings of the 1st Range Imaging Research Day, Zürich, Switzerland, pp. 97–108 (2005)
16. Lewicki, M.S., Sejnowski, T.J., Hughes, H.: Learning overcomplete representations. Neural Computation 12, 337–365 (2000)
17. Labusch, K., Barth, E., Martinetz, T.: Sparse Coding Neural Gas: Learning of Overcomplete Data Representations. Neurocomputing (2009) (in press)

Multiple Kernel Learning of Environmental Data. Case Study: Analysis and Mapping of Wind Fields

Loris Foresti, Devis Tuia, Alexei Pozdnoukhov, and Mikhail Kanevski

Institute of Geomatics and Analysis of Risk,
University of Lausanne, Switzerland
{Loris.Foresti,Devis.Tuia,Mikhail.Kanevski}@unil.ch
National Centre for Geocomputation,
National University of Ireland, Maynooth, Ireland
Alexei.Pozdnoukhov@nuim.ie
http://www.unil.ch/igar

Abstract. The paper presents the Multiple Kernel Learning (MKL) approach as a modelling and data exploratory tool and applies it to the problem of wind speed mapping. Support Vector Regression (SVR) is used to predict spatial variations of the mean wind speed from terrain features (slopes, terrain curvature, directional derivatives) generated at different spatial scales. Multiple Kernel Learning is applied to learn kernels for individual features and thematic feature subsets, both in the context of feature selection and optimal parameters determination. An empirical study on real-life data confirms the usefulness of MKL as a tool that enhances the interpretability of data-driven models.

Keywords: Multiple Kernel Learning, Support Vector Regression, Feature Selection, Wind Speed Mapping.

1 Introduction

Machine Learning Algorithms, as non linear adaptive models, are of great importance in studies of geo- and environmental spatio-temporal data [1]. The present research deals with spatial prediction of the long term average wind speed which is fundamental for natural resources evaluation and for planning the correct location of wind farms and single turbines, for climatological analysis and for a better understanding of the local topography-related patterns of wind speeds. The complex non-linear relations with topography makes spatial wind speed prediction a challenging case study for data-driven statistical methods.

Topographic features can be computed from the digital elevation models of the terrain and directly used in predictive learning regression models, such as Artificial Neural Networks and Support Vector Machines to predict the wind speed. In such studies, however, it is equally important for a successful predictive model to serve as an exploratory tool. For example, it is essential to determine the relevance of particular features or groups of features to yield a better understanding of the problem at hand. Feature selection techniques aim

C. Alippi et al. (Eds.): ICANN 2009, Part II, LNCS 5769, pp. 933–943, 2009.

at this task. Other benefits of feature selection are related to the reduction of computational time (for the final prediction model when applied operationally) and hopefully to the enhancement of performance as a result of noisy features elimination. There are two groups of feature selection techniques [2]: *filter* methods rank the features according to a predefined relevance criteria and *wrapper* methods involve the predictor as a part of the selection process by analyzing the predictive power of features. Among Support Vector Machine wrappers, the recursive feature elimination [3] method is the most used.

In this paper we present the use of Multiple Kernel Learning (MKL) scheme [4] based on the recently proposed SimpleMKL [5] as a wrapper method for detecting the subsets of important features. This scheme uses the Support Vector Regression (SVR) [6] as a predictive model. Performances of the conventional SVR and its MKL extension are compared, and the use of MKL as a feature exploratory tool is analyzed.

The paper is organized as follows. Section 2 presents the methods and the algorithms. The classical methodology for wind speed mapping is reviewed and the proposed MKL-based scheme is discussed in relation to it and the approached case study. Section 3 describes data preparation and Section 4 presents the results of the experiments carried out with SVR and MKL.

2 Learning with Kernel Methods

2.1 Support Vector Regression

Support Vector Regression is a non-linear robust method for regression estimation [6]. SVR controls the complexity of the model and provides accurate results when dealing with high-dimensional and noisy data by building sparse kernel models. Using the kernel trick, the data are mapped into a higher dimensional feature space where linear regression is achievable. The model is based on the use of an ϵ-insensitive loss function, which is responsible for its sparseness. SVR model is given by the linear expansion of kernel functions $K(\mathbf{x}, \mathbf{x}_i)$. It is a function encoding the dot products in the high-dimensional feature space between $\mathbf{x}$ and the support vectors $\mathbf{x}_i$ found by the model and for which their weights are non-zero. The weights are found by quadratic programming, hence due to the convexity of the problem its solution is unique.

2.2 Learning with Multiple Kernels

Common closed-form kernels such as the polynomial or the RBF are rigid representations of the data, that may be replaced by more flexible and data-adapted kernels. The use of multiple kernels can enhance the performance of the model and, more importantly, the interpretability of the results. A multiple kernel in the sense of [7] is built by using a convex combination of basis kernels. Thus, the simple kernel function $K(\mathbf{x}, \mathbf{x}')$ can be replaced by

$$K(\mathbf{x}, \mathbf{x}') = \sum_{m=1}^{M} d_m K_m(\mathbf{x}, \mathbf{x}') \quad \textit{with} \quad d_m \geq 0 \quad \textit{and} \quad \sum_{m=1}^{M} d_m = 1 \qquad (1)$$

where d_m are the weights associated to each kernel. For a given weight vector $\mathbf{d}$, the associated feature space is the vector product of all feature spaces $\mathcal{H}_1, ..., \mathcal{H}_M$ for which $d_m > 0$. Multiple kernel learning aims at optimizing simultaneously the SVM weights α and $\mathbf{d}$.

This formulation is very flexible and can be used in a variety of situations: the features passed to each kernel being defined by the user, combination of features spaces accounting for different features, different scales (same features, but different kernel parameters) or both can be considered. In this sense, a MKL algorithm also acts as a kernel (and therefore feature) selection method, because the kernels associated to zero weights are discarded from the final model.

2.3 SimpleMKL for Support Vector Regression

SimpleMKL algorithm [5] is a recently proposed efficient method for optimizing the weighted combination of kernels of Eq. (1). Similarly to [4], SimpleMKL wraps an SVM solver considering the kernel of Eq. (1) as a fixed single kernel. A gradient descent on the SVR's objective function $J(\mathbf{d})$ in the space of kernel coefficients $\mathbf{d}$ is iterated then. The model proposed in [5] is a general one, and its applications to several kernel methods are illustrated. In the case of SVR, the multiple kernel adaptation of the problem can be stated as

$$J(\mathbf{d}) = \begin{cases} \underset{f_m, b, \xi_i}{Min} & \frac{1}{2}\sum_m \frac{1}{d_m} \|f_m\|^2_{H_m} + C\sum_i(\xi_i + \xi_i^*) \\ s.t. & y_i - \sum_m f_m(x_i) - b \leq \epsilon + \xi_i \quad \forall i \\ & \sum_m f_m(x_i) + b - y_i \leq \epsilon + \xi_i^* \quad \forall i \\ & \xi_i \geq 0, \xi_i^* \geq 0 \quad \forall i \end{cases} \tag{2}$$

which is basically the usual formulation of the SVR, apart from the function $f(\mathbf{x})$, which has been replaced by the linear combination of sub functions $\sum_m f_m(\mathbf{x})$. The dual formulation of the problem of Eq. (2) is derived by using Lagrangian multipliers:

$$J(\mathbf{d}) = \begin{cases} \underset{\alpha, \beta}{Max} & \sum_i(\beta_i - \alpha_i)y_i - \epsilon\sum_i(\beta_i + \alpha_i) \\ & -\frac{1}{2}\sum_{i,j}(\beta_i - \alpha_i)(\beta_j - \alpha_j)\sum_m d_m K_m(\mathbf{x}_i, \mathbf{x}_j) \\ with & \sum_i(\beta_i - \alpha_i) = 0 \\ & 0 \leq \alpha_i \leq C, 0 \leq \beta_i \leq C \quad \forall i \end{cases} \tag{3}$$

The derivatives of $J(\mathbf{d})$ for gradient descent in $\mathbf{d}$-space are computed as

$$\frac{\partial J}{\partial d_m} = -\frac{1}{2}\sum_{i,j}(\beta_i^* - \alpha_i^*)(\beta_j^* - \alpha_j^*)K_m(\mathbf{x}_i, \mathbf{x}_j) \tag{4}$$

This gradient gives the direction for updating $\mathbf{d}$. The updating scheme is $\mathbf{d} \leftarrow \mathbf{d} + \gamma\mathbf{D}$, where γ is the step size and $\mathbf{D}$ is the descent direction computed using the reduced gradient algorithm, which allows to respect the equality and the positiveness conditions over the d_m, as follows:

$$D_m = \begin{cases} 0 & if \quad d_m = 0 \quad and \quad \frac{\partial J}{\partial d_m} - \frac{\partial J}{\partial d_\mu} > 0 \\ -\frac{\partial J}{\partial d_m} + \frac{\partial J}{\partial d_\mu} & if \quad d_m > 0 \quad and \quad m \neq \mu \\ \sum_{\nu \neq \mu, d_\nu > 0}(\frac{\partial J}{\partial d_\nu} - \frac{\partial J}{\partial d_\mu}) & for \quad m = \mu \end{cases} \tag{5}$$

where μ is the component showing the highest non-zero value. The first condition enforces the constraint on the positivity of the d_m in the case when there is an index m such that $d_m = 0$ and the reduced gradient is greater than 0, i.e. the coefficient is zero, but $J(\mathbf{d})$ still decreases. Once the directions vector $\mathbf{D}$ is computed, the descent direction is adjusted by modifying $\mathbf{d}$. The adjustment is repeated as long as $J(\mathbf{d})$ decreases. This is done to avoid computing the full gradient at each iteration and to decrease the computational burden of the algorithm. Algorithm 1 summarizes SimpleMKL.

Algorithm 1. Simple MKL (adapted from [5])

initialize the weights $d_m = \frac{1}{M}, \quad \forall m$
compute the objective value $J(\mathbf{d})$ according to Equation (3).
repeat
  compute the reduced gradient and find the descent direction $\mathbf{D}$.
  Set $\mu = \arg\max d_m$
  repeat {Descent direction update}
    find the component $\nu = \arg\min -d_m/D_m$
    find maximum admissible step size $\gamma_{\max} = d_\nu/D_\nu$
    update $\mathbf{d} = \mathbf{d} + \gamma_{\max}\mathbf{D}$, set $D_\mu = D_\mu - D_\nu$, $D_\nu = 0$ and normalize $\mathbf{d}$
    compute the new $J(\mathbf{d})$
  until $J(\mathbf{d})$ stops decreasing
  line search along $\mathbf{D}$ to find the optimal γ
until a stopping criterion is met.

2.4 Classical Methodology for Wind Speed Mapping

The state-of-the-art statistical model for mean wind speed mapping in the Alps [8] is built using geostatistical kriging interpolation and several ad hoc corrections as follows:

$$W_{xyz} = W_{x,y,z} + C^{rc}_{x,y} + C^{bv}_{x,y} + C^{sv}_{x,y} + C^{flat}_{x,y} + C^{sea}_{x,y} \tag{6}$$

where $W_{x,y,z}$ is the wind speed after kriging interpolation with linear slope and altitude corrections on the measurements. $C^{rc}_{x,y}$, $C^{bv}_{x,y}$, $C^{sv}_{x,y}$, $C^{flat}_{x,y}$, $C^{sea}_{x,y}$ are topographic corrections to reproduce higher wind speed over ridges, big valleys, flat regions, offshore areas and lower wind speed on canyons and narrow valleys. Corrections are based on linear regressions with respect to specific features. This empirical approach relies on prior physical knowledge for the generation of the relevant terrain features. The length scales used to compute the terrain features are set up heuristically.

3 Data Preparation with Feature Generation

The mean annual wind speed at 50m above ground (period 1987-2006) is obtained from a set of permanent and temporary weather stations [9] resulting in a total of 148 measurements. A test set of 48 measurements was reserved to carry out the models assessment throughout the study. Model selection was performed by controlling the 10-fold cross-validation root-mean-squared error (RMSE) on the training set of 100 measurements.

The first three predictive features used in this study are (X,Y,Z) coordinates. Topography-related features were extracted from DEM using convolutional filters. Gaussian smoothing filters were used to create the first subset of features. By subtracting the two smoothed DEM surfaces obtained with different smoothing bandwidths one highlights ridges and canyons, as shown in Fig. 1. These features are referred to as Differences of Gaussians, DoG. The set of DoGs is generated by gradually increasing the widths of the smoothing kernels. The resulting set of features describes terrain convexity at different spatial scales. The second set of features is computed by evaluating directional derivatives at a number of different scales as well. The third set of features is generated by computing terrain slopes with the same principles.

Finally, the resulting dataset is composed of 57 features and 1 target variable: [X,Y,Z — 17 DoG — 21 Directional Derivatives — 16 Slopes — Wind Speed]. Consecutive features are correlated since they are computed at similar spatial scales. MATLAB software was used for the experiment. SVR and simpleMKL codes can be downloaded from [10].

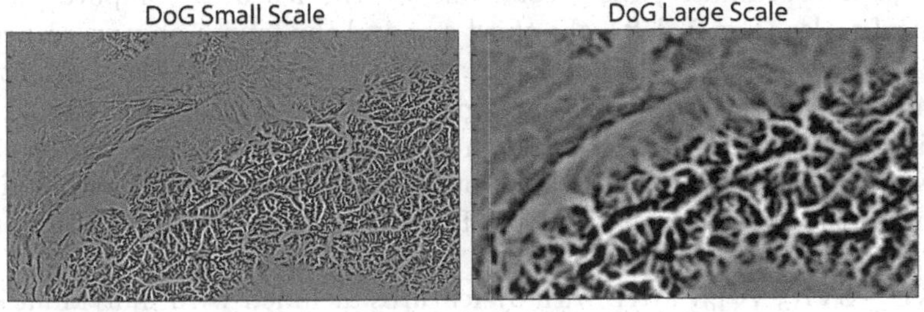

Fig. 1. Example of features computed at different scales (differences of Gaussians)

4 Experiments and Results

The data-driven method for the task of wind speed mapping has to be interpretable in describing the obtained result and exploring the spatial length scales of the terrain features. In the next sections, SVR is applied for wind speed prediction and the appropriateness of Multiple Kernel Learning SVR in providing the means for feature selection and exploratory analysis is investigated.

4.1 Wind Speed Prediction with SVR

Comparisons between SVR trained using only X,Y,Z coordinates ($SVR_{x,y,z}$) and SVR trained with the complete set of 57 features ($SVR_{x,y,z+features}$) is given in Table 1. In all the experiments of this paper, the Gaussian RBF kernel was used. The parameters were tuned according to the minimum of 10-fold cross-validation RMSE, resulting in: $C = 1$, $\epsilon = 0.7$, $\sigma = 0.6$ for the $SVR_{x,y,z}$ and $C = 10$, $\epsilon = 0.3$, $\sigma = 5$ for the $SVR_{x,y,z+features}$. An improvement in prediction performance is observed when using the full set of features. The presence of non-linearities in the wind-topography relationships can be observed by the low performance of the linear SVR (linear kernel).

4.2 Analysis of Multiple Kernel Learning SVR

In this case study, the use of MKL-SVR as a predictive and feature selection method is explored by applying it to the sets of features of increasing size. There are two basic directions to expand the baseline kernel model using MKL. First, it can be done in terms of the number of features (or the groups of features) included separately through an individual kernel. Second, each feature can be included to the model through several kernels of different parameters (for example, bandwidths of Gaussian RBF). In the first case, MKL acts purely as a feature selection method. In the second case, MKL is expected to select the features and an optimal kernel parameter from the fixed set.

The following cases are considered in this study:

1. $\mathrm{MKL} - \mathrm{SVR}_{4gr\times 1\sigma}$. MKL with 4 groups of features: first group is composed of X,Y,Z coordinates, the second one by the features computed at small spatial scales (Fig. 1, left), the third one and the fourth one are respectively the feature subsets at medium and large spatial scales (Fig. 1, right). The bandwidth of the kernel for each group is then tuned with 10-fold cross-validation, resulting in a grid search in 4-dimensional parameter space.
2. $\mathrm{MKL} - \mathrm{SVR}_{4gr\times 3\sigma}$. MKL with 4 groups of features as above, with 3 different bandwidths allowed for each group. Here MKL is tested for its ability to find appropriate bandwidth to avoid the costly cross-validation.
3. $\mathrm{MKL} - \mathrm{SVR}_{57f\times 1\sigma}$. MKL with 57 features included with individual kernels of fixed bandwidths. Here MKL is tested for its usefulness as a feature selection method.
4. $\mathrm{MKL} - \mathrm{SVR}_{57f\times 4\sigma}$. MKL with 57 features included with individual kernels of 4 bandwidths of choice. MKL is tested as a method for simultaneous feature and hyper-parameters selection.

Table 1. Performances of $SVR_{x,y,z}$ and $SVR_{x,y,z+features}$

Method	#Kernels	Eff. #Features	cv-RMSE ($\frac{m}{s^2}$)	test-RMSE ($\frac{m}{s^2}$)
$\mathrm{SVR}_{x,y,z}$	1	3	1.1141	0.9653
$\mathrm{SVRlinear}_{x,y,z+features}$	1	57	1.3346	1.0078
$\mathrm{SVR}_{x,y,z+features}$	1	57	1.0586	0.7146
$\mathrm{MKL} - \mathrm{SVR}_{4gr\times 1\sigma}$	4	39	1.0051	0.7116
$\mathrm{MKL} - \mathrm{SVR}_{4gr\times 3\sigma}$	12	16	1.0383	0.7977
$\mathrm{MKL} - \mathrm{SVR}_{57f\times 1\sigma}$	57	31	1.1203	0.9868
$\mathrm{MKL} - \mathrm{SVR}_{57f\times 4\sigma}$	228	27	1.1111	1.1141
$\mathrm{SVR}_{31features}$	1	31	1.0475	0.6487

The obtained results are summarized in Table 1. We describe below some particular interesting findings in more details.

MKL for Selecting Groups of Features and Optimal Parameters. The models $\mathrm{MKL}-\mathrm{SVR}_{4gr\times1\sigma}$ and $\mathrm{MKL}-\mathrm{SVR}_{4gr\times3\sigma}$ showed that MKL is successful in finding appropriate hyper-parameters for the groups of features. With the first model, the bandwidths are found with extensive grid search by cross-validation as σ=0.5 for spatial location, $\sigma = 3$ for small scale features, $\sigma = 3$ for medium scale features and $\sigma = 7$ for large scale features. Fig. 2 (left) shows the importance of each feature subset (weights d_m provided by MKL). The large scale features kept as a whole could be discarded.

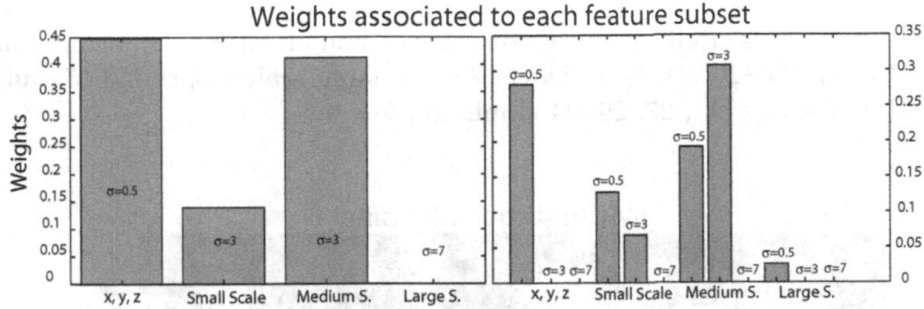

Fig. 2. Weights associated to feature subsets; location and medium scale features are the most relevant ones

SimpleMKL was able to appropriately identify the optimal bandwidth from the set (library) of *sigma*={0.5, 3, 7} for each subset. Fig. 2 (right) presents the distribution of weights d_m. The results coincide well for location features, medium and large scale features (which are almost neglected) with the ones obtained with cross-validation.

MKL for Feature Selection. Approaching individual feature selection is possible if one provides at least one single kernel to each feature. This situation can be further extended by providing a library of kernels to each feature, i.e. a set of possible kernel widths. Two models are considered, $\mathrm{MKL}-\mathrm{SVR}_{57f\times1\sigma}$ and $\mathrm{MKL}-\mathrm{SVR}_{57f\times4\sigma}$. The results obtained are presented in Table 1. The first method needs the optimization of 57 weights corresponding to the number of features in the dataset. The second method has to optimize $57 \cdot l$ weights, where l is the number of possible kernels per feature, resulting in 228 kernels.

The weights d_m after using the first method ($\mathrm{MKL}-\mathrm{SVR}_{57f\times1\sigma}$) are shown in Fig. 3 (left). A total of 31 of 57 features are clearly selected. As cross-validation is not feasible in the space of 57 parameters, MKL was applied for the latter task as well. This second experiment was carried out by building a library composed of 4 kernels per each feature, $\sigma = \{0.25, 0.5, 0.75, 1.5\}$. Features with 0 weights in all their kernels were discarded. The experiment was repeated for 100 different

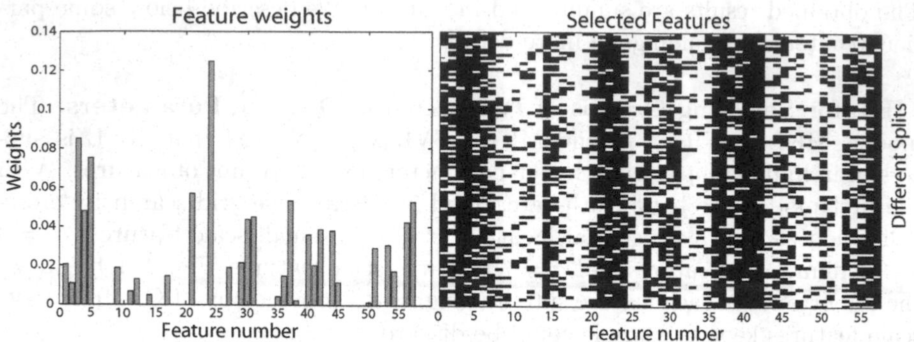

Fig. 3. Weights associated to each feature in a single run (left) and results of selecting features over 100 runs (right). Spatial coordinates and altitude (1:3), small scale DoG (4:7), medium DoG (20:24), large DoG (36:42), large scale slopes (53:57) and some directional derivatives (14, 27, 29:31) dominate.

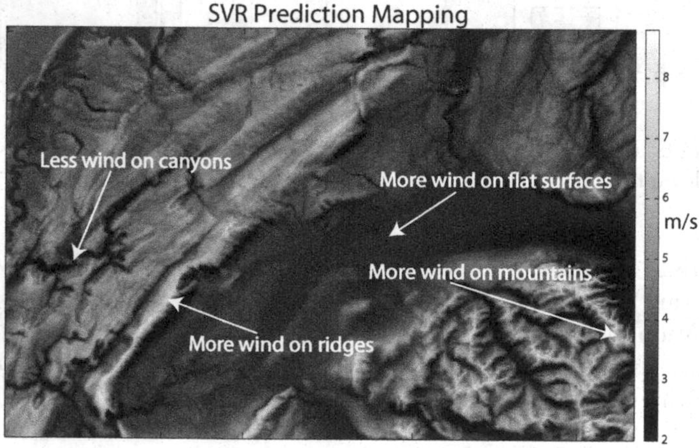

Fig. 4. SVR prediction mapping with the subset of 31 features; visual inspection of wind patterns with respect to topography is also important

training-test splits of data to analyze the stability of the set of selected features. It was observed that MKL is consistent in selecting the features and their kernel parameters (right side of Fig. 3).

One can notice significant decrease in model performance in terms of test RMSE (Table 1). It is due to the high number of kernels compared to the number of training examples. MKL gives insights about the importance of single features, but it cannot be applied as final prediction model because of the high ratio between the number of weights and the number of training examples. In this extreme limit of using 228 kernels for 100 training samples MKL fails to provide

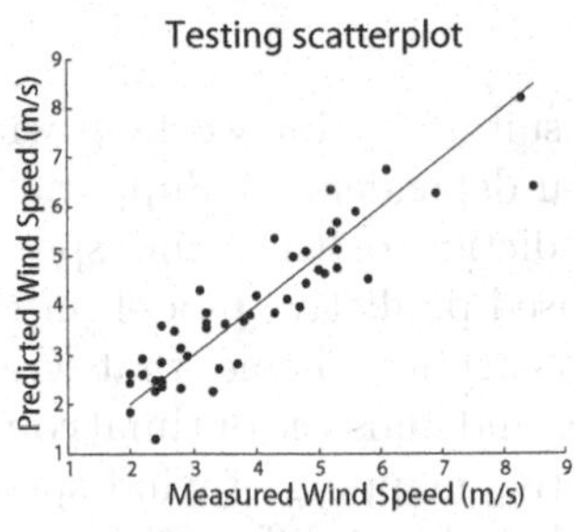

Fig. 5. Testing scatterplot for $\mathrm{SVR}_{31features}$

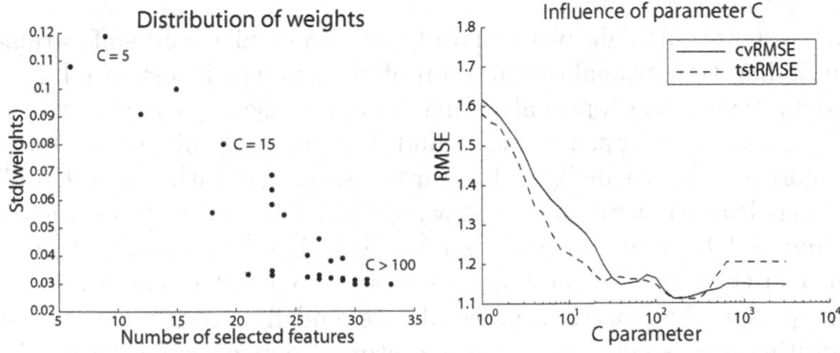

Fig. 6. Behaviour of MKL while varying trade-off parameter C

predictive model but is stable and consistent in selecting the relevant features. Surprisingly, the use of SimpleMKL as a filter method followed by the standard SVR applied to the set of non-zero weighted features (31 features) gives best results with $cvRMSE = 1.0475$ and $tstRMSE = 0.6487$. Final prediction map for the successful prediction model $\mathrm{SVR}_{31features}$ is shown in Fig. 4 and the corresponding testing scatterplot is shown in Fig. 5.

MKL and Trade-off Parameter C. MKL method was found to be very sensible to the choice of parameter C, which is known to be the parameter penalizing the misfit to the training samples. Fig. 6 presents the results of experiments illustrating the behaviour of $\mathrm{MKL} - \mathrm{SVR}_{57f \times 4\sigma}$ model, that is, an MKL scheme applied for simultaneous feature and parameter selection with 228 subkernels. For small values of C, MKL tends to select the single or few features from the available set. With increasing C, the number of selected features increases and they receive equal weights (the variance of the distribution of weights decreases as shown in Fig. 6, left). The trade-off between the number of features (interpretability of the model) and its predictive performance can be controlled with cross-validation. These findings are confirmed by low testing error, Fig. 6 (right).

5 Conclusions

Due to its robustness and suitability for working with high-dimensional input data for modelling non-linear dependencies, Support Vector Regression provided good results in spatial prediction of the wind speeds. To enhance the interpretability of this kernel-based predictive model, in this paper we explored the use of the Multiple Kernel Learning scheme, that wraps an SVR trained with a linear combination of kernels and finds the optimal combination of input features. We applied it to the predictive mapping of wind speed aiming at detecting the optimal characteristic length scales of different topographic features influencing the phenomenon.

The empirical studies of the real-life data provided interesting insights about its use for feature selection. MKL scheme was found to be successful both in detecting meaningful features and suitable kernel parameters. The sensitivity to hyperparameters (particularly, the data fit vs. complexity trade-off parameter of SVR) in finding the optimal distribution of weights was investigated.

The definition of the optimal kernel library (based currently on the prior knowledge) remains an open question and it is currently one of the limitations of the algorithm. Irrelevant kernel libraries associated with difficult and small datasets may lead to overfitting as shown empirically in [11]. Since the distance metrics induced by real processes is often variable over the input space, the non-stationarity of kernel functions is also an important research issue. Future promising perspectives for environmental data modelling concern the use of MKL for integrating multisource data from monitoring networks, for the modeling of joint multiscale physical processes, which need different types of kernels, and for automatic feature selection.

Acknowledgements. The research is funded in part by the Swiss National Science Foundation projects "GeoKernels: kernel-based methods for geo- and environmental sciences (Phase II)" (No 200020-121835/1) and "ClusterVille. Urbanisation Regime and Environmental Impact: Analysis and Modelling of Urban Patterns, Clustering and Metamorphoses" (No 100012-113506). A. Pozdnoukhov acknowledges the support of Science Foundation Ireland under the National Development Plan, particularly through Stokes Award and Strategic Research Cluster grant (07/SRC/I1168).

References

1. Kanevski, M. (ed.): Advanced Mapping of Environmental Data. ISTE Wiley, Chichester (2008)
2. Guyon, I., Gunn, S., Nikravesh, M., Zadeh, L.A. (eds.): Feature Extraction: Foundations and Applications. Springer, Heidelberg (2006)
3. Guyon, I., Weston, J., Barnhill, S., Vapnik, V.: Gene Selection for Cancer Classification using Support Vector Machines. Machine Learning 46, 389–422 (2002)
4. Sonnenburg, S., Schaefer, G., Rötsch, G., Schölkopf, B.: Large Scale Multiple Kernel Learning. Journal of Machine Learning Research 7, 1531–1565 (2006)

5. Rakotomamonjy, A., Bach, F.R., Canu, S., Grandvalet, Y.: Simple MKL. Journal of Machine Learning Research 9, 2491–2521 (2008)
6. Smola, A.-J., Schölkopf, B.: A Tutorial on Support Vector Regression. Technical Report (1998)
7. Gert, R., Lanckriet, G., De Bie, T., Cristianini, N., Jordan, M., Noble, W.: A Statistical Framework for Genomic Data Fusion. Bioinformatics 20, 2626–2635 (2004)
8. Schaffner, B., Remund, J., Rihm, B., Cattin, R., Kunz, S.: The Alpine Space Wind Map. In: European Wind Energy Conferences & Exhibitions (2006)
9. The Swiss Wind Power Data Website, `http://www.wind-data.ch/index.php`
10. Canu, S., Grandvalet, Y., Guigue, V., Rakotomamonjy, A.: SVM and Kernel Methods Matlab Toolbox, Perception Systèmes et Information, INSA de Rouen, Rouen, France (2005), `http://asi.insa-rouen.fr/enseignants/~arakotom/toolbox/index.html`
11. Lewis, D.P., Jebara, T., Noble, W.S.: Support Vector Machine Learning from Heterogeneous Data: an Empirical Analysis using Protein Sequence and Structure. Bioinformatics 22, 2753–2760 (2006)
12. Vapnik, V.: The Nature of Statistical Learning Theory. Springer, Heidelberg (1995)
13. Schölkopf, B., Smola, A.: Learning with Kernels. MIT Press, Cambridge (2001)

Contributor Diagnostics for Anomaly Detection*

Alexander Borisov[2], George Runger[1], and Eugene Tuv[2]

[1] Industrial and Systems Engineering, Arizona State University, Tempe, AZ
[2] Intel, Chandler, AZ

Abstract. Anomaly detection in data streams requires a signal of an unusual event, but an actionable response requires diagnostics. Consequently, an important task is to isolate to the few key attributes that contribute to the signal from among a large collection. We introduce this contributor problem to the machine learning community and present a solution for monitoring in modern systems (with nonlinear reference conditions, high dimensions, categorical attributes, missing data, and so forth). The objective is to identify attributes that contribute to a signal, for both individual and multiple anomalies, or from several anomaly groups. Although related to the feature selection problem, the extreme sparseness of anomalies leads to scores that are designed specifically for the contributors problem. Statistical criteria are provided to quantitatively address decision rules and false alarms and the method can be computed quickly. Comparisons are made to traditional contribution plots.

1 Introduction

The importance of anomaly detection has grown from manufacturing to include systems such as environmental, security, health, supply chains, transportation, etc.. The goal is to monitor a data stream from a system to detect an unusual event, with a minimum of false alarms. For modern systems, one often monitors a large number of attributes that might be mixed (numerical and categorical), missing, redundant with complex relationships, etc. This makes it more difficult to diagnose a signal for an effective response. However, an anomaly often manifests itself through only a fraction of the full set of attributes. Consequently, an important task is to filter the large set of attribute to those that contribute to the signal. These attributes can be further studied in more detail to complete the diagnosis. The objective of this work is to identify these contributors among a large collection of attributes. For example, suppose one monitors the RPM of a pump, the pressure differential across the pump, and numerous other process variables. A clog in the line could manifest itself through a high RPM but without the expected pressure differential. These two attributes are expected to contribute to a signal (possibly others could also) and should be a focus for further diagnosis.

* This material is based upon work supported by the National Science Foundation under Grant No. 0743160.

C. Alippi et al. (Eds.): ICANN 2009, Part II, LNCS 5769, pp. 944–953, 2009.

This diagnosis component is known in the statistical literature as a contributor problem and it occurs daily in diverse industries. The problem has been handled previously primarily though principal components analysis (and further comments appear below). Here we introduce a machine learning solution for this important problem and a promising direction for further research from the machine learning community. The objective here is to handle the complexity in data from modern systems and still provide effective methods to identify the contributors. The diagnosis task is related to feature selection, but typically a signal contains only one (or a few) instances of data and this makes the contribution problem more difficult. It is from such a very sparse sample that the contributors need to be identified.

We consider anomalies that consist of either individual or small groups of instances in unsupervised learning problems. These anomalies are assumed known prior to the contributor calculations, but we summarize our approach to detect anomalies in a following section. Attributes can be detected as contributors whether they affect the signal directly or through interactive effects. We focus on the high-dimensional problem with numerous attributes and a method that can be computed quickly. Furthermore, we provide a statistical criterion to quantitatively evaluate attributes for contributors.

Although the focus here is on contributors we briefly summarize our approach to detect anomalies in Section 2. Section 3 describes traditional methods for the contributor problem and Section 4 presents our new approach. Section 5 provides illustrative examples and compares to traditional contribution plots and Section 6 provides conclusions.

2 Anomaly Detection with Artificial Contrasts

An interesting approach for anomaly detection is to transform the problem to supervised learning. Our method was presented in detail by [1] and summarized here. Let S_0 denote a set of reference data that represents the normal operating conditions of a system. Supplement S_0 with simulated, artificial instances, denoted as S_1, that are generated to be structureless in order to contrast with any structure in S_0. Several methods to generate the artificial data are possible. For every numerical attribute in S_0 generate uniformly distributed data in S_1 that covers the range of the attribute. For categorical attributes we usually randomly permute the actual attribute values (and this could also be used for numerical attributes). For each attribute, the artificial data is generated independently so that structureless data is created in S_1. Then a class attribute y is created and defined as $y_i = 1$ $x_i \in S_0$ and zero otherwise.

A suitable learner $f(\boldsymbol{x})$ is trained to distinguish the class 0 and 1 data. Given a future instance $\boldsymbol{x}_0$ if the learner assigns it to class 1 it can be distinguished from the reference data in S_0 and $\boldsymbol{x}_0$ is flagged as an anomaly. These results can generalize monitoring in several directions (such as arbitrary, nonlinear reference conditions, fault knowledge, and categorical variables). High-dimensional problems can be handled with an appropriate learner. Extensions to tune the algorithm were presented by [2]. A related approach for anomaly detection was presented by [3]. An alternative method such as a density estimate would attempt to learn the entire density of the S_0 data. Instead the supervised learner focuses on the differences between the S_0 and S_1 classes to detect changes from S_0. Anomaly detection is a only a preliminary step and the focus of the work here is work is to determine the contributors to the signal.

3 Traditional Methods for Contributors

The contributor problem has been addressed in the statistical literature through principal components analysis (PCA). PCA and the related partial least squares (PLS) approach are the only methods that have been been applied to the contributor problem. Only PCA is summarized here because, as discussed by [4], the optimization used in PLS often results in a solution similar to PCA. Although PCA is widely applied, it is rooted in normally distributed assumptions and uses transforms to numerical attributes. PCA computes latent variables with maximum variance subject to the constraint that the variables are orthogonal. Also, the method is sensitive to the scale for attributes, and and therefore requires that an appropriate scale be selected. In most application the attributes are standardized (zero mean and unit standard deviation) and for simplicity we assume this here.

Given training data with N instances and M attributes, let $\mathbf{X}$ denote the $N \times M$ matrix of standardized data. The covariance matrix calculated from this data is

$$\mathbf{S} = \mathbf{X}^T\mathbf{X}$$

The data in $\mathbf{X}$ is summarized by derived (also called latent) attributes that are computed from an eigenvalue and eigenvector decomposition of $\mathbf{S}$. For most data sets with sufficient sample size the eigenvalues are unique and greater than zero. Assume that the eigenvalues of $\mathbf{S}$ are $\lambda_1 > \lambda_2 > \ldots > \lambda_p > 0$. The kth latent variable is $\boldsymbol{Z}_k = \mathbf{X}\boldsymbol{v}_k$ for $k = 1, 2, \ldots, p$, where $\boldsymbol{v}_k$ is eigenvector corresponding to the λ_k (the kth largest eigenvalue). The proportion of variability explained by the kth latent attribute is defined to be the variance of $\boldsymbol{Z}_k$ divided by the sum of the variances of all $\boldsymbol{Z}_k$'s. This can be shown to equal

$$\frac{\lambda_k}{\lambda_1 + \lambda_2 + \ldots + \lambda_p} = \frac{\lambda_k}{\mathrm{tr}S}$$

Often $K < p$ latent attributes are used to summarize the full data. Here K might be selected so that the total proportion of variability explained by the first K latent variables exceeds 80%.

Two common statistics are used to monitor for anomalies. A reduction to K latent variables is based on the proportion of variability explained. Let $\{Z_{01}, Z_{02}, \ldots, Z_{0K}\}$ denote the first K latent scores for a data instance $\boldsymbol{x}_0$. Hotelling's T^2 statistic [5] is applied to these latent variables as follows

$$T_0^2 = \sum_{k=1}^{K} \frac{Z_{0k}^2}{\lambda_k} \tag{1}$$

Here T^2 measures the (Mahalanobis) distance of data instance $\boldsymbol{x}_0$ from the centroid $\bar{\boldsymbol{x}}$ after a projection to the subspace defined by the first K latent variables. An anomaly is signaled if this distance is too large.

Because Hotelling's T^2 statistic is not sensitive to anomalies that are far from the subspace of the latent variables a second statistic is used that is sensitive to the distance from this subspace. The squared prediction error (SPE) is

$$SPE_0 = (\boldsymbol{x}_0 - \hat{\boldsymbol{x}_0})^T(\boldsymbol{x}_0 - \hat{\boldsymbol{x}_0}) \tag{2}$$

where $\hat{\boldsymbol{x}_0} = \sum_{k=1}^{K} Z_{0k}\boldsymbol{v}_k$ is the projection of $\boldsymbol{x}_0$ to the subspace spanned by the first K eigenvectors $\boldsymbol{v}_1, \boldsymbol{v}_2, \ldots, \boldsymbol{v}_K$.

Given a signal from an anomaly detection algorithm the PCA contribution score of $\boldsymbol{x}_0$ to T^2 was discussed by [6]. The kth PCA score for the $\boldsymbol{x}_0$ instance is $Z_{0k} = \boldsymbol{x}_0^{\prime}\boldsymbol{v}_k$ and term for attribute j is $x_{0j}v_{jk}$. The PCA contribution score of attribute j for data instance $\boldsymbol{x}_0$ to T^2 as defined by [7] is

$$C(T^2, x_0) = \sum_{k=1}^{K} \frac{Z_{0k}(x_{0j}v_{jk})}{\lambda_k} \tag{3}$$

and this can be interpreted as using the term $x_{0j}v_{jk}$ in the T^2 statistic in (1). Similarly, the PCA contribution score of attribute j for $\boldsymbol{x}_0$ to SPE is calculated from the jth term of (2) as

$$C(SPE, x_0) = (x_{0j} - \hat{x}_{0j})^2 \tag{4}$$

We also applied a different type of scoring function in previous work [8] but the method proposed here is much different and easily scales to multiple attributes.

4 Contributors Algorithm

Suppose we have already dealt with anomaly detection problem using an appropriate learning approach. Then a supervised method can be applied to estimate contributions to the chosen group or several groups of anomalies (identified by the user or by an automatic clustering procedure). Let S_0 denote the reference data from normal operations and let S_i, $i = 1, \ldots, K$ denote the ith selected group among K groups of anomalies. We define a categorical multi-class response (which reduces to binary if there is only one group of anomalies), where class 0 corresponds to samples in S_0, and class i to samples from S_i, $i = 1, \ldots, K$. In this supervised framework contributors can be related to important attributes in a feature selection problem, but with some important changes. In our previous work with the ACE feature selection algorithm [9,10,11] a hybrid of parallel and serial tree ensembles was used. Attribute scores in a tree are often computed from the decrease in an impurity measure (such as the Gini index). For the contribution problem the attribute importance scores in a random forest (RF) tree ensemble [12] are modified.

Given an RF ensemble, each instance in S_i follows a path through nodes (for each tree in the ensemble) that ends at a leaf node. For each node a specific attribute generates the split based on the change in impurity from the parent to child node. In each tree node, instead of using the Gini index impurity reduction wrt a multiclass response, we use the Gini index impurity wrt a binary response corresponding to the selected outlier group. Furthermore, if the splitting attribute is numerical, this index is weighted by difference of the splitting attribute's means in the child nodes. This allows us to target a particular group of anomalies and factor distance into account at the same time (so that more distant anomalies have higher contribution scores, as opposed to the plain Gini index that only reflects separation of anomalies from other instances).

The details of the algorithm follow.

1. Define a target attribute as $y_i = i \; x_i \in S_i, \; i = 1, \ldots, K$ and zero otherwise.
2. Build an RF model for this target. Let $T_k(x_i)$ denote the predicted target from the k-th tree for instance x_i. The prediction from the ensemble is a vote from the trees. That is, each tree generates response a $T_k(x_i)$ and the class that has the maximum number of votes is taken as the prediction from the ensemble.
3. Compute the contributions of an attribute to a selected group of anomalies S_k. The contribution calculation is based on the tree ensemble. Select an attribute X_i. For each node T in each tree, where the split variable is X_i, and the left and right child nodes are T_L, T_R correspondingly, the contribution score of X_i is increased by the following term:

$$Gini_k(T) \times \frac{|\bar{x}_i(T_L) - \bar{x}_i(T_R)|}{std(X_i)} \tag{5}$$

   where $\bar{x}_i(T_L), \bar{x}_i(T_R)$ are mean values of X_i variable in the child nodes (computed from non-missing values), and $std(X_i)$ is the standard deviation of X_i calculated over the 5%-95% percentile range on the whole sample. Here $Gini_k(T)$ is the Gini index impurity reduction in node T wrt the binary response for group S_k with equal priors. More exactly, if we denote counts of samples in and out of the S_k group in node T as $n_{k0}(T), n_{k1}(T)$, and proportions of those samples in the whole data set S, $|S| = N$ as $p_{k0} = |S_k|/N$, $p_{k1} = 1 - |S_k|/N$, we can define impurity for group S_k in node T as

$$I_k(T) = \frac{2n_{k0}(T) \cdot n_{k1}(T)/(p_{k0}p_{k1})}{n_{k0}(T)/p_{k0} + n_{k1}(T)/p_{k1}}$$

   then define $Gini_k(T)$ as

$$Gini_k(T) = |T| \cdot I_k(T) - |T_L| \cdot I_k(T_L) - |T_R| \cdot I_k(T_R) \tag{6}$$

   For a categorical variables the scaling term $|\bar{x}_i(T_L) - \bar{x}_i(T_R)|/std(X_i)$ is replaced by 1.
4. Incorporate statistical criteria to identify the statistically significant contributors. Contribution scores from actual variables are compared to the

corresponding scores for artificially constructed attributes that do not contribute to the anomaly groups. For each actual attribute X_j we generate an artificial attribute, denoted as X_j^* through a random permutation of the values of X_j. This artificial attribute has the same distribution as the original attribute but no predictive power for the anomaly groups. Then we build R small RF ensemble models G_r, $r = 1, \ldots, R$ on input data containing both the original and artificial attributes. Typical values for R is 20 and for the number of trees in the ensemble is $20 - 50$. This can be made larger if the number of attributes is large and our typical heuristic for the number of trees in the ensemble is $min(30, \sqrt{M})$ where M is number of original attributes. After we compute the contribution score C_{kjr} for each outlier group S_k, each attribute X_j, $j = 1, \ldots, M$, and the tree ensemble G_r and the corresponding scores C^*_{kjr} for artificial attributes, we apply a statistical test to select contribution values significantly greater than the contribution from the artificial attributes.

To achieve this goal, we calculate an α-quantile $C^{*\alpha}_{kjr}$ (typically $80\% \leq \alpha \leq 100\%$) of the contribution scores $C^{*\alpha}_{kr}$, $j = 1, \ldots, M$ from all the artificial attributes. The differences $C_{kjr} - C^{*\alpha}_{kr}$, $r = 1, \ldots, R$ for each attribute X_j and group S_k are compared to zero through a Student's t-test and the p-value (significance level). A significant p-value (< 0.05 for example) is used to identify an attribute that contributes to a separation of group S_k. The replicates require an attribute to be a consistently stronger contributor than the artificial attributes. Furthermore, any standard method can be applied to the p-values to control the false alarms. We often use the Bonferroni adjustment and for M actual variables we require p-values smaller than $0.05/M$.

On large data sets where number of samples N and number of attributes M is large the ACE contribution method can be considerable faster than methods rooted in linear regression like PCA. For example PCA has computational complexity $O(NM^2 + N^3)$ if $N > M$. However, most time consuming operation of our method is building RF models. For 20 series of RF models with 30 trees, where each tree has time complexity $N \cdot \sqrt{N} \cdot \sqrt{M}$ we get $O(N \cdot \sqrt{N} \cdot \sqrt{M})$ complexity. Even if we use above-mentioned heuristic for number of trees in ensemble $Ntrees = min(30, \sqrt{M})$, we have $O(N \cdot \sqrt{N} \cdot M)$ complexity.

5 Experiments

The existing competitive methods for the contribution problem are PCA (or the similar PLS approach). These can be effective for linear models and low dimensions. Consequently it is not as useful to investigate a large collection of data sets as it is to focus to two scenarios of interest: nonlinear models and high dimensions with many noisy attributes. We illustrate the limitations of PCA in these cases.

5.1 50-Dimensional Independent Data

Reference data is simulated from 50 uncorrelated attributes with 500 instances (rows). Each is normally distributed with mean zero and standard deviation one. Anomalies are generated from uncorrelated normal distributions with means zero and larger standard deviations (equal to 5). The anomalies are generated from two of the 50 attributes and the goal is to recover these attributes. Figure 1 shows a projection of the data to the plane defined by the two contributor attributes for 10 cases (each case has different pair of contributors). There are two anomalies instances (shown in red).

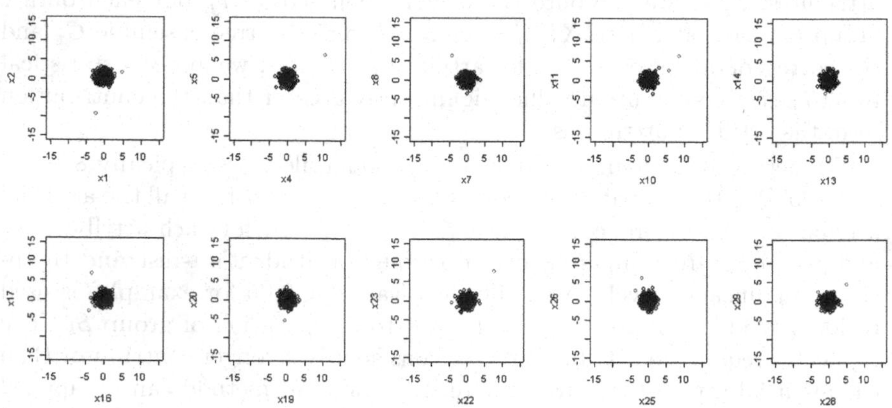

Fig. 1. Scatter plots in the plane of the two contributor attributes for 10 cases. Each case shows the two anomalies as open (red) circles and the reference data as the solid central group.

Figure 2(a) shows plots of contribution scores from the ACE contribution algorithm for the 10 cases. The bars in red are scores that are significant. The two contributors were correctly detected except for all cases; in case 7 it happened that only one (correctly detected) attribute separates both outliers. False alarms also occurred in cases 3,5, and 7 but the contribution scores were still relatively small in these cases. Among the 10 cases $48 \times 10 = 480$ attributes did not contribute to the anomalies so that our false alarm rate is low.

Figures 2(b) and 2(c) show the contribution scores obtained from the PCA statistics T^2 and SPE, respectively, for the experiments with two contributors. Sometimes the PCA statistics can detect one of the two contributors, but in general the accuracies are low and without a quantitative criterion it is difficult to identify the contributors in many of these experiments (even though the data are normally distributed).

5.2 Nonlinear Data

Another experiment considers two groups of equal size with nonlinear separation. Two attributes (X_1 and X_2) contribute to the separation and 10 other random

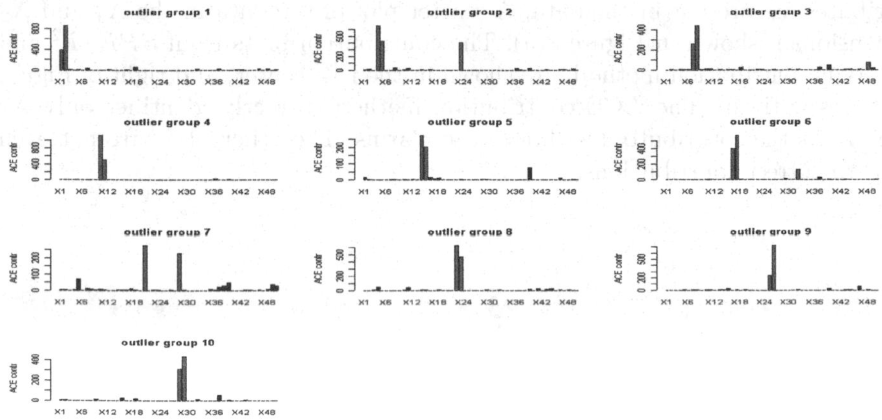

(a) Contributor plots from the ACE algorithm. The bars in red are scores that are significant.

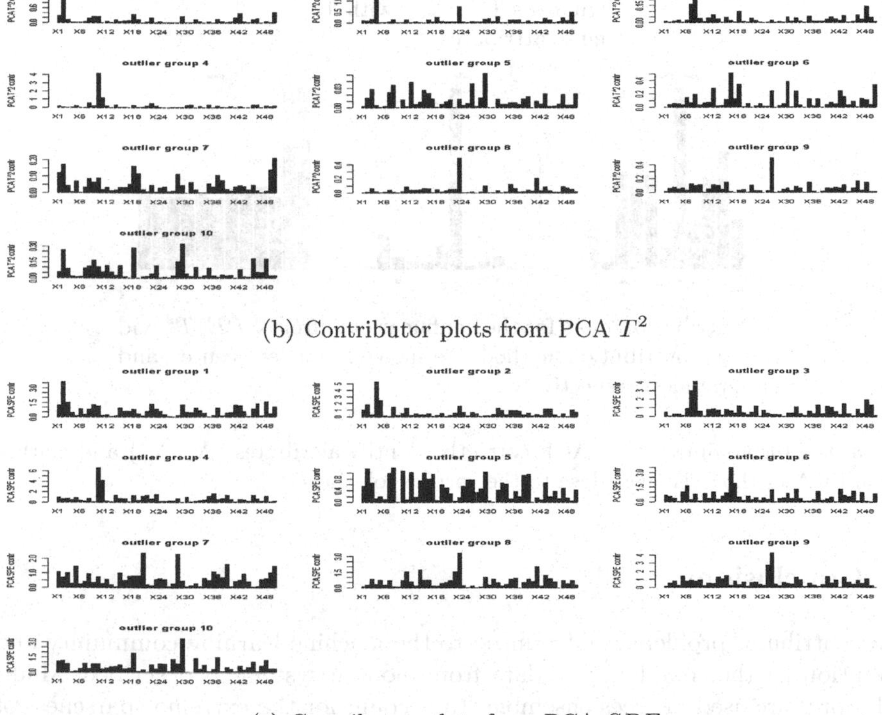

(b) Contributor plots from PCA T^2

(c) Contributor plots from PCA SPE

Fig. 2. Contributor plots from ACE, PCA SPE and T^2 for experiments with exactly two contributors among 50. Actual contributors are shown in the Figure 1. ACE correctly detects contributes in most cases while PCA statistics can detect one of the two contributors, but in general the accuracies are low and without a quantitative criterion it is difficult to identify the contributors in many of these experiments.

attributes are present in the data. A scatter plot of the data in the X_1 and X_2 dimensions is shown in Figure 3(a). The contribution plots from SPE, T^2 and the ACE contribution method, are shown in the left, center, and right of Figure 3(b), respectively. The ACE contribution method correctly identifies only X_1 and X_2 as the contributors without false alarms. The other plots are not even close to correct contributions.

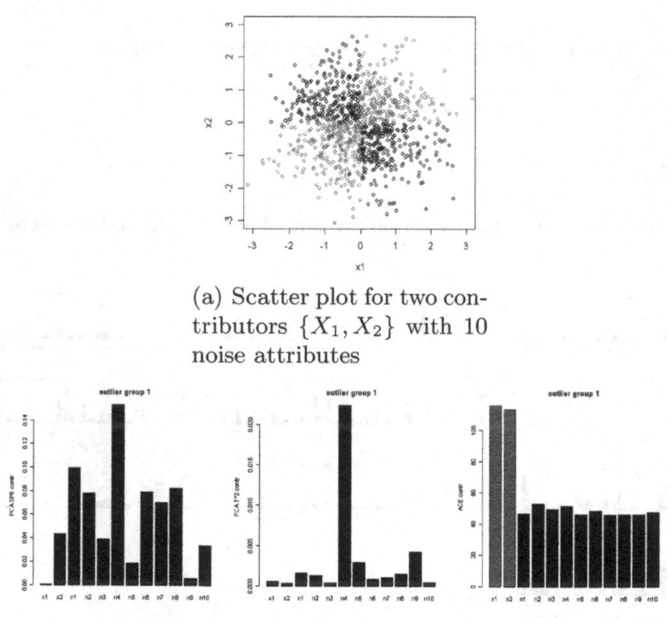

(a) Scatter plot for two contributors $\{X_1, X_2\}$ with 10 noise attributes

(b) Contribution plots for the nonlinear data from SPE, T^2 and the ACE contribution method, are shown in the left, center, and right graphics, respectively

Fig. 3. For the nonlinear data ACE correctly identify attributes $\{X_1, X_2\}$ and neither of the PCA methods is even close to the correct solutions

6 Conclusions

The contributor problem is introduced to the machine learning community and a solution for the complexity of data from modern systems is described. Modified scores are used in trees ensembles to account for the extreme sparseness of anomalies. Linear methods based on principal components often fail to detect contributors in anomaly detection, especially in the presence of noise. The linear methods are sensitive to scaling. Furthermore, linear methods rarely work when groups are separated by a non-linear boundary. The principal component methods only rank the attributes in terms of their separation power and do not indicate which are irrelevant. The proposed method works equally well for linear and non-linear cases in terms of diagnostics. One can identify (and rank) only attributes that contribute to the separation of the groups. It is insensitive to noise and identifies contributions for any anomaly group including a single (or a few data points) using a robust, nonparametric technique with statistical criteria to select contributors.

References

1. Hwang, W., Runger, G.C., Tuv, E.: Multivariate statistical process control with artificial contrasts. IIE Transactions 39(6), 659–669 (2007)
2. Hu, J., Runger, G.C., Tuv, E.: Tuned artificial contrasts to detect signals. International Journal of Production Research 45(23), 5527–5534 (2007)
3. Fei, T.L., Ting, K.M., Zhou, Z.H.: Isolation forest. In: 2008 Eighth IEEE International Conference on Data Mining (ICDM), pp. 413–422 (2008)
4. Hastie, T., Tibshirani, R., Friedman, J.: The Elements of Statistical Learning. Springer, Heidelberg (2001)
5. Hotelling, H.: Multivariate quality control-illustrated by the air testing of sample bombsights. In: Eisenhart, C., Hastay, M.W., Wallis, W.A. (eds.) Techniques of Statistical Analysis, pp. 111–184. McGraw-Hill, New York (1947)
6. Miller, P., Swanson, R., Heckler, C.: Contribution plots: A missing link in multivariate quality control. Applied Mathematics and Computer Science 8(4), 775–792 (1998)
7. Nomikos, P., MacGregor, J.F.: Multivariate spc charts for monitoring batch processes. Technometrics 37(1), 41–59 (1995)
8. Hu, J., Runger, G., Tuv, E.: Contributors to a signal from an artificial contrast. In: Informatics in Control, Automation and Robotics II, pp. 71–78. Springer, Heidelberg (2007)
9. Tuv, E.: Ensemble learning and feature selection. In: Guyon, I., Gunn, S., Nikravesh, M., Zadeh, L. (eds.) Feature Extraction, Foundations and Applications. Springer, Heidelberg (2006)
10. Tuv, E., Borisov, A., Torkkola, K.: Feature selection using ensemble based ranking against artificial contrasts. In: Proceedings of the International Joint Conference on Neural Networks (IJCNN) (2006)
11. Tuv, E., Borisov, A., Runger, G., Torkkola, K.: Best subset feature selection with ensembles, artificial variables, and redundancy elimination. Journal of Machine Learning Research (2008) (to appear)
12. Breiman, L.: Random forests. Machine Learning 45(1), 5–32 (2001)

Indoor Localization Using Neural Networks with Location Fingerprints

Christos Laoudias[1], Demetrios G. Eliades[1], Paul Kemppi[2], Christos G. Panayiotou[1], and Marios M. Polycarpou[1]

[1] KIOS Research Center for Intelligent Systems and Networks
Department of Electrical and Computer Engineering, University of Cyprus
Kallipoleos 75, P.O. Box 20537, 1678, Nicosia, Cyprus
{laoudias,eldemet,christosp,mpolycar}@ucy.ac.cy
[2] VTT Technical Research Centre of Finland
Vuorimiehentie 3, P.O. Box 1000, FIN-02044, Espoo, Finland
Paul.Kemppi@vtt.fi

Abstract. Reliable localization techniques applicable to indoor environments are essential for the development of advanced location aware applications. We rely on WLAN infrastructure and exploit location related information, such as the Received Signal Strength (RSS) measurements, to estimate the unknown terminal location. We adopt Artificial Neural Networks (ANN) as a function approximation approach to map vectors of RSS samples, known as location fingerprints, to coordinates on the plane. We present an efficient algorithm based on Radial Basis Function (RBF) networks and describe a data clustering method to reduce the network size. The proposed algorithm is practical and scalable, while the experimental results indicate that it outperforms existing techniques in terms of the positioning error.

Keywords: Localization, WLAN, Fingerprinting, Received Signal Strength, Radial Basis Function Networks.

1 Introduction

Localization techniques are used in order to determine the position of people, mobile devices and equipment. The provision of reliable location estimates is a challenge, especially indoors where satellite-based positioning is infeasible. Positioning accuracy is the key issue to effectively support indoor location aware services. Indicative applications include in-building guidance, asset tracking in hospitals or warehouses and autonomous robot navigation.

A wide variety of localization techniques have been discussed in the literature and can be categorized according to the type of measurements employed in the underlying positioning algorithm. Location is estimated with angle, timing or signal strength measurements from a number of transmitters with known locations by utilizing radio signal propagation models [1]. However, the presence of non line-of-sight paths between the receiver and the transmitter can cause accuracy degradation. Especially indoors, where multipath conditions are prevalent,

C. Alippi et al. (Eds.): ICANN 2009, Part II, LNCS 5769, pp. 954–963, 2009.

model inaccuracies may lead to high positioning errors; see [2] for an overview of technologies for wireless indoor location systems.

Localization performance in indoor environments can be improved by utilizing a premeasured map of Received Signal Strength (RSS) measurements. In this case, a set of predefined locations is associated with vectors containing RSS values from neighboring transmitters. These vectors, referred to as location fingerprints, are collected offline and stored in a database followed by the location coordinates. The unknown location can then be estimated on line from the current RSS fingerprint by finding the best match in the database. Matching is based on a distance measure between the current and collected fingerprints or on probability distributions [3,4,5].

Artificial Neural Networks (ANN) have been proposed as a solution to the location determination problem [6,7,8,9,10]. We adopt ANNs in a function approximation scheme to map RSS fingerprints in the high dimensional input signal space to locations in the physical space. In the envisioned indoor localization system data collected offline is used to train the ANN. Subsequently, when a mobile device running a location-based application enters a building, it receives the parameters of the trained ANN and is enabled to localize itself by using the currently measured RSS fingerprints. As a result, it is desirable to have an ANN-based algorithm that is computationally efficient and requires a small set of parameters in order to keep the communication cost low. Moreover, the ANN needs to be easily retrained in case the information in the database is outdated or new data become available. In this context, we evaluate different ANN models namely the Multi Layer Perceptron (MLP), Radial Basis Function (RBF) and Generalized Regression Neural Network (GRNN) for the implementation of the localization method.

The rest of this paper is structured as follows. In Section 2 the problem is defined and we briefly describe the related work in this area. In Section 3 we present the WLAN experimental setup used to conduct the measurements. The ANN designs and the proposed method based on RBF are detailed in Section 4. In Section 5 the positioning accuracy results are presented and we discuss the advantages and drawbacks of the ANN implementations. Finally, Section 6 provides concluding remarks and discusses future work.

2 Indoor Localization Overview

2.1 Problem Formulation

We introduce the theoretical framework for localization techniques based on fingerprints, assuming a WLAN infrastructure and availability of RSS measurements from neighboring Access Points (AP). Let $\mathbb{D} \subset \mathbb{R}^2$ be a 2-dimensional physical space denoting the area of interest. We define the finite set of locations $\mathcal{L} \subset \mathbb{D}$ known as reference points, where $\mathcal{L} = \{\ell_i \in \mathbb{D} | \ell_i = (x_i, y_i),\ i = 1, \cdots, L\}$. At each location $\ell_i \in \mathcal{L}$ a mobile device is used to collect RSS measurements from n neighboring APs. Thus, we form an n-dimensional input space denoted by $\mathcal{S}$. A reference fingerprint $s \in \mathcal{S}$ is a vector of RSS measurements collected at location ℓ_i, i.e. $s = [s_1, \cdots, s_n]^T$ and s_j denotes the RSS value related to the j-th AP. The reference points, can be placed over a uniform grid to cover the entire area with the desired resolution. However, the grid is usually non uniform and sparse due to building walls, furniture and other objects that limit the area where measurements can be performed. At each reference point $\ell_i \in \mathcal{L}$ we collect a series of fingerprints $s(\ell_i, m)$, $m = 1, \cdots, M$ and thus the database contains $R = L \cdot M$ fingerprints followed by the respective location coordinates. We also define the mean value fingerprint $\overline{s}(\ell_i) = \frac{1}{M}\sum_{m=1}^{M} s(\ell_i, m)$. During localization the goal is to obtain an estimate denoted as $\widehat{\ell}$, given a fingerprint $s' = [s'_1, \cdots, s'_n]^T$ that is measured at the unknown location.

2.2 Fingerprinting Techniques

Several approaches have been discussed for indoor localization using RSS fingerprints that are briefly described next.

In the **deterministic** approach, $\widehat{\ell}$ is obtained by minimizing a given norm of the difference between s' and the reference fingerprints. The Nearest Neighbor method introduced in [3], assumes the Euclidean distance as the optimization criterion and thus $\widehat{\ell}(s') = \arg\min_{\ell_i} \|s' - \overline{s}(\ell_i)\|^2$. In this case $\widehat{\ell} \in \mathcal{L}$. The K Nearest Neighbors (KNN) variant [3] determines $\widehat{\ell} \in \mathbb{D}$ as the centroid of the K locations with the shortest distances between s' and the mean value fingerprints. Weighted versions of the KNN algorithm have also been proposed.

From a **probabilistic** point of view, location is determined by calculating the conditional probabilities $p(\ell_i|s'), \forall \ell_i \in \mathcal{L}$. Then, the estimated location $\widehat{\ell} \in \mathcal{L}$ may be obtained by $\widehat{\ell}(s') = \arg\max_{\ell_i} p(\ell_i|s')$, as in [4]. Alternatively, authors in [5] calculate the expected value of the location variable ℓ, i.e. $\widehat{\ell}(s') = \mathbf{E}[\ell|s'] = \sum_{i=1}^{L} \ell_i p(\ell_i|s')$, in order to obtain the Minimum Mean Square Error (MMSE) estimate $\widehat{\ell} \in \mathbb{D}$. By application of Bayes rule the problem reduces to calculating $p(s'|\ell_i)$. Assuming that RSS measurements from neighboring APs are independent we get $p(s'|\ell_i) = \prod_{j=1}^{n} p(s'_j|\ell_i)$. Different methods have been proposed to estimate $p(s'_j|\ell_i)$ by utilizing the fingerprints in the database, namely the Kernel and Histogram methods [4,5]. In general, probabilistic techniques achieve higher positioning accuracy compared to the deterministic ones, at the expense of increased computational complexity.

2.3 Artificial Neural Network Approaches

In the context of ANNs, localization can be viewed as a classification problem. Each reference point defines a class and in this case the output of the ANN is one of the reference points $\ell_i \in \mathcal{L}$. Authors in [6] extend this approach by using a $L \times 1$ vector output for the network. The vector output provides the probability of s' belonging to each class and $\widehat{\ell} \in \mathbb{D}$ is obtained as the weighted mean of the respective reference points.

Alternatively, estimating current location can be viewed as a function approximation problem. The objective is to find a mapping $\mathbf{F}(s) : S \to \mathbb{D}$ of RSS fingerprints onto locations in the physical space. Recently, RBF networks have been discussed for localization in Wireless Sensor Networks (WSN). In [7] distance measurements from three beacon nodes, instead of RSS fingerprints, are utilized to evaluate the performance of RBF networks and compare to MLP and Recurrent Neural Networks (RNN) architectures. In [8] location is estimated with a RBF network using RSS measurements, however the number of neurons in the hidden layer is decided experimentally and can be very high thus increasing the computational cost. Positioning techniques based on ANNs have also been applied to areas where WLAN infrastructure is available. Authors in [9] propose a MLP architecture with a single hidden layer to perform localization using RSS measurements from three WLAN APs. A GRNN architecture, which is a RBF-type network with slightly different output layer, is proposed in [10] to determine location using RSS values from three transmitters.

In this paper we evaluate MLP, RBF and GRNN architectures using RSS measurements from ten WLAN APs. We focus on RBF networks and discuss a clustering method to reduce the size of the hidden layer and improve the computational complexity. In addition, this approach alleviates some of the overtraining problems of standard RBF networks. Experimental results indicate that the proposed clustered RBF design outperforms the deterministic approach and provides higher level of accuracy compared to MLP and GRNN.

3 Experimental Setup

The localization trial was carried out in a typical modern office environment at the premises of VTT Technical Research Centre in Espoo, Finland. The measurement campaign was conducted in the second floor of the 3-storey building, where $n = 10$ Cisco Aironet APs that use the IEEE802.11b/g standard are installed. We developed a Site Survey software that utilizes a floorplan map to mark $L = 107$ distinct reference points located 2-3 meters apart from each other in order to cover all public spaces and meeting rooms; see Fig. 1. RSS samples were collected with 1dBm resolution by using a WLAN-enabled smart phone. This resolution, though accurate enough for some applications, it introduces some "quantization" error since two locations that are very close to each other cannot be distinguished. This resolution depicts the lower bound of the error that any localization technique (using RSS measurements) can achieve. Typical RSS values range from -101dBm to -34dBm in close proximity to an AP. In case an AP was not hearable at a reference point, a small constant (-110dBm) was used to handle the missing RSS values in the fingerprints.

We have measured 30 fingerprints per reference point and selected randomly $M = 25$ out of these, corresponding to a total of $R = 2675$ fingerprints, which are stored in the database. This is our training set, while the remaining 5 fingerprints per reference point are kept as a test set for the performance evaluation of the ANN architectures described in Section 4. Additional fingerprints were also collected independently of the training set during a separate measurement campaign to form a second test set by following a predefined route that consists

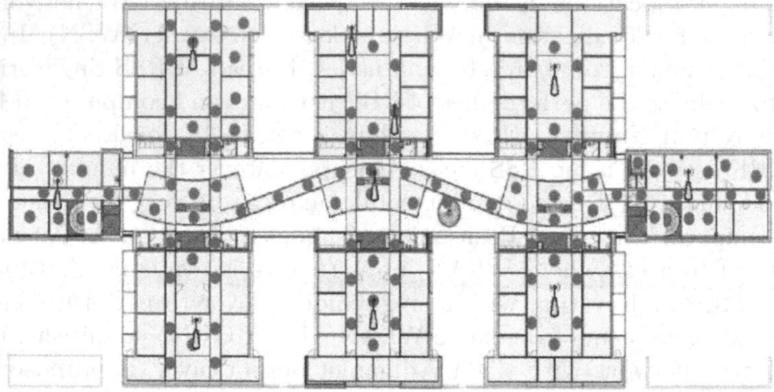

Fig. 1. Floorplan map with the reference points and AP locations

of 192 locations. One fingerprint was recorded at each location and the same route was sampled 3 times.

4 Artificial Neural Network Architectures

4.1 Multi Layer Perceptron (MLP)

The fully connected MLP network has ten inputs, corresponding to the RSS measurements from all available APs, while the output linear layer has two neurons representing the location coordinates (x, y). We use the sigmoidal transfer function for neurons in the single hidden layer. The size of the hidden layer was decided experimentally trying to keep it as small as possible, while preserving an adequate level of positioning accuracy. Specifically, we reserved 20% of the training fingerprints as a validation set and the network that achieved the best performance on this set was selected. The synaptic weights w were determined with the standard back propagation algorithm and the validation set was used as an early stopping method to avoid overfitting the training data. We have also investigated the use of a separate single output MLP network for each coordinate, x and y. We employed the same validation procedure in order to decide the network sizes, however the performance of this combination of two MLP designs on the validation set was degraded. The MLP architecture considered in Section 5 has 20 and 2 neurons in the hidden and output layers, respectively. We point out that training of the MLP is rather time consuming and the network must be retrained in case new data becomes available.

4.2 Radial Basis Function (RBF)

We examine a fully connected RBF network to approximate $\mathbf{F}(s) : S \rightarrow \mathbb{D}$ and use the normalized Gaussian function for neurons in a single hidden layer. The network has ten inputs and two outputs. Given a fingerprint s', the estimated location $\widehat{\ell}$ is given by

$$\widehat{\ell}(s') = \mathbf{F}(s') = \sum_{k=1}^{C} w_k u(\|s' - c_k\|) = \sum_{k=1}^{C} w_k \frac{\varphi(\|s' - c_k\|)}{\sum_{j=1}^{C} \varphi(\|s' - c_j\|)} \tag{1}$$

where $\varphi(\|s' - c_k\|) = \exp\big(-\beta\|s' - c_k\|^2\big)$. The number of neurons in the hidden layer is C, c_k is the 10-dimensional center for neuron k, and w_k are the 2-dimensional weights for the linear output layer. The value of β must be appropriately selected to ensure that the Gaussian basis functions are wide enough and the resulting RBF architecture implements a smooth approximation $\mathbf{F}(s)$.

In the GRNN architecture each reference fingerprint defines a center c_k, i.e. $C = R$. The weights w_k in (1) are set equal to the coordinates of the respective reference points and in that sense $\widehat{\ell}$ is the weighted average of the reference points whose fingerprints are closest to s'.

However, the weights w_k can be determined in order to optimize the fit between $\mathbf{F}(s)$ and the reference data. Thus, one may select the centers c_k and the width β and then form the following set of equations

$$(x_i, y_i) = \sum_{k=1}^{C} w_k u(\|s(\ell_i, m) - c_k\|),\ i = 1, \cdots, L \text{ and } m = 1, \cdots, M \tag{2}$$

In the standard RBF network (sRBF) each reference fingerprint defines the center of a neuron. In this case, the system of linear equations based on (2) can be written in matrix form as $\mathbf{U}\mathbf{w} = \mathbf{d}$, where $\mathbf{U} = \{u(\|s_j - c_i\|)|(j, i) = 1, \cdots, R\}$ and $u(\cdot)$ is the normalized Gaussian basis function given in (1). Matrix $\mathbf{d}$ contains the coordinates of the reference points and the weights are easily obtained by $\mathbf{w} = \mathbf{U}^{-1}\mathbf{d}$.

The number of neurons in the hidden layer can be reduced dramatically by application of a clustering technique on the reference fingerprints. In the clustered RBF (cRBF) architecture each center is set equal to the mean value fingerprint $\overline{s}(\ell_i)$. Thus, $C = L$ and the weights are calculated in a least squares sense by solving the overdetermined system of equations based on (2). The minimum-norm solution for the weight vector is $\mathbf{w} = \mathbf{U}^{+}\mathbf{d}$, where $\mathbf{U}^{+}$ is the *pseudoinverse* of matrix $\mathbf{U}$ defined as $(\mathbf{U}^{\mathbf{T}}\mathbf{U})^{-1}\mathbf{U}^{\mathbf{T}}$.

The sRBF design guarantees exact fitting for reference data at the expense of increased hidden layer size. Moreover, it is well known that sRBF is prone to overfitting and exhibits inadequate generalization capabilities. In [7] it is reported that sRBF outperforms cRBF for the localization problem and the positioning error is in the order of few cm. However, in that work the evaluation was conducted in a small scale (3×3 m) experimental test bed with line-of-sight conditions using low noise distance measurements from 3 beacon nodes. Under realistic propagation conditions the accuracy of sRBF is degraded. We consider the sRBF design in our evaluation to verify that when noisy RSS measurements collected in a real-life WLAN environment are utilized, its performance is poor compared to the clustered counterpart.

5 Results and Discussion

The MLP, sRBF, cRBF and GRNN designs are compared in terms of the positioning error, defined as the Euclidean distance between the actual and estimated location. We have implemented the deterministic KNN localization method [3] and use it as baseline for our evaluation. In our experimental setup, the value $K = 2$ provides the lowest positioning error and is therefore selected for the rest of the experiments.

5.1 Test Case 1

The first test set comprises 535 RSS fingerprints in total, i.e. 5 test vectors per reference point and has the same statistical distribution of positions as the training set. Table 1 summarizes the accuracy results. The MLP, GRNN and the proposed cRBF architectures are equivalent regarding the mean and median positioning error. The results also indicate that the error in half of the location estimates derived with the KNN algorithm is below 1.9m. This is lower compared to MLP, GRNN and cRBF, however a considerable fraction of the KNN estimates exhibit error higher than 10m leading to the same level of accuracy as far as the mean error is concerned. As expected the sRBF design achieves the highest level of accuracy for the given test set. Note that during the data collection process the RSS level in some reference points may not vary much for certain APs and duplicate fingerprints are recorded. Therefore, there is a high probability that exactly the same fingerprint is present in both the training and test sets. The sRBF design guarantees exact fitting for training data and for this reason the median error is zero. However, sRBF is prone to overfitting and its poor generalization capabilities are depicted in the maximum positioning error; even moderate deviation from the training fingerprints leads to significant accuracy degradation.

5.2 Test Case 2

The RSS fingerprints in the second test set are measured by walking inside the area of interest. Note that most of the unknown locations do not coincide with any reference point. Location estimates obtained with the MLP network for a single route are depicted in Fig. 2 (dots), while the black line denotes the actual route. The estimated locations for the same route using the sRBF and cRBF designs are illustrated in Fig. 3 and Fig. 4, respectively. The increased number of neurons in the sRBF network results in worse localization performance and the estimates do not reflect the traveled route. The sRBF network is overtrained and has essentially learned the noise in the reference fingerprints. Even when a smaller value is used for β, in order to increase the width of the Gaussian function and create a smoother approximation $\mathbf{F}(s)$, sRBF fails to accurately locate the user when new fingerprints are presented to the network. GRNN location estimates are depicted in Fig. 5 and higher accuracy is achieved compared to MLP and sRBF networks.

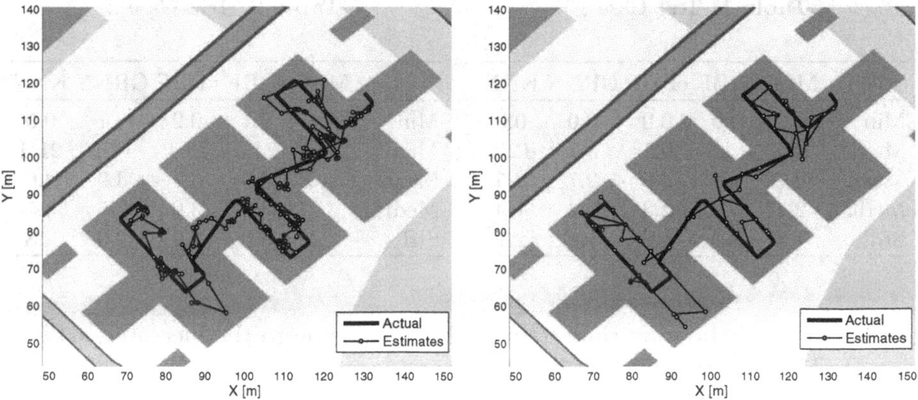

Fig. 2. Location estimates with MLP

Fig. 3. Location estimates with sRBF

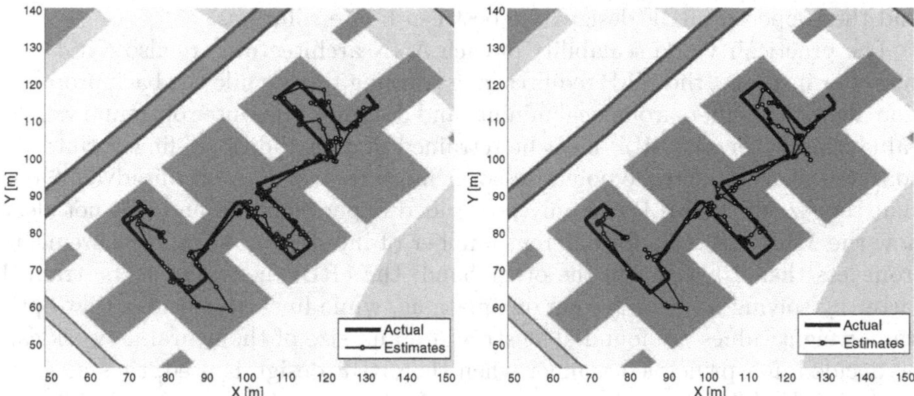

Fig. 4. Location estimates with cRBF

Fig. 5. Location estimates with GRNN

Positioning error statistics pertaining to the second test set that contains all fingerprints collected after sampling the same route 3 times are tabulated in Table 2. The cRBF design has the best localization performance according to the mean and median error. The standard deviation (Std) of the error is also low and cRBF is the only network that outperforms the KNN algorithm. This is followed by the GRNN architecture that achieves the same level of accuracy as KNN. The sRBF design provides less accurate location estimates compared to cRBF, GRNN and KNN. Finally, the MLP network exhibits the worst performance for the given test set and the maximum error is surprisingly high.

The MLP has very low memory requirements for storing the network weights and biases and essentially the fingerprint database is compressed into a small set of parameters. Moreover, the MLP is the least computationally intensive, due to the small number of neurons. In the sRBF and GRNN designs, the weights and all reference fingerprints are required to perform localization, while they

Table 1. Test Case 1

	MLP	sRBF	cRBF	GRNN	KNN
Min	0.2	0.0	0.0	0.0	0.0
Max	10.0	14.3	9.1	8.2	12.2
Mean	2.7	1.9	2.6	2.7	2.5
Median	2.4	0.0	2.3	2.4	1.9
Std	1.7	3.0	1.8	1.8	2.1

Table 2. Test Case 2

	MLP	sRBF	cRBF	GRNN	KNN
Min	0.1	0.1	0.2	0.1	0.0
Max	29.4	24.0	13.1	17.2	21.4
Mean	5.3	4.6	3.4	3.9	4.0
Median	4.2	3.6	3.0	3.5	3.5
Std	4.4	3.6	2.2	2.5	2.8

exhibit longer estimation time compared to MLP due to the increased network size. Problems related to storage memory and localization time can be alleviated by adopting the cRBF architecture. Nowadays, the memory and computational overhead of all these ANN architectures can be well handled by high-end mobile devices. However, the transmission overhead to communicate the ANN parameters to the device through the WLAN is significant, thus rendering the MLP and the proposed cRBF designs the best candidate solutions.

The practicality and scalability of each ANN architecture are also critical issues. For instance, the MLP requires long training time, while the back propagation algorithm suffers from local minima and does not guarantee optimum weight values. Moreover, the MLP must be retrained in case additional fingerprints are collected at new reference points to cover more rooms. Another disadvantage is that the size of the MLP can only be decided experimentally and it is not clear how the MLP will scale for different number of inputs, e.g. using measurements from less than 10 APs. On the other hand, the cRBF network can be trained faster by solving a linear system of equations, while linearity ensures that optimum weight values are found. The structure and size of the neural network can be decided in a principled manner when the cRBF design is used, thus increasing its applicability to other environments. In case new reference fingerprints are available the size of the cRBF network is easily decided, while retraining time can be greatly reduced by using appropriate matrix operations.

6 Conclusions

We have evaluated several ANN designs to perform indoor localization by exploiting RSS fingerprints collected in a typical office environment. We rely on WLAN infrastructure to minimize the deployment cost since no specialized equipment is required. The proposed cRBF algorithm is a promising solution to the location determination problem that can be easily scaled and applied to other indoor environments with WLAN coverage. The mobile device needs to receive only a small number of parameters through the WLAN in order to start locating itself inside the building. Moreover, experimental results indicate that the cRBF achieves higher level of accuracy compared to the sRBF, MLP and GRNN designs, as well as the deterministic KNN algorithm.

Future work will focus on further improving the cRBF approach by using a variable selection procedure in order to limit the area where the user may reside and determine which APs to use in the localization process. We also plan to use an appropriate network regularization method and variable β values in the Gaussian basis functions, based on the distribution of centers in the multidimensional signal space, to achieve higher accuracy.

Acknowledgments. This work is partly supported by the Cyprus Research Promotion Foundation under contract ENIΣX/0506/59 and the European Science Foundation (ESF) in the framework of the Middleware for Network Eccentric and Mobile Applications (MiNEMA) activity.

References

1. Stuber, G., Caffery, J.: Radiolocation Techniques. In: The Communications Handbook. CRC Press, Boca Raton (2002)
2. Pahlavan, K., Li, X., Makela, J.: Indoor geolocation science and technology. IEEE Communications Magazine 40(2), 112–118 (2002)
3. Bahl, P., Padmanabhan, V.: RADAR: an in-building RF-based user location and tracking system. In: Proceedings IEEE INFOCOM, vol. 2, pp. 775–784 (2000)
4. Youssef, M., Agrawala, A., Udaya Shankar, A.: WLAN location determination via clustering and probability distributions. In: IEEE International Conference on Pervasive Computing and Communications (PerCom), pp. 143–150 (2003)
5. Roos, T., Myllymaki, P., Tirri, H., Misikangas, P., Sievanen, J.: A probabilistic approach to WLANuser location estimation. International Journal of Wireless Information Networks 9(3), 155–164 (2002)
6. Saha, S., Chaudhuri, K., Sanghi, D., Bhagwat, P.: Location determination of a mobile device using IEEE 802.11b access point signals. IEEE Wireless Communications and Networking (WCNC) 3, 1987–1992 (2003)
7. Shareef, A., Zhu, Y., Musavi, M.: Localization using neural networks in wireless sensor networks. In: Proceedings of the 1st international conference on MOBILe Wireless MiddleWARE, Operating Systems, and Applications, ICST, pp. 1–7 (2007)
8. Lee, H., Wicke, M., Kusy, B., Guibas, L.: Localization of mobile users using trajectory matching. In: Proceedings of the first ACM international workshop on Mobile entity localization and tracking in GPS-less environments, San Francisco, California, USA, pp. 123–128 (2008)
9. Battiti, R., Nhat, T., Villani, A.: Location-aware computing: a neural network model for determining location in wireless LANs. University of Trento, Trento, Italy, Tech. Rep. DIT-02-0083 (2002)
10. Nerguizian, C., Despins, C., Affes, S.: Indoor geolocation with received signal strength fingerprinting technique and neural networks. In: Telecommunications and Networking - ICT 2004, pp. 866–875 (2004)

Distributed Faulty Sensor Detection in Sensor Networks

Xuanwen Luo and Ming Dong

Research and Development, Sindvik Mining and Construction
13500 NW CR 235, Alachua, Florida 32615, USA
Department of Computer Science, Wayne State University
5143 Cass Avenue, Detroit, Michigan 48202, USA
xuanwen.luo@sandvik.com, mdong@wayne.edu

Abstract. In wireless sensor networks, faulty sensors may produce incorrect data and transmit the data to other sensors. They may cause inappropriate data fusion. Furthermore, they would consume the limited energy and bandwidth of sensor networks. In this paper, we propose a distributed faulty sensor detection scheme, in which we assume that the sensor fault probability or reliability is unknown and data to be sensed has Gaussian distribution with unknown parameters. In the proposed method, each sensor obtains a global convergency data through data fusion and makes a local 3-level decision by hypothesis testing against the global convergency data. A final decision about the sensor is then obtained by fusing the decisions of its neighbors. The detection is carried out in a distributed fashion as each sensor only communicates with its neighbors in the entire process. Experiment results demonstrate that the proposed algorithm is able to achieve higher detection accuracy than existing methods even without the knowledge of sensor reliability and parameters of data distribution.

Keywords: Consensus value, data fusion, distributed faulty sensor detection, hypothesis testing, sensor fault, wireless sensor networks.

1 Introduction

Recent advancement in wireless communications and electronics has enabled the development of low-cost wireless sensor networks. A wireless sensor network usually comprises of a large number of small sensor nodes, which consist of sensing, data processing and communication components. The unique features of a sensor network, for example, random deployment in inaccessible terrains and cooperative effort, offer unprecedented opportunities for a broad spectrum of civilian and military applications, such as military tactical surveillance, target detection and tracking, environment monitoring, modeling and remote sensing, etc.

Sensor networks are usually deployed in an uncontrolled, harsh, even hostile environments. Due to the low cost, it is not uncommon that the sensor nodes become faulty. These faulty sensors must stop functioning to ensure the network

C. Alippi et al. (Eds.): ICANN 2009, Part II, LNCS 5769, pp. 964–975, 2009.

quality of service. To detect the faulty sensor is not trivial because of the limited energy and communication bandwidth. Sensor nodes are powered by battery, which may not be replaced or recharged after deployment. It is too costly for the base station to collect data from all nodes and detect the faulty sensors in a centralized mode. A distributed faulty sensor detection scheme is preferred in wireless sensor networks.

The basic idea of distributed detection [1] is to have each of the independent sensors make a local decision and then combine these decisions at a fusion sensor to generate a global decision. Optimal distributed designs have been sought under both the Bayesian and the Neyman-Pearson performance criteria [2]. Statistically, the distributed detection could be modeled as a hypothesis test problem: n sensors observe an unknown hypothesis; the sensor observations are independent and identically distributed, given the unknown hypothesis; each sensor transmits its decision over a multiple access channel to a fusion sensor; based on the received sensor decision, the fusion sensor makes the final decision regarding the unknown hypothesis. In many data centric applications of sensor networks, the nearby sensors are likely to have similar measurements. In distributed detection, we exploit the fact that sensor faults are likely to be stochastically unrelated, while sensor measurements are likely to be spatially correlated.

Distributed averaging or average consensus in wireless sensor network is studied in [3,4]. Their studies show that if all sensor measurements follow the same data distribution $\mathcal{N}(\mu, \sigma^2)$, the maximum-likelihood estimate of μ is the average of all sensor measurements. Each sensor can obtain the maximum-likelihood estimate or average consensus by communication with its neighbors. Since the consensus value are common to all sensors, it is actually global information. In this work, without knowing (μ, σ^2), each sensor makes a local 3-level decision by hypothesis testing against the consensus value, and its final decision is made by fusing the decisions from its neighbors. Our simulation results show that the proposed algorithm has better performance than existing ones even without knowledge about the sensor and the data to be sensed.

In this work, our major contribution includes 1) a distributed algorithm for faulty sensor detection without knowing sensor fault probability and parameters of sensor data distribution; 2) the threshold in our proposed algorithm is completely independent on the sensor fault probability, which makes it very flexible in applications. The remainder of the paper is organized as follows. In Section 2, we state the problem and related works. In Section 3, we show how each sensor obtains the global information through communicating with its neighbors. The model of sensor fault is defined in Section 4. The detail of the faulty sensor detection scheme is described in Section 5. In Section 6, we present our simulation results. We conclude in Section 7.

2 Statement of the Problem and Related Works

N sensor nodes are deployed over a interested region to perform data collection of their environment. We assume that each sensor can only communicate directly

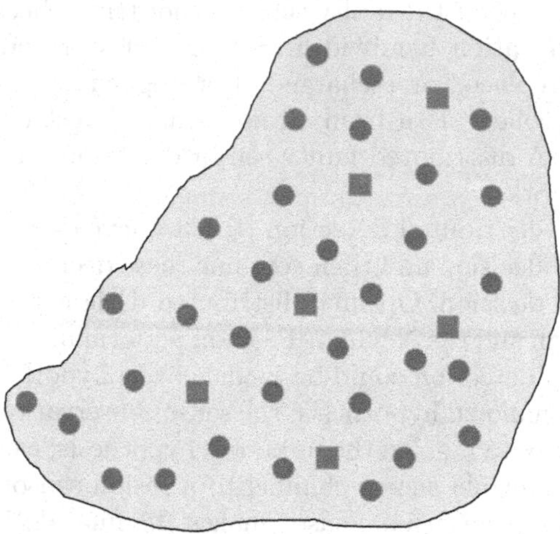

Fig. 1. A sensor network. Dot denotes normal sensors and square denotes faulty sensors

with its neighbor sensors (within its communication radius) and indirectly with other sensors through its neighbors. With broadcast/acknowledge protocol, each sensor node is also able to locate the neighbors within its communication range.

A sensor node could make certain decision independently based on its own measurement. The network considered is also likely to contain faulty sensor nodes due to harsh environment and manufacturing reasons. Normally, each sensor, may be good or faulty, communicates its own measurement with other sensor in data fusion. It is obvious that sensor faults make the fusion unreliable, i.e., a fusion data may exceed a tolerant range. Fig. 1 shows a sample scenario of the sensor network. The "dot" denotes a healthy sensor while the "square" denotes a faulty sensor. Each sensor can communicate with other sensors within its communication radius.

Faulty sensor detection in wireless sensor networks is studied explicitly and implicitly in the literatures [5,6,7,8]. In [5], distributed Bayesian algorithm was proposed to detect both faulty sensors and event regions. Each sensor collects local binary decisions from its neighbors and final decision is made based on majority-voting. The algorithm can correct over 85% – 95% of the total faults when the fault rate is as high as 10% in the entire network. An energy efficient fault-tolerant detection scheme is studied in [6] by considering both noise related measurement error and sensor fault. Two local binary decision schemes are studied under Bayesian and Neyman-Pearson frameworks. Each sensor makes a final decision by fusing the binary decisions from its neighbors with k-out-of-n rule. In [8], Ding et al. proposed a localized faulty sensor identification algorithm, in which a single reading at a sensor is compared with the median of its neighbors's readings. If the difference is large, the sensor is likely to be faulty. The algorithm is purely localized and requires low computational overhead, so that it can be easily scaled to large sensor networks. Chen et al. [7] developed a distributed fault detection algorithm by comparing difference of two sensors' current measurements with the difference of their historical measurements. If the difference becomes significantly large over the time, a sensor is more likely faulty. In above studies, the data distribution or sensor fault probability or both are assumed known. In [5,6,8], the detection threshold depends on the sensor fault probability: when the probability changes, the threshold must be changed accordingly, while the threshold in [7] strongly depends on the distribution of normal sensor data and faulty sensor data.

In many applications, such as density estimation of environment [9], the parameters of data distribution is not available for sensor network in advance. Actually, to identify the parameters is a difficult task itself. Furthermore, the sensor fault probability is difficult to predict for sensors deployed in a harsh environment. All these greatly limit the applicability of the previously proposed schemes for faulty sensor detection.

3 Global Information: Distributed Average Consensus in Networks

Distributed consensus or averaging has been extensively studied recently [10,3,4]. Let the reading of sensor node i be $x_i = \mu + v_i \sim \mathcal{N}(\mu, \sigma^2)$, $i = 1, \ldots, n$, where μ is a scalar to be estimated, and the noise v_i is i.i.d. Gaussian $v_i \sim \mathcal{N}(0, \sigma^2)$. In this case, the maximum likelihood (ML) estimation is the average of the measurements x_i at all sensors, i.e., $\hat{\mu}_{ML} = \frac{1}{n}\mathbf{1}^T x$, where $\mathbf{1}$ denotes the vector whose components are all ones. The associated mean-square error is $\frac{1}{n}\sigma^2$. The distributed linear iteration can be expressed in the form,

$$x_i(t+1) = W_{ii}x_i(t) + \sum_{j \in N_i} W_{ij}x_j(t),\ i = 1, \ldots, n. \tag{1}$$

where $t = 0, 1, \ldots,$ is the discrete time index. In the vector form, (1) can be rewritten as $x(t+1) = Wx(t)$. Define a t-step transition matrix,

$$\Upsilon(t) = W(t-1)\ldots W(1)W(0) \tag{2}$$

We have

$$x(t) = \Upsilon(t)x(0) \tag{3}$$

The weight matrices exist such that the states at all nodes converge to $\hat{\mu}_{ML} = \frac{1}{n}\mathbf{1}^T x$, i.e.,

$$\lim_{t \to \infty} x(t) = \hat{\mu}_{ML}\mathbf{1} = \left[\frac{1}{n}\mathbf{1}^T x(0)\right]\mathbf{1} \tag{4}$$

which is equivalent to $\lim_{t \to \infty} \Upsilon(t) = \frac{1}{n}\mathbf{1}\mathbf{1}^T$. Equation (4) should hold for any $x(0) \in \mathcal{R}^n$. Since all sensors in the network have the $\hat{\mu}_{ML}$, it is actually global information about the sensor network. The rule to choose the weights can be expressed as [4],

$$W_{ij}(t) = \begin{cases} \frac{1}{1+\max\{d_i(t),d_j(t)\}}, \text{ if } \{i,j\} \in E(t) \\ 1 - \sum_{\{i,k\}\in E(t)} W_{ik}(t), \text{ if } i = j \\ 0, \text{ otherwise} \end{cases} \quad (5)$$

where $d_i(t) = \|N_i(t)\|$ is the degree (number of neighbors) of node i and E is the set of existing connections among sensor nodes in the network.

4 Sensor Fault Model and Fusion Error

As the most common sensor fault is offset bias [13], at a given sensor, we assume that the faulty sensor measurement might have an offset β_0. The fault mode we consider can be written as $\mathcal{N}(\beta_0 + \mu, \sigma^2)$. Offset bias alters the sensor readings uniformly by a certain value. The bias model is a drift fault model, where the correct value is subject to alternation that is a time invariant function of the correct value [14]. Assuming that γ of n sensors have offset bias, modeled by $\mathcal{N}(\beta_0 + \mu, \sigma^2)$, we have

$$\frac{1}{n}\sum_{i=1}^{n} x_i \sim \mathcal{N}(\mu + \gamma\beta_0, \sigma^2/n) \quad (6)$$

The mean square error due to the sensor fault is $e_{err} = E[(\gamma\beta_0)^2]$ for $n \to \infty$. It is clear that when $\gamma \neq 0$, $e_{err} \neq 0$ asymptotically. Obviously, the error depends on both γ and β_0. The average consensus method previously discussed cannot remove the error from the sensor fault, leading to a low fusion accuracy.

5 Faulty Sensor Detection

The average consensus discussed in the previous section is the global maximum likelihood estimate of the sensor network and will be used for faulty sensor detection in this section. We consider data with Gaussian distribution but with unknown mean and variance.

We first introduce a concept of $\varepsilon-$consensus. Given an initial value $x(0)$ of a network at $t = 0$ and the consensus value $\mu = \frac{1}{n}\mathbf{1}\mathbf{1}^T x(0)$, we define $x(t_\varepsilon) = \Upsilon(t_\varepsilon)x(0)$ the $\varepsilon-$consensus value at $t = t_\varepsilon$ as

$$Pr_{t\geq t_\varepsilon}\left(\frac{|\mu - x(t)|}{\mu} < \varepsilon\right) = 1 \quad (7)$$

where $\varepsilon \in (0,1)$. Once sensor network reaches the $\varepsilon-$consensus $\mu_\varepsilon = x(t_\varepsilon)$ after time t_ε, we use hypothesis testing to detect faulty sensors. Notice that each sensor i has the same $\varepsilon-$consensus value $\mu_{i,\varepsilon}$. Thus, we use μ_ε instead of $\mu_{i,\varepsilon}$ for convenience in the following discussion.

5.1 Hypothesis Testing

Suppose that random samples are selected from a normal distribution with unknown parameter $\theta = \{\mu, \sigma^2\}$. The null hypothesis H_0 specifies that θ lies in a particular set of possible values: $\Omega_0 = \{\mu_\varepsilon, \sigma^2\}$; the alternative hypothesis H_1 specifies that θ lies in another set of possible values: $\Omega_a = \{\mu, \sigma^2 : \mu > \mu_\varepsilon\}$, which does not overlap with Ω_0. Let $\Omega = \Omega_0 \bigcup \Omega_a$, i.e., $\Omega = \{\mu, \sigma^2 : \mu \geq \mu_\varepsilon\}$, and $L(\hat{\Omega}_0)$ denote the likelihood function with all unknown parameters replaced by their maximum-likelihood estimators, subject to the restriction that $\theta \in \Omega_0$. Similarly, let $L(\hat{\Omega})$ be obtained the same way, but with the restriction that $\theta \in \Omega$. Suppose that, at sensor i, $x_i(1), x_i(2), \ldots, x_i(l)$ constitute random samples from a normal distribution with unknown mean μ and unknown variance σ^2. We drop the index i for convenience in the following discussion. We want to test (Test I):

$$H_0 : \mu = \mu_\varepsilon \text{ versus } H_1 : \mu > \mu_\varepsilon \tag{8}$$

For the normal distribution, we have

$$L(\Omega) = L(\mu, \sigma^2) = \\ = \left(\frac{1}{\sqrt{2\pi}}\right)^l \left(\frac{1}{\sigma^2}\right)^{l/2} \exp\left[-\sum_{i=1}^{l} \frac{(x(i)-\mu)^2}{2\sigma^2}\right] \tag{9}$$

Restricting μ to Ω_0 implies that $\mu = \mu_\varepsilon$. The value of σ^2 that maximizes $L(\mu_\varepsilon, \sigma^2)$ is $\hat{\sigma}_\varepsilon^2 = \frac{1}{l}\sum_{i=1}^{l}(x(i) - \mu_\varepsilon)^2$. With simple plug in, $L(\hat{\Omega}_0)$ can be expressed as,

$$L(\hat{\Omega}_0) = \left(\frac{1}{\sqrt{2\pi}}\right)^l \left(\frac{1}{\hat{\sigma}_\varepsilon^2}\right)^{l/2} \exp^{-l/2} \tag{10}$$

We now turn to find $L(\hat{\Omega})$. The unrestricted maximum-likelihood estimator of μ is $\bar{x} = \frac{1}{l}\sum_{i=1}^{l} x(i)$. Therefor, for θ restricted to Ω, the maximum-likelihood estimator of μ is $\hat{\mu} = \max(\bar{x}, \mu_\varepsilon)$. Notice that if the actual maximum of L is outside the region Ω, the maximum within Ω occurs at the boundary point μ_ε [15]. Similarly, $L(\hat{\Omega})$ can be expressed as,

$$L(\hat{\Omega}) = \left(\frac{1}{\sqrt{2\pi}}\right)^l \left(\frac{1}{\hat{\sigma}^2}\right)^{l/2} \exp^{-l/2} \tag{11}$$

where $\hat{\sigma}^2 = \frac{1}{l}\sum_{i=1}^{l}(x(i) - \hat{\mu})^2$. Thus,

$$\lambda = \frac{L(\hat{\Omega}_0)}{L(\hat{\Omega})} = \begin{cases} \left[\frac{\sum_{i=1}^{l}(x(i)-\bar{x})^2}{\sum_{i=1}^{l}(x(i)-\mu_\varepsilon)^2}\right]^{l/2}, & \text{if } \bar{x} > \mu_\varepsilon \\ 1, & \text{if } \bar{x} \leq \mu_\varepsilon \end{cases} \tag{12}$$

Notice that $\lambda \leq 1$. Thus, the rejection region is $\lambda < k < 1$, where k is the threshold to be computed (see Equation (15)).

Assuming the alternative hypothesis H_2: θ lies in another set of possible values: $\Omega_a = \{\mu, \sigma^2 : \mu < \mu_\varepsilon\}$, we can test (Test II):

$$H_0 : \mu = \mu_\varepsilon \text{ versus } H_2 : \mu < \mu_\varepsilon \tag{13}$$

Similarly, we have,

$$\lambda = \frac{L(\hat{\Omega}_0)}{L(\hat{\Omega})} = \begin{cases} \left[\frac{\sum_{i=1}^{l}(x(i)-\bar{x})^2}{\sum_{i=1}^{l}(x(i)-\mu_\varepsilon)^2}\right]^{l/2}, & \text{if } \bar{x} < \mu_\varepsilon \\ 1, & \text{if } \bar{x} \geq \mu_\varepsilon \end{cases} \tag{14}$$

After mathematical manipulation, the threshold k can be computed based on t-statistic. Let α be the designated significant level, $\nu = l - 1$ the degree of freedom, we can find $t_{\alpha,\nu}$ in the t-Table and get

$$k = \left(\frac{l-1}{t_{\alpha,\nu}^2 + l - 1}\right)^{l/2}, \tag{15}$$

which is independent on sensing data and sensor reliability.

5.2 Faulty Sensor Detection Scheme and Algorithm

Assume that each sensor has l observations $x(1), x(2), \ldots, x(l)$ and a ε-consensus value μ_ε after certain iterative steps with Equation (3). We first discuss the local decision in each sensor. For $\bar{x} > \mu_\varepsilon$, sensor i starts Test I. Let $u_i = 1$ if rejected, which indicates that mean value of the measurements in sensor i is great than μ_ε. Otherwise, let $u_i = 0$. That is, the mean value of the measurements in sensor i is equal to μ_ε. Similarly, for $\bar{x} < \mu_\varepsilon$, sensor i starts Test II. Let $u_i = -1$ if rejected, which indicates that the mean value of the measurements in sensor i less than μ_ε. Otherwise, let $u_i = 0$. For $\bar{x} = \mu_\varepsilon$, let $u_i = 0$ directly. The local decision rule can be expressed as,

$$u_i = \begin{cases} 1, & \text{if } \bar{x}_i > \mu_\varepsilon \\ 0, & \text{if } \bar{x}_i = \mu_\varepsilon \\ -1, & \text{if } \bar{x}_i < \mu_\varepsilon \end{cases} \tag{16}$$

The sensor nodes collect the decisions from their neighbors and makes the decision fusion u_0 by,

$$u_0 = \begin{cases} 1, & u_1 + \cdots + u_m > 0 \\ 0, & u_1 + \cdots + u_m = 0 \\ -1, & u_1 + \cdots + u_m < 0 \end{cases} \tag{17}$$

where m is the number of neighbors of each sensor. The faulty sensor detection rule can be expressed as,

$$\text{sensor } i = \begin{cases} \text{faulty}, & \text{if } u_i \neq 0, u_i \neq u_0 \\ \text{good}, & \text{if } u_i = 0 \text{ or } u_i \neq 0, u_i = u_0 \end{cases} \tag{18}$$

The entire algorithm of distributed faulty sensor detection is summarized as follows.

```
Input: threshold k, number of measurements l
Output: faulty sensor detection result

Each sensor i obtains l measurements and μ_ε
IF x̄ > μ_ε THEN
Do Test I based on k
    IF Rejected THEN
    u_i = 1
    ELSE u_i = 0
    END IF
ELSE IF x̄ < μ_ε THEN
Do Test II based on k
    IF Rejected THEN
    u_i = −1
    ELSE u_i = 0
    END IF
ELSE u_i = 0
END IF

Each sensor collects u_i from its m neighbors
IF u_1 + ··· + u_m > 0 THEN
u_0 = 1
ELSE IF u_1 + ··· + u_m = 0 THEN
u_0 = 0
ELSE u_0 < 0
END IF

IF u_i ≠ 0 and u_i ≠ u_0 THEN
Sensor i is faulty
ELSE IF u_i = 0 OR u_i ≠ 0 and u_i = u_0 THEN
Sensor i is good
END IF
```

6 Simulation and Discussion

In this section, we present our simulation results for the proposed algorithm of faulty sensor detection. The sensor network contains 100 sensors in a 10 by 10 area. The communication radius for each sensor is $\sqrt{2}$. Normal sensor readings are drawn from $\mathcal{N}(\mu, \sigma_1^2)$ and faulty sensor reading are drawn from $\mathcal{N}(\mu+\beta_0, \sigma_2^2)$. In the simulation, we choose $\mu = 10$, $\sigma_1 = 1$ and two type of sensor faults are considered:

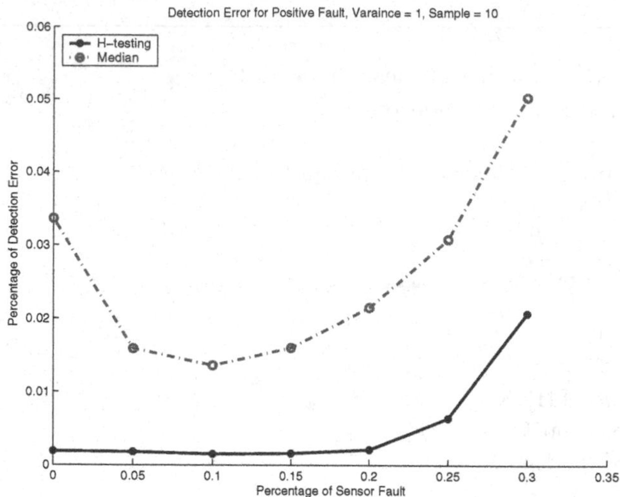

Fig. 2. Detection Error Rate of Sensor Network with Positive Fault

1. positive fault: $\beta_0 = 20$, $\sigma_2 = 1$.
2. random fault: β_0 is chosen randomly from $\beta_1 = 20$ and $\beta_2 = -20$. $\sigma_2 = 1$.

Notice that the mean and variances can be chosen arbitrarily as long as β_0 is sufficiently large when compared with σ_1 and σ_2. For each type of sensor fault, the percentage of faulty sensors r is set at 0%, 5%, 10%, 15%, 20%, 25%, 30%. We choose $\varepsilon = 0.05$ in obtaining μ_ε. Each sensor takes $l = 10$ samples from its environment and the corresponding threshold $k = 0.0421$ is computed from Equation (15) with $t_{\alpha,\nu} = 2.821$, where $\alpha = 0.01$ and $\nu = l - 1$. We repeat the experiment with each sensor fault type 200 times. The averaged detection error and false alarm rate are reported.

For sensor fault type 1 ($\beta_0 = 20$, $\sigma_2 = 1$), the consensus values corresponding to each sensor fault rate are given in Table 1. The Average Detection Error Rate (ADER) and Average False Alarm Rate (AFAR) with fault type 1 are shown in Figures 2 and 3, respectively.

In these two figures, we have compared our hypothesis testing (H-testing) method with the method proposed in [8] (Median), in which the reading at a sensor is compared with its neighbor's median reading - if the difference is large,

Table 1. Consensus Value μ_ε v.s. Sensor Fault Rate r for $\varepsilon = 0.01$

r	0%	5%	10%	15%	20%	25%	30%
μ_ε	10.05	11.18	12.11	13.05	14.06	15.03	16.02

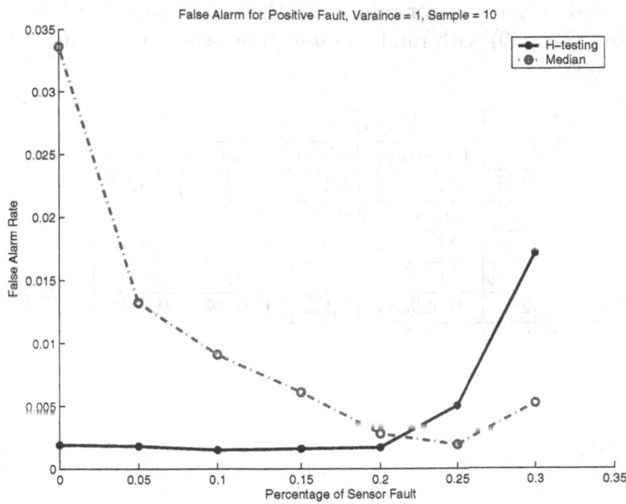

Fig. 3. False Alarm Rate of Sensor Network with Positive Fault

the sensor is likely to be faulty. The number of neighbors of each sensor we choose is $m = 20$. Clearly, we can see that ADER (solid line) of H-testing is less than ADER (dashed line) of Median. The AFAR of H-testing is also less than AFAR of Median when sensor fault rate is between $0 \sim 20\%$. The ADER of H-testing is below 0.2% with sensor fault rate up to 20% and 2.1% with sensor fault rate up to 30%.

For sensor networks with random fault, i.e., some sensors with positive bias and some sensors with negative bias, the ADER and AFAR of both the H-testing and the Median methods are reported in Table 2. ADER of H-testing method is below 0.23% with sensor fault rate up to 30% except for the case of sensor fault rate at 5%. Similarly, the H-testing method has lower AFAR than the Median method when the sensor fault rate is below 25%. For random sensor fault rate within $0 \sim 5\%$, both methods have relative large ADER and AFAR. This is because the mean value of good sensor readings is so close to the consensus value with the fault rate. That creates some confusion in detection. However, H-testing has much better performance than Median method in this fault rate range. This will be discussed in our next research report.

In general, the H-testing method has smaller ADER and AFAR than the Median method for both fault types. Notice that when the sensor fault rate is 0%, the Median method has high ADER and AFAR, which are above 3% for each fault type, while H-testing method achieves less than 0.3% rates. For both methods, ADER and AFAR of networks with random faults are smaller when compared to that with only positive faults. The reason is that, with random faults, the median in the Median method and the consensus value in the H-testing method can better reflect sensing data in the network.

Table 2. Averaged detection error rate (ADER) and average false alarm rate (AFAR) of sensor network (10×10) with random fault. The sensor fault rate r = 0%, 5%, 10%, 15% 20%, 25%, 30%.

	ADER		AFAR	
r	H-testing	Median	H-testing	Median
0%	0.23%	3.52%	0.23%	3.52%
5%	1.058%	2.05%	0.59%	1.52%
10%	0.07%	2.57%	0.06%	1.23%
15%	0.02%	2.37%	0.02%	0.61%
20%	0.02%	2.23%	0.02%	0.36%
25%	0.05%	2.27%	0.04%	0.16%
30%	0.13%	2.32%	0.09%	0.06%

7 Conclusion

We proposed a distributed fault sensor detection algorithm, where sensor fault probability or reliability is unknown and data to be sensed has Gaussian distribution with unknown parameters. We assume that each sensor has a consensus value of the sensed environment by communicating with its neighbors. The proposed algorithm uses the consensus value to make a local hypothesis testing for a local decision. The final decision about a sensor is made by fusion of the local decisions from its neighbors. The detection threshold of proposed algorithm is completely independent on the sensor fault probability and sensor data. Our simulation results show that the proposed faulty sensor detection algorithm has lower detection error even without knowledge of sensor reliability and parameters of sensed data.

References

1. Tsitsiklis, J.N.: Decentralized detection. Adv. Statist. Signal Process 2, 297–344 (1993)
2. Varshney, P.K.: Distributed Detection and Data Fusion. John Wiley & Sons, Inc., New York (1991)
3. Xiao, L., Boyd, S.: Fast linear iterations for distributed averaging. Systems and control letters 53(1), 65–78 (2004)
4. Xiao, L., Boyd, S., Lall, S.: A scheme for robust distributed sensor fusion based on average consensus. In: Proceedings of the International Conference on Information Processing in Sensor Networks (IPSN), April 25-27, pp. 63–70 (2005)
5. Krishnamachari, B., Iyengar, S.: Distributed bayesian algorithms for fault-tolerant event region detection in wireless sensor networks. IEEE Trans. on Computers 53(3), 241–250 (2004)
6. Luo, X., Dong, M., Huang, Y.: On distributed fault-tolerant detection in wireless sensor networks. IEEE Tran. on Computers 55(1), 58–70 (2006)
7. Chen, J., Kher, S., Somani, A.: Distributed fault detection of wireless sensor networks. In: DIWANS 2006, Los Angeles, California, USA, September 25, pp. 65–71 (2006)

8. Ding, M., Cheng, D., Xing, K., Cheng, X.: Localized fault-tolerant event boundary detection in sensor netwroks. In: IEEE INFOCOM 2005 (2007)
9. Nowak, R., Mitra, U.: Boundary estimation in sensor networks: Theory and methods. In: Proc. of IPSN, pp. 80–95 (2003)
10. Olfati-Saber, R., Shamma, J.S.: Consensus filters for sensor networks and distribited sensor fusion. In: Proceedings of the 44th IEEE Conference on Decision and Control, ans the European Control Conference 2005, Seville, Spain, December 12-15, pp. 6698–6703 (2005)
11. Scherber, D.S., Papadopoulos, H.C.: Locally constructed algorithms for distributed computations in ad-hoc networks. In: Proceedings of the 3rd Intermaational Symposium on Information Processing in Sensor Networks, Berleley, CA, April 2004, pp. 11–19 (2004)
12. Boyd, S., Diacoins, P., Xiao, L.: Fastest mixing markov chian on a graph. In: Proceedings of the International Conference on Information Processing in Sensor Networks (IPSN), vol. 46, pp. 667–689 (2004)
13. Balzano, L.: Fault in sensor networks. In: NESL Technical Report: TR-UCLA-NESL-200606-01 (2006)
14. Koushanfar, F., Potkonjak, M., Sangiovanni-Vincentelli, A.: On-line fault detection of sensor measurements. In: 0-7803-8133-5/03 IEEE, pp. 974–979 (2003)
15. Wackerly, D.D., Mendenhall, W., Scheaffer, R.L.: Mathematical statistics with applications, 5th edn. Duxbury Press, Boston (1996)

Detection of Failures in Civil Structures Using Artificial Neural Networks

Zhan Wei Lim[1], Colin Keng-Yan Tan[1],
Winston Khoon-Guan Seah[2], and Guan-Hong Tan[3]

[1] School of Computing, National University of Singapore
[2] Institute for Infocomm Research, A*STAR
[3] SysEng. (S) Pte. Ltd.

Abstract. This paper presents an approach to failure detection in civil structure using supervised learning of data under normal conditions. For supervised learning to work, we would typically need data of anomalous cases and normal conditions. However, in reality there is abundant of data under normal conditions, and little or none anomalous data. Anomalous data can be generated from simulation using finite element modeling (FEM). However, every structure needs a specific FEM, and simulation may not cover all damage scenarios. Thus, we propose supervised learning of normal strain data using artificial neural networks and make prediction of the strain at future time instances. Large prediction error indicates anomalies in the structure. We also explore learning of both temporal trends and relationship of nearby sensors. Most literature in anomalies detection makes use of either temporal information or relationship between sensors, and we show that it is advantageous to use both.

1 Introduction

Civil infrastructures are an important part of society and a country's economy. With the recent advances in sensing technology, the care taker of civil structures nowadays are able to obtain huge amounts of real-time data from numerous sensors, such as strain gauges installed on their structures. The structures are being monitored constantly by the sensors, at a high time-resolution of up to one reading every ten minutes. However, the abundance of data poses a problem. With the constant influx of this huge amount of real-time data, it becomes harder for humans to analyse make sense of. As the data gathered is complex, the use of simple threshold limits for triggering an alert is inaccurate. False alarms waste precious time and effort and increasingly degrade the users' confidence on the monitoring system [1].

This paper proposes the use of an artificial neural networks to detect alarm conditions in sensor data. To detect alarm condition that requires attention is equivalent to detecting anomalous behaviour of the structure. Anomalous structural behaviours manifest in anomalous sensor readings. The voluminous amount of sensor data provides much information that can be extracted. Using a machine learning approach, the problem would have been a binary classification

C. Alippi et al. (Eds.): ICANN 2009, Part II, LNCS 5769, pp. 976–985, 2009.

problem of either anomalous or non-anomalous. However, to make the binary classification, we would need training data that belongs to each of the class. In reality, there are much lesser anomalous training data than non-anomalous. Furthermore, a lot of human effort is required to tag each training instance as either anomalous or non-anomalous. While it is possible to simulate the structure using finite element modeling to generate failure cases[2][3], a specific model has to be simulated for each civil structure and the simulation may not cover all damage cases. Having plenty of data under normal conditions, we propose an approach that predicts the normal behaviour. A reading is anomalous if it deviates significantly from the prediction.

In the next section, we will present the objective of structural health monitoring and an overview of state-of-the-art techniques in structural health monitoring. In the third section, we look at the usage of artificial neural networks for fault detection and diagnosis in various domains. In section four, we will introduce our method using artificial neural networks for structural health monitoring. Finally, we will present experimental results using our method on real world strain data with anomalous condition, to validate our system.

2 Structural Health Monitoring

Structural Health Monitoring (SHM) refers to the continuous monitoring of the structure's state properties, in order to identify anomalous structural behaviour. The monitoring at its simplest can be done by visual inspection of the structure or manual physical measurements of different parts of the structure. The current state-of-the-art includes monitoring via an array of strain or optics sensors continuously feeding data to a management system that will analyse the data and send alerts automatically to relevant personnel [4][5].

Anomalous structural behaviour may be due to construction events such as post-tensioning, concreting during construction, or random events such as heavy traffic, changes in weather, rainfall, etc. These are expected loads during the lifetime of a structure and will not affect the integrity of the structure. On the other hand, anomalies can also be caused by deterioration in the material [6] and damages resulting from ground movements due to nearby constructions. These are hazardous situations and the relevant personnel must be alerted as soon as possible.

Various techniques have been proposed to detect structural fault from sensor data. One approach is the discrete wavelet transform (DWT). DWT is applied on raw strain data to filter the signal into high and low frequency components. The coefficients in the highest frequency component of the transform are then used to identify abrupt changes in the strain values which indicate likely occurrence of anomalous events on the structure [7].

Artificial neural networks were used as pattern classifiers to detect structural damage in a few studies. In one study, the damaged patterns of a bridge were generated using simulation and the neural network was trained using the generated patterns. The neural network was able to detect the damage location and damage level accurately when given a simulated pattern [3]. Artificial neural networks also improve on traditional vibration-based damage identification (VBDI) techniques. It is able to detect the damage even in presence of simulated measurement errors of a finite element model of a real bridge, which are undetected by VBDI technique.[2]. Support Vector Machine (SVM) is another pattern classifier that has been proposed to detect structural damage. In another study, SVM was able to detect damage locations from simulated streaming data that are compressed via wavelets [8].

3 Artificial Neural Networks in Fault Detection

The detection of fault from various measurements can be viewed as a pattern recognition task that artificial neural networks are very good at. Artificial neural networks have been used for detection of damage in aerospace material structures [9] and also in various other domains for fault detection and diagnostic such as manufacturing, chemical plant, power generation and nuclear plant [10][11][12].

Auto-associative neural networks (AANNs) have been used for novelty detection to diagnose damage in a simple simulated lumped-parameter mechanical system[9]. An AANN attempts to reproduce its input at its output nodes. It is a feedforward network with a "bottleneck" hidden layer that has lesser hidden nodes than input nodes. This prevents it from simply copying the input to output nodes, and to force it to extract meaningful information from the signals[11]. A damage to the lumped-parameter system will change the stiffness of the structure. The AANN that has learnt the signals under normal conditions will reproduce a different set of signals due to changes in stiffness when the system is damaged. Thus, the new signal that differs from the original normal signal signifies damage to the structure.

In the nuclear power plant setting, damage to the plant infrastructure was simulated using a virtual earthquake testbed and the neural network was trained to infer the damage from simulated data[12].

4 Using Artificial Neural Networks for Fault Detection in Civil Structures

4.1 Patterns in Strain Data of Civil Structure

Concrete structures expand and contract due to daily temperature fluctuations. The strain sensors on a structure also fluctuate in a manner that correlates to temperature fluctuations. In a cluster of sensors positioned close to one another, the fluctuation in strains of the sensors are usually highly correlated to one another as well. However, when there is damage or underlying soil movement, the sensors in the cluster will be affected differently due to difference in sensor locations. When structural changes occur, the relationship between the sensors is likely to change due to displacement or cracks in the concrete structure.

By learning the relationship between the sensors and trends in strain data due to temperature fluctuations, we can make a sensible forecast of the next instance of sensor readings assuming the structure remains the same as before without any damage or structural changes. Thus, if the actual sensor readings differ significantly from the forecast of what the readings would be under normal circumstances, it would be a strong suggestion of anomalies in the structure.

4.2 Learning the Patterns with Artificial Neural Networks

There are two kinds of information to be derived from the voluminous historical strain data: 1) the relationship between the sensors that are spatially close together, and 2) the temporal trends in strain fluctuations and relation to temperature fluctuations. An artificial neural network, being a universal function estimator, is a good candidate to learn these two kinds of information from historical data. We use a neural network with supervised learning for this purpose. The learnt information in a trained network would exist in the weights of the network connections.

Auto-associative neural networks are well known to be able to reduce measurement noise from raw data, and learn a correlation model of its input[11]. An auto-associative neural network is a multilayer perceptron neural network that approximates an identity function. Instead of simply copying its input to its output, it has hidden layers with nodes lesser than the input or output nodes that act as a bottleneck. The input is transformed into a lower dimension space in the hidden layers and then transformed back to original space at the output layer. It is possible to use AANN to learn the relationship between the strain sensors through supervised learning of the AANN with strain data under normal conditions.

To learn the trend in sensor strain fluctuations, we need a neural network that is able to model the trend and predict the next instance values using the model. A neural network with time-delayed data as input performs well at the task of making prediction or forecasting of time series data in many domains. The neural networks have been successful in major statistical time series forecasting competition[13]. Neural networks that are fed with historical data are able to learn the underlying "rules" of currency exchange rates through supervised learning, and then make predictions using the trained neural network[14].

To learn both the trend and inter-sensor relationship, we propose combining auto-associative neural networks with time-delayed data. We will show the details of the neural network structure, its training, and experimental results of this artificial neural network in the next two sections.

4.3 Training

To enable fault detection, we aim to build an artificial neural network that is able to generalizes well on unseen data under normal circumstances, but does not generalizes for anomalous condition. We train a Multilayer Perceptron (MLP) neural network configured as an AANN with historical strain data of the structure under normal circumstances, with a standard backpropagation algorithm. We exclude portions of the historical data where it is known to display anomalies. The neural network is trained to predict the vector of sensor readings for the next time instance, given an input vector for the current time instance. The training data consist of tuples of (v_{t-1}, r_t), where v_{t-1} is the input vector corresponding to time instance $t-1$ and r_t is a vector of actual sensor readings at time t.

The input and output of the neural network are:

$$\text{Input:}\quad S_{0,t-1}, S_{1,t-1}, ...S_{n,t-1}, S_{0,t-2}, S_{1,t-2}, ...S_{n,t-k}, T_t$$

$$\text{Output:}\quad S_{0,t}, S_{1,t}, ...S_{n,t}$$

where $S_{j,t}$ is the sensor j's reading at time t, T_t is temperature at time t, k, $(k > 0)$ is the number of delayed values to use (delay number) and sensors 0 to n are close together.

The input consists of three kinds of data: sensor readings of a cluster of sensors, the sensors time-delayed values and temperature readings. The temperature reading at each time instance is fed into the neural network as well to provide more information. A linear regression of temperature against strain is unable to remove temperature effect from strain data (Figure 1). Patterns still exist in the residues of such a regression. Thus, it is left to the neural network to extract any linear or non-linear relation between strains and temperature.

The delay number is the number of delayed sensor readings to be used in the input. When the delay number is zero, the neural network is simply an auto-associative neural network that has no temporal information. In the experimental result section, we will show the relationship between the delay number and the neural network's ability to detect anomalous events. Higher delay number provides more information to the neural network with diminishing return. While the MLP neural network is able to ignore irrelevant information, a higher

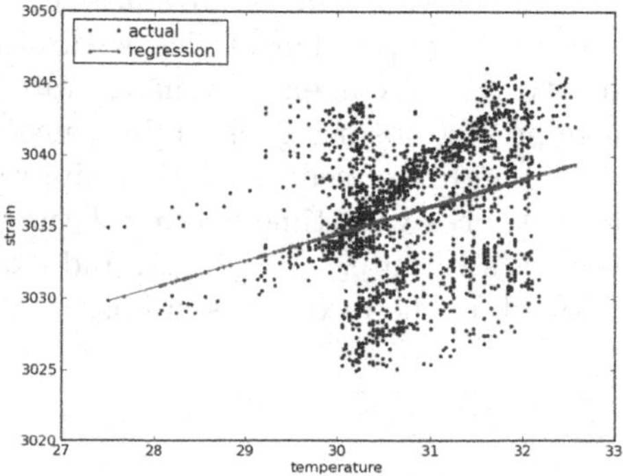

Fig. 1. Plot of regression line and actual strain values. Observe a solid streak of *dots* that lies above the regression line.

delay number results in longer training time. We find that delay number $d = 8$ is sufficient.

Through experimentation, we find that one hidden layer is sufficient. Additional hidden layers neither increase accuracy nor generalization ability on unseen normal data. We use linear units at the input and output layers, and sigmoid units at the hidden layer. We also find that the $\lfloor I/3 \rfloor$ number of hidden units serves well to learn the trends and relationship, where I is number of input variables. As the sensors' readings and their time delayed readings are highly correlated, the number of hidden nodes should be smaller than the number of input nodes to force the neural network to extract meaningful information from the high redundancy data.

Typical values of strain data are in the order of thousands, and have small deviations in order of tens. These values are numerically ill-conditioned. Large changes in the network weights may have little effect and this may result in slow training with high inaccuracies[15]. We re-scale the input and target variables to be in the range of 0.1 to 0.9 so that smaller changes in the weights are needed during the training. For each input or target variable x, we apply

$$x'_i = \frac{x_i - x_{min}}{x_{max} - x_{min}} * 0.8 + 0.1$$

where x is the input variable and x_{max} and x_{min} are the maximum value and minimum values of x respectively.

The MLP neural network is trained under generalized delta rule with weight updates using Scipy(Scientific Tools for Python)'s fast conjugate gradient optimizer for 50 training iterations. Conjugate gradient minimizer is faster than standard steepest gradient descent training and requires less training iterations. In most cases, the MLP neural network has no significant improvement in accuracy after 50 training iterations.

4.4 Monitoring

After the neural network is trained, it can be used to forecast the sensor readings under the assumption of normal circumstances. The input vector for the neural network is assembled as soon as the latest sensor readings are received. The input vector is fed into the neural network, and the neural network makes prediction of the readings. If the deviation between the prediction and actual reading is significant, anomaly is said to be observed. There can be a few ways to quantify a vector of deviations to be significant. Its deviation is significant, and thus anomalous if the more than half the sensors' predictions have error larger than the largest error encountered in the training data. Alternatively, statistical hypothesis testing can be used to quantify if the errors are large enough to be anomalous. In addition to instantaneous errors, cumulative errors over a time period can be used to detect gradual changes, for example, gradual buildup of stress in the structure.

5 Experimental Result

The artificial neural network described above is tested on strain data collected from an overhead (elevated) expressway in Singapore. There are real life occurrences of anomalous events in the test data, and the artificial neural network demonstrated its ability to pick up these real anomalous occurrences.

5.1 Strain Data from Bridge

The strain gauges are installed on the piers of the bridge. There are eight sensors per pier and 12 piers of the bridge were measured and recorded. The topological arrangement of the eight sensors is shown in Figure 2.

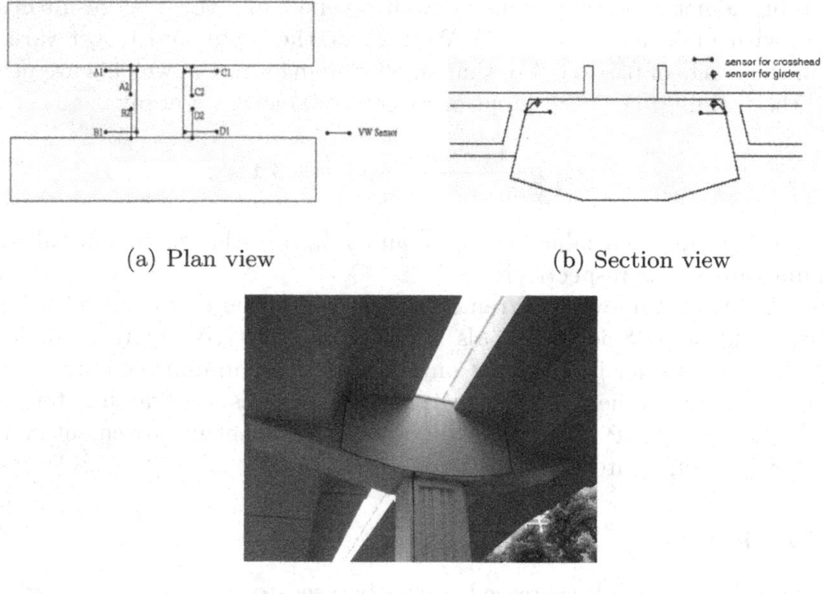

(a) Plan view

(b) Section view

(c) Photo of sensors on bridge

Fig. 2. Layout of 8 sensors on a bridge pier

There are temperature sensors installed along the strain sensors to record the environmental temperature. All sensors are read at an interval of ten minutes.

5.2 Neural Network Prediction on Normal Condition

To show the ability of our neural network for prediction in normal conditions, we tested the neural network across a period of 10 days without anomalous occurrence. We train the neural network using data from day D-5, D-4, ... D-1, and tested it on the period of D-5 to D+4. The result (Figure 3(a)) shows that there is no significant prediction error throughout the period.

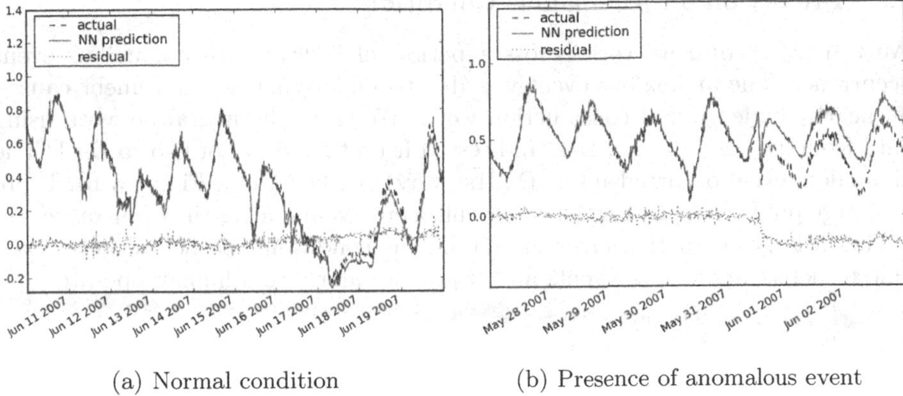

(a) Normal condition

(b) Presence of anomalous event

Fig. 3. Neural network prediction for normal condition and anomalous condition. In Figure 3(b), the anomalous event occurs at 31st May 13:00. Significant prediction error is observed from then onwards.

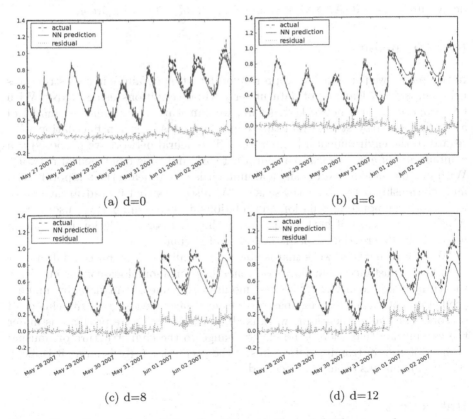

(a) d=0

(b) d=6

(c) d=8

(d) d=12

Fig. 4. Neural network prediction for different values of d. The anomalous event appears most obvious when d=*12*.

5.3 Prediction on Anomalous Condition

We test the neural network across a period of 7 days with anomalous event occurrence. The anomalous event was due to underlying soil movement caused by nearby underground construction work. We train the neural network using data from day D-5, D-4, ... D-1, and tested it on the period of D-5 to D+1. The anomalous event occurred at day D (31st May) at 13:00 on the Figure 3(b). There is a large prediction error when the event occurred and after the occurrence. By having a large prediction error on anomalous condition, the neural network is able to detect anomalous events and trigger an alarm to a human operator.

5.4 Temporal Information

Auto-associative neural network alone is able to detect some anomalous conditions through its prediction error. With the addition of time delay information, the neural network is more sensitive to anomalous conditions, showing greater error and thus greater differentiation between non-anomalous and anomalous cases. We show the result for one sensor with slight anomalous behaviour when delay number $d = 0$ (AANN), $d = 6$, $d = 8$, and $d = 12$ in Figure 4.

6 Conclusion

In this paper, we presented an approach for detection of failure in civil structures using supervised learning of data under normal conditions using ANNs. With the proposed neural network structure, we can learn the correlation model of a cluster of nearby sensors and the temporal trends in strain fluctuation with regard to the environmental temperature. The neural network we proposed is a combination of auto-associative neural network and time delay neural network. With just one hidden layer of sigmoid units, it is able to learn the temporal trend and relationship between nearby sensors. We also show that it is advantageous to learn both temporal information and relationship between nearby sensors using ANN. The presence of both temporal and inter-sensor correlation information helps in the differentiation of anomalous and normal cases.

We built a neural network that is able to generalize well on unseen data under normal circumstances, but does not generalize for anomalous condition. Using the trained neural network, we can monitor realtime strain data by making predictions of future readings based on time delayed readings of the cluster of sensors. The trained neural network is able to make accurate predictions under the assumption that there is no major change to the civil structure or unusual events not seen in training data. Thus, if the prediction error is large, it is likely that an anomalous event has occurred.

References

1. Tan, G.H., Hooi, T.H.: Causes of false alerts in real time monitoring and alert systems in deep excavation sites. In: Underground Singapore, Singapore (2007)

2. Xu, H., Humar, J.: Damage detection in a girder bridge by artificial neural network technique. Computer-Aided Civil and Infrastructure Engineering 21(6), 450–464 (2006)
3. Yan, L., Fraser, M., Elgamal, A., Conte, J.P., Fountain, T.: Applications of neural networks in structural health monitoring. Technical report, Department of Structural Engineering, University of California, San Diego (La Jolla, CA 92093-0085, USA)
4. Brownjohn, J., Tjin, S., Tan, G., Tan, B., Chakraboorty, S.: A structural health monitoring paradigm for civil infrastructure. In: 1st FIG International Symposium on Engineering Surveys for Construction Works and Structural Engineering, Nottingham, United Kingdom, pp. 1–15 (2004)
5. Tan, G.H., Ng, T.G., Brownjohn, J.: Real Time Monitoring and Alert Systems for Civil Engineering Applications Using Machine-to-Machine Technologies. In: International Conference on Structural and Foundation Failures, Singapore (2004)
6. Moyo, P., Brownjohn, J.: Application of box-jenkins models for assessing the effect of unusual events recorded by structural health monitoring systems. Structural Health Monitoring 1(2), 149–160 (2002)
7. Moyo, P., Brownjohn, J.: Detection anomalous structural behaviour using wavelet analysis. Mechanical Systems and Signal Processing 16(2-3), 429–445 (2002)
8. Bulut, A., Singh, A.K., Fountain, J.H., Yan, L., Elgamal, A.: Real-time nondestructive structural health monitoring using support vector machines and wavelets. In: Advanced Sensor Technologies for Nondestructive Evaluation and Structural Health Monitoring, San Diego, CA, USA (2005)
9. Worden, K.: Structural fault detection using a novelty measure. Journal of Sound and Vibration 201(1), 85–101 (1997)
10. Xu, X., Hines, W.J., Uhrig, R.E.: Sensor validation and fault detection using neural networks. Technical report, Maintenance and Reliability Center, The University of Tennessee, Knoxville, TN 37996-2300 (1997)
11. Kramer, M.A.: Autoassociative neural networks. Computers & Chemical Engineering 16(4), 313–328 (1992)
12. Shu, Y.: Structural health assessing by interactive data mining approach in nuclear power plant, pp. 332–345 (2007)
13. Hill, T., O'Conner, M., Remus, W.: Neural network models for time series forecasts. Management Science 42, 1082–1092 (1996)
14. Tan, C.L., Yao, T.: A case study on using neural networks to perform technical forecasting of forex. Neurocomputing 34, 79–98 (2000)
15. Reed, R.D.: Neural Smithing: Supervised Learning in Feedforward Artificial Neural Networks (Bradford Book). MIT Press, Cambridge (1999)

Congestion Control in Autonomous Decentralized Networks Based on the Lotka-Volterra Competition Model*

Pavlos Antoniou and Andreas Pitsillides

Networks Research Lab, Computer Science Department,
University of Cyprus, Nicosia, Cyprus
{paul.antoniou,andreas.pitsillides}@cs.ucy.ac.cy

Abstract. Next generation communication networks are moving towards autonomous infrastructures that are capable of working unattended under dynamically changing conditions. The new network architecture involves interactions among unsophisticated entities which may be characterized by constrained resources. From this mass of interactions collective unpredictable behavior emerges in terms of traffic load variations and link capacity fluctuations, leading to congestion. Biological processes found in nature exhibit desirable properties e.g. self-adaptability and robustness, thus providing a desirable basis for such computing environments. This study focuses on streaming applications in sensor networks and on how congestion can be prevented by regulating the rate of each traffic flow based on the Lotka-Volterra population model. Our strategy involves minimal exchange of information and computation burden and is simple to implement at the individual node. Performance evaluations reveal that our approach achieves adaptability to changing traffic loads, scalability and fairness among flows, while providing graceful performance degradation as the offered load increases.

Keywords: autonomous decentralized networks, congestion control, lotka-volterra.

1 Introduction

Rapid technological advances and innovations in the area of autonomous systems push the vision of Ambient Intelligence from concept to reality. Networks of autonomous sensor devices offer exciting new possibilities for achieving sensory omnipresence: small, (often) inexpensive, untethered sensor devices can observe and measure various environmental parameters, thereby allowing real-time and fine-grained monitoring of physical spaces around us. Autonomous decentralized networks (ADNs) as for example, Wireless Sensor Networks (WSNs) [1], can be used as platforms for health monitoring, battlefield surveillance, environmental observation, etc.

* The research leading to these results has received funding from the European Community's Seventh Framework Programme (FP7/2007-2013) under grant agreement no. 224282 (the GINSENG project) and from the Cyprus Research Promotion Foundation under grant agreement no. TPE/EPIKOI/0308(BE)/03 (the MiND2C project).

C. Alippi et al. (Eds.): ICANN 2009, Part II, LNCS 5769, pp. 986–996, 2009.

Typically, WSNs consist of small (and sometimes cheap), cooperative devices (nodes) which may be constrained by computation capability, memory space, communication bandwidth and energy supply. The uncontrolled use of the scarce network resources is able to provoke congestion. Thus, there is an increased need to design novel congestion control strategies possessing self-* properties like self-adaptability, self-organization as well as robustness and resilience, which are vital to the mission of dependable WSNs. Biological processes which are embedded in decentralized, self-organizing and adapting environments, provide a desirable basis for computing environments that need to exhibit self-* properties. In addition, their constrained nature necessitates simple to implement strategies at individual node level with minimal exchange of information.

Simple mathematical biology models [2] which aim at modeling biological processes using analytical techniques and tools are often used to study non-linear systems. Population dynamics has traditionally been the dominant branch of mathematical biology which studies how species populations change in time and space and the processes causing these changes. Information about population dynamics is important for policy making and planning and in our case is used for designing a congestion control policy. In this study, nature inspired models are employed to design a scalable and self-adaptable congestion control algorithm for streaming media in WSNs. **Based on the Lotka-Volterra (LV) competition model, a decentralized approach is proposed that regulates the rate of every flow in order to prevent congestion in WSNs**. The LV-based congestion control (LVCC) mechanism is targeted for dependable wireless multimedia WSNs [3] involving applications that require continuous stream of data.

Based on analytical evaluations performed in [4], the LVCC model guarantees that the equilibrium point of the system ensures coexistence of all flows, with stability and fairness among active flows when some conditions (presented below) are satisfied. In this paper, the validity of the analytical results is further investigated by simulating complex scenarios that cannot be formally tested. Performance evaluations are based on simulation studies conducted in Matlab and in the network simulator NS2 [5], and focus on scalability, graceful performance degradation, fairness and adaptability to changing conditions. Results have shown that the LVCC approach provides adaptation to dynamic network conditions providing scalability, fairness and graceful performance degradation when multiple active nodes are involved.

The remainder of this paper is organized as follows. Section 2 deals with the problem of congestion in ADNs and discusses previous work. Section 3 presents the analogy between ADNs and ecosystems. Section 4 proposes our bio-inspired mechanism. Section 5 evaluates the performance of our mechanism in terms of stability, scalability and fairness. Section 6 draws the conclusion and future work.

2 Congestion in AD Networks

There are mainly two types of congestion in WSNs: (a) **queue-level congestion** and (b) **channel-level congestion**. Traditionally, either high queue occupancy or queue overflow (queue drops) were considered to be key symptoms of congestion (queue-level congestion). However, simulation studies conducted by [10] and [11] revealed that in WSNs where the wireless medium is shared using Carrier Sense Multiple Access (CSMA)-like protocols, wireless channel contention losses can dominate queue drops and increase quickly with offered load. The problem of channel losses (channel-level congestion) is worsened around hot spot areas, as for example, in the area of an event, or around the sink. In the former case, congestion occurs if many nodes report the same event concurrently, while in the latter case congestion is experienced due to the converging (many-to-one) nature of packets from multiple sending nodes to a single sink node. These phenomena result in the starvation of channel capacity in the vicinity of senders, while the wireless medium capacity can reach its upper limit faster than queue occupancy [12]. Queue-level congestion is mainly attributed to the constrained nature of nodes consisting an autonomous decentralized network (e.g. limited memory and computation power), whereas channel-level congestion can be influenced by the broadcast nature of wireless networks as well as traffic variations.

Congestion causes energy waste, throughput reduction, increase in collisions and retransmissions at the medium access control (MAC) layer, increase of queueing delays and even information loss leading to the deterioration of the offered QoS and to the decrease of network lifetime. Also, under traffic load, multi-hop networks tend to penalize packets that traverse a large number of hops, leading to large degrees of unfairness.

Congestion control (CC) policies in ADNs are fundamentally different than in the traditional TCP/IP Internet, which is based on source-destination pair with reliable communication model, also involving retransmission of lost packets. This reliable end-to-end principle is tightly coupled to the client-server model of TCP/IP communication. However, this model is not very effective for ADNs, where delivery of data to a gateway (sink), without retransmission of any lost packets, is the normal objective. Their constrained and unpredictable nature provokes increased latency and high error rates that may result in reduced responsiveness e.g. for end-to-end congestion detection, leading to higher energy consumption (e.g. very high packet loss during long periods of congestion). These problems drive the need for decentralized CC approaches adopting a hop-by-hop model where all nodes along a network path can be involved in the procedure. Each node should make decisions based only on local information since none of them has complete knowledge of the system state.

Previous work on CC involving mathematical models of population biology was proposed for the Internet on the basis of either improving the current TCP CC mechanism [6] or providing a new way of combating congestion [7]. The study of [6] couples the interaction of Internet entities that involved in CC mechanisms (routers, hosts) with the predator-prey interaction. This model exhibits fairness and acceptable throughput but slow adaptation to traffic demand. Recent work by [7] focuses on a new TCP CC mechanism based on the LV competition model [8], [9] which is applied to the congestion window updating mechanism of TCP. According to the authors, remarkable results in terms of stability, convergence speed, fairness and scalability are exhibited. However, these approaches are based on the end-to-end model of the Internet, which is completely different from the hop-by-hop nature of ADNs. **The novelty of our approach lies in the fact that the LV model is applied to WSNs in a hop-by-hop manner**.

3 Autonomous Decentralized Networks: An Ecosystem View

An ADN (Fig. 1) is considered to be analogous to an ecosystem. An ecosystem comprises of multiple species that live together and interact with each other as well as the non-living parts of their surroundings (i.e. resources) to meet their needs for survival and coexist. Similarly, an autonomous network consists of a large number of cooperative nodes. Each node has a buffer in order to store packets and is able to initiate a traffic flow. All traffic flows compete with each other for available network resources in an effort to reach one or more sink nodes by traversing a set of intermediate nodes forming a multi-hop path. Just as in an ecosystem, *the goal is the coexistence of flows.*

To investigate the decentralized and autonomic nature of our approach, a network is divided into smaller neighborhoods called sub-ecosystems. Each sub-ecosystem involves all nodes that send traffic to a particular one-hop-away node. The traffic flows initiated by those nodes play the role of competing species and the buffer (queue) capacity of the receiving node can be seen as the limiting resource within the sub-ecosystem.

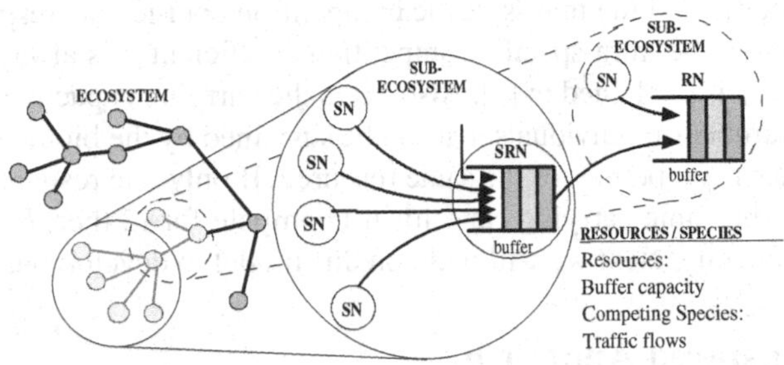

Fig. 1. Competition in AD networks

Within a virtual ecosystem, participant nodes may perform different roles. In particular, each node is able to either initiate a traffic flow i.e. is a source node (SN), or serve as a relay node (RN) for multiple other flows, or perform both roles being a source-relay node (SRN). Source nodes are basically located at the edges of a network (e.g. leaf nodes) while relay nodes are internal nodes (e.g. backbone nodes). Our strategy provides hop-by-hop rate adaptation by regulating the traffic flow rate at each sending node. *Each node is in charge of self-regulating and self-adapting the rate of its traffic flow i.e., the rate at which it generates or forwards packets. All flows compete for available buffer capacity at their one-hop-away receiving node.* Each sending node is expected to regulate its traffic flow rate in a way that limiting buffer capacities at all receiving nodes along the network path towards the sink are able to accommodate all received packets. The sending rate evolution of each flow will be driven by variations in buffer occupancies of relay nodes along the network path towards the sink. Due to the decentralized nature of our approach, each node will regulate its traffic flow rate using local information (i.e. from neighbors). The number of bytes sent by a node within a given period refers to the population size of its flow. From an ecosystem perspective, the population size of each traffic flow (i.e. of each species) is affected by interactions among competing flows (species) as well as the available resources (buffers) capacities.

The proposed strategy is based on *a deterministic competition model which involves interactions among species that are able to coexist, in which the fitness of one species is influenced by the presence of other species that compete for at least one limiting resource*. Competition among members of the same species is known as intra-specific competition, while competition between individuals of different species is known as inter-specific competition. One of the most studied mathematical models of population biology, the LV competition model [8], [9], exhibits this behavior. The generalized form of an n-species LV system is expressed by a system of ordinary differential equations:

$$\frac{dx_i}{dt} = r_i x_i \left[1 - \frac{\beta_i}{K_i} x_i - \frac{1}{K_i} \left(\sum_{j=1, j \neq i}^{n} \alpha_{ij} x_j \right) \right], \tag{1}$$

for $i = 1, ..., n$, where $x_i(t)$ is the population size of species i at time t ($x_i(0) > 0$), r_i is the intrinsic growth rate of species i in the absence of all other species, β_i and α_{ij} are the intra-specific and the inter-specific competition coefficients respectively. In the classical LV model, the intra-specific competition coefficient β is always equal to one. The reason for this is explained in [4]. Also K_i is the carrying capacity of species i i.e., the maximum number of individuals that can be sustained by the biotope in the absence of all other species competing for the same resource. If only one resource exists and all species (having the same carrying capacity K) compete for it, then K can be seen as the resource's capacity. Next we will build on this model to develop our strategy.

4 Nature-Inspired Approach

This section distinguishes the roles of the different entities (i.e., SN, RN, and SRN) involved in the congestion avoidance mechanism along the path towards a sink.

Source Node (SN): Pure source nodes (SNs) are end-entities (Fig. 2) which are attached to the rest of the network through an downstream node e.g., a relay node (RN), or a source-relay node (SRN) located closer to the sink.

Each SN is expected to initiate a traffic flow when triggered by a specific event. The transmission rate evolution of each flow is regulated by the solution of Eq. 1 (see Eq. 2)

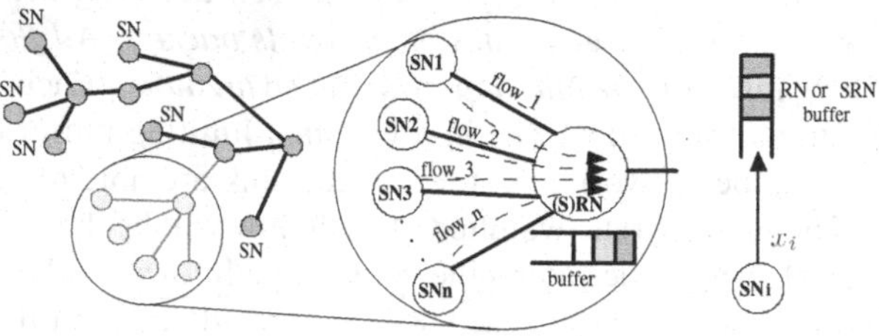

Fig. 2. Source nodes competing for a limiting resource at their downstream node

that gives the number of bytes sent x_i by flow i. In order to be able to solve Eq. 1 for a single node i, it is necessary to be aware of the aggregated number of bytes sent from all other nodes $\sum_{j=1,j\neq i}^{n} x_j$ which compete for the same resource. This quantity is denoted by C_i. In decentralized architectures, the underlying assumption of C_i-awareness is quite unrealistic. However, each SN can indirectly obtain this information through a small periodic backpressure signal sent from its downstream SRN/RN (father node) containing the total number of bytes sent from all father's children, denoted by BS. Each node can evaluate its neighbors' contribution C_i by subtracting its own contribution x_i from the total contribution BS as expressed by: $C_i = \sum_{j=1,j\neq i}^{n} x_j = BS - x_i$. Thus, Eq. 1 becomes:

$$\frac{dx_i}{dt} = rx_i \left(1 - \frac{\beta}{K} x_i - \frac{\alpha}{K} C_i\right), \; i = 1, ..., n. \tag{2}$$

To obtain x_i Eq. 2 is integrated:

$$x_i(t) = \frac{wx_i(0)}{\beta x_i(0) + [w - \beta x_i(0)]\, e^{-\frac{wr}{K}t}}, \quad w = K - \alpha C_i \tag{3}$$

The validity of Eq. 3 is based on the assumption that $K - \alpha C_i > x_i$. If we set $\alpha = 1$ then, according to the inequality, the number of bytes sent from each node i (i.e. x_i) must not exceed the empty space left on the upstream node's buffer ($K - C_i$) so as to prevent buffer overflows. If we let K be a constant, the larger the value of α the smaller the value of x_i compared to the available buffer capacity of the upstream node.

According to [4], a network (ecosystem) of flows (species) that compete for a single resource while the populations of bytes sent are regulated by Eq. 3 has a global non-negative and asymptotically stable equilibrium point when inter-specific competition is weaker than intra-specific competition i.e., $\beta > \alpha$ $(\alpha, \beta > 0)$. Under this condition, the series of values generated by each SN converges to a global and asymptotically stable *coexistence solution* given by Eq. 4. For a detailed proof of this concept refer to [4].

$$x_i^* = \frac{K}{\alpha(n-1) + \beta}, \; i = 1, ..., n. \tag{4}$$

In order to avoid buffer overflows, it needs to be ensured that when a system of n active nodes converges to the coexistence solution, each node i will be able to send less than or equal to K/n bytes. This is satisfied by Eq. 4 when $\alpha(n-1) + \beta \geq n$ or $\beta - \alpha \geq n * (1 - \alpha)$. If we set $\alpha \geq 1$ and require $\beta > \alpha$ (equilibrium stability condition), then the aforementioned inequality is always satisfied.

Each SN evaluates Eq. 3 in an iterative manner. By iterative, we mean, roughly, that Eq. 3 generates a series of values which correspond to number of bytes sent every period T. The iterative form of Eq. 3 is expressed by:

$$x_i((k+1)T) = \frac{w(kT)x_i(kT)}{\beta x_i(kT) + [w(kT) - \beta x_i(kT)]\, e^{-\frac{w(kT)r}{K}T}} \tag{5}$$

Relay Node (RN): Pure relay nodes (RNs) are internal entities which do not generate any packets, but forward packets belonging to several flows traversing themselves which

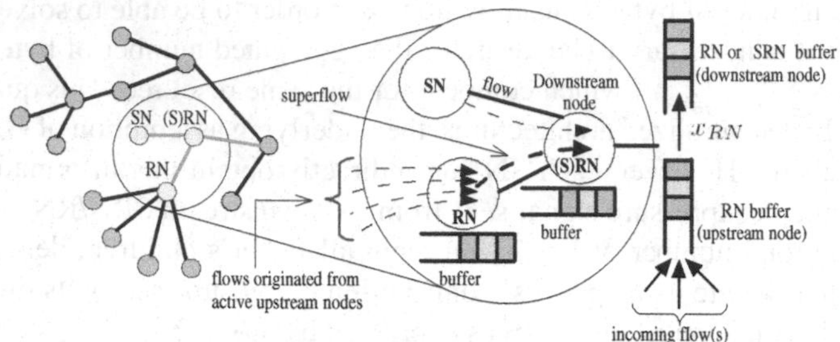

Fig. 3. Relay node creates a superflow which competes for downstream node's buffer

compete for their resources. The main function of a RN is to combine (or multiplex) all incoming flows into a superflow and relay it to the dedicated downstream node (SRN or RN) as shown in Fig. 3. However, the superflow competes with other flows destined to the same downstream node (e.g., the flow originating from SN in Fig. 3). Hence, each RN is in charge of acting on behalf of all active upstream nodes whose flows are passing through it when evaluating the transmission rate of the superflow (i.e. number of bytes sent from RN within period T). As shown in Fig. 3, each one of the four flows of the superflow as well as the flow originating from SN should be able to allocate equal share of the downstream node's limiting resource. Thus, each RN allocates resources for its active upstream nodes based on a slightly modified expression of Eq. 5 as follows:

$$x_{RN}((k+1)T) = m\left(\frac{w(kT)H(kT)}{\beta H(kT) + [w(kT) - \beta H(kT)]\, e^{-\frac{w(kT)r}{K}T}}\right), \quad (6)$$

where $H(kT) = \frac{x_{RN}(kT)}{m}$, $w(kT) = K - \alpha C^*_{RN}(kT)$ and m is the total number of active upstream nodes which belong to the tree having RN as root. The number of bytes sent from a superflow within a period kT, namely $x_{RN}(kT)$, is equal to the aggregated number of bytes sent from m RN's upstream source nodes which compete for RN's buffer. Each RN can calculate the number (m) of its active upstream nodes by examining the source id field of each packet traversing itself. $C^*_{RN}(kT)$ reflects the total number of bytes sent (BS) to the downstream node ((S)RN in Fig. 3) from all competing children nodes subtracting the contribution of a single flow belonging to the superflow. $C^*_{RN}(kT)$ can be expressed as $C^*_{RN} = BS - \frac{x_{RN}(kT)}{n}$.

Source-Relay Node (SRN): A source-relay node (SRN) acts as both source and relay node, having both functions concurrently operated as described above.

5 Performance Evaluation

Simulation studies were used to investigate how parameters affect the performance of our mechanism in terms of sensitivity to parameters, scalability and global fairness.

As discussed above, the rates of all flows converge to a global and asymptotically stable solution when $\beta > \alpha$ ($\alpha, \beta > 0$). There is no upper limitation on β but as it becomes larger, the steady state traffic rate (Eq. 4) decreases. In this case, each node will have to transmit data at a lower rate leading to lower quality of the received streams at the sink. As far as r is concerned, the system of Eq. 1 has a stable equilibrium point for any value of $r > 0$ [4], [14]. An upper bound for r is not analytically known, thus can be experimentally explored. The mathematical analysis of our model gives a general understanding of the system's behavior on the basis of stability as function of the α and β. However, the complexity of an ADN necessitates simulation evaluation using plausible scenarios that cannot be formally tested. The analytical study serves as the basis for the simulations.

In order to supplement the analytical results, some simulation experiments were conducted both in Matlab and in NS2. We considered a wireless sensor network consisting of 25 nodes which are deployed in a cluster-based topology (Fig. 4). Our mechanism was evaluated in a static and failure-free environment. All nodes were assumed to have the same buffer capacity $K = 35$KB. The time period T between successive evaluations of the number of bytes sent by each SN, as well as the time between backpressure signals was set to 1 sec. It was assumed that nodes 5, 6, 10, 14, 16 and 20 were activated at $1T$, $50T$, $150T$, $300T$, $450T$, $600T$ and 900 respectively. Node 14 was deactivated at $750T$. **Stability and Sensitivity:** Based on the analytical study of our model [4], the

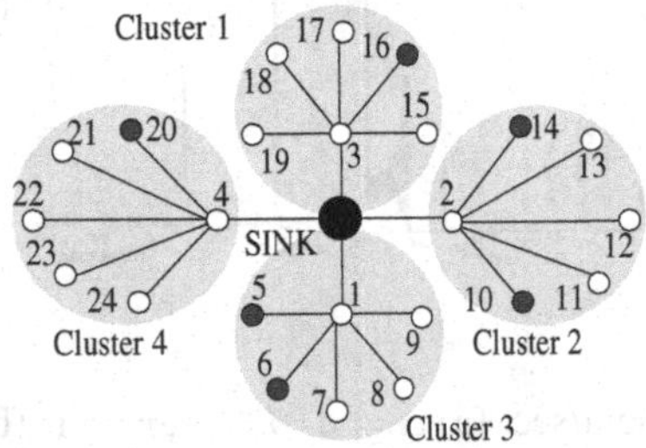

Fig. 4. Experimental cluster-based topology (all links are wireless)

satisfiability of some conditions contributes to system's stability. Their validity was further investigated by simulating complex scenarios that cannot be formally tested. It has been mathematically proved that if $\beta > \alpha$, then all sending rates converge to a stable equilibrium value $\forall r$ (detailed proofs in [4]). Initially, α and r were set equal to 1 while the value of β varied.

Fig. 5(a) depicts the estimated number of bytes that can be sent per T from each active node when $\beta = 2$. As can be observed, the system was able to re-converge to a new stable point after a change in network state (node activation). However, fluctuations in sending rates arose when (previously inactive) downstream nodes were not prepared to accommodate the increasing incoming traffic before Eq. 6 converged. This behavior was exhibited by flows initiated from nodes 10, 16 and 20. These flows were not well behaved but exhibited some oscillatory behavior after changes in network state. Also, some fluctuations occurred when the flow of node 14 was deactivated. Note that buffer

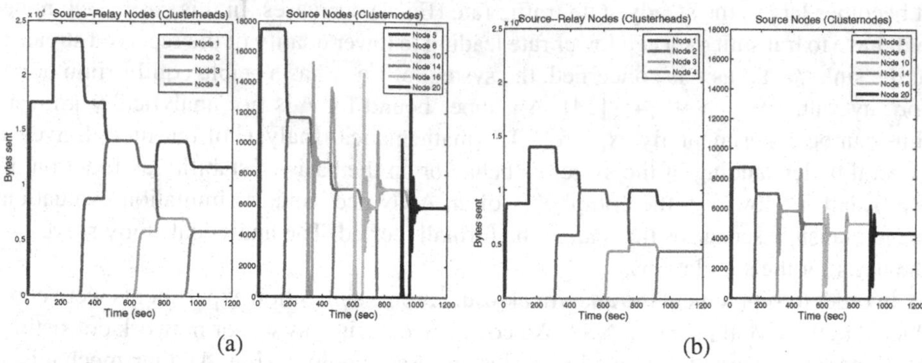

Fig. 5. Estimated bytes sent/sec: (a) $\beta = 2$, (b) $\beta = 4$, when $\alpha = 1$ and $r = 1$

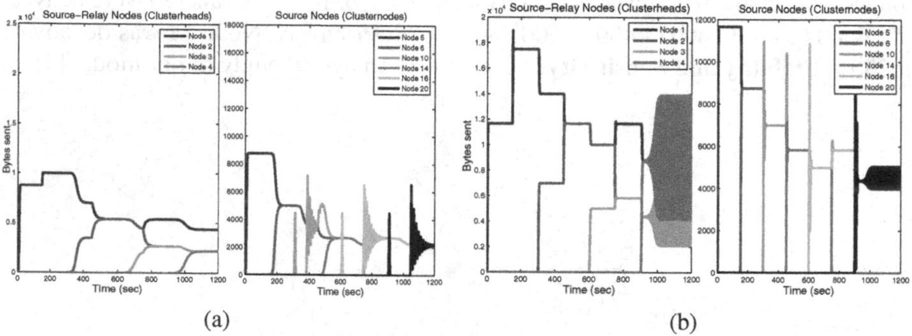

Fig. 6. Estimated bytes sent/sec: (a) $\alpha = 3, \beta = 4, r = 1$, (b) $\alpha = 1, \beta = 3, r = 4$

overflows never occurred since the amount of traffic that was sent by each flow was small compared with the downstream node's buffer capacity.

When β increased to 4 (Fig. 5(b)) all flows became well-behaved while some small oscillations occurred as a result of changes in network state. Even though there is no upper bound for β value, it is worth pointing out that as β increases, the equilibrium value decreases (see Eq. 4) and the quality of the received data at the sink may be reduced. Increasingly, the results of Fig. 5(a) and (b) suggest that β should be greater than α but greater enough (this may depends on n) such that each node can allocate much less than K/n. This observation is supported by Fig. 6(a) ($\alpha = 3, \beta = 4$). When β is much greater than α, high buffer utilization is prevented, while smooth and stable response of traffic flows is achieved. In all the previous scenarios, the parameter r was set to 1. Further simulation studies were carried out in order to study the influence of r on stability. Results showed that the stability of traffic flows rates depends on r but a different behavior was observed with the change in parameters α and β. In general, it was shown that the flow sending rates converged when $r \leq 2.5$ for quite a large number of combinations of α and β values. Therefore, r could not grow unboundedly but smooth network operation could be preserved in low r values (≤ 2.5). Fig. 6(b) illustrates large fluctuations in flow sending rates occurred for $\alpha = 1, \beta = 3, r = 4$.

Scalability and Fairness: The system proved to be adaptable against changing traffic load and achieved scalability by sharing buffer capacity of nodes to their active upstream nodes. For example in Fig. 5(b), in the presence of one sender (node 5) the stable equilibrium point of the system given by Eq. 4 was 8750 bytes/T (clusterhead node 1 transmitted at the same rate). When node 6 became active, each sender obtained 7000 bytes/T, while the downstream node 1 (clusterhead) was able to accommodate both senders by increasing its rate using Eq. 6. When the number of senders scaled up, all senders could be supported by the system by diminishing the sending rate per node, thus offering graceful degradation. Fairness was also achieved having the available buffer capacity of each node equally shared among all activated flows.

Further simulations were conducted using the discrete event based simulator NS2 in order to evaluate the performance of the LVCC mechanism under more realistic network conditions (in terms of packet loss and delay) when multiple users are involved. Performance was measured in terms of the packet delivery ratio (PDR), which is defined as the ratio of the total number of packets received by the sink to the total number of packets transmitted by source nodes. The following table presents the combinations of α and β values ($r = 1$) that achieved the highest transmission rates (bytes sent per T) and the highest mean PDR for different number of active nodes. It is worth pointing out that only the scenarios where traffic flows of all active nodes converged to stable solutions were taken into consideration.

Table 1. Performance evaluations for realistic network conditions using NS2 [5]

α	β	*No. of Active Nodes*	*Mean Packet Delivery Ratio*
1.6	3.3	3	0.99
1.5	4.5	5	0.99
1.7	6.0	7	0.88
1.4	6.2	10	0.70
1.6	6.5	15	0.64
1.9	6.5	20	0.62

The results of Table 1 support previous results obtained from Matlab simulations. It can be seen that as the number of active nodes scaled up, stable response of traffic flows was achieved with the increase of parameter β. On the other hand, α remained from 1.5 to 1.9 regardless of the number of active nodes. In addition, the mean PDR decreased below 70% when more than 10 active nodes were concurrently activated in the topology of Fig. 4. This is due to the fact that the network resources (e.g. wireless channel capacity) were incapable of sustaining such a large number of active nodes, resulting in high packet losses.

6 Conclusions and Future Work

This study investigates how nature inspired models can be employed to prevent congestion in ADNs. Inspiration from biological processes is drawn where global properties e.g., self-adaptation and scalability are achieved collectively without explicitly programming them into individual nodes, using simple computations at the node level.

Motivated by the famous LV competition model, a rate-based, hop-by-hop CC mechanism (LVCC) was designed which aims at controlling the traffic flow rate at each sending node. Simulations were performed to understand how the variations of the model's parameters influence stability and sensitivity. Simulation studies validated the correctness of analytical results of [4] and showed that our model achieves scalability, graceful performance degradation, adaptability and fairness. Realistic scenarios of network operation were also taken into consideration. However, for future work, further simulations for generalized network cases are required. Also a study of the behavior of our mechanism is needed when dynamic network conditions in terms of offered traffic load and node failures are considered.

References

1. Akyildiz, I., Su, W., Sankarasubramaniam, Y., Cayirci, E.: Wireless sensor networks: a survey. Computer Networks 38, 393–422 (2002)
2. Brauer, F., Chavez, C.: Mathematical Models in Population Biology and Epidemiology
3. Akyildiz, I., Melodia, T., Chowdhury, K.: A survey on wireless multimedia sensor networks. Computer Networks 51, 921–960 (2007)
4. Antoniou, P., Pitsillides, A.: Towards a Scalable and Self-adaptable Congestion Control Approach for Autonomous Decentralized Networks. In: NiSIS 2007 (2007)
5. The Network Simulator - ns-2, `http://www.isi.edu/nsnam/ns/`
6. Analoui, M., Jamali, S.: A Conceptual Framework for Bio-Inspired Congestion Control in Communication Networks. In: Proc. of the 1st BIMNICS, pp. 1–5 (2006)
7. Hasegawa, G., Murata, M.: TCP Symbiosis: Congestion Control Mechanism of TCP based on Lotka-Volterra Competition Model. In: ACM Intl. Conf. Proc. Series, vol. 200 (2006)
8. Lotka, A.: Elements of Physical Biology. Williams & Wilkins, Baltimore (1925)
9. Volterra, V.: Variations and fluctuations of the numbers of individuals in animal species living together, translation. In: Chapman, R. (ed.) Animal Ecology, pp. 409–448. McGraw Hill, New York (1931)
10. Wan, C., Eisenman, S., Campbell, A.: CODA: Congestion Detection and Avoidance in Sensor Networks. In: Proc. of the 1st Int. Conf. on Embedded Net. Sensor Systems, pp. 266–279 (2003)
11. Hull, B., Jamieson, K., Balakrishnan, H.: Mitigating Congestion in Wireless Sensor Networks. In: Proc. of the 2nd Int. Conf. on Embedded Net. Sensor Systems, pp. 134–147 (2004)
12. Vuran, M.C., Gungor, V.C., Akan, O.B.: On the Interdependence of Congestion and Contention in WSNs. In: Proc. of ICST SenMetrics (2005)
13. Ee, C., Bajcsy, R.: Congestion Control and Fairness for Many-to-One Routing in Sensor Networks. In: Proc. of the 2nd Int. Conf. on Embedded Net. Sensor Systems, pp. 148–161 (2004)
14. Takeuchi, Y., Adachi, N.: The existence of Globally Stable Equilibria of Ecosystems of the Generalized Volterra Type. Journal of Math. Biology 10, 401–415 (1980)

Author Index

Abdel Hady, Mohamed Farouk I-121
Abdullah, Rudwan I-198
Abe, Shigeo I-854
Aida, Shogo I-767
Aihara, Kazuyuki I-306, II-325, II-563, II-573
Aknin, Patrice II-416
Alexandre, Luís A. I-1015
Allinson, Nigel M. II-445
Almeida, Leandro M. I-485
Alonso-Betanzos, Amparo I-824
Alzate, Carlos II-315
Andreou, Panayiotis C. I-874
Antonelo, Eric A. I-747
Antonic, Radovan II-475
Antoniou, Pavlos II-986
Apolloni, Bruno I-449
Araujo, Aluizio F.R. I-515
Arcay, Bernardino II-378
Atiya, Amir F. II-275
Austin, Jim I-728

Ban, Tao I-913
Barbosa, Alexandre Ormiga G. II-495
Barth, Erhardt II-923
Bassani, Hansenclever F. I-515
Bassis, Simone I-449
Bayer, Justin II-755
Belle, Vanya Van I-60
Bertol, Douglas W. II-845
Bhaya, Amit I-668
Blekas, Konstantinos II-145
Bodyanskiy, Yevgeniy I-718
Bogdan, Martin I-181
Borisov, Alexander II-944
Bors, Adrian G. II-245
Bouchain, David I-191
Boutalis, Yiannis II-875
Brdys, Mietek A. II-823
Bumelienė, Skaidra I-618
Byadarhaly, Kiran V. I-296

Cacciola, Matteo II-455
Calcagno, Salvatore II-455
Canals, Vincent I-421
Cardin, Riccardo II-105
Carvajal, Gonzalo I-429
Castellanos Sánchez, Claudio II-745
Cecchi, Guillermo I-587
Čerňanský, Michal I-381
Cernega, Daniela II-865
Chairez, Isaac II-552
Chambers, Jon I-198
Charalambous, Chris I-874
Chen, H. I-401, I-410
Chen, Huanhuan II-185
Chen, Ning II-426
Chenli, Zhang II-345
Cherkassky, Vladimir I-932
Cheung, Yiu-ming I-10
Chinea, Alejandro I-952
Chouchourelou, Arieta I-678
Christodoulidou, K. II-145
Christodoulou, Chris I-737
Christodoulou, Christodoulos S. II-885
Christodoulou, Manolis II-875
Chudacek, Vaclav II-485
Chung, Kwok Wai II-534
Ciccazzo, Angelo I-449
Cleanthous, Aristodemos I-737
Cobb, Stuart I-229
Côme, Etienne II-416
Conti, Marco I-325
Contreras-Lámus, Victor I-277
Corbacho, Fernando II-235
Costa, Ivan G. II-20
Cpałka, Krzysztof II-435
Ćulibrk, Dubravko II-633
Cutsuridis, Vassilis I-229
Cuzzola, Maria II-10

Dafonte, Carlos II-378
Dai, Wuyang I-932
Dal Seno, Bernardo II-515
da Silva, Ivan I-807
Daucé, Emmanuel I-218
Davoian, Kristina I-111
De Brabanter, Jos I-100
De Brabanter, Kris I-100

Debruyne, Michiel I-100
De Carvalho, Francisco A.T. I-131
de Carvalho, Luís Alfredo V. II-65
de las Rivas, Javier II-195
de M. Silva Filho, Telmo I-799
De Moor, Bart I-100, II-315
Dendek, Cezary I-141
Denœux, Thierry II-416
de Paúl, Ivan I-421
De Pieri, Edson R. II-845
de Souto, Marcilio C.P. II-20
de Souza, Renata M.C.R. I-799
Diaz, David Ronald A. II-495
Direito, Bruno II-1
Dolenko, Sergey I-373, II-295, II-397
Donangelo, Raul II-65
Dong, Ming II-964
do Rego, Renata L.M.E. I-515
Dorronsoro, José R. I-904
Doulamis, Anastasios II-715, II-895
Doulamis, Nikolaos II-895
Dourado, António II-1
Doya, Kenji I-249
Drchal, Jan II-775
Duch, Włodzisław I-151, I-789
Ďuračková, Daniela I-363
Dutoit, Pierre I-277

Eggert, Julian II-804, II-855
El-Laithy, Karim I-181
Eliades, Demetrios G. II-954
Embrechts, Mark II-175, I-1015
Erwin, Harry I-208
Espinoza, Marcelo II-315

Fernandes, João L. I-356
Fernandez, Emmanuel I-296
Fiasché, Maurizio II-10
Figueroa, Miguel I-429
Filgueiras, Daniel I-515
Flauzino, Rogerio I-807
Fontenla-Romero, Oscar I-824
Foresti, Loris II-933
Franco, João M. II-1
Fuentes, Rita II-552

Galatsanos, Nikolaos I-50, I-942
García, Rodolfo V. I-259
García-Sebastián, Maite II-725
Gastaldo, Paolo II-684
Gáti, Kristóf I-698
Gavalda, Arnau I-525
Ge, Ju Hong II-534
Georgoulas, George II-485
Gianniotis, Nikolaos I-567
Gomez, Faustino J. II-765
González, Jesús I-259
Gopych, Petro II-54
Gori, Marco II-653
Goryll, Michael II-265
Gouko, Manabu I-628
Graham, Bruce P. I-229
Graña, Manuel II-725
Grim, Jiří II-165
Grochowski, Marek I-151
Gross, Horst-Michael II-735, II-804
Guan, XuDong II-30
Guazzini, Andrea I-325
Guillén, A. II-215
Guillén, Alberto I-1, II-406
Gurney, Kevin I-198
Gutmann, Michael II-623
Guzhva, Alexander I-373, II-295, II-397

Hajek, Petr II-505
Hajnal, Márton A. I-315
Haker, Martin II-923
Hans, Alexander I-70
Hara, Kazuyuki I-171
Haraguchi, Yuta II-325
Haralambous, Haris II-335
Hartley, Matthew II-40
Hasegawa, Osamu I-864, II-345
Hasler, Stephan II-663
Hatakeyama, Takashi I-767
Hauser, Florian I-191
He, Shan II-185
Hellbach, Sven II-804
Heng, Pheng Ann I-995
Hernández, Carmen II-725
Hernández-Lobato, Daniel I-90
Herrera, Luis Javier II-215, II-406
Hilbers, Peter A.J. II-305
Hobson, Stephen I-728
Höffken, Matthias I-757
Honda, Hidehito I-678
Honkela, Timo II-305
Hora, Jan II-165
Horváth, Gábor I-698
Huang, Tingwen II-583

Huang, Yu II-534
Hubert, Mia I-100
Huptych, Michal II-485
Hussain, Amir I-198
Hyvärinen, Aapo II-623, II-704

Iacopino, Pasquale II-10
Ikeguchi, Tohru I-306, II-325, II-563, II-573
Iordanidou, Vasiliki II-368
Ishii, Shin I-608
Ishikawa, Masumi I-546
Ito, Koji I-628

Joshi, Prashant I-239
Jovic, Mate II-475

Kadobayashi, Youki I-913
Kakemoto, Yoshitsugu I-688
Kanevski, Mikhail II-933
Kapral, Ondrej II-775
Kasabov, Nikola II-10, I-913
Kasai, Wataru I-864
Kaszkurewicz, Eugenius I-668
Kato, Hideyuki I-306
Kato, Yuko II-544
Kawakami, Hajimu II-673
Kayastha, Nagendra II-387
Kemppi, Paul II-954
Kencana Ramli, Carroline Dewi Puspa II-85
Kirstein, Stephan II-663
Kirvaitis, Raimundas I-618
Kiyokawa, Sachiko I-678
Kleihorst, Richard I-439
Knee, Peter II-265
Kobos, Mateusz II-125
Koene, Ansgar I-267
Kolesnik, Marina I-757
Kollia, Ilianna II-465
Konnanath, Bharatan II-265
Körner, Edgar II-663, II-804
Korytkowski, Marcin I-817
Koskela, Markus I-495
Kosmatopoulos, Elias B. I-638
Kottas, Thodoris II-875
Kounoudes, Anastasis II-913
Koutník, Jan II-775
Kovacevic, Asja II-475
Kovacevic, Darko II-475
Kryzhanovsky, Vladimir I-844
Kukolj, Dragan II-633
Kůrková, Věra I-708
Kuroe, Yasuaki II-673
Kuroiwa, Jousuke II-544
Kurokawa, Hiroaki I-834
Kwon, Jun Soo I-345

Laaksonen, Jorma I-495, II-694
Laganá, Filippo II-455
Lagaris, I.E. II-145
Lago-Fernández, Luis F. II-235
Lang, E.W. II-115
Laoudias, Christos II-954
Lawniczak, Anna T. II-358
Lee, John A. II-95
Lendasse, Amaury I-1, II-305
Lhotska, Lenka II-485
Li, Xiaoli II-185
Lió, Pietro I-325, II-358
Likas, Aristidis I-50, I-942, II-205
Lim, Zhan Wei II-976
Lindh-Knuutila, Tiina II-305
Lippe, Wolfram-M. I-111
Litinskii, Leonid I-161
Liu, Jindong I-208
Loengarov, Andreas II-613
Loizou, Christos P. II-885
López, Jorge I-904
Lu, Chih-Cheng I-401, I-410
Ludermir, Teresa I-30, I-131, I-485, I-557
Luo, Xuanwen II-964

Magdi, Yasmine M. I-894
Maggini, Marco II-653
Maguire, Liam I-335, II-794
Mańdziuk, Jacek I-141, II-125
Manolis, Diamantis I-638
Manteiga, Minia II-378
Marakakis, Apostolos I-942
Marotta, Angelo I-449
Martín-Merino, Manuel II-195
Martínez-Muñoz, Gonzalo I-90
Martínez-Rego, David I-824
Martinetz, Thomas II-923
Martinez, Régis II-75
Martins da Silva, Rodrigo I-475
Martins, Nardênio A. II-845
Martzoukos, Spiros H. I-874

Maszczyk, Tomasz I-789
Mathew, Trupthi II-265
Matsuda, Yoshitatsu II-135
Matsuka, Toshihiko I-678
Matsushima, Fumiya I-767
Matsuura, Takafumi II-563, II-573
Matsuyama, Yasuo I-767
Matteucci, Matteo II-515
Mc Bride, Michael F. II-794
McGinnity, T.M. I-335, II-794
Megali, Giuseppe II-455
Meggiolaro, Marco Antonio II-495
Meijer, Peter B.L. I-439
Melacci, Stefano II-653
Mercier, David II-255
Mesiano, Cristian I-449
Michalopoulos, Kostas II-368
Miche, Yoan II-305
Michielan, Lisa II-105
Milis, Marios II-913
Minai, Ali A. I-296
Mirikitani, Derrick Takeshi I-975
Mitsuishi, Takashi I-598
Miyoshi, Seiji I-171
Moore, Roger K. II-784
Morabito, Francesco Carlo II-10, II-455
Mori, Takeshi I-608
Moro, Stefano II-105
Morro, Antoni I-421
Motohashi, Shun II-563, II-573
Mourelle, Luiza de Macedo I-475
Müller, Marco K. II-643
Murray, Victor II-885
Mykolaitis, Gytis I-618

Nagaya, Shigeki II-345
Nakagawa, Masahiro II-593
Nakagawa, Masanori I-779
Nakamura, Yuichi II-593
Nakano, Ryohei II-155
Nakano, Takashi I-249
Nakasuka, Shinchi I-688
Nakayama, Yoichi I-171
Nascimento, André C.A. II-20
Nasios, Nikolaos II-245
Necoara, Ion II-835
Nedjah, Nadia I-475
Ni, He I-577
Nicolaides, Andrew II-885
Nishida, Youichi I-767
Nowicki, Robert I-817
Ntalianis, Klimis II-895

Oberhoff, Daniel I-757
Obornev, Eugeny II-397
Ochiai, Nimiko I-767
Odaka, Tomohiro II-544
Ogura, Hisakazu II-544
Oja, Erkki I-20, II-305
Okada, Masato I-171
Olej, Vladimir II-505
Ordóñez, Diego II-378
Osana, Yuko I-505
Ott, Thomas II-525
Ouarbya, Lahcen I-975
Oukhellou, Latifa II-416

Paechter, B. II-215
Palm, Günther I-121, I-191, I-894
Panayiotou, Christos G. II-954
Pang, Shaoning I-913
Pantziaris, Marios II-885
Papadopoulos, Harris II-335
Papageorgiou, M. I-638
Parviainen, Elina II-225
Passarella, Andrea I-325
Patan, Krzysztof I-80
Patan, Maciej I-80
Pattichis, Constantinos S. II-885
Pattichis, Marios S. II-885
Paugam-Moisy, Hélène II-75
Pavlou, Maria II-445
Pazos, Fernando A. I-668
Pelckmans, Kristiaan I-60, I-100
Pellicanó, Diego II-455
Perdoor, Mithun I-296
Perez-Gonzalez, David I-208
Perkiö, Jukka II-704
Perrig, Stephen I-277
Persiantsev, Igor I-373, II-295, II-397
Piekniewski, Filip II-603
Pitsillides, Andreas II-986
Pokrić, Maja II-633
Polycarpou, Marios M. II-954
Pomares, Héctor II-215, II-406
Popa, Cosmin Radu I-459
Popov, Sergiy I-718
Porto-Díaz, Iago I-824
Poulsen, Thomas M. II-784
Pozdnoukhov, Alexei II-933

Poznyak, Alexander II-552
Poznyak, Tatyana II-552
Prasad, Shalini II-265
Prentis, Philip I-495
Prescott, Tony J. I-267
Pribacic, Nikica II-475
Prieto, Alberto I-259
Prudêncio, Ricardo B.C. I-30, II-20
Psillakis, Haris I-648, II-815

Ralaivola, Liva I-884
Rallo, Robert I-525
Ramos, João P. II-1
Rao, A. Ravishankar I-587
Raschman, Emil I-363
Raychaudhury, Somak I-567
Rebrova, Olga II-435
Redi, Judith II-684
Rees, Adrian I-208
Refaat, Khaled S. II-275
Ribeiro, Bernardete I-923
Rigutini, Leonardo I-40
Rinaudo, Salvatore I-449
Rizzo, Riccardo I-536
Rodríguez, Roberto I-259
Rojas, Fernando I-259
Rojas, Ignacio I-1, II-215, II-406
Rosselló, Josep L. I-421
Rubio, Ginés I-1, II-215, II-406
Ruiz-Llata, Marta I-467
Runger, George II-944
Rutkowski, Leszek II-435

Sakkalis, Vangelis II-368
Sánchez-Montañés, Manuel II-235
Sánchez Orellana, Pedro L. II-745
San Román, Belén I-259
Santos, Jorge M. II-175
Särelä, Jaakko II-285
Sattigeri, Prasanna II-265
Schaffernicht, Erik II-735
Schaul, Tom I-1005
Scherer, Rafał I-817
Scherer, Stefan I-894
Schmidhuber, Jürgen I-964, I-1005, II-755, II-765
Schrauwen, Benjamin I-747, I-985
Schüle, Martin II-525
Schwenker, Friedhelm I-121, I-894
Seah, Winston Khoon-Guan II-976
Shaposhnyk, Vladyslav I-277
Shibata, Tadashi I-391
Shidama, Yasunari I-598
Shimada, Yutaka II-325
Shimelevich, Mikhail II-397
Shirai, Haruhiko II-544
Shirotori, Tomonori I-505
Shrestha, Durga Lal II-387
Shugai, Julia II-295
Silva, Catarina I-923
Simou, Nikolaos II-465
Sittiprapaporn, Wichian I-345
Sjöberg, Mats I-495
Šnorek, Miroslav II-775
Soares, Rodrigo G.F. I-131
Solanas, Agusti I-525
Solea, Razvan II-865
Solomatine, Dimitri P. II-387
Sorjamaa, Antti I-1
Souza, Jefferson R. I-485
Spaanenburg, Lambert I-439
Spanias, Andreas II-265
Sperduti, Alessandro II-105
Sportiello, Luigi II-515
Spreckley, Steve I-567
Stafylopatis, Andreas I-942, II-465
Stamou, Giorgos II-465
Stempfel, Guillaume I-884
Stephan, Volker II-735
Stoop, Ruedi I-618, II-525
Stylios, Chrysostomos II-485
Suárez, Alberto I-90
Suard, Frédéric II-255
Suykens, Johan A.K. I-60, I-100, II-315
Suzuki, Takayuki II-573

Takahashi, Norihiro I-391
Tamaševičius, Arūnas I-618
Tamaševičiūtė, Elena I-618
Tan, Colin Keng-Yan II-976
Tan, Guan-Hong II-976
Tanahashi, Yusuke II-155
Tanscheit, Ricardo II-495
Taylor, John G. II-40
Tehrani, Mona Akbarniai I-439
Teixeira, Ana R. II-115
Terai, Asuka I-779
Tereshko, Valery II-613
Theodosiou, Zenonas II-905, II-913
Thornton, Trevor II-265

Tiňo, Peter I-567
Titov, Mykola I-718
Tobe, Yutaro I-864
Togelius, Julian II-755, II-765
Tomé, Ana Maria II-115
Tomi, Naoki I-628
Torreão, José R.A. I-356
Toshiyuki, Hamada II-544
Tran, Vu Nam II-823
Trentin, Edmondo I-40
Triesch, Jochen I-239
Tsaih, Rua-Huan I-658
Tsapatsoulis, Nicolas II-895, II-905, II-913
Tuia, Devis II-933
Tuv, Eugene II-944
Tzikas, Dimitris I-50
Tzortzis, Grigorios II-205

Udluft, Steffen I-70
Ueno, Kosuke I-546
Unkelbach, Jan I-964
Urso, Alfonso I-536

Valderrama-Gonzalez, German D. I-335
Valença, Ivna I-557
Valenzuela, Olga I-259
Valenzuela, Waldo I-429
Valpola, Harri II-285
van Heeswijk, Mark II-305
Van Huffel, Sabine I-60
Vasa, Suresh I-296
Vasiljević, Petar II-633
Vassiliades, Vassilis I-737
Vehtari, Aki II-225
Velázquez, Luis I-259
Vellasco, Marley Maria B.R. II-495
Ventura, André II-1
Verleysen, Michel II-95
Verma, Anju II-10
Versaci, Mario II-455
Verstraeten, David I-985
Victer, Silvia M.C. I-356
Viitaniemi, Ville II-694
Villa, Alessandro E.P. I-277

Wan, Wang-Gen II-426
Wan, Yat-wah I-658
Wang, Jun II-583
Wedemann, Roseli S. II-65
Wei, Hui II-30
Wermter, Stefan I-208
Wersing, Heiko II-663
Wickens, Jeff I-249
Wierstra, Daan II-755
Wu, QingXiang I-335
Würtz, Rolf P. II-643

Xie, Shengkun II-358
Xu, Jian II-534

Yamaguchi, Kazunori II-135
Yang, Zhirong I-20
Yao, Xin II-185
Yébenes-Calvino, Mar I-467
Yi, Sun I-964
Yi, Zhang I-287, I-995
Yin, Hujun I-577
Yli-Krekola, Antti II-285
Yonekawa, Masato I-834
Yoshimoto, Junichiro I-249
Yu, Jiali I-287

Zeng, Hong I-10
Zeng, Zhigang II-583
Zervakis, Michalis II-368
Zhang, Chen II-855
Zhang, Haixian I-287
Zhang, Lei I-995
Zlokolica, Vladimir II-633
Zunino, Rodolfo II-684